Fundamentals of Thermochemical Biomass Conversion

Fundamentals of Thermochemical Biomass Conversion

Edited by

R. P. OVEREND

National Research Council of Canada, Ottawa, Canada

T. A. MILNE

Solar Energy Research Institute, Colorado, USA

and

L. K. MUDGE

Battelle Pacific Northwest Laboratories, Washington, USA

ELSEVIER APPLIED SCIENCE PUBLISHERS
LONDON and NEW YORK

ELSEVIER APPLIED SCIENCE PUBLISHERS LTD
Crown House, Linton Road, Barking, Essex IG11 8UJ, England

Sole Distributor in the USA and Canada
ELSEVIER SCIENCE PUBLISHING CO., INC.
52 Vanderbilt Avenue, New York, NY 10017, USA

British Library Cataloguing in Publication Data

International Conference on Fundamentals of Thermochemical Biomass Conversion (*1982: Estes Park, Colo.*)
Fundamentals of thermochemical biomass conversion.
1. Fuelwood 2. Biomass energy
I. Title II. Overend, R. P.
III. Milne, T. A. IV. Mudge, L. K.
662'.6 TP324

ISBN-13: 978-94-010-8685-1 e-ISBN-13: 978-94-009-4932-4
DOI: 10.1007/978-94-009-4932-4

WITH 539 ILLUSTRATIONS AND 216 TABLES

Softcover reprint of the hardcover 1st edition 1985

Preface

Throughout the world many projects have been underway to investigate the conversion of renewable biomass into energy and synthetic fuels by thermochemical methods such as combustion, pyrolysis, gasification and liquefaction. While many of these represent prior art used during the early 20th century, the recent decade since the 1970s oil shock has immeasurably increased the knowledge base for such processes. Much of the new knowledge has been gained by persons who were not trained in classical wood chemistry and there have not yet been many attempts to synthesize the knowledge into a corpus of systematic information.

To bring this about the International Energy Agency's Forestry Energy collaboration, the Gas Research Institute, the National Research Council of Canada and the US Department of Energy jointly sponsored a conference on the Fundamentals of Thermochemical Biomass Conversion in Estes Park, Colorado which was held on October 18–22, 1982.

The Conference, which was structured around invited plenary papers and contributions from researchers, served as the basis for the papers in this volume which reflect the substantial conclusions of the Conference. During the planning for the Conference, it was realized by the editors in their capacity as Co-chairmen that a major problem in biomass research was the lack of reproducibility between reported experiments and their intercomparison on account of the heterogeneity of biomass materials. A well-known wood chemist, George M. Barton, was persuaded to prepare a survey of thermochemical researchers' interests in specified standard biomass materials. The results of this were presented at Estes Park and a panel discussion on the topic held which the editors believe has made an advance in improving the data base for all biomass research.

This volume brings together, in one place, the state-of-the-art in the many disciplines of thermochemical biomass conversion ranging from wood ultrastructure studies, wood chemistry, through to reaction kinetics. In addition, the conference's contribution to the evolution of standard reference materials research has been made the subject of the final chapter. The International Organizing Committee for the Conference consisted of:

Michael Antal, USA
Dan Asplund, Finland
Anthony Bridgwater, UK
John Cox, USA
Simon Friedrich, USA
Nils Lindman, Sweden
Oskar Zaborsky, USA

who join with the editors in hoping that this contribution to the science of Thermochemical Biomass Conversion will be of continuing use to scientists in this field.

During the final editing of this book, the editors were saddened to hear of the death of Professor Fred Shafizadeh, one of the senior contributors to this volume. His contribution to the understanding of the pyrolysis of biomass materials had been immense and he will be sorely missed by his scientific colleagues.

R. P. Overend
National Research Council of Canada

T. A. Milne
Solar Energy Research Institute

L. K. Mudge
Battelle Pacific Northwest Laboratories

Contents

List of Contributors

F. G. van den Aarsen
Laboratory of Chemical Reaction Engineering, Twente University of Technology, PO Box 217, 7500 AE Enschede, The Netherlands

B. Åhgren
Royal Institute of Technology, Department of Chemical Technology, S-100 44 Stockholm, Sweden

M. Aho
Technical Research Centre of Finland, PL 221, 40101 Jyväskylä 10, Finland

M. J. Antal Jr
Department of Mechanical Engineering, University of Hawaii at Manoa, 2540 Dole Street, Honolulu, Hawaii 96822, USA

D. Asplund
Technical Research Centre of Finland, Domestic Fuel Laboratory, PL 221, 40101 Jyväskylä 10, Finland

E. Avni
Department of Chemical Engineering, University of Connecticut, Storrs, Connecticut 06268, USA

D. W. Bacon
Department of Chemical Engineering, Queen's University, Kingston, Ontario, K7L 3N6, Canada

E. G. Baker
Fuels and Mineral Technology Section, Battelle Pacific Northwest Laboratories, PO Box 999, Richland, Washington 99352, USA

G. M. Barton
530 Bayview Road, PO Box 308, Lion's Bay, British Columbia, V0N 2E0, Canada

A. A. C. M. Beenackers
Laboratory of Chemical Reaction Engineering, Twente University of Technology, PO Box 217, 7500 AE Enschede, The Netherlands

M. A. Bergougnou
Faculty of Engineering Science, University of Western Ontario, London, Ontario, N6A 5B9, Canada

A. Bergström
Royal Institute of Technology, Department of Chemical Technology, S-100 44, Stockholm, Sweden

E. Björnbom
Royal Institute of Technology, Department of Chemical Technology, S-100 44, Stockholm, Sweden

P. Björnbom
Royal Institute of Technology, Department of Chemical Technology, S-100 44, Stockholm, Sweden

W. Blackadder
Fuel Technology Department, Studsvik Energiteknik AB, S-611 82 Nyköping, Sweden

T. Bradley
Colorado School of Mines, Golden, Colorado 80401, USA

D. Brouillard
Department of Chemical Engineering, Université de Sherbrooke, Sherbrooke, Quebec, J1K 2R1, Canada

A. G. Buekens
University of Brussels, Pleinlaan 2, Brussels B-1050, Belgium

M. CATHONNET
Centre National de la Recherche Scientifique, Centre de Recherches sur la Chimie de la Combustion et des Hautes Températures, 1C Avenue de la Recherche Scientifique, 45045, Orleans Cedex, France

B. DE CAUMIA
Department of Chemical Engineering, Université de Sherbrooke, Sherbrooke, Quebec, J1K 2R1, Canada

W.-C. R. CHAN
Department of Chemical Engineering, University of Washington, Seattle, Washington 98195, USA

S. M. CHERN
Department of Chemical Engineering, Durland Hall, Kansas State University, Manhattan, Kansas 66506, USA

E. CHORNET
Département de Génie Chimique, Faculté des Sciences Appliquées, Université de Sherbrooke, Sherbrooke, Québec, J1K 2R1, Canada

H. L. CHUM
Solar Energy Research Institute, 1617 Cole Boulevard, Golden, Colorado 80401, USA

R. W. COUGHLIN
Department of Chemical Engineering, University of Connecticut, Storrs, Connecticut 06268, USA

R. CUELLO
Battelle Pacific Northwest Laboratories, PO Box 999, Richland, Washington 99352, USA

H. G. DAVIS
Energy and Environment Division, Lawrence Berkeley Laboratory, One Cyclotron Road, University of California, Berkeley, California 94720, USA

J. L. DAVIS
Department of Forestry, University of Wisconsin, 1630 Linden Drive, Madison, Wisconsin 53706, USA

F. DAVOUDZADEH
Department of Chemical Engineering, University of Connecticut, Storrs, Connecticut 06268, USA

W. F. DEGROOT
Wood Chemistry Laboratory, University of Montana, Missoula, Montana 59812, USA

J. DIEBOLD
Solar Energy Research Institute, 1617 Cole Boulevard, Golden, Colorado 80401, USA

J. DOWNIE
Department of Chemical Engineering, Queen's University, Kingston, Ontario, K7L 3N6, Canada

C. W. DRAKE
Western Regional Research Center, US Department of Agriculture, 800 Buchanan Street, Berkeley, California 94710, USA

R. L. EAGER
University of Saskatchewan, Saskatoon, Saskatchewan, S7N 0W0, Canada

M. A. EAMES
Energy and Environment Division, Lawrence Berkeley Laboratory, One Cyclotron Road, University of California, Berkeley, California 94720, USA

R. EDRICH
Colorado School of Mines, Golden, Colorado 80401, USA

H. EKLUND
Bergbau-Forschung GmbH, Franz-Fischer-Weg 61, D-4300 Essen 13, Federal Republic of Germany

C. EKSTRÖM
Royal Institute of Technology, Department of Chemical Technology, S-100 44 Stockholm, Sweden

D. C. ELLIOTT
Battelle Pacific Northwest Laboratories, PO Box 999, Richland, Washington 99352, USA

D. EUGÈNE
Département de Génie Chimique, Faculté des Sciences Appliquées, Université de Sherbrooke, Sherbrooke, Québec, J1K 2R1, Canada

P. T. FALLON
Process Sciences Division, Brookhaven National Laboratory, Department of Energy and Environment, Upton, New York 11973, USA

L. T. FAN
Department of Chemical Engineering, Durland Hall, Kansas State University, Manhattan, Kansas 66506, USA

C. M. FERNANDEZ
Department of Chemical Engineering, Arizona State University, Tempe, Arizona 85287, USA

C. FIGUEROA
Energy and Environment Division, Lawrence Berkeley Laboratory, One Cyclotron Road, Berkeley, California 94720, USA

R. R. GANSLEY
Energy and Environment Division, Lawrence Berkeley Laboratory, One Cyclotron Road, Berkeley, California 94720, USA

W. G. GLASSER
Department of Forest Products, 210 Cheatham Hall, Virginia Polytechnic Institute and State University, Blacksburg, Virginia 24061, USA

M. S. GRABOSKI
Colorado School of Mines, Golden, Colorado 80401, USA

R. G. GRAHAM
Forintek Canada Corporation, 800 Montreal Road, Ottawa, K1A 3Z5, Canada

K. S. GREGORSKI
US Department of Agriculture, Western Regional Research Center, 800 Buchanan Street, Berkeley, California 94710, USA

R. T. HALLEN
Battelle Pacific Northwest Laboratories, PO Box 999, Richland, Washington 99352, USA

B. W. HATT
Department of Chemical Engineering, University of Aston in Birmingham, Gosta Green, Birminghan B4 7ET, UK

G. HEATH
Department of Chemical Engineering, Arizona State University, Tempe, Arizona 85287, USA

V.-P. HEISKANEN
Technical Research Centre of Finland, Domestic Fuel Laboratory, PL 221, 40101 Jyväskylä 10, Finland

W. E. HILLIS
Chief Research Scientist, Division of Chemical Technology, CSIRO, PO Box 56, Highett, Victoria 3190, Australia

C. HÖRNELL
Royal Institute of Technology, Department of Chemical Technology, S-100 44 Stockholm, Sweden

D. A. HOVELAND
Department of Chemical Engineering, Durland Hall, Kansas State University, Manhattan, Kansas 66506, USA

J. B. HOWARD
Energy Laboratory and Department of Chemical Engineering, Massachusetts Institute of Technology, Cambridge, Massachusetts 02139, USA

J. C. HSU
Department of Chemical Engineering, Queen's University, Kingston, Ontario, K7L 3N6, Canada

J. S. Hudson
Western Regional Research Center, US Department of Agriculture, 800 Buchanan Street, Berkeley, California 94710, USA

E. R. Huff
Agricultural Engineering Department, University of Maine, Orono, Maine 04469, USA

J. Huotari
Technical Research Centre of Finland, Domestic Fuel Laboratory, PL 221, 40101 Jyväskylä 10, Finland

P. J. Iredale
Department of Chemical Engineering, University of Aston in Birmingham, Gosta Green, Birmingham B4 7ET, UK

G. A. Irlam
Department of Chemical Engineering, University of Aston in Birmingham, Gosta Green, Birmingham B4 7ET, UK

S. T. Jack
217 Academic Projects Building, Pennsylvania State University, University Park, Pennsylvania 16802, USA

O. Jönsson
Lund Institute of Technology, Box 740, 220 07 Lund 7, Sweden

G. Karlsson
Royal Institute of Technology, Department of Chemical Technology, S-100 44 Stockholm, Sweden

O. Karlsson
Royal Institute of Technology, Department of Chemical Technology, S-100 44 Stockholm, Sweden

G. G. Karsner
Energy and Environmental Science Division, WTC-1B34, Weyerhaeuser Company, Tacoma, Washington 98477, USA

M. Kelbon
Department of Chemical Engineering, University of Washington, Seattle, Washington 98195, USA

M. T. Klein
Department of Chemical Engineering, University of Delaware, Newark, Delaware 19711, USA

V. R. Koch
Covalent Associates, Inc., 52 Oragon Court, Woburn, Massachusetts 01801, USA

B. B. Krieger
Department of Chemical Engineering, University of Washington, Seattle, Washington 98195, USA

J. M. Krochta
Western Regional Research Center, US Department of Agriculture, 800 Buchanan Street, Berkeley, California 94710, USA

J. M. Kryla
Forintek Canada Corporation, 800 Montreal Road, Ottawa, Ontario, K1G 3Z5, Canada

J. L. Kuester
Department of Chemical Engineering, Arizona State University, Tempe, Arizona 85287, USA

H. I. de Lasa
Faculty of Engineering Science, University of Western Ontario, London, Ontario, N6A 5B9, Canada

W. H. Lee
Department of Chemical Engineering, University of Idaho, Moscow, Idaho 83843, USA

L. Liinanki
Royal Institute of Technology, Department of Chemical Technology, S-100 44 Stockholm, Sweden

N. Lindman
Royal Institute of Technology, Department of Chemical Technology, S-100 44 Stockholm, Sweden

E. S. Lipinsky
Battelle Columbus Laboratories, 505 King Avenue, Columbus, Ohio 43201, USA

A. E. Lipska-Quinn
Lawrence Livermore National Laboratory, Livermore, California 94550, USA

J. P. Longwell
Energy Laboratory and Department of Chemical Engineering, Massachusetts Institute of Technology, Cambridge, Massachusetts 02139, USA

K. M. McGee
University of California at Davis, Davis, California 95616, USA

P. C. Malte
Department of Mechanical Engineering, University of Washington, Seattle, Washington 98195, USA

M. Markson
Solar Energy Research Institute, 1617 Cole Boulevard, Golden, Colorado 80401, USA

J. F. Mathews
Department of Chemistry and Chemical Engineering, University of Saskatchewan, Saskatoon, Saskatchewan, S7N 0W0, Canada

H. Ménard
Department of Chemistry, Université de Sherbrooke, Sherbrooke, Quebec, J1K 2R1, Canada

T. A. Milne
Thermochemical and Electrochemical Research Branch, Solar Energy Research Institute, 1617 Cole Boulevard, Golden, Colorado 80401, USA

M. Modell
MODAR Inc., 14 Tech Circle, Natick, Massachusetts 01760, USA

L. K. S. Mok

Faculty of Engineering Science, University of Western Ontario, London, Ontario, N6A 5B9, Canada

P. M. Molton

Battelle Pacific Northwest Laboratories, PO Box 999, Richland, Washington 99352, USA

T. R. Mon

Western Regional Research Center, US Department of Agriculture, 800 Buchanan Street, Berkeley, California 94710, USA

L. K. Mudge

Battelle Pacific Northwest Laboratories, PO Box 999, Richland, Washington 99352, USA

S. P. Nandi

Institute of Gas Technology, 3424 South State Street, Chicago, Illinois 60616, USA

D. A. Nelson

Battelle Pacific Northwest Laboratories, PO Box 999, Richland, Washington 99352, USA

T. R. Nunn

Energy Laboratory and Department of Chemical Engineering, Massachusetts Institute of Technology, Cambridge, Massachusetts 02139, USA

B. Olsson

Royal Institute of Technology, Department of Chemical Technology, S-100 44 Stockholm, Sweden

M. Onischak

Institute of Gas Technology, 3424 South State Street, Chicago, Illinois 60616, USA

R. P. Overend

Energy Project, Task Coordinator Renewable Energy, National Research Council of Canada, Ottawa, K1A 0R6, Canada

A. E. Pavlath

Western Regional Research Center, US Department of Agriculture, 800 Buchanan Street, Berkeley, California 94710, USA

J. M. Pepper

University of Saskatchewan, Saskatoon, Saskatchewan, S7N 0W0, Canada

J. Peters

Department of Chemical Engineering, Queen's University, Kingston, Ontario, K7L 3N6, Canada

W. A. Peters

Energy Laboratory and Department of Chemical Engineering, Massachusetts Institute of Technology, Cambridge, Massachusetts 02139, USA

F. P. Petrocelli

Department of Chemical Engineering, University of Delaware, Newark, Delaware 19711, USA

R. Pettersson

Royal Institute of Technology, Department of Chemical Technology, S-100 44 Stockholm, Sweden

K. W. Ragland

Department of Mechanical Engineering, University of Wisconsin-Madison, Madison, Wisconsin 53706, USA

T. B. Reed

Solar Energy Research Institute, 1617 Cole Boulevard, Golden, Colorado 80401, USA

E. Rensfelt

Fuel Technology Department, Studsvik Energiteknik AB, S-611 82 Nyköping, Sweden

J. J. Reuther

Fuels and Combustion Laboratory, 320 Steidle Building, Pennsylvania State University, University Park, Pennsylvania 16802, USA

J. R. RICHARD
Centre National de la Recherche Scientifique, Centre de Recherches sur la Chimie de la Combustion et des Hautes Températures, 1C Avenue de la Recherche Scientifique, 45045 Orleans Cedex, France

G. ROSE
Institute of Gas Technology, 3424 South State Street, Chicago, Illinois 60616, USA

J. P. ROUAN
Centre National de la Recherche Scientifique, Centre de Recherches sur la Chimie de la Combustion et des Hautes Températures, 1C Avenue de la Recherche Scientifique, 45045 Orleans Cedex, France

C. ROY
Department of Chemical Engineering, Université de Sherbrooke, Sherbrooke, Quebec, J1K 2R1, Canada

J. A. RUSSELL
Battelle Pacific Northwest Laboratories, PO Box 999, Richland, Washington 99352, USA

J. SCAHILL
Solar Energy Research Institute, 1617 Cole Boulevard, Golden, Colorado 80401, USA

L. L. SCHALEGER
Energy and Environment Division, Lawrence Berkeley Laboratory, One Cyclotron Road, Berkeley, California 94720, USA

M. D. SCHEER
Chemical Kinetics Division, National Bureau of Standards, Room A241, Building 222, Washington DC 20234, USA

J. G. SCHOETERS
University of Brussels, Pleinlaan 2, Brussels, B-1050, Belgium

H. A. SCHROEDER
Department of Forest and Wood Sciences, Colorado State University, Fort Collins, Colorado 80523, USA

L. J. Sealock Jr
Battelle Pacific Northwest Laboratories, PO Box 999, Richland, Washington 99352, USA

The late F. Shafizadeh
Wood Chemistry Laboratory, University of Montana, Missoula, Montana 59812, USA

R. N. Shand
Department of Chemical Engineering, University of Aston in Birmingham, Gosta Green, Birmingham B4 7ET, UK

H. H. Sheena
Department of Chemical Engineering, University of Aston in Birmingham, Gosta Green, Birmingham B4 7ET, UK

G. M. Simmons
Department of Chemical Engineering, University of Idaho, Moscow, Idaho 83843, USA

W. W. Simmons
Department of Mechanical Engineering, University of Wisconsin-Madison, Madison, Wisconsin 53706, USA

S.-O. Sjöberg
Royal Institute of Technology, Department of Chemical Technology, S-100 44 Stockholm, Sweden

E. L. Smith
Department of Chemical Engineering, University of Aston in Birmingham, Gosta Green, Birmingham B4 7ET, UK

Y. Solantausta
Technical Research Centre of Finland, Fuel and Lubricant Research Laboratory, Espoo SF-02150, Finland

M. N. Soltys
Thermochemical and Electrochemical Research Branch, Solar Energy Research Institute, 1617 Cole Boulevard, Golden, Colorado 80401, USA

D. W. Sopher
Department of Forest and Wood Sciences, Colorado State University, Fort Collins, Colorado 80523, USA

M. Steinberg
Process Sciences Division, Brookhaven National Laboratory, Department of Energy and Environment, Upton, New York 11973, USA

E. Ström
Royal Institute of Technology, Department of Chemical Technology, S-100 44 Stockholm, Sweden

M. S. Sundaram
Process Sciences Division, Brookhaven National Laboratory, Department of Energy and Environment, Upton, New York 11973, USA

W. P. M. van Swaaij
Laboratory of Chemical Reaction Engineering, Twente University of Technology, PO Box 217, 7500 AE Enschede, The Netherlands

O. Theander
Department of Chemistry and Molecular Biology, Swedish University of Agricultural Sciences, S-750 07, Uppsala, Sweden

M. M. Thornton
Department of Chemistry, University of Washington, Seattle, Washington 98195, USA

J. H. P. Utley
Department of Chemistry, Queen Mary College, University of London, Mile End Road, London E1 4NS, UK

W. P. Walawender
Department of Chemical Engineering, Durland Hall, Kansas State University, Manhattan, Kansas 66506, USA

P. L. Walker Jr
Department of Materials Science and Engineering, Pennsylvania State University, 320 Steidle Building, University Park, Pennsylvania 16802, USA

T.-C. WANG
Department of Chemical Engineering, Arizona State University, Tempe, Arizona 85287, USA

W. WANZL
Bergbau-Forschung GmbH, Fran-Fischer-Weg 61, D-4300 Essen 13, Federal Republic of Germany

D. W. WATT
Energy and Environment Division, Lawrence Berkeley Laboratory, One Cyclotron Road, University of California, Berkeley, California 94720, USA

W. A. WILCOX
Battelle Pacific Northwest Laboratories, PO Box 999, Richland, Washington 99352, USA

R. A. YOUNG
Department of Forestry, University of Wisconsin, 1630 Linden Drive, Madison, Wisconsin 53706, USA

R. ZABRANSKY
Institute of Gas Technology, 3424 South State Street, Chicago, Illinois 60616, USA

S. H. ZERONIAN
University of California at Davis, Davis, California 95616, USA

1

Wood and Biomass Ultrastructure

W. E. Hillis
CSIRO, Division of Chemical Technology, Highett, Victoria, Australia

1. INTRODUCTION

The increasing populations of mankind with their expectations of rising living standards will cause shortages of energy and resources in the future. The shortages of large trees for conversion to wood for constructional and high quality uses will become a crucial problem if present patterns of forest use continue.[1] Serious shortages of fuelwood already exist in some global regions. Greater availability and more efficient utilization of trees and other biomass will be required to meet the demands.

The amount of small-sized and low-quality trees is significantly greater in some countries than has been documented by the traditional methods of forest inventory.[2] In other countries unrecognized and uncharacterized species exist and need to be considered for utilization. Thermochemical conversion could assist a more effective management and utilization of these renewable resources because species, size and the nature of trees and other biomass are not of major importance in this process.

From the viewpoint of energy conservation, wood will become increasingly important since it can be produced, processed and utilized at low energy costs and more cheaply than many of the alternative materials such as aluminium, steel and plastics. Consequently, it is prudent in the long run to recover as much of the fibre from biomass as is possible for these purposes before considering the use of biomass as a source of energy and chemicals. Integrated systems of processing will enable the selection of materials to be processed into the higher priority products, and those for energy and chemicals will be derived from the reject material.

Besides wood there are also lignocellulosic materials such as bamboo, cane and straw which can be used for pulp and building materials but their use for thermochemical conversion is less likely to be as important as wood. Seaweeds which contain cellulose and carbohydrates of a different nature to those in land plants but lack lignin could also be used for thermochemical conversion.

An understanding of the ultrastructure of renewable resources is necessary for effective use. In this paper ultrastructure refers to the anatomy, structure and the composition of biomass, which vary for different parts and species of living plants. The ultrastructure of the material derived from agriculture and forestry residues will frequently be a modification of that described in most references concerned with wood and plants. Because this type of material will be the raw material for thermochemical conversion emphasis will be placed on biologically modified wood, fast-grown wood, extraneous materials and heartwood.

2. FORMATION OF BIOMASS

Leaves are the primary organs of the plant, which perform two basic functions—first, the transpiration of water and the translocation of essential materials throughout the plant and, second, the manufacture of food materials. The rates of these basic functions are dependent on the amount of solar energy received. In the case of agricultural crops about 10% of this energy is converted into end-products whereas in trees about 1% is converted into woody materials.[3]

Leaves contain varying amounts of chlorophyll and carotenoids, residual carbohydrates, organic materials specific to their metabolic pattern, mineral salts and oils. Removal of leaves from the growth site can, in the long term, reduce the quality of the soil because essential nutrients are lost.

Large changes do not occur in the shape and thickness of leaves of dicotyledons after primary growth has ceased. On the other hand the basal meristem (with cell-division activity) of many monocotyledonous leaves (such as the grasses and reeds) continues for a long time after the distal portions have matured.

The primary stems of plants are covered with an epidermis often very similar to that of the leaf of the same species. The cortical tissues which may contain chloroplasts form the next layer, and some cells act as a physiological boundary between the cortex (outer layers) and the stele (central core). The vascular bundles translocating the sap form a ring just to the inner side of the cortex, and, to the inside of them, there is frequently a parenchymatous centre within the stem which in some plants breaks down to leave a canal (as found in wheat stem). Most dicotyledonous stems have nodes where leaves are attached, some of which may develop into branches.

The central ground tissue or pith in a number of plants is composed of cells that are usually parenchyma (living cells). The walls may be thin and composed largely of cellulose or thickened with lignin. This matrix of cells can contain specialized cells such as the latex cells found in members of certain botanical families.

Many monocotyledons and some dicotyledons have, to the inner side of the cortical tissues, a well-developed cylinder of sclerenchyma (thick-walled cells, often lignified, whose principal function is mechanical) in which are embedded the vascular bundles. Most dicotyledons lack such a cylinder; their vascular bundles are arranged in a ring of cambium, which is continuous in secondary growth (Fig. 1). The cells which make up meristems divide and produce new cells throughout the life of the tree. The apical meristem of the stem increases the height of the tree or length of the branches. The lateral meristem, or vascular cambium, completely encircles the stem or branches and is responsible for increases in diameter.

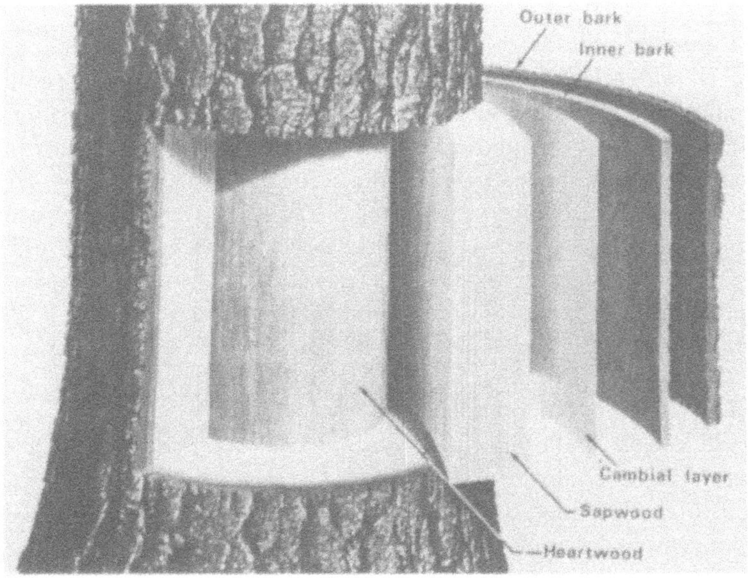

FIG. 1. Generalized structure of a tree stem showing orientation of major tissues including outer bark, inner bark, cambium, sapwood, and heartwood. (Courtesy of St. Regis Paper Co.)

During each growing season, wood cells are formed on the inside of the cambial layer and a smaller number of bark cells are formed on the outside of the cambium at the same time (Fig. 1). Immediately after cell division, the wood cell begins to enlarge in length and diameter, and the cell wall is formed. Finally, lignification commences at the cell corners, progresses through the middle lamella between the cells and then to the primary and the secondary walls (see Section 3). The entire development of fibres or tracheids from division to death takes 14–21 days. The parenchyma cells remain alive. There is a basic similarity in the arrangement of cells and cell wall structure in wood and most other land plants.

3. THE CELL WALL OF WOOD FIBRES

The cell wall is made up of a twisted skein of cellulose molecules bunched into elementary fibrils and ordered into microfibrils which are embedded in a matrix of hemicelluloses and lignins. These major components as well as the extractives and inorganic materials in cell walls and their lumens (cell cavities) have different thermal properties.[4]

3.1. Elementary Fibrils

There is evidence that each elementary fibril consists of 40 cellulose molecules held in close association by hydrogen bonding.[5–7] The cross-sectional dimensions are about 3·5 nm by a multiple of 3·5 or 4·0 nm (3·5 or $4{\cdot}0 \times 10^{-9}$ m) with the multiple varying according to the authority. In the elementary fibril, ordered crystalline regions of 300 nm length are separated by 'amorphous' regions where the molecules become less orderly (Fig. 2). It is at these amorphous regions where water may move between the elementary fibrils of cellulose molecules and where degradation by acids or by heat can occur.

3.2. The Nature of Microfibrils

The cellulose elementary fibrils of about 3·5 nm width are embedded in a matrix of hemicelluloses and lignins. (A range of lignins exists with compositions differing according to cell wall location and plant species. The conventional singular form of 'lignin' will be used here.) Different theories exist concerning the arrangements of the components of the fibre cell wall and the situation can be exemplified by the proposal of Fengel[6] (Fig. 2). In this model a cellulose elementary fibril is surrounded by a few layers of hemicelluloses which have considerably less crystallinity, lower

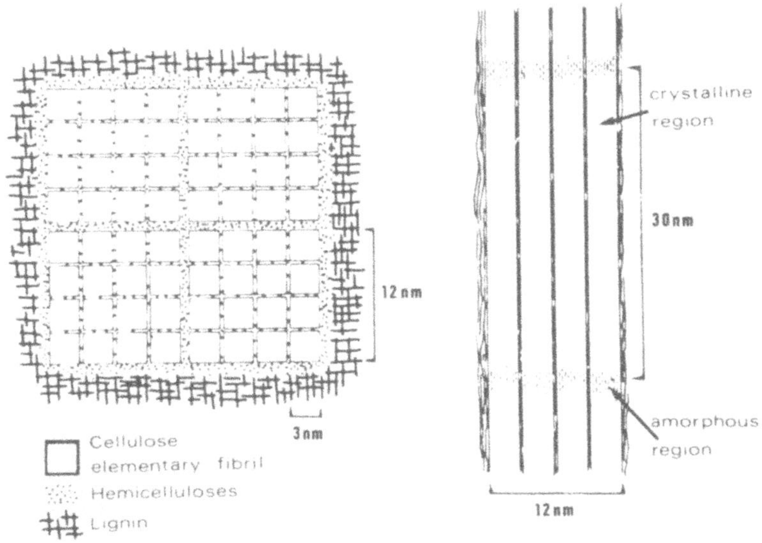

FIG. 2. Model of the ultrastructural organization of a microfibril in wood; cross-section on the left, longitudinal section on the right (after Fengel[6]). (Courtesy of John Wiley, Inc.)

molecular sizes and degree of orientation than cellulose. The hemicelluloses tend to align along the microfibril axis. The elementary fibrils are arranged in groups of 16, then into microfibrils containing a total of 64 elementary fibrils which are surrounded by hemicelluloses and lignin. The lignin is unaligned, and probably present in a range of molecular sizes as spherical microgels.

There are indications that in many cases the elementary fibrils are essentially crystals of cellulose I. However, in the conifers a small amount of a glucomannan hemicellulose seems to be associated with these crystals. It is assumed that the movement of water within the fibre wall takes place mainly at the interfaces containing hemicelluloses. At the surface of the microfibrils the hemicelluloses are in intimate contact with the lignin and it is in these places that chemical bonds between the two groups of compounds are to be found. The lignin hinders the enzymic degradation of the cellulose and hemicelluloses to sugars and alcohols.

3.3. Arrangement of Microfibrils in the Cell Wall

Plant cell walls are both chemically heterogeneous, being composed of several different substances, and physically heterogeneous as they are built up of layers. The wood cell wall consists of three morphologically distinct layers: the middle lamella, the primary wall and the secondary wall. The thin middle lamella, which is an amorphous cellulose-free layer between the primary walls of adjacent cells, becomes heavily encrusted with lignin. During the growth of the cell the thin primary wall is a flexible structure composed largely of cellulose but towards the end of growth it becomes considerably lignified. Towards the end of the growing phase the thick secondary wall begins to form in two or three layers in a process which continues after cell growth has stopped (Fig. 3).

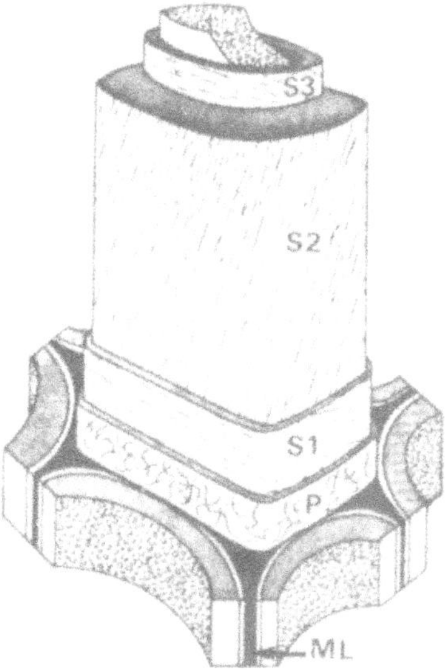

FIG. 3. The ultrastructure of the wood cell wall (ML = middle lamella; P = primary wall; S1, S2 and S3 = layers of secondary wall (after Côté[8]). (Courtesy of Springer Verlag.)

In the primary wall (about $0{\cdot}06 \times 10^{-6}$ m thick) the microfibrils are arranged in a multinet structure, and towards the end of the growth phase they tend to be oriented increasingly in a longitudinal direction. The secondary cell wall in wood fibres consists of microscopically distinct outer (S1), middle (S2) and inner (S3) layers. The middle layer is by far the thickest being $1{\cdot}5$–8×10^{-6} m with the other two about $0{\cdot}1 \times 10^{-6}$ m thick. The microfibrils in all layers are helically arranged but the pitch of the helices differs from layer to layer. In the S1 layer there are two or more helices of microfibrils which make large angles with the cell axis. The concentric lamellae of microfibrils arranged in helices in the S2 layer make a small angle with the axis. The inner S3 layer, which is sometimes absent, has a shallow pitched helical organization with several lamellae.[8]

Wood cellulose has a density of about 1·55, the amorphous hemicelluloses 1·50, and lignin a density of only 1·33[9,10] and results in a wall density of 1·474 for *Abies balsamea*.[11]

3.4. Porosity of Cell Wall

The framework of microfibrils in the cell wall in sapwood encompasses 'free' space containing water and other materials. The hydrated S2 layer of the tracheid cell walls in *Pinus resinosa* contains 25% free space.[12] The water in the capillaries of the wet walls of other woods has a volume of more than 40% (on a dry wood basis) with about half the water being in the free state. The free space in wet cell walls is in capillaries having cross-sections from 16 to 60 × 10^{-10}.[12–14] Consequently, molecules of significant size can enter and act as fillers or modify to a considerable extent some properties such as stability, durability and pulping characteristics. Inorganic salts penetrate the free spaces and remain as recognizable structures when the wood is carefully ashed. A dimensionally stable wood such as *Sequoia sempervirens* contains up to three-quarters of its extractives in the cell wall and penetration of extractives can influence the amount of absorbed water that is hydrogen-bonded to the microfibrils.[15,16] This water is additional to the free water in the capillaries which collapse when plant tissues are dried.

The small voids in the dry cell wall affect its packing density, which is a more relevant indicator of the rate of penetration than density. Packing density is higher in high density woods than in low density woods and higher in late and mature wood than in early and juvenile wood; packing density also appears to be characteristic for a species.[17,18]

3.5. Distribution of Constituents in the Cell Wall

Although the concentration of lignin in the middle lamella and primary wall can be over 50% by mass, depending where the measurement is made,[19] only 20–30% of the total lignin is found there; the remainder is in the much thicker secondary walls[20] where it is present in amounts of 20–25% (Fig. 4). The amount of lignin in the walls of ray cells and vessels is high.[21,22] For pine, the inorganic material is less concentrated in the S2 layer than elsewhere in the cell wall.[23]

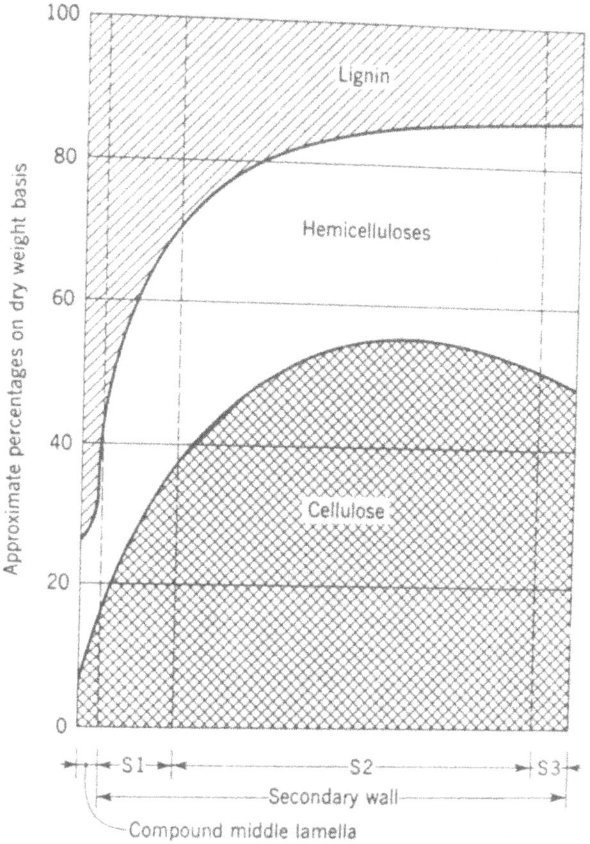

FIG. 4. Distribution of the principal chemical constituents within the various layers of the cell wall in conifers. From *Textbook of Wood Technology*,[28] Vol. I, by A. J. Panshin and C. de Zeeuw. © 1970. Used with permission of McGraw-Hill Book Co.

4. GROWTH RINGS

A new layer of wood is formed during each growing season between the bark and the stem of trees over the entire length of stem. The demarcation of these growth layers or growth rings results from differences in temperature, moisture availability, the growing and non-growing seasons and also in tree species. In most cases growth rings are formed annually (Fig. 5). In tropical and sub-tropical regions, the growth rings can be

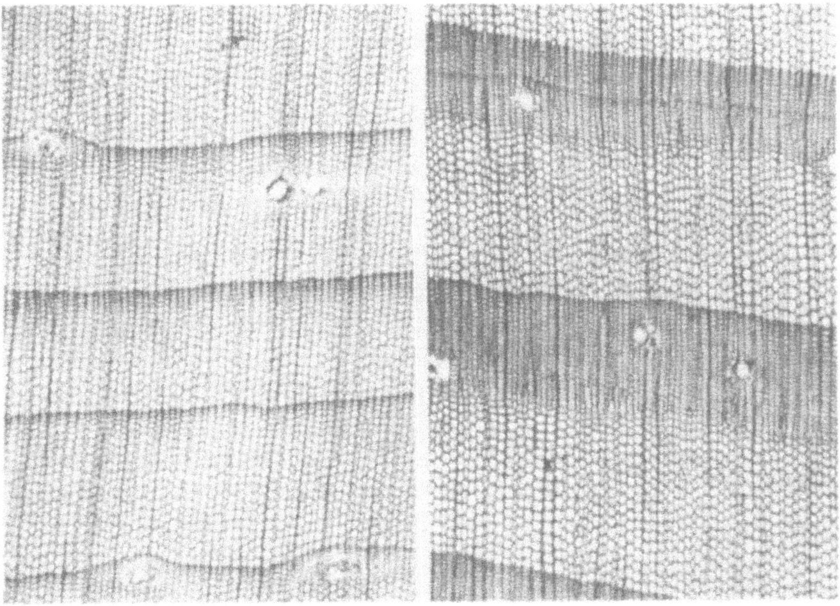

FIG. 5. Variations in transition from earlywood to latewood in *Pinus* spp. White pine (*P. strobus*, left) and loblolly pine (*P. taeda*, right) (about × 30). Note large diameter, vertical resin canals.

indistinct. In dry regions with relatively high temperatures and intermittent rainfall several rings can be formed annually.

Growth rings are usually distinct due to differences in the wood produced early in the growing season (i.e. earlywood or springwood) as opposed to that formed late in the growing season (latewood or summerwood). Earlywood has the thinnest cell walls and large lumens whereas latewood with its small lumens and thick walls may appear as a dark zone. There are variations in the contrast between earlywood and latewood, the relative amounts of them and the abruptness in which earlywood stops forming. These variations depend on the species, the differences in thickness of the cell wall, and the age of the tree when the growth ring was formed.

Dry latewood has a higher specific gravity than dry uncollapsed earlywood. However, in undried sapwood, the earlywood contains large volumes of water sometimes up to 4 or more times greater than that in the latewood due to the large lumens in the earlywood.

Some fast growing species, such as *Anthocephalus* spp., produce high volumes of wood but contain large proportions of cells with thin walls and large lumens which, when occurring in sapwood, are filled with water. Consequently water must be removed from materials such as this by drying or compression in order to reduce transport and conversion costs.

5. SOFTWOODS AND HARDWOODS

5.1. Description

The two major groups of trees—softwoods (conifers, gymnosperms) and hardwoods (angiosperms)—are distinguished by the presence or absence of a cover on the seed. Whereas there are only about 40 genera and 600 species of softwoods there are thousands of genera of hardwoods with widely differing anatomical structures. Frequently many species of hardwoods grow together in forests with a considerable proportion of the species being commercially under-utilized. The average oven-dry specific gravity of softwoods is about 0·36 with a range of 0·25–0·60 compared with 0·50 and a range of below 0·30 to above 0·80 for hardwoods.

5.2. Distribution

Much of the wood available globally is from hardwood species. In the USA the softwood volume exceeds the hardwood volume with almost 75 % of the volume of softwoods being produced in the western states.[24] The hardwoods are concentrated in the eastern and southern states where, in many areas, low quality hardwoods are increasing in volume.

5.3. Major Anatomical Differences between Softwoods and Hardwoods

The fibre is the major single structural element in wood. In softwoods, fibres serve also as the major translocation route for sap, and are interconnected by openings known as bordered pits. Because of this characteristic the fibre element of softwoods is defined as a tracheid. In hardwoods the fibre elements are usually thick-walled with simple and slit-like pits. The characteristics of these elements vary within the tree, the species and between species. Tracheids and fibres are considerably longer than they are wide, with ratios of sometimes 100 to 1. The length of tracheids range from about 1·5 to 9·5 mm and usually in commercial softwoods it is 3–4 mm. The diameters range from about 20 to 80 × 10^{-6} m. The length of hardwood fibres ranges from about 0·5 to 1·8 mm and the diameter from about 10 to 25 × 10^{-6} m. The density of wood is strongly influenced by the average fibre or tracheid wall volume, namely $(F^2 - L^2)/F^2$ where F = fibre diameter and L = lumen diameter for fibres circular in cross-section.

FIG. 6. A softwood Douglas-fir (*Pseudotsuga menziesii*) as viewed with the scanning electron microscope (about ×80), showing earlywood and latewood tracheids and rays. (Courtesy of W. A. Côté.) Scale distance 100×10^{-6} m.

The major anatomical difference between softwoods (Fig. 6) and hardwoods is the presence of vessels (or pores) in the latter (Fig. 7). The vessels, through which the sap is conducted in the living tree, have a very wide range of size, shape and arrangement. In ring-porous hardwoods the vessels in the earlywood portion of a growth ring have different sizes from those in the latewood. In diffuse-porous woods the vessels have the same size at all locations in the ring. A vessel consists of individual vessel elements stacked one on top of the other, and it 'wanders' about the vertical

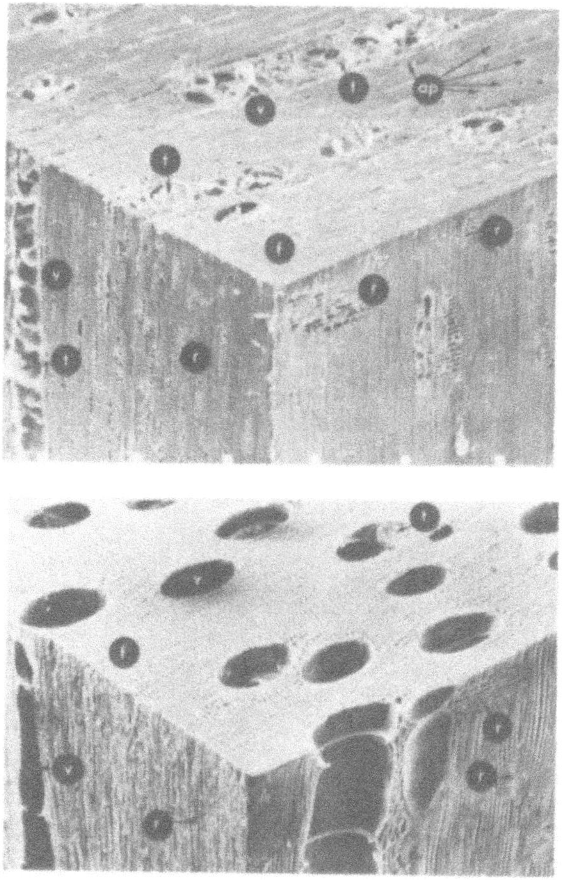

FIG. 7. Scanning electron micrograph of a small cube of the hardwood *Eucalyptus citriodora* (top) and *E. regnans* (bottom) showing rays (r), fibre tracheids (f), vessels (v), tyloses (t), axial parenchyma (ap), (unit distance 300×10^{-4} m).

axis of the stem, for a length up to about 3 m. The length of the vessel elements varies from 0·18 to 1·3 mm and the width from 20 to over 250×10^{-6} m. For a number of hardwoods the vessel diameter increases and vessel frequency decreases as the distance from the pit increases. The frequency of vessels per mm^2 varies according to the species, the size of the vessels, the position in the growth ring, the distance from the pith and growth conditions. The walls of vessels are thin (about $3–4 \times 10^{-6}$ m), relatively unchanged in thickness from pith to bark.

5.4. Parenchyma

On the cross-sections of woody stems, the ray parenchyma extend as lines from the pith outwards to the cambium and are seen as lines between the fibres. In conifers and some hardwoods, the rays are 1–2 cells wide and 10–20 cells high. In other hardwoods the rays can be up to about 30 cells wide, sometimes hundreds of cells high and are tapered on their upper and lower edges.

Vertical parenchyma like the rays are also thin-walled cells and are interspersed between the fibres and arranged diffusively or surround the vessels in various patterns. In contrast to the fibres, the ray and vertical parenchyma (see Figs 6 and 7) remain alive across the sapwood zone. In the sapwood they translocate and store primary metabolites such as starch or fats; in the heartwood they frequently contain large amounts of extractives.

5.5. Resin Canals

Coniferous wood contains resin canals (Fig. 6) of about the same size as some vessels. They exist as anastomosing networks surrounded by epithelial cells which form and exude resin into the canal. Frequently, the resin is under pressure so that when the canals are cut in the living tree the resin exudes.

5.6. Pits and Tyloses

When the pits, or openings in the cell wall, between the ray parenchyma and vessels are more than 10×10^{-6} m diameter, the parenchyma can form thin-walled, balloon-like extrusions—the tyloses—into the vessels (Fig. 7). When the pit diameters are smaller the rays secrete extractives. These changes occur when heartwood is formed or when sapwood is injured. In conifers the bordered pits also aspirate during the transition of sapwood to heartwood. All these changes hinder the movement of fluids either into or out of the wood.

5.7. Proportion of Non-Fibrous Elements

Whereas the proportion by volume of tracheids in conifers is usually over 90%, the proportion of fibres to the thin-walled elements in hardwoods is much lower (Table I). The proportion of the latter affects the water content, the density and the compressibility of the material. In addition, the lignin content of the thin-walled, non-fibrous elements is higher than that of the fibres or tracheids.[25,26]

TABLE I

PROPORTION OF TISSUE ELEMENTS IN VARIOUS WOODS*

Species	*Percentage of total volume*			*Vessels*
	Longitudinal fibres/tracheids	*Parenchyma*		
		Ray	*Axial*	
Betula papyrifera	75·7	11·7	2·0	10·6
Erythrina vespertilio	16	80		4
Eucalyptus camaldulensis	58	25		17
Juglans nigra	48·7	16·8	13·5	21·0
Liquidamber styraciflua	26·6	18·3	0·2	54·9
Picea abies	94·1	5·9	—	—
Pinus strobus	93·0	6·0	—	—
Quercus rubra	41·3	15·9	23·4	19·5
Salix nigra	54·4	6·1	2·2	38·1
Ulmus americana	34·7	11·3	6·0	48·0

* Selected from refs 27 and 28.

5.8. Chemistry

The cellulose content of both softwoods and hardwoods is normally in the range of 42 ± 2 % (on extractives-free basis). The hemicelluloses and lignin are found in complementary proportions and the woods with a high lignin content have a low hemicelluloses content. Generally, hardwoods (such as poplar) contain less (16–24 %) lignin than softwoods (24–33 %). Variable amounts of acid-soluble lignin affect the method of analysis so that the total lignin content of hardwoods can be 19–28 %. Ipil-ipil (*Leucaena leucocephala*), which is a fast growing tropical hardwood with a specific gravity of 0·54, consists of lignin 23 %, polysaccharides 72 % and ash 0·8 %.[29]

In addition, the composition of these major components differs between and within the two types of wood so that the predominant hemicellulose in hardwoods is a glucuronoxylan and in softwoods a galacto-glucomannan. For example, the polysaccharides of the extracted heartwood of the hardwood cottonwood (*Populus trichocarpa*) consist of 44·1 % cellulose, 20·8 % 4-O-methyl glucuronoxylan and 9·7 % glucomannan.[30] The major basic aromatic component isolated from softwood lignin is vanillin whereas hardwood lignin yields varying proportions of vanillin, syringaldehyde and *p*-hydroxybenzaldehyde. Different lignins occur in the fibres, vessels and ray cells of hardwoods.[11,31] Also, the lignin in the secondary wall of softwoods contains twice as many phenolic hydroxyl groups as in the middle lamella.[32]

The heating value of lignin is about 1·5 times that of cellulose and hemicelluloses, and the heating value of some extractives is even higher. The degree of inflammability is affected by the morphology of the wood as well as its composition.[33]

6. WOOD FORMED UNDER ABNORMAL CONDITIONS

Responses to abnormal growing conditions result in reaction wood with ultrastructures markedly different from those of normal wood. Reaction wood is frequently formed in leaning trees and bent stems. Because of the low density of juvenile wood and the flexible nature of fast-grown trees under windy conditions, there is frequently a higher proportion of reaction wood in the central portions of these stems when compared with the outer layers of mature trees.[34] In addition, stressed wood can be formed along the periphery of trees, particularly hardwood trees. This results when the crown seeks the optimum opening in the forest canopy so that the tree leans in a certain direction.

Leaning trees and branches that are transversely eccentric in outline often contain reaction wood in the regions of accelerated growth. In softwoods this is found in the underneath side of the trunk in the form of compression wood whereas in hardwoods it is on the upper side and is known as tension wood.

Stressed wood, and particularly reaction wood, has inferior properties for conversion into, or for use as, solid wood products. Also its presence reduces the yield of high quality pulps. Where possible these woods are rejected from high quality uses and would form the raw material for the thermal and chemical industries.

6.1. Macroscopic and Microscopic Descriptions

6.1.1. Compression Wood

This wood is readily detected in freshly cut log cross-sections (Fig. 8). Its colour varies from a pale brown to a dark reddish brown, but is generally much darker than the colour of normal wood of the same species. Unlike tension wood, pronounced compression wood can occur over large areas and contain no normal tracheids. The most characteristic anatomical features are the typical rounded appearance of the tracheids in cross-section, the presence of numerous intercellular spaces between individual cells and numerous radial checks in the cell wall. Compared with normal wood, compression wood contains shorter fibres with thicker walls, and a

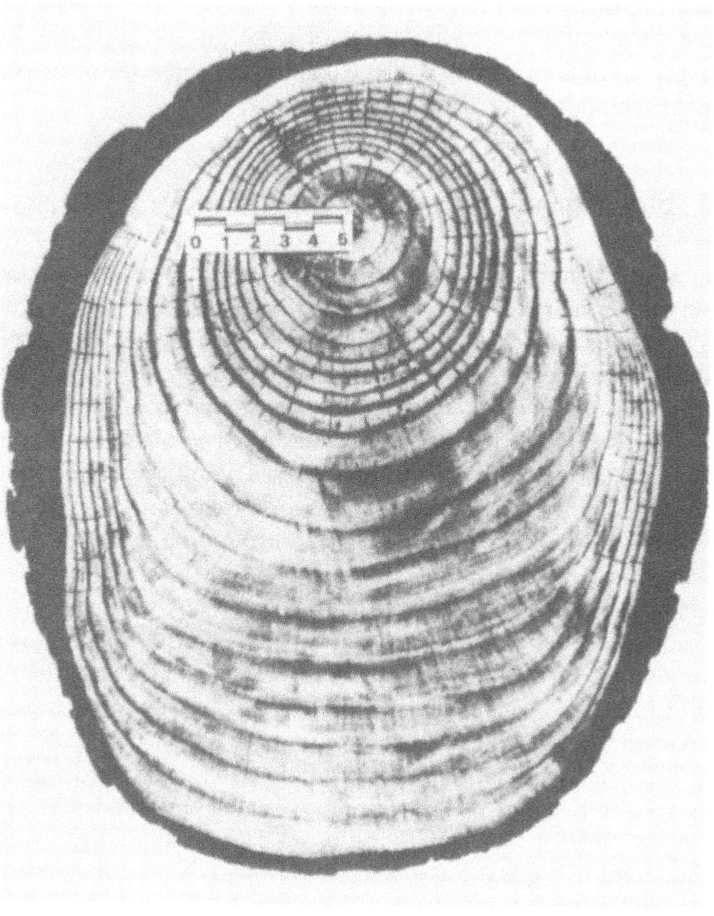

FIG. 8. Cross-section of a log of *Pinus radiata* taken from a tree that had been blown over after approximately 10 years of normal growth—note the excessive development of compression wood on the lower side (scale in cm).

greater proportion of latewood in growth rings. Compression wood tracheids have only two layers in the secondary wall, an S1 and a modified S2, with the microfibrillar angle being much larger than in normal wood.[11]

6.1.2. *Tension Wood*

Tension wood is not as conspicuous as is compression wood in log cross-sections. However, eccentricity of growth and wider growth rings are often indications of its appearance. Tension wood often forms in bands which, compared with neighbouring normal wood, are usually darker in colour, harder and denser. The main anatomical features are fibres with a light-refractory, inner layer often termed the gelatinous layer. The layer is thick in beech (*Fagus* spp.) and thin in poplar (*Populus* spp.). There is also a marked reduction in size and number of vessels, and a higher average wood density. A number of layers exists in the cell wall of tension wood, but the number varies according to growth conditions: the S1 is always present, and the S2 is usually present but its thickness depends on the severity of tension wood formation. The gelatinous layer, which is usually loosely attached to the wall, is highly crystalline cellulose with microfibrils running parallel to the cell axis; it is not lignified. The presence of the gelatinous layer can increase the specific gravity of tension wood up to 30 % above that of normal wood.[35,36]

6.1.3. *Wood Formed under Load Stress*

Peripheral growth stress can reach high levels in significant portions of the outer sapwood of hardwoods. The wood formed in those regions has an increased thickness of cell wall, a greater microfibrillar angle in the S2 layer and a higher intensity of lignification.[37,38] The high peripheral stress results in high compressive stresses in the interior of the tree so that compression failures occur in the fibres and the wood becomes brittle ('brash').

6.2. Chemistry

There is a general pattern for both hardwoods and softwoods in that the wood on the lower side of any cross-section of a leaning or sloping stem has a higher lignin content and a lower cellulose content. There are gradations in composition and properties between severe reaction wood and normal wood within the one tree. The wood on the opposite side of the stem to reaction wood also exhibits abnormal physical properties and chemical composition. It has a lower lignin content and higher cellulose content than the side wood, which is regarded as 'normal' wood and used for comparison.

6.2.1. *Compression Wood*

This wood can have substantially higher lignin (30–40 %) and galactan (10 %) contents and lower cellulose (20–25 %) contents than those in comparable normal wood with 30 %, trace and 42 % respectively. The cellulose in compression wood has a lower degree of crystallinity than that in normal wood. The galactoglucomannan content is only 9 % and half that found in normal wood. Also, the chemistry of the compression wood found in juvenile wood appears to be different from that found in adult wood.[11]

Because of the higher proportion of lignin, the cell wall density of compression wood is less than that of normal wood. On the other hand, because of the small proportion of earlywood and the thick tracheid walls, the gross density of compression wood is much higher and almost twice that of normal wood. The specific gravities, calculated on the basis of green volume and dry weight, of normal and compression woods for different conifers are 0·29 and 0·63 for *Abies balsamae*, 0·36 and 0·64 for *Picea* spp., 0·32 and 0·61 for *Pinus sylvestris*, 0·36 and 0·61 for *Pseudotsuga menziesii*, and 0·35 and 0·58 for *Tsuga canadensis*.[11]

The extra lignin occurs in the outer part of the S2 layer in amounts of 54–70 %. About 90 % of the lignin in compression wood tracheids occurs in the secondary wall, compared with 75–80 % in normal wood.[11] The lignin of compression wood has a low degree of methylation.[39]

6.2.2. Tension Wood

Tension wood contains higher percentages of cellulose (up to 50 % more) and lower percentages of lignin and pentosans than normal wood. The nature of the lignin is the same in both normal and tension wood.[40] The amounts of glucomannan, galactan and acetyl groups in tension wood differs from that in normal wood.[40] The cellulose in the gelatinous layer is more highly crystalline[35] and has a larger microcapillary system than cell walls in normal wood.[36]

7. SAPWOOD AND HEARTWOOD

7.1. Volume of Sapwood

Sapwood is the outer zone and 'the portion of the wood that, in the living tree, contains living cells and reserve material (e.g. starch)'.[41] Fibres and tracheids die as soon as the lignification phase is completed. On the other hand, the parenchyma remain viable for many years or until heartwood is formed or the wood is damaged.

The amount of sapwood in a mature tree can vary considerably, occupying the whole tree or only the outermost growth rings which surround an approximately conical shaped central core of dead heartwood cells. This variation in width occurs regardless of genera, species and location in the tree. Sapwood forms, in red oak (*Quercus rubra*), a band of constant width of 1 cm or 5–6 growth rings around the heartwood, in *Pinus* spp. 15–50 growth rings, and, of varying widths, in maple (*Acer* spp.) over

100 growth rings. The width of sapwood in a cross-section of a stem of a tree is not always uniform and the number of sapwood rings can be greater at the stump than higher in the tree. Also with some trees of mature or over-mature age the width of the sapwood, and particularly the number of growth rings in it, is much greater than in young trees of the same species.[42]

7.2. The Contents of Sapwood

Most sapwood tissues in the living tree are involved in the translocation or storage of sap. The free water content of sapwood is determined by the lumen area of the elements after allowing for stored metabolites. The water content of sapwood can be high and over 200 % (on dry weight basis) in some cases; for softwood sapwood it is usually between 110 and 170 %, whereas for hardwood sapwood it is between 60 and 110 %.[43] Usually, the water content is maintained across the sapwood but decreases abruptly at the heartwood periphery. In woods such as beech (*Fagus sylvatica*) a steady loss of water takes place across the stem and at the sapwood–heartwood boundary 60 % of water is present.

Starch, fatty acids or esters of fatty acids are formed exclusively in the parenchyma, where they are stored as reserve material for growth or for extractives formation. Examples of starch containing trees are maple (*Acer* spp.), chestnut (*Castanea* spp.), beech (*Fagus* spp.) and oak (*Quercus* spp.). Fats are the storage material in the sapwood of trees of pine (*Pinus* spp.), spruce (*Picea* spp.) and poplar (*Populus* spp.). Fats and starch are found in birch (*Betula* spp.), cherry (*Prunus* spp.) and willow (*Salix* spp.). The amounts of these stored metabolites depend on the species, the previous growth history of the tree, growth stage and location in the sapwood. In some species the amount of starch may reach levels of over 5 %.

The amounts of proteins and other organic materials are low. The nitrogen content seldom exceeds 0·3 % and is usually less than 0·1 % of the dry weight of the wood. The nitrogen content decreases progressively from the cambium to the heartwood periphery. The pith contains more nitrogen by weight than heartwood and in some cases than sapwood.[44]

Inorganic inclusions of different types are common and in most cases are present in larger amounts than in the heartwood. Spherical conglomerates of silica are formed in the middle of the sapwood of *Syncarpia glomulifera* and other species.[45] Silica is present in a range of species[46] and usually in amounts of under 2 %. Calcium oxalate crystals can sometimes have characteristic shapes and are often found in hardwoods but rarely in softwoods. Calcium carbonate deposits are found in the vessels of a number of hardwoods.

7.3. Heartwood

In most growing trees or shrubs the inner layers of the stem, branches and roots cease to contain living cells and the reserve metabolic materials are no longer present. This heartwood zone contains higher, sometimes much higher, amounts of extractives than in the sapwood and their composition is usually different. The zone is generally dark with a colour range between light yellow and dark brown. Once the zone begins to form it is spatially continuous.[47]

Less moisture is usually found in heartwood than in sapwood, and the decrease at the transition zone can be abrupt and considerable, coinciding approximately with the change in colour. This decrease in moisture content is characteristic of softwood species and most hardwoods. However, the heartwood can contain more moisture than the sapwood in some hardwood species such as *Populus* (poplar), *Ulmus* (elm), *Fraxinus* (ash), *Carya* (hickory) and *Eucalyptus*[43] and in some cases this may be due to the formation of wetwood (see Section 9.2). The moisture content in the heartwood of softwoods is usually between 33 and 95% whereas in hardwoods it is usually 60–100%.[43]

Together with loss of water on heartwood formation, a marked aspiration of the pits between tracheids restricts the transport of fluids. When heartwood is formed in many hardwoods tyloses appear in the lumen of vessels (Fig. 7). The number of vessels blocked by tyloses depends on the species and genera. Aspiration of pits and tyloses formation also takes place in sapwood in response to wounding and invasion by microorganisms and to drying.

Extractives are responsible for the distinctive colour of heartwood and are present in the parenchyma, vessels and some other cells, but also in the cell wall of the tissues. Extractives can hinder the movement of aqueous fluids. The amount of extractives increases from the pith to heartwood periphery, and abruptly decreases in the sapwood. Large amounts can be present in the outer heartwood, e.g. over 30% of polyphenolic extractives in eucalypts or up to 29% of an arabinogalactan in *Larix* spp.[47]

The factors affecting the amounts of extractives present include the lumen volume in the wood, the amount of photosynthates and metabolites remaining from primary activities and the activity of the parenchyma at the time these are available. Usually, fast-grown trees of the same species contain significantly lower amounts of extractives than do the slow-grown counterparts. The maximum amount of extractives that can be formed in trees of certain species appears to be under genetic control. The composition of extractives in the heartwood depends on the genus. In some instances the composition is specific even at species level and can be used taxonomically for characterization of genera and species.[47]

Part of the inorganic constituents of wood is located in the cell wall but larger amounts are translocated from the heartwood during its formation. A partial selection of certain metallic ions occurs during this translocation. For example, when heartwood is formed in *Eucalyptus grandis* the phosphorus content drops from 0·0103 to 0·0007% and the potassium content from 0·1087 to 0·0238%.[48] In woods of the temperate zones, the ash content seldom exceeds 1% or drops below 0·2%. In tropical species, the ash content can be 3–4% or higher. The amount in a particular species can vary between sites of different nutritional status, and most of the ash is calcium carbonate.[49]

The total mineral content increases from unblemished to discoloured wood and again to decayed wood. Potassium concentrations can increase 400–4000%, calcium significantly and other elements in a less pronounced manner.[50-52]

The major classes of organic extractives in heartwood are:

(1) Polyphenols which are present in all woods although in some cases the amount may be very low. The flavonoids form the major class and are largely present in polymerized forms. Other important classes include those based on gallic acid and lignans.

(2) Resins are present mainly in the softwoods and include terpenes, resin acids (which are found also in birch, *Betula* spp.) and sterols.

(3) Fats exist as free fatty acids and their esters usually in softwoods but also in hardwoods such as poplars (*Populus* spp.), birch (*Betula* spp.), blackgum (*Nyssa* spp.) and sweetgum (*Liquidamber styraciflua*).

(4) Tropolones. Members of the Cupressaceae such as western red cedar (*Thuja plicata*) and other cedars (e.g. *Chamaecyparis* spp.) are the only tree species containing extractives based on this seven carbon ring structure. They readily react with copper and iron.[53]

8. JUVENILE WOOD, BRANCH AND ROOT WOOD

Natural stands and plantations of both pines and hardwoods are being harvested at increasingly younger ages, and, in addition, greater use is being made of the formerly non-merchantable tops and branches. Also, there is the development and use of short-rotation, intensively grown crops of fast-growing species. These raw materials contain increasing amounts of reaction wood and juvenile wood.

The age of the cambium at the point of wood formation determines whether juvenile or mature wood is formed and the former extends 6 to 15 rings outwards from the pith. At these rings the juvenile wood has gradually graded into mature wood. It comprises the major portion of the wood formed in the first 20 years of growth at a particular level in the tree.[34] Juvenile wood also contains a high proportion of earlywood and with increasing distance from the pith the fibre length, density and cell wall thickness increase steeply in the first few years and more slowly in the mature wood. The fibre length and wall thickness of topwood is intermediate between that of juvenile and mature woods.

The moisture content varies widely with tree age and with location of the sample in the tree. The considerable drop of moisture content with age is due to an increase in density and to the presence of heartwood. The increase with height in the tree occurs with both juvenile wood from 110% to 145% at the 30 ft level and mature wood from 74% to 117% over the same increase in height.[34] The amount of extractives is lower in juvenile wood.

The few reports of the differences in ultrastructure of branch wood and root wood in comparison with stem- or topwood indicate a higher proportion of reaction wood and a lower proportion of heartwood.

Compared with the stem, roots have higher bark-to-wood ratios, a smaller number of fibres in the xylem and in the gymnosperms, wider and longer tracheids, larger vessels (when present) of a more uniform size but sometimes smaller in number, and a higher proportion of sapwood.[54]

9. IRREGULAR FEATURES

9.1. Knots

Various types of knots affect the quality of wood in different ways, and knots are often rejected. The knots of conifers become highly resinified whereas with hardwoods the branch stubs are not filled with extractives so that an entry point for insects and fungi is provided, leading to widespread discoloration and decay.

9.2. Wetwood

In freshly cut stems of some species, the small or large zones of wetwood (when present) are usually recognized by a darker or wetter appearance. They contain more water than the sapwood zone, and sometimes a gas, largely carbon dioxide or methane, can be present. The pH and density are higher than those of the surrounding tissues, and the occurrence of mineral stains or deposits, often of calcium carbonate, are frequently greater. Wetwood frequently results from the action of microorganisms which enter through wounds or branch stubs.

9.3. Discoloured and Decayed Woods

Some types of discoloured woods can be confused with heartwood because of their location and colour. They are, however, part of a sequence of damage to the living tree which proceeds eventually to decayed wood. The zones are irregular in shape, colour intensity and location within the stem; sometimes they occur as coloured tubes within the stem. Often they occur longitudinally for a short distance as zones on one side of the tree. If the tree is vigorous, and the wounds are not severe the sequence to decay is not completed. In other situations the different stages to decay take place when different microorganisms cause further changes. When this occurs along with discoloration of the tissues, the pH and the moisture content increase and inorganic materials accumulate selectively, so much so that a 9-fold increase in calcium has been recorded. With decay there can be either destruction of the carbohydrate or the lignin or both portions of the cell wall.[47,55,56]

In addition to deterioration of wood in the living tree, loss of wood substance can occur on storage of undried wood chips at 50 °C for several months and for shorter periods as the temperature rises. There is an increase of hot water soluble material (from about 5 to 10·5 %), a decrease in pH (from about 6 to 3·2) and loss of acetyl content.[57] When chip piles become affected by microorganisms even greater changes take place.[58]

9.4. Resinified Wood

The interiors of old pine trees frequently contain resin-enriched areas. Similar areas can be formed by injection into the sapwood of the herbicide 'Paraquat' (1,1′-dimethyl-4,4′-bipyridinium ion), or ethylene-releasing compounds or both.[59]

9.5. Exudates

Cambial injury of many trees, particularly during the growing season, can result in the formation of organic materials which are confined to cavities in the wood or which exude on to the bark of the tree. The amount of the exudate depends on the nature of species and its conditions of growth. Conifers exude resins of the terpenoid type, eucalypts exude kino composed of flavonoids, the *Prunus*, *Acacia* etc. genera yield carbohydrate gums.

10. BARK

The stems of woody plants are protected from injury and desiccation by an outer layer of bark (rhytidome) consisting of areas of dark, porous tissue (dead phloem). This layer, which in some trees is shed regularly, covers the narrow, light-coloured region, the phloem, next to the vascular cambium; the phloem contains the living cells translocating and storing food. The number of phloem cells formed during a given year is much smaller than that of the xylem or woody cells. The inner and outer bark commonly form 10–20 % of the cross-sectional area of trees about 50 cm in diameter.[60] This proportion varies throughout the tree, with age of tree, and between species and genera. The bark thickness of southern pines varies according to log diameter, locality and height in the tree for which relationships have been derived.[61]

Although bark structure can be characteristic for trees belonging to one genus it sometimes varies greatly within the genus. The outer layer or periderm in conifers consists of three layers of cells: the phellogen which produces phellem or cork cells to the outside and phelloderm cells to the inside. The phellem cells are generally highly impregnated with suberin, the phelloderm are thin- or thick-walled and blend into the thin-walled parenchyma of the phloem. In softwoods the sieve cells are similar in size and shape to wood tracheids. In hardwoods the sieve tube elements resemble small-diameter and relatively long vessel elements. The sieve cells and tube elements are usually thin-walled and collapse by the time they are incorporated into the rhytidome. The bark fibres are thick-walled, long and straight and, in some eucalypt barks, consititute one-quarter of the bark volume. There are also short sclereids often called stone cells and these as well as the fibres become lignified. Other cells generally remain unlignified and collapse when compressed radially.[61–67]

Considerable variation exists in the composition of barks with varying compositions of carbohydrates and 'lignin'. Polyphenols can exist in large quantities in some barks such as those of some pines (*Pinus radiata*), wattle (*Acacia mearnsii*) and mallet (*Eucalyptus astringens*). A range of other extractives such as resins and fats can also be present. The bark of western hemlock (*Tsuga heterophylla*) contains 20·5 % cellulose, 16·2 % hemicelluloses, 15·2 % lignin, 47·8 % total extractives and 0·3 % ash.[68]

The moisture content of fresh bark varies according to its nature, proportion of phloem, position in the tree, between trees and possibly with the time of the year. For example, the inner bark of southern pines have moisture contents between 150 and 300 % whereas the outer bark contains 18 to 63 %.[61]

Even though the bark can constitute 25 % of stem weight, it may contain nearly 70 % of the inorganic nutrients in the stem.[60] The amount of inorganic material in the bark varies significantly within and between species and genera, particularly in young trees. In some cases the concentration increases with age and most elements reach maximum concentration in the fall.[69,70] The amounts of various elements vary between species and genera, and calcium salts predominate. In the *Eucalyptus* genus, those species with decorticating and smooth barks have, after incineration, higher ash levels of 6·0–12·7 % than those with fibrous barks with 0·5–1·2 % ash.[49]

11. BIOMASS FROM SOURCES OTHER THAN TREES

The biomass produced by the gymnosperms and the angiosperms with their life spans of many years is more conveniently handled industrially than the less developed plants, with their bulk, and difficulties in storage. There are, however, very large volumes of biomass in plants such as the dicotyledonous vines, herbaceous dicotyledons and the herbaceous monocotyledons.

Those materials receiving attention are the quick-growing kenaf, elephant grass or Napier grass (*Pennisetum purpurem*), bagasse (the fibrous residue remaining after crushing sugar cane), bamboo (various species of *Bambusa*) found in extensive areas in tropical regions, and cereal straw.

11.1. Kenaf

Kenaf (*Hibiscus cannabinus*) is a herbaceous annual grown in regions of warm temperatures and good rainfall and reaching a height of 3–6 m with stem diameters of 15–25 mm (with a biomass production of 15–34 tonnes (dry wt)/ha/yr). The stalks have a solids content of only about 20 % with a fibrous bark (about 35 % by weight) and a wood that is largely made up of partly-lignified parenchymatous cells. The plant contains significant amounts of oils, proteins and other materials which are readily expressed on compression.[71,72]

11.2. Cereal Straws

The strands of conducting tissues (vascular bundles) in the monocotyledonous stems of the Gramineae are arranged in two basic ways: first, the vascular bundles are arranged in two circles with the outer circle consisting of thin bundles and the inner of thick bundles. Such an arrangement is found in the cereals *Triticum* (wheat), *Hordeum* (barley),

Avena (oats) and *Oryza* (rice); second, the vascular bundles are scattered through the entire cross-section of the stem as in *Saccharum officinarum* (sugar cane), *Zea* (maize) and *Bambusa* (bamboo). The individual bundles are collateral and usually surrounded by a supporting tissue of fibres.

The parenchyma cells of the pith of many cereal species often mature early and stop growing while the surrounding tissues continue to enlarge longitudinally and in circumference, with the result that a hollow stem forms between the nodes.[64]

The chemical compositions of straws differ from those of wood (Section 5). Wheat straw contains about 40% cellulose, 28% hemicelluloses, 17% lignin, 11% extractable material and 7% ash. Rice straw contains these materials in the following percentages 30, 24, 12, 18, 16 respectively.[4]

11.3. Bamboo

Bamboo groves in the tropical and sub-tropical regions are mainly *Phyllostachys*, *Dendrocalamus* and *Bambusa* spp., and there are also dwarf bamboos of *Pleioblastus*, *Sasa* and other spp. growing in the colder regions. The bamboo culm consists of about 50% parenchyma cells, in which the 40% fibres and 10% conducting tissue (xylem and phloem) are

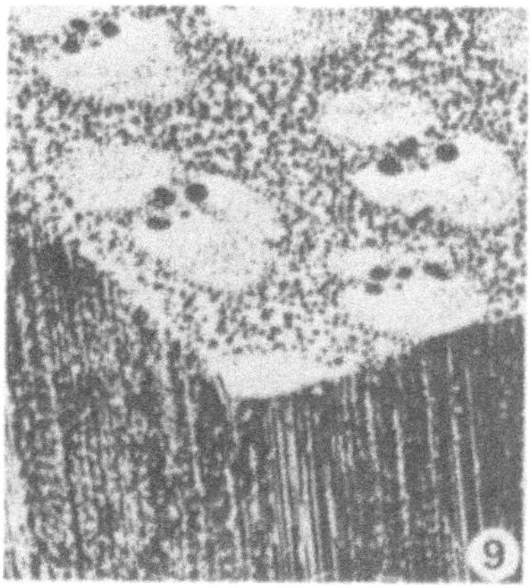

FIG. 9. Scanning electron microscope of an internode of *Cephalostachyum pergracile* showing vascular bundles and parenchyma distribution (×35) (from ref. 73). (Courtesy of Springer Verlag.)

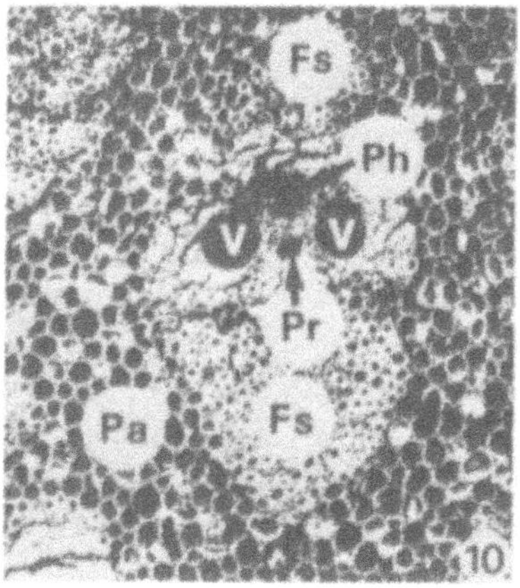

FIG. 10. Vascular bundle of *Melocanna bambusoides*. Metaxylem vessels (V), protoxylem (Pr), phloem (Ph) and fibre strands (Fs) embedded in ground parenchyma (Pa) (×90) (from ref. 73). (Courtesy of Springer Verlag.)

embedded (Figs 9 and 10). (Sugar cane has a similar arrangement.) The major portion of the fibres, which occur as bundles or isolated strands, do not possess any obvious layers in their walls. However, thick-walled fibres are characterized by lamellated walls with narrow and broad zones with different orientation of microfibrils (Fig. 11). The narrow lamella have higher lignin contents than the broad lamella and also the middle lamella between the fibres has a high lignin content particularly at the cell corners.

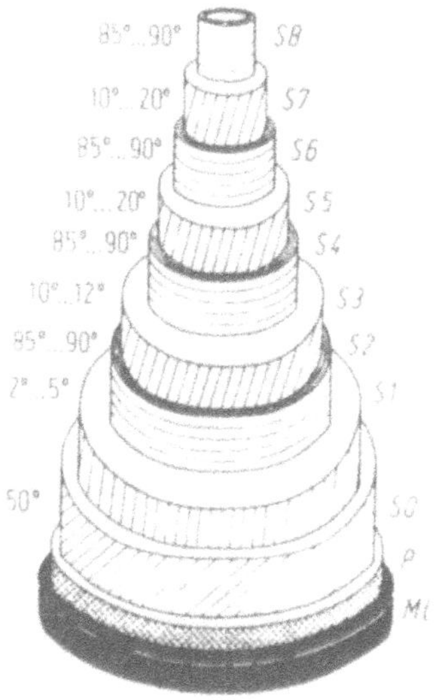

FIG. 11. Model of the polylamellate structure of a thick-walled bamboo fibre. Figures on the left indicate fibril angle, letters on the right the terminology of wall lamellae (from ref. 73). (Courtesy of Springer Verlag.)

The vertically elongated parenchyma cells with laminated lignified cell walls are interspersed with shorter cube-like cells which are mostly unlignified. The elements of the protoxylem have lignified cell walls composed of cellulose microfibrils arranged parallel to the middle lamella in a concentric manner. The two large metaxylem vessels in each vascular bundle have two lamella in their walls.

The phloem consists of large thin-walled sieve tubes and smaller companion cells. The wall of the sieve elements is composed of cellulosic microfibrils parallel to each other in a concentric manner around the cell and perpendicular to the cell axis, without obvious signs of lignification.[73,74]

11.4. Palm 'Wood'

The coconut palm (*Cocos nucifera*) and the oil palm (*Elaeis guineensis*) exemplify the commercially important palms found in the tropical countries. There are also other 'woody' monocotyledons such as the estuarine palm (*Nypa*) and scandent palms such as rattans (e.g. *Calamus*). As palms do not possess cambium their 'wood' is tissue consisting of primary vascular bundles embedded in parenchymatous ground tissue. The presence of wide vessels results in low densities. There is little known about the ultrastructure of the xylem or fibres in palms, but their arrangement and the fine structure of their cell walls is similar to that of bamboo. There is a common association of silica-containing cells with the fibres which are usually lignified.[75]

12. AQUATIC PLANTS

The algae (seaweeds) contain a large volume of biomass and are not greatly used commercially. Suitable harvesting techniques could be devised when their potential has been evaluated. Seaweeds grow in communities and their ultrastructure is not well developed and is less understood than that of the land plants. They are macroscopic members of the divisions *Chlorophyta* (green alga), *Rhodophyta* (red alga) and *Phaeophyta* (brown alga) and generally grow attached to solid substrata.

The *Phaeophyta* includes only multicellular forms which grow between and below the tide marks. Seaweed forests can grow to 30 m in six months and, thus, more quickly than terrestrial forests. The storage products of the *Chlorophyta* division is starch as in higher plants, with cellulose in the cell walls together with hydroxyproline and glycosides. The *Rhodophyta* have a Floridean starch as storage material, with cellulose, xylans and pectins in the cell wall. The *Phaeophyta* have laminarin and mannitol as storage materials and, in the cell wall, cellulose, alginic acid and sulphated polysaccharides. Most algae have an α-cellulose content between 13 and 41%.[5]

The algal walls are of two types. The first or *Valonia* type has cellulose fibrils primarily in two orientations at 90° to one another; the two orientations occur in different lamellae. In the second type of arrangement the microfibrils (cellulose, mannans or xylans) are oriented almost transversely like hoops around the cell periphery. During cell differentiation, a thicker cell wall is deposited, within which the microfibrils are oriented almost longitudinally. In this type the fibrils are usually dispersed at random in a mucilage of complex composition such as alginic acid or sulphated polysaccharides. The absence of lignin facilitates conversion of seaweeds to alcohol and other compounds.[76]

13. ULTRASTRUCTURE IN RELATION TO UTILIZATION

In most cases growing plantations solely for energy or chemicals is less attractive than using residues from other operations. Because biomass is a dispersed source, the economics of its collection and storage determines usage patterns. Transportation would be aided if the bulk could be reduced by removal of the water in the cell lumens. Also the 30–40 % water in the saturated fibre walls and the volatile materials could be removed by carbonization.

13.1. Compaction of Biomass

The amount of water in biomass varies considerably down to the equilibrium moisture content of air-dried material of 10–15 %. Moisture in wood greatly affects the calorific value[77] and its ignition. The energy required to express mechanically sap or water from fresh undried wood chips is only a small portion of that which would be needed to vaporize the liquid when the undried material is burnt.[78] However, the energy losses in the provision of mechanical energy still require attention. Reduction of moisture content in wood, for example from 150 % or more (on the dry wood basis) to 40 %, would reduce transport costs.[79] Although there has been some study of the basic relationships between tissue structure and fluid flow on compression, there is a lack of mathematical models or data.[80,81]

13.2. Carbonization

There is an interaction between microstructural features present in wood and its observed behaviour on charring. However, the chemical heterogeneity, species differences and cellular morphology, as well as the physical properties of the char layer which develops, makes this interaction difficult to characterize. Various aspects of the problem have been studied separately.[82]

The fine structure of wood is retained when charcoal is formed. Because the thermal conductivity of wood is low there is consistent development of a thin transition zone between thermally degraded and non-degraded wood. This zone appears to be critical in crack propagation and fissure formation in the char, both of which are initiated between 200 °C and 270 °C and in turn facilitate combustion. These fissures terminate only in the earlywood. Thermal degradation commences at lower temperatures for the carbohydrates (the origin of volatile products) than for lignin (the major source of char).[30] The deformations of the structure of the outer layers of the char resulting from volumetric loss of water and volatilization of cell wall constituents could be due to the softening of lignin in the degraded cell walls which no longer have the cellulose reinforcing networks.[82]

REFERENCES

1. Hillis, W. E. (1982). The future demand and supply of forest products. *Outlook on Agriculture* **11**(3), 114–22.
2. McLure, J. P., Saucier, J. R. and Biesterfeldt, R. C. (1981). Biomass in Southeastern forests. USDA For. Serv. Res. Pap. SE-227, 38 p. Southeast. For. Exp. Stn, Asheville, N.C.
3. Nautiyal, J. C. (1979). The place of forestry in the energy question. *Can. J. Forest Res.* **9**, 68–75.
4. Shafizadeh, F. (1982). Introduction to pyrolysis of biomass. *J. Anal. Appl. Pyrolysis* **3**, 283–305.
5. Preston, R. S. (1974). *The Physical Biology of Plant Cell Walls*. Chapman & Hall, London.
6. Fengel, D. (1971). Ideas on the ultrastructural organization of the cell wall components. *J. Polymer Sci. Part C* No. 36, 383–92.
7. Parameswaran, N. and Liese, W. (1982). Ultrastructural localization of wall components in wood cells. *Holz als Roh-u. Werkstoff* **40**, 145–55.
8. Cote, W. A. (1981). Ultrastructure—Critical domain for wood behaviour. *Wood Sci. Technol.* **15**, 1–29.
9. Stamm, A. J. and Sanders, H. T. (1966). Specific gravity of the wood substance of loblolly pine as affected by chemical composition. *Tappi* **49**, 397–400.
10. Stamm, A. J. (1969). Correlation of structural variations of lignins with their specific gravities. *Tappi* **52**, 2498–502.
11. Timell, T. E. (1982). Recent progress in the chemistry and topochemistry of compression wood. *Wood Sci. Technol.* **16**, 83–122.
12. Berlyn, G. P. (1969). Microspectrophotometric investigation of free space in plant cell walls. *Am. J. Bot.* **56**, 498–506.
13. Stone, J. E. (1973). The fibre saturation point, wall density and other characteristics of wood. Proc. IUFRO Division V Conference, South Africa, pp. 988–99.
14. Hillis, W. E. (1971). Distribution, properties and formation of some wood extractives. *Wood Sci. Technol.* **5**, 272–89.
15. Tarkow, H. and Krueger, J. (1961). Distribution of hot-water soluble material in cell walls and cavities of redwood. *For. Prod. J.* **11**, 228–9.
16. Wangaard, F. F. and Granados, L. A. (1967). The effect of extractives on water-vapor sorption by wood. *Wood Sci. Technol.* **1**, 253–77.

17. Perng, W. R. and Tajima, T. (1981). The packing density of wood. *Mokuzai* Gakkaishi **27**, 237–45.
18. Siau, J. F. (1971). *Flow in Wood.* Syracuse Univ. Press, New York.
19. Goring, D. A. I. (1971). Microscopic patterns in lignin removal during chemical pulping, in *The Physics and Chemistry of Wood Pulp Fibres. Tappi* Series **8**, 107–14.
20. Berlyn, G. P. and Mark, R. E. (1965). Lignin distribution in wood cell walls. *For. Prod. J.* **16**(3), 140–1.
21. Hardell, H. L., Leary, G. J., Stoll, M. and Westermark, V. (1980). Variations in lignin structure in defined morphological parts of spruce. *Svensk Papperstidn* **83**, 44–9.
22. Hardell, H. L., Leary, G. J., Stoll, M. and Westermark, V. (1980). Variations in lignin structure in defined morphological parts of birch. *Svensk Papperstidn* **83**, 71–4.
23. Zickerman, J. B. and Thomas, R. J. (1971). Distribution of loblolly pine ash residue. *Tappi* **54**, 1727–30.
24. Cliff, E. P. (1973). Timber the renewable material. National Commission on Materials Policy, USA.
25. Harada, H. and Wardrop, A. B. (1960). Cell wall structure of ray parenchyma cells of a softwood (*Cryptomeria japonica*). *Mokuzai Gakkaishi* **6**, 34–41.
26. Fergus, B. J. and Goring, D. A. I. (1970). The distribution of lignin in birch wood as determined by ultraviolet microscopy. *Holzforschung* **24**, 118–24.
27. Dadswell, H. E. and Wardrop, A. B. (1960). Some aspects of wood anatomy in relation to pulping quality and to tree breeding. *J. Aust. Pulp Paper Ind. Tech. Assoc.* **13**, 161–73.
28. Panshin, A. J. and de Zeeuw, C. (1980). *Textbook of Wood Technology* 4th edn. McGraw-Hill, New York.
29. Anon. (1977). Leucaena—Promising forage and tree crop for the tropics. Nat. Acad. Sciences, Washington, DC.
30. Shafizadeh, F. and McGinnis, G. D. (1971). Chemical composition and thermal analysis of cottonwood. *Carbohydrate Res.* **16**, 273–7.
31. Goring, D. A. I., quoted by Timell (1982).[11]
32. Yang, J. M. and Goring, D. A. I. (1980). The phenolic hydroxyl content of lignin in spruce wood. *Can. J. Chem.* **58**, 2411–4.
33. Doat, J. and Valette, J.-C. (1981). Discussion at IUFRO XVII Congress, Japan.
34. Zobel, B. and Blair, R. (1976). Wood and pulp properties of juvenile wood and topwood of the southern pines. Applied Polymer Symposium No. 28, 421–33.
35. Wardrop, A. B. and Dadswell, H. E. (1955). The structure and properties of tension wood. *Holzforschung* **9**(4), 97–103.
36. Scurfield, G. (1967). The ultrastructure of reaction wood differentiation. *Holzforschung* **21**, 6–13.
37. Nicholson, J. E., Hillis, W. E. and Ditchburne, N. (1975). Some tree growth—property relationships of eucalypts. *Can. J. Res.* **5**, 424–32.
38. Boyd, J. D. (1977). Relationship between fibre morphology and shrinkage of wood. *Wood Sci. Technol.* **11**, 3–22.
39. Bland, D. E. (1961). The chemistry of reaction wood III. The milled wood lignins of *Eucalyptus goniocalyx* and *Pinus radiata*. *Holzforschung* **15**(4), 102–6.

40. Timell, T. E. (1969). The chemical composition of tension wood. *Svensk Papperstidn* **76**, 173–81.
41. International Association of Wood Anatomists (1957). International glossary of terms used in wood anatomy. Tropical Woods No. 107, 1–35.
42. Anon. (1981). Comparative properties of old- and young-growth giant sequoia of potential significance to wood utilisation. Div. Agric. Sci., Univ. Calif. Bull. 1901.
43. Peck, E. C. (1953). The sap or moisture in wood. USDA For. Prod. Lab., Madison, Rpt No. D768.
44. Merrill, W. and Cowling, E. B. (1966). Role of nitrogen in wood deterioration: Amounts and distribution of nitrogen in tree stems. *Can. J. Bot.* **42**, 1555–80.
45. De Silva, D. and Hillis, W. E. (1979). Inorganic extraneous constituents of wood. *Holzforschung* **33**, 47–53.
46. Scurfield, G., Anderson, C. A. and Segnit, E. R. (1974). Silica in woody stems. *Aust. J. Botany* **22**, 211–29.
47. Hillis, W. E. (1977). Secondary changes in wood, in *Recent Advances in Phytochemistry*, F. A. Loewus and V. C. Runeckles (Eds). Vol. II, Plenum Press, New York, pp. 247–309.
48. Bamber, R. K. (1976). Heartwood, its function and formation. *Wood Sci. Technol.* **10**, 1–8.
49. Lambert, M. J. (1981). Inorganic constituents in wood and bark of New South Wales forest tree species. For. Comm. N.S.W. Res. Note No. 45.
50. Shigo, A. L. and Sharon, E. M. (1970). Mapping columns of discolored and decayed tissues in sugar maple *Acer saccharum*. *Phytopath.* **60**, 232–7.
51. Tattar, T. A., Shigo, A. L. and Chase, T. (1972). Relationship between the degree of resistance to a pulsed electric current and wood in progressive stages of discoloration and decay in living trees. *Can. J. For. Res.* **2**, 236–43.
52. Safford, L. O., Shigo, A. L. and Ashley, M. (1974). Gradients of cation concentration in discolored and decayed wood of red maple. *Can. J. For. Res.* **4**, 435–40.
53. Hillis, W. E. (1962). *Wood Extractives*. Academic Press, New York.
54. Esau, K. (1965). *Plant Anatomy*. John Wiley, New York.
55. Bauch, J. (1980). Natural variation of wood properties. Mitterlungen Bundesforsch Forst. u. Holzwirtschaft. Hamburg-Reinbek.
56. Shigo, A. L. and Hillis, W. E. (1973). Heartwood, discolored wood, and microorganisms in living trees. *Ann. Review Phytopath.* **11**, 197–222.
57. Hillis, W. E. (1975). The role of wood characteristics in high temperature drying. *J. Inst. Wood Science* **7**, 60–7.
58. Greaves, H. (1975). Microbiological aspects of wood chip storage in tropical environments. *Aust. J. Biol. Sci.* **28**, 315–22.
59. Lightwood Res. Coord. Council. Proceedings Annual Mtgs. 1975, 1976, 1977, 1978.
60. Young, H. E. (1971). Preliminary estimates of bark percentages and chemical elements in complete trees of eight species in Maine. *For. Prod. J.* **21**(5), 56–9.
61. Koch, P. (1972). Utilization of the southern pines. US Dept of Agric. Agric. Handbook. No. 420, Vol. I, 467–533.

62. Howard, E. T. (1971). Bark structure of the southern pines. *Wood Sci.* **3**, 134–48.
63. Martin, R. E. and Crist, J. B. (1970). Elements of bark structure and terminology. *Wood and Fiber* **2**, 269–79.
64. Fahn, A. (1974). *Plant Anatomy*. Pergamon Press, Oxford.
65. Roth, L., Saeger, G., Lynch, F. J. and Weiner, J. (1960). Structure, extractives and utilisation of bark. Bibliograph. Series 191, 446 pp. Inst. Paper Chem. Appleton, Wis., USA.
66. Roth, L. and Weiner, J. (1968). ibid. ref. 68. Suppl. 1, Bibliog. Ser. 191, 184 pp. Inst. Paper Chem. Appleton, Wis., USA.
67. Weiner, J. and Pollock, J. (1973). ibid. ref. 68. Suppl. 2, Bibliog. Ser. 191.
68. Herrick, F. W. and Hergert, H. L. (1977). Utilization of chemicals from wood, in *Recent Advances in Phytochemistry*, F. A. Loewus and V. C. Runeckles (Eds). Vol. II, Plenum Press, New York, pp. 443–515.
69. Wright, T. W. and Will, G. M. (1958). The nutrient content of Scots and Corsican pines growing on sand dunes. *Forestry* **31**, 13–25.
70. White, J. D., Alexander, L. T. and Clark, E. W. (1972). Fluctuations in the inorganic constituents of inner bark of loblolly pine with season and soil series. *Can. J. Bot.* **50**, 1287–93.
71. Wood, I. M. and Angus, J. F. (1976). Kenaf versus forests as a source of paper pulp. *Aust. For.* **39**(1), 23–39.
72. White, G. A., Cummins, D. G., Whiteley, E. L., Fike, W. T., Greig, J. K., Martin, J. A., Killinger, G. B., Higgins, J. J. and Clark, T. F. (1970). Cultural and harvesting methods of kenaf. US Dept. of Agr. ARS Prod. Res. Rep. No. 113.
73. Parameswaran, N. and Liese, W. (1976). On the fine structure of bamboo fibres. *Wood Sci. Technol.* **10**, 231–46.
74. Parameswaran, N. and Liese, W. (1981). Bamboo production and utilisation. Proc. Congress Group 5.3A, T. Higuchi (Ed.), IUFRO XVII World Congress, Japan, pp. 178–83.
75. Parthasarathy, M. V. and Klotz, L. H. (1976). Palm 'wood' I. Anatomical aspects. *Wood Sci. Technol.* **10**, 215–29. II. Ultrastructural aspects of sieve elements, tracheary elements and fibres. *Wood Sci. Technol.* **10**, 247–71.
76. Chapman, A. R. O. (1979). *Biology of Seaweeds*. University Park Press, Baltimore.
77. Ince, P. J. (1977). Estimating effective heating value of wood or bark fuels at various moisture contents. USDA For. Ser., Gen. Tech. Rept. FPL 13.
78. Haygreen, J. G. (1981). Potential for compression drying of green wood chip fuel. *For. Prod. J.* **31**(8), 43–54.
79. Hassan, A. E. D. (1977). Compaction of wood chips—energy cost. *Trans. Am. Soc. Agric. Engineers* 839–42.
80. Murray, C. R. and Holt, J. E. (1967). *The Mechanics of Crushing Sugar Cane*. Elsevier Publ. Co., Amsterdam.
81. Wingate-Hill, R. (1982). Personal communication.
82. Zickerman, J. B. and Williamson, R. B. (1981). Microstructure of wood char. Part I. Whole wood. *Wood Sci. Technol.* **15**, 237–49.

2

Cellulose, Hemicellulose and Extractives

OLOF THEANDER

Department of Chemistry and Molecular Biology,
Swedish University of Agricultural Sciences, Uppsala, Sweden

1. INTRODUCTION

According to the organizers, this lecture should place emphasis on cellulose, hemicellulose, and other important polysaccharides and on extractives in lignocellulose-rich materials, such as wood and straw and related agricultural residues, and not include the often more costly starch-rich and sugar-rich materials. The latter materials are important sources for biotechnical and organo-chemical conversion, rather than thermo-chemical biomass conversion. The lignocellulose polymers cellulose, hemicellulose and lignin represent the major portion of the enormous amount of biomass annually produced on land.

I will start by presenting the main polysaccharides and types of extractives in the wood of forest trees and then move on more briefly to two other parts of the tree which are of technical interest, namely the bark and the foliage. There has been intensive research activity in wood chemistry, particularly concerning species of industrial interest, where the work has been directed both at the main fibre constituents (cellulose, hemicellulose and lignin) as well as extractives of technical and of taxonomic interest, such as in connection with selection of plus trees. In the scope of my presentation, I have to be rather brief, giving only a few literature references to original investigations, but mainly to more reviewing articles. Straw and other main agricultural crop residues, which also mainly belong to the *Gramineae* family, will be discussed more briefly after the forest raw materials, as they have not been as thoroughly studied (this refers particularly to the extractives) and because their major components are related to those present in wood.

Peat is another large, low-cost organic matter resource of great interest for thermochemical conversion. Although it is considered as a fossil source, it is still slowly renewed, and as an average it contains substantial amounts of unaltered or slightly modified plant materials. As a comparison to the main lignocellulosic biomass resources, it will therefore be discussed here very briefly.

Studies made in the author's laboratory on conversion of carbohydrates to phenols and other cyclic non-carbohydrate compounds will be summarized at the end of the paper.

2. WOOD POLYSACCHARIDES

Plant cell walls contain three types of structural polysaccharides, namely cellulose, hemicelluloses and pectic polysaccharides. In wood—depending on species and location in the wood—there are also larger or smaller numbers of other types than those which are part of the cell wall structure, such as storage polysaccharides and gums.

2.1. Cellulose

Cellulose (Fig. 1) is the most abundant renewable organic molecule in nature and the main constituent of wood. Approximately 40–45% of the dry matter in most wood species is cellulose, and it is predominantly located in the secondary cell wall. Although it is apparently a simple chemical structure—a linear polymer composed of up to about 10 000 β-1,4-linked D-glucopyranose units—its supermolecular state, including many structural problems, is still not completely solved. The cellulose chains have a

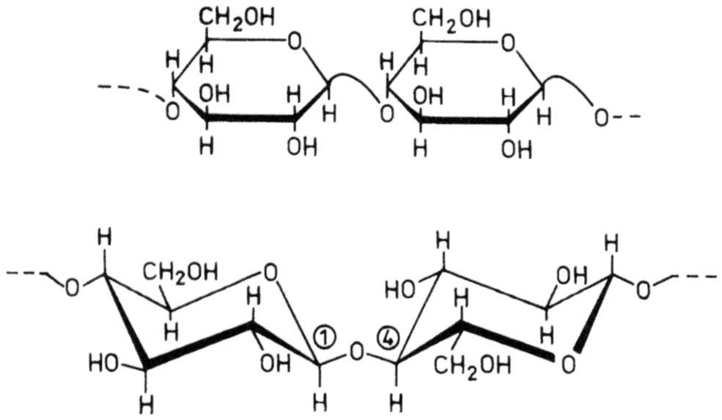

FIG. 1. Cellulose, upper in the Haworth formula and lower in the conformation formula.

conformation (see the lower part of the formula) which strongly favours the formation of intra- and intermolecular hydrogen bonds. Therefore, the cellulose molecules are aggregated together into bundles in the form of microfibrils with highly crystalline order (alternating with less ordered regions) which build up the cellulose fibres of high mechanical strength. This structure is also characterized by its insolubility in most types of solvents and its low accessibility towards acid and enzymatic hydrolysis. Because of its conformation, cellulose differs strongly in its physical and chemical properties from that of starch—also built up of 1,4-linked glucopyranose units, but glycosidically linked as α-anomers.

2.2. Hemicelluloses

Hemicellulose polysaccharides are chemically related to cellulose, having backbones of 1,4-β-linked major sugar units, and being morphologically strongly associated with cellulose in the plant cell walls as well as to lignin in lignified cell walls—to the latter also linked with different types of covalent linkages. These polysaccharides are generally heterogeneous, built up of different hexoses (C_6-sugars) and pentoses (C_5-sugars), sometimes in addition to uronic acids. They have a lower degree of polymerization than cellulose (100–200 units), are largely soluble in alkali, and also more easily hydrolysed. They usually amount to between 20% and 30% of the dry matter of the wood.

2.3. Softwood Hemicelluloses

In softwoods, galactoglucomannans (Fig. 2) are the major hemicelluloses, amounting to about 20%. They have a backbone of 1,4-β-linked D-glucopyranose and D-mannopyranose sugar units, to which are linked single α-D-galactopyranose units at some of the C-6 positions. One can usually fractionate these polysaccharides into two types having different galactose contents. It is important that these polysaccharides are partially

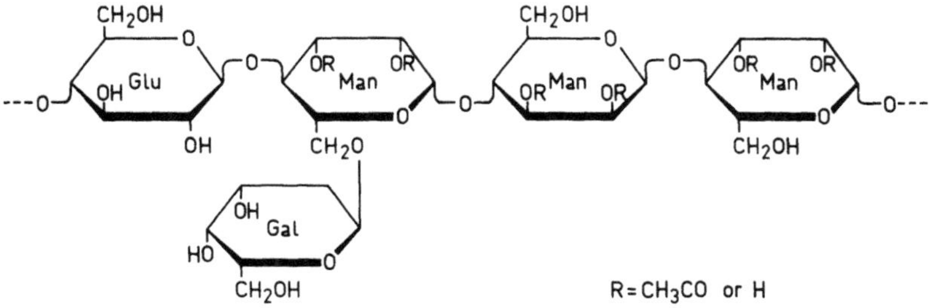

FIG. 2. Galactoglucomannans.

substituted by acetyl groups—easily cleaved by alkali—i.e. one group per 3–4 hexose units, on average, and linked at the C-2 and C-3 positions.

In addition to galactoglucomannans, softwoods contain 5–10% arabinoglucuronoxylans (Fig. 3) where the backbone contains β-1,4-linked β-D-xylopyranose units, partially substituted by C-2-arabinofuranose units (as an average about 1 unit per 10 xylose units) and 4-O-methyl-α-D-glucuronic acid units (as an average 2 residues per 10 xylose units). In technical processes it is important that arabinose is furanosidically linked and therefore very readily hydrolysed by acid. This type of hemicellulose component is generally simply called xylans or pentosans, C_5-sugars being the major constituents.

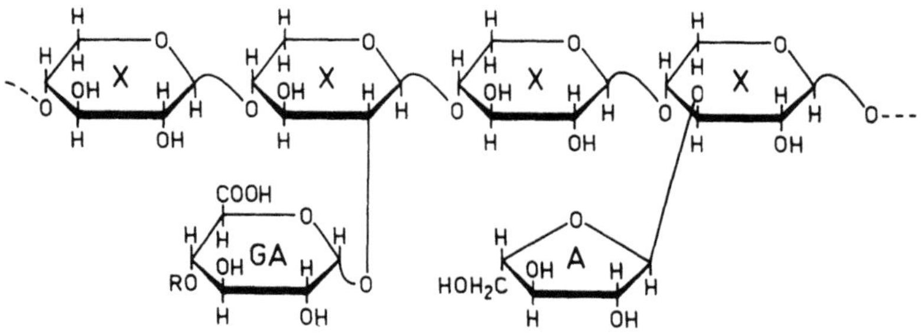

FIG. 3. Arabinoglucuronoxylans.

2.4. Hardwood Polysaccharides

In hardwood species, glucuronoxylan is the main hemicellulose component, 15–30% of the dry wood. It has the same kind of xylose-polymeric backbone as in the softwoods (compare Fig. 3) and also with C-2-linked 4-O-methyl-α-D-glucuronic acid as a significant constituent (about 1 unit per 10 xylose units). On the other hand, this type of xylan does not contain branches of arabinose but considerable amounts of acetyl groups at C-2 and C-3 positions (about 7 units per 10 xylose units). Linkages between uronic acid and sugar are very resistant towards acid hydrolysis. Recent studies on birch xylan[1] have revealed that the unit next to the reducing xylose end unit is D-galacturonic acid, linked to an L-rhamnose unit through the C-2 position. The latter unit is connected to the xylose chain by its C-3 position.

For a review of wood hemicelluloses, see Timell.[2]

TABLE I

CHEMICAL COMPOSITION OF VARIOUS WOOD SPECIES[a] [3]

Species	*Common name*	*Total extractives*	*Lignin*	*Cellulose*	*Glucomannan*[b]	*Glucuronoxylan*[c]	*Other polysaccharides*	*Residual constituents*
Softwoods								
Abies balsamea	Balsam fir	2·7	29·1	38·8	17·4	8·4	2·7	0·9
Pseudotsuga menziesii	Douglas fir	5·3	29·3	38·8	17·5	5·4	3·4	0·0
Tsuga canadensis	Eastern hemlock	3·4	30·5	37·7	18·5	6·5	2·9	0·5
Juniperus communis	Common juniper	3·2	32·1	33·0	16·4	10·7	3·2	1·4
Pinus radiata	Monterey pine	1·8	27·2	37·4	20·4	8·5	4·3	0·4
Pinus sylvestris	Scots pine	3·5	27·7	40·0	16·0	8·9	3·6	0·3
Picea abies	Norway spruce	1·7	27·4	41·7	16·3	8·6	3·4	0·9
Picea glauca	White spruce	2·1	27·5	39·5	17·2	10·4	3·0	0·3
Larix sibirica	Siberian larch	1·8	26·8	41·4	14·1	6·8	8·7	0·4
Hardwoods								
Acer rubrum	Red maple	3·2	25·4	42·0	3·1	22·1	3·7	0·5
Acer saccharum	Sugar maple	2·5	25·2	40·7	3·7	23·6	3·5	0·8
Fagus sylvatica	Common beech	1·2	24·8	39·4	1·3	27·8	4·2	1·3
Betula verrucosa	Silver birch	3·2	22·0	41·0	2·3	27·5	2·6	1·4
Betula papyrifera	Paper birch	2·6	21·4	39·4	1·4	29·7	3·4	2·1
Alnus incana	Gray alder	4·6	24·8	38·3	2·8	25·8	2·3	1·4
Eucalyptus camaldulensis	River red gum	2·8	31·3	45·0	3·1	14·1	2·0	1·7
Eucalyptus globulus	Blue gum	1·3	21·9	51·3	1·4	19·9	3·9	0·3
Gmelina arborea	Yemane	4·6	26·1	47·3	3·2	15·4	2·5	0·9
Acacia mollissima	Black wattle	1·8	20·8	42·9	2·6	28·2	2·8	0·9
Ochroma lagopus	Balsa	2·0	21·5	47·7	3·0	21·7	2·9	1·2

[a] All values are given as % of the dry wood weight.
[b] Including galactose and acetyl in softwood.
[c] Including arabinose in softwood and acetyl in hardwood.

2.5. Other Polysaccharides

There are also other types of polysaccharides present in wood—generally in lower quantities. Thus, the cell walls of wood, and probably also of most plants, contain pectic substances or what are simply called pectins. This is a complex group of polysaccharides in which the D-galacturonic acid is the principal constituent in the main polymers. Many examples are reported where L-rhamnose residues have been found at intervals along the galacturonan backbone and often other neutral sugars like L-arabinose, D-xylose, D-galactose and L-fucose are attached as side chains. Some of the uronic acid units may also be present as methyl esters. Neutral arabinans, galactans and arabinogalactans are often associated with these acidic polymers. The heartwood of larches contains exceptionally large amounts of a water-soluble arabinogalactan (20–30%). It has a backbone of β-1,3-linked D-galactopyranose units. Almost every unit has branches at C-6 containing β-1,6-linked D-galactopyranose units and also L-arabinose units. Also a few D-glucuronic acid units are present in the polymer. Compression wood of softwoods contains about 10% of a 1,4-β-linked D-galactan and 2–5% of a 1,3-β-linked D-glucan. Table I summarizes the composition of polysaccharides as well as the amounts of lignin and extractives of a series of softwoods and hardwoods.[3]

3. WOOD EXTRACTIVES

Extractives are a complex mixture of organic compounds which can be removed from the wood by solvents such as petrol ether, ether, dichloromethane, benzene, acetone, ethanol and water. One way of dividing these compounds has been into primary and secondary metabolites, the primary ones being those occurring in essentially all plants and being intermediates in various metabolic processes. Representatives of this group are low-molecular weight sugars, inositols, amino acids, simple fats and carboxylic acids. Secondary metabolites are generally more complex, have often a restricted taxonomic distribution, and their formation is probably essentially irreversible. They often have important physiological functions for the plant as protecting agents against parasites and so on. Representatives of this group are terpenes and phenolic compounds.

Another, and technically more useful, way is to divide the extractives into lipophilic ones soluble in non-polar solvents, such as petrol ether, ether and benzene, and hydrophilic ones which are water-soluble but also generally soluble in solvents such as ethanol and acetone. There are, however, compounds, such as some phenols, on the borderline of these two groups. The scheme in Fig. 4 illustrates how one may fractionate the different types of extractives from tree sources (compare ref. 4; studies on pine needles). It also indicates methods for their subfractionation and analytical determination such as column chromatography (CC), thin-layer chromatography (TLC), gas–liquid chromatography (GC) and high-pressure liquid chromatography (HPLC). Figure 5 gives an excellent example of analysis of both fatty and resin acids from Norway spruce wood[5] on the same run by capillary gas–liquid chromatography.[6] Now, a brief presentation will be given of the different groups with a few examples of individual compounds.

As seen in Table I, the amount of extractives generally constitutes 1–5% of the dry matter of woods, but there are also examples of species with higher amounts. Thus, some trees can have up to 30% tannins, and larches, as discussed above, have large amounts of water-extractable arabinogalactans.

An important lipophilic group containing several types of organic compounds is the so-called resins. Thus, softwood resin contains resin acids, fats and terpenes, as major constituents. Resins located in the resin canals (oleoresin) are mainly composed of resin acids and monoterpenes but also contain higher terpenoids and fatty acid esters. These resins protect the wood against biological damage. The physiological resin, located in the ray parenchyma cells, is rich in fats, waxes and sterols, and serves as a supply of reserve food. Hardwoods contain only this type of resin. Heartwood, which forms when the living parenchyma cells die, has a very different chemical pattern to that of the surrounding sapwood. Formation of phenolic fungicides is thus typical of pinewood, and so also is a substantial increase in the heartwood of the total amount of extractives—up to 12–14% in pine species.[7]

The content of extractives as well as their chemical pattern can vary greatly between different wood species and also between different parts of the same tree. The lipophilic extractives are usually divided into aliphatic compounds (mainly fats and waxes), terpenoids, steroids and phenolic compounds (for some examples, see Fig. 6). Some phenolic compounds like tannins and phenolic glycosides are rather hydrophilic and water-soluble. Besides some polysaccharides, one can also find small amounts of low-molecular weight sugars, inositols, proteins, salts and other constituents among the water-soluble extractives.

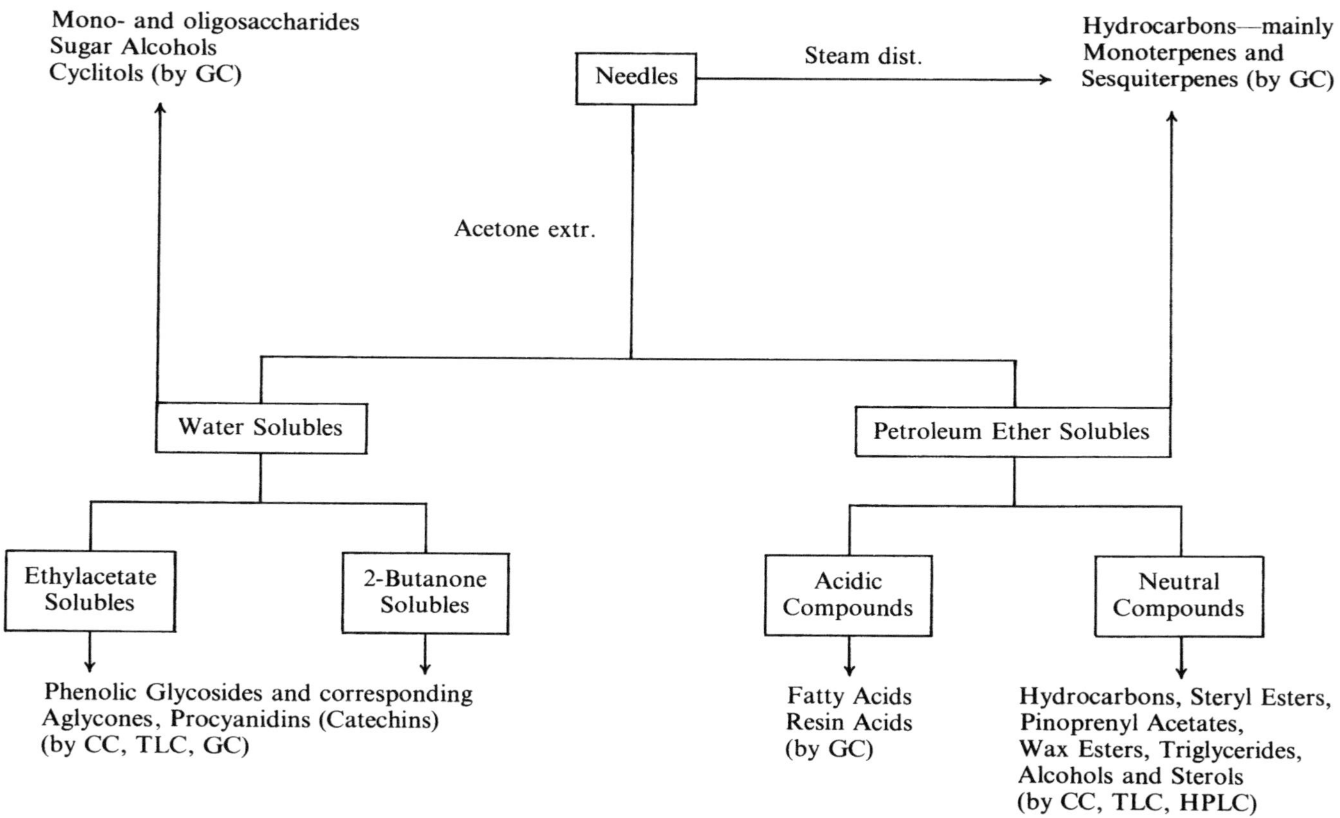

FIG. 4. Fractionation of extractives from pine needles—main steps.

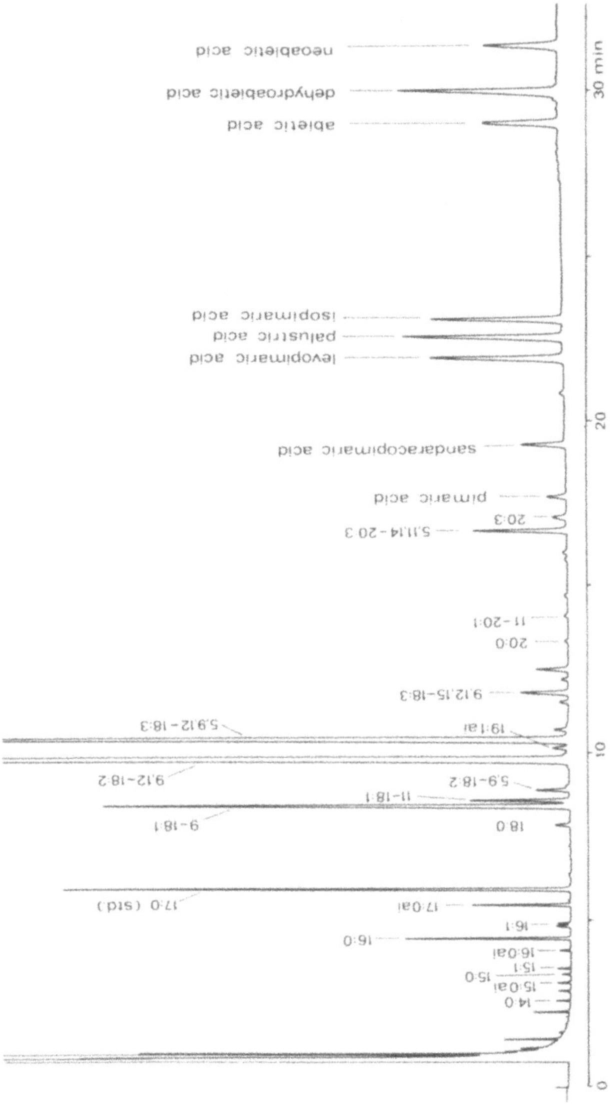

FIG. 5. Analysis of fatty and resin acids in Norway spruce wood as methyl esters on a glass capillary column: BDS, 30 m × 0·3 mm.[5]

Aliphatic components

n-Alkanes	$CH_3-(CH_2)_n-CH_3$	n = 8–30
Fatty alkohols	$CH_3-(CH_2)_n-CH_2OH$	n = 16–22
Fatty acids	$CH_3-(CH_2)_n-COOH$	n = 10–24
Fats (glycerol esters)	CH_2-OR / $CH-OR'$ / CH_2-OR''	R, R' and R'' can be fatty acid residues
Waxes (esters of other alcohols)	$RO-(CH_2)_n-CH_3$; RO – sterol; RO – terpene alcohol	R is fatty acid residue

Terpenoids and steroids

Monoterpenes

α-Pinene β-Pinene Δ^3-Carene

Diterpenes

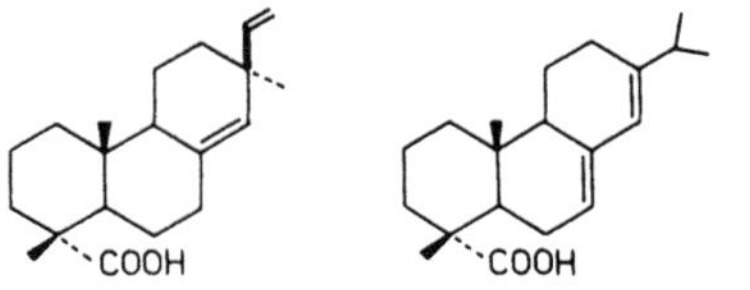

Pimaric acid Abietic acid

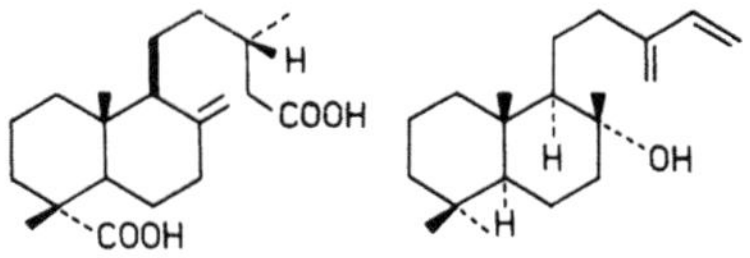

Pinifolic acid Isoabienol

Polyprenols

$$H-[CH_2-\overset{\displaystyle CH_3}{\overset{|}{C}}=CH-CH_2]_{6-9}-OH$$

Betula prenols

Sterols

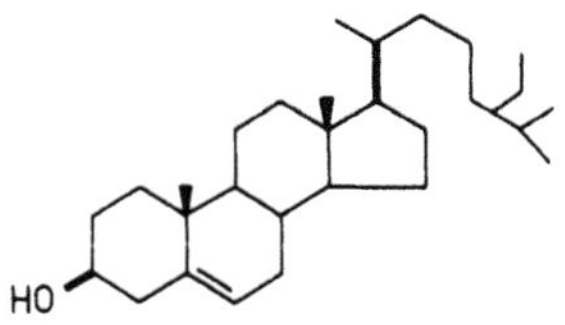

β-Sitosterol

FIG. 6. Examples of lipophilic extractives in trees.

Phenols

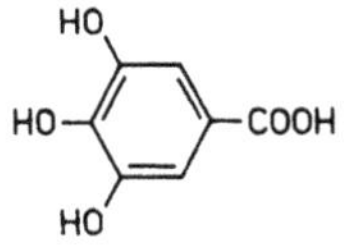

Gallic acid

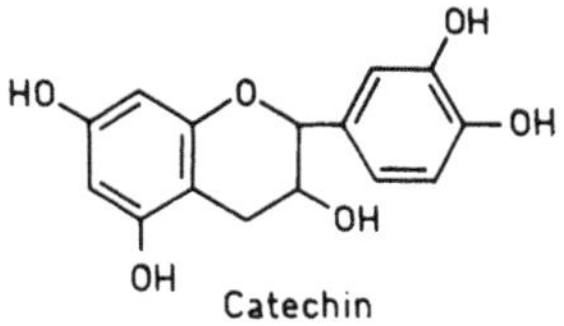

Catechin

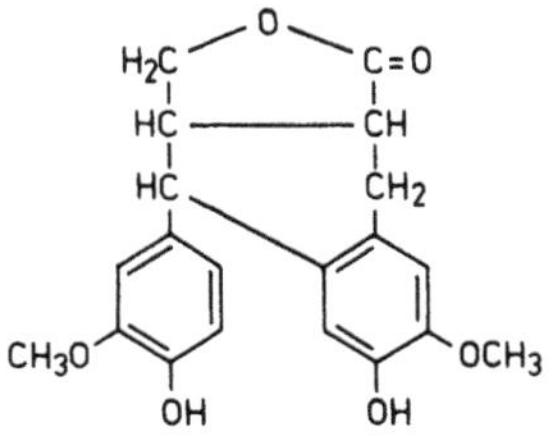

Conidendrin

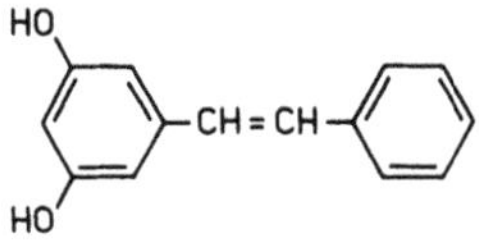

Pinosylvin

Tropolones

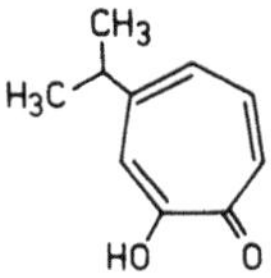

β-Thujaplicin

FIG. 6—*contd.*

3.1. Aliphatic Components

A large number of aliphatic compounds are present in wood, of which alkanes (hydrocarbons) and alcohols are generally in small amounts only. Arachinol (C_{20}), behenol (C_{22}) and lignocerol (C_{24}) represent the major alcohols. The fatty acids, the main constituents of the parenchyma resins in both softwoods and hardwoods, occur in mainly esterified form. The esters with glycerol (mono-, di- or tri-acylglycerols) are called fats and, together with fatty alcohols, terpenoids or steroids, are called waxes. The main fatty acids in wood, which can be saturated or contain one or more double bonds, are of similar type as in vegetable fats in general, e.g. palmitic, stearic, oleic, linoleic and linolenic acids.

3.2. Terpenoids and Steroids

Terpenoids, as also the related steroids, belong to the important group of natural products called isoprenoids—formally biosynthetically built up of isoprene units (an unsaturated, branched C_5-hydrocarbon). Thus monoterpenes are C_{10}-, sesquiterpenes C_{15}-, diterpenes C_{20}-, and triterpenes C_{30}-compounds. Mono-, sesqui- and diterpenes are the major terpenoids in softwoods—in particular the mono- and diterpenes—and triterpenes and polyprenoid compounds in hardwoods. Conifer oleoresin contains about 25% volatile components (turpentine) and the rest consist mainly of resin acids. In kraft pulping a turpentine fraction is recovered, with monoterpenes as main components and a tall-oil fraction with diterpenoid resin acids and fatty acids as main components.

In Fig. 6 α-pinene, β-pinene and 3-carene are given as important representatives of conifer monoterpenes and the resin acids pimaric and abietic acids as examples of diterpenoids. We have found pinifolic acid—a dicarboxylic acid—as a major diterpenoid component (1–2% of the dry weight) in the needles of Scots pine but it is not found in the wood;[8] Ekman *et al.*[9] have recently shown that isoabienol, previously erroneously characterized as abienol, is the principal diterpene alcohol in these needles. Although triterpenes are typical for hardwoods and the related steroids for softwoods, both groups of isoprenoids have been found in wood and bark from both groups of trees. Here, betulinol and β-sitosterol respectively may be mentioned as abundant representatives. Betulaprenols from birch are given, finally, as examples of more high-molecular isoprenoids.

3.3. Phenolic Components and Tropolones

This is a large, heterogeneous class of compounds which can be found both among lipophilic extractives, soluble in non-polar solvents, and hydrophilic extractives—in the latter case often as glycosides (compare the foliage section). Figure 6 gives gallic acid, which is a constituent of the hydrolysable tannins and catechin, as an example of the large group of plant phenols called flavonoids, but which is also an important constituent of the so-called condensed tannins. Lignans—conidendrin is here given as an example—which are formed by oxidative coupling of two phenylpropane units, are common in small amounts in conifers. In the reaction zone separating sound sapwood in Norway spruce from wood decayed by *Fomes annosus*, however, the amount of these compounds is strongly increased.[10,11]

Stilbenes, here exemplified by pinosylvin, having a conjugated double bond system, are very reactive and occur frequently in wood and bark of conifers, but have also been reported in *Eucalyptus* species. They are known to protect the heartwood of pines against fungal attack but also to be effective inhibitors of delignification by acid sulphite pulping. Tropolones, common in the cypress family, are related chemically to phenols and are characterized by an unsaturated seven-membered carbon ring. They form strong complexes with ferric and other ions and therefore cause corrosion problems in pulping. In Fig. 6 they are exemplified by the isopropyl derivative β-thujaplicin.

For a review of wood extractives see Hillis;[12] for the extractives in eastern US hardwoods see Rowe and Conner.[13]

4. BARK

Bark, which is the layer outside the cambium of a tree, amounts to 10–15% of the tree. It is thus a large renewable resource but so far is used technically to a very minor extent, except as fuel. The chemical composition of bark is complicated and varies with tree species as well as with the morphological parts involved. Characteristic for most barks, compared to wood, are the exceptionally high contents of extractives. The amount of polysaccharides is much lower than in wood. Minerals and certain organic components such as tannins and other phenolic compounds, pectins and so-called suberins, are often found in high contents. Suberins, which are present in the cork cells in the outer bark, are polyestolides, composed of ω-hydroxy-monobasic acids linked together by ester links. Also present are α,β-dibasic acids, which are esterified with bifunctional alcohols as well as with phenolic acids.

Table II,[14] which gives the yields of sugars after hydrolysis of bark from

TABLE II

REDUCING SUGARS OBTAINED AFTER HYDROLYSIS OF BARK[a] [14]

Species	*Glucose*	*Mannose*	*Galactose*	*Arabinose*	*Xylose*	*Total*
Softwoods						
Abies amabalis	29·0	5·4	2·3	4·1	3·2	45·3
Larix occidentalis	31·7	5·1	1·8	2·8	4·1	46·0
Picea abies	23·2	3·3	1·7	3·6	4·4	37·2
Pinus sylvestris	15·7	1·8	1·7	0·8	3·8	24·0
Pinus contorta	16·5	2·0	2·3	8·6	2·6	32·9
Tsuga canadensis	22·3	4·3	1·0	2·7	2·3	33·3
Hardwoods						
Alnus rubra	20·5	0·4	1·1	2·3	12·9	38·0
Betula papyrifera	16·0	0·3	0·6	1·8	10·8	30·1
Fagus sylvatica	16·7	0·7	2·3	5·4	8·1	34·1
Populus tremuloides	23·0	0·4	0·8	2·0	11·9	39·0
Quercus robur	28·1	0·8	1·2	0·8	9·6	40·6

[a] Figures given after extraction with neutral solvents, and expressed as % of unextracted bark weight.

a series of softwoods and hardwoods (after removal of extractives with neutral solvents), provides an indication of the amounts and composition of bark polysaccharides, glucose mainly originating from cellulose and the other sugars from hemicellulose polysaccharides.

For a recent review of bark, see Hemingway.[15]

5. FOLIAGE

The utilization of foliage—defined not only as pure leaves and needles from trees but also including varying amounts of twigs and small branches—for fodder supplement and for various technical products has been extensively studied in the USSR. In recent years, however, the interest in this biomass resource has been accelerating, especially in North America and Finland. The growing activity in plantations of short rotation forestry is one reason for this. The often heard Russian term *muka*, meaning foliage that has been dried and ground, is produced in the USSR as a high quality component for animal feeds in a quantity of over 100 000 tonnes/year.

Because foliage, unlike other tree components, is so strongly associated with the photosynthetic activity of the tree, it contains many unique chemical compounds, such as chlorophyll and carotenoids. In addition, the main part of the nitrogen content of the tree (around 1 %), in the form of

TABLE III
NON-POLAR EXTRACTIVES FROM SCOTS PINE (*Pinus sylvestris*) TECHNICAL FOLIAGE[a] [16]

Constituent	*Yield*
Triglycerides	2·5
Free acids	2·4
Sterylesters, sterols	1·1
Pinoprenyl acetates	0·9
Diterpene alcohols	0·6
Mono- and sesquiterpenes (essential oils)	0·5
Wax esters	0·2
Diterpene aldehydes	0·1
Secondary fatty alcohols	0·1

[a] The yields are given as % on dry weight basis.

proteins and other components, is found here. The amount of extractives in foliage is much higher than in wood. Table III[16] shows the composition of the non-polar, lipophilic extractives from the foliage of Scots pine. The major components are triglycerides, free acids (mainly resin acids) and steroid compounds (with β-sitosterol predominating). The essential oils (terpenes) are also technically interesting components.

In our laboratory we have studied the hydrophilic extractives of the needles of Scots pine and Norway spruce—for a recent review see Theander.[17] From the former source glucose, fructose, sucrose, pinitol and shikimic acid were each isolated in a yield of 1·5–3% of the dry weight.

TABLE IV

CHEMICAL COMPOSITION OF SCOTS PINE NEEDLES[a4]

	A	*B*
Extractives and residues		
Light petroleum	12·2	8·4
Ethyl acetate	5·0	6·5
2-Butanone	4·0	3·1
Water	19·9	9·2
Residue	65·4	73·2
Major groups of identified lipophilic extractives		
Hydrocarbons	0·3	0·2
Steryl esters	1·2	0·6
Triglycerides	2·8	0·7
Free acids	1·1	1·2
Alcohols and sterols	0·9	0·7
Some hydrophilic extractives		
Sugars and inositols (neutral)	7·7	1·6
Components in the residue		
Ash	2·3	2·3
Crude protein	5·3	2·4
Acid polysaccharide constituents	3·4	4·5
Neutral polysaccharide constituents, calculated as:		
Glucans	25·5	19·9
Mannans	4·4	5·9
Galactans	1·8	2·8
Arabinans	2·6	4·0
Xylans	1·3	2·1
Klason lignin	14·8	22·6

A = green needles; B = brown needles, from the same tree.

[a] All figures are given as % on dry weight basis of unfractionated sample.

Glucosides from Scots pine (those marked with ⋆ also from spruce):

G = β-D-glucopyranoside

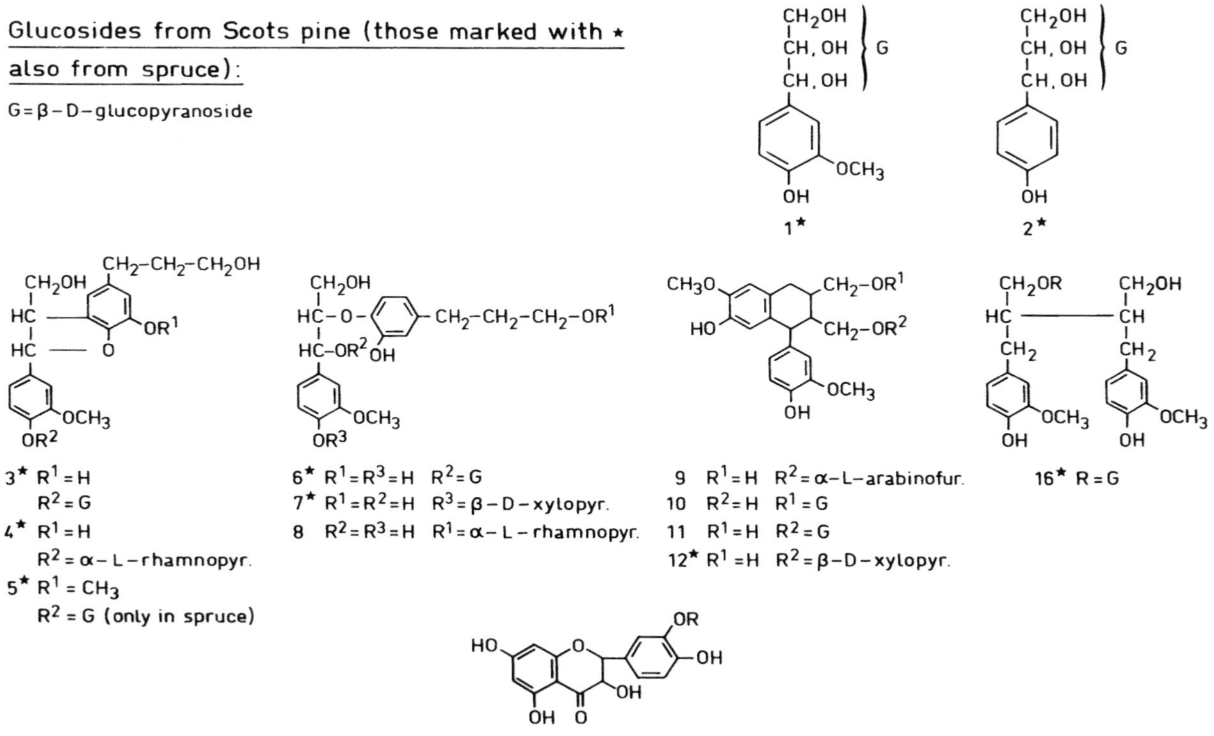

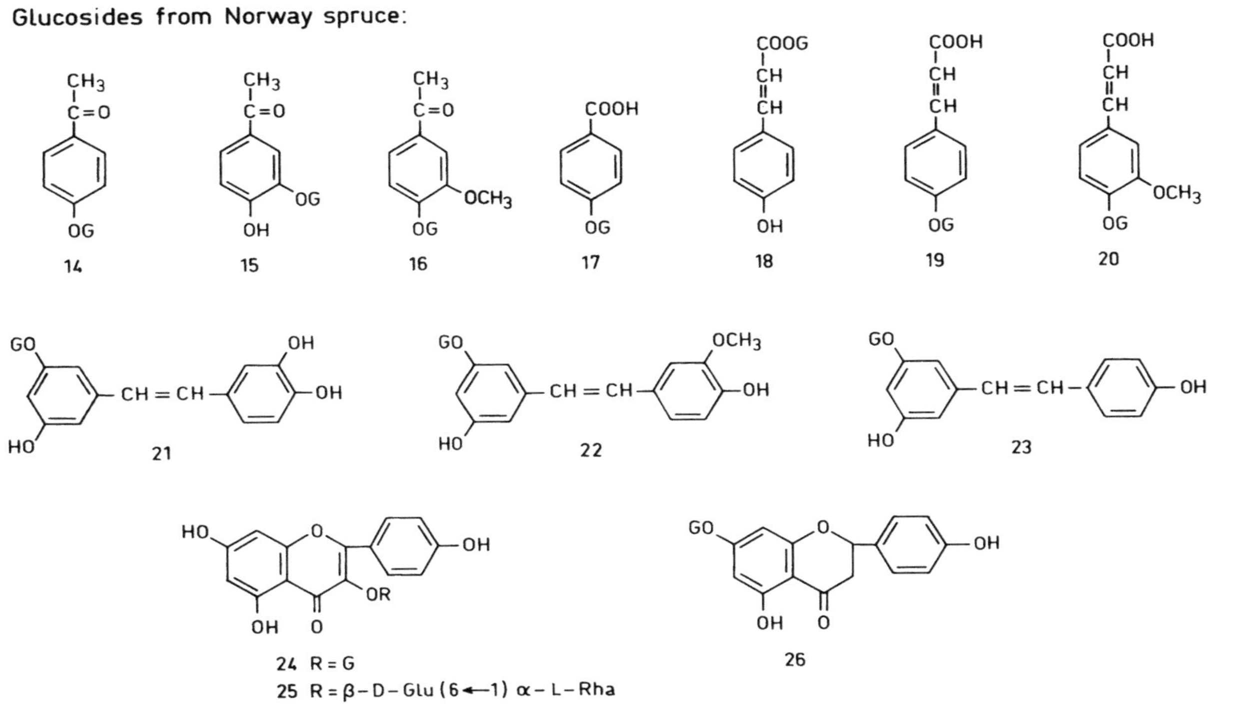

FIG. 7. Phenolic glycosides isolated from needles from Scots pine and Norway spruce.[17]

Arabinose, rhamnose, mannitol, melibiose, raffinose, myo-inositol and sequoitol were isolated in the range of 0·1–1 %. In recent years we have studied extensively the phenolic constituents of the needles, which are mainly present as glycosides, which have totally amounted to 3–5 % (see Fig. 7). Sometimes the taxifolin glucoside (formula **13**) is found in pine needles in over 2 % of the dry weight. Rather unexpectedly, we did not find this common conifer compound in our spruce needle sample. Many types of glycosides such as the monocyclic **1**–**2** (with glucose present in the different positions in the side chain) and several dilignol glycosides (such as **3**, **4**, **6**, **7**, **12** and **16**) seem to be common in both types of needles. Substantial amounts of acetophenone and stilbene glucosides (**14**–**16** and **21**–**23** respectively), however, seem to be particularly characteristic for the spruce. In connection with the Swedish Coniferous Forest Project, where we studied the changes of the chemical composition of the pine needle litter for a period of five years, we also compared the composition of green and brown needles from the same tree.[4] Since a high proportion of the needles on the tree are brown, it is of interest to obtain knowledge about the composition also of that part of the foliage. The constituents of the different groups of extractives were analysed in detail but here only a summary is given of the major groups of extractives and other components (Table IV). Pinifolic acid and isoabienol (see Fig. 6) are two principal components in both green and brown needles.

For a recent review of foliage, see Barton.[18]

6. STRAW AND OTHER AGRICULTURAL CROP RESIDUES

Of the around 2000 million tons of agricultural crop residues produced in the world per year, the straw from corn (maize), wheat, rice, barley, oats, millet and sorghum represents the major part. Sugar cane residues are also produced in large amounts. The chemical components of straw have not been so intensively investigated as those of wood—which particularly applies to lignin and extractives. Straws are rich in lignocellulose but low in storage carbohydrates (as sugars and starch) and proteins. The mineral content as well as compounds extractable with organic solvents may sometimes be quite high.

Table V gives the cellulose, hemicellulose and Klason lignin values for some different Swedish straws.[19] We determined the different neutral sugar constituents of the polysaccharides by gas–liquid chromatography and the uronic acid constituents by a decarboxylation method. The lignin

TABLE V

CHEMICAL COMPOSITION OF STRAW[a] [19]

Cultivar	*Hemicellulose*	*Cellulose*	*Klason lignin*
Barley			
Cilla	27	29	20
Ingrid	24	30	22
Senat	23	28	24
Särla	27	28	22
Wing	26	33	23
Oats			
Titus	22	30	23
Winter wheat			
Holme	21	27	21
Spring wheat			
Drabant	21	27	19
Rye			
Petkus II	23	37	20

[a] Figures are given as % on dry weight basis.

content—as Klason (sulphuric acid) lignin—is, as in most studies on lignocellulosic materials, determined gravimetrically as the residue left after removal of extractives by organic solvents, and the polysaccharides by strong sulphuric acid. This fraction often overestimates the true lignin content as, for instance, tannins and some nitrogenous matter may contribute to its weight. In a type of material like straw, calculations of the cellulose content can be made from the glucose value, and the hemicellulose content from the sum of the other sugar values plus the uronic acid value. The table shows that there are large differences between species and cultivars of the three main components. Of the hemicellulose figures, the uronic acid part varied between 2 and 3·7%. As seen in Table VI, arabinoxylans are the predominant hemicellulose polysaccharides and those containing galactose and mannose only minor ones. Structural studies on xylans from straw, cereals and grasses have been reported in several investigations[20] and Fig. 3 gives an illustration of the structural features generally found—a β-1,4-linked xylan backbone with arabinofuranosyl units and some glucuronic acid and/or its 4-O-methyl ether

TABLE VI

COMPOSITION OF NEUTRAL SUGARS IN HYDROLYSATES OF 80% ETHANOL-EXTRACTED STRAW (RELATIVE %)

Cultivar	*Arabinose*	*Xylose*	*Mannose*	*Galactose*	*Glucose*
Barley					
Cilla	7·2	35·4	0·5	1·9	55·0
Ingrid	5·9	32·7	0·5	1·3	59·6
Senat	7·0	32·1	0·3	1·7	58·9
Särla	7·3	35·6	0·4	1·9	54·8
Wing	6·1	33·9	0·3	1·4	58·3
Oats					
Titus	7·2	29·8	0·2	1·6	61·2
Winter wheat					
Holme	6·1	32·2	0·3	1·4	60·0
Spring wheat					
Drabant	6·5	32·3	0·5	1·8	58·9
Rye					
Petkus II	4·7	28·8	0·4	1·2	64·9

TABLE VII

CHEMICAL COMPOSITION OF SOME AGRICULTURAL CROP RESIDUES[a] [21]

Sample	*Cell walls*	*Hemicellulose*	*Cellulose*	*Lignin*
Barley straw	81	27	44	7
Oat straw	73	16	41	11
Paddy straw	79	26	33	7
Wheat straw	80	36	39	10
Sorghum stover	74	30	31	11
Chickpea straw	62	20	30	10
Lucerne straw	69	19	38	11
Sugarcane bagasse	82	29	40	13
Sugarcane trash	80	26	36	10
Paddy hulls	86	14	39	11
Cottonseed hulls	91	15	59	13

[a] Figures are given as % on dry weight basis.

linked to xylose units. We also find 1–2 % acetyl groups in straws, probably mainly linked to the xylans.

Table VII gives some values for the content of cell walls, cellulose, hemicellulose and lignin from straws and some other agricultural crop residues.[21] These figures are based on a gravimetric detergent fibre analysis scheme.

From the straws in Table V the amount of material extractable with 80 % ethanol varied between 5·2 and 13·7 % of the dry weight of which 1·6–3·9 % represented lipophilic extractives soluble in chloroform. Incidentally the ash content was in the range 3·2–11·8 %, which is much higher than generally found for woods. There is very little reported about the chemical composition of straw extractives. We found that low-molecular weight carbohydrates in the nine straws discussed above amounted to 0·3–1·3 %, fructose, glucose, sucrose and the sugar alcohols arabinitol and mannitol being the main constituents.[19] Streibl *et al.*[22] reported 0·2–0·5 % waxes in rye straw.

7. PEAT

The magnitude of the enormous amount of peat present in the world is only imperfectly known, but figures in the range 190–542 billions of tonnes (dry basis) have been estimated. See the recent book of Fuchsman,[23] which is recommended for further reading of the industrial chemistry and technology of peat. Regions with particularly large resources are Northern America, USSR and the Scandinavian countries.

The chemical composition of peat naturally varies enormously with botanical origin, location and age (degree of humification). One usually differentiates between low-moor peats with mineral contents of 3–14 % mainly originating from woody plants and sedges, and the high-moor peat, which is more acidic and low in mineral content, with *Sphagnum* mosses, cottongrass and heath plants as the parent sources. The former type of peat decomposes more rapidly than the latter and gives a residue relatively richer in carbon, which has made the low-moor peats attractive for fuels and as raw materials for methane production. The higher polysaccharide content of *Sphagnum* peats, on the other hand, has made them more suitable for hydrolysis to yeast cultures. Peats are also used for agricultural fertilizers. Humic acids, the chemically complex products formed in increasing amounts by the continued modification and decomposition in the peat bog of the original plant components, have been extracted by alkali and used as surface active agents.

Because of the different botanical origin of peat and dependent on the decomposition—which particularly in the low-moor type of peat will partly produce some microbial carbohydrates—the amount and pattern of carbohydrates, as expected, varies a great deal.[23] Some results from an earlier investigation by the author,[24] before the modern gas chromatographic methods were introduced, are presented in Table VIII. Given are the amounts of extractives (stepwise extraction by benzene, methanol and water), material removed from the extracted samples by oxidative solution (stepwise treatment with acidified sodium chlorite at 60–65 °C) and the

TABLE VIII

FRACTIONS FROM *Sphagnum* MOSSES AND SOME *Sphagnum* PEAT SAMPLES[a] [24]

Fraction	*Sphagnum moss*	*Sphagnum peat*	
		Age: *900–1 200 years*	*4 300–4 800 years*
Ash	1·7	1·7	1·7
Benzene extract	1·7	5·9	6·7
Methanol extract	5·4	4·6	4·9
Water extract	4·1	1·6	1·1
Losses by oxidative solution (lignin, humic acids, etc.)	3·7	43·9	57·1
Holocellulose	85·1	44·0	30·2

The *Sphagnum* moss figures are average values of *Sph. fuscum* and *imbricatum*.
[a] All figures are given as % on dry weight basis.

residual holocellulose from *Sphagnum* moss and two of the peat samples studied. The material removed by the oxidative solution should represent lignins, humic acids and other noncarbohydrate components. Holocellulose preparation by a mild delignification, such as by acidified chlorite solution, is a classical method to get an estimate of the cellulose and hemicellulose (and partly also pectins) content in woods, pulps and other plant materials.[25]

The results show that the *Sphagnum* polysaccharides in high-moor peatland are degraded rather slowly. Fractionation of the holocellulose fractions and analytical studies on their hydrolysates indicate that both in the moss and peat samples cellulose as well as hemicellulose and pectins were major constituents. In particular, the large amounts of uronic acids (proved to be both galacturonic acid and methoxylated acids) present were notable. We found that the amount of lignin or lignin-like phenolic components (characterized by a predominance of p-hydroxyphenyl units) was very low (around 1%)[26]—compare also the low amount of material removed from the moss by oxidative solution. The author is therefore of the opinion that, at least in high-moor *Sphagnum* peat, the main formation of humic acids is by chemical transformation of carbohydrates via reactive phenolic intermediates. Our later results, showing that in particular pentoses and uronic acids under slightly acidic conditions react in that way (see Section 8) give strong support for this hypothesis. Modern studies including the use of ^{13}C NMR (see ref. 27 and references therein) has proved the presence of p-hydroxyphenyl residues in *Sphagnum* moss, but suggests that they probably are not present in a polymer, which can be defined as lignin.

As seen in Table VIII the amounts of hydrophilic extractives decrease and those of lipophilic extractives increase with increasing age and increasing huminosity in a *Sphagnum* bog. The peat components soluble in organic solvents—collectively called bitumens—have been used for production of technical products for many years and studies on characterization of their different components are extensive.[23] This fraction is thereby refined and separated into resins, waxes and asphalts. The extracted peat can be pyrolysed to phenol-containing tars. In particular the peat waxes, which have shown a considerable similarity in their main chemical components from widely varying peats, have been used for many technical purposes—including as a replacement of montan wax, obtained from brown coal.

For an extensive analytical study of the lipophilic components in *Sphagnum* peat of various ages, see Ekman and Ketola.[28]

8. CONVERSION OF CARBOHYDRATES TO PHENOLS AND OTHER NON-CARBOHYDRATE COMPOUNDS

It has long been known that thermal treatment of hexoses in aqueous acidic solution gives hydroxymethylfurfural as the main reaction product and that pentoses and hexuronic acids give furfural. In connection with studies at the Swedish Forest Products Research Laboratory on the yellowing and degradation of cellulosic products (highly refined and almost free of lignin) during storage, use and heat treatment, we discovered that accelerated colour formation occurred under the prevailing slightly acidic conditions. We found that active intermediates in this colour formation were reductic

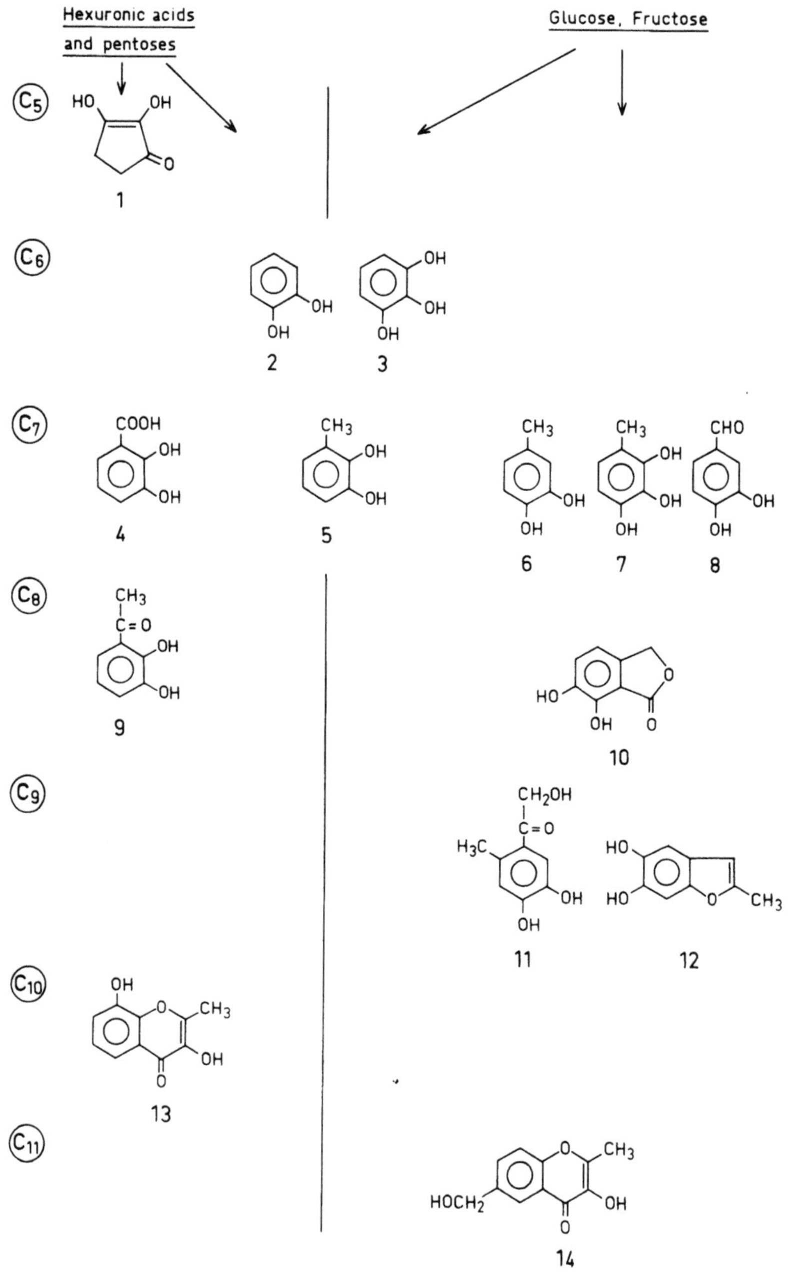

FIG. 8. Examples of products from thermal treatment of hexuronic acids, pentoses and hexoses in slightly acidic solution.[29,30]

acid but also phenols as catechols, acetophenones and chromones—all formed by degradation of pentoses and hexuronic acids, released by hydrolysis of the xylans.[29] Some examples of the compounds isolated are shown in Fig. 8, where formula **1** is reductic acid and **2–5**, **9** and **13** represent the catechols, acetophenones and chromones respectively. Under the slightly acidic conditions used the formation of these types of compounds was more pronounced than that of furfural and uronic acids were more reactive than pentoses. By later studies it was found also that glucose and fructose are converted to various phenolic compounds under similar conditions, but at a lower rate.[30] Some examples are shown in Fig. 8.

We studied later the conversion of other sugars, the formation of phenolic and other non-carbohydrate compounds under neutral and alkaline conditions and the still more complicated system of reactions which occur between sugars and amino compounds (the Maillard reaction). In the latter case, besides furans and phenols, also pyrroles, pyridines and other nitrogenous compounds are formed. It was also found that phenols formed by sugar conversion can react with the excess of sugar present to C-glycosides. For a summary of these recent investigations see ref. 31.

The conversions of carbohydrates discussed here are certainly of importance not only for discoloration but also for the formation of toxic and inhibiting compounds in the preparation of sugar solutions for fermentative processes and of phenolic and other products in liquefaction and pyrolysis processes. At present we are studying the mechanisms of the carbohydrates to phenols conversion in order to increase the yields of some phenols in an order of technical interest. We have so far increased the yield of the main chromone from pentoses and uronic acids (formula **13** in Fig. 8) to 10–15%.

REFERENCES

1. Johansson, M. H. and Samuelson, O. (1977). *Wood Sci. Technol.* **11**, 251–63.
2. Timell, T. E. (1965). *Adv. Carbohydr. Chem. Biochem.* **19**, 247–302; **20**, 409–83.
3. Janson, J., Haglund, P. and Sjöström, E. (1981). *Wood Chemistry*, E. Sjöström (Ed.). Academic Press, New York, p. 208.
4. Berg, B., Hannus, K., Popoff, T. and Theander, O. (1980). *Ecol. Bull. (Stockholm)* **32**, 391–400.
5. Ekman, E., private communication.
6. Holmbom, B. (1977). *J. Am. Oil Chem. Soc.* **54**, 289–93.

7. Assarsson, A. (1969). In Lindgren, B. and Norin, T., *Sven. Papperstidn.* **72**, 143–55 (in Swedish).
8. Enzell, C. and Theander, O. (1962). *Acta Chem. Scand.* **16**, 607–14.
9. Ekman, R., Sjöholm, R. and Hannus, K. (1977). *Acta Chem. Scand.* **B31**, 921–2.
10. Shain, L. and Hillis, W. E. (1971). *Phytopathology* **61**, 841–5.
11. Popoff, T., Theander, O. and Johansson, M. (1975). *Physiol. Plant.* **34**, 347–56.
12. Hillis, W. E. (Ed.) (1962). *Wood Extractives and their Significance to the Pulp and Paper Industries.* Academic Press, New York.
13. Rowe, J. W. and Conner, A. H. (1979). Forest Products Laboratory, Forest Service, US Dept. of Agric., Madison, General Technical Report FPL 18.
14. Chang, Y. P. and Mitchell, R. L. and Dietrichs, H. H., Garves, K., Behrensdorf, D. and Sinner, M., respectively in ref. 15.
15. Hemingway, R. W. (1981). In *Organic Chemicals from Biomass*, I. S. Goldstein (Ed.). CRC Press, Boca Raton, Florida, pp. 189–248.
16. Hannus, K. and Pensar, G., in ref. 18.
17. Theander, O. (1982). *Sven. Papperstidn.* **85**, R64–R68.
18. Barton, G. M. (1981). In *Organic Chemicals from Biomass*, I. S. Goldstein (Ed.). CRC Press, Boca Raton, Florida, pp. 249–80.
19. Theander, O. and Åman, P. (1978). *Swedish J. Agric. Res.* **8**, 189–94.
20. Wilkie, K. C. B. (1979). *Adv. Carb. Chem. and Biochem.* **36**, 215–64.
21. Jackson, M. G. (1977). *Anim. Feed Sci. Technol.* **2**, 105–30.
22. Streibl, M., Konecny, K., Trka, A., Ubik, K. and Pazlar, M. (1974). *Collect. Czech. Chem. Comm.* **39**, 475–9.
23. Fuchsman, C. H. (1980). *Peat—Industrial Chemistry and Technology.* Academic Press, New York.
24. Theander, O. (1954). *Acta Chem. Scand.* **8**, 989–1000.
25. Browning, B. L. (1967). *Methods of Wood Chemistry.* Vol. II, Interscience, New York.
26. Lindberg, B. and Theander, O. (1952). *Acta Chem. Scand.* **6**, 311.
27. Nimz, H. N. and Tutschek, R. (1977). *Holzforschung* **31**, 101–6.
28. Ekman, R. and Ketola, M. (1981). *Kemia-Kemi* n:o 7–8.
29. Popoff, T. and Theander, O. (1972). *Carbohydr. Res.* **22**, 135–49; *Acta Chem. Scand.* **B30**, 705–10.
30. Popoff, T. and Theander, O. (1976). *Acta Chem. Scand.* **B30**, 397–402.
31. Olsson, K., Pernemalm, P.-Å. and Theander, O. (1981). *Prog. Fd Nutr. Sci.* **5**, 47–55.

3

Lignin

WOLFGANG G. GLASSER
Department of Forest Products,
and Polymer Materials and Interfaces Laboratory,
Virginia Polytechnic Institute and State University,
Blacksburg, Virginia, USA

1. INTRODUCTION

Lignin is deposited by woody plants for the purpose of providing the growing plant with mechanical support, of sealing a water conducting system which links roots with leaves (needles),[1-3] and for protecting wood against degradation by weathering[4] and biodegradation.[5] Lignified plants contain between 20 and 30% lignin, depending on species, age, juvenile vs. mature wood, sapwood vs. hardwood, and normal vs. reaction wood.[1,6] The 20–30% lignin content by weight corresponds to approximately 35–45% by energy content since lignin is the polymeric cell wall component with the highest energy content, consisting of more than 60% carbon and around 30% oxygen. This is contrasted by hemicellulose and cellulose with carbon contents of less than 50% and oxygen contents of nearly 50%.[7]

This paper is to review the formation of lignin in terms of its chemical and some of its physiological aspects; the separation of lignin from carbohydrates; and some utilization prospects for lignin in non-fuel applications.

2. CHEMISTRY OF FORMATION: BASIC MECHANISMS

Lignin is synthesized exclusively from p-hydroxy cinnamyl alcohol precursors by dehydrogenative polymerization.[8] These lignin-forming, phenylpropanoid (C_9) precursors may be partially methoxylated, and they are known as p-coumaryl alcohol (with a p-hydroxy phenyl moiety as aromatic ring), coniferyl alcohol (with a 3-methoxy-4-hydroxyphenyl, 'guaiacyl', moiety as aromatic ring), and sinapyl alcohol (with a 3,5-dimethoxy-4-hydroxyphenyl, 'syringyl', moiety as aromatic ring).[9] The ratio between the three different types of cinnamyl alcohols depends on genetic factors,[10] environmental conditions[6] and the particular physiological situation of the growing plant cell at the time of lignification.[11]

Lignin precursors are synthesized from monosaccharides via the shikimic acid pathway, and this has been reviewed by Neish[12] and by Sarkanen.[13] Precursors arise from the phenylalanine and tyrosine pools by sequential (a) deamination with the enzyme(s) ammonia-lyase, (b) aromatic hydroxylation with a 4-hydroxylase and a 3-monoxygenase, (c) methylation with 3-O-methyl-transferase, and (d) reduction of the intermediary coenzyme-A thioesters with an oxidoreductase to the aldehyde, and with a dehydrogenase to the primary alcohol.[14] Since p-hydroxy cinnamyl alcohols are susceptible to (premature) oxidation leading to polymerization and termination of the cell's life functions, the living cell seems to possess what amounts to a defense mechanism providing it with the ability to control lignin formation. This is based on glucosidation of the phenolic sites resulting in (a) improved precursor stability against oxidation, (b) improved precursor solubility in water and transportability in aqueous solutions, and (c) reduced toxicity.[15] The resulting glucosides, such as coniferin, can be transported without risk of inadvertent polymerization to the site of lignification. The translocation of lignin precursors in the form of their glucosides from the cambial zone into lignifying tissue has been suggested by Freudenberg.[16] Recently, however, this hypothesis has become doubtful since (a) 'it has been concluded that lignin precursors most probably arise within the lignifying cell and not in adjacent meristems such as the cambium',[17] and (b) since 'the enzymes involved in the biosynthesis of lignin precursors are tissue-specific and are located predominantly or exclusively in the xylem in which lignification occurs'.[14] In any event, liberation of the p-OH cinnamyl alcohol from its glucoside, and its subsequent dehydrogenation by peroxidase, concludes the involvement of enzymes in the process of lignification. The enzymatic generation of phenoxy radicals triggers a polymerization process which is entirely controlled by thermodynamic principles. The chemical complexity of lignin is a result of the many different types of linkages that exist between individual cinnamyl alcohol units. This variety, in turn, is a consequence of the distribution of free π-electron spin densities throughout the C_9 (phenylpropane) molecule which represents lignin precursors.[2]

The relative distribution of free π-electron spin densities in various

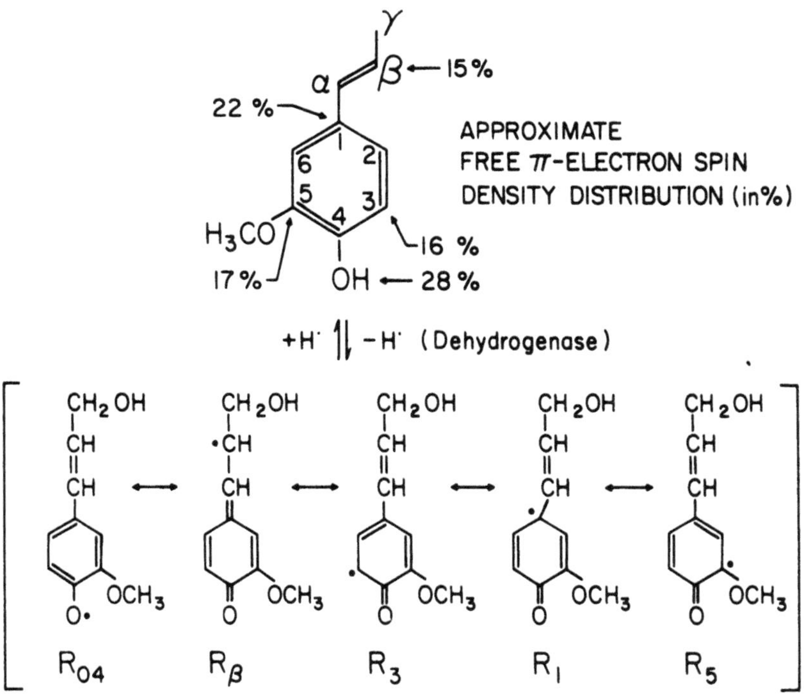

FIG. 1. The coniferyl alcohol phenoxy radical and its mesomeric forms.

lignin-like monomeric and dimeric model compounds has been studied by Martensson and Karlsson,[18] and by Remko *et al.*[19] The simplified results of some of these efforts are summarized in Fig. 1. This illustration suggests that free π-electron spin densities decline in the order of phenolic OH position (04 position), position 1, 5, 3 and β. All other positions have nil or negligible free electron concentrations.

Polymerization of phenoxy radicals involves coupling processes leading to the formation of intermediary, covalently bound reaction products, $[R_X\text{–}R_Y]$ (Fig. 2), which may either revert to their starting materials (phenoxy radicals, $R_{\dot{X}}$ and $R_{\dot{Y}}$), or which may stabilize by internal rearrangement. The extent to which these intermediates revert to the free radical form or to the stable product must depend on steric arguments, the nature of substituents, and a variety of other factors. The presence of a

$$R_{\dot{X}} + R_{\dot{Y}} \rightleftharpoons [R_X - R_Y] \rightarrow R_X - R_Y$$

FIG. 2. Basic phenoxy radical coupling reaction.

substituent other than a hydrogen atom at the free radical site, such as in positions 1 and 3 of the coniferyl alcohol radical (Fig. 1), must be suspected to result in reversion being favored. Thus, the relative frequency with which the other mesomeric phenoxy radical species—those with unpaired electrons in positions 04, β and 5—are involved in radical coupling reactions must be upgraded proportionately. Table I illustrates the products from phenoxy radical coupling and their expected frequencies if it is assumed that only 1 of 2 R_1-radicals (Fig. 1) and no R_3-radicals lead the formation of stable combinations. A total of eight interunit linkages are

TABLE I

RELATIVE THEORETICAL FREQUENCIES OF COUPLING MODES OF PHENOXY RADICALS FROM p-OH CINNAMYL ALCOHOLS

Mesomeric radical species[a]	R_{04}	R_β	R_5	R_1	R_3
R_{04}	—	0·092 β–0–4	0·101 4–0–5	0·046 1–0–4	—
R_β	0·092 β–0–4	0·048 β–β	0·053 β–5	0·024 β–1	—
R_5	0·101 4–0–5	0·053 β–5	0·058 5–5	0·012 1–5	—
R_1	0·046 1–0–4	0·024 β–1	0·012 1–5	?	—
R_3	—	—	—	—	—

Estimated from data by Martensson and Karlsson.[18]
[a] cf. Fig. 1.

ranked according to their relative expected frequency of occurrence based on free electron spin density concentrations throughout the phenylpropane unit of lignin. The results of this linkage projection based on quantum mechanics are compared in Table II to those predicted on the basis of a simulation approach known as the SIMREL technique.[20,21]

SIMREL constitutes a modeling attempt which correlates accepted biosynthetic principles and primary and secondary experimental observations with a distinct structural scheme of lignin. Where the initial simulation studies were based on published secondary experimental data,[20,21] more recent studies interfaced the SIMREL program with such primary analysis data-generating devices as gas and liquid chromatographs and spectrophotometers.[22,23] This technique permits the immediate correlation of

TABLE II
APPROXIMATE LINKAGE DISTRIBUTION OF A SOFTWOOD LIGNIN BY QUANTUM MECHANICS AND BY COMPUTER SIMULATION (SIMREL)

Linkage type	*Linkage frequency by*	
	Quantum mechanics	*Computer simulation*
β–0–4	27	41
β–β	7	3
β–5	16	13
4–0–5	30	3
5–5	9	16
β–1	7	8
Others	4	17

analysis data obtained by degradative analysis and spectroscopy of derivatives (acetates etc.) with hypothetical polymer structures. Figure 3 represents such a simulated hardwood lignin structure with a total of 98 phenylpropane units and a molecular weight of 19 000 D. This structure is in good agreement with most analytical observations made with mildly isolated hardwood lignin preparations of a syringyl-to-guaiacyl ratio of 0·7.

Where quantum mechanics predicts a greater frequency of 4–0–5 linkages followed by β–0–4, β–5, 5–5, β–β and β–1 combinations (Table II), structure analysis according to SIMREL records greater concentrations of β–0–4 linkages which are followed distantly by 5–5, β–5, β–1, 4–0–5 and β–β bonds. Several linkages based on non-dehydrogenative polymerization mechanisms, not involving free phenoxy radical coupling reactions, account for the majority of linkages listed as 'others' in Table II. The differences between the results predicted on the basis of quantum mechanics and those suggested by SIMREL are considerable, and they reflect the fact that the quantum mechanical predictions were made on the basis of monomeric radical coupling reactions only, which in itself constitutes the least likely case.[24,25] Lignification in woody tissue amounts less to a combination of monomers with each other, but more to the gradual addition of monomeric phenoxides to oligomers and a growing polymer. Using a phenylpropanol with saturated sidechain as model for a lignin-like polymer involved in growth, Ericksson and Miksche[25] succeeded in demonstrating that β–0–4 ethers are the predominant linkages

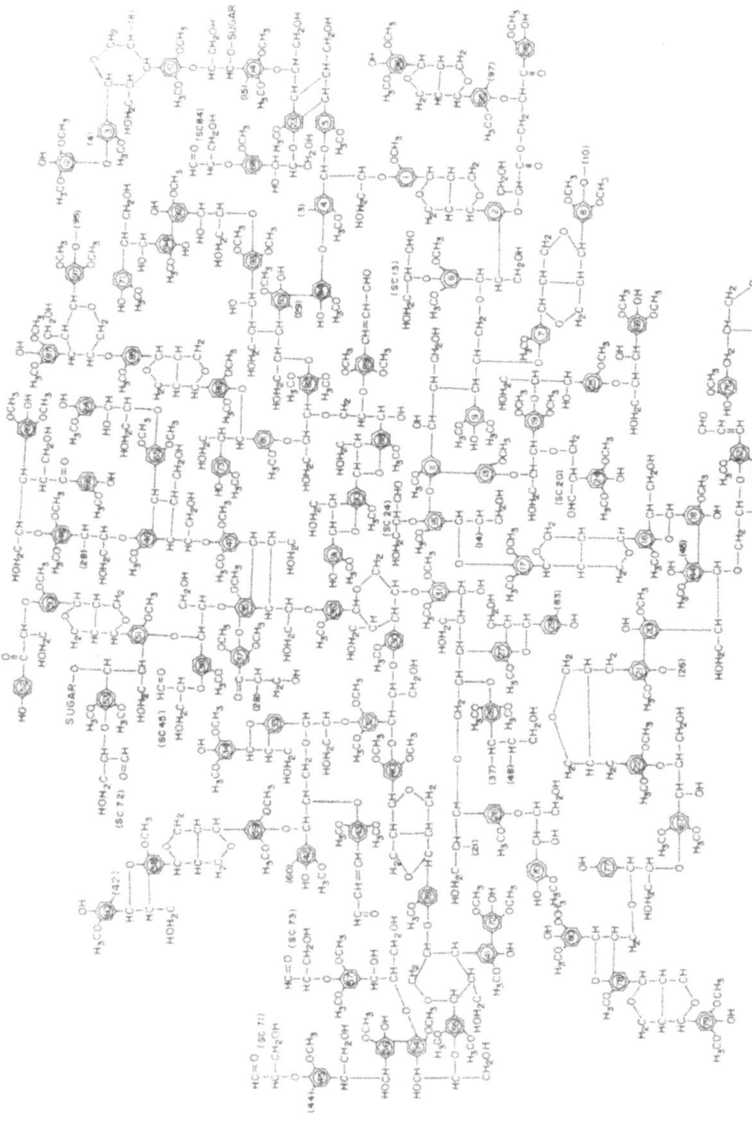

FIG. 3. Schematic formula for a hardwood lignin with syringyl-to-guaiacyl ratio of 0·7.

formed upon the gradual addition of coniferyl alcohol in the presence of peroxidase/H_2O_2. The fact that this 'cross-coupling' between two different phenolic species was favored over dimerization of any one phenol was explained with concentration effects of the phenol with the lower oxidation potential.[25] Thus, the formation of alkyl-aryl ethers (β–0–4) during lignification can be expected to be more predominant than is suggested on the basis of quantum mechanical calculations of monomers. In addition, 4–0–5 type linkages (diaryl-ethers) must represent an improbable coupling mode because of the requirement for two aromatic rings to come extremely close to each other. The lower concentration of β–β linkages in polymeric lignins than projected by quantum mechanics (Table II) must be attributed to low concentrations of phenols with unreacted β-positions.

Lignin has been known to be structurally variable, not only between softwoods and hardwoods,[10] but also with respect to morphological location within one plant.[26] The precise extent of this variability remains the subject of much speculation in the current literature. An image of surprisingly little variability results from the data presented in Fig. 4. These data were obtained by the structural analysis of milled wood lignin preparations isolated from red alder and Douglas fir sapwood and NSSC pulp preparations.[27,28] High yield, neutral sulfite pulps were prepared in a

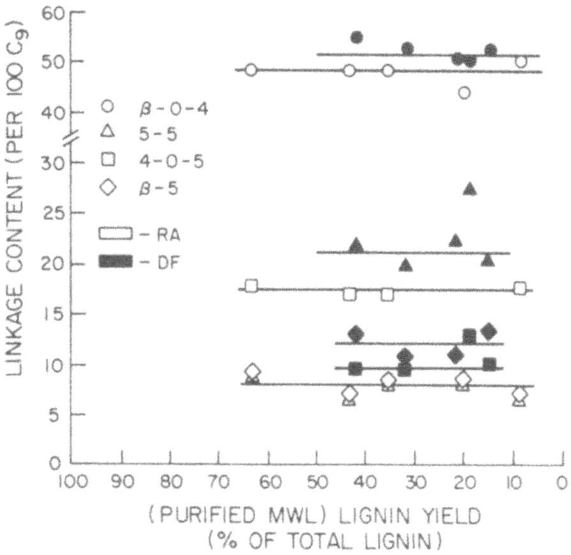

FIG. 4. Relationship between linkage distribution and lignin isolation yield of red alder and douglas fir softwood and NSCC pulp MWL preparations.

laboratory digester in cooks ranging between 5 min and 8·5 h in cooking time, and producing pulps in 59 to 95% yield. This study has been the subject of prior reports elsewhere.[29] The isolated lignin preparations were analyzed with regard to functionality and frequency of prominent interunit linkages, and these results were related to yield of purified lignin based on total lignin content of the sapwood (Fig. 4). The yield of isolated lignin preparations ranged between <10% and >60% of total lignin in wood. The results (Fig. 4), obtained with 10 different lignin preparations, demonstrate that linkage content remains unaffected by lignin yield over the entire yield range. This suggests a surprising lack of variability in lignin structure in relation to yield of isolation. By contrast, a recent study on the classification of various lignins according to their alkyl aryl ether content[30] concluded that there exists a distinct relationship between the syringyl-to-guaiacyl ratio and the alkyl aryl ether content. This is illustrated in Fig. 5. Data obtained from MWL preparations from 15 different syringyl–guaiacyl (hardwood-type) lignins demonstrated the existence of a significant relationship between the alkyl aryl ether content and the syringyl-to-guaiacyl ratio with a correlation factor (R) of 0·81. Thus, it appears that although lignin structure in terms of interunit linkages is not subject to variation with regard to yield of isolation, it is indeed subject to variation with regard to syringyl-to-guaiacyl ratio and related functionality.

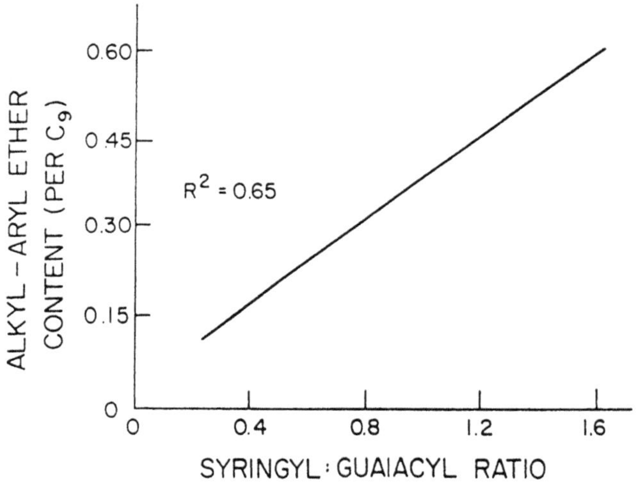

FIG. 5. Relationship between alkyl-aryl ether content and syringyl-to-guaiacyl ratio of milled wood lignins.

3. PHYSIOLOGY OF FORMATION

The lignification hypothesis established and advanced by Freudenberg and his associates and accepted throughout the lignin chemistry field for many years involves the production of lignin precursor glucosides by the cambium cells of a growing (woody) plant, their transportation across approximately 10-cell layers to the expanding xylem portion of the plant, the enzymatic release of the aglycones from the glucosides (by β-glucosidase), and the subsequent oxidation of the phenolic precursors leading to oxidative polymerization and lignin formation.[8] Differences in lignin structure in relation to morphological location of lignin,[26,31] and other results,[14,17] led to a general questioning of the intercellular transport hypothesis. Extensive work by Wardrop[17,32] and others[14,33] seems now to have established the fact that 'lignin precursors are formed in the cytoplasm of the lignifying cell'.[32] This observation, in turn, makes it possible and likely for the plant to respond to any form of stress by means of influencing and controlling the lignin formation process. Thus, it seems possible that lignin formation and lignin structure become subject to changes in the physiology of the living cell. Variation with regard to two factors, lignin content and lignin structure, become worthwhile targets for research dealing with the biochemistry of lignin formation.

Attempts to reduce the amount of lignin formed by a growing plant—lignin content manipulation—has focused on controlling, reducing and generally impairing the activities of enzymes involved in the biosynthesis of lignin precursors and of lignin itself.[34-36] Boron-,[34] calcium-,[35] and zinc-containing[36] minerals were found to be of influence for various enzyme activities, and they were shown to impair the process of lignin formation. In addition, phenylalanine ammonia-lyase (PAL)[37] was found to be inactivated by the addition of several chemical compounds. The absence of PAL activity leads to reduced p-OH cinnamyl alcohol formation by preventing diamination of phenylalanine, an important precursor of lignin formation.

Lignin structure was found to be subject to physiological manipulation by several means. The rate of precursor supply to the site of lignification was found to be affected by the temperature in which lignification took place.[38,39] Low temperatures resulted in a reduced rate of precursor supply, which in turn resulted in higher alkyl aryl ether contents and lower biphenyl (5–5) and diaryl ether (4–0–5) contents. Increased polyether nature and reduced 'condensed', carbon-to-carbon linked nature of lignin can be expected to influence thermal and reactivity properties of lignins.

The activity of peroxidase enzymes was found to vary with the pH of the aqueous medium, and the presence of auxins.[39,40] Elevated concentrations of growth hormones (auxins) were found to raise the activity of peroxidases, which in turn resulted in increased degrees of condensation. Growing stresses were found to lead to increased auxin concentrations in cambial cell regions.

Structural differences between lignins deposited when plants were exposed to light vs. dark conditions were explained with differences in the transportation of ferulate and sinapate precursors.[35,41] These seemed to be related to the permeability of certain vesicles, which appeared to be sensitive to lighting conditions. Darkness produced syringyl lignin exclusively; and lignin with high syringyl contents can be expected to be rich in alkyl aryl ether linkages.[30]

It can be concluded that current research efforts in plant physiology are demonstrating that plants retain more control over the process of lignification than was to be expected on the basis of Freudenberg's hypothesis regarding the need for intercellular transport of glucosidic lignin precursors. Plants seem to be capable of controlling lignification with regard to content as well as structure. Three distinct control mechanisms appear to offer opportunity for manipulating lignification; these are control over enzyme activities, rate of precursor delivery to the site of lignification, and nature of the composition of the precursor mixture, i.e. ratio of syringyl to guaiacyl precursors.

4. SEPARATION FROM LIGNOCELLULOSE

The separation of lignin from lignocellulosic materials in some type of water- or organic solvent-soluble form requires the elimination of all lignin to carbohydrate bonds, and sufficient depolymerization of the lignin gel to achieve solubility. The option to separating soluble lignin from insoluble, carbohydrate-rich fractions exists in the form of separation by dissolving carbohydrates away from undissolved lignin, even though this may be necessary in the form of mono- and oligosaccharidic degradation products. Although industrially practised during World War II, the latter route seems to be a less appealing technological option to lignocellulose fractionation than the former, probably because of low yields encountered during wood saccharification.[42,43] Thus, most recent research has focused on fractionating lignocellulosic resources into soluble lignin fractions and insoluble, although partially hydrolytically depolymerized, carbohydrate fractions.[44-46]

Since lignin-to-carbohydrate bonds seem to be exclusively of ether (and possibly ester) nature,[47-49] and since lignin depolymerization is most conveniently achieved by hydrolytic means, some type of alkaline or acidic hydrolysis scheme appears to hold greatest promise for achieving the fractionation goal. The reactions of lignin and lignin model compounds under acidic and alkaline conditions have recently been reviewed by Gierer.[50] Prominent intermediates in this depolymerization pathway are quinonemethides and benzyl carbonium ions of lignin.

Extent and completeness of fractionating lignocellulose into distinct lignin and carbohydrate fractions are limited by recombination reactions which have been suspected to cause the retention of residual lignins in commercial (unbleached) pulps. Such *in situ* recombinations between lignin and carbohydrates may involve any one of three different reaction mechanisms. These are based on reactions with free radicals,[51] on nucleophilic addition reactions to quinonemethide intermediates,[29,52] and on acid promoted condensation reactions within lignin and between lignin and carbohydrates.[53]

Free radicals on lignin were reported to be responsible for the reaction between maleic anhydride and toluene in the presence of sawdust,[54] for traces of toluene grafted on to lignin as a consequence of ballmilling sawdust with steel balls in the presence of toluene (Bjorkman lignin isolation procedure),[55] and for the retention of dioxane by lignin following ballmilling and solvent (aqueous dioxane) extraction.[22] That phenoxy radical coupling reactions may lead to the formation of new lignin-to-carbohydrate linkages has been pointed out previously.[29]

The frequency with which quinonemethides appear as intermediates of lignin depolymerization under alkaline and neutral conditions creates ample opportunities for combining nucleophilic centers of carbohydrates with electron deficient carbon atoms of quinonemethides.[50,52,56] This nucleophilic addition reaction has previously been held responsible for the retention of lignin by high-yield pulps even after extended periods of time.[29]

The formation of carbon-to-carbon bonds internally in lignin as a consequence of acid treatment has long been known to be responsible for the formation of insoluble and unreactive lignin, such as Klason lignin.[1] Reactions leading to similar C–C connected condensation products under alkaline reaction conditions have also been reported.[57] Internal recombination reactions of this type may lead to the regeneration of lignin gels inaccessible to further depolymerization and dissolution in organic solvents, water or alkali. Although this recombination may only rarely be responsible for creating linkages between polysaccharides and lignin, low molecular weight degradation products of carbohydrates, such as furfural and hydroxymethyl furfural, may indeed be involved in condensation reactions with benzyl carbonium ions.[53] These reactions also may lead to lignins of low solubility, low reactivity, and high crosslink density, which are trapped inside cell wall matrices.

Thus, it can be concluded that recombinations of lignin internally, and between lignin and carbohydrates, may involve the formation of dialkyl ether and carbon-to-carbon linkages under neutral and alkaline conditions, and the formation of carbon-to-carbon bonds internally in lignin and between lignin and furans under acidic conditions. In either case, a less soluble, less reactive, more highly cross-linked network polymer is formed from lignin, which is either chemically or physically trapped in the cell walls of cellulosic fibers.

5. UTILIZATION PROSPECTS

Research into the utilization of lignin is among the most challenging and promising areas of wood chemistry.[46,58,59] Blessed with an overabundance of lignin residues co-produced during pulp and papermaking, with values ranging between less than 0 and approximately 4¢/1b (fuel equivalent), the total amount of lignosulfonates and kraft lignin together outweigh the sum of all manmade polymers in the United States.[3] Formidable problems facing lignin utilization include natural variability and inhomogeneity, a complex chemical and macromolecular structure posing an exceedingly difficult task of quality control, and an almost insurmountable resistance to any type of degradation to low molecular weight chemicals.[3,58] It is no wonder, then, that industrial lignin utilization has taken the route of using lignin as a cheap extender and polymeric substitute in a variety of end products, most of which involve surface-active properties and the ability of lignin and its derivatives to disperse aqueous suspensions.[60,61] Of the approximately 1·5 billion pounds of lignin sold annually for non-fuel purposes, most sells for just slightly above its fuel value.[58] However, the fact that some (isolated and dried) lignin products sell for 10–20 times their fuel value seems to suggest that there exist utilization schemes which warrant further exploration.

Opportunities for lignin utilization obviously lie within its enormous abundance and its inherent design features as an aromatic, phenolic macromolecule designed for the purpose of mechanically supporting a fiber composite material, and as an antioxidant. The prospect of producing ethanol from biomass by process technology generating as much as 7–10 lb of lignin coproducts per gallon of ethanol poses a significant challenge to scientists and engineers to make better use of this material. Quality control and continued analytical efforts to better describe the raw material characteristics of coproduced lignins must eventually lead to fruitful results if biomass conversion to chemicals is to become industrial practice.[45,46,62]

The use of lignin as a binder for cellulosic fibers and wood particles seems to offer a natural route to lignin utilization.[63] This approach is limited by the intractable nature of polymeric lignin, by its high thermal (glass) transition temperature, and by its low solubility and consequent reactivity in solvents other than aqueous alkali.[3] The 'new lignins' from steam explosion and organosolv delignification processes are different from their pulp and paper counterparts in terms of both chemical and thermodynamic characteristics.[30,45,46,62,64–67] In addition, oxyalkylated lignin derivatives with lignin contents in excess of 60% were found to be by and large thermoplastic and solvent soluble materials.[68,69] Cross-linking these aromatic prepolymers with suitable cross-linking agents, such as isocyanates,[70,71] phenol-formaldehyde resins,[72,73] and others[74,75] seems to offer a promising avenue for utilizing large quantities of lignin in thermosetting network systems. Already advances have been made and are documented in the patent literature which suggest that water-soluble lignin derivatives (ligninsulfonates) are capable of serving as contributing copolymers to polymeric isocyanates in the manufacture of particleboards.[71] Limitations in terms of water sorption properties and brittleness, no doubt, are inherent to a particular class of pulping-derived lignin products. Additional research must focus on finding methods for circumventing these drawbacks.

6. CONCLUSIONS

1. The chemical structure of lignin is well understood, both in terms of the mechanism of biochemical formation and in terms of its chemistry. This has been modeled successfully by computer techniques. The methodology exists for analyzing and for quantitatively describing the chemistry of isolated lignin preparations.
2. The process of lignification is controlled physiologically by the lignifying cell via control over activities of enzymes involved in the production and supply of precursors for lignification. Various

stresses have been recognized to influence lignification both with regard to lignin content and lignin structure.

3. The separation of lignin from carbohydrates by hydrolytic depolymerization, or by other means, is understood as well. It is recognized that the efficiency of separating lignin from its origin in and between cellulosic fibers is limited by recombination reactions both internally in lignin and between lignin and carbohydrates. These recombination reactions may involve free radical mechanisms, nucleophilic addition to quinonemethide intermediates, and acid-promoted secondary condensation reactions. These recombination reactions may occur over a wide pH range.
4. Lignin utilization offers great opportunities. Failures have been reported for the production of low molecular weight chemicals by some type of depolymerization, and successes are being recorded with polymeric products in which lignin plays a role similar to that of its original purpose. Lignins and lignin derivatives with superior solubility and thermal characteristics are expected to be useful prepolymers for a multitude of structural polymeric materials.

REFERENCES

1. Sarkanen, K. V. and Ludwig, C. H. (Eds) (1971). *Lignins—Occurrence, Formation, Structure and Reactions*. Wiley-Interscience, New York.
2. Glasser, W. G. (1980). *Pulp and Paper Chemistry and Chemical Technology*, 3rd edn. Vol. I. J. P. Casey (Ed.), Wiley, New York, pp. 39–111.
3. Falkehag, S. I. (1975). *Appl. Polym. Symp.* **28**, 247–57.
4. Chang, S. T., Hon, D. N.-S. and Feist, W. C. (1982). *Wood and Fiber* **14**(2), 104–17.
5. Vance, C. P., Kirk, T. K. and Sherwood, R. T. (1980). *Ann. Dev. Phytopathol.* **18**, 259–88.
6. Timell, T. E. (1981). *Proc. Interntl. Symp. Wood and Pulping Chem.*, Vol. I. Stockholm, June 9–12, pp. 99–147.
7. Browning, B. L. (Ed.) (1963). *The Chemistry of Wood.* Interscience Publishers, New York.
8. Freudenberg, K. (1968). In *Constitution and Biosynthesis of Lignin*, K. Freundenberg and A. C. Neish (Eds). Springer-Verlag, New York.
9. Sarkanen, K. V. and Ludwig, C. H. In ref. 1, pp. 10–16.
10. Sarkanen, K. V. and Hergert, H. L. In ref. 1, pp. 43–94.
11. Higuchi, T. (1980). *Wood Research* No. 66, 1–16.
12. Neish, A. C. In *Constitution and Biosynthesis of Lignin*, K. Freudenberg and A. C. Neish (Eds). Springer-Verlag, New York.
13. Sarkanen, K. V. In ref. 1, pp. 95–115.
14. Grisebach, H. (1977). *Naturwissenschaften* **64**, 619–25.
15. Heitefuss, R. and William, P. H. (1976). *Physiological Plant Pathology*. Springer, Berlin.

16. Freudenberg, K. (1965). *Science* **148**, 595.
17. Wardrop, A. B. (1976). *Appl. Polym. Symp.* No. 28, 1041–63.
18. Martensson, O. and Karlsson, G. (1969). *Ark. Kem.* **31**(2), 5.
19. Remko, M. and Polcin, J. (1978). *Zeitschrift fur Physikalische Chemie Neue Folge* **110**, 229–35.
20. Glasser, W. F. and Glasser, H. R. (1974). *Macromolecules* **7**(1), 17–27.
21. Glasser, W. G. and Glasser, H. R. (1974). *Holzforschung* **28**(1), 5–11.
22. Glasser, W. G. and Glasser, H. R. (1981). *Paperija Puu* **1981**(2), 71–80.
23. Glasser, W. G., Glasser, H. R. and Morohoshi, N. (1971). *Acta. Chem. Scand.* **25**, 647–62.
24. Larsson, S. and Miksche, G. E. (1971). *Acta. Chem. Scand.* **25**, 647–62.
25. Erickson, M. and Miksche, G. E. (1972). *Acta. Chem. Scand.* **26**, 3085–96.
26. Goring, D. A. I. (1981). *Proceedings of the International Symposium on Wood and Pulping Chemistry*, Stockholm, June 9–12, Vol. 1, pp. 3–10.
27. Glasser, W. G. and Barnett, C. A. (1979). *Holzforschung* **33**(3), 78–86.
28. Glasser, W. G. and Morohoshi, N. (1979). *Tappi* **62**(12), 101–5.
29. Glasser, W. G. (1981). *Sv. Papperstidn.* **84**(6), R25–R33.
30. Glasser, W. G., Barnett, C. A. and Sano, Y. (1983). *Appl. Polym. Symp.* No. 37, 441–60.
31. Musha, Y. and Goring, D. A. I. (1975). *Wood Sci. Technol.* **9**(1), 45.
32. Wardrop, A. B. (1981). *Proceedings of the International Symposium on Wood and Pulping Chemistry*, Stockholm, June 9–12, Vol. 1, pp. 44–51.
33. Grisebach, H., Luderitz, T. and Schmid, G. (1981). *Proceedings of the International Symposium on Wood and Pulping Chemistry*, Stockholm, June 9–12, Vol. 3, pp. 51–4.
34. Acerbo, S., Kastori, R., Sochtig, H., Harms, H. and Haider, K. (1973). *Zeitschrift fur Pflanzenphysiologie* **69**, 306.
35. Boudet, A. M., Grand, C., Ranjeva, R., Alibert, G. and Marigo, G. (1981). *Proceedings of the International Symposium on Wood and Pulping Chemistry*, Stockholm, June 9–12, Vol. 3, pp. 55–9.
36. Wyrambik, D. and Grisebach, H. (1979). *European J. Biochem.* **97**, 503.
37. Amrhein, N. (1981). *Zeitschrift fur Physiologische Chemie* **362**, 3.
38. Tominura, Y., Yokoi, T. and Terashima, N. (1979). *Mokuzai Gakkaishi* **25**(11), 743–8.
39. Tominura, Y., Yokoi, T. and Terashima, N. (1980). *Mokuzai Gakkaishi* **26**(1), 37–42.
40. Gross, G. G. In *Recent Advances in Phytochemistry*, Vol. 12. *Biochemistry of Plant Phenolics*, T. Swain *et al.* (Eds). Plenum Press, New York.
41. Grand, C., Ranjeva, R., Boudet, A. M. and Alibert, G. (1979). *Planta* **146**, 281.
42. Worthy, W. (1981). *C&E News*, Dec. 7, 35–42.
43. Goldstein, I. S. (1976). *Materials: Renewable and Non-Renewable Resources.* AAAS, Science Compendium No. 4, P. H. Abelson and A. L. Hammond (Eds), pp. 179–84.
44. Marchessault, R. H. and St.-Pierre, J. (1980). In *Future Sources of Organic Raw Materials—CHEMRAWN I*, L. E. St.-Pierre and G. R. Brown (Eds). Pergamon Press, New York, pp. 613–25.
45. Myerly, R. C., Nicholson, M. D., Katzen, R. and Taylor, J. M. (1981). *CHEMTECH* 1981 (March), 186–92.
46. Bungay, H. R. (1982). *Science* **218** (4573), 643–6.

47. Ericksson, O., Goring, D. A. I. and Lindgren, B. L. (1980). *Wood Sci. Technol.* **14**, 267–79.
48. Lundquist, K., Simonson, R. and Tingsvik, K. (1980). *Sv. Papperstidn.* **83**(16), 452–4.
49. Koshijima, T., Yaku, F. and Tanaka, R. (1976). *Appl. Polym. Symp.* **28**, 1025–39.
50. Gierer, T. (1982). *Holzforschung* **36**, 43–51 and 55–64.
51. Kleinert, T. N. and Morton, J. R. (1962). *Nature* **196**, 334.
52. Leary, G. J. (1980). *Wood Sci. Technol.* **14**, 21–34.
53. Chua, M. G. S. and Wayman, M. (1979). *Can. J. Chem.* **57**(10), 1141–9.
54. Glasser, W. G. and Sandermann, W. (1970). *Holzforschung* **24**, 73–6.
55. Larsson, S. and Miksche, G. E. (1969). *Acta Chem. Scand.* **23**(10), 3337.
56. Ohara, S., Hosoya, S. and Nakano, J. (1980). *Mokuzai Gakkaishi* **26**(6), 408–13.
57. Gierer, T., Imsgaard, F. and Petterson, I. (1976). *Appl. Polym. Symp.* **28**, 1195–211.
58. Glasser, W. G. (1981). *For. Prod. J.* **31**(3), 24–9.
59. Kringstad, K. (1980). In *Future Sources of Organic Raw Materials—CHEMRAWN I*, L. E. St.-Pierre and G. R. Brown (Eds). Pergamon Press, New York, pp. 627–36.
60. Glasser, W. G., Hsu, D. H.-H. and Nakano, J. (1976). *Sv. Papperstidn.* **79**(16), 527–30.
61. Allan, B. R., Cousins, M. J. and Pierce, G. E. (1980). *Pretreatment Methods for the Degradation of Lignin.* Final Report to NSF: Battelle Columbus Laboratories: Columbus, OH, pp. 1–176.
62. Glasser, W. G. (1983). *Lignin's Potential Contribution to the Feasibility of Biomass Conversions to Ethanol.* First Annual Report to USDA/SEA, Energy and Alcohol Fuels Research Grants Program; VPI & SU, Blacksburg, VA, pp. 1–135.
63. Glasser, W. G., Saraf, V. P. and Newman, W. H. (1982). *J. Adhesion* **14**, 233–55.
64. Fengel, D., Wegener, G. and Feckl, J. (1981). *Holzforschung* **35**(2), 51–7.
65. Meier, D., Faix, O. and Lange, W. (1981). *Holzforschung* **35**(5), 247–52.
66. Marchessault, R. H., Coulombe, S. and Morikawa, H. (1982). *Can. J. Chem.* **60**, 2372–82.
67. Glasser, W. G., Barnett, C. A., Muller, D. C. and Sarkanen, K. V. (1983). *J. Agric. Food Chem.* **31**, 921–30.
68. Glasser, W. G., Hsu, O. H.-H., Reed, D. L., Forte, R. C. and Wu, L. C.-F. (1981). In *Urethane Chemistry and Applications*, K. Edwards (Ed.). ACS Symp. Ser. No. 172: pp. 311–38.
69. Glasser, W. G., Wu, L. C.-F. and Selin, T.-F. (1983). In *Wood and Agricultural Residues: Research on Use for Feed, Fuels and Chemicals*, J. Soltes (Ed.). Academic Press, New York, pp. 149–66.
70. Hsu, O. H.-H. and Glasser, W. G. (1976). *Wood Sci.* **9**(2), 97–103.
71. Lambuth, A. L. (1981). U.S. Patent 4,279,788.
72. Forss, K. and Fuhrmann, A. (1979). *For. Prod. J.* **29**(7), 39–43.
73. Dolenko, A. J. and Clarke, M. R. (1978). *For. Prod. J.* **28**(8), 41–6.
74. Nimz, H. H. and Hitze, G. (1980). *Cell. Chem. Technol.* **14**, 371–82.
75. Shen, K. C. (1977). *For. Prod. J.* **27**(5), 32–8.

4

Pretreatment of Biomass for Thermochemical Biomass Conversion

EDWARD S. LIPINSKY

Battelle Columbus Laboratories, Columbus, Ohio, USA

Developers of microbiological processes to convert biomass into chemicals and fuels recognized long ago that pretreatments were needed prior to fermentation or enzyme reactions, if high rates and yields are to be achieved.[1,2] Microorganisms and enzymes have difficulty in depolymerizing crystalline polymers, such as cellulose, and hydrophobic molecules, such as lignin and triglycerides. Furthermore, biomass constituents frequently conspire to form complexes that are difficult for microbes or enzymes to break apart. Because of these problems, numerous pretreatment methods have been developed to aid in microbiological processing by disrupting the lignocellulose complex, facilitating the depolymerization of polymers, and removing protective extractives.

Developers of thermochemical processes have observed that yields have been below expectations, the product mix from thermochemical conversion frequently is not appropriate to the end use, and deleterious biomass ash effects have been observed in gasifiers and combustors. Biomass inputs frequently are variable in composition. The typical reaction has been to complicate further the thermochemical gasification or liquefaction process. However, recently thermochemical conversion researchers have begun to apply the concept of pretreatment to give the central processes (e.g. liquefaction) a better chance for success.[3,4]

It is the purpose of this paper to provide a prolegomena to the use of pretreatment in thermochemical conversion. It should be clear at the outset that a direct lateral transfer of pretreatment processes from the microbiological field will only occasionally be appropriate. Thermochemical processing deals with the facilitation of heat transfer and mass transfer primarily of much smaller entities than are encountered in microbiological processes. This is not to say that thermochemical pretreatments are simple compared with microbiological treatments. The tendency of thermally generated species to form high molecular weight tars and chars can provide just as great a challenge as is found with the lignin challenge in microbiological processing.

1. BIOMASS MORPHOLOGY

As shown in Fig. 1, biomass has a hierarchical organization. Although the tree is used as an illustration, each type of biomass has a similar morphology. Many conventional processes are designed to use chips as feedstock because of their ease in handling. The chip has a substantial amount of water, both in free and bound form. This water can act as a cooling agent and as a chemical reactant in a thermochemical conversion process. The lignocellulose complex that holds this water is in the form of fibers, vessels, and other anatomical parts of the tree from which it was obtained. Because of this porous structure, reaction products do not leave the wood chip in random fashion but may travel through pathways or vestiges of pathways. Size reduction beyond the chip level (e.g. sawdust) yields a feedstock with quite different heat transfer and mass transfer characteristics.

The lignocellulose complex is comprised actually of three polymers—lignin, hemicellulose and cellulose. These polymers are interlocked intimately enough that pyrolysis of the whole complex does not yield the same reaction products as is obtained by independent pyrolysis of the constituents.[5] Nor does pyrolysis of the monomers that comprise the respective polymers yield the same product as does pyrolysis of the

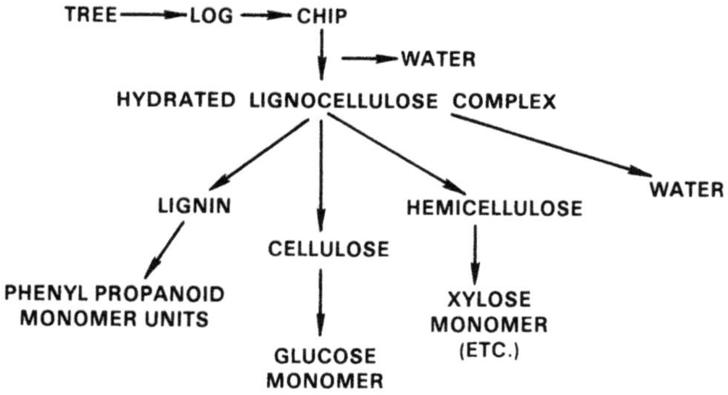

FIG. 1. Hierarchical organization of biomass.

polymers. Thus, biomass analyses in terms of higher heating values or percentage of carbon, hydrogen and oxygen fall far short of revealing the complexity of biomass as a starting material. Biomass has the advantage of greater reactivity than coal but successful research on thermochemical conversion processes requires that the reactivity be understood at the molecular level so that the reaction pathways can be guided into desirable products.

2. FEEDSTOCK SPECIFICATIONS

Consider the biomass feedstock to be used in a thermochemical process. To what degrees does this feedstock need to meet specifications and what specifications should be provided and why? The following specifications provide starting points for the development of pretreatment methods.

—Specific biomass types, e.g. all hardwoods or pine, but not Douglas fir or trees, but not wheat straw.
—Anatomical parts, e.g. wood chips, but not bark, sugarcane pith, but not rind.
—Size, e.g. standard pulpwood chips, sawdust, powder to pass through twenty mesh screen.
—Size distribution, e.g. largest particles one hundred times the size of the smallest or same size within 10%.
—Moisture content, e.g. bone dry or air dry or green or 20% plus or −2%.
—Extractives, e.g. as received or removed by extraction.
—Status of lignocellulose complex, e.g. intact or steam exploded or acid treated.
—Status of lignocellulose polymers, e.g. hemicellulose removed or cellulose partially depolymerized with acid.

Both the mechanical and thermochemical properties of biomass can vary substantially, depending on these feedstock considerations. One purpose of pretreatment is to provide some uniformity to the feedstock employed in several chemical conversions.

3. PRETREATMENT ALTERNATIVES

Given the range of variables that could be controlled by pretreatment, some discussion of pretreatment alternatives is in order. The degree to which any specific pretreatment is desirable will depend on the details of the thermochemical conversion process that follows the pretreatment. The feedstock specifications will determine which pretreatments are desirable candidates.

3.1. Anatomical Separation

Although wood chips appear homogeneous, they come from all portions of the log—heartwood, sapwood, juvenile wood, etc. Similarly, bagasse includes both the pith and the rind and corn stover includes material from the stalk and the cob. Opportunities exist for anatomical separations, some of which are likely to greatly influence the outcome of thermochemical conversion and others of which are not likely to do so.

The Canadian roll splitter,[6] based on the TVA fiberizer,[7] is an illustration of a device for pretreatment of biomass that provides a means of separating the bark from a tree and of providing strands that can be chopped in a way to separate different portions of the tree (Fig. 2). The Tilby cane separator (Fig. 3) provides rind and pith fractions.[8] Both of these processes provide lignocellulose in strand form which opens fresh opportunities for innovation in feeding of feedstocks for thermochemical conversion. The novelty of strands also provides opportunities for colossal problems in materials handling.

3.2. Comminutions

Although other aspects of pretreatment may have been ignored to a greater or lesser degree, developers of thermochemical conversion processes have

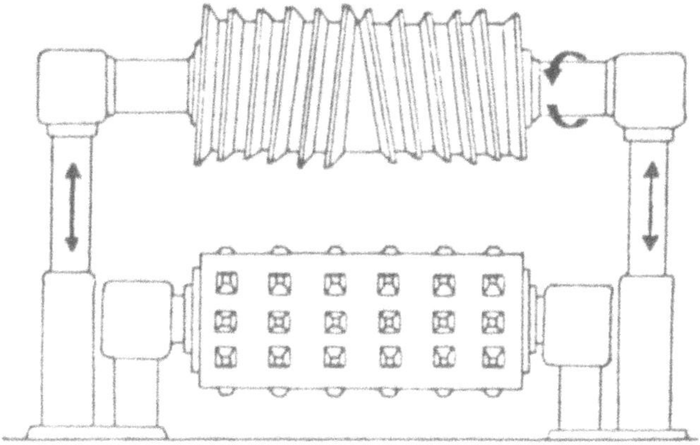

FIG. 2. Roll splitter (Source: Ref. 6).

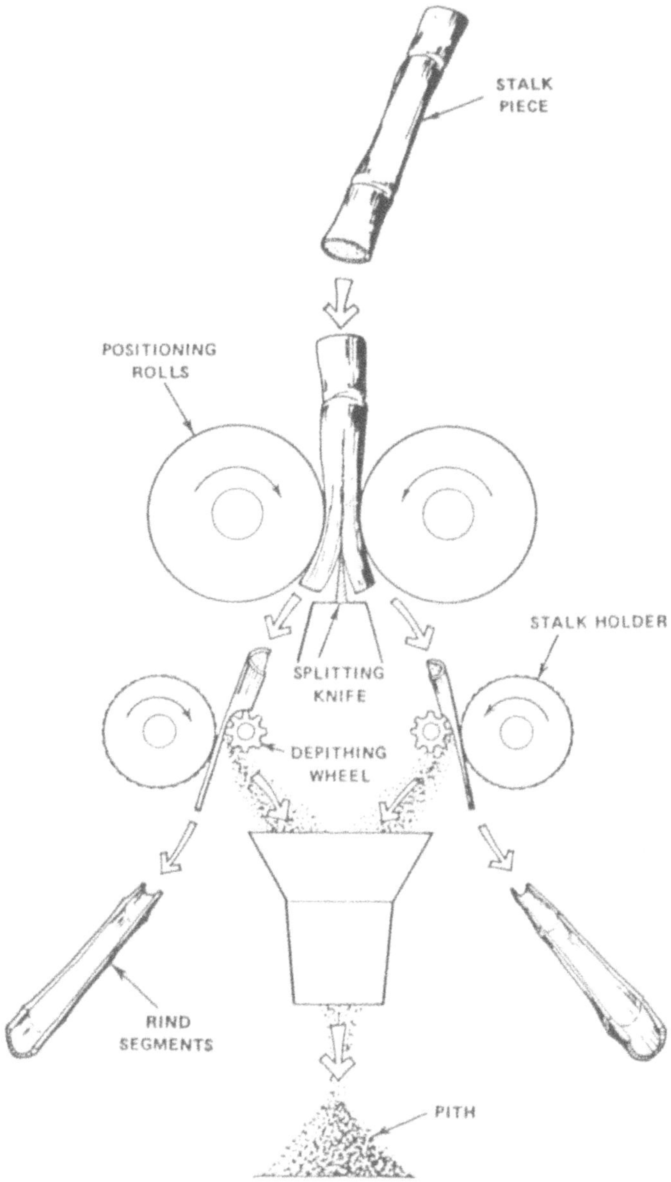

FIG. 3. Tilby separator process.

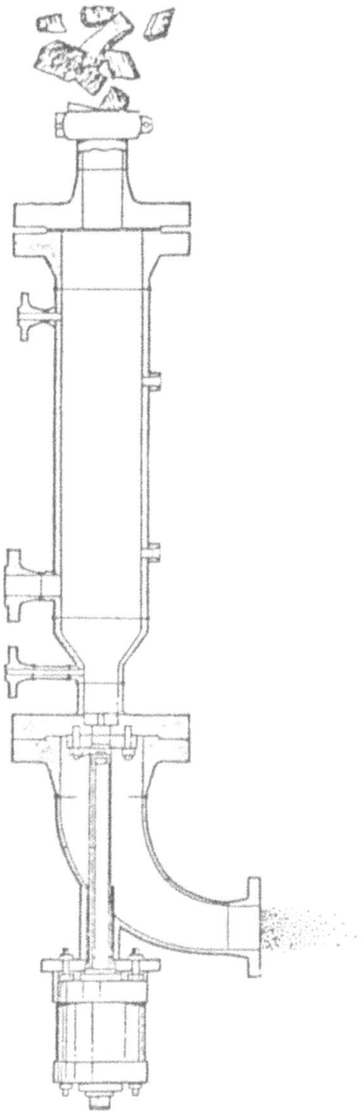

FIG. 4. Masonite gun for steam explosion process. (Adapted from Marchessault, R. H. *et al.* in *International Development Research Center Manuscript Reports*, p. 14, 1983).

been forced to face the comminution problem.[4] There has been an emphasis on chips because of their convenience but some fast rate processes require sawdust or fine powders. Steam explosion (Fig. 4) provides opportunities to convert chips to fibers or powders relatively inexpensively. However, steam exploded wood has a moisture level of approximately 60 % when made in a conventional fashion.[9] Developers of this technology have tended to emphasize steam explosion for microbiological uses where this moisture content is not disadvantageous. Wenger Manufacturing[10] is adapting a corn dry milling extrude (Fig. 5) to perform autohydrolysis of wood without use of added steam.[11] Thus, the moisture level of the product is reduced.

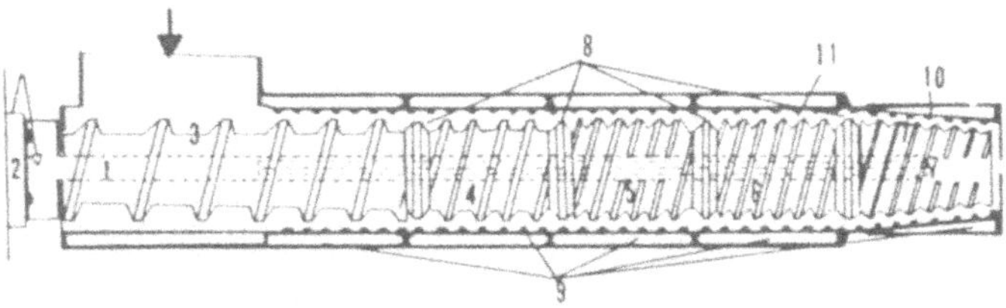

FIG. 5. Wenger extruder (Source: Ref. 11).

The production of strands by the roll splitting or Tilby processes is a form of comminution in which a high surface area is generated without resorting to grinding or water treatments. Compared with chip handling, strand handling technology is at an early stage of evolution.

The use of acid embrittlement to make such powders as Eco-fuel or agri-fuel provides a means of making relatively fine powders with high surface areas.[12,13] Where acid costs are not prohibitive, the gain in reactivity may be quite worthwhile.

3.3. Control of Particle Shape and Size

Pretreatment must be designed with thermochemical conversion in mind. Usually there are optimum transfer rates for heat going into the particle and gaseous and/or liquid products coming out of the particle. If the particle size is too large, secondary reaction can occur within the particle that may lead to excessive char or tar formation. If the particle size is too small, it may not be retained in the reaction zone long enough.

In some thermochemical processes, particles that are too small compared with the average will become overcooked and those that are too large will be undercooked, leading to reduced yields. Thus, it is not just absolute size, but also the size distribution that is important.

Densification of biomass may aid in thermochemical conversion by controlling mass transfer in the reactor and by simplifying feeding of the raw material. Typical densification processes[14,15] involve substantial chemical changes, in addition to densification.

3.4. Disruption of Lignocellulose

Steam explosion not only reduces wood chips to particles in the sawdust size range; it also disrupts the lignocellulose by means of autohydrolysis. If run under appropriate conditions, the lignin becomes quite tacky and can be used as an adhesive in a consolidated product. This self-adhesive property is exploited in the production of Masonite® wallboard. Part of the hemicellulose is converted to simple sugars and, under some conditions, to furfural. The molecular weight of the cellulose is also decreased. Thus, steam explosion is a complex process that makes changes at many hierarchical levels at once. Methods under development to achieve steam explosion by methods that minimize moisture content of the final product were mentioned previously. Such approaches could be quite desirable in the preparation of biomass feedstocks for thermochemical conversion. However, the alkaline biomass liquefaction processes make use of aqueous media and could accept steam exploded biomass produced by the present processes.[3]

3.5. Fractionation of Lignocellulose

Thermochemical processing usually has as its goal commodity fuels that are either gaseous or liquid. One can envision large-scale lignocellulose fractionation facilities in which high value products (e.g. chemical cellulose) are obtained to satisfy polymer markets. The other constituents of the lignocellulose might be converted into fuels by a gasification or liquefaction process. This concept is similar to that of the petroleum refinery in which the more valuable products are removed for sale in chemical or lubricant markets and the rest are converted into fuels.

Typical lignocellulose fractionation processes include the organosolv process in which ethanol is employed as a solvent to remove lignin.[16] The Battelle Geneva process (Fig. 6) employs aqueous phenol to achieve a simultaneous separation of cellulose, hemicellulose, and lignin.[17,18]

Like steam explosion, fractionation of lignocellulose with solvents leads to multiple, deep seated changes. The lignin becomes soluble in simple organic solvents. This type of lignin has been shown by Johansson and

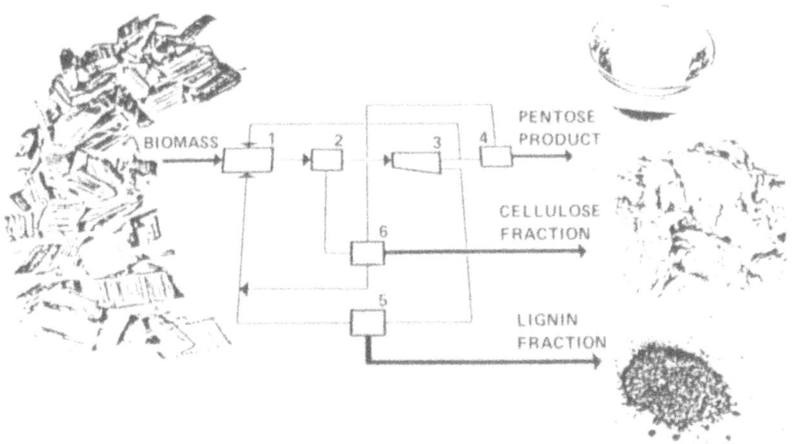

FIG. 6. Battelle-Geneva lignocellulose fractionation process (Courtesy of J. P. Sachetto, Battelle, Geneva).

coworkers[17] to be easily converted to phenol by thermochemical hydrocracking reactions.

3.6. Depolymerization

The use of steam or acid or base on lignocellulose polymers can reduce their molecular weight through depolymerization reactions. The reduced molecular weight may lead to higher rates of reaction in thermochemical conversion. Simpler pyrolytic reactions paths may also occur. The motivation for depolymerization may originate in the desire to improve the viscosity of biomass slurries.[19]

Shafizadeh[5] has shown that glucose polymers can be cracked to levoglucosan by thermal treatment. Thus, depolymerization does not have to be a hydrolytic process.

3.7. Moisture Control

Entire libraries could be filled with books on the science and technology of biomass moisture control. Water is notorious for its protection against rapid thermal processes. A flammable open-cell polyester foam containing water can be held over a bunsen burner for a considerable period of time because of refluxing effects. Comparable cooling effects occur in wood, especially if the particle size is relatively large.

Water also plays a role in hydrolysis of lignocellulose and thus in lignocellulose disruption and depolymerization reactions.

Opportunities exist to use mechanical methods to remove part of the water from lignocellulose, as illustrated by the Stake Technology dewatering device[20] and by moisture lost during roll splitting.[6] Such processes do reduce moisture content but not to the very low levels frequently desired by developers of thermochemical conversion processes.

4. CONCLUSIONS

The special needs of thermochemical conversion processes for uniform feedstock with respect to size, size distribution, shape, and moisture content differ from process to process. Therefore, it is difficult to generalize about pretreatment processes. The author strongly suspects that developers of the new processes have been so intent on what happened inside the thermochemical reactor that opportunities to facilitate the principal reactions by making use of appropriate pretreatments are being overlooked. Another factor that has reduced the role of pretreatment in thermochemical conversion is the desire to use standard chip (or sawdust) materials handling systems.

Pretreatments that lead to particles that are smaller than chips may be very advantageous in many thermochemical conversion processes. These smaller sizes may be achieved by use of steam explosion methods that are specially tailored for thermochemical (not microbiological) purposes. Strand production methods merit greater attention, based on work conducted by the Tilby group and by K. C. Jones.

Pretreatment research and development should be integrated with the development of specific thermochemical conversion processes. Pretreatments must not be solutions looking for problems.

REFERENCES

1. Millett, M. A., Baker, A. J. and Satter, L. D. (1976). Physical and chemical pretreatments for enhancing cellulose saccharification, in *Biotechnology and Bioengineering Symposium No. 6*, E. L. Gaden, Jr., M. H. Mandels, E. T. Reese and L. A. Spano (Eds). Wiley, New York, pp. 125–53.
2. Lipinsky, E. S. (1979). Perspectives on preparation of cellulose for hydrolysis, in *Hydrolysis of Cellulose: Mechanisms of Enzymatic and Acid Catalysis*, R. D. Brown, Jr., and L. Jurasek (Eds). American Chemical Society, Washington, D.C., pp. 1–23.
3. Bouvier, J. M., Gelus, M., Lamy, P. Y. and Roux, F. (1982). Wood autohydrolysis as pretreatment process—Application to liquefaction, in *Book of Abstracts—Energy from Biomass*. Commission of the European Communities, Berlin, p. 187.

4. Miles, T. R. (1982). Preparation of biomass for gasification, in *Proceedings, Biomass-To-Methanol Specialists' Workshop*. US Department of Energy, Durango, Colorado, pp. 95–118.
5. Shafizadeh, F. (1980). Introduction to pyrolysis of biomass, in *Specialists' Workshop on Fast Pyrolysis of Biomass Proceedings*. Solar Energy Research Institute, Copper Mountain, Colorado, pp. 79–103.
6. Jones, K. C. (1982). Development and testing of a roll splitter, in *Fourth Bioenergy R&D Seminar—Proceedings*. National Research Council of Canada, Winnipeg, Manitoba, pp. 177–82.
7. Marra, A. A., Hausknecht, W. A. and Day, R. F. (1975). Low density composites from high density hardwoods. Massachusetts Agricultural Experiment Station Bulletin No. 610.
8. Lipinsky, E. S. and Kresovich, S. (1982). Sugar crops as a solar energy converter, *Experientia* **38**, 13–17.
9. Nguyen, Q. and Noble, G. (1980). Preparation of powdered feedstock from biomass with steam, in *Specialists' Workshop on Fast Pyrolysis of Biomass Proceedings*. Solar Energy Research Institute, Copper Mountain, Colorado, pp. 365–74.
10. Wenger, J. (1964). US Patent 3117006.
11. Rokey, G. (1982). Personal communication to E. S. Lipinsky, October 14.
12. Ghazee, A. and Hecht, N. (1980). The embrittlement treatment of wheat straw for conversion to a powder, in *Speciliasts' Workshop on Fast Pyrolysis of Biomass Proceedings*, Solar Energy Research Institute, Copper Mountain, Colorado, pp. 375–92.
13. Hasselreiss, F. (1980). The eco-fuel II process: Producing a storable, transportable fuel for co-combustion with oil, in *Thermal Conversion of Solid Wastes and Biomass*, J. L. Jones and S. B. Radding (Eds). American Chemical Society, Washington, D.C., pp. 143–50.
14. Reed, T. and Bryant, B. (1981). Densified biomass: A new form of solid fuel. Report from Solar Energy Research Institute, Report No. SERI-35.
15. Miles, T. R. and Miles, T. R., Jr. (1980). Densification systems for agricultural residues, in *Thermal Conversion of Solid Wastes and Biomass*, J. L. Jones and S. B. Radding (Eds). American Chemical Society, Washington, D.C., pp. 179–93.
16. Sarkanen, K. V. (1980). Acid-catalyzed delignification of lignocellulosics in organic solvents, in *Progress in Biomass Conversion*, Vol. 2, K. V. Sarkanen and D. A. Tillman (Eds). Academic Press, New York, pp. 127–44.
17. Johansson, A., Sachetto, J. P. and Roman, A. (1982). A new process for the fractionation of biomass, unpublished manuscript available from Battelle-Geneva.
18. Anon. (1982). Process yields marketable biomass fractions, *C&EN* **60**(22), 38.
19. Ergun, S. (1981). Review of biomass liquefaction efforts. Report from Lawrence Berkeley Laboratory to the US Department of Energy, Report No. LBL-13957.
20. Brown, D. (1980). Design, construction, and testing of a pilot scale continuous dewatering device. Report from Stake Technology Ltd to the Forest Engineering Research Institute of Canada, ENFOR Project C-6.

5

A Kinetic Isotope Effect in the Thermal Dehydration of Cellobiose

MILTON D. SCHEER
Chemical Kinetics Division, National Bureau of Standards, Washington, D.C., USA

The thermal dehydration kinetics of cellobiose have been studied previously in the 170–250 °C temperature range.[1] A comparison with the dehydration kinetics of sucrose suggested that the eliminated water is responsible for its liquefaction at 238 °C and that this so-called melting process is a high temperature aqueous dissolution of the partially dehydrated cellobiose. The kinetics of dehydration appear to be strongly influenced by the fact that each hydroxyl hydrogen atom is hydrogen bonded, either inter- or intramolecularly, to a neighboring hydroxyl oxygen atom.[2] A typical dehydration site can be depicted as follows (where -- represents a hydrogen bond):

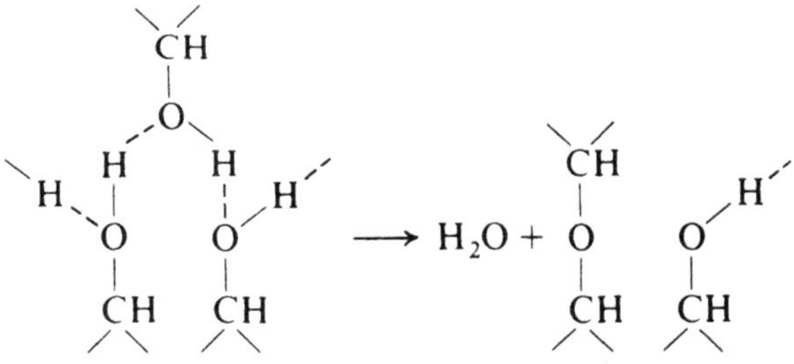

(1)

The elimination of water requires the breaking of a >CH–OH, a >CHO–H, and two hydrogen bonds, while an H–OH is formed from an H--OH hydrogen bond and a >CH–O–CH< ether linkage ends up bridging the reaction site. Using well-known values for the bond energies,[3,4] one can readily estimate that the elimination of water from a disaccharide molecule is about 45 kJ/mol endothermic. The activation energy for dehydration was found to be about three times this endothermicity.[1] It is the purpose of this communication to report the effect of deuteration of the hydroxyl hydrogens in cellobiose upon its dehydration activation energy and liquefaction temperature. It will be shown that the lowering of the zero point energy in the O–H bond (when D is substituted for the H) only accounts for a small part of the observed increase in activation energy and decrease in the rate of water elimination. An increase in the liquefaction temperature of the deuterated cellobiose, however, is quantitatively accounted for by the decrease in the rate of water elimination.

1. EXPERIMENTAL

The hydroxyl hydrogen atoms in simple sugars are known to be quite labile in aqueous solution so that the dissolution of cellobiose in D_2O will result in an exchange between the H atoms of the hydroxyl groups in the sugar molecule and the D atoms in the heavy water. Cellobiose was dissolved in an excess of D_2O (99·5 D atom %) and then vacuum evaporated to dryness at 65 °C. This process was repeated three times, after which the deuterated cellobiose was dissolved in the minimum amount of D_2O and recrystallized by the addition of C_2H_5OD. After filtering and drying at 100 °C, the

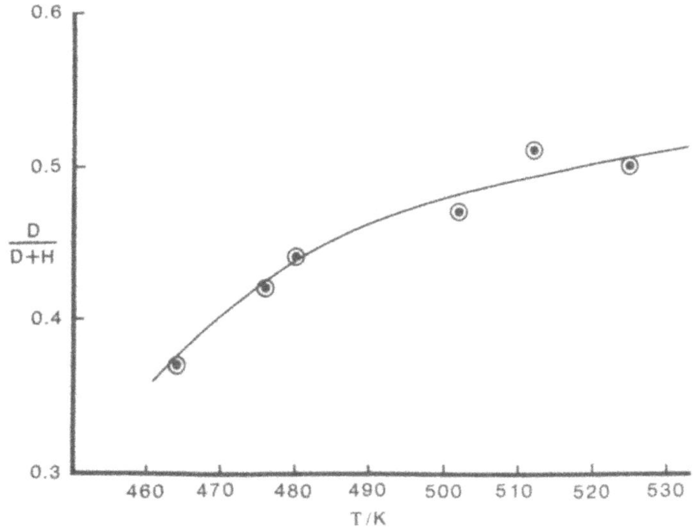

FIG. 1. A plot of D/(D + H) for the water produced by the dehydration of partially deuterated cellobiose in the 460–530 K temperature range.

deuterated cellobiose was thermally dehydrated on a heated tungsten filament that was mounted in front of the ionization chamber of a quadrupole mass filter. Details of the apparatus and its application to the measurement of thermal dehydration rates have been described previously.[1] The intensities of the mass spectral peaks with m/Z values of 18, 19, and 20 yielded the isotropic composition of the eliminated water. Figure 1 shows a plot of D/(D + H) of the eliminated water as a function of the dehydration temperature. It is seen that the above procedure yields a product in which about 50% of the hydroxyl hydrogens are replaced by deuterium atoms.

2. RESULTS AND DISCUSSION

The logarithm of the rates of water elimination for both ordinary and partially deuterated cellobiose is plotted as a function of $1/T$ in Fig. 2. The open circles, representing the elimination of $D_2O + HDO + H_2O$ from the deuterated sample, lie below the closed circles which represent the elimination of water from ordinary cellobiose. A linear regression for the latter data yielded the following:

$$\ln r_H = 10{\cdot}46 - 14\,540/T \tag{2}$$

Since a kinetic isotope effect tends to become smaller as the temperature increases, one may assume that it becomes vanishingly small as $1/T$ goes to zero. The pre-exponential Arrhenius factors are therefore assumed to be identical for both the deuterated and ordinary cellobiose and a least-squares fit of the open circle data in Fig. 2 is

$$\ln r_D = 10{\cdot}46 - 14\,645/T \tag{3}$$

The temperatures at which the dehydration rates are the same for both the ordinary and deuterated species may be obtained from refs 2 and 3 by setting $r_H = r_D$ or $T_D = 1{\cdot}007 T_H$. Since the so-called sugar melting process is assumed to be the dissolution of the decomposition products in the eliminated water, the increase in the 'melting' temperature ($\Delta T = T_D - T_H$) arising from the 50% deuteration of the hydroxyl hydrogens is given by $\Delta T = 0{\cdot}007 T_H$, where T_H is now the liquefaction temperature of ordinary cellobiose (511 K). The liquefaction temperature of the deuterated cellobiose was found to be 513 K. This agrees with the calculated value (515 K) when account is taken of the experimental uncertainties in the Arrhenius parameters.

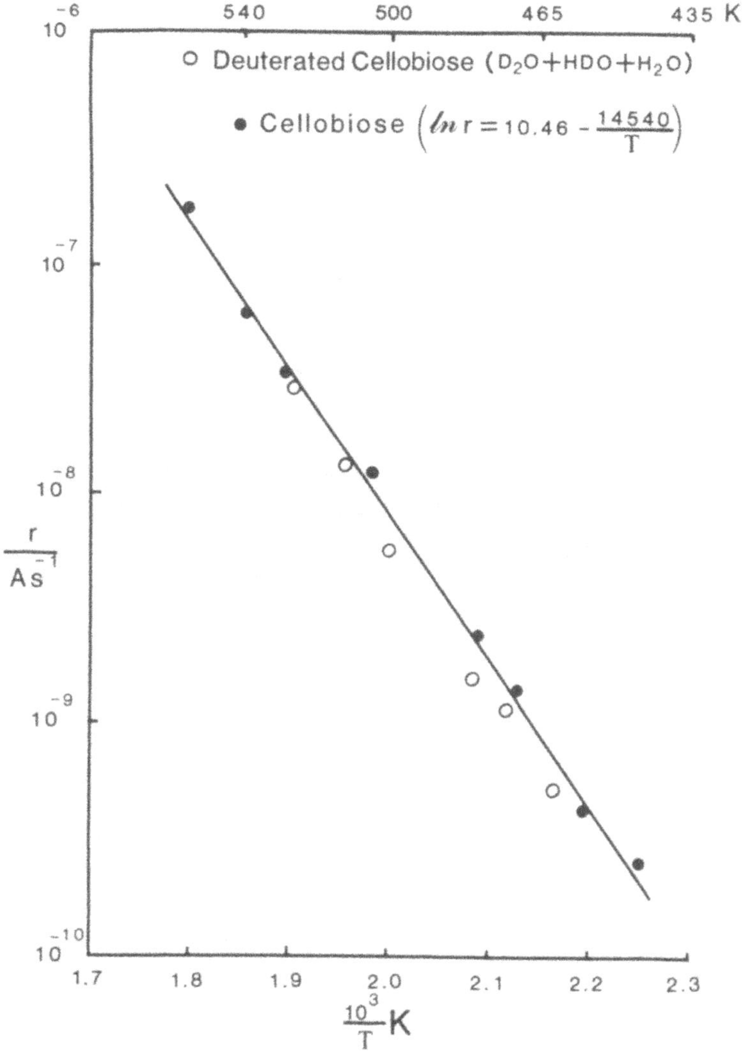

FIG. 2. An Arrhenius plot of the rate of water elimination from ordinary (●) and partially deuterated (○) cellobiose.

TABLE I
THE RATES OF H_2O, HDO AND D_2O ELIMINATION FROM PARTIALLY DEUTERATED CELLOBIOSE

T/K	$r(H_2O)/pAS^{-1}$	$r(HDO)/pAS^{-1}$	$r(D_2O)/pAS^{-1}$
462	223	213	107
473	402	522	255
480	495	745	348
502	1 776	2 616	1 515
511	3 678	5 978	3 908
525	8 875	12 685	8 637

As seen in Fig. 2, the dehydration of the partially deuterated cellobiose was studied at six different temperatures. Table I gives the individual rates of formation of H_2O, HDO and D_2O at these temperatures. The Arrhenius parameters for the isotopic rate ratios are given by

$$\ln [r(D_2O)/r(H_2O)] = 5{\cdot}41 - 2790/T \quad (4)$$

and

$$\ln [r(D_2O)/r(HDO)] = 2{\cdot}34 - 1440/T \quad (5)$$

with correlation coefficients greater than 0·9. These results correspond to activation energy differences of 23 and 12 kJ/mol for the elimination of D_2O relative to H_2O and HDO, respectively.

One component of the observed activation energy differences for the elimination of the three isotopic water molecules must be associated with the different vibrational zero point energies of the two isotopic hydroxyl bonds that are ruptured in the dehydration of the partially deuterated cellobiose. The O–H vibrational stretching frequency (ω), modified by the influence of strongly coupled neighboring hydrogen bonds, is known to be about 3000 cm^{-1} so that the zero point energy E_Z ($= hc\omega/2$) is 18 kJ/mol.[5] The ratio of the zero point energies for the two isotopically different hydroxyls is

$$E_Z(\text{O–D})/E_Z(\text{O–H}) = [\mu(\text{O–H})/\mu(\text{O–D})]^{1/2} \quad (6)$$

where μ is the reduced mass of the oscillator, so that the difference in zero point energies may be written as

$$\begin{aligned}\Delta E_Z &= E_Z(\text{O–H}) - E_Z(\text{O–D}) \\ &= \{1 - [\mu(\text{O–H})/\mu(\text{O–D})]^{1/2}\}E_Z(\text{O–H}) \quad (7)\end{aligned}$$

The quantity $\mu(\text{O–H})/\mu(\text{O–D}) = 0{\cdot}53$ so that ΔE_Z is about 5 kJ/mol. As seen in eqn (1), the elimination of HDO must involve the breaking of an

O–D and an O–H bond, one of which is hydrogen bonded. It is unlikely that there is a significant zero point energy difference between the two isotopically different hydroxyl hydrogen bonds so that no contribution is made to the kinetic isotope effect. Since the D and H atoms can be assumed to be randomly mixed in the 50% deuterated cellobiose, one half of the eliminated HDO will involve the breaking of one O−−D and one O–H, while the remaining HDO will be formed by the breaking of one O–D and one O−−H bond. Therefore, the net ΔE_Z for the elimination of D_2O relative to HDO ought to be about one half of the calculated value, or 2·5 kJ/mol. By the same reasoning, the ΔE_Z for the elimination of D_2O relative to H_2O is the full 5 kJ/mol. These values are less than 0·25 of the observed activation energy differences.

The data in Table I clearly show that as the temperature is lowered, the $r(H_2O)/r(D_2O)$ and $r(H_2O)/r(HDO)$ ratios increase more than that predicted by the zero point energy differences. A reasonable interpretation of these results is that there is a significant contribution from the wave mechanical tunnel effect,[6,7] which predicts that the light hydrogen atoms tunnel through the reaction activation barrier far more readily than do the heavier deuterium atoms. This process, unlike the Arrhenius relation, is temperature insensitive and therefore can account for the anomalously large isotope effect observed at the lowest experimental temperatures.

ACKNOWLEDGEMENT

The author would like to thank Dr J. R. McNesby of the University of Maryland for a number of useful discussions of the kinetic isotope effect.

REFERENCES

1. Scheer, M. D. (1983). *Int. J. Chem. Kinet.* **15**, 141.
2. Brown, G. M. and Levy, H. A. (1973). *Acta Cryst.* **B29**, 790.
3. Stull, D. R., Westrum, E. F. Jr. and Sinke, G. C. (1969). *The Chemical Thermodynamics of Organic Compounds*. Wiley, New York.
4. McMillen, D. F. and Golden, D. M. (1982). *Ann. Rev. Phys. Chem.* **33**, 498.
5. Herzberg, G. H. (1945). *Infrared and Raman Spectra*. D. Van Nostrand Co., New York, pp. 535–6.
6. Johnston, H. S. (1966). *Gas Phase Reaction Rate Theory*. The Ronald Press Co., New York, Chapter 13.
7. Melinder, L. (1960). *Isotope Effects on Reaction Rates*. The Ronald Press Co., New York.

6

Gasification and Liquefaction of Forest Products in Supercritical Water

MICHAEL MODELL

MODAR, Inc., Natick, Massachusetts, USA

1. INTRODUCTION

As one of the ultimate renewable resources, forest products will inevitably become a major source of fuels and organic chemicals. Any technological process for converting biomass to fuels must involve the break-down of complex organic structures and stabilization of the decomposition products. The first step in conventional physicochemical routes to biomass gasification and/or liquefaction usually involves severe decomposition of the feed at temperatures above 600 °C (e.g. pyrolysis, partial oxidation, steam reforming). In pyrolysis, high temperatures are required to provide sufficiently rapid kinetics. However, under such conditions, an appreciable fraction of the feed is converted to char, which complicates materials handling and decreases yield of fluid fuels.

Char can be minimized or eliminated by steam reforming or partial oxidation under high temperatures. However, such processes suffer from an inherent limitation because the high temperature decomposition is endothermic. For example, in the formation of SNG from biomass via steam reforming, the overall reaction can be simplified as a summation of three steps (usually carried out sequentially):

Reforming:

$$2CH_2O = 2CO + 2H_2 \qquad \text{high } T\text{; endothermic} \tag{1}$$

Water-gas shift:

$$CO + H_2O = CO_2 + H_2 \qquad \text{moderate } T\text{; thermoneutral} \tag{2}$$

Methanation:

$$CO + 3H_2 = CH_4 + H_2O \qquad \text{moderate } T\text{; exothermic} \tag{3}$$

Overall:

$$2CH_2O = CH_4 + CO_2 \tag{4}$$

While the overall reaction is exothermic, the first step requires the addition of high temperature heat, whereas the third step produces lower temperature heat. The net effect is a Second Law inefficiency. As a consequence, part of the feed material is usually oxidized to CO_2 to supply the high temperature heat and this heat is rejected from the overall system at a lower temperature.

The ideal biomass gasification process would be one which converts the feed material directly to CH_4 and CO_2 by reaction (4). However, this reaction is thermodynamically favorable only under low to moderate temperatures. To illustrate this point, the equilibria for the system of dextrose, H_2O, CH_4, CO_2, H_2 and CO is shown in Fig. 1, along with the net enthalpy of reaction. Note that below 500 °C, the equilibrium favors formation of CH_4 and CO_2 as the main products; below 600 °C, the overall reaction is exothermic.

The long-range objective of this work has been to develop a moderate temperature route for reforming forest products to liquid and gaseous fuels and organic chemicals. It is believed that such a process can provide a

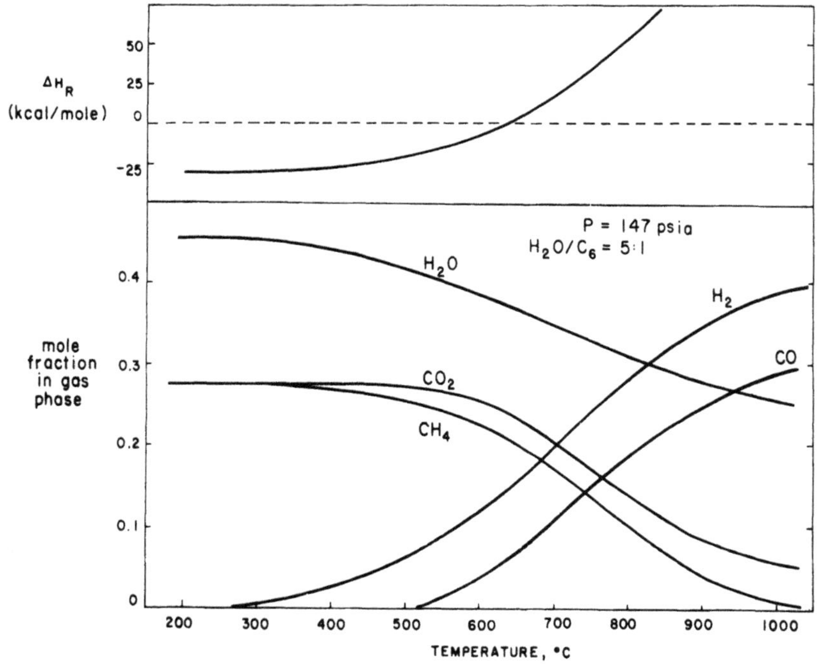

FIG. 1. Effect of temperature on equilibrium for steam reforming of D-glucose.

means for improving efficiency of biomass conversion above that of processes currently under development.

In view of the above discussion, the desired temperature range for processing biomass is below 500 °C. A number of studies on the degradation and pyrolysis of wood and wood components have shown that such materials are chemically reactive at temperatures as low as 150 °C. Hemicellulose pyrolysis has been observed to occur at 150–350 °C; cellulose at 275–350 °C; and lignin at 250–500 °C.[1] DTA and TGA experiments with wood and its components clearly indicate chemical reactivity in the range of 250–400 °C.[2,3]

Although it has been clearly established that forest products are chemically reactive with respect to degradation reactions in the range of temperatures below 500 °C, nevertheless, industrial processes for biomass gasification and liquefaction are usually designed for temperatures above 600 °C. The main reason for operating at these higher temperatures is the fact that char is a major product below 600 °C; higher temperatures are required to attain appreciable rates of char conversion via the heterogeneous reactions of char with gaseous components (e.g. H_2O, CO_2, O_2).

Prior work in our laboratory has established that char formation can be avoided in the range 375–380 °C by processing carbohydrates and wood in supercritical water.[4–6] The supercritical water serves two purposes: (1) it acts as a carrier fluid and solvent for the feed material and reaction products; and (2) water reacts with organic materials, via reforming reactions, to liberate hydrogen which, in turn, stabilizes reactive intermediates.

2. PROPERTIES OF SUPERCRITICAL WATER (SCW)

The supercritical region can best be visualized with the aid of a temperature-density diagram, as shown in Fig. 2. The critical point (CP), which lies on the vapor–liquid saturation dome, occurs at 374 °C, 218 atm and 0·3 g/cm^3. The supercritical region refers to temperatures above 374 °C and densities from 0·2 to 0·7 g/cm^3. In this region, density varies very rapidly with small changes in temperature at constant pressure.

Insight into the structure of the aqueous fluid in this region has been obtained from measurements of the static dielectric constant, shown in Fig. 3.[7,8] In the near critical (NC) region, the dielectric constant of normal

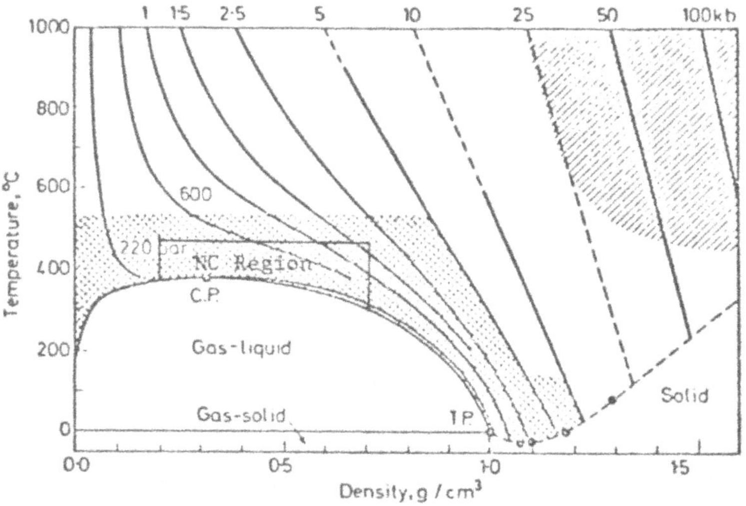

FIG. 2. Temperature/density diagram of water. Full lines: measured isobars. Dashed lines: interpolated isobars. Cross-hatched zone: viscosity and dielectric constants determined. Single shading: ionic conductance determined. (Source: ref. 8.)

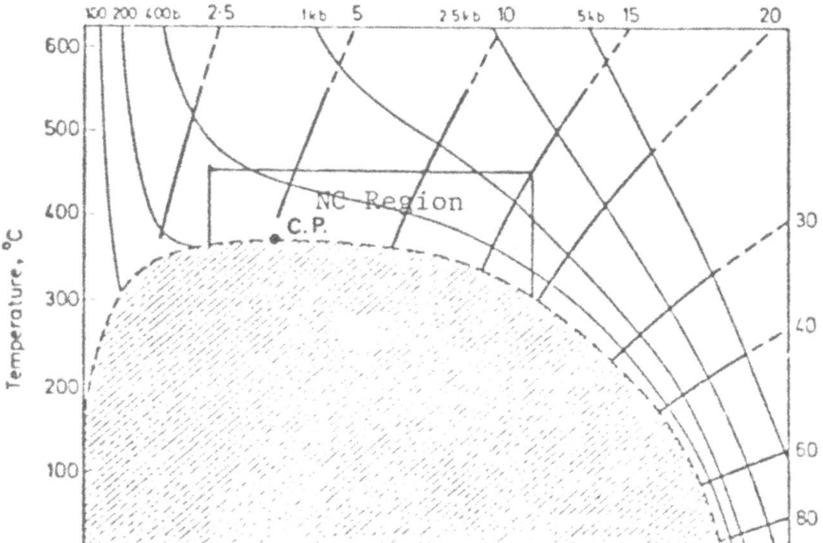

FIG. 3. Dielectric constant of water as a function of temperature and density. ——: measured; – –: calculated; ——: isobars. (Source: ref. 8.)

liquid water (80) is the result of strong hydrogen bonding. The lower values in the supercritical regime are indicative of a greatly reduced extent of hydrogen bonding. Since hydrogen bonds are very short-range forces, and since the supercritical fluid has a significantly lower density than the normal liquid, it is not surprising to find a large reduction in hydrogen bonding.[9]

Supercritical water behaves very much like a polar organic solvent. Organic materials, which are only sparingly soluble in water at room

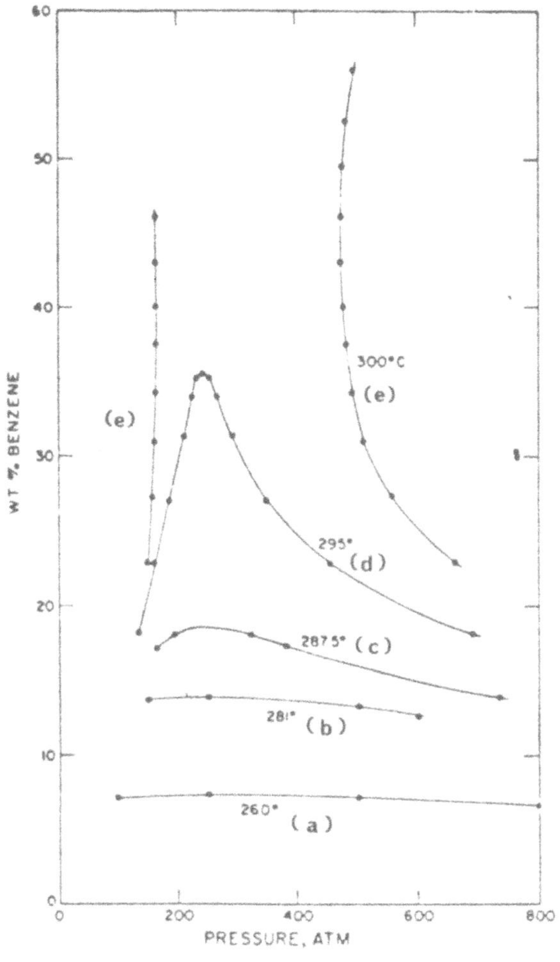

FIG. 4. Solubility of benzene in water.[10] Reprinted with permission from *Journal of Chemistry and Engineering Data*, Vol. 11, p. 13 (1966). © 1966 American Chemical Society.

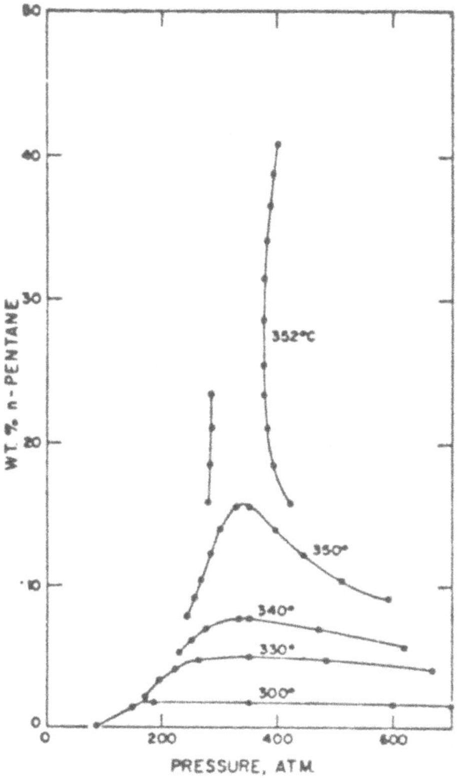

FIG. 5. Solubility of *n*-pentane in water.[10] Reprinted with permission from *Journal of Chemistry and Engineering Data*, Vol. 11, p. 13 (1966). © 1966 American Chemical Society.

temperature, become completely miscible with water in the supercritical regime. For example, the solubilities of benzene, pentane, heptane and 2-methyl pentane in water are shown in Figs 4–7.[10] Consider the case of benzene (Fig. 4). At 260 °C (curve a), the solubility is fairly independent of pressure at about 7–8 wt % benzene, which is already far higher than that at 25 °C (0·07 wt %). Above 260 °C, the solubility increases with temperature; at 287·5 °C (curve c), a maximum appears in the range 200–250 atm. The maximum is very pronounced at 295 °C (curve d). At 300 °C the solubility curve splits into two branches (curve e); between these two branches at 200–500 atm, *benzene and water are miscible in all proportions*. The lowest temperature at which complete miscibility first occurs is called the critical

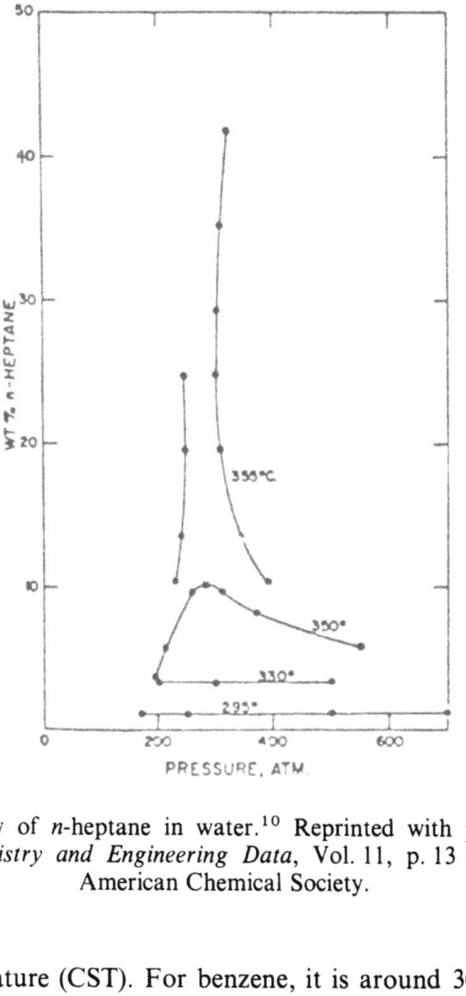

FIG. 6. Solubility of *n*-heptane in water.[10] Reprinted with permission from *Journal of Chemistry and Engineering Data*, Vol. 11, p. 13 (1966). © 1966 American Chemical Society.

solution temperature (CST). For benzene, it is around 300 °C.[10,11] The other aliphatic hydrocarbons, shown in Figs 5–7, require somewhat higher temperatures for complete miscibility: 351–353 °C.[10]

Another way to demonstrate miscibility is by examining the critical curve in a pressure–temperature–concentration diagram. The schematic shown in Fig. 8 is typical of many organic–water binary systems, such as benzene–water (see, for example, ref. 11). The critical point of water is presented by CP_h and that of benzene by CP_l. The dashed curve in Fig. 8 beginning at CP_h is the upper branch of the critical curve; to the left of this branch, the system separates into two fluid phases, whereas to the right of it (at higher temperatures) a homogeneous phase exists at all temperatures

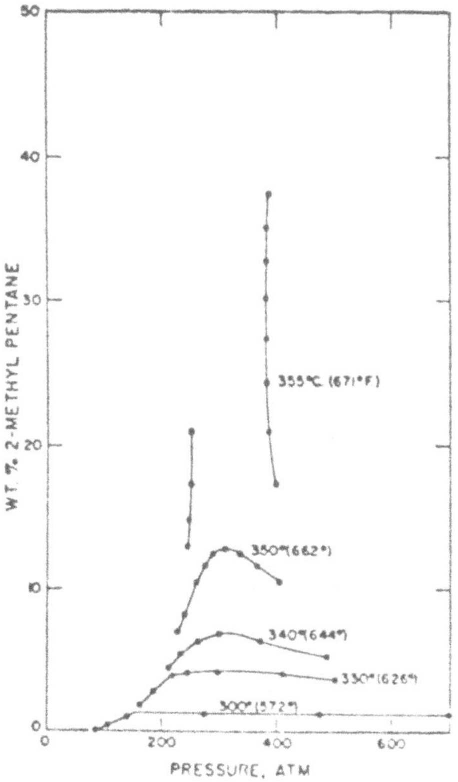

FIG. 7. Solubility of 2-methyl pentane in water.[10] Reprinted with permission from *Journal of Chemistry and Engineering Data*, Vol. 11, p. 13 (1966). © 1966 American Chemical Society.

and concentrations. In other words, *to the right of the critical curve, the components are completely miscible in all proportions.* The projection of the critical curve, in pressure–temperature space, is shown in Figs 9 and 10 for benzene–water and naphthalene–water, respectively. In both cases, it can be seen that *in supercritical water, the organics are completely miscible with water!* Although the number of organic–water systems that have been studied in this detail is very limited, it is clear that water is an excellent solvent for some of the least soluble organics at room temperature. We would anticipate that other organics would behave similarly. Indeed, we have found that substances as complex as cellulose wood can be completely solubilized in SCW, as described below.

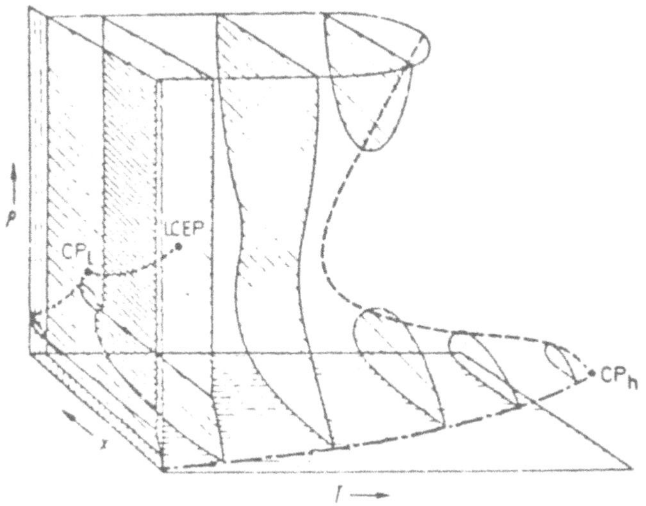

FIG. 8. Schematic pressure–composition–temperature diagram of a binary system with interrupted critical curve. CP_l and CP_h: critical points of the lower and higher boiling pure components. LCEP: lower critical end point. Branches of the critical curve: – – –.[8]

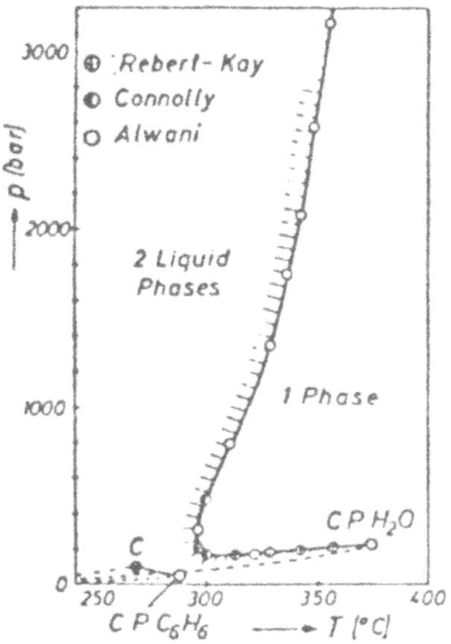

FIG. 9. P–T projection for the benzene–H_2O system.

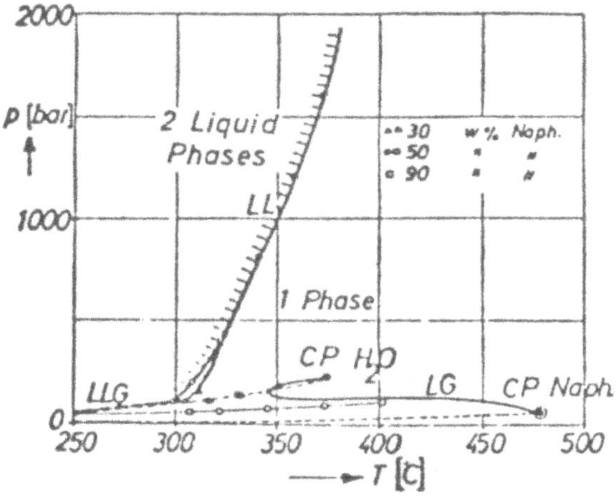

FIG. 10. *P–T* projection for the naphthalene–H_2O system.

3. APPARATUS AND PROCEDURE

A schematic of the apparatus is shown in Fig. 11. A standard 1-liter, 316 stainless steel Autoclave Engineers Model AB-1005 magnetically stirred autoclave was used as the reactor. An iron constantan thermocouple, located in the thermowell fitted on the head of the autoclave, was used to monitor the reactor temperature and to actuate a proportional controller which drove an electrically heated furnace. During a typical 1 h run, the temperature varied by ± 3 °C from the set point.

The autoclave was charged with 250–300 cm^3 distilled water and the reactor was leak-tested with helium at 600 atm. The reactor was then heated to 150 °C and vented to remove residual helium and air. Less than 1 cm^3 of water was purged during the venting. The reactor was then heated to about 10 °C above the desired operating temperature. The organic feed, maintained in a reservoir between valves A and B (Fig. 11), was flushed into the reactor with water.

When operating under critical or supercritical conditions, enough water was added so that the total water in the reactor was 310 cm^3. Thus, the contents were brought to the critical density of water (0·31 g/cm^3). As the addition of cold feed tends to cool the reactor contents, the reactor was heated above the desired operating temperature prior to feed addition so as to ensure that the temperature did not drop significantly below the critical during feed addition.

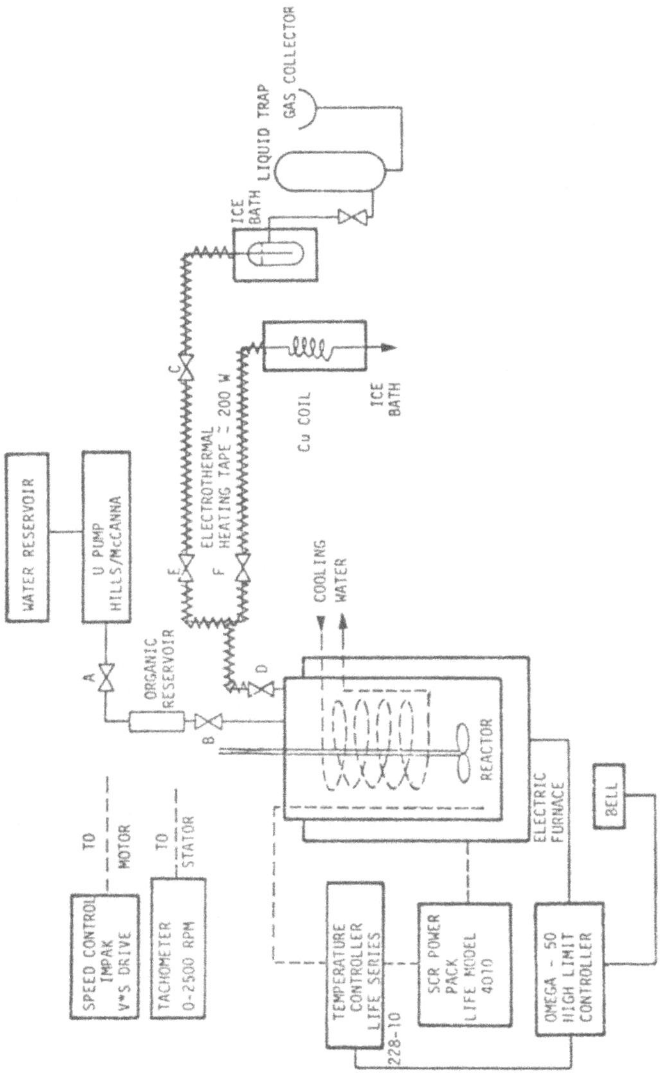

FIG. 11. Experimental apparatus. Valves: A, B, D, E, F—straight flow through; C—micrometering. All lines and fittings 316 S/S. Lines 1/8 in (except organic inlet—1/4 in). Working pressure, 11 000 psi.

At the conclusion of a run, a sample of the reactor contents was obtained by cracking valves E and C. The sample passed through electrically heated lines to a liquid trap immersed in an ice bath. Non-condensable gases were collected in a gas reservoir downstream of the liquid trap. The liquid and gas samples were typically 20 and 100 cm^3, respectively. The reactor was then quenched by passing water through the internal cooling coils. The reactor was inspected for the formation of char or tar.

The gas samples were immediately analyzed for H_2, CO, CH_4, CO_2 and C_2+ (i.e. C_2H_4 and C_2H_6) using a Beckman GC-5 gas chromatograph fitted with a 6·1 m, 0·23 cm ID Porapak Q (80/100 mesh) column. The sample size was 0·10–0·25 cm^3. A known calibration gas mixture was used to determine response factors for the thermal conductivity detector.

Liquid samples were extracted with one half their volume of methylene chloride, concentrated by mild evaporation of the solvent, and analyzed on a Hewlett-Packard 5710A gas chromatograph fitted with a 1 m, 0·23 cm ID OV-17 column with a flame ionization detector. The temperature was programmed from 70 to 250 °C at 8 °C/min. Over 60 peaks were found. Selected liquid samples were analyzed by GC/MS in an attempt to identify some of the liquid components.

4. RESULTS AND DISCUSSION

4.1. Glucose at Subcritical Conditions

A series of experiments were made with a glucose feed under subcritical conditions.[12] The disappearance of glucose and the distribution of carbon as solid, liquid and gaseous products were studied as a function of temperature. The distribution of the carbon for 2 h runs at various operating conditions is shown in Table I. The liquid and vapor traps were analyzed for percent carbon and the total amount of carbon in each phase was obtained from the known amount of water in the liquid and vapor phases at the operating temperature and pressure. The amount of carbon present as a solid product was obtained by recovering the solids and analyzing them for carbon.

The carbon balances closed to within 20 %. The error in closing the balance could be attributed to the difficulty in removal and recovery of all of the solid products from the reactor. From the results, it can be seen that about 35 % of the glucose is degraded in 2 h at 150 °C and that practically all the glucose is decomposed in 2 h at 300 °C. About 40 % of the carbon charged at 300 °C results in a solid product.

TABLE I

RESULTS AT SUBCRITICAL CONDITIONS; GLUCOSE FEED

Run no.[a]	*Temperature (°C)*	*Pressure (atm)*	*Carbon in glucose initially (g)*	*Products*					
				% original carbon in glucose	*Liquid*	*Vapor*		*Solid*	*Carbon closure (%)*
						Condensable % carbon	*Volatile*[b]		
12	150	5·1	3·36	64·88	0·85	0·003	<0·000 1	Trace	89
17	150	5·0	3·36	67·56	0·63	0·001	<0·001	Trace	86
13	200	13·6	3·36	20·54	1·63	0·008	<0·001	1·0	96
16 C	200	13·6	3·36	26·49	1·18	—	—	—	—
18 C	250	40	4·0	0·8	2·97	0·011	0·006	—	—
14 (14A) C	300	83	3·36	2·08	1·59	0·033	0·01	0·95	80
15 (15A)	300	83	3·36	1·18	1·14	0·03	0·01	1·31	95
25	350	165	4·0	1·0	—	—	—	0·44	—

Reaction time: 2 h.
Run 25 for 1 h only.
[a] C denotes catalyst (3 g, C11-2S-04).
[b] H_2, CO, CO_2, CH_4, CH_3OH identified. No details given because of insignificant gasification.

The amount of carbon in the vapor increased with temperature. However, this gas-phase carbon only accounts for about 1·5% of the total carbon at 300°C. Gasification in the form of volatile products even at 300°C is extremely low (~0·3% of original carbon).

The amount of solids formed increased with increasing temperature up to 300°C, but then decreased at 350°C. This last result may be due to a different reaction mechanism, yielding less 'char' producing intermediates or because of the decomposition of some char formed initially. From the data alone, it is not clear whether the solid products formed are a result of glucose condensation reaction or whether intermediates are formed which then polymerized to 'char'.

The distribution of carbon as glucose, liquid products and solid (obtained by difference) as a function of time and temperature is shown in Figs 12–14. Errors in measurement of amount of carbon in liquid and solid products were about ±10%, while that of glucose is approximately ±5%.

At 150°C (Fig. 12), the rate of solid formation is low initially but increases with time. At 200°C (Fig. 13), the carbon in the liquid product passes through a maximum. At 300°C (Fig. 14), practically all the glucose disappears in the first 5 min of the reaction. About 40% of the glucose is

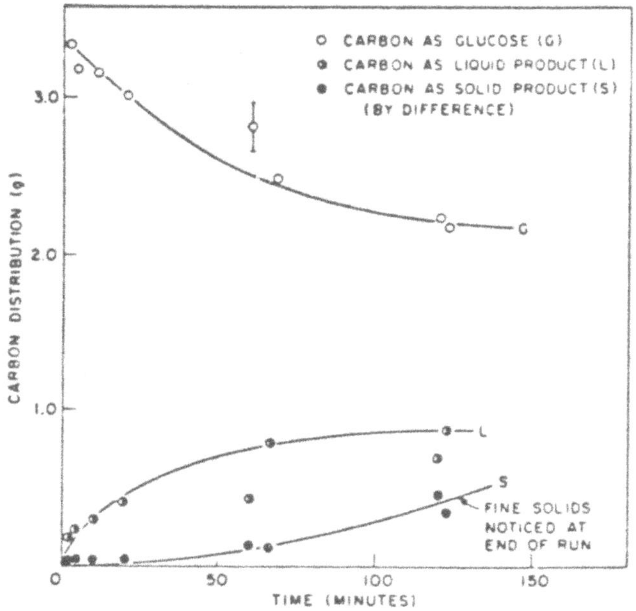

FIG. 12. Product distribution at 150°C.

converted into a solid product. Liquid products are formed which do not react further to form solids. In order to test the rate at which glucose disapppears in the first 5 min, a run was conducted in which the reactor was quenched immediately (about 1·5 min) after glucose addition. On opening the reactor, 25 % of the original carbon was found as a solid product. The datum point is shown in Fig. 14 and lies on the steep line showing solid formation. Another run was conducted at 300 °C, and the reactor quenched, after 1 h of reaction, by a rapid vapor vent. About 38 % of the carbon charged was recovered as a solid product, as shown in Fig. 14.

There is little information available in the literature on the degradation of aqueous solutions of glucose at temperatures above 150 °C. Singh *et al.*[13] showed that the dehydration of glucose at 145 °C (for 16 % glucose solution, pH range 1·6–6·0) led to the formation of hydroxy methylfurfural (HMF):

$$C_6H_{12}P_6 \rightarrow C_6H_6O_3(\text{HMF}) + 3H_2O$$

The mechanism hypothesized by Wolfrom *et al.*[14] involves the opening of the pyranose ring, development of the carbonyl group, followed by

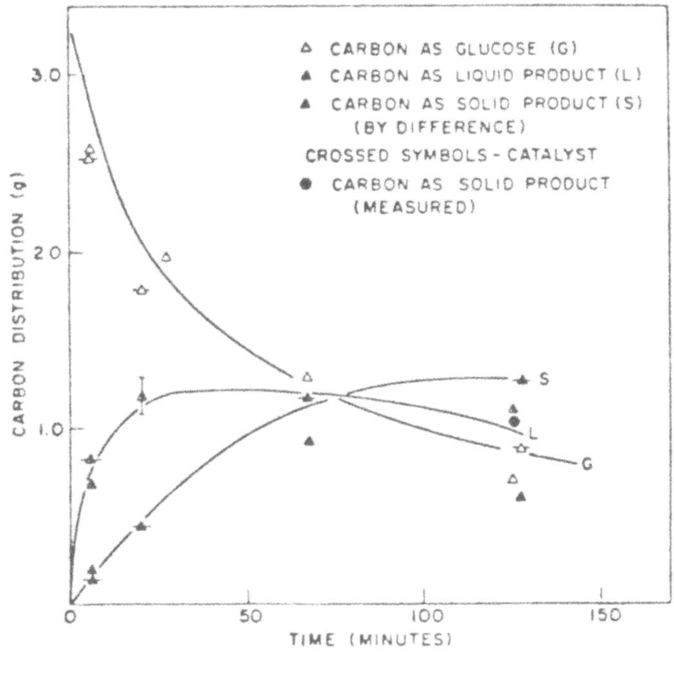

FIG. 13. Product distribution at 200 °C.

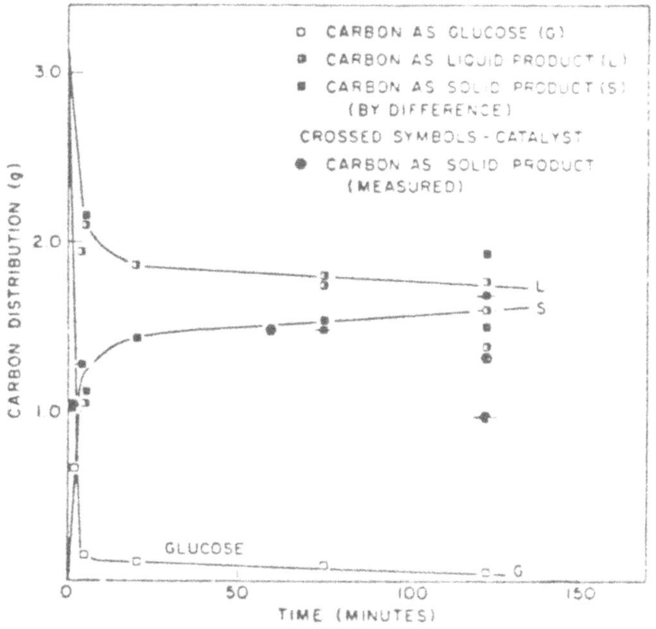

FIG. 14. Product distribution at 300 °C.

dehydration. On continued heating of HMF, levulinic and formic acids are formed.[15]

Ploos van Amstel and Rietenia[16] report that, for glucose decomposition,

$$\text{glucose} \longrightarrow \text{HMF} \longrightarrow \text{levulinic} + \text{formic acids}$$
$$\longrightarrow \text{polymerization products}$$

We conducted a run in which HMF solution was used as feed instead of glucose. At 300 °C, about 35 % of the HMF reacted to form a solid product (60 min duration). In order to determine whether levulinic acid was stable, it was also used as a reactant. No solid products were formed. About 4 % of the acid charged was decarboxylized to CO_2. GC–MS analysis of the liquid samples indicated the presence of both HMF and levulinic acid with a variety of furan and furfural derivatives. These results are consistent with the mechanism proposed in ref. 16.

4.2. Reforming under Supercritical Conditions with Catalysts

In the subcritical experiments with glucose, it was noted that the amount of char formed at 350 °C was substantially less than that at 300 °C (see Table I, runs 15 and 25). Consequently, experiments were conducted at the critical conditions of water.

At 374 °C and 218 atm, a rather startling discovery was made: *no char was formed.* The products were entirely liquid organics and a gaseous mixture of CO, H_2, CO_2, CH_4 and light hydrocarbons. The discovery is the basis of US Patent No. 4,113,446.[6]

Results of a series of experiments at 374 °C are given in Table II. In run 23, with glucose feed and a reaction time of 60 min, over 8 % of the feed carbon was converted to gaseous products. Note that the products are rich in H_2 and CO, while the CH_4 concentration is small. This composition is far from that predicted by chemical equilibrium calculations (see Fig. 1). Nevertheless, the presence of H_2 and CO indicates that we have achieved considerable reforming at moderate temperatures.

A commercial nickel catalyst was added to the feed in run 21. The results obtained were not significantly different from that of run 23 (see Table II). In run 28, 2 g each of seven commercially available catalysts were charged to the reactor. Five of the catalysts were nickel reforming (percent Ni (wt) varying from 11 to 28) and two of the remaining were Co/Mo (3·2/15·2 % CoO/MoO_3) and Pt (0·61 %) of alumina. In a reaction time of 10 min, 11 % of the feed carbon was gasified. In comparing runs 23 and 28, it is clear that these mixtures exhibited a significantly enhanced gasification rate: more gas was produced in run 28 even though the reaction time was reduced by a factor of 6.

The same catalyst mixture was used in the other runs given in Table II. At 30 min reaction time (runs 26 and 35, which are duplicates), carbon gasification increased to 20–23 %. Also note that the gas composition is significantly different from both run 23 (60 min, no catalyst) and run 28 (10 min, with catalyst): the H_2 concentration has increased at the expense of CO. The CH_4 concentration has also increased, although it is still much lower than predicted from thermodynamic equilibrium.

In order to determine if the lack of char production was caused by the increased temperature alone, run 34 was conducted at the critical temperature but at a subcritical pressure of 150 atm. Under these conditions, the density of water is less than 0·1 g/cm^3. A sizable amount of char was formed, indicating that temperature alone is insufficient; the density of water must be maintained close to the critical density in order to avoid char production. In run 36, the glucose feed concentration was increased by a factor of 3, while other conditions were the same as in runs 26 and 35. A small quantity of a tar-like solid was obtained. This material, which exhibited a low melting point ($\sim$40 °C), is very likely soluble in SCW and precipitates out upon cooling the reactor.

TABLE II

RESULTS AT THE CRITICAL CONDITIONS (374 °C, 218 atm)

Run no.	Catalyst	Reaction time (min)	Carbon initially (g)	% of original carbon fed: Final liquid phase	Final gas phase	Solid phase	Carbon closure (%)	(C:H:O) ratio Solid	H_2	CH_4	CO_2	CO	C_2^+
Glucose													
23	None	60	4	77·7	8·2	—	85·9	—	25·8	1·3	34·4	38·5	—
21	Ni	60	4	86·8	10·0	—	96·8	—	30	1·5	42	37	—
28	M	10	4	70·0	11·0	—	81·0	—	25·7	1·0	31·6	41·6	—
26	M	30	4	65·0	20·2	—	85·2	—	45·1	3·2	38·5	12·5	0·7
35	M	30	4	70·8	23·3	—	94·1	—	43·13	2·9	40·6	12·6	0·74
34[a]	M	30	4	27·75	3·4	(68·85)[b]	—	$CH_{0\cdot8}O_{0\cdot2}$	23·21	0·5	65·1	11·24	—
36	M	30	12	60·4	9·5	8·5	78·4	$CH_{1\cdot12}O_{0\cdot3}$	20·38	1·1	42·48	36·07	—
Cellulose													
27	M	30	1·42	77·4	18·3	—	95·7	—	14·5	1·5	19·7	64·2	0·13
Hexanoic acid													
32	M	30	5·75	82·6	2·8	—	85·4	—	67·61	0·8	31·6	≤5·0	—
Polyethylene													
37	M	30	3·22	25·6	2·3	62·4	90·3	$CH_{2\cdot0}$	26·6	2·54	65·87	≤5·0	—
39	M	180	3·22	27·0	4·0	62·7	93·7	$CH_{2\cdot0}$	35·75	3·73	54·43	≤5·0	1·2

[a] Operating pressure was 15·2 MN/m^2.
[b] By difference.
Ni C11-2S-04 nickel catalyst (Catalysts and Chemicals). Carbon closure = [(carbon in products)/(original carbon)]100. M ≡ catalyst mix. 2 g each of: G91B, G90B, C11-2S-04, 22-6HA, 11-3, Co/Mo, Pt/Al_2O_3.

Cellulose was substituted for glucose as the feed material in run 27. Once again, no char was formed. The conditions in run 27 were the same as those in runs 26 and 35. The amount of carbon gasified was similar to that of glucose (18 % vs. 20–23 %), although the gas composition was richer in CO and leaner in H_2.

Appell and co-workers[17,18] studied the degradation of cellulose slurries in a batch autoclave at 250–350 °C. They noted that addition of H_2 and/or CO significantly reduced the amount of char produced. In our experiments, CO and H_2 are produced *in situ*, with only carbohydrate and water as feed. Our current findings may provide a major simplification for processing of forest products.

A number of other organic materials were tested under SCW conditions. The results for hexanoic acid and polyethylene are shown in Table II. The only other cases in which solid was observed were in runs 37 and 39 with polyethylene feed. Those solids, which analyzed as CH_2, appeared to be porous beads of unreacted polyethylene.

4.3. Char Formation

In experiments conducted in the critical region, no char was formed. In most cases, a thin film of acetone-soluble material was deposited on the autoclave baffle and on the walls where liquid water was present after cooling the reactor. A very fine black powder was observed to form in the liquid product upon sitting overnight. This precipitate may be due to subsequent polymerization of reactive or unstable intermediates which formed during reaction. Alternatively, the precipitate may have been present as a dissolved solute during reaction and may slowly crystallize out from a supersaturated state resulting from the cooling down to room temperature. In either case, the evidence indicates that char is not present *during* the reaction in the critical region.

From the results of the subcritical experiments, it is not clear whether the critical conditions prevent char formation or whether char is formed but subsequently reformed rapidly to other liquid and gaseous products. In one experiment, the valve sealing the organic feed reservoir from the reactor (valve B, Fig. 11) leaked. As a result, glucose solution dribbled into the reactor during heat-up. In this case, a char was observed around the inlet port at the conclusion of the run. This char undoubtedly formed under subcritical conditions. The fact that the char remained at the end of the run (i.e. after subsequently being exposed to critical conditions) and that it had not dispersed throughout the reactor is a clear indication that the char is not soluble in supercritical water. Thus, the lack of char formation under critical conditions must be due to competing chemical reactions which prevent or preclude char formation. The implication for introducing feed is significant. Char would undoubtedly form if the feed were present in the reactor during heat-up. The organic must be brought to the critical conditions without being exposed for a significant time to the subcritical temperatures in which char formation is favored.

4.4. SCW Reforming of Glucose Without Catalyst

A number of runs with glucose feed were repeated to determine the reproducibility of the results. These data are shown in Table III. The gas compositions are, in general, fairly reproducible; H_2 shows somewhat more scatter than the other components. The percent of the carbon in the feed that is gasified is shown in the last column. The variation is largest in runs G-1 to G-4, in which it is $\pm 3\%$. This scatter is significant; it is not clear if the differences between the average percent gasification at different operating conditions are real.

Data obtained with the three feed materials are presented in Table IV. Where repeated experiments were run, only the averages are shown therein. The last column is the total amount of water collected at the end of a run. In

TABLE III
REPRODUCIBILITY OF RESULTS; GLUCOSE FEED AND 60 min RESIDENCE TIME

Run No.	*Carbon in feed (g)*	*Av. temp. (°C)*	*Gas composition*					*Carbon gasified (%)*
			H_2	CO	CH_4	CO_2	C_2^+	
G-1	2	377	16·3	67·1	1·2	14·8	0·6	16·9
G-2	2	377	10·2	70·8	1·5	17·5	—	16·6
G-3	2	377	14·4	63·0	1·9	17·8	0·9	21·3
G-4	2	377	11·5	70·8	1·5	15·5	0·7	22·9
Av.	2	377	13·1	68·4	1·5	16·4	0·3	19·5
G-5	2	371	19·9	53·5	2·3	23·3	1·1	25·9
G-6	2	371	8·5	54·6	0·8	34·8	1·2	25·0
Av.	2	371	14·2	54·1	1·6	29·1	1·1	25·5
G-8	4	371	12·3	43·9	0·9	41·7	1·7	10·7
G-9	4	371	25·8	38·5	1·3	34·4	—	8·2
Av.	4	371	19·1	41·2	1·1	38·1	0·9	9·5

TABLE IV

PRODUCT DISTRIBUTION IN THE CRITICAL REGION

Feed (Run no.)	*Reaction time (min)*	*Carbon in feed (g)*	*Av. temp. (°C)*	*Gas composition*					*Carbon gasified (%)*	*Recovered water (cm^3)*
				H_2	CO	CH_4	CO_2	C_2^+		
Glucose										
G-1-4	60	2	377	13·1	68·4	1·5	16·4	0·3	19·5	266
G-5,6	60	2	371	14·2	54·1	1·6	29·1	1·1	25·5	280
G-7	150	2	371	12·3	42·8	1·2	41·4	2·3	28·9	271
G-8,9	60	4	371	19·1	41·2	1·1	38·1	0·9	9·5	265
Levulinic acid										
L-1	60	4·82	377	4·8	87·9	0·6	6·6	—	10·8	251
Maple sawdust										
M-1	5	0·59	377	8·3	79·2	2·3	10·1	—	16·8	241
M-2	15	0·96	377	17·9	69·4	3·7	9·1	—	18·1	258
M-3	30	0·10	377	15·9	65·5	5·1	13·0	0·5	88·3	249
M-4	30	0·67	377	5·0	82·4	2·9	9·6	0·1	33·1	252
M-5	60	0·29	377	16·8	57·2	5·6	19·6	0·9	39·5	258

each run, approximately 310 cm^3 of water was charged to the reactor. We cannot account for the loss of 10–20 % of the water charged by any explanation other than leaks through the seals in the autoclave. As noted earlier, the reactor was leak-tested with helium at 60 atm and room temperature prior to each run. However, the demands on the seals at the critical conditions of water are severe.

Although the batch autoclave apparatus could not be pushed far beyond the critical conditions of water, a small temperature excursion was made in the hope that the reaction had a high activation energy. Comparing the results shown in the first two rows of Table IV, it appears that an increase in temperature from 371 to 377 °C decreased the gasification from 25·5 to 19·5 %. This result was quite unexpected; we can offer no mechanistic explanation at the present time. It is conceivable that the difference is due to a fortuitous change in some parameter other than temperature. For example, the loss of water in the runs at 377 °C was more severe than that at 371 °C. The resulting variation in density of the reaction medium may be an important factor.

Comparing the second and fourth rows of Table IV, increasing feed concentration by a factor of 2 led to a significant reduction in the percent gasification. The reaction(s) leading to gaseous products appears to be close to zero order in feed concentration.

Comparing the second and third rows in Table IV, an increase in residence time by a factor of 2·5 did not significantly alter the percent gasification. It is conceivable that the reactions leading to gaseous products are relatively rapid. For example, some liquid intermediates which reform rapidly may be formed in the early states of the reaction. Other intermediates may only reform more slowly. It is conceivable that some intermediates are relatively stable and do not subsequently reform to an appreciable extent.

At subcritical conditions, we found substantial amounts of levulinic acid [$CH_3CO(CH_2)_2COOH$] present in the liquid phase product from a glucose feed. Under subcritical conditions, levulinic acid was used as a feed and was found to be relatively stable. A run with levulinic acid as a feed was also made under the critical conditions to determine if it was a stable intermediate from glucose. The results, shown in Table IV, indicate that it is as reactive as glucose for gaseous product formation, although the gas composition is leaner in hydrogen.

4.5. SCW Reforming of Maple

A logical sequence in studying the reforming of organics at the critical conditions would be to start with glucose and then proceed to more complex materials, such as cellulose and wood. In an exploratory run with cellulose (see Table II, run 27), it was found that the gasification rate was very similar to that of glucose. However, great difficulty was encountered in feeding cellulose solutions into the reactor. A 10% slurry of cellulose in water will solidify within a half-hour.

As an alternative, sawdust was explored as a feed. Wood was tested for its swelling characteristics in water. Five grams of softwood (pine) absorbed 21 g of water; 5 g of hardwood (maple) absorbed 14 g of water. Additions of pine and maple to the reactor were tested. Pine clogged in narrow apertures, resulting in limited addition of reactant. Maple sawdust (in 3 mm × 1/2 mm needles) was successfully flushed into the reactor, although the quantity fed could not be controlled accurately.

Maple has the advantage of a low ash content. Wenzl[19] reports the composition as 44·8% α-cellulose, 18·9% lignin, 0·2% ash, the remainder (36·1%) being rosin and β-cellulose.

As shown in Table IV, five runs were made with maple sawdust at constant temperature and residence times varying from 5 to 60 min. The feed concentrations varied by a factor of 10 because an unpredictable fraction of the feed was left in the organic reservoir.

In all of the runs with maple sawdust, the feed material was completely solubilized; the product once again did not contain a char. These results are very significant in the light of the complex nature of the feed material.

At a residence time of 5 min, the percent of carbon gasified (16·8%) was substantially higher than that of glucose (without catalyst) at 60 min residence time (8·2%; Table II, run 23). Increasing residence time of maple reforming increased gasification significantly. At 30–60 min, over a third of the feed carbon appears as gaseous products. (The percent gasification in run M-3 is suspect because the amount of feed was very small and, therefore, the calculated results are subject to a large uncertainty.)

It should be emphasized that the gaseous products in SCW reforming of maple are formed very rapidly; substantial gasification was noted after 5 and 15 min. It is also interesting to note that the gas composition tends to be leaner in H_2 and CO_2 and richer in CO and CH_4.

5. CONCLUSIONS

The lack of char formation under SCW conditions is by far the most significant result of this study. The high solvent power of supercritical water for organics most probably prevents char formation by keeping intermediates highly solvated and well-dispersed. The rate of char-forming polymerization reactions is thereby inhibited. The presence of hydrogen, a product of reforming, can aid in stabilization of the intermediates.

The production of H_2 and CO at the moderate temperatures of SCW reforming can have major implications for producing gaseous and liquid fuels from forest products. There lies the potential of subsequently converting the CO and H_2 to CH_4 or liquids by the well-developed routes of Fisher–Tropsch syntheses. Since the synthesis reactions are conducted in the same temperature range as SCW reforming, it may be possible to use the exothermic heat of the synthesis reactions to provide part or all of the heat required in the endothermic reforming reactions.

ACKNOWLEDGEMENT

This work was supported by the Environmental Control Branch of the Ames Research Center, NASA, and their assistance is gratefully acknowledged. The experimental results were obtained by Dr S. Amin and Mr G. Woerner, as part of their degree dissertations. The early work was co-supervised by Professor R. C. Reid of the M.I.T. Department of Chemical Engineering.

REFERENCES

1. Shafizadeh, F. and Chin, P. S. (1976). *ACS Symposium Ser. 43*, I. S. Goldstein (Ed.).
2. Shafizadeh, F. and DeGroot, W. F. (1976). In *Thermal Uses and Properties of Carbohydrates and Lignins*, F. Shafizadeh, K. V. Sarkanen and D. A. Tilman (Eds). Academic Press, New York, pp. 1–16.
3. Fang, P. and McGinnis, G. D. (1976). In *Thermal Uses and Properties of Carbohydrates and Lignins*, F. Shafizadeh, K. V. Sarkanen and D. A. Tilman (Eds). Academic Press, New York, pp. 37–47.
4. Amin, S. I. (1975). Sc.D. thesis dissertation, M.I.T. Department of Chemical Engineering, August.
5. Woerner, G. A. (1976). M.S. thesis dissertation, M.I.T. Department of Chemical Engineering, August.
6. Modell, M., Reid, R. C. and Amin, S. (1978). Gasification process. US Patent 4,113,446, September 12.
7. Quist, A. S. and Marshall, W. L. (1965). Estimation of the dielectric constant of water to 800°. *J. Phys. Chem.* **69**, 3165.

8. Franck, E. U. (1970). Water and aqueous solutions at high pressures and temperatures. *Pure Applied Chem.* **24**, 13.
9. Franck, E. U. (1976). Properties of water, in *High Temperature, High Pressure Electrochemistry in Aqueous Solutions* (NACE-4), p. 109.
10. Connelly, J. F. (1966). Solubility of hydrocarbons in water near the critical solution temperature. *J. Chem. Eng. Data* **11**, 13.
11. Rebert, C. J. and Kay, W. B. (1959). The phase behaviour and solubility relations of the benzene–water system. *AIChE Journal* **5**, 285.
12. Amin, S., Reid, R. C. and Modell, M. (1975). Reforming and decomposition of glucose in an aqueous phase. Intersoc. Conf. on Environ. Systems, San Francisco, July 21.
13. Singh, B., Dean, G. R. and Cantor, S. M. (1948). *Journal of the American Chemical Society* **70**, 517.
14. Wolfram, M. L., Schultz, R. D. and Cavalieri, L. F. (1948). *Journal of the American Chemical Society* **70**, 514.
15. Kirk, R. E. and Othmer, D. F. (1965). *Encyclopedia of Chemical Technology*. McGraw-Hill, New York, Vol. 6, p. 920.
16. Ploos van Amstel, J. J. A. and Rietenia, K. (1970). *Chem. Eng. Tech.* **42**, 981.
17. Appell, H. R., Wender, I. and Miller, R. D. (1970). Technical Progress Report, 25, Bureau of Mines, May.
18. Appell, H. R. *et al.* (1971). RI 7560, Bureau of Mines.
19. Wenzl, H. F. I. (1970). *The Chemical Technology of Wood*. Academic Press, New York, pp. 253–96.

7

Thermochemical Fractionation and Liquefaction of Wood

RAYMOND A. YOUNG and JAMES L. DAVIS
Department of Forestry, University of Wisconsin, Madison, Wisconsin, USA

1. INTRODUCTION

Approximately 40 % of the wood harvested in the US today is for pulp and paper production.[1] This high grade use will certainly continue in the years to come. Thus our approaches to biomass utilization have emphasized systems which complement cellulose production and incorporate schemes for total utilization of the woody stem. This means more efficient utilization of wood industry wastes and byproducts. Our plan is oriented to both the energy and materials crises.

The Cellulose–Particleboard–Synfuels complex shown in Fig. 1 serves as a model for our integrated approaches to biomass research in the Department of Forestry, University of Wisconsin. Newer approaches for production of cellulose pulp from wood are an important segment of the scheme. In addition, synfuels and adhesives can be obtained from pulp and paper industry wastes and byproducts.[2,3] The synfuels and octane boosters are ethanol, methanol, ethylacetate and other esters, terpenes, aldehydes and phenols. The adhesives are formulated from the renewable wood polymers and monomers, after suitable modification. We are developing strong, water-resistant adhesives based on carbohydrate derivatives, sugars and lignin waste products.

A complete discussion of the research involved in the proposed scheme is beyond the scope of this paper and will be dealt with in succeeding publications.[44] The work reported here is part of our efforts to improve the yield of biomass liquid fuels and chemicals from woody biomass, through thermochemical fractionation and liquefaction of wood. Our supposition is that wood, as a unique polymer composite, can be fractionated into a

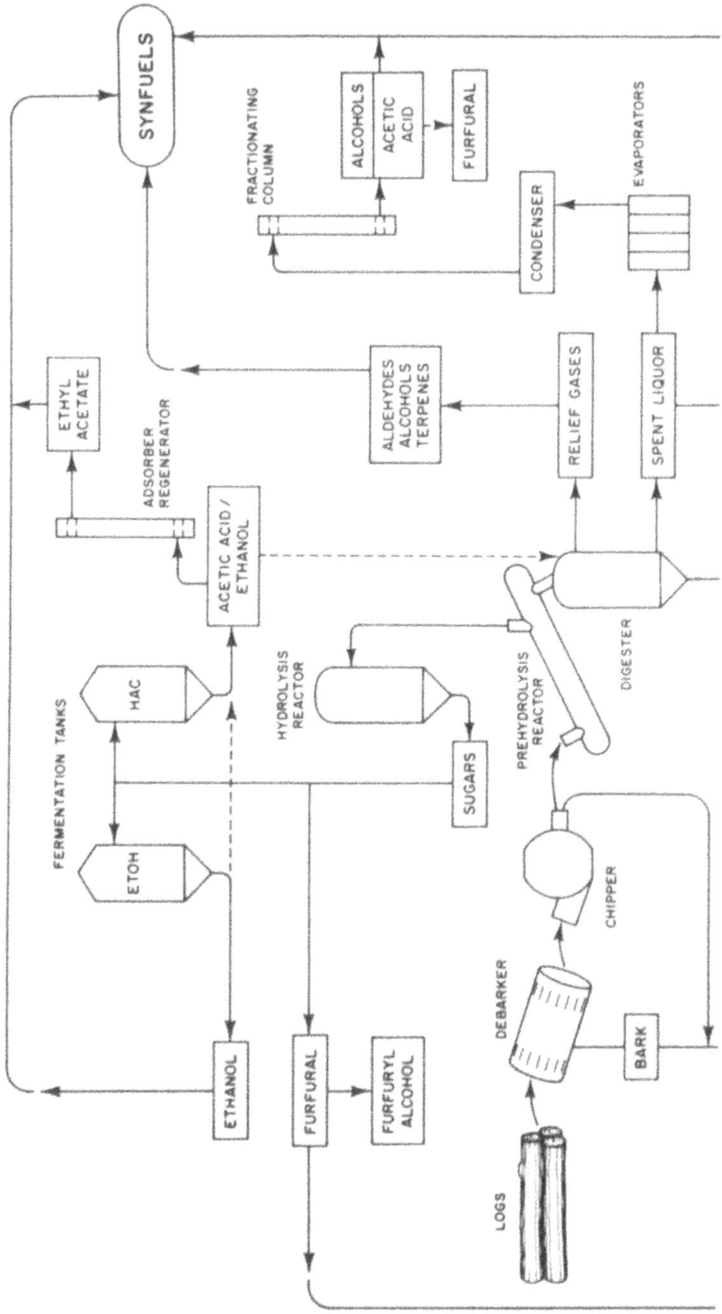
SYNFUELS
FRACTIONATING COLUMN
ALCOHOLS
ACETIC ACID
FURFURAL
CONDENSER
EVAPORATORS
ETHYL ACETATE
ADSORBER REGENERATOR
ALDEHYDES ALCOHOLS TERPENES
RELIEF GASES
SPENT LIQUOR
ACETIC ACID / ETHANOL
FERMENTATION TANKS
HAC
ETOH
HYDROLYSIS REACTOR
PREHYDROLYSIS REACTOR
DIGESTER
SUGARS
CHIPPER
DEBARKER
BARK
ETHANOL
FURFURAL
FURFURYL ALCOHOL
LOGS

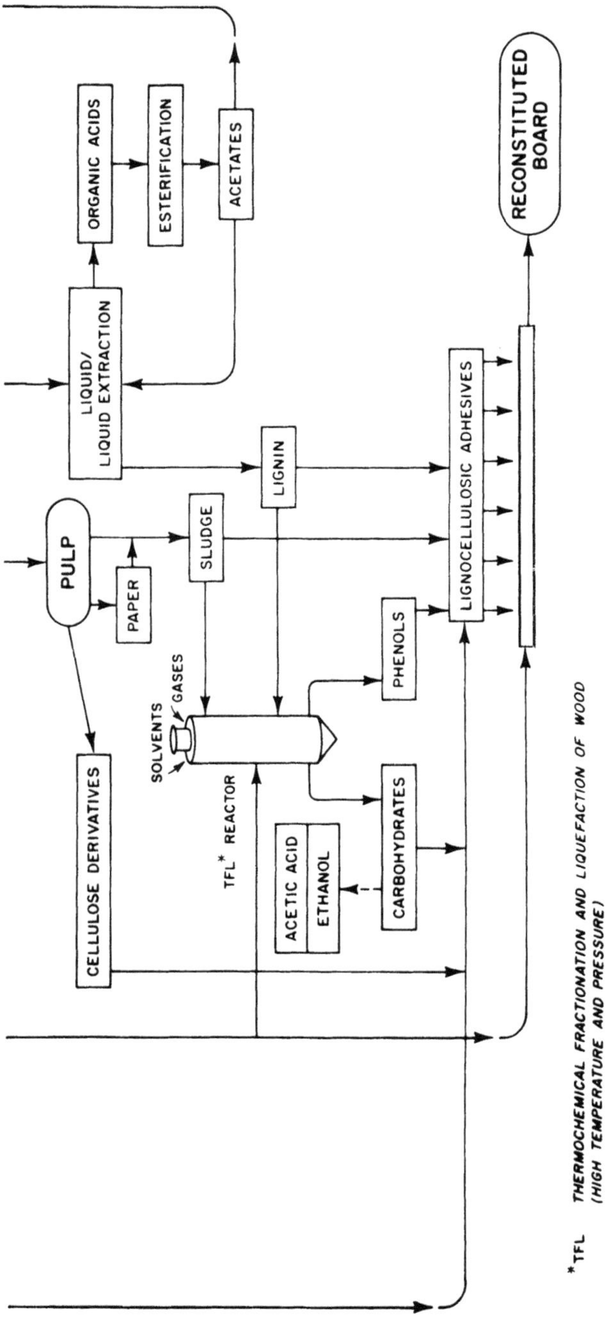

FIG. 1. Proposed Cellulose–Particleboard–Synfuels Complex Model.

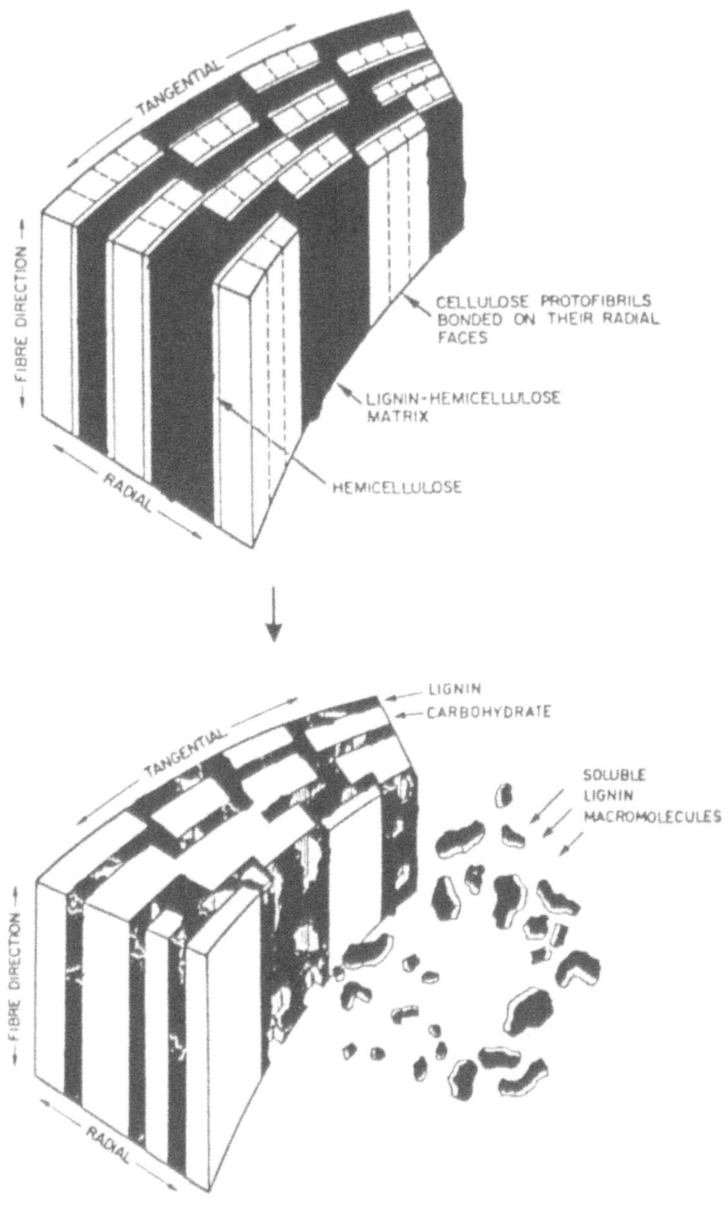

FIG. 2. Schematic of the proposed interrupted lamella model for the ultrastructural arrangement of lignin and carbohydrate in the wood cell wall and breakdown and solution of lignin during chemical pulping according to Goring.[9]

limited number of high grade fuels and chemicals by thermochemical techniques through proper control of the critical variables of gas type and pressure, temperature, duration of treatment, solvent type and catalyst. The fractionated wood could then be more efficiently converted to higher quality and more directly useable commodities. Previously described wood liquefaction methods[5-8] result in hundreds of diverse low molecular weight chemicals, the oils from which have only low grade utility.

For efficient utilization of woody biomass, a viable process must provide for clean separation of the major wood components, provide efficient saccharification of cellulose and provide efficient conversion of the components to fuels and chemicals. Wood is composed of three primary constituents: cellulose, hemicellulose and lignin in the rough proportions of 2:1:1, respectively. The wood fibers are held together by a middle lamella which is almost pure lignin and represents about 30% of the total lignin in wood. The remaining 70% of the lignin is in the cell wall with the polysaccharide components. Goring[9] has proposed an interrupted lamella model for the ultrastructural arrangement of lignin and carbohydrates in the cell wall (Fig. 2). The major wood components probably exist as domains in a complex array. Lignin removal is shown in the diagram to represent a pulping reaction. However, it should also be possible to isolate preferentially the other components as well. Clearly to avoid admixing of the components and get a clean separation, considerable specificity of the reagents must be realized.

The wood polymers can be roughly classified as follows: hemicelluloses are amorphous polysaccharides with susceptible glycosidic bonds; cellulose is a crystalline polysaccharide with susceptible but less accessible glycosidic bonds; and lignin is an amorphous cross-linked aromatic polymer with some resistant carbon-to-carbon bonds and greater heat stability.

Obviously wood is predominantly carbohydrate in nature (about 70% of the dry weight) and these polysaccharide fractions can be further distinguished by their susceptibility to acids. The relative hydrolysis rates for methyl pyranosides and disaccharides are shown in Table I.[10] It can be seen that the highest rate of acid hydrolysis is given by xylobiose, a prototype of xylan hemicellulose, and the slowest rate by cellobiose, as a cellulose model. Even if cellulose were not crystalline, it would be expected to hydrolyze slower than xylan.

Another important consideration is the rates of decomposition of the sugars in acid solutions once they are formed. As shown in Table II,[10] the xylose decomposes the fastest. Thus, xylose is the most easily produced

TABLE I
RELATIVE HYDROLYSIS RATES[a]

Methyl pyranoside		*Disaccharide*	
β-D-glucose	1·0	Cellobiose	1·5
β-D-mannose	3·0	Mannobiose	2·7
β-D-galactose	4·8	Lactose	3·6
β-D-xylose	5·8	Xylobiose	11·0

[a] Data from Ref. 10.

sugar and the most easily degraded. This suggests a short reaction time and a quick removal of the pentose sugar from the system.

Lignin, in contrast, is a complex three-dimensional polymer of phenylpropane units interlinked by a variety of bonds.[1] The β–O–4 ether linkages account for about one-half of the total interunit bonds while the much more resistant carbon-to-carbon bonds account for another 30 % of the interunit linkages.[11] Indeed, lignin is more resistant to temperature than the two polysaccharide components of wood and this could serve as an important feature in the separation design. Lignin is relatively stable up to 300–350 °C while the other wood components decompose at lower temperatures.[12,13]

The differences in the chemical and physical nature of the three major wood components were taken into account in the proposed process shown in Fig. 3. Three stages are shown which represent the sequential isolation of aspen wood for subsequent conversion to fuels and chemicals. In general, the conditions become harsher as the material moves from Stage I (mildest) to Stage III (harshest).

TABLE II
RELATIVE DECOMPOSITION RATES[a]

Sugar	*Rate constant* ($\times 10^{-2}$)
Glucose	2·4 (lowest)
Galactose	2·6
Mannose	3·6
Arabinose	4·2
Xylose	7·2 (highest)

[a] Data from Ref. 10.

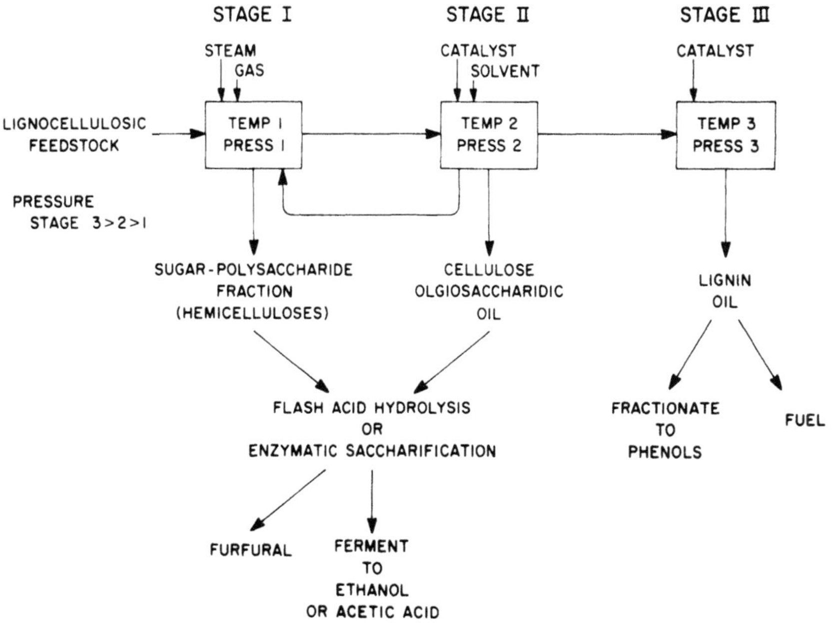

FIG. 3. Proposed three stage biomass utilization scheme. Temperature and pressure are increased in the consecutive stages.

The first stage (I) is a mild treatment for removal of the hemicellulose fraction. Moderate temperatures and pressures are utilized in this stage with the hemicellulose removal either acid or base assisted. The simultaneous isolation, decrystallization and partial saccharification (SIDS) of cellulose is the second stage (II) of the proposed sequence. The second stage reactions are performed at higher temperatures and pressures with the intent of obtaining a fermentable oil. The third and final stage (III) is designed to convert lignin to a phenolic oil. Critical factors in this stage are proper choice of catalyst and higher temperatures where lignin is more rapidly degraded.

The concept of producing an oil from wood is not new and there is considerable literature available on oil production from wood and cellulose.[5–8] However, selective fractionation of wood to mostly fermentable oils is unique and the preliminary work on this process is described here. An important aspect of the investigation was characterization of the changes occurring with wood in aqueous and organic solvents in the temperature range of 100–350 °C.

2. RESULTS AND DISCUSSION

2.1. Stage I: Prehydrolysis

Prehydrolysis of wood has been studied previously by a number of different investigators.[14] In practically all of these investigations mineral acids such as sulfuric and hydrochloric were utilized to remove the hemicelluloses from wood. Alternate methods for prehydrolysis have not been adequately treated to date. Organic acids such as acetic acid offer advantages compared to the mineral acids. For example, acetic acid:

1. is an organic acid which can be readily burned if necessary where mineral acids cannot;
2. is produced in chemical pulping, representing a large percentage of waste pulping liquors and thus is a potentially abundant byproduct of the pulp and paper industry;[14]
3. is recoverable by newer technology and can be readily converted to ethyl acetate for use as a fuel additive or octane booster;[2,3]
4. contains no chlorine or sulfur which makes it environmentally desirable;
5. can be used as an agent in subsequent organic acid pulping of wood.[15]

Carbonic acid also offers advantages as an organic acid since it can be produced *in situ* through an overpressure of carbon dioxide gas in the reaction vessel. Results obtained using acetic acid and carbonic acid for prehydrolysis of aspen are discussed below.

Figure 4 shows the effect of time of immersion on the percent soluble yield with acetic acid and carbon dioxide gas (10:1 liquor/wood ratio). Increased immersion time and higher acetic acid concentration (no CO_2 pressure) both increase the soluble fraction. The actual reaction times would be less than indicated in Fig. 4 because the reaction cylinders required about 20 min to heat to maximum temperature. The different shape of the distilled water curve reveals the low level of hydrolytic activity generated by water prehydrolysis (autohydrolysis). Some acetic acid is generated *in situ* due to hydrolysis of acetyl groups from the 4-O-methylglucuronoxylan hemicellulose in aspen.

A high yield of soluble product (24·6 %) was obtained with 20 % acetic acid at 60 min immersion time. There appears to be no advantage to using higher concentrations of acetic acid in the prehydrolysis stage. When the vessel was pressurized with carbon dioxide gas (200 psi) the concentration of acetic acid could be reduced to 5 % with about the same yield of soluble

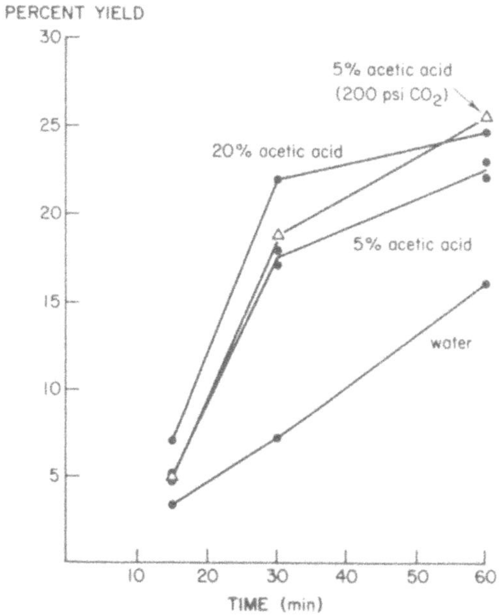

FIG. 4. Percent yield versus reaction time for aspen prehydrolysis in aqueous acetic acid at 160 °C.

product (25·3 %) (Fig. 4). The combination of pressure and additional carbonic acid probably accounts for the high percent yield in this case. However, the use of CO_2 pressure with just distilled water gave essentially the same curve as shown for the distilled water at ambient pressure. Apparently the catalytic activity of the added acetic acid is enhanced in the CO_2 pressurized vessel. Thus prehydrolysis in pressurized CO_2 vessels may offer special advantages.

A higher liquor/wood ratio also provided an increase in the soluble fraction as shown in Fig. 5. A maximum yield of 25 % soluble product was obtained with 5 % acetic acid at a liquor/wood ratio of 25:1. A high liquor/wood ratio may be necessary to improve mass transfer since the cylinder contents were not agitated during the reaction.

The composition of the prehydrolysis liquors is shown in Fig. 6. Clearly the total yield of soluble products is not only hemicelluloses. Ultraviolet spectroscopic analysis of the liquors confirmed the presence of aromatic lignin structures in the non-carbohydrate fraction. The liquors probably also contained a small percentage of extractives as well.

The greatest selectivity in prehydrolysis was obtained with 20 % acetic

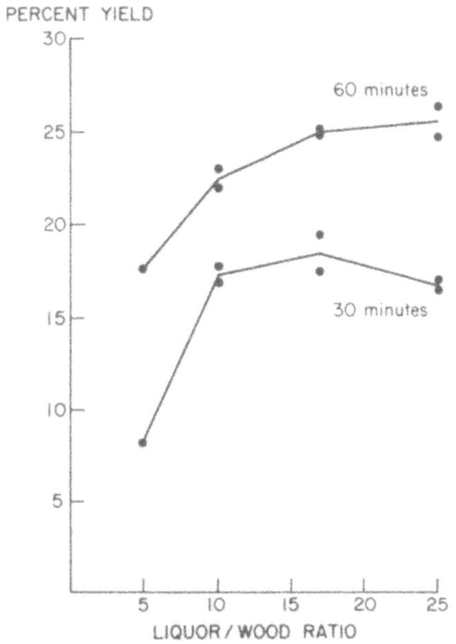

FIG. 5. Percent yield versus liquor-to-wood ratio for aspen prehydrolysis in 5% acetic acid at 160°C.

acid. At this concentration, xylose represented over 90% of the soluble carbohydrate and over 80% of the total xylose in the wood. The liquors from the acetic acid–carbon dioxide runs were not analyzed for sugars.

These results indicate that a significant portion of xylan present in wood can be removed with aqueous acetic acid. In addition, Connor[16] has also shown that very short reaction times are possible with acetic acid prehydrolysis. He was able to recover about 85% of the xylan from aspen in about 2 min at 220°C with aqueous acetic acid. Pressurized vessels should give even better results.

2.2. Characterization of Thermochemical Reactions of Wood

To ascertain proper reaction conditions for Stages II and III, further experiments were designed for general characterization of the thermochemical reactions of whole wood. The independent variables for the study were reaction temperature and time, reaction solvent, gas composition and catalyst concentration, while the dependent variables were total recovered mass, soluble yield and composition of the soluble fraction and residue.

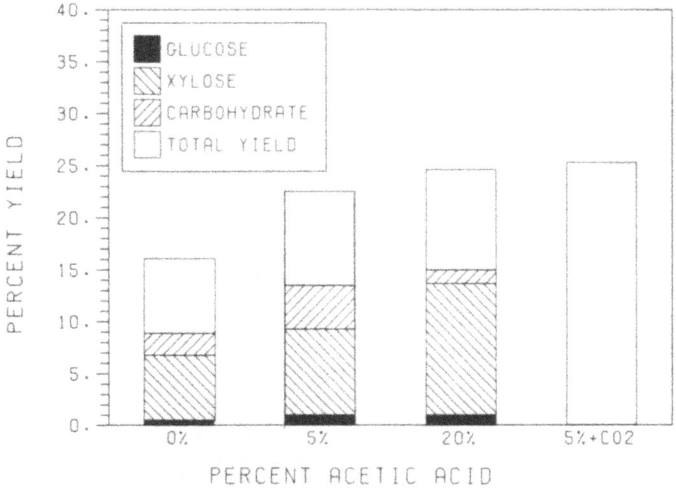

FIG. 6. Liquor composition for aspen prehydrolysis in aqueous acetic acid at 160 °C.

Carbohydrate, sugar and lignin analyses were used to characterize the composition of the reaction products.

Selection of the initial reaction conditions were based on earlier results by Lloyd and Ralph[17] who utilized homogeneous catalysts in organic solvents for liquefaction of wood. Figure 7 shows the thermochemical reactions of aspen in 95 % ethanol with palladium chloride catalyst under

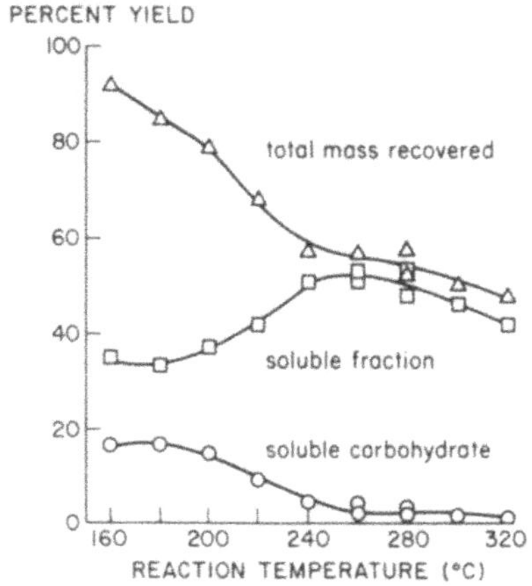

FIG. 7. Liquefaction of ground aspen in 95 % ethanol containing 6 % palladium chloride catalyst under hydrogen/carbon monoxide at various temperatures.

hydrogen/carbon monoxide gas. The total recovered mass decreases rapidly from 160 to 240 °C and then at a decreased rate from 240 to 320 °C. At 320 °C only 48 % of the wood charged to the reactor is recoverable, the rest being lost as gases. A control run without the palladium chloride catalyst yielded only 14 % soluble product at 200 °C.

The soluble fraction reaches a maximum (54 %) at 260 °C and decreases at higher temperatures. At 260 °C over 90 % of the mass recovered is as soluble product. The soluble carbohydrate decreases steadily at temperatures above 160 °C, and at 260 °C has essentially disappeared. The soluble carbohydrate decreases by 62 % in the temperature range 200–220 °C and again by 50 % from 200 to 240 °C.

The carbohydrate content of the residue was also analyzed and summative results are shown in Fig. 8. At 160 °C there is very little loss of carbohydrate, 67·5 % recovered out of the original 70 % present in untreated aspen. Most of the carbohydrate is still in the residue (51·0 %) and only about 19 % of the soluble fraction is glucose (78 % xylose). At 200 °C the total recovered carbohydrate decreased by 21 % compared to results at 160 °C. The xylose probably degraded at 200 °C because the glucose portion increased to about 55 % of the soluble carbohydrate while the total soluble fraction (15 %) was less than at 160 °C. It appears that the amorphous, xylose-rich hemicelluloses dissolve readily at 160 °C; while a

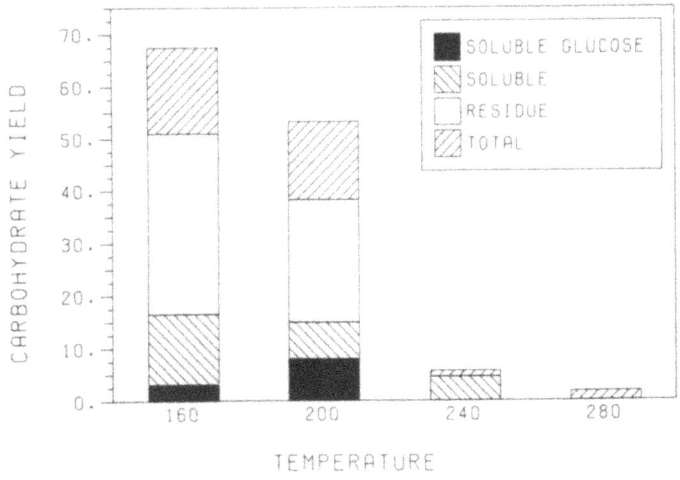

FIG. 8. Carbohydrate yield and composition for the soluble fraction and residue for reaction of aspen at selected temperatures. Conditions identical to those described in Fig. 7.

higher temperature (> 200 °C) is needed to disrupt the crystalline structure and hydrolyze the cellulose. The majority of the carbohydrate is still retained in the residue at 200 °C. We are presently characterizing the molecular weight of the polysaccharides remaining in the residue.

A dramatic change occurs around 220–240 °C. In this temperature range the majority of the carbohydrates are lost with a recovery of only 5·7 %, most of which is in the soluble fraction (4·6 %), at 240 °C. Obviously to ensure retention of the wood carbohydrates, the temperature must be kept below approximately 220 °C for the systems studied.

Fourier transform infrared spectroscopy (FTIR) was also utilized to characterize the reaction products. Differences between the soluble products formed at 200 °C (Fig. 9a) and those formed at 240 °C (Fig. 9b) are clearly visible in the difference spectrum (product at 240 °C – product at 200 °C) shown in Fig. 9c. The absorbance at 1050 cm^{-1} in the lower temperature product is probably associated with a C–O–C stretching vibration while the absorbance at 1710 cm^{-1} is unquestionably a carbonyl stretching vibration.[18]

The disappearance of the C–O–C stretching absorbance at the higher reaction temperature is consistent with the decline in soluble carbohydrate mentioned above. However, the production of carbonyl functions was unexpected. Eager *et al.*[7] reported that as the reaction temperature is increased, the water soluble products from aspen liquefaction in aqueous sodium carbonate under carbon monoxide showed a shift from ketones to alcohols. Analysis by high pressure liquid chromatography (HPLC) showed that neither furfural nor hydroxymethylfural were present in our soluble reaction product. However, there appears to be a correlation between the appearance of the carbonyl group in the FTIR and the formation of an oil in the reaction product. A possible mechanism for formation of carbonyl compounds through reaction of carbohydrates with a platinum or palladium catalyst is shown below.[19]

$$RCH_2OH + Pt \rightarrow RCHO + Pt\text{–}H_2$$

Lignin analyses of the soluble products indicated that about 75 % of the original lignin dissolved at 160 °C. As the reaction temperature increased above 160 °C, the retention of the lignin in the residue remained constant. Thus, the increase in the soluble fraction with temperature is primarily due to increasing carbohydrate dissolution.

Trials with α-cellulose samples provided more specific information on the behavior of this predominant polysaccharide (Fig. 10). The total mass recovered was greater with cellulose compared to wood (dashed lines) up to

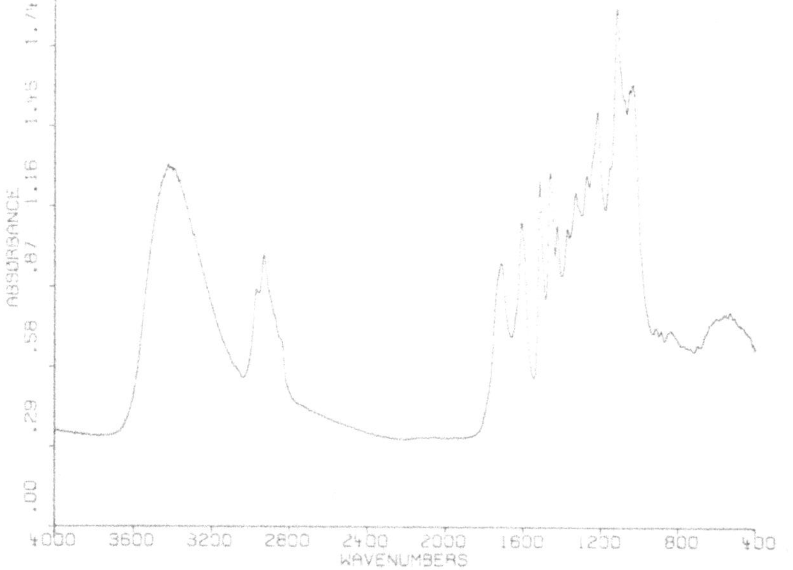

FIG. 9.a FTIR spectrum of acetone soluble product from liquefaction of aspen in 95% ethanol with palladium chloride catalyst at 200 °C under hydrogen/carbon monoxide.

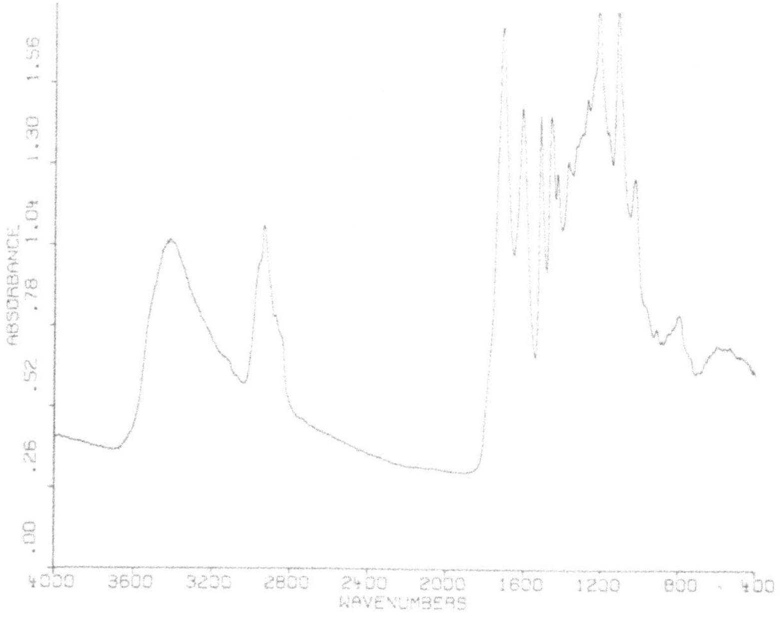

FIG. 9.b FTIR spectrum of acetone soluble product from liquefaction of aspen in 95% ethanol with palladium chloride catalyst at 240 °C under hydrogen/carbon monoxide.

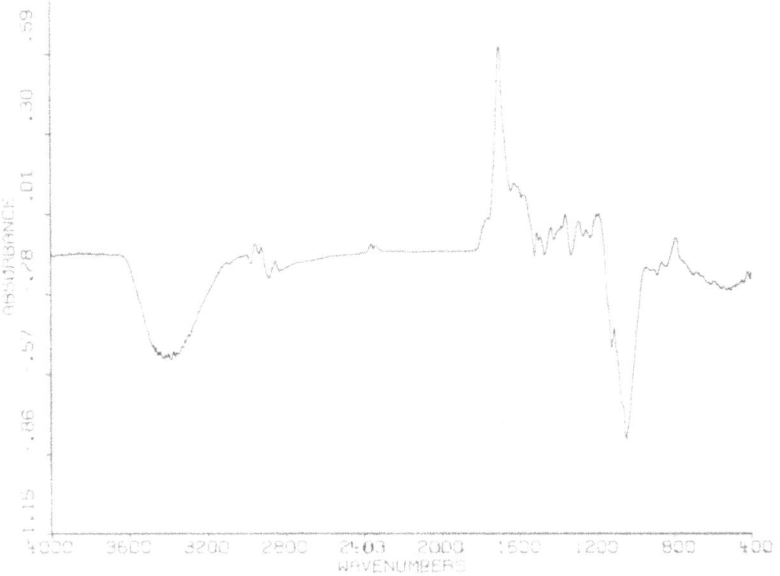

FIG. 9.c Difference spectrum of aspen liquefaction product at 240 °C (Fig. 9(b)) minus aspen liquefaction product at 200 °C (Fig. 9(a)).

240 °C where the curves overlap. The soluble fraction was much less than with wood at 160 °C but about the same as with wood at 200 °C. This difference is probably due to the greater ease of hemicellulose (and lignin) solubilization at the lower temperatures.

The soluble carbohydrate from α-cellulose, however, was the greatest at 200 °C. This most likely reflects the differences in reactivity between the more reactive, amorphous hemicelluloses versus the crystalline cellulose and confirms the previous conclusion, based on sugar analyses, that the hemicelluloses are solubilized and lost at lower temperatures, while the cellulose requires higher temperature for solubilization.

The FTIR spectra of the cellulose liquefaction products revealed the same transition for the cellulose-derived product that was observed for the wood-derived material, between the temperature of 200 and 240 °C. Thus the major changes that occur in the composition of the liquefaction products in this temperature range are primarily the result of carbohydrate degradation.

2.3. Effect of Reaction Solvent

The reaction solvent has an impact on wood liquefaction that is second only to the reaction temperature. The solvents tested in addition to 95%

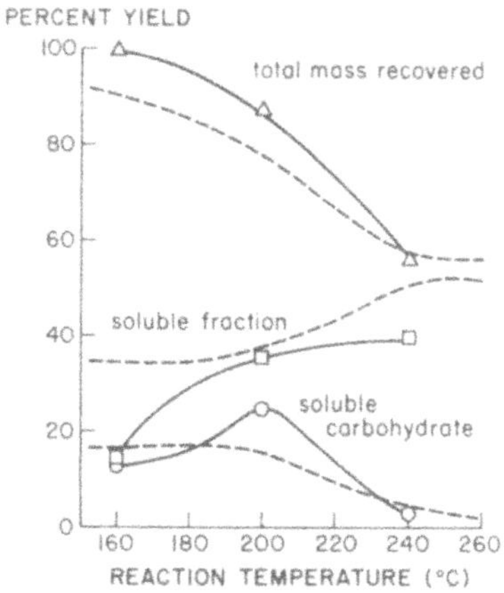

FIG. 10. Liquefaction of α-cellulose in 95% ethanol containing 6% palladium chloride catalyst under hydrogen/carbon monoxide at various temperatures. The dashed lines are the results for whole wood (Fig. 7).

aqueous ethanol were 50% ethanol, 5% ethanol, butanol, ethylene glycol and toluene. As shown in Fig. 11, the highest total recovered mass at 200°C was with 95% ethanol and butanol, both close to 80%. However, ethylene glycol produced the most soluble material (53·0%) with toluene the least, only 8·2%. The large soluble fraction from ethylene glycol may be related to the high boiling point (198°C) of this solvent.

The amount of lignin remaining in the residue was very low (4–5%) for the three solvents, 95% ethanol, butanol and ethylene glycol (Group I), but much higher with 5% ethanol, 50% ethanol and toluene (Group II). The lignin present in the residue with the 5% ethanol (37·5%) and toluene (48·5%) far exceeds the amount in the untreated wood. This is probably due to condensation of carbohydrate degradation products with lignin degradation products and subsequent precipitation of this polymer on to the residue. The insolubility of such a condensation polymer in the 72% sulfuric acid used in the Klason procedure leads to high 'lignin' content in the residue. Obviously the reactions are quite different for the two groups of solvents.

A further indication of the effect of the reaction solvents is given in Fig. 12. The Group I solvents show high yields of carbohydrate both in the residue and in the liquor, while the Group II solvents yield less than 4% total carbohydrate. Apparently with the Group II solvents the carbohydrates are rapidly degraded to products which condense and show up as

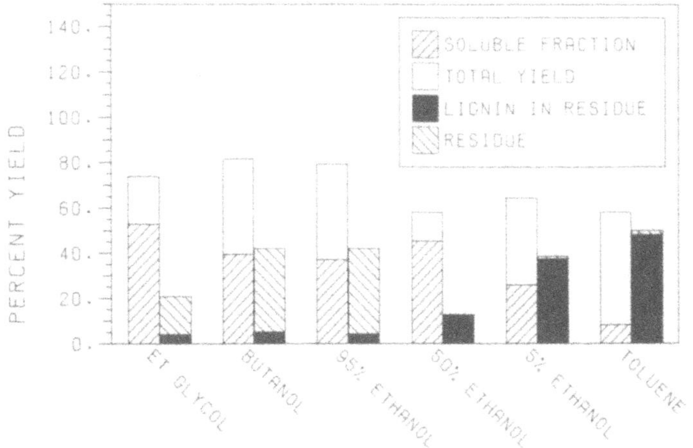

FIG. 11. Composition of reaction products from liquefaction of aspen at 200 °C with palladium chloride under hydrogen/carbon monoxide in various solvents.

'lignin'. The preferred solvent for carbohydrate retention at 200 °C would probably be 95 % ethanol which gave a total carbohydrate yield of 53·3 % and a soluble carbohydrate yield of 15·0 %.

The FTIR spectra of the liquefaction products obtained with representative solvents from the two groups are shown in Figs 9(a) and 13. With 95 % ethanol (Fig. 9(a)) the strongest absorption is in the

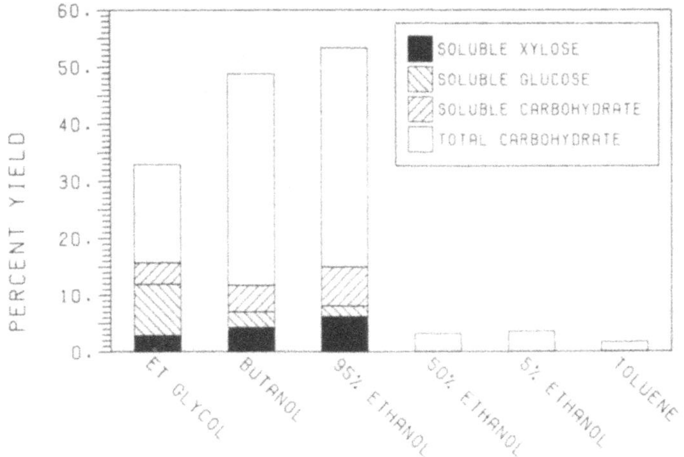

FIG. 12. Effect of solvent on carbohydrate yield for aspen liquefaction. Palladium chloride catalyst under hydrogen/carbon monoxide at 200 °C.

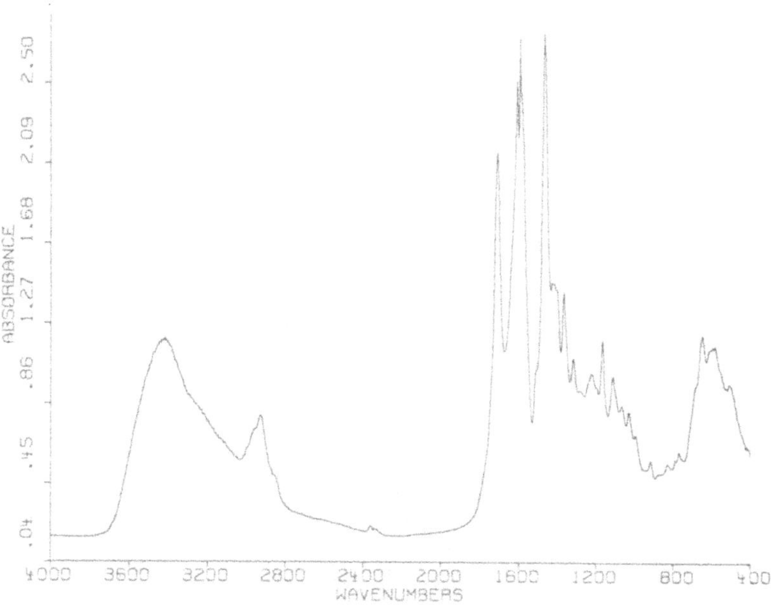

FIG. 13. FTIR spectrum of acetone soluble product from liquefaction of aspen in 5% ethanol containing palladium chloride under hydrogen/carbon monoxide at 200 °C.

1000–1200 cm^{-1} region, characteristic of C–O–C stretching vibrations. In contrast, with 5% ethanol (Fig. 13) the maximum absorption is near 1600 cm^{-1}, characteristic of aromatic skeletal vibrations.[11] The strong carbonyl absorption near 1710 cm^{-1} is obvious in both the spectra. Clearly the reaction solvent plays a major role in the course of thermochemical wood degradation reactions. Vuorinen and Sjöström[20] also noted a pronounced effect of alcohol solvents in their investigation of oxygen oxidation of polysaccharides. The alcohol solvent may affect the gas transfer and the chemical reaction rates.

2.4. Effect of Reaction Gases

Different reaction gases have been used by various investigators for liquefaction of wood and lignin.[4-8,11] Hydrogen, carbon monoxide and mixtures of the two have been commonly employed. Few studies have looked specifically at the effect of the reaction gas.

To study the influence of the reaction gas on product yield, a series of trials were run in 95% ethanol at 280 °C with 6% palladium chloride

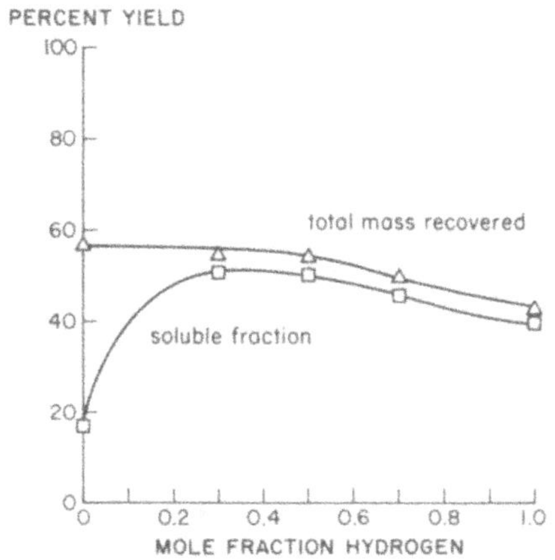

FIG. 14. Effect of gas composition (ratio of hydrogen/carbon monoxide), on aspen liquefaction at 280 °C.

catalyst. The initial pressure was held constant, at approximately 1075 psi, and the partial pressures of hydrogen and carbon monoxide varied. The temperature 280 °C was selected because this is where the maximum soluble products were obtained and thus changes should be maximized.

Figure 14 shows that the total recovered mass declines with increasing mole fraction of hydrogen, but the soluble fraction reaches a maximum when hydrogen is 30–50 % of the reaction gas. Carbon monoxide and hydrogen would have distinctly different effects on the liquefaction process and it is not clear whether the variation shown in Fig. 14 results from the increasing partial pressure of hydrogen or the decreasing partial pressure of carbon monoxide. However, it is obvious that the gas composition can also affect the liquefaction process.

3. MATERIALS AND METHODS

3.1. Prehydrolysis

Milled aspen (*Populus tremuloides*) was charged to 100 ml stainless steel cylinders with the appropriate concentration of aqueous acetic acid (w/w). The cylinder was placed upright in a silicone oil bath at 160 °C. At the end of the designated immersion time the cylinders were immediately cooled in cold water. They were then opened and flushed with distilled water, and filtered and washed on a medium porosity fitted glass filter. The oven dry weight of the residue and a 20 ml aliquot of the liquor was determined.

3.2. Liquefaction

Milled aspen was charged to a 100 ml glass-lined high pressure block reactor. The appropriate solvent was added and carbon monoxide and hydrogen gas (generally 1/1) introduced. The furnace rocking motor and furnace controller were then turned with a controlled heat-up time of 1–2½ h. When the reactor reached the desired temperature, the furnace was shut off and allowed to cool to 50 °C. This provided about 30 min at the maximum temperature. The reaction product was washed and filtered with three volumes of acetone. The dry weight of the residue and soluble fraction were then determined.

3.3. Analytical Methods

Lignin contents were determined by the Klason method[21] and total sugar by the method of Somogyi and Nelson.[22] The relative proportions of the individual sugars were determined on a Waters HPLC with a Biorad HPX87P cation exchange column (500 psig and 85 °C) equipped with a refractive index detector. All FTIR spectra were obtained with a Nicolet 7199 instrument as KBr pellets.

ACKNOWLEDGEMENTS

Appreciation is expressed to Marilyn Effland and other members of the analytical department of the USDA Forest Products Laboratory for assistance with carbohydrate, lignin and HPLC analyses. This study was made possible through grants from the Bio-Energy Council, the US Department of Agriculture (Special Grants Program), and the College of Agriculture and Life Sciences, University of Wisconsin.

REFERENCES

1. Young, R. A. (Ed.) (1982). *Introduction to Forest Science.* Wiley, New York.
2. Baierl, K., Young, R. A., Sell, N. and Goldsby, A. (1981). Production of synfuels from waste biomass seminars. University of Wisconsin—Green Bay, March 6 and October 6.
3. Baierl, K. (1979). Adsorber–regenerator. US Patent 4,155,849, May 22.
4. Young, R. A. (1982). In *Proceedings of NATO Advanced Study Institute on Biomass Utilization*, Alcabideche, Portugal, September 26–October 8.
5. Appell, H. R. *et al.* (1970, 1971, 1975). US Bur. Mines Rept. 7560 (1971); 8013 (1975); TP25 (1970); RI7560 (1971).

6. Ergun, S. (1980). Biomass liquefaction efforts in the United States. US Dept. Energy Rept. LBL-10456, February.
7. Eager, R. L., Mathews, J. F., Pepper, J. M. and Zohdi, H. (1981). *Can. J. Chem.* **59**, 2191.
8. Boocock, D. G. B., Mackay, D. and Patterson, S. (1980). *Biomass Liquefaction in Canada: A Review of the Current Studies.* Proceedings of a Workshop, Univ. of Toronto, August.
9. Goring, D. A. I. (1977). In *Cellulose Chemistry and Technology*, J. C. Arthur, Jr. (Ed.). ACS Symposium Series No. 48.
10. Harris, J. F. (1975). *Appl. Polym. Symp.* **28**, 131.
11. Sarkanen, K. V. and Ludwig, C. H. (Eds) (1971). *Lignins: Occurrence, Structure, Formation and Reactions.* Wiley-Interscience, New York.
12. Roberts, A. F. (1972). *Combustion and Flame* **14**, 261.
13. Tang, W. K. (1967). US Forest Serv. Res. Pap. FPL 71, Forest Prod. Laboratory, Madison, January.
14. Rydholm, S. A. (1965). *Pulping Processes.* Wiley, New York.
15. DeHaas, G. G. and Lang, C. J. (1971). US Patent 3,553,076, January 5.
16. Connor, A. H. (1981). Acetic acid prehydrolysis of hardwoods. Paper presented at 181st ACS National Mtg., Atlanta, Georgia, March.
17. Lloyd, J. A. and Ralph, J. (1977). Wood Chemistry Repts. 16 and 17, New Zealand Forest Serv., Forest Res. Inst., Rotorua.
18. Tipson, R. S. (1968). Infrared spectroscopy: A review of the literature. Nat. Bur. Std. Monogr. 110.
19. Heyns, K. and Paulsen, H. (1962). *Adv. Carb. Chem.* **17**, 169.
20. Vuorinen, T. and Sjöström, E. (1982). *J. Wood Chem. Technol.* **2**(2), 129.
21. Effland, M. J. (1977). *Tappi* **60**, 143.
22. Somogyi, M. (1952). *J. Biol. Chem.* **195**, 19.

8

The Pyrolysis and Gasification of Wood in Molten Hydroxide Eutectics

P. J. IREDALE and B. W. HATT

Department of Chemical Engineering, University of Aston in Birmingham, Birmingham, UK

1. INTRODUCTION

In order to avoid problems caused by solids handling, char formation and tar formation during the gasification of wood and cellulosic materials, the use of a low temperature molten salt system was investigated. This approach was adopted in order to achieve dissolution of the biomass material in the salt prior to pyrolysis, thus enabling the pyrolysis to be carried out in the liquid phase throughout.

The extensive literature on the wood pulp industry was investigated with a view to identifying a salt which is commonly used to solubilize cellulose in aqueous solution and which might also dissolve cellulose in the molten state: NaOH was identified as such a pulping medium in aqueous solution. A consideration of the physical properties of the NaOH–KOH eutectic systems (51 mole % KOH, 49 mole % NaOH) which melts at about 190 °C was thought to provide a convenient temperature range for the dissolution of wood. As fusions of NaOH with organic compounds are well known in analysis of trace inorganic compounds, it appeared that dissolution in the molten salt should be possible.

There is some literature on the fusion of wood in NaOH in order to produce low molecular weight organic acids on a commercial scale[1,2] but these processes generally use an atmosphere of either air or oxygen. There is no literature on the fusion of wood in strong anhydrous alkalis under inert or reducing atmospheres. The use of molten salts for the pyrolysis of pulping waste and domestic garbage has also been investigated[3] but in all these cases there is no attempt at dissolution of the cellulosic material before pyrolysis. Molten salts have also been used in the pyrolysis of tyres to produce a 'crude' oil.[4]

Having decided that the above molten salt system might prove of interest, it was logical to suppose that the chemical reactions which might occur during pyrolysis could be related to the degradation of cellulose in concentrated aqueous alkaline solutions. This degradation in aqueous solutions takes place in three ways:

(i) In dilute solutions of NaOH (up to around 0·5 molar) the formation of a complex is thought to occur.[3,5] If such solutions are heated to around 100 °C, the reducing end units of the cellulose molecule are removed to form D-glucoisosaccharinic acid,[6-8] this degradation continuing until about 50 glucose units per cellulose molecule are removed.

(ii) As the concentration of the hydroxide solution is increased to around 1 molar and temperature to around 170 °C, the glycosidic linkages are split to give 1,2-epoxides[9] of the form:

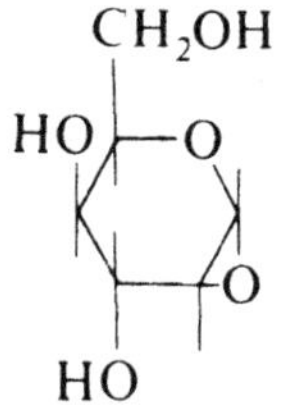

(iii) At higher temperatures than in (ii) above and in the presence of air or oxygen, the formation of low molecular weight organic acids has been noted,[1,2] although the mechanism of this reaction is apparently unknown.

The mechanism of the alkaline degradation of lignin in wood has been much studied[12] and is extremely complex. Methanol is a major component of the product of this degradation at 200 °C in concentrated aqueous alkalis leaving a residue with an increased C-to-H ratio. This degradation is also dependent on the concentration of hydroxyl groups but, again, it is not clear how the composition of the atmosphere affects these reactions.

2. EXPERIMENTAL WORK

During preliminary work the solubility of wood in the eutectic melt system was observed to be complete and rapid, and concentrations of wood in the melt of 5–10 % (w/w) were achieved. At 200 °C degradation of wood leading to gas evolution was observed, but it was extremely slow.

A number of experimental runs were undertaken in a simple apparatus consisting of a stainless steel lined glass reaction vessel capable of holding 50–100 ml of melt. This was stirred using a vibrating stirrer. The melt temperature was measured using a thermocouple in a stainless steel sheath. The system was purged with either nitrogen or an Ar/He mixture. Purge gases were purified to remove CO_2, O_2 and H_2O. The purge was required to remove gases from the reaction zone and also to ensure that O_2 was excluded from the system.

The gas was collected either in a polythene gas-bag (for total gas volume measurement and chemical analysis) or over saturated and acidified brine at different times during the experiment. The gas was analysed either by mass spectrometry or by gas chromatography with both katharometer and FID detectors.

Liquid distilling out of the system was collected in a small ice-cooled trap and was analysed by HPLC. The cold, solid melt was dissolved in water (still under inert atmosphere) and this solution was analysed for both hydroxide and carbonate concentrations by pH titration against HCl, under conditions capable of detecting significant concentrations of organic acids. The dissolved melt was also titrated against $K_2Cr_2O_7$ solution in order to detect any organic compounds remaining in the melt.

The runs themselves were of three types:

(a) Isothermal. In these experiments the wood was added to the melt at different temperatures (between 200 and 375 °C).

(b) Successive increments. The wood was added at 200 °C to the melt, and the melt kept at this temperature for 5 min. The thermostat set-point was then increased by 25 °C and the gases collected until 5 min after gas evolution had visibly subsided. The sequence was then repeated.

(c) Continuous ramp. This was found to be the most successful of the heating techniques used. The wood was first added at 200 °C and the thermostat was set to 400 °C; then, because of the thermal lag of the melt and container, the temperature was found to rise approximately linearly over the range 200–350 °C. The run was terminated 10 min after the final temperature was achieved.

The wood samples were all of *Pinus silvestri* of about 14 years old and were taken from the main trunk of the tree. All the bark was removed and the wood allowed to air dry at room temperature for about 6 weeks. The wood was then sawn into fine sawdust (knots being discarded), further dried at 95 °C for 3–4 h and then kept in a stoppered bottle. The

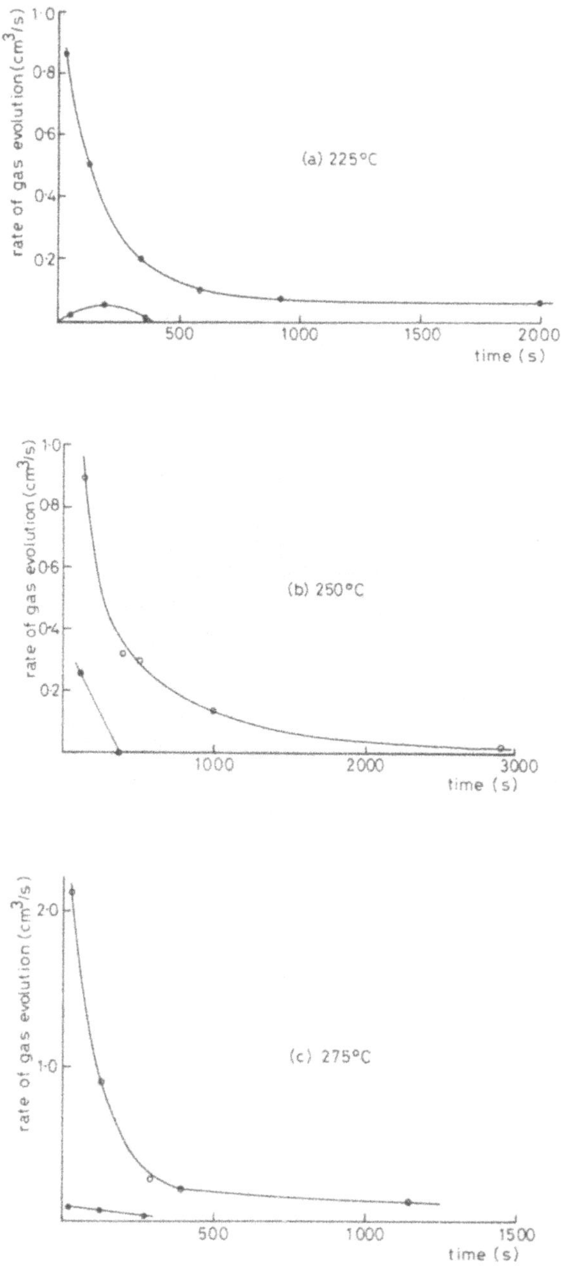

FIG. 1. Isothermal runs. ○ Hydrogen, ● methane.

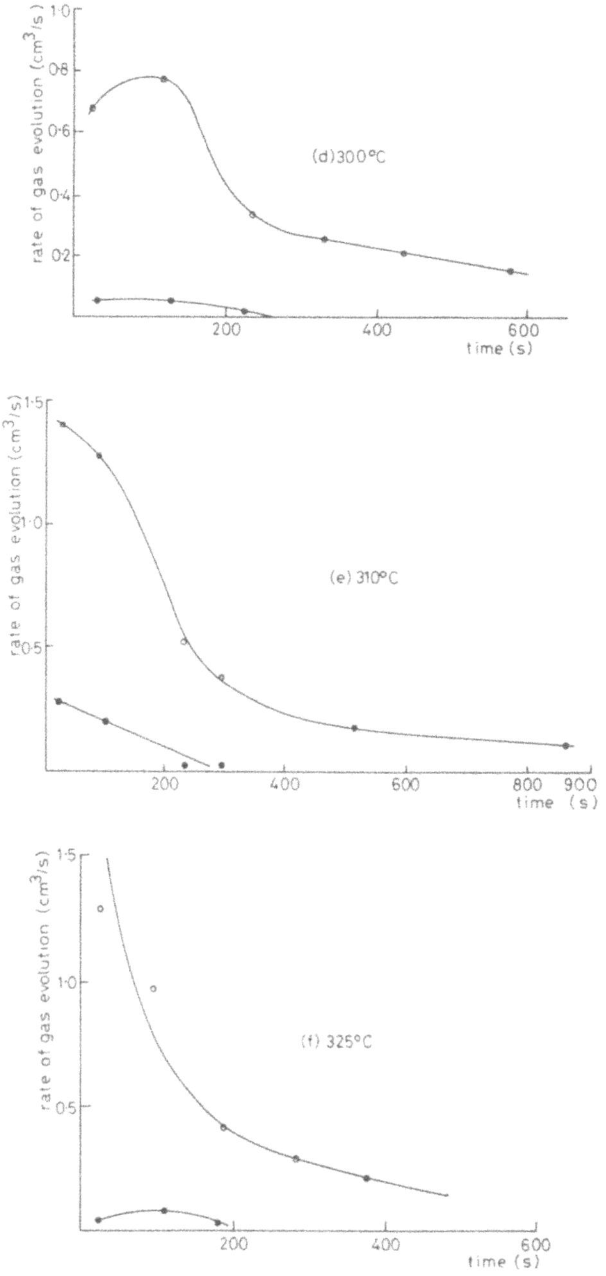
rate of gas evolution (cm^3/s)
1·0
0·8
0·6
0·4
0·2
(d) 300°C
200
400
600
time (s)
rate of gas evolution (cm^3/s)
1·5
1·0
0·5
(e) 310°C
200
400
600
800
900
time (s)
rate of gas evolution (cm^3/s)
1·5
1·0
0·5
(f) 325°C
200
400
600
time (s)

FIG. 1.—*contd.*

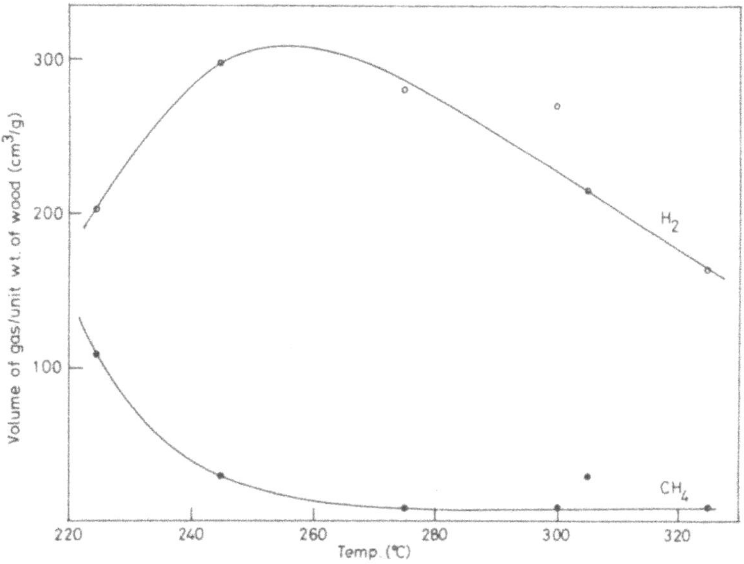

FIG. 2. Yield of gas against temperature (isothermal runs).

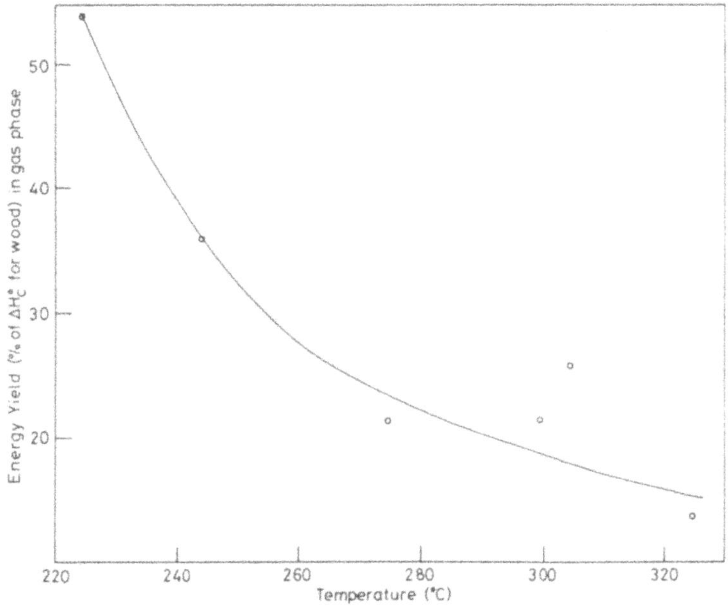

FIG. 3. Energy yield against temperature of isothermal run.

composition was found to be $C_6H_{9.07}O_{4.29}$ by analysis. Sample sizes used in the experimental work were between 1 and 5 g.

3. RESULTS

As described above, the wood was found to dissolve completely in the melt at 200 °C with very little foaming. Using 1 g of sawdust, dissolution was complete within 10–30 s, depending on the rate of stirring. Adding wood to the melt at temperatures above 250 °C resulted in the charring of the wood before it dissolved: this probably accounts for the low yields of gas from the isothermal runs at high temperatures.

3.1. Isothermal Runs

These runs were carried out at various temperatures between 200 and 375 °C. The results for the concentrations of CH_4 and H_2 in the product gases as a function of time after dissolution of the wood are shown in Fig. 1. Only H_2 and CH_4 are shown as these were by far the major components of the product gases; neither CO_2 nor CO were detected.

The total volumes of H_2 and CH_4 given off during a run against temperature are plotted in Fig. 2, and the yield in energy represented by these volumes is shown in Fig. 3.

3.2. Run using Successive Increments

The result is shown in Fig. 4.

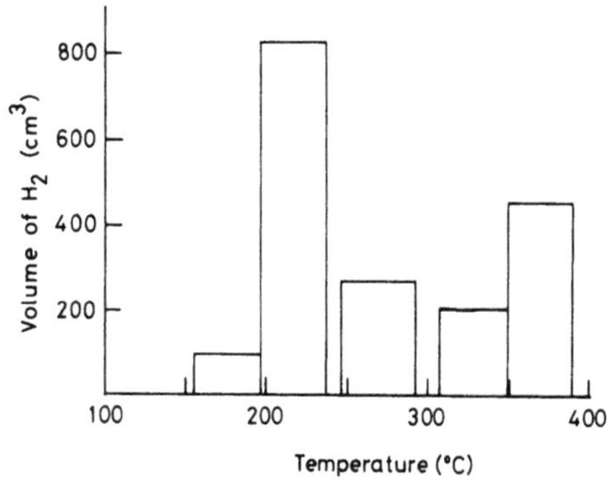

FIG. 4. Volume of H_2 given off in successive increments of temperature.

TABLE Ia

TYPICAL MASS BALANCE

In	*Out*		*Residue*	
Wood	H_2O	$9{\cdot}3 \times 10^{-2}$ mol	OH^-	0·96 mol
C $3{\cdot}84 \times 10^{-2}$ mol	$(CH_3)_2CO$	$3{\cdot}8 \times 10^{-4}$ mol	CO_3^{2-}	4×10^{-2} mol
H $3{\cdot}33 \times 10^{-2}$ mol	H_2	$7{\cdot}0 \times 10^{-2}$ mol	$CH_3CO_2^-$	$<10^{-3}$ mol
O $6{\cdot}98 \times 10^{-2}$ mol	CH_4	$8{\cdot}9 \times 10^{-4}$ mol	Total organics	
	C_2H_6	$2{\cdot}7 \times 10^{-4}$ mol		$<10^{-2}$ mol
Melt	Higher H/Cs			
OH^- 1·12 mol		$<3 \times 10^{-4}$ mol		
	CO	$<1 \times 10^{-4}$ mol		
	CO_2	$<1 \times 10^{-4}$ mol		

3.3. Continuous Ramp (200–400 °C)

The use of a continuous ramp together with the collection and analysis of all the evolved gases, the used melt and the liquid distillate enabled a total mass balance to be made (see Tables Ia and Ib). This mass balance enabled a theoretical product energy yield to be estimated: for a 200–370 °C ramp at a heating rate of 3·5 °C/min this was 95 ± 5 % (expressed as a percentage of ΔH_c° of wood).

4. DISCUSSION AND CONCLUSIONS

Important conclusions from the experimental data are summarized below.

(1) A yield (in energy of the wood) of 100 % into volatile compounds, i.e. acetone and gases (H_2, CH_4 and higher alkanes/alkenes), is possible using the 'continuous ramp' approach. This high energy yield should not be confused with the mass yield as the carbon present in the wood remains in the melt in the form of CO_3^{2-} ions and so the mass yield is in fact lower than the high energy yield might lead one to expect.

(2) Any large-scale production system incorporating the process must be able to operate continuously and to cope with the build-up of CO_3^{2-} in the melt. Some preliminary calculations from thermodynamic data (see Fig. 5) indicate that a partial pressure of steam in the reaction vessel should regenerate the hydroxide via

$$2NaOH_{(l)} + CO_{2(g)} = Na_2CO_{3(l)} + H_2O_{(g)}$$

The fact that the solids may be dissolved in a separate vessel and the melt

TABLE Ib
FULL MASS BALANCE

Material	*Wood + melt*		*Spent melt*		*Liquids*		*Gases*		*Losses*
	Wt	%	*Wt*	%	*Wt*	%	*Wt*	%	*Wt*
Wood	1·1	1·9							
Na^+	21·2	38·6	21·2	39·8					
K^+	13·6	24·9	13·6	25·5					
OH^-	19·0	34·6	16·3	30·7					
CO_3^{2-}			2·1	3·9					
$CH_3CO_2^-$			0·1	0·1					
H_2O					1·70	98·7			
$(CH_3)_2CO$					0·02	1·3			
CO_2							0·004	2·5	
CO							0·003	1·5	
CH_4							0·018	9·8	
C_2H_6							0·008	4·5	
Higher H/Cs							0·008	4·4	
H_2							0·140	77·3	
Total	54·9	100·0	53·3	100·0	1·72	100·0	0·181	100·0	
Losses									0·3
	Atom	%	*Atom*	%	*Atom*	%	*Atom*	%	
C	0·049	1·4	0·038	1·2	0·008	2·5	0·003	2·0	
H	1·19	33·8	0·960	30·3	0·200	66·6	0·144	95·4	
O	1·16	33·0	1·061	33·4	0·097	30·9	0·004	2·6	
Na	0·589	16·7	0·589	18·5	0·000	0·0	0·000	0·0	
K	0·529	15·1	0·529	16·6	0·000	0·0	0·000	0·0	
Total	3·517	100·0	3·180	100·0	0·314	100·0	0·151	100·0	

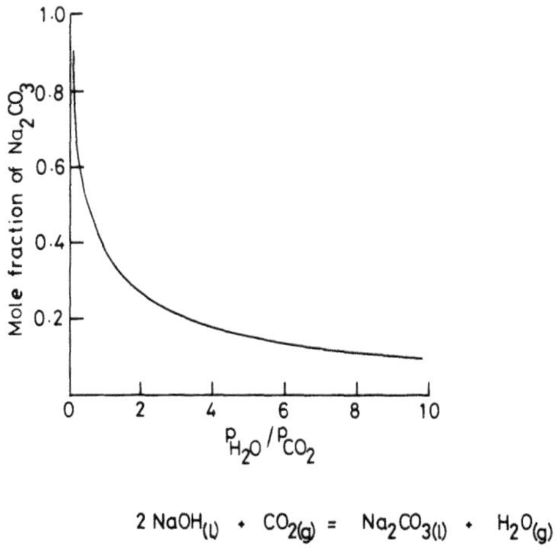

FIG. 5. Ratio of partial pressures (H_2O:CO_2) against Mole fr. of Na_2CO_3 at 500 K.

pumped to reaction and regeneration vessels is an advantage of this process. A further advantage, apart from that of energy yield, is that the complete process could be operated at 1 atm pressure.

(3) The product gas is free from poisonous compounds, especially CO and, unless regeneration were to be carried out *in situ*, the product gases could be made free from CO_2.

5. POSSIBLE CHEMICAL MECHANISMS

There are two main ways in which one might approach a mechanistic study of this system. One is to assume that the degradation is largely chemical and can be based on an extrapolation of the aqueous chemistry of cellulose (lignin and the other components of the wood will be ignored for the sake of brevity in this discussion). The second approach is to assume that the same level of degradation of cellulose will occur in molten hydroxides as in aqueous solutions at high temperatures and that the species thus formed will further degrade due to high temperatures rather than because of the strongly alkaline conditions in the melt.

A reaction scheme taking into account the above is shown in Fig. 6. In this scheme the cellulose is assumed to degrade into 6-carbon fragments (either saccharinic acids or 1,2-epoxides or possibly even further degraded species). Once formed these 6-carbon fragments could conceivably degrade

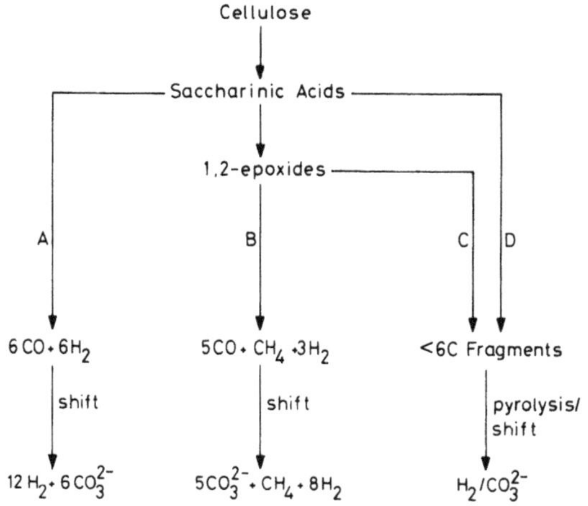

FIG. 6. Possible reaction scheme.

in several ways, but as described above these can be classified into two types.

A and B can be thought of as 'pyrolysis' routes in which the 6-carbon fragments are pyrolysed without further chemical attack from the hydroxide melt. This degradation would presumably lead to (stoichiometrically) either $6CO + 6H_2$ from the saccharinic acids or $5CO + CH_4 + 3H_2$ from the 1,2-epoxide. The CO produced would be expected to 'shift' via the reaction

$$CO + 2OH^- = H_2 + CO_3^{2-}$$

Thus this mechanism would account for the presence of CO_3^{2-} and lack of CO in the gas, as well as the presence of CH_4 (if 1,2-epoxides are involved). If this is indeed correct the ratio of H_2 to CH_4 in the product gas should be around 8 to 1 by volume, which is close to the value actually found.

C and D can be considered to represent 'chemical' degradation routes involving a purely chemical attack from the hydroxide in the melt. This degradation would be expected to produce small fragments (formaldehyde, formic acid etc.) which would be reasonably stable to pyrolysis at lower temperatures (around 200 °C), but they might either pyrolyse at slightly higher temperatures or be converted 'chemically' to hydrogen and CO_3^{2-} directly, no intra-molecular pyrolysis reactions being involved. These routes, however, would not necessarily be expected to produce any CH_4. Nevertheless, this lack of methane could possibly be explained by considering the pyrolysis of the lignin (ignored in this simplistic approach).

Conventional wood pyrolysis is thought to produce some CH_4, but this applies also to the above discussion on routes A and B.

The degradation might conceivably take place by any of the above routes—A, B, C or D—either singly, in combination, or together with methane production from the lignin.

Although the postulated mechanisms are consistent with the experimental data, it must be remembered that the chemistry discussed represents an extrapolation of the aqueous chemistry of cellulose combined with much speculation. It is by no means certain that the molten salt chemistry of hydroxides can be developed in this way and it is important to note that the degradation of cellulose may occur in completely novel ways in molten salts. The second major criticism is that lignin and hemicelluloses have been ignored. Although the hemicelluloses might be expected to behave in much the same way as cellulose, reactions involving lignin could also provide an explanation for the production of the methane; it may be that instead of forming CH_3OH, as in aqueous alkaline media, the lignin is degraded to CH_4 in the melt. Against this, it could be argued that the residue of the lignin might be expected to form chars and/or tars very readily after removal of CH_4, followed by dehydration to remove the rest of the hydrogen as H_2O; significantly, no char formation was observed during the experiments except where dissolution of the wood was incomplete.

6. FURTHER WORK

The chemistry of the system is obviously poorly understood and much work is required in order to understand the mechanism of the pyrolysis. This may be aided by examination of the behaviour of pure cellulose in order to clear up the doubt as to whether the lignin or the cellulose is providing the methane. Further studies with model compounds from cellulose (1,2-epoxides and the saccharinic acids concerned) might also help to identify the mechanisms, as might the electrochemistry of the melt system and analysis of the quenched melt.

Work with a larger, continuous system will be necessary to ensure that the suggested process can be scaled up. A study of the economics of the process is also planned.

Lastly, other molten salts which are known to dissolve wood could be investigated. The use of catalysts to either speed up the existing reactions or encourage completely new reactions might also be explored.

ACKNOWLEDGEMENTS

The authors wish to thank the Science and Engineering Research Council of the UK for provision of a research grant.

REFERENCES

1. Richards, G. N. (1963). *Methods in Carbohydrate Chemistry*, R. L. Whistler (Ed.). Academic Press, New York, pp. 154–63.
2. Wenzel, H. F. J. (1970). *Chemical Technology of Wood*. Academic Press, New York, pp. 290–3.
3. Birk, J. R. and Huber, D. A. (1973). US Pat. 3,708,270, January 2. Abstract in *Chemicals from Pulp and Wood Waste*, G. T. Maloney (Ed.). Noyes Data Corp., New Jersey (1978).
4. Domino, F. A. (Ed.) (1979). *Energy from Solid Wastes*. Noyes Data Corp., New Jersey, p. 53.
5. Viemeg, W. (1907). *Ber.* **40**, 3876.
6. Miller-Moskau (1907). *Ber.* **40**, 4903; *CA*. **2**, 1186.
7. Sugihara, J. M. (1953). *Adv. Carbohyd. Chem.* **8**, 1.
8. Rendlem, J. A. Jr. (1966). *Adv. Carbohyd. Chem.* **21**, 209.
9. Ref. 1, p. 155.
10. Whistler, R. L. and Be Miller, J. M. (1958). *Adv. Carbohyd. Chem.* **13**, 289.
11. Lloyd-Evans, W. (1942). *Chem. Rev.* **31**, 537.

9

Influence of Alkali Carbonates on Biomass Volatilization

R. T. HALLEN, L. J. SEALOCK, JR and R. CUELLO
Battelle Pacific Northwest Laboratories, Richland, Washington, USA

1. INTRODUCTION

Several thermochemical methods for converting biomass into chemicals, fuels and synthetic gases are currently under development. Various reactor configurations, catalysts and reaction conditions are employed depending primarily upon the desired product. Steam gasification is one method being developed for the production of synthetic gases from biomass.

Steam gasification involves contacting biomass at elevated temperatures with water vapor. The overall conversion of biomass to gases can be represented by two stages. During the first stage—the rapid volatilization stage—biomass reacts rapidly to produce gases, condensible liquids and a char residue. In the second stage the char residue is slowly converted to gases by pyrolysis and the carbon–steam reaction ($C + H_2O \rightarrow CO + H_2$).

Alkali carbonates have been employed as steam gasification catalysts by directly contacting them with the sample before gasification. They have been found to increase both the rate and overall conversion of carbon to gas.[1] The catalytic effect of alkali carbonates on the carbon–steam reaction has been studied using various forms of carbon. McKee has prepared a recent review of catalyzed gasification reactions of carbon in which he discusses possible mechanisms of catalyzed steam gasification.[2] The effect of alkali carbonates on volatilization has been studied relative to combustion and flame proofing of wood.[3] This paper reports laboratory investigations on the effect of alkali carbonates on volatilization relative to biomass gasification. Gas-phase experiments were performed with model compounds to develop a fundamental understanding of the influence of alkali carbonates on specific reactions occurring during volatilization.

2. EXPERIMENTAL

Gasification experiments were performed in a ball-valve batch reactor which has been discussed in earlier publications.[1,4] The gas-phase experiments were performed in a quartz reactor system illustrated in Fig. 1. The system is equipped with multiple gas and liquid inlets. The reactor contained crushed quartz glass above the catalyst bed to vaporize the liquids and heat the gases before they passed over the catalyst. A thermocouple shielded in a glass tube monitored reaction temperature in the catalyst bed. The liquid products leaving the reactor were condensed in an ice water trap and then a dry ice trap. The gaseous products were collected in a sample bag.

The gaseous productions were analyzed with a Carle AGC-S gas chromatograph. Integration was performed by a Spectra-Physics SP4000 central processor. Liquid products were analyzed with a Hewlett-Packard

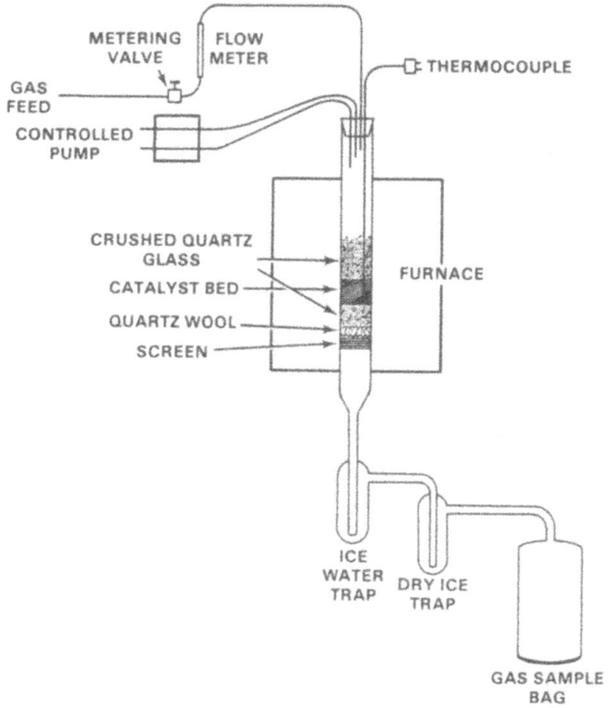

FIG. 1. Schematic of the quartz reactor system used to conduct gas-phase experiments.

5830A gas chromatograph with a HP 18850A GC terminal. Standard samples were used to determine GC response factors relative to an internal standard. A known weight of the internal standard was then added to the liquid product collected to determine product yields.

The organic chemicals used were reagent quality or better and were used without further purification. Fresh deionized distilled water with a specific resistance of 7 MΩ/cm or better was used. Degassing the solutions to remove oxygen had no detectable effect on results so the majority of the solutions were not degassed. Aldrich gold label Na_2CO_3, Aldrich 99+ % K_2CO_3 and Argent 99·9 % cesium carbonate were used.

3. RESULTS AND DISCUSSION

Gasification experiments were performed in the absence and presence of alkali carbonates (Na_2CO_3, K_2CO_3 and Cs_2CO_3) to develop a fundamental understanding of their role in increasing the production of synthetic gases from biomass. Two different catalyst contacting methods, dry-mix and solution impregnation, and two catalyst concentrations, $1{\cdot}5 \times 10^{-3}$ and $3{\cdot}0 \times 10^{-3}$ mole alkali/g sample, were evaluated relative to gas production. Experimental results for steam gasification of cellulose, holocellulose, kraft lignin and Douglas fir samples have been reported.[1] The alkali carbonates were found to have similar activities. The gas production rate and total gas produced varied only slightly for the two catalyst concentrations tested. Both solution impregnation and dry-mix were effective catalyst contacting methods.

The percent carbon converted to gas for steam gasification of Douglas fir in the absence and presence of Na_2CO_3 is shown in Fig. 2. At 550 °C the volatilization stage of gasification is complete at approximately 7 min of reaction. The carbon–steam reaction at this temperature is slow and has little effect on the total carbon converted to gas during the volatilization stage. With Na_2CO_3 present at 850 °C both volatilization and the carbon–steam reaction occur rapidly and are indistinguishable. The presence of Na_2CO_3 at temperatures of 650 °C and higher appears to decrease the carbon conversion to gas during volatilization but increases the rate and total gas produced later.

The volatilization of biomass was examined without steam present (pyrolysis). Douglas fir samples with and without Na_2CO_3 were pyrolyzed in a flowing nitrogen atmosphere. The percents carbon converted to gases versus times for the pyrolysis experiments are shown in Fig. 3. At 650, 750

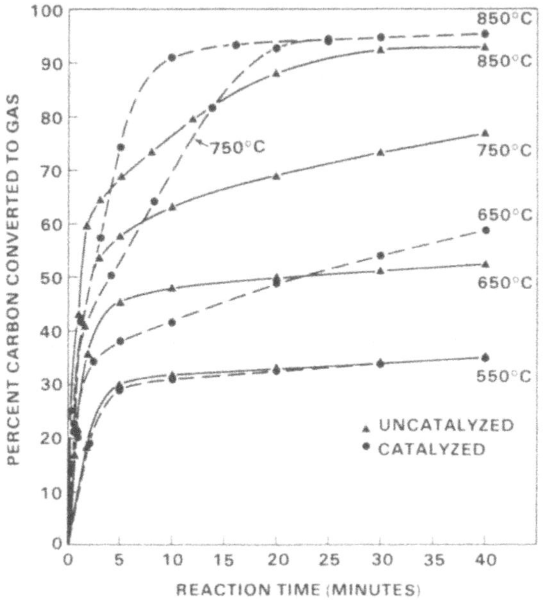

FIG. 2. Carbon conversion to gas for uncatalyzed and dry-mix sodium carbonate catalyzed steam gasification of Douglas fir.

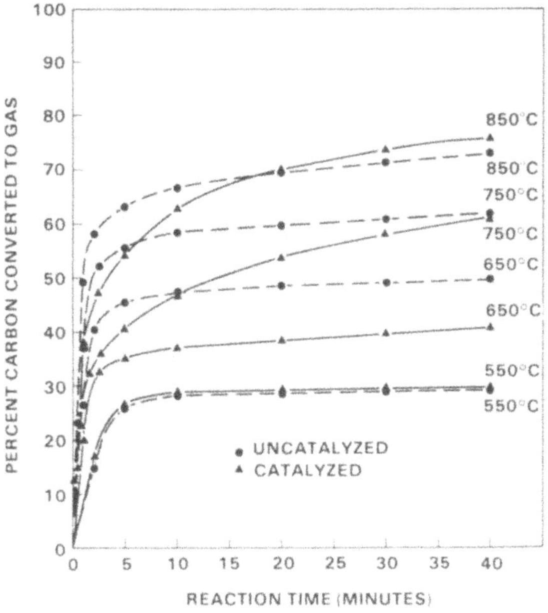

FIG. 3. Carbon conversion to gas for the pyrolysis of Douglas fir samples in the absence and presence of sodium carbonate.

and 850 °C the presence of Na_2CO_3 decreases the carbon converted to gas initially. However, at 750 and 850 °C after the initial rapid volatilization, pyrolysis continued at a faster rate for the catalyzed cases until at 40 min of reaction the carbon converted to gas was the same or better than in the uncatalyzed cases.

The gas velocity in the reactor during the pyrolysis experiments was much less than during the steam gasification experiments. The conversion of condensible liquids to gases and char has been shown to be increased by longer residence times in the heat zone of the reactor.[5] Experiments were performed at 550 and 650 °C with a steam flow rate of 0·06 g/min, which is equivalent to the nitrogen purge rate. The carbon converted to gas during the volatilization stage was the same as at the higher steam flow rate. The only significant difference was at 650 °C and 40 min of reaction with Na_2CO_3 present. The carbon conversion to gas was increased from 48 % at the lower steam rate to 60 % at the higher steam rate.

The influence of Na_2CO_3 on formation of char from Douglas fir was determined at various times during steam gasification. Table I presents the char yields for steam gasification of 2·0 g samples of Douglas fir with and without dry-mixed Na_2CO_3. Rapid volatilization was occurring at 3 min of reaction and complete by 10 min. The presence of Na_2CO_3 increased the yield of char during the volatilization stage and decreased the char during the slower second stage of gasification. Shafizadeh has reported the influence of inorganic salts on charring of cellulose relative to flame retarding.[3] The presence of Na_2CO_3 during the pyrolysis of cellulose increased the char yield to 13 % compared to 7 % without. The combustible carbon in the volatiles was decreased by Na_2CO_3 to 39·5 % compared to 76·2 % for pure cellulose.

TABLE I

CHAR YIELDS FROM STEAM GASIFICATION OF 2·0 g OF DOUGLAS FIR WITH AND WITHOUT SODIUM CARBONATE PRESENT AT 650 °C

Reaction time (min)	*Uncatalyzed char*		*Catalyzed char*[a]	
	Weight (g)	*Carbon weight (g)*	*Weight (g)*	*Carbon weight (g)*
3	0·34	0·29	0·77	0·48
10	0·29	0·25	0·72	0·44
25	0·28	0·25	0·54	0·25
40	0·27	0·24	0·46	0·16

[a] 0·32 g Na_2CO_3 added to wood samples.

Contacting alkali carbonates with biomass has been shown to increase the formation of char during the volatilization stage of gasification. Experiments were performed with the alkali carbonate not in direct contact with the wood. Dual sample baskets were used with the wood sample inside the inner basket and Na_2CO_3 inside the outer basket. The carbon conversion to gases for uncatalyzed and Na_2CO_3 catalyzed dual basket experiments are compared in Fig. 4. The increase in carbon conversion to

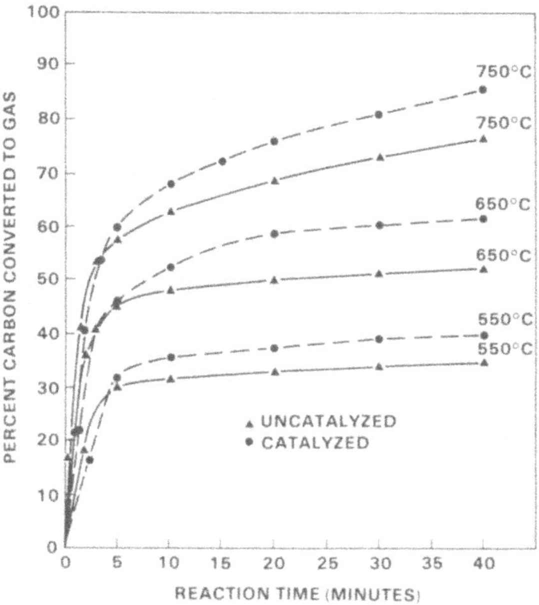

FIG. 4. Carbon conversion to gas for sodium carbonate dual basket and uncatalyzed experiments with Douglas fir.

gas with Na_2CO_3 present results from conversion of condensible liquids to gases. This could occur by direct conversion of the liquids to gases and/or by condensing the liquids to non-volatile compounds which further pyrolyze and gasify to produce gases. Baker *et al.* have used alkali carbonates as a fixed bed catalyst in a continuous feed biomass reactor.[6] At 750 °C alkali catalysts reduced the tar yield as effectively as various nickel catalysts without deactivation by carbon deposition.

The research described above demonstrates that alkali carbonates influence reactions important in biomass gasification in addition to their noted catalytic effect on the carbon–steam reaction. Gas-phase reactions of gasification products and model compounds were investigated in the quartz reactor system to develop a fundamental understanding of reaction pathways and sequences influenced by alkali carbonates. Simple hydrocarbon products from gasification, methane, ethane and ethylene did not react when contacted with alkali and steam at 600 °C. Methane reforming occurred only when the reactor reached 800 °C. The water–gas shift reaction was found to be catalyzed by Na_2CO_3 at 600 °C. Table II

TABLE II
THE INFLUENCE OF SODIUM CARBONATE ON THE WATER–GAS SHIFT REACTION AT 600 °C

Catalyst	*Flow rates*		*Gas composition (vol%)*		
	CO (ml/min)	*H_2O (g/min)*	*CO*	*CO_2*	*H_2*
None	24	0·18	98·2	1·3	0·5
Na_2CO_3	24	0·18	46·3	27·2	26·6
None	24	0·38	98·4	1·0	0·5
Na_2CO_3	24	0·38	60·7	19·5	19·8
None	52	0·18	98·7	1·0	0·3
Na_2CO_3	52	0·18	80·1	9·4	10·5
None	52	0·38	99·1	0·4	0·5
Na_2CO_3	52	0·38	71·5	13·5	15·0

presents results from experiments at various flows of CO and steam. Experiments were also performed with model compounds that have been reported as pyrolysis products from cellulose.[7,8] The carbon conversion to gas and gas composition of experiments performed with various compounds and equal weights of steam at 600 °C are reported in Table III. The results suggest that sodium carbonate increases the decomposition of alcohols and carboxylic acid but has little effect on aldehydes and ketones. Acetaldehyde was a major product from the reaction of ethanol with Na_2CO_3 and was not detected in the uncatalyzed experiment. This suggests that alkali carbonates catalyze the dehydrogenation of alcohols. The smaller effect of Na_2CO_3 on lactic acid decomposition is most likely due to autoesterification of lactic acid to form lactide or polymer.[9] The decomposition of lactic acid to CO, not CO_2, suggests the ester forms are present.

TABLE III
GAS COMPOSITION AND CARBON CONVERSION TO GAS FOR MODEL COMPOUNDS REACTED WITH STEAM AT 600 °C IN THE ABSENCE AND PRESENCE OF SODIUM CARBONATE

Compound	*Catalyst*	*Feed rate (g/min)*	*Gas composition (vol%)*					*Carbon converted to gas (%)*
			H_2	CO	CO_2	CH_4	C_2^+	
Methanol	None	0·46	75·8	11·2	7·0	6·0	0·0	0·8
Methanol	None	0·92	81·2	5·2	6·8	6·9	0·0	0·3
Methanol	Na_2CO_3	0·46	74·0	10·4	15·3	0·3	0·0	19·2
Methanol	Na_2CO_3	0·92	75·1	10·6	13·6	0·7	0·0	12·5
Methanol	Na_2CO_3	a	72·3	10·7	17·0	0·0	0·0	64·0
Ethanol	None	a	44·2	19·6	0·5	22·6	13·2	1·8
Ethanol	Na_2CO_3	a	56·3	11·7	13·4	16·3	2·3	45·7
Acetaldehyde	None	a	1·1	52·0	0·4	45·8	0·8	9·4
Acetaldehyde	Na_2CO_3	a	36·2	18·1	20·0	23·6	2·0	38·3
Acrolein	None	0·42	4·5	57·6	14·2	2·4	21·4	1·1
Acrolein	None	0·84	6·1	55·9	12·5	2·3	23·4	0·4
Acrolein	Na_2CO_3	0·42	34·4	24·6	27·9	2·9	10·1	2·9
Acetone	None	0·46	10·2	10·8	21·7	53·1	4·2	0·4
Acetone	Na_2CO_3	0·46	15·0	4·2	60·7	20·4	0·0	0·1
Acetic Acid	None	0·53	44·4	14·2	30·8	10·3	0·2	7·2
Acetic Acid	None	1·05	47·0	15·5	28·6	8·8	0·2	2·6
Acetic Acid	Na_2CO_3	0·53	2·6	0·8	65·7	30·8	0·1	53·9
Acetic Acid	Na_2CO_3	1·05	2·2	0·8	69·2	27·9	0·0	22·7
Lactic Acid	None	0·48	4·2	73·0	14·0	8·1	0·7	39·3
Lactic Acid	Na_2CO_3	0·48	22·6	52·3	13·6	9·8	1·2	57·9

[a] Preliminary experiments with varied feed rates.

TABLE IV

THE PERCENT REACTION AND YIELD OF ACETONE AND GASES FROM REACTION OF 2-PROPANOL AT 600 °C IN THE PRESENCE OF ALKALI CARBONATES

Feed rate (g/min)	*Catalyst*	*Percent reaction*	*Percent yield*[a]	
			Acetone	*Gases*
0·29	Na_2CO_3	68·2	79·7	11·6
0·39	Na_2CO_3	77·3	88·7	6·8
0·79	Na_2CO_3	49·5	83·4	7·0
0·39	K_2CO_3	72·6	81·3	11·1
0·79	K_2CO_3	64·1	84·9	6·9
0·39	Cs_2CO_3	88·1	68·9	23·1
0·79	Cs_2CO_3	86·9	77·0	15·1

[a] Based on carbon.

The increase in decarboxylation of carboxylic acids by alkali carbonates was expected due to the thermal instability of carboxylic acid salts.[10] An experiment was performed in which glacial acetic acid was contacted with Na_2CO_3 at 600 °C. 97 % of the acetic acid reacted converting 42 % of the carbon to gas and 55 % of the carbon to acetone. The production of acetone accounts for the lower yield of CH_4 compared to CO_2 reported in Table III.

The large increase in reaction of alcohols by Na_2CO_3 was unexpected. Further investigations employed 2-propanol which, upon dehydrogenation, produces acetone which was found to be relatively stable under our reaction conditions. Table IV reports percent reaction and product yields from the reaction of 2-propanol with Na_2CO_3, K_2CO_3 and Cs_2CO_3 at 600 °C. The alkali carbonates are selective and highly active for alcohol dehydrogenation. No reference has been found in the literature describing the high activity of alkali carbonates as alcohol dehydrogenation catalysts.

4. CONCLUSIONS

Alkali carbonates enhance the gasification of biomass. This results from catalysis of the carbon–steam reaction and increase in gas production at the expense of condensible liquids. The decrease in production of volatiles by the presence of alkali carbonates increases the production of char. Results from gas-phase experiments suggest that the mechanism of increased char formation by alkali carbonates involves dehydrogenation and decarboxylation reactions. The condensation of products from dehydrogenation and decarboxylation reactions is also likely to be catalyzed by alkali carbonates.

ACKNOWLEDGEMENTS

This work was funded through the Basic Research Department of the Gas Research Institute (GRI). The authors wish to thank Dr John L. Cox and Dr Charles R. Nelson of GRI for their support and encouragement of our research.

REFERENCES

1. Sealock, L. J. Jr. *et al.* (1981). Kinetics and catalysis of producing synthetic gases from biomass. Annual Report for December 7 1980–December 6 1981 (GRI-80/0116), NTIS.
2. McKee, D. W. (1981). *Chemistry and Physics of Carbon*, Vol. 16, P. L. Walker, Jr. and P. A. Thrower (Eds). Marcel Dekker, New York, p. 1.
3. Shafizadeh, F., Chin, P. S. and DeGroot, W. F. (1975). *J. Fire Flamm./Fire Retardant Chemistry* **2**, 195.
4. Mudge, L. K., Sealock, L. J. Jr. and Weber, S. L. (1979). *J. Analytical and Applied Pyrolysis* **1**, 165.
5. Shafizadeh, F. (1977). Utilization of biomass by pyrolytic methods. *Proceedings of the TAPDI 1977 Joint Forest Biology/Wood Chemistry Meeting*, Madison, Wisconsin, 191.
6. Baker, E. G., Mudge, L. K. and Wilcox, W. A. (1981). This book, Chapter 47, pp. 863–74.
7. Schwenker, R. F. Jr. and Beck, L. R. Jr. (1963). *J. Polym. Sci.: Part C* **2**, 331.
8. Shafizadeh, F., McGinnis, C. D. and Philpot, C. W. (1972). *Carbohyd. Res.* **25**, 23.
9. Hurd, C. D. (1929). *Pyrolysis of Carbon Compounds*. Chemical Catalog Co., New York, p. 807.
10. Morrison, R. T. and Boyd, R. N. (1973). *Organic Chemistry*, 3rd edn. Allyn & Bacon, Boston, p. 583.

10

Flash Pyrolysis of Biomass with Reactive and Non-reactive Gases

MUTHU S. SUNDARAM, MEYER STEINBERG and PETER T. FALLON

Process Sciences Division, Department of Energy and Environment, Brookhaven National Laboratory, Upton, New York, USA

1. INTRODUCTION

The precarious nature of the energy supply has encouraged research on renewable sources of energy. The conversion of biomass and organic wastes to gaseous and liqud fuels shows good potential. A Biomass Conversion Program was initiated at Brookhaven National Laboratory early in 1981 with the main objective of obtaining process chemistry and design information on the production of liquid and gaseous fuels from wood via flash pyrolysis. Flash pyrolysis is a fast, gas-phase decomposition of carbonaceous materials over a wide range of process conditions in the presence of an inert gas.

This work is based on the observation that very rapid pyrolysis of carbonaceous materials, specifically coal, has yielded amounts of aromatic and aliphatic hydrocarbons significantly higher than those obtained by slow pyrolysis. Flash pyrolysis in a hydrogen atmosphere (flash hydropyrolysis) has been found to produce more valuable lighter liquid and gaseous hydrocarbons. Much significant information about the process chemistry can also be obtained with other gases, both reactive and inert. The BNL biomass program is therefore directed to widely ranging studies of flash pyrolysis in a variety of atmospheres comprising gases generally classified as reactive or non-reactive gases. The non-reactive gases chosen for study include He, N_2 and Ar, in order to cover a range of gaseous thermal transport and diffusional properties. The selection covers low and high molecular weight monoatomic and diatomic gases, and gases having low and high specific heat and thermal conductivity. Since flash pyrolysis is a non-equilibrium process, it will be possible to determine whether the thermal and mass transport and diffusional properties of these gases have any effects on the yields of liquid and gaseous hydrocarbons.

The reactive gases chosen include H_2, CO_2, CO, CH_4 and steam because they are readily available and inexpensive and they are found in the product gases of both biomass and coal pyrolysis. These gases will be recycled in the process and the various reactivities of the elements H, C, O and their compounds with the biomass substrate need to be determined.

The BNL biomass program is divided into four phases:

Phase I:

Task 1: Adaptation of the existing coal pyrolysis reactor for experiments with Douglas fir wood in hydrogen, helium and methane environments.

Task 2: Determination of yields and distributions of reaction products—C_1–C_2 gases, CO/CO_2 gases, BTX, tar, H_2O and char; isolation of the tar and identification of its components.

Phase II:

Task 1: Correlation of product yields to process conditions.

Task 2: Kinetic analysis of the data and development of reaction models.

Phase III:

Task 1: Identification of new process potentials.

Task 2: Economic analysis, and reactor and process design.

Phase IV:

Task 1: Pyrolysis of Douglas fir wood in other reactive and non-reactive gaseous environments.

Task 2: Pyrolysis of other types of biomass.

This paper describes the results of Tasks 1 and 2 of Phase I and Task 1 of Phase II.

2. DESCRIPTION OF THE EXPERIMENTAL APPARATUS

A very versatile, highly instrumented, down-flow entrained tubular reactor was used in the pyrolysis studies (Fig. 1). Its design, construction and operation have been described in detail.[1–3] This apparatus is ideal for performing non-equilibrium, short residence time, flash pyrolysis experiments with a variety of feedstocks. The unit consists of (i) a compressor with gas storage cylinders and manifold, (ii) a pyrolysis reactor with preheater and sample taps, (iii) a product recovery system with char trap and product condensers, and (iv) a control console with ancillary

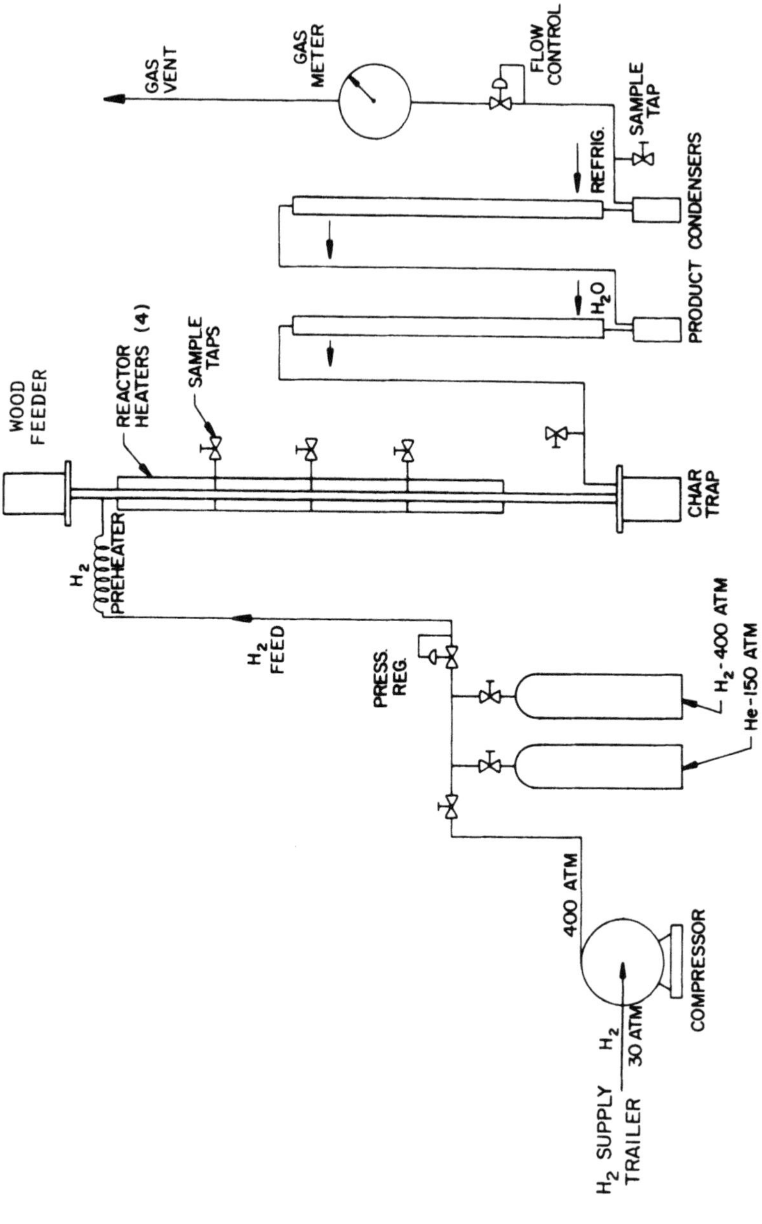

FIG. 1. Schematic flowsheet of entrained tubular reactor experiment.

equipment and an on-line gas chromatograph. The wood particles are fed by gravity at the top of the reactor in a free-fall mode. The gaseous feed is preheated in an electrical resistance tube and entrains the wood particles, which then flow concurrently down the 8-ft long, 1-in ID tube where they are heated at a rate > 10 000 °C/s to temperatures of 750–1000 °C at reactor pressures up to 4000 psi. The solid residence time in the reactor is usually about 1–10 s. Direct representative samples of light gases and BTX can be taken every 2 ft along the length of the reactor and analyzed automatically by the on-line gas chromatograph (GC). At the exit of the reactor tube, the reaction products are rapidly quenched to less than 300 °C and the residual solids are separated from the gases in an externally heated char trap. The heavier liquids are collected in a water-cooled condenser and the lighter liquids in a Freon-cooled condenser.

3. EXPERIMENTAL PROCEDURE

Pyrolysis experiments were done under various conditions:

Entraining gas	H_2, CH_4, He
Pressure (psi)	50, 200, 500
Temperature (°C)	800, 900, 1000
Solids residence time	0·5–4 s

A seasoned sawdust of Douglas fir wood (C:H:O wt % = 50:6:44) was used, which had been size-separated and vacuum dried for at least 12 h at ~70 °C. To prevent agglomeration and to provide good free-fall in the tubular reactor, Cab-O-Sil[4] was mixed with the wood particles (300–1000 μm) in an amount equivalent to 15 % by weight of the total feed. Cab-O-Sil is a fumed silicon dioxide marketed by Cabot Corp., Boston, MA, and is a fine particulate with a high surface area of 200–400 m^2/g. Its chain-forming tendency is considered ideal for reducing the agglomeration between heated wood particles. The system was flushed with helium and leak-tested via the on-line GC before each run. Usually, a steady reactor temperature was achieved within 3–4 min after starting to feed wood particles into the reactor, and the series of runs was continued for 60–75 min. Product gas samples were analyzed via the on-line GC for CH_4, C_2H_4, C_2H_6, CO, CO_2 and BTX at different points along the reactor corresponding to different wood residence times. Thus, the concentrations of these reaction products in the effluent stream can be monitored for four different residence times during a single run. However, the process water, heavy liquids and residual char are collected only at the end of each run, and their yield is reported for the longest wood residence time during a run. The wood residence time was obtained from the sum of the particle free-fall velocity calculated from Stoke's law and the velocity of the entraining gas.

4. RESULTS AND DISCUSSSION

Results from the pyrolysis of Douglas fir wood in helium, methane and hydrogen environments were reported recently.[4,5] The results of short residence time experiments (usually about 1 s) are discussed in this paper.

Almost complete (98 %) gasification of carbon in Douglas fir wood was achieved at 900 °C and 500 psi H_2 pressure in 1 s residence time. At this pressure, increasing the reaction temperature from 800 to 1000 °C had only a small effect on the total carbon conversion to hydrocarbon products and oxides of carbon. However, the individual product yields and product gas composition changed with reaction temperature. Figure 2 shows the extent

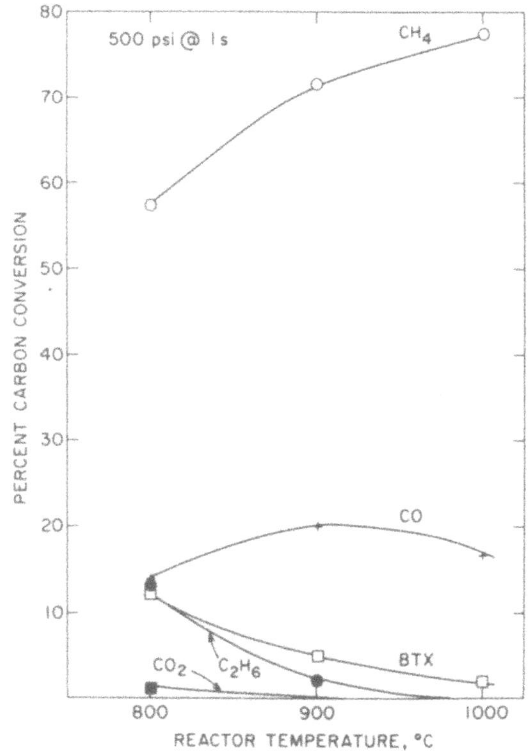

FIG. 2. Product yields vs. temperature at 500 psi and 1 s wood residence time.

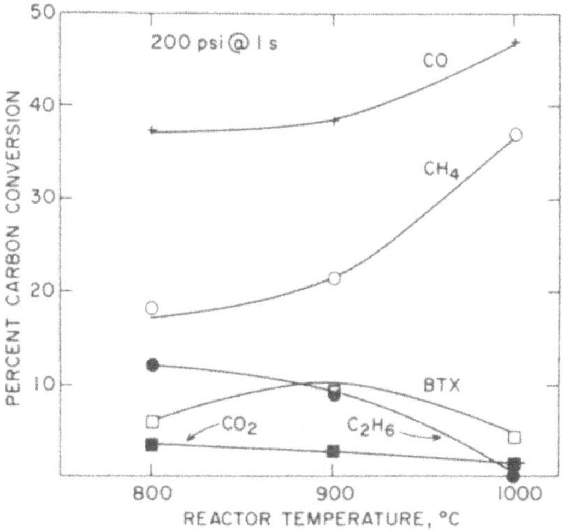

FIG. 3. Product yields vs. temperature at 200 psi and 1 s wood residence time.

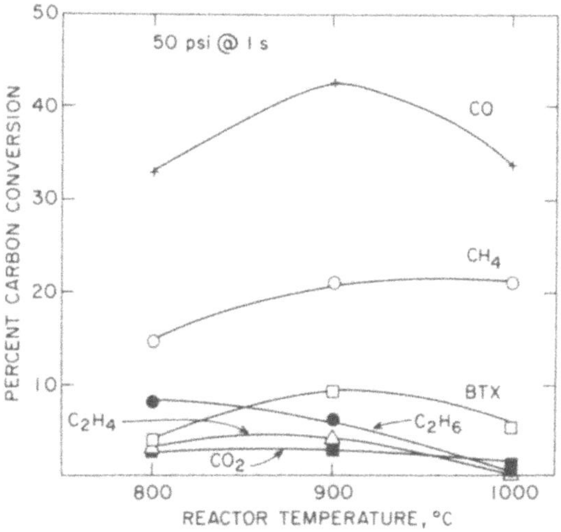

FIG. 4. Product yields vs. temperature at 50 psi and 1 s wood residence time.

of conversion of carbon contained in the feed wood, at 500 psi, to CH_4, C_2H_6, BTX, CO and CO_2 as a function of reaction temperature. Olefinic hydrocarbons, particularly ethylene, were not detected at this pressure. The conversion to methane exceeded that to all other products and increased with increasing temperature. CO yield was considerably less than that of methane and showed a broad maximum at 900 °C. As expected, the yields of ethane, BTX and CO_2 decreased with temperature.

The product distributions at other operating conditions (200 and 50 psi at 800–1000 °C) are shown in Figs 3 and 4. Comparison of Figs 2 through 4 shows a three- to fourfold increase in methane production as the hydrogen pressure was increased from 50 to 500 psi. The extremes in methane production (highest, 78·2 % C, vs. lowest, 14·5 % C conversion) correspond to the extremes in process conditions (most severe, 1000 °C and 500 psi vs. least severe, 800 °C and 50 psi).

At 50 psi, the methane yield increased only a little with temperature (6·5 % C from 800 to 1000 °C), and, as seen in Fig. 4, above 900 °C it leveled off (21·0 % C conversion), and CO yield dropped from 42·5 % C to 32·6 % C with no increase in the yield of any other measured product. This means that, at 50 psi, CO is partially removed at temperatures above 900 °C via carbon deposition reactions, possibly the following:

$$CO + H_2 \rightarrow C + H_2O \quad (1)$$

$$2CO \rightarrow C + CO_2 \quad (2)$$

$$2CO \rightarrow 2C + O_2 \quad (3)$$

Since no increase in water production was observed, and the CO_2 yield actually decreased from 2·7 % C at 900 °C to 1·2 % C at 1000 °C, reaction (3) is the probable route of CO yield reduction.

At 200 and 500 psi H_2 pressure, the increase in methane yield between 800 and 1000 °C were greater (20 %C conversion), because the hydrogenation of thermally produced fragments is made easier by the increased availability of hydrogen at higher hydrogen pressures. The dramatic increase in methane yield at 500 psi at the expense of CO (Fig. 2) is due to the following methanation reaction:

$$CO + 3H_2 \rightarrow CH_4 + H_2O \quad (4)$$

Equilibrium considerations suggest that high partial pressures of hydrogen favor the production of methane.

Ethylene production was 3–4 % C at 800–900 °C and 50 psi. At higher temperatures and pressures, ethylene was not detected in the product

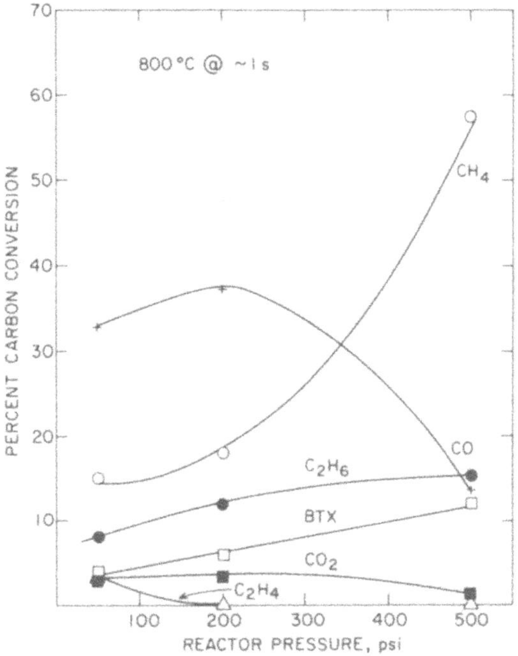

FIG. 5. Product yields vs. pressure at 800 °C and 1 s wood residence time.

stream; its absence at 200 and 500 psi is due to its hydrogenation to ethane, as evidenced by the higher ethane yields. At 1000 °C, ethane decomposition is accelerated, and its yield is reduced to zero.

An important difference in product distribution was noted in comparing the 500 psi data with the 50 and 200 psi data. At 500 psi, the product accounting for the highest total carbon conversion is methane, whereas at 50 and 200 psi it is carbon monoxide (Figs 5 through 7). As the temperature was increased from 800 to 1000 °C, the pressure at which this yield reversal occurred became lower, going from ~350 psi at 800 °C to 300 psi at 900 °C and 250 psi at 1000 °C. Thus, a 100 °C increase in reaction temperature causes about a 50 psi decrease in the pressure at which the yield reversal occurs. Also, as the reaction temperature is raised, the yield of CO and CH_4 at the point of reversal becomes higher.

At all pressures, the yields of CO_2 and ethane decreased with temperature because of secondary reactions. Considerable quantities of ethane and BTX were produced at the lowest temperature investigated, 800 °C. With increasing hydrogen pressure, the yield of BTX decreased at

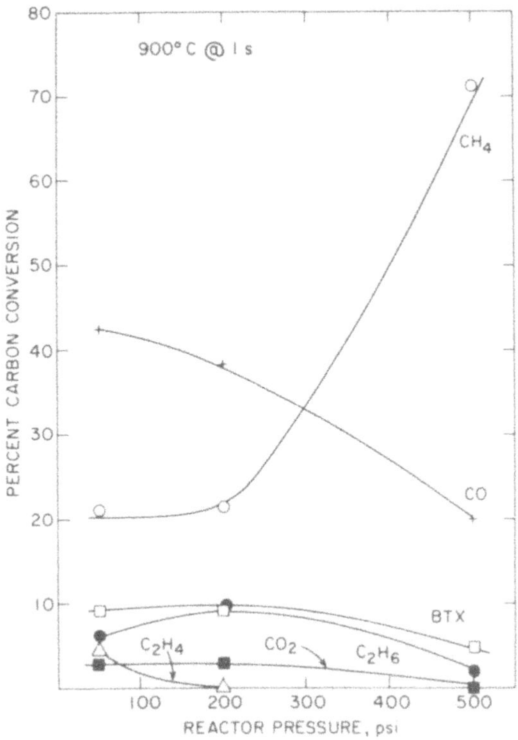

FIG. 6. Product yields vs. pressure at 900 °C and 1 s wood residence time.

900 and 1000 °C, but it increased at 800 °C, probably because of slow cracking of tar molecules. As the temperature is increased, enhanced decomposition of BTX reduces its yield.

The final yields of hydrocarbon gases (methane + ethane + ethylene) and CO were affected significantly by the residence time of the wood particles in the reactor. At all temperatures investigated, increasing the residence time decreased the gaseous hydrocarbon yield and increased the CO yield, although not to an equal extent. Figure 8 shows this variation at 800 °C and 500 psi.

The available data indicate that the primary devolatilization and attendant hydrogenation are complete within 1 s, and that increasing the residence time causes an array of secondary reactions. The flash pyrolysis of wood appears to produce mainly methane and water at residence times of 1 s or less. On the basis of calculated H and O balances, starting with

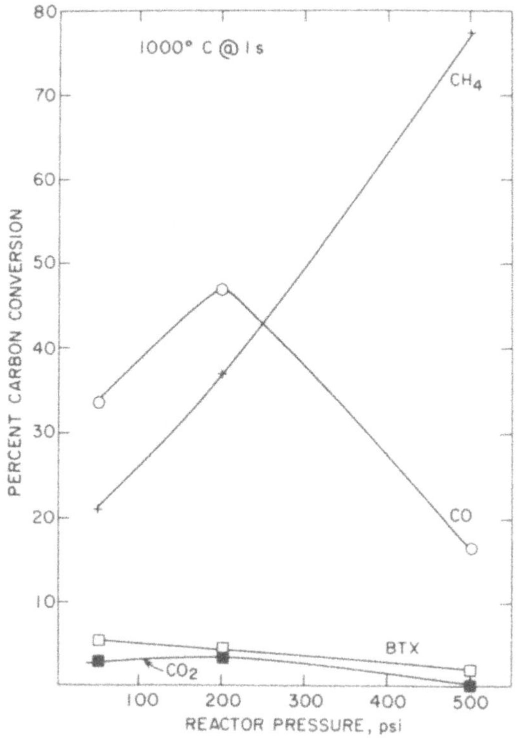

FIG. 7. Product yields vs. pressure at 1000 °C and 1 s wood residence time.

the initial composition of wood, the following stoichiometric equation approximates the experimental results:

For <1 s residence times,

$$CH_{1 \cdot 4}O_{0 \cdot 7} + 2H_2 \rightarrow CH_4 + 0 \cdot 7H_2O$$

As the reactants and products (wood plus hydrogen, methane and water) proceed further down the tubular reactor and the residence time becomes longer, the methane appears to react with the water in a reforming type of reaction which increases the amount of CO and reduces the methane and water in the product gases:

$$CH_4 + H_2O \rightarrow CO + 3H_2$$

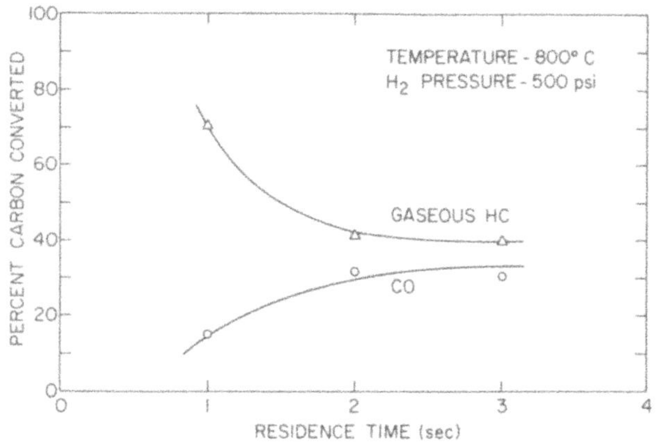

FIG. 8. Yields of CO and hydrocarbon gases vs. wood particle residence time at 800 °C and 500 psi hydrogen pressure.

The stoichiometry of the overall reaction for longer residence times can be approximated as follows:

For >4 s residence time,

$$CH_{1\cdot4}O_{0\cdot7} + 0{\cdot}5H_2 \rightarrow 0{\cdot}5CH_4 + 0{\cdot}5CO + 0{\cdot}2H_2O$$

The results of experiments done under similar conditions show that the product state is greatly altered by the presence of moisture in the feed wood. For example, at 900 °C the yield of gaseous hydrocarbons from undried wood (containing 28 % moisture) was only one-third of that from dried wood, but the yield of light oils increased from 7 % to 18 %.

To study the effect of different gaseous environments on the amounts of various reaction products obtained, the Douglas fir wood was flash pyrolyzed in an inert helium atmosphere and a reactive methane atmosphere.[5] Data from selected runs at 50 psi are shown in Table I, with hydropyrolysis data included for comparison.

Under the conditions investigated, CO emerges as the predominant species in the product gas mixture in all cases. The initial increase in CO production during helium pyrolysis in Fig. 9 indicates continued formation of CO from the wood up to 900 °C; up to this temperature the percentage oxygen conversion to water was relatively constant at 16·2 %. Above 900 °C, the CO yield decreased because of secondary reactions. Increasing the temperature from 900 to 1000 °C decreased the yields of all other products except BTX, and decreased the total carbon conversion by 6 %.

TABLE I

PERCENT CARBON CONVERSION TO PRODUCTS IN FLASH PYROLYSIS OF DOUGLAS FIR WOOD AT 50 psi

Conditions									
Temperature	*800 °C*			*900 °C*			*1 000 °C*		
Entraining gas	H_2	He	CH_4	H_2	He	CH_4	H_2	He	CH_4
Gas, lb h	1·12	1·89	4·97	1·12	1·82	4·21	1·01	2·10	4·89
Wood, lb h	0·8	0·72	0·92	0·90	0·76	0·89	0·98	0·86	0·92
Gas/wood ratio	1·40	2·63	1·23	1·24	2·39	1·03	1·03	2·44	5·31
Res. time, s	0·9	0·7	0·7	0·9	0·7	0·9	0·9	0·6	0·8
Product									
CH_4	14·6	9·2	ND	21·0	11·0	ND	21·0	10·1	ND
C_2H_4	3·1	9·2	8·4	4·0	11·9	13·0	0·0	9·5	20·7
C_2H_6	8·0	0·0	2·2	6·0	0·0	2·2	0·0	0·0	2·0
CO	32·6	40·0	41·7	42·6	45·8	47·7	33·6	41·9	48·1
CO_2	2·7	1·9	2·5	2·7	3·9	2·5	1·2	3·6	3·3
BTX	3·7	2·8	1·9	9·0	4·7	5·8	5·3	5·4	12·7
Total	64·7	63·1	56·7*	85·3	77·3	71·2*	61·1	70·5	86·8[a]

ND = Not determined.
[a] Excluding methane produced.

Pyrolysis in methane gave higher CO yields than did that in helium, probably because the reforming reaction of methane with pyrolytic water is thermodynamically favored at higher temperatures.

The methane yield was higher in hydropyrolysis experiments than in pyrolysis experiments with helium, as expected (Fig. 10). The actual amount of methane produced during pyrolysis in a methane environment could not be analytically determined, but the percent carbon conversion to methane was calculated from the monitored mass flow conditions to be ~7 %, which is only slightly less than that in pyrolysis experiments with helium.

The ethylene yields were comparable (~10 % conversion) for pyrolysis in helium and methane up to 900 °C (Fig. 11). With the helium experiments taken as the base case, it is seen that (i) in hydropyrolysis experiments, a large part of any ethylene produced initially is hydrogenated to ethane, and (ii) methane is almost inert with respect to product formation up to 900 °C. At 1000 °C, however, there are large differences between the product yields in different gases, particularly the yields of ethylene, CO and BTX. The lower CO yield in flash hydropyrolysis than in pyrolysis with helium was explained via reaction (4). At high methane partial pressures, the

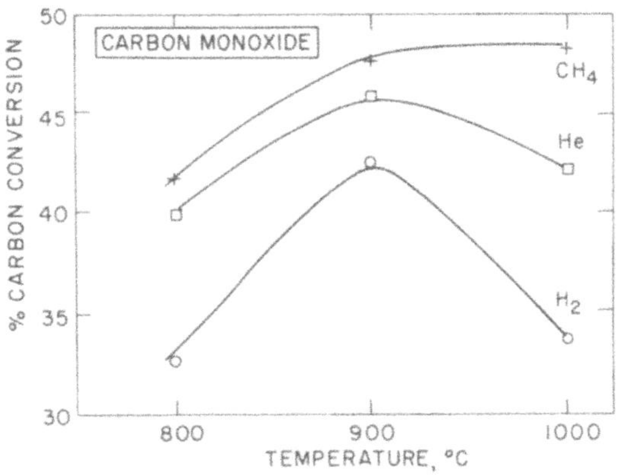

FIG. 9. Yield of CO in different gaseous environments vs. temperature at 50 psi gas pressure and 1 s wood residence time.

equilibrium shifts towards the reverse reaction favoring the production of CO.

At 1000 °C and 50 psi, 9·5 % C in the feed was converted to ethylene during pyrolysis in helium, and 20·7 % C during pyrolysis in methane. Several experiments were done to find out why, including pyrolysis of (a) methane only, (b) methane with Cab-O-Sil additive, and (c) methane with coconut char. No ethylene or any other product was detected in any of these 'blank' experiments, and it was concluded that (i) reactor walls do not catalyze decomposition of methane, (ii) Cab-O-Sil is inert towards

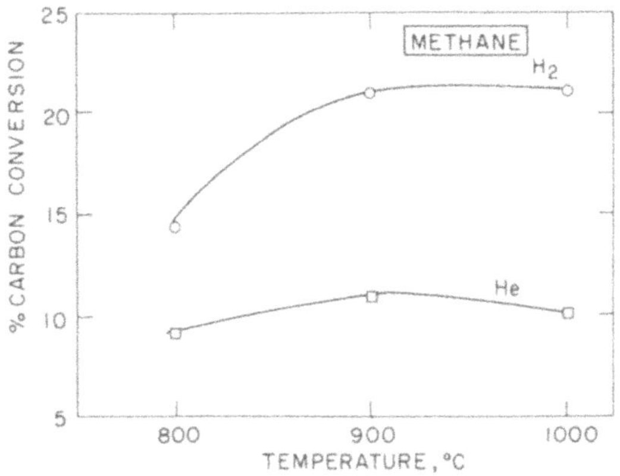

FIG. 10. Yield of methane in different gaseous environments vs. temperature at 50 psi gas pressure and 1 s wood residence time.

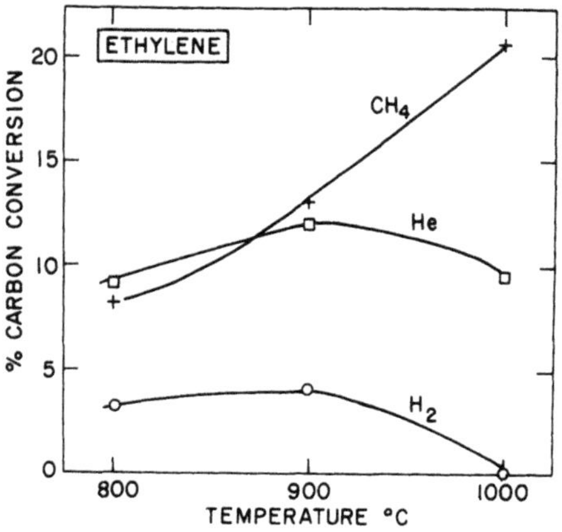

FIG. 11. Yield of ethylene in different gaseous environments vs. temperature at 50 psi gas pressure and 1 s wood residence time.

methane, and (iii) the chars do not provide active surface for methane decomposition. The enhanced ethylene production at 1000 °C and 50 psi methane pressure can probably be explained by free-radical reactions that can take place in the presence of biomass. Flash heating of biomass particles produces free radicals (thermal fragments), which are stabilized by hydrogen abstraction from methane:

$$R-R \rightarrow R\cdot + R\cdot \qquad \text{(chain initiation)}$$

$$R\cdot + CH_4 \rightarrow RH + CH_3\cdot \qquad \text{(chain propagation)}$$

Ethylene is produced from methyl free radicals:

$$CH_3\cdot + CH_3\cdot \rightarrow C_2H_6 \qquad \text{(chain termination)}$$

$$C_2H_6 \rightarrow C_2H_4 + H_2 \qquad \text{(dehydrogenation)}$$

The presence of ethane—which undergoes cracking at temperatures above 900 °C—in the product stream at all temperatures investigated (800–1000 °C) provides additional evidence. The concentration of thermal free radicals from biomass increases with the maximum temperature to which the particles are heated. This increases the demand for hydrogen, which results in a higher concentration of methyl radical intermediates and hence a higher ethylene yield. The gas-to-wood ratio was higher in the pyrolysis runs with methane than in those with helium, but the data obtained at 800 and 900 °C indicated that this had little influence on the product yields.

In the pyrolysis runs with helium, which is inert, the hydrogen necessary for free radical stabilization comes from the biomass substrate itself. Thus, in such runs, ethylene is produced from the methylene linkages by reactions of the type

$$R\text{–}CH_2\text{–}CH_2\text{–}R \rightarrow R\cdot + \cdot CH_2\text{–}CH_2\text{–}R$$

$$R\cdot + \cdot CH_2\text{–}CH_2\text{–}R \rightarrow R\cdot + \cdot CH_2\text{–}CH_2\cdot + R\cdot$$

$$\downarrow$$

$$CH_2{=}CH_2 + R\text{–}R$$

Another important deviation from the usual trend is the yield of BTX from the pyrolysis runs with methane. When the temperature was raised from 800 to 1000 °C, the BTX yield increased from 1·9 % to 12·7 % C. The decomposition of BTX at temperatures above 900 °C has been reported by several laboratories including our own.[3,6] The run with the lowest BTX yield (800 °C, 50 psi CH_4 and 1·9 % C BTX) also gave the most tar (11·4 % C). On the other hand, at 1000 °C the tar yield was only 0·3 %. The yields of BTX plus tar from these runs were almost the same (13·3 % C at 800 °C vs. 13·0 % C at 1000 °C), indicating that the high yield of BTX at 1000 °C is due to tar cracking reactions. The same result could be obtained by increasing the residence time of the tar molecules in the reactor at lower temperatures. Since only negligible amounts of tar were produced when Douglas fir wood was pyrolyzed in hydrogen or helium, it appears that methane promotes free radical condensation reactions at lower temperatures (800 °C). The elucidation of the exact mechanism warrants further investigation.

It is concluded that flash pyrolysis of wood in a methane atmosphere is a valuable method for producing the chemical feedstocks ethylene and benzene—basic raw materials for the vast polymer and plastic markets. Direct production of these products from biomass would be a valuable renewable source of supply. In addition, the large CO yield from wood pyrolysis can be converted to methanol, which has been proven to be usable as an alternative transportation fuel.

REFERENCES

1. Fallon, P. T. and Steinberg, M. (1977). The design, construction, operation, and initial results of a flash hydropyrolysis experimental unit (BNL 22519). Presented at ACS 173rd Nat. Meet., New Orleans, March.
2. Steinberg, M. and Fallon, P. T. (1976). *Safety Analysis Report in the Brookhaven Coal Flash Hydropyrolysis Experiment*. Informal Report BNL 21919, October.

3. Sundaram, M. S., Steinberg, M. and Fallon, P. T. (1982). *Flash Hydropyrolysis of Coal for Conversion to Liquid and Gaseous Fuels: Summary Report.* DOE/METC/82-48, May.
4. Steinberg, M. and Fallon, P. T. (1981). Flash hydropyrolysis of biomass (BNL 30235). Presented at 12th Biomass Thermochemical Contractor's Meet. Richland, WA, February.
5. Steinberg, M. and Fallon, P. T. (1981). *Flash Pyrolysis and Hydropyrolysis of Biomass.* Informal Report BNL 30263, October.
6. Finn, M. J., Fynes, G., Ladner, W. R. and Newman, J. D. H. (1979). The formation of BTX by the hydropyrolysis of coals. *ACS Fuel Div. Preprints* **24**(3), pp. 99–110.

11

Pyrolytic Reactions and Products of Biomass

The late FRED SHAFIZADEH
Wood Chemistry Laboratory, University of Montana, Missoula, Montana, USA

1. INTRODUCTION

The biomass or commonly available lignocellulosic materials could be converted to different types of fuel and chemical feedstock by a variety of thermochemical processes. Each of these processes involves two highly significant and inter-related general aspects: first, material and energy transformations that could be explored and understood through the discipline of chemistry; second, material transport and heat transfer that could be investigated and designed through the disciplines of process engineering. Variations of the reaction conditions and the processing design, which are closely related, are often investigated in a sporadic or empirical manner in order to define the optimum conditions that provide high yields of operational efficiency.

In this presentation I would like to discuss the role and significance of the chemical variables as a basis for understanding the mechanism and kinetics of pyrolytic reactions and their relationship with the expected yields and efficiency of the thermochemical conversion processes. The major variables are composition of the substrate, heating conditions, ambient atmosphere and catalysis.

These variables profoundly affect the sequence and kinetics of the reactions and ultimately yields of the products.[1] The products are broadly classified as char, tar, volatile (liquids) and gases, which are separated and analyzed for the individual compounds, functional groups or structures present. The pyrolysis conditions could be controlled in order to reduce the side reactions and enhance the formation of the desirable products, if the influence of the above variables on the mechanism and kinetics of the pyrolytic reactions were known or could be anticipated.

2. MAJOR VARIABLES

2.1. Composition of the Substrate

The biomass or commonly available cellulosic substrates are generally complex and heterogeneous raw materials derived from different types of dead and living plant cells, the structure and composition of which varies for different parts and species of the plants. The green leaves contain considerable amounts of proteins in their protoplasm, large amounts of extractives, and less cellulosic materials. The woody tissues contain mainly cell wall materials composed of cellulose (30–40 %), hemicelluloses (25–35 %) and lignin (12–30 %) as shown in Table I. Each of these compounds has different thermal properties and yields different products.

TABLE I
APPROXIMATE ANALYSIS OF SOME BIOMASS SPECIES

Species	*Total ash* (%)	*Solvent sol.* (%)	*Water sol.* (%)	*Lignin* (%)	*Hemicellulose* (%)	*Cellulose* (%)
Softwood	0·4	2·0	—	27·8	24·0	41·0
Hardwood	0·3	3·1	—	19·5	35·0	39·0
Wheat straw	6·6	3·7	7·4	16·7	28·2	39·9
Rice straw	16·1	4·6	13·3	11·9	24·5	30·2
Bagasse	1·6	0·3	—	20·2	38·5	38·1

At temperatures above 300 °C the polysaccharides break down at the glycosic linkage to provide tarry pyrolyzates containing lower molecular weight anhydro sugar derivatives and relatively little char, whereas the lignin is mainly condensed to a carbonaceous char and gives smaller amounts of a pyrolyzate containing phenolic compounds. However, the thermal properties of the aggregate, in general, reflect the sum properties of its components. These phenomena have been demonstrated by thermal analysis of cottonwood and its components as shown in Figs 1 and 2.[2] Among these components the thermal properties of cellulose have been more thoroughly investigated and are better known because of its relative abundance and structural homogeneity.

The heterogeneity of biomass presents a major obstacle to chemical utilization due to proportionately lower yields of individual products that can be derived from each component. It also affects the yield of charcoal or other types of fuel that could be obtained by straight pyrolysis.

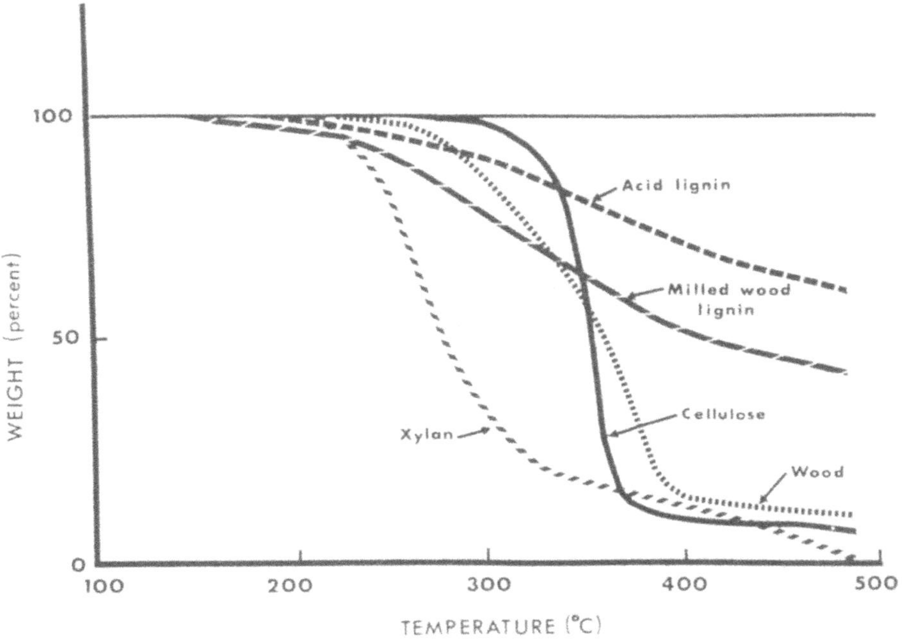

FIG. 1. Thermogravimetry of wood and its components.

2.2. Heating Conditions

The duration and intensity of heating designated as TTP (time–temperature profile) affect not only the rate and the extent of the pyrolytic reactions, but also the sequence of these reactions and compositions of the products. The pyrolytic reactions proceed over a wide range of temperatures and the products formed earlier undergo further transformation and decomposition in a series of consecutive reactions. Furthermore, as these materials are heated, the availability of sufficient energy of activation, at higher temperatures for alternatives to the minimum energy pathway, results in various side reactions, which give a variety of products. Long heating periods allow the sequence of consecutive reactions to take their course, while rapid or flash pyrolysis could reduce the secondary reactions and further degradation of the products. For this reason it is preferred for analytical pyrolysis, where too many degradation products are detrimental.

According to the employed conditions there could be a substantial difference between the temperature of the reactor and that of the substrate, because of the difference in the rate of heat transfer and the rates of heat

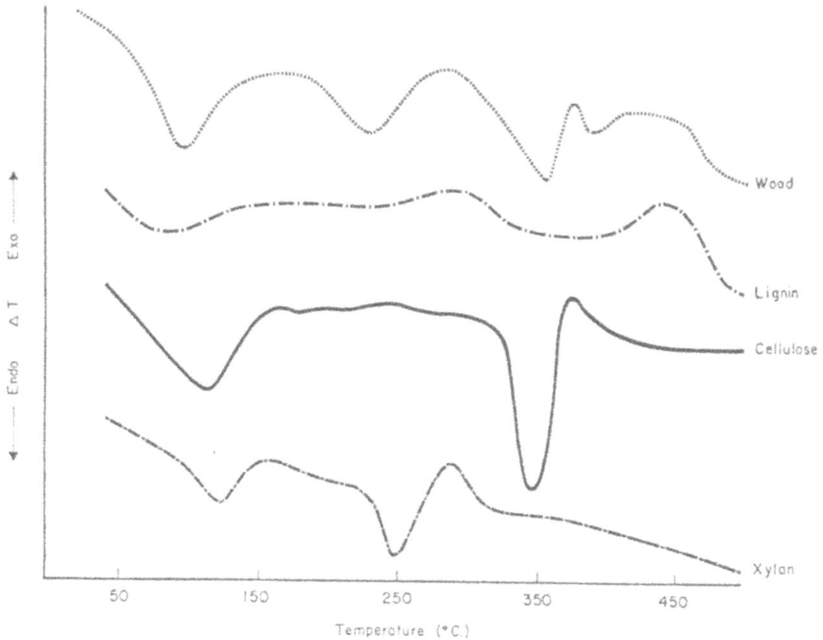

FIG. 2. Differential analysis of wood and its components.

loss due to heat capacity, heats of transition and the heat of reaction.[3] The difference is greater when rapid dynamic heating does not allow for establishment of an equilibrium condition, when larger mass-to-surface area impedes heat transfer and when rapid evaporation of the pyrolyzate cools the substrate. Consequently, at higher temperatures when rapid ablation takes place, the rate of reaction may be controlled by the rate of heat transfer rather than chemical kinetics of the pyrolytic reactions.

2.3. Ambient Atmosphere

Other than the problems of heat transfer, the ambient atmosphere affects the extent and nature of the secondary reactions according to whether the reaction is carried out in a vacuum, inert or reactive atmosphere. Under vacuum, the primary pyrolysis products are rapidly removed or thinned out in the gas phase, thus reducing their chances for further recombination and decomposition. A case in point is pyrolysis of cellulose to levoglucosan which in turn could recombine to form oligosaccharides and decompose to char and gaseous products. As shown in Table II the yield of levoglucosan

TABLE II

ANALYSIS OF THE PYROLYSIS PRODUCTS OF CELLULOSE AT 300 °C UNDER NITROGEN

Condition	*Atm. pressure* (%)	*1·5 mm Hg* (%)	*1·5 mm Hg 5% $SbCl_3$* (%)
Char	34·2[a]	17·8[a]	25·8[a]
Tar	19·1	55·8	32·5
levoglucosan	3·57	28·1	6·68
1,6-anhydro-β-D-glucofuranose	0·38	5·6	0·91
D-glucose	trace	trace	2·68
hydrolyzable materials	6·08	20·9	11·8

[a] The percentages are based on the original amount of cellulose.

is lower and the char formation is greater when the reaction is carried out under nitrogen.[4]

Presence of air increases the rate of decomposition and pyrolysis by oxidation of the substrate at the lower temperatures. At the higher temperatures oxidation of the volatile pyrolysis products results in flaming combustion and oxidation of the char leads to smoldering or incandescence combustion according to the rate of reaction.[5]

Presence of water and steam is also known to speed up the breakdown and degradation of the molecules by hydrolysis of the substrate and rearrangement of the intermediate products.[1] This process, known as hydrothermolysis, may be catalyzed by acids or alkali reagents as discussed later. At higher temperatures the steam plays a more direct and better known role as a reactant in the gasification of char and related processes. At the higher temperatures, CO_2 and H_2 could also function as reactants. In order to understand the nature of these reactions it is important to define structure, functionality and the reactivity of the pyrolytic char or the carbonaceous residues prevailing at these temperatures.

2.4. Catalysis

It has generally been known that presence of inorganic materials either as additives or natural ash content strongly affects the pyrolysis of cellulosic materials. The influence of the inorganic materials has been extensively investigated, primarily in connection with the problem of flame-proofing and fire prevention and more recently in connection with thermochemical conversion of biomass to fuel and chemicals. The catalytic effect of the inorganic materials is particularly pronounced with alkaline compounds

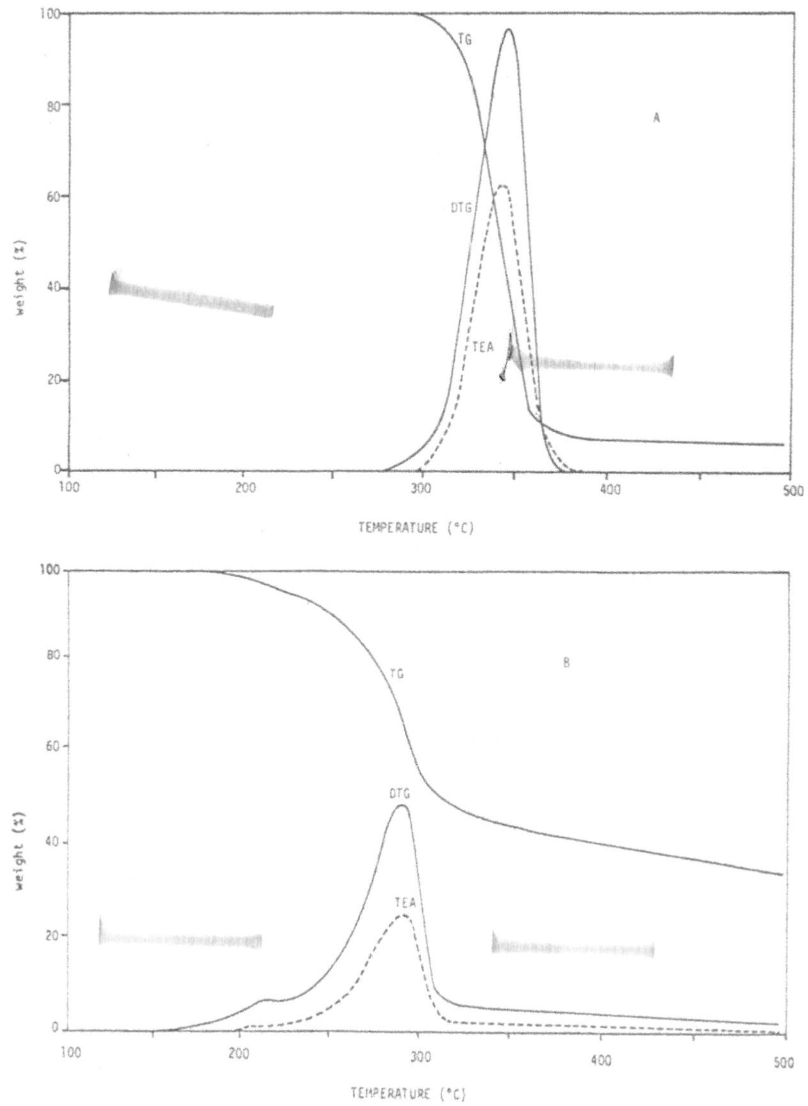

FIG. 3. Thermal analysis curves of pure cellulose (A) and cellulose containing $+5\%$ $(NH_4)_2HPO_4$(B).

and acidic reagents. This includes Arrhenius and Lewis acids as well as materials that could generate acidic or alkaline reactive species on heating, such as diammonium phosphate. However, even the natural impurities and ash content can produce a profound effect. This effect is manifested by lowering of the pyrolysis temperature and increased char formation. That could easily be assessed by thermogravimetry (TG) of the treated substrate as shown in Fig. 3.[6] The increased char formation could start with the abstraction of a hydroxyl group or proton leading to dehydration, decarboxylation, condensation and ultimately enrichment of the carbon content of the solid residue. In this respect the general effect of these additives could also be assessed by the thermal evolution analysis (TEA), the heat of combustion of the gaseous pyrolysis product (EGA) and differential thermal analysis (DTA) indicating the heats of reaction and evaporation. Thus, thermal analysis methods indicate that influence on lowering the temperature of decomposition, the relative amounts of char and volatiles formed, the related heat of evaporation and the heat of combustion of the gaseous products.[6-8]

More specifically, acidic catalysts in trace amounts catalyze the transglycosylation reactions which result in the formation of levoglucosan from cellulose at lower temperatures.[8,9] In larger amounts, catalytic dehydration of the anhydro sugar accompanied by hydrolysis and rearrangement reactions produces levoglucosenone, 5-(hydroxymethyl)-2-furaldehyde, 2-furaldehyde, 1,4;3,6-dianhydro-α-D-glucose, levulinic acid and other dehydration products (see Fig. 4) depending on the amount of water present.[1,10] Acidic catalysts also enhance condensation of the

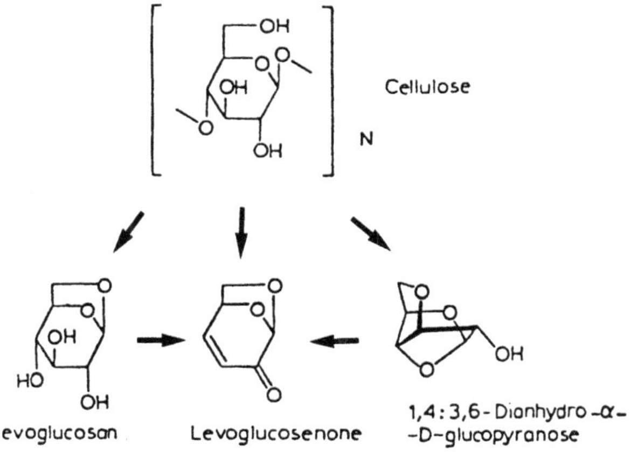

FIG. 4. Formation of levoglucosenone by dehydration reactions.

intermediate compounds and affect some of the properties of the resulting char.

Alkali catalysts affect the hydrolysis, fission (by dealdolization),[11] rearrangement to saccharinic acids, and quantity and properties of the char, particularly its rates of oxidation and gasification by steam and carbon dioxide.[12] The products again depend on the reaction conditions, particularly the TTP and the presence of steam, acid and alkali catalysts. The degradation of wood and carbohydrates in water at temperatures of up to about 200 °C, which may be considered as specific cases of hydrothermolysis, have been extensively investigated and are better known.

2.5. The Combined Effects

Because of these variables and their profound effects on the course, mechanism and kinetics of pyrolysis, the quantitative data obtained on this subject at different laboratories have often been different and even controversial.[13] This is particularly true of the kinetic data, which may range from modelling of the general process to global kinetics and chemical kinetics of specific reactions depending on whether it involves a general process, a combination of poorly defined reactions or specific chemical reactions.

Despite these variations and controversies, the kinetic studies can be highly informative provided that the employed conditions are specified and their significance understood. It is particularly helpful to distinguish between the primary and secondary reactions and recognize the changes which take place in the cellulosic substrates as the pyrolysis proceeds. In the past pyrolysis has been almost synonymous with volatilization, devolatilization or destructive distillation emphasizing the production and properties of the volatile products and neglecting the properties and reaction of the solid materials. Since the former aspects of pyrolysis have been covered in several previous reviews and presentations[14-18] and in view of the significance of the char and the solid phase reactions in processes of gasification and the smoldering or incandescent combustion, this article provides a review of the pyrolytic reaction at the solid phase as affected by the above variables.

3. FORMATION OF VOLATILE PRODUCTS

The general pathways for pyrolysis of cellulose, leading to production of char as well as gaseous and volatile products, are shown in Fig. 5.[1] The

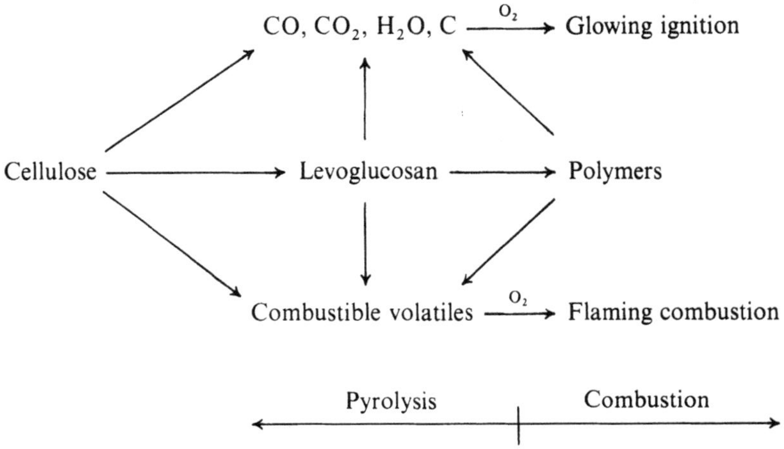

FIG. 5. The pyrolysis and combustion of cellulose.

global kinetics for isothermal evolution of volatile pyrolysis products from purified cotton linter cellulose, within the temperature range 275–310 °C, has been studied in air and nitrogen.[19] The Arrhenius plot of the results based on first order kinetics is shown in Fig. 6. These data gave an activation energy of 17 kcal/mole for the weight loss due to the overall pyrolytic reaction of cellulose in air and 37 kcal/mole in a nitrogen

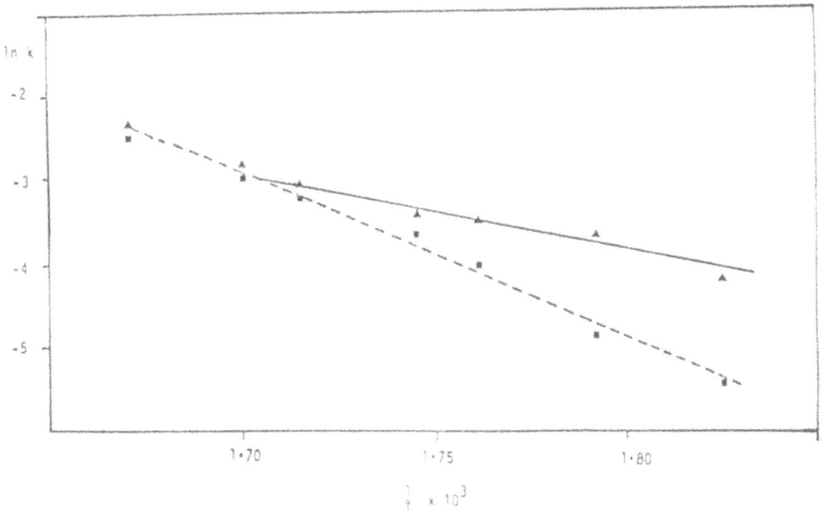

FIG. 6. Arrhenius plots for first order reactions in the isothermal degradation of cellulose in air (■) and nitrogen (▲).

atmosphere. Figure 6 indicates a transition at *ca*. 300 °C which reflects the existence of two different pathways. As seen in Fig. 7, the rate of pyrolysis followed by weight loss under isothermal conditions shows an initial period of acceleration and proceeds much faster in air than in the inert atmosphere. As the pyrolysis temperature is increased, the initiation period and the difference between pyrolysis under nitrogen and air gradually diminish and disappear at 310 °C when pyrolysis by the second pathway takes over.

3.1. First Pathway

The reactions in the first pathway, which dominates at lower temperatures, involve reduction in molecular weight or DP by bond scission, appearance of free radicals, elimination of water, formation of carbonyl, carboxyl and hydroperoxide groups (especially in air), evolution of carbon monoxide and carbon dioxide, and, finally, production of a charred residue. These reactions, which contribute to the overall rates of pyrolysis of cellulosic materials, have been individually investigated. Reduction in the degree of

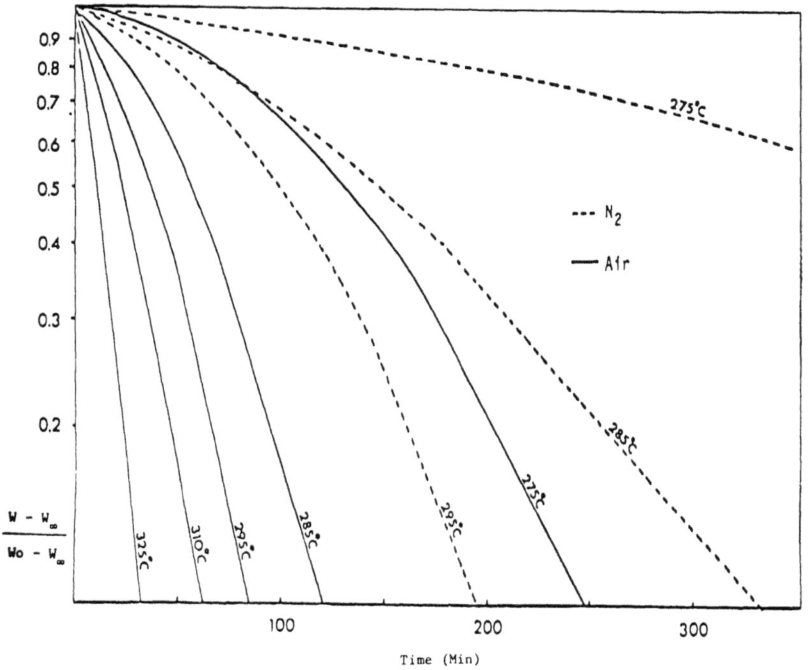

FIG. 7. First order plot for the residual cellulose weight (normalized) versus time. Plots at 310 °C and 325 °C for air and nitrogen are similar.

polymerization of cellulose on isothermal heating in air or nitrogen at a temperature within the range 150–190 °C has been measured by the viscosity method (see Fig. 8). The resulting data have been correlated with rates of bond scission given in Table III and used for calculating the Arrhenius plot shown in Fig. 9. These calculations give an activation energy of 21 kcal/mole for bond scission in air and 27 kcal/mole in nitrogen and indicate that at low temperatures a larger number of bonds are broken in air than in nitrogen.

The rates of production of carbon monoxide and carbon dioxide at 170 °C in air and nitrogen are shown in Fig. 10. The rates for evolution of these gases are much faster in air than in nitrogen and, furthermore, accelerate on continued heating. It is instructive to compare the initial linear rates for the evolution of these gases with the rates of bond scission obtained for depolymerization at 170 °C that were discussed earlier. As can

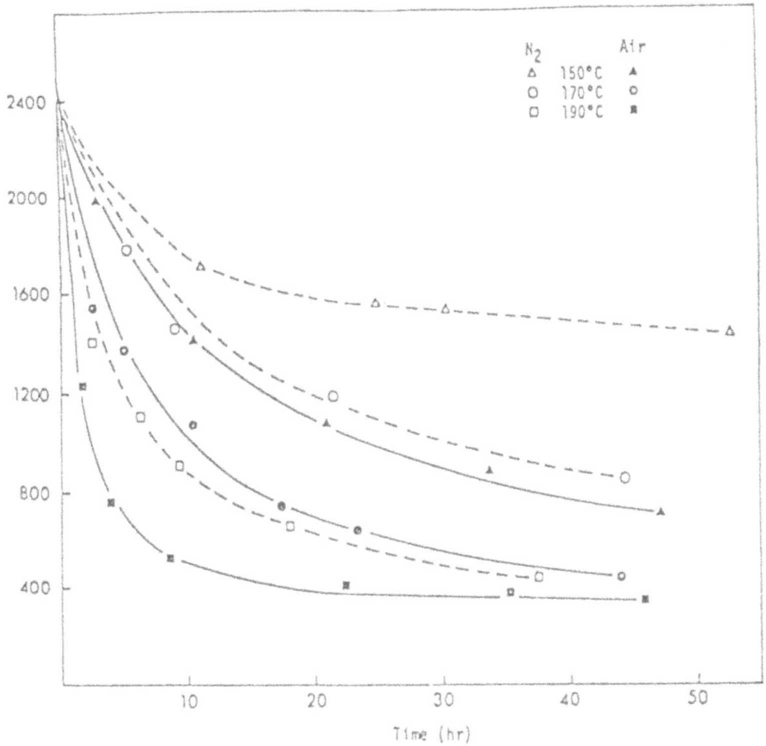

FIG. 8. Viscosity average degree of polymerization (P_v) of cellulose heated in air or nitrogen at 150, 170 and 190 °C.

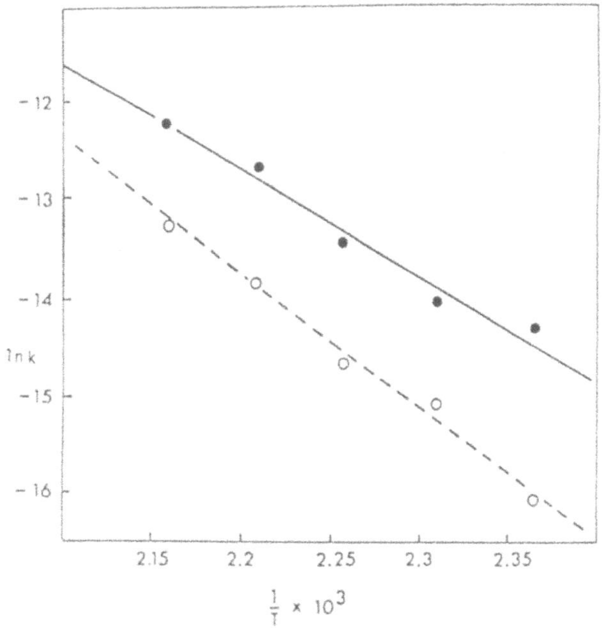

FIG. 9. Arrhenius plot for the rate of bond scission in air (●) and nitrogen (○).

be seen in Table IV, the rate of bond scission in air approximately equals the rate of production of carbon dioxide plus carbon monoxide in moles per glucose unit. In nitrogen, however, the rate of bond scission is greater than the rates of carbon monoxide and carbon dioxide evolution combined.

It is assumed that carbon dioxide and carbon monoxide are formed by decarboxylation and decarbonylation, respectively. The significance of the

TABLE III

RATE CONSTANTS FOR THE DEPOLYMERIZATION OF CELLULOSE IN AIR AND NITROGEN

Temperature (°C)	*Conditions*	$k_0 \times 10^7$ *mole/162 g min*[a]	*Conditions*	$k_0 \times 10^7$ *mole/162 g min*
150	N_2	1·1	Air	6·0
160	N_2	2·8	Air	8·1
170	N_2	4·4	Air	15·0
180	N_2	9·8	Air	29·8
190	N_2	17·0	Air	48·9

[a] 162 g represents 1 mole of monomer unit.

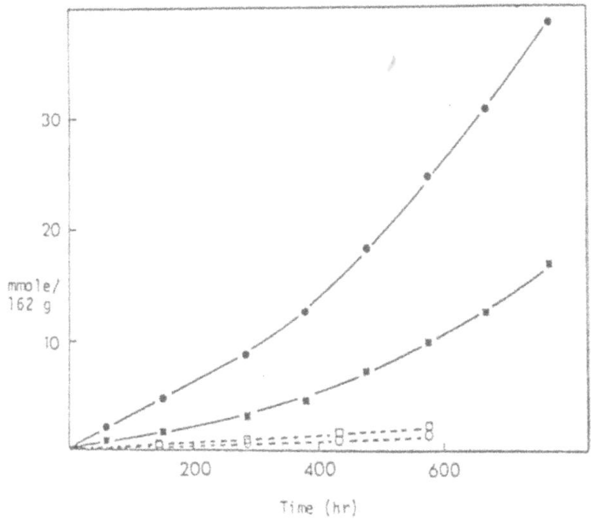

FIG. 10. Yields of CO and CO_2 from heating cellulose at 170 °C: ○ CO_2 in N_2; □ CO in N_2; ● CO_2 in air; ■ CO in air.

former reactions was determined by measuring the net rate of accumulation of carboxyl and carbonyl groups in cellulose upon heating in air at 190 °C. The results shown in Fig. 11 indicate an almost linear rate of formation on heating for 50 h. On heating for longer periods, the rate of accumulation of carboxyl groups falls off, and the rate of accumulation of carbonyl groups is increased. There was also a very small increase in the number of these functions on heating in nitrogen, formed by dehydration and rearrangement as shown for model compounds.[18,20] In order to check the extent of the decarboxylation at the lower pyrolysis temperatures,

TABLE IV

INITIAL RATES OF GLYCOSIDIC BOND SCISSION OF CARBON MONOXIDE AND CARBON DIOXIDE FORMATION AT 170 °C

Reaction	*Rate × 10^5 in N_2 mole/162 g h*	*Rate × 10^5 in air mole/162 g h*
Bond scission[a]	2·7	9·0
CO evolution[b]	0·6	6·4
CO_2 evolution[b]	0·4	2·1

[a] Calculated from the rate constants in Table I.
[b] Calculated from initial linear portion of plots in Fig. 8.

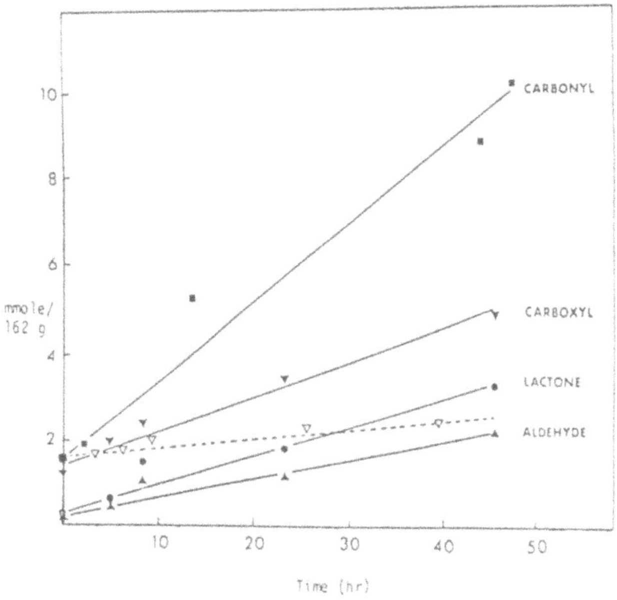

FIG. 11. Incorporation of oxidized groups in cellulose at 190 °C. ▽ represents number of lactone and carboxyl groups formed in nitrogen at 190 °C.

samples of carboxylcellulose having a low degree of substitution, with carboxyl groups at C1, C2, C3 and C6, were prepared and pyrolyzed. The results generally were not conclusive, except for the sample oxidized at C2 and C3, which showed a definite reduction in carboxyl-group content.

The thermal degradation of cellulose, similar to that of synthetic polymers, may involve a free-radical mechanism. As discussed later, in air it was difficult to observe these radicals, but it was possible to monitor the formation of hydroperoxide groups on heating cellulose in air. The hydroperoxide functions are simultaneously formed and decomposed, and their concentration rapidly climbs until a steady state is reached. The decomposition of the hydroperoxide function appeared to follow first-order kinetics with a rate constant of $2{\cdot}5 \times 10^{-2}$ min^{-1} at 170 °C from the steady-state concentration of $3{\cdot}0 \times 10^{-5}$ mole/162 g min. When compared with the initial rate of bond scission in air of $1{\cdot}5 \times 10^{-6}$ mole/162 g min at 170 °C (Table IV), it is apparent that hydroperoxide formation could make a significant contribution to bond scission.

Based on the above considerations, it is proposed that three stages are involved in the low temperature pathway: initiation of pyrolysis,

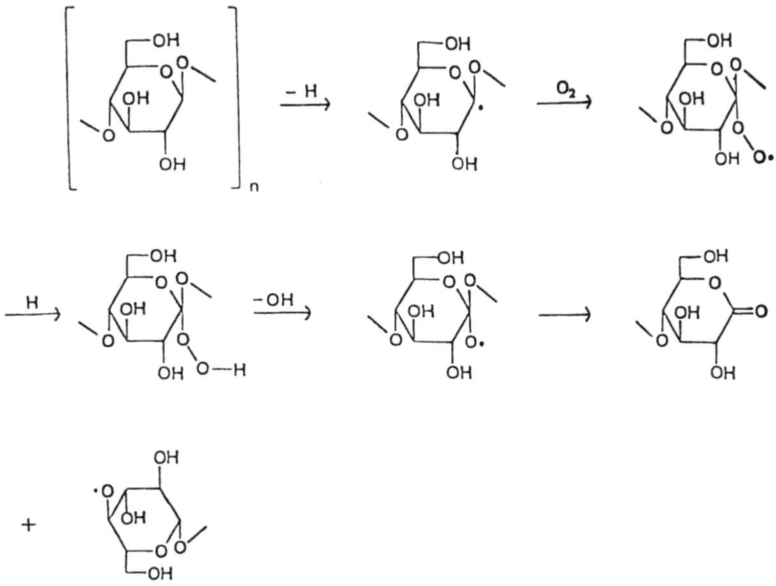

FIG. 12. Possible mechanism of formation and decomposition of cellulose hydroperoxide formed thermally in air.

propagation and product formation. The initiation period apparently involves the formation of free radicals facilitated by the presence of oxygen or inorganic impurities. Subsequent reactions of the free radicals could lead to bond scission, oxidation and decomposition of the molecule, producing char, water, carbon monoxide and carbon dioxide. Figure 12 shows the nature of the initiation, propagation and decomposition reactions involved in the thermal decomposition of cellulose by this pathway in air. In an inert atmosphere, a lactone could be formed by rearrangement and decomposition as shown with model compounds.[18,20]

3.2. Second Pathway

At temperatures of approximately 300 °C, the second pathway gradually takes over and dominates. The primary reaction in this pathway involves depolymerization by transglycosylation. This reaction takes place when the molecule has gained sufficient flexibility (activation) and produces levoglucosan (1,6-anhydro-β-D-glucopyranose), its furanose isomer (1,6-anhydro-β-D-glucofuranose) and randomly linked oligosaccharides as shown in Fig. 13.[4] The inter- and intramolecular transglycosylations shown in this figure are accompanied by dehydration and followed by

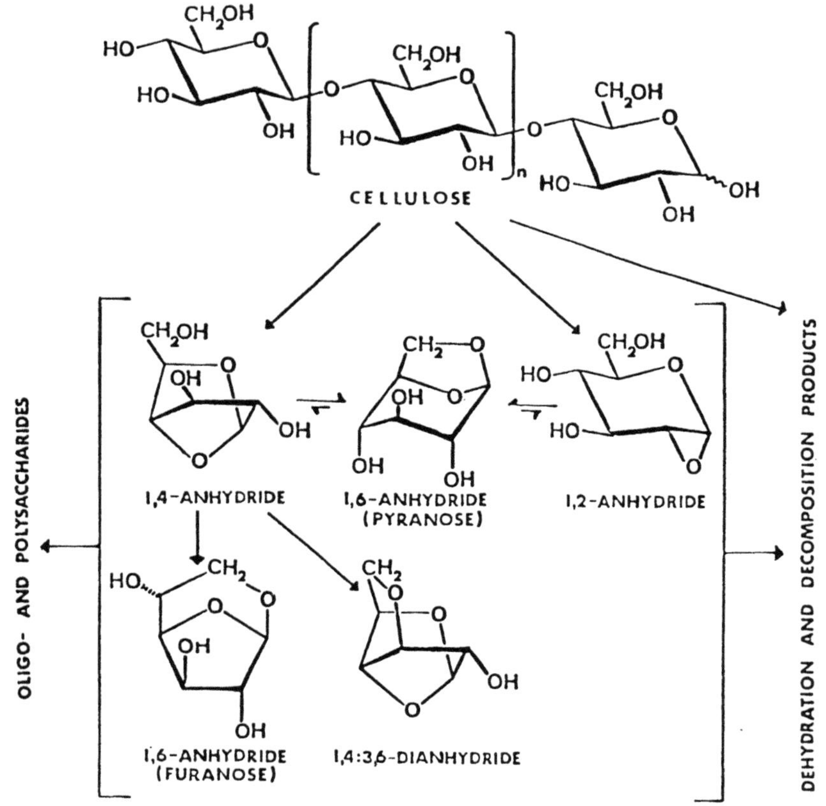

FIG. 13. Pyrolyzate of cellulose to anhydro sugars.

fission and disproportionation reactions in the gas phase and further decomposition and condensation of the solid phase, producing a mixture of gases and volatile products and a 'stable' carbonaceous char.[16–18]

On raising the temperature, the tar-forming reactions accelerate rapidly and overshadow the production of char and gases. Figure 14 shows the production of diminishing amounts of char and increasing amounts of tar (containing anhydro sugar derivatives that could be hydrolyzed to reducing sugar) as the oven temperature is raised from 300 to 500 °C.[8] At this time, it should be pointed out that evaporation of levoglucosan and the volatile pyrolysis products is highly endothermic. Thus, the increased oven temperature could raise the rate of heat transfer but not necessarily the temperature of the ablating substrate which is cooled by the heat of evaporation, especially under vacuum. As shown in Fig. 15,[21] the oven

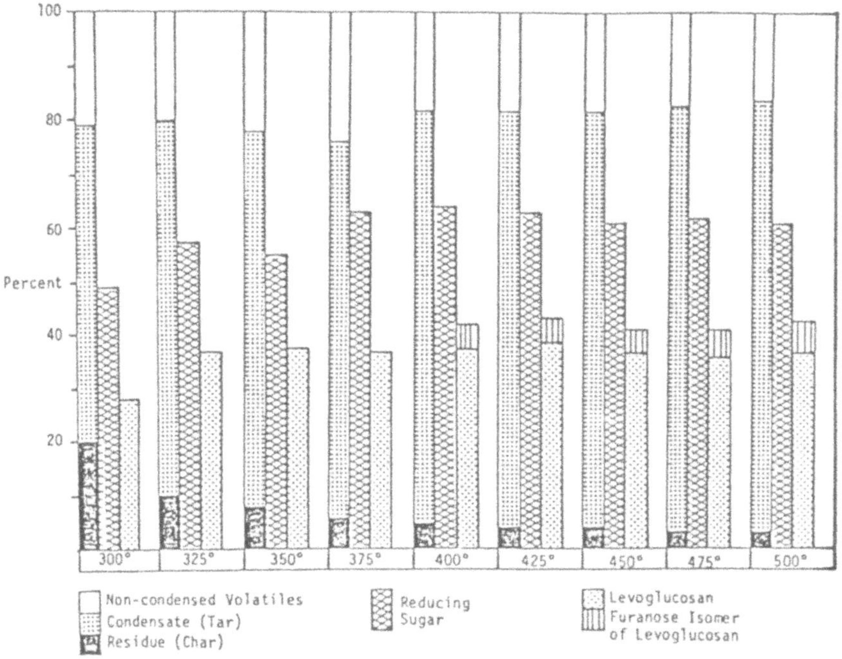

FIG. 14. Products from pyrolysis of cellulose powder under vacuum.

temperature is reached when the rapid evaporation is over. In other words, at the higher temperatures, the pyrolysis process may be controlled by the rate of heat transfer rather than the kinetics of the chemical reaction. Material transport presents another major obstacle to the investigation of chemical kinetics, because if the products of primary reactions are not removed, they can undergo further decomposition reactions. Table II shows the difference between the yield of different pyrolysis products in vacuum, which removes the primary volatile products, and in nitrogen at atmospheric pressure which allows more decomposition of the anhydro sugars.[4] It also shows the effect of inorganic catalysts in changing the nature of the reactions and products. In view of these considerations, the chemical kinetics of cellulose pyrolysis have been investigated within the limited temperature range of 260–340 °C and under vacuum in order to obtain chemically meaningful data. Under these conditions, the chemical kinetics of cellulose pyrolysis could be represented by the three reaction model shown in Fig. 16.[21] In these models it is assumed that the initiation reactions discussed earlier lead to the formation of an active cellulose,

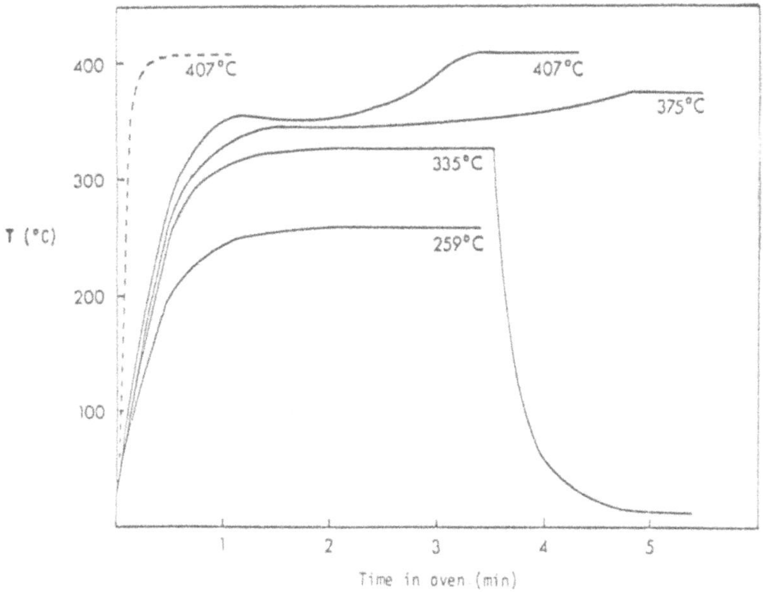

FIG. 15. Temperature of pyrolysis as measured by a thermocouple in cellulose sample. Final temperatures, T_f, are indicated. (----) Temperature measured in empty boat. The cooling curve follows removal of a sample from the oven at 335 °C.

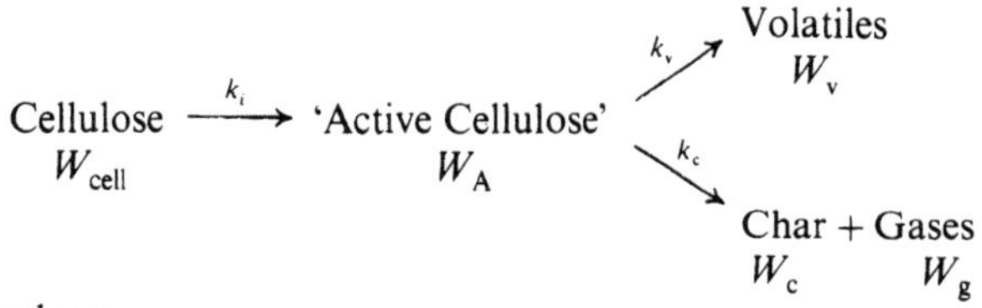

where

$$\frac{-d(W_{cell})}{dt} = k_i[W_{cell}]$$

$$\frac{d(W_A)}{dt} = k_i[W_{cell}] - (k_v + k_c)[W_A]$$

$$\frac{d(W_c)}{dt} = 0{\cdot}35k_c[W_A]$$

$$k_i = 1{\cdot}7 \times 10^{21} \exp[-(58\,000/RT)] \text{ min}^{-1}$$

$$k_v = 1{\cdot}9 \times 10^{16} \exp[-(47\,300/RT)] \text{ min}^{-1}$$

$$k_c = 7{\cdot}9 \times 10^{11} \exp[-(36\,000/RT)] \text{ min}^{-1}$$

FIG. 16. Kinetic model for pyrolysis of pure cellulose under vacuum.

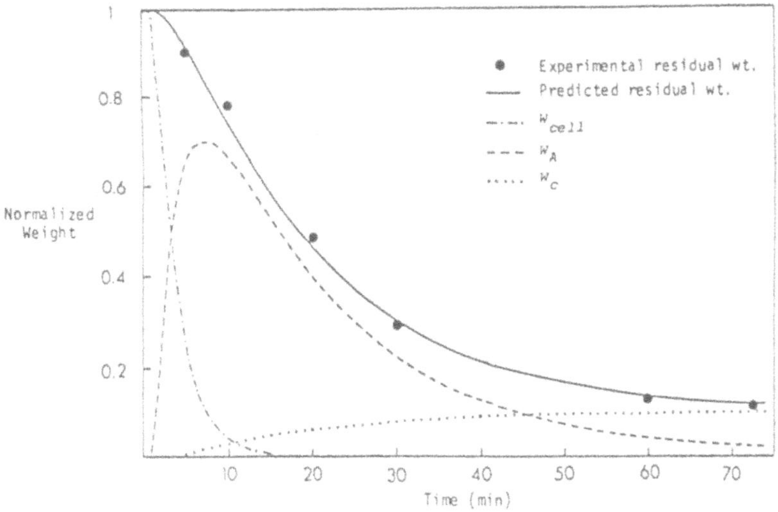

FIG. 17. Comparison of experimentally measured residual weight to that predicted from reaction model at 312 °C. W_{cell}, W_A and W_c are predicted weights.

which subsequently decomposes by two competitive first order reactions, one yielding anhydro sugars (transglycosylation products) and the other char and a gaseous fraction. Figure 17 shows a comparison between predicted and experimental results.

4. FORMATION AND PROPERTIES OF CHAR

The devolatilization reactions leave a solid residue, which is neither intact substrate nor pure carbon, but a material at various stages of charring and carbonization. The intermediate chars are characterized by the functional groups present, including aromatic and olefinic structures, a high concentration of free spins trapped in a rigid structure or stabilized by aromatic and olefinic structures, a large surface area, and a high degree of reactivity, which depend on progression of the secondary reaction in the solid phase. Development of these structures and functionalities could be investigated by several chemical and physical methods and related to the pyrolysis conditions, particularly the heating and presence of inorganic catalysts, as well as the reactivity of the products.

The cross-polarization magic-angle spinning (CP/MAS) ^{13}C-NMR spectra of cellulose before and after heating at 350 and 500 °C are shown in

Fig. 18. As can be seen in this figure, the spectrum of the unheated cellulose has sharp peaks corresponding to the aliphatic carbon atoms of the polymer units. After heating at 350 °C, four new peaks appear which are associated with aliphatic (0–60 ppm), olefinic, aromatic (100–110 ppm), carboxylic and ester (160–180 ppm) and carbonyl carbons (190–220 ppm). After heating at 500 °C, the peaks for olefinic and/or aromatic carbons are enhanced and those for the other carbons are reduced.

Development of aromaticity can be investigated by permanganate oxidation of the char which gives benzene polycarboxylic acids derived from the aromatic nuclei present, as shown in Fig. 19. These compounds

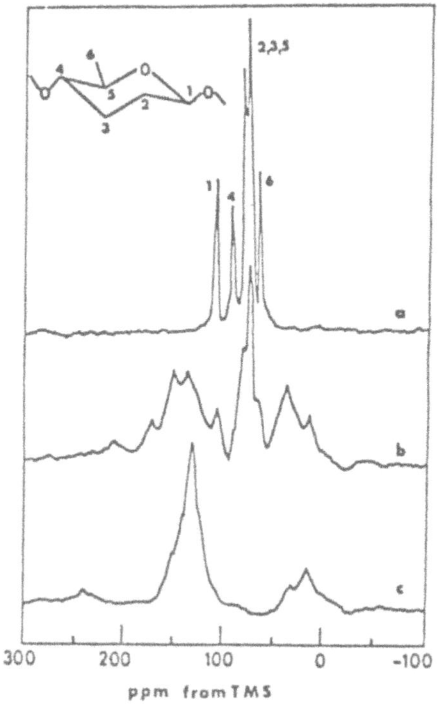

FIG. 18. CP/MAS ^{13}C-NMR spectra of cellulose after different heat treatments. The spectrum for untreated cellulose (a) gives sharp peaks numbered 1 to 6 corresponding to the carbon atoms in the polymeric units. On heating after 5 min at 350 °C (b), four new peaks associated with: aliphatic carbons located at 0–60 ppm; olefinic and aromatic carbons at about 100–160 ppm; carboxylic acid and ester carbons at about 160–180 ppm; and carbonyl carbons at 190–220 ppm are observed. On heating at 500 °C (c), olefinic and aromatic carbons increase whereas the other three types decrease.

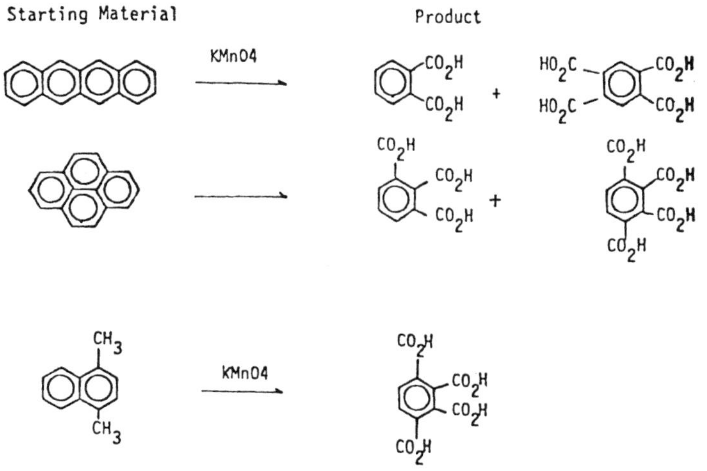

FIG. 19. Oxidation of aromatic compounds.

can be methylated by diazomethane, and analyzed by GLC. These products, shown in Fig. 20, are grouped by the number of the carboxylic acid substituents and abbreviated as B2C–B6C. Analysis of the products indicates the concentration and degree of substitution or condensation of the aromatic units. However, since the permanganate oxidation leaves only the central aromatic rings, the resulting data represent only a minimum amount of the aromatic structures present in the char.

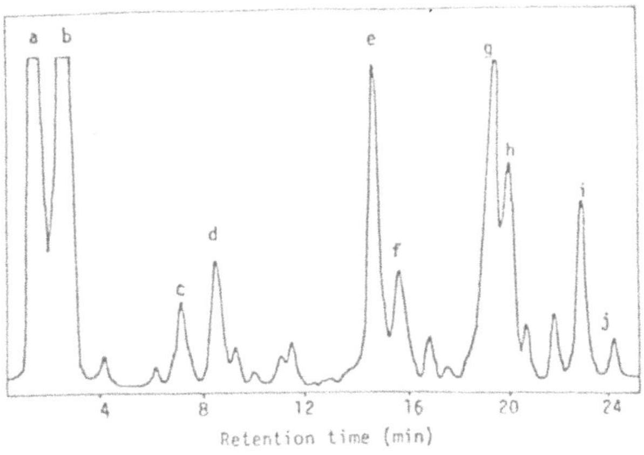

FIG. 20. Typical gas chromatogram of oxidation product. a,b: solvent; c: benzene dicarboxylic acid; d: benzene monohydroxy dicarboxylic acid; e: benzene tricarboxylic acid; f: unknown; g: benzene tetracarboxylic acid; h: unknown; i: benzene pentacarboxylic acid; j: benzene hexacarboxylic acid.

The yields, elemental composition and the empirical formulas of chars obtained by isothermal heating of cellulose samples for 5 min within the temperature range 300–500 °C are given in Table V.[22] The total aromatic carbon content of these chars and the aromatic carbon content of each benzene carboxylic acid compound as a percent of the total carbon content of the char are given in Fig. 21. These data show that the reduction in the char yield, or the increase in the weight loss on increasing the heat treatment temperature (HTT), is accompanied by the increased aromatization of the char as measured by the aromatic carbon content of the benzene polycarboxylic acids (see Fig. 22). Furthermore, the increased aromaticity of the char is accompanied by a reduction of the hydrogen-to-carbon (H/C) ratio (see Fig. 23). Figure 24 shows the yields of aromatic carbon contents calculated on the basis of the carbon content of the

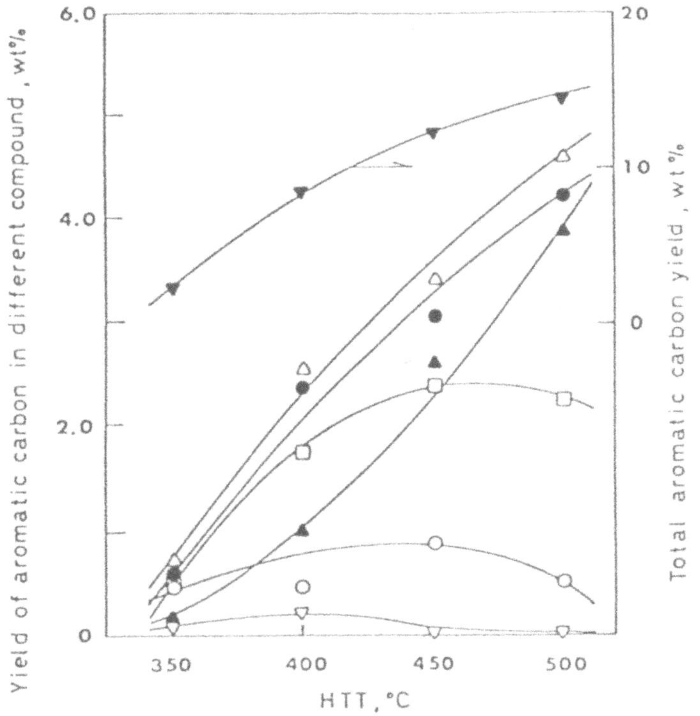

FIG. 21. Aromatic carbon content of the char from benzene dicarboxylic acid, ○; benzene tricarboxylic acid, □; benzene tetracarboxylic acid, △; benzene pentacarboxylic acid, ●; benzene hexacarboxylic acid, ▲; and total yield of hydroxy benzene carboxylic acid, ▽.

TABLE V
RESIDUAL CHAR YIELD AND ITS ELEMENTAL ANALYSIS

Condition	*Char yield (wt %)*	*Composition*			*Formula (Ref. to C_6)*	*H/C*
		C	*H*	*O (diff)*		
Untreated		42·8	6·5	50·7	$C_6H_{10}O_5$[a]	1·7
300 °C for 5 min	89·4	44·0	6·4	49·6	$C_6H_{10\cdot4}O_{5\cdot0}$	1·7
325 °C for 5 min	63·3	47·9	6·0	46·1	$C_6H_{9\cdot0}O_{4\cdot3}$	1·5
350 °C for 5 min	31·8	59·9	5·2	34·9	$C_6H_{6\cdot2}O_{2\cdot6}$	1·0
400 °C for 5 min	16·6	76·5	4·7	18·8	$C_6H_{4\cdot4}O_{1\cdot1}$	0·7
450 °C for 5 min	10·5	78·8	4·3	16·9	$C_6H_{3\cdot9}O_{1\cdot0}$	0·6
500 °C for 5 min	8·7	80·4	3·6	16·0	$C_6H_{3\cdot2}O_{0\cdot9}$	0·5

[a] Theoretical value.

original cellulose rather than the carbon content of the char, as a function of HTT. In this figure the yield of aromatic carbons reaches to 2·5 per 100 carbons of the original cellulose at 400 HTT and then levels off, while the aromatization, as indicated by drastic reduction of the H/C ratio (see Fig. 23) and the increased formation of B5C and B6C (see Fig. 21), continues at HTTs above 400 °C. Thus, these data suggest that the

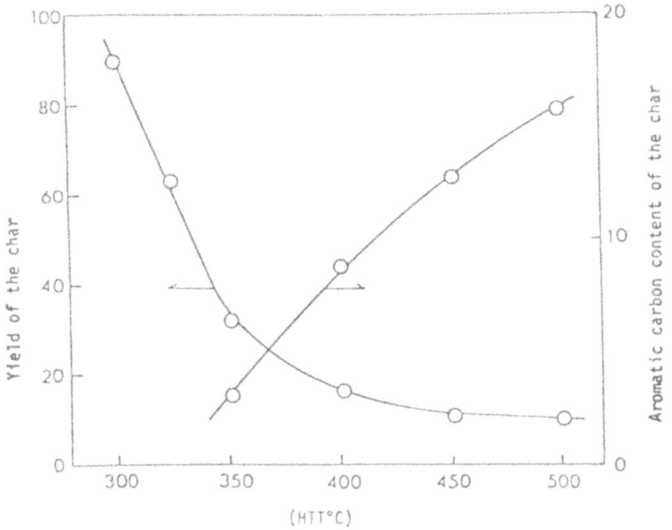

FIG. 22. Changes in the yield and aromatic content of the char as a function of HTT.

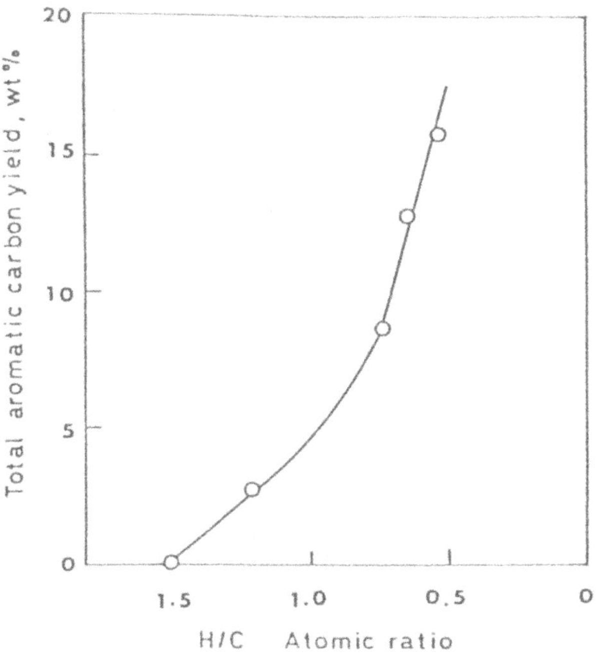

FIG. 23. Relation between total aromatic carbon and H/C ratio of char.

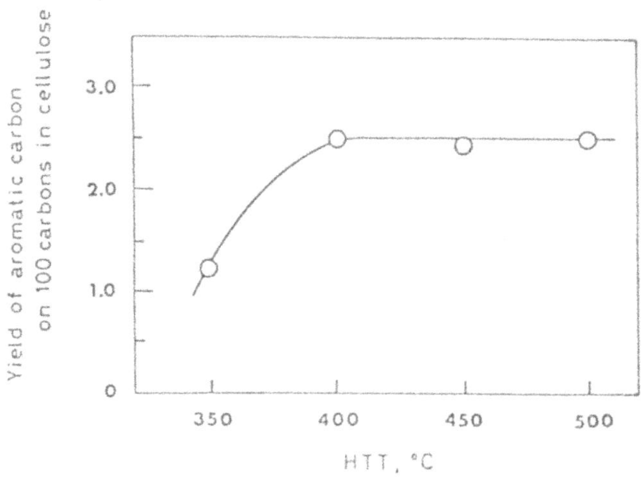

FIG. 24. Number of aromatic carbons in 100 carbons of original cellulose.

aromatization process involves nucleation and development of isolated aromatic structures at temperatures between 350 and 400 °C. Above 400 °C the number of aromatic clusters which are oxidized to polycarboxylic acids remains constant, but the aromatization continues through condensation and the growth of the individual aromatic clusters, resulting in lower H/C ratios. Furthermore, the weight loss in the 'stable' char at temperatures above 400 °C, although relatively small, must take place by elimination of aliphatic substituents accompanied by some dehydrogenation and condensation or fusion of the products. These processes apparently take place through homolytic cleavage and condensation of the resulting free radicals, evident from the measurement of the trapped free spin concentration in char that, as discussed later, are affected by HTT and other variables.

5. THE EFFECT OF INORGANICS

Inorganic impurities, additives or catalysts exert profound effects on the rate of production, yield and properties of the charred residue. The transglycosylation reactions are catalyzed by trace amounts of inorganic acids, which lower the temperature of pyrolysis and increase the yields of the volatile pyrolysis products. However, higher concentrations of acid and the presence of inorganic materials in general catalyze the dehydration and condensation of the intermediate products, resulting in higher yields of char. The effect of small amounts of sulfuric acid on production of char, tar and levoglucosan from Douglas fir lignocellulose is shown in Fig. 25.[9] Similar results have also been obtained on pyrolysis of cellulose[8] and phenyl glucoside used as a model compound.[23] Higher concentrations of acid (1–5 %) result in dehydration of the sugar units as evidenced by the formation of levoglucosenone, furan derivatives, and 1,4:3,6-dianhydro-glucose, and condensation of the intermediate compounds to provide more char.[10,16] The effects of inorganic materials on the formation and reactivity of chars have been investigated, particularly in relation to problems of flaming combustion, which is suppressed by increased formation of char at the expense of the flammable volatiles, and smoldering or glowing combustion which is related to the reactivity of the char. The flame retardants and inorganic materials in general lower the temperature of decomposition and enhance the char formation. This effect can be readily detected by the thermal analysis of cellulose, particularly TG, which shows the amount of volatiles and residue. Figure 3 shows that pyrolysis of pure cellulose

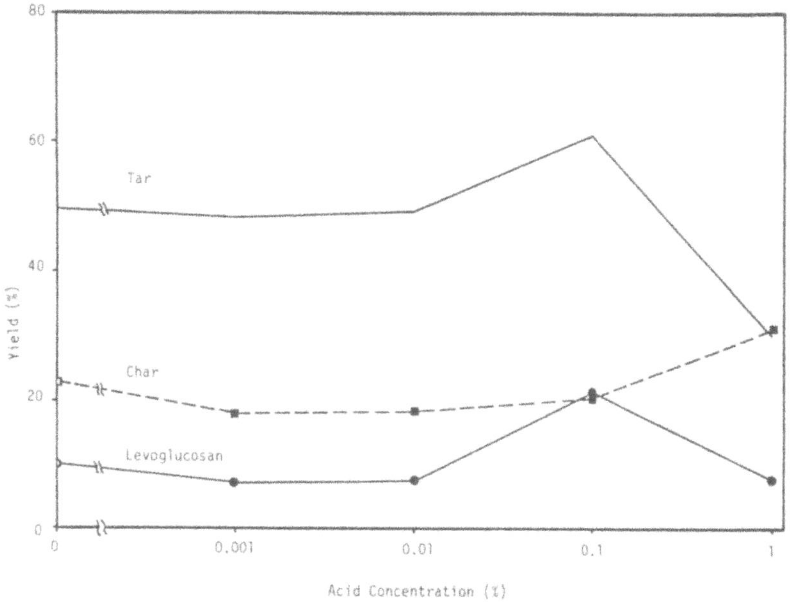

FIG. 25. Saccharification of Douglas fir heartwood by prehydrolysis, pyrolysis and tar hydrolysis. Percentages are weight percent.

proceeds rapidly at about 300–375 °C and leaves little char, whereas in the presence of diammonium phosphate cellulose is gradually decomposed at a wider and lower temperature range of 150–350 °C, and leaves substantial amounts of char. Table VI shows the TG data for some flame retardants and related inorganic compounds, ranked according to their effectiveness in enhancing the char formation. Compounds like diammonium phosphate are effective flame retardants not only because they produce more char and less volatiles but also because the volatiles formed contain more carbon dioxide and water and consequently have lower rates of combustion. Also, the rate of heat release is drastically reduced.

In order to gain further information about the mechanism of these interactions, cellulose was treated with these additives and charred at 400 °C. The chars obtained from samples of treated and untreated cellulose were oxidized with permanganate as described before to obtain the results shown in Table VII. As expected, the char yield was slightly higher for the sodium chloride treated sample (17·5 %) and substantially higher for the sample containing diammonium phosphate (28·9 %), as compared to the yield from the untreated sample (15·3 %). Furthermore, the increased char

TABLE VI

THE EFFECT OF SOME FLAME RETARDANTS AND RELATED INORGANIC COMPOUNDS ON THERMOGRAVIMETRIC ANALYSIS OF CELLULOSE

Flame retardant	*Weight loss*				*Order of effectiveness*
	Range (°C)	*Fast rate (°C)*	*Max. rate (°C)*	*Total at 500 °C (%)*[a]	
H_3PO_4	170–500	170–290	265	61	1
$(NH_4)_2HPO_4$	170–500	225–310	290	66	2, 3
$NH_4H_2PO_4$	190–500	250–325	300	66	
$ZnCl_2$	200–500	200–500	265, 320	74	4
NaOH	50–500	150–350	300	79	5
H_3BO_3	275–500	300–425	335	81	6
NaCl	275–500	275–400	375	82	7
$CaCl_2$	260–500	260–400	355	83	8
$NaHSO_4$	150–500	150–400	300	84	9, 10
$SbCl_3$	75–500	275–350	320	84	
$NaVO_3$	225–500	300–380	355	85	11
$SbCl_3$ Hydrate[b]	100–500	275–390	330	86	12, 13
$(NH_4)_2SO_4$	200–500	200–400	310	86	
Na_2CO_3	225–500	225–380	355	87	14, 15
$NaH_2PO_4 \cdot H_2O$	255–500	255–350	325	87	
$Na_2B_4O_7 \cdot 10H_2O$	240–500	300–370	350	89	16
$Na_2WO_4 \cdot 2H_2O$	240–500	300–380	360	90	17
Na_3PO_4	220–500	300–380	360	91	18
$NaHCO_3$	250–500	290–380	335	92	19
NH_4Cl	150–500	300–375	350	93	
$(NH_4)_2CO_3$	260–450	300–380	340	93	20, 21, 22
Neat	290–500	290–370	345	93	

[a] Based on the total weight of the sample.
[b] Contains 90 % water.

formation was accompanied by increased aromaticity, in terms of the yield of the aromatic carbon in the char and the original cellulose molecules. The pattern in the distribution of polycarboxyl benzene derivatives was also very interesting, because they showed that the increased aromaticity was distributed among the highly substituted compounds (B4C–B6C) and the actual aromatic content of these chars could be much higher.

These data closely correlate with the general mechanism of the flame retardants and seem to indicate that increased char formation is due to the increased condensation and cross-linking of the carbon skeleton, in addition to any role that inorganic flame retardants play in dehydration of

TABLE VII

EFFECT OF INORGANIC ADDITIVES ON AROMATIC FORMATION IN CHAR PREPARED AT 400 °C FOR 5 MIN[a]

Additives	Char yield (wt%)	Char formula (ref. to C_6)	Aromatic carbon yield in char (wt%) determined in each benzene carboxylic acid						Number of aromatic carbons in 100 carbons of original cellulose
			B2C	B3C	B4C	B5C	B6C	Total	
None	15·3	$C_6H_{4\cdot3}O_{1\cdot3}$[b]	0·7	3·1	4·5	3·8	1·6	13·7	3·5
NaCl	17·5	$C_6H_{4\cdot1}O_{1\cdot5}$	0·6	3·1	6·0	5·0	2·1	16·8	5·0
$(NH_4)_2HPO_4$	28·9	$C_6H_{4\cdot3}O_{1\cdot2}$	0·7	3·4	6·2	5·8	2·5	18·6	8·7

[a] A new pyrolyzer was employed for these runs. Thus the char yield and aromatic content of the sample with no additive differed from the 400 °C run quoted in Table I and Fig. 1.

[b] Oxygen content was obtained by difference, neglecting weight of inorganics.

the glucose units in the cellulose molecule. The effect of inorganic additives on smoldering combustion is more complex and related to the reactivity of the char as discussed below.

5.1. Char Reactivity

It has been shown that freshly prepared char is highly reactive and pyrophoric. This reactivity is closely related to smoldering combustion which has been extensively investigated and shown to involve the formation of reactive char by pyrolysis, chemisorption of oxygen on this product, evolution of CO and CO_2, and generation of new reactive sites.

Figure 26[24] shows the weight increase and heat of reaction due to chemisorption of oxygen on fresh char determined by TG and DSC (differential scanning calorimeter). In the low density, fibrous cellulosic materials where the heat loss is restricted and yet oxygen could penetrate by diffusion, the heat flux from chemisorption could play a significant role in the ignition of the active char and initiation of the combustion. Figure 27 shows the differential heat of chemisorption at 118 °C which is gradually reduced as the more reactive sites are occupied.

The chemisorption data determined gravimetrically were analyzed according to Elovich kinetics.[25] These kinetics are based on the assumption that the rate of chemisorption (dq/dt) declines as the more reactive sites are quenched according to the equation:

$$dq/dt = a\exp(-bq)$$

which is integrated over time between the limits of $-t_0$ and t to give

$$q = (1/b)\ln ab + (1/b)\ln(t + t_0)$$

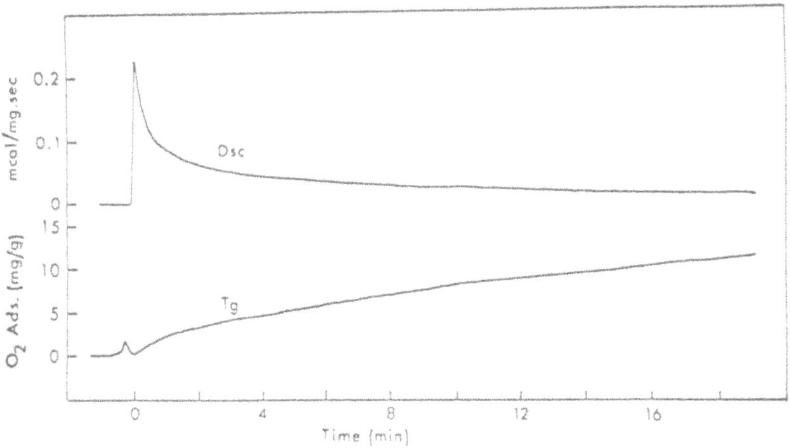

FIG. 26. Differential scanning calorimetry and thermogravimetry of oxygen chemisorption on cellulose char at 118 °C: analysis was carried out on 2·5 mg samples in aluminum pans using a Cahn R-100 Electro-balance and a DuPont calorimeter cell, attached to a DuPont Model 990 thermal analyzer, and nitrogen and oxygen gas flows (60 ml/min, dried by passing through H_2SO_4) were rapidly interchangeable for DSC and TG.

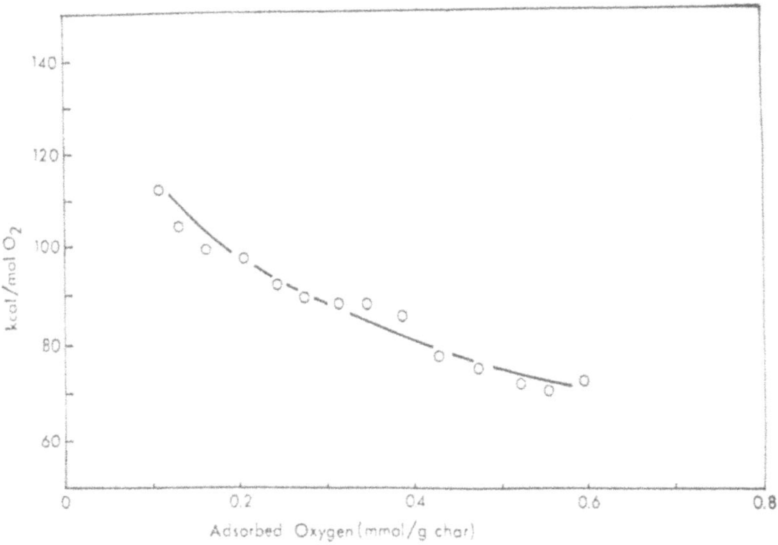

FIG. 27. Differential heat of chemisorption, as a function of the amount of oxygen adsorbed, on cellulose char at 118 °C.

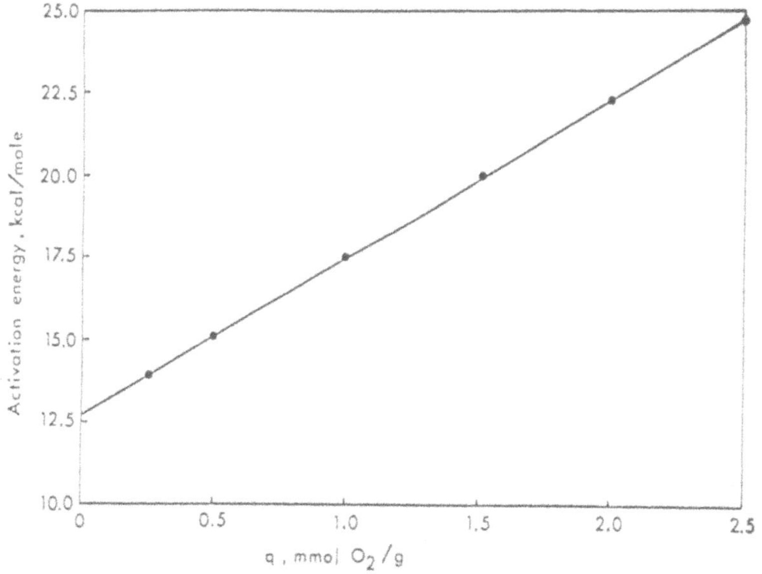

FIG. 28. Variation of activation energy for oxygen chemisorption with surface coverage.

where q is the amount of oxygen chemisorbed at time t and a and b are constants. The constant t_0 allows for an initial 'pre-Elovichian' period of chemisorption which is generally more rapid than that predicted by Elovich kinetics. The value of t_0 was determined by an iterative method to give the best linearity of plots of q vs. $(t + t_0)$.

The activation energy for oxygen chemisorption on a typical char varied linearly from 13 to 22 kcal/mole with surface coverage of zero to 2·5 mmole O_2/g char (Fig. 28). These data show the significance of nascent or pyrophoric char, which becomes harder to ignite as the low activation energy sites are occupied and the heat of chemisorption (see Fig. 27) is reduced.

The carbon oxides produced by the chemisorption process finally decompose or further react to produce new reactive sites and CO and CO_2.

The mechanism and kinetics of the gasification reactions providing a mixture of CO and CO_2 are highly significant because, as shown below, at 500 °C the heat release associated with CO_2 formation is about four times greater than that of CO formation:

$$C + O_2 \longrightarrow CO_2 \qquad H_{500} = -88{\cdot}5 \text{ kcal/mole}$$
$$C + \tfrac{1}{2}O_2 \longrightarrow CO \qquad H_{500} = -22{\cdot}9 \text{ kcal/mole}$$

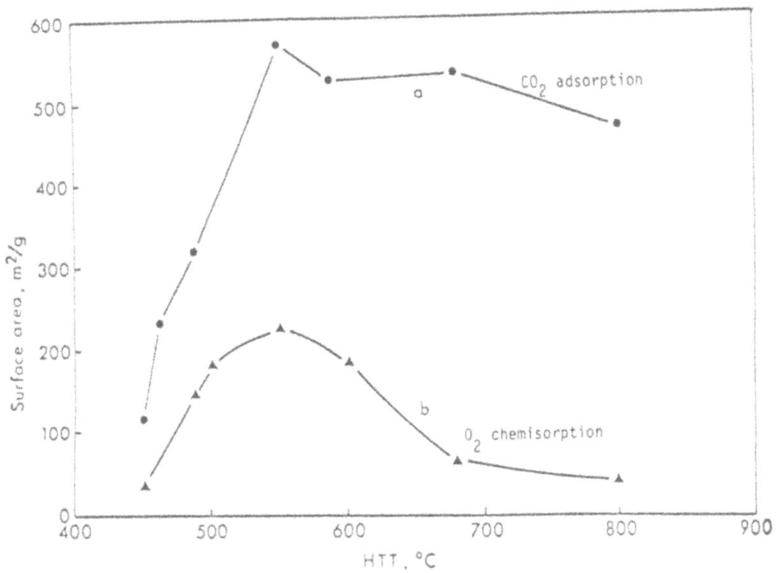

FIG. 29. Comparison between total surface area calculated from CO_2 adsorption at 25 °C (curve a) and that occupied by surface oxides calculated from oxygen chemisorption at 230 °C (curve b).

The surface areas of chars prepared from cellulose samples at different HTTs were determined by application of the Dubinin–Polanyi equation to CO_2 adsorption at room temperature and compared with the area occupied by surface oxides calculated from oxygen chemisorption at 230 °C. The results shown in Fig. 29 indicate that cellulosic chars have large surface areas which vary according to the HTT, and peak at about 550 °C. The surface oxides formed by chemisorption occupy only a portion of the total surface area, and the chemisorption also shows a peak for chars formed at about 550 °C, corresponding to the temperature of smoldering combustion.

Inorganic salts can produce different effects on smoldering combustion of cellulosic materials, ranging from total inhibition to enhancement. This effect is caused by variation in the rates of production of CO and CO_2 and the corresponding rates of heat release, which provides the driving force for this process. Table VIII[5] shows the empirical correlation between the effect of additives on the smoldering combustion and the rates of production of CO and CO_2 on oxidation of the char. A major question is whether the additives affect the properties of the char and formation of the reactive sites, or simply catalyze its oxidation.

TABLE VIII

EFFECT OF PRE-PYROLYSIS ADDITIVES ON RATE OF CHAR OXIDATION AND CO/CO_2 RATIO AT 450 °C IN AIR

Additive	*Smoldering behavior of char* (√, *smolders*/ ×, *does not*)	$(d(CO+CO_2)/dt)$max *at 450 °C* (*mmole g char*$^{-1}$ *min*$^{-1}$)	CO/CO_2 *yield at 450 °C*
—	√	5·7	1·01
HBO_3	×	4·6	1·69
$Na_2B_4O_7$	√	6·1	1·03
$(NH_4)_2HPO_4$	×	2·9	1·71
$(NH_4)_2SO_4$	×	3·9	1·43
$ZnCl_2$	×	5·1	0·90
$AlCl_3$	×	5·3	0·95
NaCl[a]	√	13·9	0·70
Na_2SO_4[a]	√	15·2	0·86

[a] Ignition occurred initially in these samples in CO, CO_2 measurements. Ignition peak prevented ΔH determination by DSC.

The effect of inorganic additives on the physical properties and chemical reactivity of chars was investigated by measurement of the surface areas by the CO_2 adsorption method as discussed before and the rates of oxygen chemisorption at 228 °C by the thermogravimetric method. These experiments were carried out with char samples prepared from treated and untreated cellulose pyrolyzed at 550 °C for 1·5 min (corresponding to the maximum surface area expected from Fig. 29). Table IX shows the surface area and the kinetic parameters of the chemisorption derived from the Elovich equation. These data show that the surface areas are remarkably

TABLE IX

ELOVICH KINETIC PARAMETERS FOR OXYGEN CHEMISORPTION AT 228 °C AND SURFACE AREA OF CHARS (HTT 550 °C) PREPARED FROM TREATED COTTON FABRIC

Additive	*T* (°C)	*a*	*b*	t_c (*min*)	t_{max} (*min*)	q_{max} (*mmole* O_2/g)	*Surface area* (m^2/g)
Untreated	228	0·74 ± 0·19	2·22 ± 0·20	0	62	2·1	359
NaCl	228	1·24 ± 0·45	2·90 ± 0·18	0	50	1·8	371
$Na_2B_4O_7$	228	1·14 ± 0·26	3·49 ± 0·20	0	48	1·6	352
H_3BO_3	228	0·44 ± 0·08	2·43 ± 0·18	0	110	1·8	324
$(NH_4)_2HPO_4$	228	0·39 ± 0·08	1·77 ± 0·04	0	120	2·5	351

similar and inorganic additives have no effect on the physical structure of the char, which appears to be a function of HTT. However, in contrast to the surface area, the kinetics of the chemisorption are quite different for the different additives and closely correlate with their effect on the smoldering combustion of the substrate. In view of this correlation, it was interesting to learn if the inorganic compounds produced a similar effect when added to the char after it is formed by the pyrolysis of cellulose rather than before. In such treatment, however, only boric acid had a noticeable effect.

These experiments suggested that the effect of inorganic materials may be due to their influence on the functionality of the char developed during pyrolysis. Thus, combustion enhancers such as sodium chloride could catalyze the formation of 'active sites'.

Measurement of the free spin concentration (free radicals) indicate a major dependence on the pyrolysis conditions rather than the presence of different additives. As shown in Fig. 30,[26] the free spin concentration is

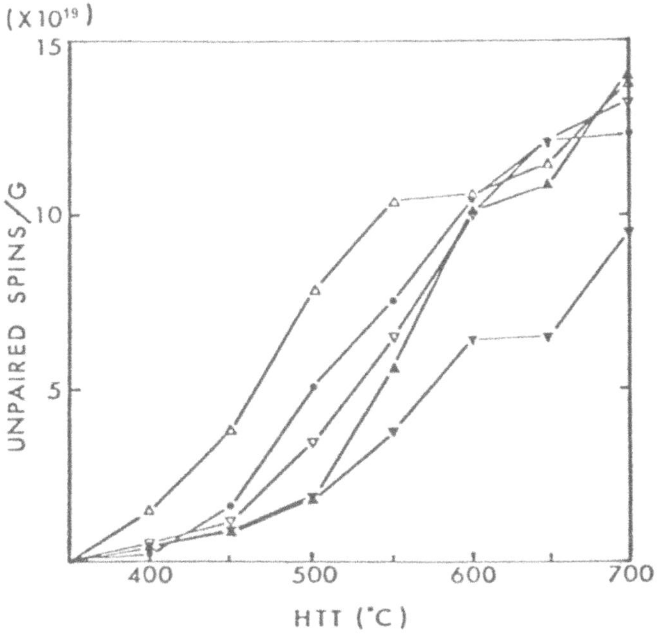

FIG. 30. Unpaired spin concentration in chars prepared from untreated cellulose (●) and cellulose treated with sodium chloride (△), borax (▲), diammonium phosphate (▽) and boric acid (▼), as measured under vacuum.

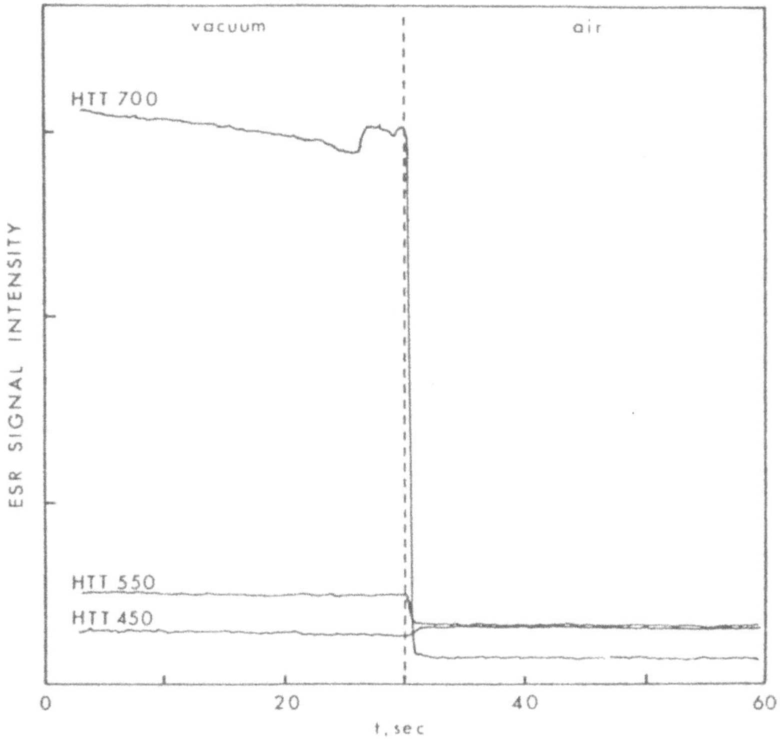

FIG. 31. Variation in relative intensities of derivative ESR curves with time for untreated cellulose char before and after exposure to air.

very high for nascent char under vacuum and increases with HTT. However, on exposure to air at room temperature the concentration of free spins drops precipitously (see Fig. 31). This effect is apparently due to physical adsorption, because it is reversible. On the other hand, if the char is heated in air the reduction is no longer reversible at room temperature unless the char is reheated under nitrogen. This phenomenon can be explained by the gasification of surface oxides formed by chemisorption and regeneration of new free radicals, as successive stages of the combustion process. These data indicate that free radicals play a significant role in oxidation of char. However, combustion behavior and catalysis of the char cannot be explained strictly in terms of changes in the free radical concentration, and the active sites must include other functionalities besides free radicals.

ACKNOWLEDGEMENTS

The author is pleased to acknowledge the contribution of the postdoctoral fellows, research associates and graduate students who are named in the following list of references, and the Colorado State University Regional NMR Center, funded by the National Science Foundation, Grant No. CHE 78-18581, for the ^{13}C-NMR. He is also grateful to the National Bureau of Standards Center for Fire Research for their interest and support of this program under Grant No. NB81NADA2066.

REFERENCES

1. Shafizadeh, F. (1968). *Advan. in Carbohydr. Chem.* **23**, 419.
2. Shafizadeh, F. and McGinnis, G. D. (1971). *Carbohydr. Res.* **16**, 273.
3. Susott, R. A., DeGroot, W. F. and Shafizadeh, F. (1975). *J. Fire Flamm.* **6**, 311.
4. Shafizadeh, F. and Fu, Y. L. (1973). *Carbohydr. Res.* **29**, 113.
5. Shafizadeh, F., Bradbury, A. G. W., DeGroot, W. F. and Aanerud, T. W. (1982). *Ind. Eng. Chem., Prod. Res. Develop.* **21**, 97.
6. Shafizadeh, F., Chin, P. S. and DeGroot, W. F. (1975). *J. Fire Flamm./Fire Retardant Chemistry* **2**, 195.
7. Susott, R. A., Shafizadeh, F. and Aanerud, T. W. (1979). *J. Fire Flamm.* **10**, 94.
8. Shafizadeh, F., Furneaux, R. H., Cochran, T. G., Scholl, J. P. and Sakai, Y. (1979). *J. Appl. Polym. Sci.* **23**, 3525.
9. Shafizadeh, F. and Stevenson, T. T. (in press). *J. Appl. Polym. Sci.*
10. Shafizadeh, F., Furneaux, R. H. and Stevenson, T. T. (1979). *Carbohydr. Res.* **71**, 169.
11. Shafizadeh, F. and Lai, Y. Z. (1972). *J. Org. Chem.* **37**, 278.
12. DeGroot, W. F. and Shafizadeh, F. *Fuel*, submitted for publication.
13. Antal, M. J. (1981). *Combust. Sci. and Technol.* **21**, 141.
14. Shafizadeh, F. (in press). Proceedings of the Ninth Cellulose Conference special edition. *J. Appl. Polym. Sci.*
15. Shafizadeh, F. (in press). *J. Pure and Appl. Chem.*
16. Shafizadeh, F. (1982). *J. Anal. and Appl. Pyrolysis* **3**, 283.
17. Shafizadeh, F. (1981). *Biomass Conversion Processes for Energy and Fuels*, Plenum Press, New York, p. 103.
18. Shafizadeh, F. (1975). *Appl. Polym. Symp.* **28**, 153.
19. Shafizadeh, F. and Bradbury, A. G. W. (1979). *J. Appl. Polym. Sci.* **23**, 1431.
20. Shafizadeh, F. and Lai, Y. Z. (1975). *Carbohydr. Res.* **42**, 39.
21. Bradbury, A. G. W., Sakai, Y. and Shafizadeh, F. (1979). *J. Appl. Polym. Sci.* **23**, 3271.
22. Shafizadeh, F. and Sekiguchi, Y. *Carbon*, submitted for publication.
23. Lai, Y. Z. and Shafizadeh, F. (1974). *Carbohydr. Res.* **38**, 177.
24. Bradbury, A. G. W. and Shafizadeh, F. (1980). *Combustion and Flame* **37**, 85.
25. Bradbury, A. G. W. and Shafizadeh, F. (1980). *Carbon* **18**, 109.
26. DeGroot, W. F. and Shafizadeh, F. (in press). *Carbon.*

12

Product Formation in the Pyrolysis of Large Wood Particles

WAI-CHUN RICKY CHAN, MARCIA KELBON and BARBARA B. KRIEGER
Department of Chemical Engineering, University of Washington, Seattle, Washington, USA

The potential for energy and chemicals production from renewable resources has re-established interest in thermal conversion of biomass. If the devolatilization of a single wood particle can be predicted accurately and optimized, process design and scale-up of biomass conversion processes can benefit from the methodology developed for heterogeneous catalysis (e.g. ref. 1). Although using small particles might be advantageous for certain gasification processes, size reduction of fibrous wood chips is difficult and expensive.[2] Therefore, large particles or a distribution of particle sizes will likely be employed. Knowledge of the effects of particle size, particle thermal properties and wood anisotropy on pyrolysis product distribution and volatiles release rate will be important in preparing an optimal biomass feedstock for any thermal conversion process.

Studies of combustion and pyrolysis of thermally thick wood particles have been conducted primarily for purposes of fire safety.[3–5] (Thermally thick is intended to imply a spatially non-uniform temperature profile in the solid.[6]) Although the effects of physical variables on heat transfer behavior during pyrolysis and combustion have been extensively investigated and modeled,[6–8] inclusion of multiple chemical reactions into such models has been avoided because of the lack of intrinsic kinetic data and mathematical complexity.[9,10] Owing to experimental complexity, time-resolved biomass pyrolysis product distributions have not often been measured,[11] although ultimate yields as functions of reaction conditions have been determined.[12–15]

The scope of this investigation is to determine experimentally the detailed product distribution (tar, gases, char and components therein) as a function of time from the rapid pyrolysis of a single, well-characterized biomass pellet using a Box–Behnken statistical experimental design for

TABLE I
LEVELS OF FACTORS IN EXPERIMENTAL DESIGN

Variable (factor)	*Factor levels*		
	Low (−)	Medium(0)	High(+)
(1) External heat flux ($cal/cm^2/s$)	2	4	6
(2) Pellet length (cm)	0·5	1·0	1·5
(3) Wood grain direction (relative to heat flux)	∥	⊥∥[a]	⊥

[a] This direction corresponds to a chord in a circular cross-section.

physical and chemical variables. The independent variables reported on in this article are: external heat flux intensity; pellet length; and wood grain orientation which affects thermal properties (see Table I). The dependent variables are generally the product compositions as functions of time and the independent variables (see Table II). Temperature and density profiles as functions of time appear elsewhere.[16] Experimental results are compared to a mathematical model.

TABLE II
TABULATION OF RESULTS FOR DESIGN CENTERPOINT[a]

Response Ultimate yield ($t = \infty$, wt %)	*Replicate No.* 1	2	3	*Standard error (s)*
Char	22·5	22·1	23·6	0·8
Measured tar (+H_2O)	52·4	43·5	42·9	5·0
Tar by difference	59·1	58·6	60·3	0·9
Tar (−H_2O; by difference)	44·4	43·9	44·8	0·4
Gas	18·4	19·3	16·1	1·6
CH_4	1·1	1·2	0·8	0·2
C_2H_2	0·006	0·04	0·0	0·02
C_2H_4	0·4	0·5	0·25	0·1
C_2H_6	0·24	0·27	0·18	0·05
CO	8·9	9·3	7·6	0·9
CO_2	7·8	7·7	7·3	0·3
H_2O	14·6	14·7	15·6	0·5

[a] Experimental conditions for centerpoint are (0, 0, 0) corresponding to 4 $cal/cm^2/s$ heat flux, 1·0 cm pellet, with ⊥∥ grain direction.

1. MODEL DESCRIPTION

At high heat fluxes, intraparticle heat transfer during biomass pyrolysis is one to two orders of magnitude slower than mass transfer (volatiles outflow) or chemical reaction using an analysis similar to Russel *et al.*[17] A mathematical model reflecting this is briefly presented here and detailed elsewhere.[16,18] It predicts volatiles release rate and composition as a function of biomass physical properties, and simulates the experiments described herein in order to assess which kinetic models are adequate to predict individual product formation rates. A diagram of the modeled phenomena appears in Fig. 1.

A one-dimensional approach is used to describe transport processes and reaction in a cylindrical pellet heated on one face by a time-varying radiation source. Convective cooling by the helium carrier gas and radiative heat loss at both faces are also treated. Within the pellet, heat is

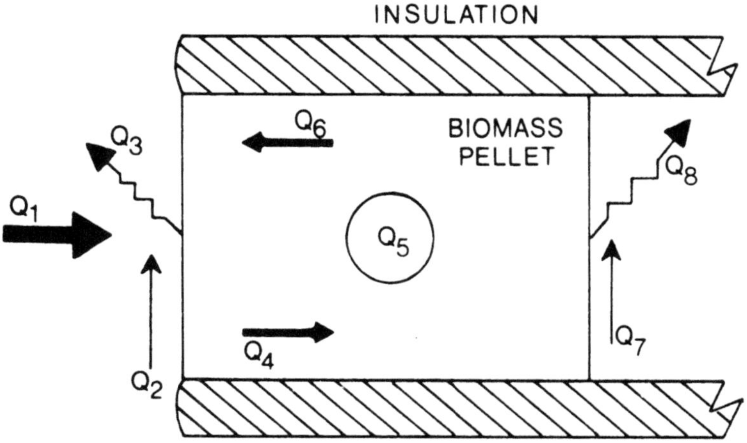

FIG. 1. Schematic of model—unsteady state 1-D energy balance for use with mass balance and kinetics. $Q_1(t)$: cylinder heated by constant radiation on 1 face; net heat to surface = f(time) owing to volatiles absorption; absorption is calibrated and used as model input. Q_2: carrier quenches volatiles; cools surface. Q_3: radiation loss from surface. Q_4: net flux is conducted to cylinder interior. Q_5:heat of reaction released according to devolatilization kinetics. Q_6: outflowing volatiles are in thermal equilibrium and cool layer behind reaction front; 1 direction flow only. Q_7: convective cooling. Q_8: radiation loss from unheated face. Thermal conductivity varies linearly with density (porosity). Heat capacity is a function of temperature and density. Porosity varies from that of wood to that of char. Heat of reaction is not well known; treated as a parameter. Unsteady mass and energy equations solved by finite difference method and Gear package.[21]

transferred by conduction and absorbed or released by reaction. The volatiles from chemical reaction are assumed to be in thermal equilibrium with the solid, and cool it during volatiles outflow. They are assumed to flow toward the heated face only, which has been experimentally verified to be a valid approximation. This analysis extends previous models[7-9] by including additional effects, more extensive treatment of variable thermal properties and different boundary conditions. Most important, series–parallel reactions are included because kinetic parameters have recently become available[11] which potentially describe intrinsic reaction rates. With the introduction of complex kinetics into this realistic heat transfer model, the effects of mass transfer (owing to wood grain and porosity) on product distribution can be assessed and compared to experiment.

For thermal conversion process design, volatiles release rate is most important to predict adequately. For the present, it is computed as the instantaneous value of the change in density integrated over the pellet length, or as a mass transfer resistance within the wood.[19] The heat of reaction for wood is not well known[20,12] and depends on the extent of secondary reaction within the wood (thus also dependent on particle size[5]). In this model, 80 cal/g (endothermic) is used or it is treated as a parameter.

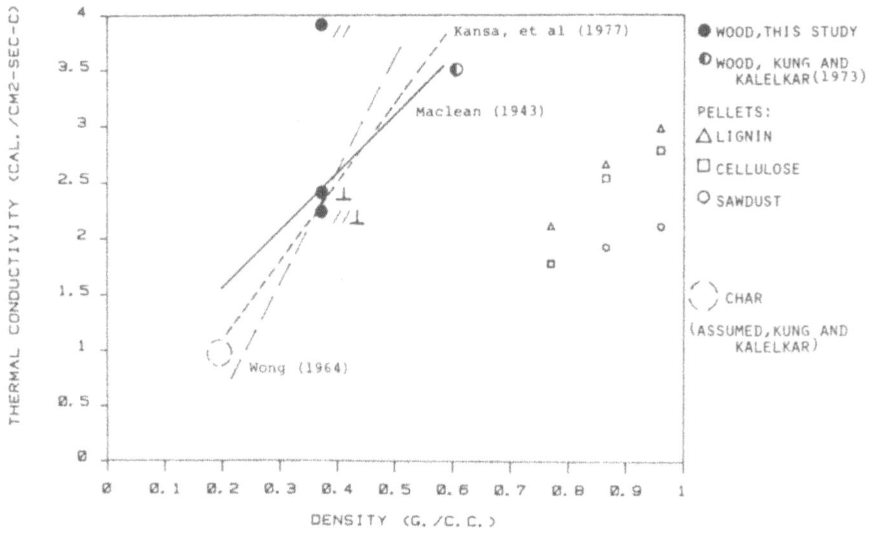

FIG. 2. Biomass thermal conductivity presented as a function of density for wood, cellulose, lignin and sawdust pellets. Parallel grain (||) and perpendicular grain (⊥) thermal conductivities are shown.

Initial thermal conductivity and thermal diffusivity are assumed to be linear functions of the density and are evaluated[16] using the experimental results in Fig. 2. The porosity and heat capacity are linear functions of their initial and final values. The equations are solved in dimensionless form using the method of finite differences.[21]

2. EXPERIMENTAL APPARATUS AND PROCEDURE

The experimental apparatus consists in part of a single pellet reactor schematically shown in Fig. 3. A high intensity xenon arc lamp provides a constant external heat flux (4–12 cal/cm^2/s) to one face of a well-characterized wood pellet which is carefully positioned in the reactor. This

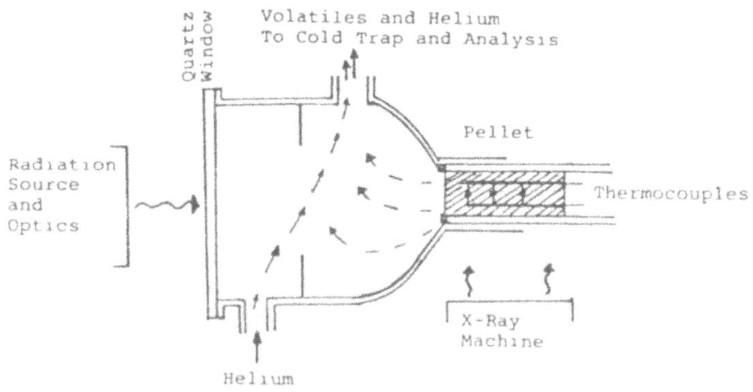

FIG. 3. Schematic diagram of the experimental apparatus mounted on optical bench—pyrometer not shown.

heat flux range is judged to include combustion level fluxes according to the comments of Shivadev and Emmons[22] and Wesson *et al.*[23] The reduction in net radiation arriving at the pellet surface during high rates of volatiles outflow is quantified using a heat flux gauge.[19,24] The experimentally determined time variation of the net heat flux is also used as input to the model.

The 1 cm dia. pellet of Oregon lodgepole pine is carefully lathed from a uniform slab of knot-free wood with attention to the grain direction. The pellet is dried, resulting in an initial pellet density of 0·37 g cm^3. Thermocouples (chromel-alumel, 0·005 in. dia.) are placed at three depths, and the wood is inserted in the reactor. The heated face temperature is measured by an infrared pyrometer calibrated with thermocouples at the pellet surface. The unheated face has a slight positive pressure applied to ensure uni-directional volatiles outflow. The evolving gases are swept by helium carrier gas through a glass reactor designed to prevent tar condensation on the window facing the lamp. Tracer studies on this reactor show a residence time distribution (RTD) characterized by a Peclet number of 3–5.[1]

During an experiment, volatile products evolving from the pyrolyzing pellet are rapidly quenched by a high helium flow rate and a dry ice–acetone trap. Uncondensed gases are sampled above the trap as a function of time by two 34-port programmable gas chromatographic (GC) sampling valves in series or in parallel, which also trigger an X-ray machine. The time and spatial variation in density of the pellet is measured by an X-ray technique.[25] X-ray photographs show no evidence of pellet fracture. After the experiment is completed, the gas samples are sequentially analyzed by gas chromatography. Tar samples and reactor washings are analyzed by a gas chromatograph coupled to a mass spectrometer and/or liquid chromatography. A slight modification of the procedure in ref. 18 is used. Normally absent from the product slate are hydrogen and light tars for reasons described in ref. 19. Char and unreacted portions are weighed and analyzed.

Mass balances on the experiment are generally 'close' to within 5 %. Tar is plotted as water-containing condensables. Gas yields and char yields are more accurate than tar yields, as seen in Table II. The major difficulty occurs in quantifying the tar that condenses on the reactor walls before the cold trap. The experimentally measured tar in Table II is that weighed from the cold trap. When reactor washings (not done quantitatively) are combined with this, the component state is complete, but the tar fraction composition is probably less quantitative than other fractions.

Since product analysis is lengthy and complex phenomena are occurring simultaneously,[23] the experiments are performed according to an incomplete factorial experimental design[26] for which preliminary model runs indicated suitable factor levels. These are shown in Table I. This design is ideally suited to determination of non-linear effects that groups of experimental variables may have on pyrolysis product yields. An additional advantage of larger experimental designs is hidden replication, although for incomplete factorials not every combination of independent variables is run. Replication is sufficient, however, to determine empirical model coefficients and predict trends. The trend lines shown in Figs 5 and 6 represent regressions of all the 15 experimental results in the design. The standard errors of the product yields from the centerpoint replicates are presented in Table II as typical results.

3. EXPERIMENTAL RESULTS AND DISCUSSION

The experimental temperature profiles at three depths in the pellet are contrasted in Fig. 4 for heat fluxes of 2 and 6 cal/cm^2/s. Data contrasting the effect of particle length show smaller temperature profile differences between 0·5 cm and 1·5 cm pellets, for a similar tripling of the independent variable. The reduction in the surface temperature at about 20 s is a result of outflowing volatiles reducing the incoming radiation as well as convectively cooling the pellet surface. Inspection of the temperature at several depths at a given time indicates that the temperature gradient becomes less steep as the reaction front moves into the pellet. After approximately 100(+) s, the temperatures are nearly constant with time. While most kinetic studies generally use temperature as an independent

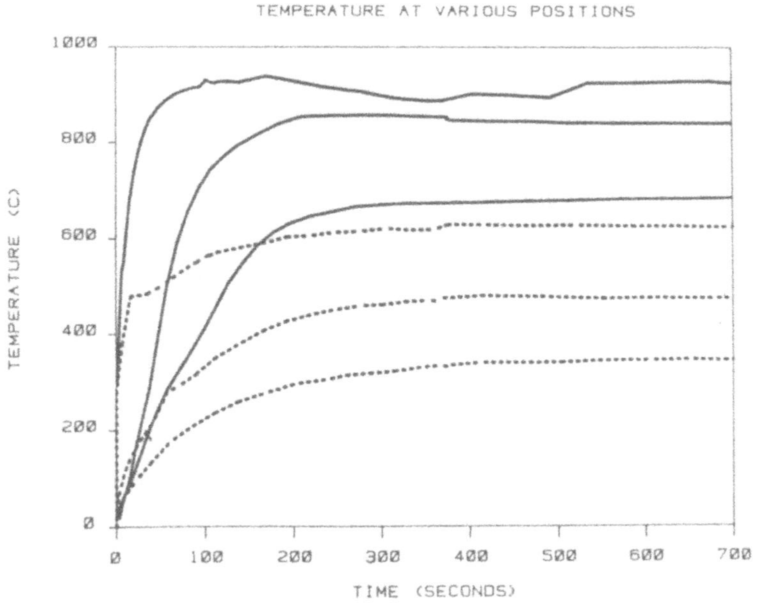

FIG. 4. Experimental temperature variation as a function of time at several depths in the pellet heat flux: 6 cal/cm^2/s (——); 2 cal/cm^2/s (---). Temperatures at the surface, 0·2 cm and 0·4 cm are presented while the data at 0·6 cm, although available, was omitted for clarity.

variable, Fig. 4 demonstrates that for these studies it is a function of time and position for any given experiment, as well as a function of applied heat flux, particle length and thermal properties.

If one ignores moisture content (dried samples are used), then grain direction and density determine the initial thermal diffusivity and conductivity of the wood particle. For the results presented in Figs 5, 6, 8 and 9, the grain is parallel to the applied heat flux (minimal mass transfer resistance) and the initial sample density is that of oven-dried lodgepole pine, 0·37 g/cm^3. The ultimate pyrolysis gas, tar and char yields (time > 12 min) are plotted in Figs 5 and 6 as functions of incident initial heat flux and pellet length respectively.

As seen from heat transfer analyses such as Murty and Blackshear[27] or Wesson *et al.*,[23] incident heat flux q, particle 'thermal' thickness L and, to a lesser extent, wood thermal properties determine ignition time or mass loss rate during combustion. In the former investigation, simple ratios of physical variables were used for data correlation. In the latter, however, an error function and fractional powers of variables were employed to correlate ignition time. Figures 5 and 6 also show a non-linear dependence of individual *products*, i.e. gas, tar and char, on both q and L. However, the

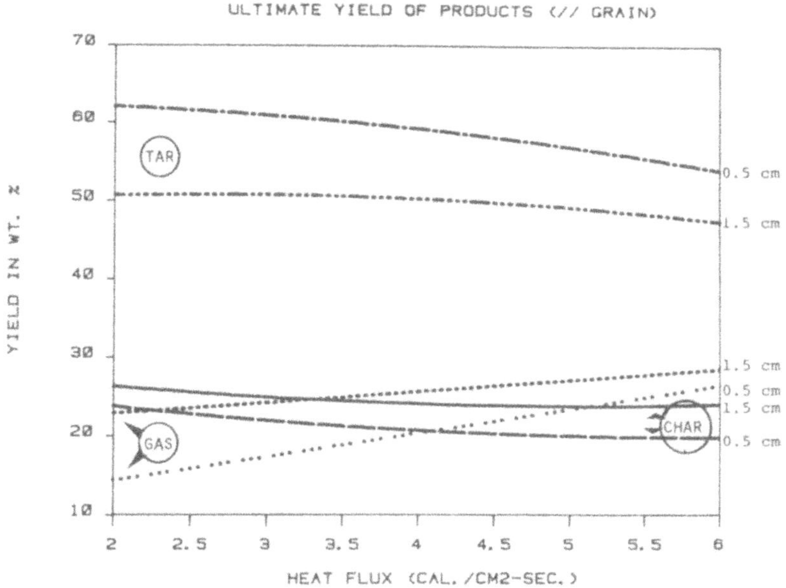

FIG. 5. Experimental ultimate gas, tar and char yields as a function of heat flux for two pellet lengths.

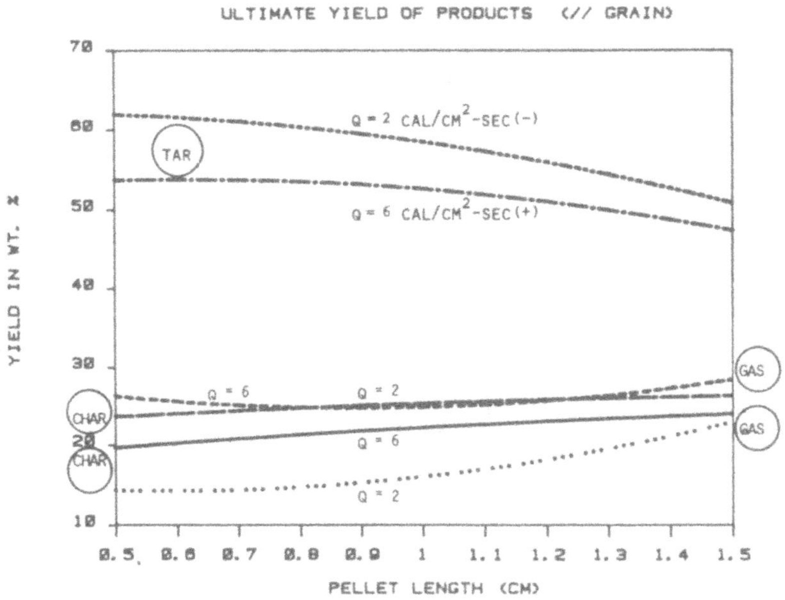

FIG. 6. Experimental ultimate gas, tar and char yields as a function of pellet length for two incident initial heat fluxes.

figures demonstrate that the same functional form does not apply to both gas and tar, although both are volatiles and thus contribute to the mass loss rate correlated by previous investigators. As observed by others, the tars behave in a fashion consistent with being chemical intermediates in series–parallel reactions during devolatilization.

Initially, a candidate function for correlation of pyrolysis yields from large particles was q/L, akin to an effective volumetric heating rate. However, for quantitative correlations, q varies with time and position in the pellet as does temperature. Similarly, it is difficult to identify the proper quantitative characteristic length for correlation purposes, since the char layer depth, reaction zone thickness and initial particle thickness all determine the local heat flux and, therefore, the local temperature. These characteristic lengths also vary with time and position. Nevertheless, initial incident heat flux divided by pellet length provides a ranking scale for comparing large particle heating rates to small particle heating rates and assessing trends in product yields. When this is done *qualitatively*, the large particle reaction product data we have obtained follow the trends described in ref. 28 moderately well. However, *quantitative* correlations indicate some product yields depend on $q^n L^m$ where $n > m > 0$. In light of the mathematical model results, current efforts focus on correlation of yield data with both time-averaged and length-averaged temperature in the pellet as well as with semi-empirical functions of the experimental controllables. Satisfactory correlations are sought for ultimate pyrolysis yields as well as time-dependent product concentrations.

While the externally applied heat flux and pellet dimensions certainly influence the ultimate product distribution, it can be seen in Fig. 7 that they also alter the instantaneous products or rate of volatiles release as a function of time. This figure compares the influence of parallel (‖) and perpendicular (⊥) heating relative to wood grain direction for two heating rates and an intermediate pellet length of 1 cm. Some of the 'structure' or relative minima and maxima in the experimental volatiles release rate curves is due to intermittent sampling. Since the more pronounced 'structure' in the experimental release rate curves varies for different components and conditions, it is judged to be real.

In Fig. 7 wood grain perpendicular to the heat flux results in lower effective thermal conductivity (Fig. 2) and increased resistance to volatiles

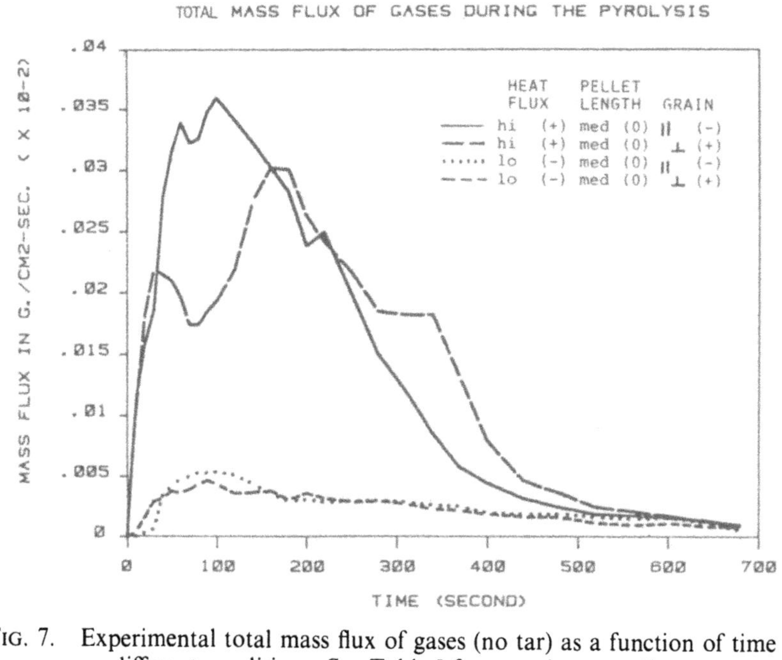

FIG. 7. Experimental total mass flux of gases (no tar) as a function of time for different conditions. See Table I for experiment coding.

outflow. Consistent with heat transfer arguments, the lower thermal conductivity causes slightly higher surface temperatures and steeper temperature gradients. Thus the initial volatile release rate in Fig. 7 rises faster and the maximum of mass flux is lower as a result of lower interior temperature for heating perpendicular to the grain. There is about 20% shrinkage of the char layer which has been observed after the pyrolysis of perpendicular grain wood. The thinner char layer allows more rapid heat transfer to the unreacted wood substrate. Consequently, an additional volatiles surge occurs at a later stage. Heating parallel to the grain presents less resistance to volatiles outflow and therefore the volatiles peak appears earlier relative to the profile for heating perpendicular to the grain in Fig. 7. At lower heat fluxes (2 cal/cm^2/s) the effect of grain direction on volatiles release rate is small.

The temporal behavior of individual gas species is presented to illustrate its potential for optimizing product distribution. Figures 8 and 9 present the instantaneous production rates of CH_4 and C_2H_4 respectively for experiments in which the wood grain was oriented parallel to the incident heat flux and the heating rate and pellet length were varied. Similar plots of CO and higher hydrocarbon production rates appear in ref. 18. A tripling

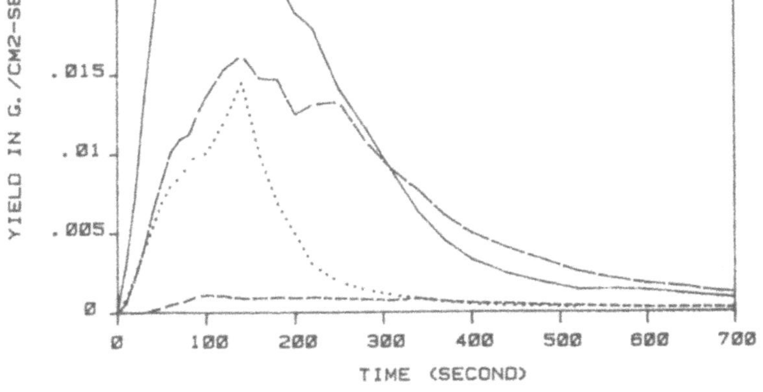

FIG. 8. Experimental CH_4 production (evolution) rate as a function of time for different conditions (see Table I).

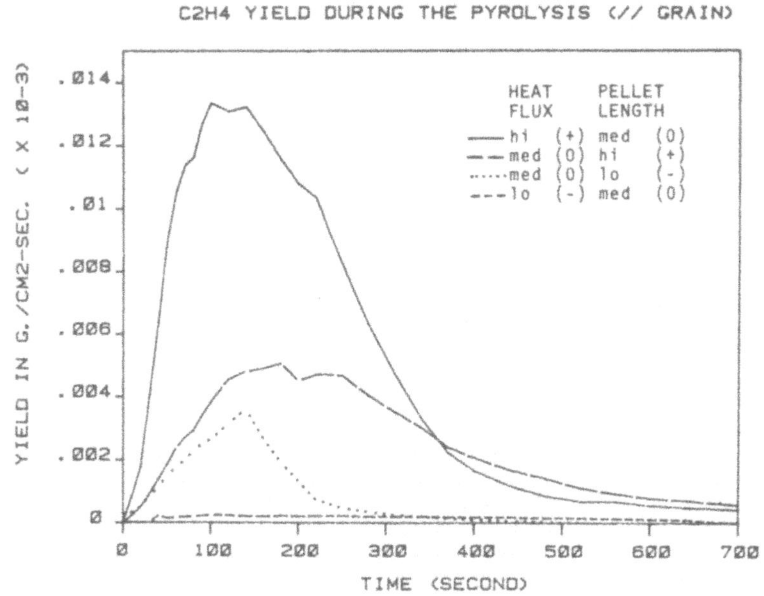

FIG. 9. Experimental C_2H_4 evolution rate as a function of time for different conditions (see Table I).

of the pellet length (Table I) is seen to extend the period over which both CH_4 and C_2H_4 are devolatilized. While division of the time scale by the normalized pellet length brings the CH_4 release rates almost into coincidence both in amplitude and time, this is not true for C_2H_4. The average level of ethylene production is enhanced more than proportionately in the longer pellet. Examination of Fig. 4 reveals the temperature profiles rise to an almost constant level after about 100–200 s. At 100–200 s, when the ethylene production rate for both pellet thicknesses heated at 4 cal/cm^2/s (medium q) is maximum, the surface temperature is no higher than earlier in the pyrolysis. The volatiles outflow from the interior keeps the growing char layer at relatively constant temperature. Thus the increased ethylene production rate cannot be attributed to a phenomenon involving higher surface temperatures for longer pellets, but to either increased residence time of product gas from the reaction zone in the pellet or to contact with the char layer as the gases flow outward.

A tripled initial heat flux at constant pellet length (similar to uppermost and lowermost curves in Figs 8 and 9 for total mass flux, but not presented here) causes a more-than-proportionate increase in the peak production rate of both CH_4 and C_2H_4, although little change in time at which the peak occurs is seen. This supports the concepts put forth in ref. 28 concerning heating rate in processes for chemicals production from biomass.

4. MODEL PREDICTIONS AND DISCUSSION

A successful model of wood pyrolysis will include appropriate physical phenomena as well as employ realistic values of variable physical properties and reaction rate parameters. The heat transfer phenomena included in the current mathematical model adequately predict the experimental temperature profiles such as those in Fig. 4[16] as functions of time and the independent variables. The initial thermal properties used are realistic since they are measured values as functions of density (Fig. 2) or consistent with those used by others.[8,29,30] However, the predicted temperature histories are sensitive to the magnitude of the char thermal properties which, to our knowledge, have not been measured. Attempts at measurement were unsuccessful since the char was too fragile to handle and the properties varied with position. Thus, for modeling purposes, char thermal conductivity from Kung[8] and that of charcoal were employed. However, as seen in Fig. 2, thermal conductivity depends on density (porosity) which in turn depends temporally on the weight loss kinetics used in the model. Birkebak and Ozil[31] have indicated a T^3 dependence of thermal conductivity on temperature when radiation is the predominant heat transfer mode in porous media. Thus, in the simulation there is a strong coupling or 'feedback'. Small errors in the char thermal properties cause large deviations in the temperature at the pellet interior for long reaction times, which in turn causes large deviations in the chemical reaction rates, amplifying further the errors in the density/thermal property functions. Comparisons of predicted and experimental density profiles measured as functions of time using X-rays were made.[19] However, favorable density comparisons between model and experiment only validate the choice of apparent weight loss rate parameters. The apparent weight loss is still an algebraic sum of the reactions to produce the volatile species and the char forming reactions.

Determination of the simplest, yet appropriate, kinetic mechanism to predict product formation rates presents a challenge. Not only do the reaction rates and mechanism postulated in the model alter the physical properties, but they have a dramatic effect on the volatiles production rate as a function of time. In an earlier article[18] it was demonstrated that a single weight loss reaction using realistic rate coefficients does not adequately predict the shape of the volatiles production rate as a function of time (e.g. Figs 7–9). An alternative kinetic mechanism for the volatiles' reaction is shown in Fig. 10. The predicted shape and amplitude compare favorably with the experimental data. The second maximum is attributed to an inappropriate value of the char thermal conductivity as described by Kung.[8] Efforts to model the thermal properties more realistically are

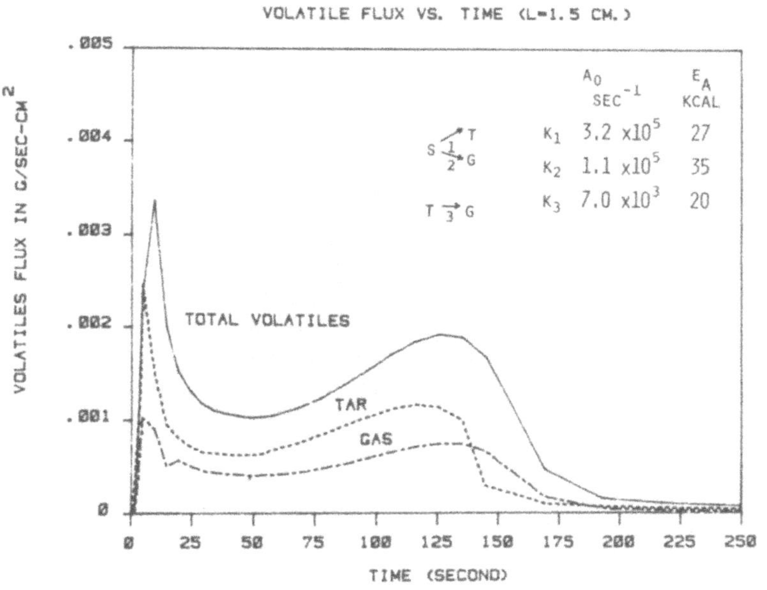

FIG. 10. Simulated evolution rate of volatiles using a complex series–parallel reaction mechanism. Presented in dimensional form.

discussed below. It should be noted that while total volatiles are plotted in Fig. 10, only the curve marked 'gas' should be compared to experimental data. Unfortunately, tars, though volatile, are collected batchwise in a cold trap, making only integrated experimental tar yields available for comparison to the model. Thus, the predicted early sharp tar volatiles peak in the model is not time-resolvable in our apparatus.

Further refinement is needed for the current kinetic mechanism and thermal property variation with temperature in the model. One approach used by others is to fix the final char density (thus fixing the char thermal properties) by using experimental char yields. This avoids the feedback problem described earlier. However, char yields and density surely depend on reaction conditions (e.g. ref. 32) and thus each model run requires prior experimental data. A more useful approach is to predict char yield in spite of the lack of evaluated kinetic data, albeit with the mathematical difficulties described above. An additional reaction ($S \rightarrow C$) can be added to the mechanism in Fig. 10 and rate parameters estimated from the sparse data available. This kinetic approach is discussed in detail in ref. 19. Although the four-reaction mechanism is the simplest, realistic lumped parameter mechanism suitable for comparison to our data, actual data from small particles judged to represent intrinsic kinetics are not often available over the complete temperature range in this study. While the data of Hajaligol *et al.*[11] are extensive, their model assumes parallel reactions form all products and is not suited to our purposes. Thus, our current modeling efforts focus on arriving at an adequate kinetic mechanism that will lend insight into the important topic of char formation rate.

A phenomenon not currently modeled is forced convection[7] and/or a mass transfer resistance to volatiles outflow which have potentially important effects on a series–parallel chemical mechanism involving tar. This extension is planned. However, while this phenomenon might predict enhanced production of gas volatiles at the expense of tar, the data on heating perpendicular to the wood grain (Fig. 7) illustrate the greater importance of predicting correctly the heat transfer and temperature profile which affect the product formation rates in an exponential way.

5. SUMMARY AND CONCLUSIONS

The thermal properties, heat flux and pellet length are shown to interact in a non-linear way to alter the ultimate yields and instantaneous rate of gas production from a rapidly pyrolyzing large pellet of biomass. For purposes of predicting ignition behavior of wood, other studies have adequately explained initial volatiles release based on heat transfer arguments. Experimental data presented here indicate that the later volatiles release can differ in composition depending on particle size and thermal properties. It is felt that further model improvement will result from adequate chemical kinetic representation of the char production rate rather than from further refinement of the transport phenomena in a large, pyrolyzing or burning wood particle. These comments apply to the one-dimensional model discussed here.

As pellet length and heat flux are varied in experiments, the gas product contains changing proportions of chemicals such as CH_4 and C_2H_4 as functions of time. This indicates a potential for optimization of product distribution from thermally thick biomass particles in the event that costly size reduction must be avoided.

ACKNOWLEDGEMENTS

The authors would like to acknowledge the Solar Energy Research Institute and the Department of Energy for financial support of this project. In addition the Department of Chemical Engineering, University of Washington, provided facilities and financial aid. The model calculations were performed with the aid of computer codes developed by Professor Bruce Finlayson whose suggestions are also gratefully acknowledged.

REFERENCES

1. Carberry, J. J. (1977). *Chemical and Catalytic Reaction Engineering*. McGraw Hill, New York.
2. Bain, R. (1979). Beneficiation of biomass for gasification and combustion, a survey of biomass gasification, Vol. II. Principles of Gasification, SERI/TR-33-239.
3. Kung, H. C. (1974). The burning of vertical wooden slabs. 15 Comb. Symp. 243–53.
4. Kanury, A. M. (1973). Rate of charring combustion in a fire. 14th Symp. (Intl.) on Combustion, The Combustion Institute.
5. Roberts, A. F. (1971). Problems associated with the theoretical analysis of the burning of wood. 13th Symp. (Intl.) on Combustion, Combustion Institute, 893.
6. Kanury, A. M. and Blackshear, P. L. Jr (1970). Some considerations pertaining to the problem of wood burning. *Comb. Sci. and Tech.* **1**, 339.
7. Kansa, E. J., Perlee, H. E. and Chaiken, R. F. (1977). Mathematical model of wood pyrolysis including internal forced convection. *Comb. Flame* **29**, 311.
8. Kung, H. C. (1972). A mathematical model of wood pyrolysis. *Comb. and Flame* **18**, 185–95.
9. Panton, R. L. and Rittman, J. G. (1970). Pyrolysis of a slab of porous material. 13th Symposium on Combustion (Intl), Combustion Institute, 881.
10. Fan, L. (1978). Volume reaction model for pyrolysis of a single solid particle accompanied by a complex reaction. *Can. J. Engng.* **56**, 603.
11. Hajaligol, M. R., Peters, W. A., Howard, J. B. and Longwell, J. P. (1980). Product compositions and kinetics for rapid pyrolysis of cellulose. *Proc. Specialist's Workshop on Fast Pyrolysis of Biomass.* SERI, Golden, Col. SERI/CP-6221096, 215–36.

12. Antal, M. J., Friedman, H. L. and Rogers, F. E. (1980). Kinetics of cellulose pyrolysis in N_2 and steam. *Comb. Sci. and Tech.* **21**, 141.
13. Milne, T. (1979). Pyrolysis—the thermal behavior of biomass below 600 °C. *A Survey of Biomass Gasification*, Vol II. SERI TR-33-239.
14. Hileman, F. D., Wojcki, L. H., Futrell, J. H. and Einhorn, I. N. (1976). Comparison of the thermal degradation products of cellulose and Douglas fir under inert and oxidative environments. *Thermal Uses and Properties of Carbohydrates and Lignins Symposium*, F. Shafizadeh *et al.* (Eds). Academic Press, New York, pp. 49–71.
15. Shafizadeh, F. (1980). Introduction to pyrolysis of biomass. *Proc. Specialists' Workshop on Fast Pyrolysis of Biomass*, SERI, Golden, Colo., October, SERI/CP-622-1096, p. 79 (and references cited therein).
16. Chan, W. C. R. (1982). Analysis of physical and chemical processes in the pyrolysis of a single large pellet of biomass. Ph.D. thesis, Chemical Engineering, University of Washington.
17. Russel, W. B., Saville, D. A. and Greene, M. I. (1979). A model for short residence time hydro-pyrolysis of single coal particles. *A.I.Ch.E. J.* **25**, 65.
18. Chan, R. C. and Krieger, B. B. (1982). Analysis of chemical and physical processes during devolatilization of a single large particle of wood. 7th Intl. Symposium on Chemical Reaction Engineering, Boston, October 4–6, 1982, published as *ACS Symposium Series.*
19. Chan, R. and Krieger, B. B. (1983). Modeling of physical and chemical processes during pyrolysis of a large biomass pellet with experimental verification. ACS Fuel Div. Preprints, **28**(5), presented at ACS Meeting, Washington, D.C., August 29–September 3.
20. Kung, H. C. and Kalelkar, A. S. (1973). On the heat of reaction in wood pyrolysis. *Comb. and Flame*, **20**, 91.
21. Finlayson, B. A. (1981). *Non-linear Analysis.* Wiley, New York.
22. Shivadev, U. K. and Emmons, H. W. (1974). Thermal degradation and spontaneous ignition of paper sheets in air by irradiation. *Comb. Flame* **22**, 223.
23. Wesson, H. R., Welker, J. R. and Sliepcefich, C. M. (1971). The piloted ignition of wood by thermal radiation. *Comb. Flame* **16**, 303–10.
24. Kashiwagi, T. (1981, 1964). Private communication, National Bureau of Standards, Washington D.C. (1981); *Fire Safety J.* May 1964, 185–200.
25. Lee, C. K., Chaiken, R. F. and Singer, J. M. (1976). Charring pyrolysis of wood in fires by laser stimulation. 16th Symposium on Combustion, Comb. Institute, 1459.
26. Box, G. E. P. and Behnken, D. W. (1960). Some new 3-level designs for the study of quantitative variables. *Technometrics* **2**(4), 455.
27. Murty, K. A. and Blackshear, P. L. Jr (1967). Pyrolysis effects in the transfer of heat and mass in thermally decomposing organic solids. 11th Symp. (Intl.) on Combustion.
28. Diebold, J. P. and Scahill, J. (1982). Progress in the entrained-flow, fast-ablative pyrolysis of biomass. Annual Report, October 80–September 81, Solar Energy Research Institute Report, SERI/PR-234-1456, p. 6.
29. Wong, P. T. Y. (1964). *Forest Products J.* May.

30. McClean, M. C. (1943). *Wood Handbook*. US Forest Service and Forest Products Laboratory, Madison, WI.
31. Birkebak, R. C. and Ozil, E. (1979). Theoretical consideration of radiative conductivities in fibrous insulation, *Studies in Heat Transfer*. McGraw Hill, New York.
32. Mackay, D. M. and Roberts, P. V. (1982). The influence of pyrolysis condition on yield and microporosity of lignocellulosic chars. *Carbon* **20**(2), 95–104.

13

The Pyrolysis under Vacuum of Aspen Poplar

CHRISTIAN ROY, BRUNO DE CAUMIA, DOMINIQUE BROUILLARD and HUGUES MÉNARD*

*Departments of Chemical Engineering and *Chemistry, Université de Sherbrooke, Sherbrooke, Québec, Canada*

1. INTRODUCTION

Pyrolysis is a reaction common to all thermochemical conversion reactions of biomass. When conducted under reduced pressure conditions, the thermal decomposition reactions are of fundamental interest since this permits one to get an insight into the primary chemical reactions which occur during the initial breakdown of the biomass macromolecules.

The theory and practice of the carbonization of wood and its major components have been well covered in the literature a number of years ago.[1–6] The chemistry of vacuum pyrolysis reactions of biomass has also been studied in the past in other laboratories. Such work led, for instance, to useful information about the mechanism of thermal decomposition of cellulosic materials.[7] The approach used by the authors in the present study was more phenomenological and focused on the role of a few parameters relevant to thermal decomposition reactions of biomass under a low vacuum. The practical importance of this study relied on the potential of the vacuum pyrolysis approach as a means of considerably enhancing the yields and quality of condensable vapours relative to the gaseous and solid products; this application aspect of our work has been treated elsewhere.[8]

2. EXPERIMENTAL

The data presented in this paper were obtained in a bench scale, low vacuum pyrolysis reactor. A schematic of the experimental set-up is depicted in Fig. 1.

The vacuum pyrolysis experiments were conducted in a batch mode in a

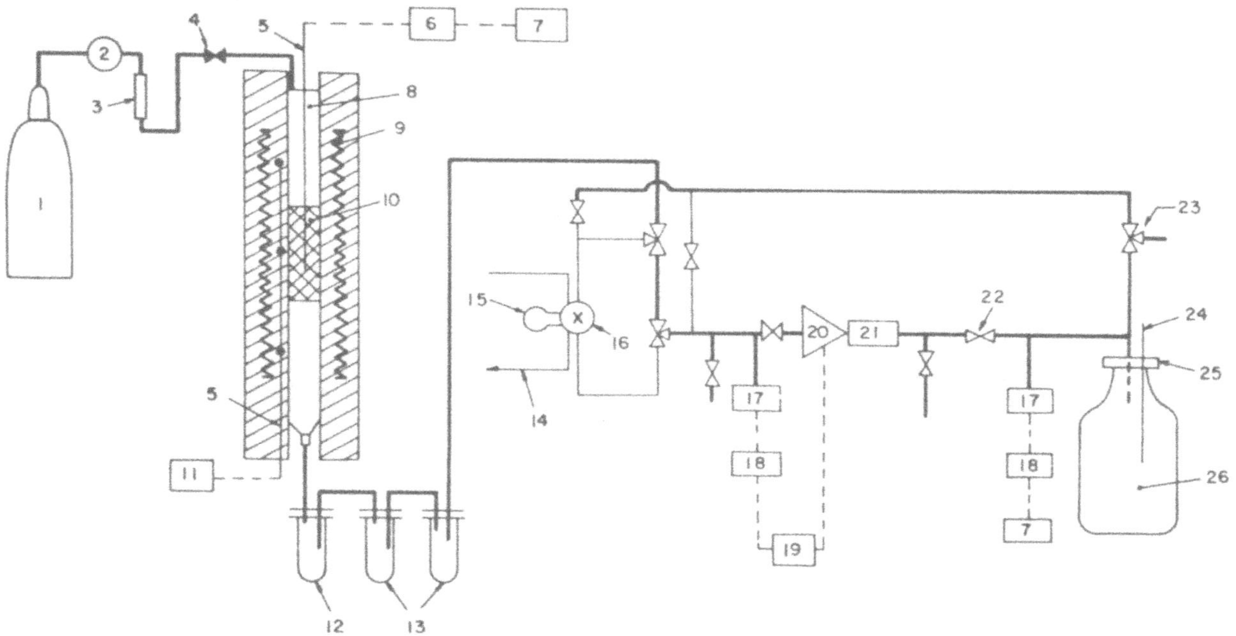

FIG. 1. Schematic of bench scale vacuum pyrolysis unit. (1) Nitrogen cylinder; (2) pressure gauge; (3) rotameter, (4) micrometric valve; (5) thermocouple; (6) thermocouple meter; (7) recorder; (8) retort; (9) three zone electric furnace; (10) sample; (11) temperature programmer; (12) cooling trap (ice and water); (13) cooling traps (dry ice and acetone); (14) carrier gas to chromatograph; (15) sampling loop; (16) gas sampling valve; (17) electronic manometer head; (18) digital readout; (19) pressure controller; (20) automatic vacuum exhaust valve; (21) vacuum pump; (22) two-way manual valve; (23) three-way manual valve; (24) thermometer; (25) vacuum seal; (26) gas collecting bottle. Reproduced from ref. 8 by permission of the Institute of Gas Technology.

cylindrical stainless steel retort equipped with an inner quartz liner 50 mm in diameter to minimize contact between the reactants and products and the hot metallic walls of the reactor. The feedstock material typically weighed 60 g and occupied a volume of about 350 cm^3 in the retort. The air-dry biomass feedstock (5–8 % moisture) was prepared in the form of chips, pins or flour depending on requirements. The retort was placed in a Lindberg vertical three-zone electric furnace equipped with a temperature programmer for close control of the decomposition temperature up to the final desired temperature level. The rate of temperature rise was varied between 2 and 14 °C min^{-1} for the experiments herein reported. The latter value corresponded to the upper limit of the furnace. For each selected heating rate the temperature rise was uniform throughout the whole section of the fixed bed.

The reduced pressure in the system was monitored by an MKS Baratron Type 252A Exhaust Valve Controller which was attached to the MKS Baratron manometer digital readout and the MKS control valve as indicated in Fig. 1. For the experiments herein reported the total absolute pressure varied between 2 and 400 mm Hg. When required, nitrogen was used to maintain the desired pressure level in the system. The residence time of vapours and gases produced in the reactor was controlled through the combination of a pressure gauge, rotameter and micrometric valve as indicated in the figure. This variable ranged between 2 and 20 s.

Thermal breakdown of the solid feedstock took place then, releasing gases and vapours which were rapidly pumped off the reactor chamber through a series of cold water and dry ice–acetone condensers and a filter-trap where both the condensable oils and the pyrolytic water were recovered while the non-condensable gases were pumped away and stored in glass bottles at room temperature. The solid residues were left behind in the retort. Water in the pyroligneous liquor was determined by Karl Fischer titration using an automatic amperometric titremeter.[9] For all experiments herein reported the material balances closed between 98 and nearly 100 % of the initial charge. Losses were assumed to be gases. Other details regarding the procedure and the equipment used will be found elsewhere.[10]

The investigated material was aspen poplar (*Populus tremuloides*). Its elemental composition was 49·5 % C, 6·2 % H, 0·3 % N and 44·0 % O. The original tree from which the samples were taken was approximately 20 years old. The trunk was used for preparing the samples. It included the bark portion for a series of runs while some experiments were performed without the bark fraction. When bark was included the ash content of the whole sample reached 1·3% (moisture-free basis); ash content of the bark-free wood sample was 0·5%.

3. FACTORS INVOLVED IN VACUUM PYROLYSIS OF BIOMASS

The complexity of thermal decomposition reactions is reflected by the number of variables involved. This includes temperature, heat pretreatment ('bertinization'), heating rate, contact time of reactives and products in the reactor, pressure, particle size and extent of secondary reactions (e.g. thermal or catalytic cracking, repolymerization). Some of these parameters have interaction effects one with the other. The feedstock itself and its constituents (cellulose, lignin, hemicelluloses and extractives) significantly determine the yields and types of pyrolytic products obtained.

It is not our present objective to make a detailed discussion of each factor governing the thermal decomposition reactions of biomass. Such an exercise has been done elsewhere (for instance in the case of peat see ref. 11). This paper sheds some light on the relative importance of a few major factors involved in vacuum pyrolysis of biomass.

3.1. Temperature

By far, temperature is the most important factor involved. Figure 2 shows the evolution of the major stream products obtained in the low vacuum pyrolysis of aspen poplar flour (40/80 Tyler Sieves) at low heating rates (10 °C min^{-1}). The total pressure in the system varied between 0·3 and 2·3 mm Hg. The whole wood was used, exclusive of the bark portion.

It has long been known that the rate of decomposition of wood increases rapidly with increasing temperature. It can be said that this rate approximately doubles for each 10–14 °C of elevation.[12] Our experiments in the vacuum pyrolysis of aspen wood indicated that the more active decompositions occurred between 240 and 300 °C. The bulk of pyrolytic oils was obtained at a temperature as low as 350 °C, the yield reaching 58% w/w in this case. This is in sharp contrast to atmospheric pressure carbonization where it has been reported that the liquid products are of great variety according to the material carbonized, and their expulsion is complete at about 600 °C.[13] The pyrolytic oil yields themselves are considerably higher than that obtained in conventional carbonization of aspen wood.[14] In Fig. 2, char yields correspondingly dropped quite steeply after 190 °C and stabilized to 15% w/w at 450 °C. As compared to

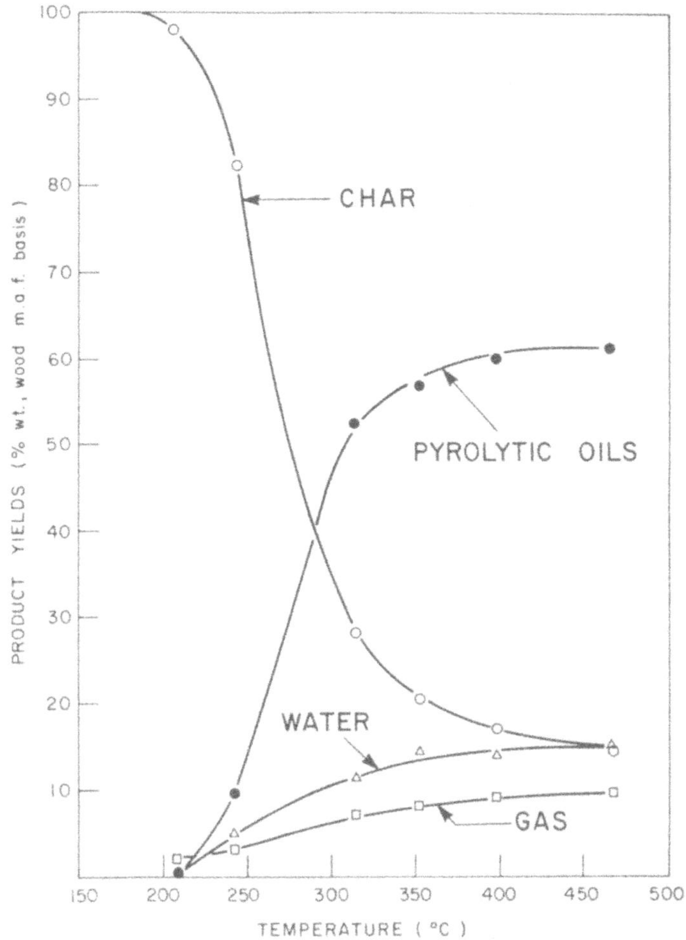

FIG. 2. Thermal decomposition of aspen poplar under vacuum. Product yields as function of temperature. Reproduced from ref. 8 by permission of the Institute of Gas Technology.

conventional pyrolysis, the yield of charcoal is about half that at the same temperature.[15] In this case increasing the temperature after 450 °C substantially decreased the yield of charcoal and this was still higher than 20 % w/w at 1600 °C. The type of wood used for the carbonization had no effect on these data. In Fig. 2, evolution of the non-condensable gases was almost linear between 200 and 400 °C; it practically ceased at 450 °C. Pyrolytic water was mainly formed between 210 and 300 °C.

3.2. Particle Size

The effect of particle size is well demonstrated in Table I. A decrease in size of the particles enhanced very significantly the formation of pyrolytic oils due to reduced mass transfer limitations. Correspondingly, there was a decrease in yield of charcoal.

Varying the particle size also influenced the concentration of gaseous products obtained as seen from the table. It is noteworthy that more oxygen was removed via H_2O and CO_2 when using wood chips as compared to pin chips or wood flour. No firm inference can be drawn at the present stage regarding the yields of carboxylic acids.

3.3. Heating Rate

In fast pyrolytic reactions, it may be assumed that the original material is brought up to the operation temperature almost instantaneously. In slow

TABLE I

EFFECT OF PARTICLE SIZE ON VACUUM PYROLYSIS OF ASPEN POPLAR UP TO 450 °C (%wt, wood m.a.f.)

Product	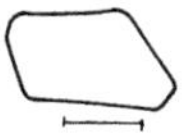 *1 cm chips*	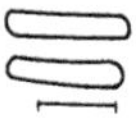 *1 cm pins*	*Flour*	*Standard deviation*
Oils	50·7	56·9	60·8	0·2
Water	16·3	13·9	12·8	0·2
Char	20·4	18·1	15·8	0·1
Gas	12·6	11·1	10·6	0·2
	100·0	100·0	100·0	
CO_2	7·87	6·79	6·32	0·2
CO	3·58	3·40	3·58	0·1
CH_4	0·28	0·27	0·23	0·05
H_2	0·01	0·01	0·02	0·01
C_2H_4	0·04	0·04	0·04	0·01
C_2H_6	0·06	0·05	0·04	0·01
Others	0·76	0·53	0·37	0·1
Acetic acid	4·7	3·8	3·6	0·4
Formic acid	2·3	2·3	2·6	0·2
Glycolic acid	0·4	0·4	0·2	0·2
Others	0·6	0·3	0·7	0·2

Feedstock: ~20 years old tree; bark was removed.
Absolute pressure: 5 mm Hg (constant).
Final temperature reached: 450 °C. Heating rate = 10 °C min^{-1}.
Reproduced from ref. 8 by permission of the Institute of Gas Technology.

pyrolysis, the heating rate of the feedstock can be used to calculate the nominal residence time of the solids.

In fast pyrolysis of biomass, contact time of the residual solids and the vapour–gas mixtures is of importance in determining the extent of gas–solid reaction. At slower heating rates, contact time still remains an important parameter, but the course of the reaction is significantly influenced by even a slight variation of the heating rate as indicated in Table II. Relatively low heating rates reduce the rate of generation of volatile matter from the solids, particularly in the temperature range 200–400 °C, and this in turn simultaneously influences the contact time of the vapour–gas mixtures and the residual solids in the reactive zone. One practical illustration of the relevance of the heating rate in pyrolytic processes will be found, for instance, in the manufacture of peat coke briquettes used in the metallurgical industry. Such processes require that

TABLE II

EFFECT OF HEATING RATE ON VACUUM PYROLYSIS OF ASPEN WOOD UP TO 450 °C (%wt, wood m.a.f.)

Product	*2 (°C min^{-1})*	*8 (°C min^{-1})*	*14 (°C min^{-1})*	*Standard deviation*
Oils	50·0	50·6	50·9	0·2
Water	15·4	15·4	15·5	0·2
Char	22·8	21·3	20·8	0·1
Gas	11·8	12·7	12·8	0·2
	100·0	100·0	100·0	
CO_2	7·60	8·11	8·33	0·2
CO	3·51	3·90	3·88	0·1
CH_4	0·28	0·21	0·18	0·05
H_2	0·02	0·02	0·01	0·01
C_2H_4	0·04	0·04	0·05	0·01
C_2H_6	0·04	0·06	0·07	0·01
Others	0·31	0·36	0·28	0·1
Acetic acid	4·7	3·7	3·9	0·4
Formic acid	2·4	1·9	1·9	0·2
Glycolic acid	0·2	0·2	0·2	0·2
Others	0·7	0·8	0·6	0·2

Feedstock: ~20 years old tree; bark was removed; 0·371 in./4 mesh, Tyler.
Absolute pressure: 5 mm Hg (constant).
Final temperature reached: 450 °C.
Reproduced from ref. 8 by permission of the Institute of Gas Technology.

the heating rate of the original material be 1·2 °C min^{-1}, or even lower, in order to prevent breaking of the carbonized structures caused by the escape of the volatile matters from the solids. Sundgren and Ekman[16] reported that when carbonization of some types of peats was carried out at 3 °C min^{-1}, the peat briquettes did not remain intact, and breakages were common.

Comparative experimental studies have shown that the chemical reactions which occur under fast heating conditions differ substantially from those conducted under slow heating, with respect to both the yields and the composition of the pyrolytic products.[11] In Table II, an increase in the rate of temperature rise was accompanied by a greater deoxygenation of the carbonizing material. Such treatment eliminated part of the oxygen content of wood as CO_2, CO and H_2. The preliminary heat treatment of biomass at low temperatures (200–450 °C) is defined as the *bertinization* process. This is generally conducted under fast pyrolysis conditions since slow heating makes the bertinization less complete: less oxygen forms pyrolytic water and more goes into tars.

It is difficult to obtain a firm cut-off conclusion for the changes in yields of carboxylic acids in view of the distribution of results. In atmospheric pressure slow pyrolysis of birchwood (maximum temperature 400 °C) the formation of acetic acid has been found to be practically independent of the rate of carbonization, while that of formic acid followed a progressive decrease.[14]

3.4. Residence Time, Pressure

The extent of interaction among solid residues, steam, organic vapours and other gaseous products is largely determined by the *contact time* or the *residence time* spent by such products in the pyrolysis reactor. Gas–solid and liquid–solid interactions as well as secondary reactions in the gaseous phase are important in pyrolysis since they may lead to recondensation or repolymerization of the broken organic macromolecules, secondary thermal cracking, catalytic cracking and oxidative or reductive reactions.

Two significant factors in such secondary reactions are the *heating rate* of the original material as discussed before and *pressure* of gases and vapours in the system. Heating rate is important mainly at the solid phase level where recondensation of the primary products may occur under slow heating rate conditions. Total pressure of the system, and especially partial pressures of the released organic vapours which can be diluted by a carrier gas, are of great importance in pyrolysis since the number of gas phase collisions between the molecules increases proportionally to the square of the pressure. However, it is difficult to distinguish clearly between contact time effects and those attributable to the frequency of gas phase collisions.

As a consequence, heating rate, pressure and residence time of products in the reactor are not wholly independent variables. It would be wise, then, to devise a scheme which would permit the study of the interaction effects between these parameters. Such an attempt is described in the following paragraphs.

Workers in the field of fast pyrolysis of biomass currently agree that a contact time of the order of 0·1 s distinguishes between so-called fast and slow pyrolysis.[17] Such short contact time of the reactants and products in the reactor is possible when very high heating rates of the solids, typically of the order of 100 °C s^{-1} or higher, are used in the processes. Flash pyrolysis involves even higher heating rates. The shortest contact time used in our experiments was approximately 2 s. According to Diebold's definition, this is truly a slow pyrolysis process.

The data presented in Figs 3–6 and Table III were obtained from a factorial design of experiments. This technique estimates the statistical significances of primary effects and their levels of interactions. The primary effects considered here were the heating rate ($\Delta T/\Delta t$), the total pressure (P) in the system and the residence time of gases and vapours (τ) in the reactor. Heating rate was varied between 2 and 13 °C min^{-1}. Residence time extended from the lower limit of about 2 s to about 20 s. The absolute pressure was kept in the range 10–400 mmHg; the higher value was *a priori* considered to be desirable process vacuum condition, a hypothesis which

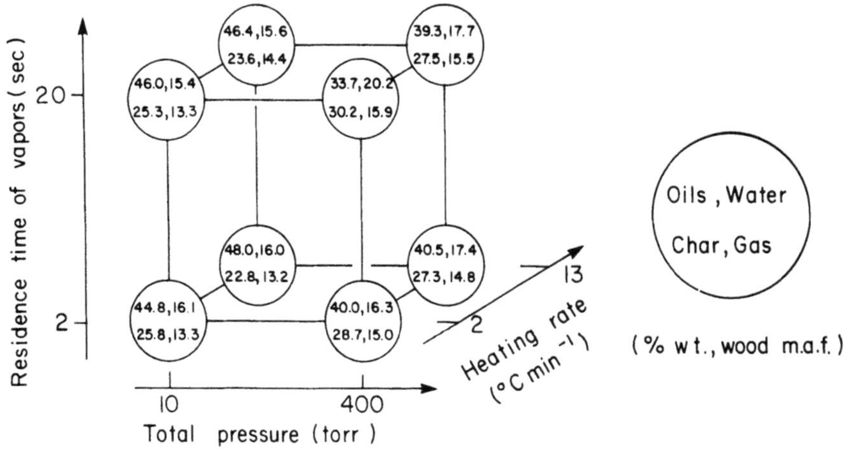

FIG. 3. Factorial design of experiments. Results for product yields. Reproduced from ref. 8 by permission of the Institute of Gas Technology.

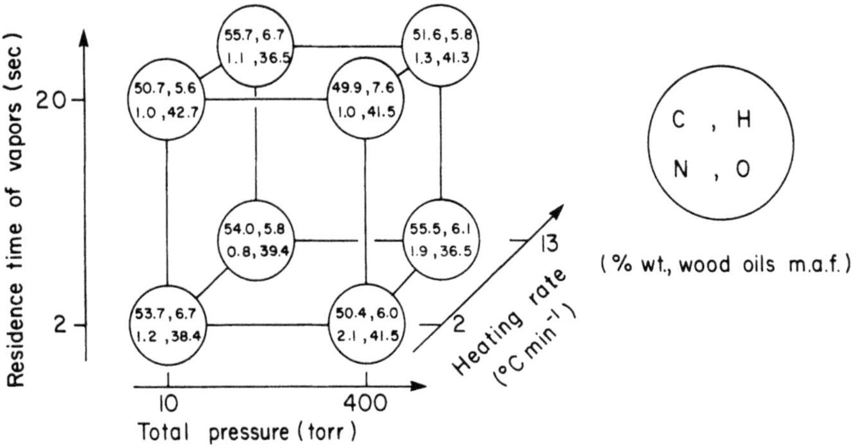

FIG. 4. Factorial design of experiments. Results for elemental composition of wood-derived oils.

will require substantiation via economic analysis later on. No control, however, of the partial pressure of the vapours generated in the reactor was possible in this unit. It is likely that the latter varied during the course of the reactions since evolution of the volatile matters changed with rising temperature while the pressure and inlet flow rate of the nitrogen carrier gas was kept constant for each experiment. This eventually resulted in

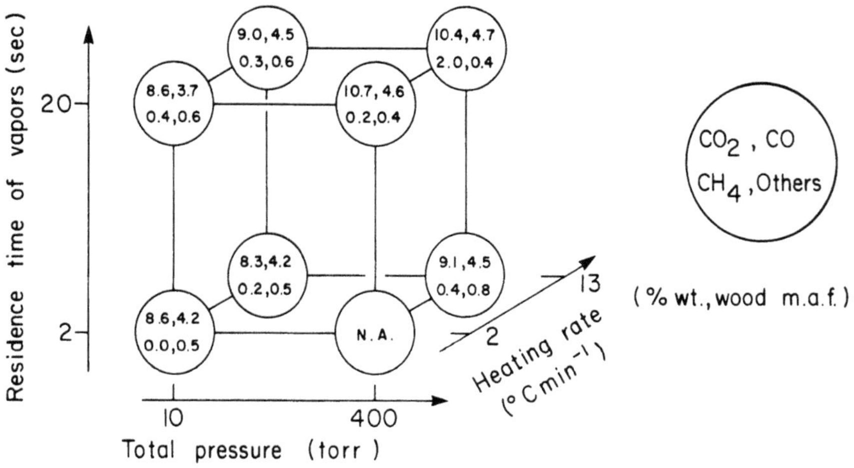

FIG. 5. Factorial design of experiments. Results for final gas phase composition.

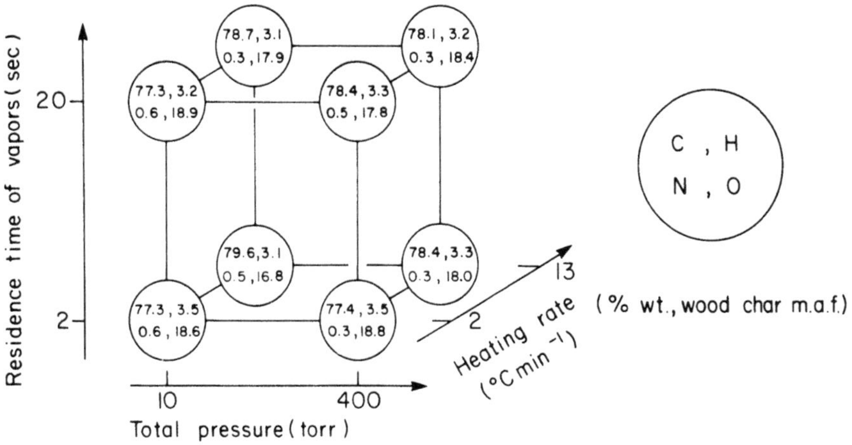

Fig. 6. Factorial design of experiments. Results for elemental composition of charcoal.

experimental noise which was reflected in some of the results presented in Table III.

The feedstock used for this series of experiments was a sample of aspen wood containing bark. It was reduced in the form of pin chips and sieved to 0·371 in./4 mesh, Tyler. The maximum temperature was fixed at 450 °C.

Results are summarized in Figs 3 to 6 for pyrolytic product yields, elemental composition of wood-derived oils, final gas phase composition and elemental composition of charcoals.

The results for oil, water, char and gas yields are presented in Fig. 3. As expected, a set of experimental conditions with a low pressure, short residence time and high heating rate positively influenced the pyrolytic oil yield; the yield of charcoal was correspondingly low. In the opposite direction, a higher pressure, longer residence time and lower heating rate influenced the char formation process at the expense of oils.

The data can be expressed in the form of polynomial models in order to estimate the statistical significances of primary effects and their levels of interactions. Models have been determined for char, gas and oil yields as well as oxygen content in pyrolytic oils. Those variables are of interest particularly for a potential vacuum pyrolysis process. The models which are included in Table III were found to be well adjusted, with good prediction at the central point. They are valid only for the exact ranges of experimental conditions which have been specified. The levels of significance are indicated in the table.

TABLE III

POLYNOMIAL MODELS DERIVED FROM FACTORIAL DESIGN OF EXPERIMENTS[a]

Variable	*Model*[b]	*Test at central point*	*Statistical significance*
Char yield	$Y_{char} = 26{\cdot}4 - 1{\cdot}1\left(\frac{\Delta T}{\Delta t}\right) + 2{\cdot}0(P)$	26·5	No lack-of-fit at $\alpha = 1\%$
Gas yield	$Y_{gas} = 14{\cdot}4 + 0{\cdot}9(P) + 0{\cdot}4(\tau)$	15·1	No lack-of-fit at $\alpha = 10\%$
Oil yield	$Y_{oil} = 42{\cdot}3 + 1{\cdot}2\left(\frac{\Delta T}{\Delta t}\right) - 4{\cdot}0(P) - 1{\cdot}0(\tau) - 0{\cdot}9(P\tau) + 1{\cdot}0\left(\frac{\Delta TP\tau}{\Delta t}\right)$	41·1	No lack-of-fit at $\alpha = 1\%$
% O in oil	$Y_{O} = 39{\cdot}7 - 1{\cdot}3\left(\frac{\Delta T}{\Delta t}\right) + 1{\cdot}5\left(\frac{\Delta TP\tau}{\Delta t}\right)$	38·0	No lack-of-fit at $\alpha = 5\%$

[a] All yields are on a % wt, wood m.a.f. basis.
[b] Based on F distribution.

The yield of charcoal was adequately modelled by two single parameters which excluded residence time. The model predicted that both a higher pressure and a lower heating rate increased the yield of char and, conversely, decreased the yield of volatile matter ($Y_{VM} = 73{\cdot}6 + 1{\cdot}1\,\Delta T/\Delta t - 2{\cdot}0P$). The non-condensable gas yield was well represented by two primary effects, pressure and residence time. The model predicted that an increase in both P and τ positively influenced the gas formation. The model is particularly well adjusted in this case ($\alpha = 10\%$). On the other hand the formation of pyrolytic oils was significantly influenced by each primary variable investigated and also by interaction variables. The unexpected three-factor interaction involving P, τ and $\Delta T/\Delta t$ could mean that experimental noise appeared during the course of experiments. The model indicated, however, that the net production of oils was more sensitive to pressure than contact time, as observed from the coefficients for each term in the corresponding expression.

The ultimate analysis of oils was performed (Fig. 4). These oils contained substantial amounts of oxygen, ranging from 36·5 to 42·7 % w/w. This is indicative that little increase in energy was achieved. The oil oxygen content was represented by an equation with one primary factor ($\Delta T/\Delta t$) and a three-factor interaction (Table III). The model predicted that the largest difference in oxygen content of the oils was 5·6 % over the wide range of experimental conditions used.

The results for gas phase composition are shown in Fig. 5. These data were useful for calculating partial mass balances for C, H and O which were distributed in the different pyrolytic products. Elemental mass balances for C and O always closed within 9 % for this series of experiments.

The elemental composition of charcoals was also reported (Fig. 6). Based on Pohl's study[15] on the elemental composition of charcoals produced in the carbonization of wood at 450 °C, the oxygen content of vacuum char is much higher as compared to atmospheric pressure char while the carbon content is lower. The largest difference in carbon content of vacuum charcoal was rather small (2·3 %).

4. THE ROLE OF WOOD COMPONENTS IN THERMAL DECOMPOSITION

The yields of charcoal, oils, water and gas may vary within wide limits with different woods and different conditions of carbonization. This makes it difficult to compare the majority of studies reported in the literature. For instance, the vacuum pyrolysis under identical experimental conditions of two samples of aspen wood of variable age gave substantially different results, as noted in Table IV. This raised the question of the role of the various wood components on thermal decompositions under a vacuum. This subject will be treated briefly in the following paragraphs.

The thermal decomposition of different wood constituents can be studied either by carbonizing a number of woods with greatly varying chemical composition or by carbonizing isolated wood fractions. Most

TABLE IV
RESULTS ON VACUUM PYROLYSIS OF TWO SAMPLES OF ASPEN POPLAR HAVING DIFFERENT AGES

Component[a]	*Mature*	*Young*
Extractives	11·5	5·3
Holocellulose	70·0	76·9
Lignin	18·0	15·2
Ash	0·5	0·6
Pyrolytic product[b]		
Oils	61·0	54·9
Char	16·0	18·0
Water	15·5	15·7
Gas	9·7	11·4

[a] %wt, anhydrous and bark-free basis. ASTM Standards were used except for holocellulose which was determined by the chlorite method.
[b] %wt, wood m.a.f. basis. Experimental conditions: wood flour (40/80 Tyler Sieves); 10 °C min^{-1}, up to 450 °C; absolute pressure <2 mm Hg.

studies have been carried out with isolated fractions. The thermal decomposition of different wood components can also be studied by successive treatment of wood with different reagents and by carbonization of the original wood and the residues remaining after each treatment. The thermal decomposition yields of the extracted fractions can then be calculated by difference. This approach was followed in our laboratory. Properties like the *participation degree* of a wood component and the *degree of thermal decomposition* were defined. The participation degree of a wood component was defined as the ratio between weight loss of the component and the weight loss of the wood itself; the degree of thermal decomposition was defined as the ratio between weight loss of the wood component and the original content of this component in wood. These studies, which are currently being conducted for the range of temperatures 150–450 °C, will provide estimates of the thermal stability of the various wood components. It will also permit one to answer whether in its response to thermal treatment under a vacuum wood behaves as if it were a mixture of its major constituents.

It has been reported that yields of the various products of wood carbonization are usually about what would be obtained by carbonizing proportional amounts of the isolated cellulose, hemicellulose and lignin components.[18] Lebeau and co-workers[19,20] have published data in 1935 and 1937 on the pyrolysis of woods and their principal components. The materials were heated, under vacuum, to 1000 °C in 100 °C stages, and maintained for 1 h at each stage. From poplar were isolated 12·3 % hemicellulose, 23·9 % lignin and 57·8 % cellulose. These were carbonized separately and the results, when proportioned according to their content in the wood and totalled, were compared with results from the wood itself. The agreement was found to be quite good. However, the interpretation was based on the charcoal volatile matter contents. Such data give no information on the distribution of tar, water and non-condensable gases following treatment of the various fractions.

Our results for the vacuum pyrolysis up to 315 °C of aspen poplar wood flour and the residues remaining after extraction of the wood components are summarized in Tables V and VI.

The relatively large amounts of lignins required in these experiments made it impractical to produce and isolate it by the usual laboratory techniques. The lignin used was prepared from aspen poplar wood

TABLE V
PRODUCT YIELDS ON VACUUM PYROLYSIS UP TO 315 °C OF ASPEN WOOD CONSTITUENTS[a]

Fraction	*Oils*	*Char*	*Water*	*Gas*
Wood (*P. tremuloides*)	52·8	27·7	12·3	7·2
Extractive-free wood	46·7	41·5	7·6	4·2
Holocellulose	40·0	29·6	16·5	13·9
Lignin[b]	17·0	72·2	—	10·8

[a] Yields are on a feedstock m.a.f. basis.
[b] Lignin from STAKE process, washed to remove ashes.

TABLE VI
GAS PHASE COMPOSITION ON VACUUM PYROLYSIS UP TO 315°C OF ASPEN WOOD COMPONENTS

Fraction[a]	CO_2	CO	CH_4	C_2	C_3	C_4	C_5	C_6	*Others*[b]
Wood	5·3	1·9	<0·1	<0·1	<0·1	—	—	—	<0·1
Extractive-free wood	2·5	0·9	—	<0·1	<0·1	0·1	0·1	0·1	0·4
Holocellulose	9·0	4·2	<0·1	<0·1	0·6	<0·1	<0·1	<0·1	<0·1
Lignin (STAKE)	9·3	0·6	—	—	—	—	—	0·2	0·7

[a] Yields are on a feedstock m.a.f. basis.
[b] Identified gases: methanol, acetone, ethanol.

containing 50% moisture (waste from the Eddy Match Company, Pembroke, Ontario), autohydrolysed in the STAKE cattle feed plant in Maine with steam at 200°C for 5 min and then extracted with sodium hydroxide, and the lignin reprecipitated with 2% w/w sulphuric acid at 72–80°C at the Herty Laboratories in Savannah, Georgia. The fibre was separated with an Impco Vari-Nip press. The lignin was spray dried in the Dairy Laboratory of the Experimental Farm in Ottawa, Ontario. The process is that described and patented by Lora and Wayman, of the University of Toronto, Ontario.[21]

The as-received lignin sample contained, however, 32% ash due to the presence of sodium sulphate which formed during the reprecipitation process. Ash was removed by washing the solids several times at pH 4·5. This treatment reduced the ash content of the lignin sample to 0·4%.

The results included in the tables bring useful information about the role of the various wood components during pyrolysis. It is observed, for instance, that lignin significantly contributed to the production of charcoal during vacuum pyrolysis of wood. These results ought to be interpreted, however, in the light of the forthcoming data which will be obtained at various carbonization temperatures. Future work along this line will also involve determination of the source of the major chemicals in the liquid phase and the contribution of each wood component in the formation of such compounds.

5. OIL CHARACTERISTICS

The typical aspen wood-derived oil product is a low viscosity, red oil containing small clear droplets which are apparently water. It is highly oxygenated, as mentioned before. Analyses performed by Battelle Pacific

TABLE VII

ORGANIC COMPOUNDS FORMED BY VACUUM PYROLYSIS OF ASPEN POPLAR

Compound	*Formula*
Methanol	CH_4O
Acetaldehyde	C_2H_4O
Acetic acid	$C_2H_4O_2$
Ethanol	C_2H_6O
Acetone	C_3H_6O
2-Propen-1-ol (Allyl alcohol)	C_3H_6O
Propionaldehyde	C_3H_6O
Methyl acetate	$C_3H_6O_2$
Propionic acid	$C_3H_6O_2$
Methylal (dimethoxymethane)	$C_3H_8O_2$
Crotonaldehyde	C_4H_6O
Butyrolactone	$C_4H_6O_2$
Crotonic acid	$C_4H_6O_2$
2-Butanone	C_4H_8O
Isobutanoic acid	$C_4H_8O_2$
Butanoic acid	$C_4H_8O_2$
Furfural	$C_5H_4O_2$
2-Furan carboxylic acid	$C_5H_4O_3$
2,3-Pentadione	$C_5H_8O_2$
2-Methyl 2-butenoic acid	$C_5H_8O_2$
2-Pentanone	$C_5H_{10}O$
3-Methyl-2-butanone	$C_5H_{10}O$
Phenol	C_6H_6O
2-acetylfuran	$C_6H_6O_2$
5-Methyl furfural	$C_6H_6O_2$
Hexanoic acid	$C_6H_{10}O_2$
Methylcyclopentane	C_6H_{12}
3-Hexanone	$C_6H_{12}O$
Acetal (acetaldehyde)	$C_6H_{14}O_2$
3-Methylphenol	C_7H_8O
2-Methylphenol	C_7H_8O
o-Methoxyphenol	$C_7H_8O_2$
2-Ethylphenol	$C_8H_{10}O$
2,5-Dimethylphenol	$C_8H_{10}O$
2,4-Dimethylphenol	$C_8H_{10}O$
2,6-Dimethoxyphenol	$C_8H_{10}O_3$
Octanoic acid	$C_8H_{16}O_2$
Nonanoic acid	$C_9H_8O_2$
2,4,6-Trimethylphenol	$C_9H_{12}O$
Ethyl methyl phenol	$C_9H_{12}O$
2-Isopropylphenol	$C_9H_{12}O$
4 ethyl-2-methoxy phenol	$C_9H_{12}O_2$
4-allyl-2-methoxyphenol (eugenol)	$C_{10}H_{12}O_2$
2-Methoxy-4-propyl phenol	$C_{10}H_{14}O_2$

Northwest Laboratories, Richland, Washington, indicated that light acids, aldehydes, ketones, furans, phenols and methoxylated phenols constituted about 34 % w/w of the oils on an anhydrous basis. This type of pyrolysis oil was found to bear very little resemblance to the type of oil produced at the Albany facility in Oregon.[8]

Water was the major pure compound in the liquid phase product. Carboxylic acids from C_2 to C_7 constituted the largest group of organic compounds in the pyroligneous liquors. It was found that acids amounted to nearly 19 % w/w of the as-received liquid phase.[22] Acetic, formic and glycolic acids constituted the bulk of these acids. The high yields of acetic acid determined in this series of experiments are in line with that reported in the literature.[14] The source of acetic and formic acids from wood has also been discussed in this literature.

A detailed analysis of the pyroligneous liquor derived from aspen poplar has also been performed in our laboratories. These results are available.[23] The approach which was used involved liquid phase preparative chromatography combined with GC–MS and analytical HPLC techniques. A list of organic compounds which have been detected in the liquid phase product is presented in Table VII.

6. CONCLUSION

The role of a few key parameters in vacuum pyrolysis reactions of aspen wood has been reviewed. These variables include temperature, particle size, heating rate, total pressure and residence time of gases and vapours in the reactor. Short residence times and low pressures characterize decompositions under vacuum; the formation of pyrolytic oils, however, was found to be more sensitive to pressure than residence time for the range of experimental conditions investigated. The role of wood components appears to be of primary importance in the formation of the various pyrolytic products; more research is still needed in this area.

The fundamental investigation which was reported is being extended with practical application aspects. The technique of vacuum pyrolysis of wood and biomass has been found to produce high yields of liquid organic products under relatively moderate operating conditions. The vacuum pyrolysis process which is now being contemplated has many advantages of commercial significance related to its insensitivity to the type and form of feedstocks—wood chips are an adequate feed, lower energy and equipment requirements to prepare the fibre, the immediate separation of the liquids from solid residuals within the reactor and the possibility which is also envisaged of using the reactor of primary fractionation of the liquids at each reactor zone.[8]

ACKNOWLEDGEMENTS

The collaboration and technical assistance of Dr Esteban Chornet, Department of Chemical Engineering, Université de Sherbrooke, and of Mr Douglas C. Elliott, Battelle Pacific Northwest Laboratories, Richland, Washington, has been greatly appreciated. Thanks are due to Dr Morris Wayman, Department of Chemical Engineering, University of Toronto, for providing the lignin samples.

This work has been supported by the Ministry of Environment, Canadian Forestry Service, ENFOR programme.

REFERENCES

1. Hawley, L. F. (1923). *Wood Distillation.* Chemical Catalog Company, Inc. (Reinhold Publishing Corp.), New York.
2. Bunbury, H. M. (1926). *The Destructive Distillation of Wood.* Van Nostrand Co., New York.
3. Klar, M., translated by Rule, A. (1925). *The Technology of Wood Distillation.* D. Van Nostrand Co., New York.
4. Wise, L. E. and Jahn, E. C. (1952). *Wood Chemistry,* Vol. 2. Reinhold Publishing Corp., New York.
5. Stamm, A. J. and Harris, E. E. (1953). *The Chemical Processing of Wood.* Chemical Publishing Co., New York.
6. Wenzl, H. F. J., translated by Brauns, F. E. and Brauns, A. D. (1970). *The Chemical Technology of Wood.* Academic Press, New York.
7. Shafizadeh, F., Furneaux, R. H., Cochran, T. G., Scholl, J. P. and Sakai, Y. (1979). Production of levoglucosan and glucose from pyrolysis of cellulosic materials. *J. Appl. Polymer Sci.* **23**: 3525–39.
8. Roy, C., de Caumia, B., Plante, P. and Ménard, H. (1983). Production of liquids from biomass by vacuum pyrolysis—Development of data base for continuous process. *Symposium Papers, Energy from Biomass and Wastes VII.* Lake Buena Vista, Florida, January 24–28. Institute of Gas Technology, Chicago.
9. Roy, C., Chauvette, G., de Caumia, B. and Ménard, H. (in preparation). The determination of water in wood-derived oils.
10. Roy, C. (1980). Study of the vacuum thermal decomposition of peat doped with metallic additives and kinetic modelling of evolution of the noncondensable gases. Ph.D. Thesis, Université de Sherbrooke, Sherbrooke, Québec, Canada (in French).

11. Roy, C., Chornet, E. and Fuchsman, C. H. (in press). The pyrolysis of peat: a comprehensive review of literature. Part I: Fundamentals. *J. Anal. Appl. Pyrolysis.*
12. Stamm, A. J. (1958). Thermal degradation of wood and cellulose. *Ind. Eng. Chem.* **48**: 413–7.
13. Milner, G., Spivey, E. and Cobb, J. W. (1943). *J. Chem. Soc.* **1943**: 578.
14. Wenzl, H. F. J., *op. cit.*, Chapter V.
15. Pohl, W., in Ullmann, *Enzyklopadie der Techn. Chemie*, Vol. 8. Urban & Schwarzenberg, München-Berlin, p. 585.
16. Sundgren, A. and Ekman, E. (1968). Experiments on the cokability of various compressed peat products. *Second Intern. Peat Congress Transactions.* Edinburgh, HMSO, **2**: 863.
17. Diebold, J. (1980). Workshop summary. *Proceedings, Specialists' Workshop on Fast Pyrolysis of Biomass.* Copper Mountain, Colorado, October 19–22. Available: NTIS Report No. SERI/CP-622-1096.
18. Wise, L. E. and John, E. C. *op. cit.*, p. 840.
19. Lebeau, P., Marmasse, P., Michel, R. and Viel, G. (1935). *Ann. combustibles liquides* **10**: 1027–76.
20. Lebeau, P., Marmasse, P., Michel, R., Viel, G. and Voyer, M. (1937). *Annales combustibles liquides* **12**: 651–709, 929–58.
21. Lora, J. (1977). MASc. Thesis, Department of Chemical Engineering and Applied Chemistry, University of Toronto, Toronto, Ontario.
22. Ménard, H., Gaboury, A., Bélanger, D. and Roy, C. (1982). Analysis of carboxylic acids in pyroligneous liquors using HPLC. *J. Anal. Appl. Pyrolysis.* Submitted for publication.
23. Ménard, H. and Roy, C. (1982). *Caractérisation des huiles pyrolytiques provenant du peuplier faux-tremble.* Rapport final, projet ENFOR C-256. Disponible: Approvisionnements et Services Canada, Dossier No. 41SS. KN107-1-4315.

14

Simulation of Kraft Lignin Pyrolysis

FRANCIS P. PETROCELLI and MICHAEL T. KLEIN*
Department of Chemical Engineering, University of Delaware, Newark, Delaware, USA

1. INTRODUCTION

The profitable conversion of lignin to low-molecular-weight products is hindered by the complexity of both its chemical structure and also its conversion product spectra. These complexities not only mask the actual fundamental events occurring during lignin processing but render design, scale-up and optimization difficult. Each of these tasks and also potential catalyst screening and future research guidance would be facilitated by the development of a fundamentals-based conversion simulation model. The present communication reports on our attempt at such a description of Kraft lignin pyrolysis reactions.

Our work followed the *a priori* simulation approach advanced by Klein and Virk,[1] where native spruce lignin pyrolysis was predicted using reaction parameters obtained from model compound pyrolyses; no actual lignin pyrolysis information was used. Their simulation is hereafter referred to as the MIT model.

The MIT model's interpretation of the lignin polymer was as an ensemble of single-ring phenolic units. Derived from one of the three coniferyl, sinapyl and coumaryl alcohol monomers, each phenolic unit contained a methoxyphenol (in the '1' and '2' position) and propanoid side chain (in the '4' position) substituent, and were joined by one or more of eight types of interunit linkages. The Freudenberg spruce lignin structural schematic was used to ascertain the types and proportions of both substituents and interunit linkages.

Both lignin structure and lignin pyrolysis were described probabilistically. The probability of the occurrence of a given aromatic unit in lignin

* To whom correspondence should be addressed.

was equal to the product of the two independent probabilities of the occurrence of the implied methoxyphenol and propanoid side chain substituents; the latter probabilities were extracted from the Freudenberg structure. Aromatic rings were conserved during simulated pyrolysis and the independent reactions of the methoxyphenol and propanoid side chain substituents described the lignin pyrolysis. Model compound pyrolysis reaction pathways and kinetics were used to deduce the alterations of a given aromatic unit's methoxyphenol and side chain substituent. The probability that a given pyrolysis product formed, then, was equal to the product of the independent probabilities that the implied alteration of the methoxyphenol and propanoid side chain substituents occurred.

The objective of the present work was to extend the MIT modeling approach toward simulation of Kraft lignin thermolysis. Kraft lignin is the residue obtained in industrially important Kraft pulping operations and is quite different in chemical structure from native lignin. Simulation of its pyrolysis thus seemed relevant.

In summary of what follows, the first of the remaining sections describes the development of this Kraft lignin pyrolysis model. We then delineate model predictions and consider them in the light of relevant experimental data. Our discussion section concerns the accuracy, inherent limitations and usefulness of the present model, and we follow this by listing important conclusions emanating from this work.

2. KRAFT LIGNIN PYROLYSIS SIMULATION MODEL DEVELOPMENT

The substantial differences between the development of the Kraft and milled wood lignin pyrolysis simulation models were largely on account of the lignin structural differences caused by Kraft pulping. Although Kraft lignin could also be interpreted as an ensemble of single-ring aromatic units connected by a limited number of interunit bonds, the utility of considering only one methoxyphenol (i.e. '1' position) and one 3-carbon side chain (i.e. '4' position) substituent was lost. This is because Kraft pulping changes not only the types but also the bonding patterns of interunit linkages. Thus, whereas '4'–'4' side chain links are plentiful in Kraft lignins, only methoxyphenol-propanoid side chain (i.e. '1'–'4') interunit linkages were included in the MIT spruce lignin model, save the one obvious pinoresinol '4'–'4' link; also, Kraft lignins contain '3'–'6' and more labile '4'–'6' linkages. Note that whereas α and β ethers are predominant in native lignins, these are cleaved during pulping and stilbene, diphenylethane and diphenylmethane linkages are important in Kraft lignins.

The foregoing led us to characterize Kraft lignin's single-ring aromatic units by the simultaneous occurrence of substituents on each of the six ring positions. The Marton[2] structure was used to ascertain the types and distributions of these substituents, which included terminal side chains as well as interunit linkages. Since simulated pyrolysis would not fragment the aromatic rings, Kraft lignin pyrolysis could be followed by monitoring the reactions of the six ring substituents.

Kraft lignin substituents were as follows. All aromatic rings had an oxygen substituent at their '1' position that was either a free phenolic hydroxyl or involved in a β aryl alkyl ether bond. All '2' positions were one of hydride-, hydroxyl- or methoxyl-substituted. All '3' and '5' positions were hydride-substituted. The '4' positions either contained a terminal side chain substituent or were involved in an interunit linkage, and the '6' positions were either hydride-substituted or involved in an interunit linkage. Aside from the '1'–'4' β-ether bonds, interunit linkages were either '4'–'4', '6'–'6', or '4'–'6' and were akin to bibenzyl, stilbene and diphenylmethane molecules.

A probabilistic interpretation of lignin structure was formulated to account for the likelihood that Kraft lignins contain more types of aromatic units than only those fourteen shown in the Marton structure. Substituent types and frequency distributions, taken from the Marton structure, were matched so as to define the probability of the occurrence of each Kraft lignin aromatic unit as the probability of the simultaneous occurrence of six substituents, i.e. $P^j_{AU} = P^{1j}_1 \times P^{2j}_2 \times P^{3j}_3 \times P^{4j}_4 \times P^{5j}_5 \times P^{6j}_6$, where P^j_{AU} is the probability of a given aromatic unit j with ring substituents $1j$ through $6j$ on positions '1' through '6', respectively. Each ring position thus had a substituent probability vector, $\vec{P}_i$, defined in eqn (1), which depicted the types and proportions of a given ring position's substituents:

$$\vec{P}_i = \sum_{j=1}^{N_i} P^{ij}_i \tag{1}$$

i = ring positions '1' through '6'
j = substituent type counter
N_i = number of different substituent types at ring position i
P^{ij}_i = proportion of substituents of type j at ring position i.

Pyrolysis was modeled by following the temporal variations of the

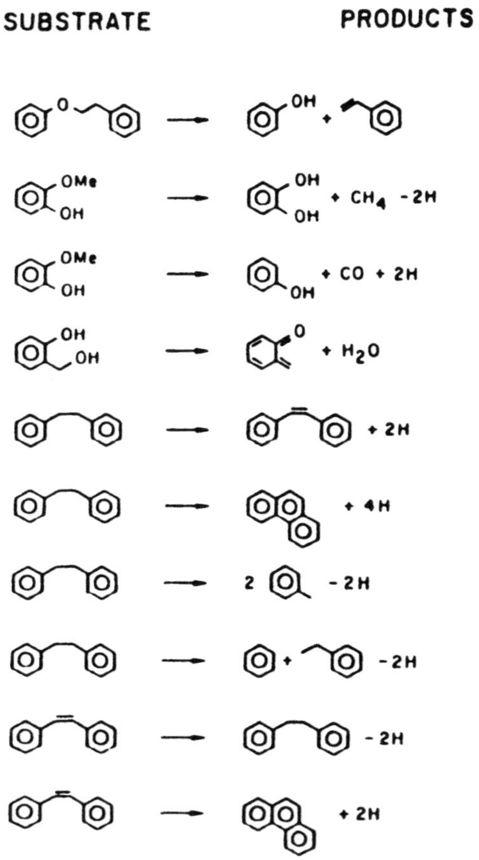

PATHWAY NUMBER	$-\log_{10} k_{400°C}$ (s^{-1})	REFERENCE
1	3.51	10,14
2	3.29	8,10,15,16,17,18
3	3.89	
4	-2.56	10, 19, 20,
5	4.85	4
6	5.91	4
7	4.53	4
8	5.62	4
9	4.97	4
10	7.39	4

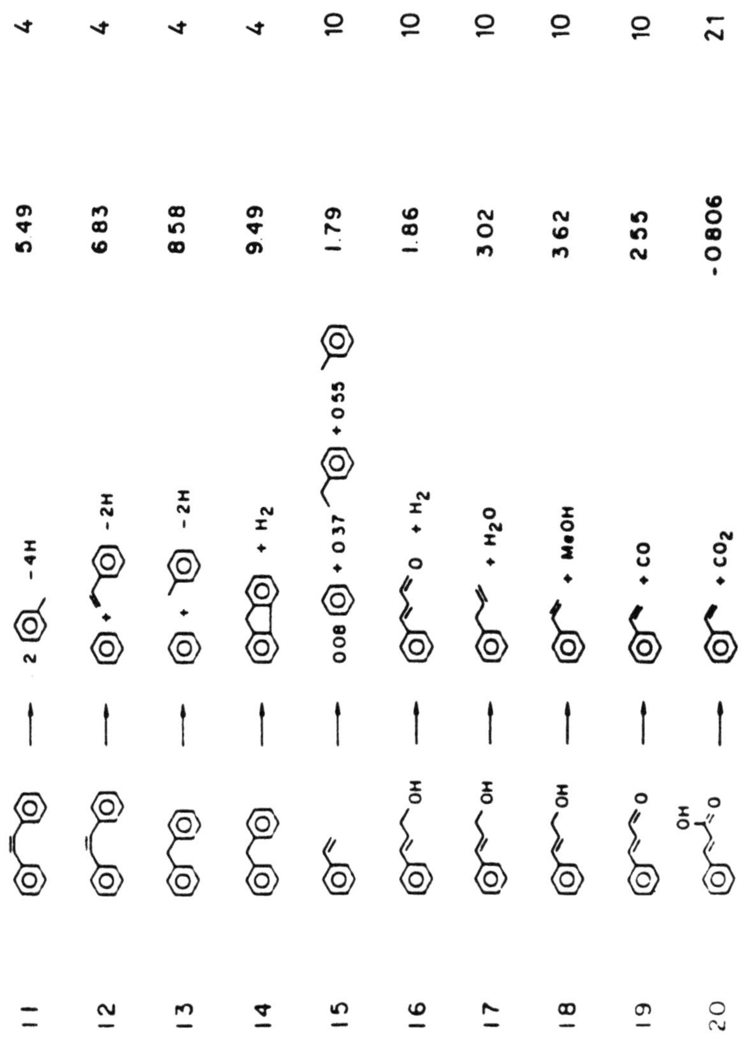

FIG. 1. Simulation model compound pyrolysis data base.

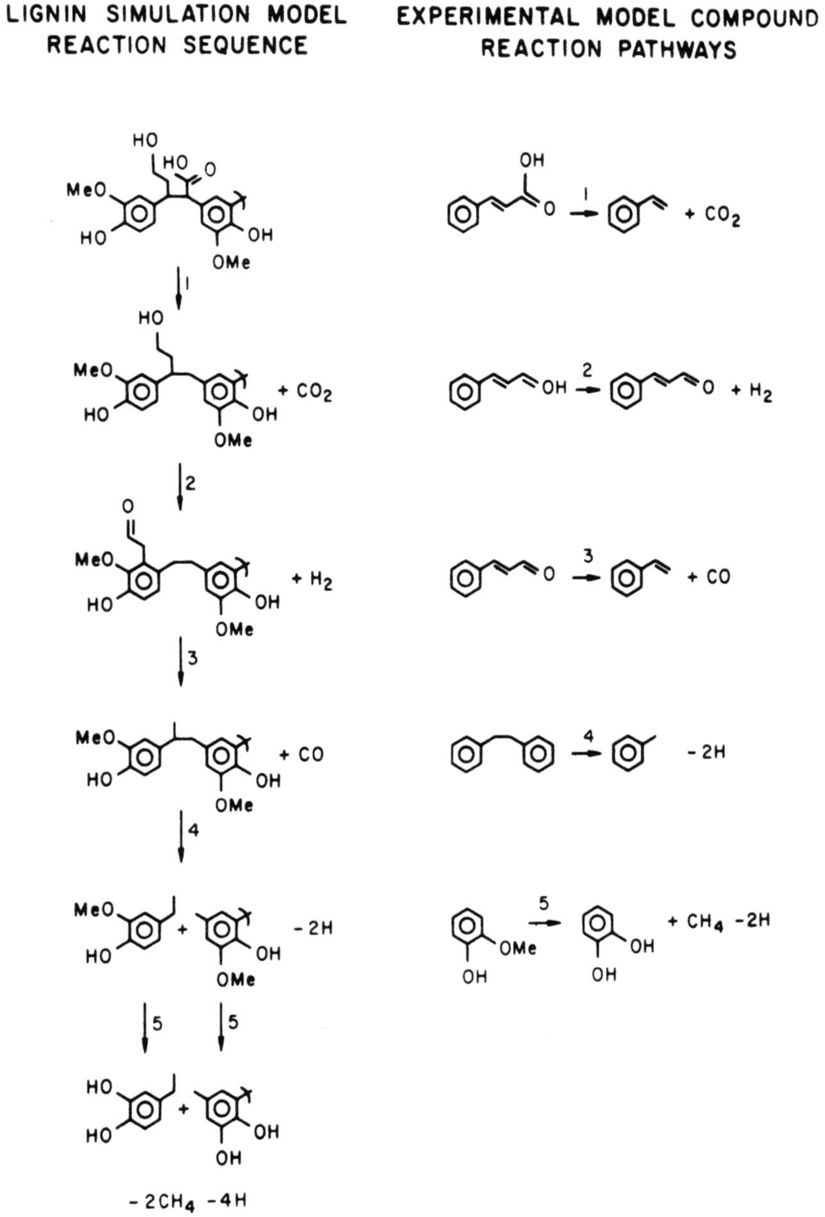

FIG. 2. Example of the application of model compound results to Kraft lignin pyrolysis.

substituents' proportions in the six substituent probability vectors; pyrolysis reactions enlarged N_i for each of the six substituent vectors. Pyrolysis products occurred as the probability of the occurrence of six substituents, i.e. $p^j_{PR} = \prod_{i=1}^{6} P_i^{ij}$ where j indicates the substituent type required for the given product (different in each vector i) and P_{PR} is the yield of that product.

Reaction pathways and kinetics for the entries in each substituent probability vector were based on the experimental model compound pyrolysis results summarized in Fig. 1. Our application of model compound results in the simulation of Kraft lignin pyrolysis is illustrated in Fig. 2, where one of many possible reaction sequences for a typical Kraft lignin moiety is shown. In Fig. 2 the column on the left depicts the model's pyrolysis reactions of the lignin moiety, and the column on the right lists the experimental model compound pathways from which the model's reactions were derived. The pyrolysis of each Kraft lignin moiety and reactive intermediate was modeled similarly. Complete simulation of Kraft lignin pyrolysis was achieved by accounting for the pyrolysis reactions of each lignin moiety. Since all reaction pathways and kinetic parameters incorporated into the simulation model were derived solely from model compound pyrolyses, model predictions were thus truly *a priori* with regard to actual lignin pyrolysis.

The simulation was of a constant volume, batch pyrolysis, and accounting for substituent changes amounted to the solution of six sets of ordinary differential rate equations. Numerical integration of these was accomplished using a prepackaged IMSL subroutine, DGEAR, based on the work of Gear[3] and available on the University of Delaware's DEC-10 computing system.

Details are available.[4]

3. RESULTS

Kraft lignin pyrolysis was simulated at 300, 400 and 500 °C for reaction times up to 8000, 5000 and 1000 s, respectively. Simulation predictions are presented at 400 °C, for the most part, to allow for effective comparison with the experimental Kraft lignin pyrolyses of Iatridis and Gavalas,[5] Chan and Krieger,[6] Graef *et al.*,[7] Connors *et al.*[8] and Domburg *et al.*[9]

Simulated Kraft lignin pyrolysis products were grouped into one of a permanent gas, light liquid, single-ring phenolic, multiring phenolic and carbonaceous residue fraction. The gas fraction consisted of carbon monoxide, carbon dioxide, methane and hydrogen, and light liquids comprised water and methanol. A total of 126 single-ring phenolics, all derivatives of guaiacol, catechol and phenol, were produced. Multiring phenols were lumped according to both the number of oxygen substituents they contained and also their type of interatomic unit linkage. These products were oxygen-substituted bibenzyl, stilbene, diphenylmethane, phenanthrene and fluorene derivatives. The carbonaceous residue was calculated as the difference between the original lignin weight and the weight of the gas, liquid, single-ring phenolics and multiring phenolic products. Yields are reported as percent of original lignin weight.

3.1. Permanent Gases

Evolution of carbon dioxide was rapid and in yields of 7·6% for all temperatures and times considered. This compares with a value of 5·1% reported by Chan and Krieger[6] and a value of 7·2% found by Iatridis and Gavalas[5] at 650 °C and 120 s.

Simulated carbon monoxide and methane yields at 400 °C and 500 °C are shown in Fig. 3. These plots show carbon monoxide to be the predominant product gas, surpassing even the carbon dioxide yield at 400 °C and 1000 s pyrolysis time.

The model prediction for CO yield at 400 °C and 120 s holding time was 1·95%, whereas Iatridis and Gavalas reported 1·2% at the same

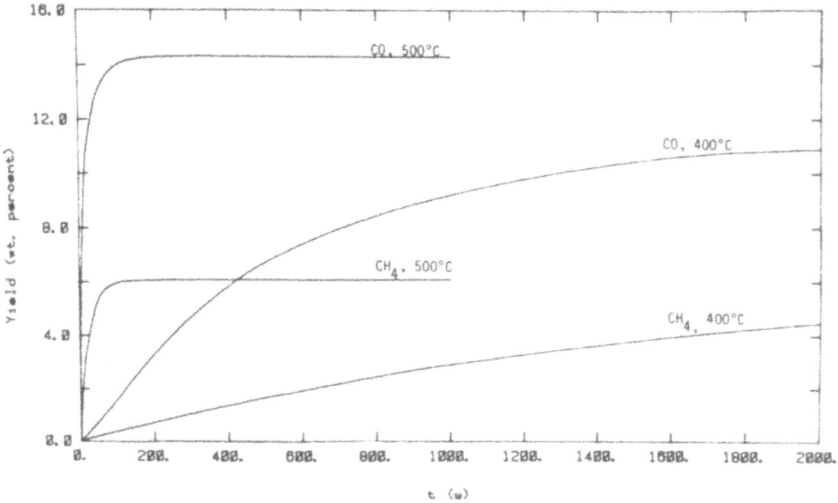

FIG. 3. Simulated carbon monoxide and methane yields.

conditions. Agreement was less favorable at 500 °C and 60 s, where the simulated yield of 13·3 % far exceeded the experimental yield of 2·05 %. This likely occurred, in part, because the simulation allowed for secondary guaiacol pyrolyses, whereas the reactor of Iatridis and Gavalas emphasized primary reactions. It is noteworthy that Iatridis and Gavalas reported a yield of 9·2 % at 650 °C and 120 s and that Chan and Krieger observed a yield of 21·9 % from microwave pyrolysis at temperatures in excess of 650 °C.

The simulation prediction for methane release at 400 °C and 120 s was 0·45 %, and at the same conditions Iatridis and Gavalas reported a yield of 0·52 %. At 500 °C and 60 s simulated and experimental yields were 5·55 % and 2·21 %, respectively; again this is likely due, in part, to the secondary guaiacol pyrolyses allowed in the simulation. The maximum simulation yield was about 6·1 % at both 400 and 500 °C, while Iatridis and Gavalas reported 4·83 % at 650 °C and 120 s, and Chan and Krieger found 4·4 % from the microwave pyrolysis.

The simulation predicted maximum hydrogen gas yields of about 0·3 % at both 300 and 400 °C. Chan and Krieger found hydrogen in 0·8 % yield.

3.2. Light Liquids

Maximum simulated water yields at each of 300, 400 and 500 °C were only 1 %, whereas the 8·6 % yield found by Chan and Krieger is typical of experimental yields of water.

Simulated methanol yields were also underpredicted; the maximum model yield of 0·06 % was far less than experimental yields, e.g. about 2 % reported by Iatridis and Gavalas.

3.3. Single-Ring Phenols

3.3.1. Guaiacols

The most prevalent single-ring guaiacols produced in simulated Kraft lignin pyrolyses were guaiacol (G), 4-methylguaiacol (MG) and 4-ethylguaiacol (EG). Temporal variations of these products' yields at 400 °C are illustrated in Fig. 4. At low conversions these products represented the largest fraction of single-ring phenolics; the maximum in yield attained by each is indicative of secondary decomposition to phenols and catechols.

Quantitative comparison with the literature is possible in only a few instances. At 400 °C and 120 s simulated yields were G:MG:EG = 0·69 %:3·21 %:0·77 %, whereas Iatridis and Gavalas found G:MG:EG = 1·1 %:0·76 %:0·38 %. At 500 °C and 60 s these yields were G:MG:EG = 0·18 %:0·36 %:0·14 % from the simulation and 0·98 %:1·01 %:0·16 %

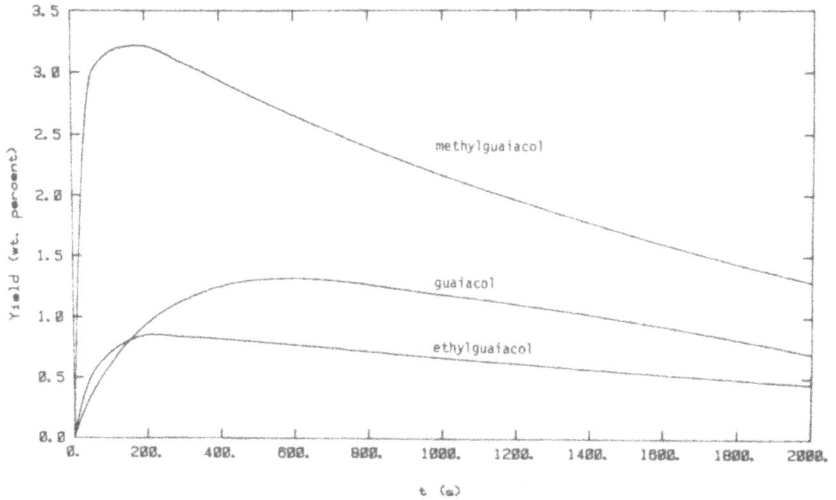

FIG. 4. Simulated single-ring guaiacols yields, 400 °C.

from the experimental work. Thus at 400 °C simulated yields were generally higher than experimental yields, whereas at 500 °C the simulated yields were lower. This was likely because the reactor used by Iatridis and Gavalas was designed to emphasize primary reactions while simulated pyrolysis allowed secondary guaiacol decomposition. Connors *et al.* found G:MG:EG:PG (propylguaiacol) = 0·7 %:0·89 %:0·98 %:0·35 % for pyrolysis at 400 °C and 900 s, whereas the simulation predicted yields of 1·19 %:2·17 %:0·67 %:0·002 % at 400 °C and 1000 s.

3.3.2. Catechols

Temporal variations of the yields of important catechol products, namely, catechol (C), 4-methylcatechol (MC) and 4-ethylcatechol (EC) are shown in Fig. 5 for simulated pyrolysis at 400 °C. Since the Marton structure contains only one catecholic aromatic unit, these products were formed mainly through secondary decomposition reaction of guaiacols. The model incorporated no catechol decomposition reactions, so the yields of these products always increased with pyrolysis time.

Comparison with the literature is informative. Neither Chan and Krieger nor Iatridis and Gavalas reported catechols as products from Kraft lignin pyrolyses in reactors designed to emphasize primary reactions. Domburg *et al.* reported yields of C:MC:EC = 0·80 %:1·10 %:0·50 % for pyrolysis at 475 °C, whereas simulated yields at 500 °C and 20 s holding time were 1·21 %:2·13 %:0·69 %, respectively. Connors *et al.* found C:MC:EC in

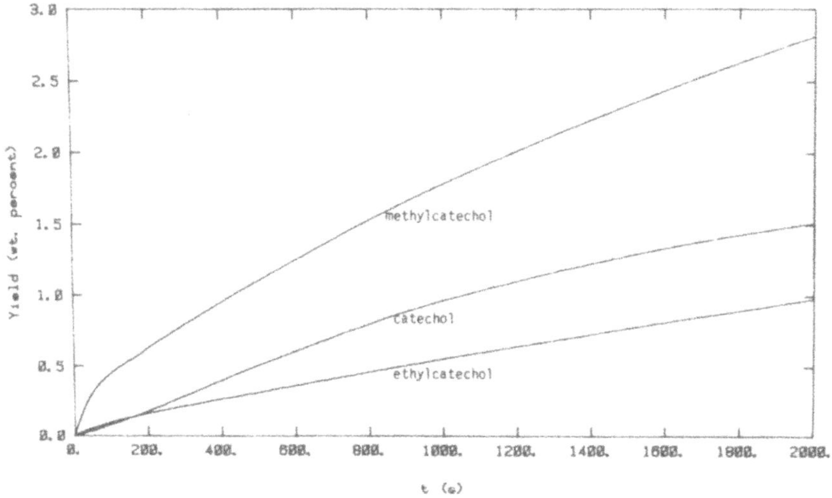

FIG. 5. Simulated single ring catechols yields, 400 °C.

yields of 0·99 %:1·11 %:0·80 % at 400 °C and 900 s from Kraft lignin pyrolysis in tetralin in a batch autoclave reactor. Simulation predictions at 400 °C and 1000 s of 0·97 %:1·79 %:0·56 % are in good agreement.

3.3.3. Phenols

The major monohydroxy phenolics ('phenols') predicted in simulated pyrolysis were phenol (P), p-cresol (PCR) and 4-ethylphenol (EP). Yields of these were negligible for pyrolysis at 300 °C and holding times up to 8000 s, but became appreciable at 400 °C. Phenol yields were always lower than catechol yields primarily because guaiacol pyrolysis reaction selectivity favors catechol over phenol.[10]

Domburg *et al.* found P:TCR(total cresols):EP yields were 0·10 %: 0·50 %:0·10 % from pyrolysis at 475 °C, which compares favorably with model predictions of P:PCR:EP = 0·29 %:0·52 %:0·17 % at 500 °C and 20 s. Iatridis and Gavalas report phenol in 0·10 % yield at both 400 °C, 20 s and 500 °C, 60 s, conditions at which the model predicted 0·008 % and 0·48 %, respectively. Connors *et al.* reported a product spectrum including P:PCR:EP = 0·48 %:1·39 %:0·49 % at 400 °C and 2700 s, and predicted yields at 400 °C and 3000 s were 0·32 %:0·64 %:0·24 %. Model predictions are thus in reasonable accord with available experimental information. Simulated and experimental total single-ring phenolics yields agree rather well at 400 °C. The model prediction of 8·1 wt % at 400 °C and 1000 s compares with the experimental values of 7·0% at 400°C and 900 s and 6·2% at 400°C and 1200 s found by Connors *et al.*[8] and Jegers,[11] respectively. Experimental single-ring phenolics yields decreased at higher pyrolysis times due, presumably, to secondary catechol degradation.

3.4. Multiring Phenolics

Simulated multiring phenolic products accounted for as much as 15% of the original lignin weight at 500°C and 1000 s. Individual products were not identified, but rather multiring phenolics were lumped as follows. First, products were categorized as either condensed or uncondensed. In the former category fell phenanthrene and fluorene derivatives, whereas the latter comprised aromatics like bibenzyl, stilbene and diphenylmethane. Second, entries in both categories were subdivided on the basis of the number of oxygen atoms (includes hydroxyls and methoxyls) that each contained.

The simulation predicted yields of bibenzyl:stilbene:diphenylmethane: phenanthrene:fluorene derivatives to be 4·72%:1·57%:4·59%:0·018%: $3{\cdot}8 \times 10^{-6}$% for pyrolysis at 400°C and 3500 s; 90% of these multiring products contained either three or four oxygen-containing substituents. Bibenzyl links were reactive and stilbene and diphenylmethane links were more stable at 500°C and 1000 s where these same products were predicted in yields of 0·95%:3·77%:9·72%:0·42%:$4{\cdot}6 \times 10^{-3}$%; 93% of these compounds contained either three or four oxygen-containing substituents.

The foregoing results show that even the lightest multiring phenolics contained three to four (oxygen-containing) substituents, and thus had molecular weights in the range 250–300. Recovery and gas-chromatographic analysis of products like these is difficult, and it is not surprising that very little experimental information concerning these products is available.

Chan and Krieger reported 0·091% fluorene and 0·143% anthracene from their microwave pyrolysis, whereas Iatridis and Gavalas found dimers and larger fragments in their tar fractions. Their tar yields of approximately 10% at 400°C (60 and 120 s) and 12% at 500°C (60 s) agree rather well with multiring phenolic model predictions of 7·9% at 400°C (60 and 120 s) and 11·2% at 500°C (60 s).

3.5. Carbonaceous Residue

Simulated carbonaceous residue yields at 300, 400 and 500°C are shown in Fig. 6. Inspection of Fig. 6 shows that asymptotic yields were attained

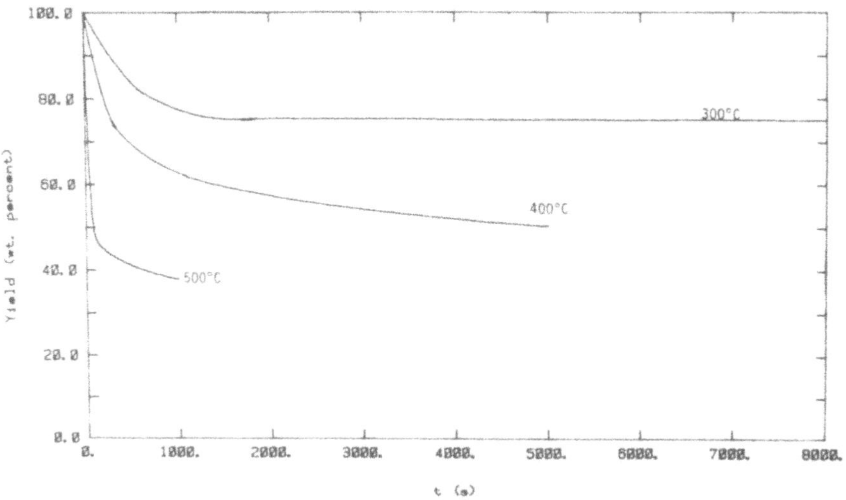

FIG. 6. Simulated carbonaceous residue yields.

quickly at 300 and 400 °C, whereas the residue yield continued to fall at 500 °C. This suggests that scissile β-ether and bibenzyl bonds were cleaved rapidly at 300 and 400 °C, and that the more refractory biphenyl and diphenyl ether bonds were stable; at 500 °C the reactivity of these latter bonds was more appreciable.

Quantitative comparison with experiment is available in only one instance. The temporal variations of lignin weight loss reported by Iatridis and Gavalas at 400 and 500 °C can be used to infer residue yields. Residue yields of 79 % (400 °C, 60 s), 72 % (400 °C, 120 s) and 53 % (500 °C, 60 s) agree well with model predictions of 74·6 %, 73·3 % and 51·2 % at the same conditions.

4. DISCUSSION

The present discussion concerns the model's accuracy, inherent limitations and likely applications.

4.1. Accuracy of Model Predictions

Kraft pulping renders a residual lignin that is quite different in chemical structure than wood's native lignin. Because of this, our Kraft lignin pyrolysis model encompassed significant changes from previous native lignin models: the phenolic ring substituents and inter-ring linkages shown in the Marton structure are unlike those in the Freudenberg structure; also, the concept of identifiable methoxyphenol and propanoid side chain substituents on each ring was lost, and fully all six ring substituents needed to be considered.

In this light it is interesting that Kraft lignin pyrolysis model predictions were not strikingly different from native lignin pyrolysis model predictions. Both Kraft and native lignin pyrolysis simulation models predicted similar yields of methane, carbon monoxide, methanol, water and phenolics. These similarities likely occurred because the α and β ether bonds that are cleaved during native lignin dissolution during Kraft pulping are also thermally scissile. Fused-ring products, however, like phenanthrene and fluorene, were important Kraft lignin pyrolysis products because of the prevalence of bibenzyl and diphenylmethane bonds in this lignin. These bonds are less prevalent in native lignins.

The inclusion of pathways for the formation of fused-ring products also affects the pyrolysis reaction hydrogen balance. Thus, for example, hydrogen formally evolved in the reaction of stilbene to phenanthrene can balance the hydrogen formally consumed in conversion of bibenzyl to two toluene molecules. This pathway for formal hydrogen liberation was not present in the MIT native lignin model.

As regards the accuracy of model predictions, it is reasonable to conclude that except for the two obvious examples of water and methanol yields, model predictions were not only qualitatively correct but in near quantitative agreement with previous experimental pyrolyses. Decisive evaluation is difficult because the deviations among previous experimental pyrolyses are as large as the differences between model predictions and any one experiment. While predicted water yields are quite likely lower than experimental water yields because (at least in part) of physical water associated with the experimental lignin substrates, methanol under-predictions must be considered a legitimate failure of the model as currently formulated.

4.2. Model Limitations

The simulation focuses on chemical aspects of lignin pyrolysis. Although actual lignin thermolysis is likely a complicated solid-, gel- or liquid-phase reaction, simulated pyrolysis essentially comprises an independent set of gas and liquid phase pyrolysis pathways. No account of the lignin macrostructure or the influence of transport limitations was made. Delineation of operative pyrolysis mechanisms will be helpful since, for example, whether a lignin fragment decomposes via a free-radical chain or a concerted pericyclic mechanism has profound modeling implications.

The underprediction of water and methanol yields points to the incompleteness of the simulation's model compound data base. The selectivity of propanoid side chain decomposition to carbon monoxide and water, recognition of the experimentally observed decomposition of catechols,[11] inclusion of the reaction paths of pinoresinols, lactones and phenylcoumaran lignin moieties, and the chemical effects of substituents are all areas where the current model requires attention.

4.3. Model Utility

The distinguishing feature and ultimate utility of the pyrolysis model derives from the lignin structural analyses and experimental model compound reactions. The model comprises fundamental reaction pathways with interpretable kinetic parameters and, in a few favorable cases, mechanisms. Thus, the model is truly *a priori* with respect to lignin pyrolysis and can be used for extrapolation and prediction of: pyrolysis at different reaction conditions; pyrolysis of hardwood lignins (by accounting for the additional methoxyl groups and side chain redistributions); the likely effect of a promising solvent or reactor configuration; and optimization, design and scale-up.

The model should also be useful for guidance of future research. Thus, the potential process benefit of, for example, a hundred-fold increase in catalyst activity can be tested before considerable effort and expense is devoted toward its development. Also, experimental phenolic product spectra obtained from lignin pyrolyses are well known to be complex and often different from one investigation to another. The simulation model can be used to suggest possible phenolic products and thus help in GC or GC–MS identification of the components of the experimental phenolic product spectra.

Although the model is critically dependent upon lignin structural information, it is not limited to the Marton or Freudenberg structures. Other and, importantly, newly acquired lignin structural information[12,13] can be used, and the modeling approach can also be used to describe coal or other macromolecule pyrolyses.

Finally, it is interesting to compare the kinds of information available from the model predictions and actual lignin pyrolyses. The model can be 'integrated' from a microscopic, chemical bonding description of a macromolecule's pyrolysis to predict overall gas, liquid, tar and char fractions. On the other hand, experimental observations of weight loss or overall gas, liquid, tar and char yields cannot be 'differentiated' to reveal fundamental reaction paths, kinetics and mechanisms. This motivates our continued interest in the present modeling approach.

5. CONCLUSIONS

1. A statistical interpretation was proposed of lignin, lignin pyrolysis and pyrolysis product formation coupled with experimental model compound reaction pathways to allow *a priori* mathematical simulation of Kraft lignin pyrolysis.
2. Predicted pyrolysis products, including methane, carbon monoxide, carbon dioxide, hydrogen, water, methanol, single-ring phenolics (phenols, guaiacols and catechols), multiring phenolics (bibenzyl, stilbene, diphenylmethane, fluorene and phenanthrene derivatives) and char were those observed in experimental pyrolyses. Quantitative agreement between predicted and experimental product yields and weight loss was favorable save water and methanol yields; methanol prediction was the most conspicuous model failure.
3. The model comprises fundamental reaction pathways and rate parameters that can be used not only to help explain experimental pyrolysis results but also to predict pyrolyses effected under untested reaction conditions, i.e. various temperatures, pressures, solvents and catalyst types. The model should also be a useful tool for the guidance of future lignin conversion research.

ACKNOWLEDGEMENT

The authors gratefully acknowledge the financial support of the University of Delaware Research Foundation.

REFERENCES

1. Klein, M. T. and Virk, P. S. (1982). Modelling of lignin thermolysis. *Fuel* (accepted).
2. Marton, J. (1971). In *Lignin*, Sarkanen and Ludwig (Eds). Wiley, New York.
3. Gear, C. W. (1971). *Numerical Initial Value Problems in Ordinary Differential Equations*. Prentice-Hall, Englewood Cliffs, N.J.
4. Petrocelli, F. P., M.Ch.E. Thesis, University of Delaware, in preparation.

5. Iatridis, B. and Gavalas, G. R. (1979). *Ind. Eng. Chem. Prod. Res. Dev.* **18**(2), 127.
6. Chan, R. W. and Krieger, B. B. (1981). *J. Appl. Polym. Sci.* **26**, 1.
7. Graef, M., Allan, G. G. and Krieger, B. B. (1979). *ACS Div. Petroleum Chem. preprints* **24**(2), 432.
8. Connors, W. J., Johanson, L. N., Sarkanen, K. V. and Winslow, P. (1980). *Holzforschung* **34**, 29.
9. Domburg, G. E., Sergeeva, V. N. and Kalnish, A. I. (1972). Thermal Analysis, Vol. 3. *Proc. 3rd ICTA Davos [1971], Burkhaeuser, Basel*, pp. 327–40.
10. Klein, M. T. (1981). Sc.D. Thesis, M.I.T.
11. Jegers, E. H. (1982). M.Ch.E. Thesis, University of Delaware.
12. Glasser, W. G., Glasser, H. R. and Morohoshi, N. (1981). *Macromolecules* **14**, 253.
13. Adler, E., Becker, H. D., Ishihara, T. and Stamvik, A. (1966). *Holzforschung* **20**(1), 3.
14. Domburg, G. E. (1974). Thermal Analysis, *2. Proc. 4th ICTA*, Budapest.
15. Friedlin, L., Balandin, A. A. and Nazarova, N. M. (1949). *Izvest. Akad. Nauk S.S.S.R. Otdel Khim. Nauk* No. 1, 102.
16. Kravchenko, M. I., Kiprianov, A. I. and Korotov, S. (1970). *Nauch. Tr. Leningrad Lesotekh. Akad.* **135**(2), 60.
17. Obolentsev, R. D. (1946). *J. Gen. Chem. (USSR)* **16**, 1959.
18. Shaposhnikov, U. K. and Kosyukova, L. V. (1965). *Khim. Pererabotka Drev., Ref. Inform.* No. 3, 6.
19. Cavitt, S. B., Sarrafizadeh, R. H. and Gardner, P. D. (1962). *J. Org. Chem.* **27**, 1211.
20. Gardner, P. D., Sarrafizadeh, H. R. and Brandon, R. L. (1959). *J. Am. Chem. Soc.* **81**, 5515.
21. Juntgen, H. and van Heek, K. H. (1968). *Fuel* **47**, 103.
22. Benjamin, B. M., Raaen, V. F., Maupin, P. H., Brown, L. L. and Collins, C. J. (1978). *Fuel* **57**, 69.

15

Kinetics of Wood Gasification by Carbon Dioxide and Steam

WILLIAM F. DEGROOT and the late FRED SHAFIZADEH
Wood Chemistry Laboratory, University of Montana, Missoula, Montana, USA

1. INTRODUCTION

Gasification of wood and related biomass fuels provides a means of converting these renewable resources to an energy form that can be stored, transported and burned more cleanly and efficiently. The availability of biomass materials, particularly in industries such as the food processing and forest products industries, and their lower nitrogen, sulfur and ash content compared to conventional fuels such as coal, offer distinct advantages for the use of biomass as gasification substrates. Furthermore, gasification can be used in combination with other thermal processes such as pyrolytic saccharification of wood[1] to recover the fuel value of the energy-rich char which is a byproduct of this process.

The development of technically and economically feasible systems for the conversion of biomass to energy by thermal processes requires a fundamental understanding of the complex transitions involved. There is also a general need for more thorough characterization of substrates which will allow the results of one study to be extended to predict the behavior of other substrates on the basis of their chemical composition. While the structure and reactivity of coal have been, and continue to be extensively studied with respect to the production of gaseous fuels, relatively little attention has been focused on biomass fuels. Through research in this and other laboratories we have just begun to understand the processes leading to the structure of biomass chars and their reactivity.[2-5] In this paper we present the results of recent research directed toward a chemical description of the factors controlling gasification of biomass by steam and carbon dioxide and the effect of catalysts on these processes.

2. EXPERIMENTAL

Samples of Douglas fir heartwood (*Pseudotsuga menzeisii*) and cottonwood sapwood (*Populus trichocarpa*) were ground and seived to 20/40 mesh. The samples were analyzed for Klason lignin and ash content by standard methods. The naturally occurring inorganic elements in the wood were determined by flame emission spectroscopy. The elements analyzed were chosen for potential catalytic activity.

Wood samples were treated with inorganic salts by placing the ground wood in an aqueous solution of the salt, removing excess solution by filtration with suction, and air drying the treated wood. Solution strengths were adjusted to give 0·1 mole of the active element per kg of wood, and the treated samples were analyzed for the active element of the added salt by a commercial laboratory. Levels of treatment from 0·05 to 0·2 moles/kg were considered satisfactory for further analysis. Chars were prepared at a heat

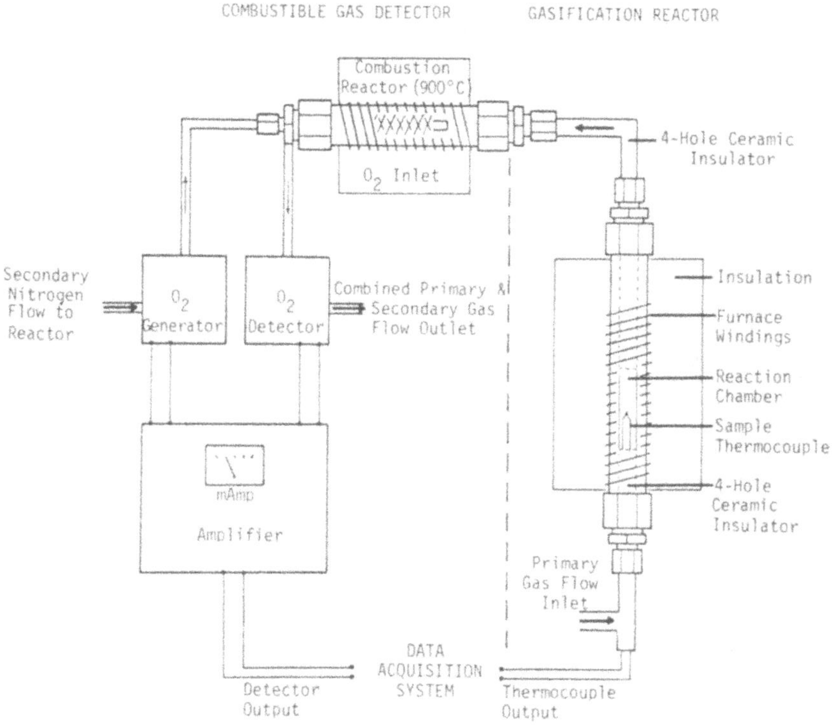

FIG. 1. Schematic diagram of gasification reactor and detection system.

treatment temperature (HTT) of 1000 °C by pushing a porcelain boat containing 300–500 mg of the wood into the heated zone of a well purged, preheated tube furnace. The sample was heated for 10 min under flowing nitrogen (100 ml/min). Traces of oxygen were removed from the nitrogen stream by passing it through a high capacity 'Oxy-Trap' (Alltech Associates) and then through copper turnings heated to 800–900 °C. Chars were cooled for at least 15 min under nitrogen before being removed from the tube furnace. The chars were stored under reduced pressure of nitrogen.

The rate of gasification of chars was determined by measuring the rate of evolution of combustible gases. Gasification temperatures were limited to 950 °C to minimize competition between the concurrent pyrolysis and gasification reactions. Combustible gases were detected by evolved gas analysis (EGA) using the reaction coulometer detector[6] placed at the outlet of the gasification reactor, as shown in Fig. 1. This detector is specific for combustible gases and provides a quantitative measure of the stoichiometric quantity of oxygen required for combustion of evolved gases. Thus, for the detection of H_2 and CO, the primary products of the reactions of interest in this study, the detector produces 0·5 mole of O_2 per mole of product gas. The detector does not differentiate between the two, but provides a quantitative determination of the combined yields of both. This in turn provides an accurate measure of the energy content of the product gas.[6] The performance of the detector system was checked daily by applying a known current to a hydrogen generator placed in line with the gasification reactor nitrogen purge gas and comparing the detector response with the theoretical value. In all cases the response was within 3 % of the theoretical value.

The gasification reactor was composed of a $\frac{1}{4}$ in. OD × $\frac{1}{8}$ in. ID impervious, pure alumina ceramic tube (99·8 % Al_2O_3, Omega Engineering) wound with nichrome wire. The reaction chamber was 2·0 cm long and was bounded at either end by a 4-hole ceramic insulator ($\frac{1}{8}$ in. OD, 0·02 in. ID) inserted into the alumina tube, which allowed gases to flow into and out of the reaction chamber. In extended isothermal gasification experiments, where depletion of the char particles could lead to entrainment in the gas stream, the upper insulator was replaced by one having four holes of 0·016 in. ID. The pressure inside the reactor was approximately 2 % greater than atmospheric with either outlet tube. The lower insulator also supported a Chromel-Alumel thermocouple which extended 1·0 cm into the reaction chamber and was used for furnace temperature control as well as temperature data. Char samples (8–12 mg) were placed in the reaction chamber and the sample tube was purged with

30 cm^3/min of nitrogen or a combination of nitrogen and the reactant gas. The reactor heater was controlled to within 1 °C of the set point by a proportioning controller.

Reactant gases were added to the high purity nitrogen purge gas using linear multiport valves positioned near the reactor for rapid switching. For CO_2 gasification the nitrogen purge gas was the combination of 10 and 20 cm^3/min flows, and 10 cm^3/min of dried CO_2 (Matheson, Coleman Inst. Grade, 99·99% min) replaced the equivalent flow of N_2 during the gasification. For steam gasification the purge gas was diverted and bubbled through water heated to 41·0 ± 0·5 °C to give a partial pressure of H_2O of 58 ± 2 mm Hg. The switching valve and transfer lines between the water bath and gasification reactor were heated to >50 °C to prevent condensation. Periods during which the char was exposed to the reactive gas were accurately timed using a stopwatch. Chars were weighed before and after gasification using a Perkin-Elmer AD-2 microbalance sensitive to 0·1 μg.

The output of the reaction coulometer detector and the sample thermocouple millivoltage were monitored with an interactive micro-computer-based data aquisition system using a Hewlett Packard 3497A Data Aquisition Unit. The detector output, sample thermocouple and digital clock were each read at 4·0 s intervals during the experimental runs. The thermocouple millivoltage was converted to temperature during data aquisition using a pair of quartic equations.[7] This conversion was accurate to within 0·1 °C in the temperature range required for gasification.

3. RESULTS AND DISCUSSION

3.1. Analysis of Wood and Chars

Analysis of the wood and char samples used in this study is given in Table I. The major difference in the two species is the higher lignin content in Douglas fir. This in turn gives rise to the higher char yield from Douglas fir because of greater thermal stability of the aromatic lignin structure relative to that of the carbohydrate fraction. The cottonwood sample is higher in ash content, which is important in terms of the ability of the naturally-occurring inorganic constituents to catalyze gasification reactions in the char. The surface areas of unreacted char are very high, indicating high availability of carbon for reaction. This may be due to the high temperature at which these chars were prepared, and the correspondingly high degree of

TABLE I

ANALYSIS OF WOOD AND CHAR SAMPLES (HTT 1 000 °C)

Species	*Percent ash*	*Klason lignin (%)*	*Char yield (%)*	*Surface area (m^2/g)*
Cottonwood	0·41	20·3	9·2 ± 0·3	797
Douglas fir	0·10	27·4	11·7 ± 0·3	891

pyrolytic gasification leading to their formation. By comparison chars from pure cellulose (HTT 550 °C)[3] have surface areas of 300–400 m^2/g.

The analysis of the inorganic fraction of the two samples is shown in Table II. The higher overall ash content in cottonwood is reflected in significantly higher quantities of potassium, magnesium and calcium in that sample. These differences are expected to be significant in terms of char reactivity since all of these elements are known to be good catalysts for the gasification of coal and graphite by either steam or CO_2. Other elements which are known to be good catalysts, such as sodium and iron, are not present in sufficient quantities to have a strong effect on the gasification of these samples.

The effect of addition of inorganic additives on the yields and elemental analysis of chars is shown in Table III. The level of active species added is less than 1 % in all cases and varies from 0·08 to 0·19 mmole/g. This level of addition represents an eight-fold increase over the natural level of potassium in the cottonwood sample and an increase of nearly 200 times the potassium level naturally present in Douglas fir. The effect of additives on char yield is quite consistent in both species. The addition of alkali metals results in a substantial increase in the yield of char, while the addition of magnesium, iron and nickel salts results in a slight increase in the char yield in Douglas fir and a slight decrease in cottonwood.

The elemental analysis of the chars indicates an approximate empirical

TABLE II

ANALYSIS OF INORGANIC CONTENT OF WOOD SAMPLES

Species	*Percent ash*	*Elemental analysis (μg/g wood)* Li	Na	K	Mg	Ca	Sr	Mn	Fe
Cottonwood	0·41	< 1	18	668	324	869	< 5	< 5	5
Douglas fir	0·10	< 1	3	32	53	314	< 3	20	< 3

TABLE III

YIELD AND ELEMENTAL ANALYSIS OF CHAR FROM UNTREATED AND TREATED WOOD

Species	*Additive*	*Additive concentration* (*Percent*)	(*mmole/g*)	*Percent char*	*Percent ash in char*[a]	*Elemental analysis* *C*	*H*	*O*[b]
Cottonwood	Untreated	(0·41 % ash)[c]		9·2	4·5	90·38	0·78	4·3
	Sodium carbonate	0·30 % Na	0·13	16·1	2·5			
	Potassium carbonate	0·56 % K	0·14	15·9	4·2			
	Magnesium acetate	0·20 % Mg	0·08	9·5	2·2			
	Ferric citrate	0·46 % Fe	0·08	8·1	5·8			
	Nickel acetate	0·55 % Ni	0·09	7·8	8·0			
Douglas fir	Untreated	(0·10 % ash)[c]		11·7	0·9	94·69	0·57	3·8
	Sodium carbonate	0·39 % Na	0·17	18·8	2·8	91·91	0·78	4·5
	Potassium carbonate	0·76 % K	0·19	21·0	4·4	88·50	0·80	7·9
	Magnesium acetate	0·21 % Mg	0·09	13·6	1·7	91·03	1·08	6·2
	Ferric citrate	0·50 % Fe	0·09	14·1	3·6	94·23	0·77	1·4
	Nickel acetate	0·68 % Ni	0·12	12·6	6·1	93·16	0·61	0·1

[a] Calculated as the weight of inorganic material in the original wood (calculated as the oxides of the lowest common oxidation number of added inorganic material in the case of treated samples).
[b] Organic oxygen by difference.
[c] For analysis of ash see Table II.

formula of $C_{28}H_{2.9}O$ for cottonwood char and $C_{33}H_{2.5}O$ for Douglas fir char, suggesting a high degree of carbonization and unsaturation in these chars. Additive treatment has a less pronounced effect on the elemental composition than on the char yield. Sodium, potassium and magnesium all reduce the carbon content in the char slightly, even though the effect of magnesium on the char yield is very different from that of the alkali metals.

3.2. Isothermal Gasification of Chars by Carbon Dioxide and Steam

The behavior of chars prepared from untreated cottonwood and Douglas fir under conditions of isothermal gasification by CO_2 are shown in Figs 2 and 3, respectively. In both cases the extent of gasification of the chars was greater than 95%. In order to achieve comparable rates of gasification between the two samples, the gasification of the Douglas fir char had to be run 50 °C higher, presumably because of the higher level of catalytic minerals naturally present in the cottonwood sample.

Figure 2 shows the rate of gasification of cottonwood char, first under pyrolytic conditions under N_2 up to 1000 °C, then under 33% CO_2 in N_2 at 900 °C. The reaction coulometer detector output is converted to the equivalent rate of CO production in this and other figures, although other

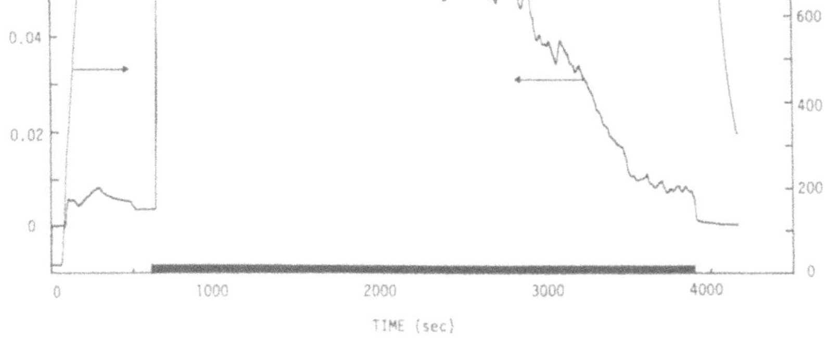

FIG. 2. Isothermal CO_2 gasification of char (HTT 1000 °C) prepared from untreated cottonwood. Heavy line on time axis indicates period during which char was exposed to 1:2 (v/v) mixture of CO_2 in N_2.

gases are also formed. The erratic rate behavior in the latter stages of the reaction is unexplained at this time, but was occasionally seen in other samples as well, although generally when additives were present. It is obvious from Fig. 2 that the gasification does not proceed in a manner consistent with a kinetic expression which is first order with respect to char. Instead there is an apparent 'induction period' during which the rate increases. This period could be explained either by increasing the available surface area of the char during the initial stages of reaction or by the accumulation of a steady state level of surface oxides. After this induction period the rate of gasification decreases, although a small fraction of the char remains resistant to gasification. The gasification was terminated before the char was completely gasified. Repeated isothermal experiments of the type shown in Fig. 2 indicated that this type of rate behavior was characteristic of cottonwood, although the sharp peak in the reaction rate at the outset of the gasification was sometimes much more pronounced, and the rate decreased from this initial peak and reached a constant level at the end of the induction period before decreasing again.

The gasification of Douglas fir char, shown in Fig. 3, also occurs with an induction period, and the rate begins to decrease when one-third to one-half of the sample has been gasified. This rate behavior during isothermal

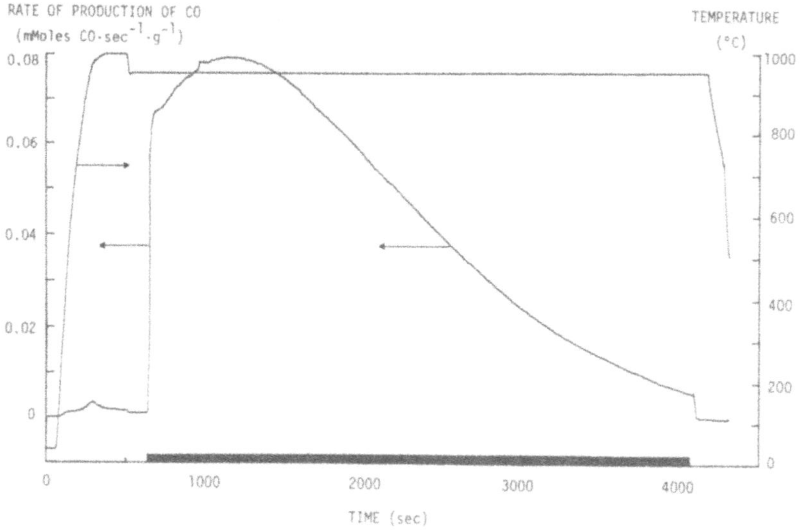

FIG. 3. Isothermal CO_2 gasification of char (HTT 1000 °C) prepared from untreated Douglas fir. Heavy line on time axis indicates period during which char was exposed to 1:2 (v/v) mixture of CO_2 in N_2.

gasification was again quite reproducible. The behavior of cottonwood and Douglas fir differed primarily in the presence of the rate maximum at the outset of the gasification in cottonwood.

The isothermal gasification of these samples by steam proceeded at approximately the same rate and exhibited the same rate behavior with time with two exceptions: (1) the rate was generally more constant over time, with a less pronounced increase in the rate during the induction period and a less pronounced decrease thereafter; and (2) in the gasification of cottonwood char there was consistently a sharp rise in the rate when the gasification was approximately 70 % complete, as shown in Fig. 4. The char remaining after this maximum was always much less reactive, and was not gasified in the experiment shown.

3.3. Kinetic Analysis of the Gasification of Chars from Treated and Untreated Wood

In view of these unique variations in gasification rate with time, it is very difficult to formulate a meaningful kinetic expression which will predict the gasification rate throughout the reaction. Therefore, for the purposes of

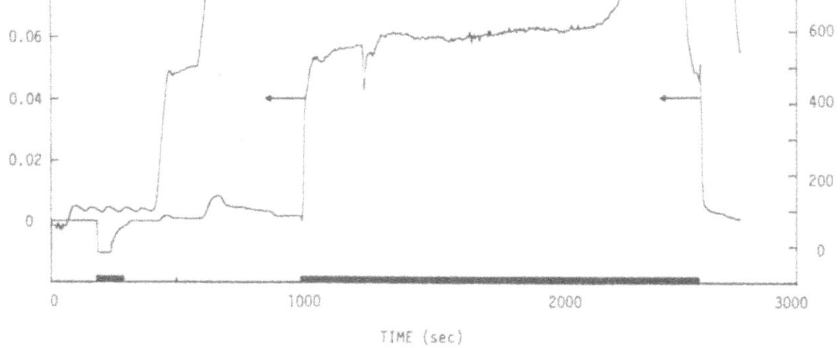

FIG. 4. Steam gasification of char (HTT 1000 °C) produced from untreated cottonwood. Heavy lines on time axis indicate periods during which char was exposed to mixture of 58 torr of water vapor in N_2.

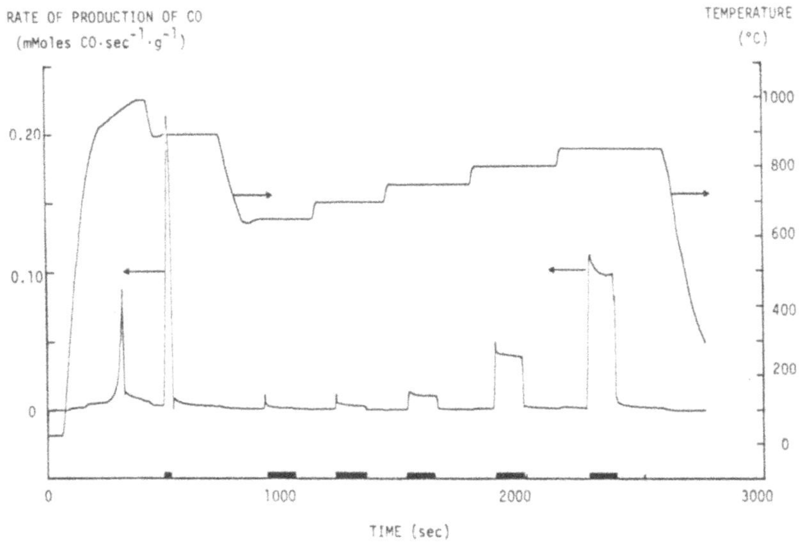

FIG. 5. Temperature dependence of the rate of gasification of char (HTT 1000 °C) produced from Douglas fir treated with sodium carbonate. Heavy lines on time axis indicate periods during which char was exposed to 1:2 (v/v) mixture of CO_2 in N_2.

this study, the following zero order kinetic expression was used to predict the gasification rate:

$$\text{Rate} = d[CO]/dt = k W_0$$

where k is the experimentally determined rate constant and W_0 is the initial weight of the char. The rate constant for this expression was determined in the initial stages of the gasification, prior to the induction period. Therefore, the rate expression is applicable to the average rate of gasification during the period of reaction in which most of the char (60–80 %) is gasified. The time required for complete gasification of the char would be up to two times greater than that predicted by the rate expression.

A typical gasification experiment for the determination of Arrhenius kinetic parameters is shown in Fig. 5. The first gasification peak is due to pyrolysis of the char. This was generally higher for treated samples such as the one shown, and usually peaked before the maximum pyrolysis temperature was reached. Since the chars had been previously treated at 1000 °C for 10 min, we assume that this peak is due to surface oxides which may have formed during storage and handling. The initial pyrolysis at

1000 °C was included to remove these surface species. The temperature was then reduced to 900 °C (950 °C for untreated samples) and the reactant gas was valved into the reactor for 30 s. This 'pregasification' step was included to eliminate particularly reactive sites such as those producing the initial rate maximum in Fig. 2. The gasification rate for this peak was not included in the kinetic analysis. The temperature was then lowered and gasification rates were determined for temperature intervals of at least 50 °C up to, but not exceeding, the temperature of 'pregasification'. Before each gasification step, the temperature was allowed to stabilize and the baseline corresponding to the pyrolysis rate at that temperature was allowed to come to a constant level. After replacing the reactive gas with pure N_2 the char was held at the same temperature for 1–4 min, depending on the size of the preceding peak, to allow any surface species formed during gasification to be desorbed and for a new pyrolysis baseline to be established. The rate at each temperature was determined by digital integration of the peak over time from a lower limit corresponding to a point on the pyrolysis baseline just prior to the gasification peak to an upper limit corresponding to the last point on the pyrolysis baseline before the temperature setting was changed. The rate was then determined by dividing the integral by the time period that the char was exposed to the reactive gas. In this way the gasification due to reaction with the reactive gas was separated from that due to pyrolysis, and any surface intermediates which were not immediately desorbed were included in the rate determination. After the final gasification peak the reactor temperature was reduced to establish a detector baseline free of pyrolytic gasification. In a typical kinetic experiment 80–90 % of the gasification was due to the oxidative gasification reaction being studied, and the remainder due to pyrolysis.

Rates determined in this way were then analyzed according to zero-order Arrhenius kinetics to determine the apparent activation energy, E_a and frequency factor, A. Arrhenius plots for gasification of cottonwood and Douglas fir are shown in Figs 6 and 7, respectively. In these plots data from several kinetic experiments were pooled and a single activation energy and frequency factor were calculated from the pooled data.

Arrhenius parameters calculated in this way for the gasification of chars by CO_2 are given in Table IV. Estimated standard errors were calculated from pooled data according to standard statistical methods. This treatment is not strictly valid, since the values from separate kinetic experiments are not independently variable. However, Arrhenius parameters calculated by averaging the results of individual experiments agreed very well with those for the pooled data, although the standard deviations were up to two times

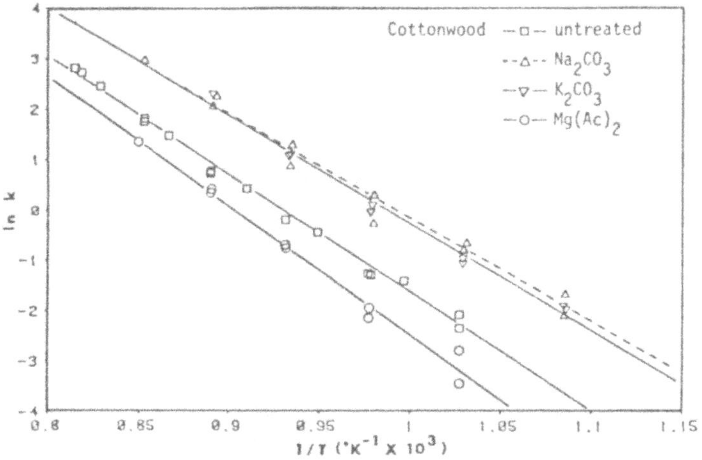

FIG. 6. Arrhenius plots for the CO_2 gasification of chars (HTT 1000 °C) prepared from untreated and treated cottonwood (33 % CO_2 in N_2).

higher. Thus, the deviations given in Table IV should be considered as indicators of the relative deviations in the Arrhenius parameters and should not be used to evaluate statistically differences in them. It is clear from the data given in Table IV that the activation energies for gasification of Douglas fir chars by CO_2 with alkali metal salts is lower than the activation energy for the uncatalyzed reaction. Because of the catalysis by the natural

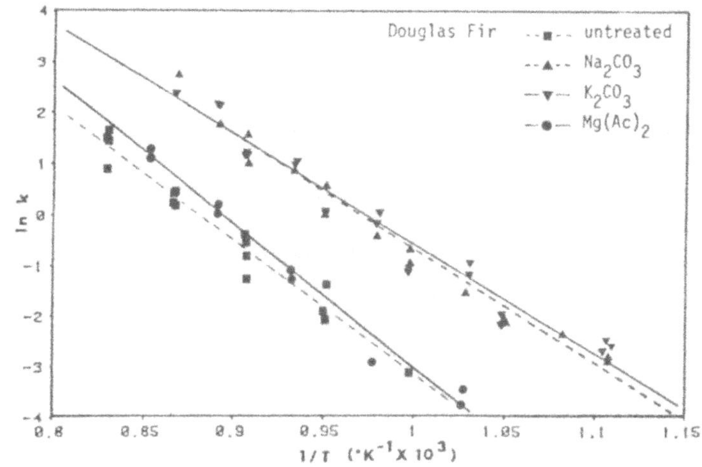

FIG. 7. Arrhenius plots for the CO_2 gasification of chars (HTT 1000 °C) prepared from untreated and treated Douglas fir (33 % CO_2 in N_2).

TABLE IV

ARRHENIUS KINETIC PARAMETERS AND CORRESPONDING ZERO ORDER RATE CONSTANTS FOR THE CO_2 GASIFICATION OF CHARS PREPARED FROM TREATED AND UNTREATED WOOD (HTT 1 000 °C)

Species	*Treatment*	*Arrhenius parameters* E_a	ln A	*Calculated rate constants*[a] *(mmole CO min*$^{-1}$ *g*$^{-1}$*)* $k_{700\,°C}$	$k_{900\,°C}$
Cottonwood	Untreated	46·9 ± 0·9	22·9 ± 0·4	0·11	6·6
	Na_2CO_3	40·9 ± 1·9	20·4 ± 0·9	0·47	17·4
	K_2CO_3	42·9 ± 1·8	21·3 ± 0·9	0·41	18·1
	$Mg(Ac)_2$	51·4 ± 2·4	23·5 ± 1·1	0·05	4·3
Douglas fir	Untreated	52·8 ± 2·8	23·4 ± 1·3	0·02	2·1
	Na_2CO_3	44·8 ± 1·8	21·8 ± 0·9	0·25	13·2
	K_2CO_3	42·6 ± 1·8	20·8 ± 0·9	0·29	12·5
	$Mg(Ac)_2$	55·9 ± 2·7	25·1 ± 1·3	0·02	3·1

[a] Units of weight refer to the initial weight of the char.
Gasification is by 33 % CO_2 in N_2.

mineral components of cottonwood, the differences in activation energies for the catalyzed and uncatalyzed reactions are less pronounced. However, the data strongly indicate a reduction in the activation energy by added catalysts. Another interesting feature of the results given in Table IV is the higher apparent activation energy in cottonwood char with magnesium acetate. This treatment had no apparent effect on the gasification of Douglas fir char. The increase in the activation energy for cottonwood char is probably due to the effects of ion exchange during treatment of the wood, which would replace some of the effective naturally occurring catalytic ions with the less effective magnesium ions, resulting in a reduction of the gasification rate and an increase in the apparent activation energy.

The Arrhenius parameters were used to calculate rates of gasification by CO_2 for chars from treated and untreated samples at 700 and 900 °C. These data, which are also given in Table IV, indicate the pronounced effect of added catalysts in enhancing the gasification reaction. For instance, addition of alkali metal salts to Douglas fir increases the gasification rate by a factor of 10–15 at 700 °C and a factor of 6–7 at 900 °C.

Arrhenius plots for gasification of Douglas fir chars (HTT 1000 °C) by 58 torr of steam are shown in Fig. 8. Because the results obtained in the kinetic analysis of gasification by CO_2 indicated that there were no significant differences between the species in terms of gasification behavior apart from the catalytic effect of the natural mineral component,

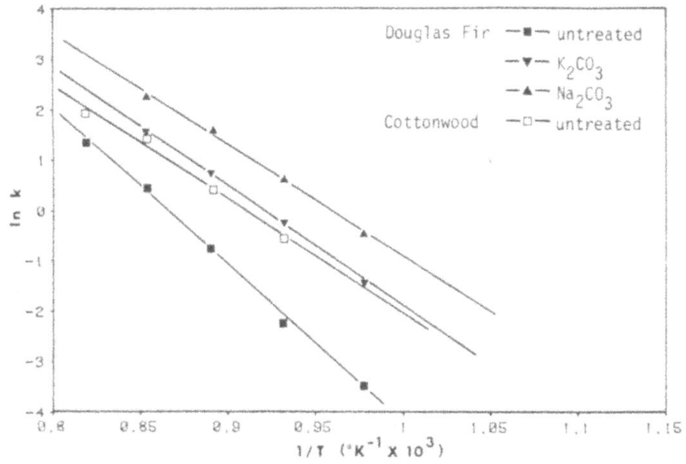

FIG. 8. Arrhenius plots for the steam gasification of chars (HTT 1000 °C) prepared from untreated cottonwood and untreated and treated Douglas fir (58 torr H_2O in N_2).

gasification kinetics were not determined for treated cottonwood chars. The effect of catalysis by magnesium on the gasification of char with steam was excluded from this phase of the study on the basis of its ineffectiveness in catalyzing gasification by CO_2 and preliminary results with steam gasification which indicated similar results. The Arrhenius kinetic parameters for the gasification of chars by steam are given in Table V. It is clear that catalysis by the alkali metals significantly lowers the activation

TABLE V

ARRHENIUS KINETIC PARAMETERS AND CORRESPONDING ZERO ORDER RATE CONSTANTS FOR THE STEAM GASIFICATION OF CHARS PREPARED FROM UNTREATED COTTONWOOD AND UNTREATED DOUGLAS FIR (HTT 1 000 °C)

Species	*Treatment*	*Arrhenius parameters* E_a	ln A	*Calculated rate constants*[a] *mmole* $[CO/H_2]$ min^{-1} g^{-1} $k_{700\,°C}$	$k_{900\,°C}$
Cottonwood	Untreated	45·0	20·6	0·07	3·7
Douglas fir	Untreated	61·7	26·9	0·01	1·5
	Na_2CO_3	43·9	21·2	0·22	10·7
	K_2CO_3	47·3	21·9	0·08	5·0

[a] Units of weight refer to the initial weight of the char.
Gasification is by 58 torr H_2O in N_2.

energy in Douglas fir chars and that the natural mineral component in cottonwood provides significant catalysis of this reaction. The activation energies are similar to those for gasification by CO_2 although rates calculated at both 700 and 900 °C are somewhat lower for the gasification by steam. The lower rate reflects the lower partial pressure of steam in the reactor gas stream; this should have no effect on the activation energies, however.

The apparent activation energies determined in this study for gasification of wood chars are relatively low on the spectrum of activation energies for coals and chars reported in published reviews,[8,9] which range from 40 to 90 kcal/mole, and they are comparable to those reported recently for coal chars[10] and even some graphitizing carbons.[11] The primary effect of the catalysts used in this study was to reduce the apparent activation energy of gasification. This finding differs from the results of McKee *et al.* for catalysis of coal chars, in which the rate enhancement was found to be due to an increase in the pre-exponential factor of the Arrhenius kinetic expression.[12] This implies a mechanism of catalysis due to increased reactivity of reactive sites as opposed to an increased number of reactive sites. The increase in apparent activation energy of CO_2 gasification of chars prepared from magnesium-treated cottonwood also supports this mechanism, since the reduction in reactivity was accompanied by an increase in the activation energy. As mentioned previously, this decreased reactivity is apparently due to ion exchange of magnesium for more active catalytic species, probably calcium. Thus, it appears that wood contains a relatively constant number of active catalytic sites which are associated with inorganic species and that prepyrolysis addition of gasification catalysts results in a change in the reactivity of these sites.

3.4. Gasification Catalysis by Transition Metals

Figure 9 shows repeated thermal cycles for the CO_2 gasification at 650 °C of a char prepared from cottonwood treated with ferric citrate. The initial gasification rate is nearly as great as that of a char prepared from untreated cottonwood gasified at 900 °C (see Fig. 2), but the rate declines very rapidly. When the sample is heated under nitrogen to higher temperatures, the catalytic activity of the iron is renewed, apparently due to reduction to the active form by the carbon. The reproducible two-step transition accompanying this process results in at least partial gasification of the carbon involved. When steam was used as the reactive gas, similar behavior was observed, although the catalyst deactivation was much more rapid.

With nickel as the added catalyst the initial gasification rate is greater

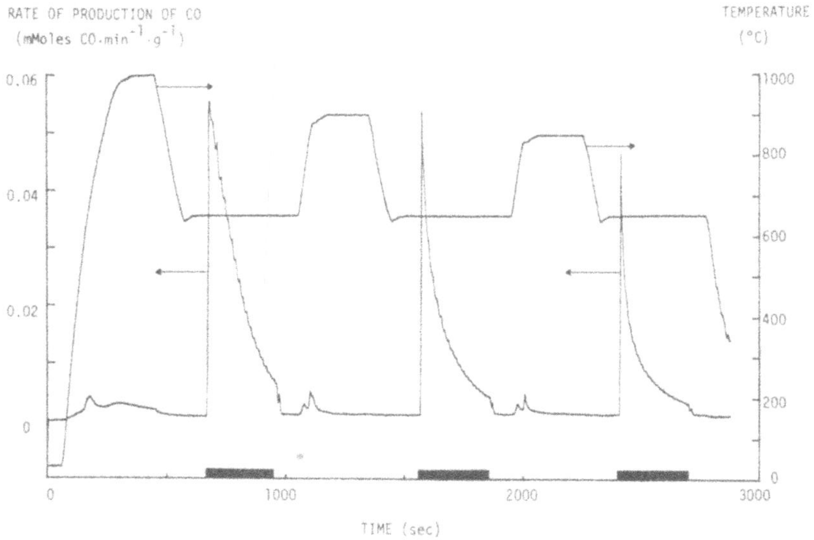

FIG. 9. Repeated gasification and reduction cycles in the CO_2 gasification of char (HTT 1000 °C) prepared from cottonwood treated with ferric citrate. Heavy lines on time axis indicate periods during which char was exposed to 1:2 (v/v) mixture of CO_2 in N_2.

and the deactivation of the catalyst is less rapid, as shown in Fig. 10. The initial rate of gasification at 650 °C of char prepared from Douglas fir treated with nickel acetate is approximately three times higher than that of untreated Douglas fir gasified at 950 °C. More than 60 % of the char was gasified before the gasification rate fell to 10 % of its original value. However, in contrast with the iron catalyst, the nickel was not reactivated when heated to 900 °C under nitrogen. Catalysis by nickel was much less effective for steam gasification than for CO_2 gasification. The initial gasification rate at a given temperature was much lower, and the rate of deactivation was greater.

Because of the deactivation of catalysts with time during gasification, no kinetic analysis of gasification catalysis by transition metals was attempted. It is clear that these metals are very active catalysts, particularly for the CO_2 gasification of wood chars. However, obtaining satisfactory catalysis with these metals would require that the catalysts be maintained in an active state, possibly in a reducing atmosphere or by reaction at temperatures other than those employed in this study. The results shown in Fig. 10 for CO_2 gasification of chars containing nickel indicate that it would be a

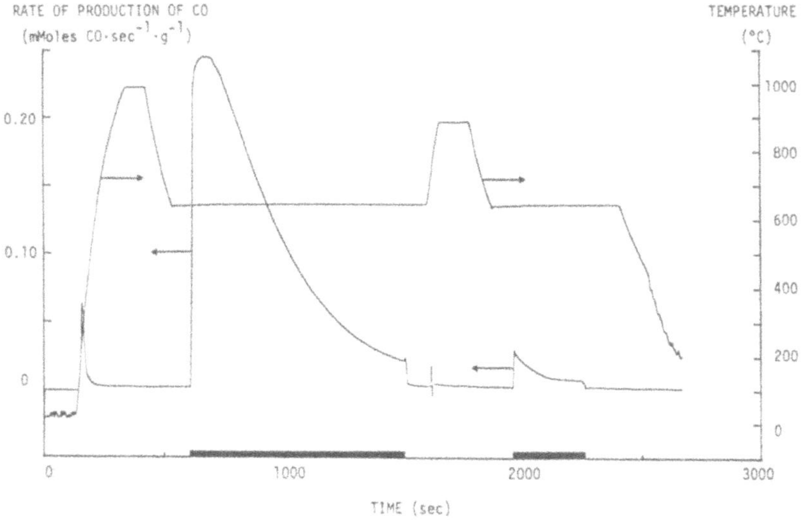

FIG. 10. Gasification by CO_2 of char (HTT 1000 °C) prepared from Douglas fir treated with nickel acetate. Heavy lines on time axis indicate periods during which char was exposed to a 1:2(v/v) mixture of CO_2 in N_2.

useful catalyst under the conditions employed, especially in combination with a second catalyst, e.g. one from the alkali metal group, which would provide catalysis when the nickel became inactive.

4. CONCLUSIONS

Gasification of wood chars by either steam or carbon dioxide is not a simple first order process, but goes through an induction period early in the reaction. After the induction period the reaction rate decays in a manner intermediate between zero and first order behavior. The overall process can be approximated by a zero order expression in which the rate is dependent primarily on the initial mass of the sample. In this study the rate of gasification by 225 torr of CO_2 was approximately the same as that obtained with 58 torr of steam. At equal partial pressures of the reactant gas the rate of steam gasification would no doubt be higher.

The gasification of wood chars by both steam and carbon dioxide is influenced by the natural mineral fraction, and high levels of ash can provide significant catalysis of gasification. Prepyrolysis addition of low levels of inorganic salts to the wood from solution is an effective means of further catalyzing these reactions. Elements of the alkali metal group were the most effective of the catalysts used in this study. Transition metals provided a very high level of catalysis initially, but their activity was lost relatively rapidly. For gasification by CO_2 nickel was a better catalyst than iron because of its slower deactivation, although it was less effective for catalysis of steam gasification. The addition of catalysts appears to involve ion exchange of the added ions for those naturally present in the wood. This mechanism would allow for very high dispersion of inorganic species in the resulting chars and correspondingly high levels of catalysis.

ACKNOWLEDGEMENT

The authors are pleased to acknowledge the interest and support of the Gas Research Institute through Contract No. 5080-363-0385.

REFERENCES

1. Shafizadeh, F., Cochran, T. G. and Sakai, Y. (1979). *AICHE Symposium Series* No. 184, Vol 75, 24.
2. Bradbury, A. G. W. and Shafizadeh, F. (1980). *Carbon* **18**, 109.
3. DeGroot, W. F. and Shafizadeh, F. (in press). *Carbon.*
4. Earl, W. L. (1982). In *Residential Solid Fuels: Environmental Impacts and Solutions*, J. A. Cooper and D. Malek (Eds). Oregon Graduate Center, Beaverton, Oregon.
5. McKay, D. M. and Roberts, P. V. (1982). *Carbon* **20**, 105.
6. Susott, R. A., Shafizadeh, F. and Aanerud, T. W. (1979). *J. Fire Flamm.* **10**, 94.
7. Powell, R. L. *et al.* (1974). National Bureau of Standards Monograph No. 125, US Government Printing Office, Washington, D.C.
8. Walker, P. L. Jr., Shelef, M. and Anderson, R. A. (1968). In *Chemistry and Physics of Carbon*, Vol 4, P. L. Walker, Jr. (Ed.). Marcel Dekker, New York.
9. McKee, D. W. (1981). In *Chemistry and Physics of Carbon*, Vol 16, P. L. Walker, Jr. and P. A. Thrower (Eds). Marcel Dekker, New York.
10. McKee, D. W., Spiro, C. L., Kosky, P. G. and Lamby, E. J. (1982). Preprints of *Fuel Chem. Div., ACS 183rd Mtg.* **20**(1), 74.
11. Marsh, H. and Taylor, D. W. (1977). *Carbon* **15**, 265.
12. McKee, D. W. (1980). *Fuel* **59**, 308.

16

Studies of the Rapid Pyrolysis of Sweet Gum Hardwood

THEODORE R. NUNN, JACK B. HOWARD, JOHN P. LONGWELL and WILLIAM A. PETERS*

Energy Laboratory and Department of Chemical Engineering, Massachusetts Institute of Technology, Cambridge, Massachusetts, USA

1. INTRODUCTION

Many previous studies on biomass pyrolysis, including several kinetic investigations, are reviewed by: Molton and Demmitt,[1] Lewellen *et al.*[2] and Hajaligol[3] (cellulose); Allan and Mattila[4] and Klein[5] (lignin); Roberts[6] and Wenzl[7] (wood); and Milne[8] and Antal[9] (different forms of biomass). Much remains to be learned about the quantitative effects of temperature, heating rate, solids residence time, volatiles residence time, pressure, gaseous atmosphere, sample dimension and biomass type on product yields, compositions and rates of production. Improved understanding of the role of primary decomposition and volatiles secondary reactions in pyrolysis is also needed. Better information in these areas would advance basic understanding of biomass thermal conversion pathways, and facilitate development of models of the thermochemical conversion and combustion of biomass.[2]

This paper presents recent results from a quantitative study of the effect of temperature on the yields, composition and rates of evolution of major products from the rapid pyrolysis of sweet gum hardwood under 5 psig of helium. It also describes preliminary findings from an investigation of methods for simulating the pyrolysis behavior of whole biomass from knowledge of the pyrolysis kinetics of its major constituents. The latter work draws on studies in this laboratory on the rapid pyrolysis of filter paper cellulose[3,10] and on milled wood lignin.[11,12]

* To whom correspondence should be addressed.

2. EXPERIMENTAL SECTION

2.1. Reactor

Pyrolysis of the wood occurred in a captive sample electrical screen heater reactor described in detail previously.[3,11,10] All products except H_2 are collected to allow direct measurement of material and elemental balances. Sample time–temperature histories are recorded to allow kinetic parameters to be derived from the measured integral product yields. Freshly formed volatiles rapidly exit the high temperature region around the hot stage, so that their secondary reactions are minimized (but not totally eliminated), and primary decomposition of the sample is more closely interrogated. Collaborative effects on pyrolysis of heating rate, sample residence time, sample size, extent of sample dispersion and sample temperature can be independently studied.

2.2. Experimental Procedure

The experimental procedure is described in more detail by Nunn.[11] In brief, powdered sweet gum hardwood was dry sieved to give particles in the 45–88 μm size range. This material was dried over silica gel for at least 4 weeks. Its elemental composition is given in Table I. In a typical run approximately 100 mg of this sample were spread as a thin layer on the reactor heater screens, and then pyrolyzed under 5 psig of helium by

TABLE I

ELEMENTAL COMPOSITION OF TARS AND CHARS FROM PYROLYSIS OF SWEET GUM HARDWOOD

Feedstock or product	*Peak temp. (K)*	*Yield (wt %)*[a]	*Percentage of element*		
			C	*H*	*O*[b]
Sweet gum wood	—	—	49·5	6·1	44·6
Tar	770	22·9	52·6	6·1	32·3
Tar	895	52·5	53·9	5·9	37·1
Tar	1 355	50·2	55·0	6·2	32·3
Char	610	91·0	50·1	6·2	42·2
Char	810	53·2	51·5	6·1	39·7

[a] Based on dry wood.
[b] Oxygen elemental analyses were obtained with a Coulometrics carbon dioxide coulometer.[32]
Reprinted with permission from ref. 13. Copyright 1984 American Chemical Society.

TABLE II

ELEMENTAL, TOTAL MASS AND ENERGY BALANCES FOR PYROLYSIS OF SWEET GUM HARDWOOD

Component	*Approx. ultimate yield (wt%)*	*Percentage of element*			*Heat of combustion*[b] *(Btu/lb)*	*Percent of wood energy in component*	*Approx. peak temp.*[c] *(K)*
		C	*H*	*O*			
Wood	—	49·5	6·1	44·6	8 450	100·0	—
Char	7·0	7·0	—	—	14 660	12·1	1 000
Tar	46·0	26·0	3·7	16·3	11 470	62·4	1 300
CO	17·0	7·3	—	9·7	4 340	8·7	1 200
CH_4	2·3[d]	1·7	0·6	—	23 860	6·5	1 520[d]
CO_2	6·1	1·7	—	4·4	—		950
C_2H_4	1·3[d]	1·1	0·2	—	21 630	3·3	1 520[d]
C_2H_6	0·2	0·16	0·04	—	22 300	0·5	950
H_2O	5·1	—	0·6	4·5	—	—	900
HCHO	2·0	0·8	0·1	1·1	8 190	1·9	900
C_3H_6	0·4	0·3	0·1	—	21 000	1·0	950
CH_3OH	1·5	0·6	0·2	0·7	9 770	1·7	[e]
CH_3CHO	1·4	0·8	0·1	0·5	11 400	1·9	900
Butane + ethanol	0·6	0·3	0·1	0·2	12 780	0·9	[e]
Acetone + furan	0·9	0·6	0·1	0·2	13 280	1·4	[e]
Acetic acid	1·5	0·6	0·1	0·8	6 270	1·1	[e]
Misc. oxygenates	0·7	0·6	0·1	—	18 020	1·4	[e]
Total	94·0	49·5	5·9	38·6		104·8	
Closure[a] (%)	94	100	97	86		105	

[a] Asymptotic yields of molecular hydrogen were estimated to be <0·2 wt % and were ignored in this analysis.
[b] All heats of combustion are from ref. 16, except for wood, char and tar which are calculated from eqn (1) after ref. 17.
[c] Approximate temperature at which the product yield becomes constant.
[d] Yield still increasing as temperature increases.
[e] Insufficient data to determine.

heating to preselected final (peak) temperatures between 600 and 1400 K at rates of 1000 K/s. Sample cooling, at an average nominal rate of 200 K/s, was then immediately begun with no holding time spent at the peak temperature. All light liquid and gaseous products, except H_2, were collected downstream of the reactor and analyzed by gas chromatography. Tar is operationally defined as heavy liquid material: (a) condensed within the reactor vessel; or (b) not recoverable from a glass wool trap downstream of the reactor by warming to 100 °C.[10] The yields of this product and of char were determined gravimetrically. Tar and char were further characterized by elemental analysis. The collection, recovery and analysis procedures were similar to those presented by Hajaligol and co-workers.[3,10] Some modifications were introduced by Nunn[11] to improve the precision of the wood data. Material, and C and H elemental balances were usually closed to within ±5–10% (see Table II).

3. RESULTS AND DISCUSSION

3.1. Product Yields and Volatiles Compositions

Yields are expressed as a percent by weight of the dry wood and plotted as a function of peak temperature. In Figs 1–3 points are laboratory data and curves are free-drawn trend lines.

Figure 1 presents the results for char, tar and gas (including water), and shows that for the above time–temperature history decomposition of the wood is first observed at about 600 K. Weight loss increases with temperature until 93% conversion to volatile material is attained at 950 K. Above 950 K, the char yield remains constant at 7 wt%. Most of the devolatilization occurs between 700 and 900 K.

Tar and gas are initially evolved at the same rate (at 600 K), but the tar yield markedly exceeds the gas make as the peak temperature rises above 700 K. With further temperature increases, tar yield goes through a maximum of about 55 wt% at 850–950 K and attains an essentially asymptotic yield of 46 wt% at 1300 K. Secondary cracking to light volatiles is believed to be responsible for the decrease in tar yield above 950 K.

Secondary cracking of biomass tars to form light volatiles was also observed in rapid pyrolysis of cellulose in this apparatus.[10] In the sweet gum measurements tar secondary cracking declines dramatically, but (based on CH_4 and C_2H_4 high temperature yield data—see below) not completely, by 1300 K. One interpretation of this behavior is that primary pyrolysis of sweet gum produces two broad classes of tarry materials, one

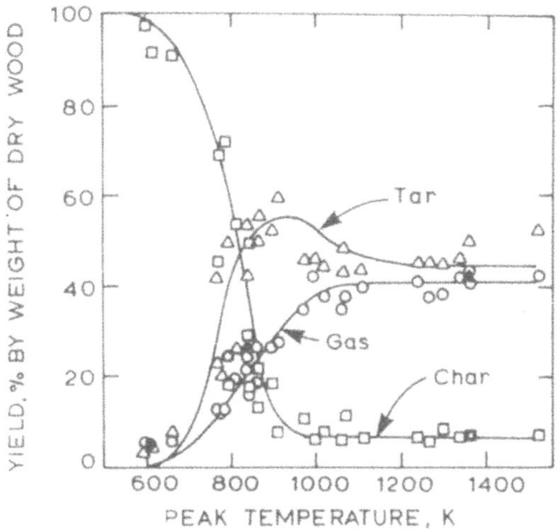

FIG. 1. Effect of peak temperature on yields of char, tar and gas (including water), from pyrolysis of sweet gum hardwood powder. Points: experimental data; curves: trendlines; pressure = 5 psig (helium); nominal heating rate = 1000 K/s; powder diameter = 45–88 μm.

distinctly thermally labile under the present conditions, the other essentially unreactive. As temperature increases above 950 K, most of the reactive tar is converted to either light volatiles or additional unreactive tar by 1300 K.[13] Formation of pyrolytic tars of significantly different thermal stability is implied by other biomass pyrolysis studies.[10,7,14] An alternative view is that some or all of the 'unreactive' tar is material that was formed during the earlier (i.e. lower temperature) stage of sample heatup and which escaped the environs of the hot stage before its temperature became high enough to cause secondary reactions.

Figure 2(a) shows that CO is first observed at about 750–800 K and its yield increases monotonically with temperature to an asymptote of 17 wt % near 1200 K. Since almost half of this CO is produced at temperatures above 950 K, where weight loss (i.e. 100-char yield, see Fig. 1), and hence primary decomposition of the sample, has essentially ceased, it is concluded that at higher pyrolysis temperatures CO is also a major product of secondary cracking of the tar. The magnitude of this increase (about 8 wt %) accounts for most of the 9 wt % decrease in tar yield between these two temperatures. On a weight basis, CO dominates the gaseous products at temperatures above 900 K.

Methane production, shown in Fig. 2(b), starts at around 800 K,

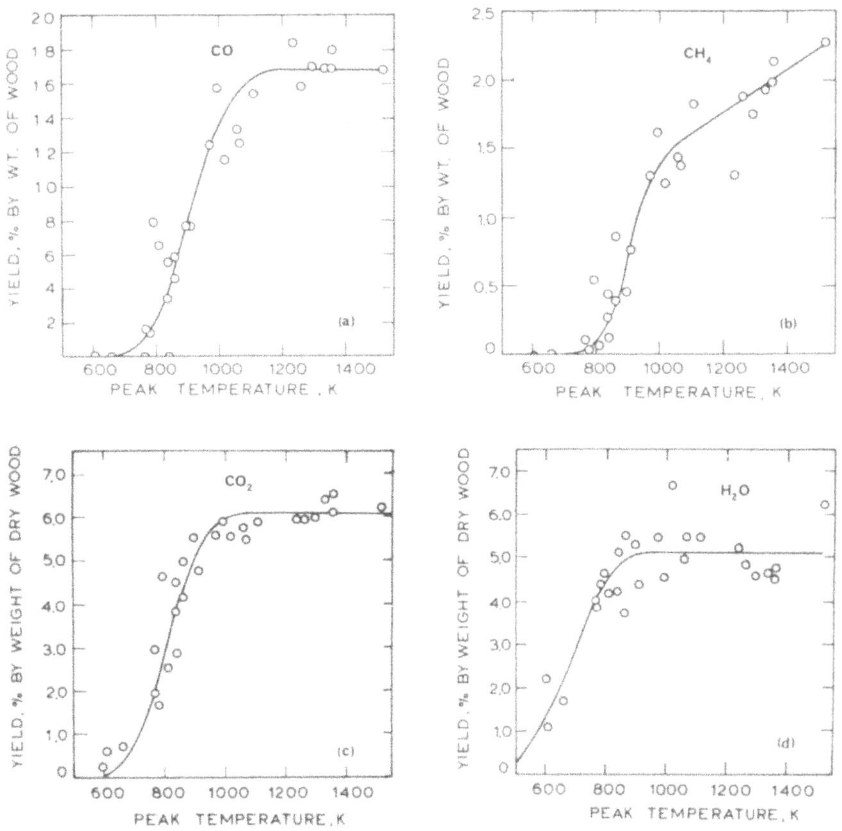

FIG. 2. (a) Effect of peak temperature on the yield of carbon monoxide from pyrolysis of sweet gum hardwood powder. Points: experimental data; curve: trendline; pressure = 5 psig (helium); nominal heating rate = 1000 K/s; powder diameter = 45–88 μm. (b) Effect of peak temperature on the yield of methane from pyrolysis of sweet gum hardwood powder. Points: experimental data; curve: trendline; pressure = 5 psig (helium); nominal heating rate = 1000 K/s; powder diameter = 45–88 μm. (c) Effect of peak temperature on the yield of carbon dioxide from pyrolysis of sweet gum hardwood powder. Points: experimental data; curve: trendline; pressure = 5 psig (helium); nominal heating rate = 1000 K/s; powder diameter = 45–88 μm. (d) Effect of peak temperature on the yield of water from pyrolysis of sweet gum hardwood powder. Points: experimental data; curve: trendline; pressure = 5 psig (helium); nominal heating rate = 1000 K/s; powder diameter = 45–88 μm. Reprinted with permission from ref. 13. Copyright 1984 American Chemical Society.

increases rapidly with temperature to about 1·3 wt % at 950 K, and then increases more slowly with temperature to a non-asymptotic value of 2·3 wt % at 1520 K. Comparison of Fig. 2(b) with Fig. 1 shows that the incremental CH_4 production between 1200 and 1520 K (0·5 wt % of wood) falls within the uncertainty in the tar yield curve. These observations imply that high temperature (>950 K) CH_4 production arises from secondary cracking of tar.

Figure 2(c) shows that measurable quantities of CO_2 are found at temperatures as low as 600 K. Its yield increases with temperature, leveling off at 6 wt % near 950 K. This behavior suggests that CO_2 is derived mainly from direct degradation of the wood and that the secondary cracking reactions contributing to CO and CH_4 production at higher temperatures do not significantly impact CO_2 evolution.

The yields of chemical (non-moisture) water are shown in Fig. 2(d). The data exhibit somewhat more scatter than many of the lighter products. This is believed to arise from severe tailing of the water peak during the GC analysis.[15] Despite this scatter the broad trends in yield behavior are readily discernable. Water is evolved immediately after decomposition of the wood begins at ~600 K. This observation is consistent with the postulate that the major pathways for water formation from lignocellulosics are dehydration and depolymerization reactions, which can occur at low temperaures.[3] The water yield increases monotonically with temperature and plateaus at 5·0 wt % at 900 K.

Figure 3(a) shows that ethylene (C_2H_4) evolution is first observed at 800 K and increases steadily with temperature. At around 950 K the slope of its yield curve decreases, and with further temperature increases its yield increases more slowly to a maximum measured value of 1·4 wt % at 1520 K. This does not appear to be an asymptote, however. This behavior indicates that the incremental production of C_2H_4 between 950 and 1520 K (~0·6 wt % of wood) arises from secondary cracking of tar as discussed above for CH_4.

Figure 3(b) shows that measurable yields of ethane (C_2H_6) are seen at 800 K, with production increasing with temperature to a plateau of 0·17 wt % at 950 K. Approximately 60 % of this yield occurs above 900 K where significant secondary cracking of the tar is occurring (Fig. 1), suggesting that this pathway is an important reaction channel for C_2H_6 production during rapid pyrolysis.

As shown in Fig. 3(c), propylene (C_3H_6) yield behavior closely parallels that of ethane. Propylene production begins at a peak temperature near 800 K, increases strongly with temperature between 850 and 900 K, and

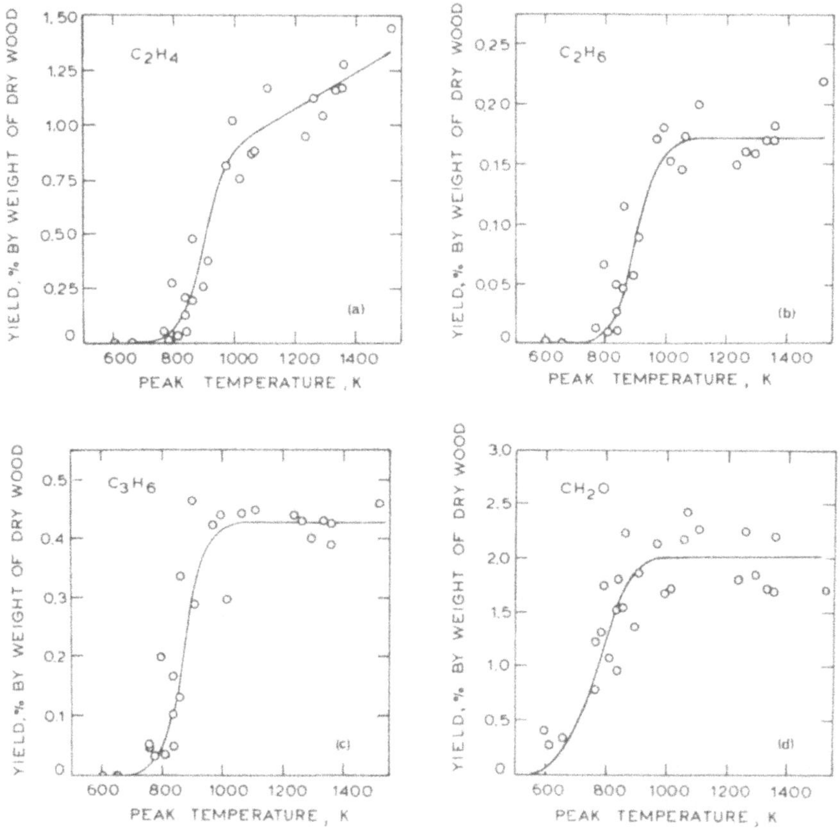

FIG. 3. (a) Effect of peak temperature on the yield of ethylene from pyrolysis of sweet gum hardwood powder. Points: experimental data; curve: trendline; pressure = 5 psig (helium); nominal heating rate = 1000 K/s; powder diameter = 45–88 μm. (b) Effect of peak temperature on the yield of ethane from pyrolysis of sweet gum hardwood powder. Points: experimental data; curve: trendline; pressure = 5 psig (helium); nominal heating rate = 1000 K/s; powder diameter = 45–88 μm. (c) Effect of peak temperature on the yield of propylene from pyrolysis of sweet gum hardwood powder. Points: experimental data; curve: trendline; pressure = 5 psig (helium); nominal heating rate = 1000 K/s; powder diameter = 45–88 μm. (d) Effect of peak temperature on the yield of formaldehyde from pyrolysis of sweet gum hardwood powder. Points: experimental data; curve: trendline; pressure = 5 psig (helium); nominal heating rate = 1000 K/s; powder diameter = 45–88 μm. Reprinted with permission from ref. 13. Copyright 1984 American Chemical Society.

levels off at its ultimate yield of 0·42 wt % near 950 K. The propylene data exhibit some scatter which is probably the consequence of occasional water interference on the GC. Secondary tar cracking probably contributes to C_3H_6 evolution above 900 K.

The yield data for many of the light oxygenated compounds, including methanol, acetaldehyde, ethanol (+butene), acetone + furan and acetic acid, exhibit such a high degree of scatter that it is difficult to discern temperature effects reliably. The uncertainties arise from water interferences during GC analyses, and contamination due to a methanol/acetone solvent mixture used for tar recovery.[11,13] Fortunately some of the high temperature data have a sufficiently low scatter to allow apparent ultimate yields of these products to be estimated with confidence. The results are discussed below and summarized in Table II.

The formaldehyde (HCHO) data generally exhibited less scatter than those of the other light oxygenates listed above. Figure 3(d) shows that this product is observed at about 600 K and its yield increases monotonically with temperature to a reasonably well-defined asymptote of 2·0 wt % above 900 K. The appearance of HCHO at temperatures well below those where tar cracking is significant suggests that primary decomposition of the wood contributes to its production.

3.2. Material, Elemental and Energy Balances

Elemental analyses of the wood, and of product tars and chars obtained at different peak temperatures, are presented in Table I. The chars show increasing carbon and decreasing oxygen content with increasing preparation temperature. This suggests that oxygen, rather than carbon/hydrogen functionalities, are more thermally labile in the pyrolyzing wood. The maximization of tar oxygen content with increasing temperature suggests that oxygen-containing compounds are significant constituents of tar formed at intermediate temperatures, and that some of these compounds are preferentially removed by cracking at higher temperatures.

Table II presents elemental, total mass and energy balances for pyrolysis of this wood. Because of the large amount of scatter in the yield data for some of the light volatiles the analysis used asymptotic yields rather than results from a specific experimental run. The table includes estimates of the peak temperature where the asymptotic yield, or highest observed yield, occurs. For these calculations, the butene and ethanol, acetone and furan, and miscellaneous oxygenated compound fractions were assumed to be ethanol, acetone and benzene, respectively. Since no elemental analysis was available for high temperature char, this product was assumed to be 100 %

carbon. Heats of combustion for the individual gaseous products were taken from ref. 16. Those for wood, char and tar were calculated using a correlation from ref. 17:

$$Q = 146{\cdot}58(\mathrm{C}) + 568{\cdot}78(\mathrm{H}) - 51{\cdot}53(\mathrm{O}) \tag{1}$$

where Q is the gross heating value in Btu/lb on a dry basis and (C), (H), and (O) respectively are the contents of elemental carbon, hydrogen and oxygen in the sample, in weight percent.

The balances for total mass, carbon and hydrogen are very good. The oxygen balance is somewhat low, perhaps due to the assumptions that the char and miscellaneous oxygenated compounds contain no oxygen, or as a result of errors in the elemental analyses.

The energy balance shows that for high temperature pyrolysis only 12·1 % of the wood energy content is retained in the char, while tar accounts for over 62 %. The product gases account for another 30 %. Carbon monoxide and methane together contain more than half of the heating value of the gases. Although on a weight basis the asymptotic CO yield is over seven times greater than that of methane, the latter accounts for almost as much of the energy of the wood (6·5 %) as does CO (8·7 %).

The total energy content of the measured products exceeds that of the wood by under 5 %, which is within the uncertainty of the present analysis. The high temperature thermal decomposition of sweet gum hardwood would thus appear to be approximately thermal neutral.

3.3. Kinetic Modeling

Previous models of the high temperature pyrolytic decomposition of coal, biomass and other organic solids have ranged in descriptive and mathematical sophistication from simple unimolecular chemical models to complex analyses which incorporate physical transport and chemical reaction mechanisms.[18]

The present data on yields of several individual products (except tar) were correlated with a single-reaction, first-order decomposition model. This model is simple, useful in engineering calculations, and has a history of successful utilization for condensed phase substances.[10,19-21,2]

In this model the rate of formation of a product i in yield V_i at time t is given by the expression

$$\frac{\mathrm{d}V_i}{\mathrm{d}t} = (V_i^* - V_i)k_{0i}\exp\left(-\frac{E_i}{RT}\right) \tag{2}$$

where k_{0i} and E_i are respectively the pre-exponential factor and apparent activation energy for component i in the standard Arrhenius rate constant. The quantity V_i^* is the ultimately attainable yield of i, i.e. the yield at high temperature and long residence times. The other symbols have their usual meaning.

Best fit values for the kinetic parameters k_{0i}, E_i and V_i^* (i.e. those that minimize the sum of squared errors between calculated and observed yields) are derived by comparing integral product yields $V_i(T')$, measured in non-isothermal experiments of duration t', to different peak temperatures T', with corresponding values calculated using measured laboratory time–temperature histories, a non-linear least squares regression code, and the integrated form of eqn (2):

$$\ln\left[\frac{V_i^* - V_i(T')}{V_i^*}\right] = -\int_0^{t'} k_{0i} \exp\left(-\frac{E_i}{RT}\right) dt \qquad (3)$$

Thus the fitting procedure takes account of contributions to conversion during both heatup and cooldown of the sample, i.e. t' is the total time of the temperature program, not just the time to attain T'. The results for several products are shown in Table III along with the standard error of the fitting procedure. The tar evolution kinetics were not analyzed with this model because single-reaction first-order kinetics alone cannot predict a maximum in yield. Several of the light oxygenates were omitted from the kinetic analysis because of large uncertainties in their yield data.

The derived kinetic parameters were used to calculate curves showing yield as a function of peak temperature. The results for total weight loss, total gas, CH_4 and CO are compared to the laboratory data in Fig. 4(a–d, respectively). In general the calculated curves fit the laboratory data quite well and similar behavior was found for the other products listed in Table III.[11] These findings and the generally low values for the standard error in the estimate (Table III) show that the single-step, first-order kinetic model provides statistically satisfying correlations of the laboratory data. This model is thus a useful tool for comparing different data sets, and performing engineering calculations. However, this does not imply that these products are formed by simple mechanisms. The present yield vs. temperature data indicate an important role for multiple chemical pathways in the production of CO, CH_4, C_2H_4 and other light gases, and in the tar evolution at higher temperatures. Further, except for C_3H_6, the activation energies in Table III are much lower than those usually expected for unimolecular thermal decomposition reactions (30–70 kcal/mole[21]), again suggesting that multiple reaction pathways contribute to the observed yields.[18]

TABLE III

KINETIC PARAMETERS FOR SWEET GUM PYROLYSIS

Product	E_i (*kcal/g-mole*)	$\log_{10} k_{0i}$ (s^{-1})	V^* (*wt %*)	*Standard error of estimate*[a] (*wt %*)
Weight loss	16·5	4·53	92·97	7·66
Total gases	11·8	2·88	41·01	2·54
CO	14·6	3·36	17·05	2·15
CH_4	16·6	3·79	1·91	0·21
CO_2	14·3	3·77	5·97	0·51
C_2H_4	19·2	4·41	1·17	0·12
C_2H_6	23·7	5·87	0·17	0·02
H_2O	11·5	3·35	5·14	0·65
HCHO	12·9	3·51	1·99	0·27
$H_2O + HCHO$	11·5	3·26	7·13	0·74
C_3H_6	42·8	11·20	0·41	0·06
CH_3CHO	21·3	5·80	1·40	0·34

[a] Defined as

$$\sum_{j=1}^{n} (V_{j,\text{model}} - V_{j,\text{exper.}})^2/(n-3)$$

where n is the number of data points.

Table IV and Fig. 5 compare the kinetic parameters for total weight loss from sweet gum hardwood from this work with those for other woods presented in refs 20, 22 and 23. In the regions of overlapping temperature, the present findings predict higher rates and extents of overall devolatilization. Some of these disparities may be from differences in wood type. Further, the data of refs 22 and 23 were obtained using samples of larger dimensions, and may reflect a significant contribution from intraparticle char-forming reactions which would reduce the overall rates and extents of weight loss. However, it has also been argued that larger sample dimensions promote intra-sample autocatalytic reactions which increase overall rates of weight loss.[6] Further research on the role of secondary reactions in wood pyrolysis is clearly needed.

Figure 5 also implies that: (a) extrapolating kinetic data outside the range of temperatures over which they were obtained may cause serious errors; and (b) derivation of reliable kinetic parameters from wood pyrolysis experiments, and successful utilization of those parameters in

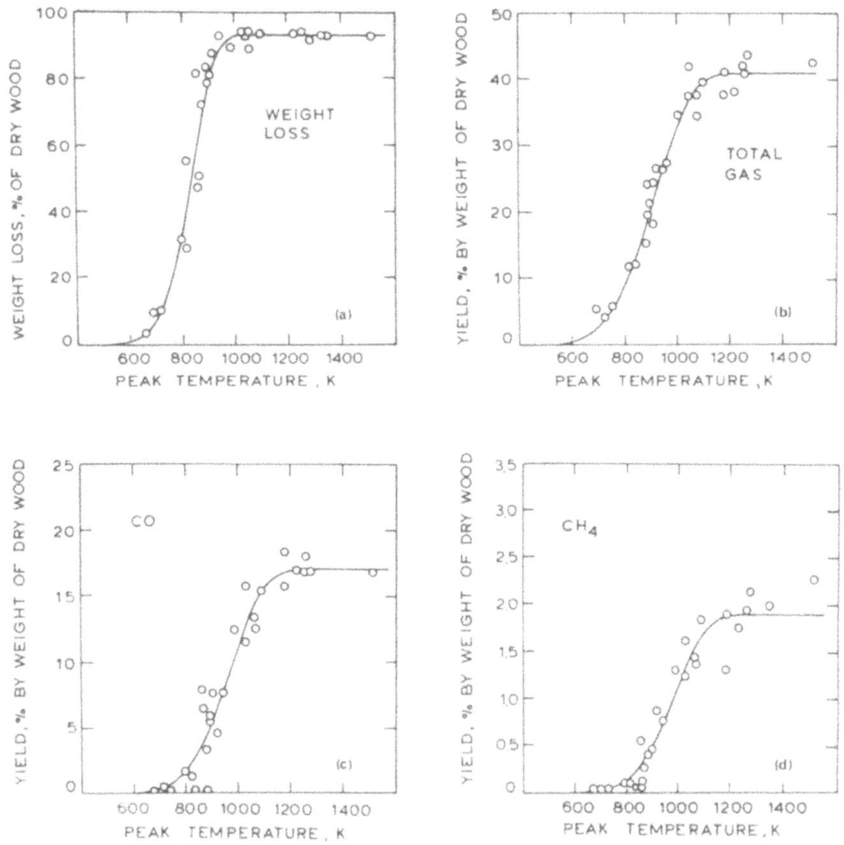

FIG. 4. (a) Experimental data, ○, and model-generated curve for total weight loss from pyrolysis of sweet gum hardwood powder. Pressure = 5 psig (helium); nominal heating rate = 1000 K/s; powder diameter = 45–88 μm. (b) Experimental data, ○, and model-generated curve for total gas production from pyrolysis of sweet gum hardwood powder. Pressure = 5 psig (helium); nominal heating rate = 1000 K/s; powder diameter = 45–88 μm. (c) Experimental data, ○, and model-generated curve for carbon monoxide production from pyrolysis of sweet gum hardwood powder. Pressure = 5 psig (helium); nominal heating rate = 1000 K/s; powder diameter = 45–88 μm. (d) Experimental data, ○, and model-generated curve for methane production from pyrolysis of sweet gum hardwood powder. Pressure = 5 psig (helium); nominal heating rate = 1000 K/s; powder diameter = 45–88 μm. Reprinted with permission from ref. 13. Copyright 1984 American Chemical Society.

TABLE IV

SINGLE-REACTION, FIRST-ORDER KINETIC PARAMETERS FOR TOTAL WEIGHT LOSS IN THE PYROLYSIS OF SEVERAL WOODS

Reference	*Wood type*	*Sample dimensions* (μm)	*Temp. range studied* (K)	E ($kcal/g\text{-}mole$)	$\log_{10} k_0$ (s^{-1})	*Ultimate yield*, V^* ($wt\%$)
Reactor type: bath of molten metal						
23	Sitka Spruce Veneer	1 588 × 25 400 × 149 200	440–573	29·8	7·45	12
Reactor type: tube furnace						
22	Beech	Cylinder 20 000–21 400 OD × 130 000–145 000	626–778	15	3·18	72
Reactor type: tube furnace						
20	Missouri Oak	Sawdust surface mean size 615	573–665	25·4	5·87	70
Reactor type: screen heater						
This work	Sweet gum	Powder 45–88	573–1 373	16·5	4·53	93

Reprinted with permission from ref. 13. Copyright 1984 American Chemical Society.

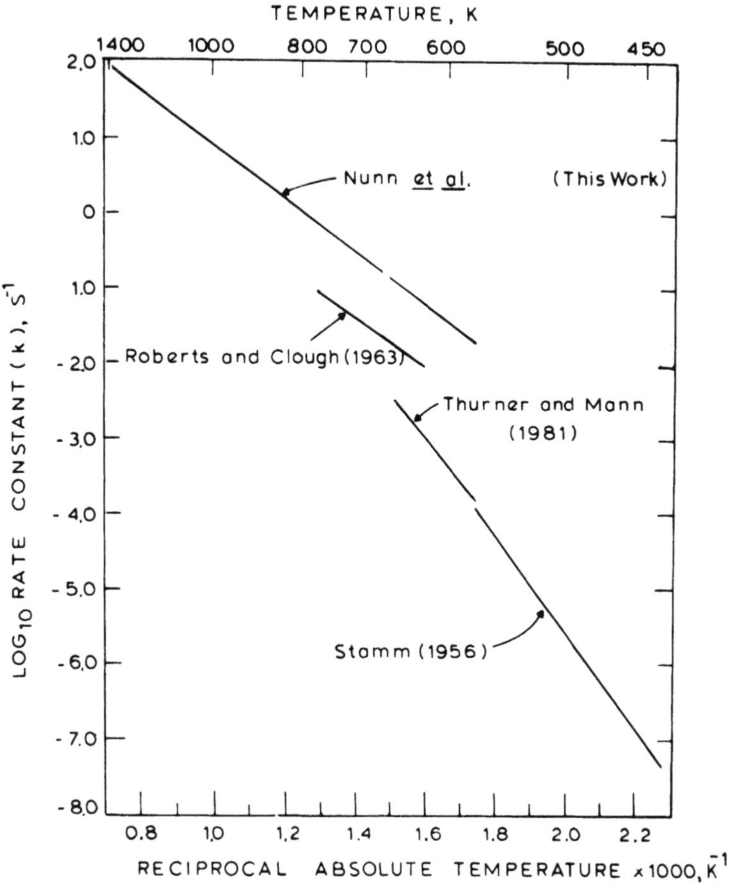

FIG. 5. Kinetic data for total weight loss from the pyrolysis of several woods. Sample types and experimental details are given in Table IV. Reprinted with permission from ref. 13. Copyright 1984 American Chemical Society.

reactor engineering calculations, requires that secondary reactions be accounted for.

3.4. Preliminary Approach to Simulating Whole Wood Pyrolysis

Because of the many different types of wood and other biomass, there is interest in determining the viability of predicting their pyrolysis behavior, under a given set of reaction conditions, from corresponding information on their constituents. Many forms of biomass consist primarily of different amounts of three major molecular components—cellulose, hemicellulose and lignin. These building blocks may be chemically bonded and physically intermixed. Detailed molecular structures, particularly for the latter two, can vary among different biomass types.

A preliminary pyrolysis simulation model is proposed. It is assumed that the simulated integral yield $Y_{i,w}(t)$ of a given product i from pyrolysis of this wood up to time t is given by

$$Y_{i,w}(t) = \sum_{j=1}^{3} W_j Y_{i,j}(t) \tag{4}$$

where W_j is the weight fraction of constituent j (i.e. cellulose, hemicellulose or lignin) in the wood, and $Y_{i,j}(t)$ is the integral yield of product i from pyrolysis of constituent j up to time t. Values for W_j were based on constituent determinations for sweet gum reported by Andrews.[24] Values for $Y_{i,j}(t)$ can be calculated from kinetic models of the pyrolysis of each constituent.

The sweet gum and samples of α-cellulose, milled wood lignin and hemicellulose (xylan and glucomannan) derived from it were each specially prepared for this investigation.[25] The necessary kinetic studies on the hemicellulose and α-cellulose had not been completed at the original time of

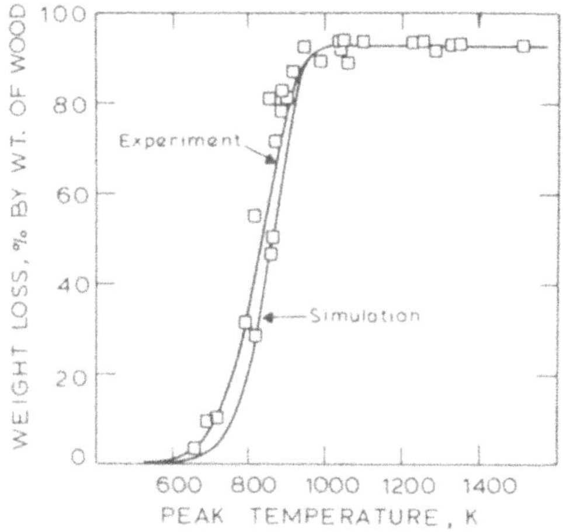

FIG. 6. Comparison of experimental weight loss from sweet gum wood pyrolysis with simulation given by sum of predicted weight losses from wood components pyrolyzed alone. Points: experimental data. The simulation assumes the pyrolysis behavior of hemicellulose and cellulose are the same—see text. Reprinted with permission from ref. 31. Copyright Butterworths, Borough Green, Sevenoaks, Kent, England.

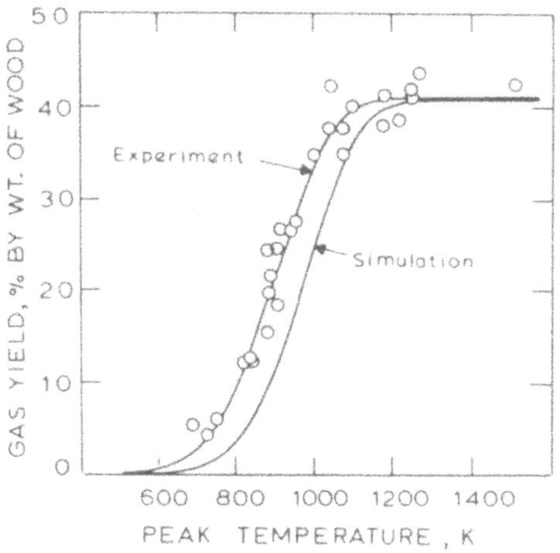

FIG. 7. Comparison of experimental gas yield from sweet gum wood pyrolysis with simulation given by sum of predicted weight losses from wood components pyrolyzed alone. Points: experimental data. The simulation assumes the pyrolysis behavior of hemicellulose and cellulose are the same—see text. Reprinted with permission from ref. 31. Copyright Butterworths, Borough Green, Sevenoaks, Kent, England.

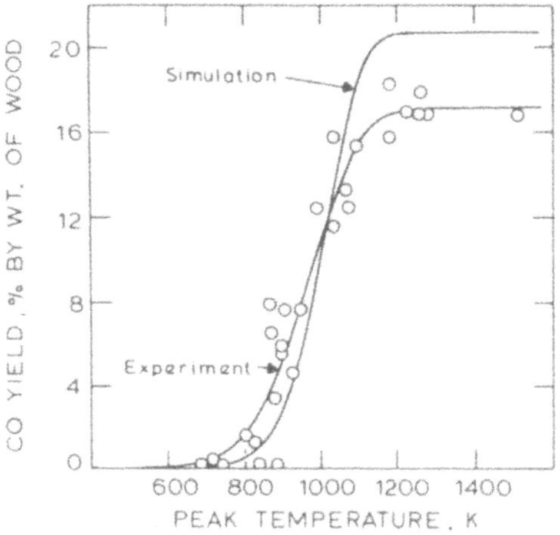

FIG. 8. Comparison of experimental carbon monoxide yield from sweet gum wood pyrolysis with simulation given by sum of predicted weight losses from wood components pyrolyzed alone. Points: experimental data. The simulation assumes the pyrolysis behavior of hemicellulose and cellulose are the same—see text. Reprinted with permission from ref. 31. Copyright Butterworths, Borough Green, Sevenoaks, Kent, England.

writing. It was therefore assumed that in this wood: (1) kinetic data on filter paper cellulose[3,10] adequately represented the behavior of cellulose; and (2) the hemicellulose behaved the same as the cellulose. The requisite kinetic information on milled wood lignin was available[11,12] and was employed in eqn (4).

Results are shown in Figs 6–8. Total weight loss (Fig. 6) was well simulated by this method. However, total gas yield (Fig. 7) and yields of individual products[11] including CO (Fig. 8) could not be predicted by this model. Work is now in progress[26] to determine if the discrepancies arise from interactions between products from pyrolysis of the different constituents and/or from failure of the assumption of similar in-wood pyrolysis behavior of cellulose and hemicellulose.

4. CONCLUSIONS

The following conclusions on the rapid pyrolysis behavior of sweet gum hardwood can be drawn from this study.

(1) The captive sample electrical screen heater reactor gives reliable information of product yields, compositions and formation kinetics in the rapid pyrolysis of wood under the present conditions. Good to excellent material and C and H elemental balances have been achieved and product energy values can be readily estimated.

(2) High conversions of the wood to volatile products can be obtained in very short solids residence times (≤ 1 s). An asymptotic weight loss of 93 % occurs above 1100 K.

(3) Tar is the major pyrolysis product at temperatures above 800 K. It exhibits maximum and asymptotic yields of 55 (900 K) and 46 (1300 K) wt % respectively. These yields respectively represent around 75 % and 62 % of the energy content of the parent wood.

(4) Secondary cracking of this tar contributes significantly to the yields of CO, CH_4, C_2H_4 and most of the other light gaseous products at temperatures above 900–950 K.

(5) Total gas production (including chemical water) increases monotonically with increasing temperature to an ultimate yield of 40 wt % above 1300 K, representing 30 % of the wood energy content.

(6) On a weight basis, CO is the dominant gaseous product above 850 K, reaching an ultimate yield of 17 wt % above 1300 K. Below 850 K, CO_2 and chemical water dominate the gas make, and attain close to asymptotic yields of 5·7 and 5·1 wt % at 1000 and 900 K respectively. The CH_4 yield

(2·3 wt %) at 1520 K accounts for almost as much of the parent wood energy (6·5 %) as does the 17 wt % asymptotic CO yield (8·7 %).

(7) Although their yield increases with temperature, light hydrocarbon gases (CH_4, C_2H_4, C_2H_6, C_3H_6) are relatively minor products with their combined yield at 1300 K remaining below 5 wt %.

(8) Light oxygenated liquids, such as formaldehyde, methanol and acetaldehyde, cumulatively account for approximately 9 wt % of the products collected above 1300 K.

(9) The rates of evolution of most major products except tar were well fitted by a single reaction first order decomposition kinetic model. However, the detailed mechanisms for products generation are in general more complex than single step pathways. This simple model should therefore be viewed only as a convenient tool for correlating data and comparing results from different biomass materials.

(10) Rates of total weight loss were higher than those reported by other investigators, probably because of reduced contributions from apparatus-specific char-forming secondary reactions in the present equipment, and differences in wood composition.

(11) A linear simulation of whole wood pyrolysis behavior, based on kinetic data for the pyrolysis of milled wood lignin and filter paper cellulose, predicts total weight loss quite satisfactorily, but breaks down for individual products and total gas yield.

Much remains to be learned about the fundamental chemical and physical processes governing the pyrolytic decomposition of biomass. However, the present results show that for heating rates of 1000 K/s, primary decomposition dominates the rapid pyrolysis of sweet gum below 900 K, with tar being a major product. Above 900 K, secondary cracking of tar contributes significantly to the product spectrum.

From a practical viewpoint, reaction conditions (such as lower temperatures) could thus be selected to minimize secondary reactions and obtain large yields of volatile organic liquids and tar. Elevated temperatures and extended volatiles residence times would enhance secondary reactions leading to higher yields of CO, CH_4 and other light gases. It is expected that many practical scale biomass thermochemical conversion reactors will be designed to process biomass particles substantially larger than those studied here. In the absence of mechanically assisted removal of tars and other volatiles from the reacting sample (so-called ablative pyrolysis[27]) such particles would offer increased opportunity for secondary, char-forming reactions. Thus the low char yields (high total weight loss) above

1000 K observed in the present investigation probably represent an upper limit on sample conversion to potentially valuable gaseous and liquid products, that may not be attainable for larger particles in non-ablative reactors. The present observations can thus be viewed as identifying reasonable expectations for upper bounds on pyrolysis yields in several types of larger scale equipment.

The present results also confirm the need to obtain quantitative understanding of the separate contributions of primary decomposition and of volatiles secondary reactions to the overall thermal conversion behavior of biomass. Such information would offer the best hope for developing reliable models capable of predicting biomass pyrolysis behavior in practical scale reactors where secondary char forming and volatiles cracking reactions cannot be totally arrested.

ACKNOWLEDGEMENTS

We gratefully acknowledge the financial support for this work provided by the United States Department of Energy under Grant No. DE-FG02-79ET00084 and by the Edith C. Blum Foundation of New York, NY. The DOE grant is administered by the Solar Energy Research Institute, and Drs A. Kotch and T. A. Milne have served as technical project officers. Samples of sweet gum hardwood, milled wood lignin and other wood constituents were specially prepared for this work by Professor H.-M. Chang and his colleagues in the Department of Wood and Paper Science at North Carolina State University, and are gratefully acknowledged. We also acknowledge important contributions to this work by P. Bhadha, R. Caron, J. Curme, H. D. Franklin, J. Ghosh, M. R. Hajaligol, P. Houghton, J. Rau and S. G. Taliaferro.

REFERENCES

1. Molton, P. M. and Demmitt, T. F. (1977). Reaction mechanisms in cellulose pyrolysis, a literature review. Report No. BNWL-2297, Battelle Pacific Northwest Laboratories, Richland, WA.
2. Lewellen, P. C., Peters, W. A. and Howard, J. B. (1977). *Proceedings, Sixteenth Symposium (International) on Combustion.* The Combustion Institute, Pittsburgh, p. 1471.
3. Hajaligol, M. R. (1980). Ph.D. Thesis, Department of Chemical Engineering, Massachusetts Institute of Technology, Cambridge, MA.

4. Allan, G. G. and Matilla, T. (1971). In *Lignins Occurrence, Formation, Structure, and Reactions*, K. V. Sarkanen and C. H. Ludwig (Eds). Wiley Interscience, New York.
5. Klein, M. T. (1981). Sc.D. Thesis, Department of Chemical Engineering, Massachusetts Institute of Technology, Cambridge, MA.
6. Roberts, A. F. (1970). *Combustion and Flame* **14**, 261.
7. Wenzl, H. F. J. (1970). *The Chemical Technology of Wood*. Academic Press, New York.
8. Milne, T. (1979). Chapter 5 in *Principles of Gasification*, Vol. II of *A Survey of Biomass Gasification*, T. B. Reed (Ed.). Report No. SERI/TR-33-239, The Solar Energy Research Institute, Golden, Colorado.
9. Antal, M. J. (1980). Chapter II in Vol. III, Part C of *Energy from Biological Processes*. Office of Technology Assessment, United States Congress, Report No. PB-81-134793, September.
10. Hajaligol, M. R., Howard, J. B., Longwell, J. P. and Peters, W. A. (1982). *Ind. Eng. Chem. Process Des. Dev.* **21**, 457.
11. Nunn, T. R. (1981). M.S. Thesis, Department of Chemical Engineering, Massachusetts Institute of Technology, Cambridge, MA.
12. Nunn, T. R., Howard, J. B., Longwell, J. P. and Peters, W. A. (1984). Product compositions and kinetics in the rapid pyrolysis of milled wood lignin. *Ind. Eng. Chem. Process Des. Dev.* Publication expected.
13. Nunn, T. R., Howard, J. B., Longwell, J. P. and Peters, W. A. (1984). Product compositions and kinetics in the rapid pyrolysis of sweet gum hardwood. *Ind. Eng. Chem. Process Des. Dev.* Publication expected.
14. Stamm, A. J. and Harris, E. E. (1953). *Chemical Processing of Wood*. Chemical Publishing Co., New York.
15. Cosway, R. G. (1981). M.S. Thesis, Department of Chemical Engineering, Massachusetts Institute of Technology.
16. CRC (1976). *Handbook of Chemistry and Physics*, 57th edn, R. C. Weast (Ed.). CRC Press.
17. Mason, D. M. and Gandhi, K. (1980). *ACS Div. Fuel Chem. Preprints* **25**(3), 235.
18. Howard, J. B. (1981). Chapter 12 in *Chemistry of Coal Utilization, Second Supplementary Volume*, M. A. Elliott (Ed.). Wiley, New York.
19. Franklin, H. D., Peters, W. A., Cariello, F. and Howard, J. B. (1981). *Ind. Eng. Chem. Process Des. Dev.* **20**, 670.
20. Thurner, F. and Mann, U. (1981). *Ind. Eng. Chem. Process Des. Dev.* **20**, 482.
21. Suuberg, E. M., Peters, W. A. and Howard, J. B. (1978). *Ind. Eng. Chem. Process Des. Dev.* **17**, 37.
22. Roberts, A. F. and Clough, G. (1963). *Proceedings, Ninth Symposium (International) on Combustion*. Academic Press, New York, p. 158.
23. Stamm, A. J. (1956). *Ind. Eng. Chem.* **48**, 413.
24. Andrews, E. K. (1980). M.S. Thesis, Department of Wood and Paper Science, North Carolina State University, Raleigh, NC.
25. Chang, H.-M. (1980). North Carolina State University, Raleigh, NC, Personal Communication.
26. Ghosh, J. (1983). M.S. Thesis, Department of Chemical Engineering, Massachusetts Institute of Technology, Cambridge, MA.

27. Diebold, J. P. and Scahill, J. W. (1984). This volume, p. 539.
28. Diebold, J. P. (Ed.) (1981). *Proceedings Specialists Workshop on Fast Pyrolysis of Biomass.* Report No. SERI/CP-622-1096, The Solar Energy Research Institute, Golden, CO.
29. Howard, J. B., Hajaligol, M. R., Longwell, J. P., Nunn, T. R. and Peters, W. A. (1982). Pyrolysis of biomass, in *Proceedings, Joint Meeting of Chemical Engineering Chemical Industry and Engineering Society of China.* American Institute of Chemical Engineers, Beijing, People's Republic of China, Sept. 19–22, Vol. II, pp. 638–52.
30. Klein, M. T. and Virk, P. S. (1981). Report No. MIT-EL 81-005, Energy Laboratory, Massachusetts Institute of Technology, Cambridge, MA.
31. Peters, W. A. (1984). Studies of the pyrolysis behavior of condensed phase fuels with applications to fuel conversion technology. *Proceedings, Fifth International Symposium on Analytical Pyrolysis.* Vail, Colorado, Sept. 26–30, 1982. K. J. Vorhees *et al.* (Eds). Butterworths, London.
32. Raines, D. (1981). Huffman Laboratories, Inc., Wheatridge, CO., Personal Communication.

17

Pyrolysis and Hydropyrolysis of Peat at High Heating Rates

H. EKLUND* and W. WANZL

Bergbau-Forschung GmbH, Essen, Federal Republic of Germany

1. INTRODUCTION

For the utilization of solid fuels at a low coalification stage, such as peat and biomass, processes are in development which in a number of cases involve rapid heating, short solids residence times and high pressure. For the simulation of the above process conditions in the laboratory different types of microreactors have been developed.[1,2]

In the work described here heating with the Curie-point technique has been applied to studies of pressurized pyrolysis. The Curie-point technique is well established as a method for finger-print analysis in connection with pyrolysis gas chromatography.[3-5] The choice of the Curie-point technique as a heating principle in this investigation is mainly due to the ability of the method to obtain very high heating rates and simultaneously a self-stabilizing temperature control at a relatively low level.

2. EXPERIMENTAL

2.1. Curie-Point Method

Inductive heating with the Curie-point technique is accomplished by introducing a ferromagnetic conductor into a high frequency magnetic field. The conductor will absorb energy until the Curie-temperature is reached, when the conductor turns paramagnetic and absorbs no more

* On leave from Lund University, Sweden; present address: Norsk Hydro AS Research Center, N-3901 Porsgrunn, Norway.

energy. The temperature will thus be automatically controlled and stabilized at the Curie-temperature.

To obtain a high heating rate and simultaneously a good temperature stabilization, the magnetic permeability must have a highest possible value but must also show a very sharp drop at the Curie-temperature. These properties are in principle fulfilled for pure iron and certain iron/nickel and iron/cobalt alloys having Curie-points in the temperature range of technical interest for biomass and coal pyrolysis, i.e. 600–1000 °C. Besides the chemical composition of the conductor, geometric factors, reactor pressure and the operation load of the high frequency generator are of importance in obtaining the desired heating characteristics.

To obtain proper heating conditions, i.e. defined and reproducible heating profiles, it was necessary to follow a proposal by Bühler[6] and Oertli[7] and to operate the generator in a two-pulse mode. With single-pulse operation it was not possible to obtain a high heating rate and simultaneously a good temperature stabilization at the Curie-temperature: a high heating rate was followed by a considerable overheating of up to 400 °C above the Curie-point. Pre-loading of a large condenser gave the high energy demand for the heating phase followed by an appropriate continuous lower energy supply for the isothermal phase.

2.2. Experimental Set-up

The experimental set-up, which is shown in Fig. 1, has the following main components:

—pressurized reactor with induction coil and sample holder;
—high frequency power generator;
—temperature and pressure recorders;
—sampling and analysis system for the pyrolysis gas.

The pyrolysis experiments are carried out in a stationary gas atmosphere with batch samples of the solid fuel in the milligram range. The reactor is designed to allow for a rapid quench of the volatiles released, e.g. the bulk of the gas atmosphere in the reactor is maintained at ambient temperature during the pyrolysis run. With this arrangement secondary reactions of the primary pyrolysis products are diminished.

The reactor is made of stainless steel and designed for a maximum pressure of 30 MPa. The reactor chamber together with the connected pipes and pressure transmitters are all minimized in volume to give the highest possible concentrations of the pyrolysis products for the product analysis. The sample holders used are made from thin-walled tubes of

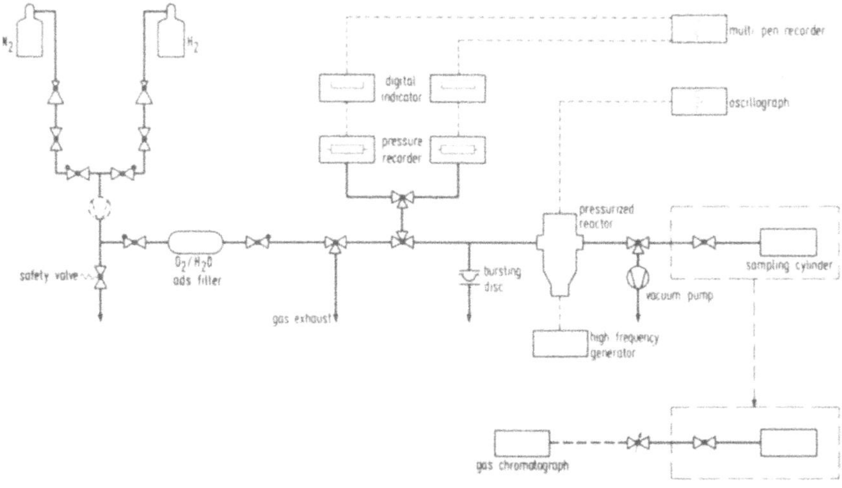

FIG. 1. Flow sheet of experimental set-up for high pressure Curie-point pyrolysis.

ferromagnetic alloys having Curie-points of 600, 700, 800 and 950 °C. The tubes are 50 mm in length, 1·2 mm in diameter with a wall thickness of 0·1 mm. The sample holder and its position inside the induction coil is shown in Fig. 2.

In the reactor the tube is kept in position by a support rod which also acts as one pole—the positive—in a thermocouple together with a 0·1 mm diameter constantan wire welded to the tube surface—the negative pole. The thermoelectrical potential from the sample holders were 'pseudo'-iron

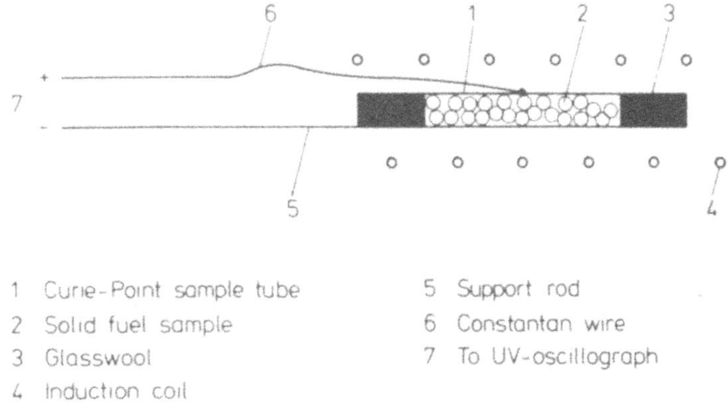

FIG. 2. Curie-point pyrolyser sample tube.

constantan thermocouples and were calibrated against a Pt/PtRh-standard in a furnace. By welding six constantan wires along the tube length, the sample holders were measured to be isothermal on about 60% of the length.

The registration of the time/temperature profiles was made on a high performance UV-oscillograph which in addition allowed the recording of the power output from the HF-generator.

In the experiments the sample holders are filled with 3–6 mg of the fuel which is fixed in position with a plug of quartz wool in each end of the tube.

2.3. Data Evaluation

The pyrolysis yields, e.g. total weight loss, gas and tar production, are determined by weighing the sample before and after the run and from a gas analysis. Non-condensable gases H_2, CO, CO_2, CH_4, C_2H_4 and C_2H_6 are analysed with a GC by flushing a fraction of the reactor gas volume into an evacuated sample bottle as indicated in Fig. 1.

The tar yield m_T is calculated from a mass balance:

$$m_T = \Delta m - m_G \tag{1}$$

The value m_T defined in this way will include also moisture from the original fuel and water chemically formed during the pyrolysis reaction. One possible way to correct the tar yield will be to make an oxygen mass balance according to the scheme below:

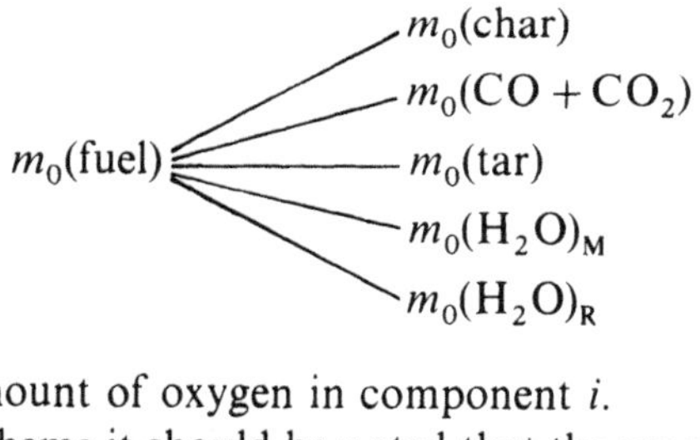

where $m_0(i)$ = amount of oxygen in component i.

In the above scheme it should be noted that the amount of moisture in the original fuel $m_0(H_2O)_M$ and the yields of CO and CO_2 could be experimentally determined in each individual run. However, the small sample sizes used do not allow for an elementary analysis of the residual char and the tar formed. Test runs in a thermobalance with larger samples of peat in the pressure range of interest but at lower heating rates[8,9] provide a data base on the oxygen content of residual peat char and tar which has been used for the calculation of a corrected tar yield. The amount of chemically formed water has been calculated iteratively by eqns (2) and (3) by inserting the results from eqn (2) into (3) until convergence is obtained.

$$(m_{H_2O})_R = \tfrac{18}{16}\{m_0(\text{fuel}) - [m_0(\text{char}) + m_0(\text{CO} + \text{CO}_2) + m_0(\text{tar}) + m_0(\text{H}_2\text{O})_M]\} \quad (2)$$

$$m_T' = \Delta m - m_G - (m_{H_2O})_R - (m_{H_2O})_M \quad (3)$$

The influence of the chemically formed water was found to be at a maximum 15% of the tar yield. This result was obtained for the lowest pyrolysis temperatures; at higher temperatures it was considerably lower.

3. RESULTS

3.1. Analysis Data of Feed Material, Char and Tar

The peat samples used were obtained from the Hasselfors Bruk AB in Sweden with analysis data according to Table I. For comparison the analysis of the sub-bituminous coal Westerholt is included in the table.

A set of analysis data for peat char and tar produced under pyrolysis conditions which approach those of the Curie-point reactor are given in Table II. A detailed characterization of the tar indicated an oxygenated

TABLE I
ANALYSIS DATA FOR COAL WESTERHOLT AND HASSELFORS PEAT

	Coal Westerholt (*wt %*)	*Hasselfors Peat* (*wt %*)
Proximate analysis		
Moisture (ar)	3·1	13·0
Ash (mf)	3·8	1·1
Volatile matter (mf)	37·1	75·3
Char	61·0	21·6
Ultimate analysis (mf)		
C	81·4	50·1
H	5·2	5·45
N	1·28	0·51
S	3·33	0·18
Cl	0·15	0·05
Ash		1·1
O_{diff}	6·8	42·6

TABLE II

ANALYSIS OF PEAT TAR AND CHAR OBTAINED BY HYDROPYROLYSIS

Reaction conditions: T = 848 K
p = 10 MPa H_2
Heating rate = 26 K/s
Total reaction time = 600 s

	%*C*	%*H*	%*N*	%*O*	%*S*	*H*/*C*	*VM*% (*maf*)
Tar	77·5	6·7	0·19	9·5	0·46	1·04	—
Char	94·0	3·4	0·8	1·5	0·16	0·43	6·4

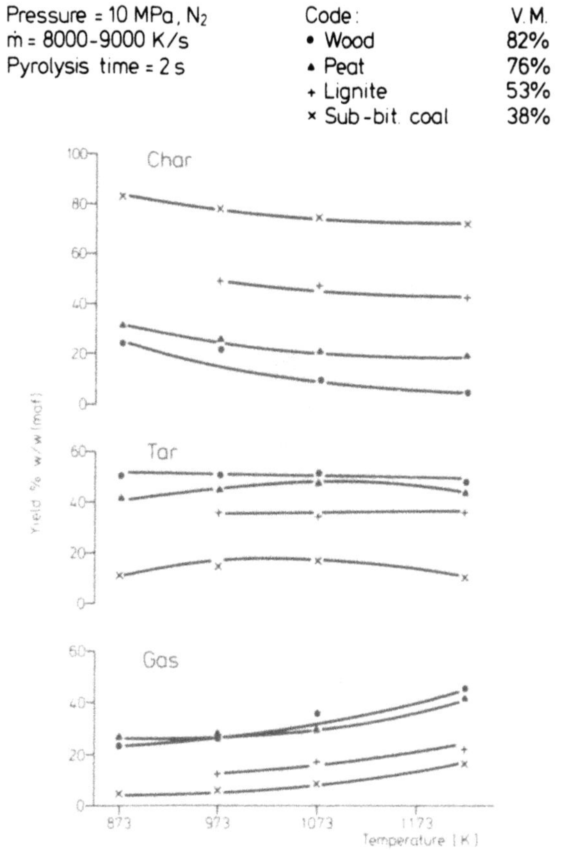

FIG. 3. Char, tar and gas yields as a function of pyrolysis temperature for different solid fuels.

material with a high content of carboxylic acids, carbonyl compounds and phenols but also long aliphatic chains which are either free or substituted to an aromatic structure.[9] The char has a low volatile matter content of about 6% and a high calorific value of 36·5 MJ/kg.[9]

3.2. Influence of Pyrolysis Temperature

The devolatilization of peat is a rapid process. Above 800 °C the reaction is finished within 2 s at a pressure of 10 MPa. There is virtually no difference in the total weight loss of the peat sample for experiments in nitrogen or hydrogen atmosphere. In nitrogen a very slight decrease in the tar yield and a corresponding increase in the gas yield was observed at 950 °C. This cracking of the tar is outweighed by hydrogenation reactions in hydrogen atmosphere where both tar and gas yields are constant above 800 °C.

In Fig. 3 the yields of char, tar and gas are shown as a function of isothermal pyrolysis temperature in N_2 for fuels of different coalification

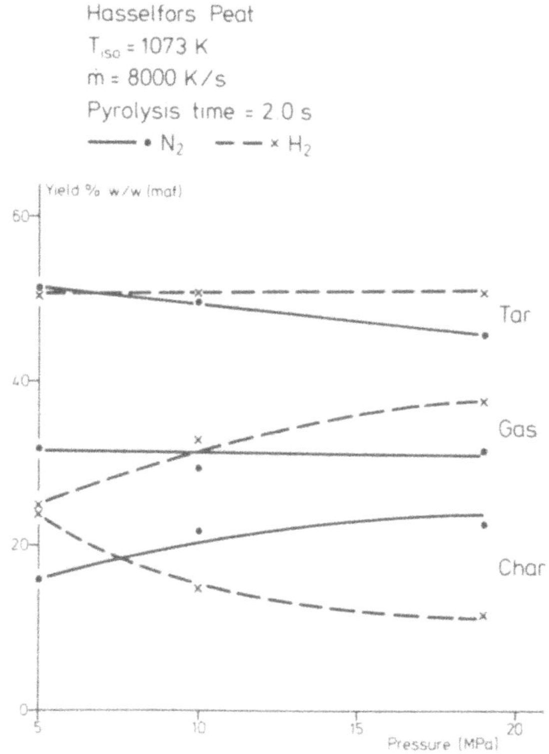

FIG. 4. Pyrolysis yields in N_2 and H_2 as a function of pressure for Hasselfors Peat.

stages. The yields reflect the volatile matter content and elementary composition of the different materials giving a conversion series

wood > peat > lignite > sub-bituminous coal

The figure also illustrates the tar cracking at higher temperatures mentioned above. The effect is more marked for sub-bituminous coal than for peat and wood. It is likely that the coal-derived tar is heavier compared with tar from peat and thus more easily cracked already within the particle.

3.3. Influence of Pressure

Figures 4 and 5 display the pyrolysis yields obtained as a function of pressure in N_2 and H_2 atmosphere. In N_2 the tar yield is decreasing with

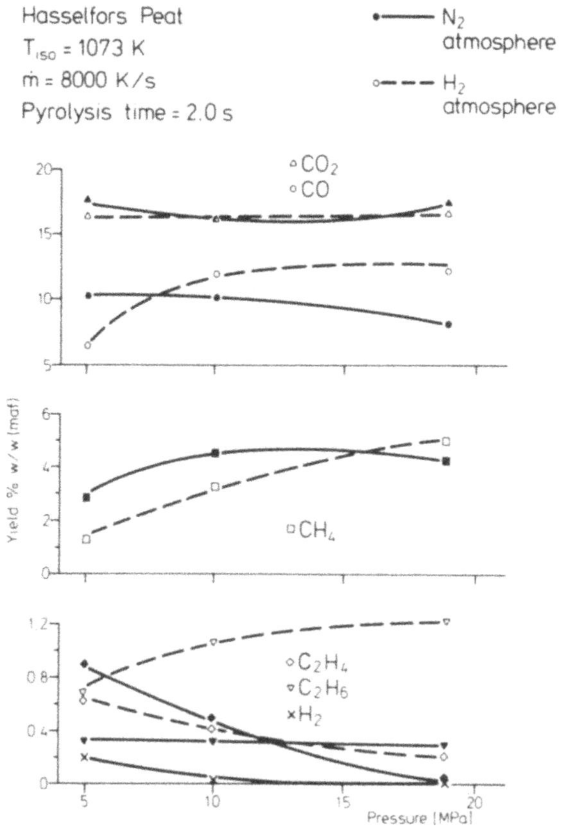

FIG. 5. Gas yield in N_2 and H_2 as a function of pressure for Hasselfors Peat.

increasing pressure giving a higher char yield. The same tendency, but more marked, is also found for the sub-bituminous coal as shown in Figs 6 and 7. In the coal case the diffusion rate of tar within the particle during the plastic phase and the evaporation from the particle surface will be reduced with increasing pressure leading to increased cracking of the tar into char and light gases. For the peat there is no increase in the gas yield—for some components there is even a slight decrease as seen in Fig. 5. Since the peat, in contrast to coal, has no defined pore structure and also no caking properties the pattern of cracking of the volatiles will be different. The data in Figs 4 and 5 indicate that at higher pressures the devolatilization is suppressed, with a small extent of secondary cracking, yielding a more oxygenated char compared with lower pressures.

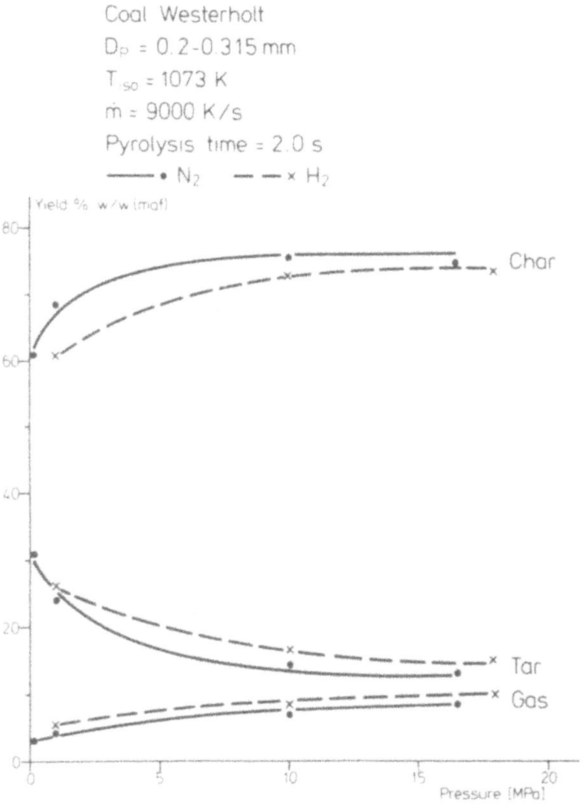

FIG. 6. Pyrolysis yields in N_2 and H_2 as a function of pressure for the sub-bituminous Coal Westerholt.

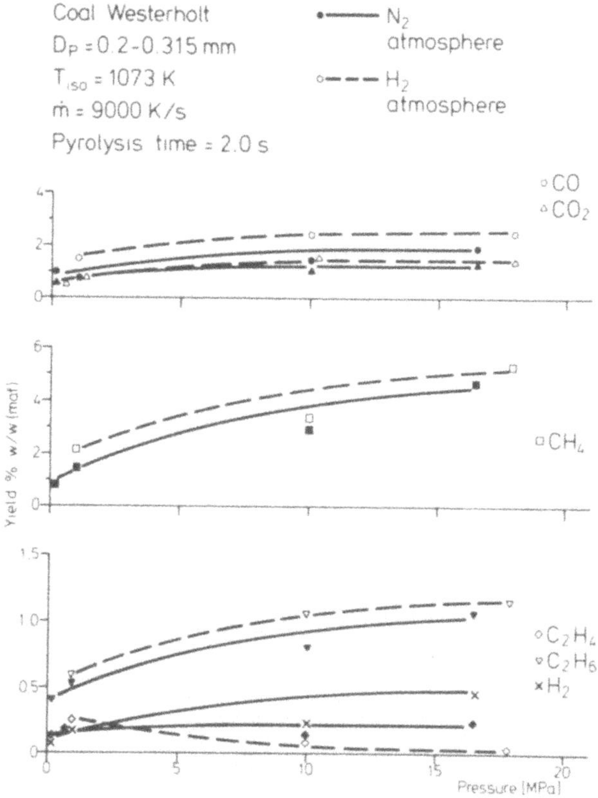

FIG. 7. Gas yield in N_2 and H_2 as a function of pressure for the sub-bituminous Coal Westerholt.

When looking at the yields obtained in hydrogen (dashed lines in the Figs 4–7), the hydrogenation reactions appear to have a greater impact on peat than on coal at comparable conditions. In contrast to coal, the char yield from peat decreases with increasing pressure giving a constant tar and increasing gas yield. This could be explained by the more open structure of the peat giving a better access of hydrogen due to lower diffusion resistance. In addition the high content of heteroatoms and less aromatic structure for the peat could create more favourable hydrogenation conditions.

3.4. Influence of Heating Rate

In general it is known that rapid heating in inert atmosphere yields more and lighter volatiles compared with slow heating as in a coke oven.[1,10,11]

TABLE III

INFLUENCE OF HEATING RATE BY PYROLYSIS OF PEAT IN N_2, $T_{iso} = 1073$ K, $p = 10$ MPa

Peat	*Heating rate 500 K/s*	*Heating rate 10 000 K/s*
Char, wt %	24	19
Gas, wt %	23	30
Tar, wt %	50	49
Gas yield % w/w		
H_2	below det. limit	0·22
CO	7·0	10·0
CO_2	13·2	15·7
CH_4	2·6	3·4
C_2H_4	below det. limit	0·1
C_2H_6	0·1	0·3

Table III summarizes the effects of heating rate for peat in nitrogen. As seen, there is a considerable increase in the gas yield at the higher heating rate. There is also an indication for the influence of the heating rate on the gas composition, with an increased production of CO and H_2 at an increased heating rate. This effect was also stated by Suuberg[12] in a study of lignite decomposition. In H_2 the pyrolysis followed much the same pattern as for nitrogen atmosphere as shown in Table IV. This is again in

TABLE IV

INFLUENCE OF HEATING RATE BY HYDROPYROLYSIS OF PEAT, $T_{iso} = 1073$ K, $p = 10$ MPa

Item	*Heating rate 500 K/s*	*Heating rate 10 000 K/s*
Char, wt %	24	11
Gas, wt %	21	28
Tar, wt %	52	60
Gas yield % w/w		
CO	5·6	11·9
CO_2	8·3	10·8
CH_4	6·4	5·4
C_2H_4	0·3	0·3
C_2H_6	0·4	0·3

TABLE V

INFLUENCE OF HEATING RATE BY HYDROPYROLYSIS OF COAL WESTERHOLT, $D_P = 0{\cdot}2$–$0{\cdot}315$ mm, $T_{iso} = 1073$ K, $p = 10$ MPa

Item	*Heatin rate 800 K/s*	*Heating rate 10 000 K/s*
Char, wt %	73·3	81·8
Gas, wt %	6·7	6·5
Tar, wt %	18·2	10·3
Gas yield % w/w		
CO	1·45	1·55
CO_2	0·92	0·69
CH_4	3·32	3·08
C_2H_4	0·09	0·15
C_2H_6	0·92	1·07

contrast to hydropyrolysis of hard coal where an increase in heating rate gives a considerably lowered tar yield: the tar yield from hard coal in Table V shows a decrease of almost 40 % when increasing the heating rate from 800 to 10 000 K/s. A similar effect has been observed also by changing the heating rate from 10^2 to 10^3 K/s.[13] A possible explanation is that at lower heating rates the release of reactive species will have a longer duration giving more favourable conditions for the hydrogen to react with and stabilize the tar radicals before they recombine, polymerize or decompose. With peat the structural differences from coal pointed out above give other reaction conditions by diminishing the intra-particle diffusion for both the tar and the hydrogen.

4. CONCLUSIONS

The above work has shown the possibility of using the Curie-point technique as the heating method in the study of pressurized pyrolysis of solid fuels. With an appropriate design of induction coil, sample holder and high frequency generator reproducible time–temperature profiles and high heating rates could be obtained.

The experimental results from the pyrolysis of peat and hard coal under similar conditions pointed out different devolatilization patterns which are related mainly to different chemical constitution and physical structure of the fuel.

ACKNOWLEDGEMENT

This work has been carried out within the framework of the Pyrolysis Project of the International Energy Agency (IEA) with funding from the governments of the Federal Republic of Germany, United Kingdom and Sweden.

REFERENCES

1. Arendt, P. and van Heek, K. H. (1981). *Fuel* **60**(9), 779–87.
2. Hajaligol, M. R., Howard, J. B., Longwell, J. P. and Peters, W. A. (1982). *Ind. Eng. Chem. Process Des. Dev.* **21**, 457.
3. Oertli, C., Bühler, C. and Simon, W. (1973). *Chromatographica* **6**, 499.
4. Schmid, J. P., Schmid, P. P. and Simon, W. (1976). *Chromatographica* **9**, 597.
5. Jones, C. E. R. and Cramers, C. A. (Eds) (1977). *Analytical Pyrolysis*. Elsevier, Amsterdam, p. 203.
6. Bühler, C. (1971). Ph.D. Thesis, ETH Zürich.
7. Oertli, C. (1974). Ph.D. Thesis, ETH Zürich.
8. Herod, A. A., James, R. G. and Snape, C. E. (1980). Internal Report TR06/80 of the IEA-Pyrolysis Project.
9. Fynes, G. and Newman, J. O. H. (1982). Internal Report TR12/82 of the IEA-Pyrolysis Project.
10. Peters, W. and Bertling, H. (1965). *Fuel* **44**, 317.
11. Howard, J. B. (1981). Chapter 12 in *Chemistry of Coal Utilization*, Second Supplementary Volume, Elliott (Ed.). Wiley, New York.
12. Suuberg, E. M., Peters, W. A. and Howard, J. B. (1978). *Ind. Eng. Chem. Process Des. Dev.* **17**, 37.
13. Arendt, P. (1980). Ph.D. Thesis, TH Aachen.

18

Flash Pyrolysis of Lignin

EITAN AVNI,* FAEGH DAVOUDZADEH AND ROBERT W. COUGHLIN
Department of Chemical Engineering, University of Connecticut, Storrs, Connecticut, USA

1. INTRODUCTION

Lignin, a major component of biomass found in all vascular plants, is an important potential source of fuels and chemicals. Pyrolysis of lignin is important in itself for the potential production of tars and liquids which can contain valuable chemicals. In addition, understanding pyrolysis has a broad importance based on the assumption that thermal decomposition is an important initial step in the combustion, hydrocracking or donor-solvent liquefaction of lignin, just as in the case of coal.

The model structure of lignin is generally assumed to include phenylpropane repeating units as well as functional groups such as carboxyl, methoxy, carbonyl and ether. Thermal cracking initially severs relatively weak aliphatic bonds releasing large fragments (tar). Some of these fragments are highly reactive free radicals which can undergo a variety of secondary reactions such as cracking and repolymerization. Simultaneously, the various functional groups are thought to decompose to form low-molecular-weight gaseous species, e.g. CO_2 from carboxyl, H_2O from hydroxyl, CO from carbonyl, methanol from methoxy and light hydrocarbons from aliphatic moieties. At higher temperatures ($>500\,°C$) the aromatic rings are thought to rearrange and condense, releasing hydrogen in the process; CO probably also arises from ether linkages at higher temperatures. Based on similar reasoning for coal pyrolysis,[1,2] most aliphatic and alicyclic hydrogen is probably removed from the lignin at lower temperature in the form of cracked hydrocarbons whereas the

* Present address: Department of Chemical Engineering, Florida State University, Tallahassee, FL 32306, USA.

condensation of aromatic rings is believed to contribute molecular hydrogen as a product.

Generally, it is known that overall pyrolysis reactions are strongly influenced by pressure, temperature, the rate and duration of heating and the nature of the ambient atmosphere. Primary pyrolytic reactions probably depend on far fewer variables. More information about the effects of such variables can not only contribute to a better understanding of lignin conversion processes but also has utility for the rational design of pyrolysis processes. We are investigating the course of lignin pyrolysis by measuring weight loss and following the evolution of volatile products as a function of time and reaction conditions.

2. EXPERIMENTAL

Pyrolysis was conducted in four types of apparatus: (a) a Du Pont Model 951 thermogravimetric analyzer; (b) an ohmically heated screen for rapid 'flash-pyrolysis' within a quartz tube as shown schematically in Fig. 1; (c) a specialized pyrolysis apparatus (CDS Pyroprobe Model 120) which communicates directly with a gas chromatograph as illustrated in Fig. 2; and (d) ohmically heated vacuum pyrolyzer in line with a Fourier-Transform Infrared Analyzer (FTIA) as shown in Fig. 3.

In the flash-pyrolysis apparatus (b), the lignin is placed on a stainless steel grid contained within a quartz tube. The grid is ohmically heated rapidly to the desired final temperature using a regulated power supply while the temperature is monitored continuously using a recorder. In this system, flash pyrolysis can be conducted at high heating rates (up to 500 °C/s) in environments of different flowing gases in order to study the effect of gaseous environment on tar yield. In general, the variables studied included pyrolysis temperature, ambient gases, sample particle size and time of pyrolysis. Tar products were collected and analyzed separately by a gas chromatograph–mass spectrometer system (Hewlett Packard GC-MS Model 5985).

In the Pyroprobe flash pyrolysis system (c), lignin pyrolysis products are analyzed by an associated, dedicated gas chromatograph. The calibrated platinum element in the probe serves as temperature sensor, heater and sample holder; it is incorporated as one leg of a Wheatstone bridge circuit which is electrically balanced by the Pyroprobe control module. Consequently, the time/temperature profile applied to the sample is precisely reproducible. In this system, pyrolysis can be conducted at very high

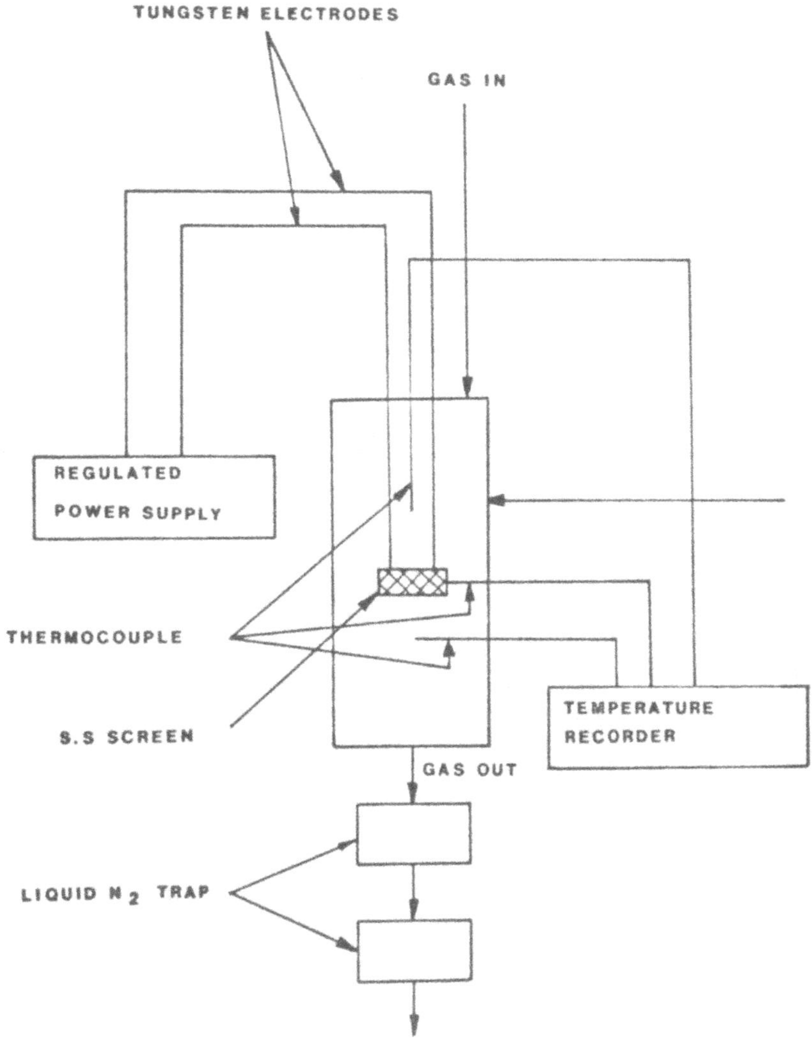

FIG. 1. Schematic diagram of pyrolysis apparatus.[6]

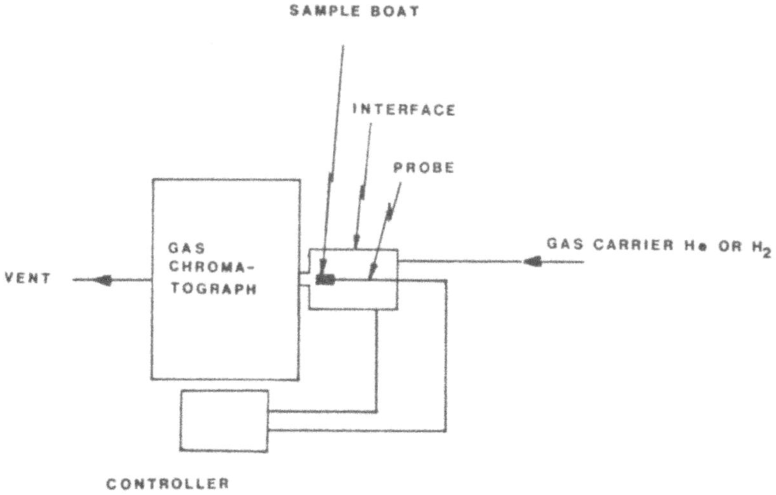

FIG. 2. Schematic diagram of pyrolysis apparatus (c), showing the associated gas chromatograph.

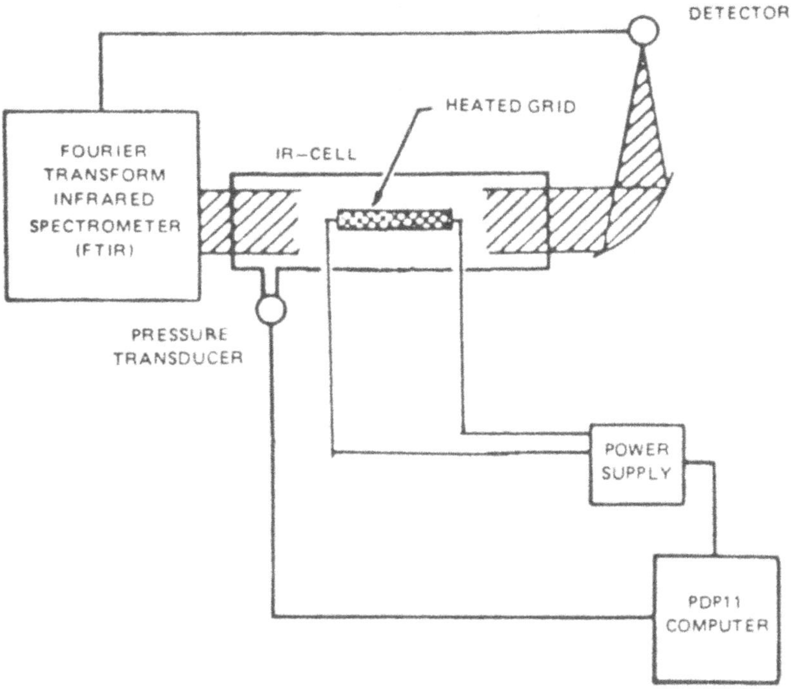

FIG. 3. Schematic diagram of pyrolysis apparatus (d), FTIR.

heating rate (up to 20 000°C/s) under a flow of different gases, although the sample temperatures themselves probably lag somewhat behind the recorded thermocouple temperatures. Different GC columns are used to separate the different volatile fractions: a Porapak Q column for gases, a Porapak T column for liquids and a DC-200 column for tar.

In the FTIR flash pyrolysis system (d), product evolution is measured continuously by an in-line FTIR Analyzer (Nicolet Model 7199). Here the sample is placed on an ohmically heated metal grid and pyrolysis conducted under an initial vacuum of 0·1 mm Hg. With the FTIR the release of pyrolysis products can be monitored directly by rapid measurement of gas concentrations during the process. The FTIR permits low-resolution analysis at 2-s intervals, as shown in Fig. 4, followed by high-resolution analysis for selected wavenumbers at the completion of pyrolysis by which the time-course of evolution of individual gaseous species can be displayed as illustrated in Fig. 5. Hydrogen is estimated by difference. Figures 4 and 5 are presented here to show the type of information that can be obtained by such experiments. Most of the data on evolution of individual products are not presented in this paper; they are currently being

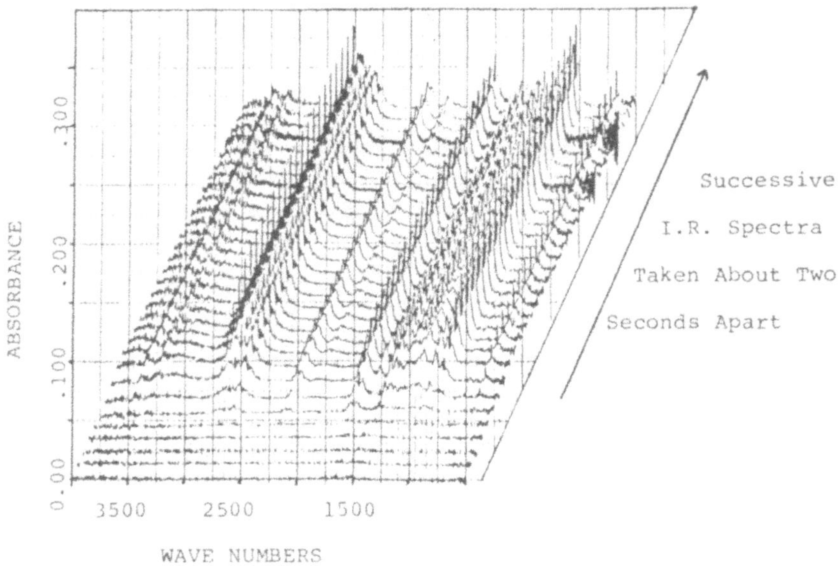

FIG. 4. Low resolution FTIR analysis of accumulated pyrolysis gases at 2-s intervals. Pyrolysis of alkali lignin from steam exploded aspen; static pyrolysis at 500°C.

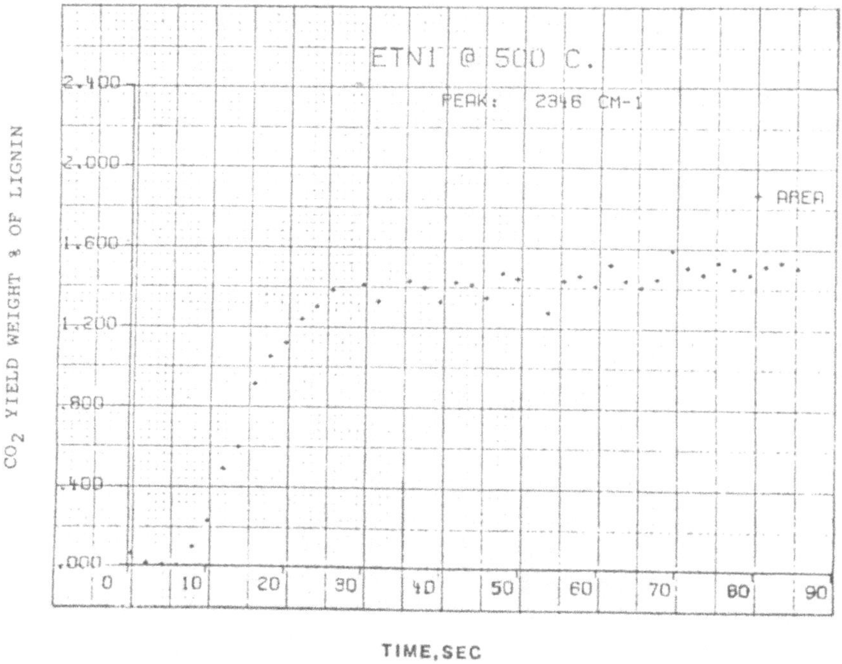

FIG. 5. Time behavior of cumulative CO_2 production (area of CO_2 absorption peak). Pyrolysis of alkali lignin from steam exploded aspen; static pyrolysis at 500 °C.

evaluated and used to develop kinetic models of the pyrolysis reactions. Such results and information will be reported in a future publication.

The du Pont 951 TGA (System (a)) is used with the electronic programming and recording features of the du Pont 990 thermal analyzer module. The instrument is capable of providing different specified heating rates in the range of 0·1 °C/min to 100 °C/min. The recorder plots weight loss and its rate versus time (static pyrolysis) or versus temperature (dynamic pyrolysis).

During pyrolysis there is a heat-up period, an isothermal period and a cool-down period. In the discussion below we refer to a dynamic temperature range or 'dynamic pyrolysis' to mean pyrolysis during the heat-up period. Dynamic pyrolysis incorporates transient heating only to a specified temperature—then the experiment is over. The term static temperature or static pyrolysis refers to pyrolysis during the isothermal period. Only the du Pont TGA apparatus (System (a)) could be operated in either the dynamic mode (slow, steady heating) or the static mode (a prolonged, constant, final temperature following a rapid heat-up to final temperature). Each of the experimental systems (b), (c) and (d) operated essentially isothermally—the heat-up to final static temperature was rapid and the pyrolysis was conducted at the final static temperature for a period that was long compared to the heat-up period.

The term conversion or weight loss is defined as $(W_0 - W/W_0) \times 100$, reaction rate is $-1/W_0(\mathrm{d}W/\mathrm{d}t)$ where W is weight at any time t and W_0 is the initial weight.

3. RESULTS AND DISCUSSION

Figure 6 shows pyrolysis yield data using the FTIR system (d) and plotted versus the final, static temperature. Each such final temperature corresponds to a different initial heating rate, values of which ranged up to 600 °C/s for the experiments of Fig. 6.

Figure 7 compares tar and char yields from static pyrolysis in the following systems:

(1) System (d) (FTIR) at 0·1 torr,
(2) System (b) under 1 atm N_2.

Results for system (c), also under 1 atm of N_2, were essentially identical

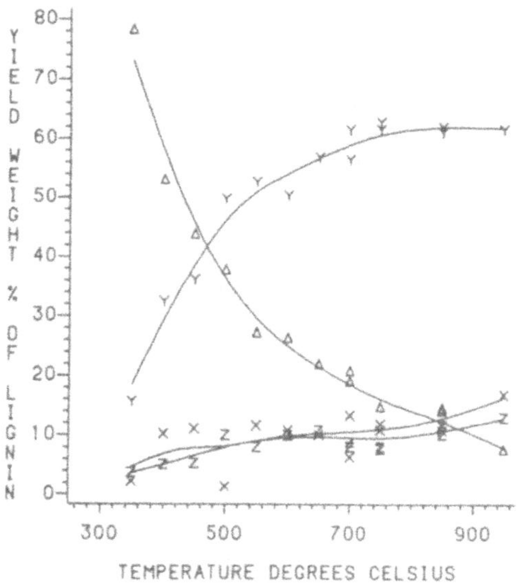

FIG. 6. Flash pyrolysis of alkali lignin from steam exploded aspen under vacuum in FTIR system (d). Heating rates up to 600 °C/s. (Y) Tar; (Δ) char; (Z) gas; (X) mass unaccounted for.

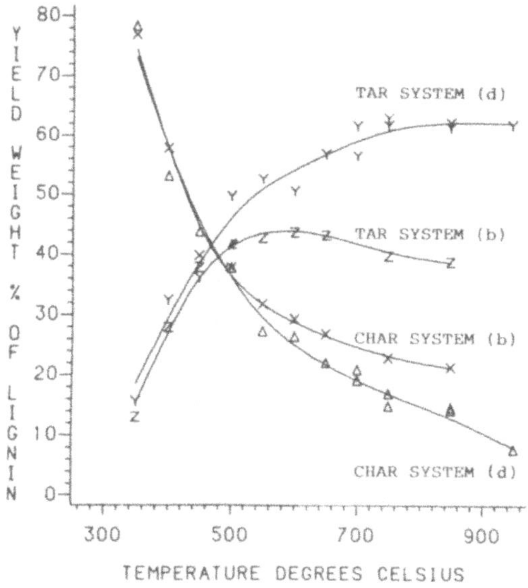

FIG. 7. Flash pyrolysis of alkali lignin from steam exploded aspen in systems (d) and (b). (Y) Tar yield, system (d), FTIR, heating rates to 600 °C/s, 0·1 torr. (Δ) Char yield, system (d), FTIR, heating rates to 600 °C/s, 0·1 torr. (Z) Tar yield, system (b), heating rates to 600 °C/s, 1 atm N_2. (X) Char yield, system (b), heating rates to 600 °C/s, 1 atm N_2.

to those for system (b). Figure 7 shows the important feature that pyrolysis under vacuum gives significantly greater tar yields and low char yields compared to pyrolysis under 1 atm of N_2 but with otherwise similar conditions. Under 1 atm of N_2, the tar yield reached a maximum value in the 550–600 °C range and then declined; Tyler[7] observed a similar trend for pyrolysis of ten bituminous coals in a fluidized bed. Unlike the results at one atmosphere, the lignin tar yields obtained under vacuum do not show any definite decrease with temperature in the range 300–800 °C (Fig. 7).

Figure 8 compares the conversion (weight loss) for the following systems:

(1) System (b) under 1 atm N_2 and a heating rate which ranged up to 600 °C/s,

(2) System (a) TGA under 1 atm N_2 and a heating rate of 10 °C/min.

The higher heating rates gave higher conversions (weight loss) and lower char yield. At 10 °C/min in the TGA, the maximum conversion was about 62 wt % and the corresponding char yield 38 wt %; at heating rates up to 600 °C/s (System (b)) the maximum conversion (weight loss) was 79 wt %

and the corresponding char yield 21 wt %. For flash pyrolysis of coal, Wen and Tone[3] observed similar results, namely higher conversion to volatile material at higher heating rates in the 100–800 °C range; moreover, they got more tar and liquids at higher rates than at lower heating rates. In contrast Suuberg[4] found no change in the specific gas or tar yields from coal as heating rates increased from 270 °C/s to 10 000 °C/s. Allen *et al.*[8] and Fletcher *et al.*[5] obtained pyrolysis results for lignin at low heating

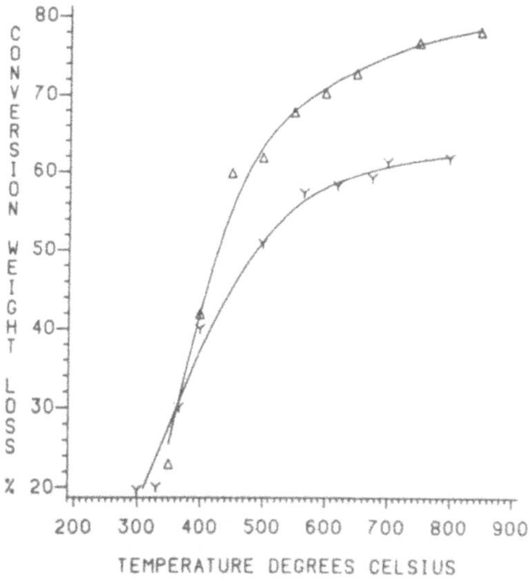

FIG. 8. Pyrolysis of alkali lignin from steam exploded aspen in systems (a) TGA and (b). (Δ) Conversion (weight loss), system (b), heating rates to 600 °C/s, 1 atm N_2. (Y) Conversion (weight loss), system (a), heating rate of 10 °C/min, 1 atm N_2.

rates similar to those reported here: 35–45 % weight loss conversion at 400–450 °C.

Gases evolved during flash pyrolysis were analyzed either by GC or FTIR depending on the experimental apparatus employed. Figure 9 shows that the total gas yield increased as the static temperature was increased; it reached 12 wt % at a static temperature of 800 °C during vacuum pyrolysis. In contrast, under 1 atm of N_2 (System (c)) the total gas yield was 38 wt % at a static temperature of 800 °C; results for system (b), also under 1 atm of N_2, were essentially identical to those of system (c). Figures 10 and 11 show

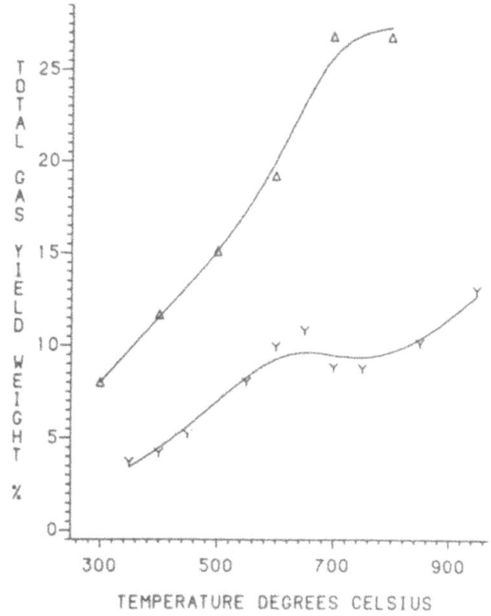

FIG. 9. Flash pyrolysis of alkali lignin from steam exploded aspen in systems (d) and (c). (Δ) Gas yield, system (c), heating rates to 600 °C/s, 1 atm N_2. (Y) Gas yield, system (d), FTIR, heating rates to 600 °C/s, 0·1 torr.

the individual gas yields as a function of static temperature for the vacuum FTIR system (d) and Pyroprobe system (c), respectively.

It should be noted that the FTIR system (but not the Pyroprobe) can detect H_2 (Fig. 10) whereas the Pyroprobe (but not the FTIR) can detect CH_3OH and acetone (Fig. 11).

Dynamic pyrolysis of lignin was carried out using (System (a)) the TGA at a heating rate of 100 °C/min in different gaseous environments; weight loss conversions versus temperature are shown in Fig. 12. The data at a total pressure of 1 atm in Fig. 12 show that the presence of H_2 (up to 30 %) in the N_2 has essentially no effect. On the other hand adding O_2 to N_2 shows a substantial effect, as expected, because of oxidation (with ignition evident at about 500 °C). Figure 12 also shows that in pure N_2 lowering the pressure from 750 to 15 torr increased the pyrolysis weight loss.

It is interesting to consider why lowering pyrolysis pressure increases the yields of tar (Fig. 7), the weight loss conversion (Fig. 12) and lowers the gas yield (Fig. 9). The corresponding char yield is decreased by lowering pressure (Fig. 7) as expected from mass-balance considerations. Lowered

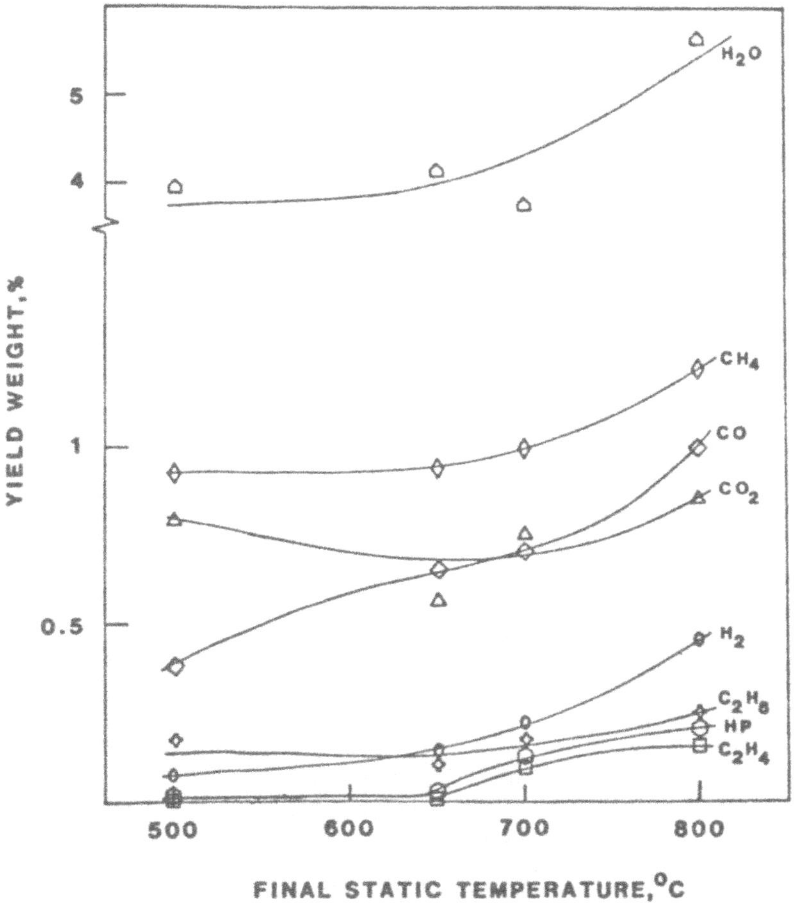

FIG. 10. Flash pyrolysis of alkali lignin from steam exploded aspen in system (d), FTIR; heating rates to 600 °C/s, 0·1 torr. (HP = higher paraffins.)

pressure is expected to enhance the transport of volatile species away from the decomposing solid mass. Furthermore, such increased rates of transport would also be expected to reduce secondary reactions between highly reactive volatile fragments and the residual solid material. Pyrolyzing thin layers of lignin (as opposed to deep beds) would be expected to produce effects similar to lower pressures by reducing the amount of interaction between reactive volatile fragments and the residual reactive solid. Similar low-pressure and thin-layer effects have been observed[9] for coal pyrolysis. Suuberg[6] explained these effects of pressure on coal pyrolysis by means

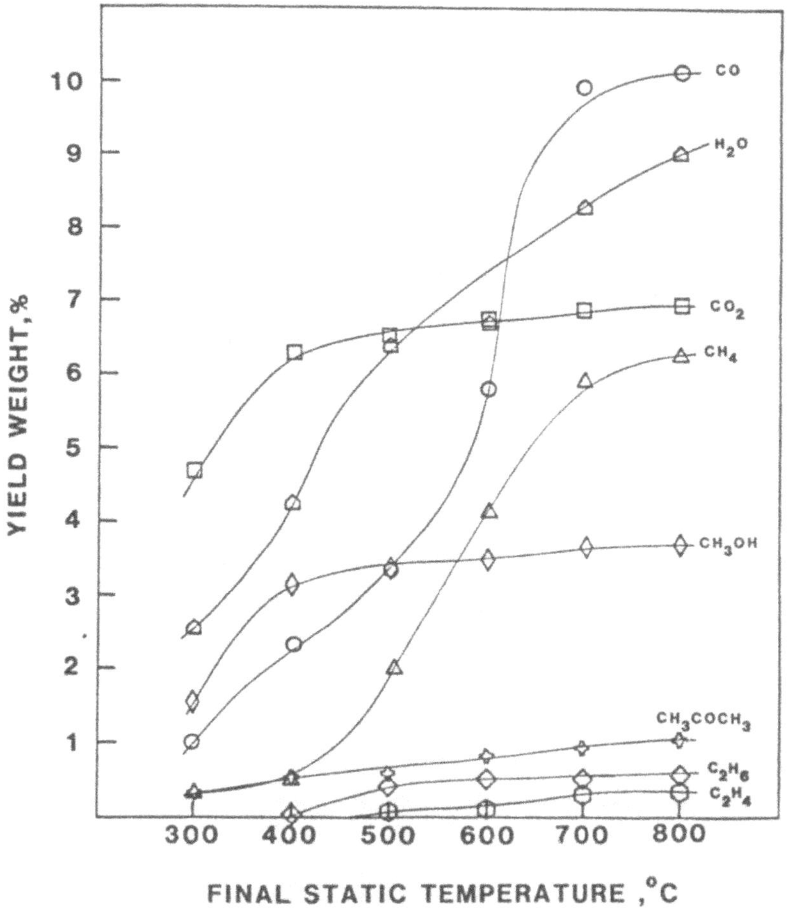

FIG. 11. Flash pyrolysis of alkali lignin from steam exploded aspen in systems (c); heating rates to 600 °C/s, 1 atm N_2.

of a diffusion model that incorporated a molecular diffusivity that varied inversely with pressure. Such mass-transfer phenomena are probably most important for the large (tar) fragments produced during the early stages of pyrolysis and bituminous coals showed[6] greater sensitivity to particle diameter and pyrolysis pressure than did lignite which forms less tar.

Solomon[2] has envisioned the pyrolysis of coal to involve a competition for aliphatic hydrogen between tar molecules and lighter species—when this hydrogen is used up the evolution of tars and aliphatic gases ceases. Solomon[9] observed that thick beds and higher pressures appear to inhibit

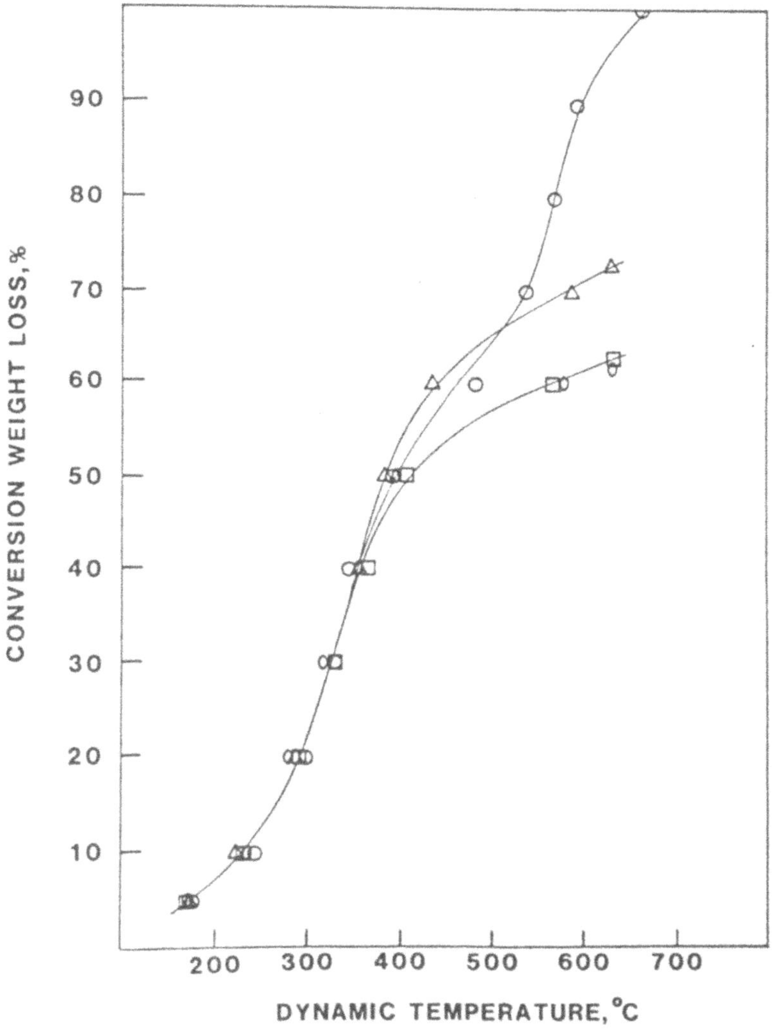

FIG. 12. Dynamic pyrolysis of alkali lignin from steam exploded aspen in system (a), TGA; heating rate of 100 °C/min; different gaseous environments were used. (○) 14·7 V% O_2 in N_2, 1 atm; (△) 15 torr of N_2, (O) 30% (by volume) of H_2 in N_2, 1 atm; (□) 1 atm of N_2.

the evolution of tar while still permitting lighter species to escape, taking with them a major portion of the aliphatic hydrogen. It is likely that related but quite different processes take place during donor-solvent liquefaction wherein the solvent retains, transfers and distributes the hydrogen between different molecules or even between different parts of the same molecule. Figure 12 shows that the effects of pressure do not appear during lignin pyrolysis until about 350 °C, the approximate temperature at which tar formation begins. It is also interesting to note from the data of this figure that the presence of molecular hydrogen does not appear to affect the extent of weight loss; this suggests that the hydrogen transferred between pyrolysis fragments is more reactive than molecular hydrogen under these conditions. It will be interesting to learn more about how the presence of molecular hydrogen may affect the yields of individual volatile products from lignin pyrolysis.

Experiments currently in progress are exploring the effects of the reactive atmosphere (e.g. hydrogen or air) on the product yields. Heating rates ranging from 1 °C/min to 20 000 °C/s are also under investigation. The following six different lignins are under study: (1) a caustic lignin extracted from steam exploded aspen using NaOH solution (the lignin used in the experiments reported here from Iotech Corp., Ltd,); (2) a Kraft lignin from Westvaco Corp.; (3) alkali lignin from autohydrolysis of aspen (from Stake Tech., Ltd.); (4) H_2SO_4-lignin from steam exploded aspen; (5) HCl-lignin from steam exploded aspen; (6) ethanol-lignin from steam exploded aspen. The last three lignins were extracted in our laboratories. Experiments are also being conducted to explore the effects of downstream catalytic reaction to convert further the pyrolytic products after their formation.

ACKNOWLEDGEMENT

We thank the National Science Foundation for support of this research under Grant No. CPE 8014523.

REFERENCES

1. Fitzgerald, D. and Van Krevelen, D. W. (1959). *Fuel* **38**, 17.
2. Solomon, P. R. (1981). *Fuel* **60**, 3.
3. Wen, C. Y. and Tone, S. (1978). *Chem. Res. Eng. Dev.* ACS Symp. Ser 72. San Francisco, CA, p. 18.

4. Suuberg, E. M., Peters, W. A. and Howard, J. B. (1978). *Ind. Eng. Chem. Process, Res. Dev.* **17**(1), 37.
5. Fletcher, T. L. and Harris, E. E. (1952). *Tappi* **35**, 536.
6. Suuberg, E. M. and Unger, P. E. (1981). *18th Symposium (International) on Combustion*. The Combustion Institute, Pittsburg, PA, 1203.
7. Tyler, R. J. (1980). *Fuel* **53**, 218.
8. Allan, G. G. and Matilla, T. (1971). High energy degradation, in *Lignins*, K. Sarkanen and C. H. Ludwig (Eds). Wiley-Interscience, New York, Chapter 14.
9. Solomon, P. R. (1979). ACS Div. of Fuel Chemistry, Preprints **24**(2), 184.

19

Devolatilization of Maple Hardwood

GERARD ROSE and ROBERT ZABRANSKY
Institute of Gas Technology, Chicago, Illinois, USA

1. INTRODUCTION

The thermal decomposition behavior of biomass, and in particular cellulose, has been the subject of experimental investigations for a considerable time. Most of the early work, and current pertinent publications, examine pyrolysis of biomass with respect to the propagation or retardation of flames. A great deal of useful kinetic information has thus resulted; but the models and rate constants so obtained primarily describe the behavior of cellulose at conditions of temperature and pressure less severe and gas compositions much simpler than are likely to exist in a gasifier. Hence, there is a need to obtain kinetic information which describes the behavior of the entire biomass sample when it is subjected to temperatures and pressures conducive to the production of usable gaseous fuels such as methane and syngas mixtures.

The present investigation under the sponsorship of DOE was undertaken to determine the effects of temperature, pressure, gas composition and solids residence time upon the devolatilization behavior of maple hardwood chips. For the investigation an isothermal laminar flow reactor, in which each of the above mentioned parameters can be accurately controlled, was utilized. The experiments were run over a temperature range of 1000 °F (538 °C)–1600 °F (871 °C), a pressure range of 114·7 psia (790·8 kPa)–314·7 psia (2169·7 kPa), and solids residence times of 1·5–4 s, in either pure helium or a mixture of 5 % H_2/95 % He.

2. EXPERIMENTAL

A schematic of the isothermal laminar flow reactor utilized in this investigation is shown in Fig. 1. The main flow of the devolatilization medium,

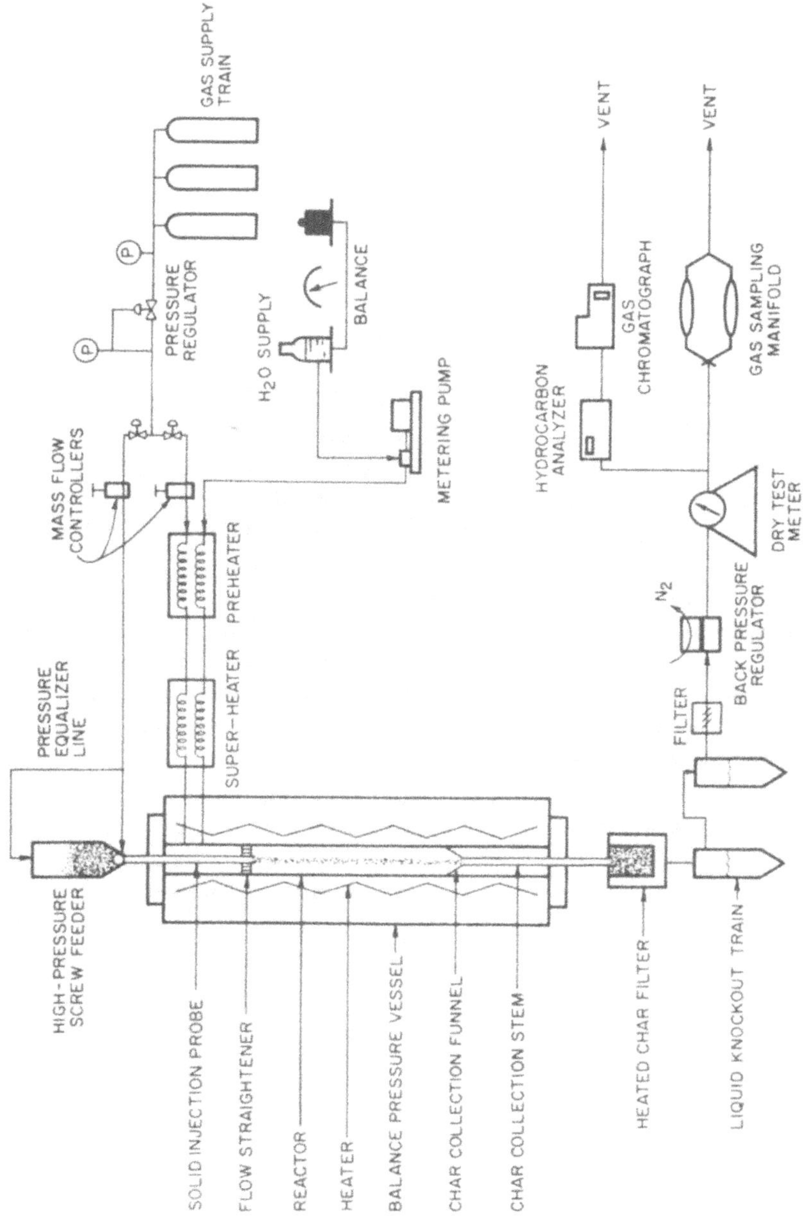

FIG. 1. Schematic flow diagram of the laminar-flow reactor system.

either pure helium or a 5% H_2/95% He mixture, was delivered at the desired reaction temperature through the annular section surrounding the solids injection probe. Maple hardwood solids having a size of $-60+80$ mesh which were dried at ~100°F (~40°C) for approximately 16 h, were screw fed through a water cooled solid injection probe to prevent any premature reaction. A small cold gas flow was maintained over the screw to improve the feeding characteristics. The particles then flowed concurrently with the gas in free-fall fashion through the hot reaction zone. The length of the reaction zone, hence solids residence time, could be varied by adjusting the height of the char collection funnel. (In this paper char represents the total solids collected in each experiment which may contain partially reacted feed.) The temperature was controlled by seven independent heater zones. For economy, ease of operation and safety a balanced pressure design was employed for the reactor. The char funnel includes an outside jacket through which either a water or a gas cooling medium could be passed. Temperatures low enough to prevent any further reaction yet high enough to sustain the liquid products in the vapor phase were maintained in the section of the probe below the funnel.

The gas and char stream first pass through a hot sintered metal filter (5 μm size) which retained the solid products and allowed the liquids (as vapors) and gaseous products to pass. The liquid products were collected in two liquid traps maintained at ice-water temperatures. Finally, the aerosols which form were trapped in an in-line fiber filter which was weighed before and after each run.

After expansion to atmospheric pressure the complete product gas stream was metered and passed through a sample train where gas chromatographic analysis could be employed or samples taken for subsequent mass spectral analysis. On-line infrared analyzers for CO and CO_2 were used to follow the course of the reaction and to monitor steady state operation. Steady state operation generally was achieved within 15 min and maintained for the balance of the experiment, usually about 4 h.

For this investigation mass spectrometry was used for all gas analyses. Feed and char samples were analyzed using standard ASTM ultimate analysis methods. Representative wood samples were obtained by riffling a large quantity of prepared feed to 100 g samples.

Three reaction zone lengths, 18, 30 and 48 inches, corresponding to solids residence times of 1·5, 2·5 and 4 s respectively, were used. These residence times were calculated using standard equations for the free fall of a particle in a static system and an empirical relation developed by Pennsylvania State[1] for a particle falling in a system where the fluid is not stationary. calculated heat-up rates indicate that even at the lowest heat-up rate, the particle should be at temperature after being in the hot zone less than 0·1 s.

3. RESULTS AND DISCUSSION

The effects of temperature and gas composition are summarized in Table I. Plots of devolatilization yields and product gas composition as a function of temperature are shown in Figs 2 and 3 respectively. Increasing temperature increased the amount of feed converted to gaseous products at the expense of both the char and liquid yields. However, the decrease in liquid yields from 34% (5% H_2/95% He) at 1000 °F (538 °C) to 7% at 1600 °F (871 °C) is three times the decrease in observed char yields, indicating that predominantly the gas yields increase at the expense of the liquids. Similar results were obtained for the helium devolatilizations. These results are consistent with the results reported earlier by Shafizadeh,[2] where he has shown that heating cellulose at temperatures above 932 °F (500 °C) causes the glucose polymers to depolymerize to anhydrosugars, primarily levoglucosan, which further react to form a tarry pyrolysate. Secondary reactions of this product such as dehydration, disproportionation and decarboxylation result in the observed mixture of low molecular weight gaseous and volatile products.

The observed decreases in the char yields with increasing temperature may be the result of several concurrent and competing reactions. Increasing temperature not only increases the rate of decomposition reactions with higher activation energies and initiates those reactions which are thermodynamically limited at the lower temperatures, but also affects the char yields by increasing the heat-up rates. It has been demonstrated that char yields can be diminished[3-7] by increasing heat-up rates.

Char yields may also be affected by the occurrence of several alternate reactions:

$$H_2O + C \rightleftarrows CO + H_2 \quad (1)$$

$$2H_2 + C \rightleftarrows CH_4 \quad (2)$$

$$CO_2 + C \rightleftarrows 2CO \quad (3)$$

where in these reactions C represents char. Thermobalance studies with biomass chars indicate that char gasification reactions are relatively slow at 1500 °F even in environments where there are high concentrations of the

TABLE I

DATA FOR THE DEVOLATILIZATION OF MAPLE HARDWOOD IN THE LAMINAR FLOW REACTOR

Run No.	*MHC 17*	*MHC 18*	*MHC 31*	*MHC 33*	*MHC 19*	*MHC 20A*	*MHC 21*	*MHC 22*	*MHC 22A*	*MHC 32*	*MHC 23*	*MHC 24A*	*MHC 38*
Temperature, °F	1 000	1 300	1 300	1 300	1 500	1 600	1 000	1 300	1 300	1 300	1 500	1 600	1 600
Pressure, psig	300	300	300	300	300	300	300	300	300	300	300	300	300
Gas medium	He	He	He	He	He	He	H_2/He[a]	H_2/He	H_2/He	H_2/He	H_2/He	H_2/He	H_2/He
Run time, min	200	300	292	210	300	270	152	333	299	250	284	360	210
Amount fed, g (dry)	30·95	56·44	47·80	35·61	51·55	51·02	17·81	55·25	59·92	40·81	47·87	78·13	28·06
% feed moisture	3·1	3·1	3·1	3·1	3·1	3·1	3·1	3·1	3·1	3·1	3·1	3·1	3·1
Reactor length, in.	30	30	30	30	30	30	30	30	30	30	30	30	30
Experimental devolatilization yields, % dry feed													
Gas	45·8	68·3	66·8	70·5	72·5	72·2	48·4	65·7	64·8	72·2	79·7	80·7	79·6
Liquids	25·0	11·4	11·6	14·2	8·0	5·4	33·4	15·1	14·2	15·0	9·1	6·3	6·6
Char	12·4	10·3	7·9	9·7	9·0	6·6	13·6	4·7	4·8	6·3	3·5	6·5	6·4
Gas composition, mole %													
CO	50	48	50	47	44	45	58	54	56	54	51	57	53
CO_2	15	10	9	11	10	11	14	8	9	9	8	9	7
H_2	12	16	16	17	23	24	—	—	—	—	—	—	—
CH_4	14·0	19·3	18·7	19·7	19·2	19·3	17·8	27·1	25·0	26·1	37·8	32·5	37·6
C_2H_6	2·6	1·5	1·6	1·2	0·3	—	4·1	7·9	7·6	7·5	0·7	—	0·3
C_2H_4	3·5	3·7	3·5	3·5	1·6	—	4·1	1·4	1·4	1·7	0·7	—	0·3
C_3H_6	—	—	—	—	—	—	—	—	—	—	—	—	—
C_3H_8	1·8	—	—	—	—	—	1·4	—	—	—	—	—	—
C_6H_6	—	0·7	1·2	0·8	1·6	0·7	—	0·9	1·0	1·7	1·7	1·4	1·9
C_3H_6O	0·9	—	—	—	—	—	1·4	—	—	—	—	—	—

[a] 5% H_2/95% He mole %.

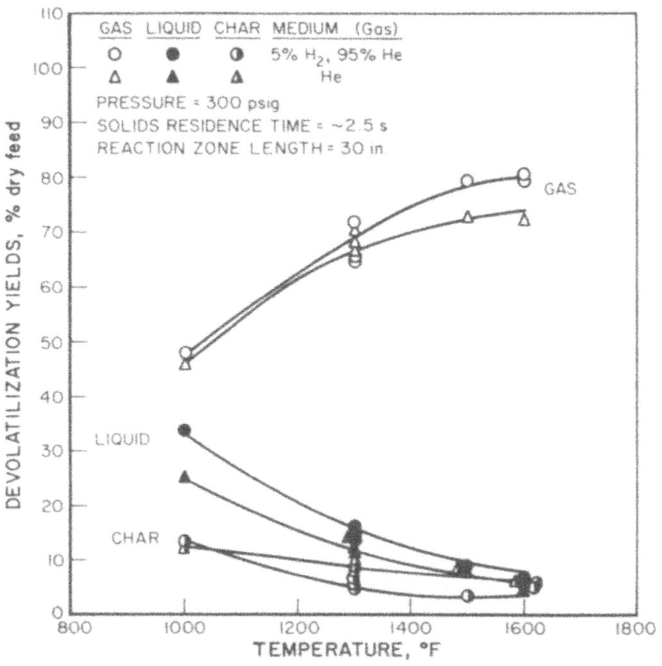

FIG. 2. Devolatilization yields of maple hardwood as a function of temperature.

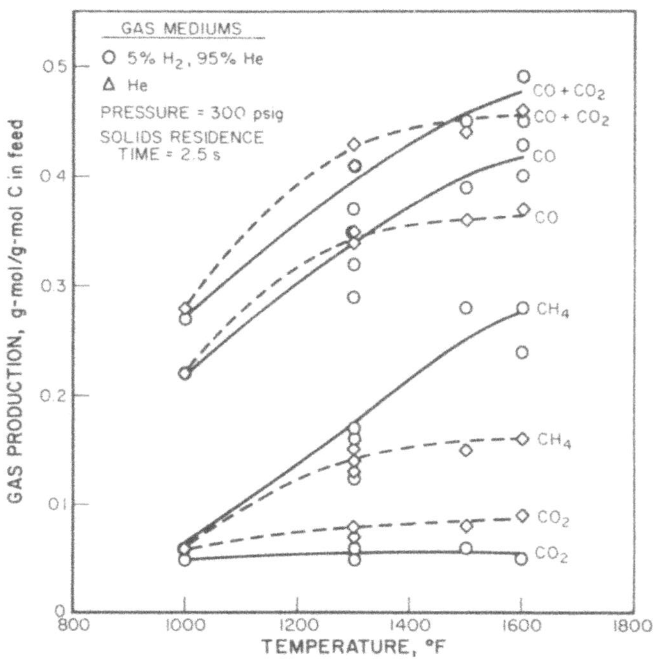

FIG. 3. Devolatilization gas product composition, maple hardwood.

reactive gases (i.e. CO, CO_2, H_2O and H_2). In the laminar flow experiments under discussion the average concentration of these species, with the exception of H_2, is less than 1 % and H_2 concentration was only about 5 %. Therefore, the combined effect of a small reaction rate constant under reaction conditions coupled with low reactive gas concentrations and short reaction times, *ca.* 2·5 s, make it reasonable to assume that none of the above reactions contribute significantly to char conversions in these experiments.

It is more reasonable to assume that the presence of hydrogen in the devolatilization atmosphere promotes the secondary reactions of the primary liquid devolatilization products to gaseous products. In agreement with such a conclusion is the hypothesis presented by Broido and Nelson,[8] who maintain that pyrolysis gas evolution will prevent significant gas–solid contact. Under these circumstances the observed beneficial effects on product yields and gas composition result from secondary reactions of primary pyrolysis with hydrogen and not the effects of hydrogen on the primary feed decomposition reactions. It is known that thermal cracking reactions of primary liquid products can also produce char. It may be possible, then, that the presence of hydrogen in the reactive environment may prevent, or at least diminish the effect of the secondary char forming reactions and thereby cause reduced char yields when hydrogen is present. Although thermobalance experiments clearly indicate the high reactivity of chars with hydrogen, this type of experiment is conducted with completely devolatilized char, and thus no pyrolysis gases are evolved and no inhibiting effect of gas evolution is present.

The major gaseous products (CO, CO_2 and CH_4) account for from 73 % to 99 % of the total gaseous products. Their production, relative to the amount of carbon in the feed, as a function of temperature and for the two gas compositions (pure helium and 5 % He/95 % H_2) is shown in Fig. 3. Increasing the temperature increased both the CO and CH_4 concentration with very little effect upon the CO_2 concentration. These increased concentrations, as discussed previously, result from secondary reactions of the primary liquid products through disproportionation and decarboxylation[2] type reactions. It is particularly interesting to note the effect of added hydrogen at temperatures above 1300 °F. When the hydrogen results are compared with those obtained in pure helium the CO yield increases while the CO_2 yield decreases. This suggests that the concentration of these components is being affected by the water gas shift reaction.

$$CO + H_2O \rightleftarrows CO_2 + H_2 \tag{4}$$

TABLE II

LAMINAR FLOW DEVOLATILIZATION DATA, BALANCES AND CONVERSION

Run No.	*MHC 17*	*MHC 18*	*MHC 31*	*MHC 33*	*MHC 19*	*MHC 20A*	*MHC 21*	*MHC 22*	*MHC 22A*	*MHC 32*	*MHC 23*	*MHC 24A*	*MHC 38*
Temperature, °F	1 000	1 300	1 300	1 300	1 500	1 600	1 000	1 300	1 300	1 300	1 500	1 600	1 600
Pressure, psig	300	300	300	300	300	300	300	300	300	300	300	300	300
Gas medium	He	He	He	He	He	He	H_2/He[a]	H_2/He	H_2/He	H_2/He	H_2/He	H_2/He	H_2/He
Run time, min	200	300	292	210	300	270	152	333	299	250	284	360	210
Amount fed, g (dry)	30·95	56·44	47·80	35·61	51·55	51·02	17·81	55·25	59·92	40·81	47·87	78·13	28·06
% feed moisture	3·1	3·1	3·1	3·1	3·1	3·1	3·1	3·1	3·1	3·1	3·1	3·1	3·1
Reactor length, in.	30	30	30	30	30	30	30	30	30	30	30	30	30
Balances													
Carbon	73	84	80	86	85	76	81	80	71	88	90	91	94
Hydrogen	86	90	86	96	91	83	97	95	88	101	102	82	96
Oxygen	92	95	93	103	94	100	103	92	90	98	95	104	91
Overall[t]	83	90	86	94	90	84	96	85	77	94	92	93	95
Conversion[b]	85	89	91	90	90	92	86	95	94	93	96	93	93

[t] Overall material balance = [(gas + liquid + char)/feed] × 100, by weight.

[a] 5% H_2/95% He mole %.

[b] Conversion = [gas + liquid/(gas + liquid + char)] × 100, by weight.

Evaluation of the ratio $(CO_2)(H_2)/(CO)(H_2)$, using the data in Table III and an estimate of the H_2O concentration in the reactor, when compared to the equilibrium constant at these temperatures, indicated that for experiments where H_2 is present the reaction as written would proceed to the left.

The most striking difference in Fig. 3 is the effect of the added hydrogen upon the methane yields. While at the lower temperatures there is very little effect, at 1600 °F (871 °C) there is a 75 % increase in total methane yield. In the light of the overall devolatilization yields in Fig. 2 this improvement in methane yield must result from secondary reaction of hydrogen with primary decomposition liquids.

Elemental and overall material balances as well as conversion, defined as the ratio of the difference in weight of feed and char to feed weight, are shown in Table II. Overall material balances average *ca.* 90 %. Although the average hydrogen and oxygen balances are above 90 %, the average carbon balance is only 83 %. The most probable cause for this is the difficulty in collecting all the oil and tarry products. Feed measurements

TABLE III

PRODUCT GAS ANALYSIS REPRODUCIBILITY

Run No.	*MHC 18*	*MHC31*	*MHC 33*	*MHC 22*	*MHC 22A*	*MHC 32*
Temperature, °F	1 300	1 300	1 300	1 300	1 300	1 300
Pressure, psig	300	300	300	300	300	300
Gas medium	He	He	He	H_2/He[a]	H_2/He	H_2/He
Run time, min	300	292	210	333	299	250
Amount fed, g (dry)	56·44	47·80	35·61	55·25	59·92	40·81
% feed moisture	3·1	3·1	3·1	3·1	3·1	3·1
Reactor length, in.	30	30	30	30	30	30
Devolatilization yields % dry feed						
Gas	68·3	66·8	70·5	65·7	64·8	72·2
Liquids	11·4	11·6	14·2	15·1	14·2	15·0
Char	10·3	7·9	9·7	4·7	4·8	6·3
Gas composition, mole %						
CO	48	50	47	54	56	54
CO_2	10	9	11	8	9	9
H_2	16	16	17	—	—	—
CH_4	19·3	18·7	19·7	27·1	25·0	26·1
C_2H_6	1·5	1·6	1·2	7·9	7·6	7·5
C_2H_4	3·7	3·5	3·5	1·4	1·4	1·7
C_3H_6	—	—	—	—	—	—
C_3H_6O	—	—	—	—	—	—
C_6H_6	0·7	1·2	0·8	0·9	1·0	1·7

[a] 5 % H_2, 95 % He mole %.

TABLE IV

SUMMARY OF DATA ON THE EFFECT OF SOLIDS RESIDENCE TIME VARIATION

Run No.	*MHC 21*	*MHC 25*	*MHC 25A*	*MHC 28*	*MHC 22*	*MHC 22A*	*MHC 26A*	*MHC 29*	*MHC 32*	*MHC 23*	*MHC 27A*	*MHC 30*
Temperature, °F	1 000	1 000	1 000	1 000	1 300	1 300	1 300	1 300	1 300	1 500	1 500	1 500
Pressure, psig	300	300	300	300	300	300	300	300	300	300	300	300
Gas medium	H_2/He[a]	H_2/He	H_2/He	H_2/He	H_2/He	H_2/He	H_2/He	H_2/He	H_2/He	H_2/He	H_2/He	H_2/He
Run time, min	152	194	180	266	333	299	240	270	250	284	270	184
Amount fed, g (dry)	17·81	27·28	32·12	30·73	55·25	59·92	52·53	38·21	40·81	47·87	53·64	23·91
% feed Moisture	3·1	3·1	3·1	3·1	3·1	3·1	3·1	3·1	3·1	3·1	3·1	3·1
Reactor length, in.	30	18	18	48	30	30	18	48	30	30	18	48
Devolatilization yield % dry feed												
Gas	48·4	48·3	47·2	60·2	65·7	64·8	58·8	75·4	72·2	79·7	73·3	84·6
Liquids	33·4	26·3	24·6	18·0	15·1	14·2	12·1	12·9	15·0	9·1	8·5	6·7
Char	13·6	14·5	12·8	16·0	4·7	4·8	5·6	9·7	6·3	3·5	7·4	9·2
Gas composition, mole %												
CO	58	55	54	55	54	56	55	54	54	51	51	52
CO_2	14	18	20	16	8	9	10	9	9	8	9	7
CH_4	17·8	17·8	17·9	21·6	27·1	25·0	24·7	25·5	26·1	37·8	35·2	38·8
C_2H_6	4·1	2·8	3·3	4·3	7·9	7·6	7·5	8·7	7·5	0·7	2·1	0·4
C_2H_4	4·1	1·9	1·6	0·9	1·4	1·4	2·1	1·4	1·7	0·7	1·2	—
C_3H_8	1·4	0·9	0·8	0·9	—	—	—	—	—	—	—	—
C_3H_6	—	1·9	0·8	0·9	—	—	—	—	—	—	—	—
C_3H_6O	1·4	1·9	1·6	—	—	—	—	—	—	—	—	—
C_6H_6	—	—	—	0·9	0·9	1·0	0·8	1·4	1·7	1·7	1·8	1·8

[a] 5% H_2, 95% He mole %.

and char collection are quite reliable. Replicate analyses shown in Table III indicate that reproducible gas compositions were obtained. Since liquids, which have carbon as their major component, may condense before the liquid collection system, in spite of the precautions taken to prevent it, they, therefore, are probably the primary cause of the relatively low carbon balance.

A preliminary evaluation of the effects of residence time and pressure was undertaken. A summary of the results of the variation in residence times and temperature is shown in Table IV and presented graphically in Figs 4 and 5. Gas yields consistently increase as the solids residence time is increased. Gas velocities were maintained constant, so as the reaction zone length was increased, residence times of the gaseous and vapor products likewise increased. The longer reaction times not only allowed more time for the primary decomposition to occur, but also increased the time available for primary products to undergo secondary gas-phase reactions.

The data collected on the effect of residence time on liquid yields is

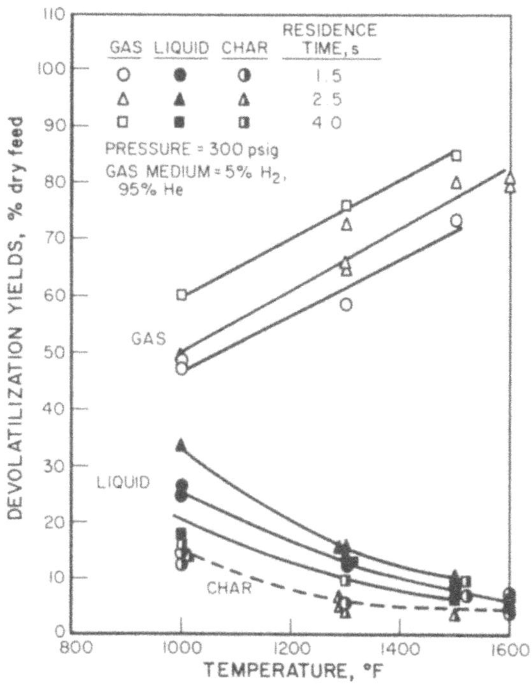

FIG. 4. Devolatilization yields as a function of temperature for various residence times.

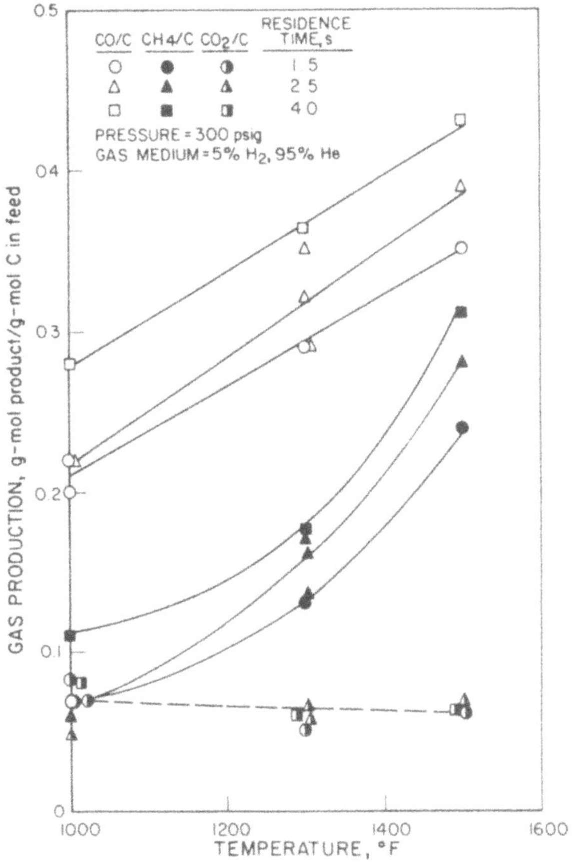

FIG. 5. Gas production as a function of temperature for various residence times.

scattered. The literature reports[2,9] that liquid yields may go through a maximum at overall conditions that significantly reduce the probability of secondary reaction of primary liquid products. The data on liquid yields tend to support this hypothesis.

Although it is known that char gasification reactions occur, the data suggest that for the residence times investigated at these pressure and temperature conditions their rates may not be fast enough to significantly affect the total char yield. As previously mentioned there are known gas-phase decomposition reactions, whose by-products may be char. The overall char yield, however, at these conditions, appears to be unaffected by residence times.

Residence time effects upon product gas composition are shown in Fig. 5. While no measurable effect was observed upon the CO_2 make, both the CO and CH_4 concentration consistently increase as the residence time is increased. Based upon the data available, it is difficult to determine the reaction which will lead to these increased methane and carbon monoxide yields. Earlier it was shown that as temperature increased there was a consistent increase in gas yield accompanied by a simultaneous decrease in both liquid and char yields (predominantly liquid); however, as shown in Fig. 4, the steady increase in gas yields with increasing temperature was accompanied by char yields which were unaffected and liquid yields which passed through a maximum.

Some preliminary experiments to define the effect of pressure in the range 100–300 psig were performed (Table V). In these preliminary experiments,

TABLE V

LAMINAR FLOW DEVOLATILIZATION OF MAPLE HARDWOOD CHIPS: EFFECT OF PRESSURE ON YIELDS

Run No.	*MHC 24A*	*MHC 34*	*MHC 35*	*MHC 38*
Temperature, °F	1 600	1 600	1 600	1 600
Pressure, psig	300	100	200	300
Gas medium	H_2He[a]	H_2/He[a]	H_2/He[a]	H_2/He[a]
Run time, min	360	300	300	210
Amount fed, g (dry)	78·13	47·29	57·01	28·06
% feed moisture	3·1	3·1	3·1	3·1
Reactor length, in.	30	30	30	30
Devolatilization yields % dry feed				
Gas	80·7	85·8	84·9	79·6
Liquids	6·3	11·4	6·6	6·6
Char	6·5	5·4	5·1	6·4
Gas composition, mole %				
CO	57	58	58	53
CO_2	9	8	7	7
H_2	—	—	—	—
CH_4	32·5	32·2	33·7	37·6
C_2H_6	—	0·1	0·4	0·3
C_2H_4	—	0·3	0·4	0·3
C_3H_8	—	—	—	—
C_3H_6	—	—	—	—
C_3H_6O	—	—	—	—
C_6H_6	1·4	1·7	1·1	1·9

[a] 5% H_2, 95% He, mole %.

no significant effect was observed on devolatilization yields (Fig. 6) and product gas composition (Fig. 7). In published reports[2,6,9-12] on the effect of pressure on biomass pyrolysis their low pressure ranges were well below atmospheric pressure. In the reported investigations they found that, when pyrolyzing at vacuum, tar yields increase and char and gas yields decrease.

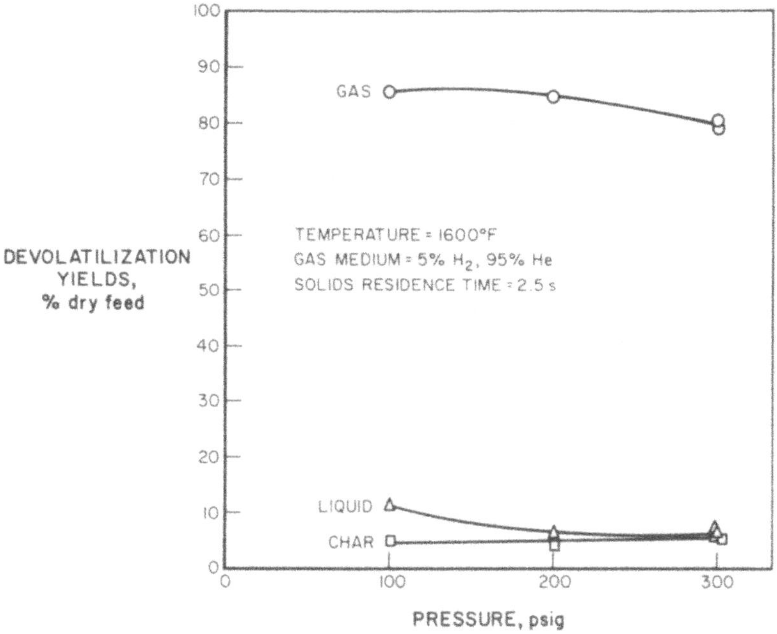

FIG. 6. Devolatilization yields from maple hardwood chips, as a function of pressure.

This has been attributed to secondary reactions of the primary decomposition products. The nature of these secondary reactions may be purely pyrolytic or they may be the result of the increased opportunity for occurrence of catalytic reaction of primary products on the char as well as on inorganic matter, ash present in the biomass samples.[13] The lowest pressure used, 100 psig, produced residence times sufficient for the catalytic or thermally driven reactions mentioned above to proceed to completion. Thus one would not expect agreement with the results reported for vacuum and atmospheric pressure pyrolysis.

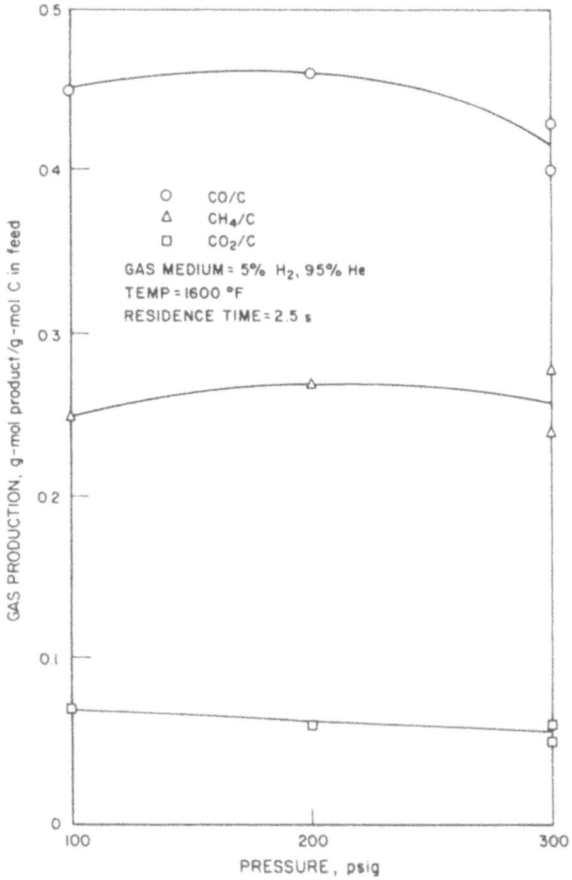

FIG. 7. Gas production from maple hardwood chips as a function of pressure.

4. SUMMARY

These devolatilization experiments are interesting and provide useful information on the effect of several important reaction variables on the behavior of maple hardwood chips under various simulated reaction conditions. Increasing reaction temperature clearly improves the amount of conversion to gaseous and liquid products and also increases the amount of feed carbon which is converted to useful gaseous products, such as CO and CH_4, while having little effect upon the CO_2 production.

The addition of even relatively small amounts of hydrogen, 5 % in these experiments, produced several desirable effects upon the global devolatilization yields as well as 'improvement in product gas composition', especially methane at the higher temperatures.

Increasing solid and gas residence times increased the amount of feed converted to gases; however, the effect upon the liquid and char yields are not readily explainable. The increased gas product yields are predominantly the result of increased CO and CH_4 production. No discernable difference was observed with CO_2 yields as a function of residence time. More information is needed on the reactivity of the basic components of biomass (i.e. cellulose, hemicellulose and lignin) and upon the whole biomass sample.

REFERENCES

1. Jones, J. H., Braun, W. G., Daubert, T. E. and Allendorf, H. D. (1966). *A.I.Ch.E. Journal* November, 1070.
2. Shafizadeh, F. (1980). Introduction to pyrolysis of biomass. *Proceedings of the Specialists' Workshop on the Fast Pyrolysis of Biomass.* SERI/CP-622-1096, Copper Mountain, CO, Oct 19–22.
3. Diebold, J. P. (1980). Ablative pyrolysis of macroparticles of biomass. *Proceedings of the Specialists' Workshop on the Fast Pyrolysis of Biomass.* SERI/CP-622-1096, Copper Mountain, CO, Oct 19–22.
4. Reed, T. B., Diebold, J. P. and Desrosiers, R. (1980). Perspective in heat transfer requirements and mechanisms for fast pyrolysis. *Proceedings of the Specialists' Workshop on the Fast Pyrolysis of Biomass.* SERI/CP-622-1096, Copper Mountain, CO, Oct 19–22.
5. Lincoln, K. A. (1965). Flash-pyrolysis of solid-fuel materials. *Pyrodynamics* **2**, 133.
6. Lleweller, P. C., Peters, W. A.. and Howard, J. B. (1976). Cellulose pyrolysis kinetics and char formation mechanisms. Sixteenth (Int.) Symposium on Combustion, The Combustion Institute, p. 1471.
7. Antal, M. J., Edwards, W. E., Friedman, A. L. and Rogers, F. E., A study of the steam gasification of organic wastes. University Grant No. R804836010 to Princeton University by U.S. EPA, IERI, Cincinnati, Ohio.
8. Broido, A. and Nelson, M. A. (1975). *Combustion and Flame* **24**, 263.
9. Hajaligol, M. R., Peters, W. A. and Howard, J. B. (1980). Product compositions and kinetics for rapid pyrolysis of cellulose. *Proceedings of the Specialists' Workshop on the Fast Pyrolysis of Biomass.* SERI/CP-622-1096, Copper Mountain, CO, Oct 19–22.
10. Madorsky, S. L., Hart, V. E. and Straus, S. (1958). *J. of Research of the NBS* **60**(4), 343, April.
11. Roberts, A. F. (1970). *Combustion and Flame* **14**, 261.
12. Stamm, A. J. (1956). *Industrial Engineering Chemistry* **48**(3), 413.
13. Shafizadeh, F.Amn, Bradburg, G. W., DeGroot, W. F. and Aanernd, T. W. (1982). *Ind. Eng. Chem. Prod. Res. Dev.* **21**, 97.

20

The Direct, Mass-Spectrometric Study of the Primary and Secondary Pyrolysis Behavior of Biomass and its Constituents

THOMAS A. MILNE and MICHAEL N. SOLTYS

Thermochemical and Electrochemical Research Branch, Solar Energy Research Institute, Golden, Colorado, USA

1. INTRODUCTION

One of the important gasification research needs identified in our comprehensive survey of biomass gasification[1] is a more detailed understanding of the molecular processes and kinetics involved in the pyrolysis of biomass and its constituents. Of particular interest is the fast pyrolysis of biomass, which can accomplish nearly charless gasification to valuable intermediates.[2] Improved understanding of the details of fast pyrolysis requires a versatile sampling and analytical technique. Such a technique should allow observations of gases, condensible species and reactive intermediates in real-time, sampled from realistic laboratory fast-pyrolysis reactors. Bench-scale studies[3,4] indicate that millisecond processes are occurring and that tars and oils are likely precursors to the final product slate of lighter hydrocarbons.

To accomplish the necessary sampling and analysis, we have chosen the well-studied technique of extractive sampling, free-jet expansion, molecular-beam mass spectrometry.[5] One of us has used such a technique in analogous studies of the early stages of pulverized coal combustion in rich flames.[6] This technique provides rapid quenching of all species sampled directly from the reactor environment. The mass spectrometer provides real-time, nearly universal detection of even condensible vapors. In the present work, the pyrolysis medium is a laminar column of hot gases, generated by either the burnt-gas region of a H_2–O_2–Ar, He flame or from a resistance-heated flow system. This combination permits a 'wall-less' system for both pyrolysis and sampling. A major additional feature is that the system can be used to observe separately primary and secondary pyrolysis events.

The system and its qualitative performance has been described in detail in two previous publications.[7,8] The next section outlines briefly the apparatus and its use.

2. APPARATUS

The sampling system, with a resistance-heated heated inert gas pyrolysis medium, is shown in Fig. 1. Gaseous mixtures or pure gases such as Ar, He, N_2, H_2 or CH_4 can be heated to temperatures of the order of 600–700 °C to produce a laminar flow of hot gases that bathes specimens to be pyrolyzed, thus simulating heating under realistic gasifier conditions. For higher temperatures, and easy creation of a steam environment, the tube system

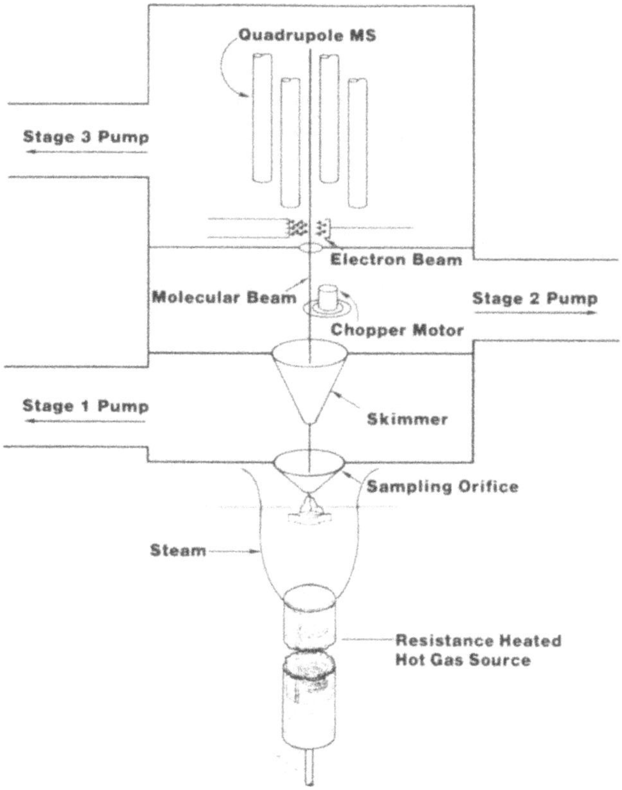

FIG. 1. Molecular beam apparatus for fast biomass pyrolysis.

shown is replaced by a flat-flame burner whose burnt-gas region provides a steam–inert gas environment at temperatures from about 900 °C to around 3000 °C, if desired. For most of the work carried out so far, we have used a stoichiometric H_2–O_2–Ar flame, producing a burnt gas of about 15 % steam and 85 % argon at an uncorrected thermocouple temperature of 900 °C.[7]

The samples to be pyrolyzed are suspended in the hot gases either as slivers of wood, strips of filter paper, or threads. Powders are contained in thin, metal shim-stock boats.

Gases emerging from the pyrolyzing material into the hot surrounding steam–argon are sampled and rapidly quenched by extraction into a sonic orifice and subsequent expansion to molecular flow. The residence time of the primary pyrolysis products in the hot gas is varied by changing the position of the sampling orifice relative to the pyrolyzing surface (or burner surface in the case of injected materials). Free-jet, molecular-beam mass spectrometry has been applied to a number of high temperature systems. Details may be found in several reviews.[5,9] Suffice it to mention here that the free-jet expansion to collisionless flow typically occurs in times of the order of microseconds, allowing the preservation of even highly condensible species, while achieving significant internal-state cooling of the sampled molecules, a phenomenon of some significance for the interpretation of mass spectra in directly coupled systems. The use of a molecular beam introduction to the ion source of the mass spectrometer allows observation of only gas that has arrived by line-of-sight from the hot environment, through the use of neutral-beam modulation and phase-sensitive detection. Details of the specifications of the sampling system and the performance and sensitivity of the mass spectrometer and data acquisition system are given by Soltys and Milne.[10,7]

The system just described has been exercised, in a semi-quantitative way, on permanent gases and biomass and its constituents in order to determine the complexity of the pyrolysis events and the sensitivity and time-resolution limitations of the present system. Typical performance and results are shown next, followed by a discussion of quantitative work now underway.

3. OPERATING PHENOMENA

The general sampling technique, for steady-state sampling of high temperature systems, can achieve a large dynamic sensitivity range. In a 1-atm

CH_4–air flame, we were able to detect the Kr isotopes in air at a level of 100–600 ppb. We were also able to observe the vaporization of zinc atoms when zinc-coated wire was passed through the hot flame gases.[7]

Previous studies had indicated significant internal state cooling of hot *n*-butane during free-jet sampling.[11] In the present work, in which steam–argon is expanded through a 0·50-mm dia. orifice, we appear to be achieving sufficient internal cooling of solute species such as propane, *n*-butane and methanol such that room-temperature fragmentation patterns are achieved when sampling these molecules from a 900 °C environment. Such cooling will greatly simplify the interpretation of mass spectra of directly-sampled species.

The sensitivity of the combined system is such that a good profile of pyrolysis products can be followed in real time in the pulsed pyrolysis of samples of the order of 10 mg in weight. Signal averaging techniques should allow a further reduction in the amount of pyrolyzed material required, and hence a reduction in the physical dimensions, or concentration, of material to be pyrolyzed—important parameters we wish to vary.

4. QUALITATIVE OBSERVATIONS OF PRIMARY PYROLYSIS PRODUCTS

We have surveyed the nature and complexity of the primary products from a number of carbonaceous materials suddenly immersed in the approximately 900 °C steam–argon combustion gases. Two kinds of results are shown. First, the integrated mass spectrum of the entire evolution of products for a complete pyrolysis, with the sampling orifice placed just a few mm above the pyrolyzing surface. In a typical pyrolysis, a 10 mg sliver of material is suspended in the unignited flame gases. The gases are lit simultaneously with the start of repetitive scans of the quadrupole mass spectrometer. In the case of wood or filter paper (cellulose), the gas evolution lasts about 8–10 s so that repetitive 1 s scans over the desired mass range can be stored in a multichannel analyzer[10] to give an integrated picture of the total pyrolysis wave. Figures 2–4 show spectra of cellulose, lignin and wood with tentative mass assignments. These spectra are taken at an ionizing electron energy of 16·5 eV to achieve a mass spectrum dominated by parent ions. In addition, these spectra are taken under focusing conditions such that the masses in the range 60–400 are presented in approximately their true proportions as judged from the NIH-EPA standard spectrum of PFTBA. Lighter species such as H_2, H_2O, CO,

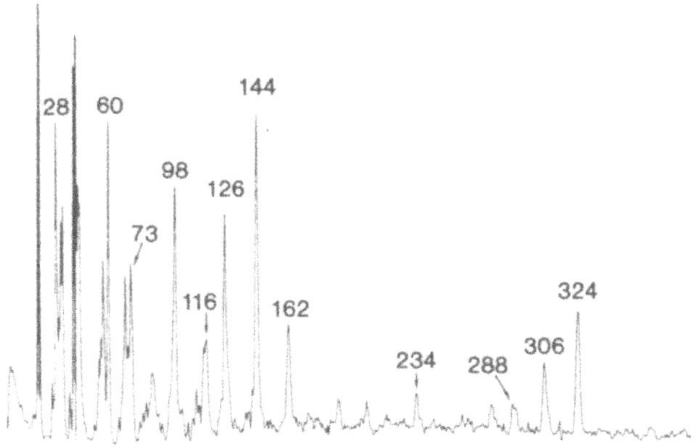

FIG. 2. Positive ion mass spectrum of the primary products of pyrolysis of filter paper (cellulose) heated in 900 °C $Ar{-}H_2O$.

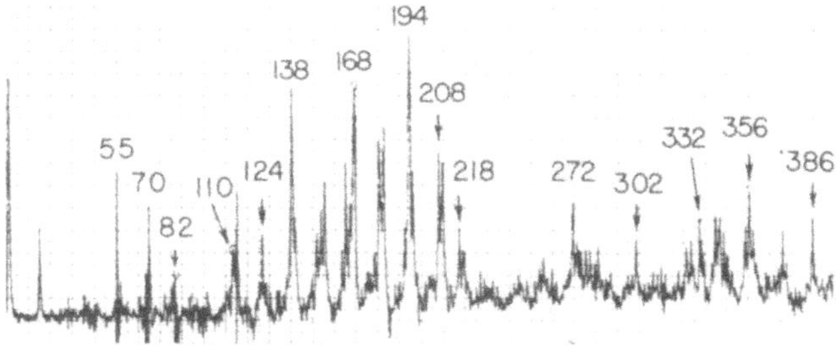

FIG. 3. Positive ion mass spectrum of the primary products of pyrolysis of ethanol-extracted, steam exploded aspen lignin heated in 900 °C $Ar{-}H_2O$.

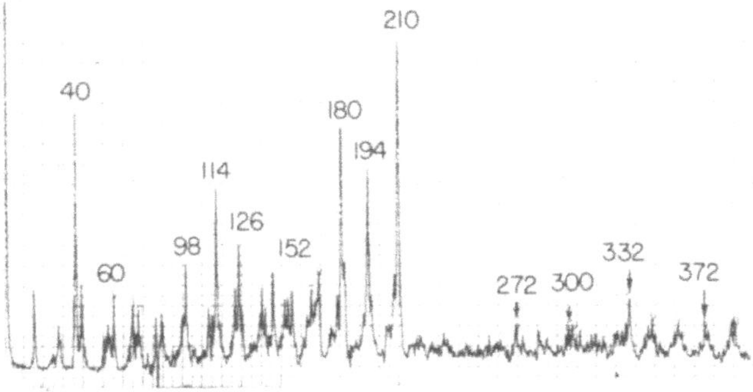

FIG. 4. Positive ion mass spectrum of the primary products of pyrolysis of a sliver of hardwood dowel.

CH_3OH, C_2H_4, CO_2, etc. are strongly discriminated against under these conditions, however, due both to the low electron energy used and to mass discrimination effects inherent in the quadrupole mass spectrometer. Separate tuning conditions can, of course, be used to study these species.

Noteworthy in Figs 2–4 is the rich array of primary pyrolysis products (oils and tars) being preserved and observed by the present techniques of sampling and detection. Equally noteworthy are the distinctive signatures of the primary pyrolysis products from the different sources. Preliminary comparison of these results with prior work using analytical pyrolysis and other techniques[12–17] allows some immediate identification of the major species coming from the biomass materials.

5. TENTATIVE IDENTIFICATION OF CELLULOSE, LIGNIN AND WOOD PRIMARY PYROLYSIS PRODUCTS

The assignment of possible neutral precursors to the prominent ions in the cellulose spectrum is greatly aided by the results from the analytical pyrolysis of glycogen carried out several years ago by Schulten and Gortz.[12] Using Curie-point pyrolysis of micrograms of glycogen heated at $>1000\,°C/s$, with fairly direct inletting of pyrolysis gases to be field-ionized in a high resolution mass spectrometer, they assigned unambiguous chemical compositions to the observed major ions. Then the reports of chemical products in the literature were correlated with observed ions. The suggested assignments in Table I are based largely on the work of Schulten and Gortz. As can be seen, our observed mass spectra from cellulose are plausible in terms of other work on cellulose degradation.

It should be noted that many of the peaks observed from cellulose are similar in proportion to those observed from levoglucosan (MW 162), a major primary product of cellulose pyrolysis.[15] Recent comparative studies of electron impact and chemical ionization spectra of levoglucosan[18] suggest that fragmentation is extensive, even at 16·5 eV, for free-jet-cooled parent levoglucosan. Thus, peaks identified in Table I as chemical species (e.g. 60, 70, 98, 126, 144) may be dominated by fragment ions. A major thrust of future quantitative work is to resolve the questions of fragment identities using techniques such as MS/MS, calibrations with pure compounds and time-dependence of evolution and secondary cracking.

The very distinctive series of ions from lignin shows great similarity to Curie-point spectra of lignin obtained by Meuzelaar *et al.*[17] Meuzelaar

TABLE I
ION PEAKS FROM CELLULOSE PYROLYSIS (16·5 eV)

Mass	*Likely parent*	*Mass*	*Likely parent*
26	C_2H_2	82	?
28	CO, C_2H_4	84	?
29	HCO^+	85	?
30	formaldehyde	86	2,3-butanedione
31	CH_3O^+	87	?
M32[a]	methanol	88	pyruvic acid
42	propene, ketene	89	?
M43	CH_3CO^+	90	glyceraldehyde, dihydroxypropane
44	CO_2, acetaldehyde,		
47	L^+	96	2-furaldehyde
55	$C_4H_7^+$	97	?
56	acrolein	M98	furfuryl alcohol, L^+
M57	L^+		
58	glyoxal, acetone	102	?
M60	acetic acid, hydroxy-acetaldehyde, L^+	110	5-methyl-2-furaldehyde
		114	?
69	?	115	?
M70	2-butenal, L^+	116	L^+
72	pyruvaldehyde, L^+	M126	levoglucosenone, 5-hydroxy-methyl-2-furaldehyde, L^+
M73	L^+		
M74	1-hydroxy-2-propanone, L^+	128	?
80	?		
81	?	M144	L^+, 1,4:3,6-dianhydro-α-D-glucopyranose
		162	levoglucosan, 1,6-anhydro-β-D-glucofuranose

[a] M = 10 major peaks. L^+ = levoglucosan fragment ion.

has identified ion series for alkyl phenols and for mono- and dimethoxy compounds, as shown preliminarily in Table II.

We have not yet obtained a satisfactory spectrum of a hemicellulosic component of wood. Experiments with deacetylated xylan show few distinctive peaks, perhaps because the pyrolysis occurred too quickly or the hot environment caused premature secondary cracking.

Figure 4 shows the spectra from a small wood splint (dowel) at 16·5 eV. Major sequences of peaks from cellulose and lignin can readily be identified and, in addition, peaks at 96 and 114 match reported hemicellulose products of 2-furaldehyde and 3-hydroxy-2,2-penteno-1,5-lactone, respectively.[19]

TABLE II

ION PEAKS FROM STEAM-EXPLODED HARDWOOD LIGNIN PYROLYSIS (16·5 eV)

Mass	*Possible parent*
94, 108, 122, 136, 150, 164, 178	Alkyl phenol series
110	Dihydroxy benzene
124, 138, 152, 164, 180	Monomethoxyphenol series
154, 168, 182, 196, 210	Dimethoxy phenols series

It thus appears that the primary products from wood pyrolysis satisfactorily reflect the contributions from its major constituents, though the lignin-related peaks seem over represented.

6. EFFECTS OF SEVERAL PARAMETERS ON CELLULOSE AND WOOD PYROLYSIS

As a preliminary to quantitative work, a number of the key parameters likely to affect pyrolysis of cellulose and wood were varied to reveal qualitative responses in the primary product state. These were: physical form, quantity and type; ionizing electron energy; temperature of pyrolyzing gas; and effect of added K^+ for cellulose and physical size for wood.

Celluloses surveyed were Whatman No. 40 ashless filter paper, commercial cotton thread, Avicel microcrystalline powder in 90 and 20 μm size, and Sigma amorphous powder. The filter paper was prepared in rolls ranging from 10 mg to 100 mg in weight; Avicel 90 powder was pyrolyzed from large and small boats containing from 20 mg to 100 mg.

The spectra of all these samples were remarkably similar when pyrolyzed in the 900 °C flame, but at 16·5 and 50 eV ionizing electron energy. Without more careful quantitative work it is not clear that there are significant differences in primary product slates, in spite of the variations in the parameters listed as well as presumed differences in crystallinity and degree of polymerization (not determined in this work but known to be variable from literature citations). Figure 5 shows several typical low-voltage spectra for five of the cellulose samples.

Ionizing electron energy likewise had only second order effects on the mass spectrum, a somewhat surprising result in view of the above-cited evidence that levoglucosan fragments extensively even under our low voltage conditions. Figure 6 shows the spectrum of several celluloses at 50 eV for comparison with the low voltage spectrum. The examples

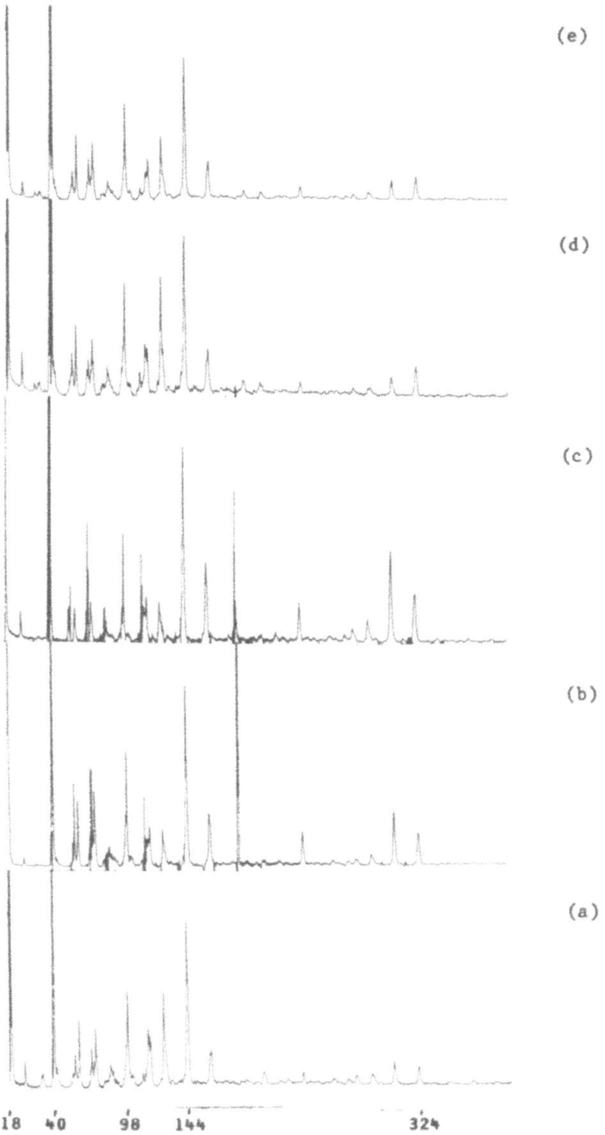

FIG. 5. Primary pyrolysis mass spectra of five different specimens heated in 900 °C $H_2/O_2/Ar$ flame gases; electron energy ~17 eV. (a) Filter paper (thin sheet); (b) cotton thread; (c) Sigma chemical amorphous powder; (d) Avicel crystalline powder, 90 μm; (e) Avicel crystalline powder, 20 μm (powders heated in small boats).

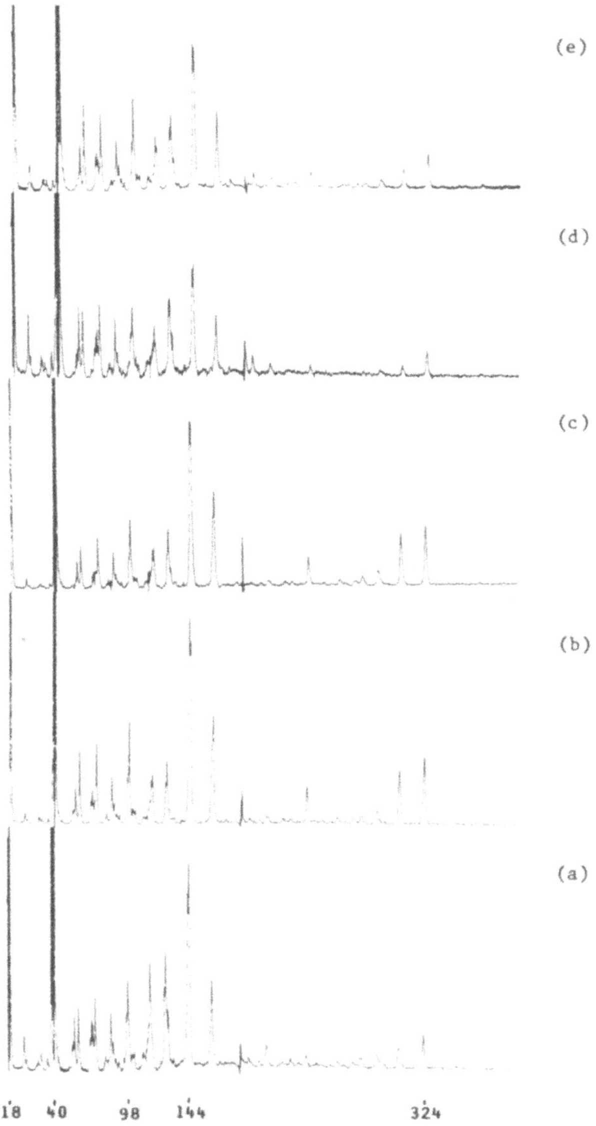

FIG. 6. Primary pyrolysis mass spectra of five different specimens heated in 900 °C $H_2/O_2/Ar$ flame gases; electron energy 50 eV. (a) Filter paper (thin sheet); (b) cotton thread; (c) Sigma chemical amorphous powder; (d) Avicel crystalline powder, 90 μm; (e) Avicel crystalline powder, 20 μm (powders heated in small boats).

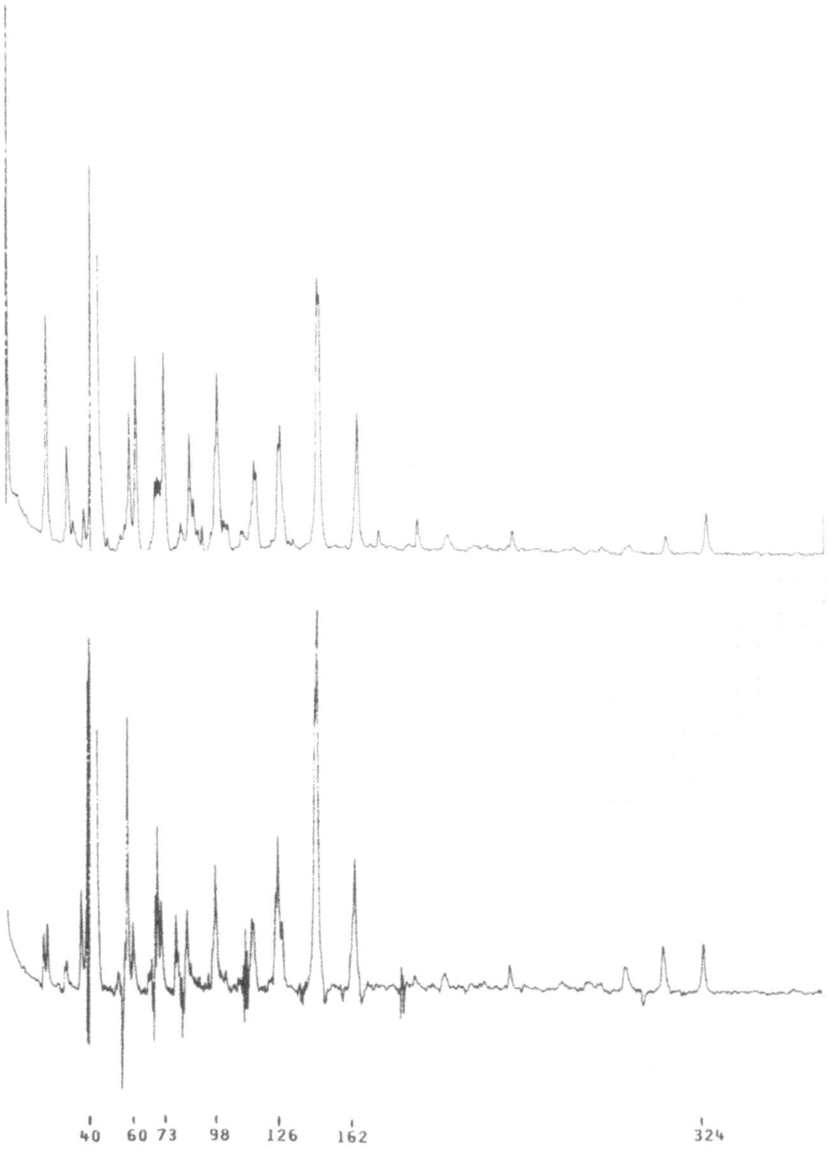

FIG. 7. Comparison of cellulose spectra in 900 °C steam/argon (upper) and in 450 °C hot argon (lower).

just shown were all taken under focusing conditions that discriminated seriously against masses below 40^+. However, under low-mass focusing conditions, there were again great similarities in cellulose spectra over type and electron energy.

Samples of filter paper were pyrolyzed in hot argon over temperature ranges from 450 to 600 °C to compare with the 900 °C steam–argon flame pyrolysis results. Once again, at 50 eV electron energy, there was great similarity in spectra from 450–600 °C. Figure 7 shows spectra at 450 and 900 °C. These spectra differ little, showing that the steam and higher temperature were not playing major roles. Caution is suggested in interpreting these results. First, the temperature varied is that of the heating gas, not necessarily the temperature of the pyrolyzing cellulose surface. To the extent that the endothermic pyrolysis causes ablative cooling, the 'true' pyrolysis temperature may not be varying that much. Heated-grid experiments are planned soon to try to control more directly the solid temperature rise. Second, we were focusing in these experiments on the heavy primary pyrolysis products (above mass 40^+). To the extent that levoglucosan and its fragments dominate the high molecular weight mass spectrum, temperature effects on true product slate will be masked.

A third comparison of cellulose pyrolysis was done to observe the effects of inorganic impurities. Potassium ion, in the form of KOH or K_2CO_3, was added to pure filter paper to achieve K levels of about 1 wt %. Such salts have been widely reported to affect pyrolysis, even at levels of 0·1 % or less.[20] The resultant total pyrolysis times and char residue amounts are shown in Table III. Equally dramatic was the effect on the primary product spectra in 900 °C steam–argon, as shown in Fig. 8. These results are consistent with reports of the effect of impurities and acid-washing on levoglucosan and tar production.[21]

TABLE III

EFFECT OF POTASSIUM SALTS ON CELLULOSE PYROLYSIS

Sample condition	*Approximate weight (g)*	*Pyrolysis time (s)*	*% Residue*
40 Whatman filter paper	0·009 6	12	2·0
With 1 % K from K_2CO_3	0·011 2	4	16·0
With 1 % K from KOH	0·011 4	4	17·0
With 10 % K from KOH	0·014 1	Plugged sampling orifice[a]	50·6

[a] Extensive foaming of char.

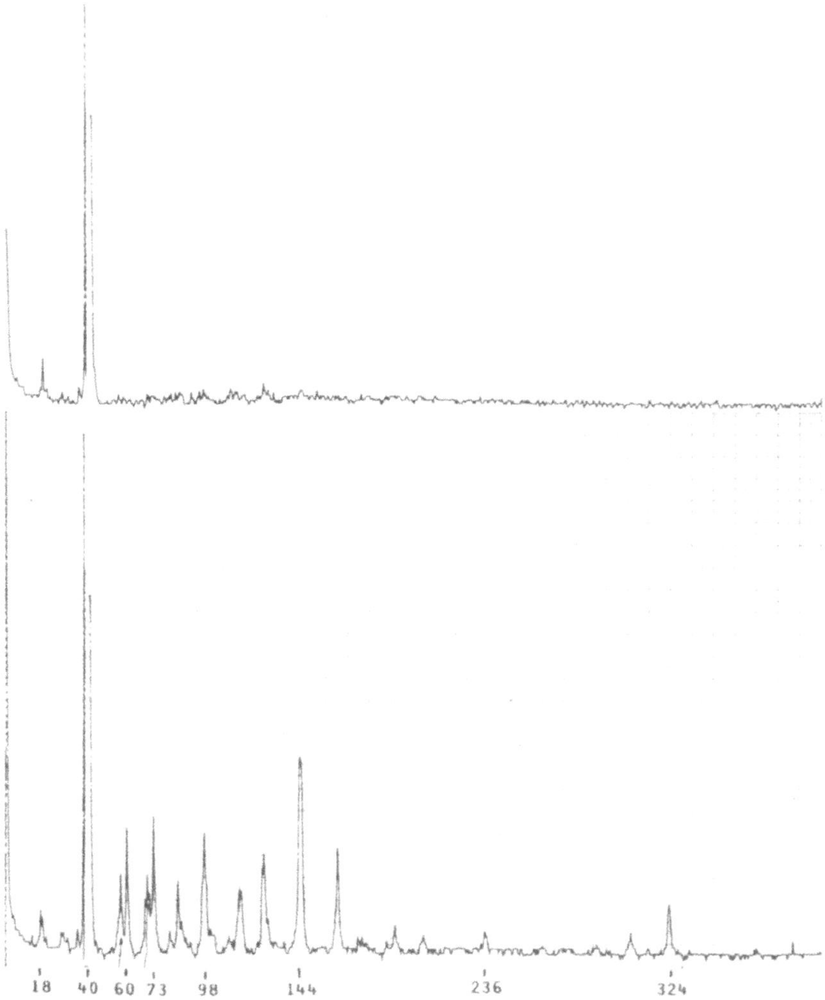

FIG. 8. Effect of 1 % K (from K_2CO_3) (upper) on pyrolysis spectrum at high masses of cellulose (filter paper) in 900 °C flame. Electron energy 50 eV.

TABLE IV
EFFECT OF SIZE OF WOODEN DOWELS ON PYROLYSIS BEHAVIOR

Approximate diameter (mm)	*Weight (g)*	*Pyrolysis time (s)*	*% Residue*
7·0	0·604	180	17·8
3·7	0·105	40	13·0
1·5	0·020 6	15	9·7
1·0	0·017 3	8	8·5

One set of experiments was done with wood splints, all from 1/4 in. (birch?) dowel to observe major effects of size of a dense specimen on integrated pyrolysis products. Table IV shows the effect of diameter on pyrolysis time and char yield in 900 °C steam–argon. The spectra for the extremes were quite similar in the tar region, with perhaps more light gases relative to tars, the larger the specimen diameter. Such an effect is expected to the extent that the pyrolysis is slower with the larger specimens, and the interior tars have the opportunity to crack as they diffuse out through a hot char layer.

In addition to the integrated spectra of pyrolysis products discussed so far, our real-time sampling permits the following of single or multiple ions throughout a pyrolysis wave. As an example, Fig. 9 shows the behavior of a single major ion, 60^+, during pyrolysis of a 17-mg piece of rolled filter paper. One curve shows evolution with the standard H_2–O_2–Ar flame. The

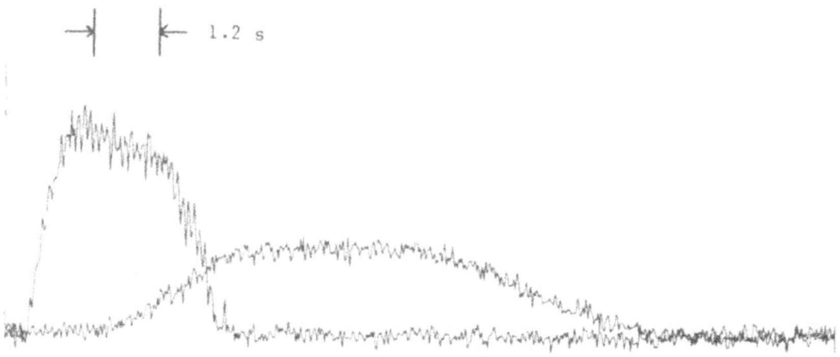

FIG. 9. Pyrolysis profiles from cellulose (60^+) heated in He/H_2/O_2 and Ar/H_2/O_2 flames of the same temperature. The helium flame profile is on the left.

other shows the effect of substituting He for Ar, which does not alter flame temperature but alters heat transfer to the filter paper considerably.

One can also carry out repetitive scans without averaging. Our present hard-wired memory in the multichannel analyzer limits us to about a 50 amu scan each second, to monitor a pyrolysis wave over about 10–20 s. An example is shown in Fig. 10, where scans over the ranges 38 to 88 and 120 to 170 are shown for small specimens of filter paper. To a first approximation, species heavier than CO_2 all evolve at the same time for these small samples.

7. QUALITATIVE OBSERVATIONS OF SECONDARY CRACKING OF PRIMARY PRODUCTS FROM CELLULOSE, LIGNIN AND WOOD

One of the prime objectives of our pyrolysis work is to observe the dependence of final product slate on the time, temperature and environment of secondary cracking of primary products such as those shown above. In our experimental approach this is easily accomplished by simply sampling at progressively greater distances from the pyrolyzing surface so that the primary products spend longer times in the hot-gas environment.

To observe qualitative behavior of secondary cracking, small samples of cellulose, powdered lignin and wood slivers have been pyrolyzed in 900 °C steam–argon and the spectrum recorded at varying distances from the sample surface. Typical results are shown in Figs 11, 12 and 13, where low voltage spectra are displayed, with key mass assignments. Because of the line source nature of the pyrolyzing material, there is considerable dilution of products with distance, even though the approximately 2 cm × 2 mm specimens and flame gases are confined in 5-cm diameter Vycor tubes. Thus, one should only compare changes in relative peak intensities within one scan.

Figure 11 shows the behavior of cellulose at an ionizing electron energy of 20 eV. From 5 to 10 mm above the surface, the spectrum is only slightly altered and the familiar cellulose spectrum is present. Between 10 and 20 mm, major cracking has occurred, mass 78^+ (benzene) is becoming prominent and some intermediate olefins are probably present. At 40 mm only a few peaks predominate: 28^+ (ethylene, CO), 44^+ (CO_2, propane), 66^+ (probably cyclopentadiene from thermal decomposition of toluene or phenol since these pure compounds give little 66^+ at low ionizing electron

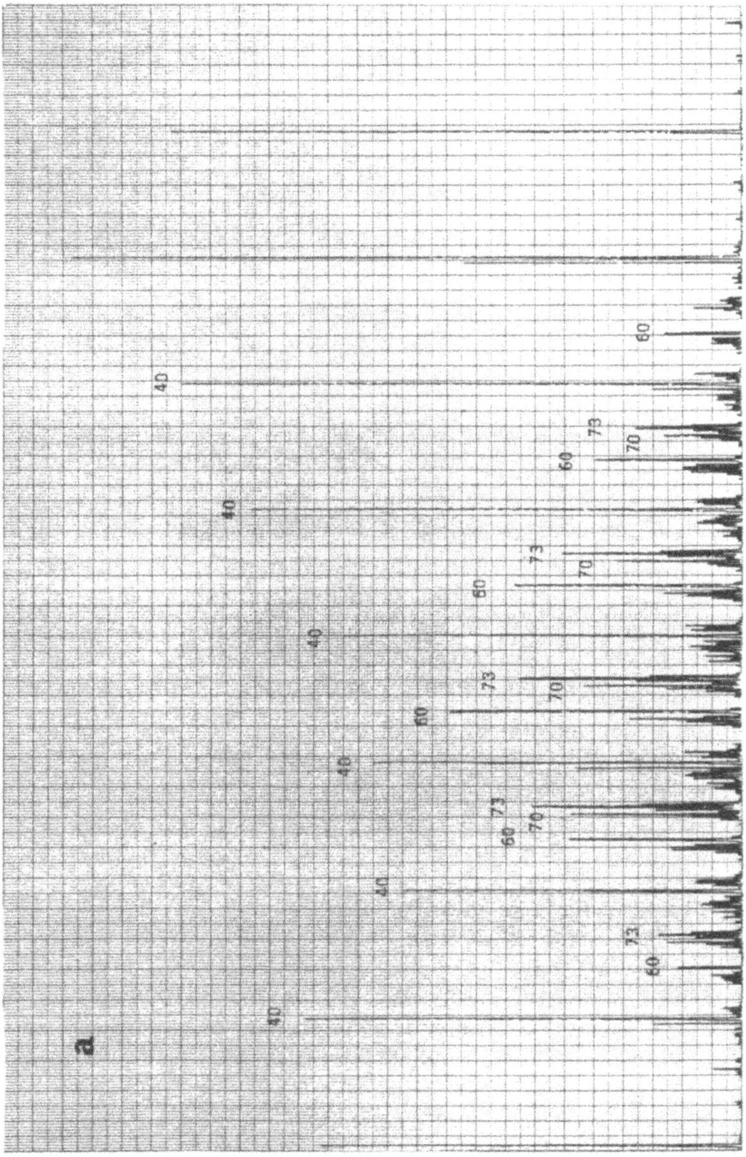
a
40
60
73
70

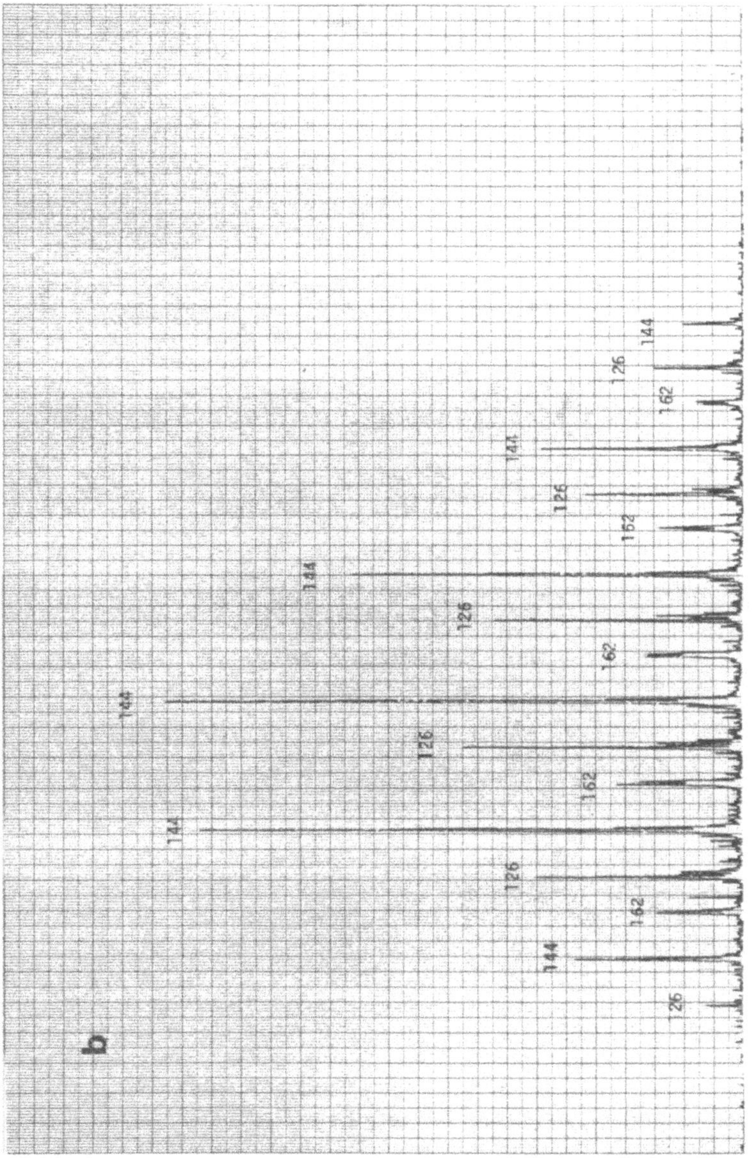

FIG. 10. (a) Repetitive one-second scans during the pyrolysis of a sliver of filter paper. Mass range ~38–88 amu. (b) Repetitive one-second scans during the pyrolysis of a sliver of filter paper. Mass range ~120–170 amu.

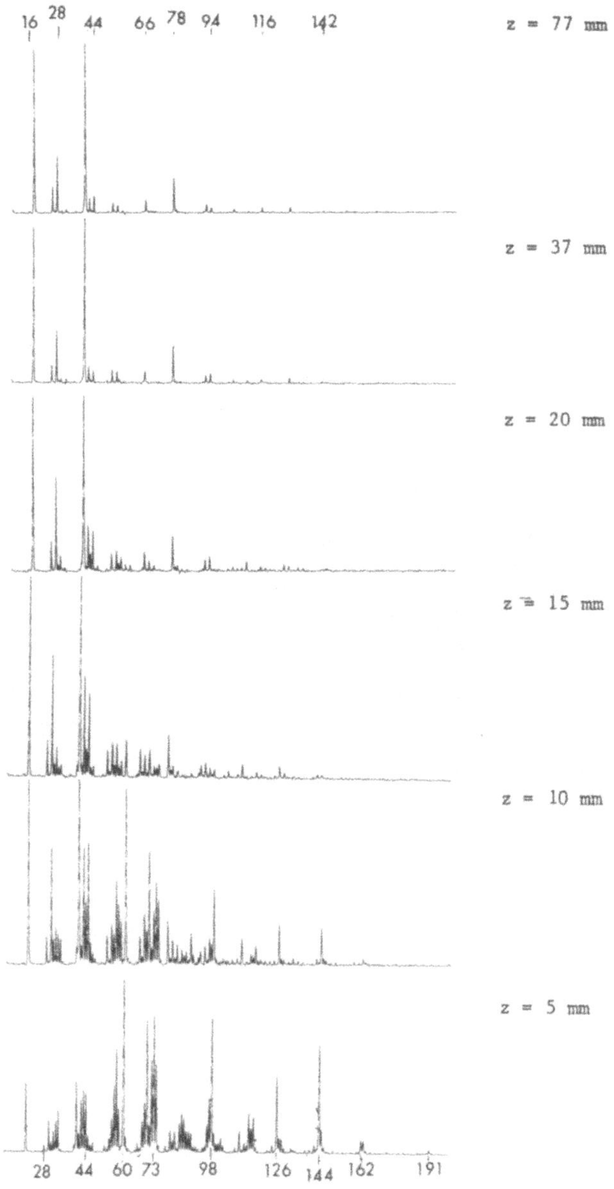

FIG. 11. Progressive secondary cracking of cellulose in 900 °C steam/argon. z is the distance of sampling orifice above the pyrolyzing surface 10 mm $\simeq$ 10 ms. Prominent primary masses indexed at bottom. Prominent secondary peaks indexed at top.

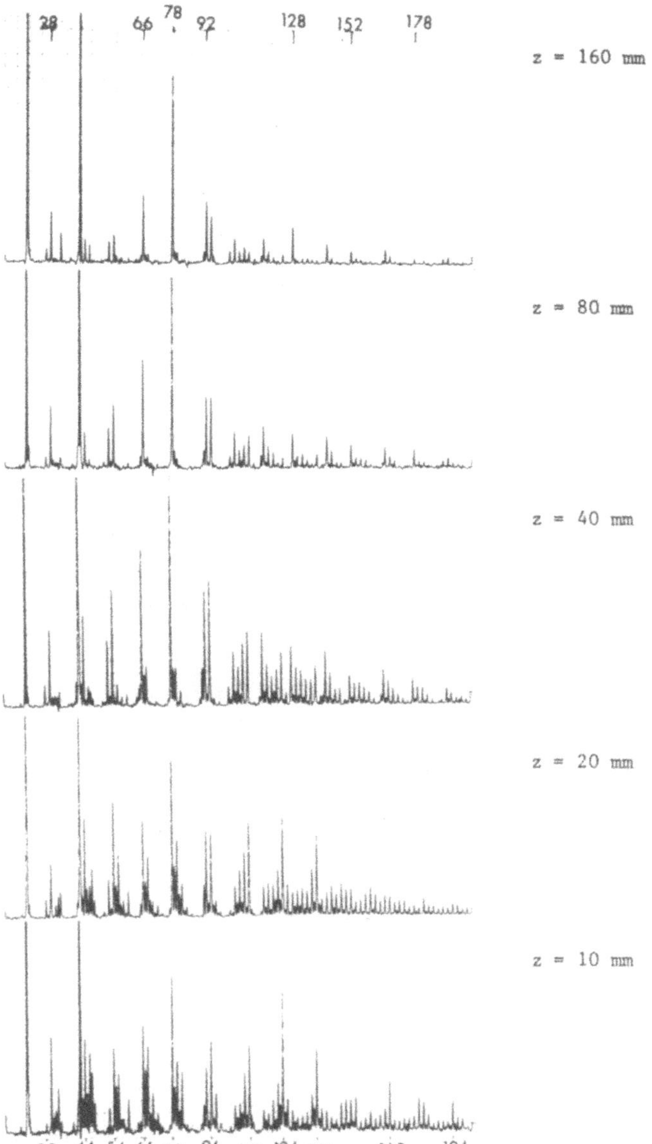

FIG. 12. Secondary cracking of lignin. Same conditions as for Fig. 11.

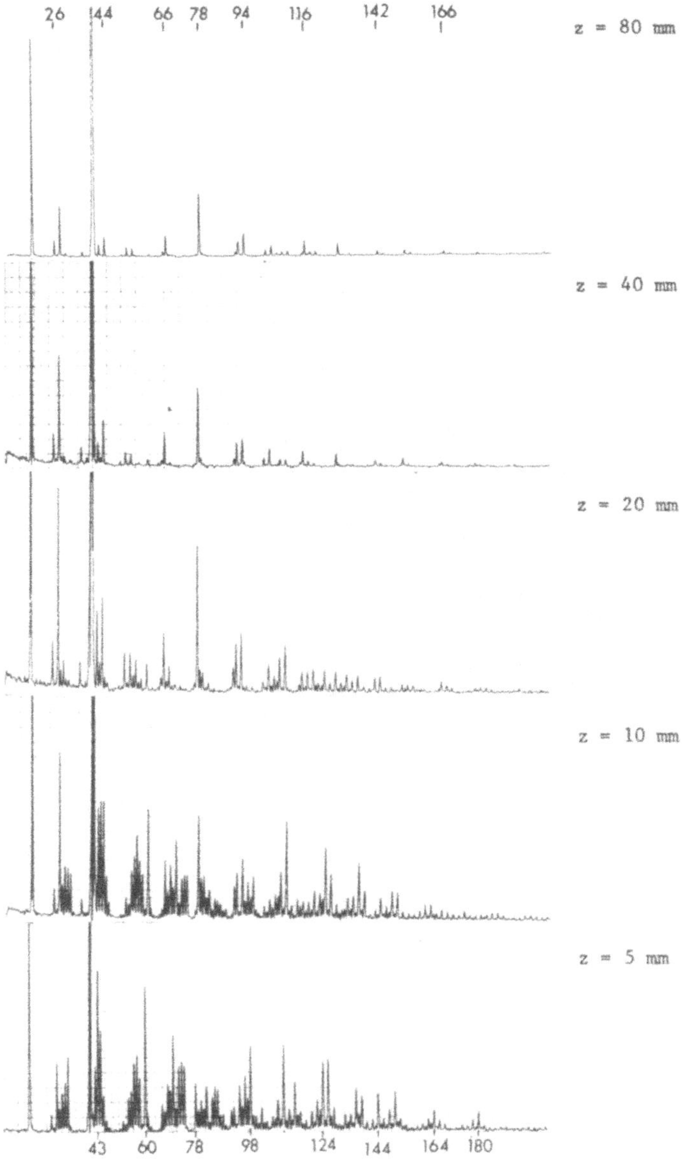

FIG. 13. Secondary cracking of wood. Same conditions as for Fig. 11.

energies), and 92^+ (toluene) and 128^+ (possibly naphthalene, which has been detected in the fast pyrolysis of wood[22]).

Figure 12 shows similar data for ethanol-extracted, steam-exploded lignin. The primary products seem more refractory and complex changes are shifting the product slate toward lower molecular weight species. Benzene is a major product ion. The prominent 66^+ peak is higher than can be attributed to the toluene (92^+) or phenol (94^+). (The reader is cautioned not to correlate ion intensities automatically with relative abundances. At these low ionizing electron energies, species with low appearance potentials can easily be over-represented in the positive ion spectrum.)

Finally, in Fig. 13, the cracking of wood is shown. Some new peaks appear to be present, perhaps due to the hemicellulose or extractives component, but the general behavior is consistent with the cellulose and lignin behavior. It appears that 900 °C and 10–100 ms time scales are matched to achieve major cracking of primary products from wood and its major components.

Gaseous compounds can be pyrolyzed and studied by introducing them into the burnt-gas region of flames. Examples of the pyrolysis of *n*-butane and ethane are given in ref. 8.

ACKNOWLEDGEMENTS

The authors wish to thank the US Department of Energy, Biomass Energy Technology Division, for support of this work, and D. Stevens, who serves as technical monitor. Helpful discussions with R. Evans, J. Diebold, T. Reed, H. Chum and H. Meuzelaar are also acknowledged.

REFERENCES

1. Reed, T. B. (Ed.) (1979–80). *A Survey of Biomass Gasification*. SERI/TR-33-29. 3 vols. Golden, CO: Solar Energy Research Institute. (See particularly Chapter 5 of Vol. II: Pyrolysis—The thermal behavior of biomass below 600 °C, by T. Milne.) Also published as T. Reed (Ed.). *Biomass Gasification: Principles and Technology*. Energy Technology Review No. 67 (Park Ridge, NJ: Noyes Data Corp, 1981).
2. Diebold, J. (Ed.) (1981). *Proceedings: Specialists' Workshop on Fast Pyrolysis of Biomass*. SERI/CP-622-1096, Copper Mountain, CO; 19–22 Oct. 1980. Golden, CO: Solar Energy Research Institute.

3. Diebold, J. and Scahill, J. (1982). This volume, Chapter 30, pp. 537–55.
4. Antal, M. J. (1982). Biomass pyrolysis—A review of the literature. Part I. Carbohydrate Pyrolysis. To appear in *Advances in Solar Energy—Annual Review of Research and Development*, Vol. 1, K. W. Boer and J. A. Duffie (Eds). ASES Publications, New York.
5. Stearns, C. A., Kohl, F. J., Fryburg, C. G. and Miller, R. A. (1978). High pressure molecular beam mass spectrometric sampling of high temperature molecules. *Tenth Materials Research Symposium on Characterization of High Temperature Vapors and Gases*. NBS Special Publication 561. National Bureau of Standards, Washington, DC, Vol. 1, p. 303.
6. Milne, T. A. and Beachey, J. E. (1977). The microstructure of pulverized coal–air flames: I. Stabilization on small burners and direct sampling techniques; II. Gaseous species, particulate and temperature profiles. *Comb. Sci. and Techn.* **16**, 123–38, 139–52.
7. Milne, T. A. and Soltys, M. N. (1983). Direct mass-spectrometric studies of the pyrolysis of carbonaceous fuels. I. A flame-pyrolysis, molecular-beam sampling technique. *J. Anal. and Appl. Pyrolysis.* **5**, 93–110.
8. Milne, T. A. and Soltys, M. N. (1982). Direct mass-spectrometric studies of the pyrolysis of carbonaceous fuels. II. Qualitative observations of primary and secondary processes in biomass. *J. Anal. and Appl. Pyrolysis* **5**, 111–31.
9. Milne, T. A. and Green, F. T. (1969). Molecular beams in high temperature chemistry. *Adv. in High Temperature Chemistry*. Academic Press, New York, Vol. 2, pp. 107–50.
10. Soltys, M. N. and Milne, T. A. (1982). *The SERI High-Pressure, Molecular Beam, Mass Spectrometric, Sampling System*. SERI/TR-622-1172, Solar Energy Research Institute, Golden, CO.
11. Milne, T. A. and Green, F. T. (1972). Study of relaxation in free jets using temperature dependence of *n*-butane mass spectra. *J. Chem. Phys.* **56**, 3007.
12. Schulten, H.-R. and Gortz, W. (1978). Curie-point pyrolysis and field ionization mass spectrometry of polysaccharides. *Anal. Chem.* **50**(3), 428.
13. Schulten, H.-R., Bahr, U. and Gortz, W. (1982). Pyrolysis field ionization mass spectrometry of carbohydrates. Part B: Polysaccharides. *J. Anal. and Appl. Pyrolysis* **3**, 229.
14. Schulten, H.-R., Bahr, U. and Gortz, W. (1981). Curie-point pyrolysis field ionization mass spectrometry of carbohydrates. Part A: Methodology. *J. Anal. and Appl. Pyrolysis* **3**, 137.
15. Shafizadeh, F., Cochran, T. G. and Sakai, Y. (1979). Application of pyrolytic methods for the saccharification of cellulose. *AIChE Symposium Series* **75**(184), 24.
16. Shafizadeh, F. (1968). Pyrolysis and combustion of cellulosic materials. *Advances in Carbohydrate Chemistry* **23**, 419.
17. Meuzelaar, H. L. C., Haverkamp, J. and Hileman, F. D. (1982). *Pyrolysis Mass Spectrometry of Recent and Fossil Biomaterials*. Elsevier Scientific Publishing Co., New York.
18. Milne, T. A., Evans, R. J. and Soltys, M. N. (1983). *Fundamental Pyrolysis Studies: Annual Report, Oct. 1, 1981–Nov. 30, 1982*. SERI/PR-234-1884, Solar Energy Research Institute, Golden, CO.

19. Ohnishi, A., Takagi, E. and Kato, K. (1976). Curie-point pyrolysis, gas–liquid chromatography of xylan. *Carbohydrate Research* **50**, 275.
20. Broido, A. (1966). *Pyrodynamics* **4**, 243.
21. Shafizadeh, F., Furneaux, R. H. and Cochran, T. G. (1979). *J. Appl. Polymer Science* **23**, 3525.
22. Diebold, J. (1982). Private communication.

21

Kinetics of Gas Formation from Cellulose and Wood Pyrolysis

G. M. Simmons and W. H. Lee
Department of Chemical Engineering, University of Idaho, Moscow, Idaho, USA

1. INTRODUCTION

The rates for decomposition of both cellulose and wood (Douglas fir sawdust) have been examined by monitoring the transient formation of gaseous species. Since no significant difference was detected in the rate constants as determined by monitoring separately all major gaseous products,[1] recent work has concentrated on measurement of only carbon monoxide, which is one of the principal gaseous reaction products over the temperature range studied. The rates obtained for cellulose decomposition by measuring carbon monoxide compare favorably with those determined by Min[2] when measuring total volatile formation. Weight loss data, obtained by Bradbury *et al.*[3] under vacuum conditions, showed faster reaction rates, indicating that pressure (in this case atmospheric) may suppress the volatiles forming reaction.

The experimental techniques employed have been described previously,[1] with the addition of an infrared gas detector for continuously monitoring the carbon monoxide produced. Briefly, a small sample (up to 20 mg) is charged to a heated reaction tube through a double set of ball valves. The sample falls to a fine mesh screen (400 mesh) where reaction takes place. Gaseous reaction products are carried away by a helium sweep gas to the infrared detector. To facilitate sample addition, powdered samples are first placed on top of an inverted thumb tack, which falls carrying the sample with it. Quantitative and fast addition of the sample to the reaction zone is possible by this method whereas powdered sample by itself falls in a mist over a large period of time.[4]

2. CELLULOSE PYROLYSIS

A kinetic scheme based on parallel solid decomposition reactions is used to analyze the experimental results. Such a reaction scheme, illustrated by Fig. 1, is consistent with that employed by Shafizadeh.[5]

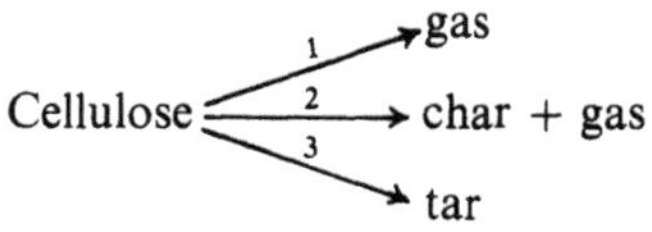

FIG. 1. Simplified reaction scheme.

A first order solid decomposition is then assumed.

$$\frac{\mathrm{d}m}{\mathrm{d}t} = -\left(\sum_i k_i\right)m \tag{1}$$

where m = mass of unreacted cellulose and k_i = first order reaction coefficient.

Integrating, we obtain

$$m = m_0 \,\mathrm{e}^{-Kt} \tag{2}$$

where m_0 = initial sample mass and $K = \sum_i k_i$.

The formation rate of carbon monoxide is then equal to the rate coefficient for carbon monoxide, k_{CO}, times the mass. Experimentally, the rate is determined by monitoring the carbon monoxide concentration and multiplying by the gas flow rate:

$$C_{CO}F = k_{CO} m_0 \,\mathrm{e}^{-Kt} \tag{3}$$

where C_{CO} = carbon monoxide concentration and F = gas flow rate.

With high sweep gas flow rates compared to the gas flow produced by the reaction products from 20 mg of sample, it is justified to make the approximation that the total flow rate F is equal to the helium sweep gas rate.

The rate coefficient for carbon monoxide, k_{CO}, is not likely to be a simple function. If, for example, we assume that carbon monoxide is formed via both reactions 1 and 2 in Fig. 1, then

$$k_{CO} = \alpha_1 k_1 + \alpha_2 k_2 \tag{4}$$

where α = stoichiometric ratios.

A plot of $\ln(C_{CO}F)$ vs. the time, t, results in a slope equal to minus the overall rate coefficient, K.

$$\ln(C_{CO}F) = \ln k_{CO}m_0 - Kt \tag{5}$$

Figure 2 shows a typical concentration vs. time curve for carbon monoxide, while Fig. 3 shows the same information on a semi-log plot.

For all practical purposes, the decomposition curve as depicted by Fig. 3 shows a single slope, indicating that the use of the reaction scheme of Fig. 1

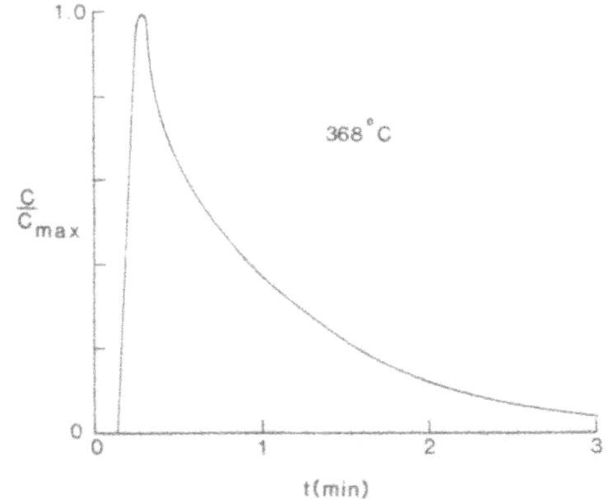

FIG. 2. Carbon monoxide concentration as a function of time for cellulose decomposition.

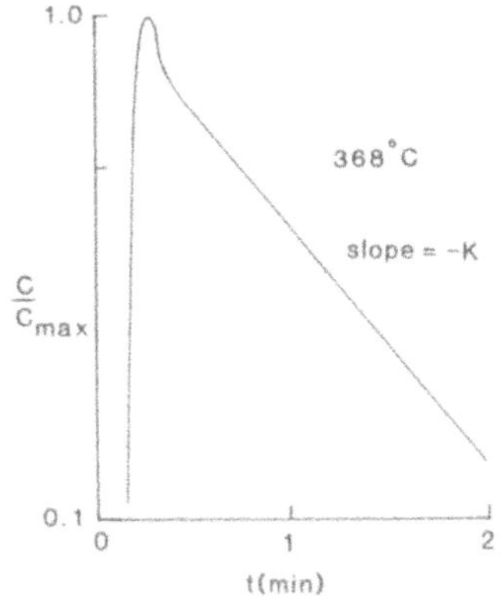

FIG. 3. Log of carbon monoxide concentrations from cellulose.

is adequate to correlate these data. This result does not, however, suggest that this reaction scheme is necessarily correct. The addition of a gas forming reaction from the tar fraction has also been suggested.[5,6] Different time functions result, depending on whether or not the tar is considered 'volatile'. If the tar is volatile, and has the same residence time as the gaseous species, then a plot of $\ln C_{CO}F$ vs. time still shows a single slope. However, a change in residence time should exhibit a change in carbon monoxide yield, which we have not observed. The following analysis illustrates these statements.

3. KINETIC MODEL WITH VOLATILE TAR

In this case, it is assumed that the tar, once formed, leaves the reaction zone and is carried out of the reactor with the gaseous products (Fig. 4). In

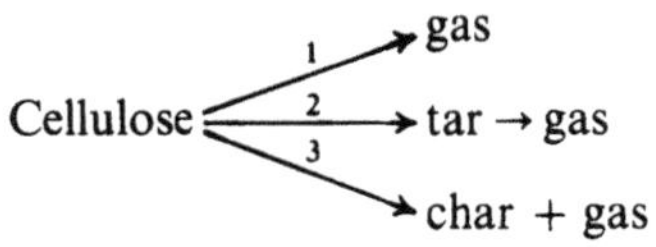

FIG. 4. Reaction scheme including gaseous production from tar.

addition, it is hypothesized that the tar further decomposes to produce gaseous products, including carbon monoxide, by a first order gas phase reaction. This secondary reaction then proceeds until the reactants are quenched at the reactor outlet. The residence time of the gaseous species in the reactor is designated τ.

For the cellulose decomposition, eqns (1) and (2) still hold. However, the generation of carbon monoxide as predicted by eqn (3) must be augmented by that formed via tar decomposition.

The formation of tar is expressed similar to eqn (3):

$$C_{\text{tar}}F = k_{\text{tar}_f} m_0 \, \mathrm{e}^{-Kt} \tag{6}$$

where k_{tar_f} = tar formation rate coefficient.

It is then assumed that the tar fraction decomposes homogeneously along the heated reactor tube to form additional carbon monoxide and other products:

$$\frac{\mathrm{d}C'_{\text{CO}}}{\mathrm{d}t} = k_{\text{CO}_t} C_{\text{tar}} \tag{7}$$

$$\frac{\mathrm{d}C_{\text{tar}}}{\mathrm{d}t} = k_{\text{tar}_d} C_{\text{tar}} \tag{8}$$

where k_{CO_t} = rate coefficient for CO formation from tar, C'_{CO} = concentration of CO formed from tar, and k_{tar_d} = tar decomposition rate coefficient.

Integrating eqn (8)

$$C_{\text{tar}} = C_{\text{tar}_0} \exp(-k_{\text{tar}} t') \tag{9}$$

where C_{tar_0} = tar concentration produced at the reaction screen, expressed by eqn (6), and t' = time coordinate as tar proceeds down the heated reactor tube.

Combining eqns (7) and (9),

$$\frac{\mathrm{d}C'_{\mathrm{CO}}}{\mathrm{d}t}=\frac{\mathrm{d}C'_{\mathrm{CO}}}{\mathrm{d}t'}=k_{\mathrm{CO}_t}C_{\mathrm{tar}_0}\exp(-k_{\mathrm{tar}_d}t')$$

and integrating:

$$C'_{\mathrm{CO}}=\frac{k_{\mathrm{CO}_t}}{k_{\mathrm{tar}_d}}C_{\mathrm{tar}_0}[1-\exp(-k_{\mathrm{tar}_d}t')] \tag{10}$$

Since the reactor exit, where the reaction mixture is quenched, is defined in time as $t'=\tau$, there results with the use of eqn (6)

$$C'_{\mathrm{CO}}=\frac{k_{\mathrm{CO}_t}}{k_{\mathrm{tar}_d}}\left(\frac{k_{\mathrm{tar}_f}}{F}m_0\exp(-Kt)\right)(1-\exp(-k_{\mathrm{tar}_d}\tau)) \tag{11}$$

The total carbon monoxide concentration is then determined by combining eqns (3) and (11):

$$C_{\mathrm{CO}}=\frac{m_0}{F}A\exp[-K(t-\tau)] \qquad t>\tau \tag{12}$$

where

$$A=\left\{k_{\mathrm{CO}}+\frac{k_{\mathrm{CO}_t}}{k_{\mathrm{tar}_d}}k_{\mathrm{tar}_f}[1-\exp(-k_{\mathrm{tar}_d}\tau)]\right\}$$

Replacing t by $t-\tau$ puts both reactions on the same time coordinate. A plot of $\ln C_{\mathrm{CO}}$ vs. t will still predict a slope equal to $-K$; however, a decreased residence time, τ, as effected by an increase in the helium sweep gas rate, could result in a decrease in the quantity of CO produced. This has not been observed in the temperature range studied.

4. KINETIC MODEL WITH ACTIVE INTERMEDIATE

In this model, as illustrated by Fig. 5, an intermediate solid is formed and stays in the solid reaction zone where further decomposition takes place.

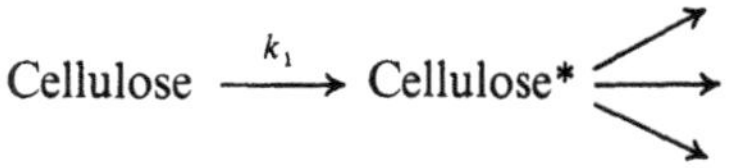

FIG. 5. Reaction scheme with active intermediate.

Such a model is proposed by Shafizadeh[5] and is consistent with his observations at lower temperatures. The formation of the active intermediate gives rise to an induction period which shortens as temperature increases. In our temperature range, this induction period is apparently quite short and indistinguishable from the time required for heating the sample to reactor temperature.

5. RESULTS

Figure 6 is an Arrhenius plot for cellulose decomposition based on CO formation. The rate constant plotted in Fig. 6 is K, which is the sum of all parallel first order reactions. The resulting overall rate coefficient, K, is lower than extrapolated values from the vacuum pyrolysis results of Shafizadeh.[5] The overall rate coefficient K is higher in magnitude than the char forming rate of Shafizadeh[5] but has nearly the same activation energy. This may indicate that the action of atmospheric pressure tends to inhibit the volatile forming reaction.

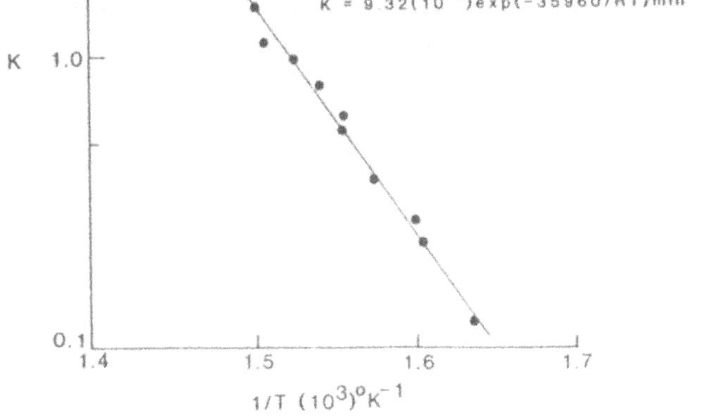

FIG. 6. Arrhenius plot for cellulose decomposition.

6. RATE COEFFICIENTS FOR CARBON MONOXIDE FORMATION

Once a value for the overall rate coefficient, K, is determined, the rate coefficient, k_{CO}, for the formation of CO from cellulose can be found. Equation (3) can be integrated and rearranged to obtain

$$k_{CO} = \frac{KF}{m_0} \int_0^{\infty} C_{CO}\, dt \tag{13}$$

The resulting activation energy is similar in magnitude to that observed for the overall rate, K, but is somewhat larger. This observation is consistent with the parallel scheme suggested by Fig. 1 and eqn (4) for relatively small values of k_2.

Figure 7 is the Arrhenius plot for k_{CO} as determined from eqn (13). No differences are observed when changing the sweep gas rate, so that a reaction scheme as illustrated by eqn (12) is not required. To reach more conclusive answers, time dependent tracking of additional products will be necessary.

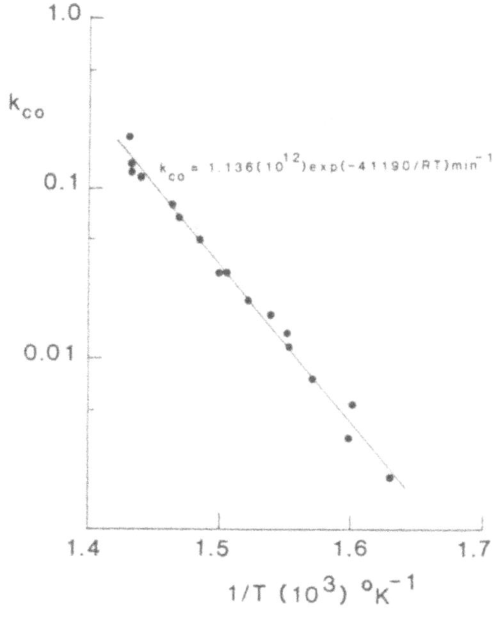

FIG. 7. Arrhenius plot for carbon monoxide formation from cellulose.

7. WOOD DECOMPOSITION

Kinetic studies involving wood decomposition by pyrolysis have also been performed in a similar manner. A typical plot of ln C_{CO} vs. t, as illustrated by Fig. 8 for fine particulate Douglas fir, now shows a change in slope during the course of the reaction. If the two slopes are associated with two rate coefficients, k_1 and k_2, Arrhenius plots can be readily constructed.

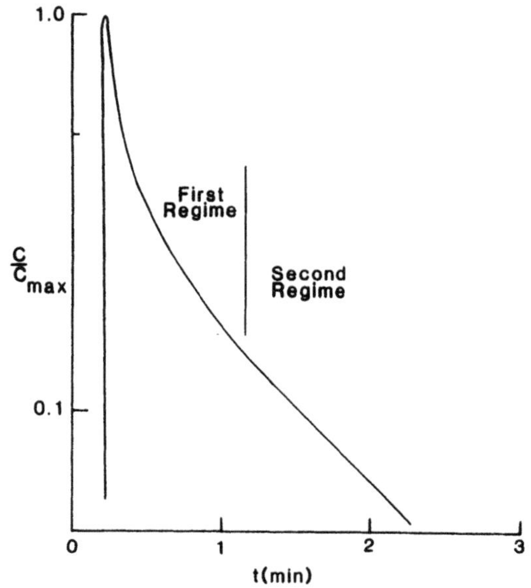

FIG. 8. Log of carbon monoxide concentration from wood decomposition.

Figure 9 shows the Arrhenius plot for the second regime rate coefficient, k_2. Also shown in Fig. 9 are those data points for overall cellulose decomposition. The close agreement between cellulose decomposition and the second rate coefficient for wood, k_2, indicates that separate decomposition of the major constituents in wood is taking place. This hypothesis is consistent with the results of Mackay and Roberts,[7] who found that char formation from wood was an additive property of the various constituents in wood. By subtracting the effects of the kinetic reaction associated with k_2, a separate concentration vs. time curve can be constructed and analyzed for k_1. Figure 10 shows the resulting Arrhenius plot for k_1. It can be hypothesized that k_1 may be related to the decomposition of hemicellulose.

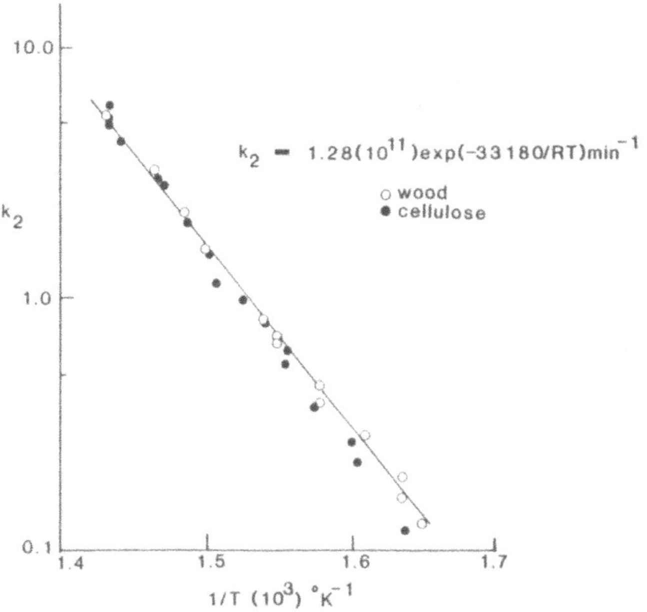

FIG. 9. Arrhenius plot for second regime of wood decomposition.

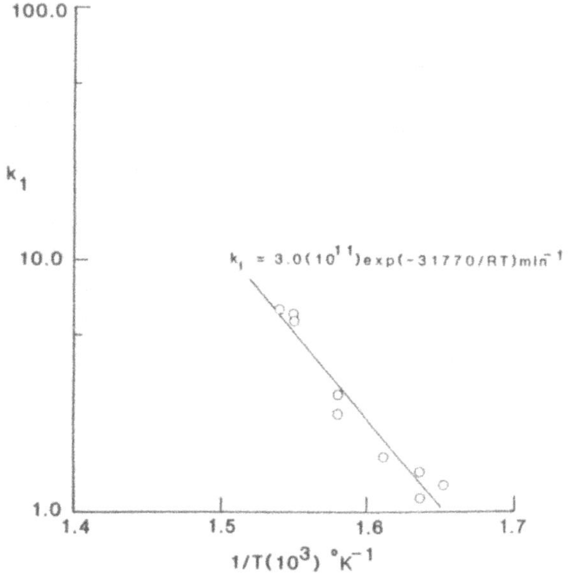

FIG. 10. Arrhenius plot for first regime of wood decomposition.

With k_1 being several times larger than k_2, the use of the second slope to calulate k_2 values is therefore justified.

The ratio of k_1 to k_2 is consistent with the results of Stamm[8] and Akita[9] for hemicellulose and cellulose, and the values obtained for k_1 are similar to those obtained by Min[2] and attributed to hemicellulose decomposition. In Min's work, all of the volatile product was combusted and the heat of combustion was measured to estimate decomposition rates, whereas in our work the production of carbon monoxide has been analyzed. We have also conducted isothermal experiments whereas Min's experiments all involved a controlled heating rate.

Reaction rate data reported previously[1] were determined by integration of eqn (3) and substitution of eqns (1) and (2) to obtain

$$C_{CO}F = k_{CO}m_0 - K\int_0^t C_{CO}F\,dt \tag{14}$$

A plot of $C_{CO}F$ vs. $\int_0^t C_{CO}F\,dt$ likewise gives a slope equal to $-K$. Only the asymptotic slope at large t was reported, which now is seen to correspond to k_2.

8. SUMMARY

Pyrolysis kinetics for cellulose, as determined by measuring the production rate of CO, have been interpreted with a kinetic scheme involving production of gaseous products via two parallel reactions from cellulose. The overall rate coefficient for cellulose decomposition was found to be $K = 9{\cdot}32 \times 10^{11}\exp(-35\,960/RT)\,\text{min}^{-1}$, while the rate coefficient for the production of CO from cellulose was found to be $k_{CO} = 1{\cdot}136 \times 10^{12}\exp(-41\,190/RT)\,\text{min}^{-1}$.

Rate data for wood pyrolysis showed two separate reaction regimes, with the second (in time) corresponding closely with rate data for pure cellulose, and the first regime corresponding to data previously reported by other investigators for hemicellulose.[2,8,9]

ACKNOWLEDGEMENT

The authors are pleased to acknowledge support for this project from the Gas Research Institute.

REFERENCES

1. Simmons, G. M. and Sanchez, M. M. (1981). High temperature gasification kinetics of biomass pyrolysis. *J. of Analytical and Applied Pyrolysis* **3**, 161.
2. Min, K. (1977). Vapour-phase thermal analysis of pyrolysis products from cellulosic materials. *Comb. and Flame* **30**, 285.
3. Bradbury, A. G. W., Sakai, Y. and Shafizadeh, F. (1979). A kinetic model for pyrolysis of cellulose. *J. of Applied Polym. Sci.* **23**, 3271.
4. Ramachandran, S. (1980). Wood pyrolysis—A kinetic study. M.S. Thesis, University of Idaho, August.
5. Shafizadeh, F. (1982). Introduction to pyrolysis of biomass. *J. of Analytical and Applied Pyrolysis* **3**, 283.
6. Chan, R. W. C. and Krieger, B. B. (1982). Analysis of chemical and physical processes during devolatilization of a single, large particle of wood. Seventh International Symposium on Chemical Reaction Engineering, Boston, Mass., October.
7. Mackay, D. M. and Roberts, P. V. (1982). The dependence of char and carbon yield on lignocellulosic precursor composition. *Carbon* **20**, 87.
8. Stamm, A. J. (1956). Thermal degradation of wood and cellulose. *Ind. and Engr. Chem.* **48**(3), 413.
9. Akita, K. (1956). *Rep. Fire Res. Inst. Japan*, **9**, 50.

22

Fast Pyrolysis (Ultrapyrolysis) of Biomass Using Solid Heat Carriers

R. G. GRAHAM,* M. A. BERGOUGNOU, L. K. S. MOK and H. I. DE LASA

* *Forintek Canada Corporation, Ottawa, Canada*

and

Faculty of Engineering Science, University of Western Ontario, London, Canada

1. INTRODUCTION

Some of the best features of conventional pyrolysis and air/oxygen gasification can be incorporated into the development of a successful fast pyrolysis process without many of the associated disadvantages. Conventional biomass pyrolysis and flash pyrolysis research, which tend to maximize the production of multi-component condensibles, has clearly illustrated the need to characterize and upgrade the liquid products, and to identify viable industrial applications prior to commercial development.

Conversely, the product end-use of air/oxygen gasification has been clearly identified in a variety of industrial sectors, and the commercial viability has already been demonstrated. However, a major consideration for air gasification product applications is their limitation to close-coupled combustion-related processes (i.e. steam generation, drying, space heating and engine shaft power). Although these are valuable and important applications, storage and piping of the low energy gas (which is diluted with nitrogen and combustion products) over any significant distance is not economically attractive. Low energy gas is not suitable as a synthesis gas for methanol or gasoline, nor is it suitable as a feedstock for the petrochemical industry.

Oxygen gasification gives a medium energy gas which may be transported over moderate distances (i.e. 50–80 km) and may be used for the synthesis of methanol and ammonia in addition to the low energy gas uses. However, the gas product cannot directly be used as a petrochemical feedstock, nor as a syngas for commercial synthetic gasoline production.

In order to commercially generate gasoline as an end-product via the oxygen gasification route, biomass must be degraded to its simplest form (CO, H_2, C), then reconstructed to methanol, and finally converted from methanol to gasoline (for instance by the Mobil synthesis process). Available energy is lost in each step.

The development of a successful fast pyrolysis process would possess the following advantages:

(1) Maximum yields of a non-condensible gaseous product (i.e. limited solid or heavy liquid by-products).
(2) A relatively homogeneous gas product (resulting in simpler beneficiation and purification).
(3) Primary product flexibility (i.e. a multiproduct facility with could produce such primary products as olefins, syngas or SNG in response to the market demand).
(4) The potential for high quality secondary products (i.e. a synthetic gasoline or a petrochemical feedstock).
(5) No oxygen plant (thus improving process economics).

The technical viability of a fast pyrolysis system depends on the development of good engineering techniques which would allow:

(1) Rapid heat transfer rates to the biomass (about 10^5 °C/s).
(2) High reactor temperatures (>600 °C).
(3) Short residence times (<0·5 s).
(4) Rapid product quenching (<30 ms).

A continuous fast pyrolysis (Ultrapyrolysis) designed to exploit the conditions listed above is currently being developed at the University of Western Ontario in cooperation with Forintek Canada Corp. and funded by the Canadian Government ENFOR programme (Environment Canada). In this paper, the objectives, approach, equipment and preliminary results of the Ultrapyrolysis project will be presented. In order to put fast pyrolysis into perspective, terms will first be defined and the chemical mechanism of fast pyrolysis will be related to conventional pyrolysis mechanisms.

2. PYROLYSIS, FAST PYROLYSIS AND GASIFICATION

Pyrolysis, flash pyrolysis, gasification, pyrolytic gasification and fast pyrolysis are terms which appear frequently in the biomass and coal thermochemical conversion literature. *Pyrolysis* may be defined, in its broadest sense, as the thermal decomposition of organic matter occurring in the absence of air or oxygen. Some authors would consider oxygen-deficient (i.e. less air or oxygen than is required stoichiometrically for complete combustion) thermal decomposition to be a true pyrolysis process as long as the primary products are liquid or solid. When the primary products of the oxygen-deficient decomposition are non-condensible fuel or synthesis gases, the process is known as *gasification.*

Conventional pyrolysis is typically characterized by a slow feedstock heating rate (less than 2 °C/s), low temperatures (less than 500 °C) and relatively long gas and solids residence times (greater than 5 s for the gas; solids residence times can be minutes, hours or days in duration). Tar and char are the primary products due to slow devolatilization of the feedstock, and the secondary coking and polymerization (recombination) reactions which are allowed to occur in the reactor after primary pyrolysis has taken place.

Flash pyrolysis is a rapid heating rate process (greater than 2 °C/s) occurring at moderate temperatures (typically 400–600 °C) whose liquid tar/oil products are maximized at the expense of char and gas. Vapour residence times are normally less than 2 s. *Fast pyrolysis* is differentiated from flash pyrolysis primarily with respect to temperature, vapour residence times and end-product. Extremely rapid heating rates (about 10^5 °C/s), high temperatures (greater than 600 °C) and short vapour residence times (less than 0·5 s) are required to maximize the production of high quality gases (i.e. olefins, other hydrocarbons) at the expense of char and condensible liquids.

Pyrolytic gasification is a term applied to any anaerobic pyrolysis process, fast or slow, which gives a non-condensible synthesis gas or medium energy fuel gas (10–20 MJ/Nm^3) as the primary product. Tar and char are viewed as undesirable byproducts.

One final term, *liquefaction*, was historically linked to hydrogenation and other high-pressure (10–30 MPa) thermal decomposition processes which employed reactive hydrogen or carbon monoxide carrier gases to produce a liquid fuel from organics at moderate temperatures (300–400 °C). However, over the past decade, liquefaction has often been used in the literature to encompass any thermochemical conversion process which primarily yields a liquid tar/oil product (pyrolysis is therefore included with hydrogenation as being 'liquefaction').

It is important to note that the pyrolysis terms used in this paper have been defined from a general consensus which was gleaned from an exhaustive literature survey. On occasion, terms such as *flash* and *fast* pyrolysis have been used interchangeably. The biomass community would avoid considerable misunderstanding and confusion should a standard nomenclature be established.

3. PYROLYTIC PATHWAYS

The current interest in fast pyrolysis and its potential for producing high quality fuel and synthesis gas from biomass is due primarily to the fundamental research performed during the past decade by Soltes,[1-3] Shafizadeh,[4,5] Broido,[6] Lewellen,[7] Milne,[8-10], Antal,[11] and others.[12] Their work has attempted to identify the major competitive and consecutive reactions which may occur during the thermal decomposition of biomass, and is summarized in an excellent review by the Solar Energy Research Institute.[12] Fundamental research of this nature is extremely valuable and necessary for the engineering of a viable continuous fast pyrolysis process and, therefore, a short discussion on pyrolysis mechanisms is given in the following paragraphs. Although much of the modelling is based on cellulose pyrolysis, these results appear to be consistent with those obtained in processes using wood and other biomass.[3]

The chemical mechanisms involved in pyrolysis (especially fast pyrolysis) are not well understood. However, an overall mechanism is presented in Fig. 1. This figure is reproduced with permission from J. Diebold of the Solar Energy Research Institute, and was formulated from a general consensus of pyrolysis specialists at a workshop sponsored by SERI.[12]

The reactions of *primary* pyrolysis are those which directly affect the biomass feedstock, and can generally be classified either as fragmentation (fission) or dehydration reactions. Dehydration is dominant at lower temperatures (below 300 °C), is typical of conventional pyrolysis, and involves a reduction in molecular weight, the evolution of H_2O, CO and CO_2, and the formation of char.[5] Fragmentation predominates at higher temperatures (greater than 300 °C) and involves the depolymerization of biomass to anhydroglucose units (i.e. levoglucosan) and other light combustible volatiles. These higher temperature reactions are of more interest with respect to the development of a flash or fast pyrolysis process.

The primary products of pyrolysis may repolymerize (recombine), undergo further fragmentation (cracking) or react with free radicals depending on their temperature, residence time and pressure. These *secondary* reactions must be better characterized in order to engineer a system which can selectively maximize one or more of the many possible products or intermediates.

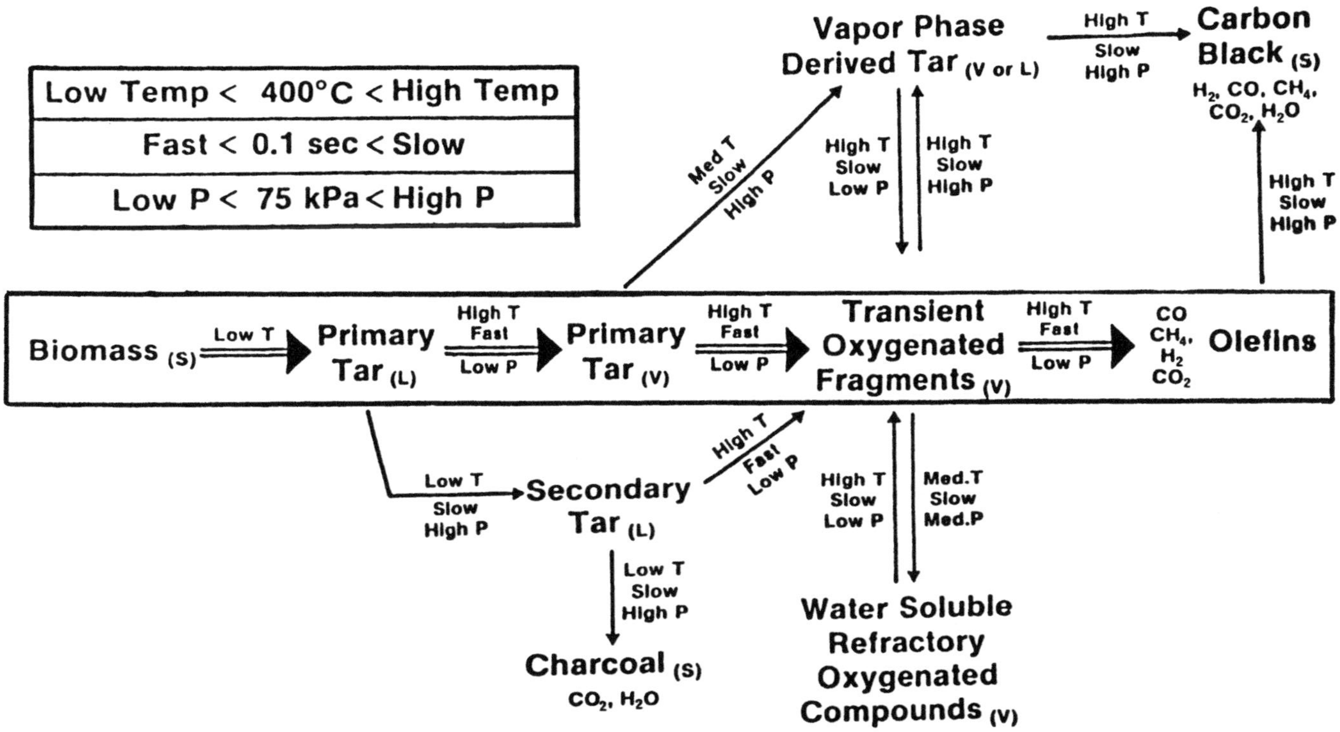

FIG. 1. Reactions competing with fast pyrolysis reactions (reproduced with the permission of J. Diebold of the Solar Energy Research Institute).

The recombination of primary products to form stable secondary tars is favoured by long vapour residence times, low to medium temperatures (200–600 °C) and high partial pressures. These conditions allow the primary molecules to be in close proximity for significant periods of time in the condensed phase, and permit frequent collisions to occur in the vapour phase. Extended reaction times at lower temperatures result in the formation of char, water and carbon oxide gases from secondary tars. Elevated temperatures favour the cracking of the secondary tar to hydrocarbons, H_2, CO, CO_2 and carbon black.

A pyrolytic pathway which clearly favours olefin production is illustrated in Fig. 1. Once the liquid primary tar is formed on the biomass surface, high temperatures, rapid heating rates and low vapour partial pressures will promote olefin formation by the following reaction sequence:[13]

(i) Primary tars are vaporized from the liquid tar surface. This precludes the formation of stable secondary tars by recombination (repolymerization).
(ii) The vaporized primary tars are thermally cracked to produce transient oxygenated fragments. These fragments can form pyroligneous acids (i.e. acetaldehyde, methanol, furan, acetic acid, acetone, etc.) and vapour-phase-derived tars (by recombination) under conditions which are not suitable for olefin production.
(iii) The transient oxygenated fragments are thermally cracked to yield olefins, other hydrocarbons, carbon oxides and hydrogen.

Extended reaction times will allow the olefins and related products to react and form carbon black, permanent equilibrium gases and condensible aromatic vapours. Therefore, to preserve the olefins, the pyrolysis pathway must be interrupted by rapid quenching of the intermediates before equilibrium can occur.

Experimental work[7,11–14] has largely substantiated the mechanisms proposed by Diebold and Antal, and has been summarized in an earlier publication.[15]

4. ULTRAPYROLYSIS PROCESS DESIGN

A continuous fast pyrolysis process (Ultrapyrolysis), designed to exploit the high temperatures, rapid heating rates, short vapour residence times and rapid quenching rates required for the production of olefins and other valuable chemicals on a commercial basis, is currently under development at the University of Western Ontario. The research is being conducted in cooperation with the Forintek Canada Corp. and under the sponsorship of the ENFOR programme of the Canadian Forestry Service (Environment Canada). Figure 2 illustrates the Ultrapyrolysis process at the laboratory scale, which employs particulate solids to carry heat to and from the reactants and products respectively. Rapid mixing and heat transfer are carried out in two vortical contactors. The first has been termed a 'Thermovortactor' and allows heat to be transferred from hot solids to the biomass.

The second is a 'cryovortactor' and allows fast quenching of the products by the transfer of heat to cold solids. In addition to high heat transfer rates by conduction and convection, there are other advantages associated with the use of solids, rather than a gas, to carry heat. First, the heat capacity of solids is higher per unit volume resulting in the design of equipment which is more compact. Second, solids are more easily separated from the product gas allowing less dilution and a corresponding higher energy density. Although a detailed description of the experimental equipment is given elsewhere in the literature,[16-18] a summary of the major components is given in the following paragraphs.

4.1 Feeders

Thermofor, cryofor and wood feeders are employed in the system. They are all fluidized bed, pneumatic transport feeders which essentially 'pump' the fluidized solids through a transport line. The thermofor feeder sends the particulate solids to a preheater coil, the cryofor sends the particulates to the cryovortactor, and the wood feeder supplies wood, cellulose or lignin powders to the thermovortactor. The solids loading and mass flow rates are controlled by manipulating the pressure in the fluidized bed, and by varying the volumetric flow of the transport gas.

4.2 Preheater Coil

The preheater coil assembly consists of two separate coils which are housed in an electrical resistance oven, and allow the transfer of heat to the thermofor (heat-carrying solid particulates). Each coil is 20 m in length and is made of 1/4 in. OD Inconel 601 tubing. The coil loops are 100 mm in diameter and 1 m in length.

4.3 Thermovortactor

Rapid mixing and heat transfer between the thermofor and the biomass feedstock occurs in the thermovortactor. The thermovortactor is a cyclonic

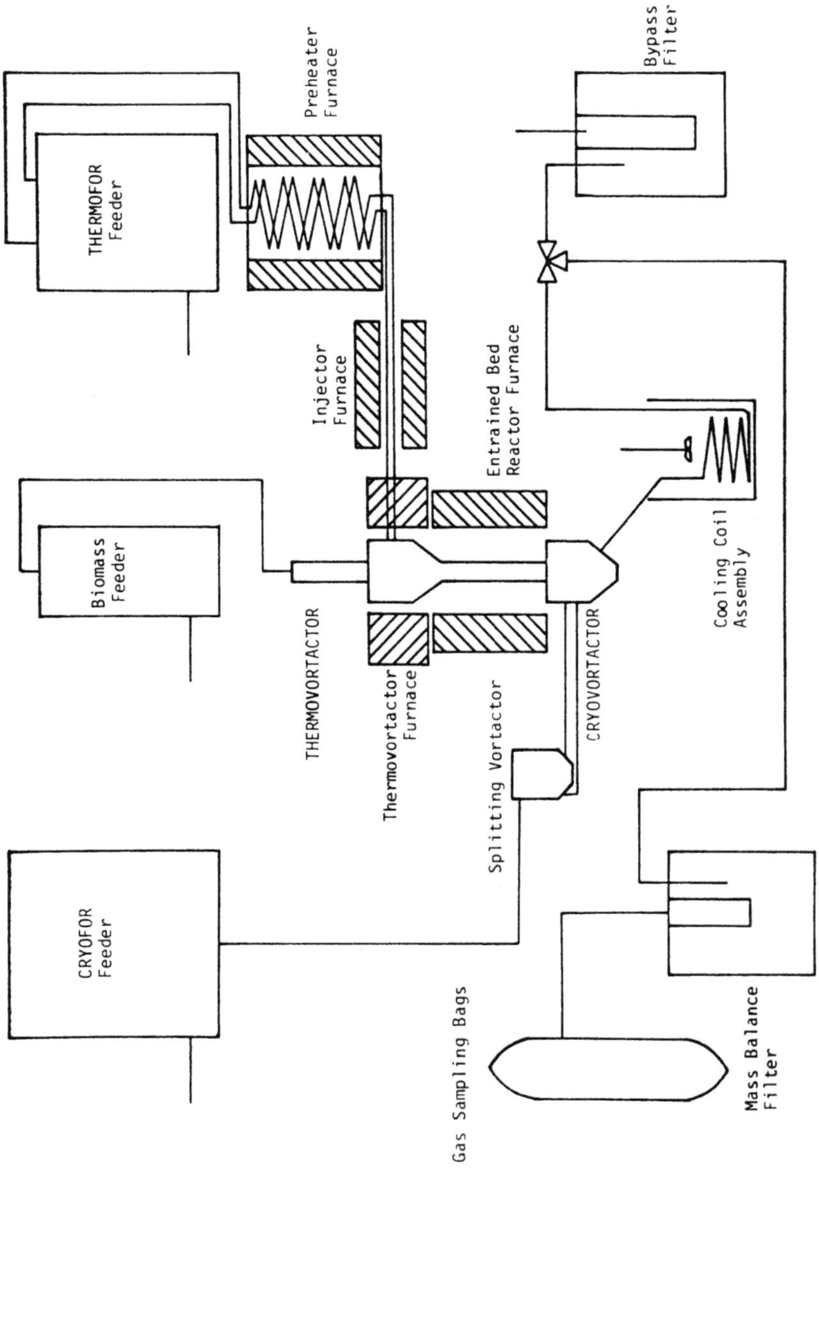

FIG. 2. Laboratory flow scheme for the Ultrapyrolysis process.

vessel with two opposing tangential inlets for the thermofor. One tangential stream effectively destroys the momentum of the other, causing severe turbulence. Biomass feedstocks are then injected from the top of the thermovortactor through a water-cooled tube into the turbulent region. The design mixing time is 20 ms.

4.4 Cryovortactor

The cryovortactor is a cyclonic contactor which serves to quench the hot gaseous products by rapidly mixing them with cold solids (cryofor). It is similar in design, but is larger than the thermovortactor. Hot gases enter through a tube from the top of the vessel, and are injected into the turbulent zone which is caused by the destruction of momentum of two opposing cold solid tangential streams. The design quenching time is 20 ms.

4.5 Entrained Bed Reactor

The fast pyrolysis of biomass commences in the thermovortactor and continues in a plug-flow entrained bed reactor. The reactor is simply a tube or coil of varying lengths (to vary the residence time) which is housed in an

TABLE I
DESIGN PARAMETERS: UWO FAST PYROLYSIS

(I)	Solids preheat temp.	1 100 °C
(II)	Reactor temperature	1 000 °C
(III)	Solids/wood mix time	~20 ms
(IV)	Reactor residence time	50–200 ms
(V)	Quenching time	~20 ms
(VI)	Heating rate	10^4–10^5 °C/s

electrical resistance oven. The mixture of hot solids and biomass passes from the thermovortactor through the entrained flow reactor to the cryovortactor. To date the total elapsed time from the heating of the biomass to the quenching of the product gas has been less than 700 ms.

A summary of the design parameters is given in Table I.

Cold model testing and construction of the equipment has been completed, and hot testing with Avicel™ cellulose commenced in June, 1982. Wood flour and lignin powder will be used upon completion of the work with cellulose.

5. PRELIMINARY RESULTS

The preliminary results using Avicel PH-102™ microcrystalline cellulose (which has an average particle size at 100 μm) are given in Table II. A typical run was carried out at steady state for 2–3 min, and all of the products and residues were collected during this time. The solid (thermofor and cryofor) and condensible fraction was removed from the cyclone/filter and weighed. Several samples from this fraction were sent to the Forintek Canada Corp. (Ottawa) for elemental and calorific analysis. The remaining sample was washed with acetone to determine the amount of condensibles in the product.

The non-condensible gas was collected in a large sample bag and subsequently analysed using a dedicated Carle refinery gas chromatograph both during and after the run. Upon completion of the gas analysis, the sample bag contents were drawn into an evacuated industrial propane tank (Superior Propane Model 123) where pressure was recorded before and after the operation. The pressure difference permitted an estimation of the total number of moles in the non-condensible gas product. Although the elemental, calorific and condensible analyses for the first experiments with cellulose have not been completed at the time of publication, the data in Table I give a general indication of the effect of temperature and residence time on the gas product. Runs 7, 9, 10 and 11 were carried out at 900 °C with residence times ranging from 170 to 600 ms. With decreasing residence time one detects an increase in the proportions (by mass and volume) of hydrogen, ethylene and acetylene, and a decrease in the proportions of carbon monoxide, carbon dioxide and methane. A comparison of runs 8 and 9 (which assumes the difference in residence time to be negligible) appears to demonstrate an increase in the proportions of ethylene and methane with increasing temperature. This trend is confirmed by a comparison of runs 6 and 7 (if one considers the difference in residence times to be negligible when compared to that of temperatures), which also indicates an increase in acetylene with an increase in temperature between 700 and 900 °C. Although the literature[12,15] substantiates these observations and suggests a maximum for ethylene and acetylene at 850 °C (800–900 °C) and 1300 °C (1200–1400 °C) respectively, it must be reiterated that the Ultrapyrolysis results presented here are preliminary. Systematic testing and analysis are currently underway to determine an accurate mass balance and the precise effects of temperature and residence time on the product gas distribution. The biomass feed rate will be determined by a load cell apparatus in future experiments. However, it is evident that maximum

TABLE II

PRELIMINARY RESULTS USING AVICEL CELLULOSE POWDER

Run number	Reactor temp. (°C)	Residence time (ms)	Product gas composition[a] H_2	C_2H_4	C_2H_6	C_2H_2	CO	CO_2	CH_4
6[b]	700	700	17·5 (1·5)	4·5 (5·3)	1·0 (1·3)	0·4 (0·5)	64·6 (78·4)	3·7 (7·1)	8·3 (5·8)
7	900	600	18·6 (1·7)	6·3 (8·0)	0·4 (0·5)	1·0 (1·2)	58·1 (73·5)	3·0 (6·0)	12·7 (9·2)
8	800	230	21·1 (1·9)	6·0 (7·7)	0·6 (0·8)	1·1 (1·3)	59·1 (75·5)	3·1 (6·2)	8·9 (6·5)
9	900	200	22 (2·1)	6·5 (8·5)	0·4 (0·6)	1·2 (1·5)	56·8 (74·1)	2·7 (5·5)	10·4 (7·8)
10	900	200	21·8 (2·0)	6·4 (8·4)	0·5 (0·7)	1·2 (1·5)	56·9 (74·2)	2·7 (5·5)	10·4 (7·8)
11	900	170	22·3 (2·1)	6·5 (8·5)	0·5 (0·7)	1·6 (1·9)	56·6 (73·9)	2·7 (5·5)	9·8 (7·3)

[a] % by volume (% by mass).
[b] Trace amount of propane was detected.

ethylene yields of 8·5 % (by mass in the product gas) and total hydrocarbon yields of 18·4 % (of ethylene, acetylene, ethane and methane by mass) have been realized in the preliminary non-optimized Ultrapyrolysis experiments.

6. CONCLUSIONS

Fast pyrolysis promises high yields of concentrated high quality intermediates, petrochemical syngas and fuel gas products from biomass. These gases include olefins and other hydrocarbons. In addition, fast pyrolysis incorporates some of the best features associated with conventional pyrolysis, flash pyrolysis and gasification without many of the related disadvantages.

The University of Western Ontario Ultrapyrolysis is employing solids, rather than a gas, to transfer heat to biomass in a continuous vortical mixing vessel to accomplish the rapid pyrolysis of cellulose, lignin and wood powder. Initial experiments are promising and have demonstrated ethylene yields of 8·5 % (by mass) in non-optimized runs.

ACKNOWLEDGEMENTS

The authors wish to acknowledge the generous funding of this research by the ENFOR programme (Environment Canada and Energy, Mines, Resources, Canada). Special gratitude is offered to W. R. Grace Co. for the supply of catalyst and to Iotech (Mr G. Noble, Ottawa), FMC (Mr H. Rosenthal, Philadelphia), Composition Materials Inc. (Mr T. Diamond, Stamford) and Westvaco (Mr J. Roberts, Charleston Heights) for their supply of alkali-extracted lignin, Avicel cellulose, birch wood flour and Indulin AT lignin, respectively. The cooperative technical supervision of Dr Ralph Overend (National Research Council of Canada) and Dr David Fung (Energy, Mines and Resources, Canada) is also gratefully acknowledged.

Portions of the theoretical discussions of this paper are taken from a paper presented by the authors in Winnipeg, Canada.[15] The experimental work has not been published previously.

REFERENCES

1. Soltes, E. J. and Elder, T. J. (1978). Thermal degradation routes to chemicals from wood. 8th World Forestry Congress, Jakarta, Indonesia, October.
2. Soltes, E. J. and Elder, T. J. (1981). Pyrolysis. In *Organic Chemicals from Biomass*, I. S. Goldstein (Ed.). CRC Press, Boca Raton, FL.
3. Soltes, E. J., Wiley, A. T. and Lin, S. C. (1981). Biomass pyrolysis—Towards an understanding of its versatility and potentials. 3rd Symposium on Biotechnology in Energy Production and Conservation. Gatlinburg, TN, May 12–15, sponsored by the United States Department of Energy.
4. Shafizadeh, F. (1968). Pyrolysis and combustion of cellulosic materials. *Advances in Carbohydrate Chemistry* **23**, 419.
5. Shafizadeh, F. (1980). Introduction to biomass pyrolysis. In *Specialists' Workshop on Fast Pyrolysis of Biomass* (see ref. 12).
6. Broido, A. (1976). Kinetics of solid phase cellulose pyrolysis. In *Thermal Uses and Properties of Carbohydrates and Lignins*, F. Shafizadeh (Ed.). Academic Press, New York.
7. Lewellen, P. C., Peters, W. A. and Howard, J. B. (1976). Cellulose pyrolysis kinetics and char formation mechanism. 16th International Symposium on Combustion, The Combustion Institute, Pittsburgh, PA.
8. Milne, T. A. (1979). Pyrolysis—Thermal behaviour of biomass below 600 °C. In *A Survey of Biomass Gasification*, Vol. II, *Principles of Gasification*. Solar Energy Research Institute. Golden, CO.
9. Milne, T. A. and Soltys, M. N. (1981). Kinetics and mechanisms of fast pyrolysis. 12th Biomass Thermochemical Conversion Contractors' Review Meeting, Washington, D.C., March 18–19, sponsored by the United States Department of Energy.
10. Milne, T. A. and Soltys, M. N. (1982). Fundamental pyrolysis studies. Quarterly Report FR-234-1537, Solar Energy Research Institute, Golden, CO.
11. Mok, W. S. and Antal, M. J. (1982). Effects of pressure on biomass pyrolysis. 14th Biomass Thermochemical Conversion Contractors' Meeting, Washington, D.C., April 23–24, sponsored by the United States Department of Energy.
12. SERI (1980). Specialists' Workshop on Fast Pyrolysis of Biomass. Report CP-622-1096, Solar Energy Research Institute, Golden, CO.
13. Diebold, J. and Scahill, J. (1981). October 1981. Update on the progress in the entrained flow, fast ablative pyrolysis of biomass. 13th Biomass Thermochemical Conversion Contractors' Review Meeting, Washington, D.C., October 27–29, sponsored by the United States Department of Energy.
14. Reed, T. B., Milne, T., Diebold, J. and R. Derosiers, (1981). Biomass gasification for production of gaseous and liquid fuels. 3rd Symposium on Biotechnology in Energy Production and Conservation, Gatlinburg, TN, May 12–15, sponsored by the United States Department of Energy.
15. Graham, R. G., Mok, L. K., Bergougnou, M. A. and de Lasa, H. I. (1982). Fast pyrolysis of biomass for the production of chemicals and fuels from wood. 4th Bioenergy R&D Seminar, National Research Council of Canada, Ottawa, Canada.

16. Mok, L. K., de Lasa, H. I., Bergougnou, M. A. and Graham, R. (1982). Continuous ultra-flash pyrolysis of wood using solid heat carriers and double jet vortical contactors (ultrapyrolysis). 4th Bioenergy R&D Seminar, National Research Council of Canada, Ottawa, Canada.
17. Mok, L. K., Graham, R. G., Bergougnou, M. A. and Lasa, H. I. (1982). Continuous ultra-flash pyrolysis of wood using solid heat carriers and vortical contactors (ultrapyrolysis). Industrial Wood Energy Forum '82 Conference, Washington, D.C., March 8–10, Forest Products Research Society, Madison, WI.
18. Bergougnou, M. A., de Lasa, H. I., Mok, L. K., Graham, R. and Hazlett, J. (1981). Flash pyrolysis of wood using solid heat carriers (ultrapyrolysis). 3rd Bioenergy R&D Seminar, National Research Council of Canada, Ottawa, Canada.

23

Calorific Value of Wood as Affected by Sample Particle Size and Radial Position in Stem

JOHN M. KRYLA
Forintek Canada Corporation, Ottawa, Ontario, Canada

1. INTRODUCTION

Standard methods of wood analyses frequently require that the material be reduced (comminuted) to a specified size range prior to testing. It is recommended that a particle size smaller than 250 μm be utilized in determinations of gross heat of combustion by adiabatic oxygen bomb calorimetry.[1] Wood and bark calorific values have been determined using various particle sizes.[2-4] In most cases, the materials are oven-dried before testing.

Recent research conducted at the University of Toronto has indicated that a decrease in particle size will achieve an increase in the gross heat of combustion of wood.[5] Comminuted juvenile hybrid poplar (2 year growth) wood and/or bark, less than 75 μm in size, yielded heating values ranging from about 8 to 20 % higher than those from corresponding particles larger than 850 μm. These results prompted discussions concerning the potential energy benefits in comminuting wood to a fine particle size. However, since these initial tests were conducted on physically and chemically heterogeneous materials which may have been subject to substantial fractionation during grinding, additional research was necessary to evaluate fully the effect of particle size.

The major objective of this study is to examine the calorific value of wood, as affected by:

(a) particle size,
(b) drying method,
(c) radial position in stem.

2. METHODS AND MATERIALS

2.1. Particle Size Experiment

2.1.1. Material Preparation and Chemical Analyses

Material for the particle size experiment was prepared from flat-sawn, kiln-dried trembling aspen (*Populus tremuloides* Michx.) and white spruce (*Picea glauca* (Moench) Voss) lumber. Selection of material was limited to defective-free sapwood portions of the lumber.

Lignin contents of the spruce and aspen were determined in duplicate on oven-dried, extractive-free 180–250 μm material using standard procedures.[6,7] Holocellulose was prepared in duplicate using the acid chlorite method.[8] One holocellulose sample was corrected for residual lignin while the other was used for determination of α-cellulose. Alpha-cellulose content was conducted in duplicate according to recommended techniques.[9]

Carbon, hydrogen and nitrogen levels of the spruce and aspen were determined using a Perkin-Elmer model 240B elemental analyser following recommended procedures.[10] Tests were conducted in triplicate on oven-dry materials of two particle sizes (250–425 μm and 38–75 μm). Ash contents, conducted in triplicate, were determined by placing crucibles containing 1·5 g oven-dry samples in a muffle furnace at about 600 °C for 7 h. Since sulphur was assumed to be negligible, oxygen was determined by difference.

2.1.2. Adiabatic Oxygen Bomb Calorimetry

The base material for the oxygen bomb tests was prepared by cutting the wood to a size ranging from 2000 to 3350 μm. This material was then comminuted in a Thomas Wiley mill fitted with a 425 μm sieve. After grinding, the materials were classified using a Ro-Tap sieve shaker into particle size classes of 250–425, 150–250, 75–150, 38–75 and less than 38 μm. The original base 2000–3350 μm wood particles were included in the testing plan to represent coarse, non-comminuted material.

From the six particle size classes, samples of approximately 0·5 g were oven-dried, weighed and stored over a saturated salt solution of potassium carbonate until a constant mass was achieved. Equilibrium moisture content (emc) over the salt solution was approximately 6·8 % (wet basis) for spruce and 5·9 % for aspen. No effect of particle size upon emc was discernible.

Calorific values were determined on loosely packed materials using a Parr model 1241 adiabatic oxygen bomb calorimeter at 3·0 MPa oxygen following prescribed methods.[1] Calorific values were expressed on a basis of original oven-dry mass. Three calorimeter tests were conducted for each of the six particle size classes.

Reference oxygen bomb tests were also conducted on five particle sizes of sucrose. An aqueous sucrose solution was crystallized on a shallow dish. After drying, the material was ground using a mortar and pestle to particle size classes of 2000–4750, 250–425, 75–250, 38–75 and less than 38 μm. Three oxygen bomb tests were conducted for each of the particle sizes.

Experimentally determined wood calorific data were compared to theoretical values calculated using the Institute of Gas Technology (IGT) correlation equation[11] described as follows:

$$CV = [146{\cdot}58\,C + 568{\cdot}78\,H + 29{\cdot}45 - 6{\cdot}58\,A - 51{\cdot}53(O + N)]/429{\cdot}917 \quad [1]$$

where CV = calorific value (MJ/kg), A = ash, C = carbon, H = hydrogen, N = nitrogen, O = oxygen, expressed as percentage of original oven-dry mass.

2.2. Drying Experiment

Sapwood portions of freshly cut white spruce and trembling aspen were selected for the drying study. The aspen and spruce were sliced into strips and partially air-dried prior to reduction into two particle size classes. A coarse segment was produced by manually cutting the strips into particles (non-comminuted) ranging from 2000 to 3350 μm. A fine fraction was produced by comminuting the samples in a Wiley mill to pass a 250 μm sieve. The two particle classes of aspen and spruce were subjected to two different methods of drying. In the first method, samples of approximately 0·5 g were oven-dried, weighed and stored over a saturated salt solution of potassium carbonate prior to oxygen bomb calorimetry. In the second method, materials were stored over the salt solution without oven-drying.

Calorific values were determined in triplicate by the previously described methods and expressed on a dry basis. For the materials which had not been oven-dried, representative moisture content samples were removed prior to oxygen bomb testing. Since sample compaction may have an influence upon degree of combustion,[1] several oxygen bomb tests were also conducted on oven-dried pelletized spruce particles ($<250\,\mu$m).

2.3. Radial Position Experiment

A stem section of 10-year old hybrid poplar clone I-262 (*Populus euramericana* (Dode) Guinier) was supplied to Forintek by the Ontario Ministry of Natural Resources, Maple, Ontario (Fig. 1). The material measured approximately 30 cm in diameter and had an average growth

FIG. 1. Hybrid poplar clone I-262 disc and test section.

ring width of 14 mm. Material for analyses was prepared by removing a test section which measured about 5 and 25 cm in the tangential and longitudinal directions, respectively, and contained all annual rings on one side of the pith (Fig. 1).

Lignin, holocellulose, α-cellulose and extractive contents were determined by the previously described methods on materials from the heartwood (ring 4 (no. from pith)) and sapwood ring 9).

Ring numbers 1, 3, 5, 7 and 10 and inner and outer bark components were removed for calorific and elemental analyses. Rings 3, 5, 7 and 10 were divided into earlywood and latewood samples by selecting material from the inner and outer 3 mm of each growth ring. For ring 1, a mixture of earlywood and latewood was utilized. All materials were comminuted to pass a 250 μm sieve. Calorific value, carbon, hydrogen and nitrogen were determined in triplicate while ash content was conducted in duplicate. The elemental analysis data were used in eqn (1) to calculate theoretical calorific values of the various hybrid poplar components.

3. RESULTS AND DISCUSSION

3.1. Particle Size Experiment

3.1.1. Constituent and Elemental Analyses

Lignin content of the aspen was found to be 17·6%, compared to 26·8% for spruce (Table I). Holocellulose content, corrected for residual lignin, was 78·7% for aspen and 69·2% for spruce. Summations of lignin, holocellulose and extractive contents yielded totals of 100·0% and 98·5% for aspen and spruce, respectively.

TABLE I

CONSTITUENT AND EXTRACTIVE CONTENTS OF TREMBLING ASPEN, WHITE SPRUCE AND HYBRID POPLAR CLONE I-262[a]

Species	*Lignin* (%)	*Holocellulose* (%)	*α-Cellulose* (%)	*Extractives* (%)
Trembling aspen	17·6	78·7	46·4	3·7
White spruce	26·8	69·2	45·0	2·5
Hybrid poplar clone I-262 sapwood	22·5	78·3	42·3	4·4
Hybrid poplar clone I-262 heartwood	23·5	78·0	42·3	5·2

[a] Values based upon oven-dry unextracted wood. Lignin and α-cellulose are average of two observations while holocellulose and extractive values are based upon one observation.

Elemental analysis did not detect significant differences between the 250–425 μm and 38–75 μm particle size classes of aspen and spruce. Average carbon, hydrogen, nitrogen and ash levels for aspen were 48·84, 6·13, 0·10 and 0·16%, respectively. Respective levels for spruce were 49·65, 6·13, 0·01 and 0·20%.

3.1.2. Calorific Values

Calorific values of the five sucrose particle sizes are shown in Table II(a). Analysis of variance (Table IIb) did not show a significant treatment effect,

TABLE II(a)

AVERAGE AND RANGE OF SUCROSE CALORIFIC VALUES[a]

Particle size (μm)	Calorific value (MJ/kg)	
	Average	*Range*
2 000–4 750	16·34	16·32–16·37
250–425	16·49	16·45–16·54
75–250	16·47	16·42–16·52
38–75	16·37	16·24–16·45
<38	16·45	16·38–16·55

[a] Average values are based upon three observations.

TABLE II(b)

ANALYSIS OF VARIANCE-SUCROSE CALORIFIC VALUES

Source of variation	*Degrees of freedom*	*Sum of squares*	*Mean square*
Replications	2	0·005 9	0·002 9
Treatments	4	0·052 4	0·013 1
Error	8	0·045 3	0·005 7

TABLE III(a)

CALORIFIC VALUES OF TREMBLING ASPEN AND WHITE SPRUCE AS AFFECTED BY PARTICLE SIZE[a]

Particle size (μm)	Calorific value (MJ/kg)			
	Trembling aspen		*White spruce*	
	Average	*Range*	*Average*	*Range*
2 000–3 350	19·56a	19·40–19·68	20·01a	19·94–20·06
250–425	19·01c	18·90–19·10	19·70b	19·51–19·80
150–250	19·23b	19·13–19·30	19·45c	19·35–19·57
75–150	19·24b	19·15–19·31	19·86ab	19·84–19·88
38–75	19·19bc	19·11–19·29	19·87ab	19·72–19·98
<38	19·52a	19·41–19·65	19·93a	19·82–20·05

[a] Values based upon oven-dry wood; averages are based upon three observations. Column means with the same letter are not significantly different at the 5% level.

TABLE III(b)
ANALYSIS OF VARIANCE—ASPEN AND SPRUCE CALORIFIC VALUES, PARTICLE SIZE EXPERIMENT

Source of variation	*Degrees of freedom*	*Sum of squares*	*Mean square*
Replications	2	0·002 7	0·001 3
Species (S)	1	2·346 1	2·346 1[a]
Particle size (P)	5	1·012 7	0·202 5[a]
S × P	5	0·251 9	0·050 4
Error	22	0·288 3	0·013 1

[a] Significant at the 1% level.

indicating that, for chemically homogeneous substances, particle size has little influence upon gross heat of combustion.

Calorific values of the aspen and spruce wood are summarized in Table III(a) and Fig. 2. Analysis of variance (Table IIIb) showed that particle size had a significant effect upon the calorific values of both species. Comparison of mean values by Duncan's multiple range test[12] showed that the highest heating values were measured for the base 2000–3350 μm and for the ultra-fine materials. Samples in the 150–250 μm and 250–425 μm particle size classes were found to have the lowest calorific values. Grinding of the wood may have expelled volatile extractives, thus reducing the calorific values of the intermediate particle sizes. Variations in calorific

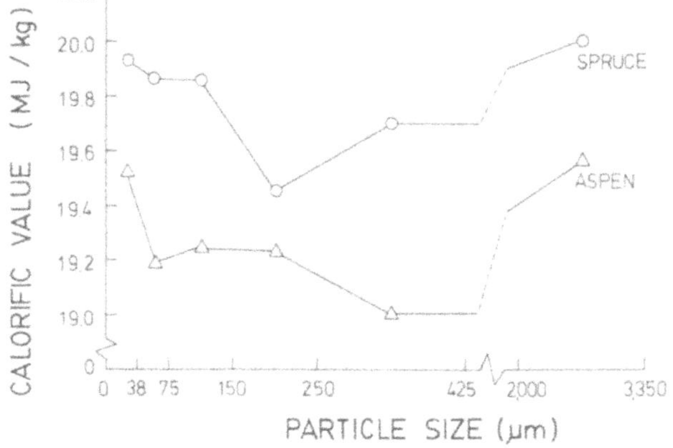

FIG. 2. Effect of particle size on calorific values of spruce and aspen wood.

values may have also resulted from differences in chemical composition due to fractionation. Harris[13] observed that during grinding of white ash, the lignin-rich middle lamella portion tended to generate more fine materials than did the other components of the cell wall. Since lignin has the highest heating value of the wood constituents, finely comminuted materials would be expected to exhibit a higher calorific value than coarsely ground particles.

Although significant, the differences between average maximum and minimum calorific values were less than 3 % for both aspen and spruce. Larger variations may be encountered for species such as ash and oak which have a less uniform cellular structure than spruce and aspen.

The average calorific value of spruce was 19·80 MJ/kg compared to 19·29 MJ/kg for aspen. The higher calorific value of spruce can be attributed to larger amounts of lignin (Table I). For trembling aspen, the IGT equation and elemental analysis data resulted in theoretical calorific values of 19·60 (error = +3·09 %) and 19·30 (error = +0·59 %) MJ/kg for particle sizes of 250–425 and 38–75 μm, respectively. For spruce, calculated heating values were 19·86 (error = +0·81 %) and 19·79 (error = 0·37 %) MJ/kg for particle sizes of 250–425 and 38–75 μm, respectively.

3.2. Drying Experiment

Calorific values of the white spruce and trembling aspen, which were dried from the green condition, are outlined in Table IV(a). Analysis of variance (Table IVb) showed that method of drying and particle size had significant effects upon calorific value. As observed with the kiln-dried wood, the average calorific value of the non-comminuted 2000–3350 μm material was higher than that of the material comminuted to less than 250 μm. Material which was air-dried had a higher heating value than wood which had been oven-dried.

Analysis of the significant interactions indicated that the effect of particle size depended upon the species being tested. As shown by Duncan's comparison of means (Table IVa), the calorific value of 2000–3350 μm spruce was higher than that of the finely ground material. For aspen, particle size by itself did not significantly affect calorific value. For both aspen and spruce, the effect of drying was dependent upon the particle size of the material. For materials less than 250 μm in size, oven-drying resulted in lower calorific values than did air-drying. Drying method did not affect the heating value of the 2000–3350 μm material. These results indicate that the potential calorific value of wood will be lowered after comminution and oven-drying. The calorific value of oven-dried finely ground spruce was

TABLE IV(a)
CALORIFIC VALUES OF FRESHLY CUT WHITE SPRUCE AND TREMBLING ASPEN AS AFFECTED BY DRYING METHOD AND COMMINUTION[a]

Species	*Calorific value (MJ/kg)*			
	Non-comminuted (particle size = 2 000–3 350 μm)		*Comminuted (particle size = <250 μm)*	
	Oven-dried	*Air-dried*	*Oven-dried*	*Air-dried*
White spruce	20·45a (6·82)	20·56a (8·02)	19·68c (6·78)	20·16b (7·04)
Trembling aspen	19·41de (6·03)	19·42de (7·45)	19·32e (6·12)	19·56cd (7·24)

[a] Values based upon oven-dry wood and are average of three observations. Values with the same letter are not significantly different at the 5 % level.
Values in parentheses represent average equilibrium moisture content over saturated solution of potassium carbonate prior to oxygen bomb test.

TABLE IV(b)
ANALYSIS OF VARIANCE—ASPEN AND SPRUCE CALORIFIC VALUES, DRYING EXPERIMENT

Source of variation	*Degrees of freedom*	*Sum of squares*	*Mean square*
Replications	2	0·010 2	0·005 1
Species (S)	1	3·725 7	3·725 7[a]
Particle size (P)	1	0·485 1	0·485 1[a]
Drying method (D)	1	0·274 8	0·274 8[a]
S × P	1	0·554 5	0·554 5[a]
S × D	1	0·044 0	0·044 0
P × D	1	0·135 6	0·135 6[a]
S × P × D	1	0·008 4	0·008 4
Error	14	0·095 6	0·006 8

[a] Significant at the 1 % level.

approximately 4·3% lower than that of air-dried 2000–3350 μm material. There was no statistical difference between the oven-dried fine spruce and air-dried fine aspen. Such possible overlapping illustrates the necessity of utilizing standardized preparatory techniques in oxygen bomb calorimetry.

The reductions in calorific values can be partially attributed to the loss of volatile extractives during comminution and oven-drying. The relatively large reduction in spruce reflects the presence of resinous volatile extractives. This phenomenon is not possible to evaluate fully since the moisture content used to adjust calorific values of air-dried material to an oven-dry basis may have been erroneous due to loss of volatiles during oven-drying. However, examination of calorific values which were not adjusted for moisture still revealed distinct differences. Average unadjusted calorific value of the finely ground spruce which was oven-dried and stored over the salt solution was 18·34 MJ/kg, compared to 18·74 MJ/kg for fine material which had not been oven-dried.

Pelletization of the fine material was found to have an effect upon calorific value. The calorific value of fine oven-dried pelletized spruce was 20·13 MJ/kg, or about 2·3% higher than respective non-pelletized material. Although this suggests incomplete combustion of the loosely packed materials, inspection of the bomb contents did not reveal differences in reaction behaviour. The calorific values of pelletized fine particles of spruce were lower than those of the non-pelletized 2000–3350 μm air-dried and oven-dried materials.

The effect of pelletization appears to be species dependent. An increase in calorific value would not be expected for pelletized poplar since, unlike the spruce data, calorific values of the fine materials were similar to those of the coarse samples.

3.3. Radial Position Experiment

3.3.1. Constituent and Elemental Analyses

Table I summarizes the chemical constituents of the heartwood and sapwood of the hybrid poplar. Both lignin and extractive concentrations were higher in the heartwood than the sapwood portion of the disc. The holocellulose content (corrected for residual lignin of 2·7%) was approximately 78% for both the heartwood and sapwood. This value was comparable to the holocellulose content of trembling aspen. The α-cellulose content was about 42% for both heartwood and sapwood, which was approximately 4% lower than for trembling aspen. This indicates that the hybrid poplar had a higher pentosan content than did the trembling aspen.

Summation of lignin, holocellulose and extractive contents resulted in values of 105·2 and 106·8 % for sapwood and heartwood, respectively. Overlapping, as was observed for the hybrid poplar, may have been caused by the presence of protein or polysaccharide degradation products in the isolated lignin or by residuals of ash in the holocellulose and lignin preparations. Although attempted, the samples of isolated constituents were too small to permit accurate determination of ash content.

Carbon, hydrogen, nitrogen and ash contents for the earlywood and latewood of annual rings 3, 5, 7 and 10 are summarized in Table V. Carbon

TABLE V
ELEMENTAL ANALYSIS OF FOUR WOOD GROWTH RINGS AND INNER AND OUTER BARK OF HYBRID POPLAR CLONE I-262[a]

Sample	*Carbon* (%)	*Hydrogen* (%)	*Nitrogen* (%)	*Oxygen* (%)	*Ash* (%)
Ring 3 E[b]	48·71b	5·83b	0·56ab	43·54	1·36a
L	49·24ab	5·99a	0·50bc	43·09	1·18a
Ring 5 E	49·91a	5·98ab	0·53ab	42·67	0·91b
L	48·86b	5·91ab	0·54ab	43·74	0·95b
Ring 7 E	49·35ab	5·92ab	0·50bc	43·53	0·70c
L	48·62b	5·96ab	4·43c	44·37	0·62cd
Ring 10 E	49·35ab	5·96ab	0·57ab	43·67	0·45d
L	48·91b	5·99a	0·60a	44·04	0·46d
Inner bark	43·23c	5·30c	0·77d	44·15	6·55e
Outer bark	50·55d	5·42d	0·33e	40·17	3·53f

[a] Values based upon oven-dry wood. Carbon, hydrogen and nitrogen are average of three observations while ash values are average of two observations; oxygen determined by difference. Column means with the same letter are not significantly different at the 5 % level.
[b] E = earlywood, L = latewood.

level did not appear to change with distance from pith; the average carbon content of the hybrid poplar was 49·12 %, However, carbon concentration was generally higher in the earlywood than the latewood portion of the annual ring. Comparison of mean values showed that the highest carbon content was measured for the earlywood of ring 5 (49·91 %). Intra-incremental studies of wood have shown that earlywood typically contains a higher lignin content than does latewood.[14]

Hydrogen and nitrogen levels did not appear to be affected by either distance from pith or position within the ring. Mean hydrogen and nitrogen contents for the hybrid poplar wood were 5·94 and 0·53%, respectively.

Wood ash content decreased with increasing growth ring number from pith. Average ash content of rings 3, 5, 7 and 10 were 1·27, 0·93, 0·66 and 0·46%, respectively. Wood ash contents of North American hardwoods and softwoods typically range from 0·2 to 0·8%,[15] while ash contents of 2–3% are not uncommon for wood of tropical species.[16] The high ash levels observed in hybrid poplar clone I-262 may be associated with factors such as fertilization levels, soil properties, site conditions and clonal attributes.

Ash contents of extractive-free heartwood (ring 4) and sapwood (ring 9) were found to be 0·23 and 0·30%, respectively (based on oven-dry unextracted wood mass). Although not tested for ash prior to extraction, it is reasonable to assume that their original ash contents were similar to those of the adjacent rings. This indicates that approximately 68% of the ash in the heartwood was removed during extraction and that about 12% of the extractives consisted of ash. For the sapwood, extraction removed about 50% of the ash which constituted approximately 5% of the total extractives mass. Thus when ash is included in summative analyses, some overlapping is possible whereby a component of the ash is reported in the extractives concentration. To eliminate this potential overlapping, it is recommended that for wood and bark with high levels of inorganics, ash content be conducted prior to and after extraction. Any reduction in ash level due to extraction should then be subtracted from the apparent extractive content. The nature of the combination of ash in wood is largely unknown[8] and further research appears needed to clarify fully this phenomenon.

Carbon, hydrogen, nitrogen and ash levels of inner bark were all significantly different from respective outer bark values. Carbon content of the outer bark was 50·55% compared to 43·23% for the inner bark. Outer bark displayed higher hydrogen and lower nitrogen levels than did the inner component. These variations can be attributed to basic differences in constituent concentrations. Analyses of hardwood barks have shown that outer bark contains more lignin and less holocellulose than does inner bark.[17,18] Ash content of the hybrid poplar inner bark was 6·55% compared to 3·53% for outer bark.

3.3.2. Calorific Values

Calorific values of the various hybrid poplar growth rings are listed in Table VI(a) and shown graphically in Fig. 3. Analysis of variance for rings 3, 5, 7 and 10 (Table VIb) showed that wood calorific value was significantly affected by growth ring position. Calorific value exhibited a cyclic pattern

TABLE VI(a)

CALORIFIC VALUES OF FIVE WOOD GROWTH RINGS AND INNER AND OUTER BARK OF HYBRID POPLAR CLONE I-262[a]

Sample	*Calorific value (MJ/kg)*			
	Experimental		*Theoretical according to IGT equation (1)*	*Error (%)*
	Average	*Range*		
Ring 1 E&L[b]	19·66a	19·56–19·74	—	—
Ring 3 E	19·66a	19·57–19·77	19·08	−2·95
L	19·39bc	19·18–19·53	19·54	+0·74
Ring 5 E	19·71a	19·59–19·79	19·80	+0·48
L	19·66a	19·61–19·72	19·22	−2·25
Ring 7 E	19·60a	19·48–19·72	19·44	−0·81
L	19·52ab	19·45–19·61	19·14	−1·95
Ring 10 E	19·19d	19·16–19·20	19·47	+1·50
L	19·20cd	19·07–19·30	19·31	+0·56
Inner bark	16·62e	16·47–16·74	16·33	−1·75
Outer bark	20·14f	20·04–20·22	19·56	−2·90

[a] Values based upon oven-dry wood. Experimental averages are based upon three observations.
[b] E = earlywood, L = latewood.

TABLE VI(b)

ANALYSIS OF VARIANCE—CALORIFIC VALUES OF RINGS 3, 5, 7 AND 10, RADIAL POSITION EXPERIMENT

Source of variation	*Degrees of freedom*	*Sum of squares*	*Mean square*
Replications	2	0·0480	0·0240
Position within ring (W)	1	0·0500	0·0500
Ring number from pith (R)	3	0·7833	0·2611[a]
W × R	3	0·0680	0·0227
Error	14	0·1430	0·0102

[a] Significant at the 1% level.

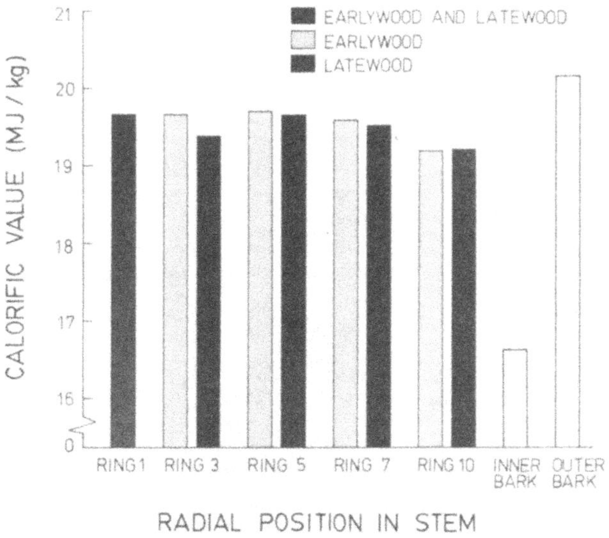

FIG. 3. Wood and bark calorific values of hybrid poplar clone I-262 (ring number represents number from pith).

with maximum values occurring in the middle growth rings. The highest calorific value was measured for ring 5 (19·68 MJ/kg) while the lowest was found for ring 10 (19·20 MJ/kg). The larger values measured in the inner rings can be attributed to higher amounts of lignin and extractives, although elemental analysis did not detect a significant change in carbon or hydrogen with distance from pith. Although significant, the calorific value difference between average minimum and maximum growth ring values was less than 3 %. The average calorific value of earlywood was approximately 0·5 % higher than that of latewood, although the difference was not significant.

Table VI(a) lists the theoretical calorific values of the hybrid poplar wood and bark, calculated from the elemental analysis data (Table V) and the IGT equation. The IGT equation provided satisfactory estimates of heating values with average and maximum absolute errors of 1·41 and 2·95 %, respectively.

The calorific value of inner bark was 16·62 MJ/kg compared to 20·14 MJ/kg for outer bark. As discussed previously, outer bark contained more carbon and hydrogen and less nitrogen and ash than did the inner bark. Utilizing the IGT equation and the elemental composition data, theoretical heating values of inner and outer bark were calculated to be 16·33 and 19·56 MJ/kg, respectively.

For hybrid poplar (Populus clone DN-30), Sastry and Anderson[4] found that the heating value of inner bark was about 5–10% lower than that of outer bark. Sandala *et al.*[18] reported that calorific values for inner bark of three native hardwoods were between 14 and 25% lower than for corresponding outer bark values.

Inappropriate sampling and preparation techniques may lead to variations in the calorific analysis of bark. Ottone and Baldwin[19] have shown that inner and outer bark will fractionate into different particle size classes during grinding. Sieving of whole-bark for calorific testing may consequently produce a sample which is not representative of the original material. Furthermore, whole-bark calorific values may vary significantly with stem diameter, since the ratio of outer to inner bark increases with increased tree diameter.

4. SUMMARY

Oxygen bomb calorimetry was conducted on (a) six particle sizes (2000–3350, 250–425, 150–250, 75–150, 38–75 and <38 μm) of trembling aspen and white spruce wood obtained from kiln-dried lumber, (b) two particle sizes (2000–3350 and <250 μm) of white spruce and trembling aspen which had been either air-dried or oven-dried from the green state, and (c) inner and outer bark and earlywood and latewood samples from various growth rings of hybrid poplar clone I-262. In (c), only one particle size (<250 μm) was utilized for calorific testing.

Lignin, holocellulose and α-cellulose contents were determined on 180–250 μm wood from the aspen, spruce and hybrid poplar. Selected wood and bark samples were analysed for carbon, hydrogen, nitrogen and ash concentrations. Experimentally determined calorific data were compared to theoretical values calculated using the IGT correlation equation. The results of this study are summarized as follows.

(a) Particle size was found to have a significant effect upon calorific value. For both spruce and aspen, the calorific values of the coarse (2000–3350 μm) and ultra-fine (<38 μm) materials were higher than those of the intermediate particle sizes. Although significant, the differences between average minimum and maximum calorific values were less than 3%. Spruce was found to have both a higher lignin content and a higher heating value than the aspen.

(b) Grinding the 2000–3350 μm material to less than 250 μm in size followed by oven-drying was found to reduce the calorific values of spruce and aspen wood. The calorific value of oven-dried <250 μm spruce was 4·3% lower than that of air-dried 2000–3350 μm material.

(c) Calorific value of hybrid poplar wood exhibited a cyclic pattern with maximum values occurring in the middle growth rings. Although significant, the difference between minimum and maximum growth ring heating value was less than 3%. The lowest calorific value was measured for the outermost growth ring. Correspondingly, the hybrid poplar sapwood had slightly lower lignin and extractive contents than the heartwood. The average calorific value of earlywood was about 0·5% higher than that of latewood, although the difference was not significant. Wood ash content decreased with increasing growth ring number from pith. The heating value of inner bark was approximately 17·5% lower than that of outer bark. Theoretical heats of combustion calculated by the IGT equation were within 3% of the experimentally determined values.

Based on the results of this study, it is recommended that a standard procedure be utilized in the preparation of wood and bark for oxygen bomb calorimetry. Grinding to less than 250 μm in size, as recommended in standard procedures, is not essential for complete combustion of wood and, when combined with oven-drying, may cause significant reductions in calorific value. The use of air-dried materials in the 2000–3350 μm range is recommended for calorific evaluation.

REFERENCES

1. Anonymous (1975). Instructions for the 1241 and 1242 adiabatic calorimeters. Parr Instrument Co, Moline, Ill. 61265, Manual 153.
2. Howard, E. T. (1973). Heat of combustion of various southern pine materials. *Wood Sci.* **5**(3), 194–7.
3. Bowersox, T. W., Blankenhorn, P. R. and Murphey, W. K. (1979). Heat of combustion, ash content, nutrient content, and chemical content of populus hybrids. *Wood Sci.* **11(1)**, 257–62.
4. Sastry, C. B. R. and Anderson, H. W. (1980). Clonal variation in gross heat of combustion of juvenile populus hybrids. *Cdn. J. Forest Res.* **10**(3), 245–9.
5. Sastry, C. B. R. (1981). Forintek Canada Corp., Ottawa, Ont. Personal communication.

6. American Society for Testing and Materials (1981).Standard method for preparation of extractive-free wood. ASTM D1105-56, Philadelphia, Pa.
7. American Society for Testing and Materials (1981). Standard test method for lignin in wood. ASTM D1106-56, Philadelphia, Pa.
8. Browning, B. L. (1967). *Methods of Wood Chemistry*, Vols 1 and 2. Interscience Publishers, Wiley, New York.
9. American Society for Testing and Materials (1981). Standard test method for alpha-cellulose in wood. ASTM D1103-60, Philadelphia, Pa.
10. Anonymous (1980). Instructions model 240B elemental analyzer. Perkin-Elmer, Norwalk, Conn.
11. Graboski, M. and Bain, R. (1979). Properties of biomass relevant to gasification. Chapter 3, *A Survey of Biomass Gasification*, Vol. II—*Principles of Gasification*. SERI/TR 33-239, Solar Energy Research Institute, Golden, Colo.
12. Steel, R. G. D. and Torrie, J. H. (1960). *Principles and Procedures of Statistics*. McGraw-Hill, New York.
13. Harris, E. E. (1953). A lignin-carbohydrate bond in wood. *Tappi* **36**(9), 402–5.
14. Wilson, J. W. and Wellwood, R. W. (1965). Intra-increment chemical properties of certain western Canadian coniferous species. *Cellular Ultrastructure of Woody Plants*, W. A. Coté, Jr. (Ed.). Syracuse University Press, Syracuse, N.Y., pp. 551–9.
15. Panshin, A. J. and De Zeeuw, C. (1970). *Textbook of Wood Technology*, Vol. 1. McGraw-Hill, New York.
16. Doat, J. (1977). Le pouvoir calorifique des bois tropicaux. *Bois et Forets des Tropiques* March–April 1977. No. 172, 33–54.
17. Binotto, A. P. and Murphey, W. K. (1975). Season and height variation in extractives and cell wall components of chestnut oak bark. *Wood Sci.* **7**(3), 185–90.
18. Sandala, M. G., Blankenhorn, P. R. and Kline, D. E. (1981). Thermal analysis of white oak, northern red oak, and black cherry bark chemical components. *Wood Sci.* **13**(3), 157–65.
19. Ottone, S. P. and Baldwin, R. C. (1981). The relationship of extractive content to particle size distribution in milled yellow-poplar (*Liriodendron tulipifera* L.) bark. *Wood and Fiber* **13**(2), 74–85.

24

The Effects of Atmosphere on Pyrolysis of Solid Fuels Produced in Finland

M. AHO and J. HUOTARI

Domestic Fuel Laboratory, Technical Research Centre of Finland, Jyväskylä, Finland

1. INTRODUCTION

Finland has considerable peat and wood resources which are extensively used as fuels and there is a need to investigate the pyrolysis of those fuels. Boilers are not perfect and their efficiency can probably be improved. Burning profiles of fuels provide information for that aim.[1]

In this research we have carried out experiments in seven different atmospheres. The methods used were thermogravimetric analysis (TG, DTG) and differential thermoanalysis (DTA). TG and DTG curves give information about weight loss profiles and allow the calculation of kinetic parameters. Differential thermoanalysis gives information about reaction heat as a function of temperature, which does not always correlate to weight loss.

2. FUELS STUDIED

Two types of peat were investigated: one with a low humification degree, H1–2 in von Post's scale (peat 1), and another with a medium humification degree, H5–6, a typical peat for use as fuel (peat 2).

Sawdust was from a softwood, mainly Scots pine (*Pinus silvestris*), and Norway spruce (*Picea excelsa*), used as chips in boiler.

Charcoal was chosen to complete the range of solid fuels. Charring increases the energy content of raw fuel. The sample was hardwood, mainly charred birch (*Betula verrucosa*).

All the experiments had been carried out after drying the samples at 105 °C for 24 h and milling in a propeller mill to a fine powder. After that

TABLE I
THE PROPERTIES OF THE SAMPLES

Sample	*C* (%)	*H* (%)	*N* (%)	*S* (%)	*O* (%)	*Ash* (%)	*Calorimetric value* (*MJ/kg*)
Peat 1[a]	46·5	5·72	0·76	0·28	43·2	2·5	19·9
Peat 2[b]	52·8	5·84	2·00	0·42	33·0	5·5	23·2
Sawdust	50·9	6·14	0·23	0·01	42·5	1·0	20·3
Charcoal	80·3	2·66	0·55	0·07	13·5	2·5	30·5

[a] From Haukineva peat production area.
[b] From Rastunsuo peak production area.

the water content of the samples was about 4 % by weight. The properties of the samples are illustrated in Table I.

3. EXPERIMENTAL CONDITIONS

3.1. Apparatus

A thermobalance (Setaram 1131) and a DTA (Setaram D80) were connected directly to the computer (Commodore 4032). The rate of increase of temperature was in all experiments 10 K/min. The gas mixer was a Teflon tube filled with glass beads.

3.2. Atmospheres

The atmospheres were selected to represent the main gas components in boilers: oxygen, nitrogen and carbon dioxide. When oxygen concentration is small, large amounts of carbon monoxide, hydrogen and many other gases, mainly hydrocarbons, can exist, too. Atmospheres used were: 20, 15, 10, 5 and 0 % O_2 in N_2, 100 % CO_2 and 10 % O_2–40 % N_2–50 % CO_2.

3.3. Calculations

The results of the kinetic parameters reported have been calculated by the Coats-Redfern method[2] with an ABC-80 minicomputer. To make sure of the values, the program also contained the Broido method,[3,4] which gives about 10 kJ/mole greater values.

4. RESULTS

Thermogravimetry, and DTA to some extent, have been used in fuel research.[1,5–7] Most of the experiments were made in air or nitrogen. In this investigation we also considered other atmospheres, and correlation between DTG and DTA information.

The effect of decreasing the oxygen content in nitrogen on DTG curves of the samples is seen in Fig. 1. When excluding charcoal the curves have two peaks, which originate from two main reaction steps. There is also weight loss between the main steps. The samples have complex chemical structures and contain very many compounds, so their pyrolysis is also a very complicated series of reactions. The shape of the curves can to some extent be explained with the aid of three main components: cellulose, hemicellulose and lignin. Cellulose has probably the strongest effect on the shape of the curve near the top of the first peak. Lignin may have the greatest effect on the shape of the second peak. The two reactions can best be observed with sawdust and peat 1. Peat 2 has an uncertain second reaction and charcoal has only one main reaction.

Reaction 2 is much more sensitive to decreasing oxygen concentration than reaction 1, when we examine the spreading to higher temperatures of the two reactions. The peak due to the first reaction becomes rounded very soon, when oxygen concentration decreases from 20 to 15 %. The spreading of the temperature range of the pyrolysis in the function of decreasing oxygen concentration follows the order: charcoal → peat 2 → peat 1 → sawdust.

The kinetic parameters are found in Table II. The values of activation energies and the frequency factor decrease in proportion to oxygen concentration in oxygen–nitrogen mixtures. Peat 1 differs slightly from that order in 20 % oxygen.

Carbon dioxide has a small reducing effect on the parameters in 10 % oxygen although the initial and final temperatures of the pyrolysis reaction are nearly the same.

Excluding the pure carbon dioxide atmosphere, the reaction order is 1 in step 2 and also in the one-step reaction of charcoal. Step 1 is more complicated. The order is roughly 1 (0·8–1·0) at an oxygen concentration ≤ 10 %. Sawdust is an exception; this may be due to rapid release of its large oxygen content (42·5 %). The reaction order has the most uncertain value at 15 % O_2, and the value is 0 at 20 % O_2.

Carbon dioxide is not an inert diluent like nitrogen. The curves and kinetic parameters differ only slightly in reaction step 1, independent of oxygen concentration. Step 2 is more sensitive. In the two atmospheres with 10 % oxygen the calculated values are nearly identical, but increase in inverse proportion to oxygen concentration. In pure CO_2 the initial temperature of the second step is 960–1040 K; in nitrogen there is no second step.

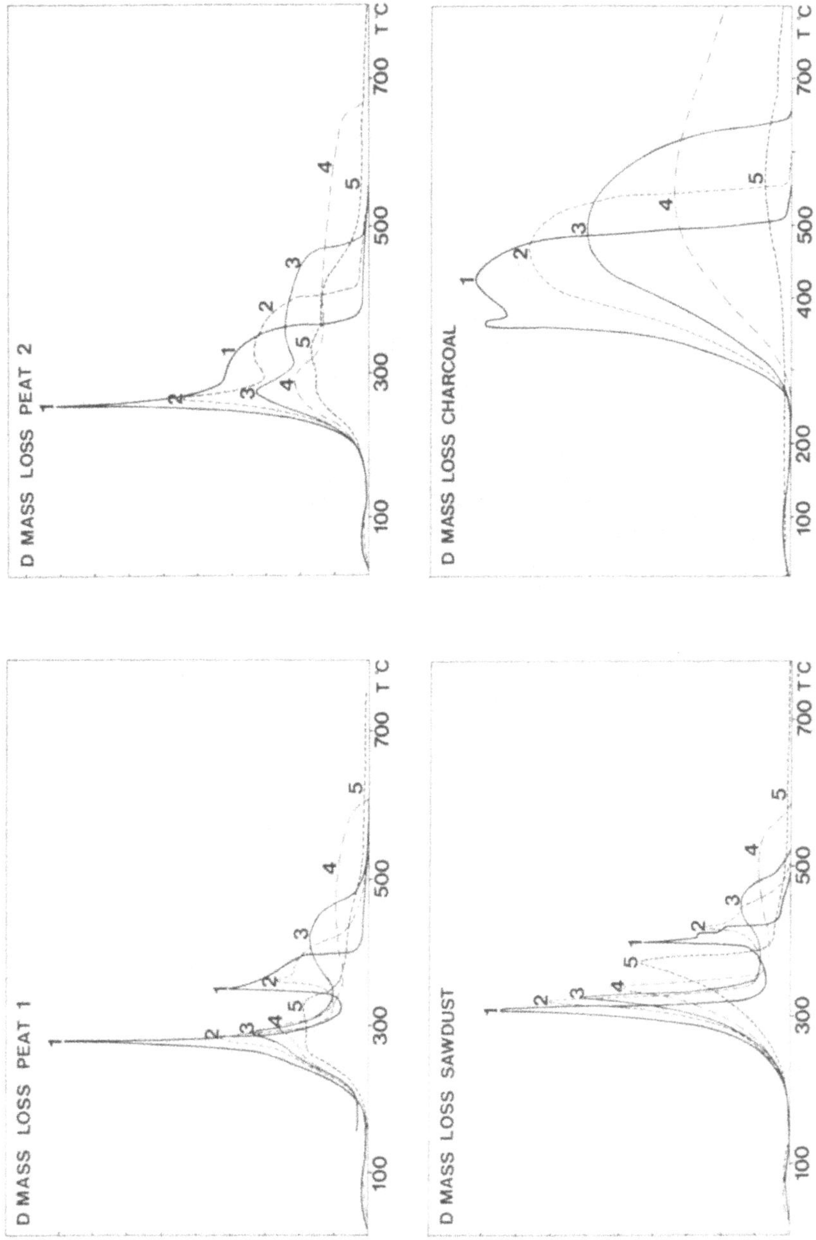

FIG. 1. DTG curves in five oxygen–nitrogen mixtures 1–5: 20, 15, 10, 5 and 0% O_2 in N_2.

TABLE II
KINETIC PARAMETERS

Reaction step	*A*	*Peat 1*			*Peat 2*			*Sawdust*			*Charcoal*		
		n	E	$\log K_0$	n	E	$\log K_0$	n	E	$\log K_0$	n	E	$\log K_0$
1	1	0·1	87	5·9	0·1	109	8·3	0·0	94	6·0	1·0	106	5·0
1	2	0·6	96	6·8	0·5	94	6·6	0·1	91	5·5	1·0	91	3·9
1	3	1·0	96	6·7	0·9	92	6·4	0·2	86	5·0	1·0	77	2·5
1	3[a]	0·8	89	6·1	0·8	80	5·1	0·2	84	4·9	1·0	84	3·1
1	4	1·0	85	5·5	1·0	71	4·0	0·4	82	4·6	1·0	51	0·0
1	5	1·0	76	4·1	1·0	59	2·5	0·5	76	3·8	—	—	—
1	5[a]	1·0	78	4·6	1·0	59	2·1	0·2	63	2·6	0·7	226	7·3
2	1	1·0	270	20	1·0	170	12	0·9	320	23			
2	2	1·0	230	16	1·0	150	10	1·0	300	20			
2	3	1·0	180	11	1·0	130	6·9	1·0	300	14			
2	3[a]	1·0	160	9·7	1·0	130	7·2	1·0	220	14			
2	4	1·0	130	6·2	1·0	88	2·8	1·0	170	9·2			
2	5	—	—	—	—	—	—	—					
2	5[a]	0·6	230	7·8	0·8	210	7·0	0·7	240	8·1			

n = reaction order, E = activation energy (kJ/mole), $\log K_0$ = logarithm of frequency factor (min^{-1}). Atmospheres (A): 1, 20% O_2 in N_2; 2, 15% O_2 in N_2; 3, 10% O_2 in N_2; 3[a], 10% O_2–40% N_2–50% CO_2; 4, 5% O_2 in N_2; 5, 100% N_2; 5[a], 100% CO_2.

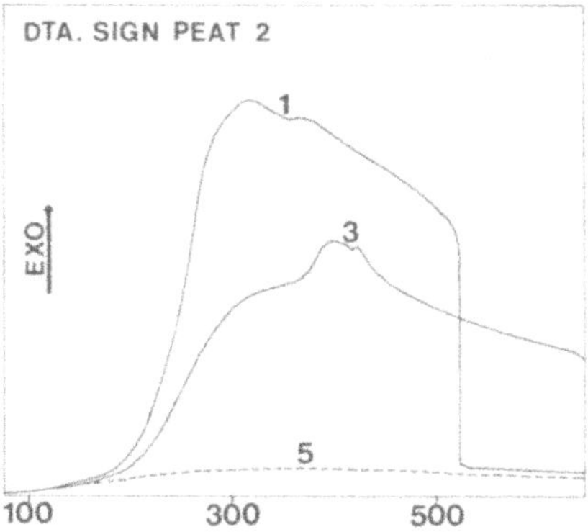

FIG. 2. DTA curves of peat 2 at 20, 10 and 0% O_2 in N_2.

DTA curves of peat 2 in three atmospheres are seen in Fig. 2. The two reaction steps can be distinguished on the curves in 10 and 20% oxygen. Also, their shifts to higher temperatures when oxygen concentration decreases can be observed.

DTA nor TG nor DTG curves alone give sufficient information about pyrolysis. The reason is seen in Fig. 3. At the temperature of maximum weight loss of the sawdust sample the reaction is only slightly exothermic. The main exothermic reaction begins after step 1. All wood samples do not behave similarly. For example, willow (*Salix alba*) has a more exothermic first weight loss step.

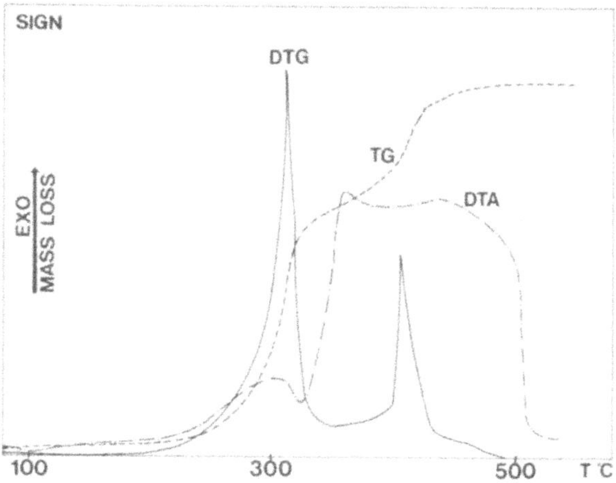

FIG. 3. TG, DTG and DTA curves of sawdust at 20% O_2.

5. SUMMARY

The solid fuels studied have different pyrolysing rates; this can be observed on DTG and DTA curves. No anomalous behaviour in the function of decreasing oxygen concentration in the oxygen–nitrogen mixture was observed. Carbon dioxide is not an inert diluent like nitrogen, when oxygen concentration is small under these conditions. DTA curves supplement TG and DTG information for use in boiler design.

ACKNOWLEDGEMENT

The experimental part of this work was carried out in France and the kinetic calculations were done in Finland. The authors are indebted to Director Delbourgo, Centre National De La Recherche Scientifique, Centre De Recherches Sur La Chimie De La Combustion Et Des Hautes Temperatures, Orleans, France, who allowed the use of equipment for thermoanalysis, and to Associate Professor Tummavuori, Department of Inorganic Chemistry in the University of Jyväskylä, Finland, who allowed the use of the computer and program for kinetic calculations.

REFERENCES

1. Vecci, S. J., Wagoneer, C. L. and Olson, G. B. (1978). Fuel and ash characterization and its effect on the design of industrial boilers. *Proceedings of the American Power Conference 40*, 850–64.
2. Coats, A. W. and Redfern, J. P. (1964). Kinetic parameters from thermogravimetric data. *Nature* **201**, 68.
3. Broido, A. (1969). A simple, sensitive graphical method of treating thermogravimetric analysis data. *J. Polymer Sci.* **A2**(7), 1761.
4. Tummavuori, J. and Suontamo, R. (1979). The thermal decomposition of copper(II) acetate, its mono-, di- and tridoro derivates and copper(I) and copper(II) chlorides. *Finn. Chem. Lett.* **6**, 176.
5. Tummavuori, J., Venäläinen, H. and Nyrönen, T. (1978). The usability of thermogravimetry in fuel research, Parts I–V. Part I, On the application in peat research, Suo 29 (1978), 33–38; Part II, On the kinetics of different peat types, Suo 29 (1978), 59–67; Part III, Comparison of the kinetics of the pyrolysis of solid fuels, Suo 29 (1978), 99–104; Part IV, The effect of the reduced pressure on the pyrolysis, Suo 30 (1979), 1–5; Part V, On the kinetics of the pyrolysis of materials suitable for fuel, Suo 30 (1979), 65–69.
6. Ranta, J. and Nyrönen, T. (1980). The simultaneous thermic analysis of solid fuels. Technical Research Centre of Finland, Fuel and Lubricant Laboratory, Report 34.
7. Patzak, W. (1972). *VDI-Forschungsheft 552*, 1.

25

Thermoanalytical Studies of Carbohydrate Pyrolysis

ATTILA E. PAVLATH and KAY S. GREGORSKI

Western Regional Research Center, US Department of Agriculture, Berkeley, California, USA

1. INTRODUCTION

Reactions of carbohydrates form practically a separate branch of organic chemistry, but the study of the pyrolysis of carbohydrates was neglected for a long time and only a few research groups have been pursuing this research area. Increasing oil prices, however, have created more interest in alternative energy resources. Carbohydrate based biomasses have the best potential to supplement our energy and chemical feedstuff need which is presently derived from fossil fuel. During the past few years, considerable efforts were made on the thermal reactions of various biomasses. Unfortunately, most of the attempts were directed toward developing reactors for special biomasses and not for the basic reactions of those carbohydrates which are the main components of biomasses. For this reason, our knowledge is quite limited on the various simultaneous and consecutive reactions involved in carbohydrate pyrolysis even for such simple compounds as hexoses and pentoses. This paper reports the first step of a mathematical approach to develop more information about these reactions.

2. THEORY

Most of the known studies on the high temperature reactions of simple carbohydrates were done at reduced pressure which allows the immediate escape of the evolving gaseous products, thus limiting secondary reactions.

Reference to a company and/or product named by the Department is only for purposes of information and does not imply approval or recommendation of the product to the exclusion of others which also may be suitable.

Even in these cases, the products are numerous, making it difficult to follow the reactions. Shafizadeh—who was the most outstanding pioneer on the research of carbohydrate pyrolysis—carried out an extensive study with ^{14}C labeled levoglucosan.[1] The formation of the latter is considered by most researchers the first step in vacuum pyrolysis of carbohydrates. Most of these studies were done on cellulose which, without any catalyst, has a narrow range of decomposition temperature. A most recent article on cellulose pyrolysis provides an excellent summary in its introduction about work done on the kinetics of cellulose pyrolysis.[2]

The decomposition of other carbohydrates, however, is more complex. The reactions are even more complicated at atmospheric pressure where the pyrolysis is to be carried out for practical purposes. Unfortunately, the identification and evaluation of such overlapping and interrelated reactions are very difficult. Baker[3] attempted to characterize the decomposition of tobacco by first measuring the rate of CO and CO_2 evolution and then curvefitting the resulting functions by a visual and empirical method which led to the determination of kinetic parameters. Thermogravimetric studies can provide similar curves, measuring the weight loss as a function of temperature. The analysis and comparison of the weight loss curves of similar carbohydrates may provide information about the relative influence of structural factors on the pyrolytic decomposition of carbohydrates.

A large number of mono-, oligo- and polymeric carbohydrate samples were heated in a thermogravimetric instrument measuring not only the weight losses, but also the rate of weight losses. Though the rate curves exhibited one or two distinct peaks, it is difficult to assume that only two reactions would occur in such a complicated high temperature reaction system. In addition, the peaks were frequently asymmetrical and the presence of other peaks could be assumed. A study of these peaks revealed that they could be well approximated by one or more Gaussian functions: $\mathrm{d}W/\mathrm{d}T = A \exp[-b(T - T_0)^2]$ where W is the weight, A is related to the absolute intensity, b determines the halfwidth while T_0 is the temperature where the rate of decomposition for a given component is the highest. For the mathematical evaluation of these thermogravimetric rate curves, it was assumed that, within this family of related carbohydrates, the same halfwidth may be used if the reaction variables such as the heating rate and sample size were kept identical. The minimum value of the halfwidth could be unlimited, but the maximum value is the one by which all similar curves may be fitted. Based on this, all the rate curves were fitted in such a way to obtain a very low weighted chi-square value. Figure 1 shows a curvefitted rate curve for the pyrolysis of glucose at a heating rate of 20 °C/min. Nine

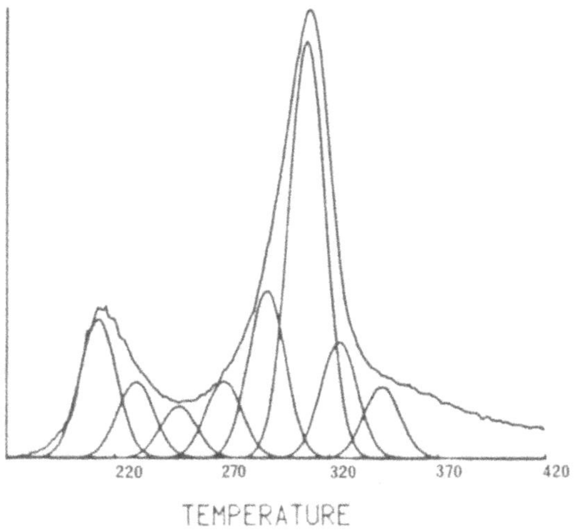

FIG. 1. Rate of weight loss curve of glucose pyrolysis (20 °C/min) curvefitted with Gaussian functions of 20 °C halfwidth.

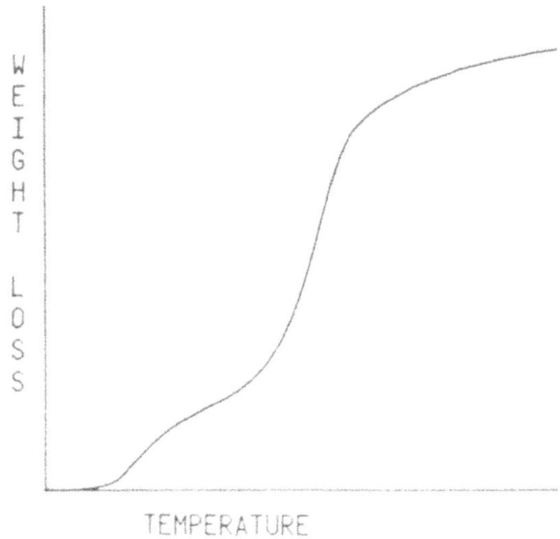

FIG. 2. Weight loss curve of glucose pyrolysis (20 °C/min).

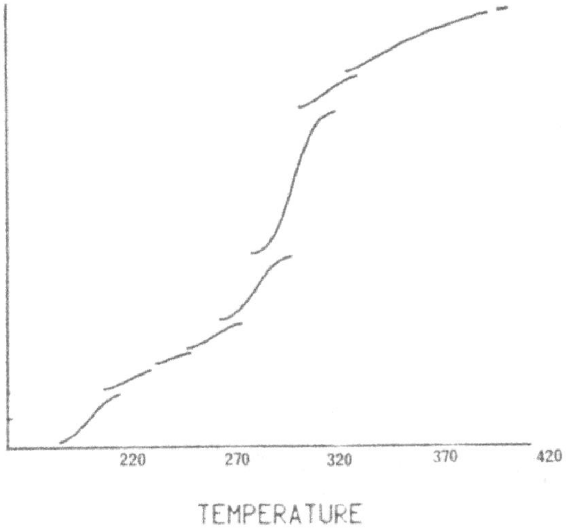

FIG. 3. Segmented weight loss curve of glucose pyrolysis in proportion of assumed rate components illustrated in Fig. 1.

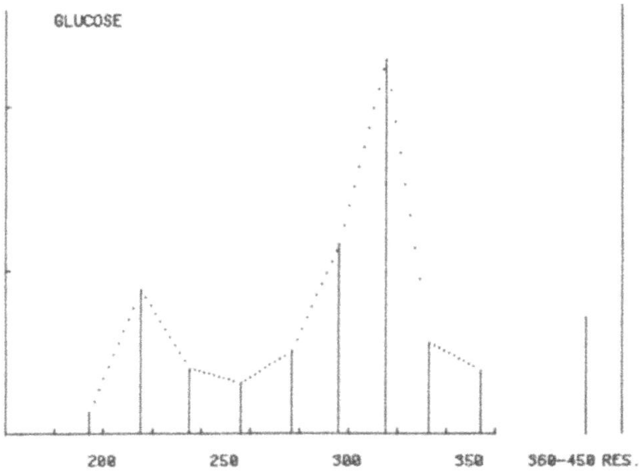

FIG. 4. Combination of amount of segmented weight loss with temperature where each rate is maximum.

Gaussian curves with a halfwidth of 22 °C provided an acceptable fit (with weighted chi-square of 2·38) for the decomposition up to 360 °C. Since the decomposition between 360 and 450 °C is very slow and without any characteristic feature, this range was not included in the curvefitting. When these assumed rate component functions were combined with the weight loss curve shown in Fig. 2, not only the rate of weight loss but the actual weight loss could be also separated into its assumed components as shown in Fig. 3. This was then graphically represented by Fig. 4, where each component is shown as a bar at the temperature where the individual rate reaches its maximum, with height proportional to the weight loss. The dotted line provides an envelope for easy visual comparison of various starting materials. For comparison, the weight loss between 360 and 450 °C and the amount of residue at 450 °C are also shown with bars of appropriate height.

3. EXPERIMENTAL

A Perkin-Elmer TGS-2 system equipped with a microprocessor controlled programmer was used in these experiments. Sample sizes were 1·5–2·0 mg. The compounds tested were of commercial origin. Altogether, 21 carbohydrates were studied: hexoses—glucose, mannose, galactose, fructose; pentoses—arabinose, lyxose, xylose; disaccharides—cellobiose, maltose, gentibiose, melibiose, lactose, turanose, sucrose, trehalose; trisaccharides—melezitose, raffinose; tetrasaccharide—stachyose; polysaccharides—amylose, cellulose, inulin. Catalysts included $ZnCl_2$, NaOH and Na_2CO_3. When the effects of catalysts were studied, the catalysts were dissolved in a small amount of water and ethanol, then mixed with the appropriate carbohydrate. The excess solvent was evaporated and the paste was dried in vacuum at room temperature. In these experiments, the catalyst-to-carbohydrate ratio was 1:10. Samples were encapsulated in large volume aluminum sample pans under nitrogen atmosphere and a small hole was made on top for the escape of volatiles. Unless otherwise indicated the analysis was carried out with a heating rate of 20 °C/min. At least two determinations were carried out in each case. An interactive Tektronix 4052 graphic computer was used for visual curvefitting.

4. RESULTS AND DISCUSSION

The comparison of the combined pyrolytic decomposition patterns of hexoses indicates that even small changes in structure could influence the

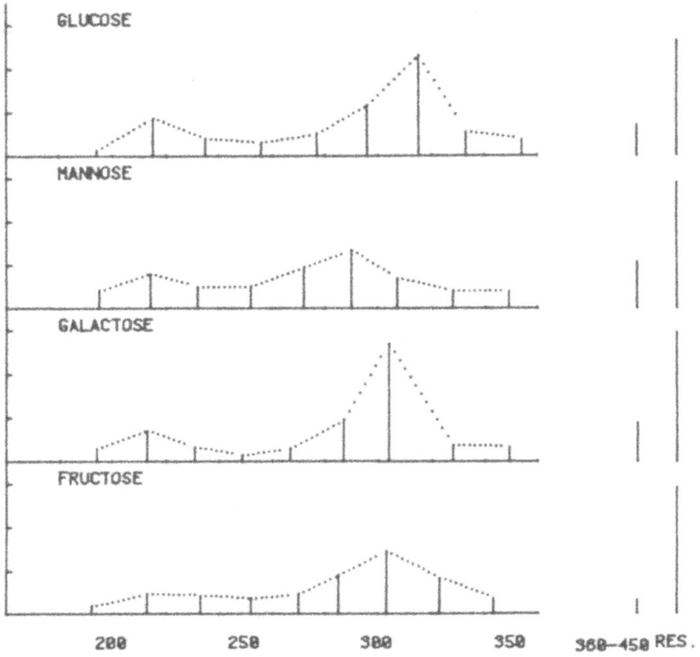

FIG. 5. Comparison of the pyrolytic pattern of hexoses.

pyrolysis (Fig. 5). The difference in the pyrolytic pattern can be observed between mannose and glucose which differ only in the steric configuration of the hydroxyl group on the second carbon atom. On the other hand, comparing glucose and galactose, the rotation on the fourth carbon atom does not alter the pyrolytic pattern of glucose considerably. The difference between glucose and fructose seems to indicate that the furanose-type structure is more easily pyrolyzed than to the pyranose-type. This appears to be further confirmed by the pyrolytic pattern shown in Fig. 6, where various pentoses with furanose structures are compared to fructose. It can also be concluded from comparing lyxose and xylose that the rotation of the hydroxyl group on the second carbon atom affects pyrolysis in the same way as with mannose and glucose.

In the case of disaccharides, the pyrolysis is influenced not only by the structure of the basic components, but also by the mode in which they are coupled. In Fig. 7, the stability of various couplings between two glucose molecules is compared. The alpha-coupling shows earlier decomposition than the beta-coupling when comparing maltose to cellobiose. However,

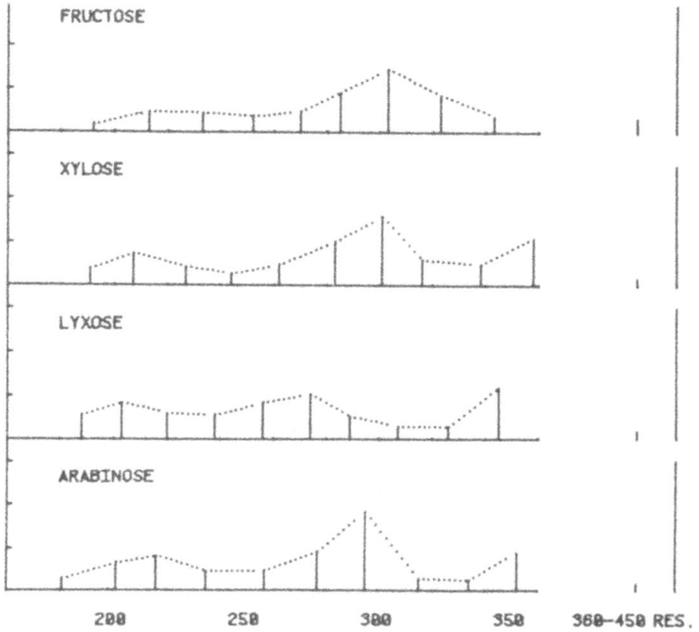

FIG. 6. Comparison of pyrolytic pattern of fructose and pentoses.

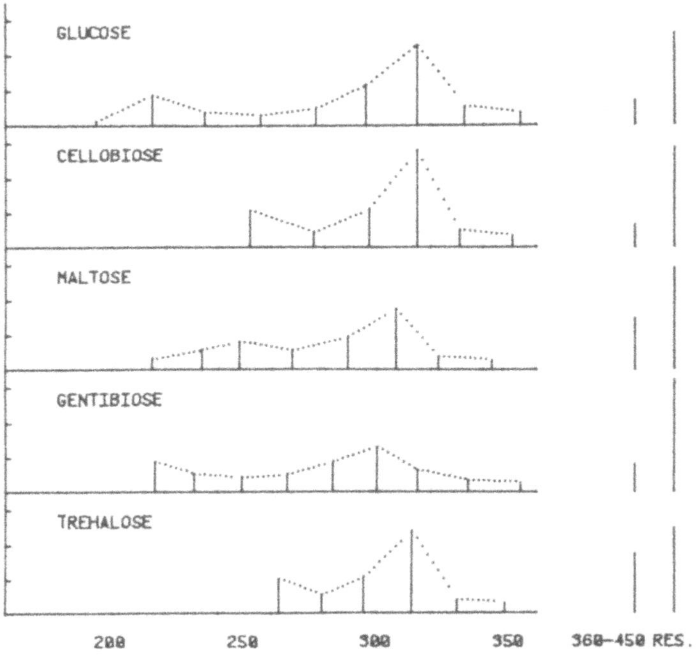

FIG. 7. Comparison of pyrolytic pattern of various disaccharides composed of glucose only.

looking at cellobiose and gentibiose, which both have beta-coupling, the 1–6 bonding also seems to induce pyrolysis at lower temperature. On the other hand, in the comparison of maltose and trehalose, which are bound together by alpha-coupling, the 1–1 bonding shows the most resistance to pyrolysis.

The number of available disaccharides with other hexoses are limited; therefore, it is difficult to make concrete deductions about the effect of individual factors. In Fig. 8, a few available galactose-containing saccharides are compared to each other. The higher reactivity of the 6-alpha-coupled mellibiose as compared to the 4-beta-coupled lactose is in agreement with the observations reported for glucose based disaccharides. In raffinose, a trisaccharide composed of fructose, glucose and galactose, the destabilizing effect of fructose seems to surface again, but it is counteracted in stachyose, a tetrasaccharide with one more galactose molecule in it. The stability of some of the 'missing links' needs to be studied to get more information about the effect of galactose on pyrolytic reactivity.

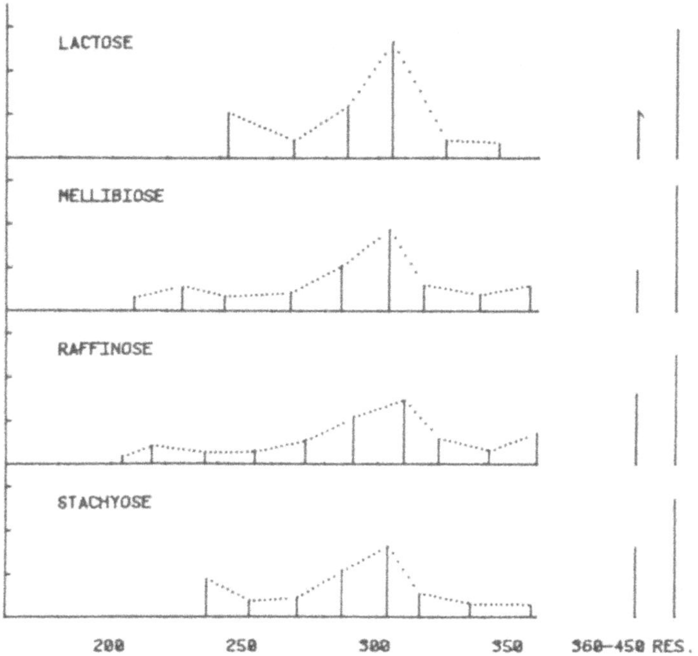

FIG. 8. Comparison of pyrolytic pattern of disaccharides containing galactose.

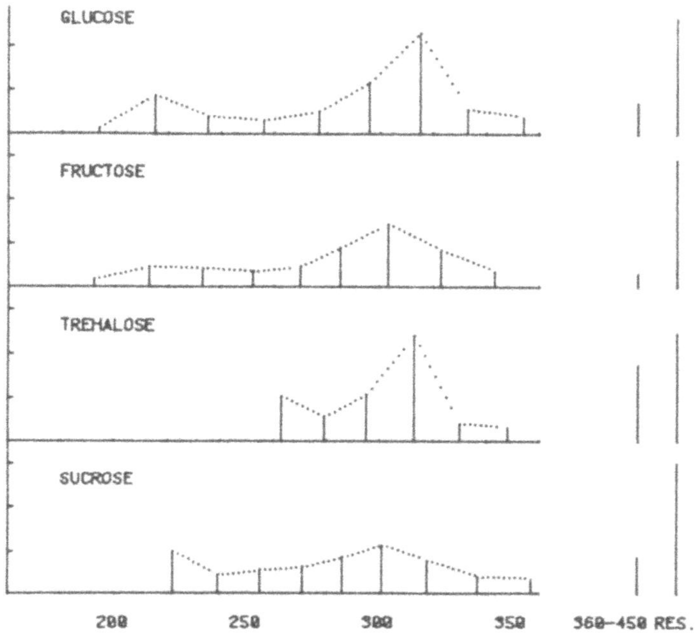

FIG. 9. Comparison of pyrolytic pattern of non-reducing sugars.

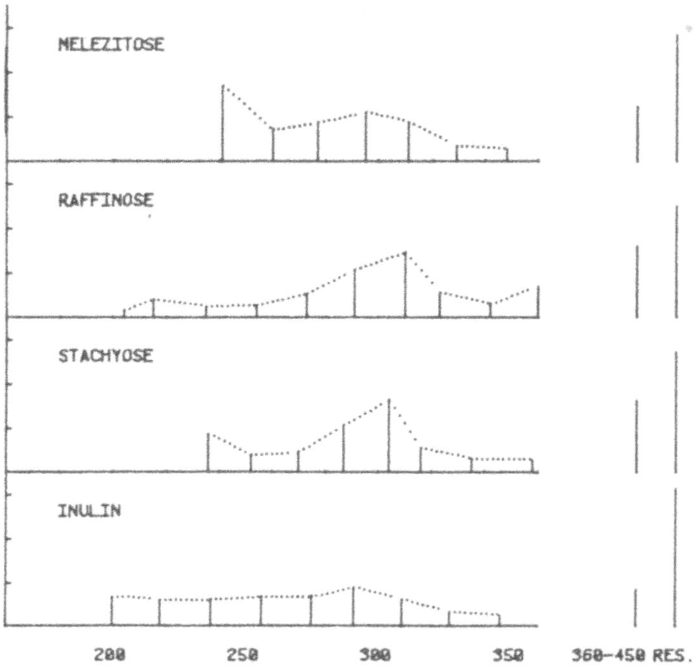

FIG. 10. Comparison of pyrolytic pattern of various saccharides based on fructose.

The influence of fructose on pyrolytic stability is evident from Fig. 9 when comparing glucose and fructose with trehalose and sucrose. In the case of these disaccharides, the bonding is through the two glucoside hydroxyl groups. The fructose-containing sucrose starts to pyrolyse almost 50 °C earlier. In Fig. 10, various fructose based saccharides are compared and similar conclusions can be made. Both melezitose and raffinose are trisaccharides; the fructose is surrounded by two glucose molecules in melezitose, while in raffinose it is at the end of the chain, thus more readily available for possible reactions. Inulin, a polyfructose, decomposes at still lower temperature. In spite of the differences in the bonding of these examples, it may be concluded that the presence of fructose in a carbohydrate initiates pyrolysis at lower temperatures.

While the polymerization of fructose appears to speed up its pyrolysis, an opposite effect can be observed in the decomposition of glucose as shown in Figs 11 and 12. Both amylose and cellulose start the decomposition at much higher temperatures than the basic glucose component, but the strength of the beta-coupling is quite evident here too. No

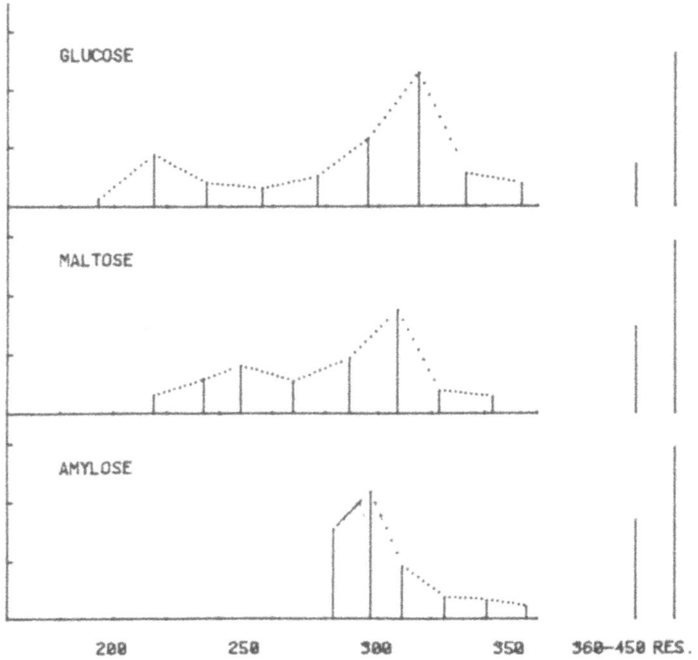

FIG. 11. Comparison of pyrolytic pattern of alpha-4-coupled glucose based saccharides.

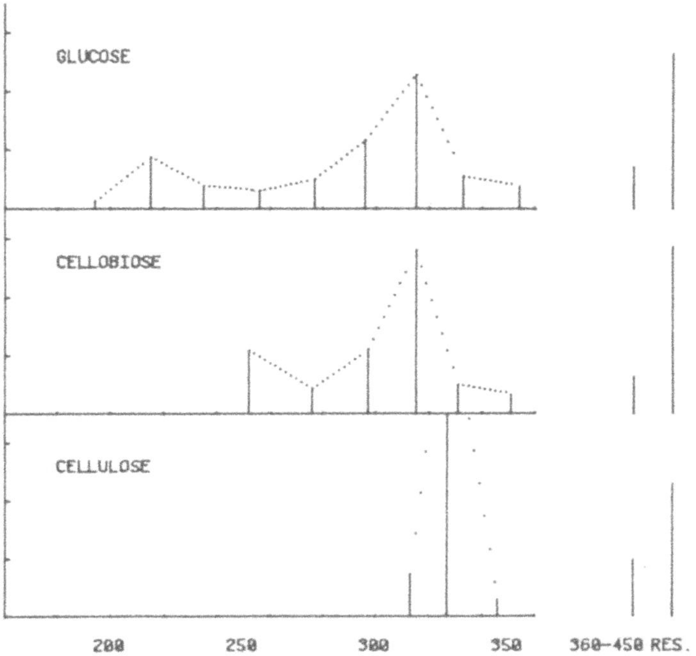

FIG. 12. Comparison of pyrolytic pattern of beta-4-coupled glucose based saccharides.

information is available on the stability of compounds with medium degree of polymerization. Especially, it would be interesting in the case of the cellulose to see whether the high stability is present already at oligomers. The strengthening is quite visible in the case of the cellobiose.

Only scattered data are available for the influence of other factors, but one may venture to make interesting speculations from the available results. Figures 13 and 14 contrast the effect of alpha- and beta-couplings. It appears from this collection, too, that beta-couplings increase the pyrolytic stability of disaccharides regardless of the components. The data for gentibiose (1–6 bonding) in Fig. 14, however, indicate that other factors could counteract this effect. Such an action can be observed in Figs 15 and 16 where the pyrolytic pattern of 1–4 coupled and 1–6 coupled carbohydrates are shown, respectively. Most carbohydrates with 1–4 bondings appear to be more stable than those with 1–6 couplings.

The effect of various factors discussed so far is limited to the thermal decomposition of carbohydrates without any additives. The situation changes and becomes much more complicated, requiring a more detailed

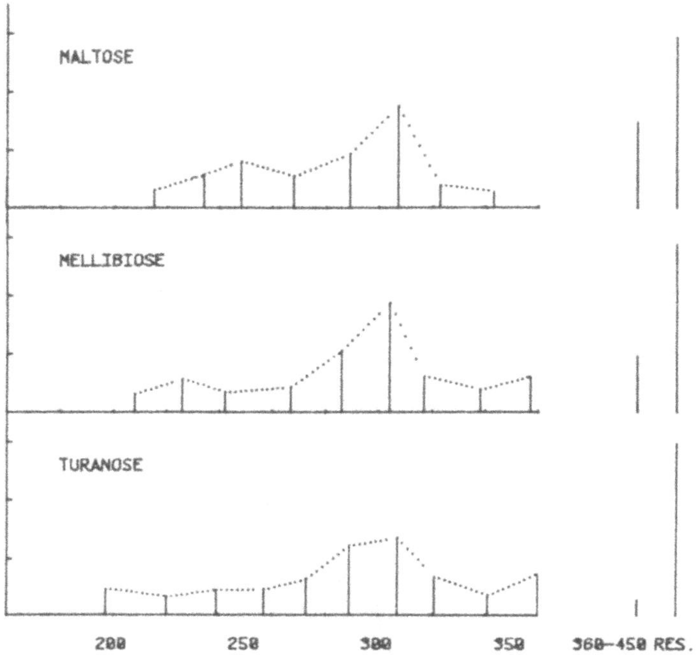

FIG. 13. Comparison of pyrolytic pattern of alpha-coupled disaccharides.

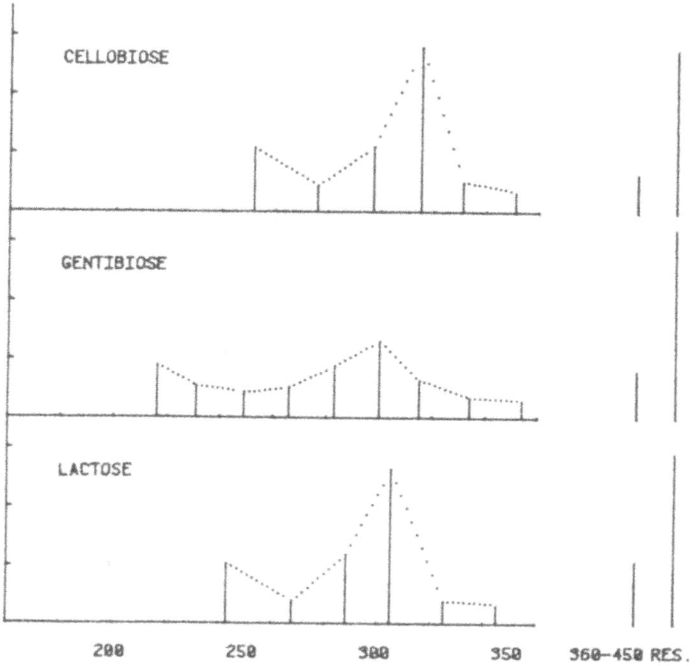

FIG. 14. Comparison of pyrolytic pattern of beta-coupled disaccharides.

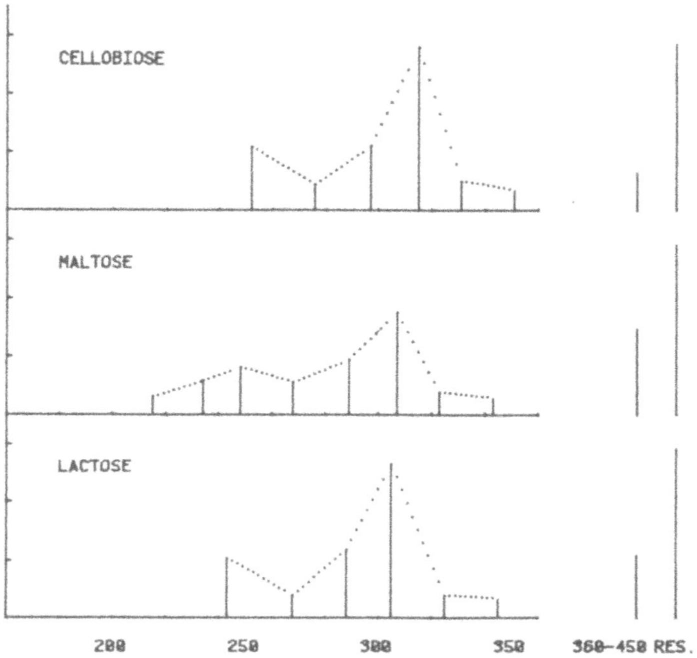

FIG. 15. Comparison of pyrolytic pattern of 1–4-coupled disaccharides.

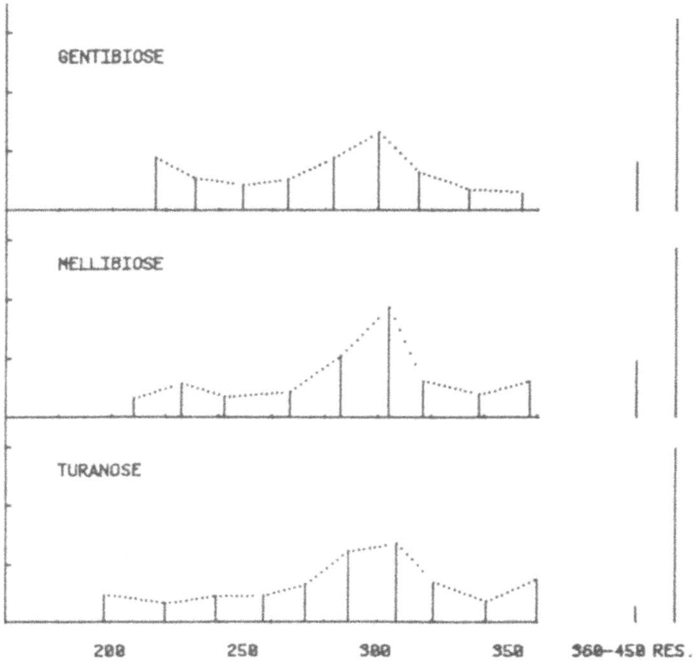

FIG. 16. Comparison of pyrolytic pattern of 1–6-coupled disaccharides.

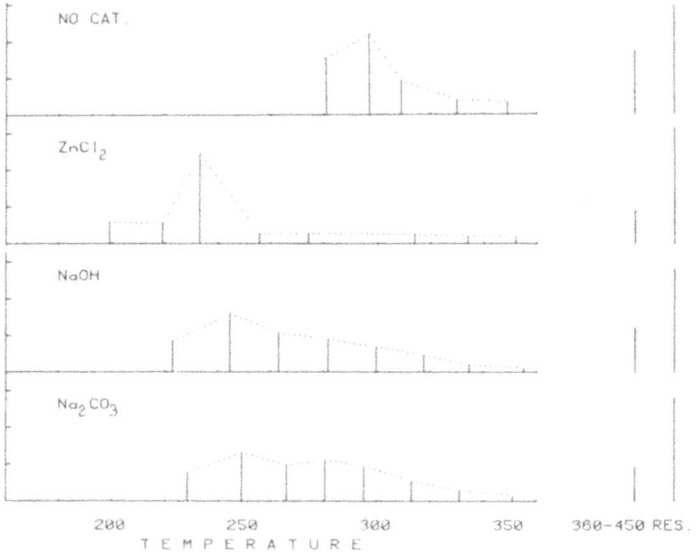

FIG. 17. Comparison of pyrolytic pattern of amylose with various catalysts.

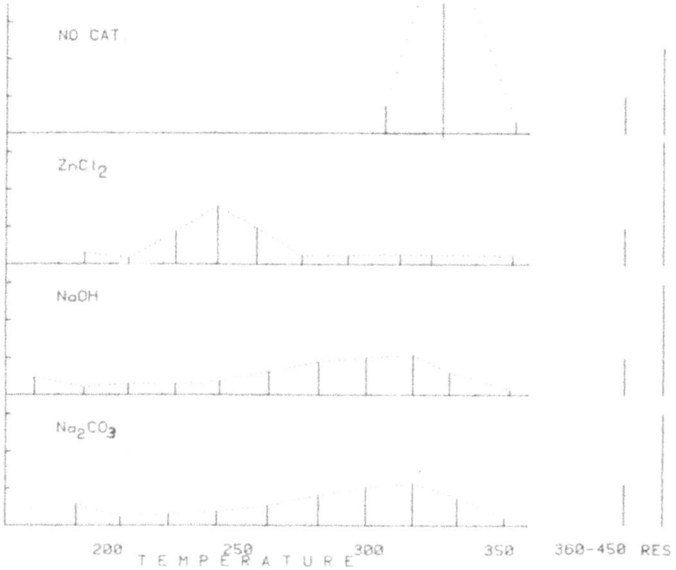

FIG. 18. Comparison of pyrolytic pattern of cellulose with various catalysts.

study, when the pyrolysis is carried out in the presence of some inorganic catalysts. A previous study on cellulose indicated that not only strong acidic and basic catalysts changed the decomposition pattern, but more or less neutral additives, e.g. LiCl or $MgCl_2$, also influenced the pyrolysis.[4] Figures 17 and 18 show the decomposition patterns of amylose and cellulose in the presence of various catalysts. The effect of $ZnCl_2$ appears to be similar on both, that is the decomposition starts at a lower temperature. However, when basic catalysts are used, it is the beta-bonding which exhibits lower stability. In the presence of NaOH and Na_2CO_3, cellulose starts to decompose earlier than amylose.

These preliminary catalytic studies raised an interesting observation seemingly in contradiction with some flame retardant studies. $ZnCl_2$ has a flame retardant action on cotton—a relatively high purity of cellulose—but this study indicates that $ZnCl_3$ catalyzes pyrolysis. This can be explained by the larger amount of residue formed when $ZnCl_2$ is used. The wide range of possible reactions in the pyrolysis of a carbohydrate may be bracketed between two extreme possibilities.

(1) Dehydration

$$C_6H_{12}O_6 \dashrightarrow 6C + 6H_2O \text{ (40\% residue)}$$

(2) Rearrangement

$$C_6H_{12}O_6 \dashrightarrow 3CH_3COOH \text{ (or } 3CH_4 + 3CO_2\text{)}$$

In the case of a high amount of residue, as with $ZnCl_2$, the volatile material is rich in water which does not support combustion. Therefore, even though $ZnCl_2$ initiates pyrolysis at lower temperature, it still could act as a flame retardant. To utilize carbohydrate based biomass as an alternative energy resource or chemical feedstuff through thermal decomposition, those reaction conditions need to be determined, where the pyrolysis starts at relatively low temperatures and the amount of residue is minimum.

5. CONCLUSION

The complex thermal decomposition of carbohydrates can be followed through thermogravimetric measurements in nitrogen environment. The curvefitting of the resulting weight loss rate curves of related carbohydrates can provide an empirical method to compare the decomposition patterns and to pinpoint the effect of structure, catalyst and other reaction variables. The components obtained during the curvefitting procedure do not necessarily represent any actual reaction. However, if sample size and heating rate are kept constant for all carbohydrates investigated, the comparison of the patterns obtained through this mathematical approach may lead to the identification of the actual reactions occurring during pyrolysis. The combination of these curvefitting data with those of other characteristics plotted as a function of temperature can provide this needed information for the utilization of biomass.

REFERENCES

1. Shafizadeh, F. and Lai, V. Z. (1972). *J. Org. Chem.* **37**, 278.
2. Antal, M. H. Jr., Friedman, H. L. and Rogers, F. E. (1980). *Combustion Science and Technology* **21**, 141.
3. Baker, R. R. (1979). *Thermochimica Acta* **28**, 45.
4. Randall, V. G., Masri, S. M. and Pavlath, A. E. (1982). *Proc. of the 7th Int. Conf. on Thermal Analysis*, Kingston, Ontario, Aug. 1982.

26

Thermal Degradation of Rice Straw and its Components

A. E. LIPSKA-QUINN,* S. H. ZERONIAN and K. M. MCGEE

* *Lawrence Livermore National Laboratory, Livermore, California, USA*

and

University of California, Davis, California, USA

1. INTRODUCTION

The disposal of rice crop residues has received increasing attention over a number of years from those both inside and outside the field of agriculture. The residues present a growing and costly management problem as grain yields increased by improved technology are accompanied by increasing straw yields. The traditional method of economical disposal, namely burning, faces growing public opposition and legal regulation, as air pollutant levels from many sources increase and state and local controls are established. To continue utilizing this practice, cleaner burning techniques must be developed. However, to alter the rice straw's burning mode effectively a better understanding must be obtained of its overall thermal degradation mechanism. Although cellulose, one of the components of rice straw, has been studied extensively by numerous researchers,[1-13] rice straw itself has not been given the same consideration.

This study examines the thermal decomposition mechanism of rice straw and its components in nitrogen and in air. The approach consists of:

- —Isolation of major components of rice straw.
- —Thermogravimetric analysis of rice straw and its components at several heating rates in air and nitrogen.
- —Differential scanning calorimetric analysis of rice straw and its components at 10 °C in air and nitrogen.
- —Identification of thermal degradation products of rice straw and its components generated at the main phase of degradation.

2. EXPERIMENTAL PROCEDURE

2.1. Materials

Rice straw from a commercially grown rice (variety 5-6) was used. It was ground in a Wiley Mill to pass through a 20-mesh screen. The purified wood cellulose was an acetate pulp obtained from ITT Rayonier.

2.2. Preparation of Extracted Rice Straw

The straw was extracted for 8 h in a Soxhlet extractor with a 1:2 by volume solution of 95% ethyl alcohol and c.p. benzene. It was next washed with alcohol to remove benzene and then allowed to dry in the ambient atmosphere.

2.3. Preparation of Holocellulose

Holocellulose was prepared from the extracted rice straw by a standard procedure.[14]

2.4. Extraction of Cellulose from Holocellulose

Alpha-cellulose was prepared by extracting holocellulose with a 17·5% NaOH solution for 45 min. The solution was then diluted to 8·3% NaOH and allowed to stand for 1 h. At that time it was filtered, washed with acetic acid, then water until neutral, and air dried. The procedure was essentially that described in ASTM D1103-60.[15]

2.5. Thermal Analysis

The unextracted 20 mesh size rice straw, holocellulose and alpha-cellulose samples were subjected to thermogravimetric (TG) and differential scanning calorimetric (DSC) analysis in dry air and nitrogen. The samples weighed between 3 and 5 mg and were heated at a rate of 5, 10, 20 and 40 °C/min from ambient to 600 °C in the TG and a rate of 10 °C/min in the DSC. All TG and DSC runs were made in triplicate and the results were compared to those obtained from wood cellulose and the extracted rice straw samples. The TG technique was applied to determine the threshold decomposition temperature, the region of active pyrolysis, the rate of pyrolysis, and the char and ash yield. The DSC technique was used to determine the temperatures for endothermic and exothermic pyrolysis reactions.

2.6. Chemical Analysis

For a better understanding of the general decomposition trend of the rice straw, the pyrolyzate generated in the active pyrolysis region was subjected to gas chromatographic/mass spectrometric (GC/MS) analysis utilizing

HP's 5985A GC/MS/data acquisition system. Samples of rice straw and its components weighing 20 mg were used to generate the degradation products in the TG analyzer. They were heated in nitrogen and in air and the resulting products were collected on the TG furnace tube at the predetermined weight loss step. After each collection, the furnace tube was cooled to room temperature and then washed with methylene dichloride in preparation for GC/MMS analyses. Each extractant was concentrated down to 0·5 ml and a 3 μl aliquot was injected on a 25 m SE-30 fused silica capillary column. The column oven was programmed from 70 °C to 220 °C at a rate of 10 °C/min, helium flow rate was maintained at 6 ml/min, and the injection port, mass spec. transfer line and jet separator were set at 250 °C. The components comprising the separated peaks were identified with the help of a computerized spectral library system. In this system, the spectra of unknown compounds are compared to approximately 40 000 compounds stored on a library disk pack, and the 10 best matched compounds are printed out. These are subjected to a spectral compare program where the masses of the sample peak and those of library hit are printed in graphic form for easier comparison of mass peak numbers and intensity.

3. RESULTS AND DISCUSSION

3.1. Composition of Tested Samples

The composition of the tested samples is presented in Table I. The composition of the rice straw is similar to that reported by other workers.

TABLE I
COMPOSITION[a] OF PRODUCTS[16]

Sample	*Moisture regain*[b]	*Ash*	*Lignin*	*Alcohol–benzene extractives*	*Holo-cellulose*	*Alpha-cellulose*
Unextracted rice straw	8·7	15·9	9·7	4·4	66·0	33·7
Holocellulose	9·8	15·3	2·07	—	—	—
Cellulose extracted from rice straw (alpha-cellulose)	5·6	9·3	0·38	—	—	—
Wood cellulose (cellulose standard)	6·3	0·10	0·20	—	—	—

[a] All values expressed as a percentage of oven-dry weight.
[b] Moisture regains are not directly comparable since samples did not have a similar history before determinations were made. Moisture regain values were determined as follows: % = [(conditioned weight − oven-dry weight)/oven-dry weight] × 100.

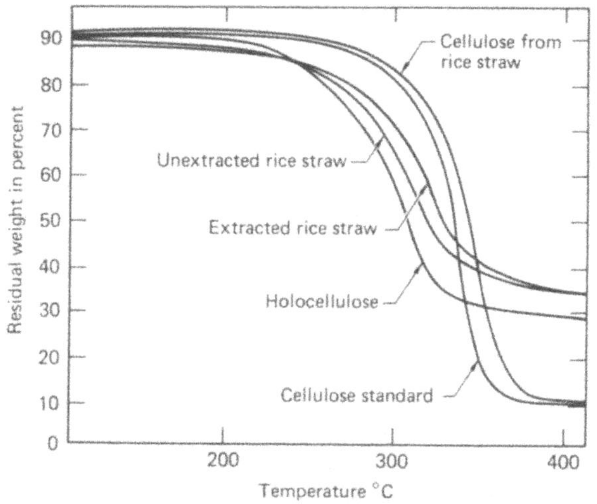

FIG. 1. Thermograms of rice straw and its components heated in nitrogen at 10 °C/min.

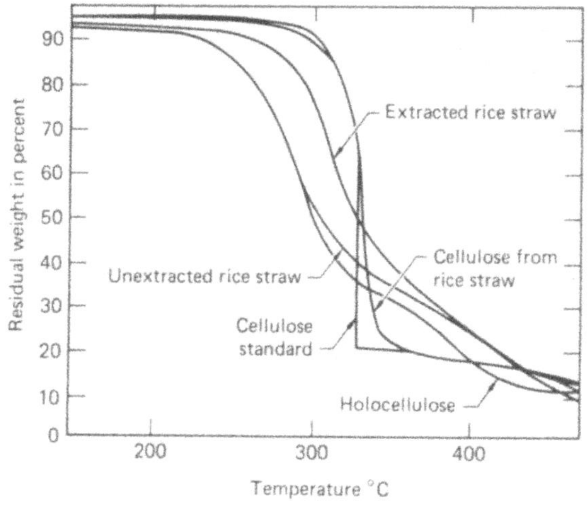

FIG. 2. Thermograms of rice straw and its components heated in air at 10 °C/min.

The high ash content of the straw should be noted. The ash content of the holocellulose and of the cellulose extracted from the rice straw remained high. In contrast, the wood cellulose had a very low ash content.

3.2. Thermogravimetric Analysis

Typical TG curves are presented in Figs 1 and 2. TG was conducted at several different heating rates in both nitrogen and air. Heating rates above 20 °C/min markedly increased the rate of degradation of all samples in both nitrogen and air (Tables II–VI). These high heating rates also markedly

TABLE II
THE EFFECT OF RATE OF HEATING ON THERMAL DEGRADATION OF UNEXTRACTED RICE STRAW

Pyrolysis environment	*Rate of heating (°C/min)*	*Rate of degradation in the active pyrolysis region (%/min)*	*Temp. of initial degradation (°C)*	*Residual weight at 500 °C (percentage)*
N_2	5	4·5	205	33
	10	8	175	33
	20	15	165	34·5
	40	29	140	16·5
Air	5	4	125	13
	10	8	120	14
	20	17	110	13·5
	40	34	110	15

lowered the initial degradation temperature of the rice straw samples and the cellulose samples. It can be deduced from the thermograms in nitrogen that the cellulose samples have the highest threshold temperature for active pyrolysis at 230–265 °C, and the lowest residual weight of 11·5–12·5 % at 500 °C. The unextracted rice straw samples required the lowest threshold temperature for active pyrolysis at 175 °C and gave a residue ranging from 16·5 % to 34·5 % at 500 °C. This high residue is partly due to the presence of silica in the rice straw samples (Table I). The extracted rice straw gave similar residue yields as the unextracted rice straw but the former sample appeared to have a slightly higher threshold temperature (cf. Tables II and III). Holocellulose had a threshold temperature intermediate between that of the rice straw and the cellulose (Table IV). The rate of degradation in the most active pyrolysis region fell in the following order: cellulose standard

TABLE III
THE EFFECT OF RATE OF HEATING ON THERMAL DEGRADATION OF EXTRACTED RICE STRAW

Pyrolysis environment	*Rate of heating (°C/min)*	*Rate of degradation in the active pyrolysis region (%/min)*	*Temp. of initial degradation (°C)*	*Residual weight at 500°C (percentage)*
N_2	5	4·35	215	35
	10	8·5	200	35·5
	20	17·0	180	33·0
	40	27·0	150	35·0
Air	5	4	230	13
	10	8	180	10
	20	17	170	10
	40	34	150	13

and cellulose from rice straw > holocellulose > extracted rice straw and unextracted rice straw.

The trend of the degradation was generally similar when TG was conducted in air but there were some notable differences. For example, the temperature of initial degradation did not appear to be affected by the ambient atmosphere at any heating rate for the holocellulose whereas the temperature of initial degradation for all the other samples appeared to be

TABLE IV
THE EFFECT OF RATE OF HEATING ON THERMAL DEGRADATION OF HOLOCELLULOSE EXTRACTED FROM RICE STRAW

Pyrolysis environment	*Rate of heating (°C/min)*	*Rate of degradation in the active pyrolysis region (%/min)*	*Temp. of initial degradation (°C)*	*Residual weight at 500°C (percentage)*
N_2	5	6	200	27
	10	11	200	30
	20	17	200	27
	40	30	200	30
Air	5	6	220	14
	10	9·5	215	14·5
	20	18	200	14·5
	40	48	200	14

TABLE V

THE EFFECT OF RATE OF HEATING ON THERMAL DEGRADATION OF CELLULOSE EXTRACTED FROM RICE STRAW

Pyrolysis environment	*Rate of heating (°C/min)*	*Rate of degradation in the active pyrolysis region (%/min)*	*Temp. of initial degradation (°C)*	*Residual weight at 500°C (percentage)*
N_2	5	10	265	10
	10	17	250	11·5
	20	30	240	10
	40	44	230	12
Air	5	16	248	6·5
	10	28	240	11·5
	20	49	205	12
	40	92	195	14·5

lower in air than in nitrogen at many heating rates, especially the higher ones. The rice straw and holocellulose samples yielded markedly lower residual weights in air than in nitrogen. The residues of the cellulose samples prepared from rice straw was less affected. Air substantially increased the rate of degradation of the two cellulose samples but the change in the rate of decomposition of the rice straw or holocellulose samples was less marked.

TABLE VI

THE EFFECT OF RATE OF HEATING ON THERMAL DEGRADATION OF STANDARD CELLULOSE PULP

Pyrolysis environment	*Rate of heating (°C/min)*	*Rate of degradation in the active pyrolysis region (%/min)*	*Temp. of initial degradation (°C)*	*Residual weight at 500°C (percentage)*
N_2	5	10	240	11·5
	10	22	240	12·5
	20	23	240	10·0
	40	63	240	9·0
Air	5	30	240	10·0
	10	43·0	240	10·0
	20	42·5	195	0
	40	100	135	0

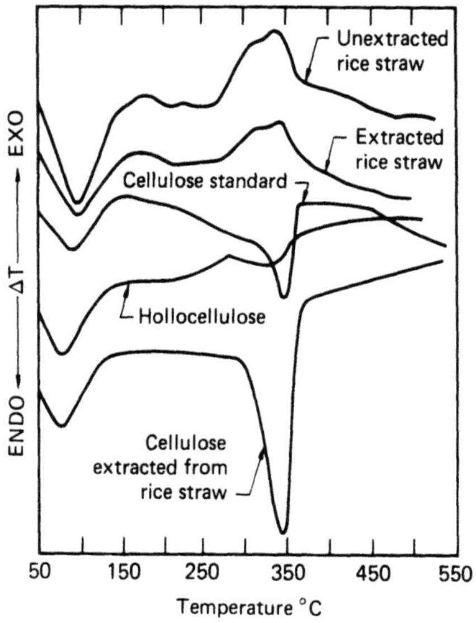

FIG. 3. DSC of rice straw and its components heated in nitrogen at 10 °C/min.

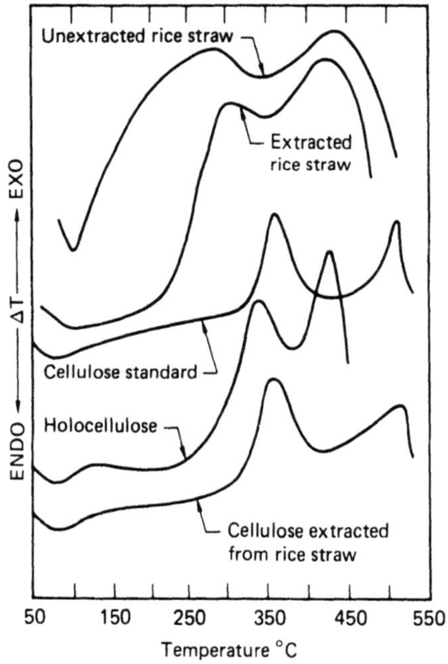

FIG. 4. DSC of rice straw and its components heated in air at 10 °C/min.

3.3. Differential Scanning Calorimetry

DSC curves of rice straw and its components as well as wood cellulose were obtained in nitrogen and in air at a heating rate of 10 °C/min. The curves are presented in Figs 3 and 4. Examining the curves for cellulose first, it can be observed that in nitrogen (Fig. 3) a water loss endotherm at 75 °C was found on the DSC curves of both celluloses. In contrast to previously reported findings,[17] no noticeable endotherm or exotherm occurred at 170 °C. A decomposition endotherm occurred at 340 °C which is attributed by some authors mainly to depolymerization and volatilization reactions.[18,19] In air, the DSC curves of both celluloses revealed a water loss endotherm and two exotherms. The first exotherm which began at 310 °C has been attributed to oxidation of the volatile products of decomposition and the second peak which began at 450 °C for standard cellulose and at 425 °C for cellulose extracted from rice straw is thought to be due to oxidation of the residual char.[17]

The curve for holocellulose in nitrogen showed a moisture endotherm at 75 °C and an indication of an exotherm at 265 °C, followed by an endotherm between 310 and 360 °C. It can be discerned also from the TG curves (Fig. 1) that the major decomposition of the holocellulose occurs between 265 and 360 °C. In air, holocellulose shows a water loss endotherm and two exotherms; the first exotherm begins at 230 °C and the second at 375 °C. Again, the first exotherm is thought to be due to oxidation of the volatiles generated during the active pyrolysis and the second exotherm due to oxidation of the char residue. A large exothermic peak with a shoulder is present on the DSC curves for both the extracted and unextracted rice straws in the presence of nitrogen, indicating that degradation occurs via two exothermic processes in the range of 275 and 350 °C. This range corresponds to the active pyrolysis range in the TGA studies (Fig. 2). In air the temperature at which the initial exothermic peak occurs is lowered but that of the second exothermic peak is raised.

3.4. Gas Chromatography/Mass Spectrometry

Results of the GC/MS analyses of the various pyrolyzates are presented in Figs 5 through 8. Peak identification tables are also included (Tables VII through X). The chromatograms include only those products which were condensed on the walls of the TG glass tube; therefore the lighter molecular weight gaseous products are excluded. In addition, products retained on the SE-30 fused silica capillary column are excluded from the identification tables.

Comparison of the pyrolyzates obtained in nitrogen from rice straw,

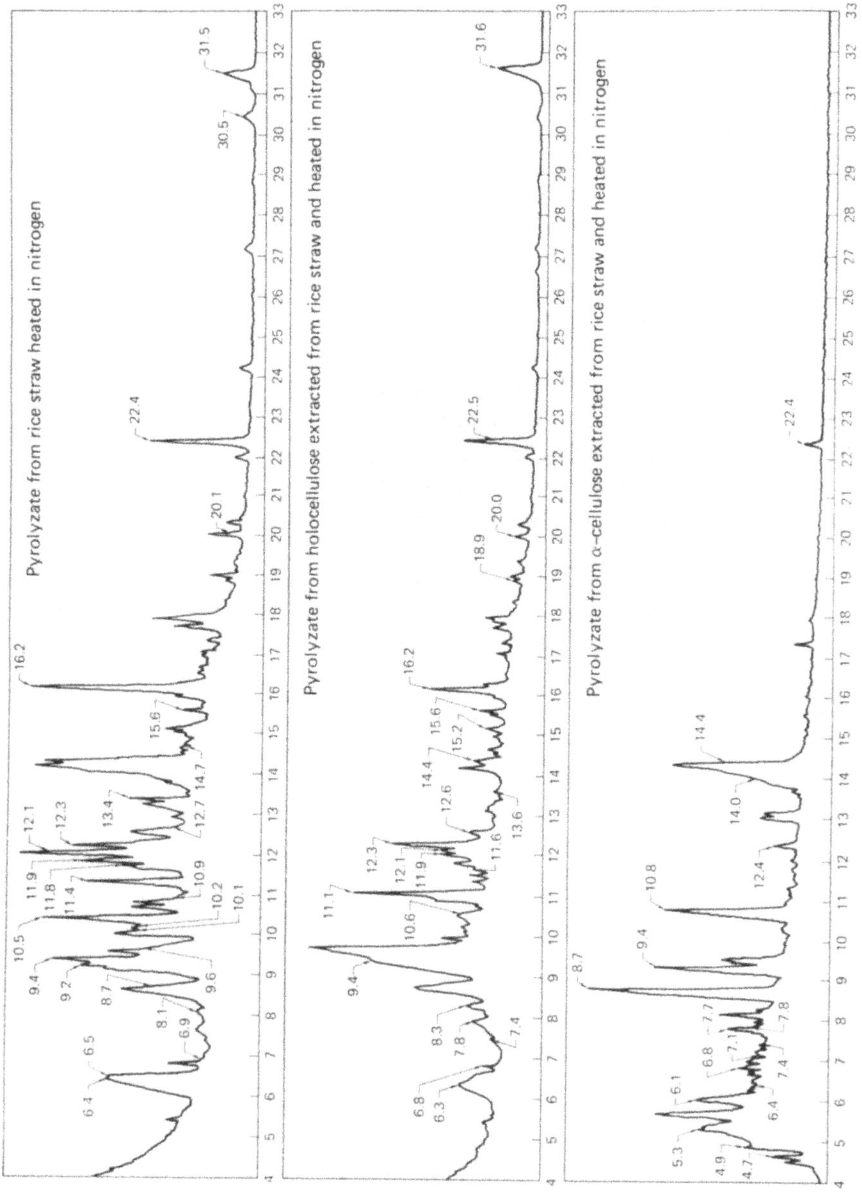

FIG. 5. Comparison of thermal degradation products generated in nitrogen from pyrolyzing rice straw and its components.

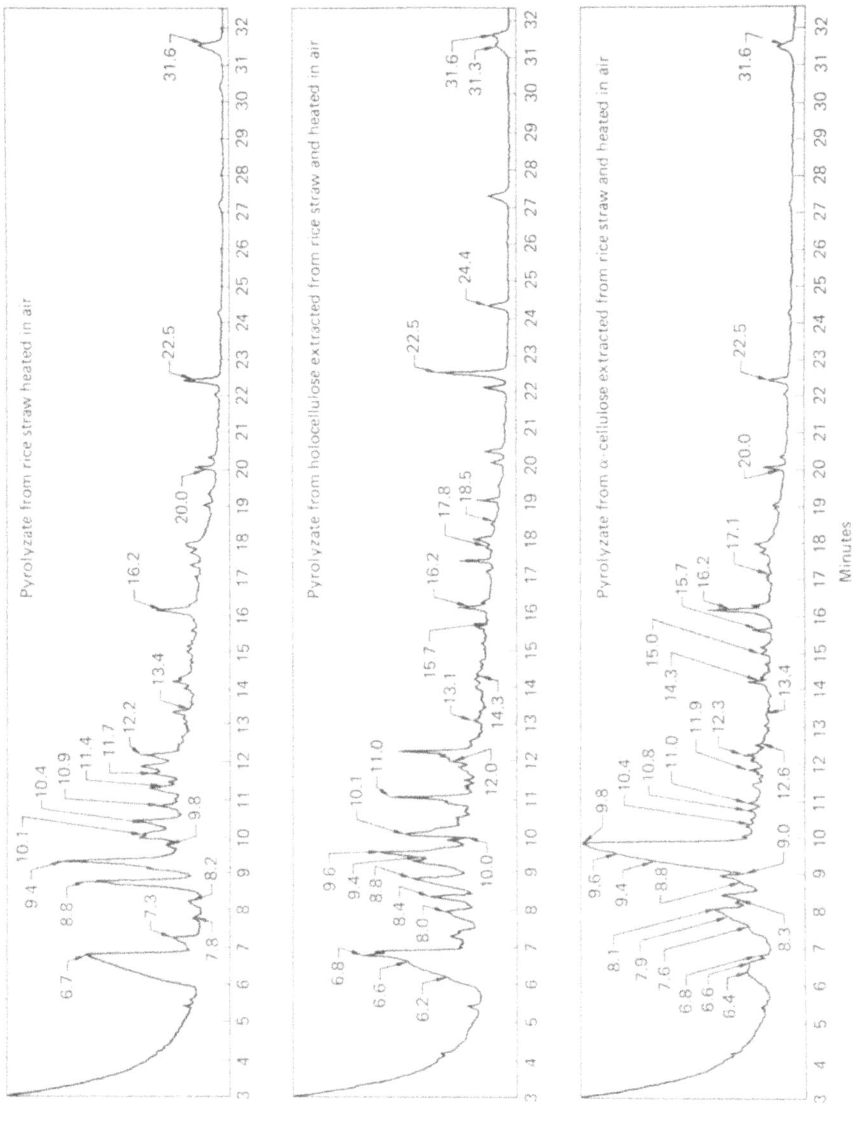

FIG. 6. Comparison of thermal degradation products generated in air from pyrolyzing rice straw and its components.

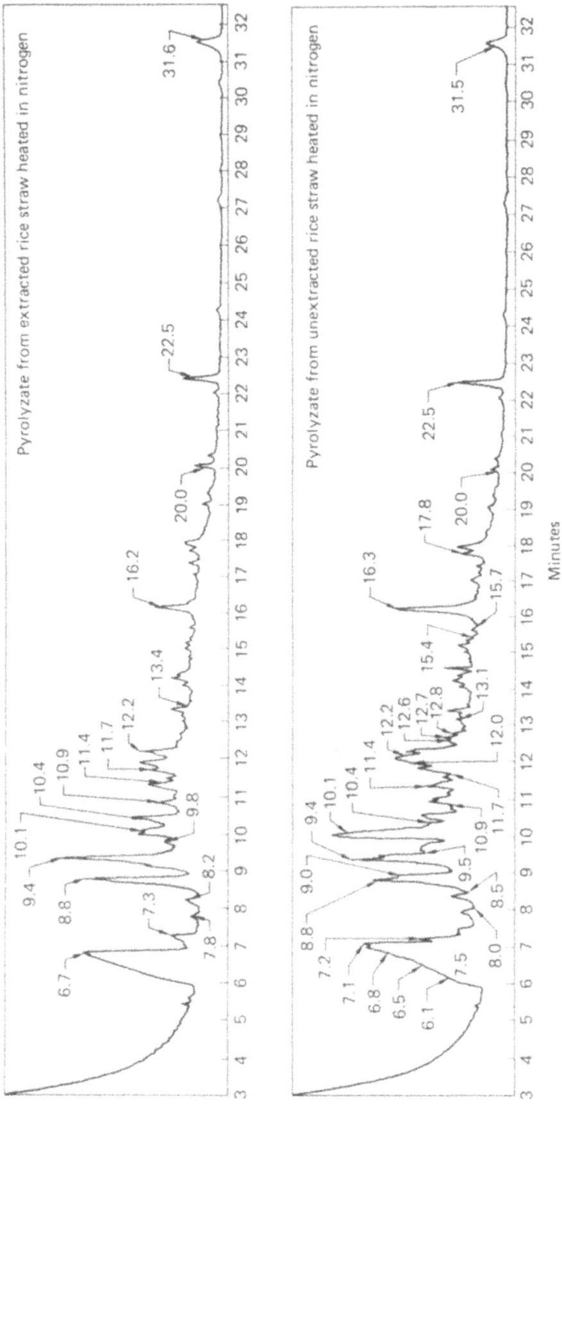

FIG. 7. Comparison of thermal degradation products generated in nitrogen from pyrolyzing extracted and unextracted rice straw.

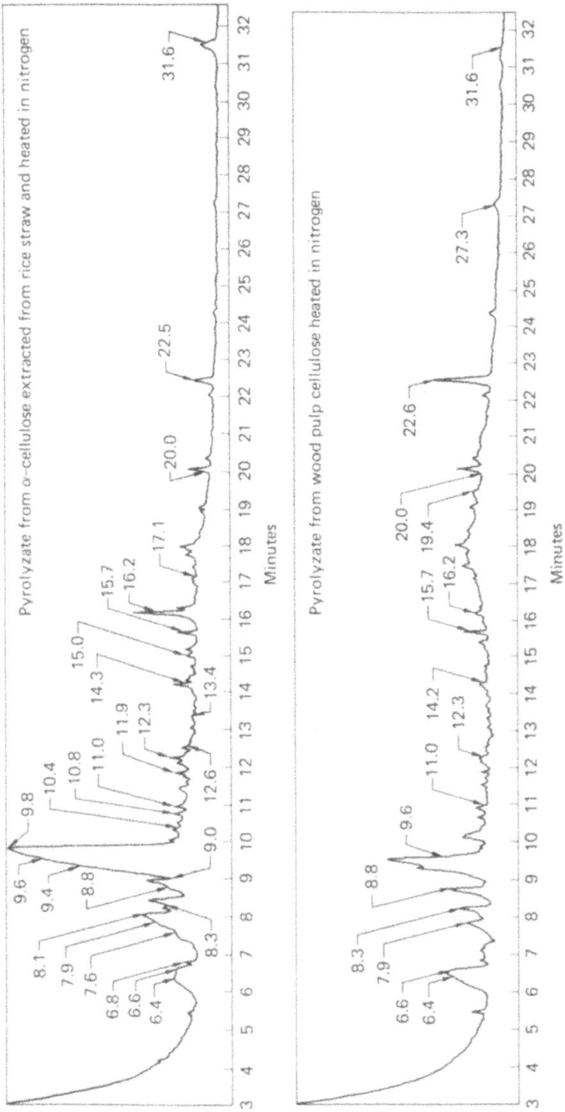

FIG. 8. Comparison of thermal degradation products generated in nitrogen from pyrolzing α-cellulose extracted from rice straw and wood pulp cellulose.

TABLE VII

PEAK IDENTIFICATION FOR FIG. 5

Retention time (min)	
4·7	3-oxo-4,6-heptadiyne
4·9	glutaric acid
5·3	3-methyl-cis-2-hexene
6·1	2,3,4-trimethylpentan-3-ol
6·3	2-methyl-3-hydroxypyrone
6·4	cyclohexanone
6·5	octane-2,7-dione
6·8	3-buten-2-ol
7·1	2-methyl-3-heptanol
7·4	4-methyl-4-hepten-3-one
7·7	2,5-dimethylhexene-1
7·8	2-methyl-3-ethylhexane
8·1	(S)-(−)-2-isopropyl gamma butyrolactone
8·3	2-propyl-4-methylfuran
8·7	3-ethyl-1-pentene
9·2	2,6-dimethoxy phenyl acetate
9·4	2,3-dimethyl-1-butene
9·6	3-ethoxy-3,4-dimethyl heptyne
10·1	pyrocatechol
10·2	2-methyl-1-thiopropyl benzene
10·5	vanillin
10·6	3-hydroxy-4-methoxybenzaldehyde
10·9	9,10-phenanthraquinone
11·1	1,4-dimethoxynaphthalene
11·4	4-hydroxy-3-methoxyacetophenone
11·6	4-ethoxycarbonyl benzo (c) cinnoline
11·8	1-heptadecanyl-3-methoxy benzene
11·9	m-hydroxybenzaldehyde
12·1	eugenol
12·3	2,3,5,6-tetramethyl benzoquinone
12·4	methyl-15,16-methylene tetracosanoate
12·6	hexacosanol-1
13·4	syringaldehyde
13·6	bis(1-ethylhexyl)azelate
14·0	nonanoic acid
14·4	4-allyl-2,6-dimethoxyphenol
14·7	unidentified
15·2	2-linolenyl-1,3-diacetin
15·6	unidentified
16·2	palmitic acid
18·9	2-(2-octyldecyl)cis bicyclo (0,3,3) octane
20·0	unidentified
22·4	unidentified
31·5	farnesol

TABLE VIII

PEAK IDENTIFICATION FOR FIG. 6

Retention time (min)	
6·2	2-propyl-5-acetoxy-5-methyltetrahydropyran
6·4	cyclohexanone
6·6	2,2-dimethyl-1-hexanol
6·7	sec-butylamine
6·8	3-buten-2-ol
7·3	methylthiofuran
7·6	hexanal
7·8	2-methyl-3-ethylhexane
7·9	decylaldehyde
8·0	4,5-dimethylnonane
8·2	2-propyl-5-acetoxy-5-methyltetrahydropyran
8·3	vinyl isopropyl ether
8·4	2-propyl-4-methylfuran
8·8	3-ethyl-1-pentene
9·0	3-methyl-cis-2-hexene
9·4	2,3-dimethyl-1-butene
9·6	4-methyl-cis-pent-2-ene
9·8	4-cyclohexyl-butan-2-one
10·0	1-tetradecanol
10·1	pyrocatechol
10·4	vanillin
10·8	3-(2,6,6-trimethyl-1-cyclohexene-1-yl)2-propen-1-ol
10·9	phenanthraquinone
11·0	1,4-dimethoxynaphthalene
11·4	4-hydroxy-3-methoxyacetophenone
11·7	1-(4-hydroxy-3-methoxy phenyl) propan-2-one
11·9	hydroxybenzaldehyde
12·0	N-hexadecyl acetate
12·2	eugenol
12·3	6-benzo-8,9-dimethylbicyclo(3,2,2)-non-6,8-diene
12·6	hexacosanol-1
13·4	4-allyl-2,6-dimethoxyphenol and phenol
14·3	n-butyl acetoacetate
15·0	glucitol-1,2,3,4,5,6-hexaacetate
15·7	unidentified
16·2	palmitic acid
17·1	3-ethyl-5-(2-ethylbutyloctadecane)
17·8	tri-(3,5,5-trimethylhexanoyl glycerol)
18·5	1-dotriacontanol
20·0	unidentified
22·5	unidentified
27·3	N-heptacosane
31·6	'farnesol

TABLE IX
PEAK IDENTIFICATION FOR FIG. 7

Retention time (min)	
6·1	2,6-dimethylpyran-4-one
6·5	octane-2,7-dione
6·7	sec-butylamine
6·8	3-buten-2-ol
7·1	3-methyl-4-heptanone
7·2	4,5-dimethyl-3-hydroxy-hept-2-ene
7·3	methylthiofuran
7·5	N-nitroso-4-methyl-piperidine
7·8	2-methyl-3-ethylhexane
8·0	4,5-dimethylnonane
8·2	2-propyl-5-acetoxy-5-methyltetrahydropyran
8·5	trans-2-heptanol
8·8	3-ethyl-1-pentene
9·0	3-methyl-cis-2-hexene
9·4	2,3-dimethyl-1-butene
9·5	3,7,7-trimethyl-bicyclo-(3,1,1)-2-heptanol
9·8	4-cyclohexylbutan-2-one
10·1	pyrocatechol
10·4	vanillin
10·9	phenanthraquinone
11·4	4-hydroxy-3-methoxyacetophenone
11·7	1-(4-hydroxy-3-methoxyphenyl)-propan-2-one
12·0	3,4-dihydroxycinnamic acid
12·2	eugenol
12·6	4-(2,6-trimethylcyclohex-1-en-1-yl)pentan-2-one
12·7	hexacosanol-1
12·8	4,8,8-trimethyl-3-oxa-bicyclo(5,4,0)undec-1,7-en
13·1	syringaldehyde
15·4	methyl palmitate
15·7	unidentified
16·3	palmitic acid
17·8	3-ethyl-5-(2-ethylbutyloctanol)
17·8	tri-(3,5,5-trimethylhexanoyl glycerol)
20·0	unidentified
22·5	unidentified
31·5	farnesol

TABLE X

PEAK ASSIGNMENT FOR FIG. 8

6·4	cyclohexanone
6·6	2,2-dimethyl-1-hexanol
6·8	3-buten-2-ol
7·9	decylaldehyde
8·1	4,5-dimethylnonane
8·3	vinyl isopropyl ether
8·8	3-ethyl-1-pentene
9·0	3-methyl-cis-2-hexene
9·4	2,3-dimethyl-1-butene
9·6	4-methyl-cis-pent-2-ene
9·8	4-cyclohexylbutan-2-one
10·4	alpha-humulene
10·8	3-(2,6,6-trimethyl-1-cyclohexene-1-yl)2-propen-1-ol
11·0	1,4-dimethylnaphthalene
11·9	hydroxybenzaldehyde
12·3	6-benzo-8,9-dimethylbicyclo(3,2,2)-non-6,9-diene
12·6	hexacosanol
13·4	2-phenyl-2,3-dihydroindene
14·3	n-butylacetoacetate
15·0	glucitol-1,2,3,4,5,6-hexaacetate
15·7	di-N-butyl-phthalate
16·2	palmitic acid
17·1	3-ethyl-5-(2-ethylbutyloctadecane)
19·5	unidentified
20·0	unidentified
22·5	unidentified
27·3	unidentified
31·5	farnesol

holocellulose and cellulose extracted from rice straw can be made from Fig. 5. The cellulose pyrolyzate consists of hydrocarbons, aldehydes, ketones, ethers, esters, acids, alcohols and their aromatic counterparts. Holocellulose (cellulose and hemicellulose) forms condensables which are composed mainly of higher molecular weight hydrocarbons, acids, esters, alcohols, pyran and furan derivatives. It has been suggested by some researchers[20] that upon pyrolysis hemicellulose forms a substantial amount of acetic acid; however, we did not detect it in our mixture at this time. In addition to compounds comprising the pyrolyzates of cellulose and holocellulose, thermally degraded rice straw forms syringaldehyde, eugenol and pyrocatechol. These are believed to be formed from the pyrolyzing lignin present in the rice straw.[21]

Comparison of the thermal degradation products generated in air from pyrolyzing rice straw and its components can be made from Fig. 6. The cellulose pyrolyzate generated in air renders more oxygen containing compounds than the pyrolyzate formed in nitrogen. The pyran derivatives seen in the pyrolyzate generated from thermally degrading holocellulose in nitrogen were not detected in the pyrolyzate mixture generated in air. The pyrolyzate from thermally degrading rice straw in air again renders an increased variety of oxygenated compounds over that found in the pyrolyzate from nitrogen pyrolyzed rice straw. In addition, eugenol, syringaldehyde and pyrocatechol were also detected in the mixture.

The chromatograms of thermal degradation products of extracted and unextracted rice straw pyrolyzed in nitrogen are presented in Fig. 7. The differences observed are in the quantity rather than in the quality of compounds formed. The pyrolyzate mixture from the unextracted rice straw shows larger quantities of all the components present in the pyrolyzate from extracted rice straw. However, this does not exclude the possibility that there are differences in the compounds comprising the lighter molecular weight fraction, not condensed on the walls of the TG glass tube and therefore not detected in our analysis.

The chromatograms of pyrolyzates of wood pulp cellulose and cellulose extracted from rice straw are compared in Fig. 8. Again there could be differences in the lighter components not condensed out on the walls of the TG glass tube or components not separated on the SE-30 column.

4. CONCLUSIONS

The preliminary findings of this study indicate the following:

(1) The mode of decomposition of rice straw depends on the pyrolysis environment, the rate of heating and the pretreatment of the sample before pyrolysis.

(2) Rice straw samples appear to have the lowest threshold temperature for active pyrolysis while cellulose samples have the highest.

(3) Once active pyrolysis commences, unextracted rice straw degrades at the lowest rate and leaves the highest residue at 500 °C, while cellulose degrades at the highest rate and leaves the lowest residue at 500 °C.

(4) All samples show lower residues in air than in nitrogen.

(5) High heating rates (20 °C/min and above) markedly increase the

rates of degradation in the active pyrolysis region of all samples with the exception of the wood cellulose. In that case the rate of degradation at heating rates of 10 °C/min and 20 °/min are similar.

(6) In the active pyrolysis temperature range in nitrogen both cellulose samples degrade through an endothermic process, while in air degradation occurs via two exothermic processes. Holocellulose degrades via an endothermic process in nitrogen and via two exothermic processes in air. Rice straw degrades via two exothermic processes both in nitrogen and in air.

REFERENCES

1. Shafizadeh, F. (1968). *Adv. Carbohydr. Chem.* **23**, 419.
2. Shafizadeh, F. (1971). *J. Polym. Sci.* **C21**.
3. Shafizadeh, F. (1975). *Appl. Polym. Symp.* **28**, 153.
4. Schwenker, R. F. and Beck, L. R. (1963). Fourth Cellulose Conference (*J. Polym. Sci.* **C2**), R. H. Marchessault (Ed.). Interscience, New York, p. 331.
5. Schwenker, R. F. and Pacsu, E. (1957). *Chem. Eng. Data* **2**, 83.
6. Madorsky, S. L., Hart, V. E. and Strauss, S. (1956). *J. Res. Nat. Bur. Stand.* **56**, 343.
7. Tang, M. M. and Neil, W. K. (1964). Thermal Analysis of High Polymers (*J. Polym. Sci.* **C6**), B. Ke (Ed.). Interscience, New York, p. 65.
8. Halpern, Y. and Patai, S. (1969). *Israel J. Chem.* **7**, 691.
9. Dollimore, D. and Holt, B. (1973). *J. of Polym. Sci: Polym. Physics Edition II* 1703.
10. Lipska, A. E. and Wodley, F. (1969). *J. Appl. Polym. Sci.* **13**, 851.
11. Fung, D. P. C., Tsuchiya, Y. and Sumi, K. (1972). *Wood Sci.* **5**, 38.
12. Zeronian, S. H. and Cabradilla, K. E. (1976). In *Thermal Uses and Properties of Carbohydrates and Lignins*, F. Shafizadeh, K. V. Sarkanen and D. A. Tillman (Eds). Academic Press, New York, pp. 73–96.
13. Zeronian, S. H. and Cabradilla, K. E. (1978). In *Modified Cellulosics*, R. M. Rowell and R. A. Young (Eds). Academic Press, New York, pp. 321–40.
14. Whistler, R. L. and BeMiller, J. N. (1963). *Methods in Carbohydrate Chemistry, III, Cellulose*, R. L. Whistler (Ed.). Academic Press, New York.
15. American Society for Testing and Materials, *1977 Annual Book of ASTM Standards*. Part 22: Wood; Adhesives. American Society for Testing and Materials, Philadelphia.
16. McGee, K. M. (1980). M. S. Thesis, University of California, Davis.
17. Ramiah, M. V. (1970). *J. of Appl. Polym. Sci.* **14**, 1323.
18. Shafizadeh, F. and Fu, Y. L. (1973). *Carbohydrate Res.* **29**, 113.
19. Shafizadeh, F., Lai, Y. Z. and McIntyre, C. R. (1978). *J. Appl. Polym. Sci.* **22**, 1183.
20. Soltes, E. J. and Elder, S. J. (1978). *Organic Chemicals from Biomass*, I. S. Goldstein (Ed.). CKC Press, Inc., Boca Raton, Florida, Chapter 6.
21. Fletcher, T. L. and Harris, E. E. (1952). *Tappi* **35**, 536.

27

Research on the Gasification of Biomass at the University of Aston in Birmingham

B. W. Hatt, P. J. Iredale, G. A. Irlam, R. N. Shand, H. H. Sheena and E. L. Smith

Department of Chemical Engineering, University of Aston in Birmingham, Birmingham, UK

1. BACKGROUND

The production of mixtures of carbon oxides and hydrogen from biomass is attractive since such 'synthesis gas' can be converted subsequently to a range of relatively high-grade fuels, such as methanol and methane, by established chemical process routes. Over the last decade, much research on the thermal processing of different types of biomass has been carried out and reports in which this work was summarized and reviewed have been prepared for the Energy Technology Support Unit of the UK Department of Energy by Ader Associates and staff at the University of Aston.[1] Following publication of these reports a programme of experimental investigations and design studies was initiated: the overall objectives of this programme were (a) to determine for specific biomass feedstocks the optimum conditions for their gasification to fuels and (b) to select and design reactor systems for assessment in a 40 kg/h test facility designed, constructed and operated by Foster Wheeler Power Products Ltd, UK. This programme has been running since the beginning of 1980 and the following sections of the paper describe some of the concepts and methods used by the Aston research team.

2. OVERALL MASS AND ENERGY BALANCES

Given a particular biomass feedstock and a range of potential fuel products, it is important to be able to predict maximum mass and energy yields. The coupling of mass and energy balances is consequently unavoidable but by using what we call the concept of 'oxygen equivalents' it is possible to make a rapid assessment of any proposal with the minimum of computational effort.

Consider, by way of example, the production of methanol from wood, for which the simplified formula $C_6H_{10}O_5$ will be used. Complete combustion of $C_6H_{10}O_5$ to CO_2 and H_2O requires 12 O-atoms and so this material has an oxygen equivalent (OE) of 12: CO_2 and H_2O cannot be further oxidized and so have OE values of zero. The desired product, CH_3OH, when fully oxidized to CO_2 and H_2O, will have reacted with 3 O-atoms and, hence, has an OE value of 3.

It is now possible to use the OE concept in an approximate, but useful, way as energy currency. If in converting $C_6H_{10}O_5$ to CH_3OH we seek to retain as much fuel energy as possible, then 12 OEs are required in the product stream: in other words 4 molecules of CH_3OH are desired. This can be summarized as follows:

	Feedstock		Desired product
	$C_6H_{10}O_5$	$\longrightarrow$	$4CH_3OH$
OE value	12		4×3

Atomic balancing must next be considered and here it must be remembered that CO_2 and H_2O can be either added or removed from the system without 'charge'—they are 'free' since they have a zero OE value. Now, in the above example, consideration of the C and H balance reveals that we need to remove 2C atoms and to add 6H atoms: H atoms can be conveniently supplied in the form of $3H_2O$ and this then makes it possible to remove $2CO_2$ molecules. The results of the atomic balancing are shown in diagrammatic form below:

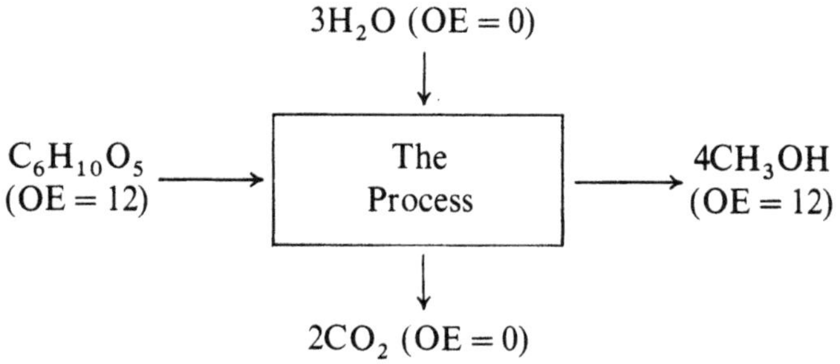

Balanced equation: $C_6H_{10}O_5 + 3H_2O = 4CH_3OH + 2CO_2$

The same type of analysis can readily be carried out for any conceivable

TABLE I

PROCESS ANALYSIS USING THE OXYGEN EQUIVALENT (OE) CONCEPT

Product Formula	*OE value*	*Product yield* Moles per mole of biomass	kg per kg of biomass	*Heat of combustion of product derived from 1 mole of biomass (kJ)*
C	2	6	0·44	−2 360
CH_4	4	3	0·30	−2 670
$(-CH_2-)_n$	3	4	0·35	−2 630
CH_3OH	3	4	0·79	−2 910
C_2H_5OH	6	2	0·57	−2 730
$C_6H_{10}O_5$ (biomass)	12	1	1·0	−2 810

Bases: 1 mole of biomass has the formula $C_6H_{10}O_5$; CO_2 and H_2O can be either added or removed during the biomass conversion process; heats of combustion based on data for 25 °C, 0.1 MPa with H_2O in liquid state.

production scheme and interesting ideas can be generated for increasing the OE value of the product relative to the feedstock. Results using this procedure for some common fuels and chemicals are given in Table I.

A more detailed analysis of gasification processes can be made using the approach of Wei.[2] He rightly points out that, for the system $C–H_2O–O_2$, overall changes and product composition are constrained by stoichiometric considerations. Further, if it is assumed that the gasification step must be thermally balanced, then it is possible to define a feasible operating range.

3. EQUILIBRIUM THERMODYNAMIC STUDIES

The methods described in Section 2 cannot be used to estimate the product gas composition from a biomass gasifier. For this purpose use must be made of thermodynamic studies as exemplified by the work of Gumz,[3] Baron *et al.*[4] and Desrosiers.[5]

At operating temperatures above about 600 °C, gas mixtures leaving many types of biomass gasifier will be close to equilibrium. Consequently thermodynamic studies can be used to provide data for assessing the performance of prototype and commercial-scale reactors.[6] The results of equilibrium calculations can also be used to guide the choice of parameter values when planning experimental programmes.

We have used the generally accepted assumption that only five gases—CO, CO_2, H_2, H_2O and CH_4—are present in significant quantities under biomass gasification conditions. In the presence of solid C and without O_2, the reactions usually considered are:

$$C + H_2O \rightleftharpoons CO + H_2$$
$$C + 2H_2 \rightleftharpoons CH_4$$
$$CO + H_2O \rightleftharpoons CO_2 + H_2$$

Equilibrium constants can be readily calculated from tabulated data for all these reactions at a given temperature and pressure. Three atom balances (for C, H and O) are also used to define completely the behaviour of the six-component system.

In situations where there is no carbon present a five-component (and five-equation) system must be considered. The reactions used then are:

$$CH_4 + H_2O \rightleftharpoons CO + 3H_2$$

and

$$CO + H_2O \rightleftharpoons CO_2 + H_2$$

At high temperatures, above say 900 °C, the amount of CH_4 is negligibly small and so the equation system can be further simplified.

The above approach can be extended to cover processes in which air rather than O_2 is used: this results in the appearance of N_2 as an inert in the product stream. By incorporation of energy balances it is also possible to explore what happens under adiabatic conditions and when energy is either added to or removed from the processing system.

The effect of using different biomass feedstocks on product gas composition can be readily accommodated since it is only necessary to know the so-called ultimate analysis of the feed. Other factors that can be studied are moisture content of the feed and input gas composition.

Development of a computational procedure has involved considerable effort and a heuristic approach has been necessary. Solution of the sets of simultaneous equations has been performed using the Newton–Raphson method coupled with the Gaussian Elimination technique for handling the associated matrices. During running of the computer program, problems have arisen (a) when iterative calculations have been performed in a region close to the carbon–no carbon system boundary, (b) when the CH_4 concentration is low, and (c) when initial guesses have been far removed from the final values. A typical set of data is presented in Fig. 1.

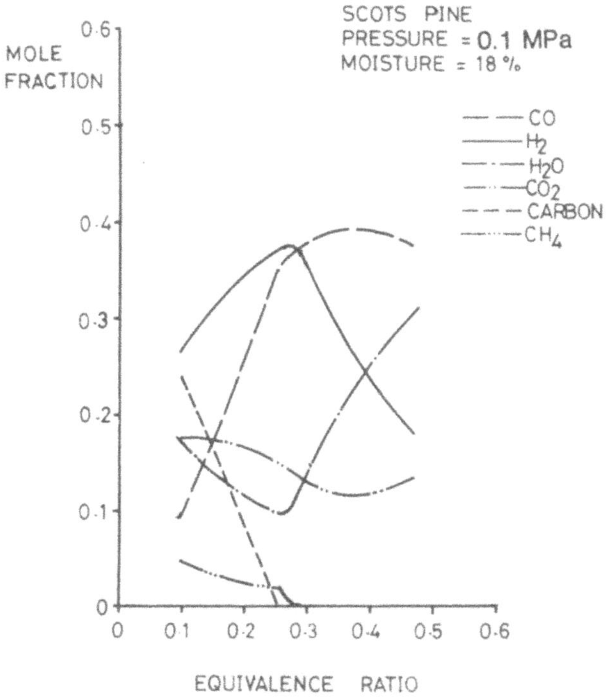

FIG. 1. Equilibrium thermodynamic studies.

4. BIOMASS PARTICLE GASIFICATION MODELS

Although product gas compositions can be estimated using thermodynamic calculations, no information is given about reaction times and the size of reactor required for a given process. To determine these particular design parameters, and to deduce temperature and phase composition profiles within a reactor, it is necessary to take account of chemical and physical rate processes.

Generally, consideration must be given to the drying of the biomass, pyrolysis leading to gas and char formation and then char gasification with CO_2, H_2O and possibly O_2. As far as the drying step is concerned, we have assumed that the moisture content of the biomass will be reduced to about 15% before it is fed to the gasifier: consequently, drying has been lumped with pyrolysis.

Pyrolysis is a complex process but for relatively dry biomass particles and high heating rates it is reasonable to assume that heat conduction into the particle is the rate-controlling step. The time required for pyrolysis can then be estimated from heat penetration times, assuming that the temperature at the centre of the particle must reach, say, 500 °C.

The development of models describing gasification of individual particles of char is a difficult task. Despite the recent flood of papers on this subject, conceptual and computational problems remain in addition to those associated with the formulation of rate expressions and derivation of rate process parameters (see, for example, Haynes[7]). We have based our work on the model proposed and tested by Groeneveld.[8] The reasons for this choice are

(a) the flexible and realistic structure of the model,
(b) the availability of details of the model, and
(c) the fact that the author had obtained his own experimental data for wood particles and from co-current moving bed gasifiers.

The Groeneveld Model assumes that the particles of char are spherical, porous and isothermal and that CO_2 and H_2O are the gasifying agents. Three reactions:

$$C + H_2O \rightleftharpoons CO + H_2$$
$$C + CO_2 \rightleftharpoons 2CO$$
$$CO + H_2O \rightleftharpoons CO_2 + H_2$$

are considered, with the last assumed to be at equivalent throughout the particle. Component mass balances are set up using the pseudo-steady state assumption to describe gaseous reactant concentrations within the particle. To solve the resulting equations, it is necessary to have expressions accounting for the rate of surface reaction, for the change in particle porosity with carbon conversion and for diffusion of gaseous components within pores. Finally, allowance is made for the decrease in particle size with time by introducing a shrinking criterion based on local carbon concentration.

The results of computations based on the model are conveniently summarized using conversion-time plots and carbon concentration profiles. Figure 2 is an example of the latter type of plot: this shows the radial concentration as a function of time under conditions where there is rapid particle shrinkage.

The single particle model can be used to gain insight into the effects of changing particle size, gas-phase composition and operating temperature and pressure on char gasification. Particle history within different reactor

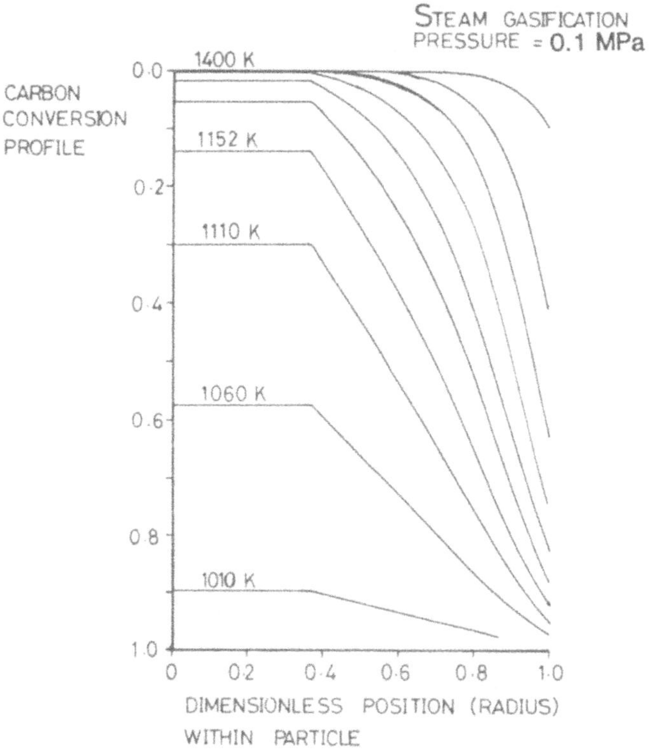

FIG. 2. Biomass particle gasification model.

configurations can also be explored: in such cases, changes in bulk gas composition and temperature must be taken into account using appropriate phase flow-rates, mass and energy balances and mass and heat transfer between phases. Our experience is limited to the use of the model for designing co-current down-draft gasifiers.

5. TG/DTA STUDIES[9]

At an early stage in the research programme a combined thermogravimetric (TG)/differential thermal analysis (DTA) technique was selected to study the effects of reaction variables on the pyrolysis of wood and on char gasification. A Linseis L81 Thermobalance was purchased for this purpose. When the balance arm is used in the horizontal mode, the equipment will take samples of about 1000 mg in weight, although simultaneous TG/DTA is only possible when using sample weights of about 100 mg. The balance can be operated at temperatures up to 1100 °C with a maximum controlled heating rate of 100 °C/min. The gaseous environment for the sample can be varied and facilities for introducing steam are available: both static and dynamic atmospheres can be maintained but high pressure operation is not possible. Evolved gas analysis (EGA) of low molecular weight volatiles is carried out using a Centronic Type 200 Mass Spectrometer.

Operating the equipment has not been without problems. To achieve reproducible results it has been necessary to

(a) standardize methods of sample preparation and presentation,
(b) dry and purify purge gases (and in particular to remove traces of O_2), and
(c) take account of mass and heat transfer limitations when varying reaction conditions.

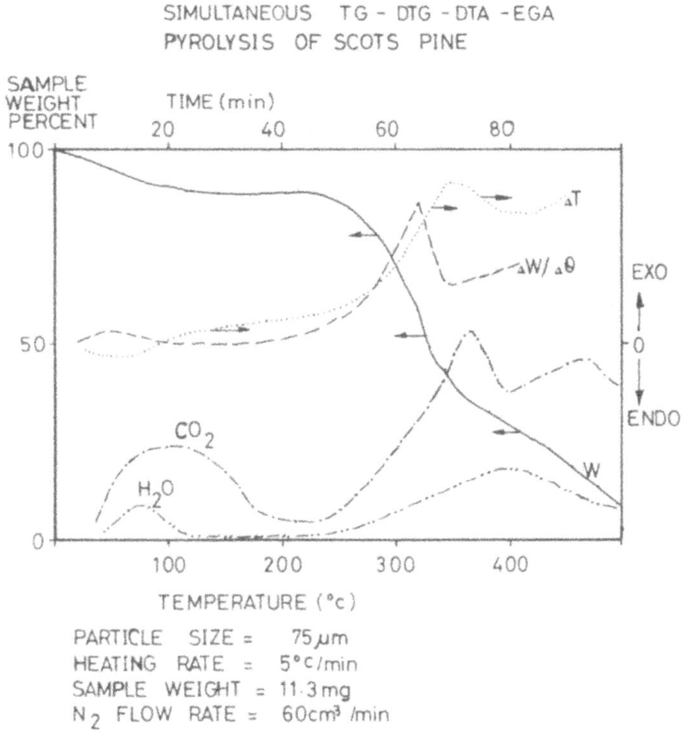

FIG. 3. Thermogravimetric differential thermal studies.

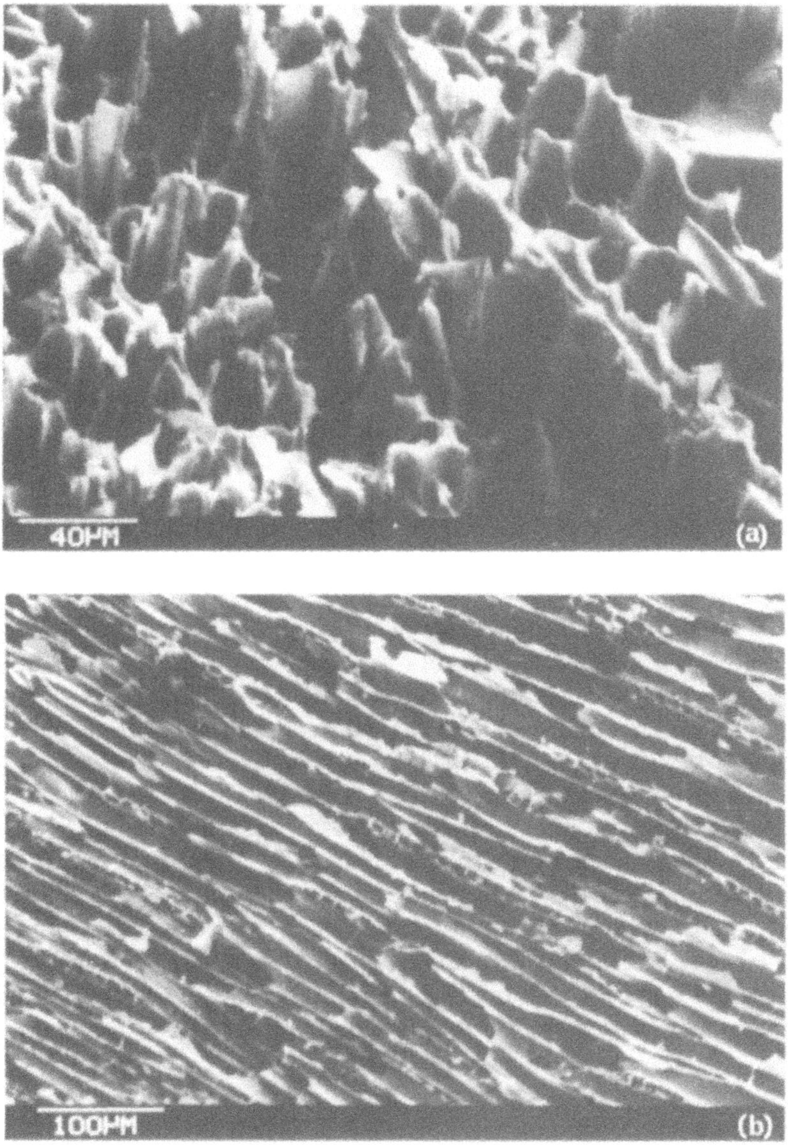

FIG. 4. (a) SEM photograph of spruce char. Section across the grain after slow pyrolysis in a nitrogen atmosphere up to 800 °C. (b) SEM photograph of spruce char. Section with the grain.

The generation of steam at low flow-rates and prevention of its condensation within the thermobalance have also caused problems. Finally, it should be noted that isothermal studies are difficult to perform because of spatial variations of temperature within the apparatus.

Some typical results are presented in Figs 3 and 4 and these compare favourably with those in the literature. Figure 3 shows TG-DTG-DTA and EGA plots for a sample of Scots pine when using 11·3 mg of particles $<75\,\mu m$ in size with a heating rate of 5 °C/min and air flow-rate of 60 cm^3/min. It will be noted that pyrolysis starts at about 250 °C and slows down at about 350 °C. Even in the presence of O_2 it is necesary to raise the temperature of the wood to about 250 °C before any significant gas evolution occurs.

A systematic study of wood pyrolysis is now nearing completion and simple kinetic models are being used to interpret the data. We are hoping that factors affecting char reactivity can be identified by examining the combustion rates of chars formed under different conditions.

6. BIOMASS GASIFICATION ON THE LABORATORY SCALE

Work on the particle gasification model described in Section 4 highlighted the need to examine the properties of large char particles. For this reason, chips and blocks of Scots pine and spruce have been pyrolysed in a stream of N_2 in a vertical tube furnace heated slowly to 800 °C. The mass yield of char from the dry feed was about 20 % whereas the decrease in volume was about 70 %: this corresponds to a decrease in linear dimensions and density of only 33 %. Microscopic examination of the char particles reveals that the skeletal structure of the original wood is very well preserved (see Fig. 4). Clearly, the structure of such chars will have an effect on the overall rate of gasification, especially at temperatures of about 1000 °C when pore diffusion is likely to be rate-controlling. Also, for particles of the order of 1 cm in size, the essentially unidirectional nature of the pores will be an important physical feature. Measurements of the porosity of chars, produced under different pyrolysis conditions, are now being made although preliminary results have been inconsistent: it is possible that structural models based on microscopic examination of specimens may aid interpretation of the data.

The tube furnace has also been used to gasify small blocks of wood with steam. Some of the results obtained at an operating temperature of 840 °C are presented in Fig. 5: the scatter in the points is due in part to operational difficulties and in part to uncertainties about char yield following initial drying and devolatilization. Char conversion values based on the single particle model described in Section 4 are also included in Fig. 5: a period of 120 s has been allowed for preheating, drying and char formation based on experimental measurements. It will be noted that a faster rate of char

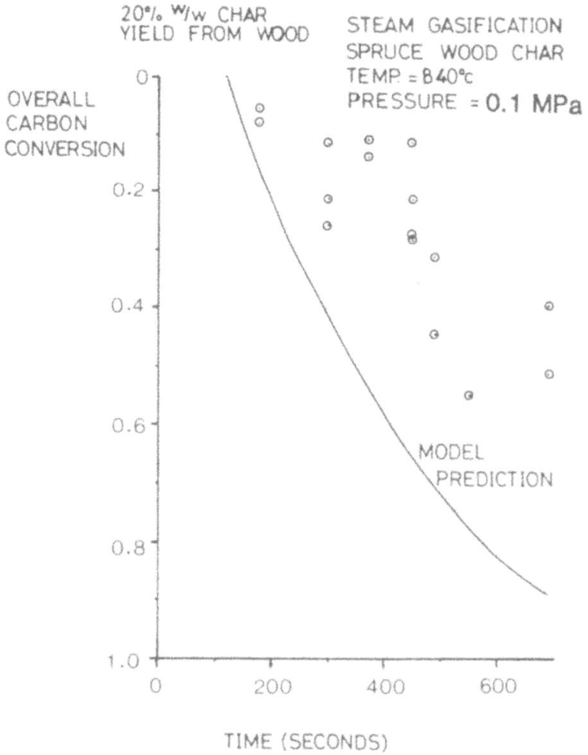

FIG. 5. Experiments with wood blocks.

gasification is predicted by the model: this is partly due to the assumption that the particles were spheres rather than cubes. At this stage in the research programme we are satisfied that the model can provide useful results for initial design purposes.

Finally, we wish to draw attention to our work using molten salts as heat-transfer and reaction media. The paper of Iredale and Hatt[10] should be consulted for further details.

ACKNOWLEDGEMENTS

The work is supported by the Energy Technology Support Unit for the Department of Energy, UK and the Science and Engineering Research Council, UK.

REFERENCES

1. Ader, G., Bridgwater, A. V. and Hatt, B. W. (1980). Conversion of biomass to fuels by thermal processes—Optimisation studies for conversion to methanol, methane and liquid fuels, Phases 2A, 2B and 2C. Energy Technology Support Unit, Department of Energy, UK, April, July and September.
2. Wei, J. (1981). Stoichiometric analysis of synthetic fuels and chemicals from coal. *Ind. Eng. Chem. Process Des. Dev.* **20**, 294–8.
3. Gumz, W. (1950). *Gas Producers and Blast Furnaces: Theory and Methods of Calculations.* Wiley, New York.
4. Baron, R., Porter, H. and Hammond, O. (1976). *Chemical Equilibria in Carbon–Hydrogen–Oxygen Systems.* M.I.T. Press, Cambridge, Mass.
5. Desrosiers, R. (1979). Chapter 6: Thermodynamics of gas–char reactions, in *A Survey of Biomass Gasification*, Vol. II. Solar Energy Research Institute, Colorado, pp. 133–75.
6. Gibbins, J. R. and Wilson, H. T. (1982). Equilibrium modelling—a cheap and effective aid to gasifier system design and analysis. International Symposium—Energy: Money, Materials and Engineering, London, 12–15 October. Institution of Chemical Engineers, London, UK.
7. Haynes, H. W. (1982). An improved single particle char gasification model. *A.I.Ch.E.J.* **28**(3), 517–21.
8. Groeneveld, M. R. (1980). The co-current moving bed gasifier. Ph.D. Thesis, Twente University of Technology, Enschede, Netherlands.
9. Hatt, B. W. and Sheena, H. H. (1982). A thermochemical study of wood pyrolysis and char gasification reactions. Second E.C. Conference on Energy from Biomass, Berlin, 20–23 September.
10. Iredale, P. J. and Hatt, B. W. (1982). This volume, Chapter 8, pp. 143–55.

28

Char Properties and Gasification

P. L. WALKER, JR

*Department of Materials Science and Engineering,
Pennsylvania State University, University Park, Pennsylvania, USA*

1. INTRODUCTION

There is increasing interest in the conversion of coal and biomass to gaseous fuels through the use of both high temperature and low temperature gasification processes. In gasification at high temperatures, the exothermic $C–O_2$ reaction supplies the heat to drive the endothermic $C–H_2O$ reaction and CO and H_2 are the primary fuel gases produced. Gasification occurs under conditions (Zone b or III in Fig. 1) where reaction rates are limited in part or completely by mass transport of reacting gas to carbon sites. Concentration of carbon active sites or the presence of gasification catalysts are not of primary importance in determining gasification rates. The extent of utilization (η) of active sites is small; thus the apparent activation energy for carbon gasification (E_a) is also small. In the limit, gasification rates are independent of starting organic precursor.

By contrast, the desire to produce CH_4 as the primary gaseous fuel, coupled with the attractiveness of enhanced thermal efficiency, has resulted in interest in low temperature gasification and the overall reaction

$$2C + 2H_2O \rightarrow CH_4 + CO_2 \tag{1}$$

This reaction, which is essentially thermoneutral, is a combination of the endothermic carbon gasification reaction

$$2C + 2H_2O \rightarrow 2CO + 2H_2 \tag{2}$$

and the exothermic water-gas shift and methanation reactions

$$CO + H_2O \rightarrow CO_2 + H_2 \tag{3}$$

$$3H_2 + CO \rightarrow CH_4 + H_2O \tag{4}$$

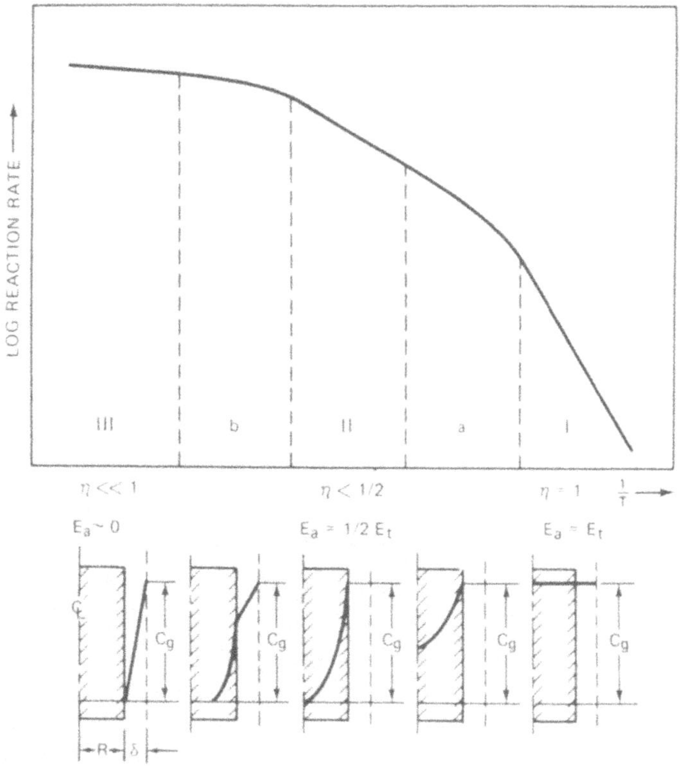

FIG. 1. Ideally, the three zones representing the change of reaction rate at a porous carbon with temperature.

Reaction (2) will be conducted at sufficiently low temperatures such that gasification takes place in Zone I, a, or II (Fig. 1); the rate is controlled, at least in part, by the concentration of active carbon sites and catalysis. This paper will be concerned with gasification conducted at lower temperatures.

There is a voluminous literature covering research on gasification of carbon derived from fossil fuels. It is the primary purpose of this paper to bring some of the main findings of this research to the attention of workers now interested in the gasification of biomass. Undoubtedly many analogies will exist in the gasification of these different organic precursors.

2. FUNDAMENTALS OF CARBON GASIFICATION

We have previously considered the fundamentals of carbon gasification insofar as reaction mechanisms[1] and catalysis[2] are concerned.

2.1. Uncatalyzed Carbon Gasification

The building block of chars produced upon pyrolysis of coals and biomass is the pseudo-graphitic building block (Fig. 2). Carbon crystallites are small and poorly aligned because of cross-linking between them. Small crystallite size and poor crystallite alignment result in a large concentration of accessible carbon sites at crystallite edges. They have been shown to be the active sites for carbon gasification. Uncatalyzed gasification occurs by: (i) the formation of an activated complex between a molecular species, say O_2, with active carbon sites, (ii) breakage of the oxygen–oxygen bond,

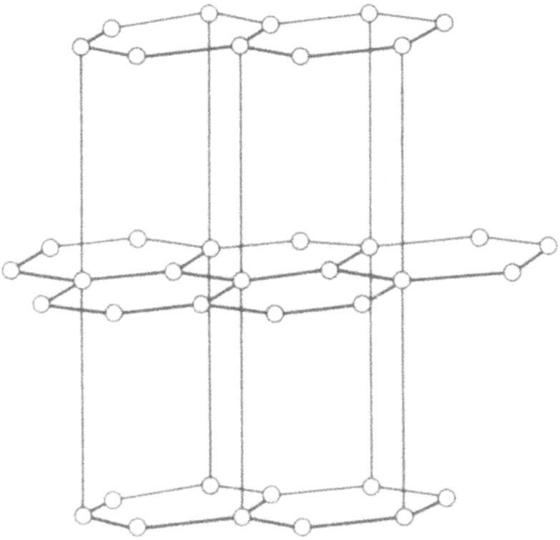

FIG. 2. The graphite crystal structure.

(iii) formation of carbon–oxygen bonds, and (iv) breakage of carbon–carbon bonds and desorption of CO into the gas phase. In short, the oxygen transfer step is followed by the carbon gasification step. Thus, the oxygen transfer step for the important gasification reactions is followed by a common gasification step as follows:

$$2C + O_2(g) \rightarrow 2C(O) \tag{5}$$

$$C + CO_2(g) \rightarrow CO(g) + C(O) \tag{6}$$

$$C + H_2O(g) \rightarrow H_2(g) + C(O) \tag{7}$$

and

$$C(O) \rightarrow CO(g) \tag{8}$$

As previously discussed,[1] the gasification of carbon by O_2 is much more rapid than its gasification by H_2O, which in turn is slightly more rapid than its gasification in CO_2.

Surprisingly, little data exist on fundamental rates for carbon gasification reactions, that is rates expressed per carbon active site. At this time, the author is aware of no data for the C–H_2O or C–CO_2 reactions. Consider data available for the C–O_2 reaction.

Hennig's development of the etch decoration technique to follow the enlargement of vacancies in the basal plane of graphite gave us a visual quantitative technique to follow gasification rates at carbon active sites.[3] It offered a direct check to previous kinetic studies in our laboratory on the C–O_2 reaction, where active surface area (ASA) was measured by oxygen chemisorption.[4] Hennig's microscopy studies on reaction of highly purified Ticonderoga natural graphite with O_2 was followed by a number of other light and electron microscopy studies, also using Ticonderoga natural graphite. It finally has enabled us to have a good picture of fundamental rates of C–O_2 reaction. Conditions under which the experiments were conducted are summarized in Table I, and rates, in different units, are given

TABLE I
STUDIES ON RATES OF CARBON GASIFICATION BY OXYGEN

Worker(s)	*Reference*	*Exptl. pressure, kPa*	*Method*
Laine *et al.*	4	$5{\cdot}2 \times 10^{-3}$	Mass spectroscopy and oxygen chemisorption
Hennig	3	1·3	TEM[a] with vacancy decoration
Thomas & Hughes	5	1·3	Light microscopy with enlargement of hexagonal pits
Baker	6	6·6	TEM with vacancy decoration
Evans & Thomas	7	1·3	TEM with vacancy decoration
Baker & Harris	8	2·6	CAEM[b] with channel enlargement
Yang & Wong	9	20·0	TEM with vacancy decoration
Baker & Chludzinski	10	0·66	CAEM with channel enlargement

[a] Transmission electron microscopy.
[b] Controlled atmosphere electron microscopy.

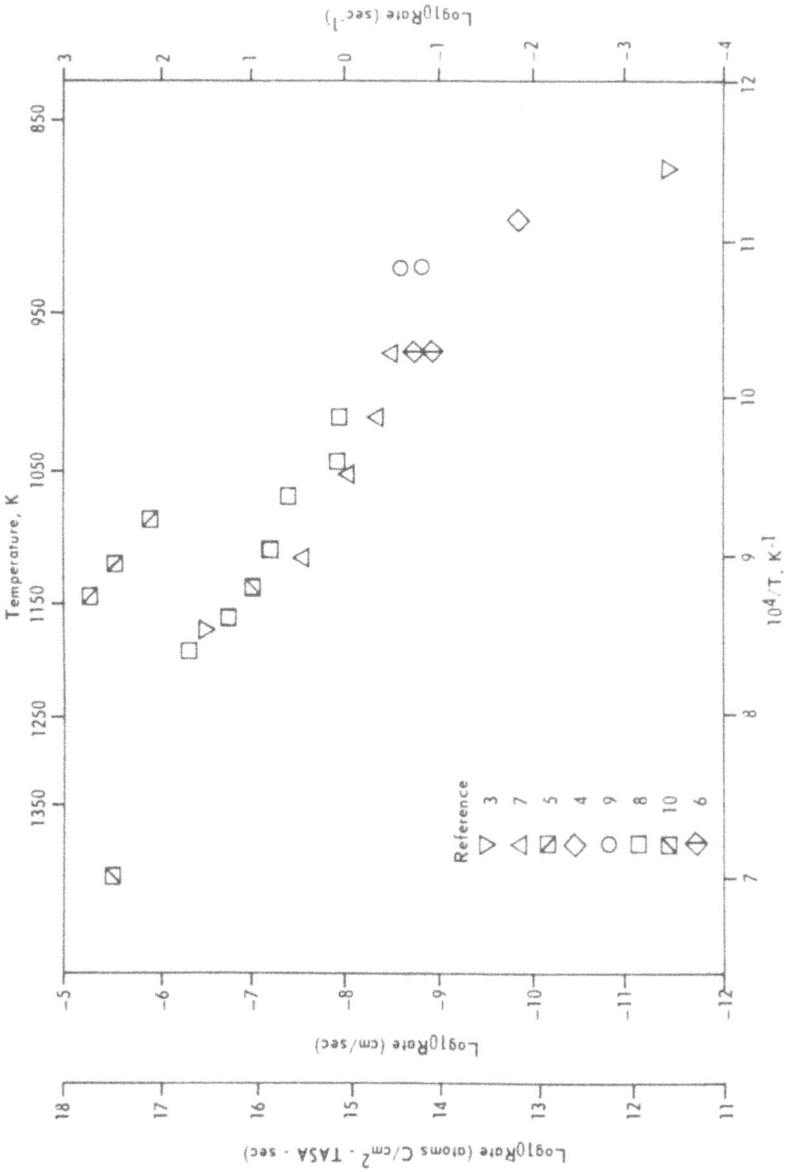

FIG. 3. Arrhenius plot of rate constants for the C–O_2 reaction at 1·3 kPa pressure.

on an Arrhenius plot (Fig. 3), at an O_2 pressure of 1·3 kPa. From microscopy studies, rates can be measured in units of cm/s, that is the recession rate of a vacancy, a pit or a channel. Rates can be converted to atom C per cm^2 of total ASA (TASA) per second, using the true density of graphite (2·268 g/cm^3). Rates can be given as 'turn-over' numbers by assuming that the area occupied by an active carbon site is 0·08 nm^2.

Let us consider the results and their possible limitations. It should first be understood that specific rates are being given in Fig. 3 and not specific rate constants. That is, Laine *et al.*, who measured the kinetics of the C–O_2 reaction on graphitized carbon black using a combination of mass spectroscopy and oxygen chemisorption, showed clearly that gasification rates depend upon the fraction of ASA uncovered by a stable oxygen complex.[4] It is only when the amount of this complex is known that a specific rate constant, based on the uncovered ASA, can be determined. To place the Laine *et al.* result on Fig. 3, their specific rate constant has been converted to a specific reaction rate by converting to a basis of TASA. Unfortunately, investigators using microscopy have not all worked at the same pressure and have failed to clearly establish the order of reaction. Hennig suggests 0·5 order;[3] this order is assumed in giving all specific rates at an O_2 pressure of 1·3 kPa in Fig. 3. It is noted, as previously pointed out by Thomas *et al.*,[11] that recession rates from multilayer steps (followed by light microscopy) are nearly a factor of 100 greater than recession rates from monolayer steps (followed by TEM). The reason for this difference is not understood at this time, but it obviously has a fundamental bearing on our understanding of carbon gasification.

Leaving out the light microscopy results of Thomas and Hughes, best fit equations for rates of gasification are given in Table II, with activation energy in units of kJ/g mole.

2.2. Catalyzed Carbon Gasification

All carbons contain at least traces of inorganic impurities. Whether these impurities result in significant catalysis of carbon gasification depends upon their concentration, their extent of dispersion and their specific

TABLE II

SPECIFIC REACTION RATES FOR THE C–O_2 REACTION AT 1·3 kPa PRESSURE

$$\text{Rate (cm/s)} = 9{\cdot}35 \times 10^3 \exp(-239/RT)$$
$$\text{Rate (atoms C/cm}^2\text{ TASA-s)} = 1{\cdot}06 \times 10^{27} \exp(-239/RT)$$
$$\text{Rate (1/s)} = 8{\cdot}82 \times 10^{11} \exp(-239/RT)$$

catalytic activity. Essentially, inorganic impurities do not reside within carbon crystallites either substitutionally or interstitially; rather they are located at crystallite boundaries. Thus their concentration (and importance) at positions where carbon gasification is occurring is magnified many times.

A number of mechanisms have been proposed for catalysis of carbon gasification.[2] Probably the most important in the context of char gasification involves oxidation–reduction cycles, with carbon reducing the metal peroxide in the case of the O_2 reaction and metal carbonate in the case of the CO_2 and H_2O reactions. McKee and coworkers conclude that this mechanism is operative in the case of catalysis by Na and K.[12–16] McKee further concludes that catalysis of the C–CO_2 and C–H_2O reactions by Ca involves a similar mechanism. Calcium is an excellent catalyst for the C–O_2 reaction, but unlike K and Na it does not form a bulk peroxide. However, it is thought that Ca has the ability to form a higher surface oxide. As we will see, Mg is a poor catalyst for the C–O_2 reaction. A bulk peroxide has not been reported; and because of the relatively small size of the Mg^{2+} cation, it is doubtful if even a surface peroxide is formed.

Not all metals which form different bulk oxidation states are active carbon gasification catalysts. To be a good catalyst, it is necessary that the rate of reduction of the higher oxide to the lower oxide by carbon be rapid. When Fe is oxidized, it is a poor catalyst for carbon gasification; however, in the reduced state it is an excellent catalyst.[2,12] Iron is an active catalyst for the dissociation of O_2, CO_2 and H_2O to produce oxygen atoms. Some fraction of these oxygen atoms either react with an active carbon site at the iron–carbon interface, leading to carbon gasification, or spill off the iron and diffuse across the carbon surface until they find an active carbon site.

Another mechanism of catalysis is attributed to electron transfer by Long and Sykes.[17,18] That is, π-electrons from carbon are transferred into the unfilled d-band of transition metals, thereby weakening surface carbon–carbon bonds and facilitating carbon gasification.

More recently, it has been suggested that catalysis involves the breakage of carbon–carbon bonds at the metal–carbon interface and rapid reaction of carbon atoms on the metal surface.[19]

3. RATES OF GASIFICATION OF COAL CHARS

3.1. Shape of Carbon Burn-off Plots

We have shown that the shapes of carbon burn-off versus time plots for most coal-derived chars in air, CO_2 and H_2O are quite similar.[20] A typical

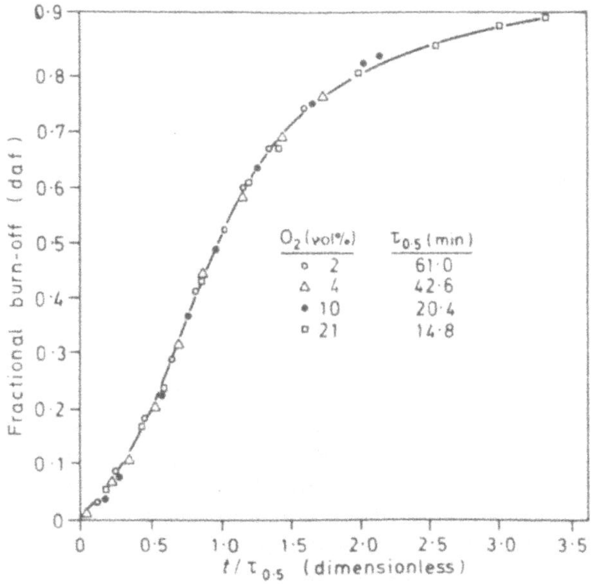

FIG. 4. Normalized plot for reaction of lignite char at 678 K in different concentrations of O_2.[20]

plot for a lignite char, reacted in different partial pressures of O_2 in O_2–N_2 mixtures having a total pressure of 0·1 MPa, is shown in Fig. 4. The abscissa is made dimensionless by dividing the experimental reaction times by the time to attain a fraction burn-off to 0·5 ($\tau_{0 \cdot 5}$). Results for most coal chars in different atmospheres can be expressed by a cubic equation of the form: burn-off $= a(t/\tau_{0\cdot5}) + b(t/\tau_{0\cdot5})^2 + c(t/\tau_{0\cdot5}),^3$ with $\tau_{0\cdot5}$ being the only variable, to a high degree of correlation. Attempts to explain the form of the empirical equation have been made.[21,22] Basically, the increasing rate at low burn-offs can be attributed to opening up of closed porosity and enlargement of pores which are already open. Eventually, gasification leads to the reacting away of pore walls, a decrease in the number of open pores, and a decrease in reactivity.

3.2. Effect of Coal Rank

Coals of all ranks, varying from anthracite to lignite, were heated at 10 K/min to 1273 K in N_2 and held for 1 h prior to studying reactivities of the chars produced. (Strictly speaking, thermal treatment of coals leads to either a char or a coke. If upon heating, the organic precursor goes through a liquid state, the resulting solid is termed a coke. If, however, negligible softening occurs, the resulting solid is termed a char. For convenience, we will use the term char throughout this paper.) Reactivity of 40 × 100 mesh chars, measured in 0·1 MPa of air at 773 K[23] and 0·1 MPa of CO_2 at 1173 K,[24] are expressed as maximum reactivity per unit initial char weight. As seen from Fig. 5, there is a wide variation in reactivity of American coal chars—a factor of 100 in going from a low volatile bituminous coal char from Pennsylvania (PSOC 127) to a lignite char from Montana (PSOC 91). Changes in coal rank have a similar effect on char reactivity in air and CO_2. A similar effect is also found for reactivity in H_2O.[25]

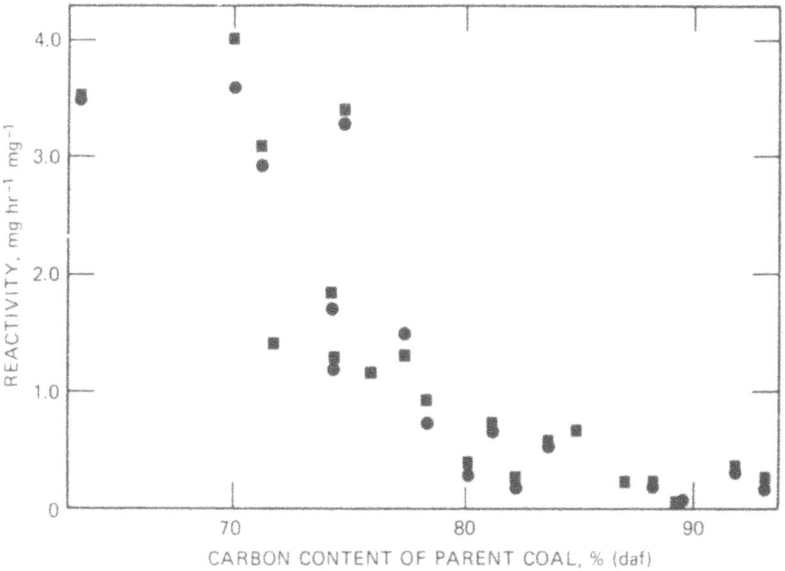

FIG. 5. Variation of reactivity of 1273 K chars produced from coals of different rank. ● 0·1 MPa CO_2 at 1173 K, ■ 0·1 MPa air at 773 K.[23,24]

Differences in reactivity can be attributed to three main factors: (i) concentration of active sites, (ii) accessibility of the reactant gas to active sites, and (iii) catalysis. As the rank of the starting coal decreases, the amount of volatile matter released upon pyrolysis increases. Enhanced release of volatiles results in the production of a more open structure or an increase in surface area as seen in Fig. 6.

We have previously shown in this laboratory that coals of lower rank have a greater volume of transitional and macropores than do coals of higher rank.[26] Even though these larger pores contribute little to surface

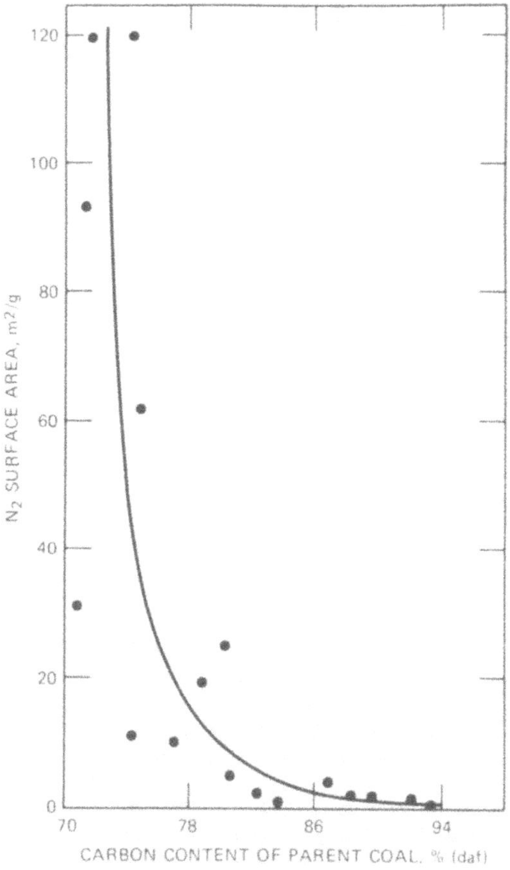

FIG. 6. Correlation between the N_2 surface areas of chars (77 K) and the carbon content of the parent coals.

area or concentration of active sites, they serve as feeder pores through which reacting molecules can rapidly diffuse to reach the micropores. Therefore, the extent of utilization (η) of active sites within the micropores is enhanced.

Lignites contain between 2 and 3 meq/g of carboxyl groups.[27] Roughly half of the carboxyl groups have undergone exchange, primarily with Ca and Mg. Upon pyrolysis, the carbonyl groups decompose liberating CO_2 and yield well dispersed metals or metal oxides, which may act as catalysts for char gasification. Calcium, for example, is an excellent catalyst. Figure 7 shows a correlation between char reactivities in air and CO_2 and CaO

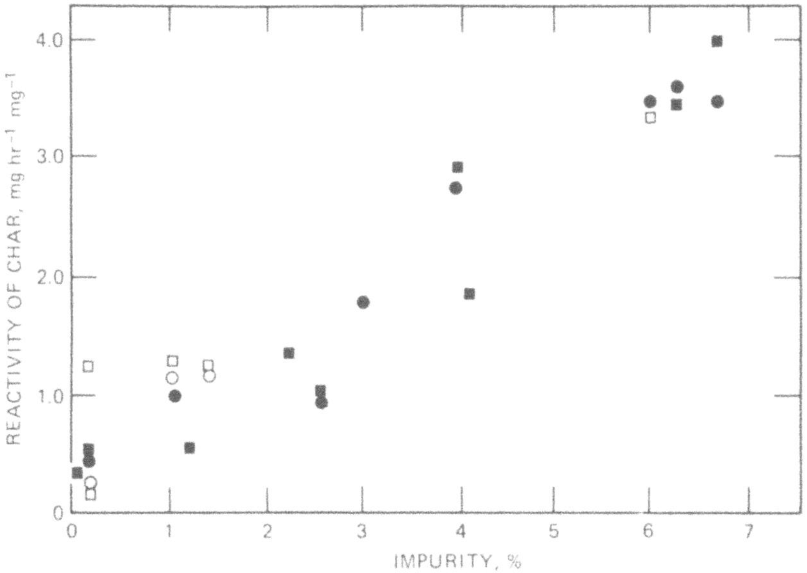

FIG. 7. Reactivities of 1273 K chars in 0·1 MPa CO_2 and air versus wt % CaO present in the chars. Closed and open symbols represent chars from original and acid washed coals, respectively. ●○ CO_2 at 1173 K, ■□ air at 773 K.[23,24]

content in the chars. Not only do chars produced from coals of lower rank contain more CaO but it is better dispersed since Ca in the higher rank coals is present primarily as relatively larger particles of calcite. It is concluded that Ca is the most important *in situ* catalyst for accelerating coal char gasification in O_2, CO_2 and H_2O. It is found that reactivities of coal chars to H_2 do not show extreme variations with changing rank of starting coal.[28] As will be seen later, this is because Ca is a poor catalyst for the hydrogasification reaction whereas Fe is a good catalyst.[28] The concentration of Fe, derived from pyrite, is highest for the bituminous coals.

Reactivities of the chars produced from 40 × 100 mesh coals can be compared with those of chars produced from −400 mesh high purity cellulose and 30 × 100 mesh lignin (1·6 wt % ash) under identical pyrolysis conditions. The cellulose had an amorphous content of 26 wt %, as estimated from water adsorption. Upon pyrolysis to 1273 K, its weight loss was 95·2 % (dry basis). The lignin was extracted from a hybrid poplar; upon pyrolysis its weight loss was 72·1 %. Reactivities of the two chars were 0·90 and 1·35 g/g h, respectively.

3.3. Effect of Pyrolysis Conditions

There are essentially an infinite number of pyrolysis conditions which can be used as an organic precursor is converted to a char. Changing pyrolysis conditions changes the structure of the char produced and catalyst activity and, therefore, char reactivity.

3.3.1. Atmosphere

In some recent results, a sub-bituminous coal was first heated to 859 K in He followed by heating to 1159 K in either He, Ar or N_2.[29] It was found that the atmosphere used had an effect on the surface area (as measured by CO_2 adsorption at 273 K) and reactivity (in CO_2 at 1109 K) of the chars produced in the order: in He > in Ar > in N_2. The authors conclude that the results are not due to variable amounts of reactive impurities in the gases. Previously we found in this laboratory that the rates of the C–O_2 and C–CO_2 reactions were affected by the presence of these same diluent gases in the following order of decreasing rate: in N_2 > in Ar > in He.[30] This phenomenon was attributed to the effect of interaction of the diluent gas with nascent active sites following a gasification event and, hence, the rate of their decay to less active sites. It was suggested that the rate of decay is expected to be: in He > in Ar > in N_2. If nascent active sites catalyze secondary cracking reactions of hydrocarbons released during coal pyrolysis, leading to carbon deposition, loss of char surface area and char reactivity, the result of Thakur and Brown on the effect of atmosphere on coal pyrolysis is to be expected.

Since removal of oxygen complex as CO and CO_2 produces nascent carbon sites, it is expected that if pyrolysis occurs in the presence of a reactive atmosphere, gasification rates are enhanced. We have confirmed this expectation when a graphitized carbon black was devolatized in a CO_2 atmosphere.[31]

3.3.2. Rate of Heating

We heated coals in N_2 up to 1073 K at rates of 10 K/min and 10^4 K/s and held at maximum temperature for less than 1 s.[32] Reactivities of chars in 0·1 MPa air at 773 K were higher for samples heated rapidly. Even rapid heating to 1073 K of a char which was previously slowly heated to 1073 K resulted in some enhancement of reactivity. The char prepared by rapid heating to 1073 K was more reactive than the char prepared by slow heating to 773 K, despite the fact that weight loss upon pyrolysis was similar in the two cases.

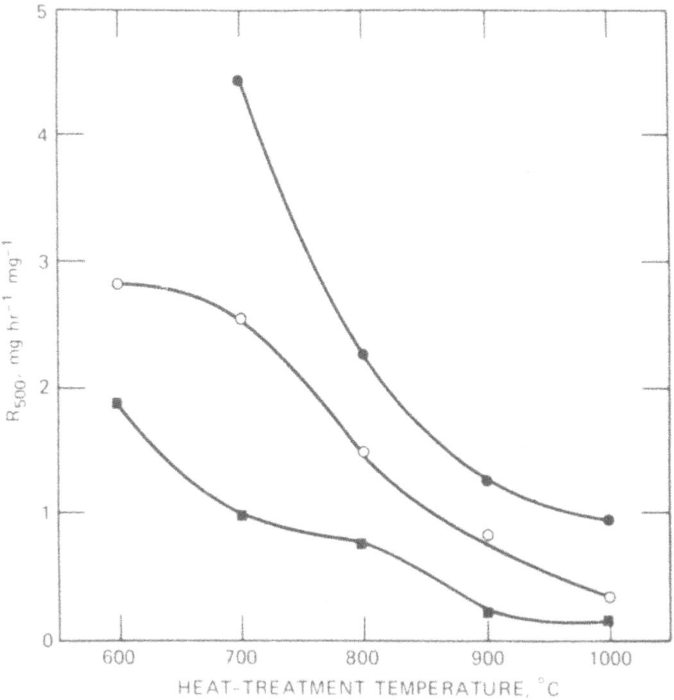

FIG. 8. Effect of pyrolysis temperature of coals in N_2 on subsequent char reactivity in 0·1 MPa air at 773 K. ● lignite, ○ HVB bituminous, □ HVA bituminous.[23]

3.3.3. Pyrolysis Temperature

Figure 8 shows typical results for the effect of heat treatment temperature (HTT) on subsequent char reactivity.[23] Three coals of different rank were heated in N_2 at 10 K/min to 1273 K, held for 2 h, and then reacted at 773 K in 0·1 MPa of air. The sharp decrease in reactivity with increasing HTT has been attributed to a progressive decrease in catalyst dispersion by sintering and a loss of carbon active sites.

3.3.4. Residence Time at Pyrolysis Temperature

In some recent studies in our laboratory, a raw lignite was treated with HCl and HF to remove the mineral matter and place all carboxyl groups in the H^+ form.[33] The demineralized coal was subsequently ion-exchanged with Ca^{2+}, using a 1·5 M solution of Ca-acetate, to add about 2·9 wt % Ca. Figure 9 shows the effect of pyrolysis residence time in N_2 on subsequent reactivity of the chars in 0·1 MPa air following either rapid (R) or slow (S)

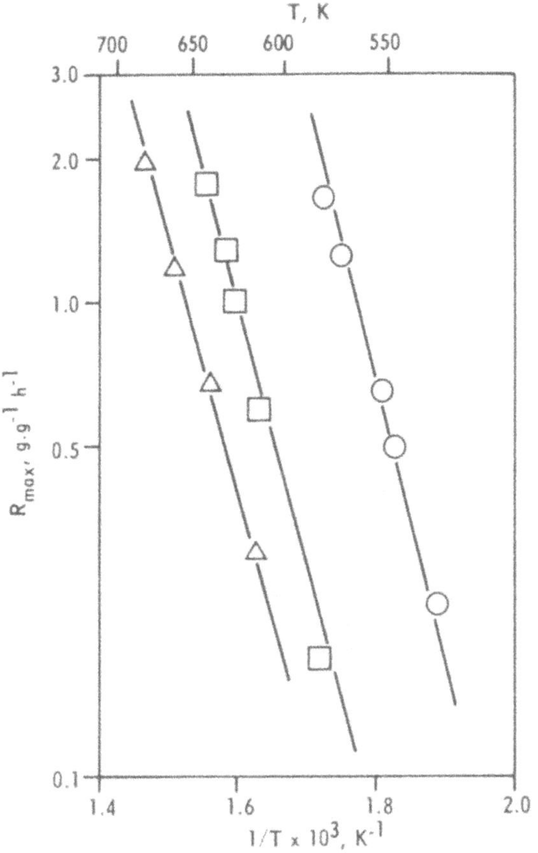

FIG. 9. Effect of pyrolysis residence time at 1275 K on the reactivities of Ca-loaded lignite chars in 0·1 MPa air. ○ R–0·3 s, □ R–5 min, △ S–1 h.

rates of heating. As residence time at 1275 K increases, char reactivity decreases. As seen in Fig. 10, longer residence times lead to the development of sharp X-ray diffraction peaks of CaO, indicating a growth in crystallite size and a sintering of CaO particles. Decrease in catalyst activity is expected to accompany decrease in catalyst dispersion.

3.3.5. Preoxidation

Coals are microporous materials.[34] If upon their pyrolysis they do not soften significantly, this microporosity and, hence, high surface area is maintained in the chars. However, most bituminous coals soften

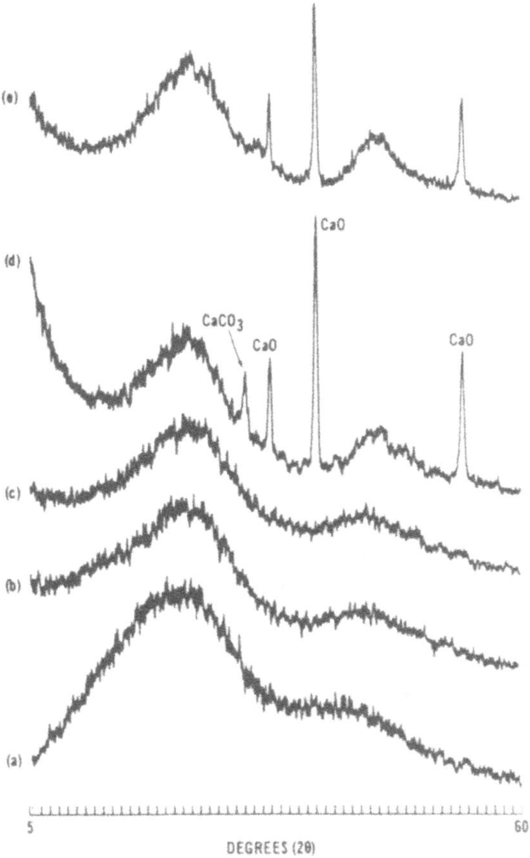

FIG. 10. X-ray diffraction patterns for Ca-loaded lignite and chars obtained at 1275 K. (a) Coal, (b) R–0·3 s, (c) R–1·5 s, (d) R–5 min, (e) S–1 h.

significantly during pyrolysis, leading to crystallite alignment in the chars and a sharp decrease in surface area. Softening can be prevented to varying degrees by exposing the coal to air at low temperatures. The oxygen added to the coal results in the formation of ether cross-links between aromatic regions upon pyrolysis. Figure 11 shows the marked effect of preoxidation of the low volatile bituminous coal (PSOC 127) on the subsequent reactivity of chars to 0·1 MPa air at 743 K. The coal was pretreated in 0·1 MPa of air at temperatures between 453 and 523 K to add varying amounts of oxygen. It is seen that for oxygen additions up to 3·0 wt %, there is a major increase in reactivity, the time required to reach a burn-off of

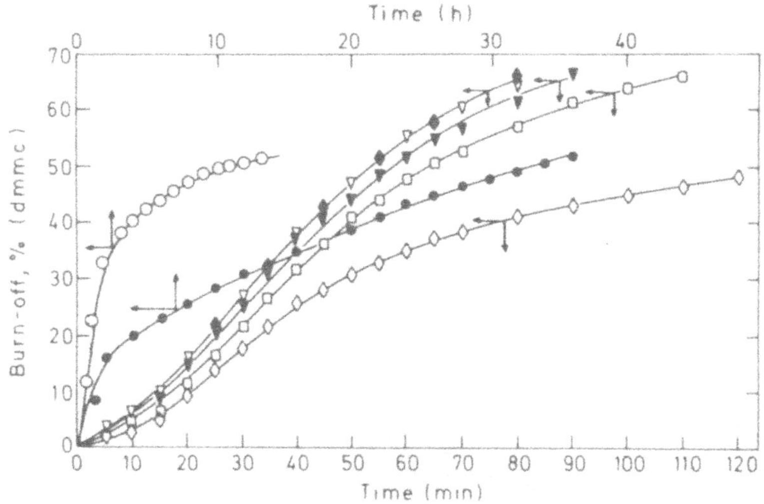

FIG. 11. Effect of different levels of preoxidation on burn-off curves in 0·1 MPa air at 743 K for chars produced from a low volatile bituminous coal (200 × 250 mesh). Weight gain during preoxidation (%): ● none, ○ 0·45, ◇ 0·67, □ 1·4, ▼ 1·9, ▽ 3·00, ◆ 4·5.[35]

50 % decreasing by a factor of 30. Since biomass is relatively high in oxygen, it behaves as a thermosetting precursor upon pyrolysis, leading to relatively high surface area chars.

3.4. Effect of Catalyst Addition

Even though chars produced from lignite coals have high reactivities, we have shown that reactivities can be further enhanced by the addition of more cations to the coal.[36] As an example, a Texas lignite was first treated with HCl and HF to place all the carboxyl groups in the H^+ form. This lignite was then exchanged with varying concentrations of Ca-acetate from 0·04 to 2 M. The loaded coals were then pyrolyzed in N_2 by heating at 10 K/min to 1073 K and holding for 2 h. Reactivities in 0·1 MPa steam at 923 K are seen to increase linearly with increasing Ca loading (Fig. 12). The reactivity of the char loaded to 12·9 wt % Ca is about 10 times higher than that for the char produced from the raw lignite.

The acid treated Texas lignite was loaded with a number of different cations to about 0·3 mmole/g, prior to its pyrolysis in N_2 to 1073 K. Reactivity plots in 0·1 MPa steam at 923 K, 0·1 MPa air at 663 K, 0·1 MPa CO_2 at 1033 K, and 0·1 MPa H_2 at 1063 K are shown in Figs 13–16. In the

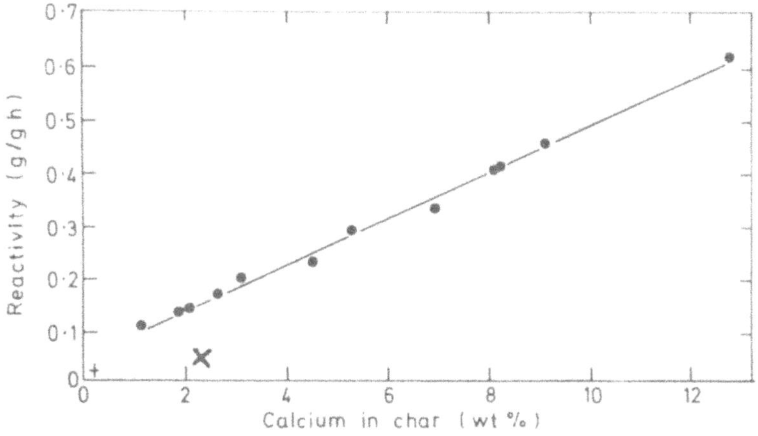

FIG. 12. Reactivities in 0·1 MPa steam at 923 K of 1073 K chars prepared from (×) raw, (+) demineralized, and (●) various Ca-exchanged lignites.[36]

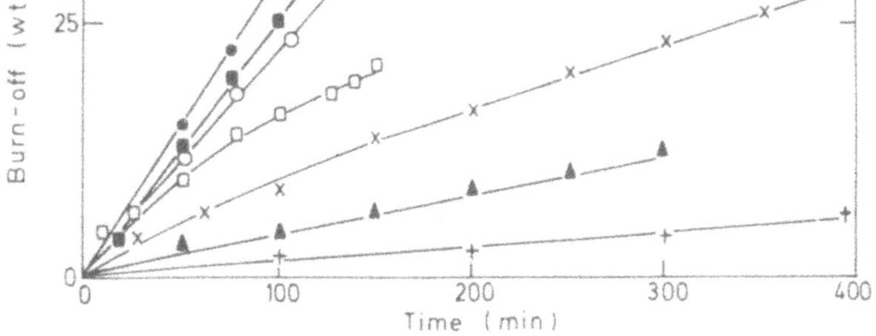

FIG. 13. Reactivities in 0·1 MPa steam at 923 K of 1073 K chars produced from raw, demineralized, and various metal-exchanged lignites. ×, Raw, +, demineralized, ● K, ○ Na, ■ Ca, □ Fe, ▲ Mg.[36]

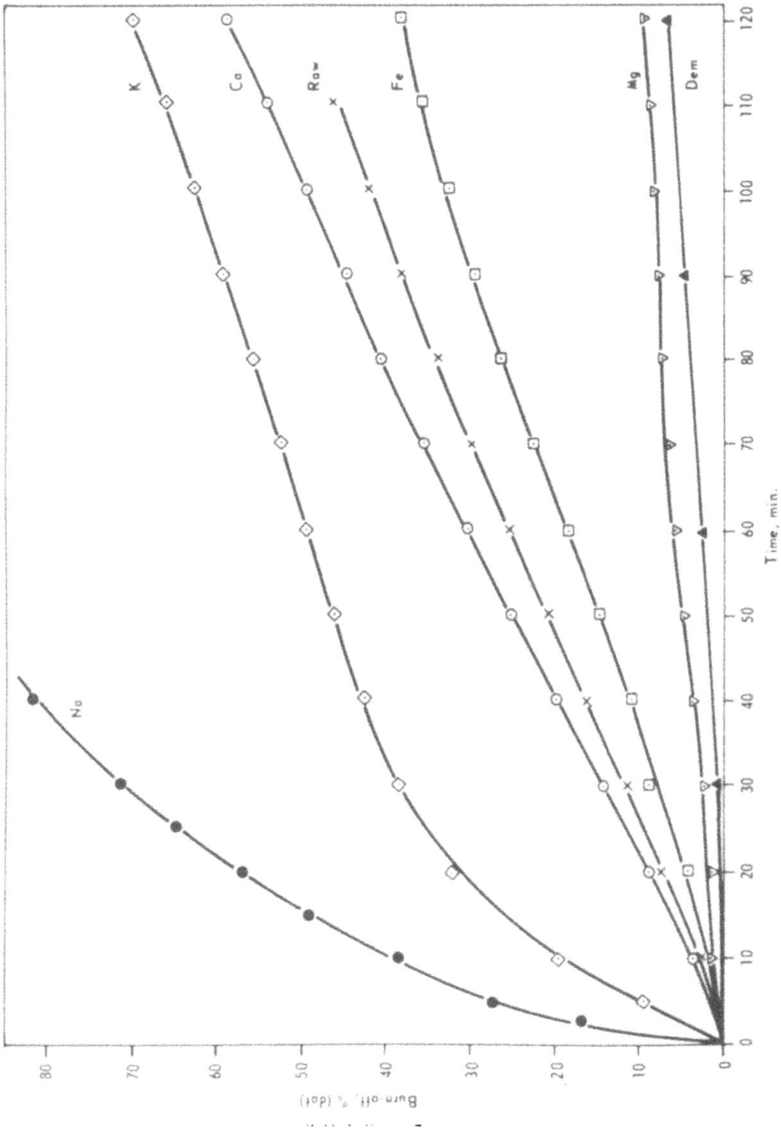

FIG. 14. Reactivities in 0·1 MPa air at 663 K of 1073 K chars.

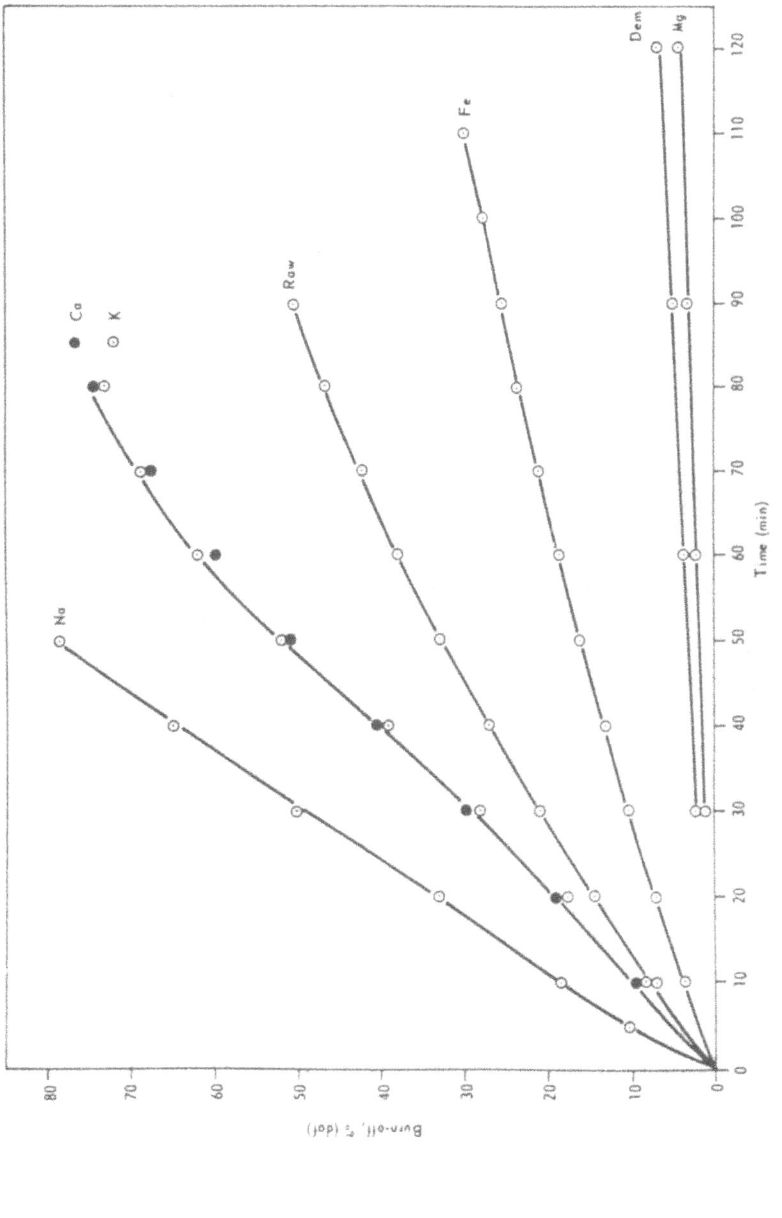

FIG. 15. Reactivities in 0·1 MPa CO_2 at 1033 K of 1073 K chars.

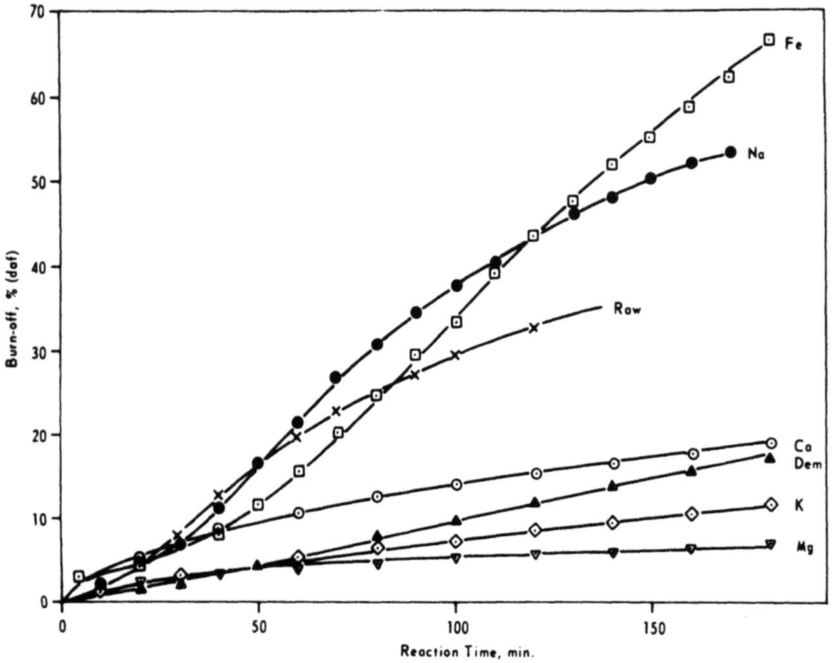

FIG. 16. Reactivities in 0·1 MPa H_2 at 1063 K of 1073 K chars.

oxygen containing gases, Ca, K and Na are excellent catalysts. Iron is an excellent catalyst at low carbon burn-offs, but it becomes oxidized and progressively loses its catalytic activity. Magnesium is a poor catalyst. Calcium, K, and Mg are poor hydrogasification catalysts; Fe is an excellent hydrogasification catalyst. Interestingly, Na also shows significant catalytic activity for the C–H_2 reaction.

3.5. Inhibitors of Steam Gasification

In the Exxon catalytic gasification process for the production of CH_4 through the use of overall eqn (1), CO and H_2 leaving the gasifier are separated from CH_4 and CO_2 and reinjected into the gasifier.[37] Potassium is added to the coal as a catalyst for the C–H_2O reaction. Therefore, it is of interest to determine the effect of H_2 and CO on the rate of char gasification by steam. An acid treated Texas lignite (PSOC-623) was loaded with 4·0 wt % K prior to converting it to a char by heating in N_2 at 20 K/min to 1123 K and holding for 30 min. Char gasification in 0·1 MPa of wet N_2 (H_2O partial pressure of 2·7 kPa) was then followed at 1123 K. As seen in Fig. 17, burn-off was essentially constant over the first 2 h or up to a

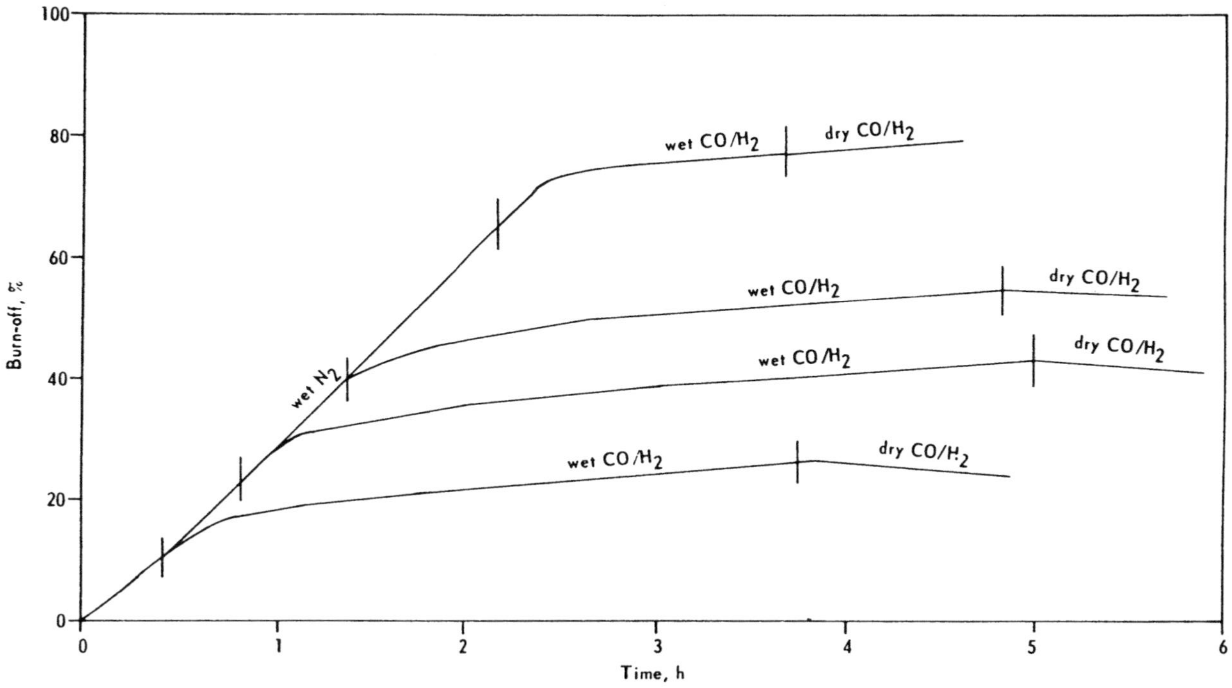

FIG. 17. Reactivity plots of K-loaded 1123 K lignite char in 2·3 kPa steam at 1123 K showing inhibiting effect of H_2/CO mixture.

burn-off of about 70%. However, if at different levels of burn-off an H_2/CO mixture ($3H_2/1CO$) was substituted for N_2, the rate of gasification was sharply reduced. If, at at longer time in the run, dry H_2/CO was substituted for wet H_2/CO, in some cases a weight gain was observed suggesting that disproportionation of CO was occurring. Gasification of chars, loaded with Na and Ca, in steam was also strongly inhibited by the presence of an H_2/CO mixture. Therefore, it is clear that a price is being paid in the Exxon catalytic gasification process, as a result of reinjecting an H_2/CO mixture into the gasifier.

It is of interest to separate the effects of H_2 and CO on char gasification by steam. Results for the Texas lignite char, loaded with different metals, are summarized in Figs 18 and 19. Gasification was conducted at 1123 K at a total pressure of 0·1 MPa and a water partial pressure of 2·3 kPa. For the

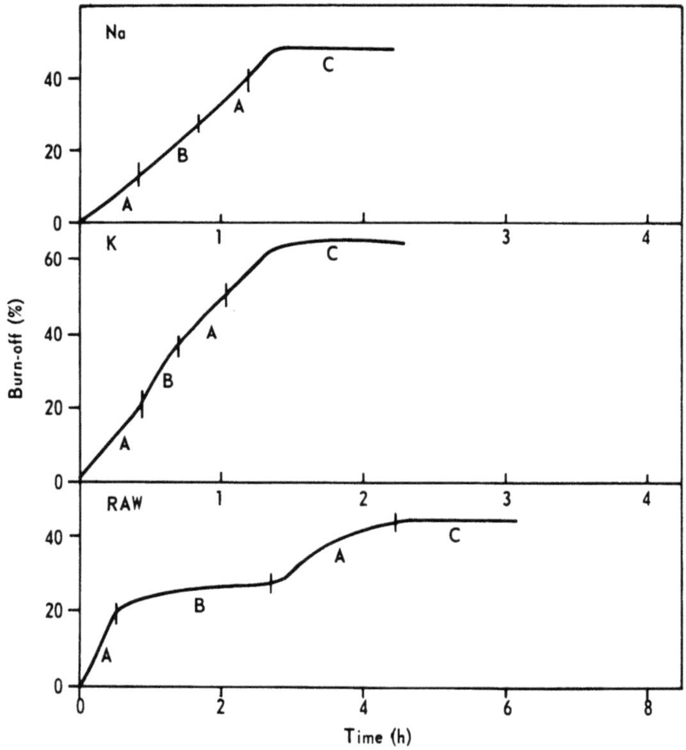

FIG. 18. Reactivity plots at 1123 K in 2·3 kPa steam for 1123 K chars prepared from raw, Na and K-loaded lignites. Wet N_2 (regimes A), wet H_2 (B) and wet CO (C).

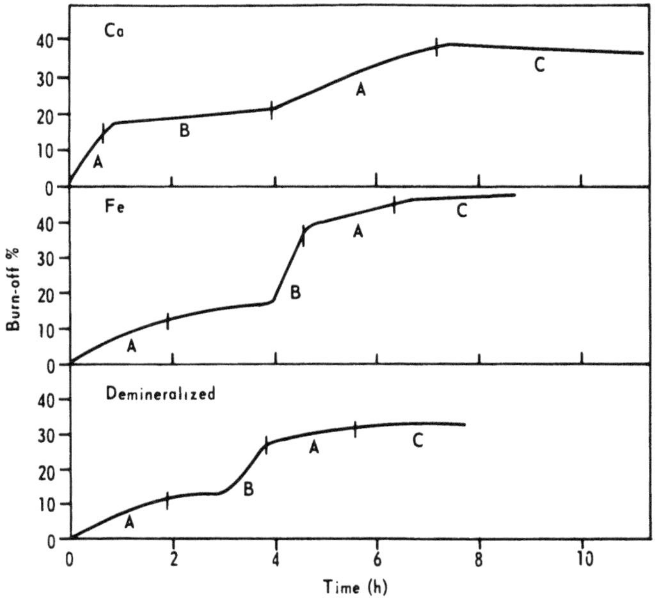

FIG. 19. Reactivity plots at 1123 K in 2·3 kPa steam for 1123 K chars prepared from demineralized, Ca- and Fe-loaded lignites. Wet N_2 (regimes A), wet H_2 (B) and wet CO (C).

K- and Na-loaded chars, gasification rates are similar in wet N_2 and wet H_2; CO, however, is a strong inhibitor of the C–H_2O reaction. For the Ca-loaded char, both H_2 and CO are strong inhibitors of the C–H_2O reaction. For the Fe-loaded char, H_2 is a strong promoter of the C–H_2O reaction once iron oxide is reduced to metallic iron. It is of interest that the behaviour of the char produced from the raw lignite resembles that found for the Ca-loaded char. This is to be expected since Ca is the predominant inorganic species in the raw lignite. On the other hand, the behaviour of the acid-treated (demineralized) lignite resembles that of the Fe-loaded char. This again is to be expected since acid treatment will not effectively remove pyrite. Upon pyrolysis of the lignite in N_2, pyrite will be converted to a pyrrhotite and, subsequently, to metallic Fe in the presence of wet H_2.

ACKNOWLEDGEMENTS

Support of our research, reported on in this paper, by the US Department of Energy under Contracts EX-76-C-01-2030, DE-AC01-79ET14882 and

DE-AC02-79ER10488 and the Gas Research Institute under Contract 5014-363-0235 is gratefully acknowledged.

REFERENCES

1. Walker, P. L., Jr., Rusinko, F., Jr. and Austin, L. G. (1959). *Advances in Catalysis*, Vol. 11, D. D. Eley, P. W. Selwood and P. B. Weisz (Eds). Academic Press, New York, p. 164.
2. Walker, P. L., Jr., Shelef, M. and Andersen, R. A. (1968). *Chemistry and Physics of Carbon*, Vol. 4, P. L. Walker, Jr. (Ed.). Marcel Dekker, New York, p. 287.
3. Hennig, G. R. (1966). *Chemistry and Physics of Carbon*, Vol. 2, Walker, P. L., Jr. (Ed.). Marcel Dekker, New York, p. 1.
4. Laine, N. R., Vastola, F. J. and Walker, P. L., Jr. (1963). *J. Phys. Chem.*, **67**, 2030.
5. Thomas, J. M. and Hughes, E. E. G. (1964). *Carbon*, **1**, 209.
6. Baker, J. J. (1970). Ph.D. Thesis, The Pennsylvania State University.
7. Evans, E. L. and Thomas, J. M. (1971). *Proceedings Third Industrial Conference on Carbon and Graphite*. London, Soc. Chem. Industry, pp. 3–9.
8. Baker, R. T. K. and Harris, P. S. (1973). *Carbon*, **11**, 25.
9. Yang, R. T. and Wong, C. (1981). *J. Chem. Phys.*, **75**, 4471.
10. Baker, R. T. K. and Chludzinski, J. J. (1981). *Carbon*, **19**, 75.
11. Evans, E. L., Griffiths, R. J. M. and Thomas, J. M. (1971). *Science*, **171**, 174.
12. McKee, D. W. (1981). *Chemistry and Physics of Carbon*, Vol. 16, P. L. Walker, Jr. and P. A. Thrower (Eds). Marcel Dekker, New York, p. 1.
13. McKee, D. W. and Chatterji, D. (1975). *Carbon*, **13**, 381.
14. McKee, D. W. and Chatterji, D. (1978). *Carbon*, **16**, 53.
15. McKee, D. W. (1979). *Carbon*, **17**, 419.
16. McKee, D. W. (1982). *Carbon*, **20**, 59.
17. Long, F. J. and Sykes, K. W. (1952). *Proc. Roy. Soc.* (*London*), **A215**, 100.
18. Long, F. J. and Sykes, K. W. (1950). *J. Chim. Phys.*, **47**, 361.
19. Holstein, W. L. and Boudart, M. (1982). *J. Catalysis*, **75**, 337.
20. Mahajan, O. P., Yarzab, R. and Walker, P. L., Jr. (1978). *Fuel*, **57**, 643.
21. Simons, G. A. (1980). *Fuel*, **59**, 143.
22. Simons, G. A. (1982). *Carbon*, **20**, 117.
23. Jenkins, R. G., Nandi, S. P. and Walker, P. L., Jr. (1973). *Fuel*, **52**, 288.
24. Hippo, E. and Walker, P. L., Jr. (1975). *Fuel*, **54**, 245.
25. Linares-Solano, A., Mahajan, O. P. and Walker, P. L., Jr. (1979). *Fuel*, **58**, 327.
26. Gan, H., Nandi, S. P. and Walker, P. L., Jr. (1972). *Fuel*, **51**, 272.
27. Morgan, M. E., Jenkins, R. G. and Walker, P. L., Jr. (1981). *Fuel*, **60**, 189.
28. Tomita, A., Mahajan, O. P. and Walker, P. L., Jr. (1977). *Fuel*, **56**, 137.
29. Thakur, S. C. and Brown, L. F. (1982). *Carbon*, **20**, 17.
30. Walker, P. L., Jr., Pentz, L., Biederman, D. L. and Vastola, F. J. (1977). *Carbon*, **15**, 165.
31. Phillips, R., Vastola, F. J. and Walker, P. L., Jr. (1970). *Carbon*, **8**, 197.

32. Ashu, J. T., Nsakala, N. Y., Mahajan, O. P. and Walker, P. L., Jr. (1978). *Fuel*, **57**, 250.
33. Radovic, L. R., Walker, P. L., Jr. and Jenkins, R. G. (1982). International Symposium on Fundamentals of Catalytic Coal and Carbon Gasification, Amsterdam, September.
34. Walker, P. L., Jr. (1981). *Phil. Trans. Roy. Soc. London*, **A300**, 65.
35. Mahajan, O. P., Komatsu, M. and Walker, P. L., Jr. (1980). *Fuel*, **59**, 3.
36. Hippo, E. J., Jenkins, R. G. and Walker, P. L., Jr. (1979). *Fuel*, **58**, 327.
37. Hirsch, R. L., Gallagher, J. E., Jr., Lessard, R. R. and Wesselhoft, R. D. (1982). *Science*, **215**, 121.

29

A Review of the Vapor Phase Pyrolysis of Biomass Derived Volatile Matter

MICHAEL JERRY ANTAL, JR
Department of Mechanical Engineering, University of Hawaii, Honolulu, Hawaii, USA

NOTATION

A	Apparent frequency factor
C	Mass fraction of carbon
D	Reactor diameter
E	Apparent activation energy
k	$= A\exp(-E/RT)$; lumped parameter rate constant
L	Reactor length
t	Time
T	Absolute temperature
U	Superficial velocity

Subscripts

g	gas
t	refractory condensable materials (tars)
v	reactive volatile materials
0	initial value entering the vapor phase reactor
∞	asymptotic (long time) value

1. INTRODUCTION

Upon heating, biomass materials undergo solid phase pyrolysis at relatively low temperatures (> 300 °C), forming reactive volatile matter, a few permanent gas species and solid char. Unlike the various coals and peats, biomass materials typically lose 70 % or more of their weight by the solid phase pyrolysis reactions. This transformation of the bulk of the biomass material from the solid to the vapor phase suggests the important role of vapor phase chemistry in the pyrolysis of biomass materials. Recognizing the highly reactive nature of the major constituents of the volatile matter, the significance of the vapor phase chemistry becomes even more apparent.

In 1976 research on the vapor phase pyrolysis of biomass derived volatile matter began in this laboratory. Since then, numerous publications were prepared describing the effects of vapor phase temperature and residence time on the pyrolysis of cellulose derived volatile matter,[1-3] lignin derived volatile matter,[3,4] and wood derived volatile matter.[5] Because laminar flow reactors were best suited for these studies, a significant effort was made to explore the ability of these reactors to provide reliable mechanistic and kinetic data on gas phase pyrolysis phenomena.[6-8] Studies of the effects of increased pressure on the vapor phase pyrolysis phenomena were also completed.[9-11] Because of the inherent complexity of these reactions, studies of the vapor phase pyrolysis of the model compounds glycerol,[12] 1,3 dioxolane[8] and 1,6 anhydro-β-D glucopyranose[8] were undertaken and are now being completed.

During the past two years, a number of the author's colleagues have requested a review of the results of this laboratory's research on gas phase pyrolysis phenomena, described in refs 1–7. As a response to these requests, this paper was prepared. A more complete review of the field must await the conclusion of ongoing research in this laboratory[8] and elsewhere.[13]

2. PREVIOUS RESEARCH

Apart from the research activities of this laboratory, no detailed studies of the effects of vapour phase temperature and residence time on the vapor phase pyrolysis of biomass derived volatile matter exist in the literature.

Houminer and Patai[14] reported studies of the low temperature (175–275 °C) pyrolysis of ^{14}C labelled D-glucose under vacuum in a closed vessel. Similarly, Shafizadeh *et al.*[15] described studies of the pyrolysis of ^{14}C labelled, 1,6 anhydro-β-D-glucopyranose at 600 °C under vacuum. Both these studies offered considerable insight into the pyrolysis mechanism, but neither included an investigation of the effects of temperature and residence time on the mechanism.

Broido *et al.*,[16] McCarter[17,18] and Baker[19] reported studies using flow reactors similar to the laminar flow reactor used in this work.

Unfortunately none of these authors used their equipment to study systematically the effects of reactor conditions on the pyrolysis products.

More recently, Deglise and his colleagues[20] described studies of the high temperature pyrolysis (1000 °C) of wood in a flow reactor. Results of these studies corroborate the gas yields reported here, but do not include a clear statement of the effects of residence time on the product distribution.

Many other studies exist which emphasize practical engineering, but typically do not differentiate between solid and gas phase conditions and phenomena. Consequently, these studies are of little interest in the context of this review.

3. EXPERIMENTAL APPARATUS AND PROCEDURES

For the experiments described below, dry Whatman No. 1 filter paper, Indulin AT Kraft Pine Lignin Polymer, Cherry and Southern Yellow Pine wood sawdust stored in desiccant jars were used as sample materials. (Because the Indulin AT Kraft Pine Lignin Polymer is highly altered, it may not be representative of native lignin. Similarly, the presence of some ash in Whatman No. 1 filter paper may cause its behavior to depart from that of pure cellulose, or native cellulose in the presence of mineral matter naturally present in biomass.) The cherry wood sawdust (HW) and the Yellow Pine wood sawdust (SW) were grab samples from industrial sources with particle sizes ranging between 75 and 200 μm. The cellulose was assumed to have the elemental composition 0·444 C, 0·062 H and 0·494 O on a mass fraction basis. Table I lists elemental compositions of the Indulin AT Lignin, the two woods, tar derived from pyrolysis of the lignin and chars derived from the cellulose lignin, and wood sample materials.

TABLE I

ULTIMATE ANALYSES OF SAMPLE AND PRODUCT MATERIALS

	C	*H*	*O*
Kraft lignin	0·635	0·059	0·251
Cherry wood	0·495	0·061	0·433
Southern yellow pine wood	0·492	0·061	0·395
Cellulose char	0·783	0·040	0·177
Kraft lignin char	0·793	0·035	0·101
Cherry wood char	0·838	0·036	0·080
Southern yellow pine wood char	0·842	0·036	0·092
Kraft lignin tar	0·743	0·056	0·185

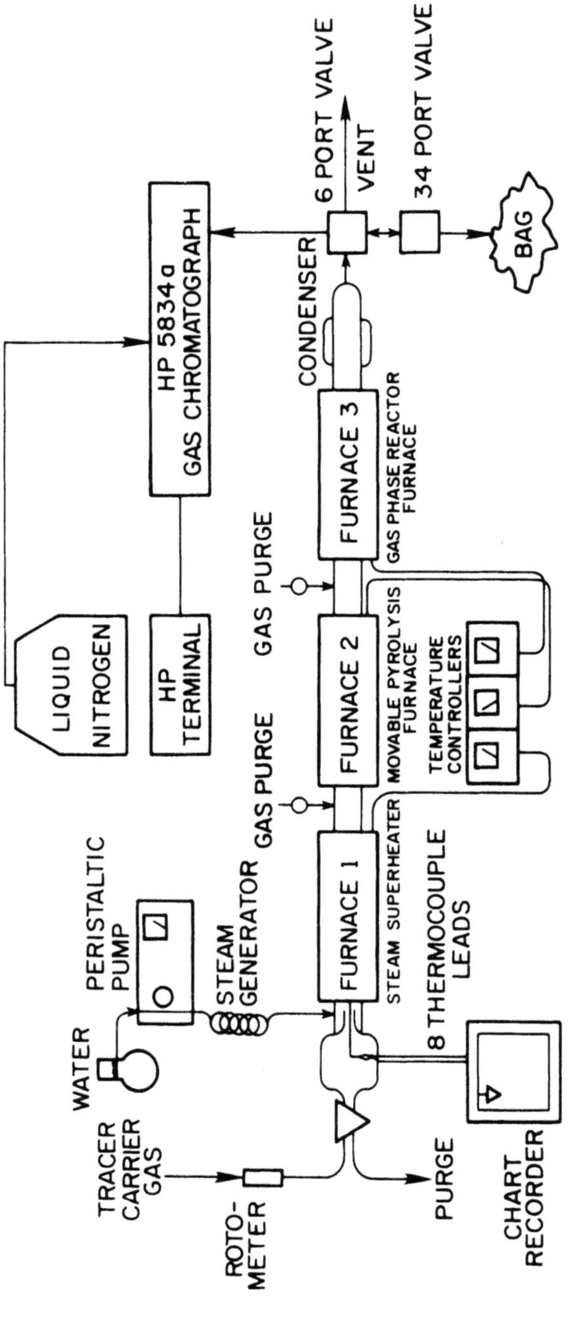

FIG. 1. Reactor schematic.

The reactor used in these studies was based on a semi batch, laminar flow design concept. This design was selected because of its simplicity, ease of operation and ability to handle the wide variety of products known to result from biomass pyrolysis. In essence, a small sample of biomass was devolatilized in a flowing steam environment within one zone of the reactor. The temperature of this zone was sufficiently low to minimize the onset of the gas phase pyrolysis reactions. The fully developed, laminar flow of steam carried the diluted volatile matter into a high temperature, 'isothermal' zone of the reactor. In this zone the gase phase pyrolysis reactions were permitted to occur for a selected time period. Product composition was then studied as a function of gas phase temperature and residence time. Some details of the apparatus are discussed below.

As shown in Fig. 1, the reactor system consisted of a long 1 in. diameter quartz tube, heated by three standard 'clam shell' furnaces. The middle furnace, used to pyrolyze the sample material, was movable and could be preheated to the desired pyrolysis temperature without heating the sample. For the experiments described here, the steam superheater zone was maintained at 350 °C, and the pyrolysis furnace at 500 °C. This latter setting caused a measured sample heating rate of about 100 °C/min. The 500 °C pyrolysis temperature was chosen because it was sufficiently high to devolatilize the sample rapidly, but too low to effect char gasification by the water gas or Boudouard reactions. Temperatures within the reactor were controlled by various Omega and Honeywell temperature controllers and monitored by Type K thermocouples with continuous recording on chart recorders.

A measured flow of steam was obtained using a peristaltic pump to force liquid water through a heated, coiled tube steam generator. The residence time of the steam and volatile matter mixture in the gas phase reactor was altered by varying the peristaltic pump's flow rate between 0·06 and 0·34 g/min, and by inserting a closed quartz cylinder into the gas-phase reactor to reduce its apparent volume. Residence times of 2–3 s were obtained using a water flow rate of 0·34 g/min with a 0·25 g sample. Shorter residence times were obtained using the quartz insert. For the cellulose experiments, longer residence times were obtained by reducing the water flow rate and cellulose sample size proportionately, thus ensuring a comparable steam dilution of the cellulose volatile matter. Because the wood and lignin experiments were not strongly influenced by variations in steam dilution, and because larger samples were required to produce sufficient gas for analysis, the wood and lignin sample sizes were maintained at about 0·3 mg for most of the experiments.

Prior to the initiation of the experiment, a small (100–300 mg) sample of cellulose, wood or lignin was placed in a quartz boat and moved to the center of the pyrolysis zone of the reactor. After purging the reactor with argon, the preheated pyrolysis furnace was moved into place around the pyrolysis zone of the reactor and flowing superheated steam carried the volatile matter evolved during pyrolysis of the sample into the preheated gas phase ('cracking') zone of the reactor. Exiting the reactor, the gaseous products were rapidly quenched in a condenser, the tars and other condensibles being collected in Teflon gas-tight sample bags. The volume of gas produced was measured by water displacement, and the gas composition was analyzed.[1–5]

Several weaknesses of laminar flow reactors impact on the interpretation of results obtained in these studies. These weaknesses include: (1) non-instantaneous heat-up of the reactants and quench of the products, (2) departures from isothermality along the reactor, (3) thermocouple temperature measurement errors, (4) departures from plug flow idealizations (i.e. laminar flow), (5) dispersion and molecular diffusion of the reactants and products, (6) variations of the concentration of volatile matter in steam due to the time dependent rate of pyrolysis of the solid sample, (7) the potential effects of wall reactions on the product distribution, and (8) the difficulty in completely collecting the tars and other condensible materials at the outlet of the reactor. As described in the following paragraphs, efforts were made to explore the extent of these weaknesses and their effects on the interpretation of our data.

References 2 and 5 detail both the calculated and experimentally measured temperature–time history of the gas flow during heat-up and quench. At the higher flow rates the gas flow traversed as much as 50 % of the high temperature cracking zone's length prior to reaching 'isothermal conditions'. Consequently, an average residence time was estimated for a given set of flow and temperature conditions. The details of this estimate are discussed in refs 2 and 5. In addition to heat-up and quench, the temperature uniformity along the length of the furnaces used in this research was typically no better than $\pm 10\,^{\circ}\mathrm{C}$. Even larger departures were evidenced near the ends of the furnace. Finally, radiative heat transfer to the thermocouple was estimated to cause a potential 13 °C overestimate of the gas temperature at 600 °C. No experimental verification of this calculated error was attempted, and temperatures reported here were those measured directly by the thermocouple in contact with the flowing gases.

Our early interpretations of the experimental data involved the idealization of plug flow within the reactor.[1–4] More recent theoretical work[6,7] established the agreement of the plug flow idealization with more realistic interpretations based on laminar flow for simple single step reactions of all orders.

Experimental studies[5] of radial and longitudinal dispersion within the reactor resulted in the measured value (see Notation)

$$D/UL \cong 0{\cdot}12$$

for the vessel dispersion number D/UL. The magnitude of this value suggests a significant departure from plug flow; however we believe that much of the dispersion occurred after the gases exited the reactor and entered the condenser system, where their residence typically exceeded 20 s. Theoretical studies[6,7] of the effects of radial diffusion on the kinetic interpretation of data derived from laminar flow reactors indicate that these effects are negligible relative to other sources of error.

For first order reactions in the gas phase, theoretical studies[2] indicated that the temporal variation in volatile matter concentration within the reactor has no effect on the kinetic interpretation of measured reaction rates. However, the effects of this variation on the interpretation of data involving non-first-order reactions is not known.

Due to the use of the insert to reduce the apparent reactor volume for the short residence time experiments, the reactor's apparent surface-to-volume ratio underwent considerable variation during the course of this research. No effects of the surface-to-volume ratio variation were noted; however, such effects could have been masked by the assumed stronger dependence of the data on temperature and residence time.

The recovery of tars from the reactor was difficult due to their solubility in water, and our inability to effect an analysis of the organic constituents of the water during the time that these experiments occurred. Carbon balances for the high temperature cellulose and lignin experiments typically exceeded 0·9; however, the experiments involving wood were often somewhat lower ($>0{\cdot}8$). In large part this difference reflected the formation of water soluble products from the woody materials, which were not included in the carbon balance.

4. RESULTS

Products of the steam pyrolysis process include a variety of light permanent gases, accompanied by smaller yields of char and condensible materials. The focus of attention of this research was the influence of reactor

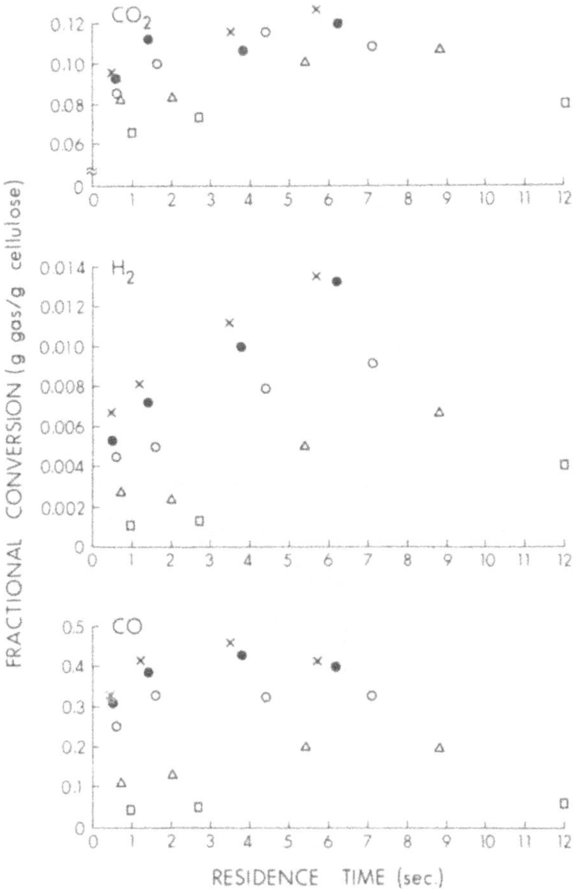

FIG. 2. Fractional conversion of cellulose to hydrogen and carbon oxides (□ 500 °C, △ 600 °C, ○ 650 °C, ● 700 °C, × 750 °C).

conditions on gas yields (especially the yields of alkenes and alkanes); however, some attention was also given to the formation of char and condensible materials.

The yields of light, permanent gases (given as a mass fraction of the feedstock) derived from the gas phase pyrolysis of cellulose, Kraft lignin, Cherry (HW) and Yellow Pine (SW) wood are displayed as a function of the gas phase temperature and residence time in Figs 2–4 (cellulose), 6–8 (lignin), and 10–16 (HW and SW). Similar displays of the influence of reactor conditions on the carbon, hydrogen, and oxygen efficiencies

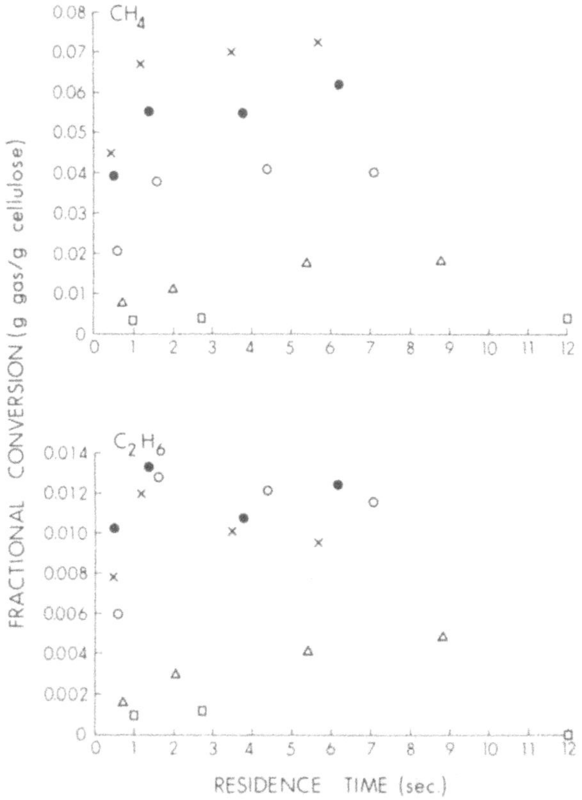

FIG. 3. Fractional conversion of cellulose to alkanes (□ 500 °C, △ 600 °C, ○ 650 °C, ● 700 °C, × 750 °C).

(carbon efficiency = carbon in gas/carbon in feedstock) are given in Figs 5, 9 and 17.

The most striking phenomenon exhibited in Figs 2–17 is the remarkably similar behavior of all the materials studied. Gaseous species yields from each feedstock display an asymptotic ('long' residence time; ~5 s) behavior that is strongly temperature dependent. Higher temperatures result in dramatic increases in the asymptotic yields of all the light permanent gases produced. One goal of this research was to develop an explanation for this striking phenomenon, held in common by all the materials studied.

Yields of the chars formed from each material, and their elemental compositions, are given in Table I. Under conditions which maximized gas formation, the yields of the condensible materials were small. Because

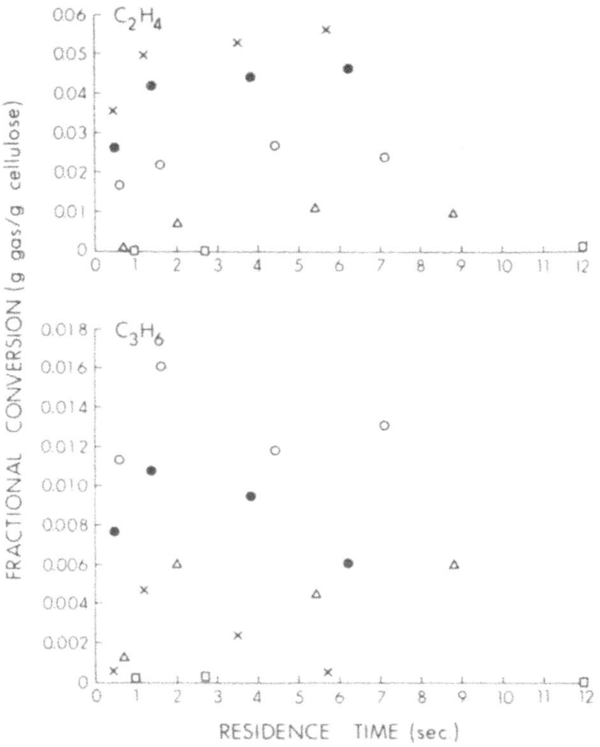

FIG. 4. Fractional conversion of cellulose to alkenes (□ 500 °C, △ 600 °C, ○ 650 °C, ● 700 °C, × 750 °C).

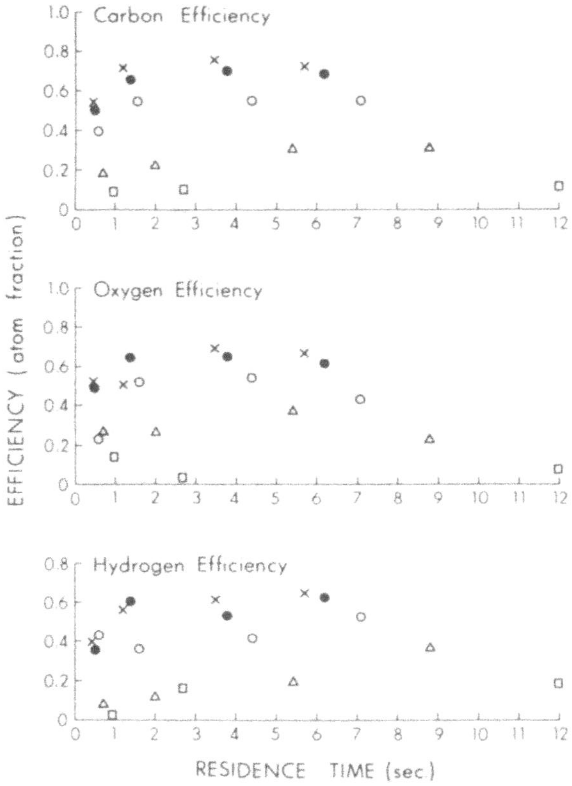

FIG. 5. Carbon, oxygen, and hydrogen efficiencies for cellulose (□ 500 °C, △ 600 °C, ○ 650 °C, ● 700 °C, × 750 °C).

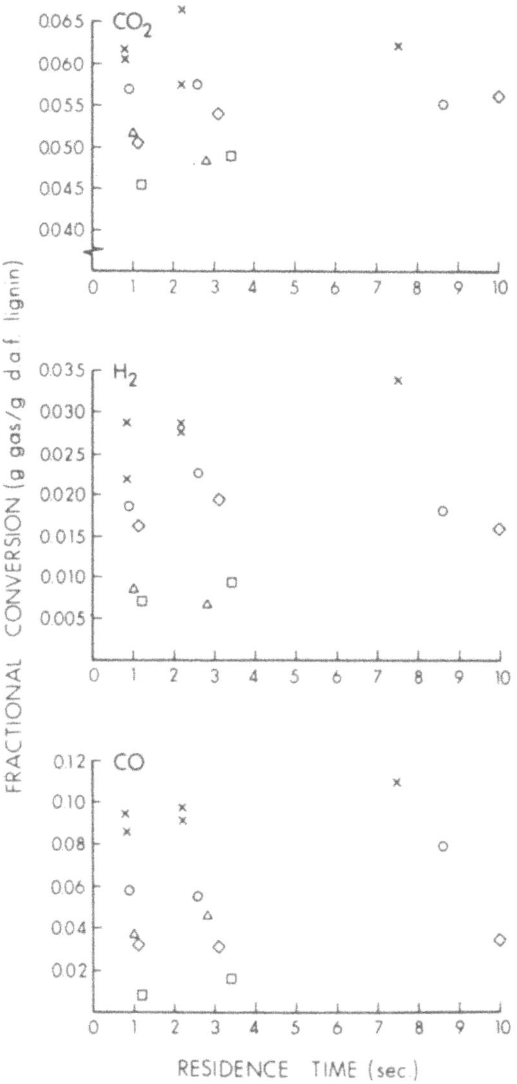

FIG. 6. Fractional conversion of lignin to hydrogen and carbon oxides (□ 500 °C, ◇ 550 °C, △ 600 °C, ○ 650 °C, × 750 °C).

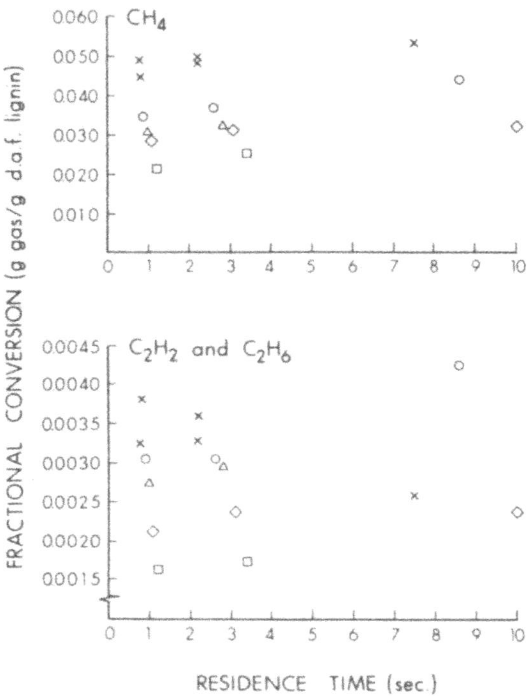

FIG. 7. Fractional conversion of lignin to alkanes and alkynes (□ 500 °C, ◇ 550 °C, △ 600 °C, ○ 650 °C, × 750 °C).

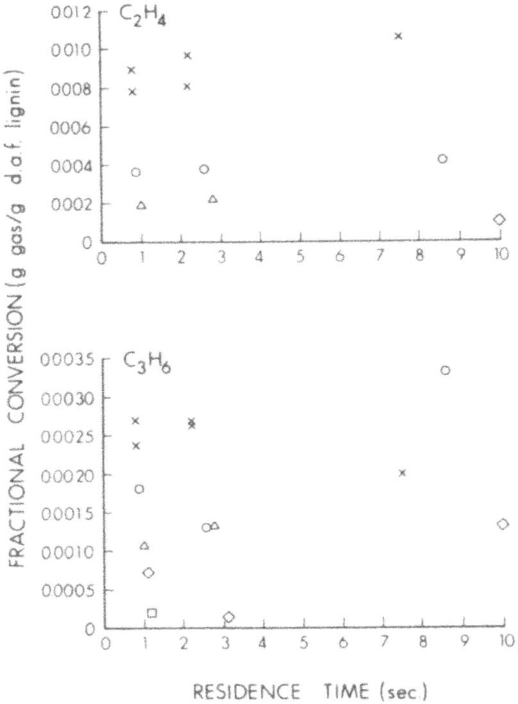

FIG. 8. Fractional conversion of lignin to alkenes (□ 500 °C, ◇ 550 °C, △ 600 °C, ○ 650 °C, × 750 °C).

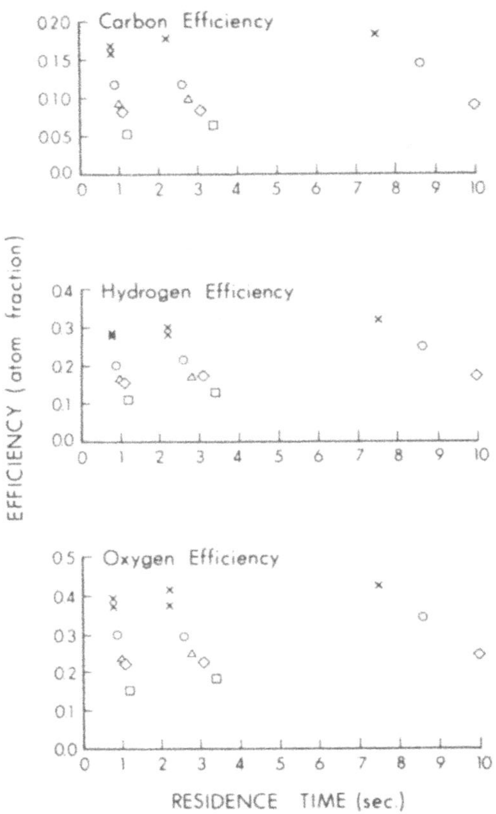

FIG. 9. Carbon, hydrogen and oxygen efficiencies for lignin (□ 500 °C, ◇ 550 °C, △ 600 °C, ○ 650 °C, × 750 °C).

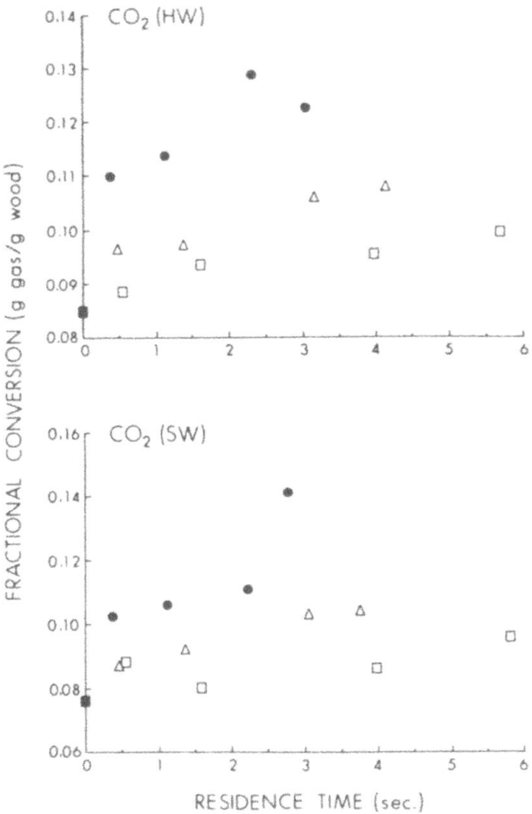

FIG. 10. Fractional conversion of HW and SW to CO_2 (■ 25 °C, □ 500 °C, △ 600 °C, ● 700 °C).

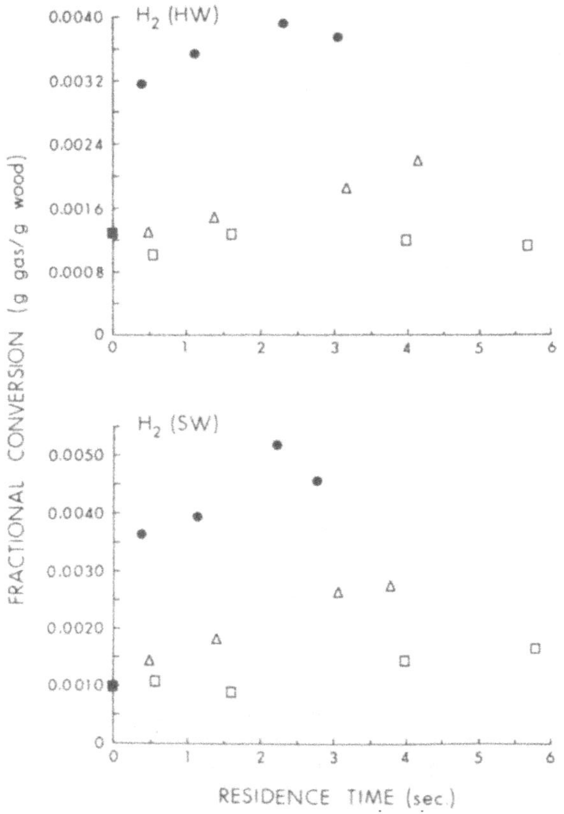

FIG. 11. Fractional conversion of HW and SW to H_2 (■ 25 °C, □ 500 °C, △ 600 °C, ● 700 °C).

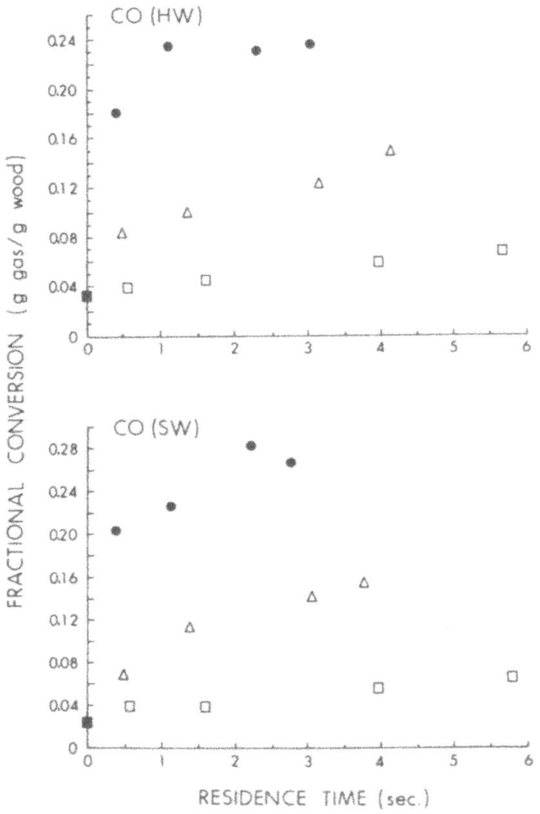

FIG. 12. Fractional conversion of HW and SW to CO (■ 25 °C, □ 500 °C, △ 600 °C, ● 700 °C).

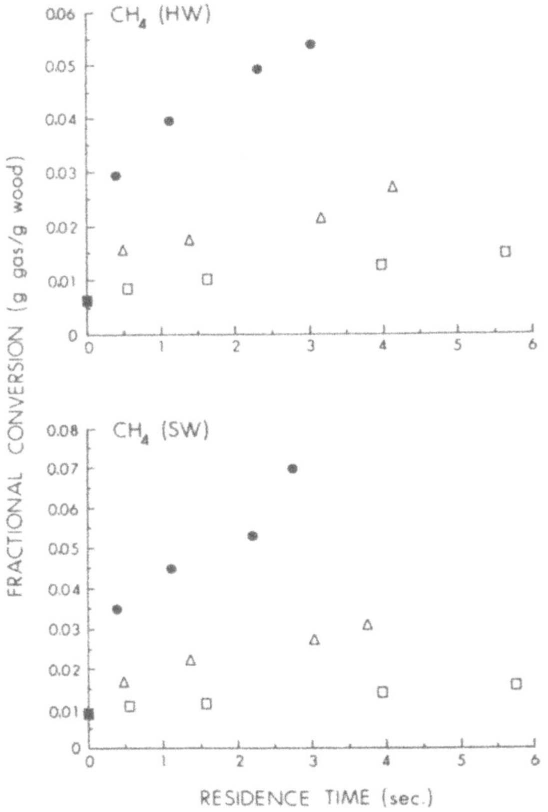

FIG. 13. Fractional conversion of HW and SW to CH_4 (■ 25 °C, □ 500 °C, △ 600 °C, ● 700 °C).

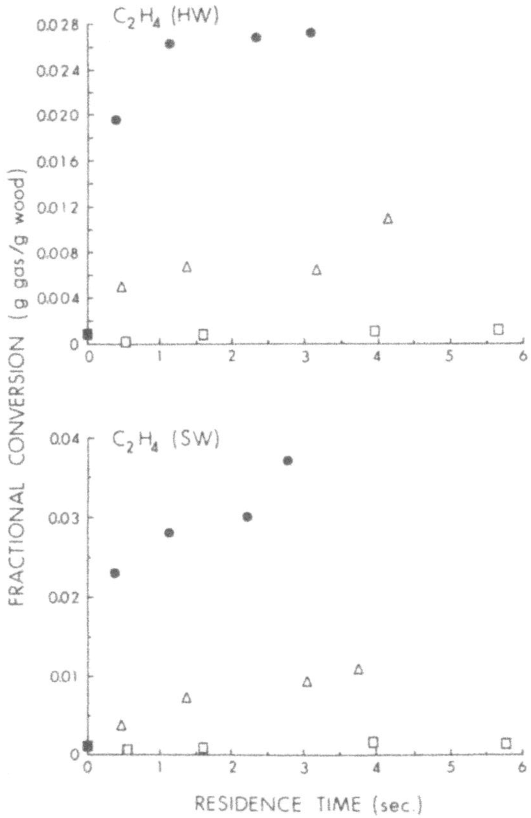

FIG. 14. Fractional conversion of HW and SW to C_2H_4 (■ 25 °C, □ 500 °C, △ 600 °C, ● 700 °C).

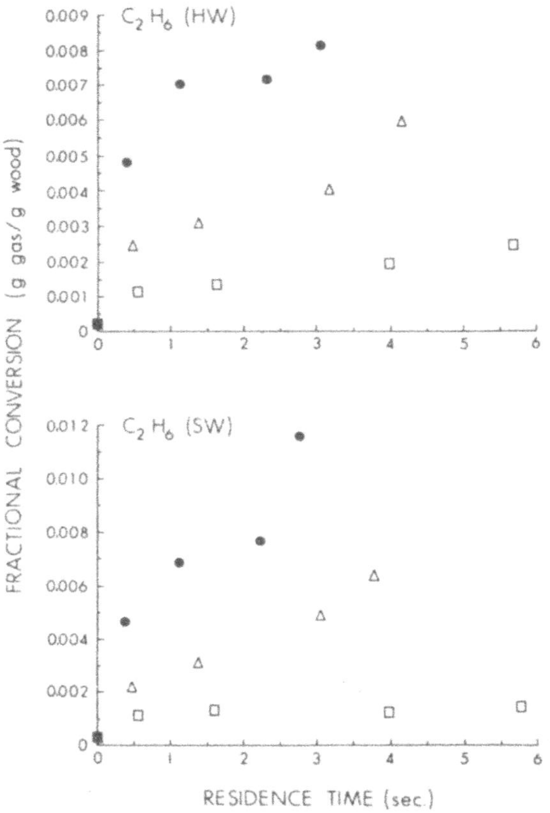

FIG. 15. Fractional conversion of HW and SW to C_2H_6 (■ 25 °C, □ 500 °C, △ 600 °C, ● 700 °C).

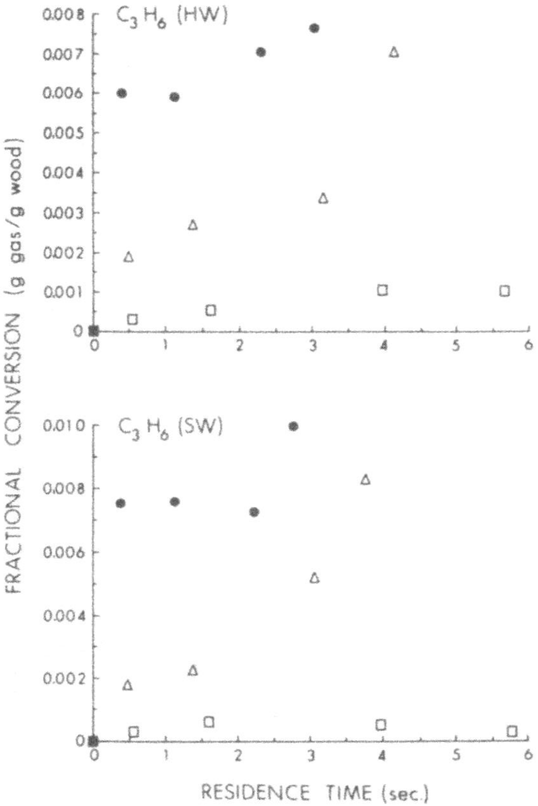

FIG. 16. Fractional conversion of HW and SW to C_3H_6 (■ 25 °C, □ 500 °C, △ 600 °C, ● 700 °C).

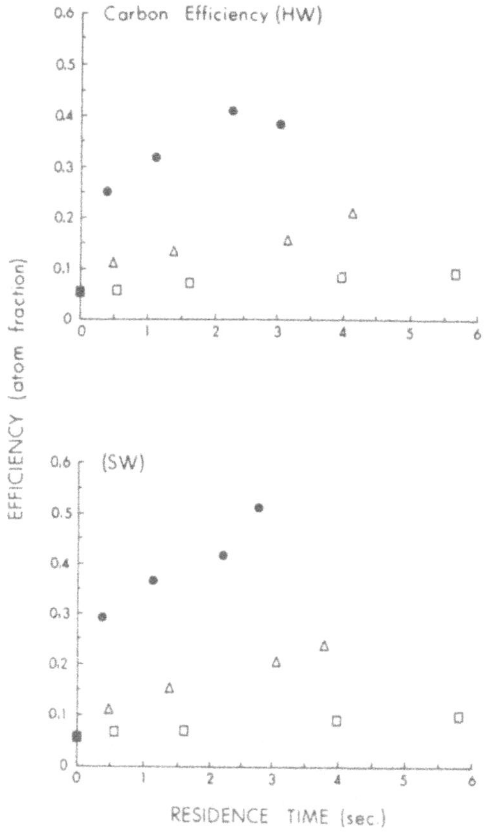

FIG. 17. HW and SW carbon efficiency (■ 25 °C, □ 500 °C, △ 600 °C, ● 700 °C).

of this small yield, and the evident dependence of both the absolute yield and its chemical composition on the gas phase reactor conditions, no attempt was made to analyze chemically the condensible products. A carbonaceous deposit on the walls of the quartz reactor was observed following experiments involving wood or lignin.[5] The amount of material formed was too small to permit analysis.

Several experiments were performed to explore the effects of steam dilution, particle size and heating rate on the product distributions displayed in Figs 2–17. Contrary to expectation, increasing the steam dilution by a factor of six reduced the methane yield from the Cherry wood by about 6 %, while the ethylene yield increased by 24 % before returning to its initial value.[5] The steam dilution had little effect on the formation of CO and char. Variations in particle size from 45 μm to 5·66 mm had little effect on the gaseous product composition. Higher solid phase pyrolysis temperatures increased the formation of secondary char, which collected on the walls of the quartz sample holder, but otherwise had little influence on the experimental results.

5. DISCUSSION OF RESULTS

An attempt was made to correlate the observed gas yields with the gas phase reactor conditions using the well known kinetic severity function KSF:[21]

$$\text{KSF} = K_5\,\mathrm{d}t = (1{\cdot}77 \times 10^{12})\exp(-55\,420/RT(t))\,\mathrm{d}t$$

where K_5 is the reaction velocity associated with *n*-pentane disappearance and the other terms are defined in the Notation section. Because temperature dependent asymptotes were still evident when gas yields obtained from the cellulose, lignin and two woods were plotted against KSF, it was concluded that the kinetic severity function commonly used to quantify reactor 'severity' has limited relevance to biomass pyrolysis.

An attempt was made to explain the gas yields obtained from the two woods using a linear programming optimization algorithm.[5] The algorithm was constructed to vary the fractional contributions of experimentally determined gas yields from cellulose, Kraft lignin, and D-mannose in such a way as best to fit the experimentally observed gas yields from the two woods. Mathematically this was accomplished by minimizing the value of a specified objective (error) function. For both woods, the best fit was obtained when only the cellulose data was used to fit the gas yields obtained from each wood. However, in both cases the value of the objective function was relatively large, indicating an unsatisfactory fit. Other, less sophisticated approaches[5] also failed to explain the wood gas species yields in terms of a linear superposition of yields obtained from the individual wood components. Although it could be argued that the cellulose, hexose and lignin samples studied did not adequately represent the behavior of the *in situ* materials, Mattocks[5] concluded that wood gasification probably does not simply reflect the independent behavior of its individual components.

The temperature dependent asymptotes evidenced in Figs 2–17 are reminiscent of the well known temperature dependent asymptotic char yields resulting from cellulose pyrolysis.[22] Broido and Nelson[22] used this phenomenon to assert the role of competitive reactions in cellulose pyrolysis. Consequently, the role of competitive reactions in the gas pyrolysis of biomass derived volatile matter was examined as a potential explanation for the temperature dependent asymptotic species yields noted in Figs 2–17.

The pyrolysis of cellulose, lignin and wood within the solid phase pyrolysis zone of the laminar flow reactor produced char, permanent gases and a reactive volatile matter. A competitive reaction mechanism was posited to consume the reactive volatile matter:

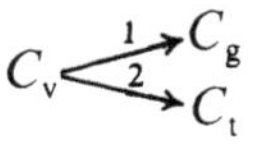

I

In Scheme **I** C_v represents the mass fraction of carbon composing the reactive volatile matter (carbon in volatile matter/carbon in sample pyrolyzate), C_g the mass fraction of carbon composing the permanent gases, and C_t the mass fraction of carbon composing the posited refractory condensible materials, including the tars. This mechanism presumes the reactive volatile matter to be concurrently consumed by two reactions: (1) a higher temperature cracking reaction forming light permanent gases, and (2) a lower temperature reaction forming refractory condensible materials, including tars. Because this research did not include extensive studies of the condensible materials, the existence of reaction 2 had to be inferred from the gas yield data. The temperature dependent asymptotes evidenced in these data essentially required the existence of reaction 2 in order to explain the disappearance of carbon atoms in the gas phase when the gas phase temperature was reduced.

The following global rate laws were assumed to govern the disappearance of C_v and C_t:

$$\frac{dC_v}{dt} = -(k_1 + k_2)C_v \tag{1}$$

$$\frac{dC_g}{dt} = k_1 C_v \tag{2}$$

$$\frac{dC_t}{dt} = k_2 C_v \tag{3}$$

where k_1 and k_2 are Arrhenius rate constants and t is the gas phase residence time. Solution of eqns (2)–(4) gives

$$C_g - C_{g0} = \frac{k_1 C_{v0}}{k_1 + k_2} \{1 - \exp[-(k_1 + k_2)t]\} \tag{4}$$

$$C_t = \frac{k_2 C_{v0}}{k_1 + k_2} \{1 - \exp[-(k_1 + k_2)t]\} \tag{5}$$

Recognizing the relationship

$$C_{v0} = (C_{g\infty} - C_{g0}) + C_{t\infty} \tag{6}$$

the following equations were derived:

$$k_1/k_2 = \frac{A_1}{A_2} \exp[-(E_1 - E_2)/RT] = \frac{C_{g\infty} - C_{g0}}{C_{t\infty}} \tag{7}$$

$$1 - \frac{(C_g - C_{g0})}{C_{v0}} (1 + k_2/k_1) = \exp[-(k_1 + k_2)t] \tag{8}$$

$$\frac{C_g - C_{g0}}{t} \cong k_1 C_{v0} \qquad (k_1 + k_2)t \ll 1 \tag{9}$$

From eqn (7) a plot of $\ln[(C_{g\infty} - C_{g0})/C_{t\infty}]$ vs T^{-1} was anticipated to be a straight line with slope $-(E_1 - E_2)/R$ and intercept $\ln(A_1/A_2)$. Values of k_1/k_2 obtained in this way were used with eqn (8) to obtain values for $k_1 + k_2$, which were then used to evaluate the individual values of k_1 and k_2. Finally, plots of $\ln k_i$ vs T^{-1} provided values for A_1, A_2, E_1 and E_2. The consistency of this approach was checked by plotting $\ln[(C_g - C_{g0})/t]$ vs T^{-1} for $(k_1 + k_2)t \ll 1$ as suggested by eqn (9). The slope $-E_1/R$ of the line obtained for values $t \ll (k_1 + k_2)^{-1}$ was required to agree with the values obtained from eqns (7) and (8) (based on $t \ll k_1 + k_2$) for the global mechanism to be meaningful.

Table II summarizes the results of these procedures for the four materials

TABLE II

KINETIC PARAMETERS FOR THE GAS PHASE PYROLYSIS OF VARIOUS BIOMASS MATERIALS

	HW	*SW*	*Cellulose*	*Lignin*
$E_1 - E_2$ (cal/mol)	19 700	22 900	34 300	13 000
E_1	23 600	24 100	48 800	NA
E_2	3 900	1 200	14 500	NA
A_1/A_2 (s^{-1})	$4{\cdot}39 \times 10^4$	$7{\cdot}50 \times 10^5$	$2{\cdot}07 \times 10^8$	$1{\cdot}05 \times 10^2$

TABLE III

APPARENT ACTIVATION ENERGIES ASSOCIATED WITH THE FORMATION OF INDIVIDUAL GAS SPECIES FROM THE GAS PHASE PYROLYSIS OF CELLULOSE AND WOOD DERIVED VOLATILE MATTER

	HW (*kcal/g mol*)	*SW* (*kcal/g mol*)	*Cellulose* (*kcal/g mol*)
CO	25	29	60
H_2	66	39	35
CH_4	23	24	67
C_2H_4	50	62	55
C_2H_6	17	14	38
C_3H_6	41	29	55
Iso-C_4H_8	18	34	—
CO_2	—	—	21

studied. The data fell on straight lines within the temperature ranges studied, and the correlation coefficients R measuring the goodness of fit often exceeded 0·99.

Reference 2 presents the derivation of an equation similar to eqn (9), which can be used to estimate the apparent activation energies associated with the formation of individual gas species. Table III lists values obtained for the two wood and the cellulose samples studied. Apart from ethylene, little agreement exists between these values of E_i for the three materials studied. The very low values obtained for the two woods are unsatisfying, but may point to the action of lignin derived volatile matter as a free radical scavenger.

The derivation of eqn (7) and its success in correlating the cellulose, lignin and wood data points to the utility of graphs which display the logarithm of a ratio of product yields vs T^{-1}. When competitive reactions are prevalent, plots of the logarithm of the ratio of asymptotic yields of individual products vs T^{-1} result in straight lines whose slopes reflect the difference in activation energies of the prevalent competing reactions. For this reason, Figs 18–21 were prepared to explore the role of competitive reactions in the formation of individual gas species from the gas phase pyrolysis of volatile matter derived from cellulose, lignin and the two woods. Essentially, this involved an exploration of the details of reaction 1 in Scheme **I**. In Figs 18–21 the gas species yield was calculated to be the difference between the asymptotic yield of the gas and the yield resulting from the solid phase pyrolysis reactions alone (i.e. ΔCH_4 = asymptotic mass fraction yield of CH_4 less the yield obtained for $T = 500\,°C$ and $t \rightarrow 0$).

Referring to Figs 18 and 19, both cellulose and lignin evidence straight

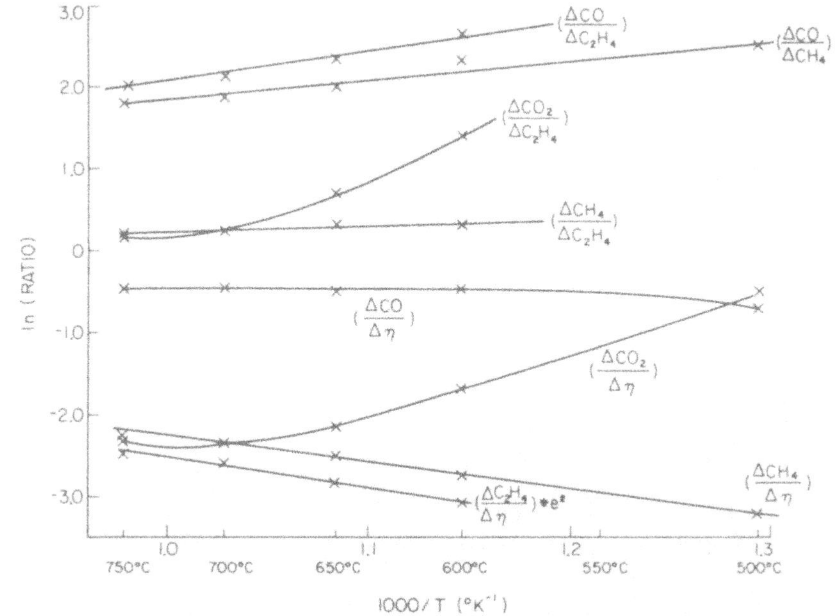

FIG. 18. Natural logarithm of the ratio of various gas yields and $\Delta\eta$ vs T^{-1} for cellulose pyrolysis.

lines for ratios involving the gases CO, CH_4 and C_2H_4, as would be expected if competitive reactions were postulated to play a role in the formation of these individual gas species. Only ratios involving CO_2 do not evidence linear behavior. Examination of Fig. 18 shows competititon to favor the formation at higher temperatures of C_2H_4 over CO, and CO_2 and CH_4 formation over CO for cellulose volatile matter pyrolysis. Similarly, higher temperatures favor the formation of C_2H_4 over CO, CH_4 and CO_2 for lignin derived volatile matter; however, no competition is observed between CO and CH_4 formation. Data from the two woods evidence similar trends, which more closely resemble the lignin than the cellulose data. Temperature has little effect on the relative yields of CO and CH_4 from the two woods. Ethylene formation competes strongly with CO and methane derived from the Cherry wood at the lower temperatures, but less strongly at the higher temperatures. In general, the wood data show more scatter than the cellulose and lignin data, and the trends are somewhat less certain.

Graphs of the natural logarithm of the ratio of gas species yield (as defined above) to the fraction of carbon $\Delta\eta$ converted to permanent gases by

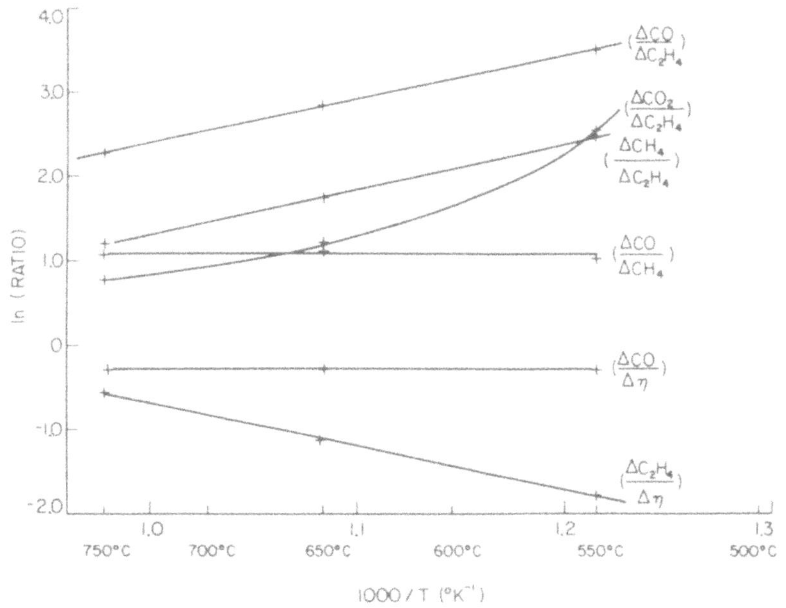

FIG. 19. Natural logarithm of the ratio of various gas yields and $\Delta\eta$ vs T^{-1} for lignin pyrolysis.

the gas phase cracking reaction vs T^{-1} also result in a relatively linear behavior for the gases CO, CH_4 and C_2H_4. Trends shown in Fig. 18 indicate that methane and ethylene together compete with carbon dioxide for cellulosic carbon atoms, with higher temperatures strongly favoring CH_4 and C_2H_4 over CO_2 formation. In marked contrast with this behavior, the fraction of carbon atoms allocated to CO formation by the gas phase cracking reactions remains constant over the temperature range 500–750 °C. For lignin, the competition appears to be between ethylene formation (alone) and CO_2 formation, since the ratios $\Delta CO/\Delta\eta$ and $\Delta CO/\Delta CH_4$ displayed in Fig. 19 appear to be independent of temperature. Both woods evidence competition of methane and ethylene for carbon atoms, with higher fractions of the total number of carbon atoms converted by the vapor phase pyrolysis reactions being dedicated to CH_4 and C_2H_4 formation at the higher temperatures. This behavior more closely resembles cellulose than lignin. Like cellulose and lignin, a fixed fraction of carbon atoms appears to be dedicated to CO formation from the wood derived volatile matter at temperatures above 600 °C.

The decreasing value of $\Delta CO_2/\Delta\eta$ for increasing temperature indicates

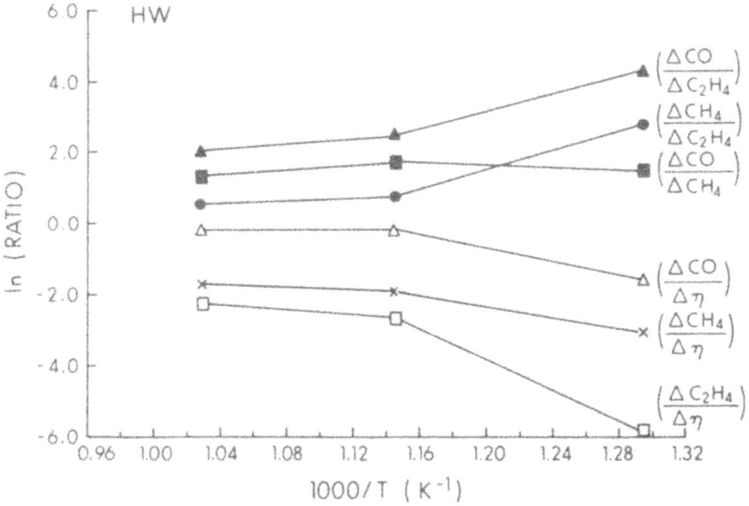

FIG. 20. Natural logarithm of the ratio of various gas yields and $\Delta\eta$ vs T^{-1} for HW pyrolysis.

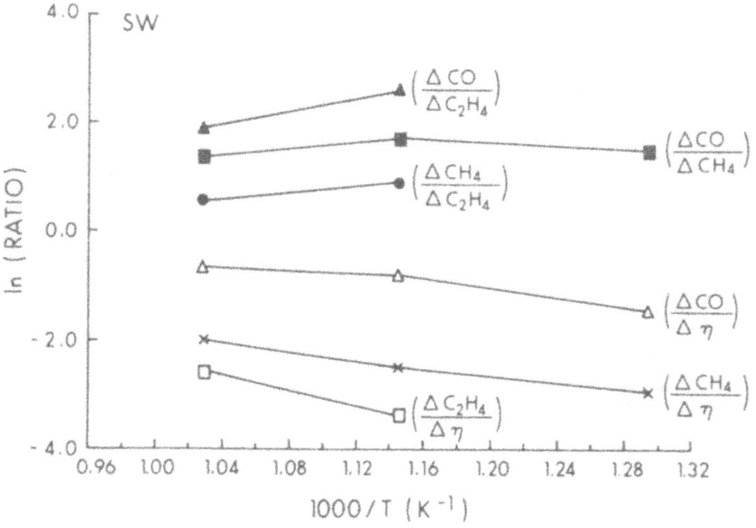

FIG. 21. Natural logarithm of the ratio of various gas yields and $\Delta\eta$ vs T^{-1} for SW pyrolysis.

the increasingly important role of dehydration reactions at the higher temperatures. As suggested by the stoichiometry shown in the following example for levoglucosan pyrolysis,

$$C_6H_{10}O_5 \rightarrow 2C_2H_4 + 2CO_2 + H_2O$$

optimal yields of olefins are obtained from carbohydrates only when both dehydration and decarboxylation reactions participate in the vapor phase pyrolysis chemistry. Thus the decreasing role of decarboxylation reactions at higher temperatures does not bode well for attempts to increase significantly the observed yields of C_2H_4 and other olefins. To secure a dramatic increase in the yields of C_2H_4 and other olefins, some means must be found to access those carbon atoms presently dedicated to CO formation, and cause the formation of CO_2 instead.

6. CONCLUSIONS

Yields of light permanent gases obtained from the gas phase pyrolysis of biomass derived volatile matter cannot be correlated to reactor conditions using the kinetic severity function. Instead, the yields are well described by a global mechanism consisting of two competing reactions. One of these reactions forms refractory, condensible materials (perhaps by polymerization or condensation reactions), whereas the second forms light permanent gases by a cracking reaction.

Attempts to represent the gas species yields derived from two woods as the simple linear superposition of yields derived from the woods' components did not succeed. This conclusion may reflect the influence of the chemical procedures used to isolate the woods' components. However, the facts that from certain perspectives pyrolytic yields of gases obtained from the two woods closely resembled those from cellulose, whereas from other perspectives the wood yields resembled those from lignin, suggest the pyrolytic behavior of wood derived volatile matter to be very complex.

A consideration of the effects of temperature on the ratios of gas yields obtained from the gas phase cracking reaction indicates the dedication of a fraction of volatile matter carbon atoms to carbon monoxide formation for all the materials studied. Higher temperatures favor the competitive formation of ethylene via dehydration reactions. For cellulose, methane and ethylene appear to be common products of the same reaction, whereas for lignin and the hardwood sample methane and carbon monoxide appear to be formed together.

Trends evidenced in this paper suggest that dramatic increases in the yields of olefins from biomass materials will be effected only by the discovery of some new influence which reduces the role of decarbonylation reactions in the gas phase pyrolysis chemistry.

REFERENCES

1. Antal, M. J., Jr (1978). Synthesis gas production from organic wastes by pyrolysis/steam reforming. In *Energy from Biomass and Wastes*, D. E. Klass (Ed.). Washington, D.C.
2. Antal, M. J., Jr (1981). The effects of residence time, temperature and pressure on a steam gasification of biomass. In *Biomass as a Non Fossil Fuel Source*, D. E. Klass (Ed.). American Chemical Society, Washington, D.C., pp. 313–34.
3. Antal, M. J., Jr. The effects of reactor severity on the gas phase pyrolysis of cellulose and lignin derived volatile matter. Submitted to I&EC Product Research and Development.
4. Antal, M. J., Jr (1980). Pyrolysis studies: Solid and gaseous pyrolysis of lignin. Final Report to SERI, Subcontract AM-9-8245-1, July.
5. Mattocks, T. W. (1981). Solid and gas phase phenomena in the pyrolytic gasification of wood. MSE Thesis, Princeton University.
6. Antal, M. J., Jr, Stein, Y. and Jones, M. (1981). Laminar flow reactor studies of the gas phase pyrolysis of glycerol. Western States Section/The Combustion Institute, Spring Meeting, Pullman, Washington.
7. Cutler, A., Antal, M. J., Jr and Jones, M. To appear.
8. Cutler, A. (1983). PhD Thesis, Princeton University.
9. Mok, W. and Antal, M. J., Jr (1981). The effects of pressure on biomass pyrolysis. In *Proceedings of the 13th Biomass Thermochemical Conversion Contractor's Meeting*. Arlington, Virginia, October 27–29.
10. Mok, W. and Antal, M. J., Jr (1982). The effects of pressure on biomass pyrolysis. In *Proceedings of the 14th Biomass Thermochemical Conversion Contractor's Meeting*. Arlington, Virginia, June 23–24.
11. Mok, W. MSE Thesis, Princeton University.
12. Stein, Y. S., Antal, M. J., Jr and Jones, M. (in press). A study of the gas phase pyrolysis of glycerol. *J. Anal. Applied Pyrolysis.*
13. Milne, T. A. and Soltys, M. N. (1981). Fundamental pyrolysis studies. Annual Report for Fiscal Year 1981, SERI/PR-234-1454.
14. Houminer, Y. and Patai, S. (1967). Thermal decomposition of D glucose labelled with ^{14}C at various positions. *Tetrahedron Letters* **14**, 1297–1300.
15. Shafizadeh, F. and Lai, Y. F. (1972). Thermal degradation of 1,6-anhydro-β-D-glucopyranose. *J. Org. Chem.* **37**(2).
16. Broido, A., Evett, M. and Hodges, C. C. (1975). Yield of 1,6 anhydro-3,4-dideoxy-β-D-glycero-hex-3-enopyranos-2-ulose (levoglucosenone) on the acid catalyzed pyrolysis of cellulose and 1,6 anhydro-β-D-glucopyranose (levoglucosan). *Carbohydrate Research* **44**, 267–74.
17. McCarter, R. J. (1973). A new technique for thermal analysis of vapor producing reactions. *J. Appl. Polym. Sci.* **17**, 1833–46.

18. McCarter, R. J. (1972). The pyrolysis of cellulose at rates approaching those in burning. *Textile Res. J.* **42**, 709–19.
19. Baker, R. R. (1975). Thermal decomposition of cellulose. *J. Thermal Analysis* **8**, 163–73.
20. Deglise, X., Richard, C., Rolin, A. and Francois, H. (1981). Fast pyrolysis/gasification of lignocellulosic materials at short residence time. EUR 1981, EUR 7091, Energy Biomass Conf., 1st, 548–53.
21. Zdonik, S. B., Green, E. J. and Haliee, L. P. (1970). *Manufacturing Ethylene.* The Petroleum Publishing Co., Tulsa, Oklahoma.
22. Broido, A. and Nelson, M. A. (1975). Char yield on pyrolysis of cellulose. *Combustion and Flame* **24**, 263–8.

30

Ablative Pyrolysis of Biomass in Solid-Convective Heat Transfer Environments

JAMES DIEBOLD and JOHN SCAHILL
Solar Energy Research Institute, Golden, Colorado, USA

1. INTRODUCTION

The heating of biomass in an oxygen-deficient situation is known as pyrolysis and has been used for centuries to produce charcoal, tars, wood alcohol and other solvents. The traditional slow heating of biomass produces about equal amounts of gases, char and tarry liquids. These tarry liquids have been promoted as boiler fuels, but they are not thought to be suitable for use in internal combustion (IC) engines. Current state-of-the-art dictates that the liquid fuels used for IC engines be either a low boiling hydrocarbon material, an alcohol, or a mixture of the two.

Under certain very rapid pyrolysis conditions, valuable products can be recovered which are not predicted by chemical thermodynamic equilibrium. These non-equilibrium products can be varied considerably depending upon the reactor conditions selected. The products can be almost entirely gaseous and contain a significant amount of very desirable unsaturated hydrocarbons such as ethylene and propylene. The chemistry of these unsaturated hydrocarbons is very well known and is the basis for a large part of the existing petrochemical industry. In the past, unsaturated hydrocarbons have been used by the petroleum industry to make gasoline and by the petrochemical industry to make alcohols, plastics, etc.

Historically, the energy and chemical industries have used the most economical feedstock available, which was wood or cellulose until the fossil fuel age arrived. With the advent of cheap petroleum and natural gas, the industry became based on low-molecular-weight compounds, such as ethylene, for use as starting materials to make rubber, plastics, alcohols, and liquid fuels. Although the highest yields of ethylene may be obtained from ethane, the increase in cost and the decrease in availability of ethane has gradually shifted the industry toward using less desirable feedstocks, even though the yields may be lower and the processing more difficult. The use of biomass as a source of ethylene via a rapid thermal reaction would have considerable merit especially in areas without assured access to petroleum sources.

The conversion of biomass to ethylene via fast pyrolysis has been demonstrated by several isolated researchers at scales ranging from a few milligrams in batch experiments to continuous flow, bench-scale experimentation at nominal flow rates of 5 kg/h. In addition to the fundamental pyrolysis studies by Milne and Soltys,[2] DOE's Division of Biomass Energy Technology has sponsored a fast pyrolysis program at SERI having three areas of effort: laboratory-scale pyrolysis of a few milligrams of material from monolithic pieces of biomass to determine optimal heat transfer mechanisms; the bringing together of the several isolated fast pyrolysis researchers to compare methods, results and conclusions as to the fast pyrolysis phenomenon;[3] and the scaling-up of the fast pyrolysis of biomass to an engineering research scale to pave the way for a pilot plant demonstration using practical, heat transfer techniques and biomass feedstocks larger than fine powders.

2. BACKGROUND

As was reported and demonstrated at the Specialists' Workshop on Fast Pyrolysis of Biomass held at Copper Mountain, CO in 1980,[1] laboratory-scale heat-transfer experiments have shown that when biomass is moved relative to a red-hot Nichrome wire, the wire will 'cut' through the biomass. The rate of cutting, or pyrolysis, can be as high as 3 cm/s when it is a very localized surface phenomenon. With this method of heat transfer, pyrolysis appears to proceed by the depolymerization, melting and vaporization of the biomass without observable char formation; the term 'ablative' seems to best describe this fast pyrolysis mechanism.

The rate of heat transfer from the red-hot metal surface to the biomass is extraordinarily high. Based on an assumed energy of primary pyrolysis of 2000 J/g,[4] the 0·025-cm diameter wire moving across the biomass at 20 cm/s and penetrating at a rate of 3 cm/s was transferring 3500 W/cm^2, which is very impressive compared to the mere 15 W/cm^2 radiated by a black body reactor wall at 1000 °C. Thus, this solid-convective approach to pyrolysis transfers energy to the biomass at rates of over two orders of magnitude greater than black body radiation at similar wall temperatures. This would imply that a pyrolysis reactor relying on solid-convective heat transfer could have over 100 times the throughput of a similarly sized reactor relying only on radiative heat transfer.

The mechanism of this solid-convective heat transfer appears to be the conduction of heat across a very thin film (thought to be about 10 μm thick, based on lubrication theory) from a nearly isothermal metal surface at 1000 °C, while the biomass depolymerizes at about 400 °C to primary tars which are wiped away and/or vaporized. Since heat conduction is proportional to this large temperature difference divided by the very small film thickness, very high heat fluxes are predicted. Because the surface regression rate is nearly the same as the thermal penetration rate, any biomass which is located more than a calculated 15 μm from the pyrolyzing surface is still at the low initial temperature and is unaffected by the ablative pyrolysis taking place. Consequently, this charless, ablative pyrolysis will proceed in a similar manner whether the biomass is a 1-cm chip or a fine, 50-μm powder.[4] Because the pyrolysis front moves so quickly through the biomass, the temperature gradient is very steep with a calculated heating rate of about 500 000 °C/s.[3]

The ability to use wood chips rather than a fine powder or dust as feedstock improves the overall process with respect to an estimated 10–15% process energy reduction, less equipment to amortize, and enhanced safety by eliminating dust explosion hazards in storage bins, transfer lines, etc.

3. SYSTEM DESCRIPTION

As shown in Fig. 1, the system is designed to pyrolyze biomass at just above atmospheric pressure and at wall temperatures up to 1000 °C in the presence of superheated steam. The high temperatures are attained through the use of electric heaters (a gas-fired boiler provides saturated steam at 1400 kPa (200 psig)). The entrained biomass particles enter tangentially into the 14-cm diameter by 70-cm long cyclonic pyrolysis reactor so that the particles are centrifuged against the red-hot, externally heated Incoloy 800H walls held at temperatures as high as 1000 °C. Incoloy 800H is a high nickel and chromium steel widely used in naphtha-cracking units to produce ethylene. The entering velocities are on the order of 100 m/s for the entraining steam and probably somewhat lower for the still accelerating biomass particles. As the particles slide and bounce on the inside of the cyclonic pyrolysis reactor, they are thought to be heated

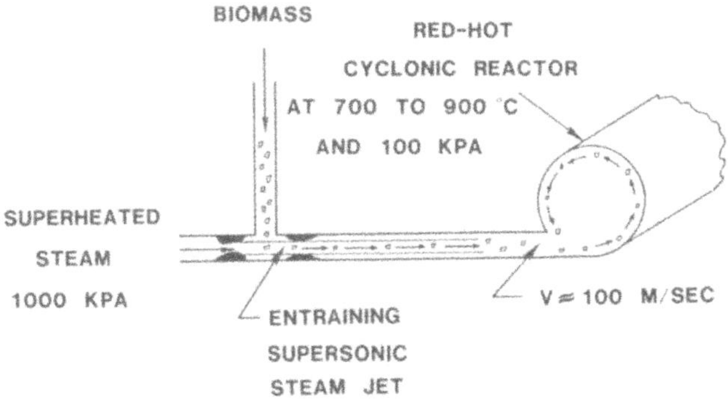

FIG. 1. Ablative macroparticle pyrolysis flow schematic.

almost entirely by a solid-convective heat transfer mechanism in a manner analogous to the hot-wire pyrolytic cutting. This causes an extremely rapid, ablative pyrolysis to occur at the interface of the biomass particle and the reactor wall. The biomass decomposes to an oil having a low viscosity at elevated temperatures or possibly to vapors at the interface which then serve both as a heat-transfer fluid and as a lubricant for the particle. If the particle is traveling fast enough, the centrifugal forces are expected to create sufficient pressure to maintain the oil in the liquid phase for the 0·1 ms or so that the oil is underneath the particle. The pyrolytic oil will probably be left behind the fast moving particle on the reactor wall in a manner similar to the trial left by a snail. The pyrolytic oil left on the reactor wall will vaporize before decomposing to char if the reactor walls are maintained at a high enough temperature. The high molecular weight pyrolytic oil vapors have a relatively high density and are thought to be held quite closely to the cyclone wall until they crack to lower density gases or reach the end of the cyclone reactor. The cyclonic reactor is electrically heated by three independently controlled furnaces.

The 5-m long by 5-cm diameter vapor cracking tube was also made of Incoloy 800H so that the wall temperature can be held as high as 1000 °C without adverse effects. Its function is to pyrolyze the tar vapors completely to gases containing the desirable olefins such as ethylene. The vapor cracker has eight evenly spaced thermowells and six independently controlled furnaces. The temperature profile used in the vapor cracker can be varied at will to produce a very fast rise in product temperature followed by isothermal conditions, a uniform temperature rise, or a delayed temperature rise. If the production of pyrolytic oils or less severe cracking is desired, a quench fluid may be added through any of the thermocouple locations to prevent further reaction or decomposition. Thus, the vapor cracker has eight kinetic 'lengths'. Gas samples may be withdrawn through any of the eight thermocouple locations to determine the optimal residence times.

After passing through the vapor cracker, the pyrolysis products pass through a char cyclone and then enter the first scrubber system which is a 38-cm diameter cyclone. The hot gaseous products enter tangentially into the cyclonic quench drum at a nominal 50 m/s. Cooled water is sprayed across the gas flow to quench the chemical reactions, to condense the tar and oil vapors, and to scrub the condensates out of the gas stream. The initial high gas velocity is used to impart the centrifugal forces ($\sim 500\,g$) to separate the fine condensate particles from the gases. The amount of cooling done by the scrubbing water can be adjusted so that only about 10% of the steam is condensed in this first scrubber. This small amount of condensed steam along with the tar condensates, non-volatile soluble organics, and any char and ash entrained from the char cyclone can be pumped to the flare for incineration. With this type of operation, the cyclone scrubber acts as a single stage evaporator to concentrate the less volatile organics (tars and water solubles) for disposal.

The cooled gas stream (at about 80 °C) then passes to the packed scrubber where it flows upward counter-currently to the cool, descending scrubbing water. Above the ceramic packing are about 1·5 m of a packed bed made up of short lengths of plastic packing resembling bottle brushes. As the temperature of the pyrolysis gas and steam mixture decreases, progressively more steam and volatile organics condense. The condensing of the steam and organic volatiles is thought to aid in the coalescing of the fine mists of water and tar droplets. The cooled gases then pass through an orifice meter before being compressed for storage or being sent to the flare for incineration.

4. SHAKEDOWN OPERATION

Initial operation of the pyrolysis system indicated that the feed was not spreading out on the inside surface of the cyclone reactor, but instead would form a narrow helical band with a rather large pitch of about 1·2 times the inside diameter. To study this phenomenon, a full-scale Plexiglas™ reactor was constructed. With nitrogen as the carrier gas at room temperature, the banding of the biomass was verified. Various changes in the geometry were attempted, but the natural direction of the solids at the wall was found to be independent of size and axial location of the outlet tube, entraining gas velocity, small changes in the angle of the entering tangential tube, and particle size of the biomass. This large helical pitch for the flow of gases has been reported in the literature relative to Ranque–Hilsch vortex tubes[5,6] and our solids at the wall appear to follow the direction of the gas flow at the wall. Unfortunately with this large helical pitch, only a small fraction of the heat transfer surface was being utilized for pyrolysis.

Extensive cold-flow testing in the Plexiglas reactor model showed that the desired tight helical path of the solids could be attained with a helical guide vane on the ID of the cyclone reactor. No other feasible physical changes to the existing reactor had any substantial effect on the solid's path on the reactor wall.[5]

The pyrolysis reactor has been modified by the addition of an internal, handmade, helical spiral of $\frac{1}{4}$-in. diameter 316SS tubing. The helix has a 12·5-cm diameter and a 3-cm pitch. This helical spiral was chosen for evaluation rather than a more ideal machine-turned spiral made of square rod, due to the relative non-availability of the square rod in a high temperature alloy. The use of the handmade spiral was to verify the effectiveness of the concept of guiding the particles at elevated temperatures in a tight spiral rather than the natural path consisting of a coarse spiral with a pitch greater than the ID of the cyclone reactor. From cold-flow studies made with the Plexiglas model, it was observed that where the handmade helix does not quite touch the reactor ID, the feed particles will take an undesirable shortcut and pass between the helix and the wall.

During pyrolysis runs made with the helical coil in place, the inside cyclone reactor surface appeared to have a uniform color (temperature) gradation from relatively dark red near the entrance to a brighter red near the exit. This uniformity suggested that the entraining steam and the entrained biomass particles were spread out uniformly on the reactor surface, whereas, without the helical coil the particles had a preferred narrow pathway which utilized only a small fraction of the total heat transfer surface, giving rise to a banded color zone on the cyclone walls.

The 10-cm diameter char cyclone has been effective in dropping out entrained char prior to entering the first scrubber. By keeping this cyclone over 125 °C, the collected char was found to be a dry powdery material (with wood flour feed) which can be weighed without the time consuming steps of drying, tar extraction in acetone, and subsequent drying operations to determine char yields. Based on the brown caramel appearance of the tars collected from the less severe cracking experiments, it appears that very little fine char or carbon black is entrained through the char cyclone.

Perhaps the very fine char ($< 10\ \mu m$), which could escape the char cyclone is preferentially oxidized due to its high surface-to-volume ratio by the hot steam environment in the vapor cracker section.

The cross-sectional area of the entraining tube at the cyclone reactor entrance has an inverse effect on the velocities of the entering carrier gas and the entrained feed particles. This effect of increased entering velocity with a smaller entraining tube was observed throughout the reactor length as evidenced by the higher particle velocities at the wall of the cold-flow Plexiglas model. An entrainment tube of 1·09 cm ID was inserted into the 2·5 cm ID entrainment tube. This smaller entrainment tube serves to increase the entering particle and gas velocities by a factor of five. The equilibrium particle velocity in the Plexiglas cyclone reactor was observed to be increased by a factor of five, due to the slow decay of the gas velocity down the cyclone length. A similar tube was placed in the high temperature reactor and successfully used.

5. PYROLYSIS EXPERIMENTATION

5.1. Preliminary Softwood Sawdust Feedstock Evaluation

Pyrolysis runs were made with minus 5 mm softwood sawdust in the pyrolysis system shown in Fig. 1 to evaluate the function of the solids-feeding system. As an example of our experimentation, Run G was made with a feed rate of 9 kg/h with a steam-to-biomass ratio of 2. The cyclone reactor heaters were set at 950 °C and the vapor cracker heaters at 850 °C. The thermocouples in the cyclone reactor indicated temperatures between 810 and 950 °C, whereas the vapor cracker thermocouples indicated temperatures very close to the set point as shown in Fig. 2. However, due to radiation and conduction effects, these indicated temperatures are partway between the wall temperatures and the true gas temperatures. The apparent gas temperature 15 in. downstream of the last vapor cracker heater was 720 °C, which is probably the best value for the final pyrolysis product temperature. The mole fraction of methane is very constant at about 10 % in the pyrolysis gases and is convenient to use to determine cracking severity parameters. With high temperatures and a long residence time due to the low volumetric throughput, the pyrolysis gases were severely overcracked with a molar ratio of methane to propylene of over 100. Less severe temperature profiles resulted in less secondary cracking and therefore a lowering of the methane-to-propylene ratio (more surviving propylene) as well as the methane-to-C_{5+} ratio (more surviving C_{5+}). The water in the first

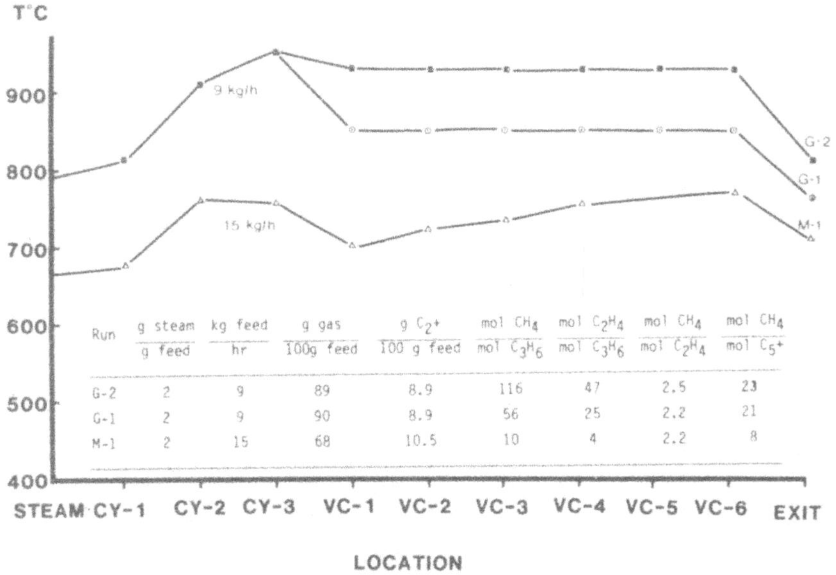

Run	g steam / g feed	kg feed / hr	g gas / 100g feed	g C_{2}+ / 100 g feed	mol CH_4 / mol C_3H_6	mol C_2H_4 / mol C_3H_6	mol CH_4 / mol C_2H_4	mol CH_4 / mol C_5+
G-2	2	9	89	8.9	116	47	2.5	23
G-1	2	9	90	8.9	56	25	2.2	21
M-1	2	15	68	10.5	10	4	2.2	8

FIG. 2. Preliminary pyrolysis results.

scrubber had a naphthalene (moth ball) odor, but was relatively clean. Most of the tar collected in the first scrubber was soluble in acetone and it is thought to contain polycyclic aromatic compounds formed by the decomposition reactions of propylene, butenes and ethylene. The char material collected in the char cyclone contained no acetone-solubles, implying that the biomass had failed to vaporize during pyrolysis before it polymerized and/or decomposed to form char.

Figure 2 shows how a lowered cracking severity (lower temperature and/or residence time) affects the product slate at a steam-to-biomass weight ratio of 2. The product slate parameters are in tabular form overlayed on the temperature profiles. The lower severity runs have lost less propylene and aromatic vapors to secondary cracking reactions which produce unwanted polycyclic aromatic tars. It is interesting to note the relative constancy of the molar methane-to-ethylene ratio at 2·2 to 2·5, whereas the highly sensitive molar methane-to-propylene ratio varies between 10 and 116. This reflects the highly reactive nature of propylene at these temperatures and the importance of cracking severity on the product slate. These temperatures were not corrected for measurement errors due to radiation or conduction effects.

TABLE I
COMPARISON OF WOOD FLOUR AND SAWDUST PYROLYSIS PRODUCTS

	Run R, wood flour (g/100 g feed)	*Run 12–4, sawdust (g/100 g feed)*
H_2	1·4	1·8
CH_4	7·5	7·3
C_2's	7·5	7·3
C_3's	1·8	1·1
C_4's	1·3	0·8
C_{5^+}	4·9	4·7
CO	54·5	48·1
CO_2	13·9	16·8
Gasification	92·6	87·9
C_{2^+}	15·3	13·9
Char	1·1	6·8
Tar	Unknown	4·8
Molar CH_4/C_3H_6	12	20
'Full volume' residence time, s	0·76	1·08
Corrected average temp., °C	718	743
Product gas partial pressure, kPa	11	13
Steam-to-biomass weight ratio	5·0	5·4

5.2. Particle Size

To evaluate the effect of particle size on pyrolysis products, pine flour smaller than 0·177 mm (minus 80 mesh) was used to compare to the normally used feedstock, a fairly granular sawdust hammermilled through 5 mm ($\frac{3}{16}$ in.) openings. The relative mesh sizes were about thirty to one. The results shown in Table I indicate that the products can be very similar with our pyrolysis system even though the flour and the sawdust were from two different sources. In actuality, the product distribution appears to be much more sensitive to operating conditions than to initial particle size, as was predicted by our concept of ablative pyrolysis.

With the −80 mesh pine flour, an apparent gasification of 93% was calculated based on gas flow measurements with only 1% char recovered. A total of 15 wt % of C_{2^+} hydrocarbons was achieved, with about 30% of them being in the C_{5^+} backflush presumed to be low molecular weight aromatics such as benzene, toluene and xylene (BTX).

5.3. Process Variable Studies

Several series of pyrolysis experiments were performed with minus 5 mm softwood sawdust to investigate different temperature profiles at steam-to-biomass ratios between 1·4 and 4·1 with inlet steam temperatures between 720 and 830 °C. In addition, injecting quench water into the vapor cracker section was demonstrated and shown to 'shorten' kinetically the vapor cracker. Attempts to correlate yields with the steam-to-biomass ratios did not result in any recognizable trends.

The experimentation with sawdust resulted in gasification yields between 64 and 92 % by weight. However, the char yield was typically only 5–7 %. By difference, this leaves about 3–30 % by weight of the products as tars and as condensible materials which passed through the char cyclone in the vapor state, but which condensed in the water scrubbing systems. These condensible materials are thought to be of two general classifications: (a) primary pyrolysis products (oxygenated fragments of biomass components); and (b) tertiary pyrolysis products (polycyclic aromatic compounds formed by the pyrolysis of the secondary pyrolysis products, i.e. gaseous olefins). The optimization of ablative pyrolysis to olefins must first deal with the pyrolysis of the biomass to the primary tar-vapor products. These primary tar products are oxygenated compounds formed by the depolymerization of the biomass, tend to be water soluble, and are thought to have molecular weights of between 50 and 400. These compounds are optimally vaporized to reduce their tendency to polymerize to higher molecular-weight tars, which can be quite stable at the reaction temperatures and have a char-like appearance. The global reaction of the primary biomass tar vapors to the olefin containing gases is apparently strongly influenced by temperature and is thought to consist of unimolecular decompositions, i.e. first-order reactions.

The kinetics of first-order reactions are well understood, i.e. the rate of change of the concentration of the reactant (C_v) is equal to the concentration of the reactant times the reaction rate constant (k_T) for the given absolute temperature (T):

$$\frac{dC_v}{dt} = k_T C_v = [A_v \exp(-E_v/RT)]C_v$$

where E_v and A_v are Arrhenius rate constants, R is the ideal gas constant, and t is time. By rearrangement and integration the following is obtained:

$$\ln\left(\frac{C_{v2}}{C_{v1}}\right) = A_v \int_{t_1}^{t_2} \exp(-E_v/RT)\,dt$$

This expression is difficult to evaluate from experimental results because it requires that the temperature history be known precisely. However, at a constant reaction temperature, the equation reduces to

$$\ln\left(\frac{C_{v2}}{C_{v1}}\right) = k_T\,\Delta t = 2{\cdot}303\log\left(\frac{C_{v2}}{C_{v1}}\right)$$

When this relationship is plotted on semi-log graph paper, a straight line is generated with a slope of $k_T/2{\cdot}303$ for a specific reaction temperature. With this brief background in kinetic theory, an examination of the experimental conditions and results is in order.

The measurement of the true gas-stream temperature in high temperature reactors is technically difficult because the thermocouples are being simultaneously heated by radiation and conduction from the externally heated reactor walls and cooled by convection by the cooler endothermically reacting gases. Thus, the thermocouples will always indicate temperatures part way between the true gas temperature and the reactor wall temperatures. With higher velocity gases, the convective heat transfer will be enhanced and the error of temperature measurement will be somewhat reduced. Using several assumptions, heat transfer theory can be used to calculate a corrected temperature, which for our conditions can be shown to be in the range of 50 to 100 °C lower than the recorded temperature. However, a more direct method which we have used is to compare the indicated temperature of the last thermocouple in the vapor cracking section (TC-18) to a thermocouple in the gas stream 38 cm downstream of the last heater (TC-20). Since the process tube walls at this location are heated only by the gas stream and are well insulated, the temperature of this thermocouple should be very close to the true gas temperature. The temperature error by these measurements (TC-18 minus TC-20) has been between 40 and 130 °C and does qualitatively decrease with an increase in gas flow as predicted by convective heat transfer concepts. To minimize these radiation and conduction induced errors in the temperature measurements in the vapor cracker, the thermocouples in the process stream were moved to the 5 cm gap between the electric heaters. This seems to have reduced these errors to only a few degrees based on the temperature differences of these last two thermocouples.

As a first approximation of the gas-temperature history in the cyclone reactor and vapor cracker, the twelve temperatures involved were all corrected by the difference between that indicated by TC-18 and TC-20. Each corrected thermocouple value was equally weighted along with the uncorrected inlet steam temperature and the uncorrected TC-20 temperature to calculate a 'corrected average' temperature. (The inlet steam

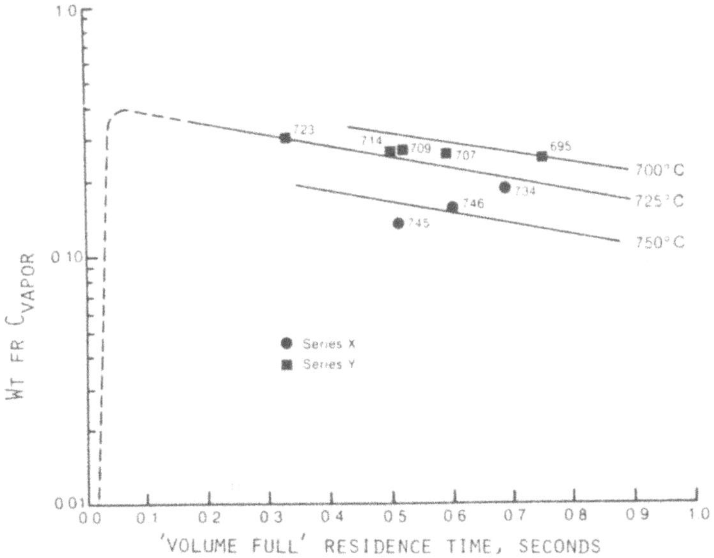

FIG. 3. Uncracked biomass vapors vs. residence time.

temperature was not corrected because the thermocouple measuring it is also outside of the furnaces.)

The volumetric gas flow was calculated based on adding the molar steamflow rate to the molar flow rate of the final product-gas. The total molar flow rate was then used with the 'corrected average' temperature and the cyclone-reactor pressure to calculate the volumetric gas flow rate using the ideal gas equation. The residence time of the gases in the reactor was found by dividing the physical internal volume of the cylone reactor and the vapor cracker by the volumetric gas flow rate. The inverse of this residence time is often referred to as the 'space velocity' in reactor design. This assumes that the product gases are formed instantaneously at the reactor entrance, which ignores the expansion of the gas volume due to temperature increases or to gasification. However, these effects lose their importance if the temperatures are fairly constant and if the overall gas composition in the reactor is primarily steam (e.g. at a steam-to-biomass ratio of 2·7, the final gross product gases are only ~20% pyrolysis gases).

Since the reactor volume used in these calculations was the full physical volume of the reactor, this residence time will be referred to as the 'full-volume' residence time. In reality, the gas flow in a cyclone or vortex tube is quite complex and experimentation by others has shown that only an outer annulus of the vortex tube (comprising two-thirds of the cross-sectional area) is actively used by the swirling gases to move toward the exit. In addition, the axial velocity was found to be a function of radial location and to be proportional to the tangential entering velocity.[6,7] The true average residence times of the gases in the pyrolysis reactor is probably $\frac{1}{3}$ to $\frac{1}{2}$ of that predicted by the full-volume simplification. A refinement in the residence time will be sought, but the full-volume residence time will be used here for discussion.

The effect of residence time on the weight fraction yield of vapors for several runs are plotted in Fig. 3. The vapors were defined as the condensible products (the feed minus the gases and char produced). The 'corrected average' temperatures are indicated at each point. Although the data are preliminary and further experimentation is needed, temperature parameter lines have been added by visual interpolation. Note that if the biomass vapor had all been formed at time equals zero, the straight parameter line should have had a common starting point at zero time of 1·0 weight fraction of vapors. However, the sawdust particles require a finite time to vaporize. Thus, one would expect the biomass vapor concentration to start at zero and rapidly rise until the vapor-cracking reactions proceed faster than the biomass vaporization, as indicated by the dashed line in Fig. 3.

The importance of temperature is shown dramatically in Fig. 3 by the significant change in the amount of biomass vapor cracked caused by only a 25 °C change. Note that linear isotherms seem to fit the data fairly well as is predicted by first-order kinetic theory. Considering the uncertainty of the residence times and temperature corrections, there seems to be a very good correlation based upon the application of first-order kinetic theory concepts to the cracking of the biomass vapors to gases. However, there are also some data which do not appear to fit the kinetic correlation too well. It is thought that a more rigorous application of kinetic theory requiring integration of several rate equations simultaneously with temperature as a variable would result in a better fit of the data to theory.

5.4. Byproduct Char

Although char is not considered to be a desirable product in our program, some knowledge of the properties of the char produced is of interest. From experimental run series X, 3·7 % by weight of the feed was recovered as char in the char cyclone for an apparent overall vaporization of 96·3 % of the feed. This char had been formed at corrected average temperatures in the 734–745 °C range and was selected arbitrarily for extensive tests. To the naked eye, this char had a particle size and shape resembling the sawdust

TABLE II

ANALYSIS OF SAWDUST FEED, CHAR AND TAR

	−5 mm sawdust feed		*Run X char*		*Run 7 tar*	
	As received	*Dry*	*As received*	*Dry*	*As received*	*Dry*
Proximate, wt %						
Moisture	6·62	0·00	2·02	0·00	15·55	—
Ash	0·44	0·47	5·67	5·79	0·37	0·44
Volatile	76·00	81·39	<0·01	<0·01	52·91	62·65
Fixes C	16·94	18·14	92·31	94·21	31·17	36·91
Ultimate, wt %						
Moisture	6·62	0·00	2·02	0·00	15·55	—
Carbon	47·60	50·97	85·35	87·11	71·24	84·36
Hydrogen	5·61	6·01	1·77	1·81		
Nitrogen	0·07	0·07	0·11	0·11	0·08	0·10
Sulfur	0·01	0·01	0·01	0·01	0·02	0·02
Ash	0·44	0·47	5·67	5·79	0·37	0·44
Oxygen[a]	39·45	42·47	5·07	5·17	7·93	9·38
High heating value,	18·6	20·0	30·8	31·4		
kJ/g (Btu/lb)	(8 024)	(8 593)	(13 256)	(13 530)	(13 975)	(14 796)
Empirical formula	$C_6H_{8\cdot5}O_{3\cdot7}$		$C_6H_{1\cdot5}O_{0\cdot3}$		$C_6H_{4\cdot9}O_{0\cdot5}$	

[a] By difference.

feed. Examination of the char at a magnification of 30 revealed that it had non-porous glossy black surfaces which had apparently been only semi-molten in that the original fibrous nature was barely evident. To determine the thermochemical nature of this char, it was submitted for heat of combustion, volatiles, proximate, and ultimate analyses. A summary of this testing is shown in Table II. For comparison, the tests were also run on a sawdust sample and are also shown in the table. The majority of the ash in the feed is recovered in the char, but apparently some goes ash on to the scrubbers. The char is a completely devolatilized material containing 92·5 % carbon, 5·5 % oxygen, and 2 % hydrogen by weight (moisture and ash free basis) with an empirical formula of $C_6H_{1\cdot5}O_{0\cdot3}$. The higher heating value was reported to be 31·4 kJ/g (13 530 Btu/lb), which reflects the low ash and low oxygen content of the char. The lower heating value was 31·1 kJ/g (13 362 Btu/lb) which is within 0·2 % of that calculated from the elemental analysis using the Dulong formula developed for use with coal.

5.5. Byproduct Tar

The amount of biomass converted to water-insoluble tar is a variable, but has been in the 2–4 wt % range. At room temperature these black tars (formed in the corrected average temperature range of 650–800 °C) appear as a high viscosity, sticky tar. However, when heated to above 70 °C, the viscosity is greatly lowered so that the tars could probably be easily pumped, atomized and burned in a furnace. When hot, a mixture of tar and water can easily be drained out of the bottom of the cyclone scrubber water reservoir. Upon cooling, the tars settle to the bottom and excess water can be decanted. The heat of combustion of these as-decanted tars was fairly high at 32·4 kJ/g (13 928 Btu/lb). Drying the tars at 107 °C to drive off residual moisture (and organic volatiles) increased the heat of combustion only to 34·2 kJ/g (14 725 Btu/lb) and increased the apparent viscosity. The results of the ultimate and proximate analyses are shown in Table II. From these analyses, the empirical formula of $C_{12}H_{10}O$ can be calculated. This empirical formula implies that the tars do not have significant amounts of either the cellulosic or lignin primary fragments due to the low oxygen content. The low hydrogen content suggests a highly aromatic material. Although we suspect that the tars are a complex mixture of organic compounds, the empirical formula suggests that the 'average' tar compound could be represented as phenyl phenol. It should be noted that the test for moisture consisted of placing the tar sample in an oven held at 107 °C which would also effectively drive off the water and volatiles such as benzene, toluene, etc., along with the water. The calculated heat of combustion of the 'moisture' driven off at 107 °C was about 22·1 kJ/g (9500 Btu/lb), suggesting that the 'moisture' was at least half organic volatiles.

5.6. Contaminated Scrubber Water

Following Run 7, samples of the cylone scrubber and packed scrubber recirculation water were taken. The primary purpose of these scrubber waters is to cool and clean the hot pyrolysis gases upon exit from the vapor cracker section. In doing so, some pyrolysis products are dissolved to form an aqueous solution. The samples were analyzed using IR and NMR techniques in an attempt to identify the major species in solution. Chloroform was used to extract the dissolved pyrolysis products from the samples before injection. As expected, the scans indicated the presence of many materials. A review of the five volumes of the Aldrich Library of NMR Spectra, with emphasis on aromatic hydrocarbons, turned up four likely species present in the cyclone scrubber water. They were *o*-ethyltoluene, *o*-cresol (trace), 2,3-dimethylindole, and acetic acid. Only *m*-xylene was recognizable in the packed-scrubber water scan among the many compounds, but the acetic acid was not present.

6. CONCLUSIONS

The process variable study is still in its early stages. The amount of gas and olefins produced appears to be controlled primarily by the severity of cracking of the biomass vapors. This reaction appears to be a first-order reaction. The effects of increasing the amount of superheated steam in the reactor appear to include a lowering of the partial pressure of the product gases, but more importantly it reduces the gaseous residence time, and increases the temperature of the tar vapors by the mixing of the hot steam with the cooler tar vapors. It appears that previously reported trends of increasing hydrocarbon yields with increasing steam-to-biomass ratios may have been primarily due to the higher temperatures achieved by the reactants in the reactor which tended to crack more completely the primary tar vapors to gases containing the hydrocarbons. If this is true, the amount of steam can be reduced to where it does not represent such a large fraction of the process energy. Using first-order kinetic reaction theory, the extent of vapours cracking to olefin containing gases is qualitatively predictable even though several simplifying assumptions were involved in calculating the values used for the average temperatures and residence times. The current results indicate that a 15 wt % (35–40 % by energy) conversion of softwood to C_2^+ hydrocarbons is feasible and will include significant amounts of aromatic liquids. This yield is lower than reported for processed municipal solid waste (ECO II Fuel)[8] and may reflect compositional differences or the effect of feedstock preparation, suggesting the potential of studies of these variables. The sawdust pyrolyzes nearly as well as the fine wood flour indicating that the larger particles are probably ablatively pyrolyzing as expected. Additional studies need to be completed before firm conclusions may be drawn.

REFERENCES

1. Diebold, J. P. (1980). Ablative pyrolysis of macroparticles of biomass. *Proceedings of the Specialists' Workshop on Fast Pyrolysis of Biomass.* Copper Mountain, CO, 20–22 October. SERI/CP-622-1096. Solar Energy Research Institute, Golden, CO.
2. Milne, T. A. and Soltys, M. N. (1982). The direct mass-spectrometric study of the primary and secondary pyrolysis behavior of biomass and its constituents. Presented at the International Conference on Fundamentals of Thermochemical Biomass Conversion, 18–22 October 1982, Estes Park, CO.
3. Diebold, J. P. (Workshop Chairman) (1981). *Proceedings of the Specialists' Workshop on the Fast Pyrolysis of Biomass*, op. cit.

4. Reed, T. B., Diebold, J. and Desrosiers, R. (1980). Perspectives in heat transfer requirements and mechanisms for fast pyrolysis. *Proceedings of the Specialists' Workshop on the Fast Pyrolysis of Biomass, op. cit.*
5. Diebold, J. P. and Scahill, J. W. (1982). Progress in the entrained-flow, fast ablative pyrolysis of biomass. Annual Report October 1980 to September 1981. Presented at DOE-BmES 13th Thermochemical Conversion Contractors' Meeting. October 17–19, 1981, Arlington, VA. SERI/PR-234-1456. Solar Energy Research Institute, Golden, CO.
6. Hartnett, J. P. and Ecker, E. R. G. (1957). Experimental study of the velocity and temperature distribution in a high velocity vortex-type flow. *Trans. ASME*, 751–8.
7. Scheller, W. A. and Brown, G. M. (1957). The Ranque-Hilsch vortex tube. *Ind. Eng. Chem.* **49**(6), 1013–16.
8. Diebold, J. P. (1980). Gasoline from solid wastes by a noncatalytic, thermal process. *Thermal Conversion of Solid Wastes and Biomass.* ACS Symposium Series 130, American Chemical Society, Washington, D.C.

31

The Gasification of Ponderosa Pine Charcoal

RICHARD EDRICH, TIMOTHY BRADLEY and MICHAEL S. GRABOSKI*
Colorado School of Mines, Golden, Colorado, USA

1. INTRODUCTION

The gasification of wood chars with CO_2 and steam is an important process step in the conversion of biomass to fuel and synthesis gases. Wood fuels can be gasified in a wide variety of sizes, shapes and densities. Commonly, chips and pellets are suitable for fixed and fluidized beds while sawdust is preferred for entrained units. In terms of fuel preparation, handling and feeding, whole tree chips and pelletized sawdust are most desirable: therefore fixed and fluid-beds are widely used with wood fuels.

Figure 1 shows two types of gasifiers used to process whole tree chips and pellets. In the updraft mode, the fuel and oxidant flow in a counter-current fashion. The charcoal produced in the upper portion of the bed therefore sees a progressively higher temperature as it is converted. A somewhat different situation exists in the downdraft or cocurrent case. Here, pyrolysis takes place in the presence of oxygen. The highest temperature observed in this type of gasifier is at the beginning of the char conversion process. Heat liberated by pyrolysis combustion is absorbed by the endothermic gasification steps thus decreasing the temperature of the exit of the reactor.

When whole chips and pellets are pyrolysed, the remaining char contains on the order of 30% of the original wood carbon and heating value. In steam and CO_2 atmospheres, the required residence time for char gasification may be 10–100 times greater than for pyrolysis for large particles at temperatures between 800 and 1100 °C. Thus in non-slagging gasifiers operating in this temperature range, the char gasification kinetics determine the size of the reactor.

* Author to whom correspondence should be addressed.

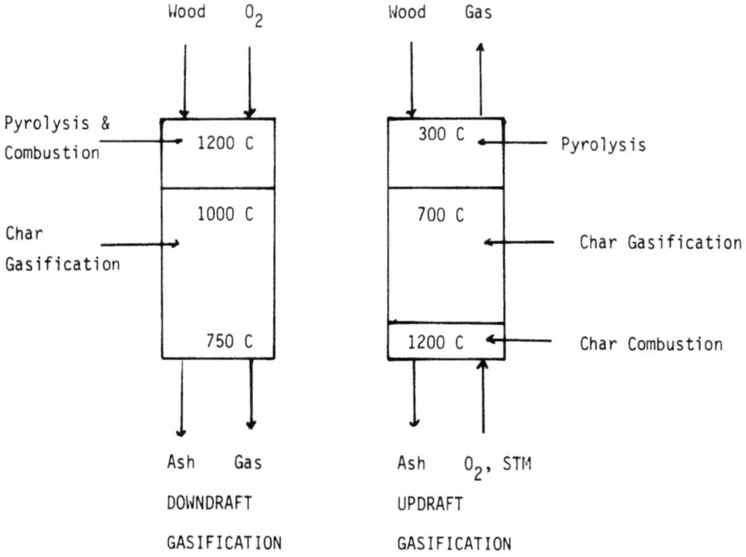

FIG. 1. Fixed bed gasification modes.

In order to obtain information on pyrolysis rates and gasification rates of wood solids of a consistency similar to those fed to commercial non-slagging gasifiers, this study funded by the Department of Energy was undertaken. The results reported in this paper represent a summary of the char gasification experiments on powder, chip and pelletized ponderosa pine feedstock.

2. APPARATUS AND EXPERIMENTAL METHOD

The gasification test apparatus is shown in Fig. 2. The reactor consists of a Cahn electrobalance fitted with a quartz reaction tube. The reaction tube is maintained at reaction temperature using a tube furnace capable of operating at temperatures up to 1200 °C. Reaction temperature is determined by a type K thermocouple located in the reaction chamber. Because of the large length-to-diameter ratio of the reaction cavity and because of the gas preheating provision, the thermocouple temperature correction is found to be negligible. Reaction gas or purge gas is fed to the reactor tube at atmospheric pressure and a rate such that about a 10/1 molar ratio of gas/solid is maintained assuring that the gas composition is known. The sample size is typically 3–30 mg of charcoal as either powder or

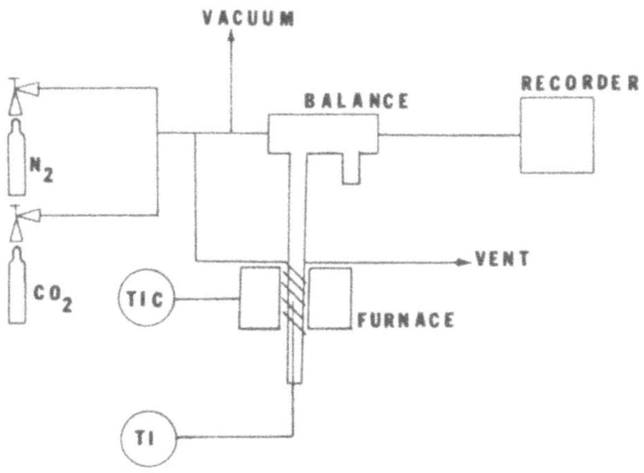

FIG. 2. Cahn electrobalance apparatus.

a large particle. Powder is gasified on a platinum pan while large particles are hung directly on the balance hang-down wire.

For powder tests, a representative sample of ponderosa pine (analysis given in Table I) was pyrolyzed in nitrogen at 800 °C in a dedicated pyrolysis reactor. The sample was cooled and ground to −200 mesh for gasification tests. Whole chips and pellets were also pyrolyzed using this apparatus. While there is generally an aging effect on char reactivity

TABLE I
PONDEROSA PINE FEEDSTOCK ANALYSIS

Proximate analysis	*Wt %*
Volatile matter	78·90
Fixed carbon	15·70
Ash	1·69
Moisture	3·71
Ultimate analysis, m.a.f. basis	*Wt %*
Carbon	50·17
Hydrogen	6·04
Oxygen	43·58
Nitrogen	0·16
Sulfur	0·05

Gross heating value, m.a.f. basis: 8647 Btu/lb.

associated with cooling and exposing the char to air, all samples were treated in the same fashion. Therefore the results reported on the importance of mass transfer are valid. The sample was charged to the reactor and brought to reaction temperature, where it was held typically for 30 min in nitrogen. The experiment was initiated by step changing the atmosphere.

3. GASIFICATION OF PONDEROSA PINE IN CO_2–CO ATMOSPHERES

Generally in the study of gas–solid reaction kinetics, the reaction rate depends on temperature, composition, particle size and pretreatment conditions. Particle size effects are particularly important in intraparticle diffusion mass transfer analysis. To obtain gasification rate data under conditions where the gross solid structure does not appreciably affect the reaction rate, all rates were determined for −200 mesh charcoals. This is an adequate criterion since it was shown that the rate for −80 mesh and −200 mesh powders were identical.

Figure 3 shows the effect of residence time on fractional conversion for a

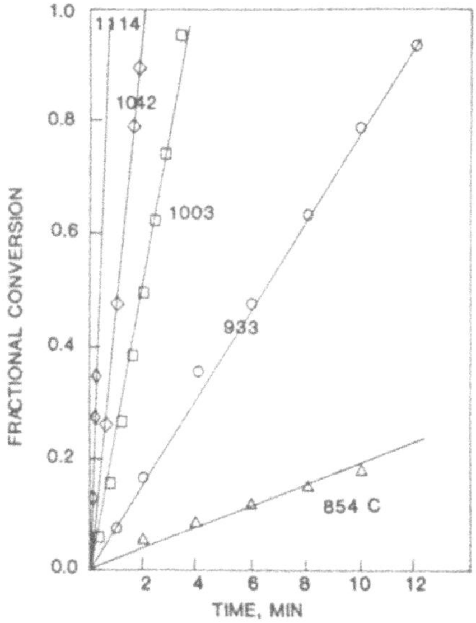

FIG. 3. Gasification rates for −200 mesh char.

series of runs where the pretreatment and reaction temperatures are the same. The general rate law which describes the data is given as follows:

$$-r_c = -\frac{1}{W_0}\frac{dw}{dt} = f(T)g(C) \tag{1}$$

where W_0 = initial carbon inventory, W = local carbon inventory, t = time, $f(T)$ = a function of temperature and $g(C)$ = a function of gas atmosphere.

Integrating eqn (1) at constant temperature and composition results in eqn (2):

$$X_A = \frac{W_0 - W}{W_0} = f(T)g(C)t \tag{2}$$

where X_A = fraction of char converted.

Equation (2) adequately describes the zero order dependency of the observed rate on carbon inventory at least up to high conversions. These data suggest that the available surface area for reaction remains essentially constant.

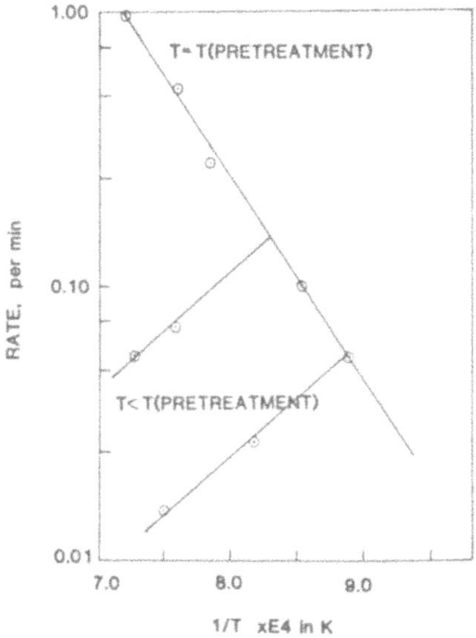

FIG. 4. The effect of temperature on char gasification rate.

Following Menster and Ergun,[4] for example, we postulate that the reaction rate follows an expression far from equilibrium of the form:

$$-r_c = \frac{k_1 P_{CO_2}}{1 + k_2 P_{CO_2} + k_3 P_{CO}} \quad (3)$$

For a pure CO_2 feed as in Fig. 3, the data may be correlated in terms of the principal activation energy. Figure 4 shows that this activation energy is 34 059 cal/mole.

Figure 4 also shows that if the pretreatment exceeds the reaction temperature, a predictable reduction in reaction rate occurs. Following

TABLE II

COMPARISON OF EXPERIMENTAL AND CALCULATED REACTION RATES FOR PONDEROSA PINE CHARCOAL AT 0·8 ATMOSPHERES TOTAL PRESSURE RATE,[a] g/g min

Temperature (°C)	*Experimental*	*Eqn (5)*	*Mole fraction CO_2*
853	0·056	0·054	1·000
1003	0·288	0·325	1·000
1042	0·529	0·485	1·000
800	0·034	0·026	1·000
1029	0·594	0·425	1·000
1114	0·961	0·953	1·000
900	0·099	0·100	1·000
852	0·011	0·003	0·310
949	0·039	0·039	0·310
1047	0·197	0·150	0·310
880	0·066	0·027	0·685
943	0·076	0·094	0·685
1002	0·157	0·208	0·685
1046	0·278	0·337	0·685
844	0·056	0·016	0·821
897	0·068	0·058	0·821
950	0·111	0·135	0·821
1007	0·263	0·270	0·821
919	0·064	0·050	0·591
863	0·050	0·010	0·544
942	0·093	0·068	0·544
1019	0·177	0·197	0·544
1013	0·253	0·202	0·594

[a] Pretreatment temperature equals reaction temperature.

Johnson,[2] we correlate this pretreatment effect in terms of a thermal annealing contribution to the apparent activation energy.

$$E = E(\text{true}) + E(\text{pretreatment}) \quad (4)$$

where E(true) = 53 870 cal/mole and E(pretreatment) = − 19811 cal/mole.

It is interesting to note that the coal pretreatment effect reported by Johnson is −9340 cal/mole. This lower value probably reflects the greater aromaticity of coal chars (more highly annealed state). The true activation energy is in good agreement with values reported for a variety of carbons.[3]

Equation (3) suggests that carbon monoxide and carbon dioxide retard the rate of reaction. The effect of gas composition was determined by varying the CO_2/CO ratio of the reaction mixture. Over the temperature range of interest, it was found that only carbon monoxide retardation was significant in the rate expression denominator. The resulting kinetic constants are as follows:

$$-r_c = \frac{f k_1 P_{CO_2}}{1 + k_2 P_{CO}}$$

$k_1 = \exp(-17\,141/T + 12{\cdot}3091)$ g/g min atm; $f = \exp[-9970(1/T - 1/T(\text{pretreatment})]$ for $T_P \geq T$; $f = 1$ for $T_P < T$; $k_2 = \exp(34\,021/T - 28{\cdot}4295)$; T = temperature in K.

Table II shows the favorable comparison between calculated and experimental reaction rates.

4. EFFECT OF PARTICLE SIZE – WHOLE CHIPS

The largest reaction rates occur for the situation where the pretreatment and reaction temperatures are the same and the reaction atmosphere is pure CO_2. Figure 5 compares rates for 5 mm chips and −200 mesh solid. Commercial hard and softwood chips generally exhibit a thickness of 2 to 6 mm ($\frac{1}{16}$ to $\frac{1}{4}$ in.) with lengths and widths many times the thickness. The data show that mass transfer effects in powders and chips are identical; therefore, the macropore structure of the charcoal does not affect the charcoal reactivity to temperatures at least as high as 1135 °C, which was the highest temperature used in this study. This does not mean that the solid reaction is not diffusion limited; the results suggest that the same micropore structure exists in the ground solid and the whole pyrolyzed chip. Furthermore, since the rates are the same, we can infer that heat

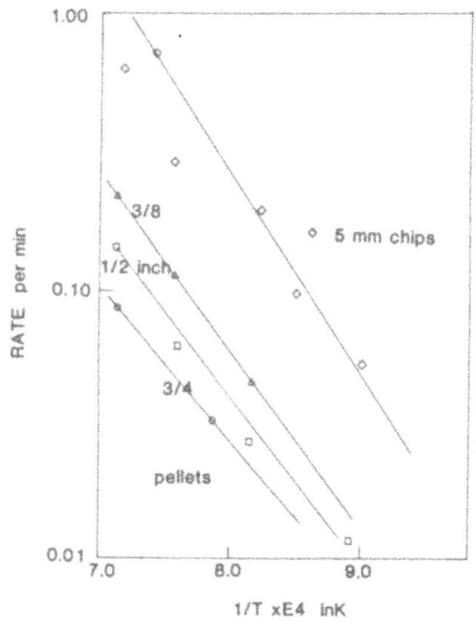

FIG. 5. The effect of size on gasification rate.

transfer is not a limiting factor. In terms of gasifier modelling at the temperature of interest, this result is obviously an important simplification.

5. EFFECT OF PARTICLE SIZE – WHOLE PELLETS

Whole pine pellets prepared by our standard laboratory technique[1] were also investigated. To obtain a benchmark for rate studies, a pyrolyzed pellet was ground to −200 mesh and gasified. The results were identical to those obtained for undensified ponderosa pine charcoal.

A series of experimental runs were made for whole pellets. The data, shown in Fig. 5, show that the reaction rate depends on particle size. However, the reaction rates exhibit an apparent activation energy not substantially different than that found for fine powders and chips. This indicates that macropore diffusion is again not limiting in pellets. In these runs, we were not able to maintain a large excess of reactant. Therefore, the gas composition, as estimated by material balance, was used for data interpretation purposes. The results are summarized in Table III. At the lowest temperatures, the gas is essentially CO_2. At higher temperatures, the gas tends to contain substantial CO. Correcting the reaction rates for the effect of CO raises the rate curve bringing the apparent activation energy

TABLE III
RATE DATA FOR PONDEROSA PINE PELLETS

Pellet size (in.)	*Reaction temperature (°C)*	*Gas, mole% CO_2*	*Rate (pellet) / rate (powder)*
1/2	950	70	0·14
1/2	1040	55	0·14
1/2	850	83	0·29
1/2	810	93	0·15
1/2	993	50	0·28
1/2	1130	32	0·22
3/8	950	75	0·19
3/8	1050	53	0·23
3/8	1130	37	0·29
3/4	1127	23	0·19
3/4	1000	41	0·15

more in line with the fine powder data. We can conclude that no additional mass transfer effects arise due to the pelletization procedure.

Figure 6 compares calculated rates based on the estimated gas composition for powders with those measured. The comparison between rate values suggests that, in most instances, the rates for pellets fall significantly outside the bounds of error for the rate model. Furthermore, since the estimated gas composition is based on the total amount of carbon converted, rather than an average, the calculated rate will always tend to be low. To determine the cause of the rate discrepancy, a series of pellets were drilled and mounted on thermocouples in the electrobalance to determine whether a heat transfer limitation exists. The maximum temperature

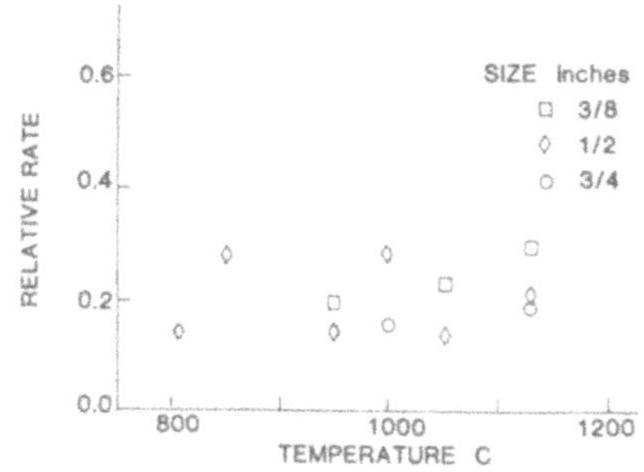

FIG. 6. Comparison of gasification rates for pellets and powder.

differential found was 50 °C for the $\frac{3}{4}$ in. particle. Since the whole solid saw reaction temperature prior to introduction of reaction gas, the true activation energy must be used to estimate rate sensitivity. At the center of the particle, this temperature drop reduces the rate by a factor of 2. Since the entire solid is at an average temperature greater than the center and since the majority of mass is located near the outer radius, heat transfer cannot be responsible for the discrepancy in rates.

The average ratio of experimental to calculated rates in Fig. 6 is $0{\cdot}25 \pm 0{\cdot}11$. Results for powders, chips and powder made from densified pine show that the micropore structure is not affected while the activation energy data show that other mass transfer resistances do not affect the rate since the gasification activation energy is not materially reduced with pellets. We propose that in the pellet making process significant external area of grains is lost in forming contact area between particles. This would result from the application of heat and pressure on the sawdust causing the lignin to melt and flow.

6. CONCLUSIONS

The gasification rates of ponderosa pine particles, chips and pellets are correlated by a single rate equation suggesting that mass transfer limitations, if they exist, are confined to the micropore structure of the solid. Pellets exhibit a rate somewhat less than chips and powder. This decrease is attributed to a loss of reactive surface which has been converted into the contact area between fused grain of pyrolyzed wood.

REFERENCES

1. Herman, D. R. (1981). The rate of pyrolysis of ponderosa pine. MS Thesis, Colorado School of Mines.
2. Johnston, J. L. (1979). *Kinetics of Coal Gasification.* Wiley, New York.
3. Lowry, H. H. (Ed.) (1963). *Chemistry of Coal Utilization.* Wiley, New York, Chapter 20.
4. Menster, M. and Ergun, S. (1973). US Bur. Mines Bull. (644).

32

Gasification of Chars Obtained from Maple and Jack Pine Woods

SATYENDRA P. NANDI and MICHAEL ONISCHAK
Institute of Gas Technology, Chicago, Illinois, USA

1. INTRODUCTION

Biomass—forestry and agricultural residues—can locally be directly utilized for combustion. However, conversion of these solid materials to gaseous products is of great importance to make transport of energy practicable.

All gasification processes, except those utilizing the agency of microorganisms, are conducted at relatively high temperatures. At elevated temperatures the constituents of biomass (mainly carbohydrates and lignin) decompose, yielding gas, liquid and char products. The relative distribution of these products of thermal decomposition (pyrolysis) depends on process conditions.

It is well-known that the rates of pyrolytic reactions are fast compared to the subsequent gasification of char by reactive gases. Therefore the overall rate of gasification will be determined by the slower rate of gasification of biomass char. Only a limited number of experimental investigations on the reactivity of biomass chars have been reported. Rensfelt *et al.*[1] conducted stream-gasification of chars produced from straw, poplar wood and bark, in the temperature range from 770 to 850 °C and at a total pressure of 1 atm (101 kPa). Sealock *et al.*[2] gasified cellulose and lignin chars in the temperature range from 658 to 850 °C and at 1 atm pressure. Ross and Fikis[3] gasified jack pine bark at a total pressure of 1 atm (steam partial pressure $\approx 7{\cdot}58$ kPa) at 650 °C and determined the rate of evolution of gases.

No data were available on the reactivity of biomass chars in any gaseous medium other than steam or at elevated pressure for pressurized process design. The present work was undertaken to determine the gasification characteristics of chars from two wood species in different reaction media in addition to steam over a range of temperatures and pressures of practical interest.

2. EXPERIMENTAL

2.1. Raw Materials

Two biomass materials—a hardwood (maple) and a softwood (jack pine)—were used in the experimental study. All measurements were conducted with $-12+16$ mesh (Tyler) size particles. The proximate and ultimate analyses of the samples are shown in Table I.

2.2. Methods

The kinetic measurements were conducted using a high pressure thermobalance with the capability of controlling various gaseous reaction media. The details of the apparatus and experimental procedure have been reported elsewhere.[4] Two series of gasification runs were conducted in the course of this study. A series of runs were conducted with 'stabilized chars'. For these runs the gasification temperature was set equal to the

TABLE I
PROXIMATE AND ULTIMATE ANALYSES

	Maple	*Jack pine*
Proximate analysis, wt %		
Moisture	5·5	6·6
Volatile matter	81·2	79·2
Fixed carbon	13·1	13·9
Ash	0·2	0·3
Total	100·0	100·0
Ultimate analysis, wt % dry		
Carbon	49·80	51·50
Hydrogen	6·05	6·38
Nitrogen	0·15	0·00
Sulfur	0·00	0·02
Oxygen (by diff.)	43·79	41·77
Ash	0·21	0·33
Total	100·00	100·00

devolatilization temperature used for char preparation. The rate of weight loss of the chars was measured under different experimental conditions.

In the other series, gasification was done with dry wood samples such that, in a typical run, 1 g of sample was taken in a wire mesh basket and introduced quickly to the reaction zone of the thermobalance maintained at chosen reaction conditions. The typical weight losses undergone by maple hardwood introduced in this way in a steam–nitrogen mixture ($H_2O/N_2 = 1$) at 2·17 MPa and at different temperatures are shown in Fig. 1. At all temperatures a very rapid weight change takes place within a

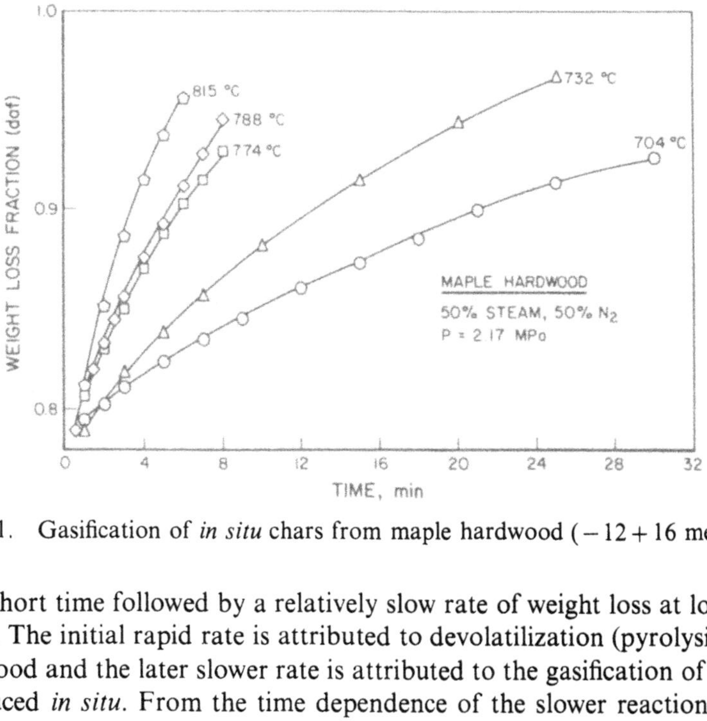

FIG. 1. Gasification of *in situ* chars from maple hardwood (−12+16 mesh).

very short time followed by a relatively slow rate of weight loss at longer times. The initial rapid rate is attributed to devolatilization (pyrolysis) of the wood and the later slower rate is attributed to the gasification of char produced *in situ*. From the time dependence of the slower reaction, the gasification characteristics of the *in situ* chars were determined. The experimental procedure outlined above is similar to that used by Sealock *et al.*[2]

The devolatilization of the biomass materials (−12+16 mesh) to produce char outside of the thermobalance was conducted in a tube furnace in a nitrogen atmosphere (101 kPa) at different temperatures from 732 to 927 °C. The heat-up rate was 11 °C/min with a soak time of 30 min at the maximum temperature. The residues left on devolatilization at different temperatures were named 'stabilized chars'. The particle sizes on the chars were not measured. By visual inspection there appeared to be only a small change in particle size of the feed due to devolatilization. The ultimate analyses of the chars are shown in Table II. It is seen that the carbon content of the stabilized chars was >93%.

Surface areas of these char samples in nitrogen and carbon dioxide were determined. The total open pore volumes of the char samples were obtained from density measurements in helium and in mercury. The apparatus and experimental procedures for these measurements have also been described elsewhere.[5]

Gasification tests in the thermobalance apparatus were performed in the temperature range from 704 to 871 °C at pressures from 0·17 to 2·17 MPa. The reaction media had the following composition:

Medium	*Composition (mole percent)*
I	$H_2O = 50{\cdot}0$ and $N_2 = 50{\cdot}0$
II	$H_2O = 50{\cdot}0$, $H_2 = 5{\cdot}0$ and $He = 45{\cdot}0$
III	$H_2O = 50{\cdot}0$, $H_2 = 10{\cdot}0$ and $He = 40{\cdot}0$
IV	$H_2O = 50{\cdot}0$, $H_2 = 25{\cdot}0$ and $He = 25{\cdot}0$
V	$H_2 = 100{\cdot}0$
VI	$H_2O = 55{\cdot}3$, $H_2 = 28{\cdot}7$, $CO = 5{\cdot}4$ and $CO_2 = 10{\cdot}6$

The rate of heat-up of the biomass samples has not been determined exactly as it involves a number of unknowns. The experimental procedure for the *in situ* char tests approximates the heat-up conditions of biomass materials such as when introduced to a fluidized bed maintained at elevated temperature and pressure.

The experimental results of gasification of maple hardwood and jack pine chars were analyzed and described with the first order rate expression

$$\frac{dX_c}{dt} = k(1 - X_c) \tag{1}$$

or

$$-\ln(1 - X_c) = kt \tag{2}$$

where X_c = weight loss fraction of char (daf), t = time (min) and K = reaction rate constant (min^{-1}).

When starting with a raw wood sample to study *in situ* char gasification the following definitions were used: X = total weight loss fraction (daf) of

TABLE II

ULTIMATE ANALYSES OF CHARS PRODUCED BY DEVOLATILIZATION IN NITROGEN (1 ATM) FROM DIFFERENT BIOMASS MATERIALS

Temperature of devolatilization °*F*	*Maple*				*Jack pine*			
	Ash	*C*	*H*	*Diff.*[a]	*Ash*	*C*	*H*	*Diff.*
				wt % dry				
1200	0·62	92·80 (93·40)[b]	1·46 (1·47)	5·12 (5·13)				
1350	0·87	93·70 (93·50)	1·09 (1·10)	4·34 (4·40)	0·77	94·10 (94·80)	1·09 (1·10)	4·04 (4·10)
1425	1·02	94·20 (95·20)	0·71 (0·71)	4·07 (4·09)	1·19	95·00 (96·10)	0·97 (0.98)	2·84 (2.92)
1500	0·96	94·70 (95·60)	0·65 (0·66)	3·69 (3·74)	1·10	95·20 (96·20)	0·82 (0·83)	2·88 (2·97)
1600	1·10	95·20 (96·20)	0·39 (0·39)	3·30 (3·41)	1·08	95·80 (96·80)	0·71 (0·72)	2·41 (2·48)
1700	1·30	95·50 (96·70)	0·33 (0·33)	2·87 (3·30)	1·62	95·20 (96·80)	0·60 (0.61)	2·58 (2·59)

[a] Diff. = (oxygen + nitrogen + sulfur) content of char.
[b] Numbers in parentheses denote weight percent on dry ash free basis.

feed, and X^* = weight loss of feed due to devolatilization. Therefore, the char carbon fraction is expressed as

$$X_c = (X - X^*)/(1 - X^*) \tag{3}$$

and from eqns (2) and (3)

$$-\ln(1 - X) = -\ln(1 - X^*) + kt \tag{4}$$

3. RESULTS AND DISCUSSION

3.1. Stabilized Char Gasification

Typical experimental results for the gasification of 'stabilized chars' are shown in Figs 2 and 3 in expected process steam–hydrogen concentration (Medium II). In these figures the fractional weight losses of maple and jack pine stabilized chars are plotted as a function of time for gasification in Medium II in the temperature range from 649 to 927 °C at a pressure of 2·17 MPa. The effect of pressure (0·79–2·17 MPa) on the rate of gasification was studied with maple chars in Medium II. Gasification of the

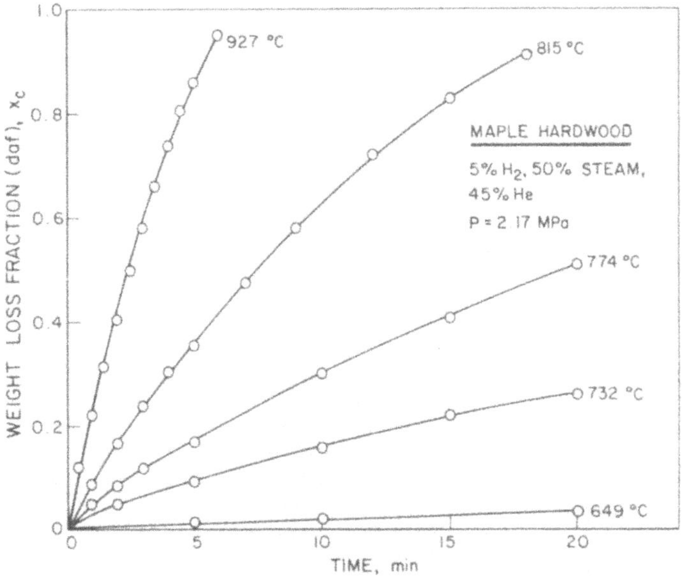

FIG. 2. Base carbon conversion as a function of time for maple hardwood stabilized chars.

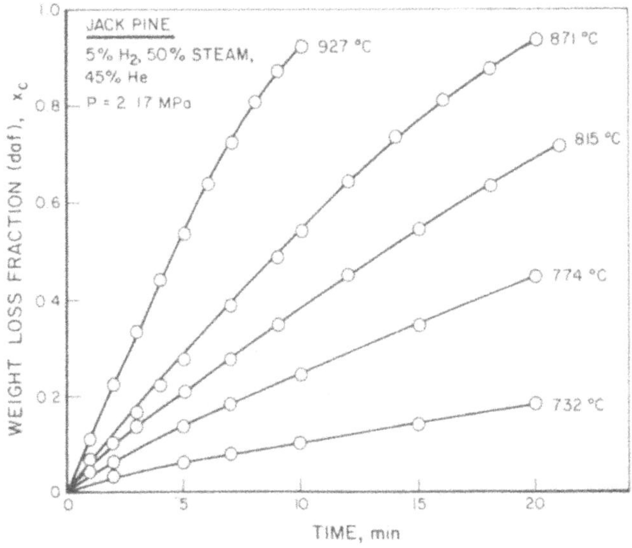

FIG. 3. Base carbon conversion as a function of time for jack pine stabilized chars.

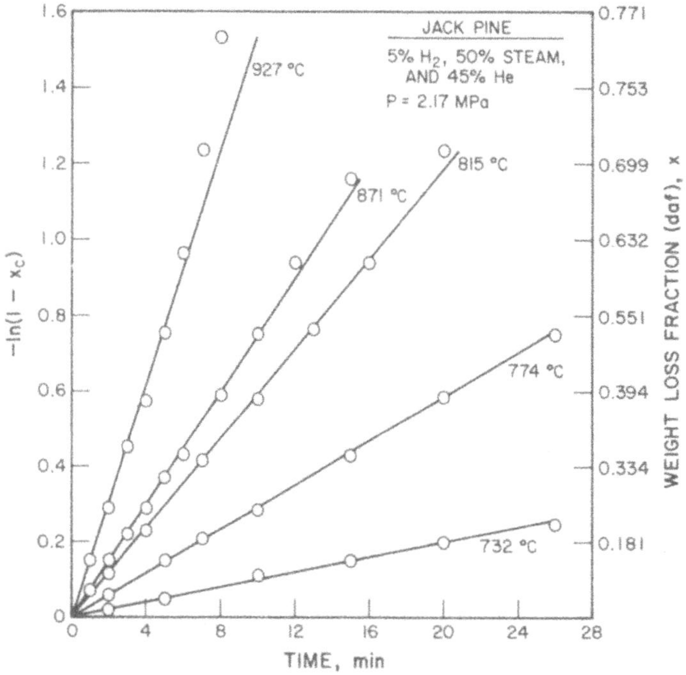

FIG. 4. Gasification of stabilized chars from jack pine—first-order plot.

stabilized chars from both maple and jack pine was also done in Medium I (steam and nitrogen) in the temperature range from 704 to 815 °C at 2·17 MPa.

The first order rate plots for the gasification of stabilized chars from jack pine in Medium II (data of Fig. 3) are plotted according to eqn (2) in Fig. 4. For gasification up to 871 °C the experimental data can be represented by straight lines passing through the origin for values of $X_c \approx 0{\cdot}7$. At 927 °C, however, the experimental data were linear to $X_c \approx 0{\cdot}5$. The gasification characteristics of maple chars are also similar to those of jack pine in the same medium. For gasification in Medium I, the rate characteristics for chars from both materials were similar to that in Medium II. The values of the overall rate constant, k, were obtained from the slope of the straight lines shown in Fig. 4 and similar plots. The values for all of the stabilized char tests are given in Table III.

In Table III, for gasification in Medium II with the maple chars it is seen that there is only a small effect of pressure on k, in the range of 0·79 to 2·17 MPa. The values of k for maple chars are higher compared to that of

TABLE III

OVERALL RATE CONSTANT FOR THE GASIFICATION OF STABILIZED CHARS FROM MAPLE HARDWOOD AND JACK PINE

Reaction Medium I (50% H_2O and 50% N_2)			*Reaction Medium II* (5% H_2, 45% He and 50% H_2O)		
Temperature, °C	*Pressure, MPa*	k, min^{-1}	*Temperature, °C*	*Pressure, MPa*	k, min^{-1}
Maple hardwood					
704	2·17	0·0122	732	0·79	0·0129
760	2·17	0·0458	732	2·17	0·0131
815	2·17	0·1079	774	0·79	0·0371
			774	2·17	0·0350
			815	0·79	0·0983
			815	2·17	0·0983
			927	0·79	0·3447
			927	1·48	0·3393
			927	2·17	0·2861
Jack pine					
704	2·17	0·0088	732	2·17	0·0108
760	2·17	0·0280	774	2·17	0·0292
760	2·17	0·0319	815	2·17	0·0593
815	2·17	0·0700	871	2·17	0·0750
			927	2·17	0·1530

jack pine. For both materials, the values of k increase with temperature. The variation has been represented by the Arrhenius equation and is shown in Fig. 5. Gasification in Medium I was conducted over a smaller temperature range compared to that for Medium II. The values of the apparent activation energy and frequency factor for char gasification in the common temperature (for Media I and II) range are given in Table IV. The rate parameters reported by Rensfelt *et al.*[1] for three biomass materials are also included in the above table for comparison. It is seen that the activation energy for gasification in Medium II was slightly higher than that for Medium I for both maple and jack pine chars. The E_a value for maple was slightly higher compared to that for jack pine for gasification in the same medium. The value of the frequency factor for maple, on the other hand, was higher compared to that of jack pine. It can also be seen that the kinetic parameters reported by Rensfelt *et al.*[1] for gasification of poplar, straw and poplar bark chars are comparable to that of maple and jack pine chars obtained in the present work.

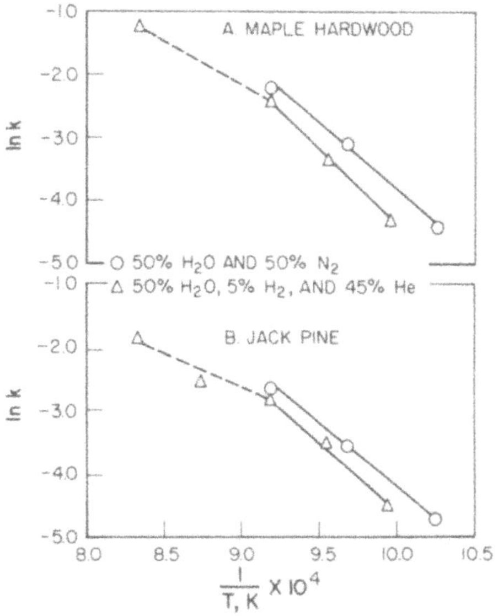

FIG. 5. Gasification of stabilized chars from maple hardwood and jack pine—Arrhenius plot.

TABLE IV

APPARENT ACTIVATION ENERGY AND FREQUENCY FACTORS FOR GASIFICATION OF STABILIZED CHARS FROM MAPLE HARDWOOD AND JACK PINE

Sample	*Gas composition, mole percent*	*Pressure, kPa*	*A, min^{-1}*	*E_a, kcal/mole*
Maple hardwood stabilized char	50% H_2O & 50% N_2	2170	$3{\cdot}329 \times 10^7$	42·27
Maple hardwood stabilized char	50% H_2O, 5% H_2 & 45% He	2170	$1{\cdot}778 \times 10^8$	46·75
Jack pine stabilized char	50% H_2O & 50% N_2	2170	$1{\cdot}031 \times 10^7$	40·73
Jack pine stabilized char	50% H_2O, 5% H_2 & 45% He	2170	$1{\cdot}732 \times 10^7$	42·50
Poplar wood char[1]	73% H_2O & 27% Ar	101	$1{\cdot}2 \times 10^8$	43·48
Straw[1]	73% H_2O & 27% Ar	101	$5{\cdot}9 \times 10^7$	43·48
Bark[1]	73% H_2O & 27% Ar	101	$9{\cdot}1 \times 10^7$	42·52

The slope of the Arrhenius plot in Fig. 5 for gasification of chars in Medium II decreases at higher temperature ($T > 815\,°C$). The change in pore structure of char on being exposed to progressively higher temperature may be a factor for this observation. To follow the change in pore structure with the increase of devolatilization temperature, the nitrogen and carbon dioxide surface areas as well as the total open pore volume of the chars were determined. The results are given in Table V.

TABLE V

SURFACE AREAS AND TOTAL OPEN PORE VOLUMES OF STABILIZED CHAR

Devolatization temperature, °C	*Surface area m^2/g*		*Apparent density,[a] g/cm^3*	*True density,[b] g/cm^3*	*Total open pore volume, cm^3/g*
	N_2	CO_2			
		Maple hardwood			
649	58·1	676	0·44	1·58	1·64
732	37·5	693	0·41	1·65	1·83
774	46·2	640	0·42	1·72	1·80
815	39·3	606	0·42	1·76	1·81
871	13·3	680	0·42	1·76	1·81
927	11·8	398	0·44	1·82	1·72
		Jack pine			
732	91·2	808	0·46	1·71	1·59
774	45·0	777	0·47	1·75	1·56
815	1·2	780	0·48	1·83	1·54
871	0·7	800	0·50	1·83	1·45
927	0·3	836	0·51	1·83	1·41

[a] By displacement of mercury at 101 kPa and at 25 °C.
[b] By displacement of helium at 25 °C.

For the maple hardwood chars, the nitrogen surface area decreases markedly from 815 to 871 °C, whereas the carbon dioxide area of the char still remains high at 871 °C. Only at 927 °C does the carbon dioxide area decrease. The open pore volume remains unchanged from 732 to 871 °C. For the jack pine chars the nitrogen area decreases continuously in the range 732 to 927 °C and the value is quite low even at 815 °C. The carbon dioxide surface area of the chars, on the other hand, remains high at all temperatures.

It has been indicated[5] that the 'carbon dioxide area' of a char is the 'total potential surface area', which includes areas contained in molecular size pores. The 'nitrogen area', on the other hand, can be assumed to be the sum of the geometric area of the particles and surface area in pores larger than $\sim 12 \times 10^{-8}$ cm in diameter. It has also been deduced[6] that the intermediate size pores (which contribute to nitrogen area) act as feeder channels to the molecular size pores. From the above arguments it follows that the decrease of nitrogen area will restrict the diffusion of the reactants to the molecular size pores. This effect will be important only at high temperature where the rate of chemical reaction is high. The decrease of the slope of the Arrhenius plot shown in Fig. 5 may be due to restricted pore diffusion which is indicated by the nitrogen surface area values.

3.2. *In Situ* Char Gasification

Typical experimental results for the gasification of '*in situ* chars' are shown in Figs 1 and 6. The effect of pressure on rate of reaction was studied in Media I and V (hydrogen) at 815 °C for the two wood samples. An estimate of the effect of pressure on the extent of devolatilization weight loss was obtained by exposing the wood sample to different pressures and temperatures in nitrogen using the same procedure as in the reactive gas-gasification tests. The experimental weight losses of the samples after 2 min at various temperatures are reported in Table VI.

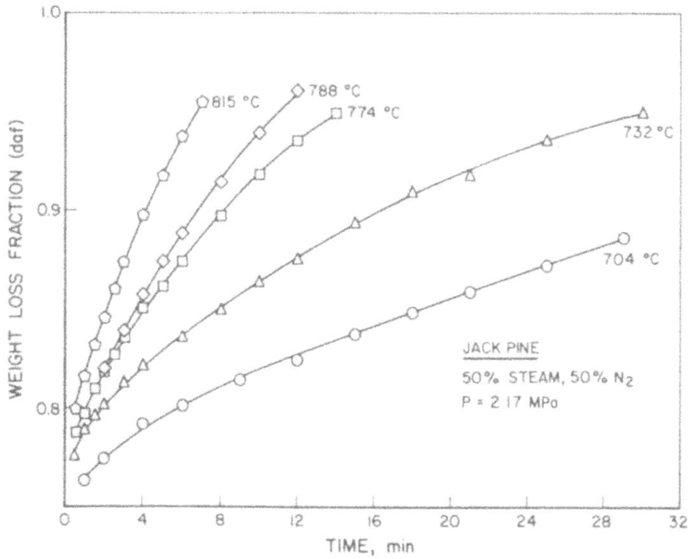

FIG. 6. Gasification of jack pine (−12+16 mesh).

TABLE VI

DEVOLATILIZATION OF BIOMASS SAMPLES IN NITROGEN

Maple hardwood (− 12 + 16 mesh)			*Jack pine (− 12 + 16 mesh)*		
Temperature, °C	*Pressure, MPa*	*Weight loss, % (daf)*	*Temperature, °C*	*Pressure, MPa*	*Weight loss, % (daf)*
732	2·17	78·7	774	2·17	79·4
774	2·17	79·8	815	2·17	80·5
815	2·17	79·9	815	0·17	84·3
815	0·79	83·0	871	2·17	81·1
815	0·17	87·3			
871	2·17	81·1			

The table shows that at 2·17 MPa the weight loss increased with the increase of the devolatilization temperature for both materials. However, the change was not large, e.g. for maple the weight loss figures changed from 78·7 % to 81·1 % for a temperature change from 732 to 871 °C. The table also shows that the weight loss at the same temperature increased with a decrease in pressure. From the standpoint of char yield the effect of pressure was considered to be significant. In addition the char yield in these devolatilization tests provides a measure of the amount of '*in situ* chars' that would be produced under comparable gasification conditions.

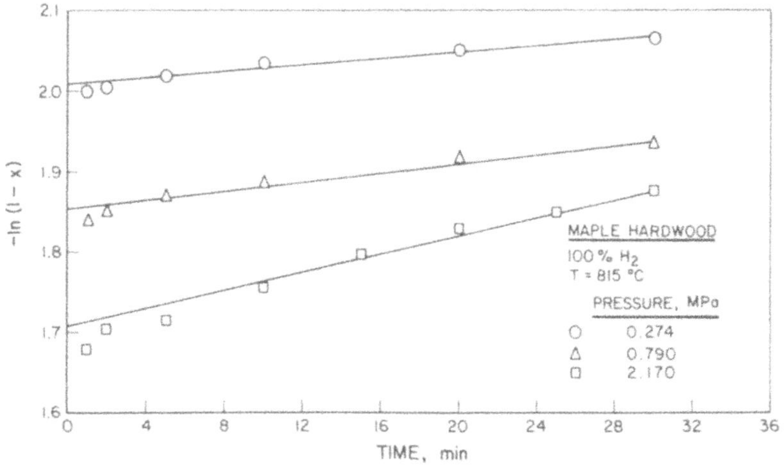

FIG. 7. Gasification of *in situ* chars from Maple hardwood (− 12 + 16 mesh)—first-order plots.

Gasification in hydrogen (Medium V) was conducted only at 815 °C. A very small weight change occurred after initial devolatilization. The results are shown in Figs 7 and 8. In these figures $-\ln(1-X)$ has been plotted against reaction time. For maple chars the values of k are 0·0020, 0·0028 and 0·0056 min^{-1} at pressures of 0·27, 0·79 and 2·17 MPa respectively. The corresponding values of the jack pine chars were 0·0002, 0·0034 and 0·0054 min^{-1}. As is discussed later, the values of k shown above are low compared to those obtained in Medium I ($H_2O/N_2 = 1$). From the intercept values in Figs 7 and 8, it can be seen that the weight loss due to devolatilization increases with decrease of pressure.

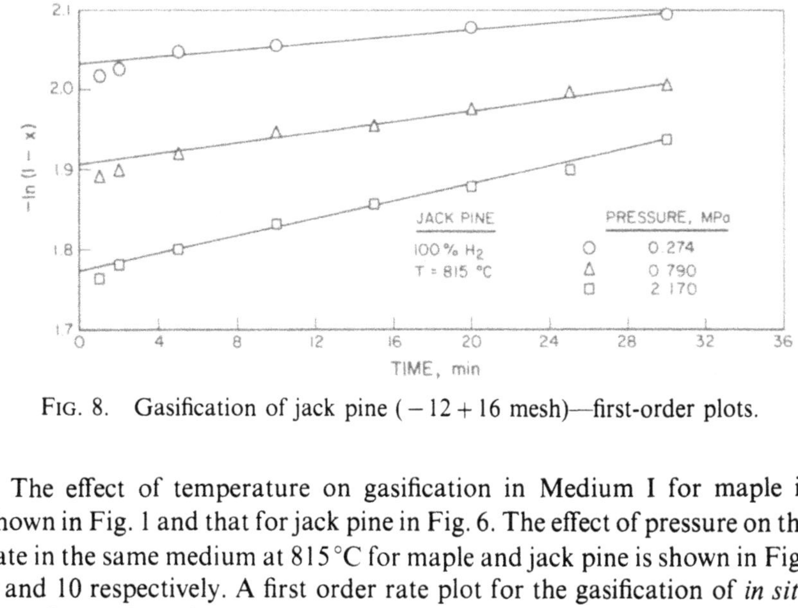

FIG. 8. Gasification of jack pine (−12 + 16 mesh)—first-order plots.

The effect of temperature on gasification in Medium I for maple is shown in Fig. 1 and that for jack pine in Fig. 6. The effect of pressure on the rate in the same medium at 815 °C for maple and jack pine is shown in Figs 9 and 10 respectively. A first order rate plot for the gasification of *in situ* chars in Medium I is shown in Fig. 11. As was the case with 'stabilized chars' the experimental data can be represented by straight lines for the period after the initial devolatilization. The straight lines here, however, do not pass through the origin as was the case for the stabilized chars. The intercept on the ordinate has been equated to the term $-\ln(1-X^*)$ in accordance with eqn (4).

The gasification rates for *in situ* chars from maple and jack pine were also measured in steam–hydrogen mixtures (Media II, III, IV). All of the experimental data were correlated with eqn (4). In addition, a few gasification runs in Medium I were done in two steps to investigate char

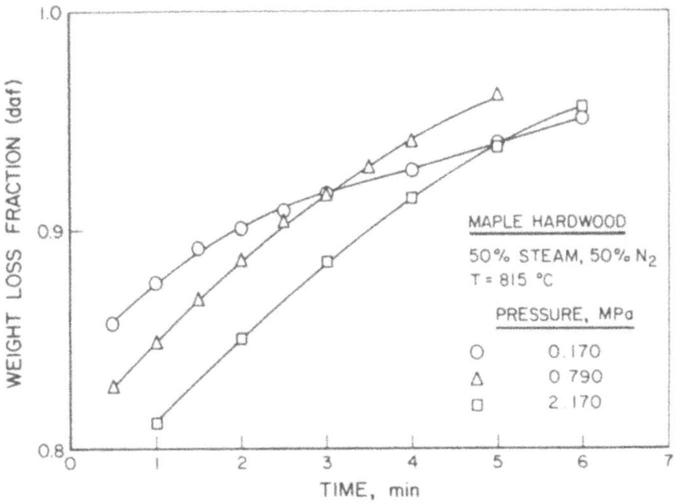

FIG. 9. Gasification of *in situ* chars from maple hardwood (−12+16 mesh).

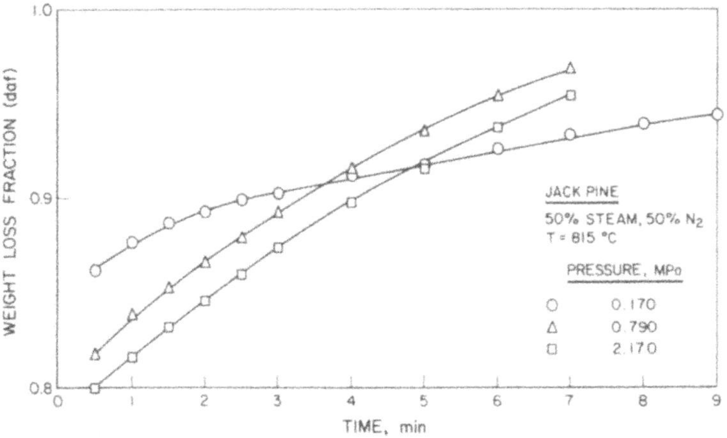

FIG. 10. Gasification of *in situ* chars from jack pine (−12+16 mesh).

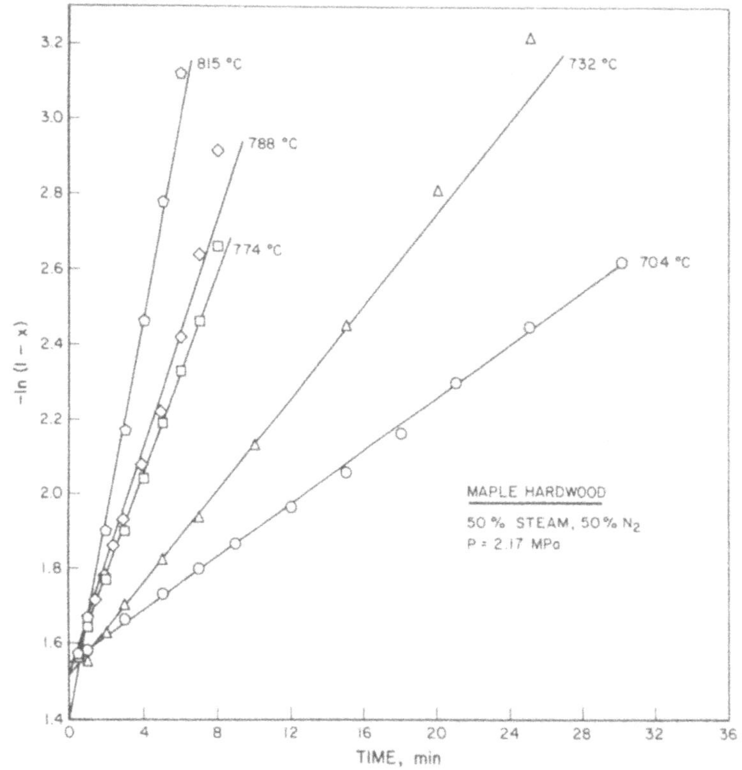

FIG. 11. Gasification of *in situ* chars from maple hardwood (−12+16 mesh)—first-order plot.

reactivity upon exposure to air. The procedure used was to gasify the wood at chosen experimental conditions for 2 min, remove the unreacted char from the reaction zone, and cool the char in helium to room temperature. The char was then exposed to air at room temperature for 24 h, after which the char sample was again reacted for a further period of time. This procedure, called 'gasification in two steps', is in contrast to the usual experiments. The 'two step' gasification tests indicated whether exposure of *in situ* chars to air at room temperature alters their reactivity.

The values of all the reaction rate constants in different reaction media along with the X^* values obtained from Medium I are given in Table VII. Note that the values of X^* in Table VII for experiments conducted at 2·17 MPa are comparable to those for pure nitrogen devolatilization weight loss data shown in Table VI. This indicates that the initial process of

TABLE VII

KINETIC PARAMETERS FOR THE GASIFICATION OF *in situ* CHARS FROM MAPLE HARDWOOD AND JACK PINE IN STEAM–HYDROGEN–NITROGEN/HELIUM MIXTURES

Temp., °C	Pressure, MPa	Medium I		Medium II k, min⁻¹	Medium III k, min⁻¹	Medium IV k, min⁻¹
		X*	k, min⁻¹			
			Maple hardwood			
704	2·17	0·788	0·036	—[a]	—	—
732	2·17	0·781	0·062	—	—	—
760	2·17	—	—	0·086	0·047	0·024
774	2·17	0·781	0·133	—	—	—
788	2·17	0·781	0·151	0·135	0·084	0·055
788	0·79	—	—	0·142	—	—
815	2·17	0·765	0·258	0·230	0·161	0·130
815	0·79	0·798	0·288	—	—	—
704[b]	2·17	0·785	0·032	—	—	—
732[b]	2·17	0·785	0·063	—	—	—
843	2·17	—	—	0·295	0·220	—
871	2·17	—	—	0·433	0·346	—
			Jack pine			
704	2·17	0·765	0·025	—	—	—
732	2·17	0·777	0·050	—	—	—
760	2·17	—	—	0·057	0·035	0·018
774	2·17	0·777	0·100	—	—	—
788	2·17	0·772	0·122	0·101	0·069	0·040
815	2·17	0·777	0·201	0·155	0·123	0·097
815	0·79	0·798	0·212	—	—	—
815	0·17	0·867	0·101	—	—	—
704[b]	2·17	0·789	0·022	—	—	—
732[b]	2·17	0·777	0·049	—	—	—
843	2·17	—	—	0·246	0·160	—
871	2·17	—	—	0·360	0·278	—

[a] Experiment not done for the particular medium at the temperatures and pressures.
[b] Experiment done in two steps.

devolatilization was unaffected by the presence of steam. The effect of total pressure on devolatilization is also similar for both nitrogen and steam. The values of k at 0·17 MPa are lower than the corresponding values at 0·79 MPa for the gasification of both materials. On the other hand, the value of k at 0·79 MPa is slightly higher than the corresponding values at 2·17 MPa for both the materials. The difference, however, is considered to be within experimental uncertainty. It is therefore concluded that the effect of steam partial pressure on the rate of gasification was insignificant for steam partial pressures above 0·395 MPa.

Comparing the k values of runs conducted in two steps with that obtained from one step gasification (at common temperatures, pressure, etc.) it is seen that the values are nearly identical. Therefore it is concluded that exposure of partially reacted *in situ* char (from maple or jack pine) to air at room temperature had no effect on its reactivity.

It is also seen from Table VII that the rate of gasification of the *in situ* chars from both woods decreased with the increase of H_2/H_2O ratio in the reaction medium. Inhibition of the rate of steam–carbon reaction by hydrogen is well known. The values of the apparent activation energy and frequency factors for the gasification of *in situ* chars in different gas mixtures were obtained from Arrhenius plots shown in Fig. 12. The values of the kinetic parameters are given in Table VIII. It is seen that both the activation energy and frequency factor increase with increase of H_2/H_2O ratio. The kinetic parameters (A and E_a) compensate each other to some extent but the effect of E_a appearing as an exponent predominates. The slope of the Arrhenius plot decreases at higher temperatures for

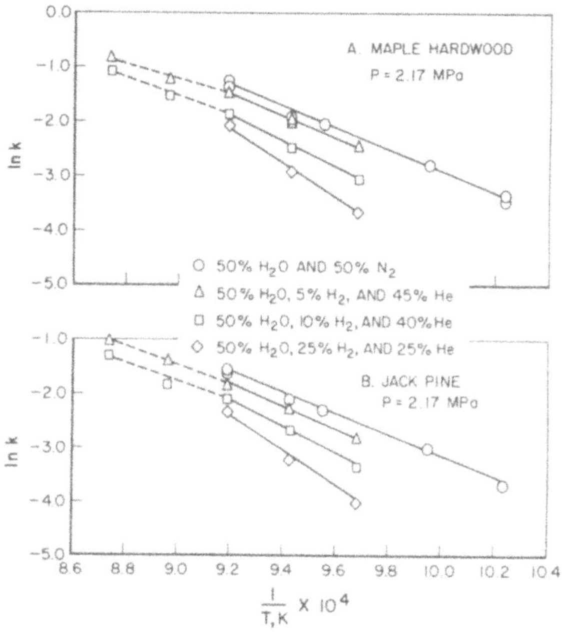

FIG. 12. Gasification of *in situ* chars from maple hardwood and jack pine—Arrhenius plot.

TABLE VIII

APPARENT ACTIVATION ENERGY AND FREQUENCY FACTORS FOR THE GASIFICATION OF *in situ* CHARS FROM MAPLE HARDWOOD AND JACK PINE IN DIFFERENT REACTION MEDIA

Sample	*Composition of reaction medium*	*A*,[a] *min*$^{-1}$	E_a,[a] *kcal/mole*
Maple hardwood *in situ* char	50 % H_2O & 50 % N_2	$2{\cdot}415 \times 10^{7}$	39·83
Maple hardwood *in situ* char	50 % H_2O, 5 % H_2 & 45 % He	$2{\cdot}187 \times 10^{7}$	40·00
Maple hardwood *in situ* char	50 % H_2O, 10 % H_2 & 40 % He	$2{\cdot}663 \times 10^{8}$	46·40
Maple hardwood *in situ* char	50 % H_2O, 25 % H_2 & 25 % He	$4{\cdot}143 \times 10^{11}$	62·80
Jack pine *in situ* char	50 % H_2O & 50 % N_2	$1{\cdot}286 \times 10^{7}$	39·10
Jack pine *in situ* char	50 % H_2O, 5 % H_2 & 45 % He	$1{\cdot}061 \times 10^{8}$	42·00
Jack pine *in situ* char	50 % H_2O, 10 % H_2 & 40 % He	$8{\cdot}000 \times 10^{8}$	49·20
Jack pine *in situ* char	50 % H_2O, 25 % H_2 & 25 % He	$4{\cdot}534 \times 10^{11}$	63·60

[a] The values of activation energy and frequency factor in the temperature range from 760 to 815 °C.

gasification Media II and III. This effect is similar to that observed with the stabilized chars.

The results of maple and jack pine *in situ* char gasification in Medium VI ($H_2O = 55{\cdot}3$, $H_2 = 28{\cdot}7$, $CO = 5{\cdot}4$ and $CO_2 = 10{\cdot}6$) is shown in Fig. 13. The first order rate behavior of the data at 815 and 843 °C for maple and jack pine in this medium is shown in Fig. 14. It may be noted that the experimental points up to 4 min do not fall on the straight lines; consequently, the rate constant obtained from the slope applies to gasification times greater than 4 min. The values of k and X^* for gasification in this medium are given in Table IX. The values of X^* obtained from the intercept are higher compared to the values of X^* that are shown in Table VIII. The kinetics of coal char gasification in synthesis gas mixtures (containing H_2O, H_2, CO and CO_2) has been discussed by Johnson.[8] Inhibition of the rate by carbon monoxide in the gasification medium has been noted. The mechanism of the inhibition is complex. The present data indicate that the gasification of biomass chars in a synthesis gas mixture would be qualitatively similar to that found with coal.

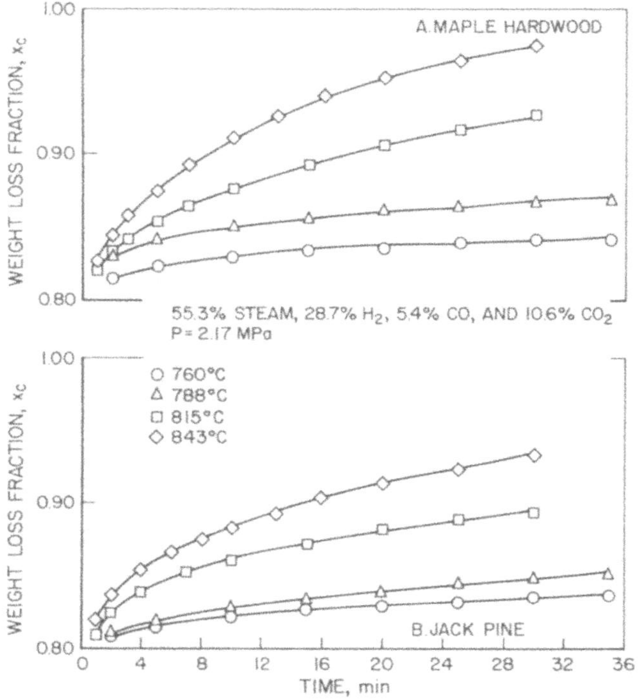

FIG. 13. Gasification of *in situ* chars from maple hardwood and jack pine.

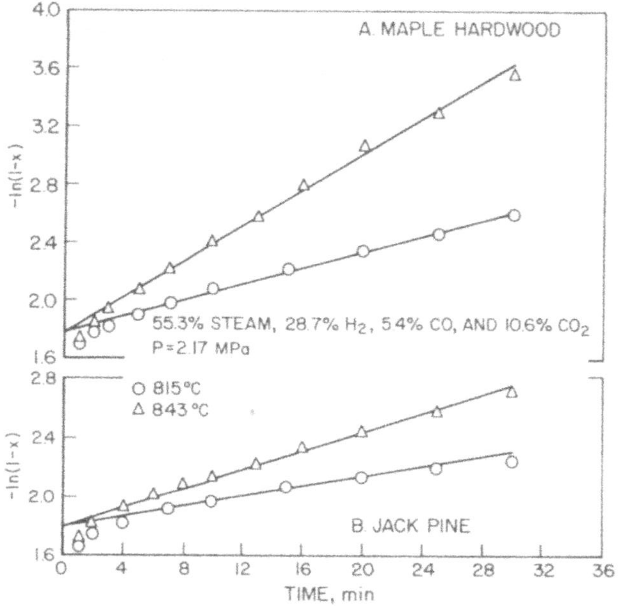

FIG. 14. Gasification of *in situ* chars from maple hardwood and jack pine—first-order plot.

TABLE IX

KINETIC PARAMETERS FOR THE GASIFICATION OF *in situ* CHARS FROM MAPLE HARDWOOD AND JACK PINE IN 55% STEAM, 29% H_2, 5·4% CO, 10·6% CO_2

Sample	*Temperature, °C*	*Pressure, MPa*	*X**	*k, min⁻¹*
Maple hardwood	815	2·17	0·835	0·027
Maple hardwood	843	2·17	0·835	0·064
Jack pine	815	2·17	0·835	0·016
Jack pine	843	2·17	0·833	0·033

Comparing the value of k at 815 °C for this medium with that for Medium IV (the $H_2/H_2O = 0{\cdot}5$ in both IV and VI) it is seen that the presence of 5·4% CO in the medium reduced the rate constant by about a factor of 4.

The values of the first order rate constant, k, for the stabilized chars of maple hardwood and jack pine were given in Table IV and the values for *in situ* chars were given in Table VIII. It is seen that the reactivity of the *in situ* chars was ~2 times as large as that for the stabilized chars for both materials. The possibility that some free radical species possibly could react with oxygen (air) at room temperature was tested by the two-step gasification scheme. The rate data showed that the reactivity of the *in situ* chars was not altered as a result of exposure to oxygen at room temperature.

Surface areas of a number of partially reacted (in H_2O/N_2) *in situ* and

TABLE X

SURFACE AREAS OF PARTIALLY REACTED CHARS FROM MAPLE HARDWOOD AND JACK PINE IN STEAM–NITROGEN MIXTURES

Sample	*Type of char*	*Gasification temperature, °C*	*Pressure, MPa*	X_c	*Surface area* CO_2, m^2/g	N_2
Maple hardwood	*In situ*	704	2·17	0·065	639	—
Maple hardwood	Stabilized	704	2·17	0·034	692	—
Maple hardwood	*In situ*	732	2·17	0·120	667	460
Maple hardwood	Stabilized	732	2·17	0·045	620	331
Jack pine	*In situ*	704	2·17	0·051	676	—
Jack pine	Stabilized	704	2·17	0·048	836	—
Jack pine	*In situ*	732	2·17	0·094	678	387
Jack pine	Stabilized	732	2·17	0·047	804	447

stabilized chars were measured. The results are shown in Table X. The value of X_c was below 0·1 except for one sample of *in situ* char. The value of the surface areas of the stabilized chars (devolatilization in nitrogen) were given in Table V. It is seen that gasification of less than 5% of char caused a significant increase in nitrogen surface area. Compare the value of the nitrogen surface area of 732 °C jack pine stabilized char, with $X_c = 0$ of 91 m^2/g, to the 447 m^2/g value for $X_c = 0{\cdot}04$ in Table X. The carbon dioxide surface areas, on the other hand, remain unchanged. However, the results in Table X clearly indicate that there is no difference in the surface area value between the partially reacted stabilized chars and the *in situ* chars. Therefore, the higher reactivity of the *in situ* chars cannot be attributed solely to any differences in total surface areas of the samples. The stabilized chars remain at temperature (in N_2 atmosphere) for at least 30 min. This prolonged exposure at relatively high temperature could, for example, bring about some annealing of defects or some order (crystallite alignment) to the disordered carbon structure.

REFERENCES

1. Rensfelt, E., Blomkvist, G., Ekström, C., Engstrom, S., Espena, B. G. and Lunanki, L. (1978). Basic gasification studies for development of biomass medium—BTU gasification processes. Symposium Papers—Energy from Biomass and Wastes, Washington, D.C. August 14–18. p. 465.
2. Sealock, L. J., Jr, Elliott, D. C., Weber, S. B. and Robertus, R. J. (1981). Kinetics and catalysis of producing synthetic gases from biomass. GRI-79/0100.
3. Ross, R. A. and Fikis, D. V. (1980). Gasification reactions of chars and modified chars from jack pine bark. *Canadian J. Chem. Eng.* **58**, 230.
4. Johnson, J. L. (1974). Kinetics of bituminous coal char gasification with gases containing steam and hydrogen. *Advances in Chemistry Series* **131**, 145.
5. Gan, H., Nandi, S. P. and Walker, P. L., Jr (1972). Nature of porosity in American coals. *Fuel* **54**, 169.
6. Dubinin, M. M. (1966). Porous structure and adsorption properties of active carbons. *Chemistry and Physics of Carbon*, Vol. 2. Marcel Dekker, Inc., New York, pp. 51–118.
7. Walker, P. L., Jr, Rusinko, F., Jr and Austin, L. G. (1959). Gas reaction of carbon. *Advances in Catalysis*, Vol. II. Academic Press, New York, p. 148.
8. Johnson, J. L. (1981). Fundamentals of coal gasification. *Chemistry of Coal Utilization*, Second Supplementary Volume, M. A. Elliott (Ed.). Wiley-Interscience, New York, pp. 1579–92.

33

Gasification of Charcoal: Influence of Water Vapor

J. R. RICHARD, M. CATHONNET and J. P. ROUAN

Centre de Recherches sur la Chimie de la Combustion et des Hautes Températures, Orleans, France

1. INTRODUCTION

Direct use of biomass as a fuel offers a limited field of applications. Conversion of biomass to gases and liquids is of great importance to make transport of energy cheaper, gases and liquids being more easily handled than solid forestry and agricultural residues. Moreover, this conversion makes the utilization of biomass energy possible in the transport sector.

Gasification of biomass occurs in three steps:

(1) pyrolysis producing volatile matter and char;
(2) secondary reactions of tars and primary gases yield mainly low molecular weight gaseous products;
(3) char gasification.

A detailed review of existing knowledge and present work in this area has been published recently.[1] This review presents scientific background information and surveys the current status of gasification activities.

Extensive literature exists on slow pyrolysis and, recently, a number of studies have appeared on fast pyrolysis. This technique gives a high yield of gases and, among them, products of high value such as olefins.

Fewer studies have been published[2] on secondary reactions of tars and primary gases evolved during slow pyrolysis.

Research described in this paper focuses on the third step of the gasification process: char gasification. Numerous studies have been published on gasification, but most of them concern coal gasification. Recent compilations of research on the fundamental aspects of coal gasification are available.[3,4]

Like coal char, biomass char is composed principally of carbon. It can be oxidized to synthesis gas through heterogeneous reactions with carbon dioxide, steam, hydrogen or oxygen. Equilibrium calculations can predict the effects of temperature, pressure and feed moisture content on product composition, but there exist kinetic restrictions which limit the quantitative significance of an equilibrium calculation. For this reason, it is necessary to perform experimental work on the reactivities of the various forms of char that arise during pyrolysis.

The main objective of this research has been to study the steam gasification of wood char to determine the experimental conditions leading to the maximum generation of synthesis gas.

Fixed bed and transverse flow, together with temperatures higher than 800 °C and pressures ranging between 1 and 20 bars with variable amounts of water vapour were the conditions chosen which could lead to noticeable reaction rates.

2. EXPERIMENTAL

The apparatus is shown in Fig. 1. The charcoal sample is positioned on a sintered quartz plate held by a quartz tube and placed inside a furnace. The gas flow passes through the sintered plate and the sample. The gases produced by the reaction between the char and the flowing water are analysed and vary according to the sample, temperature and pressure.

2.1. Experimental Details

The furnace is mobile and is regulated at the desired temperature. It can be slid along the axis holding the quartz tube containing the sample. The quartz tube is itself mobile through a hydraulic piston and can slide to be positioned inside the furnace.

The water is introduced in a vaporization vessel by a metering pump. The water vapour produced is mixed with a flow of nitrogen to increase the gas velocities, in order to eliminate the influence of mass transfer resistance. The pressure in the furnace is maintained to the desired value by a regulation valve.

The gases produced by gasification are drawn through a quartz probe and stored in glass bottles. They are later analysed by gas chromatography.

2.2. Conduct of Experiment

0·5 g of char are placed on the sintered glass plate at the bottom of the furnace. Then the gases are allowed to flow through the sample maintained

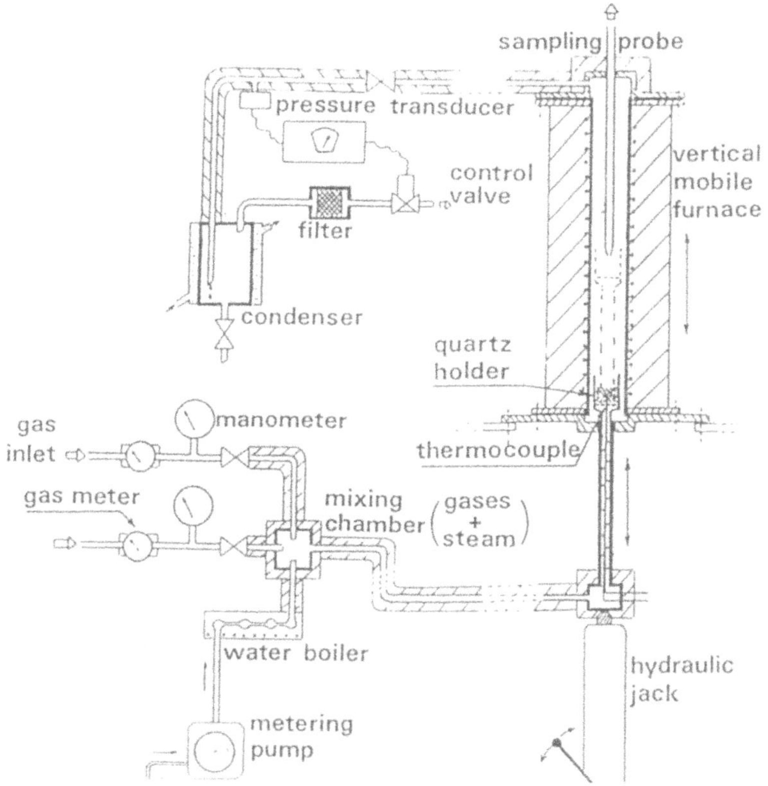

FIG. 1. Gasification system.

at 250 °C (so that no reaction takes place). When the desired pressure is attained in the apparatus, the tube is slid up into the furnace heated at the reaction temperature.

Two thermocouples constantly monitor the temperature at two points of the sample.

A probe is placed at a few centimeters above the reactor and connected with a set of sampling vessels. These are filled with discrete quantities so that the evolution of each species can be easily followed. The different gases produced are CO, CO_2, CH_4 and H_2. CO is also analysed on-line by infrared absorption.

2.3. Material

The results presented here concern char obtained by pyrolysis of fir wood particles (30 mesh) under nitrogen. The specific surface area is 180 m^2/g as measured by BET under nitrogen. Composition of dry char is C = 90 %, H = 1·2 %, O = 6 %, N = 0·3 %, ash = 2·5 %. All samples have been treated at the temperature of experiment.

3. DATA

The following results are developed:

—Variation of the gasification rate with extent of reaction, steam pressure, temperature and total pressure.
—Variation of the proportions of the different species analysed as a function of the previous parameters.

3.1. Gasification Rate

The mass flow rate of each gaseous species is calculated during the reaction. This allows one to determine the mass of carbon gasified as the reaction progresses. Figure 2 shows the time evolution of sample weight loss during gasification. From such curves, it is possible to calculate the rate of gasification.

To calculate reaction rate, many authors[3,5] use the following expression:

$$r = \frac{1}{W}\frac{\mathrm{d}W}{\mathrm{d}t} \tag{1}$$

where W is the carbon weight remaining at different values of time t, and $\mathrm{d}W/\mathrm{d}t$ represents the slope at corresponding times.

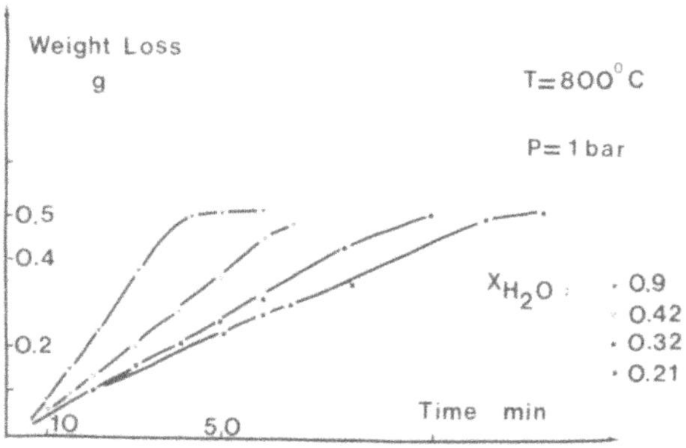

FIG. 2. Weight loss as function of water molar fraction.

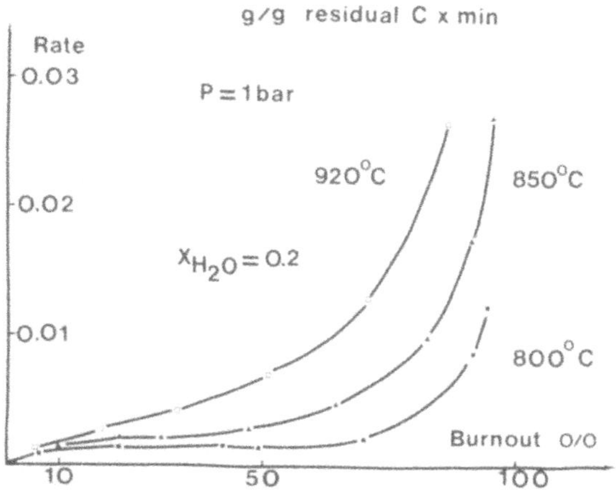

FIG. 3. Influence of burnout on gasification rate at different temperatures.

The influence of burnoff (or burnout) on gasification rate is illustrated by the curves in Fig. 3. These curves are similar to those obtained by Rensfelt *et al.*,[5] suggesting that gasification increases surface area.[1]

In eqn (1), the term W in the denominator may be replaced by W_0, the starting weight of carbon.[6] This modified equation was used to calculate the reaction rates plotted in Figs 4–8.

Figures 4, 5 and 6 show that, at low temperature or low water molar fraction, the rate is nearly constant during most of the reaction.

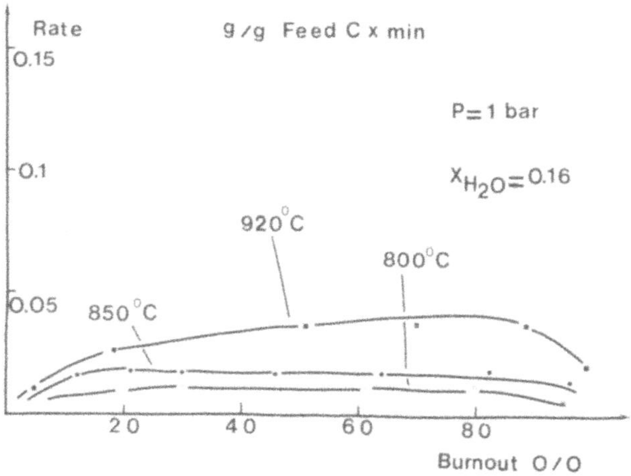

FIG. 4. Influence of burnout on gasification rate at different temperatures.

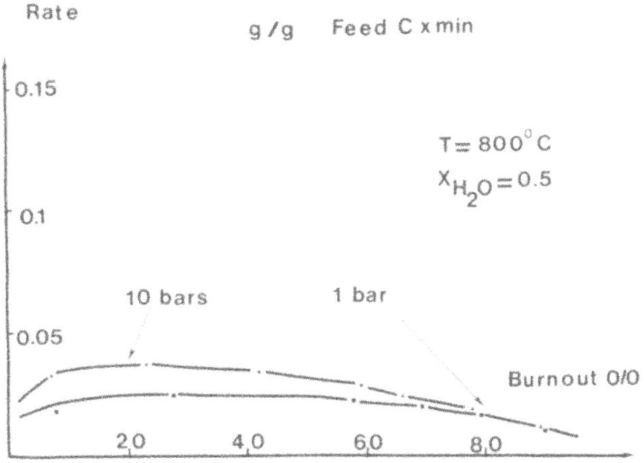

FIG. 5. Influence of burnout on gasification rate at different pressures.

The influence of steam partial pressure on gasification rate is illustrated in Fig. 7, which shows an approximately linear relationship.

The variation of gasification rate with total pressure is less evident. In Fig. 5 it is shown that for $X_{H_2O} = 0{\cdot}5$, the rate is multiplied by a factor of 2 at 800 °C when pressure is increased from 1 to 10 bars. But at 920 °C, for $X_{H_2O} = 0{\cdot}5$, the rate decreases when pressure is varied from 1 bar to 10 bars and remains nearly constant from 10 to 20 bars (Fig. 6).

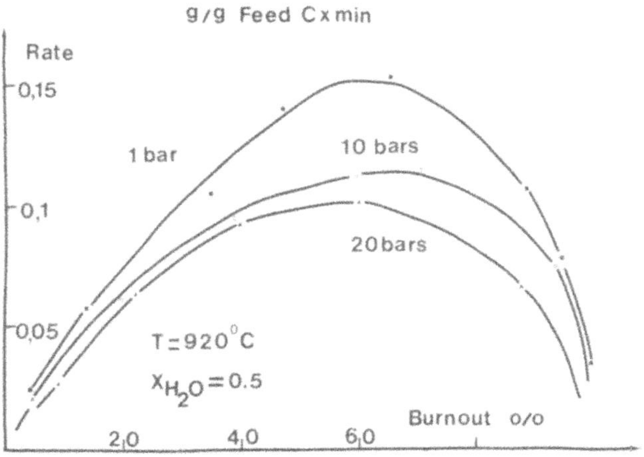

FIG. 6. Influence of burnout on gasification rate at different pressures.

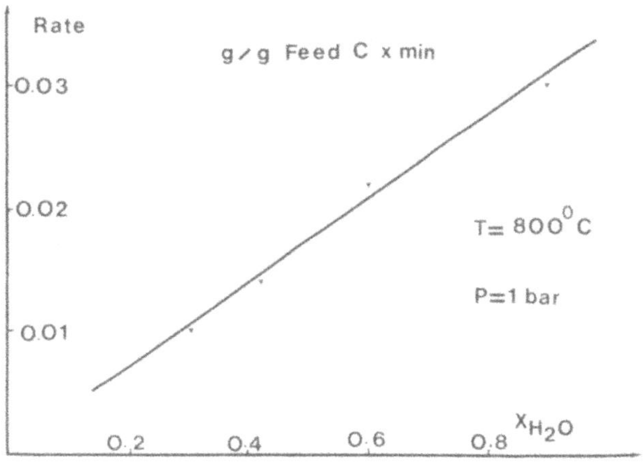

FIG. 7. Influence of water vapor molar fraction on gasification rate.

The rate dependence on temperature is shown in Fig. 4, and in Fig. 8 maximum rates have been plotted in Arrhenius form. The calculated apparent activation energy is 25 ± 2 kcal/mol, which is lower than values found by other authors.[1,5] This low value may be attributed to mass transfer limitation. According to ref. 1, external mass transfer is not important for the small particles used in this study; the only limiting process on gasification rate may be pore diffusion mass transfer.

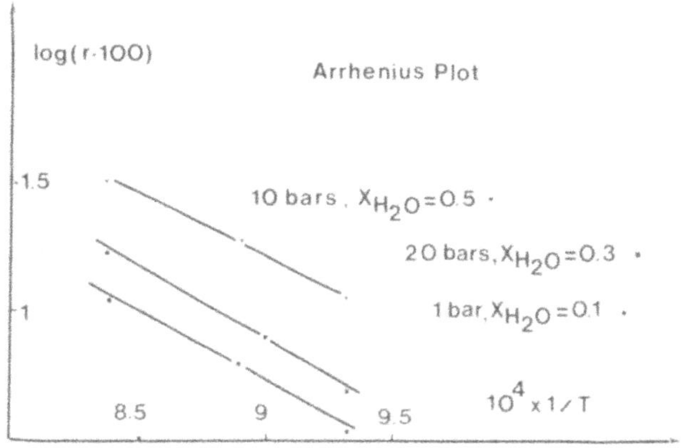

FIG. 8. Arrhenius plots for different water vapour molar fractions and pressures.

3.2. Gaseous Species Formed

In all the experiments carried out, it was noticed that the whole mass of char sample was converted during the process. The volume of CO, CO_2, H_2 and CH_4 formed in each experiment was calculated and their molar fractions were determined. H_2, CO_2 and CO are the main products. CH_4 molar fraction is very low at 1 bar and reaches a noticeable value (0·01–0·02) only when pressure is raised from 1 to 20 bars.

The influence of temperature on CO, CO_2 and H_2 molar fractions is illustrated in Fig. 9. It is shown that X_{CO} increases and X_{H_2} and X_{CO_2} decrease when temperature is raised from 800 to 900 °C.

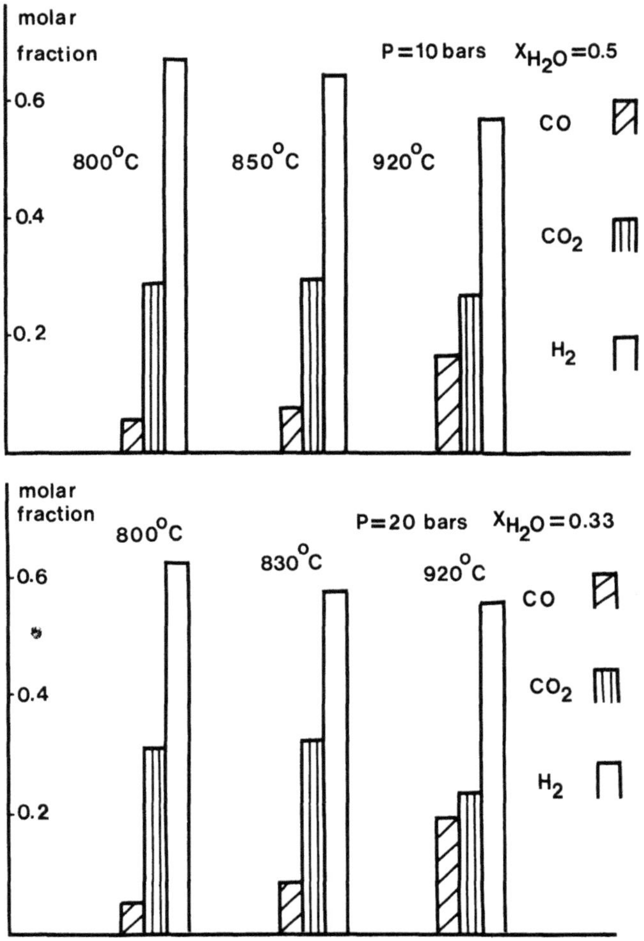

FIG. 9. Influence of temperature on CO, CO_2 and H_2 molar fractions.

The pressure dependence of CO, CO_2 and H_2 molar fractions is illustrated in Fig. 10. It is shown that X_{CO_2} increases, but X_{CO} and X_{H_2} decrease when pressure is raised from 1 to 20 bars.

The variations of H_2/CO and CO_2/CO molar ratios as a function of temperature and pressure are presented in Tables I and II.

Table I shows that for given total pressure and H_2O molar fraction, H_2/CO and CO_2/CO decrease when temperature is varied from 800 to 920 °C.

The pressure dependence is shown in Table II: for given temperature and

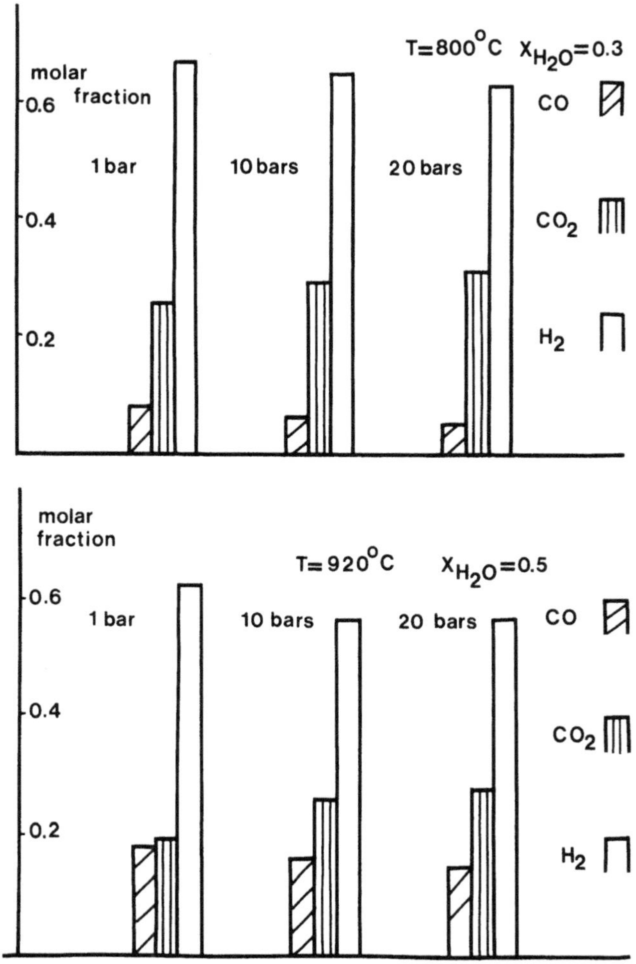

FIG. 10. Influence of pressure on CO, CO_2 and H_2 molar fractions.

TABLE I

EFFECT OF TEMPERATURE ON THE H_2/CO AND THE CO_2/CO MOLAR RATIOS

T (°C)	800	850	920	800	850	920	800	830	920
P (bars)		1			10			20	
X_{H_2O}		0·16			0·5			0·30	
H_2/CO	8·7	4·6	2·7	13·1	8·5	3·3	11·5	6·7	2·9
CO_2/CO	2·8	1·4	1·2	5·6	3·8	1·6	5·7	3·7	1·2

TABLE II

EFFECT OF PRESSURE ON THE H_2/CO AND THE CO_2/CO MOLAR RATIOS

T(°C)		800			800			920	
P (bars)	1	10	20	1	10	20	1	10	20
X_{H_2O}		0·3			0·5			0·5	
H_2/CO	8·2	9·9	11·5	9·4	13·1	19·6	3·4	3·5	3·9
CO_2/CO	3·2	4·5	5·7	3·1	5·6	7·6	1·1	1·6	1·9

H_2O molar fraction, H_2/CO and CO_2/CO increase when total pressure increases.

The H_2/CO ratios obtained in these experimental conditions are higher than the desired value for methanol synthesis ($H_2/CO = 2$). Such a value is only approached at high temperature (920 °C) and low steam and total pressures.

ACKNOWLEDGEMENTS

This work was supported by Agence Française pour la Maîtrise de l'Energie under Grant No. 80 75 092.

REFERENCES

1. Reed, T. B. (Ed.) (1981). *Biomass Gasification. Principles and Technology.* Energy Technology Review 67, Noyes Data Corporation, Park Ridge, New Jersey.

2. Antal, M. J., Jr (1981). The effects of residence time, temperature and pressure on the steam gasification of biomass. *A.C.S. Symp. Ser.* **144**, 313.
3. Laurendeau, N. (1978). Heterogeneous kinetics of coal char gasification and combustion. *Progress in Energy and Combustion Science* **4**, 221.
4. Johnson, J. L. (1981). *Kinetics of Coal Gasification*. Wiley-Interscience, New York.
5. Rensfelt, E., Blomkvist, G., Ekström, C., Engstrom, S., Espenas, B. G. and Linanski, L. (1978). Symposium papers, Energy from Biomass and Wastes. Institute of Gas Technology, Chicago, p. 465.
6. Linares-Solano, A., Mahajan, O. M. P. and Walker, P. L. (1979). *Fuel* **58**, 327.

34

Catalytic Conversion of Tars, Carbon Black and Methane from Pyrolysis/Gasification of Biomass

CLAS EKSTRÖM, NILS LINDMAN and RUNE PETTERSSON

The Royal Institute of Technology, Department of Chemical Technology, Stockholm, Sweden

1. INTRODUCTION

The use of catalysts in biomass gasification has been suggested for a long time.[1] Fung and Graham found that potassium carbonate and calcium oxide have catalytic influences on the gasification rate and the product yield.[2] This article does not, however, deal with catalysts for the primary gasification. The objective has been to find catalysts for refinement and upgrading of the gas already produced.

The use of catalysts to convert tar and methane can increase overall efficiency of biomass conversion processes by 10%.[3] This is achieved by lower oxygen consumption, better heat recovery and higher carbon conversion than in processes based on non-catalytic techniques. The necessity to convert tar is obvious, as such a high fraction of the energy in the raw material is bound up in the primary formed tar. This fraction is even increased when the high reactivity of biomass is exploited by lowering the gasification temperature. The methane content has to be minimized when syngas is produced.

Both the tar and the methane can be converted in a simple way by increasing the temperature to 1100 °C. This, however, implies an increased formation of soot and a high oxygen consumption. The potential to avoid this and to instead use catalysts without intermediate gas cleaning is given by the low content of sulphur and heavy metals in biomass compared to coal.

This article discusses the different catalytic options at hand, and presents some experimental results from tar and methane conversion with different catalytic materials.

2. EXPERIMENTAL APPARATUS AND PROCEDURE

For most of the experiments presented in this paper, an already existing LRU-apparatus for continuous flash pyrolysis was supplemented with a reforming step. A detailed description of this equipment is given by Ekström and Rensfelt.[4] A picture of this apparatus is shown in Fig. 1.

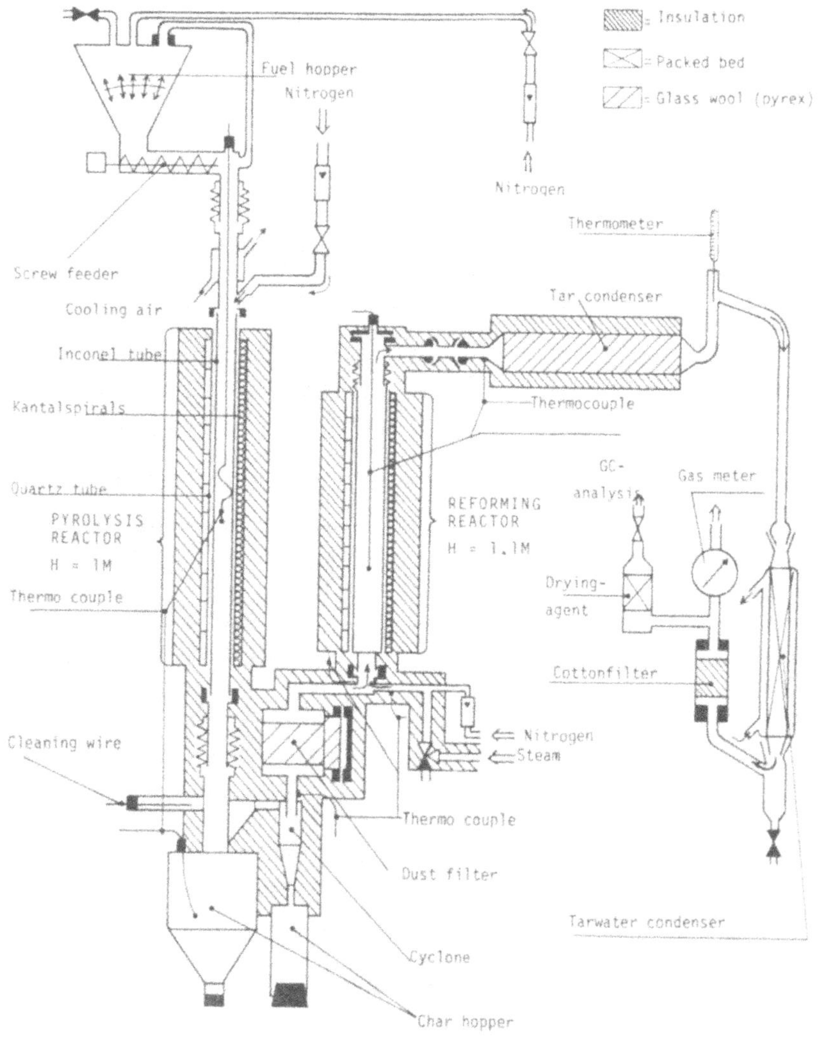

FIG. 1. Experimental apparatus.

The total amount of liquid products is calculated from the weight increase during a run of all parts in the condensing system. The relative amounts of tar and carbon black and the tar content in the water phase are estimated visually. (The carbon black is a gaseous product at temperatures above 300 °C and it is achieved as a solid together with the most high boiling tars.) Also, some water phase samples are analysed on water content. This method is very approximate and allows a precision not greater than 0·1 g of tar and carbon black.

The characteristics of the catalytic materials were studied by treating them with a mixture of steam and dust-free raw gas obtained from flash pyrolysis of the fuel at 700–750 °C and at atmospheric pressure. The amounts of different pyrolysis products entering the reforming reactor are given by the results from flash pyrolysis experiments at the same temperature in which the reformer was kept at a rather low temperature (530 °C) without steam supply. Since the experiments with Ni-based catalysts were carried out at higher reforming temperatures (900–1000 °C), the effects of thermal cracking occurring before the inlet of the catalyst bed was investigated by performing experiments with an empty reforming reactor at the same temperature level and steam supply as in the catalytic experiments.

Different space velocities were obtained by filling the reformer with different amounts of catalytic material without changing the fuel throughput.

The experiments with catalytic steam reforming of methane at pressures between 1 and 3 MPa were carried out in a bench scale unit consisting of a fluidized bed gasifier followed by a reforming step. This apparatus is described by Liinanki *et al.*[5]

3. FUEL AND MATERIAL CHARACTERIZATION

3.1. Fuels

Two types of biomass, Salix and mixed hardwood, and one type of peat were used. The experiments were carried out with substances dried at 105 °C for 24 h. All calculations are based on dry biomass, although the moisture content was about 1·5 % for both peat and biomass.

3.2. Catalytic Materials

3.2.1. Dolomite

Calcination of the dolomite before experiments at 800 °C in 4 h resulted in a weight loss of 46 %. This corresponds to a devolatilization of CO_2 resulting in a rough composition of 40% MgO, 35% CaO and 25% $CaCO_3$. A fraction with particle sizes between 2 and 5 mm, which was the main part of the dolomite, was used.

3.2.2. Char

The char used as catalyst was produced by flash pyrolysis followed by final carbonization at 700–750 °C of Salix and mixed hardwood. For Salix, the same particle size, 0·2–0·5 mm, was used as fuel and as catalyst. For mixed hardwood, the char was made from a fraction with particle sizes bigger than 2 mm while a fraction with particle sizes between 0·2 and 0·6 mm was used as fuel in the experiments.

3.2.3. Ni-catalysts

Commercially available Ni-based catalysts were used. The delivered pellets were crushed until all particles were smaller than 5 mm, and the fraction finer than 2 mm was removed.

4. CONVERSION OF TAR AND CARBON BLACK

Biomass is always decomposed to form the primary products char, tar, water and gas, when heated to temperatures above 300 °C. The distribution is very much dependent on the heating rate, residence time and the surrounding atmosphere. Above 500 °C the primary products then react further and some options to control the final product yields are presented in Fig. 2. Substantial amounts of secondary tar, or at higher temperatures

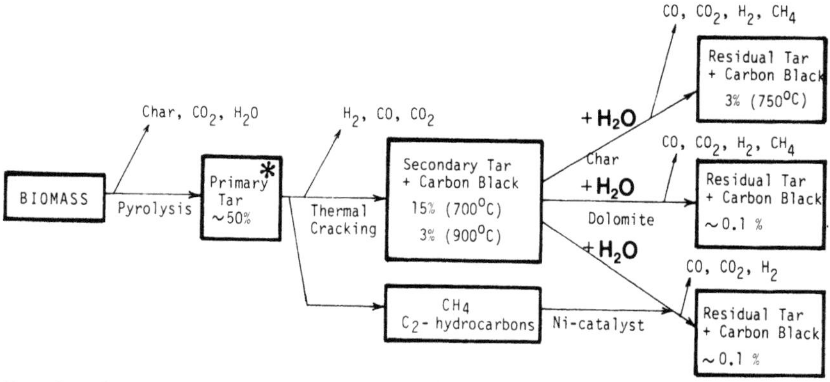

FIG. 2. Conversion routes for tars and carbon black, together with typical weight fractions, based on m.f. fuel. *Fast pyrolysis of cellulose.[6]

carbon black, are always formed in conventional gasifiers operating without catalysts. Almost all of the methane, C_2-hydrocarbons and substantial amounts of the carbon oxides and hydrogen, can be traced from decomposition or hydrogenation of the primary formed tar.[4,6,7]

No tar or carbon black can be tolerated in the product gas when it is not to be burned directly after the production. This implies that the gas has to be cleaned by means of, e.g., a scrubber or that a catalytic step has to be implemented. The catalysts to be used, char or dolomite or steam reforming catalysts, are discussed in the following sections.

4.1. Conversion over Biomass Char

It is well known that passing pyrolysis gas through a bed of char results in decreased yields of tar. This phenomenon is used in the design of downdraft gasifiers.[1] Very low tar yields from a fluidized bed peat gasifier with fuel feeding into the bed have also been reported.[8] Investigations of the effects of char and ash on the gas phase reactions were therefore initiated at the RIT.

Typical results are shown in Fig. 3 for mixed hardwood. Passing the gas phase through a char bed resulted in markedly reduced yields of tar and carbon black. At the same time, some methane, carbon dioxide and hydrogen was produced, as can be seen in Fig. 4. The high production of carbon dioxide and hydrogen is to some extent a result of gasification of the char in the bed, and from the water–gas shift reaction approaching equilibrium. Results for Salix at 730 °C were similar, although the degree of tar conversion and consequently the amounts of produced carbon dioxide, hydrogen and methane were higher. At lower temperatures (570 °C) the degree of conversion was lower.

The rather small change in total yield of tar and carbon black due to increased amounts of char in the reformer indicate that at 700–750 °C it is not possible to achieve complete conversion of tar and carbon black. However, Chembukulam *et al.*[9] have reported nearly complete conversion of tar and carbon black over semicoke and charcoal from teakwood sawdust at 950 °C. These results indicate that the temperature and the type of wood used are two important factors that determine the final degree of tar conversion. It shall, however, be noted that the teakwood used in the experiments reported by Chembukulam *et al.* had a rather high ash content, 10·5 wt % of m.f. fuel.

Pure ash from combustion of Salix was shown to have a very weak influence on tar conversion compared to char and no influence on the water–gas shift reaction, although the amount of ash used was higher than

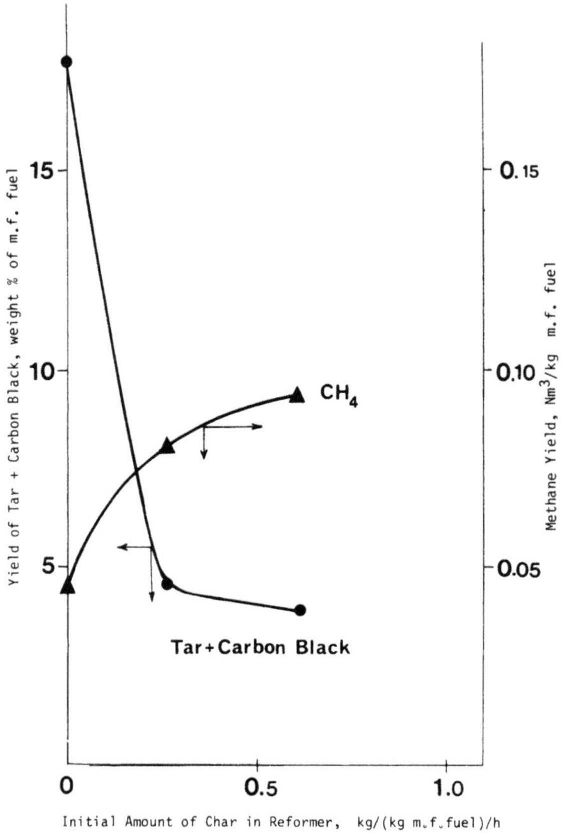

FIG. 3. Conversion of tar and carbon black over charcoal. Fuel: mixed hardwood. Char bed temperature: 745 °C. Steam supply: 0·25–0·3 kg/kg m.f. fuel.

the amount of ash in the char used as catalyst. This indicates that the ash components might not be the only active components for tar conversion over char and/or that the rather high inner surface of char compared to pure ash is an important factor. It is also possible that the activity of ash components formed during oxidizing conditions is different from that of ash components formed during reducing conditions. In that case, the ashes remaining from pyrolysis followed by steam gasification would still be active for tar breakdown.

The residual yields of tar and carbon black after conversion over char at 700–750 °C are too high for the needs for production of syngas or fuel gas. For syngas production, it is also necessary to minimize the yield of methane, and for that purpose a steam reforming catalyst is necessary. At

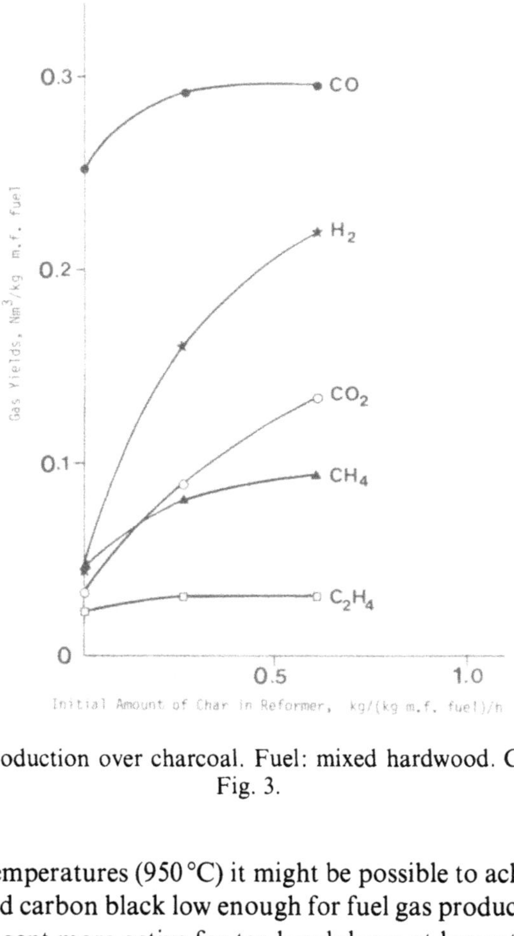

FIG. 4. Gas production over charcoal. Fuel: mixed hardwood. Conditions, see Fig. 3.

much higher temperatures (950 °C) it might be possible to achieve residual yields of tar and carbon black low enough for fuel gas production. The use of a catalytic agent more active for tar breakdown at lower temperatures, e.g. dolomite, will, however, result in increased overall energy efficiency. The catalytic effect of char is of interest mainly because of its impact on the gasifier design.

4.2. Conversion over Dolomite

A strong activity for conversion of residual oils to gases with dolomite in the temperature range of 700–900 °C has been reported by Morita.[10]

The results of our investigations showed that it is possible also to achieve a nearly complete conversion of tar and carbon black from biomass.

The results from the experiments are presented in Figs 5 and 6. From the curves in Fig. 5 it can be seen that passing of the pyrolysis gas through a

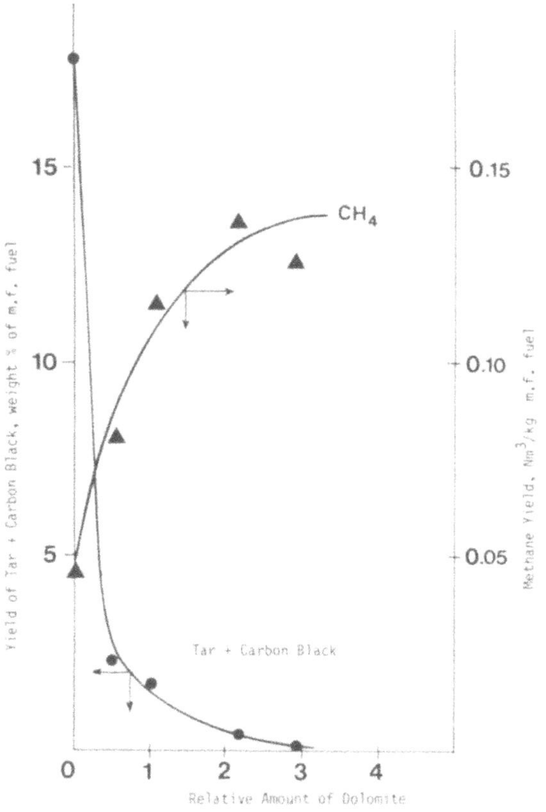

FIG. 5. Conversion of tar and carbon black over dolomite. Fuel: mixed hardwood. Dolomite bed temperature: 700–800 °C. Steam supply: 0·4–0·5 kg/kg m.f. fuel.

dolomite bed resulted in a nearly total conversion of tar and carbon black. Simultaneously, there was a markedly increased production of methane. A large part of the methane formed can be considered as a product of tar hydrogenation.[11]

The increased yields of CH_4 and C_2H_4 (Fig. 6) together with the previously discussed reduced tar yields is an indication of the cracking/reforming activity of the dolomite. Yeboah *et al.* also report the lack of activity of dolomite for methane conversion.[11]

The high increase in hydrogen and carbon dioxide yields (Fig. 6) is to some part a result of the rapid approach towards equilibrium for the water–gas shift reaction.

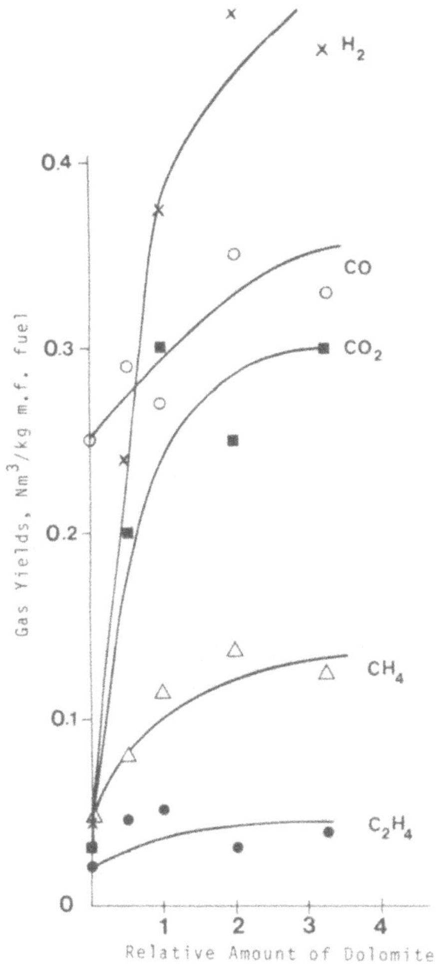

FIG. 6. Gas production over dolomite. Fuel: mixed hardwood. Conditions, see Fig. 5.

4.3. Conversion of Tars and Carbon Black over Ni-Catalysts

Syngas production by steam reforming of different hydrocarbons, such as naphtha and methane, in the presence of Ni-based catalysts has long been an established process.[12] The possibility of using a similar technique for tar and methane reforming was therefore close at hand, and research in that field was initiated at RIT in 1978. Reduced tar yields and reduced yields of methane in the presence of Ni-containing catalysts have also been reported by Weber and Mudge.[13,14]

Steam reforming of higher hydrocarbons can be regarded as an endothermic irreversible reaction:

$$C_nH_m + nH_2O \rightarrow nCO + (n + m/2)H_2$$

The outlet gas composition is then determined by the two reversible reactions (1) and (2) discussed later in connection with methane reforming.

For all investigated commercially available Ni-based catalysts nearly total conversion of tars and carbon black was achieved. Typical results for peat and hardwood are shown in Figs 7 and 8. As can be seen in Fig. 8 the main products are carbon oxide and hydrogen. The eventual activity

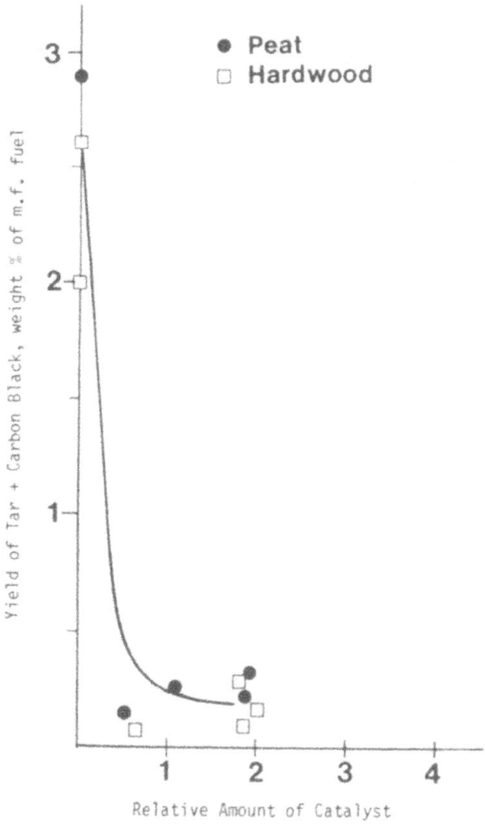

FIG. 7. Steam reforming of tar and carbon black from peat and mixed hardwood on an Ni-catalyst. Reformer temperature: 900–1000 °C. Steam supply: 0·3–0·4 kg/kg m.f. fuel.

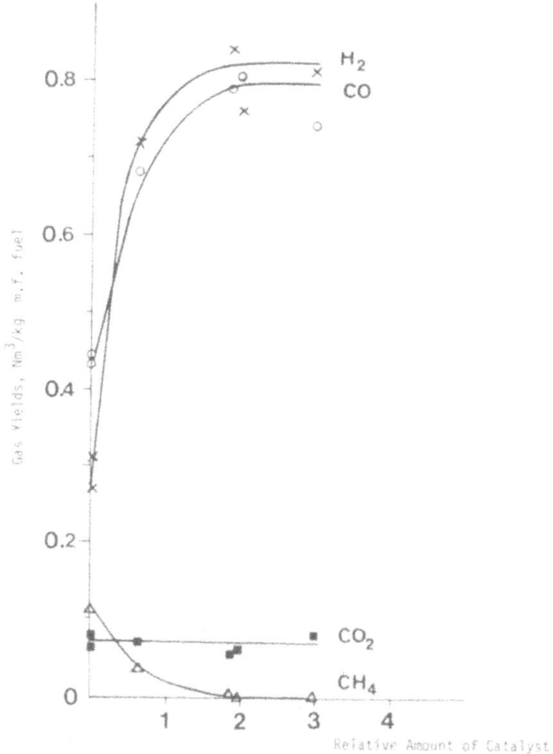

FIG. 8. Gas production from steam reforming of tar, carbon black and methane from mixed hardwood on an Ni-catalyst. Reformer temperature: 900–1000 °C. Steam supply: 0·3–0·4 kg/kg m.f. fuel.

differences between the investigated catalysts concerning tar breakdown, are all within the limits of the experimental error discussed in Section 2.

Other important factors affecting both methane and tar reforming, such as carbon deposition and steam ratio, are discussed in Section 5.1.

5. CONVERSION OF METHANE

When discussing methane conversion, it is convenient to use a separate scheme of the type shown in Fig. 9. Thermal decomposition has a very high activation energy (330–420 kJ/mol) and the reaction is very rapid at temperatures above 1100 °C, and negligible below 900–1000 °C. The methane yield in biomass gasifiers is controlled by the methane producing

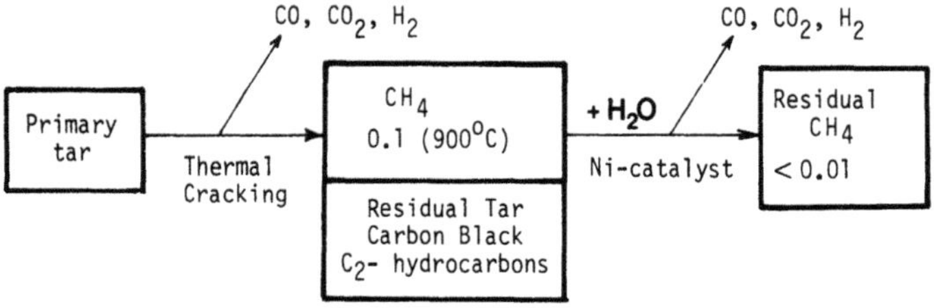

FIG. 9. Methane conversion routes, together with typical yields in N m^3/(kg m.f. fuel).

reactions at temperatures below 950 °C. The main source of the methane is the primary formed tar. It may be concluded that the conversion involves a hydrogenation step, from the strong hydrogen partial pressure dependence on the methane yield reported by Liinanki *et al.*[5] The tar is, however, also consumed in a steam reforming reaction producing no methane. This leads to two postulates:

—Methane is produced as long as there is tar present and the hydrogen partial pressure is substantial.
—Less methane is produced at higher temperatures, where the steam reforming reaction of tar is fast and methane is also consumed.

The high activation energy of the methane decomposition reaction makes it necessary to increase the temperature to approximately 1100 °C to get methane concentrations which meet the demand for, e.g., methanol synthesis. This gives a great potential for the use of catalysts to reach lower methane yields at lower temperatures.

Conversion of tar over both char and dolomite resulted in increased methane yield. Only Ni-catalysts were found to be active for methane conversion.

5.1. Steam Reforming of Methane over Ni-Catalysts

Ni-catalysts are used commercially for steam reforming of methane in natural gas. The product composition is determined by the two following reactions:

$$CH_4 + H_2O \rightleftharpoons CO + 3H_2 \tag{1}$$

$$CO + H_2O \rightleftharpoons CO_2 + H_2 \tag{2}$$

Direct conversion of methane to carbon dioxide has also been discussed:[15,16,20]

$$CH_4 + 2H_2O \rightleftharpoons CO_2 + 4H_2 \tag{3}$$

Reactions (1) and (3) are strongly endothermic and reaction (2) is slightly exothermic. Reaction (2) reaches equilibrium almost instantly, while the rates of approach to equilibrium for reactions (1) and (3) depend on partial pressures and the activity of the used catalyst, *inter alia*. Among important factors that affect the catalyst activity are the Ni-content, the Ni-crystallite size, the pore size distribution and the influence of different promotors, added to avoid carbon deposition.[17,18]

The investigated catalysts showed some slight differences in activity for methane reforming when treated with the gas phase from pyrolysis of hardwood followed by thermal cracking in the inlet of the reformer. Typical results for one of the catalysts are shown in Fig. 10.

Two main types of total kinetic expressions for the methane reforming reactions (1) and (3) have been found in the literature for higher temperatures. The mechanism is not clearly understood, and several suggestions have been made.[15] In many cases data have been correlated to simple first order kinetics with respect to methane.[12,15] A similar treatment assuming the rate-determining step to be a first order reversible reaction has also been reported.[19] For higher pressures and temperatures, Mayer and Köpsel[15,16] report satisfactory correlation of data for two expressions of Langmuir–Hinshelwood type, one for the formation of carbon oxide and one for the formation of carbon dioxide, involving partial pressures of all the reactants in reactions (1) and (3). These expressions mean that at higher pressures the reaction rate does not increase proportionally to increased total pressure. Another similar treatment reported by De Deken *et al.*[20] gives somewhat different rate equations with decreased reaction rate at increased total pressure. Preliminary experiments with pressurized steam reforming of methane at RIT have shown a very weak positive dependence of pressure for the methane conversion rate.

Considerable pore diffusion control has been reported for temperatures between 550 and 680 °C[20] and 800 and 900 °C.[12] Depending on the differences in catalyst activity, that may also play an important role during our conditions (temperatures between 900 and 1000 °C). Preliminary calculations have also shown some influence from mass-transfer between the gas phase and the catalyst particles.

During steam reforming of methane from biomass fuels, the situation might be complicated by the tar reforming, and by the presence of product gases from pyrolysis and oxidation.

5.1.1. Comparison of Different Fuels

To achieve a comparison of different fuels' influence on the catalytic

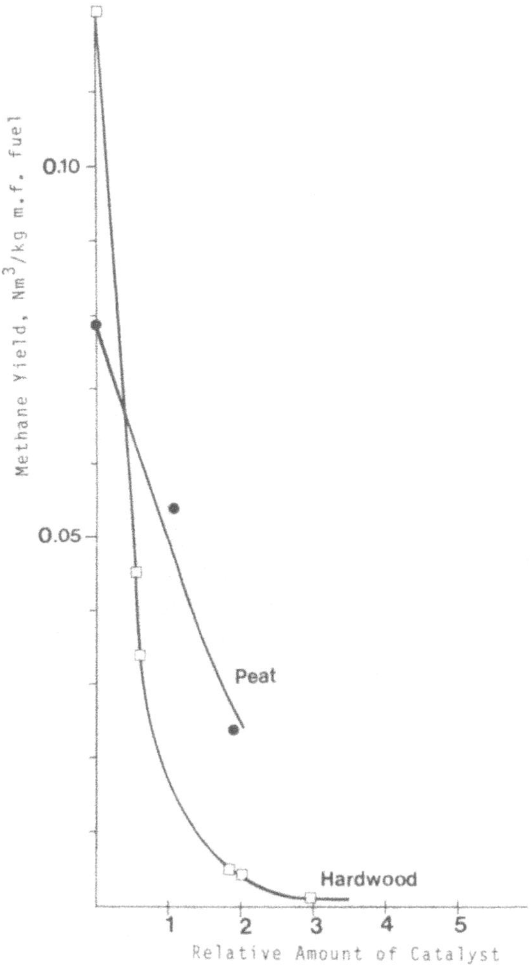

FIG. 10. Steam reforming of methane from peat and mixed hardwood on an Ni-catalyst. Conditions, see Fig. 7.

conversion, two different experimental series, one with peat and the other with mixed hardwood fuel, were performed. The results can be summarized as follows:

—Peat requires a lower space velocity for total methane conversion than does hardwood (Fig. 10).
—Total conversion of tar, carbon black and C_2-hydrocarbons was achieved for both peat and hardwood at the same space velocities.

5.1.2. The Formation and Removal of Coke from Nickel Catalysts

Two types of carbon deposition have been observed in our studies on Ni-catalysts. Between 700 and 850 °C, the catalyst was deactivated for methane reforming but not for tar reforming and the water–gas shift reaction, independent of amount of steam added. Carbon deposition was shown to be the reason for deactivation by the observed formation of carbon oxides during regeneration of the catalyst with steam (at 950 °C). Above 900 °C, the catalyst was not deactivated, but at low steam supplies, carbon deposition resulted in remarkable pressure drop enhancement. These observations correspond well to the two main types of carbon deposition occurring at higher temperatures, which are described in the literature.[21–23]

The observed lowest steam supply necessary to avoid carbon deposition above 900 °C was remarkably low, compared to what is normally used for steam reforming of higher hydrocarbons. This is probably due to the high oxygen contents in the gases and tars entering the reformer when biomass is used as fuel. Calculations have shown that the corresponding total oxygen-to-carbon ratio in the wet outlet gas was approximately the same as the minimum steam-to-carbon ratio recommended by the catalyst manufacturer for steam reforming of naphtha on the actual catalyst.

5.2. Water–Gas Shift Reaction

The water–gas shift reaction needs some catalytic material to reach thermodynamic equilibrium, at least at temperatures up to about 1000 °C, as can be seen in Table I. At higher temperatures, the equilibrium constant K_p becomes lower, at the same time as the conversion of carbon oxide increases in the experiments, which results in a very close approach to equilibrium for thermal cracking at 1150–1200 °C.

Passing the gas phase through a char bed results in achieved equilibrium for Salix at 730 °C and a change towards equilibrium for mixed hardwood at 750 °C, but no influence at all for Salix at 570 °C. This shows that both the temperature and the type of char determine the degree of approach to equilibrium.

For dolomite equilibrium was achieved for residence times above 0·3 s.

Equilibrium was rapidly achieved over Ni-catalysts with outlet temperatures between 950 and 1050 °C, which also is the case in conventional steam reformers.

It can be noted that those of the investigated catalytic materials that had a marked influence on tar breakdown also influenced the water–gas shift reaction.

TABLE I

WATER–GAS SHIFT REACTION: $CO + H_2O \rightleftharpoons CO_2 + H_2$

Gas phase treatment		*Experimental* $\left(\frac{P_{H_2} \times P_{CO_2}}{P_{CO} \times P_{H_2O}}\right) / K_p$	K_p	*Residence time, s*
Thermal cracking	740 °C	0·02	1·3	1·8
	900–950 °C	0·2	0·7	2·0
	1 150–1 200 °C	0·9	0·4	1·9
Charcoal, Salix	570 °C	0·02	3·1	1·1
	730 °C	1·0	1·3	1·3
hardwood	750 °C	0·4	1·2	1·3
Dolomite	750–800 °C	1·0	1·0	0·3
Ni-catalyst	950–1 050 °C	0·6–1·4	0·5–0·6	0·1

K_p = Equilibrium constant at reformer outlet temperature.

6. CONCLUSIONS

It can be concluded that biomass char has a catalytic influence on tar and carbon black conversion, and a nearly total elimination may be achieved at higher temperatures. This has certain implications for the design of biomass gasifiers. A long contact time between pyrolysis gas and char has to be obtained for a minimum tar yield. This is realized in a down-draft gasifier by the mode of contact and the flow direction and it can be realized in a fluid bed by feeding the fuel into the bottom of the bed.

Dolomite has been shown to possess a very high activity for tar and carbon black conversion. As dolomite is a fairly cheap raw material it may even be considered to mix it directly with the fuel in, e.g., a fluid bed. Some catalytic influence on the rate of gasification is then also to be expected as the activity of calcium oxide has been shown.[2] Another obvious application is as tar cracker after a high temperature dust filter, in fuel gas production, or as primary catalyst with an Ni-based secondary catalyst in syngas production. The risk of carbon deposition would then be reduced because of the lower contents of higher hydrocarbons entering the Ni-catalyst bed.

The potential of using more active catalysts like nickel-based steam-reforming catalysts appears to be very large from the results presented here. This is, of course, under the precondition that the potential catalysts have some tolerance to sulphur and tarry compounds over a longer period of time. There have been at least three applications for these catalysts suggested in the literature.

Mudge *et al.* have developed a system based on a fluid bed operating with mixtures of biomass and catalyst grains.[13,14] Black presents a gasification process where the catalytic reaction is performed in conventional steam-reformer tubes, fired on the outside.[24] Lindman *et al.* present the MINO-process for biomass gasification where a product gas from a fluid bed is contacted with the catalyst in an autothermal reformer where oxygen is added to the gas.[3]

It is very probable that some of the more resistant catalysts, which are under development, appear to be ideal, and can be used both for tar and methane conversion. But it is also possible that a combination of two or more types will be optimal.

ACKNOWLEDGEMENT

This work has been performed under contract from the Swedish National Board for Energy Source Development.

The authors are indebted to Professor Olle Lindström and Erik Rensfelt, by whom this research work was initiated, for their many helpful suggestions and encouraging support.

REFERENCES

1. *Generator Gas* (1979). A translation of the Swedish book *Gengas* (1950). Solar Energy Research Institute, Golden, CO, SERI/SP-33-140.
2. Fung, D. P. C. and Graham, R. (1979). The role of catalysis in wood gasification. 178th Meeting of the Am. Chem. Soc., Washington, D.C.
3. Lindman, N. *et al.* (1981). A new synthesis gas process for biomass and peat. *Energy from Biomass and Wastes V, Symp. Papers*, IGT, Lake Buena Vista.
4. Ekström, C. and Rensfelt, E. (1980). Flash pyrolysis of biomass in Sweden. *Proc. Specialists' Workshop on Fast Pyrolysis of Biomass.* SERI, Copper Mountain.
5. Liinanki, L. *et al.* (1982). On the methane yield from biomass gasification at high temperatures and pressures. To appear in *Fundamentals of Thermo-chemical Biomass Conversion, Symp. Papers*, Estes Park.
6. Peters, W. *et al.* (1980). Product compositions and kinetics for rapid pyrolysis of cellulose. *Proc. Specialists' Workshop on Fast Pyrolysis of Biomass.* SERI, Copper Mountain.
7. Rensfelt, F. *et al.* (1978). Basic gasification studies for development of biomass medium-Btu gasification processes. *Energy from Biomass and Wastes, Symp. Papers.* IGT, Washington, D.C.

8. Sundgren, A. *et al.* (1963). Manufacture of water gas from milled and powdered peat. *2nd Int. Peat Congr., Symp. Papers.* Leningrad.
9. Chembukulam, K. *et al.* (1981). Smokeless fuel from carbonized sawdust. *Ind. Eng. Chem. Prod. Res. Dev.* **20**(4), 714–19.
10. Morita, Y. (1978). Calcined dolomite. *J. Japan Petrol. Inst.* **21**(1).
11. Yeboah, Y. D. *et al.* (1980). Effect of calcined dolomite on the fluidized bed pyrolysis of coal. *Ind. Eng. Chem. Process Des. Dev.* **19**(4), 646–53.
12. *Catalyst Handbook with Special Reference to Unit Processes in Ammonia and Hydrogen Manufacture* (1970), Wolfe Scientific Books, London.
13. Weber, S. L. *et al.* (1980). Gasification of biomass in the presence of multiple catalysts for the direct production of specific products. *Energy from Biomass and Wastes IV, Symp. Papers.* IGT, Lake Buena Vista.
14. Mudge, L. K. *et al.* (1982). Catalytic generation of methanol synthesis gas from wood. *Proc. Biomass To Methanol Specialists' Workshop.* SERI, Durango.
15. Mayer, B. and Köpsel, R. (1978). Reaktionskinetik der MethanWasserdampf-Konvertierung am Katalysator GIAP 3-6N. *Freib. Forsch.-H. A 591.* Leipzig.
16. Köpsel, K. D. T. and Meyer, B. (1980). Kinetik der Dampfreformierung von Methan. *Chem. Techn.* **32**(9).
17. Bridger, G. W. (1976). Catalysts for steam reforming of hydrocarbons. *Ammonia Plant Safety* **18**, 24–31.
18. Puri, V. K. *et al.* (1975). Dispersion of active component, pore size distribution and activity of naphtha steam reforming catalyst. *Proc. Indian Natl. Science Acad.* **41A**(1), 30–5.
19. Grover, S. S. (1970). Theoretical model for hydrogen production by steam–methane reforming. *Chimie et Ind., Gén. Chim.* **103**(1), 93–8.
20. De Deken, J. C. *et al.* (1982). Steam reforming of natural gas: Intrinsic kinetics, diffusional influences, and reactor design. 7th Int. Symp. on Chem. Reaction Eng., Boston.
21. Rostrup-Nielsen, J. R. (1975). *Steam Reforming Catalysts.* Danish Technical Press Inc., Copenhagen.
22. Gruber, G. (1974). Equilibrium considerations in the methane synthesis system. 168th Meeting of Am. Chem. Soc., Div. of Fuel Chem., Atlantic City.
23. Trimm, D. L. (1977). The formation and removal of coke from nickel catalyst. *Catal. Rev.-Sci. Eng.* **16**(2), 155–89.
24. Black, J. W. (1982). The pressurized fluidized bed gasifier in the synthesis of methanol from wood. *Proc. Biomass to Methanol Specialists' Workshop.* SERI, Durango.

35

Modelling of Biomass Gasification

ALFONS G. BUEKENS and JOS G. SCHOETERS
Vrije Universiteit Brussel (VUB), Brussels, Belgium

1. SURVEY

Gasification is a complex, but versatile process. There is a wide variety in the physical and chemical processes involved, the possible reactor types and operating conditions, and the potential feedstock materials, with their different structure, properties and behaviour. A generalized treatment of the process is therefore precluded.

The modelling of a gasification process may be undertaken with different aims, such as (1) the design of an industrial gasifier, (2) the simulation of such a reactor under either steady-state or transient operating conditions, (3) the extraction of rate data from experimental work, or even (4) to advance the art of reactor modelling and of the mathematical analysis of the system of equations it generates. The level of sophistication of the model and the nature of the simplifying assumptions depend on the statement of the problem and on the final aim, and therefore may vary considerably from one case to another.

The tools available in modelling are mass, energy and momentum balances, as well as equilibrium data and rate laws. In this review paper, they are applied to the most important types of gasifier.

2. PHYSICAL AND CHEMICAL PHENOMENA IN GASIFICATION[1-3]

Fuel fed into a gasifier undergoes a number of physical and chemical changes, which cannot be entirely dissociated from each other, i.e.

—preheating,
—drying,
—devolatilization,
—gasification.

A realistic model of a gasifier should incorporate the rates and yields of each of these steps, unless the rate can be shown to be extremely high and the yield a fixed number. Whereas the physical effects, such as heating, drying and cooling, can readily be modelled on the basis of theory and available rate correlations the chemical processes are still incompletely documented and sometimes even ill-understood. Experimental evidence is required to set values for rates and yields. Since numerous biomass materials, such as green wood, manure or sewage sludge, have an appreciable or even a very high moisture content, a preliminary mechanical and thermal drying may be required, prior to gasification.

The devolatilization, also termed dry or destructive distillation, charring, coking or pyrolysis, is especially important in the gasification of biomass, because in some systems the products of pyrolysis merely join those of gasification (updraft moving bed gasifier, fluidized bed gasifier).

Upon heating, cellulosic materials generate variable amounts of gas, liquids (tars, oils) and char. The main factors of influence are

—pyrolysis temperature,
—rate of heating,
—structure, composition and size of the cellulosic particle,
—presence of extraneous catalytic matter, e.g. ash components or carbonizing catalysts.

The detailed product distribution can only be established experimentally. Moreover, the primary products may react further, e.g. by

—secondary gas phase or heterogeneous pyrolysis,
—condensation of reactive components to tars,
—partial oxidation in the presence of oxidizing gases.

TABLE 2.I
YIELD AND COMPOSITION OF CHARCOAL[4,5]

Temperature, °C	*Yield of charcoal, wt%*	*Elementary composition, wt%* *C*	*H*	*O*
Wood				
300	51·4	73·2	4·9	21·9
500	31·0	90·2	3·1	6·7
1 000	26·5	96·6	0·5	2·9
Pure cellulose				
300	20	50	5·6	44·4
500	9	90	5·4	2·7

As the devolatilization continues the char is degasified more and more completely; its carbon content increases, whereas the H– and O– contents decrease (Table 2.I).

Most gasifiers can be subdivided into several zones, e.g. a drying, pyrolysis, gasification and ash cooling zone. The relative importance of each zone depends on the moisture content and the rate of drying, the heat of pyrolysis and the rate of heat transfer, the rate of gasification and that of cooling.

The rate of drying depends on such factors as:

—the temperature, velocity and moisture content of the drying gas,
—the external surface of the particle,
—the internal diffusivity of moisture in the particle,
—the physicochemical nature of the bonding of the moisture to the particle, and
—(at high temperature only) radiative heat transfer.

The temperature inside the particle is still relatively low at the end of the drying period, generally of the order of 100–150 °C. The dimensions and physical aspect of the particle have changed fairly little; only some shrinkage may have occurred due to drying and compression.

In the zone of pyrolysis the temperature increases rapidly, because of the large temperature differential between the relatively cold particles and the

TABLE 2.II
PROXIMATE ANALYSIS DATA FOR SELECTED SOLID FUELS AND BIOMASS MATERIALS (DRY BASIS, WEIGHT PERCENT)

	Volatile matter (VM)	*Fixed carbon (FC)*	*Ash*	*Reference*
Coals				
Pittsburgh seam coal	33·9	55·8	10·3	Bituminous Coal Research[6]
Wyoming Elkol coal	44·4	51·4	4·2	Bituminous Coal Research[6]
Lignite	43·0	46·6	10·4	Bituminous Coal Research[6]
Oven dry woods				
Western hemlock	84·8	15·0	0·2	Howlett and Gamache[7]
Douglas fir	86·2	13·7	0·1	Howlett and Gamache[7]
White fir	84·4	15·1	0·5	Howlett and Gamache[7]
Ponderosa pine	87·0	12·8	0·2	Howlett and Gamache[7]
Redwood	83·5	16.1	0·4	Howlett and Gamache[7]
Cedar	77·0	21·0	2·0	Howlett and Gamache[7]

TABLE 2.III

SELECTED PYROLYSIS MODELS, DEVELOPED FOR SINGLE PARTICLES

Authors	*Topic*	*Heat balance*	*Type of reaction model*	*Postulated kinetics of decomposition*
Bamford *et al.*[8]	Pyrolysis of a wood slab	A, B, C	Volumetric	First order
Kanury[9]	Pyrolysis of a wood slab, cylinder or sphere	A, B	Volumetric	Instantaneous when local temperature > critical temperature
Kung[10]	Pyrolysis of a wood slab	A, B, C, D	Volumetric	First order
Maa and Bailie[11]	Pyrolysis of a cellulosic cylinder	A, B, D	Shrinking core	First order
Fan[12]	Pyrolysis of a slab, cylinder or sphere	A, B, C, D	Volumetric	*n*th order
Fan[13]	Pyrolysis of a slab, cylinder or sphere	A, B, C, D	Volumetric	Consecutive *n*th order reactions
Van Ginniken and Tels[14]	Pyrolysis of a slab of a waste particle	A, B, C, D	Volumetric	Parallel and consecutive reactions (Kilzer and Broido)

A, Transient heat accumulation. B, Conduction. C, Heat of reaction. D, Convection.

hot gases. The rate of temperature rise is controlled mainly by heat transfer and influenced by the occurrence of endothermic and exothermic reactions. Rapid charring, accompanied by an important reduction in volume, occurs. The structure, together with the physical and thermal properties, vary continuously.

The effects of pyrolysis become more important when the raw material has a higher content of volatile matter (VM). From Table 2.II it follows that dry biomass has a much higher content of volatile matter than fossil solid fuels, such as lignite or young coal. Anthracite contains as little as 4–10 wt % of VM!

The rate of pyrolysis of large chunks of cellulosic material is controlled by internal heat transfer. A number of investigators have derived mathematical models in order to describe the effective rate of pyrolysis (Table 2.III).

Most modelling work in gasification has been based on coal gasifiers. In this particular application drying and pyrolysis are of less consequence than in the case of biomass. A majority of the models simply disregard both steps, so that the model and the design is based on gasification only.

This assumption seems reasonable in the case of the fluid bed gasification of small particles, having a relatively large gas/solids interphase and a good rate of external mass transfer. For large particles and for reactor types with poor gas/solids heat and mass transfer rates the drying and pyrolysis sections represent a significant part of the reactor volume.

The thermal behaviour of coal during pyrolysis (plasticity, baking, swelling) may, of course, create modelling (and operating) problems in its own right, but is outside the scope of this review. The pyrolysis behaviour of cellulosic material is being treated in depth by Shafizadeh (this symposium).[16]

3. REACTOR TYPES IN GASIFICATION

3.1. Survey

Coal has been gasified in three types of reactors:

—the vertical shaft or moving bed gasifier (Lurgi),
—the fluidized bed gasifier (Winkler),
—the lean phase gasification of entrained particles (Koppers-Totzek).

Some characteristics of these three reactor types have been summarized in Table 3.I.

TABLE 3.I
CHARACTERISTICS OF COAL GASIFICATION REACTORS[17]

Gasification process		*Fixed bed*	*Fluidized bed*	*Entrained bed*
Operating pressure	kPa	Up to 3 000	105	105
	bar	30	1·05	1·05
Gas temperature				
Gasifier outlet	°C	400–600	800–900	1 400–1 600
Fuel requirements:				
—size	mm	5–30	2–8	below 0·1
—caking behaviour		must not cake	must not cake	no qualifications
—behaviour at high temperatures		must not disintegrate	must not disintegrate	no qualifications
—Ash melting behaviour		Softening point 1 000 °C	Softening point 1 000 °C	no qualifications
Byproducts yielded		gasoline, oil, tar, phenol/ NH_3-containing waste liquor	none	none

In biomass gasification only the moving bed and the fluidized bed reactor have been applied in practice. Hence most attention will be devoted to the operation and modelling of these specific reactor types.

3.2. Moving Bed

The oldest and most widespread type of gasifier is the moving bed gasifier. It consists of a vertical shaft, either refractory-lined or water cooled, with a rotary or fixed grate at the bottom. Fuel is fed on top of the reactor and slowly descends to the bottom as material is consumed. Blast gas (air, oxygen + steam) is introduced through the grate in the updraft (or countercurrent) gasifier. Upon its rise through the reactor it creates different zones (Fig. 3.1): an ash cooling zone at the bottom, a combustion zone in which almost all oxygen is consumed, a gasification zone in which the devolatilized biomass (char) is gasified with CO_2 and H_2O, a pyrolysis zone and a drying zone.

The same processes take place in the *downdraft* or cocurrent gasifier. Due to the almost complete absence of heat transfer between the gaseous stream and the descending solids the high temperature zones of pyrolysis, gasification, and cooling are much more concentrated in the immediate vicinity of the hearth.

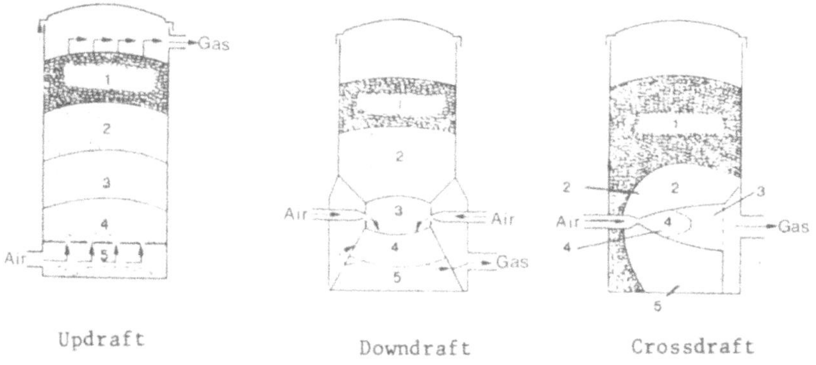

FIG. 3.1.

Various types of small moving bed units, operating on wood or charcoal, have been developed during the Second World War for driving motor vehicles. Some of these types have been adopted or slightly modified for use in developing countries.

Moving bed gasifiers are normally operated with 'dry' elimination of ash, as opposed to operation in the slagging mode. In the latter mode the gasification air is either preheated or enriched with oxygen in order to attain hearth temperatures well above the ash fusion point. Examples of this technique are the Andco-Torrax, the Union Carbide (Purox) and the Nippon Steel process for refuse gasification. The advantages and disadvantages of the various types of gasifiers have been discussed repeatedly in the literature. Tables 3.II and 3.III summarize some characteristics of moving bed gasifiers, which are relevant in biomass gasification.

3.3. Fluid-Bed Gasifiers

Although coal gasification was the first industrial application of fluid bed technology (Winkler, 1922), fluid bed gasifiers are relative newcomers in the gasification of biomass. They feature several important advantages (Table 3.IV) which explain their rapidly increasing popularity.

In contrast to moving bed reactors there are no distinct reaction zones in a fluidized bed gasifier. Drying, pyrolysis and gasification occur simultaneously over the whole reactor volume, which is almost perfectly mixed and thus isothermal.

The fluidized bed consists of inert particles, such as sand or alumina, with a typical average dimension of 0·25–2 mm. The bed is supported by a

TABLE 3.II

BASIC TYPES OF MOVING BED GASIFIERS

Type	*Updraft*	*Downdraft*	*Crossdraft*
Flow of solid and gas	Countercurrent	Cocurrent	Cross-current
Preferred fuel type and characteristics	Low tar fuel with low or moderate reactivity, e.g. coke, anthracite, charcoal	Fuel with high reactivity and volatiles content, e.g. wood, lignite, peat	Fuels with low volatiles and ash content, e.g. charcoal, coke, anthracite
	Uniform size and low ash content desirable		
Steam addition	Yes	No	No
Other characteristics	Broad oxidation zone with low temperatures, but many ignition points. Longer time for preheating. Fluctuating gas demand is met with some delay. Tar separator is often required	Narrow oxidation zone Short time for preheating. Fluctuating gas demand is rapidly met. Tar separator can often be omitted.	Rather narrow oxidation zone and preheating time. Fluctuating gas demand is rapidly met

TABLE 3.III

ADVANTAGES AND DISADVANTAGES OF MOVING BED GASIFIERS

Advantages	—simple, inexpensive units —high thermal efficiency (updraft) —high carbon conversion efficiency —low dust load of the generated gas —low tar load of the generated gas (downdraft)
Disadvantages	—limited possibilities for controlling the hearth zone (moving the grate only) —requires a calibrated feedstock for optimal operation —high tar load of the generated gas (updraft) —possible clinkering on the grate and the refractories —relatively small volumetric capacity

TABLE 3.IV
ADVANTAGES AND DISADVANTAGES OF FLUID BED GASIFIERS

Advantages	—flexible with regard to feed rate and composition —high volumetric capacity —easy control of temperature
Disadvantages	—the operating temperature is limited by the possibility of the sintering of ash —the dust load of the gas is high —the loss of carbon in the fly-ash may be high —tar is formed at low operating temperature

distributor, through which the gasifying medium is introduced. The latter maintains the bed material in a suspended, fluidized state.

The fuel particles can be introduced into the reactor by means of locks, screw conveyors or pneumatic systems. They mix rapidly with the bed material and run through the drying, thermal decomposition and gasification stages while moving through the bed.

In biomass gasification there is no specific mechanism for discharging the residue. The charcoal is gasified completely or elutriated as soon as it attains a sufficiently low particle size. In the latter case it will be collected together with dust and ash in a cyclone separator. The ash content is normally quite low.

Some crude forms of agricultural and forest wastes may also contain stones, debris, etc. Specific mechanisms for their removal have been developed.[18]

4. THERMODYNAMIC ASPECTS OF MODELLING

4.1. Pyrolysis

Under all practical circumstances the devolatilization of biomass materials proceeds as an irreversible chemical process.

The nature of the primary products can be derived from the structure of the parent molecule, the stability of which is directly related to the strength of its weakest bonds.

Primary products still undergo secondary reactions to a distinct extent, determined by their thermodynamic stability and their reactivity.

Thermodynamic considerations may thus help in sorting out possible reaction mechanisms, identifying the primary products and explaining the formation of secondary products.

4.2. Gasification

Gasification proceeds over the following set of reactions:

(a) *Heterogeneous (gas–solid) reactions*

Oxidation of carbon		H (kJ/mole)
(1)	$C + \frac{1}{2}O_2 = CO$	−110·6
(2)	$C + O_2 = CO_2$	−393·8
Boudouard reaction		
(3)	$C + CO_2 = 2CO$	172·6
Water–gas reaction		
(4)	$C + H_2O = CO + H_2$	131·4
Methane formation		
(5)	$C + 2H_2 = CH_4$	−74·93

(b) *Homogeneous (gas phase) reactions*

(6)	$CO + H_2O = CO_2 + H_2$	−41·2
(7)	$CH_4 + H_2O = CO + 3H_2$	−201·9

Reactions 1 and 2 are fast and can be considered to be irreversible at normal gasification temperatures.

On the basis of thermodynamic data, equilibrium compositions can be computed for various temperatures, pressures and compositions of the gasifying medium. Provided all reaction rates are sufficiently high the real composition of generator gas will be close to the one derived by thermodynamic computations. Since all independent variables may be varied at will it is comparatively easy to predict the effect of such variations, without having to resort to experiments. On the other hand the attainment—or not—of the computed equilibria depends on

—the operating temperature,
—the quality of the gas/solids contact,
—the reactivity of the feedstock,
—the type of reactor.

According to the reactions under scrutiny, heterogeneous or homogeneous equilibrium calculations are possible. For further details, the reader is referred to the abundant literature on the topic.[19–22]

4.3. Heterogeneous Equilibria

The composition of generator gas can be computed according to the classical treatment of Gumz[20] on the basis of the equilibrium of the heterogeneous reactions of carbon with steam, carbon dioxide and hydrogen (reactions 3, 4 and 5). It is assumed that (a) these reactions attain equilibrium, and (b) sufficient carbon is available for reaction with those gases.

In moving bed gasifiers the rate of consumption of carbon is controlled by the flow rate of the blast. Under normal operating conditions an excess of carbon will be available. Equilibrium will almost be obtained, provided the bed temperature is sufficiently high and the blast evenly distributed.

In fluid-bed and lean phase gasification, on the other hand, the flow of carbon and that of blast gas can be varied independently, so that an excess of carbon cannot be guaranteed under all operating circumstances. The validity of the second assumption should then be checked.

Some typical results of equilibrium calculations are given in Figs 4.1 to 4.3. Figure 4.1 gives the equilibrium composition of gas as a function of temperature when carbon is gasified with steam (this supposes the external

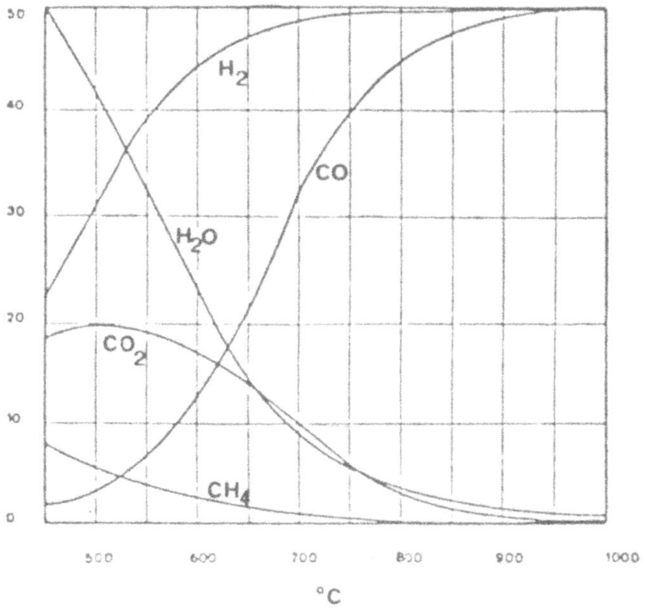

FIG. 4.1. Equilibrium composition for steam gasification of carbon (heterogeneous equilibria).

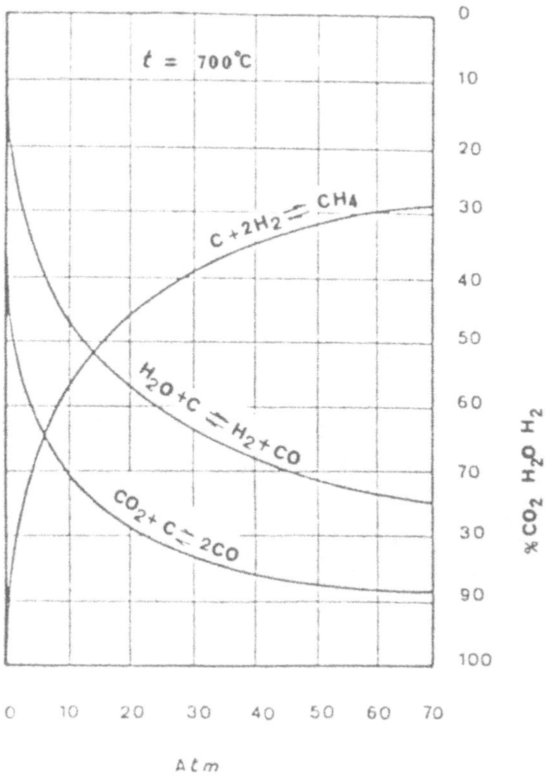

FIG. 4.2. Equilibrium composition as a function of pressure for different heterogeneous reactions.

supply of the required heat of reaction). At high temperatures (above 800 °C), the main products are H_2 and CO, whereas an almost negligible methane concentration is predicted.

Figure 4.2 gives the equilibrium percentage of carbon dioxide, steam and hydrogen in the gasification of carbon as a function of total pressure. It follows that the formation of methane is enhanced, but the equilibria of gasification with steam and carbon dioxide are less favourable.

Figure 4.3 shows the predicted carbon consumption for gasification at 850 °C with a mixture of air and steam as a function of the steam content of the blast. The specific consumption of carbon per kg gasifying medium fed increases with steam content from an initial value of 0·26 with no steam added to a final value of 0·61 when the blast consists of steam exclusively.

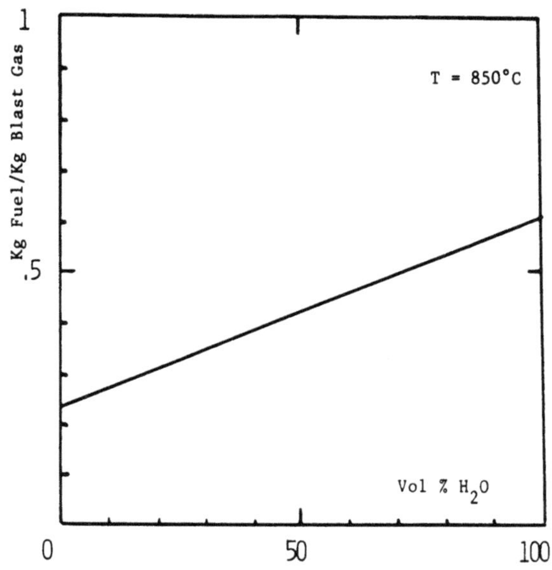

FIG. 4.3. Fuel consumption as a function of steam content of the blast for equilibrium calculation based on heterogeneous equilibria.

4.4. Homogeneous Equilibria

A theoretical gas composition can also be derived from (a) the feed rate and elementary composition of the solid (biomass) fuel, and (b) the flow rate and composition of the blast gas, using a mass balance together with homogeneous reaction equilibria. The operating temperature can either be varied as an independent parameter (which supposes that heat is either delivered to or possibly withdrawn from the system), or derived from a thermal balance over the system, supposed to operate adiabatically.

When only four components, namely H_2, H_2O, CO and CO_2, are important, equilibrium data can be derived from a single homogeneous reaction:

$$H_2O + CO = CO_2 + H_2$$

If methane is also considered, one additional reaction equilibrium has to be taken into account, e.g.

$$2H_2 + 2CO = CO_2 + CH_4$$

Above 800 °C the equilibrium content of methane is always low so that, for equilibrium calculations at atmospheric pressure, its presence can be ignored. The CH_4 present in the product gases will predominantly be formed by pyrolysis reactions.

5. GASIFICATION OF SINGLE CHAR PARTICLES

5.1. Survey

In the modelling of a gasifier there is a need for relatively simple rate expressions, to represent the rate of gasification over the range of operating conditions considered. These conditions are characterized by

—a range of temperatures,
—total pressure,
—the partial pressure of reactants and product gases,
—the possible presence of catalysts.

When all factors which influence the overall rate of reaction are taken into account, the resulting rate laws are quite complicated and incorporate numerous and sometimes experimentally unaccessible parameters. In practice several simplifications are used to obtain a more tractable rate equation.

The reactions under scrutiny are those of the devolatilized char with reactive gases, such as O_2, H_2O, CO_2 and, to a minor extent, H_2. The problem is treated in more detail in the review paper of Walker, presented at this symposium and in earlier reviews by Walker *et al.*,[23] Von Fredersdorff[24] and Grabowski.[25]

5.2. The Gasification of Char by H_2O and CO_2

5.2.1. Influence of Internal and External Mass Transfer

The intrinsic chemical rate of these reactions is relatively small. External (gas film) diffusion is not expected to be rate limiting at normal operating temperature. The importance of internal diffusion strongly depends on particle size, the intrinsic rate of the chemical reaction and the effective diffusivity.

A basic difficulty in the evaluation of the diffusion resistance is the almost complete lack of structural, physical and heat and mass transfer data obtained on biomass chars. Utilization of data derived on coal chars seems inappropriate considering the substantial differences in structure. Wood chars can be expected to be more porous, to present a lower resistance to diffusion, and to exhibit markedly anisotropic properties. The only diffusion data for biomass chars available to these authors are those of

Groeneveld,[26] suggesting the following relation for the effective diffusivity D_{eff} (m^2/s):

$$D_{eff} = D(7{\cdot}69\varepsilon - 5{\cdot}62) \qquad \text{for } 0{\cdot}75 < \varepsilon < 0{\cdot}84 \tag{5.1}$$

and

$$D_{eff} = D\varepsilon \qquad \text{for } 0{\cdot}84 < \varepsilon < 1 \tag{5.2}$$

where D is the molecular diffusivity (m^2/s) and ε the porosity of the charcoal. The importance of internal diffusion can be estimated by computing the Thiele modules on the basis of a spherical particle reacting according to first order kinetics:

$$\Phi = \frac{d_p}{6}\sqrt{\frac{k}{D_{eff}}} \tag{5.3}$$

where d_p = particle size, k = volumetric reaction rate (m^3 gas/m^3 cs).

The effectiveness factor is then given by

$$\eta = \frac{1}{\Phi}\left(\frac{1}{\tanh(3\Phi)} + \frac{1}{3\Phi}\right) \tag{5.4}$$

whereas the effective reaction rate r_G becomes

$$r_G = \eta r \tag{5.5}$$

with r the intrinsic or chemical reaction rate (moles/h).

The dependence of internal diffusion on particle size is illustrated by Table 5.I which gives the effectiveness factor as a function of d_p (on the basis of the correlation of Groeneveld for the effective diffusivity $D_{eff} = 4{\cdot}14 \times 10^{-5}$ m^2/s for $\varepsilon = 0{\cdot}75$). The intrinsic rate parameter k is varied as a parameter.

TABLE 5.I

EFFECTIVENESS FACTOR

d_p (*m*) \ k (s^{-1})	*10*	*1*	*0·1*	*0·01*
0·001	0·996	1	1	1
0·005	0·91	0·990	0·994	1
0·01	0·742	0·962	0·996	1
0·05	0·224	0·573	0·91	0·990
0·1	0·117	0·355	0·742	0·962

Plausible values of k for biomass chars vary between 0·1 and $4\,s^{-1}$ at 900 °C.[27] From Table 5.I it follows that internal diffusion is noticeable for particles larger than 5 mm. Gasification of particles larger than 10 cm will be diffusion limited.

Correlation (5.1) predicts diffusion coefficients for charcoal which are roughly one order of magnitude larger than those for coke. On the other hand, the intrinsic rates of chemical reaction are also larger for biomass chars, so that an *a priori* prediction of the relative occurrence of internal diffusion limitations is hazardous. Still, it follows from a comparison between Table 5.I and literature data that such limitations occur more readily in coke than in charcoal.

5.3. Influence of Carbon Burnoff

The rate of reaction of char with a gas is generally expressed as a function of residual carbon (or conversion), temperature and partial pressure of reactant and product gases:

$$r = F(M_c)f(T, r_i) \tag{5.6}$$

or

$$r = F(X)f(T, p_i) \tag{5.7}$$

r, r' = intrinsic reaction rate (mole C/h or gr C/h); M_c = residual carbon (kgC); X = carbon conversion $(M_{C_0} - M_c)/M_{C_0}$ and M_{C_0} = initial amount of carbon (kg).

In the case where the rate of reaction is proportional to the amount of residual carbon these equations can be written:

$$r = M_c f(T, p_i) \tag{5.8}$$

or

$$r = (1 - X)f(T, p_i) \tag{5.9}$$

Usually, however, the reaction rate varies in a more complex fashion with residual carbon, because of a variation in internal surface or in pore size distribution during burnoff. Empirical expressions for $F(X)$ have appeared in the literature (Table 5.II) but caution should be exerted in applying these expressions to other biomass chars.

5.3.1. *Influence of the Partial Pressure of Reactant and Product Gases*

It is generally accepted that the heterogeneous reaction of charcoal with H_2O or CO_2 follows a rate law of the Langmuir–Hinshelwood type, e.g.

$$f(T, p_i) = \frac{k_1 p_{H_2O}}{1 + k_2 p_{H_2O} + k_3 p_{H_2}} \tag{5.10}$$

TABLE 5.II
SELECTED EMPIRICAL EXPRESSIONS FOR $F(X)$

Author	$F(X)$
Johnson[28 a]	$(1-X)^{2/3}\,e^{-\alpha X^2}$
Dutta and Wen[29 b]	$(1 \pm 100X^{u\beta} e^{-\beta} e^{-\beta X})(1-X)$
Agarwal and Sears[30 c]	$(1 \pm AX^{u\beta} e^{-\beta X})(1-X)$
Agarwal and Sears[30 c]	$[1 - M \ln(1-X)](1-X)$

α, β, u, A and M are empirical parameters.
[a] Bituminous coal char.
[b] Various coal types.
[c] Lignitic, bituminous and sub-bituminous coal chars.

and for the carbon/CO_2 reaction

$$f(T, p_i) = \frac{k_2 p_{CO_2}}{1 + k_5 p_{CO_2} + k_6 p_{CO}} \tag{5.11}$$

These expressions have a theoretical basis; each type is linked to one or more complex reaction mechanisms, featuring elementary adsorption/desorption steps as well as chemical reactions among adsorbed species. For a further discussion refer to Von Fredersdorff.[24]

The possible inhibiting effect of the reaction products H_2 and CO upon the rate of reaction is noteworthy. Caution should therefore be exerted when extrapolating simple pseudo-first order rate data derived under conditions with negligible amount of the reaction products CO and H_2. The influence of rate hindrance by H_2 for the C/H_2O reaction has been discussed, *inter alia*, by Sundaresan and Amundson[31c] (see Section 7), and by Schoeters.[32]

The value of the actual rate and adsorption coefficients depends on complex physicochemical factors, which are influenced by the original structure of the biomass material, its thermal history and the presence of extraneous matter, such as ash, exerting a catalytic influence. For each specific application it is advisable to evaluate the rate parameters experimentally.

5.4. The Combustion of Char

The reaction between carbon and oxygen is much faster than those with H_2O or CO_2. It is generally accepted that at high temperatures (>1200 °C),

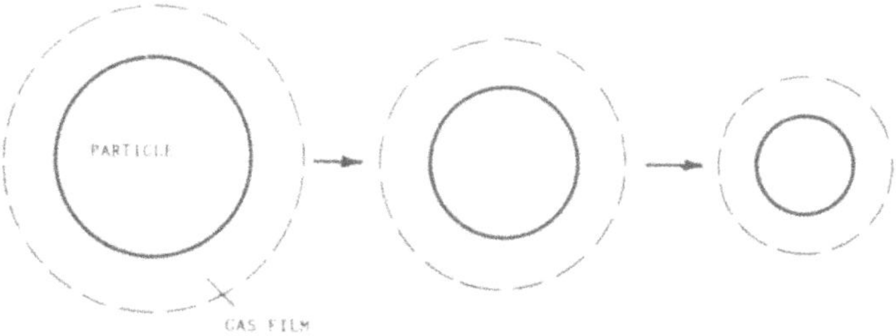

FIG. 5.1. Shrinking core reaction mechanism.

the reaction can be described by shrinking core model. According to this model the rate of chemical reaction is so fast that all oxygen diffusing to the particle is completely consumed upon reaching its external surface. Hence no reaction (with oxygen) takes place inside the particle. In biomass chars the amount of ash is negligible, so that no outer layer of ash will be formed. The particle consists of a continuously shrinking core of carbon (Fig. 5.1). The overall reaction is then controlled by external diffusion as well as by intrinsic chemical reaction at the surface. Under these assumptions, the overall rate of reaction can be expressed as

$$r = S_{ex} \frac{[O_2]}{\frac{1}{k_s} + \frac{1}{k}} \qquad \text{(mole/h)} \tag{5.12}$$

where $[O_2]$ = oxygen concentration in the bulk of the gas phase (mole/m^3); S_{ex} = external particle surface (m^2); k_s = mass transfer coefficient (m/s); k = (surface) reaction rate coefficient (m/s).

Equation (5.12) can be applied to the high temperature zone of a moving bed gasifier and possibly to fluid-bed combustors, but its use in fluid-bed gasifiers is contestable. Indeed, in a fluid-bed combustor, the temperature of the particles may be considerably higher (100–200 °C) than the average bed temperature. For a fluid-bed gasifier, however, it was shown by Rehmat[33] and Weimer and Clough[34] (Section 7) that the particle temperature is only slightly higher than that of the bed, because of the simultaneous occurrence of exothermic and endothermic reactions.

In the modelling of fluid-bed gasification a volumetric reaction model hence seems more appropriate than a shrinking core model. An important factor in most gasifier models is the relative amount of CO and CO_2 generated by the oxidation of char. Some correlations are available[35,36] but it is questionable whether they are generally applicable.

6. MOVING BED GASIFIERS

6.1. Mass and Energy Balances

Mathematical models for moving bed gasifiers are composed of a set of equations describing mass and energy balances over an elemental slice of reactor volume and incorporating rate expressions for the various heterogeneous and homogeneous reactions.

Most of the models described in literature do not consider momentum balances, which limits their field of application to quasi-isobaric reactors. It is reasonable to assume an almost constant operating pressure for a reactor with a relatively shallow bed of coarse wood slabs, but not for one operating on a feedstock containing a lot of fines. Pressure drop also increases when a layer of fine ash particles is allowed to build up on the grid.

A mass balance should be written for each component considered in the reaction mechanism. Depending on the complexity of the model these could be:

gases	CO, CO_2, CH_4, H_2, H_2O, N_2, O_2, H_2S, NH_3
vapors	tars, oils, ...
solids	fixed carbon, ash, ...

Some components, such as H_2S and NH_3, are mainly important with regard to the corrosive and polluting properties of gas. They are important for coal-derived gas only.

It is impractical to consider all physical and chemical processes, i.e. drying, pyrolysis and gasification, simultaneously. The conservation equations become quite complicated and computationally untractable. It is therefore a common procedure to split the reactor up into successive compartments, each one corresponding to a single dominant phenomenon (Fig. 6.1), which is supposed to halt abruptly at the end of the zone considered.

For an updraft gasifier these successive zones are:

(1) a *drying zone*, where the feedstock is dried by the hot upflowing gas;
(2) a *pyrolysis zone*, where volatile matter is driven off. The solid material leaving this zone consists of fixed carbon and ash;
(3) a *gasification zone*, where both heterogeneous and homogeneous reactions take place;
(4) a *combustion zone*, where the oxygen in the blast is rapidly consumed generating CO_2 and CO.

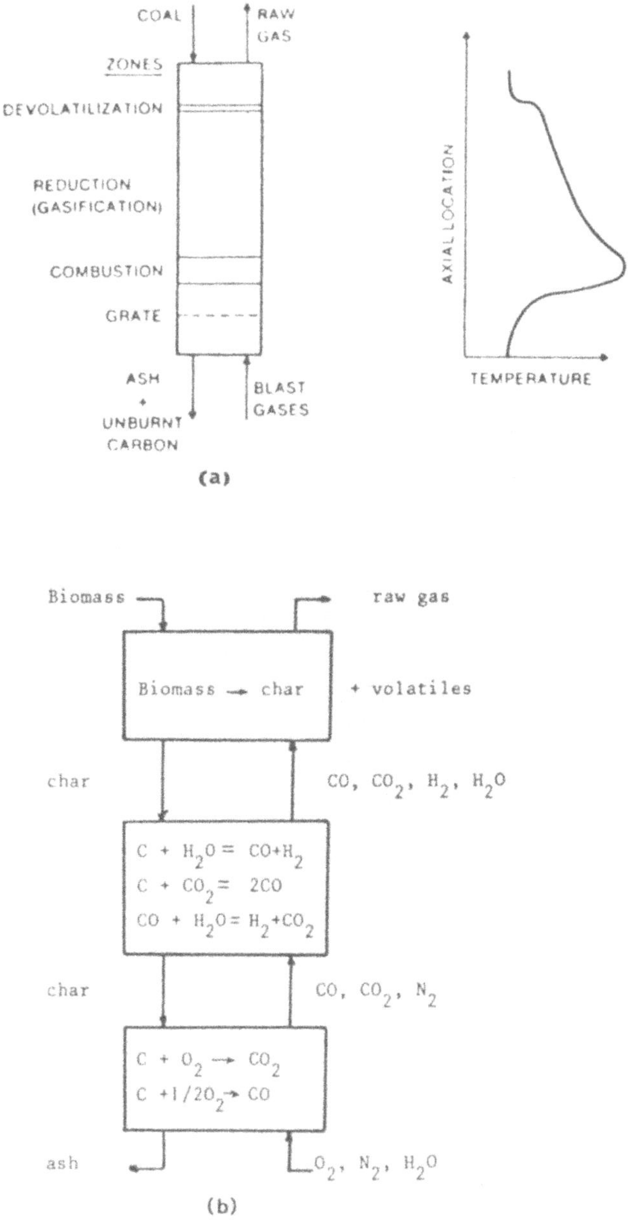

FIG. 6.1. (a) Continuous model. (b) Reaction zone model. Reproduced from Kosky, P. G. and Floess, J. K. (1980). *I & EC Process Des. Dev.* **19**, 586, with permission of the publisher. © 1980 American Chemical Society.

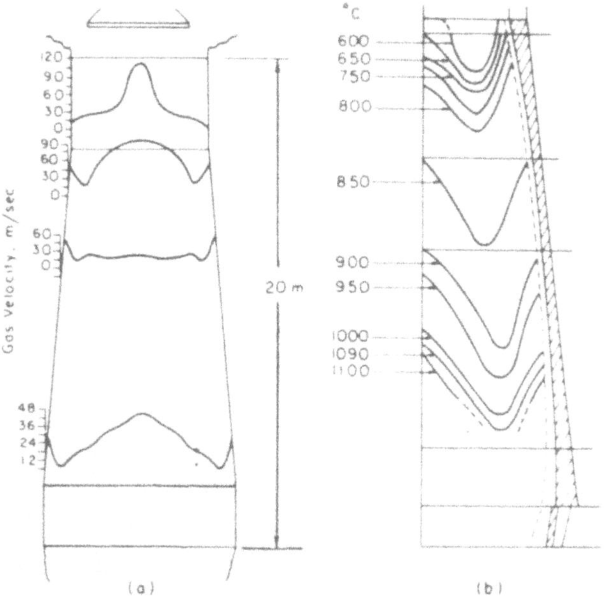

FIG. 6.2. (a) Velocity profile in a blast furnace. (b) Isotherms in a blast furnace. Reproduced from Szekerly, J., Evans, J. W. and Sohn, H. Y. (1976). *Gas–Solid Reactions*. Academic Press, New York, p. 316, with permission of the publisher.

The flow of gas through a fixed (or slowly moving) bed of identical particles is a matter of paramount importance in chemical engineering (fixed catalytic bed reactors, filtration, packed towers). The plug flow assumption is in general use. Still, radial variations in gas flow velocity are known to exist. Figure 6.2 gives a typical radial velocity and temperature profile for a blast furnace. As can be seen, local maxima of gas velocity exist both in the centre of the reactor and near the walls, where the packing is less dense. Nevertheless, radial velocity and concentration gradients are normally neglected in the modelling of moving bed gasifiers, because of fast radial diffusivity.

The flow of solids through a moving bed is not necessarily smooth and continuous. Ideally, the charge of solids should consist of uniform, dust-free, non-baking material. With a less ideal material channelling, bridging and slagging are difficult to avoid, especially for large gasifiers, of which the diameter exceeds 2 m. In blast furnaces the diameter increases towards the hearth, to promote an even flow of solids and gases.

The packing should be as uniform as possible. Near the wall the packing is looser which brings the blast and the fire to the periphery. Dense plugs of material are circumvented by the blast. They tend to remain relatively cold and yield a gas of poor quality.

Often one can observe a periodic movement in the gas output, corresponding to the formation of a large blowhole in the middle of the hearth, followed by its collapse under the pressure of the rest of the charge.

Despite those complicating factors (which have a bearing on gas output quality and operation) the plug flow assumption is also in general use for the flow of solids.

Yoon[37] calculated the Péclet number.

$$\text{Pé} = d_{\text{p}} \frac{u_{\text{z}}}{D_{\text{ax}}}$$

where d_{p} = particle size (m), u_{z} = axial velocity (m/s), and D_{ax} = axial diffusivity (m^2/s), for a Lurgi gasifier, operating at 1100 °C with particles of a diameter of 10 mm. He found a number of about 500 so that the reactor can be assimilated to approximately 250 stirred tanks in series. Hence axial dispersion is negligible.

Assuming plug flow a mass balance over a differential reactor slice can be written as follows:

$$\frac{\partial C_i}{\partial t} \pm \frac{\partial C_i u_i}{\partial x} = \sum_j \nu_{ij} R_j \tag{6.1}$$

where C_i is the concentration of species (mole/m^3); u is the absolute velocity of this species (m/h); x and t are the space and time coordinates; R_j the rate of reaction j (mole/m^2/h); ν_{ij} is the stoichiometric coefficient of species i in reaction j. The plus sign is for gas phase components, the minus sign for solids components.

The energy balance can be written for both the solid and the gas phase. For the solid phase one obtains

$$\frac{\partial C_{\text{ps}} C_{\text{s}} T_{\text{s}}}{\partial t} - \frac{\partial C_{\text{ps}} C_{\text{s}} T_{\text{s}} u_{\text{s}}}{\partial x} = -\sum_j R_j^{\text{s}} \Delta H_j^{\text{s}} + ha(T_{\text{G}} - T_{\text{s}}) \tag{6.2}$$

The s stands for total solids (ash + carbon), and G for gas. The first term corresponds to transient accumulation of heat, the second term to convective heat transport. The first term on the right hand side relates to heat generation by reactions in the solid phase; the second term represents convective heat transfer from gas to solids. Radiative heat transfer can also be incorporated.

In the previous equation C_{ps} is the average heat capacity of solids (J/K kg); C_s is the local solids concentration (kg/m^3);u_s is the absolute velocity of the solids (m/h); ΔH_j^s is the heat of reaction of heterogeneous reaction j (J/mol); h is the heat transfer coefficient (J/K h m^2); a is the specific area (1/m); T_G and T_s are the temperature of the gas and the solids (K).

Similarly, the heat balance over the gas phase becomes

$$\frac{\partial C_{pG} C_G T_G}{\partial t} + \frac{\partial C_{pG} C_G T_G u_G}{\partial x} = -\sum_j R_j^G \, \Delta H_j^G - ha(T_G - T_s) \quad (6.3)$$

In the above equations it is assumed implicitly that the temperature within the solids is homogeneous. Klose and Toufar[15] calculated the steady state internal temperature profile of a pyrolysing particle by means of the Fourier differential equations for non-stationary heat conduction and found that the hypothesis of a homogeneous particle temperature reduces the height of the pyrolysis zone by a factor of eight; this leads to an unrealistically low height of the reaction zone, as can be seen in Fig. 6.3. These results confirm that pyrolysis is really controlled by heat transfer.

In their general form the mass balance, together with the energy balance for the gas and solid phase, forms a set of coupled partial differential equations, which is difficult to solve with the current computational means.

For that reason various simplifications were introduced to make these equations easier to solve. These simplified models will be discussed in other chapters.

6.2. Types of Model

The models can be classified as follows in order of increasing sophistication:

(1) 'Equilibrium' models, in which the gas composition is calculated assuming a fixed reaction temperature and either heterogeneous or homogeneous reaction equilibrium.

(2) 'Kinetics free' models in which the reactor is subdivided into different zones. The gas composition in the gasification zone is calculated from equilibrium data whereas the reaction temperature is calculated in each zone by a separate heat balance.

(3) 'Steady-state' models using eqns (6.1)–(6.3) in which the time derivative has been neglected. The temperature of gas and solids can be identical or different.

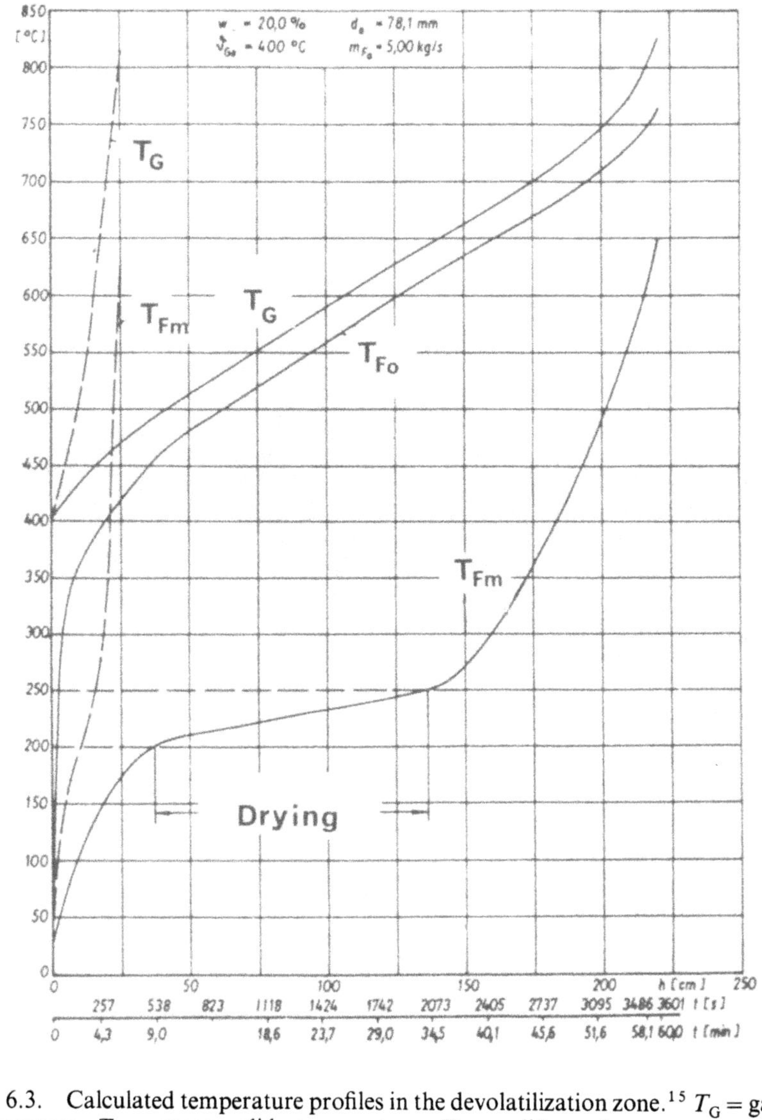

FIG. 6.3. Calculated temperature profiles in the devolatilization zone.[15] T_G = gas temperature, T_{Fm} = mean solids temperature, T_{F0} = solids surface temperature. (- - -) Homogeneous particle temperature; (——) internal temperature gradient. Reproduced from Klose, E. and Toufar, W. (1976). *Energietechnik* 26, with permission of the publisher.

(4) 'Semi-transient' models in which transients are calculated using the pseudo-steady-state assumption on the basis of the results of the steady-state models.
(5) 'Transient' models.

'Equilibrium' and 'kinetics free' models lead to algebraic equations which are easy to solve. These models allow one to predict exit gas compositions and to estimate reaction temperatures. 'Steady-state' models allow one to compute the gas composition and the temperature profile along the reactor axis. These models include rate equations for the heterogeneous and sometimes for the homogeneous reactions. These models result in a set of ordinary differential equations and algebraic equations. Mathematical software packages for their solution are readily available.

In 'semi-transient' models, a set of results from the previous steady-state model is used (for interpolation) to calculate the effect of small changes in operating conditions, such as solids feed rate or blast gas flow rate on reactor performance. 'Transient' models require the solution of a set of partial differential equations (6.1)–(6.3). Simplifications are usually made by removing the time dependency of some of the variables.

6.3. Equilibrium Models

The basis of equilibrium models has been discussed in Section 4. The potential usefulness of these models varies together with the particular type of moving bed gasifiers.

In a downdraft gasifier both the pyrolysis and the gasification products are forced through the hottest zones of the hearth. At these elevated temperatures the reaction rates are so high that equilibrium is established after a relatively brief time period. The equilibrium approach is controlled by the quality of mixing in the gas phase.

In an updraft gasifier the gasification products communicate their sensible heat to the incoming charge, and thus get cooled and mixed with the generated pyrolysis products. Equilibrium may have been established in the hearth of the gasifier, but the quantity and composition of the added products can only be predicted on the basis of separate pyrolysis data.

The real gas composition also deviates from the thermodynamic one because of imperfections in the temperature and concentration fields, caused by irregularities in the flow pattern. As a rule of thumb the deviation from ideality, expressed as an 'equilibrium approach', is of the order of 30–60 °C for industrial gasifiers operating at a temperature of 900 °C.

6.4. Kinetics Free Model

In these models, the reactor is subdivided into two or more separate reaction zones in which distinct processes occur: drying, pyrolysis, gasification, combustion. When the flow rate, composition and temperature of the solid fuel and the blast gas, together with the final carbon conversion, are given, it is possible to calculate the exit gas composition and estimate the temperature. As an example, the model developed by Kosky and Floess[41] is discussed in some detail.

The reactor is divided into two zones (Fig. 6.4). In the upper, or pyrolysis zone (II), solid reactants are introduced, dried and devolatilized. Char composed of fixed carbon, hydrogen and ash (determined from proximate and ultimate analyses) is generated here. This char then enters the lower gasification zone (I) where combustion and gasification reactions take place. The composition and temperature of this zone is supposed to be homogeneous (well mixed). The water–gas shift reaction is assumed to be in equilibrium. Methane is not considered in this zone, but is produced in the pyrolysis zone II.

The following set of reactions is considered in the gasification zone:

$$C + H_2O \rightarrow CO + H_2 \quad (1) \qquad C + \tfrac{1}{2}O_2 \rightarrow CO \quad (5)$$

$$C + CO_2 \rightarrow 2CO \quad (2) \qquad C + O_2 \rightarrow CO_2 \quad (6)$$

$$C + 2H_2 \rightarrow CH_4 \quad (3) \qquad H_2 + \tfrac{1}{2}O_2 \rightarrow H_2O \quad (7)$$

$$CO + H_2O = CO_2 + H_2 \quad (4) \qquad CO + \tfrac{1}{2}O_2 \rightarrow CO_2 \quad (8)$$

or

$$\lambda C + O_2 \rightarrow 2(\lambda - 1)CO + (2 - \lambda)CO_2 \qquad (9)$$

The mass balances for this zone are:

Carbon
$$N_c = NCO + NCO_2 \qquad (6.4)$$

Hydrogen
$$NH_2 = NH_2O^\circ - NH_2O + \delta NH_2 \qquad (6.5)$$

Oxygen
$$NH_2O^\circ + 2NO_2^\circ = NH_2O + 2NCO_2 + NCO \qquad (6.6)$$

The water–gas shift equilibrium is given by:

$$\frac{[CO_2][H_2]}{[CO][H_2O]} = K(T^x) \quad (8a) \qquad (6.7)$$

$$K(T^x) = 0{\cdot}0265 \exp\left(\frac{7860}{RT^x}\right) \quad (8b) \qquad (6.8)$$

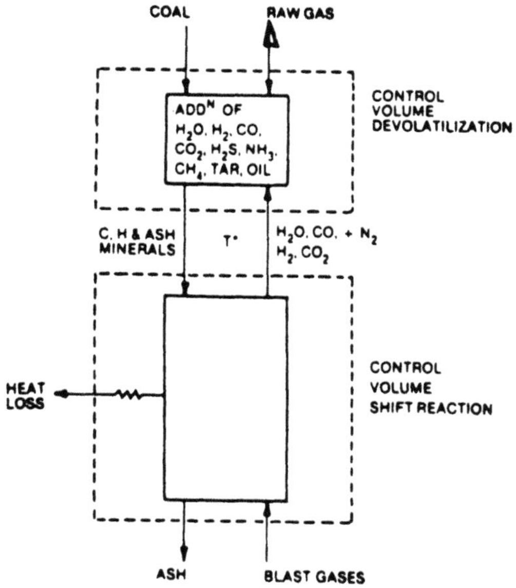

FIG. 6.4. 'Kinetics free' model representation.[41] Reproduced from Kosky, P. G. and Floess, J. K. (1980). *I & EC Process Des. Dev.* **19**, 586, with permission of the publisher. © 1980 American Chemical Society.

N_c = molar flow rate of carbon entering zone I (mole/h); Ni = molar flow rates of components leaving zone I (mole/h); $Ni°$ = molar flow rates of components in blast gas (mole/h); $\delta N H_2$ = molar flow rates of hydrogen in char (mole/h); $[i]$ = concentrations of gas composition i; $K(T^x)$ = equilibrium constant of the water–gas shift reaction; T^x = temperature of gasification (K).

Equations (6.4)–(6.8) are solved readily in combination with the corresponding energy balance yielding the temperature of gasification and the composition of the gas leaving this zone. The gases enter the pyrolysis zone where the pyrolysis gases are simply added to give the final gas composition.

The model of Denn *et al.*[44] is very similar. It provides, however, for separate combustion and gasification zones so that an estimate of the maximum temperature for this combustion zone is possible. Milne *et al.* use a model similar to that of Denn.

Kinetics free models yield exit gas compositions essentially equal to those predicted by rate models because in most cases the water–gas shift reaction is close to equilibrium (Table 6.I). The main difference is in the amount of CH_4 produced by gasification which fails in the 'kinetics free' models. For

TABLE 6.I
COMPARISON OF 'KINETICS FREE' MODELS WITH RESULTS FROM THE RATE MODEL OF YOON *et al.* FOR WYOMING COAL

	Yoon	*Denn*		*Yoon*	*Kosky and Floess*
H_2O	0·428	0·406			
H_2	0·525	0·594			
CO	0·332	0·356	CO	21·5 vol %	20·4 vol %
CO_2	0·328	0·328	CO_2	26·8	25·1
CH_4	0·025	0	H_2	42·5	47·0
T, °C	762	734	CH_4	5·8	5·8
H_2/CO	1·57	1·67	Others	1·2	1·7
T_{max}, °C	1 149	1 199	Others	1.2	

[a] Denn *et al.* (air blown).[44]
[b] Kosky and Floess (see Table 6.II).[41]

biomass gasifiers, however, which operate at atmospheric pressure, the carbon/hydrogen reaction is not important and CH_4 is mainly produced by pyrolysis reactions.

The principal drawback of 'kinetics free' models is that they require realistic input data. For certain sets of input conditions the gasification reactions are too slow to consume all of the input carbon, so that a considerable fraction of the carbon is lost in the residue. The use of these 'kinetics free' models is not to be recommended in case not a single feasible operating point is available (from experiment or from a rate model). The model can readily be used to predict variations in the gas compositions which arise due to small variations around this operating point.

6.5. Rate Models

'Rate' models consist of the mass and energy balances (6.1) to (6.3), in which the time derivatives are neglected, and of the rate expressions for the various heterogeneous reactions. The water–gas shift equilibrium is usually supposed to be established.

The rate model developed by Yoon *et al.*[40] is frequently cited in the literature. Therefore, it will be discussed in some detail.

The reactor is described as a continuous medium in which different reactions are taking place simultaneously. The temperature of gas and solid is assumed to be equal and the following reactions are considered:

Devolatilization Feedstock → char + volatiles

Combustion $\lambda C + O_2 \rightarrow 2(\lambda - 1)CO + (2 - \lambda)CO_2$ (9)

Gasification $$C + H_2O \rightarrow CO + H_2 \tag{1}$$

$$C + CO_2 \rightarrow 2CO \tag{2}$$

$$C + 2H_2 \rightarrow CH_4 \tag{3}$$

Water–gas shift $$CO + H_2O = CO_2 + H_2 \tag{4}$$

In the original model, the reactor was divided into two concentric zones: an inner adiabatic core and a surrounding annular layer, which exchanges heat with an external water jacket. The influence of this outer annular layer was small and will not be reported here.

For the gasification and combustion zone, the mass and balances can be written as:

Carbon $$\frac{NH_2O}{\Omega}\frac{dX_c}{dx} = R_1 + R_2 + R_3 + R_9 \tag{5.12}$$

Oxygen $$\frac{NO_2^\circ}{\Omega}\frac{dXO_2}{dx} = R_9 \tag{5.13}$$

Water $$\frac{NH_2^\circ O}{\Omega}\frac{dXH_2O}{dx} = R_1 \tag{5.14}$$

Methane $$\frac{NH_2^\circ O}{\Omega}\frac{dv}{dx} = R_3 \tag{5.15}$$

Ω = cross-section of the gasifier (m^2); XO_2, XH_2O = fractional conversion of O_2 resp. H_2O; X_c = fraction of carbon remaining in the char particle; v = dimensionless methane composition with respect to the molar feed rate of carbon.

The water–gas shift reaction was assumed to be in equilibrium.

The corresponding energy balance is

$$\left(\frac{N_C^\circ}{\Omega} X_c \tilde{C}_{pc} + \frac{G^\circ}{\Omega} f_{ash} C_{pash}\right)\frac{dT}{dx} - i\frac{N_i^G}{\Omega}\tilde{C}_{pi}^s\frac{dT}{dx}$$

$$= R_1\,\Delta H_1 + R_2\,\Delta H_2 + R_3\,\Delta H_3 + R_4\,\Delta H_4 \tag{5.16}$$

with N_c° = molar feed rate of fixed carbon (mol/h); $\tilde{C}_{pc}$, $\tilde{C}_{pi}^s$ = molar heat capacity of carbon resp. component i (J/mol K); G° = feed rate of coal (kg/h); f_{ash} = weight fraction of ash in coal.

The intrinsic rate of reaction is described by means of the law of mass action. Literature values are used for the rate parameters.

The C/CO_2 and C/H_2 reactions are supposed to be slow and their apparent reaction rates R_i are controlled by chemical reaction.

TABLE 6.II
OPERATING CONDITIONS USED IN SIMULATION

Feed rate	
Steam	0·591 5 $kg/m^2 s$
Oxygen	0·154 6 $kg/m^2 s$
Gas inlet temperature	371 °C
Operating pressure	25 atm ($2{\cdot}53 \times 10^6$ Pa)

The rates of C/O_2 and C/H_2O are fast and the overall rate may be limited by any of the following steps:

—gas film diffusion,
—ash layer diffusion,
—core diffusion,
—chemical reaction.

Two models are used to describe the reaction of individual particles: an ash segregating model (AS) in which any ash formed is immediately stripped off, and a shell progressive (SP) model in which the reacting core remains surrounded by an ash layer.

Devolatilization is assumed to be very fast and occurs instantaneously in the first reactor zone. The composition of the volatiles was determined in either of two ways: the first was from published devolatilization data of coals, the second from proximate and ultimate analysis of the coal and evolved tars.

The mass and energy balances were solved numerically for the conditions given in Table 6.II.

Typical temperature/height curves are shown in Fig. 6.5 for the AS model. The sharp temperature peak corresponds to the combustion front. The low reactivity fuel (Illinois) gives a higher maximum temperature because less heat is consumed by the endothermic carbon/steam reaction. Inlet conditions are chosen so that for both types of coal the maximum temperatures are at the same axial position.

Figures 6.6 and 6.7 give the residual fixed carbon and the gas composition for both types of coal using the SP model. There is a small intense combustion and gasification zone in which most of the oxygen is consumed. This zone lies much lower for the low reactivity coal. In the upper part of the reactor gas compositions and carbon consumption vary more slowly. The small change in gas composition, noticeable at the top of the reactor, is due to the pyrolysis reaction products.

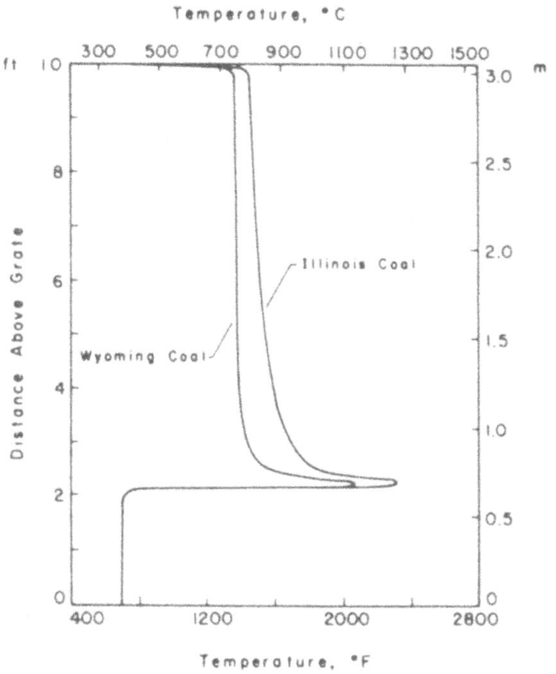

FIG. 6.5. Calculated temperature profile in a coal gasifier.[40] Reproduced from Yoon, U. *et al.* (1978). *AIChE Journal* **24**(5), 885, with permission of the publisher.

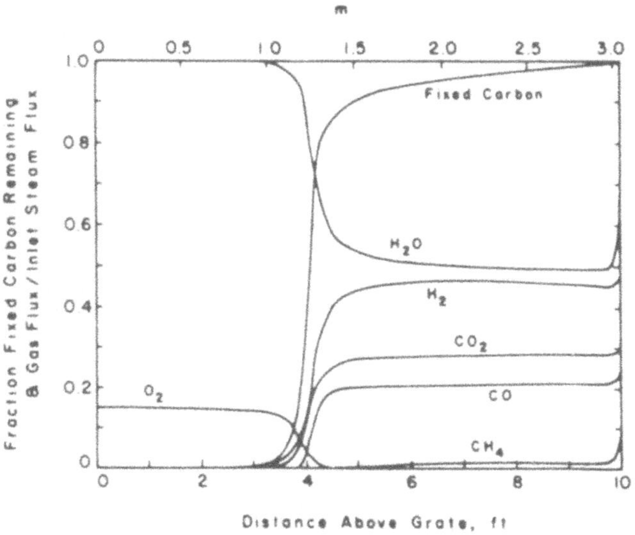

FIG. 6.6. Gas composition profiles with Wyoming coal.[40] Reproduced from Yoon, U. *et al.* (1978). *AIChE Journal* **24**(5), 885, with permission of the publisher.

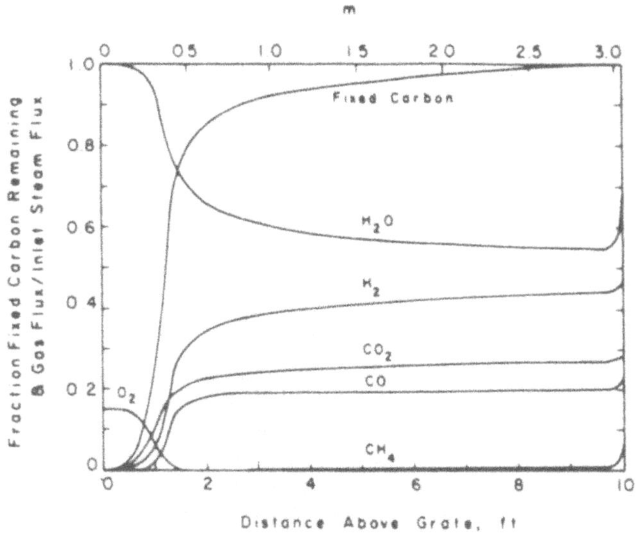

FIG. 6.7. Gas composition profiles with Illinois coal.[40] Reproduced from Yoon, U. *et al.* (1978). *AIChE Journal* **24**(5), 885, with permission of the publisher.

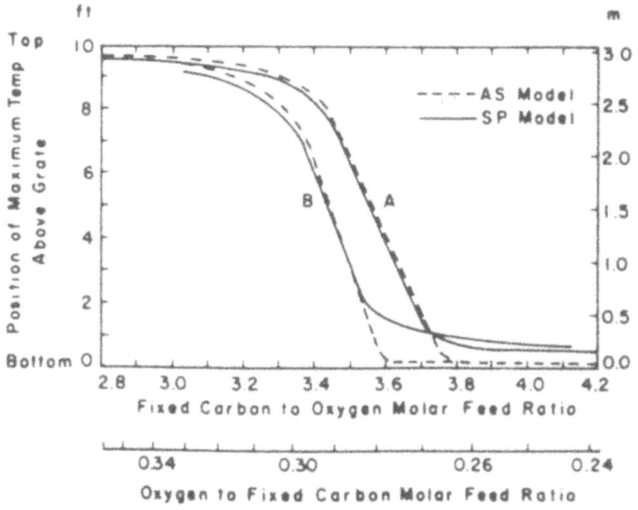

FIG. 6.8. Position of maximum temperature as a function of fixed carbon-to-oxygen molar feed ratio (Wyoming coal). Curve A: steam and oxygen feed rates as in Table 6.II. Curve B: steam and oxygen feed rates twice those in Table 6.II.[40] Reproduced from Yoon, U. *et al.* (1978). *AIChE Journal* **24**(5), 885, with permission of the publisher.

TABLE 6.III

COMPARISONS BETWEEN CALCULATED AND EXPERIMENTAL EXIT GAS COMPOSITIONS[40]

	Simulation		*Plant data*		
	Wyoming coal	*Illinois coal*			
Component			Rudolph (1972)	Hoogendorn (1973)	Moe (1974)
H_2	42·5 vol %	43·9 vol %	38 vol %	39 vol %	37·9
CO_2	26·8	26·6	28	28	28·6
CO	21·5	22·2	24	22	20·3
CH_4	8·0	6·2	10	9	11·4
Others	1·2	1·1	0	2	1·9

The model predicted a higher hydrogen and a lower methane content in the exit gas than was found in actual plant operating data (Table 6.III).

The position of maximum temperature varies as a function of the feed ratio of fixed carbon/oxygen (Fig. 6.8), the feed ratio of steam to oxygen being kept constant. For a high value of the former ratio the maximum temperature is lowered, which increases the amount of non-converted carbon in the ash. For low C/O_2 ratios the maximum temperature zone is shifted towards the top of the reactor. When this zone is too high gasification reactions will not proceed sufficiently and thermal efficiency will be lowered. Eventually oxygen breakthrough will occur.

Finally the influence of a finite rate of the water–gas shift reaction was estimated. No appreciable differences with respect to the equilibrium assumption became apparent when published rate data for this reaction were used.

Neglecting the temperature differences between solid and gas phase appears unrealistic since experimental evidence suggests that differences of approximately 700 K can exist near the combustion zone.[41]

For this reason Cho and Joseph[42] modified the model of Yoon *et al.* to incorporate this distinction. Gas solid heat transfer was described by a term

$$h\zeta a(T_s - T_G)$$

where h is the usual gas–solid heat transfer coefficient for a non-reacting particle. ζ is an empirical correction factor to account for differences occurring due to reaction in the particle. From comparison of computed results with plant data $\zeta = 0{\cdot}3$ was chosen as the best value. The influence of this factor is shown in Fig. 6.9. A value of 5·0 resulted in essentially equal gas and solid temperatures.

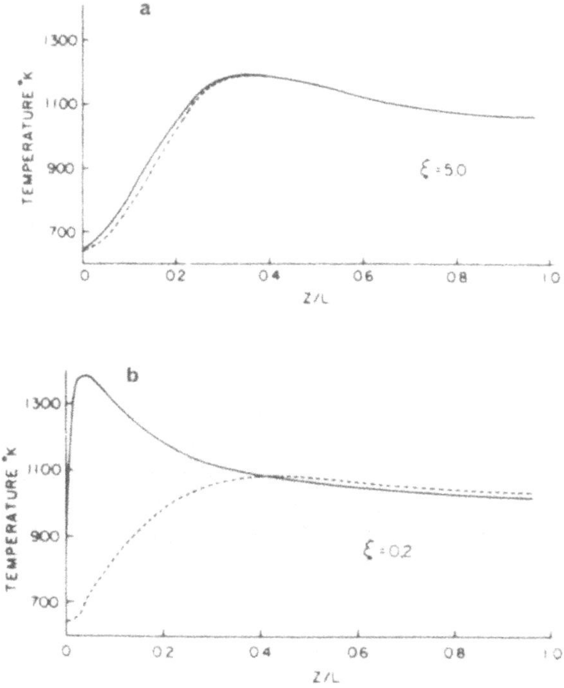

FIG. 6.9. Influence of ζ on temperature profile.[42] ——, Solid temperature; - - -, gas temperature. (a) $\zeta = 5$, (b) $\zeta = 0{\cdot}2$. Reproduced from Cho, Y. S. and Joseph, B. (1981). *I & EC Process Des. Dev.* **20**, 314, with permission of the publisher. © 1981 American Chemical Society.

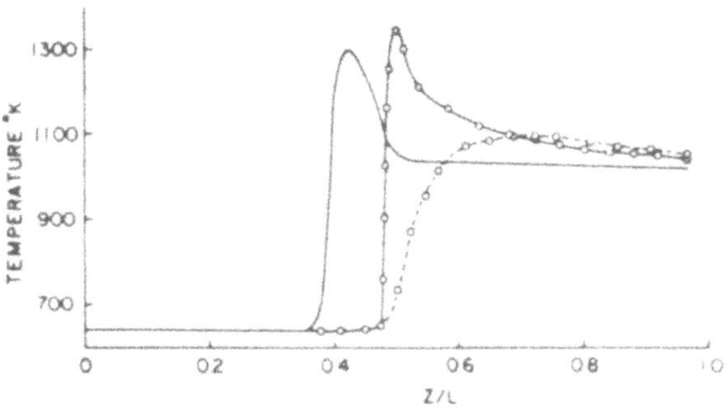

FIG. 6.10. Comparison of temperature profiles obtained by Yoon *et al.* ($T_G = T_s$) and Cho and Joseph ($T_G \neq T_s$) ($\zeta = 0{\cdot}5$).[42] Yoon's work, ——. This work: —O—, solid temperature; - -O- -, gas temperature. Reproduced from Cho, Y. S. and Joseph, B. (1981). *I & EC Process Des. Dev.* **20**, 314, with permission of the publisher. © 1981 American Chemical Society.

TABLE 6.IV

COMPARISON OF PRODUCT GAS DISTRIBUTION WITH WESTFIELD PLANT DATA FOR ILLINOIS COAL[42]

Component	*Cho and Joseph*	*Plant data*	*Yoon (1978)*
H_2	44·4 vol %	39·1 vol %	45·0 vol %
CO	19·5	17·3	14·4
CO_2	27·5	31·2	33·7
CH_4	6·5	9·4	5·8
Others	2·1	3·0	1·1

A comparison between exit gas composition calculated by Yoon and experimental data is given in Table 6.IV.

The temperature profiles obtained by Cho and Joseph are compared to those of Yoon *et al.* in Fig. 6.10. The temperature profile of the solids has the same form but the gas temperature predicted by Cho and Joseph is substantially lower.

These authors also studied the influence of the rate of the water–gas shift reaction on the carbon conversion and the H_2/CO ratio in the product gas. They found that the effect was small at rates above 10^{-3} kmol/m^3 kPa^2 h. At lower velocities, char conversion drops rapidly (Fig. 6.11). Other steady state rate models were developed by Klose and Toufar,[15] Biba *et al.*[43] and Amundson and Arri.[39]

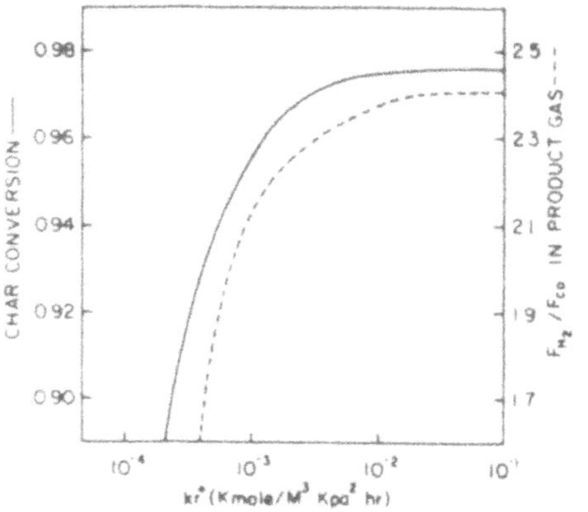

FIG. 6.11 Effect of water–gas shift reaction on char conversion and hydrogen-to-carbon monoxide ratio in product gas.[42] Reproduced from Cho, Y. S. and Joseph, B. (1981). *I & EC Process Des. Dev.* **20**, 314, with permission of the publisher. © 1981 American Chemical Society.

6.6. Transient Models

Real transient models are based on the mass balances 6.1 to 6.3. They have the disadvantage of generating a set of partial differential equations which are difficult to solve.

Yoon *et al.*[44] developed a transient model based on a pseudo steady state analysis. They used a set of results from their steady state rate model to describe the evolution of the zone of maximum temperature along the reactor axis when the feed rate of solids and gas is varied by small increments. These changes are slow phenomena taking several hours.

Figure 6.12 shows some typical results for a step change of 5, 10 and 15% in the feed rate of coal. The variation of the position of the combustion front is given for a low reactivity (Illinois) and a high activity (Wyoming) coal and for blast gas consisting of either oxygen/steam or air/steam mixtures.

Daniel neglected the time dependency of the concentrations[45] from eqn

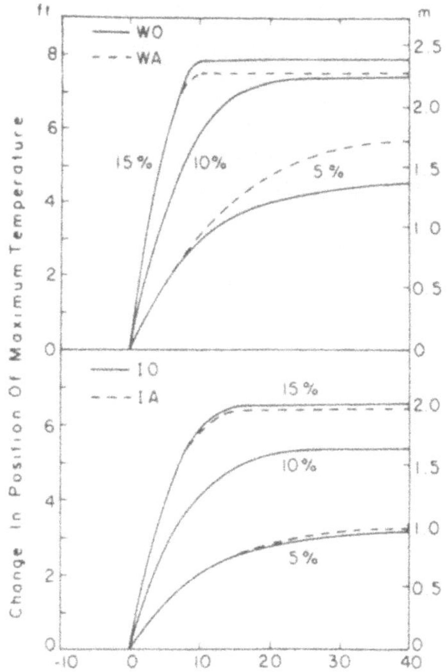

FIG. 6.12. Change in the position of the combustion zone caused by a step change in the coal feed rate in a non-slagging gasifier.[44] Reproduced from Yoon, U. *et al.* (1979). *AIChE Journal* **25**(3), 429, with permission of the publisher.

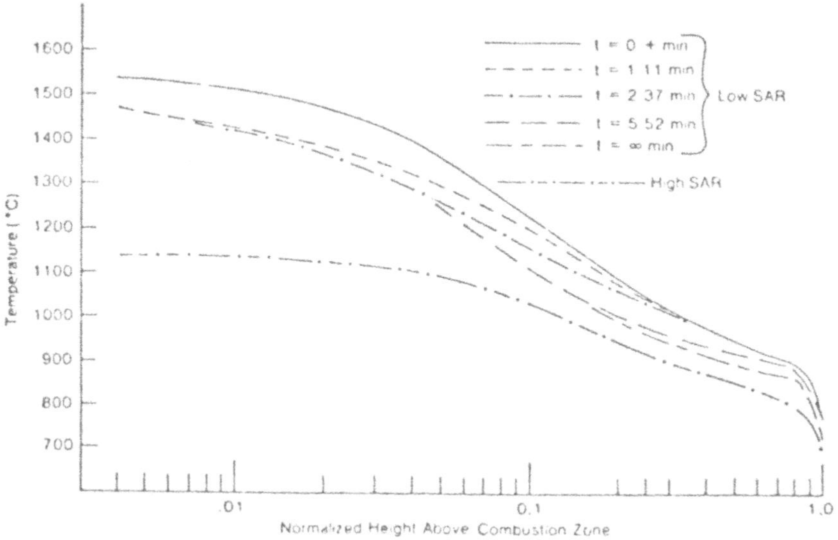

FIG. 6.13. Predicted temperature profile within the gasifier.[45] Reproduced from Daniel, K. J. (1980). Paper 52F, 83rd National AIChE Meeting, with permission of the publisher.

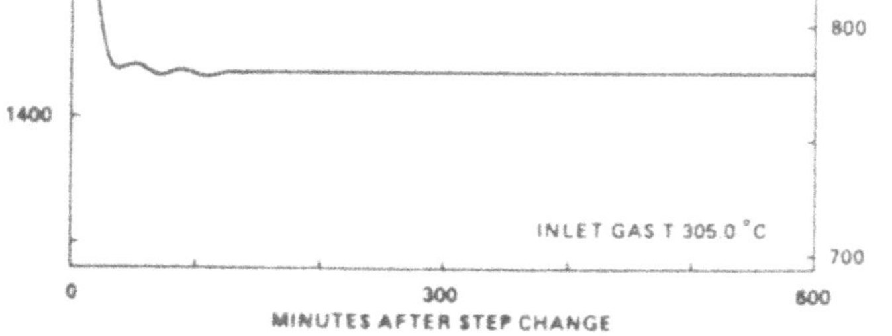

FIG. 6.14. Solids temperature at 0·3 m, inlet gas temperature reduced from 361 °C to 305 °C.[46] Reproduced from Stillman, R. (1981). ACS Symposium Series No. 168, with permission of the publisher. © 1981 American Chemical Society.

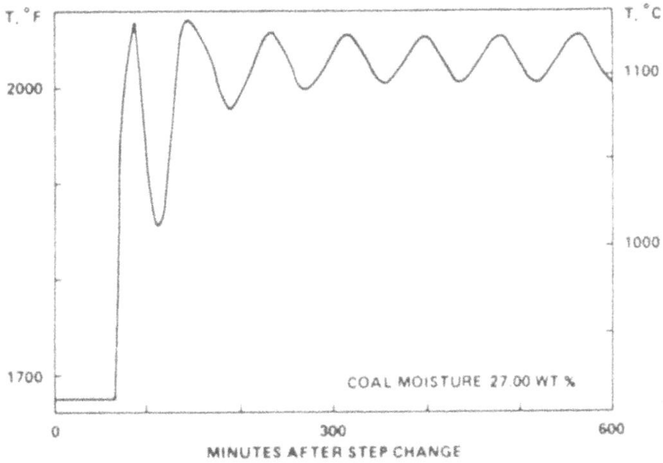

FIG. 6.15. Solids temperature at 0·3 m, coal moisture reduced from 34·7 to 27 wt %.[46] Reproduced from Stillman, R. (1981). ACS Symposium Series No. 168, with permission of the publisher. © 1981 American Chemical Society.

(6.1). This simplification results in some loss of information on rapid transient responses such as the propagation of gases through the reactor (the time of which is in the order of 10 s). The reactor is divided into separate pyrolysis, gasification and combustion zones. The transient response of the combustion zone is also neglected (duration approximately 30 s) and the combustion front is assumed to be stationary. This assumption makes it impossible to include long term transients caused by the movement of this combustion front. The analysis of Daniel is thus limited to relatively brief transients between 30 s and 10 min. Daniel also investigated the transient response of the composition of the output gas when step changes are made in the blast conditions. Solid and gas temperatures are assumed to be equal and the water–gas reaction is supposed to be in equilibrium in the gasification zone.

Figure 6.13 shows the effect of a 40 % step change in the flow rate of the blast on the temperature profiles in the gasification zone. Typically, a new steady state is reached after 10 min. Processes which have a major influence on these transients are the drying and devolatilization of the feedstock and the thermal energy stored in the bed.

Stillman[46] resolved the complete set of mass and energy balances given in 6.1 with the sole assumption that the time dependency of the concentrations and temperature in the gas phase were neglected. The set of hyperbolic partial differential equations was then solved numerically.

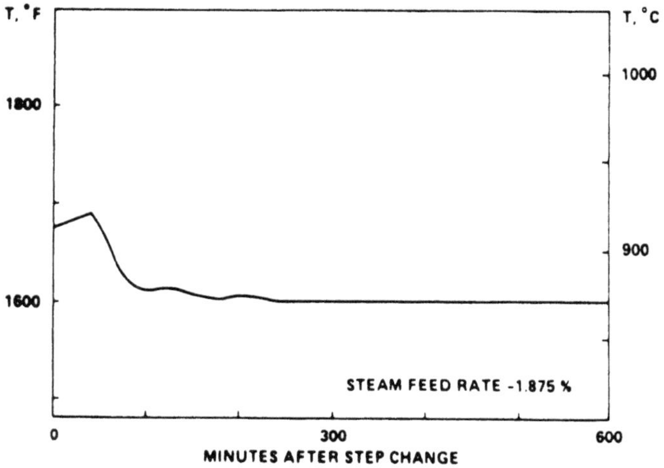

FIG. 6.16. Solids temperature at 0·3 m, steam-to-oxygen molar ratio reduced from 8·2 to 8·05.[46] Reproduced from Stillman, R. (1981). ACS Symposium Series No. 168, with permission of the publisher. © 1981 American Chemical Society.

Stillman compared the method of lines and the method of characteristics for solving these equations. The latter proved to be the more accurate.

The influence of step changes in the inlet gas temperature, the coal moisture content, the steam flow rate and the exit gas pressure were investigated. Some results are shown in Figs 6.14–6.16.

Different types of responses such as steady state convergent, underdamped, limit cycle and decaying oscillation phase limit responses were observed. Multiple steady states were also encountered when changing the exit gas pressure.

Unfortunately, as Stillman points out, 'no experimental gasifier data was found to verify any of the simulation results'. The only evidence for the observed phenomena was the presence of limit cycle responses for the pressure in combustion chambers and boilers reported by Friendly.[47]

7. FLUIDIZED BED GASIFIERS

7.1. Introduction: Flow Regimes

For a moving bed gasifier it is reasonable to assume that both the solid and the gaseous phase move in plug flow. In a fluidized bed, however, distinct flow regimes are possible, depending on the diameter of the bed particles d_p

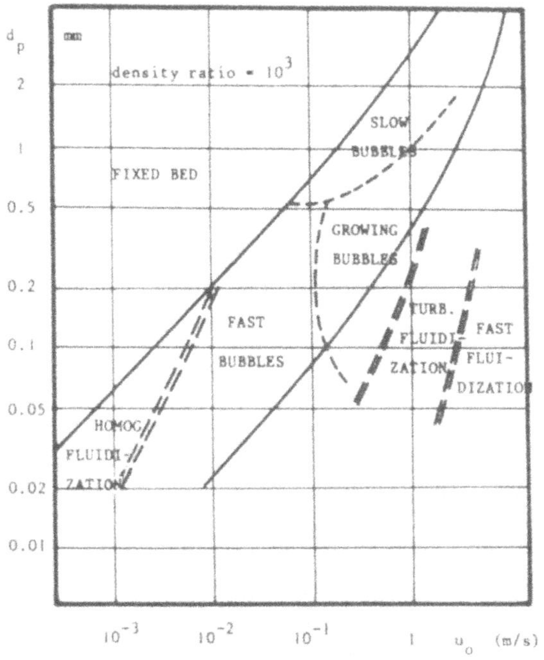

FIG. 7.1. (——) Geldart's classification.

and the gas velocity (Fig. 7.1). The latter is defined by

$$u_0 = V/\Omega$$

where u_0 = superficial fluidization velocity (m/h), V = volumetric gas flow rate (m^3/h), and Ω = cross-section of fluidized bed (m^2).

In the *fixed bed region* the particles are too large and the gas velocities too low for fluidization.

In the zone of *homogeneous fluidization* the bed is expanded and fluidized by the rising gas flow. The fluidized bed forms a homogeneous medium, without bubbles.

A regime of *fast bubbling* may be established with relatively low gas velocities, but small particles. The fluidizing gas moves through the bed under the form of rapidly rising bubbles and to a lesser extent through the continuous or emulsion phase. There is only limited mass transfer between the gas in the bubbles and the gas rising slowly through the emulsion phase. Two phase models are applicable to this zone with fast bubbles.

For the large particles and the high gas velocities typical for some coal combustion processes, the bubbles rise slower than does the surrounding

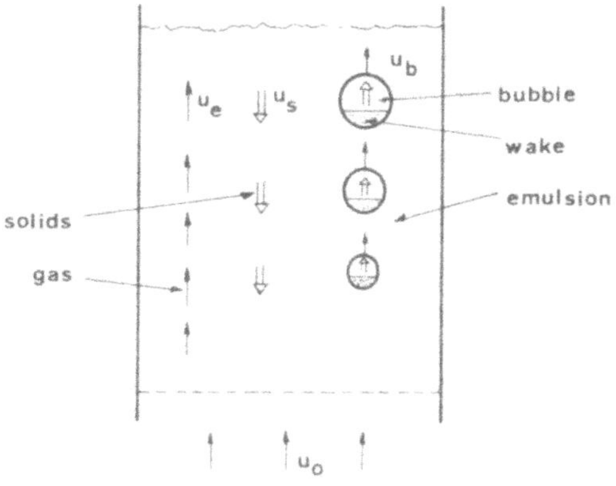

FIG. 7.2. A fluid bed in the fast bubble regime.[50]

emulsion gas. They are flushed frequently by this emulsion gas and no distinction exists between the composition of lean and dense phase. The flow regime can be approximated by plug flow in this region with *slow bubbles*.[48,49]

In a regime of *turbulent fluidization* so many solid particles are entrained that the bed level can no longer be indicated with precision. Also no clear distinction exists between bubbles and emulsion. This regime occurs in the regenerator bed of modern double-bed catalytic cracking units. In the region of *fast fluidization* large amounts of solids are entrained by the fluidizing gas. The regime is used in circulating bed reactors.

The last classic work in fluidization,[55] dating from 1971, contains more than 8000 pages and since then a wealth of additional data has appeared in literature. A detailed coverage of the phenomena in fluidization and of their modelling are obviously outside the scope of this work.

7.2. The Two Phase Theory of Fluidization

In the regime of fast bubbles, two distinct phases can be identified in a fluid bed (Fig. 7.2). In the two phase theory of fluidization it is postulated that gas flows through the emulsion phase at a rate corresponding to the minimum fluidization velocity. This assumption allows one to calculate the flow of emulsion gas through the bed.[50]

The relative velocity of gas and particles in the emulsion phase equals the minimum fluidization velocity:

$$u_e + u_s = \frac{u_{mf}}{\varepsilon_{mf}} \tag{7.1}$$

Here, u_{mf} is the minimum superficial fluidization velocity (m/h), u_e is the absolute interstitial (upward) emulsion gas velocity (m/h) and u_s is the absolute downward particle velocity in the emulsion phase (m/h). ε_{mf} is the porosity at minimum fluidization conditions.

The superficial velocity is the velocity of the gas when the fluid bed is free of solids. The interstitial velocity is the (real) velocity of the gas in the emulsion zone of the reactor.

The total flow rate of gas is then given by

$$\Omega u_0 = (\text{flow through emulsion phase}) + (\text{flow through bubbles})$$

$$u_0 = (1 - \delta - \alpha\delta)\varepsilon_{mf}u_e + (\delta + \delta\alpha\varepsilon_{mf})u_b \tag{7.2}$$

with δ = fraction of bed consisting of bubbles $= V_b/V_t$ (m^3 bubble); α = fraction of wake associated with bubbles $= V_w/V_b$ (m^{3}wake/m^3 bubbles); u_0 = superficial fluidization velocity (m/h); u_b = absolute bubble rise velocity (m/h).

The amount of solids moving upward (in the wake) must be equal to the downflow of solids in the emulsion phase:

$$\Omega(1 - \delta - \alpha\delta)u_s = \alpha\delta u_b\Omega \tag{7.3}$$

so that

$$u_s = \frac{\alpha\delta u_b}{1 - \delta - \alpha\delta} \tag{7.4}$$

Equations (7.1) to (7.4) allow δ to be calculated as:

$$\delta = \frac{u_0 - u_{mf}}{u_b - u_{mf} - \alpha u_{mf}} \tag{7.5}$$

The following parameters should be available: u_{mf}, α, u_0, ε_{mf} and u_b.

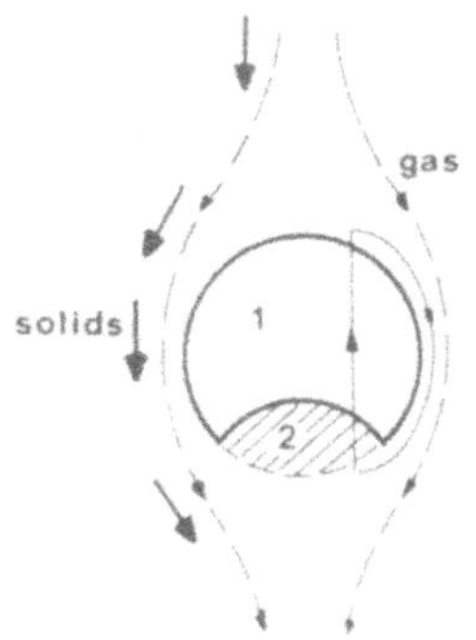

FIG. 7.3. Gas flow in the vicinity of bubbles.[50]

The minimum fluidization velocity can be estimated from correlations such as[55]

$$u_{mf} = 7{\cdot}90 \times 10^{-3} d_p^{1{\cdot}82} (\rho_s - \rho_f)^{0{\cdot}54} / \mu^{0{\cdot}88} \tag{7.6}$$

The bubble rise velocity is calculated from

$$u_b = u_0 - u_{mf} + 0{\cdot}711 \sqrt{g \,.\, d_B} \tag{7.7}$$

where d_B = bubble diameter (m) and g = gravitational acceleration (m/h^2).

The wake fraction can be estimated, e.g. from Kunii and Levenspiel.[50] Typical values are 0·2–0·4.

It is also possible to solve eqns (7.1) to (7.4) for u_e, the gas rising velocity in the emulsion phase:

$$u_e = \frac{u_{mf}}{\varepsilon_{mf}} - \left(\frac{\alpha u_0}{1 - \delta - \alpha\delta} - \alpha u_{mf} \right) \tag{7.8}$$

From this equation it follows that u_e may become negative, i.e. the gas in the emulsion phase flows down. This will occur when

$$\frac{u_0}{u_{mf}} > (1 - \delta - \alpha\delta) \left(1 + \frac{1}{\alpha \varepsilon_{mf}} \right) \tag{7.7}$$

7.3. Gas Flow Patterns

While the gas in the bubble phase is usually assumed to be in plug flow, various assumptions have been formulated concerning the flow in the emulsion phase (Fig. 7.4): a *dense or emulsion phase*, which forms the continuous phase of the bed, and a *lean phase* consisting of numerous discrete bubbles, which generally expand while rising, but may also coalesce or be split up under the action of bed internals. Each bubble has the form of a cap and is accompanied by a wake, i.e. solids moving upward together with the bubble. The bubble is surrounded by a cloud, which contains gas circulating to the bubble or back.

The most frequently used models are the well mixed flow of Davidson and Harrison (D-H) and the stagnant model of Kunii and Levenspiel (K-L), probably because of the simplicity of the resulting mass balances for individual components. The model of Kato and Wen,[53] featuring well mixed tanks in series, has also been applied to the modelling of gasifiers.

Much research work has been devoted to the selection of a suitable flow pattern in the emulsion phase. Yet the nature of the selected model does not seem to be critical unless high gas conversions in a single stage are sought. This is evidenced by results of Caram and Amundson[54] who studied the

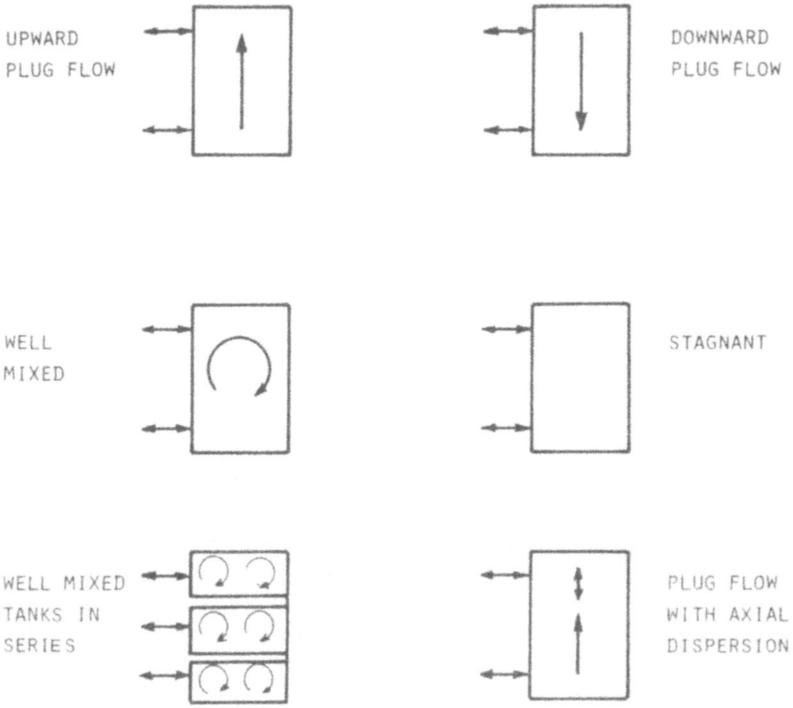

FIG. 7.4. Alternative schemes used in reactor models to represent axial dispersion of gas in the dense or emulsion phase.[52]

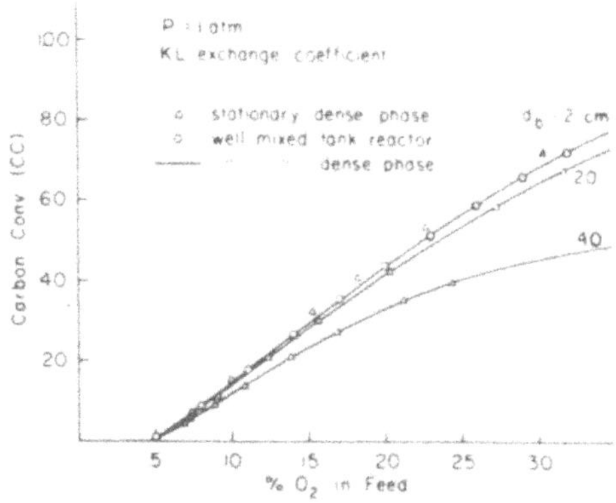

FIG. 7.5. Effect of flow patterns in the dense phase (1 atm).[54] Reproduced from Caram, H. S. and Amundson, N. P. (1979). *I & EC Process Des. Dev.* **18**, 80, with permission of the publisher. © 1979 American Chemical Society.

influence of the dense phase flow pattern by comparing well mixed emulsion phase, stationary emulsion phase and the case in which the whole reactor is treated as a continuous stirred tank reactor (CSTR), using the same gas interphase exchange coefficients. As can be seen in Fig. 7.5, only for small bubble diameters does there exist some (small) differences in carbon conversion.

7.4. Solids Mixing in Fluidized Beds

One of the main features of fluidized-bed reactors is their good mixing characteristics for solid fuels. In most of the models, the solid phase is therefore assumed to be perfectly mixed.

In practice, some provisions have to be made to ensure that this assumption is valid. The axial mixing of two types of material depends on different parameters. The most important are:

(1) particle size,
(2) particle density,
(3) ratio of actual to minimum fluidization velocity (u_0/u_{mf}).

The more the different kind of materials (on this occasion sand and biomass) differ in particle size and density, the more danger exists for segregation to take place. In this case, biomass will float on top of the sand bed. If this occurs, the possibility exists that not all the oxygen is consumed in the sand bed so that it will react with the fuel on top of the bed, causing locally high temperatures and eventually sintering of the sand.

The effect of density and size of the materials was quantified by Nienow and Rowe[79] in the formula:

$$M \sim \left(\frac{\rho_H}{\rho_L}\right)^{5/2} \left(\frac{d_L}{d_S}\right)^{1/5}$$

M is the 'degree of mixing' and varies from 0 for complete segregation to 1 for complete mixing. ρ_H and ρ_L are the density of the heavier, resp. lighter material and d_L and d_S are the particle size of the larger, resp. smaller particles. From this equation it follows clearly that the density is the more important parameter.

The obvious way of preventing particle segregation is a suitable location of the feeding point and a sufficient u_0/u_{mf} ratio. For light materials such as, e.g., wood shavings or sawdust the best feeding point is under the surface of the sand bed. The influence of the fluidization velocity on the mixing of sawdust or wood shavings and sand was investigated experimentally by Masson *et al.*[80] They found that segregation took place for ratios of u_0/u_{mf}

up to 3 for particles of 1·5 and 5·2 mm. Segregation increased with particle size and with the amount of fuel in the bed.

In order to avoid the formation of a supernatant layer of carbonized material Tels suggested the use of spouted fluidized beds.

Radial mixing is known to be a problem for large diameter reactors (1 m and more). For these reactors different feeding points are necessary.

7.5. Interphase Gas Transfer

Interphase mass transfer seems to be the most critical hydrodynamic parameter in establishing the performance of a fluid bed reactor. Unfortunately reliable quantitative data on this rate of mass transfer are still unavailable. In numerous models the interphase mass transfer coefficient is therefore treated as an empirical parameter used to fit the results computed with the model to the experimental results.

Interphase mass transfer is described by one or two transfer coefficients, depending on the model. In models with two phases (bubble and emulsion) the only resistance is situated at the bubble/emulsion interface. Mass transfer occurs here through a combination of convection and diffusion. According to Grace,[51] the correlation which shows the best agreement with experimental data for this transfer coefficient is

$$k_{bc} = 2u_{mf}/d_B + 12[D\varepsilon_{mf}\bar{u}_b/\pi d_B^3]^{1/2} \tag{7.8}$$

where k_{bc} is the bubble/cloud mass transfer coefficient based on bubble volume (m^3 gas/h m^3 bubble) and D is the molecular diffusivity (m^2/s).

In three phase models (e.g. the Kunii–Levenspiel model), an additional resistance exists at the cloud/emulsion boundary. According to Kunii and Levenspiel, mass transfer occurs only by diffusion, hence

$$k_{ce} = 6{\cdot}78[D\varepsilon_{mf}u_b/d_B^3]^{1/3} \tag{7.9}$$

k_{ce} is the cloud/emulsion mass transfer coefficient based on bubble volume (m^3 gas/h m^3 bubble).

Grace argues that this formula seriously underestimates the cloud-emulsion mass transfer. Additional mass exchange can be caused by:

(a) adsorbed gas entering and leaving the cloud together with the adsorbing particles,
(b) shedding of wake material,
(c) cloud volume changes caused by bubble interaction and coalescence.

According to the same author it is even plausible that the principal mass

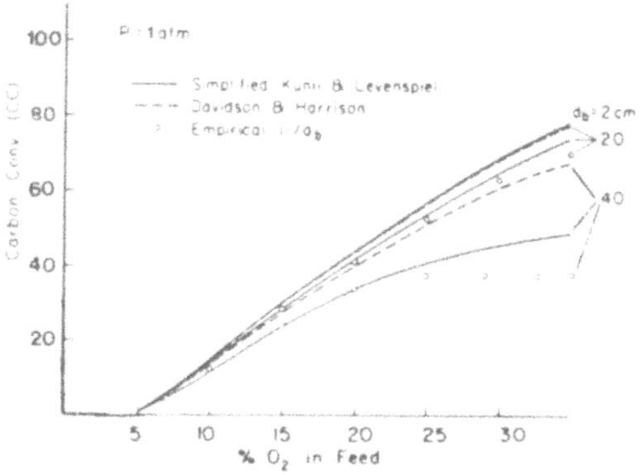

FIG. 7.6. Effect of bubble–emulsion exchange coefficient in the carbon conversion at 1 atm.[54] Reproduced from Caram, H. S. and Amundson, N. P. (1979). *I & EC Process Des. Dev.* **18**, 80, with permission of the publisher. © 1979 American Chemical Society.

transfer resistance is not at the cloud/emulsion interphase, but rather at the bubble/cloud interphase. Then it would be more sensible to include the cloud gas in the emulsion phase rather than in the bubble phase.

The influence of the interphase mass transfer coefficient on gasifier performance was also investigated by Caram and Amundson.[54] They compared the exchange coefficient predicted by the D-H model, the K-L model and an empirical exchange coefficient proposed by Kobayashi.[56]

For small bubbles, no appreciable differences exist. For large bubble sizes differences in carbon conversion are considerable (Fig. 7.6). The K-L model gives lower conversions because of the higher resistance to mass transfer predicted by this model.

7.6. Importance of Hydrodynamic Parameters

For *slow reactions*, the gas exchange between bubbles and emulsion is fast compared to the rate of reaction so that no concentration gradients will be created. This may be the case for the heterogeneous gasification of C with steam, carbon dioxide or hydrogen. The case of fast reactions, e.g. C with oxygen, will be discussed later.

Yoshida and Kunii[57] used the K-L approach for modelling a char gasifier with a comparatively slow C/H_2O reaction. The conversion of a reactant (e.g. steam) may be described by

$$u_b \frac{dC^b_{H_2O}}{dx} = K_{H_2O} C^b_{H_2O} \tag{7.10}$$

with the effective reaction rate coefficient K_{H_2O} given by

$$K_{H_2O} = k'\left[\gamma_b + \frac{1}{(k'/k_{bc}) + \{\gamma_c + [(k'/k_{ce}) + (1/\gamma_e)]^{-1}\}^{-1}}\right] \quad (7.11)$$

in which k' = first order rate coefficient for the C/H_2O reaction ($m^3\,H_2O$/m^3 carbon), γ_b, γ_c, γ_e = fractions of carbon in the bubble, cloud or emulsion phase (m^3 carbon/m^3 gas).

When $k' \ll k_{bc}$ and k_{ce}, eqn (7.11) reduces to

$$K_{H_2O} = k'(\gamma_b + \gamma_c + \gamma_e)$$

i.e. the apparent reaction rate equals the true reaction rate over the complete reactor. Hence there is no effect of the hydrodynamic parameters.

Consider the rate coefficient of Rensfelt[27] for the reaction of steam with poplar wood char at 900 °C.

$$k' = 3{\cdot}83\,\frac{m^3\,H_2O}{m_C^3\,s}$$

An estimate of k_{be} (which according to Grace could be the lower of the two interchange coefficients) is given by eqn (7.8) for particles of a size 0·36 mm and fluidized by air at 900 °C at a fluidizing velocity $u_0 = 5 \times u_{mf}$.

$$u_{mf} = 0{\cdot}045 \quad m/s \qquad \text{(from eqn (7.6))}$$

$$u_0 = 0{\cdot}225 \quad m/s$$

$$u_b = 0{\cdot}18 + 2{\cdot}23\sqrt{d_B} \quad m/s \qquad \text{(from eqn (7.7))}$$

This gives for $d = 0{\cdot}05$ m $\quad k_{bc} = 4{\cdot}72\,s^{-1}$

$d = 0{\cdot}1$ m $\quad k_{bc} = 1{\cdot}38\,s^{-1}$

$d = 0{\cdot}4$ m $\quad k_{bc} = 0{\cdot}24\,s^{-1}$

In this particular example, the reaction rate coefficient is of the same order of magnitude as the interphase mass transfer coefficient. It follows that only a two phase model is appropriate.

It appears that the rate coefficient reported by Rensfelt is rather high. Values for the less reactive coal chars are typically 100–300 times lower whereas our own results for wood charcoal are roughly 20 times lower. This example shows, nevertheless, that neglecting interphase mass transfer may lead to errors for larger sized bubbles.

For *extremely fast reactions*, on the other hand, the conversion is

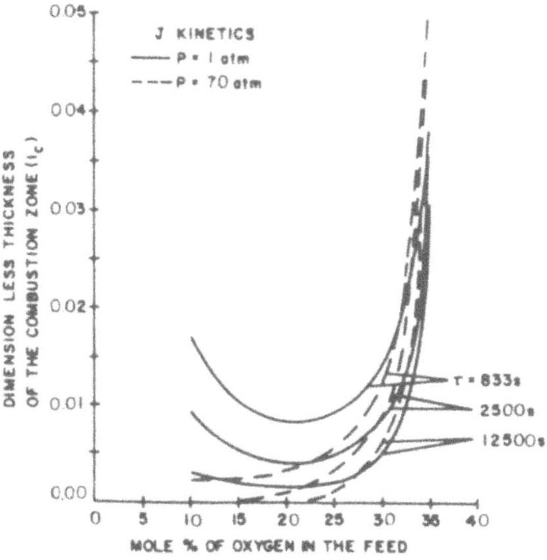

FIG. 7.7. Dependence of the height of the combustion zone on the oxygen content of the feed gas and the residence time (τ) of the solids.[31a] Reproduced from Sundaresan, S. and Amundson, N. (1979). *Chem. Eng. Science* **34**, 359–64, by permission of the publisher.

essentially completed in the inlet region immediately above the distributor. This is the basis for some models which assume that all oxygen is consumed in a small zone above the distributor.

The height of this combustion zone was evaluated by Sundaresan and Amundson[31a] (Fig. 7.7). At normal operating conditions (oxygen concentrations below 30 %, solids residence time 12 500 s = 3·5 h), its height is always below 1 % of total bed height. Weimer and Clough[58] obtained comparable results.

7.7. Bubble Size

The average size of the bubbles is a major parameter in the modelling of fluid bed reactors, because this parameter appears in numerous empirical correlations, e.g. those relating to the bubble rise velocity and the rate of gas exchange between the bubbles and the emulsion phase. Smaller bubbles improve the transfer of gas between the two phases.

Although the bubbles continuously increase in size while rising from the distributor to the freeboard zone their diameter d_B is assumed to be constant in most models. Numerous empirical correlations are available for predicting this bubble size d_B but usually their field of application is restricted to the particular range of experimental conditions for which they were determined. The rate of generation and fate of the bubbles is somewhat unpredictable since they may coalesce or be broken up by internals, such as the heat exchanger tubes of a fluidized bed boiler. Also the type of distributor is vitally important.

In numerous cases d_B is used an an empirical parameter used to fit the computed to the experimental results.

7.8. Solids Population Balance and Carbon Holdup

A basic difference between reactor operation for coal gasification and biomass gasification is the absence of an overflow for char in the bed. This is a consequence of the low ash content and the fine particulate nature of the biomass ash which strips off from a reacted particle and is removed readily as fly ash.

When the feed rate of solid fuel is altered (all other parameters remaining constant) the bed height, the carbon content of the bed and the gasifying capacity will tend to a new steady state. The rate of gasification is roughly proportional to the amount of carbon present, so that an increase in feed rate also enhances carbon holdup, rate of reaction and capacity.

There is, however, a practical limit: when the carbon content of the bed becomes too high, the characteristics of fluidization will no longer be determined by the bed material alone so that poor mixing, segregation and eventually loss of fluidization may occur. Hence, it is important to have a clear idea of the acceptable carbon content of the bed and on the relation between carbon, air and steam flow rates and carbon holdup.

The modelling of a fluidized bed gasifier rests on equations for

—mass balances for each species in each phase,
—energy balances for each phase.

Most of these equations contain reaction rate terms. As was already mentioned in Section 5, the rate of reaction can depend on the conversion (in the case of a volumetric reaction) or on particle size (in the case of a shrinking core reaction).

In a fluid bed reactor, which ideally acts as a perfect mixer for the solid particles, particles with different residence times and thus with different conversion levels and/or particle size will be present. The gas will thus contact particles with a different reactivity (dependent on particle size and/or conversion). In order to compute a weighted rate of reaction it is required to determine the size distribution of the solids in the reactor.

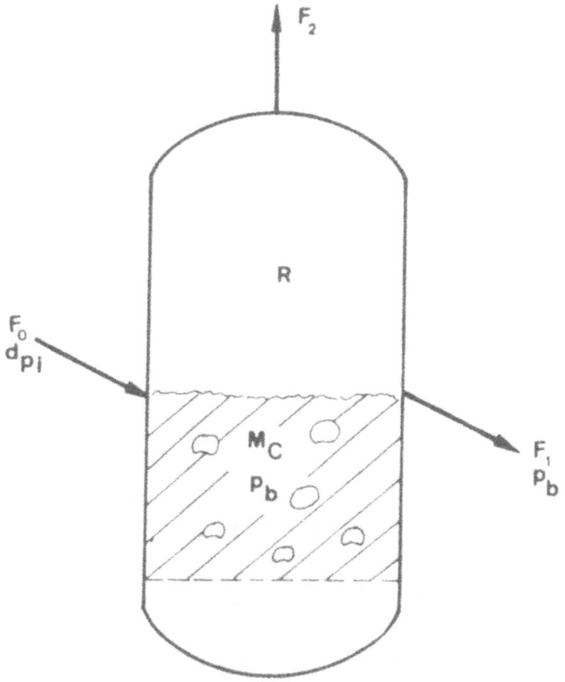

FIG. 7.8. Fluid bed with overflow and entrainment.[50]

In the case of a perfectly mixed reactor with continuous in- and outflow (e.g. by a weir or discharge pipe) and without reaction, attrition or elutriation, the particle residence time distribution is given by

$$P_b(t) = \frac{1}{\bar{t}} \exp(-t/\bar{t})$$

In this equation $P_b(t)$ is the fraction of particles having a residence time between t and $t + dt$. $\bar{t}$ is the average residence time.

When a particle reacts according to a shrinking core reaction in a reactor with in- and outflow, with particle entrainment, the particle size distribution $P_b(d_p)$ has to be known.

This theory was developed by Kunii and Levenspiel.[50,59]

For a uniformly sized feed of solids, the normalized particle size distribution $P_b(R)(1/s)$ in the bed is given by (Fig. 7.8)

$$P_b(d_p) = \frac{F_0}{W/R(d_p)} \Bigg/ \frac{d_P^3}{d_{Pi}^3} I(d_P, d_{Pi}) \tag{7.12}$$

with F_0 = the feed rate of solids (kg/h), W = the weight of reacting solids in the bed (kg), $R(d_p)$ = the rate of growth or shrinkage (m/h), d_P = the particle diameter (m), d_{Pi} = the diameter of feed particles (m).

$I(d_p, d_{Pi})$ is given by

$$I(d_P, d_{Pi}) \exp\left[\int_{d_{Pi}}^{d_P} \frac{F_1/W + K(d_P)}{R} \, \mathrm{d}.d_p\right] \tag{7.13}$$

in which F_1 = overflow (kg/h) and $K(d_P)$ = the elutriation rate constant (1/h).

Normalization of the size distribution function $P_b(R)$ gives

$$\frac{W}{F_0} = \int_0^{d_{Pi}} \frac{d_P^3}{R(d_P)d_{Pi}^3} I(d_P, d_{Pi}) \, \mathrm{d}d_{Pi} \tag{7.14}$$

Equations (7.12)–(7.14) can be solved when the reaction rate $R(d_P)$ is known. In general, this rate will depend on the concentrations of the reacting gases. Hence eqns (7.12)–(7.14) have to be solved together with the mass and energy balance over the gas phase.

The same authors derived expressions for the case of a feed of solids with a wide size distribution.

The above treatment is valid only for shrinking core-type of reactions. When the reaction proceeds homogeneously throughout the particle, the particle size stays constant and a different treatment is necessary.

The rate of reaction of a particle undergoing a homogeneous conversion can generally be expressed as

$$r = F(X)f(T_i, P_i)$$

when X is the conversion of the particle. Since X is variable, $F(X)$ should be averaged over all particles to obtain an average rate of reaction. For a homogeneous reaction model with a reaction rate proportional to the mass of residual carbon, $F(X)$ is proportional to $(1 - X)$ so that

$$\overline{F(X)} = F(\overline{X})$$

In the case where the rate of elutriation is independent of the conversion it is not necessary to establish a population balance.

In general $\overline{F(X)} \neq F(X)$ and treatments similar to the one developed by Kunii and Levenspiel should be followed on a basis of conversion rather than particle size.

An analysis in which both the particle size and the conversion vary was performed by Chen and Saxena.[60]

As was already mentioned, biomass gasifiers, in contrast to coal gasifiers, usually operate without outflow of solids.

This situation can be coped with easily in the solids population balances mentioned earlier by neglecting the outflow term F_1/W in eqn (7.13).

7.9. Survey of Literature Models

Again the models are classified in a rising order of sophistication. This gives the following sequence:

(1) thermodynamic models;
(2) single phase models based on either a well mixed (CSTR) or a plug flow (PF) reactor;
(3) two or three phase models featuring a distinct type of gas flow in the dense and the lean phase, together with some form of interphase mass transfer;
(4) transient models.

A survey of the models for fluid bed gasifiers which have appeared in recent literature is given in Table 7.I.

As can be seen, the underlying assumptions of the different models are extremely varied. This makes an evaluation quite difficult especially because here also experimental data are limited.

The majority of the models are for steady state conditions with a tendency in recent years towards the development of non-isothermal models.

The kinetics for the gas–solid reactions have different origins. The most favoured are those of Johnson.[28] His semi-empirical expression was developed for bituminous coals. It takes into account the following reactions:

$$C + H_2O = CO + H_2$$

$$C + 2H_2 = CH_4$$

$$H_2 + H_2O + 2C = CO + CH_4$$

Moreover, allowance is made for a change of reactivity with conversion according to the equation given in Table 3.II. It should be stressed here that the Johnson reaction kinetics were set up for coal and are not necessarily applicable to biomass.

Most other models also assume the water–gas shift reaction to be in equilibrium and the C/O_2 reaction to be instantaneous.

TABLE 7.I
SELECTED FLUIDIZED BED GASIFIER MODELS

Authors	*Steady state (SS) or transient (Tr)*	*Isothermal (Iso or non-isothermal (N. Iso)*	*Reaction kinetics* Gas–solid reactions	*Reaction kinetics* Water–gas shift reaction	*Combustion kinetics* Instantaneous = Inst.	*Fluid-bed hydrodynamics*
Yoshida and Kunni[57]	SS	Iso	Mayers[65]	Equilibrium	Parker and Hottel[72]	K-L
Caram and Amundson[54]	SS	N. Iso	Johnson[28] Gibson and Euker[66]	Equilibrium	Inst.	CSTR K-L D-H
Caram and Amundson[61]	SS	N. Iso	Johnson[28]	—	Inst.	CSTR
Sundaresan and Amundson (1979)	SS	N. Iso	Johnson[28]	Equilibrium	Inst.	CSTR
Sundaresan and Amundson (1979)	SS	N. Iso	Johnson[28]	Equilibrium	Inst.	DH (modified)
Sundaresan and Amundson (1979)	SS	N. Iso	Johnson[28]	Equilibrium	Finite	PF
Purdy *et al.*[62]	SS	Iso	Johnson[28]	Wen[69]	Inst.	CSTR
Weimer and Clough[58]	SS	N. Iso	Shrinking core	Karim[70] Winslow[71]	Khaustovich[73]	Two phase + jetting
Weimer and Clough[63]	Tr	Iso	—	—	—	Gas phase comp. assumed
Weimer and Clough[34]	Tr	N. Iso	Dutta Hedden Intern. diff.	—	Dutta[74]	Gas phase comp. assumed
Raman *et al.*[64]	Tr	Iso	Biba[43]	Biba[43]	No O_2 fed	Plug flow[a]

[a] Plug flow with negligible axial diffusion in bubble phase and infinite axial dispersion in emulsion phase.

7.9.1. Equilibrium Models

In a fluid bed gasifier the temperature is fairly uniform, but rather low (750–900 °C). Heterogeneous chemical equilibria will usually not be attained and can only serve to set the trend of the variations in gas composition with the operating parameters. Homogeneous equilibrium is more likely to be established; it is generally accepted that the water–gas shift reaction is essentially in equilibrium above 800–850 °C.[75] The deficiencies of equilibrium models have been discussed in the section on moving bed gasifiers.

7.9.2. Single Phase Models

The simplest possible assumption is that of a well mixed reactor. The resulting molar (or mass) balances can be written for both the reacting gas and the carbon under the form of simple algebraic equations:

$$N_i - N_{i0} = \sum_j \nu_{ij} R_j \qquad \text{(gas)}$$

$$N_{\text{carbon}} = \sum_j R_j + N_{\text{overflow}} + N_{\text{elutriated}} \qquad \text{(carbon)}$$

These equations can be solved readily with specific rate equations and, when desirable, a population balance incorporated. A heat balance can also be considered. Models of this type were used, *inter alia*, by Sundaresan and Amundson,[31b] Purdy *et al.*[62] and Caram and Amundson.[54]

Purdy *et al.*[62] used the rate equation of Johnson simplified by setting $\overline{F(X)} = F(\overline{X})$. The kinetics of the water–gas shift reactor were also taken into account.

Results computed with this model were compared to experimental data on the gasification of devolatilized bituminous coal in a pressurized fluidized bed.

The model contains three adjustable parameters used to fit the model to the experimental data. These parameters are: the reactivity of the solids, the CO/CO_2 ratio at the end of the combustion zone, and the water–gas shift reactivity.

Eventually these parameters were used to predict the conversion of the gas. The results were fairly good (Fig. 7.9). The model was also used to predict the ratio $(CO_2)(H_2)/(CO)(H_2O)$. Here the agreement was rather poor and the authors ascribe this to a departure from isothermicity in the bed, with deviations in the axial temperature as large as 50 °C (Fig. 7.10).

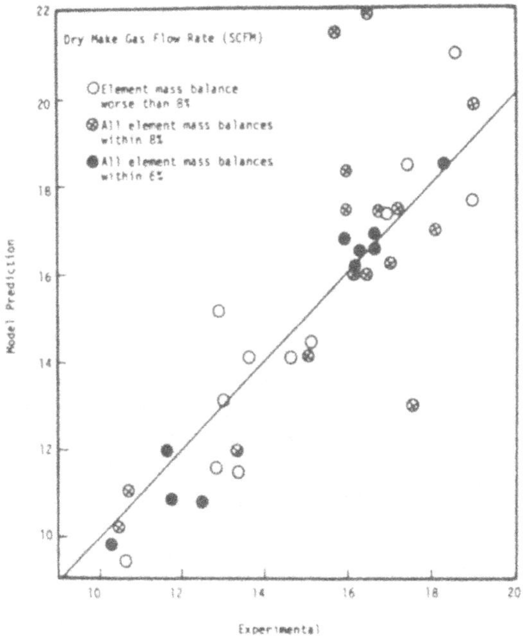

FIG. 7.9. Predicted vs. experimental dry make gas flow rate.[62] Reproduced from Purdy, M. J. *et al.* (1981). *I & EC Process Des. Dev.* **20**, 675, with permission of the publisher. © 1981 American Chemical Society.

(*a*) *Influence of kinetic models.* Sundaresan and Amundson[31b] compared results obtained using Johnson kinetics with others using a pseudo first order homogeneous model, proposed by Gibson and Euker. The model features a well mixed reactor with a population balance for the carbon particles.

Assuming that Johnson kinetics (which are based on a larger number of data) give a better representation of physical reality they conclude that Gibson/Euker kinetics are unfit for predicting reactor performance, especially at higher pressures.

Caram and Amundson[61] calculated the difference between $\overline{F(X)}$ and $F(\overline{X})$ for Johnson-type kinetics and found only small differences when the average conversion is low. The differences became appreciable for higher $\bar{X}$ values (typically $>0{\cdot}4$) but depended upon the value of X, an empirical constant used in the rate expression, and of X_0, the initial conversion of the feed.

This initial conversion, however, is only useful if staged fluidized beds are considered since in normal operation inlet conversions of fuel are zero.

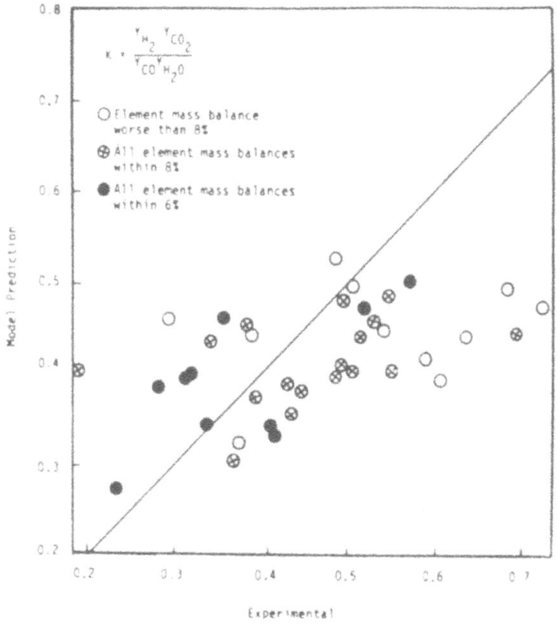

FIG. 7.10. Predicted vs. experimental K values.[62] Reproduced from Purdy, M. J. *et al.* (1981). *I & EC Process Des. Dev.* **20**, 675, with permission of the publisher. © 1981 American Chemical Society.

(*b*) *Comparison between CSTR, PF and two phase models.* Sundaresan and Amundson[31a] compared the CSTR model to plug flow, also for Johnson-type kinetics. The plug flow model predicted higher conversions of carbon (Fig. 7.11). This can be ascribed to the inhibiting effect of H_2 on gasification by steam (Section 5), which is more pronounced in a CSTR reactor. This conclusion can be generalized to all cases based on Langmuir kinetics with inhibition by the reaction products.

In another paper, Sundaresan and Amundson[31c] compared the CSTR model to a modified Davidson–Harrison model assuming a stationary dense phase and Johnson kinetics.

They found no appreciable difference in the total carbon conversion for average bubble sizes smaller than 10 cm (Fig. 7.12). For larger bubble sizes transfer of reactant gas from the bubbles to carbon in the dense phase becomes limiting which results in lower conversions of carbon. These results were obtained with a rather deep bed (2·5 m). The effect of bubble size is expected to be still more pronounced for lower bed heights.

Similar results were obtained by Mori *et al.*[76] who compared results of

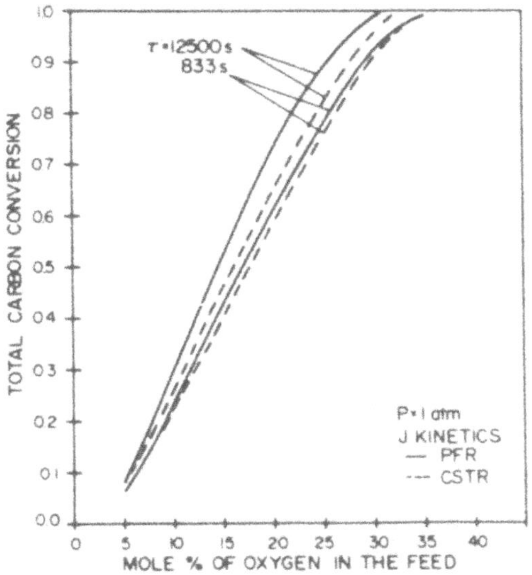

FIG. 7.11. Comparison between CSTR and plug flow.[31a] Reproduced from Sundaresan, S. and Amundson, N. (1979). *Chem. Eng. Science* **34**, 359–64, with permission of the publisher.

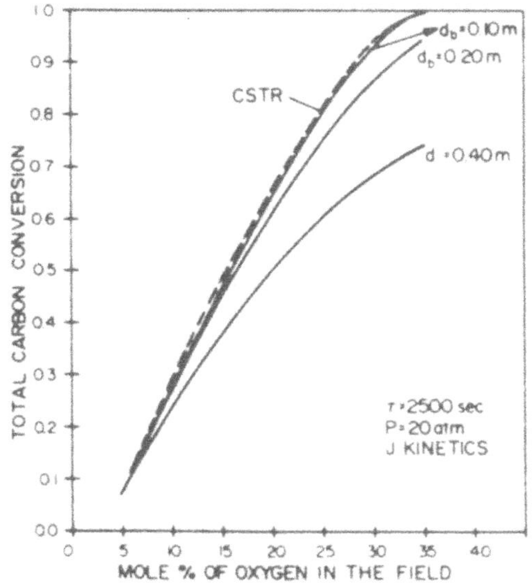

FIG. 7.12. Comparison between a CSTR and a two phase model.[31c] Reproduced from Sundaresan, S. and Amundson, N. (1979). *Chem. Eng. Science* **34**, 359–64, with permission of the publisher.

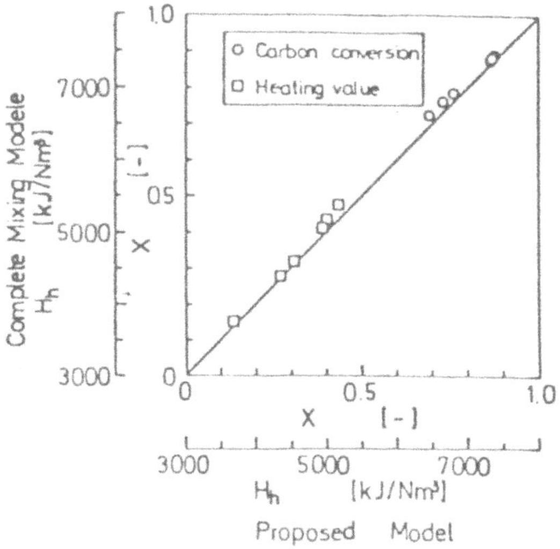

FIG. 7.13. Comparison between CSTR and two phase (bubble assemblage model).[76]

their two phase bubble assemblage model and a CSTR model. No appreciable differences were found with regard to the predicted carbon conversion and the gas heating values (Fig. 7.13).

The bubble size used in this model was fairly small and varied between 0·02 and 0·07 m with rising height.

7.9.3. Two Phase Models

Considering the large number of distinct gas flow patterns it is no longer feasible to write generalized mass and energy balances for these models.

The two models which have been used most frequently are the Kunii–Levenspiel (K-L) and the Davidson–Harrison (D-H) models.

The D-H model considers a bubble and an emulsion phase; the K-L model adds a third phase, the cloud phase. In the D-H model the dense phase is well mixed, whereas it is stationary (i.e. without in- or outflow of gas) in the K-L model. Both assume plug flow for the bubble phase. The models further differ in the correlation used for interphase mass exchange (see Section 7.4).

Yoshida and Kunii[57] applied the K-L model to the gasification of carbon with either air/oxygen mixtures or steam. Pseudo first order reaction kinetics were used for the C/O_2, C/CO_2 and C/H_2O reactions. The homogeneous water–gas shift reaction is in equilibrium. Figure 7.14 shows

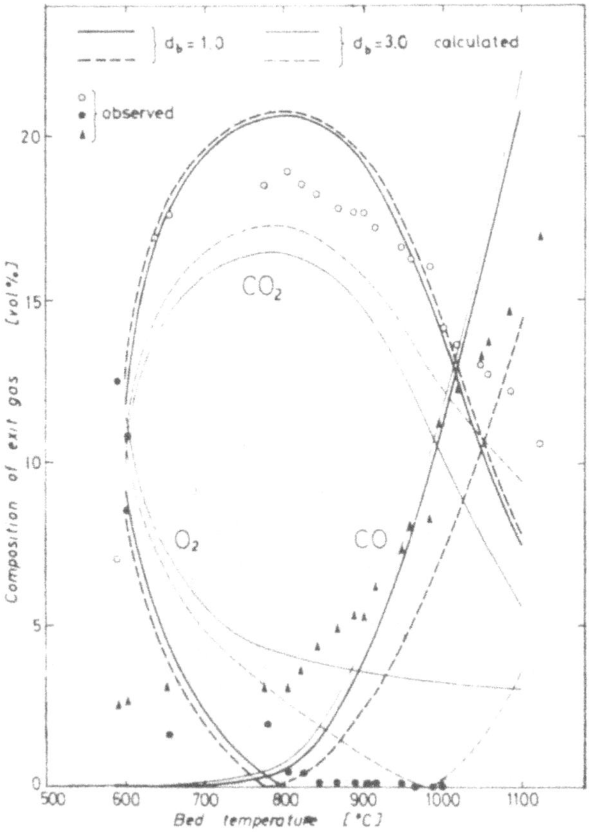

FIG. 7.14. Comparison between model calculations for gasification of carbon with CO_2. $d_t = 8{\cdot}0$ cm, $L_f = 8{\cdot}3$ cm, $d_p = 0{\cdot}771$ mm, $u_0 = 11{\cdot}8$ cm/s, $u_{mf} = 32$ cm/s, $\varepsilon_{mf} = 0{\cdot}50$, $\alpha = 0{\cdot}2$ (NTP). Reproduced from Yoshida, K. and Kunii, D. (1974). *J. Chemical Engineering Japan* 7(1), 34–9, with permission of the publisher.

a comparison of their results with experimental data. The dotted lines correspond to calculations in which CO and O_2 are not coexistent.

Weimer and Clough used a modified D-H model with a separate jetting region close to the distributor. Their model features pyrolysis, shrinking core kinetics for the heterogeneous reactions, and separate rate expressions for the catalyzed and the homogeneous water–gas shift reaction. The former takes place in the fluid bed, the latter in the freeboard zone.[58]

Allowance was made for distinct temperatures of the particles, bubble gas and the emulsion gas.

They also investigated the influence of fluidizing velocity, gas

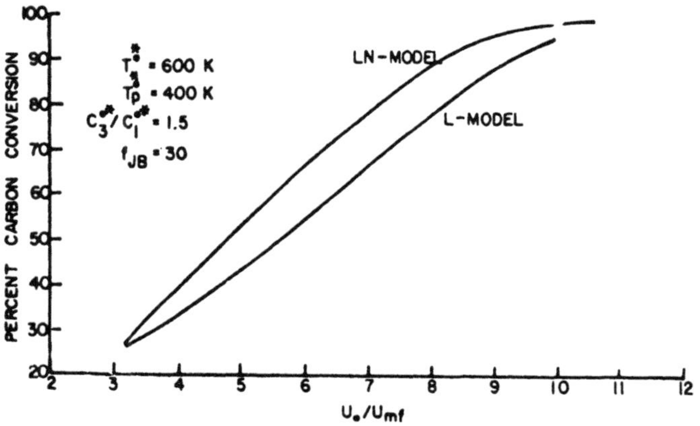

FIG. 7.15. Effect of u_0 on carbon conversion.[58] Reproduced from Weimer, A. W. and Clough, D. E. (1981). *Chem. Eng. Science* **36**, 549–67, with permission of the publisher.

composition, jetting-emulsion interchange, carbon conversion, average bed temperature and H_2/CO ratio in the product gas.

Figure 7.15 shows the influence of the fluidizing velocity u_0 upon the conversion of carbon. As expected the conversion increases with u_0 (the O_2/steam ratio is fixed). In this figure the L-model refers to computations in which O_2 is allowed to react with CO and H_2 in the gas phase, the LN-model to others in which such reactions do not occur. The L-curves give lower conversions because of lower O_2 concentrations in the emulsion phase.

The effect of the oxygen content in the feed gas (the balance being steam) on the average bed temperature is given in Fig. 7.16.

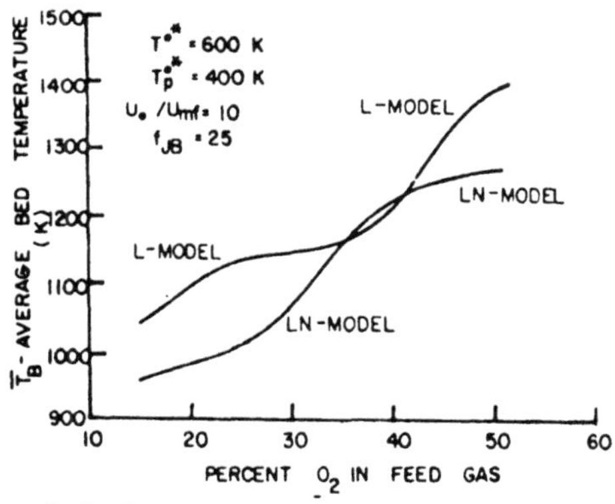

FIG. 7.16. Effect of feed gas composition on average bed temperature.[58] Reproduced from Weimer, A. W. and Clough, D. E. (1981). *Chem. Eng. Science* **36**, 549–67, with permission of the publisher.

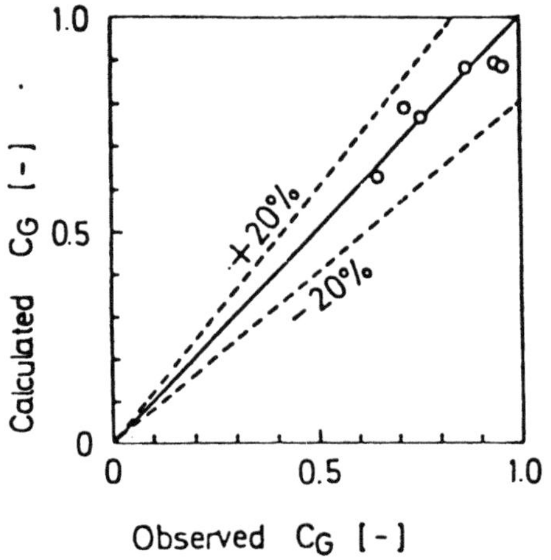

FIG. 7.17. Comparisons of calculated and observed carbon conversions.[76]

Mori *et al.*[76] used the bubble assemblage model of Kato and Wen[53] to fit their experimental data obtained in a pressurized air/steam gasifier. They used their own kinetic data for gas/solid reactions and measured the amount and composition of pyrolysis gas. The water–gas shift reaction was assumed to be in equilibrium.

In the bubble assemblage model the bed is subdivided into a number of compartments, the height of which is dependent on bubble size and jetting

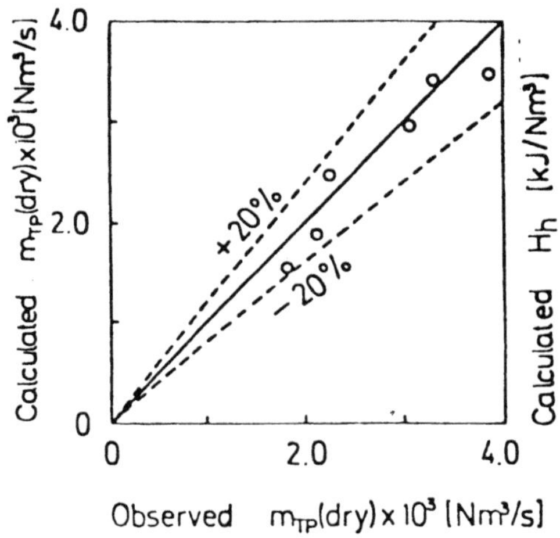

FIG. 7.18. Comparison of calculated and experimental total flow rates of dry produced gas.[76]

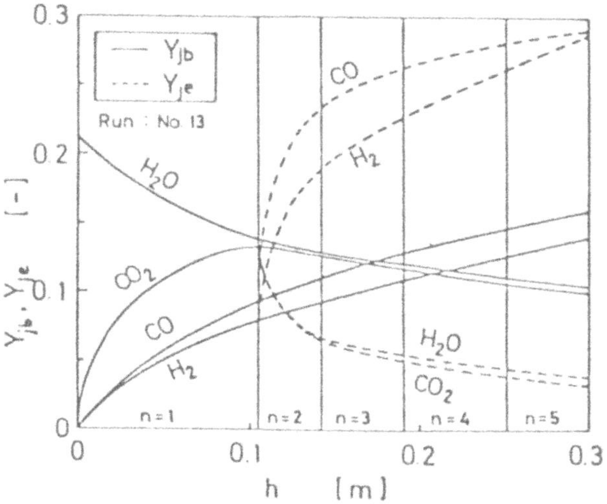

FIG. 7.19. An estimated profile of the gas compositions in the bubble and emulsion phase.[76]

height. Each compartment contains both an emulsion and a dense phase which are well mixed.

Results computed with this model show good agreement with the experimental data (Figs 7.17 and 7.18). The gas composition calculated as a function of height shows considerable differences between the bubble and the emulsion phase (Fig. 7.19). As mentioned earlier the authors found no appreciable differences for carbon conversion and gas heating value when using a CSTR model. This was attributed to the low rates of char gasification.

7.9.4. Transient Models

Weimer and Clough[63] analysed the transient behaviour of an isothermal fluid bed gasifier. They developed a general population balance for homogeneously reacting particles of a constant size.

Their model included elutriation and overflow. Not included were the hydrodynamic behaviour and the mass balances over the gas phase. The gas phase composition was varied as a parameter.

The same authors[34] later extended their model to non-isothermal systems with solids changing in both conversion and size.

The difference between particle and gas temperatures was also considered. The particle temperature was found to be 5–15 °C higher than

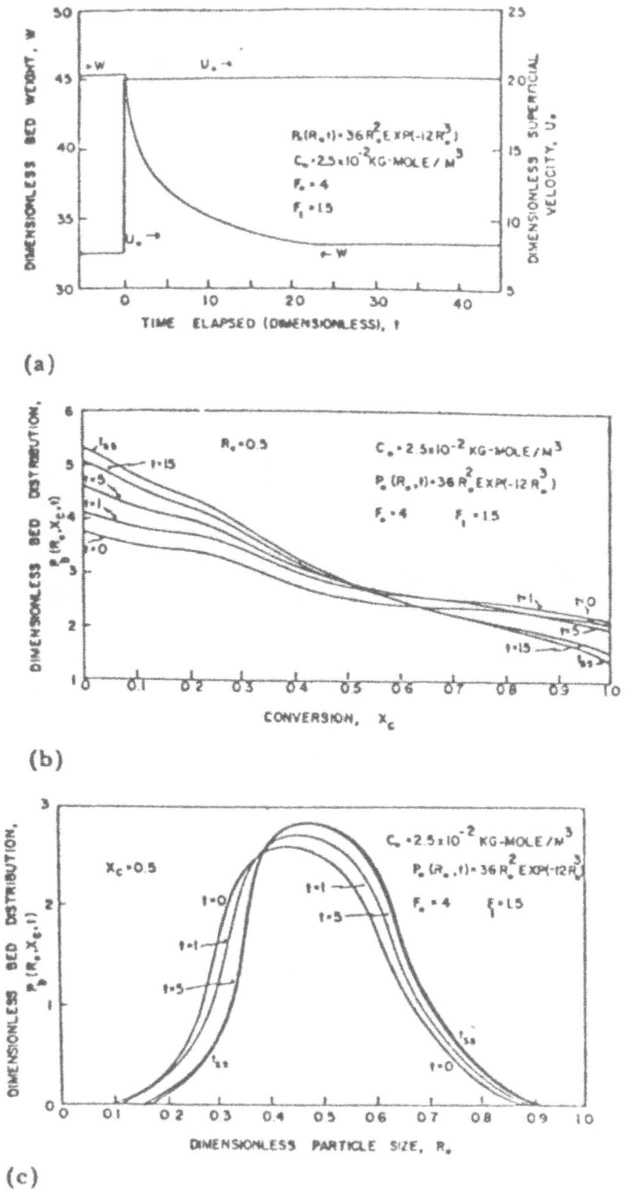

FIG. 7.20. (a) Transient response of solids holdup to a step increase in superficial gas velocity. (b) Transient response of particle size/conversion distributions to a step increase in superficial gas velocity (for $R_0 = 0{\cdot}5$). (c) Transient response of particle size/conversion distributions to a step increase in superficial gas velocity (for $X_c = 0{\cdot}5$).[58] Reproduced from Weimer, A. W. and Clough, D. E. (1980). *Powder Technol.* **27**, 85–103, with permission of the publisher.

TABLE 7.II
PARTICLE TEMPERATURE FOR $X_0 = 0{\cdot}5$[34]

Dimensionless particle size, R_0	*Temperature*, K
0·025 5	1 240·1
0·129	1 238·5
0·297	1 237·3
0·5	1 236·6
0·703	1 236·2
0·871	1 236·0
0·975	1 235·9
1·0	1 235·8

$P_0(R_0) = 36R_0^2\exp(-12R_0^3)$, $u_0 = 7{\cdot}5$, $F_0 = 4$, $W = 40$.

the average gas temperature. The temperatures of the particle increased with decreasing particle size (Table 7.II).

Their calculations further showed that internal temperature gradients were negligible for particles below a diameter of 7 mm.

The response to a 170 % increase in u_0 is shown in Fig. 7.20. The bed weight gives a rapid response. The new steady state is reached after twenty times the average residence time of the solids (W/F_0).

Elutriation losses are varying almost instantaneously.

After a time period of 5τ, the distribution of the curve has reached about 56 % of its ultimate value, while particle size distribution responds more rapidly.

The effect of a 25 % step decrease in solid flow rate is shown in Fig. 7.21.

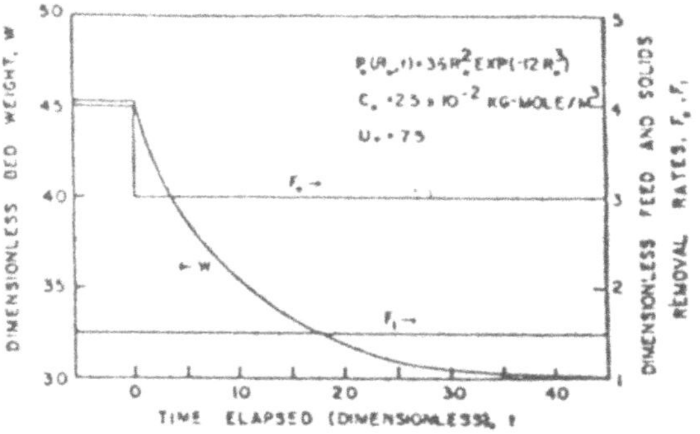

FIG. 7.21. Transient response of solids holdup to a step decrease in solids feed rate.[58] Reproduced from Weimer, A. W. and Clough, D. E. (1980). *Powder Technol.* **27**, 85–103, with permission of the publisher.

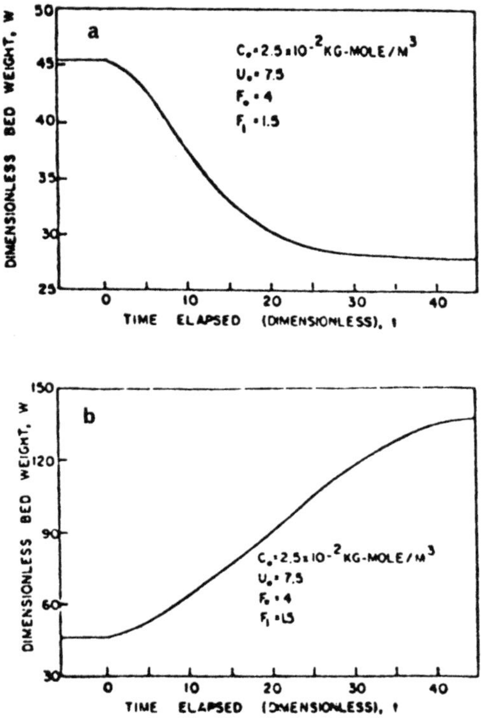

FIG. 7.22. Transient response of solids holdup to a step change in solids feed distribution.[58] (a) Smaller size feed, (b) larger size feed. Reproduced from Weimer, A. W. and Clough, D. E. (1980). *Powder Technol.* **27**, 85–103, with permission of the publisher.

The change of the carbon holdup is much slower than when changing the fluidization velocity. The effect of changing the feed size distribution towards larger and smaller particles was also studied. The effect of this factor on bed weight is shown in Fig. 7.22(a) and (b). Shifting towards a smaller particle size results in a lower bed weight because of the larger number of particles elutriated and the higher rate of reaction.

Raman *et al.*[64] proposed an isothermal transient two phase model for the gasification of feedlot manure. For computational purposes, the gas flow in dense and bubble phase were represented by plug flow with axial dispersion. The axial dispersion coefficient was selected so that the bubble phase was almost in plug flow and the dense phase almost completely mixed.

A distribution of pyrolysis products was established experimentally. Pyrolysis is instantaneous and occurs homogeneously over the fluid bed.

TABLE 7.III
COMPARISON OF CALCULATED AND EXPERIMENTAL RESULTS

	Calcd, %	*Exptl,* %	*Calcd,* %	*Exptl,* %
H_2	8·27	14·96	5·97	9·91
N_2	65·0	56·87	68·36	59·92
CH_4	1·51	2·75	2·46	3·21
CO	2·8	7·83	1·35	8·59
CO_2	22·01	15·02	21·19	16·19
C_2H_4	0·06	1·25	0·10	1·2
C_2H_6	0·34	0·21	0·58	0·39
Produced gas yield, % DAF	56·3	65·0	22·7	31·6
Liquid yield, % DAF	24·5	20·2	55·3	50·7
Char, % DAF	19·0	14·8	22·0	17·7
Operating temperature, K	983		869	

Kinetic expressions for the C/CO_2, C/H_2O, C/H_2 and the water–gas shift reaction were taken from Biba *et al.*[43]

Their model predictions were compared with experimental data (Table 7.III). Differences were ascribed to the omission in the model of secondary gas phase cracking of heavy pyrolysis products and possible differences between the char yield used in the calculations (from TGA experiments) and actual yields in a fluid bed (with higher heating rates).

No transient responses are given although the model is a transient one. The authors state that the composition of the gas reaches a steady state in about 20 s.

8. OTHER TYPES OF GASIFIERS

Apart from the moving bed and the fluidized bed gasifier several other reactor types are possible. They can be classified as follows on the basis of increasing intensity of mixing:

—rotary kiln gasifier,
—spouted bed gasifier,
—lean phase or entrained bed gasifier.

In a rotary kiln gasifier the contact between the slowly tumbling charge and the gas phase is relatively limited so that the gasification of the char remains incomplete unless very long contact times are provided. Rotary kiln gasification was applied on municipal refuse by Monsanto Co. After initial success at a pilot scale a gigantic single-train plant was installed in Baltimore, Md. The extrapolation to a much larger scale (900 tonnes/day) led to complete failure. The plant could only be operated at about half capacity and after fairly extensive revamping. After a couple of years of trial and error it was abandoned altogether. Obviously a good model was not available, or more probably it was extrapolated beyond its limits of validity!

A multiple hearth gasifier is especially adapted to a relatively fine feedstock. The contact surface between solids and gas is renewed periodically by a set of rotary rabble arms, operating on each successive floor. Normal operation is in countercurrent but the gas flow can also be organized in other ways. Garrett Energy and Research Corp.[76] is developing a gasification process featuring a multiple hearth furnace. An obvious reason for the lack of success of multiple hearth furnaces is its cost and sensitivity to thermal shock.

The spouted bed gasifier is a variant of a fluidized bed gasifier, featuring one or more spouts through which also coarse material can be introduced into the bed. Spouted bed gasifiers are capable of mixing materials of widely different density and particle size. Tels and collaborators[77] have used a spouted bed reactor in the pyrolysis of biomass.

Entrained bed gasification of coal is performed on a commercial scale in the Koppers-Totzek process. Various newly developed coal gasification processes use the same concept. Models for entrained bed coal gasifiers have been proposed by, e.g., Ubhayakar *et al.*[78] and Wen and Chaung.[79] Both used a CSTR model for the gas in the inlet nozzle region, followed by a plug flow zone. The solids were assumed to be in plug flow.

9. CONCLUSIONS

Gasifiers of solid fuels were an important tool to gas distributing companies and the chemical industry. The number of gasifiers operating on biomass (wood, charcoal) ran into many tens of thousands at the end of World War II, their main use being in traction.

During those heydays of solid fuel gasification the knowledge of the finer details of the process was insufficient to warrant serious efforts in modelling. Dr Winkler, who introduced fluid bed technology in 1922, had no idea at all of what happened inside his fluid bed gasifier, at least from a viewpoint of bubbles, wakes, clouds, or emulsion phase. People concentrated on the problems of operation, which are mechanical in nature, or deal with materials selection.

The amount of thermodynamic equilibria to compute the composition of generator gas (Gumz, 1952) was the groundbreaking effort in modelling. Since then enormous progress has been achieved in the understanding of the behaviour of single particles and of the entire reactor. Sophisticated models have been analysed by means of advanced mathematical and computer techniques. The results of computation are sometimes so detailed that their experimental verification is not yet feasible.

The art of modelling has flourished, as a fine branch of chemical engineering science, at the moment when the process of solid fuel gasification was almost abandoned due to cheap oil. The oil crisis is expected to restore a venerable technique to its former status.

REFERENCES

1. Buekens, A. (1978). *Mull und Abfall* **12**, 353–62.
2. Buekens, A. and Schoeters, J. (1980). ACS Symposium on Thermal Conversion of Solid Wastes, Residues and Energy Crops. ACS Series No. 131.
3. Buekens, A. and Masson, H. (1979). *Proc. of the Second World Recycling Conference.* Manilla, 121–38.
4. Alger, R. (1972). N.B.S. Special Publication 357, 171–84.
5. Brocksiep, H. (1971). In *Chemische Technologie*, Band III, K. Winacker and L. Kücher (Eds). Carl Hanser Verlag, 417–35.
6. Anon. (1974). Gas generator research and development, Phase II. Process and equipment development. Bituminous Coal Research, Inc., OCR-20-F; PB-235530/3GI.
7. Howlett, K. and Gamache, A. (1977). Forest and mill residues as potential sources of biomass, Vol. VI. Final Report, McLean, VA. The Mitre Corp./Metrek Division, ERDA Contract No. E(49-18)2081; MTR 7347.
8. Bamford, C. *et al.* (1946). *Proc. of Cambridge Philos. Soc.* **42**, 166–82.
9. Kanury, M. (1976). *J. Fire and Flammability* **2**, 191–205.
10. Kung, H.-C. (1972). *Combustion and Flame* **18**, 185–95.
11. Maa, P. and Baillie, R. (1973). *Combustion Science and Technology* **7**, 257–69.
12. Fan, L. *et al.* (1977). *Can. J. Chem. Engng* **55**, 47–53.
13. Fan, L. *et al.* (1978). *Can. J. Chem. Engng* **56**, 603–9.
14. van Ginniken, C. *et al.* (1979). *Recycling Berlin '79* **1**, 663.
15. Klose, E. and Toufar, W. (1976). *Energietechnik* **26**(12), 546–55.
16. Shafizadeh, F. Paper presented at this Symposium.
17. Anon. Publication by Koppers-Totzek.
18. Schoeters, J. and Buekens, A. (1981). *Rev. Gen. Thermique* **229**, 31–7.
19. Rammler, E. (1962). *Technologie und Chemie der Braunkohleverwertung.*
20. Gumz, W. (1950). *Gas Producers and Blast Furnaces.* Wiley, New York.

21. Desrosiers, R. (1979). In *A Survey of Biomass Gasification*. SERI, Chapter 6.
22. Neumann, K.-K. (1978). *Erdöl und Kohle* **31**(5), 228–33.
23. Walker, P. *et al.* (1959). In *Advances in Catalysis*, Vol. 11, D. Eley *et al.* (Eds). Academic Press, New York, pp. 133–221.
24. Von Fredersdorff, C. and Elliott, M. (1963). In *Chemistry of Coal Utilization*, Supplementary Volume, H. Lowry (Ed.). Wiley, New York, p. 982.
25. Grabowski, M. and Bain, R. (1979). In *A Survey of Biomass Gasification*. SERI, Chapter 3.
26. Groeneveld, M. and van Swaay, W. (1980). *Chem. Eng. Sci.*, **35**, 307–13.
27. Rensfelt, E. *et al.* (1979). In *Energy from Biomass Wastes*. IGT.
28. Johnson, J. (1974). *Advances in Chemistry Series*, No. 131, 145–78.
29. Dutta, S. *et al.* (1977). *Ind. Eng. Chem. Process Des. Dev.* **16**, 20–30.
30. Agarwal, A. and Sears, J. (1980). *Ind. Eng. Chem. Process Des. Dev.* **19**, 364–71.
31. (a) Sundaresan, S. and Amundson, N. (1980). *Chem. Eng. Sci.* **34**, 359–64.
    (b) *Ibid.*, 345–54.
    (c) *Ibid.*, 355–8.
32. Schoeters, J. (1983). PhD Thesis, Vrije Universiteit Brussel.
33. Rehmat, A. (1980). PhD Thesis, University of Illinois.
34. Weimer, A. and Clough, D. (1980). *Powder Technology* **27**, 85–103.
35. Arthur, J. (1951). *Trans. Faraday Soc.* **47**, 164–78.
36. Field, M. *et al.* (1967). *Combustion of Pulverized Coal*. BCURA, Leatherhead.
37. Yoon, H. *et al.* (1978). *AIChE J.* **24**, 885–903.
38. Szekely, J. *et al.* (1976). *Gas–Solid Reactions*. Academic Press, New York.
39. Arri, I. and Amundson, N. (1978). *AIChE J.* **24**, 87–101.
40. Yoon, H. *et al.* (1979). *AIChE J.* **25**, 429–39.
41. Kosky, P. and Floess, J. (1980). *Ind. Eng. Chem. Process Des. Dev.*, **19**, 586–92.
42. Cho, Y. and Joseph, B. (1981). *Ind. Eng. Chem. Process Des. Dev.* **20**, 314–18.
43. Biba, V. *et al.* (1978). *Ind. Eng. Chem. Process Des. Dev.* **17**, 92–6.
44. Denn, M. *et al.* (1979). *Ind. Eng. Chem. Fundamentals* **18**, 286–8.
45. Daniel, K. (1980). Paper presented at 88th National AIChE Meeting, Philadelphia PA, June.
46. Stillman, R. (1981). In *Chemical Reactors*. ACS Symposium Series No. 168, H. Fogler (Ed.), ACS.
47. Friendly, J. (1972). *Dynamic Behavior of Processes*. Prentice-Hall, Englewood Cliffs.
48. van Deemter, J. (1980). *Proceedings of the Symposium on Fluidization*, pp. 69–89.
49. Park, D. *et al.* (1981). *Fuel* **60**, 295–306.
50. Kunli, D. and Levenspiel, O. (1969). *Fluidization Engineering*. Wiley, New York.
51 Davidson, J. and Harrison, D. (1971). *Fluidization*. Academic Press, New York.
52. Grace, J. (1981). In *Chemical Reactors*. ACS Symposium Series No. 168, Fogler, H. (Ed.), ACS.
53. Kato, K. and Wen, C. (1969). *Chem. Eng. Sci.* **24**, 1351.
54. Caram, H. and Amundson, N. (1979). *Ind. Eng. Chem. Process Des. Dev.* **18**, 80–96.
55. Leva, M. (1959). *Fluidization*. McGraw-Hill, New York.

56. Kobayashi, H. and Arai, F. (1967). *Chem. Eng. Tokyo* **31**, 239.
57. Yoshida, K. and Kunii, D. (1974). *J. Chem. Eng. Japan* **7**, 34–9.
58. Weimer, A. and Clough, D. (1981). *Chem. Eng. Sci.* **36**, 549–67.
59. Levenspiel, O. *et al.* (1968/9). *Powder Technology* **2**, 87–96.
60. Chen, T. and Saxena, S. (1978). *AIChE Symposium Series* **74**(176), 149–61.
61. Caram, H. and Amundson, N. (1979). *Ind. Eng. Chem. Process Des. Dev.* **18**, 96–102.
62. Purdy, M. *et al.* (1981). *Ind. Eng. Chem. Process Des. Dev.* **20**, 675–82.
63. Weimer, A. and Clough, D. (1980). *Powder Technology* **26**, 11–16.
64. Raman, P. *et al.* (1981). *Ind. Eng. Chem. Process Des. Dev.* **20**, 686–92.
65. Mayers, M. (1934). *J. Am. Chem. Soc.* **56**, 1879.
66. Gibson, M. and Euker, C. (1975). Presented at 68th AIChE Annual Meeting, Los Angeles.
67. Dutta, S. and Wen, C. (1977). *Ind. Eng. Chem. Process Des. Dev.* **16**, 31–7.
68. Hedden, K. (1976). In *Alternate Energy Sources*, J. Hartnett (Ed.). Academic Press, New York.
69. Wen, C. and Tseng, H. (1979). Presented at 72nd Annual AIChE Meeting, San Francisco.
70. Karim, G. (1974). *J. Inst. Fuel* **53**, 219.
71. Winslow, A. (1976). Sixteenth Symp. on Combustion, 503.
72. Parker, A. and Hottel, H. (1936). *Ind. Eng. Chem.* **28**, 1334.
73. Khaustovich, G. (1968). *Zh. Fiz. Khim.* **42**, 1684.
74. Schoeters, J. *et al.* (1981). Presented at 2nd World Congress of Chemical Engineering, Montreal.
75. Mori, S. *et al.* (1982). Presented at the Japan–China Symposium on Fluidization, China.
76. Mikesell, R. *et al.* (1980). ACS Symposium Series, 131.
77. van Ginniken, C. and Tels, M. (1981). Presented at the 2nd International Symposium on Materials and Energy from Refuse, Antwerp, October.
78. Ubhayakar, S. *et al.* (1977). *Fuel* **56**, 281–91.
79. Nienow, A. and Rowe, P. (1976). *Proceedings of the 1st International Symposium on Materials and Energy from Refuse*. Antwerp, October.
80. Masson, H. *et al.* (1981). Presented at the 2nd International Symposium on Materials and Energy from Refuse, Antwerp, October.

36

Wood Pyrolysis and Carbon Dioxide Char Gasification Kinetics in a Fluidized Bed

F. G. VAN DEN AARSEN, A. A. C. M. BEENACKERS and W. P. M. VAN SWAAIJ

Laboratory of Chemical Reaction Engineering, Twente University of Technology, Enschede, The Netherlands

LIST OF SYMBOLS

A	heat transfer area (m^2)
a	thermal diffusivity ($=\lambda/\rho c_p$) (m^2/s)
a	specific surface area per unit weight (m^2/g)
a_0	initial specific surface area per unit weight (m^2/g)
c_p	specific heat (J/kg °C)
d_p	particle diameter (m)
E	activation energy (kJ/mol)
ΔH_r	heat of reaction (171 for carbon dioxide-char reaction) (kJ/mol)
k_0	frequency factor for pyrolysis (s^{-1})
k_0	frequency factor for gasification (m/s)
k^*	heating rate constant ($=\alpha A/\rho c_p V$) (s^{-1})
M	molecular weight (kg/kmol)
R	gas constant (J/mol K)
R	volumetric reaction rate ($mol/m^3\,s$)
R''	surface reaction rate ($mol/m^2\,s$)
S	specific surface area per unit volume (m^2/m^3)
S_0	initial specific surface area per unit volume (m^2/m^3)
T	temperature (K)
T_1	bulk gas temperature (°C or K)
T_m	mean particle temperature (°C or K)
t	reaction time (s)
V_p	particle volume (m^3)
W_t	particle weight after reaction time t (g)
W_∞	ultimate particle weight (g)
W_0	initial particle weight (g)

Greek Symbols

α overall heat transfer coefficient (W/m^2 K)
α_c convective heat transfer coefficient (W/m^2 K)
α_r radiative heat transfer coefficient (W/m^2 K)
α_i internal heat transfer coefficient (W/m^2 K)
D bulk diffusion coefficient (m^2/s)
D_k Knudsen diffusion coefficient (m^2/s)
ε particle porosity (m^3/m^3)
ε_0 initial particle porosity (m^3/m^3)
ε emission factor
λ thermal conductivity (W/m K)
η viscosity (N s/m^2)
ρ density (kg/m^3)
σ Stefan Boltzmann constant ($= 5{\cdot}75 \times 10^{-8}$) (W/m^2 K^4)
ξ conversion

1. FAST PYROLYSIS OF BEECH WOOD PARTICLES IN A BENCH SCALE FLUIDIZED BED

1.1. Introduction

Since pyrolysis is the first step in the gasification process of a solid material, rates and temperatures at which pyrolysis occur together with pyrolysis product yields are critical parameters in the design of a gasification process.

Many investigators have reported kinetics for the pyrolysis of wood and related substances (e.g. Roberts,[3] Lewellen,[1] Tran[4] which were studied under a wide variety of experimental conditions. Particle size, particle heating rate and pyrolysis temperature are the major parameters determining the ultimate product yield. Char production is maximized at a low heating rate (~5 °C/min) and relatively low temperatures (up to 500 °C), whereas at extremely high heating rates (10 000–250 000 °C/s) and temperatures up to 1000 °C, no char is produced at all.[2,1]

However, only a very limited amount of information is available for pyrolysis of small wood particles at fluidized bed conditions. This study is concerned with the interpretation of kinetic data on small particle wood pyrolysis at heating rates of 250–1000 °C/s and on the fast pyrolysis product distribution.

1.2. Experimental

Both fast pyrolysis product yields and carbon dioxide–char gasification kinetics were studied in a 30 mm diameter fluidized bed reactor (Fig. 1).

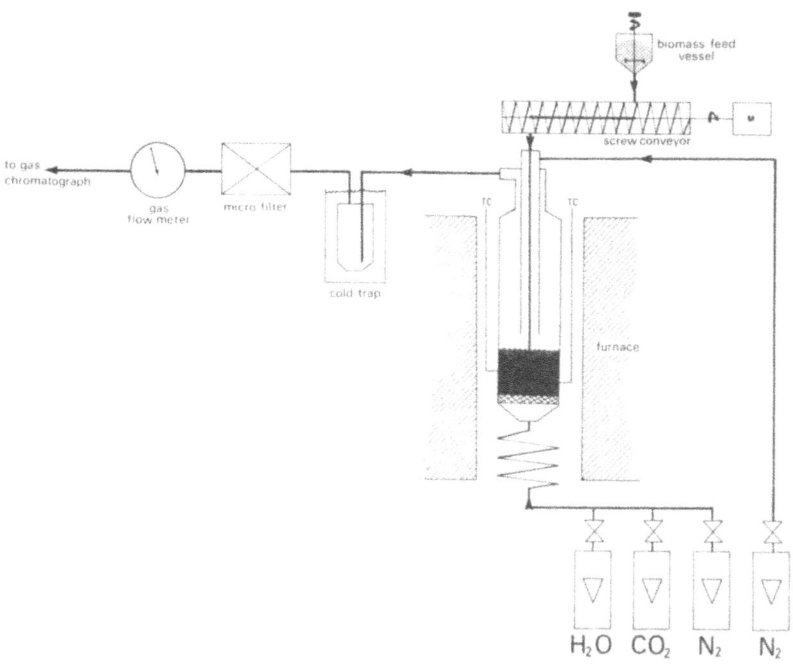

FIG. 1. Bench scale fluidized bed unit.

Beech wood particles (1–2 mm in diameter) were continuously fed into a nitrogen fluidized bed of 125 μm alumina particles, during pyrolysis experiments. Fluid bed operating temperature was varied between 700 and 1000 °C. Table I lists the range of operating variables.

Generally, a pyrolysis experiment lasted for 40 min, in which period about 10 g beech wood was fed into the bed. Product gas leaving the reactor

TABLE I
OPERATING VARIABLES, PYROLYSIS EXPERIMENTS

Reactor pressure	atmospheric
Reactor temperature	700–1 000 °C
Feed material	beech wood, 1–2 mm diameter, 3–15 wt % water
Bed material	Al_2O_3, 100–150 μm
Fluidizing agent	nitrogen
Bed diameter	30 mm
Bed height	30–60 mm
Superficial gas velocity	40–80 mm/s

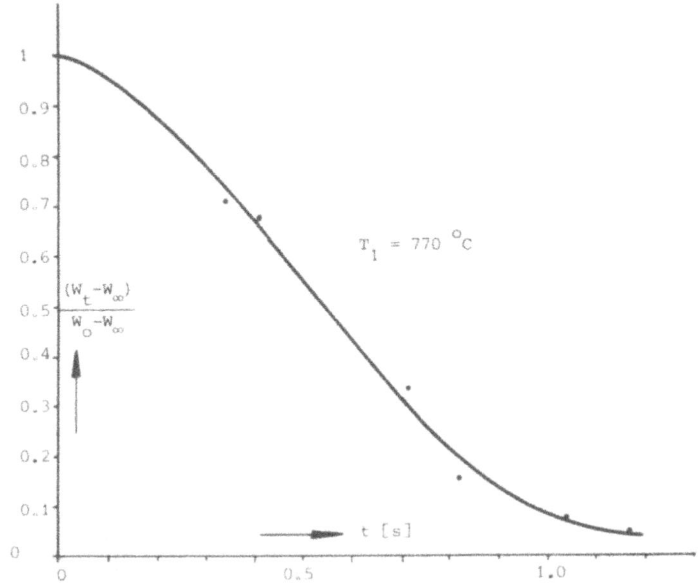

FIG. 2. Conversion of 0·4–0·6 mm wood particles (1·6 wt % H_2O) as a function of reactor residence time (data from Lindman[6]).

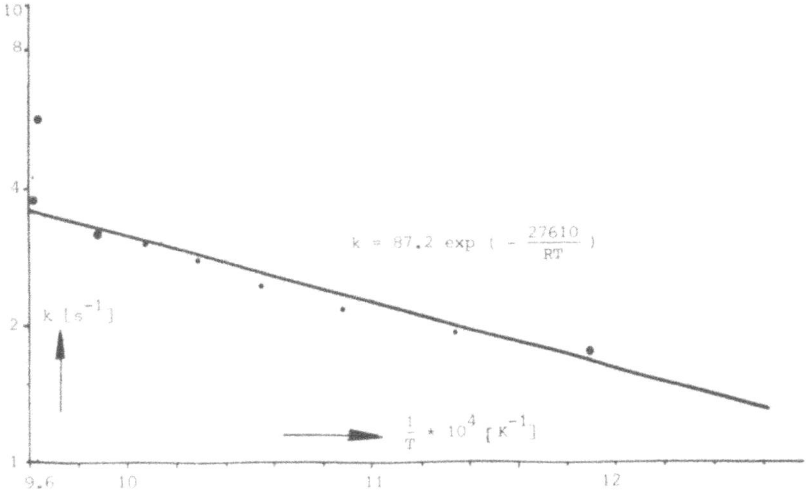

FIG. 3. Arrhenius plot for first order pyrolysis kinetics.

was cooled down, filtered and sent to the gas sampling loop after monitoring the volume produced. Gas samples were analysed using a double Varian 920 gas chromatograph.

The amount of condensate collected in the microfilter and the cold trap was determined by dissolving the tar–water mixture in a chloroform–methanol mixture from which the water content and tar content could be determined, using Karl Fischer titration and solvent evaporation. Char yield was determined by complete combustion of the resulting char with air and absorbing the produced carbon dioxide in a solution of potassium hydroxide. The elemental char analysis together with the determined carbon contents of the reactor served as a basis for calculating the char product yield.

1.3. Fast Pyrolysis Kinetics of Small Wood Particles

Lindman[6] has measured the decomposition rate of 0·5–0·6 mm wood particles in an entrained flow reactor at temperatures ranging from 650 to 1000 °C and heating rates of 500–1000 °C/s. These conditions are very much like those for a 1 mm diameter wood particle subject to pyrolysis in a fluidized bed reactor operated at elevated temperatures. Therefore, we used Lindman's data to derive a kinetic expression for the fast pyrolysis of small wood particles (Fig. 2).

It is customary in wood pyrolysis literature to assume a first order Arrhenius type rate law.[1–4,7]

$$\frac{dW_t}{dt} = k_0(W_\infty - W_t)\exp\left(-\frac{E}{RT}\right) \tag{1}$$

A plot of $[1/(W_\infty - W_t)](dW_t/dt)$ versus $1/T$ on semilog paper is expected to yield a straight line with a negative slope equal to E/R. Figure 3 gives the results.

As the wood particle is heated up during the pyrolysis process, it is necessary to know the particle temperature as a function of the pyrolysis time, for use in calculating dW_t/dt.

From a heat balance it follows that

$$\alpha A(T_1 - T_m) = \rho c_p V_p \frac{dT_m}{dt} \tag{2}$$

The heat of reaction for fast pyrolysis is taken to be zero. This assumption is based on the findings of Kanury,[9] Antal[2] and Lewellen.[1] Kanury reported exothermic heats of reaction ranging from −120 J/g to −320 J/g for slow, low temperature pyrolysis of wood.

This, as suggested by Lewellen, for example, is probably due to the fact that primary decomposition products are partly polymerized when on their way out. At very high heating rates, the residence time of volatile products is insufficient for further reaction inside the particle, which causes the reaction to appear endothermic. Antal[2] reported an endothermic heat of reaction of +360 J/g for the flash pyrolysis of wood particles at heating rates up to 250 000 °C/s.

Thus, for our case of fast pyrolysis at heating rates of 250–1000 °C/s, the heat of reaction will be negligible compared to the heat necessary to heat up the wood particle to reaction conditions.

Integrating eqn (2) yields

$$(T_1 - T_m) = (T_1 - T_0)\exp\left(\frac{-\alpha A}{\rho c_p V}\,t\right) \tag{3}$$

or

$$T_m = T_1 - (T_1 - T_0)\exp(-k^* t) \tag{3a}$$

The overall heat transfer coefficient is calculated from

$$\frac{1}{\alpha} = \frac{1}{\alpha_c + \alpha_r} + \frac{1}{\alpha_i} \tag{3b}$$

The convective heat transfer coefficient is calculated from the Ranz and Marshall equation:

$$Nu_p = 2 + 0{\cdot}6\,\mathrm{Pr}^{1/3}\,\mathrm{Re}_p^{1/2}$$

resulting in $\alpha_c = 400$ W/m^2 °C (Appendix A).

The heat transfer coefficient for radiation is calculated to be $\alpha_r = 70$ W/m^2 °C (Appendix B).

The internal heat transfer coefficient is calculated from a Fourier analysis for heat penetration into a finite medium, resulting in

$$\alpha_i = 5{\cdot}8\,\frac{\lambda_s}{d_p} \tag{4}$$

The internal heat transfer coefficient is constant for $F_0 = at/d^2 > 0{\cdot}1$, which is true in this case for $t > 0{\cdot}2$ s.

Particle properties like density, specific heat and thermal conductivity are a function of conversion and temperature. We accounted for these dependencies by taking the mean value of those properties over the conversion range considered (Appendix C). Results of these calculations

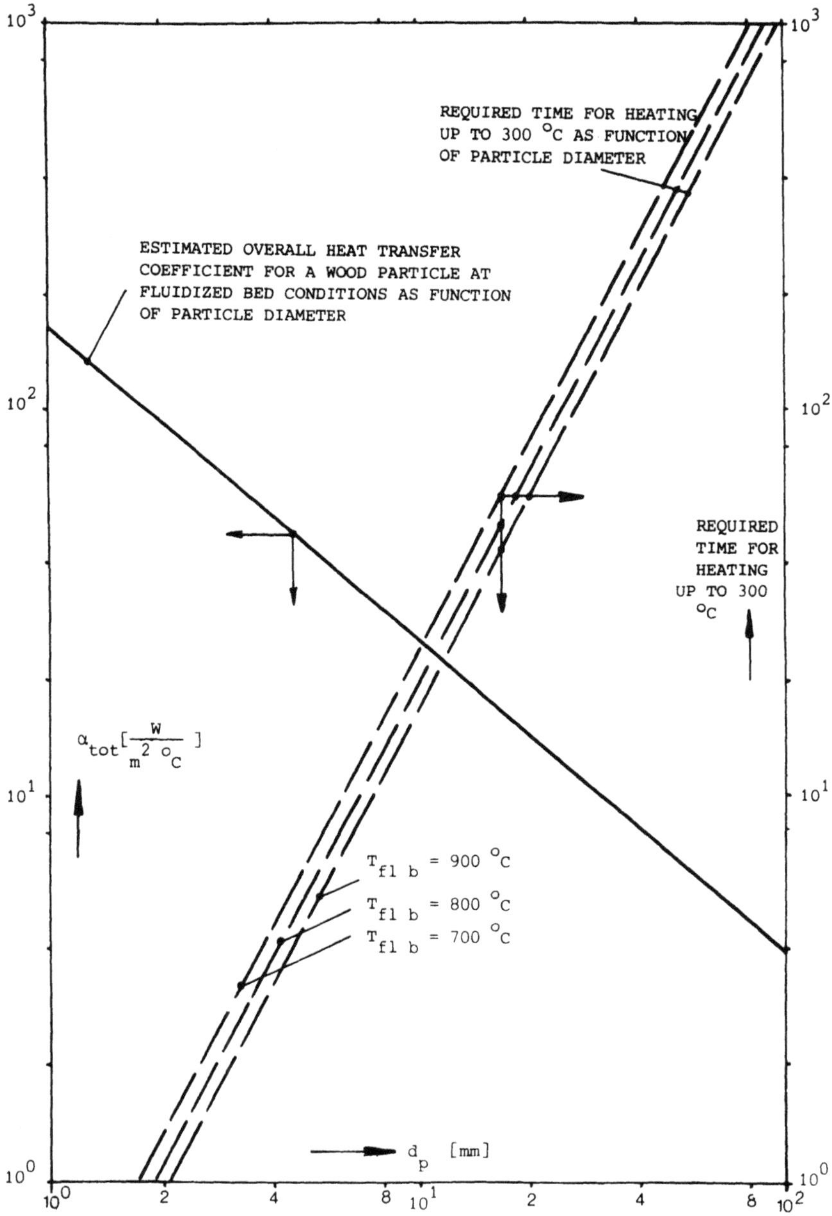

FIG. 4. Estimated overall heat transfer coefficient for pyrolising particle in a fluidized bed and required time for heating up to 300 °C as a function of particle diameter.

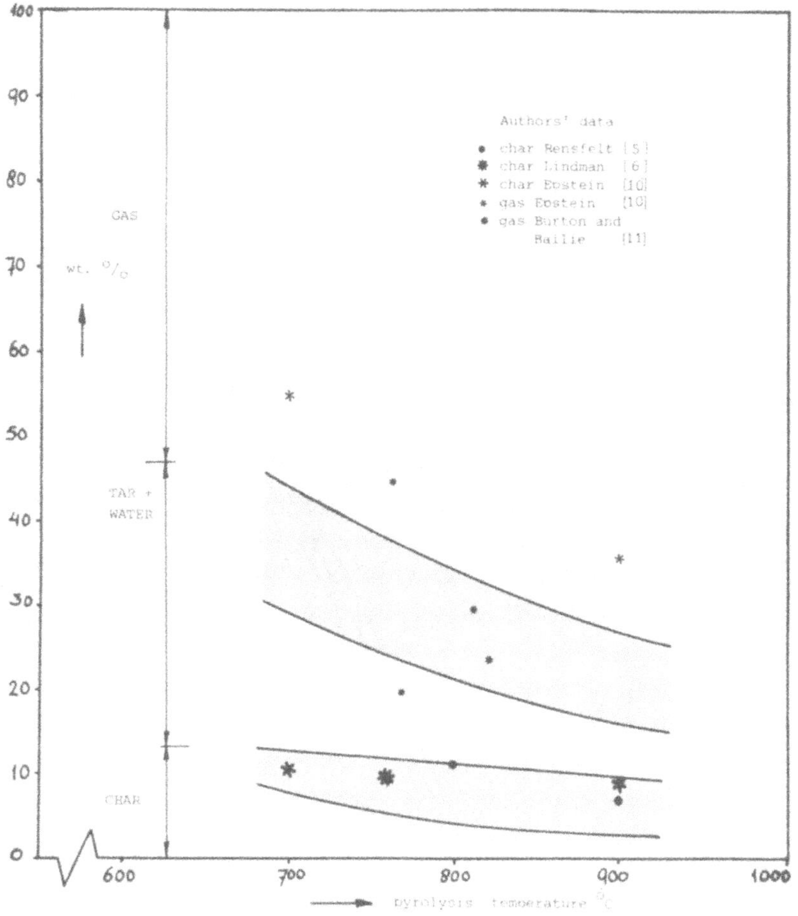

FIG. 5. Fast pyrolysis decomposition products.

(Appendix D) lead to the following expressions for the fast pyrolysis of small wood particles:

$$\frac{\mathrm{d}W_t}{\mathrm{d}t} = 87(W_\infty - W_t)\exp\left(-\frac{27\,610}{RT}\right) \tag{5}$$

From this expression we can conclude that the pyrolysis reaction is already taking place at relatively low temperatures. Therefore, the particle heating rate will play an important role in the time required for pyrolysis, dependent on particle diameter. To illustrate this, we made approximate

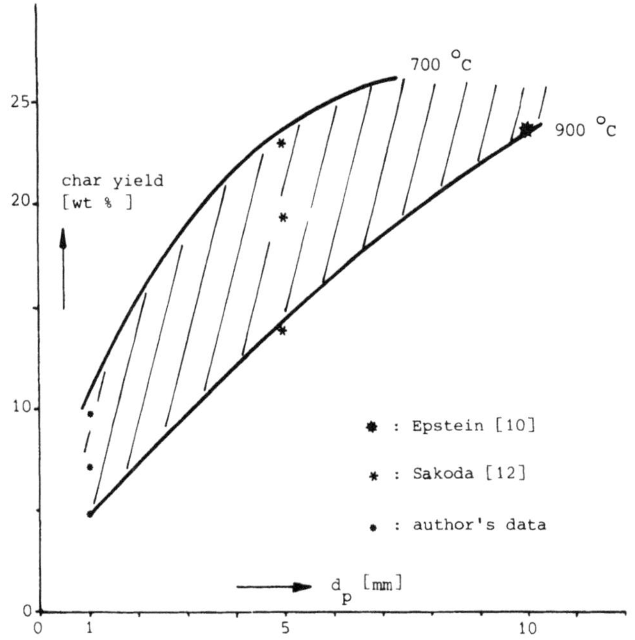

FIG. 6. Char yield in fluidized pyrolysis of wood as function of particle diameter at variable reactor temperatures.

calculations for heating rates as a function of particle diameter in fluidized bed pyrolysis (Appendix E).

Figure 4 illustrates the results and we see that the overall heat transfer coefficient to the pyrolysing particle is steadily decreasing with increasing particle diameter. Therefore, the time required to obtain 300 °C is increasing with increasing particle diameter. If a particle is kept at 300 °C it takes about 10 s to obtain 95 % conversion. Thus, we see that at a particle diameter between 5 and 10 mm, the time for heating up the particle to 300 °C is about equal to the time required for 95 % conversion of a wood particle and so at increasing diameter heat transfer will become the rate limiting step.

1.4. Fast Pyrolysis Product Yields

Product yield of the fast pyrolysis of 1 mm beech wood particles has been determined in the aforementioned experimental set-up (Fig. 1) at temperatures ranging from 700 to 900 °C. The heat transfer coefficient from the fluidizing gas and fluidized particles to the entering wood particles is estimated to be between 50 W/m^2 °C and 250 W/m^2 °C (Appendices E and

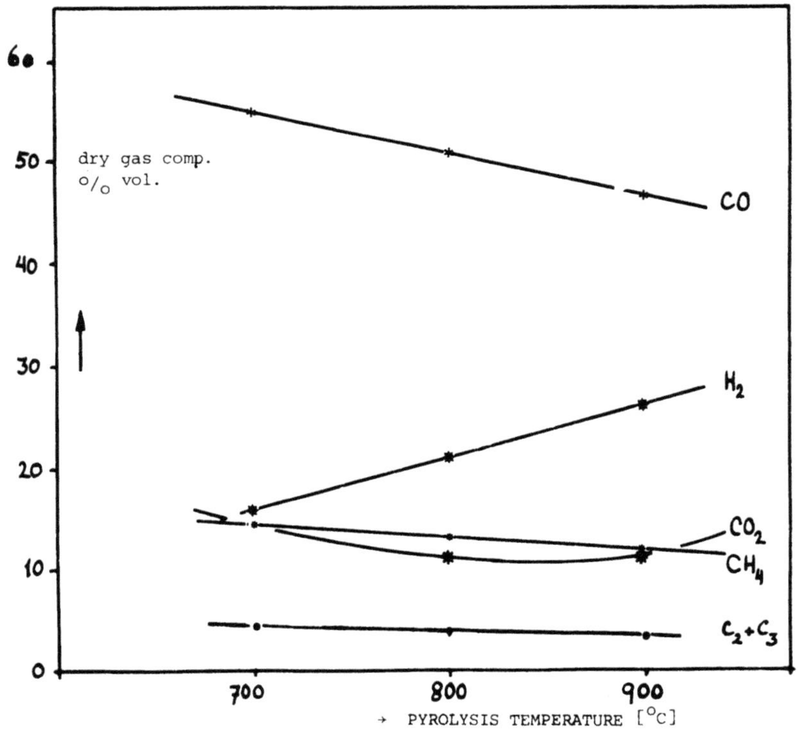

FIG. 7. Pyrolysis product gas composition.

F). In fact this heat transfer coefficient is very difficult to calculate from existing correlations for gas–solid fluidized beds operated at Reynolds numbers below 1. However, at those heat transfer coefficients, particle heating rates (applying eqn (3)) of 500 °C/s are obtained. Applying the kinetics as derived in the first section of this paper, the pyrolysis process of a 1 mm wood particle is nearly completed in 2 s.

The mass balance of the decomposition products is presented in Fig. 5 as a function of the final pyrolysis temperature.

Literature data for the same particle size range in fluidized bed or entrained flow pyrolysis fit well in this picture. It is apparent that the amount of pyrolysis char tends to be below 10 wt% for all temperatures. The use of larger particles in fluidized bed pyrolysis results in char yields which are about 50–100% higher (Fig. 6) relative to small particle pyrolysis.[10,12] These results are in agreement with earlier statements that char yield is decreased at increasing heating rates and decreasing particle diameter.

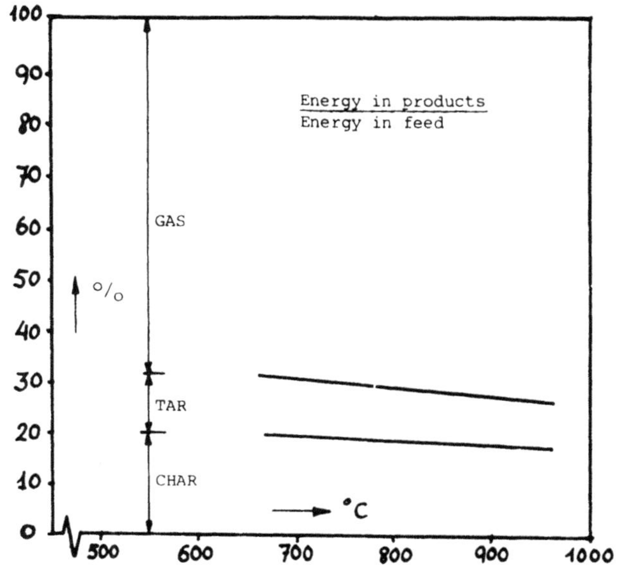

FIG. 8. Energy distribution in pyrolysis products.

Total gas yield was found to increase with temperature, at the cost of condensate, indicating a further breakdown of tar at higher temperatures.

The dry product gas composition is characterized by a very high carbon monoxide content compared to the other components (Fig. 7). The same result was found by previously mentioned investigators.[5,6,10-12] Methane, C_2^+ and carbon dioxide concentrations hardly vary but the hydrogen content increases with temperature. Decomposition reactions for wood and tar seem to determine the gas product distribution rather than gas phase equilibrium reactions.

Figure 8 shows the relative energy distribution in fast pyrolysis products. About 70% of the energy contained in the mother material is recovered in the gas phase, 20% in char and 10% in tar.

1.5. Conclusions

The pyrolysis of 1 mm wood particles in a fluidized bed at 800 °C is a very rapid process which is nearly completed within 2 s. In fluidized bed pyrolysis, char is produced in small quantities (less than 10%) and as such fast pyrolysis in a fluidized bed at heating rates of ~500 °C/s can be placed between slow pyrolysis processes (1 °C/s) and flash pyrolysis processes (2×10^4 °C/s) where 25% and 0% char are produced respectively.

2. GASIFICATION KINETICS OF THE CHAR–CARBON DIOXIDE REACTION

2.1. Introduction

Knowledge on the reactivity of char as it is present in a pilot plant fluidized bed wood gasifier is necessary for proper modelling of this gasifier. Therefore, we took char from our pilot plant reactor contents after operating stationary at a certain temperature. This char was used in the same bench scale fluidized bed as described in the previous section for studying the reactivity as a function of carbon dioxide concentration and temperature. Experimental conditions were such that external and internal mass transfer were not controlling the reaction rate. In Appendix G results of these considerations are presented. As the CO_2 gasification reaction of char is highly endothermic[13] ($\Delta H_r = +171$ kJ/mol) the particle temperature differs from the reactor bed temperature at high reaction rates; at the highest observed reaction rate a temperature correction of 17·5 °C was necessary. For the gasifying char particle itself, the temperature can be assumed to be homogeneous throughout the particle.

2.2. Char–Carbon Dioxide Gasification Fundamentals

Considerable information is available concerning the mechanism of gasification reactions. The carbon–carbon dioxide reaction is the most extensively studied gasification reaction as no other reactions occur next to gasification. Mechanistic studies indicate the reaction to be a surface reaction, with an adsorption and desorption step, resulting in Langmuir–Hinshelwood kinetics. The mechanism according to Ergun[16] is

$$(1)\quad C_f + CO_2 \underset{k_1}{\overset{k_i}{\rightleftharpoons}} C(O) + CO \qquad \text{and} \qquad (2)\quad C(O) \underset{k_2}{\rightarrow} CO + nC_f$$

Since in this case the char surface is also the reactive surface, the reaction rate can be written in terms of a specific rate multiplying the specific char surface area.

$$-\frac{dC_c}{dt} = R = S \times R'' \tag{1}$$

where R = volumetric reaction rate (mol C/m^3 s), S = specific surface area (m^2/m^3), and R'' = specific reaction rate (mol C/m^2 s).

The surface area will develop with conversion and we tried the grain model as described by Szekely and Evans[14] to relate surface area and conversion. Accordingly, the reactive area per unit particle volume is given by

$$\frac{S}{S_0} = \left(\frac{1-\varepsilon}{1-\varepsilon_0}\right)^{2/3} \tag{2}$$

where S_0 is the initial solid surface area and ε and ε_0 are the instantaneous and initial porosity respectively.

The carbon concentration in a particle is related to the porosity:

$$C_c = \frac{\rho_c}{M_c}(1-\varepsilon) \tag{3}$$

where ρ_c is the true carbon density and M_c is the molecular weight of carbon.

Carbon conversion can now also be related to the porosity:

$$\xi = 1 - \frac{C_c}{C_0} = 1 - \frac{1-\varepsilon}{1-\varepsilon_0} \tag{4}$$

Substitution and rearranging eqn (1) results in

$$(1-\xi)^{1/3} = 1 - \tfrac{1}{3}S_0 \frac{M_c}{\rho_c} \frac{R''}{(1-\varepsilon_0)} t \tag{5}$$

or, with a specific surface area per unit weight,

$$(1-\xi)^{1/3} = 1 - 4a_0R''t \tag{5a}$$

A plot of $(1-\xi)^{1/3}$ against reaction time t should give a straight line with slope $(-)4a_0R''$.

Such a plot is given for one of our experiments in Fig. 9 and the data correlate fairly well.

This enabled us to determine the gasification reaction rate from plots of $(1-\xi)^{1/3}$ against t at different conditions.

2.3. Results and Discussion

For several carbon dioxide concentrations and reaction temperatures the conversion–time relationship was determined, yielding the reaction rate. This was done for char originating from our pilot plant, having the following properties:

$$a_0 = 271\ \mathrm{m^2/g}$$
$$S_0 = 105 \times 10^6\ \mathrm{m^2/m^3}\ \text{particle volume}$$
$$\varepsilon_0 = 0{\cdot}69\ \mathrm{m^3/m^3}$$

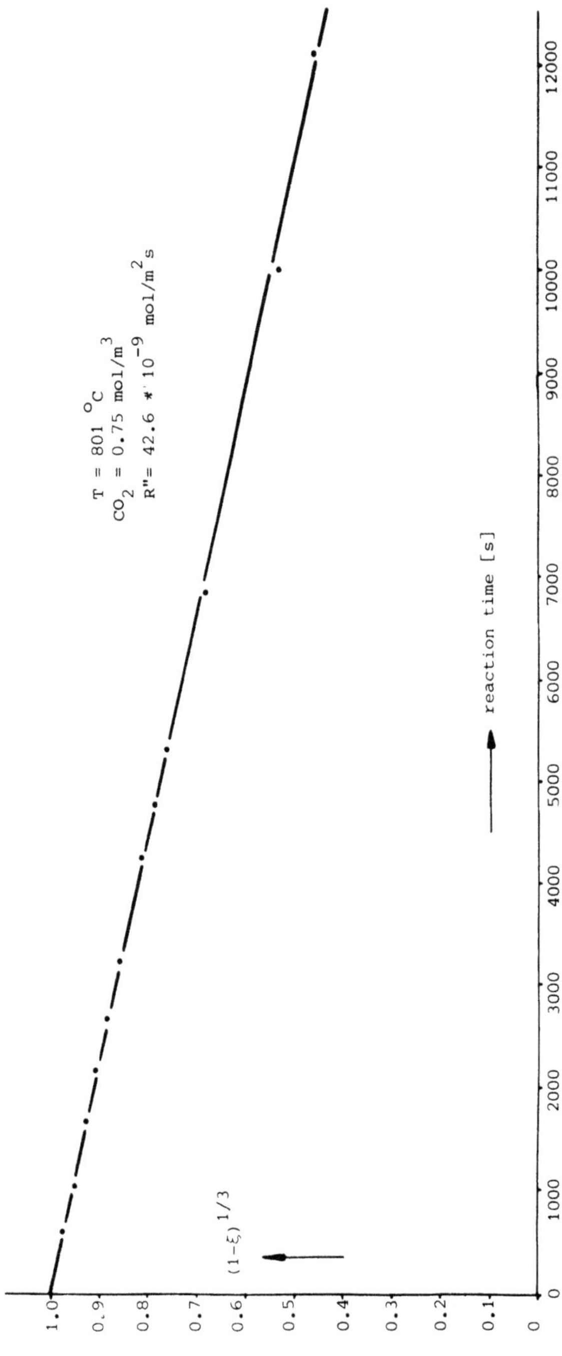

FIG. 9. Progression of conversion with reaction time.

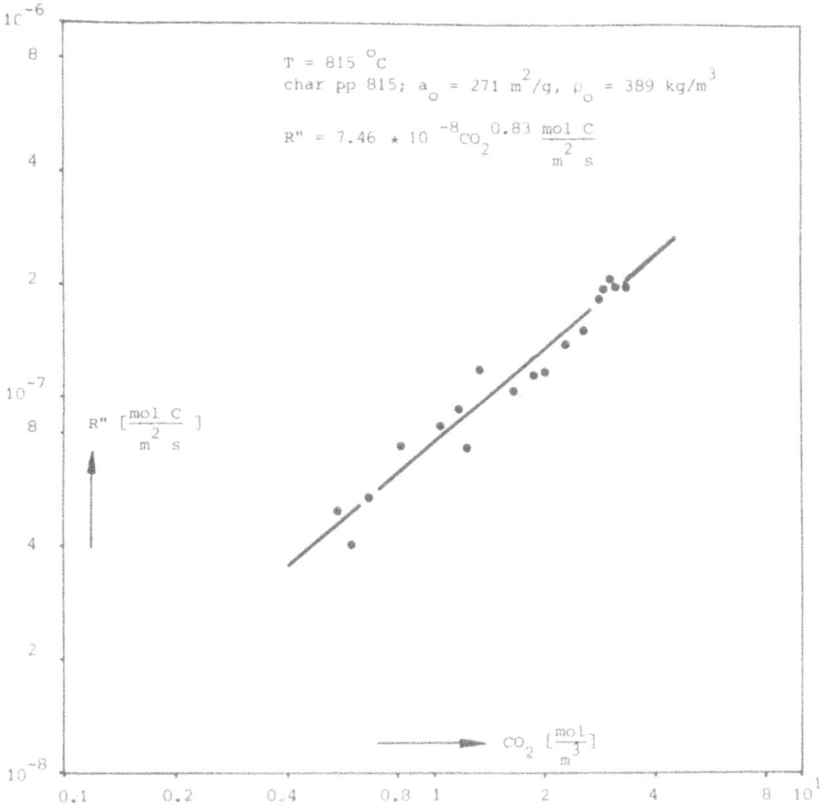

FIG. 10. Surface reaction rate of fluidized bed gasifier char as function of carbon dioxide concentration.

Figure 10 gives the dependency of reaction rate on carbon dioxide concentration, resulting in

$$R'' = k\,CO_2^{0.83} \qquad \text{mol C/m}^2\,\text{s} \tag{6}$$

The 0·83 power indicates the inhibiting effect of CO_2 and thus Langmuir–Hinshelwood kinetics. At present we are still studying true L-H kinetics on which we will publish early next year.

The dependency on reaction temperature is represented with an Arrhenius plot in Fig. 11, resulting in

$$k = 7{\cdot}2 \exp\left(-\frac{166\,156}{RT}\right) \tag{7}$$

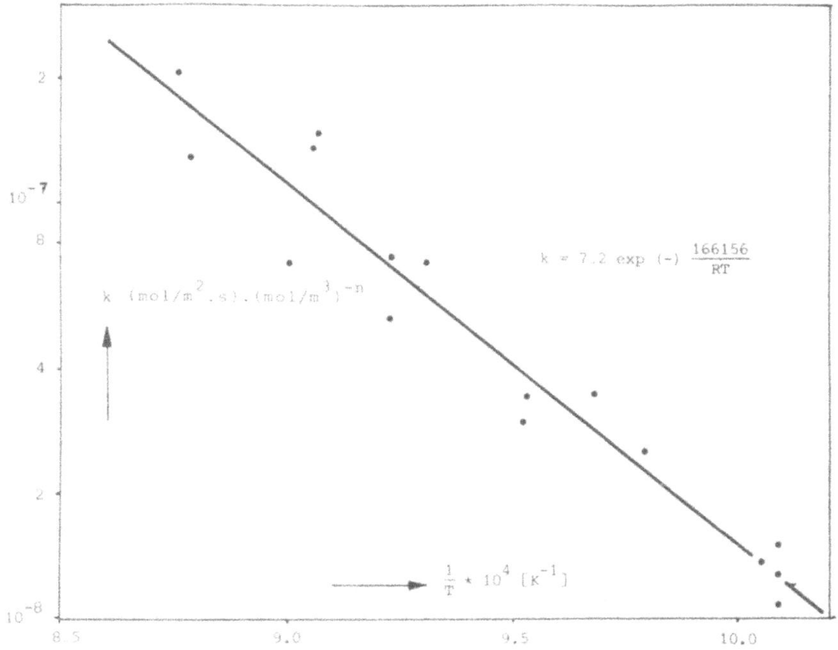

FIG. 11. Arrhenius plot for reaction rate constant of char–carbon dioxide reaction.

In Fig. 12 some reported biomass reactivities are compared with this result; kinetic parameters are given in Table II. The kinetic parameters as presented by Rensfelt[5] were measured under steam gasification conditions at $p_{H_2O} = 0{\cdot}73$ atm. To make their kinetic parameters comparable to the reactivity as found by this study, we had to correct those values by a factor $[2 \times (8{\cdot}3)^{0{\cdot}83}]^{-1}$. The factor 2 originates from the difference in reactivity between water and carbon dioxide, as reported by Rensfelt; the factor $(8{\cdot}3)^{0{\cdot}83}$ is the correction for difference in concentration, where we took the same dependency of the reaction rate on the water concentration as we have found for carbon dioxide.

The reactivity of chars with different temperature history was determined at 815 °C and 1 mol CO_2/m^3 and was found to be independent of history. Table III gives the characteristics of chars originating from our pilot plant after steady operation at 720, 780, 815, 840 and 885 °C respectively.

Surface areas as presented in Table III were determined by measuring the CO_2 adsorption isotherm according to Brunauer, Emmett and Teller. The

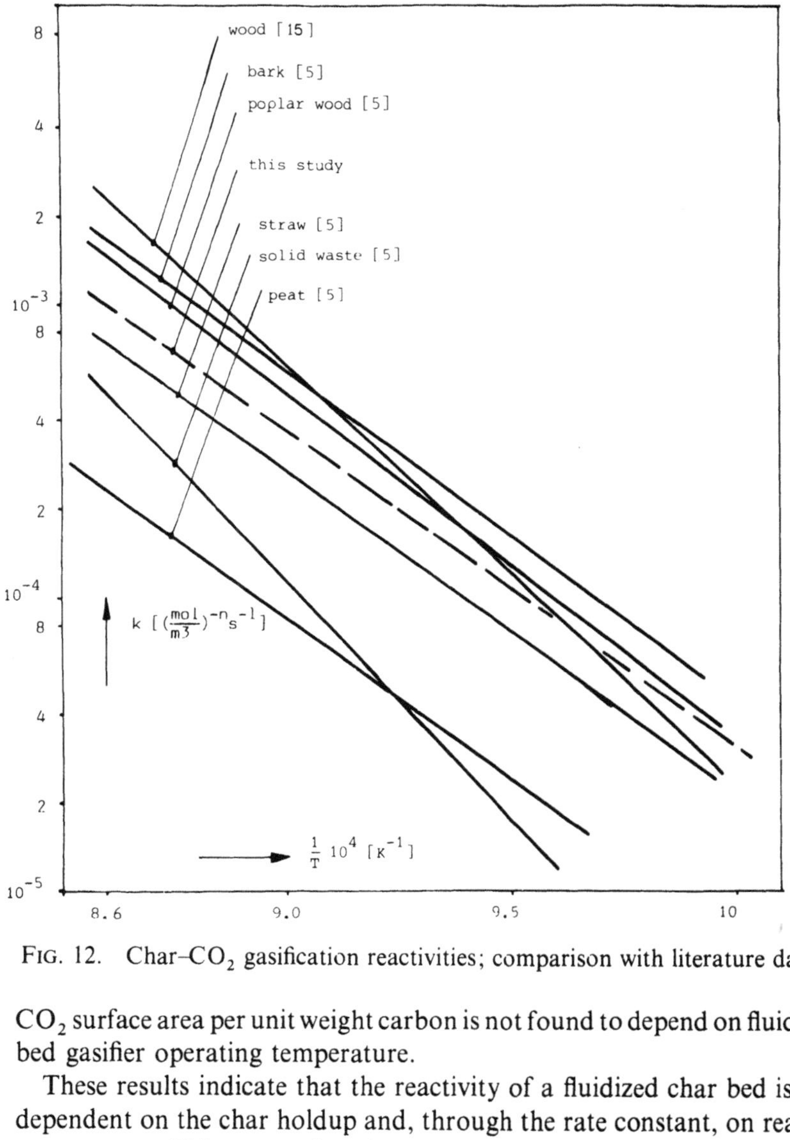

FIG. 12. Char–CO_2 gasification reactivities; comparison with literature data.

CO_2 surface area per unit weight carbon is not found to depend on fluidized bed gasifier operating temperature.

These results indicate that the reactivity of a fluidized char bed is just dependent on the char holdup and, through the rate constant, on reactor temperature. This means that the char holdup establishes itself at a given throughput and operation temperature: char holdup decreases at increasing operating temperature.

In our laboratories we are presently executing experiments in the pilot plant 50 kg/h wood gasifier to establish the holdup–temperature relationship.

TABLE II

COMPARISON WITH REPORTED BIOMASS REACTIVITIES

Author	*Fuel*	k_0 (s^{-1})	E_{act} (kJ/mol)
Rensfelt[6]	Solid waste	5.62×10^6	249
	Poplar wood	1.73×10^5	181·7
	Straw	8.47×10^4	181·7
	Bark	8.38×10^3	154·9
	Peat	7.35×10^3	169·1
Groeneveld[15]	Wood	10^7	217
This study	Beech wood	2.34×10^4	166·2

TABLE III

PILOT PLANT CHAR CHARACTERISTICS

Charring temp.	R_{815} ($mol/m^2 s$)	a_0 (m^2/g)	S_0 (m^2/m^3)	ε_c (m^3/m^3)
720	72.05×10^{-9}	372	110×10^6	0·64
785	88.5×10^{-9}	388	135×10^6	0·67
815	75.9×10^{-9}	271	105×10^6	0·69
840	—	268	138×10^6	0·75
885	76.4×10^{-9}	368	136×10^6	0·73

As the grain model was successfully applied as a basis for the conversion–time relationship, we started research on the verification of this model by measuring the development of the CO_2-BET surface area with conversion. We will report on this early next year.

2.4. Conclusions

CO_2-reactivity of char originating from a 50 kg wood/h fluidized bed gasifier can be described with an Arrhenius type surface reaction rate:

$$R'' = 7{\cdot}2 \exp\left(-\frac{166\,156}{RT}\right) CO_2^{0{\cdot}83} \qquad \frac{\text{mol C}}{\text{m}^2\,\text{s}}$$

This correlation is preliminary, as we are still studying true Langmuir–Hinshelwood kinetics. Agreement with the literature is satisfactory.

The reactivity of a fluidized bed gasifier is determined by the char hold-up and reactor temperature and not by the temperature history of the fuel.

The grain model could satisfactorily explain observed conversion–time relations; model verification is still underway.

ACKNOWLEDGEMENT

We gratefully acknowledge the financial support by the Commission of the European Communities, which made this work possible as part of a fluidized bed biomass gasification project.

APPENDIX A: EXTERNAL HEAT TRANSFER COEFFICIENT CALCULATION

$T = 770\,°C$

$p =$ atmospheric

$d_p = 5 \times 10^{-4}\,m$

Nitrogen gas flow $v_g \sim 1{\cdot}2\,m/s$

$\eta = 4{\cdot}4 \times 10^{-5}\,N\,s/m^2$

$c_p = 1200\,J/kg\,°C$

$\lambda = 0{\cdot}061\,W/m\,°C$

For a single particle falling through a heated pipe:

$$Nu_p = 2 + 0{\cdot}6\,Re^{1/2}\,Pr^{1/3} \rightarrow \alpha_c = 399\,W/m^2\,°C$$

APPENDIX B: THE MEAN RADIATIVE HEAT TRANSFER COEFFICIENT

$$\varepsilon\sigma(T_1^4 - T_p^4) = \alpha_r(T_1 - T_p)$$

with $\varepsilon = 0{\cdot}5$

$\sigma = 5{\cdot}75 \times 10^{-8}\,W/m^2\,K^4$

giving $\bar{\alpha}_r = 70\,W/m^2\,°C$

APPENDIX C: PARTICLE PROPERTIES

$\lambda_{\text{wood}} = 0{\cdot}14\ \text{W/m °C}$

$\lambda_{\text{char}} = 0{\cdot}07\ \text{W/m °C}$

$c_{p_1\ \text{wood}} = 2500\ \text{J/kg °C}$

$c_{p\ \text{char}} = 670\ \text{J/kg °C}$

$\rho_{\text{wood}} = 700\ \text{kg/m}^3$

$\rho_{\text{char}} = 150\ \text{kg/m}^3$

APPENDIX D: PYROLYSIS RATE CALCULATIONS

From the fitted curve in Fig. 2 conversion levels are read between 30 and 80 % conversion, as experimental error will be smallest in this region. These data yielded results as tabulated in Table D.

TABLE D
PYROLYSIS RATE CALCULATIONS

ξ	t	α_{c+r}	α_i	k^*	$^{\bullet}T_m$	k
0·67	0·41	470	1 493	3·25	572	1·73
0·40	0·45	470	1 511	3·41	608	1·97
0·45	0·5	470	1 469	3·61	647	2·17
0·51	0·55	470	1 417	3·76	675	2·43
0·57	0·6	470	1 393	3·92	699	2·72
0·63	0·65	470	1 371	4·13	719	3·07
0·67	0·72	470	1 339	4·44	739	3·17
0·84	0·82	470	1 297	4·92	757	5·69
0·925	1·04	470	1 238	5·70	768	3·79

APPENDIX E: HEAT TRANSFER COEFFICIENT FROM GAS TO SOLID IN PILOT PLANT FLUIDIZED BED GASIFIER

Conditions were taken to be those of a commercial fluidized bed gasifier:

$u_g = 1\ \text{m/s}$

$d_p = \text{variable}$

$T = 700 - 800 - 900\ \text{°C}$

The particle Nusselt number is calculated according to the Ranz and Marshall equation:

$$Nu_p = 2{\cdot}0 + 0{\cdot}6\,Re^{1/2}\,Pr^{1/3}$$

The inside heat transfer coefficient is calculated from $\alpha_i = 5{\cdot}8(\lambda_s/d_p)$. This will lead to conservative values of α_i, as this equation is only valid for $F_0 >$ 0·1. At large particle diameters, F_0 will remain below 0·1 for an extended period and α_i should, for example, be calculated according to the penetration theory. The resulting overall heat transfer coefficient is applied in eqn (3) to calculate the time required to attain a mean particle temperature of 300 °C.

APPENDIX F: GAS–PARTICLE HEAT TRANSFER COEFFICIENT IN FLUIDIZED BEDS AT LOW REYNOLDS NUMBERS

For a charring particle in a fluidized bed, the gas particle heat transfer coefficient might be calculated from the Ranz and Marshall equation, which holds for a single particle:

$$Nu_p = 2 + 0{\cdot}6\,Re^{1/2}\,Pr^{1/3}$$

At typical conditions in our bench scale unit:

$$T = 800\,°C$$
$$\rho_g = 0{\cdot}36\ \text{kg/m}^3$$
$$v_g = 0{\cdot}06\ \text{m/s}$$
$$\bar{d}_{char} = 10^{-3}\ \text{m}$$
$$\bar{d}_{\text{fluidized heat carrier}} = 127 \times 10^{-6}\ \text{m}$$
$$\eta_g = 4{\cdot}4 \times 10^{-5}\ \text{N s/m}^2$$
$$c_{p,g} = 1180\ \text{J/kg °C}$$
$$\lambda_g = 0{\cdot}062\ \text{W/m °C}$$

This leads to $\alpha_{p1} = 152$ W/m^2 °C, when the charring particle diameter is taken as the characteristic diameter. For small Re_p Kunii and Levenspiel[17] give:

$$Nu_p = 0{\cdot}3\,Re_p^{1{\cdot}3} \qquad 0{\cdot}1 < Re_p < 100$$

If we use $d_p = 1$ mm this results in $\alpha_{p2} = 10{\cdot}7$ W/m^2 °C; for $d_p =$ 127×10^{-6}, $\alpha_{p3} = 5{\cdot}8$ W/m^2 °C. The real value for α_p will be somewhere between α_{p1} and α_{p3}; we take $\alpha_p = 50$ W/m^2 °C.

The choice of gas velocity and conductivity in the above calculations is not straightforward in the case of heat transfer from gas and solids to a solid particle of different diameter. This heat transfer problem is at present further studied in our laboratories.

APPENDIX G: MASS AND HEAT TRANSFER RATES IN RELATION TO CHEMICAL KINETICS

In measuring chemical kinetics, one has to realize the possible influence of mass and heat transfer processes and their contribution to the measured reaction rate under experimental conditions.

G.1. Mass Transfer from Bulk Gas to the Reacting Particle

In case the fluidized bed is bubbling, the first mass transfer resistance is for gas being transferred between bubble and dense phase. We checked the possible existence of this resistance by operating the reactor bed as a fixed bed at different gas velocities. Results are tabulated in Table G.I.

TABLE G.I
OBSERVED REACTION RATE CONSTANT AT DIFFERENT GAS VELOCITIES

u_g (m/s)	$k_{observed}$ $(m/s \times 10^8)$
0·029	5·87
0·049	5·97
0·064	7·75
0·07	5·32
0·101	6·45

Experimental conditions: $T = 807\,°C$, $CO_2 = 1\ mol/m^3$, bench scale reactor operated as fixed bed.

There is no clear increase of reaction rate with gas velocity; from this observation we conclude that the mass transfer resistance from bubble to dense phase is very small and can be neglected in the interpretation of reaction rate data.

Mass transfer from gas surrounding the reacting particle in the dense phase of the fluidized bed is calculated from the Nusselt relation:[17]

$$\mathrm{Sh} = \frac{k_m d}{D} = 0{\cdot}374\,\mathrm{Re}^{1\cdot18} \qquad \text{for} \qquad 0{\cdot}1 < \mathrm{Re}_p < 15$$

The mean free path length for CO_2 diffusing inside the reacting particle is larger than the mean pore diameter. Therefore, we have to calculate with the Knudsen diffusion coefficient as the effective diffusivity:

$$D_k = \tfrac{1}{3}\bar{d}_{pore}\sqrt{\frac{8RT}{\pi M}}\,M_{CO_2}$$

with

$$\bar{d}_{pore} = 4\,\frac{\varepsilon_c}{S}$$

From the calculated values of k_m and D_k we learn that the outside mass transfer resistance is small compared to the inside mass transfer resistance. The internal mass transfer rate is compared to the observed reaction through use of the Thiele modulus:

$$\Phi = \frac{d_p}{6}\sqrt{\frac{r_{observed}}{DCO_2}}$$

For $\Phi < 0{\cdot}3$, the influence of mass transfer can be neglected.

G.2. Heat Transfer from Bulk Gas to the Reacting Particle

The heat transfer coefficient from gas to particle is estimated to be

$$\alpha_p = 50\ \mathrm{W/m^2\ ^\circ C}\ \text{(Appendix F)}$$

The internal particle heat transfer coefficient is calculated from

$$\alpha_i = 5{\cdot}8\,\frac{\lambda_s}{d_p} = 406\ \mathrm{W/m^2\ ^\circ C}$$

We calculate the particle temperature from a heat balance over the reacting particle:

$$\Delta H_r V_p R'' S_0 = \alpha A(T - T_s) \qquad \text{(G1)}$$

or

$$\Delta T = (T - T_s) = \frac{d_p}{6}\,\frac{R'' S_0}{\alpha}\,\Delta H_r \qquad \text{(G1a)}$$

The calculated ΔT is the correction on the bulk gas temperature.

In Table G.II results are gathered.

TABLE G.II

RESULTS OF MASS AND HEAT TRANSFER CALCULATIONS IN RELATION TO CHEMICAL KINETICS

T (K)	Re_p	D_{CO_2,N_2} $\left(\frac{m^2}{s} \times 10^4\right)$	$D_{K,char}$ $\left(\frac{m^2}{s} \times 10^6\right)$	k_m (m/s)	$R'' \times 10^8$ $\left(\frac{mol}{m^2 s}\right)$	Φ	ΔT $(°C)$
995	0·52	1·59	5·34	0·027	1·36	0·088	0·08
1 055	0·49	1·74	4·56	0·028	4·27	0·187	3·3
1 090	0·46	1·84	6·32	0·028	7·84	0·190	4·7
1 113	0·44	1·91	5·32	0·027	11·46	0·287	9·0
1 157	0·42	2·03	6·66	0·026	22·68	0·359	17·5

$d_p = 10^{-3}$ m; char properties as in Table III.

From Table G.II we see that mass transfer influence is small enough to be neglected, $k_m \gg D_k/d$, $\Phi < 0{\cdot}36$. The particle temperature correction varies from 0·8 to 17·5 °C at reaction conditions.

REFERENCES

1. Lewellen, P. C. *et al.* (1976). *Proceedings of the 16th International Symposium on Combustion*, August.
2. Antal, M. J. *et al.* (1980). ACS Symposium Series 130, 178th Meeting of ACS. Jones and Radding (Eds), 237–55.
3. Roberts, A. F. (1970). *Combustion & Flame* **14**, 261–72.
4. Tran, D. Q. and Rai, C. (1978). *Fuel* **57**, 293.
5. Rensfelt, E. *et al.* (1978). *Proc. I.G.T. Conference Energy from Biomass and Wastes.* August, Washington, D.C.
6. Lindman, N. (1971). *Proc. Energy from Biomass and Wastes* **5**, 571–611.
7. van Krevelen, D. W. *et al.* (1951). *Fuel* **30**, 253.
8. Shafizadeh, F. (1977). *Fuels from Waste*, L. L. Anderson and D. A. Tillman (Eds). Academic Press, New York, Chapter IX.
9. Kanury, M. (1970). *Combustion Science and Technology* **1**, 339–55.
10. Epstein, E. *et al.* (1978). I.G.T. Symposium, Energy from Biomass and Wastes. August, Washington, D.C.
11. Burton, R. S. and Bailie, R. C. (1974). *Combustion* Feb., 13–19.
12. Sakoda, A. *et al.* (1981). *The Chemical Engineering Journal* **22**, 221–7.

13. Perry, R. M. and Chilton, C. W. (Eds) (1973). Heat and free energies of formation. *Chemical Engineers' Handbook*, 5th edn. McGraw-Hill, New York, Table 3–202, p. 3–138.
14. Szekely, J. and Evans, J. W. (1970). *Chem. Eng. Sci.* **25**, 1091; *ibid.*, **26**, 1901.
15. Groeneveld, M. J. (1980). Ph.D. Thesis, Twente University of Technology.
16. Ergun, S. (1962). Bulletin 598, V.S. Bureau of Mines.
17. Kunii, D. and Levenspiel, O. (1969). *Fluidization Engineering*. Wiley, New York.

37

Modelling of Fluidized Bed Wood Gasifiers

D. W. Bacon, J. Downie, J. C. Hsu and J. Peters
Department of Chemical Engineering, Queen's University, Kingston, Ontario, Canada

1. INTRODUCTION

The world oil crisis stimulated interest in alternative sources of energy for industry, and the gasification of biomass to produce a clean fuel gas is gaining support as an attractive alternative in particular situations. In Canada several types of gasifiers at different stages of development, ranging from pilot scale to prototype units, are being developed by industry with government assistance.

Among the various types of gasifiers being developed the fluidized bed gasifier is probably the most promising. It minimizes the tar problems associated with the up-draft, and has more potential for large scale operation than the down-draft.

Fluidized beds have been in use for several decades in the chemical process industries but this is a new technology in gasification. Although a considerable amount of research has been done on fluidization the process is still not well understood, and the present report on wood gasification is limited to an operational feasibility study. In this work an attempt has been made to develop a mechanistic model to describe the gasifier performance, with the aim of providing a technical basis for the assessment of this type of gasifier.

In the development of the model, the Modified Bubble Assemblage Model is used for fluidization. Due to the lack of kinetic information on wood pyrolysis and gasification, assigned degrees of approach to equilibrium are used to describe the overall gasification process. The results of simulated calculations and sensitivity analyses on the model are presented.

2. MATHEMATICAL MODELLING

2.1. Gasification Processes

In an air gasifier, the wood undergoes the processes of heating, drying, pyrolysis, gasification and combustion. Due to the high circulation rate of sand in a fluidized bed the wood chips are rapidly distributed throughout the bed. Hence these processes occur throughout the bed with one exception—at the bottom of the bed the combustion reactions predominate due to the high concentration of O_2 from the inlet air in this region.

It is apparent from the survey paper by Milne[8] that wood gasification is a very complicated process, and that the product distribution is particularly sensitive to the rate of heating and the residence time in the reactor. The research on kinetics of wood pyrolysis is mainly based on weight loss, which is of no help in estimating product distribution. Furthermore, the experimental conditions under which the kinetic data were obtained are usually different from those expected in a fluidized bed gasifier. Thus a number of simplifying assumptions must be made in modelling this process.

The simplest approach to this problem would be to assume thermodynamic equilibrium from the reactions and to consider the reaction products to consist of only CO, CO_2, H_2, H_2O and CH_4. However, it has been recognized for some time that the CH_4 concentration in the product gas from wood gasifiers is substantially higher than would be estimated from an equilibrium calculation.[4] Therefore, a modified equilibrium approach will be used here.

The simplification of considering the reaction products to consist of only CO, CO_2, H_2, H_2O and CH_4 has some experimental support from the pilot plant results of Black *et al.*,[2] in which the tar loading of the gas was given at $<0{\cdot}1\%$. The gases in Black's paper did contain some C_2H_2, C_2H_4 and C_2H_6, in addition to CH_4, but considering these as CH_4 is unlikely to change the overall results. The simplified product gas can then be used as a first approximation, and this has obvious advantages in modelling.

Desrosiers[3] identified five principal gasification reactions, of which only three are independent. For convenience we use the following three reactions:

$$C + CO_2 = 2CO \qquad \text{(i)}$$

$$C + H_2O = CO + H_2 \qquad \text{(ii)}$$

$$C + 2H_2 = CH_4 \qquad \text{(iii)}$$

In computing the equilibrium composition, the approach of minimizing the total Gibbs free energy of the system was used.[11]

Since we adopted an approach similar to Gumz[4] for the equilibrium calculations, a modified free energy term was used. A modified equilibrium constant, K', is defined as the true equilibrium constant multiplied by the degree of approach to equilibrium, ε, giving

$$\ln K_j'/\varepsilon_j = -\sum_i \Delta G_{f,i}^{\circ}/RT \tag{1}$$

where $G_{f,i}^{\circ}$ = the free energy of formation of compound i at T. Defining

$$a_k = 1 - (RT/\Delta G_{f,k}^{\circ}) \tag{2}$$

we can write the modified equilibrium constants for the three reactions as

$$\begin{aligned} K_1' &= \exp\left[-(a_{CH_4}\Delta G_{f,CH_4}^{\circ})/RT\right] \\ K_2' &= \exp\left[-(2\,\Delta G_{f,CO}^{\circ} - a_{CO_2}\Delta G_{f,CO_2}^{\circ})/RT\right] \\ K_3' &= \exp\left[-(\Delta G_{f,CO}^{\circ} a_{H_2O}\Delta G_{f,H_2O}^{\circ})/RT\right] \end{aligned} \tag{3}$$

This is equivalent to a new set of equilibrium equations with $\Delta G_{f,k}^{\circ}$ replaced by $a_k \Delta G_{f,k}^{\circ}$.

The thermodynamic data needed in equilibrium calculations were obtained from JANAF thermodynamic tables compiled by Stull and Prophet.[12]

2.2. Fluidization Model

The fluidization model adopted in this work is the modified version of Wen's Bubble Assemblage Model.[9] The fluidized bed is considered to be composed of a number of compartments consisting of bubble and emulsion phases. The gas in the emulsion phase flows upwards at the minimum fluidization velocity, which was estimated according to Leva's equation.[7] The correlation contains a constant α_{mf}, which has a value of $9{\cdot}27 \times 10^{-3}$. Realizing that this value might not apply to the system of our interest we treated α_{mf} as an adjustable parameter.

The wood enters the fluidized bed through a feed nozzle which is usually located above the middle of the bed. Due to the vigorous circulation of the sand particles in the bed the wood is quickly distributed throughout the bed. The quantitative distribution of wood is unknown, and an empirical distribution function was used.

$$f(n) = 2{\cdot}2576 P_A^{1{\cdot}5} \exp\left[-P_A(P_B - h)\right]^2 (P_B - h)^2 \tag{4}$$

where h = the vertical distance from the bottom of the bed, P_A = an adjustable parameter, P_B = a parameter equal to $P'_B\,(U_0 = U_{mf})$ where P'_B is a constant, u_0, u_{mf} = superficial velocity and velocity at minimum fluidization respectively ($m\,s^{-1}$).

It is assumed that most of the wood pyrolysis occurs at the level of the feed nozzle, and to incorporate this in the distribution function we use the following relationship:

$$h_{max} = P_B - (1/P_A)^{0.5} \tag{5}$$

where h_{max} = the distance from the bottom of the bed to the wood feed nozzle.

The wood feed is assumed to be instantly distributed according to this function. The wood is then assumed to undergo instantaneous devolatilization. The wood and resulting char particles are assumed to exist only in the emulsion phase. However, 1 % of the sand is considered to enter the bubble phase.[5]

The total gas in the emulsion phase is considered to include the gas associated with the cloud and wake of the bubbles. The balance of the gas flow is assigned to the bubble phase. There is an interchange between the emulsion phase and the bubble phase within the compartment, due to gaseous component concentration gradients between the phases. The rate of interchange is calculated from the following equation:

$$N_j = K_{GT} A_t h_i (P/RT_i)(Y_{i,j} - X_{i,j}) \tag{6}$$

where N_j = rate of transfer of component j from the bubble phase to the emulsion phase ($mol\,s^{-1}$); A_t = cross-sectional area of the fluidized bed (m^2); h_i = height of the ith compartment (m); P = pressure (atm); R = gas constant ($atm\,m^3\,K^{-1}\,mol^{-1}$); T_i = temperature of the ith compartment (K); $X_{i,j}$ = mole fraction of component j in the emulsion phase; $Y_{i,j}$ = mole fraction of component j in the bubble phase; K_{GT} is the interchange coefficient, estimated according to Kobayashi.[10]

$$K_{GT} = 0{\cdot}11/D_B \tag{7}$$

where D_B is the bubble diameter. For the first compartment the jet effect must be considered. No quantitative correlation of the mass transfer coefficient between the jet and the emulsion phase is available. However, from the experimental data of Behie and Keho[2] the ratio of the mass transfer coefficients for the two regions is fairly constant, and is in the range 40–50.

Since it is assumed that the wood chips enter the emulsion phase only, the gas produced from the gasification and combustion will exceed the minimum fluidization velocity. This excess gas is assumed to bleed continuously into the bubble phase.

In the mass balance calculations the bed was divided into compartments. The height of a compartment is determined by the size of the local bubble. Temperature is assumed uniform throughout the compartment. However, for the mass balance and equilibrium calculations a compartment is further divided into subcompartments, in which pseudo-equilibrium is assumed for either bubble or emulsion phase.

The bubble diameter and the height of a compartment were calculated from the correlations developed by Mori and Wen,[9,10] which include the effect of bed diameter and distributor design. The use of these equations differs from that of Mori and Wen in two respects. First, the height is measured from the distributor, whereas Mori and Wen measure it from the top of the jets. Second, Mori and Wen use the combined height of jet penetration and the critical height as the height of the first compartment. In this work these are treated separately as the first and second compartments respectively.

The mass of sand in the nth compartment is given by the following equation:

$$W_{sn} = 1{\cdot}01 A_t(\Delta h_n)(1 - \varepsilon_{Bn})(1 - \varepsilon_{mf})\rho_s \tag{8}$$

where A_t = cross-sectional area of the gasifier (m^2); Δh_n = the height of the nth compartment (m); ε_{Bn} = volume fraction of bubbles in the bed in the nth compartment; ε_{mf} = void fraction in the emulsion phase at minimum fluidization velocity (ε_{mf} is assumed constant); ρ_s = density of sand (kg/m^3).

The calculations of ε_{Bn} and ε_{mf} follow the approach of Mori and Wen.[9] The coefficient 1·01 is included to account for the 1 % sand assumed to be in the bubble phase.

2.3. Energy Balance

The flows into and out of the ith compartment include wood feed W_d, total molar gas flow G_j, upward solid flow W_{mix}, and downward solid flow W'_{mix}. By assuming the temperature of the compartment to be uniform and gas ideal, the energy balance equation over compartment i can be written as follows:

$$\begin{aligned}\sum_j (G_{j,i}H_{j,i} - G_{j,i-1}H_{j,i-1}) + (W_{mix,i} + W'_{mix,i})C_{ps}(T_i - 298)\\ - W'_{mix,i+1}C_{ps}(T_{i+1} - 298) - W_{mix,i-1}C_{ps}(T_{i-1} - 298)\\ - W_{d,i}(\Delta H_v) + Q_{loss,i} = 0\end{aligned} \tag{9}$$

where C_{ps} = the heat capacity of the solid ($kJ\,kg^{-1}\,K^{-1}$) and $H_{j,i}$ = the enthalpy of component j at T_i.

The reference state used in calculating enthalpies is 298 K and 1 atm. The enthalpies of the gases are based on the constituent elements of each component.

It is assumed that the flow rate of solid upwards from a compartment, $W_{mix,i}$ is equal to the flow rate of solid downwards from the compartment above, i.e.

$$W_{mix,i} = W'_{mix,i+1} \tag{10}$$

At the bottom of the first compartment both flows are zero.

The simplified expression for the rate of solid mixing developed by Kunii and Levenspiel,[6] p. 155, is used to estimate W_{mix}.

$$W_{mix,i} = (u_0 - u_{mf})A_t f_w (1 - \varepsilon_{mf})\rho_s \tag{11}$$

where f_w = the ratio of the wake volume to the bubble volume, which can be considered as a solid mixing parameter.

This ratio, f_w, is treated as a model parameter which may be assigned a value in the range 0·1–0·3. Provision is made in the model for an additional parameter, f_{ws}, at the top of the bed where part of the solid in the wake is thrown into the freeboard.

$W_{d,i}\Delta H_v$ is the thermal energy associated with devolatilization. This is assumed to consist of two terms: one for the evaporation of the moisture in the wood, and the other for formation of the reaction products of devolatilization.

$$W_{d,i}\Delta H_v = W_{vd,i}\Delta H_{vd} + W_{H_2O,i}\Delta H_{H_2O} \tag{12}$$

where $W_{vd,i}\Delta H_{vd}$ = the thermal energy associated with devolatilization of dry wood, and $W_{H_2O,i}\Delta H_{H_2O}$ = the thermal energy associated with 'devolatilization' of moisture in the wood.

The term involving heat loss from the gasifier, Q_{loss}, is estimated from heat transfer considerations.

2.4. Computational Algorithm

Calculations begin at the bottom of the fluidized bed where the gas enters. A temperature is assigned to the first compartment at the start of the program, and as the computation proceeds the temperatures of successive compartments are estimated from the energy balance calculation. The progress of the calculations towards the top of the fluidized bed is monitored by calculating the cumulative mass of sand in the compartments

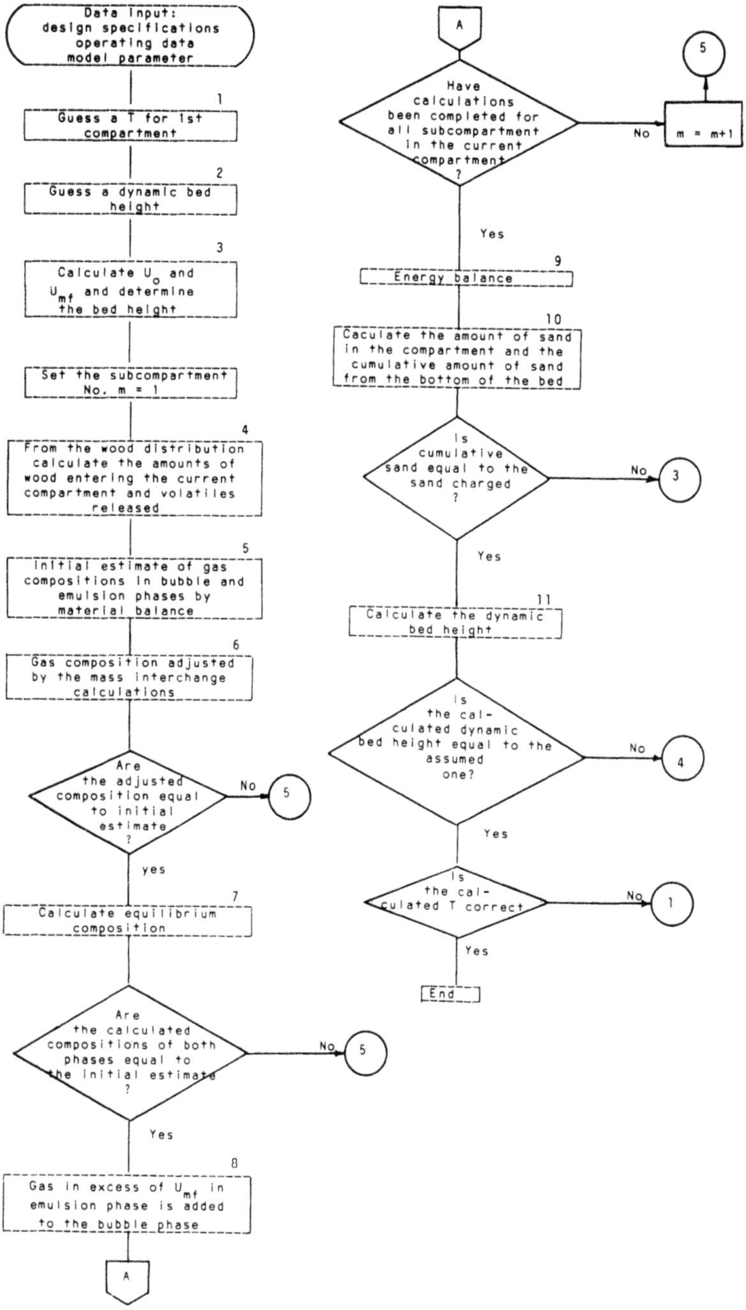

FIG. 1. Computational algorithm.

and comparing this with the total mass of sand charged to the gasifier. The computational algorithm is summarized in Fig. 1.

In addition to a guessed temperature for the first compartment, the calculations start with a guessed value of the dynamic bed height. If the calculated dynamic bed height does not agree with the guessed value, the calculations are repeated with a new dynamic height. If the calculated temperature at the top of the bed is not equal to the desired operating temperature, a new temperature for the first compartment is used and the calculations are repeated.

3. RESULTS AND DISCUSSION

The gasifier specifications and a typical set of operating conditions are shown in Fig. 2. The product gas flow rate and composition are summarized in Table I.

The temperature profiles in the bed resulting from three different temperature values being assigned to the first compartment are shown in Fig. 3. The abscissa, fractional bed height, is defined as the cumulative height of the compartments divided by the dynamic bed height.

The temperature varies in the bed by about 10 K, which is approximately the range in practice. A maximum occurs near the bottom of the bed. This variation would be expected to decrease as the circulation rate of solid in the bed increases, and this effect is shown in Fig. 3 by including three curves for different values of the solid mixing parameter, f_w.

The composition profiles for each component of the gas are shown in Fig. 4. The shapes are as anticipated, with oxygen decreasing rapidly in the

TABLE I

PRODUCT GAS FLOW RATE AND COMPOSITION: A SUMMARY OF RESULTS

Component	*Flow rate (kmol/h)*	*Mole fraction (wet basis)*	*Mole fraction (dry basis)*
CO_2	6·682	0·138	0·161
CO	5·728	0·118	0·137
H_2O	6·654	0·138	—
H_2	4·176	0·081	0·100
CH_4	2·181	0·045	0·052
N_2	23·933	0·474	0·550
Total	48·359 (1 083 SCM/h)		

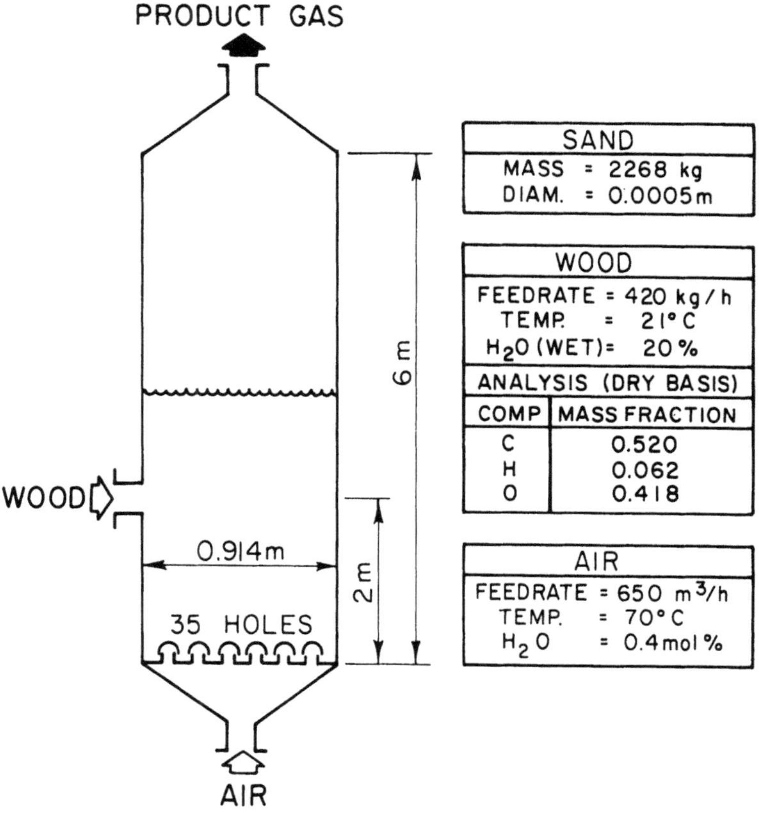

FIG. 2. Example of gasifier specifications and operating conditions.

lower section of the bed. In this section the carbon dioxide concentration increases rapidly and we have what might be described as a 'combustion zone'. In this zone the H_2, CO and CH_4 which form are reacted with the excess oxygen present. The concentrations of these 'fuel' gases increase rapidly in the upper section of the bed. When the air/dry wood ratio is reduced the CO_2 profile can show a maximum near the wood feed location.

3.1. Sensitivity Analyses

Since the model incorporates many assumptions with adjustable parameters, useful information can be derived by conducting sensitivity analyses to determine the relative importance of the parameters and variables used in the model. These analyses will clarify the implications of the model, and will focus attention on areas which should be most

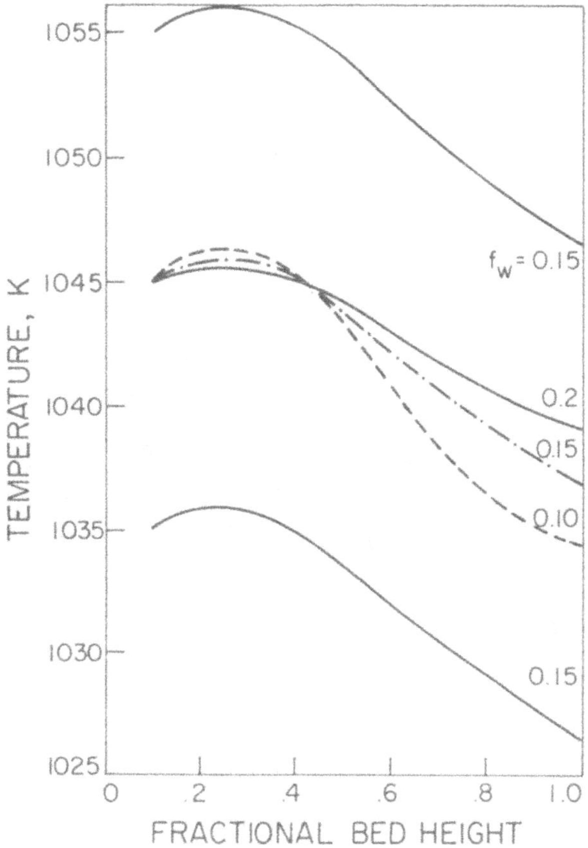

FIG. 3. Temperature profiles as a function of T_1 and the solid mixing parameter, f_w.

profitable to explore from the point of view of testing the adequacy of the model.

Three sets of sensitivity analyses were performed: one studying the influence of model parameters on overall model performance, one for operating variables, and one for design variables.

In the study of model parameters, a two-level fractional factorial design was used to identify the individual and joint effects of the 14 model parameters; a_{CH_4}, a_{CO_2}, a_{H_2O}, f_w, ε_{mf}, ΔH_{vd}, K_{GT}, P'_B, f_{ws}, α_{mf}, R_s, R_{CH}, R_{CO} and G_{rf} where G_{rf} is the correction factor for the mass interchange coefficient in the jet zone, and R_s, R_{CO} and R_{CH} are the ratios of char/C in wood, CO/CO_2 and CH_4/H_2 respectively in the volatiles. Under the reasonable assumption

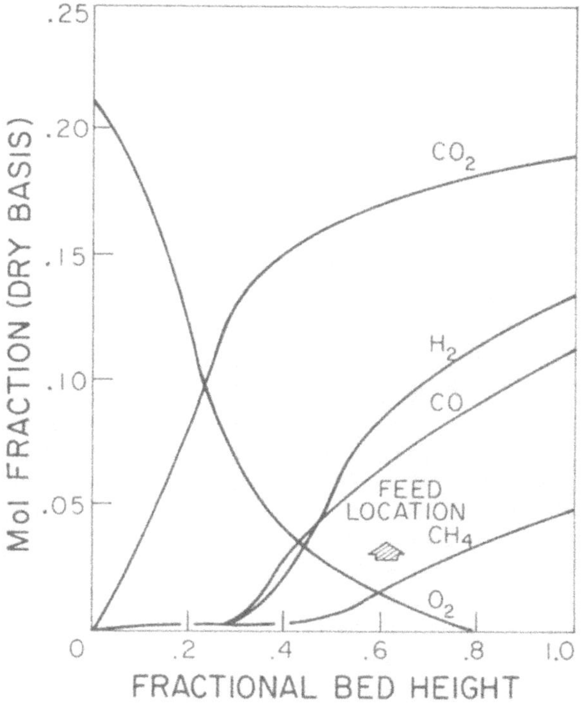

FIG. 4. Composition profiles in the fluidized bed.

that interaction effects involving more than two parameters are negligible, this design yields uncorrelated estimates of the individual effects of all 14 parameters and of 15 linear combinations of two-parameter interaction effects.

With the exception of CH_4, the composition of gas is almost solely determined by the parameters that determine the degree of approach to equilibrium. Methane, on the other hand, is affected by almost all the parameters.

The temperature at the top of the bed is significantly affected by f_w, α_{mf}, H_{vd} and P'_B and the effects of certain linear combinations involving two-parameter interactions. However, with the experimental design used here, we cannot identify the specific combinations.

Surprisingly, the calorific value is insensitive to interactions in this analysis over the ranges investigated. Although the parameter a_{CH_4} was varied from 0·95 to 1·0 (a significant variation in a_{CH_4}) in this sensitivity analysis, the resulting average change in calorific value was only 0·4 mJ/m^3.

Since the resulting gas composition varies significantly for such a perturbation in a_{CH_4} it can be concluded that the gas composition does not influence the calorific value of the product gas as long as the wood and air flow rates are fixed.

Individual and joint effects of seven operating variables were also studied. The variables selected were inlet gas temperature T_{in}, gas pressure P, volumetric gas flow rate G_v, oxygen mole fraction in gas y_{O_2}, dry basis moisture content of gas y_w, dry wood to oxygen ratio W/O, and dry basis moisture content of the wood. Once again a two-level fractional factorial design was employed and with 64 runs it was possible to obtain uncorrelated estimates of the individual effects of all 7 operating variables and of the 21 two-variable interactions involving those variables.

The significance of the effect of each operating variable and each interaction was determined using a t-statistic at the 10 % significance level. It appears that the 7 operating variables have no clear effects on the temperature profile and the bed height. Over the ranges of perturbation tested, gas inlet temperature, water vapour in the gas and moisture content of the wood all were found to have negligible effects on each of the seven selected model response variables.

Although the inlet gas temperature has little effect on the gas composition or the calorific value, it is important to the operating temperature under certain circumstances. For example, if the wood feed contains 50 % moisture on a wet basis the model indicates that the gasifier cannot be operated satisfactorily at an inlet gas temperature of 70 °C. However, satisfactory performance is predicted when the temperature of the inlet gas is raised to 400 °C.

The temperature of a fluidized bed gasifier can normally be controlled by adjusting the feed rates but there is a limit beyond which the gasifier cannot be operated. For example, according to our model, gasification reactions cannot be sustained at 1395 K without an external supply of thermal energy. Desirable operating temperatures lie in the range from 900 K to 1200 K.

Since we were not interested in high pressure operation, the pressure range studied in the sensitivity test was between 1·5 and 2·5 atm. Within that range, the calorific value and product gas composition remained essentially constant.

On the other hand, at an operating pressure of 2·5 atm the model predicts reductions of 25 % in bed height and almost 50 % in bubble size. Thus, a slight increase in pressure could have the advantage of moving the fluidized bed away from the slugging region.

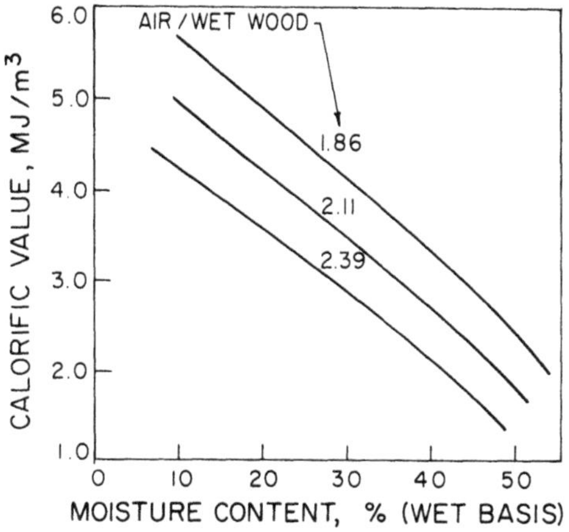

FIG. 5. Calorific value of the product gas versus moisture content of the wood (wet basis), at constant air/wet wood ratios.

The most significant variable we found in this analysis is the air/dry wood ratio. This led us to investigate this matter further. Model calculations were performed to predict the calorific values of the product gas, C_v, as a function of moisture content in wood on a wet basis for different values of the air/wet wood ratio. The results showed a strong functional relationship between C_v and moisture content as reported by others.[2] However, as C_v is plotted against the moisture content keeping air/dry wood ratio constant (Fig. 5), it is clear that C_v is not a strong function of the moisture content of wood. In contrast to this, a change in the air/dry wood ratio of approximately 15 % creates a change in C_v of over 30 %. Further evidence of the dominant effect of the air/dry wood ratio is shown in Fig. 6 where C_v is plotted against this ratio.

It is interesting to compare our results with those calculated by Desrosiers[3] based on complete equilibrium for an adiabatic system. In Desrosiers' calculation the thermal efficiency C_m defined as the calorific value of gas produced per unit mass of dry wood gasified passes through a maximum at a value of equivalence ratio in the neighbourhood of 0·25, which is equivalent to an air/dry wood ratio of 1·57. This led Desrosiers to suggest that gasification should be carried out at this ratio. For fluidized bed conditions, our model calculations show no such maximum but C_m seems to approach a constant value. Our calculated results indicate that when the ratio has a value of 1·5 or below, the gasifier cannot be operated properly.

Simulated calculations using O_2-enriched gas were also performed.

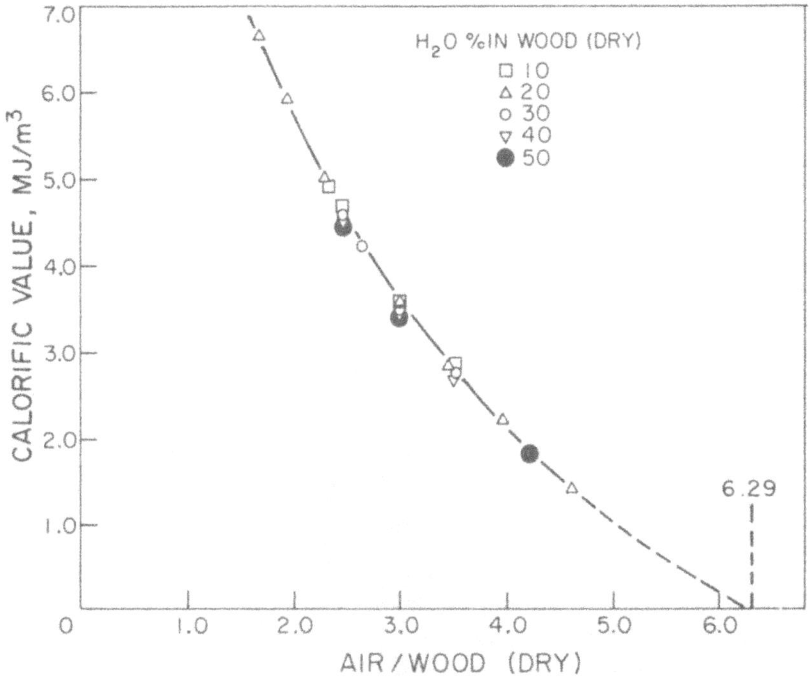

FIG. 6. Calorific value of the product gas versus air/dry wood ratio.

Results showed that C_v is an increasing function of O_2 content. The influence becomes stronger at high dry wood/O_2 ratios. The increase in C_v is a direct result of increased concentrations of CO, H_2 and CH_4. However, the relative increase in CO_2 leads to no improvement in thermal efficiency and the total calorific value of the fuel gas produced per unit mass of wood gasified remains essentially constant.

A two-level fractional factorial design was used in the sensitivity analysis on the following gasifier design variables: sand particle size, location of wood feed, insulation, sand loading and number of holes in the distributor. The analysis provides uncorrelated estimates of the individual effects of those five design variables and their ten two-variable interactions.

With the exception of those runs in which little insulation was used, the results reveal that the design of a fluidized bed has little effect on the gasification reaction. Within the tested range of design specifications, the gas composition, calorific value and thermal efficiency remained fairly constant.

The analysis on the effect of insulation indicates that if the heat loss exceeds about 10% of the total enthalpy of the gas flow with respect to the reference state, the gasification reaction cannot be self-sustaining.

Other design variables have a direct effect on how the fluidized bed behaves. Sand particle size has strong influence on the bubble size. With large particles the flow rate of gas is severely limited if slugging is to be avoided.

4. CONCLUSION

A mathematical model has been developed as a basis for assessing the performance of a sand fluidized bed wood gasifier. The model uses the Modified Bubble Assemblage Model for fluidization, and assigned degrees of approach to equilibrium for the gasification reactions.

Simulation runs were made using industrial operating conditions. The results for the composition and calorific value of the product gas, and the temperature difference across the fluidized bed, were in agreement with the limited amount of industrial data available. The assumptions used therefore appear to be reasonable. However, it must be emphasized that additional data are needed before the model can be adequately tested.

Sensitivity analyses were carried out on the model parameters, operating variables and design variables. The results show that the gas composition, with the exception of CH_4, is almost entirely dependent on the assigned degree of approach to equilibrium. The calorific value of the product gas is strongly dependent on air/dry wood ratio, but not the moisture content of wood. Using O_2 enriched gas increases the calorific value of the product gas but does not improve the overall thermal efficiency.

Within the range of design variables used in the sensitivity analysis these variables had a negligible effect on the product gas composition and calorific value. This implies that the fluidized bed may be designed solely from the standpoint of stable fluidization behaviour without the additional complication of trying to design for gasification.

ACKNOWLEDGEMENTS

We would like to acknowledge the helpful advice of Dr J. H. Verner of the Department of Mathematics and Statistics, Queen's University, on numerical analysis, and the computer programming of Mr S. Lui.

The interest and encouragement of Dr Ralph Overend of the National

Research Council of Canada was greatly appreciated. The effort of ECO-RESEARCH personnel to supply us with industrial data was also appreciated.

REFERENCES

1. Behie, L. A. and Kehoe, P. (1973). The grid region in a fluidized bed reactor. *A.I.Ch.E. Journal* **19**, 1070.
2. Black, J. W., Bircher, K. G. and Chisholm, K. A. (1979). Fluidized bed gasification of solid wastes and biomass—The C.I.L. program. Symposium on Thermal Conversion of Solid Waste and Biomass, Amer. Chem. Soc. Meeting, Washington, D.C., September 10–14.
3. Desrosiers, R. (1979). Thermodynamics of gas–char reactions. *A Survey of Biomass Gasification*, Vol. 11, Chapter 6. SERI/TR-33-239, July.
4. Gumz, W. (1950). *Gas Producers and Blast Furnaces.* Wiley, New York.
5. Kobayashi, H., Aria, F. and Chiba, T. (1965). Behavior of bubbles in a gas–solid fluidized bed. *Kagaku Kogaku* **4**(1), 147; Kobayashi, H. and Aria, F. (1966). 6th Symposium on Chem. Reaction Eng., Soc. Chem. Engrs., Japan, Nagoya, November.
6. Kunii, D. and Levenspiel, O. (1977). *Fluidization Engineering*. Robt. K. Kreiger Publ. Co., New York.
7. Leva, M. (1959). *Fluidization.* McGraw-Hill, New York.
8. Milne, T. (1979). Pyrolysis—The thermal behavior of biomass below 600 °C. *A Survey of Biomass Gasification*, Vol. 11, Chapter 5. SERI/TR-33-239, July.
9. Mori, S. and Wen, C. Y. (1975). Simulation of fluidized bed reactor performance by modified bubble assemblage model. *Fluidization Technology*, Vol. 1, D. Kearns (Ed.).
10. Mori, S. and Wen, C. Y. (1975). Estimation of bubble diameter in gaseous fluidized beds. *A.I.Ch.E. Journal* **21**, 109.
11. Smith, J. M. and Van Ness, H. C. (1975). *Introduction to Chemical Engineering Thermodynamics* 3rd edn. McGraw-Hill, New York, p. 426.
12. Stull, D. R. amd Prophet, H. (1971). *JANAF Thermochemical Tables* 2nd edn. Office of Standard Reference Data, National Bureau of Standards.

38

Thermal Cracking of Tars and Hydrocarbons by Addition of Steam and Oxygen in the Cracking Zone

O. JÖNSSON

Lund Institute of Technology, Lund, Sweden

1. INTRODUCTION

In the gasification of biomasses for production of methanol it is desirable to reduce the contents of methane and other higher hydrocarbons in the gas.

A feasible option to obtain this is by heating the gas and simultaneously injecting steam and oxygen in the cracking zone. The work under way at the Department of Chemical Engineering, Lund Institute of Technology is aimed at finding the optimal temperature and steam/oxygen ratio at which the highest conversion is obtained in such a cracking process.

For this purpose laboratory equipment consisting of a pyrolysis reactor, a cracking reactor and an analysis system has been constructed and a series of experiments has been performed.

2. THEORY OF WOOD PYROLYSIS

Pyrolysis is defined as the gas-producing heating of a material under inert atmosphere and can be observed at temperatures as low as 300 °C. At low reaction temperatures the main products from wood pyrolysis are char, carbon dioxide and water. However, a rise in temperature is followed by volatilization of components such as CH_4, C_2H_x, C_3H_x and H_2. The gas composition is strongly affected by the reaction temperature and the biomass composition.

2.1. Effects of Wood Properties on Pyrolysis

Wood consists of three different components: cellulose, hemicellulose and lignin. Traces of Na, K and other minerals are also very common as an effect of fertilizers etc. The reactivity of the three main components differs widely, with hemicellulose as the most reactive. At temperatures above 320 °C cellulose becomes the most reactive because most hemicellulose then is pyrolysed and cellulose's relative importance increases with increasing temperature.

Lignin is very hard to pyrolyse and yields *ca.* 45% solid residue when heated to constant weight. The same figure for cellulose is 15% solid residue. Tracers (Na, K etc.) are found to have a catalytic effect on the decomposition of wood. This effect is detectable at contents as low as 0·15% inorganic additive and by adding 5% $(NH_4)_3PO_4$ a 100-times increase in reaction rate has been measured.[1]

When doing experiments with pyrolysis of cellulose cylinders a certain autocatalytic effect is also observed. This effect is detected as a higher reaction rate in the centre of the cylinder than on the surface of the cylinder. This is explained by the autocatalytic effect of the products. Due to the higher concentration of products in the centre of the cylinder, the autocatalytic effect yields a higher reaction rate. From these available data the conclusions can be drawn that pyrolysis of wood is favored by

(1) high content of hemicellulose in the wood;
(2) low content of lignin in the wood;
(3) addition of inorganic salts.

2.2. Effect of Temperature on Wood Pyrolysis

As mentioned earlier the product distribution is strongly affected by the reaction temperature. This can be expressed as: the higher the pyrolysis temperature, the simpler the pyrolysis products. Thus the production of tars and other higher hydrocarbons decreases at high temperatures in favour of CO, H_2 and, to some extent, CH_4. At very high temperatures even the amount of CH_4 produced is decreased in favour of an extended production of CO and H_2. In connection with these statements it might be appropriate to point out that these conclusions are from experiment in a fixed bed reactor and might differ for other types of reactors (flash pyrolysis etc.).

The amount of gas produced from a pyrolysis reaction is also affected by the reaction temperature, while the product composition is driven towards a higher percentage of gaseous components at higher temperatures. As the density of the gas also decreases at higher temperatures the total effect is a larger volume of gas produced from the pyrolysis reactions.

2.3. The Kinetics of Wood Pyrolysis

Pyrolysis is usually defined as a large number of reactions which, in this case, involves gas production from a solid material. Because of this very complex reaction system a general kinetic model is very hard to accomplish. Another usual method is fitting the first part of the pyrolysis to a zero order expression (one author[3] has found this adequate up to 20% weight loss) and then using a first order expression for the second part of the pyrolysis. One very common expression is

$$\partial\rho/\partial t = (\rho - \rho_f)K \qquad K = k_0 e^{-E/RT}$$

Kinetic data are usually obtained as values of k_0 and E. Typical values of these constants are $10^9\,s^{-1}$ for k_0 and 146 kJ/mole for E. Other theories[4] claim that the values of k and E are changed when the pyrolysis process has reached a certain level. This level can be characterized by, e.g., $\rho/\rho_0 = 1/2$ or 1/3.

Such theories can be, and have been, explained by the difference in reactivity between cellulose and lignin and the 'level' is said to be where the pyrolysis of cellulose loses its large influence on the total reaction. The zero order reaction at the beginning of pyrolysis can be explained in a similar way by the fast pyrolysis of hemicellulose.

3. THERMAL CRACKING

Thermal cracking can be defined as a high temperature decomposition of, in this case, products from a primary pyrolysis. In the laboratory equipment constructed, the heat is supplied by means of an electric furnace but on a commercial level it will probably be necessary to supply heat by combusting some of the pyrolysis products in the cracking reactor.

3.1. Thermal Cracking of Tars

Wood pyrolysis in a fixed bed reactor produces approximately 30–50% condensable products—mainly tars. These tars are undesirable components in a product gas and can be decomposed by heating. The cracking takes place, to some extent, in the pyrolysis reactor, but to accomplish a more efficient decomposition higher temperatures and greater residence times are desirable. Some studies[2] have shown that thermal cracking does not occur if the cracking temperature is lower than the primary pyrolysis temperature. Another criterion on the cracking temperature is the products desired and also what residence time is considered reasonable.

A very high temperature will most probably drive the cracking towards very simple components such as CO and H_2. As a consequence of this the content of hydrocarbons will decrease at high temperatures. Indications of the size of a sufficient residence time for the tar cracking is quite hard to obtain from published material since this type of study seems to be rather unusual. In the present laboratory equipment residence times between 0·5 and 8 s can be tested and it is probably somewhere in this area that a commercial plant would be operated.

3.2. Thermal Cracking of Methane

Methane can be removed from a synthesis gas via three different reactions:

(1) $CH_4 + 2O_2 \rightarrow CO_2 + 2H_2O$
(2) $CH_4 + H_2O \rightarrow CO + 3H_2$
(3) $CH_4 \rightarrow C + 2H_2$

Reaction (2) is the most desirable, while this reaction produces the main components in a synthesis gas. Reaction (1) is the combustion reaction and reaction (3) produces solid carbon which yields a decreased gasification efficiency. Unfortunately, various experiments have shown that reaction (2) is the one that is the furthest from equilibrium. Reaction (3) will also be far from equilibrium, while the hydrogen content in a synthesis gas possesses an inhibiting effect on the kinetics of this reaction. This effect has been studied by (among others) Eisenberg and Bliss.[5] Their model is based on the assumption that hydrogen 'destroys' methyl radicals that otherwise are a part of the first step in the following scheme:

$$CH_4 \rightarrow CH_3 + H$$
$$CH_3 + CH_4 \rightarrow C_2H_6 + H$$
$$C_2H_6 \rightarrow 2CH_3$$
$$C_2H_6 \rightarrow H_2 + C_2H_4 \rightarrow C_2H_2 + H_2 \rightarrow 2C + H_2$$
$$H_2 + CH_3 \rightarrow CH_4 + H$$

This model also explains the positive influence of ethane on the pyrolysis rate of methane while, according to the authors,[5] the main part of the ethane is converted to methyl radicals that then 'attack' the methane molecule. Besides the influence of hydrogen and ethane on the methane pyrolysis, Eisenberg and Bliss also found that increases in temperature and methane content resulted in a higher pyrolysis rate and that a higher space/volume ratio did not influence the reaction rate.

Schneider[7] has also studied the kinetics of methane pyrolysis and found that at low conversions a first order expression is applicable. At higher conversions he recommends use of

$$\frac{d[CH_4]}{dt} = k[CH_4]\{1/(1 + [H_2]/[CH_4])\}^{1/2}$$

Another kinetic expression that better fits experimental data has been developed by Makarov and Pechik.[6] This model is based on the assumption that the reaction occurs on active centres that can be regenerated and also deactivated (by hydrogen) during the pyrolysis. The mathematical expression for this model is, however, far more complicated than the one presented in ref. 5.

All of these kinetic equations have been developed using only pure reactants (CH_4, H_2 and inert gas) which have been mixed and fed to the reactor. Therefore only very general conclusions can be applied to a system where the gas comes from wood pyrolysis and contains various impurities and catalytic material. Certain basic conclusions can, however, probably be made, such as:

(1) A higher cracking temperature gives higher decomposition rates.
(2) Hydrogen content in a gas inhibits the cracking of tars and hydrocarbons kinetically.
(3) It is possible that the concentration of methane in the gas affects the reaction rate.

4. APPARATUS (see Fig. 1)

The laboratory equipment consists of five different parts: pyrolysis reactor for wood, cracking reactor, gas cleaning, flow measuring and finally gas analysis.

4.1. Pyrolysis

The pyrolysis reactor consists of a vertical steel tube, 30 cm in length and 4 cm in diameter. The reactor is surrounded by a tube furnace which can be moved vertically by means of a motor. In this way 32 g of sawdust can be pyrolysed during 30 min. This yields a gas production of about 60 l/h. Both the pressure and the temperature in the pyrolyser can be detected: the pressure at the top of the reactor and the temperature in the middle and the bottom of the reactor.

The gas from the reactor is withdrawn through the bottom of the reactor via an electrically heated tube.

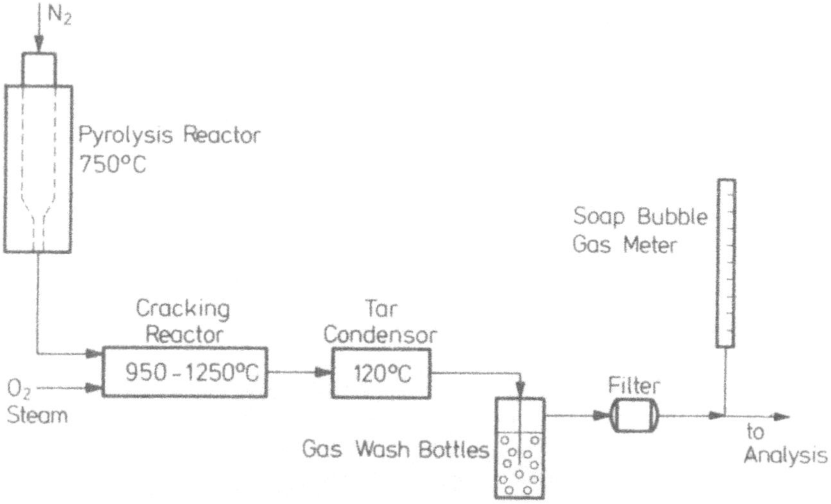

FIG. 1. Experimental set-up for cracking studies.

4.2. Cracking Zone

From the pyrolysis reactor the gas enters the cracking zone. This zone is 60 cm long and 2·2 cm in diameter and can be heated to about 1300 °C by means of a tube furnace. Tests have also been conducted with direct heating by addition of oxygen to the gas and thereby increasing the temperature in the cracking zone. The electric furnace was then only used to keep the reactor under adiabatic conditions.

4.3. Gas Cleaning

The gas from the cracker must be cleaned from tars and solids before it is fed to the gas chromatographs. This is done in three steps:

—*Tar condensor* where the tar (but not the steam) is condensed.
—*Steam condensing* bottles where the steam is condensed and the gas is washed in water.
—*Filter* made of glass wool and with water-cooled walls for further steam condensing.

After these cleaning steps the gas is led to the flow meter.

4.4. Flow Measuring

The gas is measured by means of a soap bubble meter that is intermittently connected to the gas filter.

4.5. Gas Analysis

The gas is analysed using two GCs for the determination of the content of CH_4, CO_2, H_2, CO, N_2, O_2, $C_2H_{2,4,6}$ and C_3H_6. Tars are measured by weighing the tar condensor before and after a test run.

5. EXPERIMENTAL

An experiment is started by switching on the cracker furnace and the GCs and letting them stabilize in temperature. The pyrolysis reactor is loaded with sawdust and then weighed in the same way as the tar condensor. After connecting the condensor and the pyrolysis reactor to the cracker the system is filled with nitrogen. When the gas analysis shows that there is no oxygen in the gas stream through the reactor, the nitrogen is turned off, steam and oxygen are connected to the cracker and the pyrolysis is started.

The gas production is allowed to stabilize and is then measured. When the GCs show that there is no variation in the gas composition, the composition data are registered and saved for further evaluation.

Temperature is also measured in five different spots in the equipment: in the beginning and end of the cracker tube, in the middle and bottom of the pyrolysis reactor, and on the wall of the tar condensor. When the pyrolysis furnace has reached its upper position, it is allowed to stay in that position until the gas production rapidly decreases (which means that the pyrolysis is complete) and is then turned off.

All experiments are then followed by a cooling phase after which the pyrolysis reactor and the tar condensor are weighed again to supply data on the amounts of tars and char produced, normally about 40–50% tars without cracking and 5–15% char. Besides char, tars and gas, the reactions also produce a certain amount of graphite which is deposited in layers inside the cracker tube. These layers can reach considerable thicknesses after only 10–12 runs and can be quite difficult to remove from the walls of the cracker.

After every ten runs the cracker tube is therefore heated to approximately 1400 °C and blown with pure oxygen to remove all carbonaceous materials on the walls. Influence of the graphite layers on the reactions have not been observed during runs. The experimental programme can be divided into three sub-groups:

(1) Runs without cracking to check at what temperature the pyrolysis reaction produces a gas with a tar content and composition suitable for cracking studies.

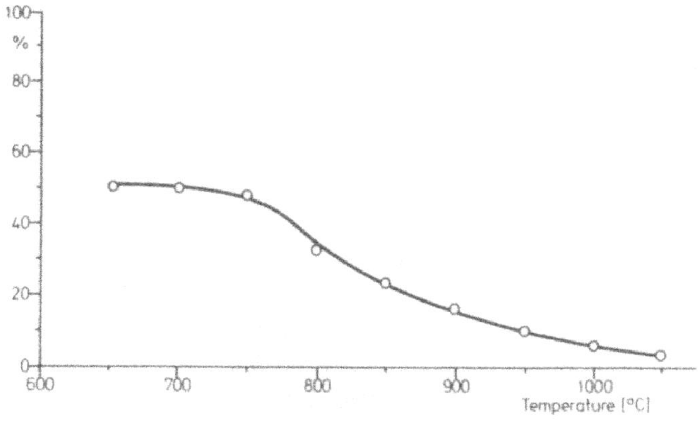

FIG. 2. Yield of condensable products vs. primary pyrolysis temperature.

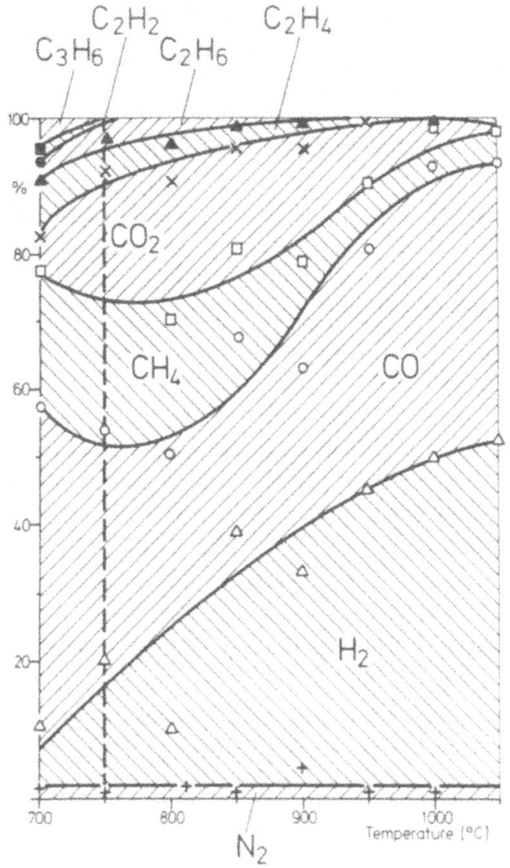

FIG. 3. Gas composition from the primary pyrolysis vs. pyrolysis temperature.

(2) Uncatalysed cracking of tars.
(3) Both steam and oxygen addition in the cracking zone.

The steam was supplied in rates of 0·2 and 0·3 g/g fuel and when oxygen was used it was supplied by 0·1 g/g fuel. At all runs the results were obtained as gas composition and production of condensable components.

6. RESULTS

The results of the first runs are presented in Figs 2, 3 and 7 where tar and gas production and gas composition are plotted vs. the pyrolysis temperature. From these data it can be seen that the tar production decreases and the production of CO and H_2 rapidly increases at temperatures above 750 °C. Therefore a pyrolysis temperature of 750 °C was chosen for further runs because of the suitable gas and tar composition at this temperature.

In the following runs the tar cracker was connected between the pyrolysis reactor and the tar condensor and six different test series were performed.

The first series without steam addition resulted in an obvious lack of steam in the reactants because very small amounts of water vapour were obtained in the product stream. Therefore 0·2 or 0·3 g steam (and 0·1 g O_2) per g fuel were added to the reactants in the succeeding runs. The oxygen addition was tested mainly to see if it was possible to use oxygen to raise the

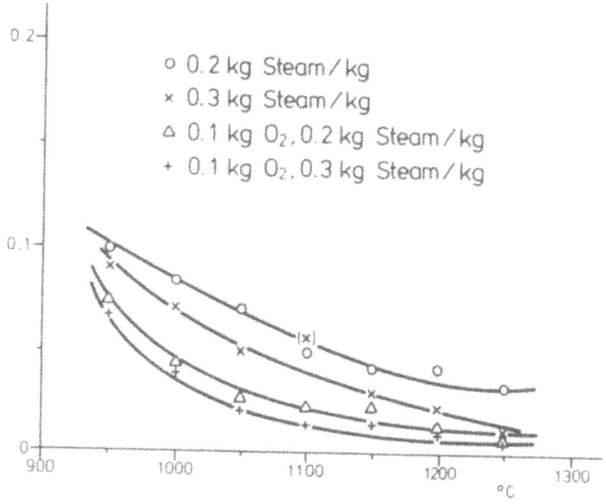

FIG. 4. Production of tar and soot (kg/kg dry feed) vs. cracking temperature.

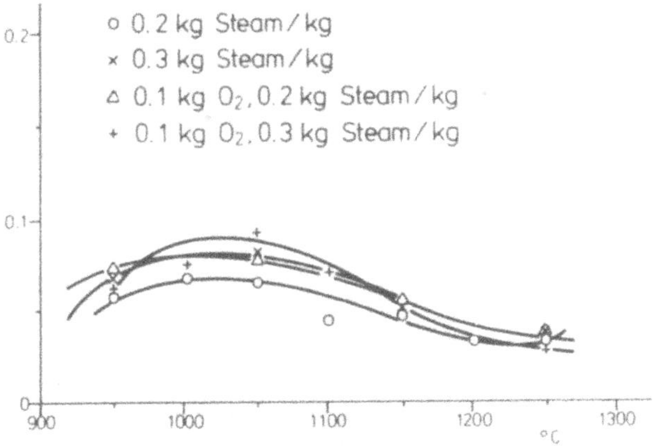

FIG. 5. Production of methane (N m^3/kg dry feed) vs. cracking temperature.

temperature in the reactor without burning off too much of the CO and H_2 content. To further increase the conversion of hydrocarbons, tests have also been performed where the gas production has been reduced to half the ordinary by filling only one half of the pyrolysis reactor with sawdust.

The results of cracking with steam and oxygen are presented in Figs 4–9 and show that:

(1) Tar conversion increases with increasing temperature and, to some extent, by increasing oxygen and steam addition.
(2) Cracking of methane and C_2-hydrocarbons increase by increasing temperature but is not affected by further moderate steam and oxygen addition.
(3) An increased residence time results in increased conversion of mainly C_2-hydrocarbons and methane.

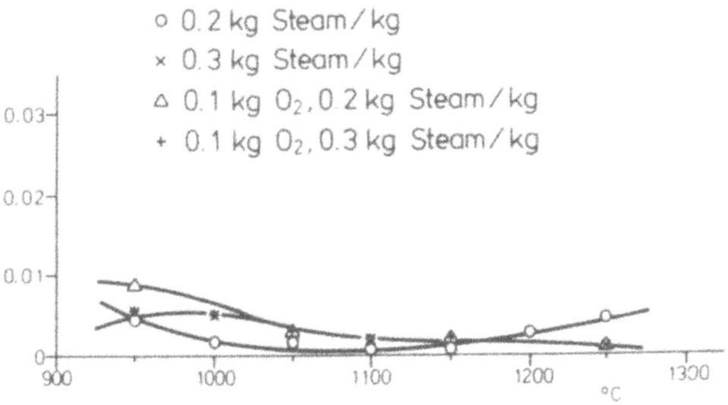

FIG. 6. Production of ethylene (N m^3/kg dry feed) vs. cracking temperature.

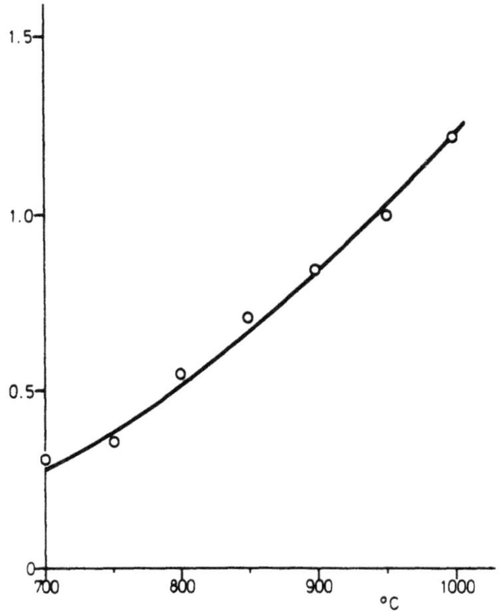

FIG. 7. Gas production after the primary pyrolysis vs. pyrolysis temperature ($N\,m^3$/kg dry feed).

7. CONCLUSIONS

All the experiments performed were aimed to find the conditions most favourable for a sufficient conversion of tars and hydrocarbons. These experiments have pointed out that the conversion of tars is sufficient at 1050 °C, 4 s residence time and 0·2 g steam plus 0·1 g O_2 added per g fuel (conversions over 90 %). The tar produced under these conditions is very different from the tar produced directly from biomass pyrolysis and consists mainly of soot. Analytic results from the National Coal Board have shown that the content which can be characterized as 'tars' is 5–7 % at cracking temperatures of 1050–1150 °C. The 'tar' content is further reduced at higher cracking temperatures.

The addition of steam and oxygen has also turned out to have an obviously positive influence on the cracking rate, and a residence time of 3·5 s can be used if only steam (and oxygen) are sufficiently supplied. Further increases in residence time then only result in a very small increase in conversion and less tars in the cracking residue. A decrease in residence time is not recommended, while the conversion rate of lower hydrocarbons,

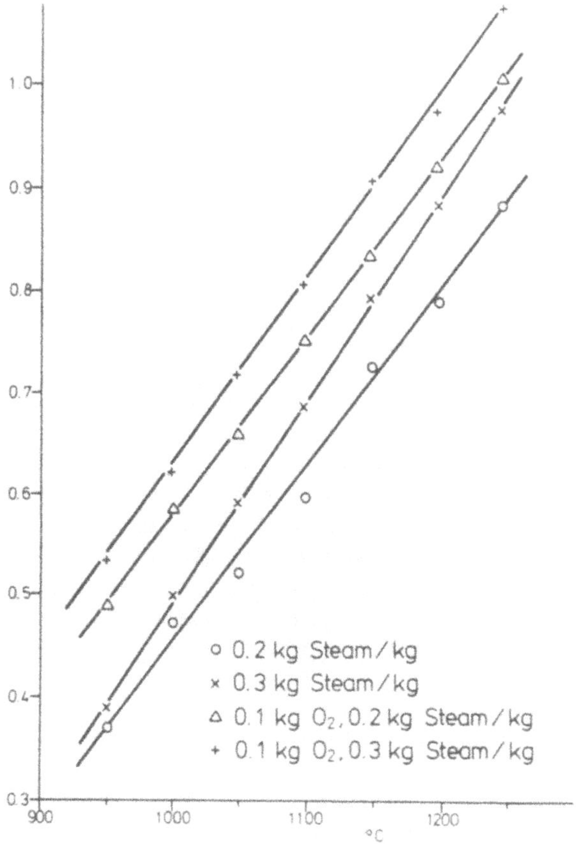

FIG. 8. Total gas production (N m^3/kg dry feed) vs. cracking temperature.

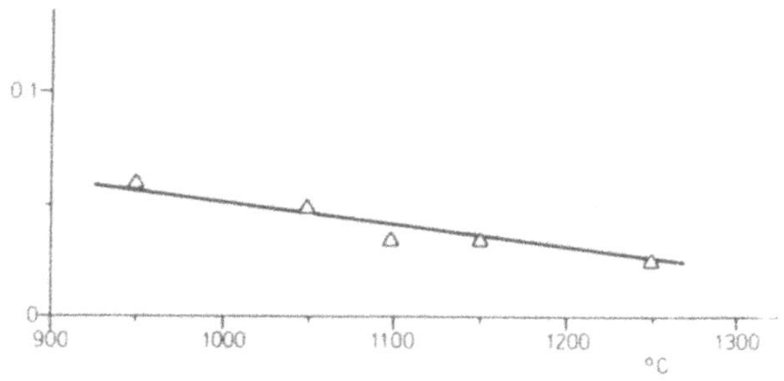

FIG. 9. The production of methane (N m^3/kg dry feed) vs. cracking temperature at 8 s residence time and 0·2 kg steam/kg dry feed.

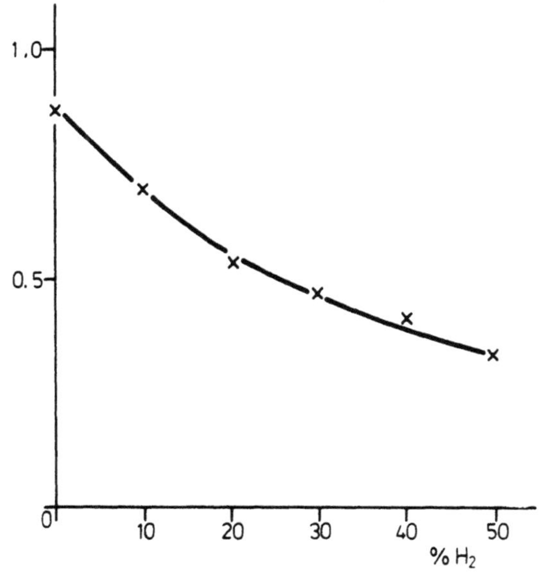

FIG. 10. Methane conversion vs. hydrogen content in the reactants; cracking temperature: 1100 °C, residence time: 4 s.

such as methane, ethane etc., is very much lower. A reduced residence time would result in lower conversions of these hydrocarbons. The most interesting component of the lower hydrocarbons is methane, while the contents of the others are very small.

The experiments performed have shown that a methane conversion of more than 60 % is hard to obtain in the experimental equipment even if the residence time is increased from 4·2 to 8 s.

The low decomposition rate can also be explained by the influence of hydrogen, which is presented in Fig. 10. There you can see that a high hydrogen content lowers the conversion down to approximately half the 'normal'. This effect is increased by the other cracking reactions which produce an increasing amount of hydrogen in the gas. This proceeds throughout the cracker and results in continuously decreasing decomposition rate.

REFERENCES

1. Roberts, F. A. (1970). A review of kinetic data for the pyrolysis of wood and related substances. *Combustion and Flame* **14**, 261–72.

2. Ekström, C. and Rensfelt, E. (1980). Flash pyrolysis of biomass in Sweden. Presented at the Flash Pyrolysis Workshop in Golden, Colorado, October.
3. Madorsky, S. L. *et al.* (1956). *J. Res. Natl. Bur. Std.* **56**(6), 343–54.
4. Tinney, E. R. (1965). *Tenth Symposium on Combustion.* The combustion Institute, Pittsburgh, p. 925.
5. Eisenberg, B. and Bliss, H. Kinetics of methane pyrolysis. *Chem. Engr. Prog. Symp. Ser., No. 72* **63**, 3–16.
6. Makarov, K. I. and Pechik, V. K. (1974). Kinetics of methane pyrolysis under conditions of pyrolytic carbon formation. *Carbon* **12**, 391–404.
7. Schneider, I. A. (1962). *Ibid.*, **220**, 119–209.

39

A Pressurized Thermo Balance for Pyrolysis and Gasification Studies of Biomass, Wood and Peat

WILLIAM BLACKADDER and ERIK RENSFELT
Studsvik Energiteknik AB, Nyköping, Sweden

1. INTRODUCTION

Since the oil crisis of 1973 there has been an ever-increasing interest in biomass, wood and peat for the production of gas (synthetic gas and fuel gas).

To improve the overall efficiency and reduce apparatus size the trend has been to pressurize many solid–gas reaction processes. Gasification of coal, biomass and peat are important examples. The need for accurate data under these conditions is rapidly increasing for proper reactor design.

When modelling the complex pyrolysis–gasification–combustion reactions it is important to know the nature of each step.

Pressurized thermo balance studies are thus a valuable tool supplementing different rapid-heating, short residence time equipment and pressurized fluidized beds.

2. GENERAL PROBLEMS WITH PRESSURIZED THERMO BALANCES

Control of the flow regime to avoid mass transfer limitations and thus to study the true kinetics is important in all kinds of TG studies. Temperature control and measurement is also a difficult problem. At high pressure the buoyancy effect will be important together with pressure control to avoid disturbances. Mechanical design for loading the sample is also complicated at high pressures.

Different technical solutions to these problems have been proposed and tested.[1–3] No design has solved all the problems because of the contradictory requirements such as large sample and good temperature control.

3. TG PROBLEMS RELATED TO BIOFUELS

Most solid fuels are rather inhomogeneous. Peat, which is one of Sweden's main energy resources, is especially difficult in this respect due to the fact that it consists of fibrous matter and fine powder, as well as larger pieces of wood. Even with more homogeneous fuels like wood it is interesting to be able to study different particle sizes.

For tests with coupled pyrolysis–gasification of biofuels, e.g. in steam mixtures, a large sample is essential due to the very low amount of char (10–20 %).

Problems with large samples are of course 'high' pressure drop and/or recondensation of volatiles; also additional problems are gas bypassing and control of the flow regime. The latter is of special importance for gasification studies where an active gas phase reactant is participating.

The combination of large sample size (1–8 g) and relative low density material means rather large volumes, which must be accommodated by the reactor. High pressure will consequently require large amounts of gas. A superheater is therefore needed to give the necessary temperature to the gas before entering the reactor. High gas flow for control of flow regime might also lead to disturbances causing vibrations and shaking of the sample holder.

Samples of fuels with poor heat transfer properties like biofuels means that the temperature measurement is of great importance. Endothermic gasification reactions and exothermic pyrolysis steps can yield important temperature deviation inside the sample.

Measuring the sample temperature has been considered almost impossible.[2] Data from atmospheric runs at the Royal Institute of Technology since 1976 have clearly demonstrated the problems as well as indicated some of the solutions to those problems.[4]

For biofuels the tars have to be transported out of the unit without condensing on the weighing device. Hanging sample holder and downflow is to be preferred over standing sample holder and downflow due to the problems with tar condensation. Upflow includes a greater risk of blowing away the sample.

4. EXPERIMENTAL APPARATUS

The thermo balance was built to work at pressures up to 5 MPa, temperatures up to 1000 °C, heating rates up to 40 °C min^{-1} and gas flow

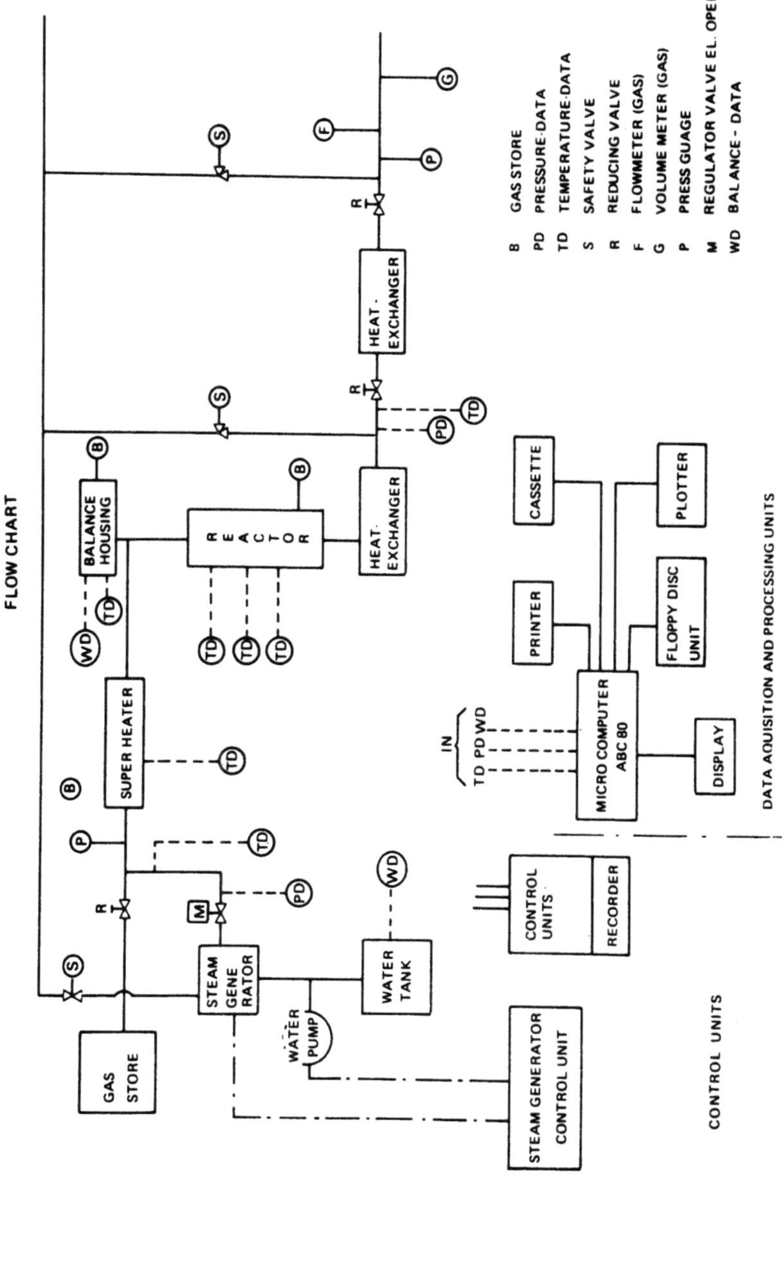

FIG. 1. Pressurized thermo balance flow chart.

FIG. 2(a). Pressurized thermo balance: pressure vessels.

rates of 4 cm s^{-1} in the reactor. One of the features of the thermo balance is that the temperature is measured in the sample during the TG analysis, but the heating rate is controlled by an outside thermocouple. Pyrolysis and gasification can be carried out in different environments: nitrogen, steam, hydrogen, carbon dioxide, etc.

A modified Sartorius 1202 balance was used, the differential transformer cell being placed in the pressurized housing with the electronics on the outside.

Figure 1 shows a flow diagram and Fig. 2(a) a photograph of the thermo balance system. A steam generator and a superheater is also included in the auxiliary equipment of the thermo balance. Figure 3 shows a cross-section of the reactor, the superheater and the balance housing. The powers of the steam generator, superheater and reactor are 8 kW, 6·5 kW and 2·5 kW respectively.

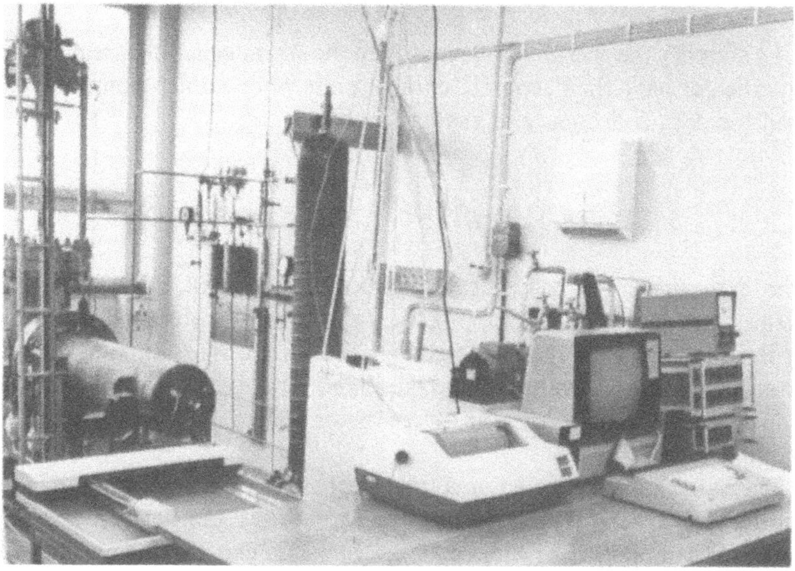

FIG. 2(b). Pressurized thermo balance: data collection system.

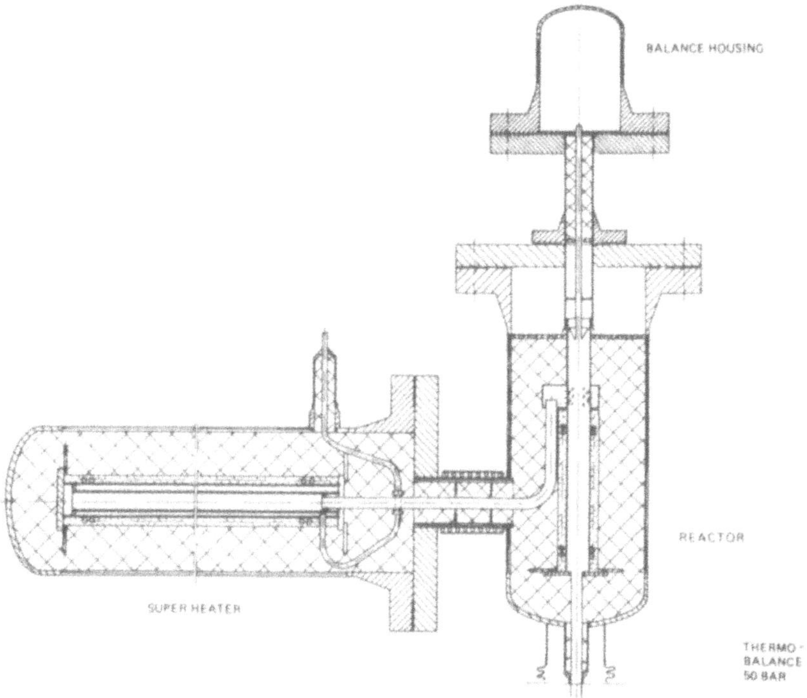

FIG. 3. Pressurized thermo balance, general drawing.

To simplify the thermo balance system the steam generator, reactor and superheater have their separate control units while a microcomputer was used for data collection and evaluation.

5. DATA HANDLING AND CORRECTIONS

The data are collected using a microcomputer (ABC 80) with a 32K memory, display, printer, floppy disk, cassette and plotter, and Assembler or Basic language may be used (see Fig. 2b). With this system the data may be collected, stored, evaluated and presented. There is also the possibility of transference of data to the central computer system if a greater capacity is required.

The data can be collected on a time (seconds) or a temperature (degrees centigrade) basis. A data collecting cycle comprises: temperature in, under and over sample, superheater, balance housing, exit gas, steam inlet and water outlet, and sample weight, remaining weight of feed water, system pressure, and time from start. The display is continually updated at the rate of collection while a printout can be given at an interval for any given number of cycles. Both the frequency of data collection and printout can be changed during the run; thus the data collection frequency can be altered depending on the change in reaction rates.

At the end of the run the data are then stored on floppy disk or they can be transferred to cassette. The data are background-corrected and corrected for change in sample displacement—this is a rather significant part of the treatment for the previously mentioned reasons. The corrected data can also be treated mathematically by applying a sliding fit third degree polynomial for generating derivation values for calculating activation energies or for plotting DTGA curves. The number of points in each fit determines the sensitivity of the derivation. The flexibility of the system is such that any combination of the collected data may be plotted.

The buoyancy effect on the balance system becomes appreciable at higher pressures. The buoyancy effects are not only pressure- but temperature-dependent. Reference runs were therefore carried out to determine the background corrections which could then be made using the computer. It was found for this system that the reference curves taking into account pressure, temperature and gas velocity could be solved as a logarithmic equation as a function of temperature (see Fig. 4). The buoyancy effect must also be considered with regard to displaced sample volume which changes during pyrolysis or gasification. An added source of

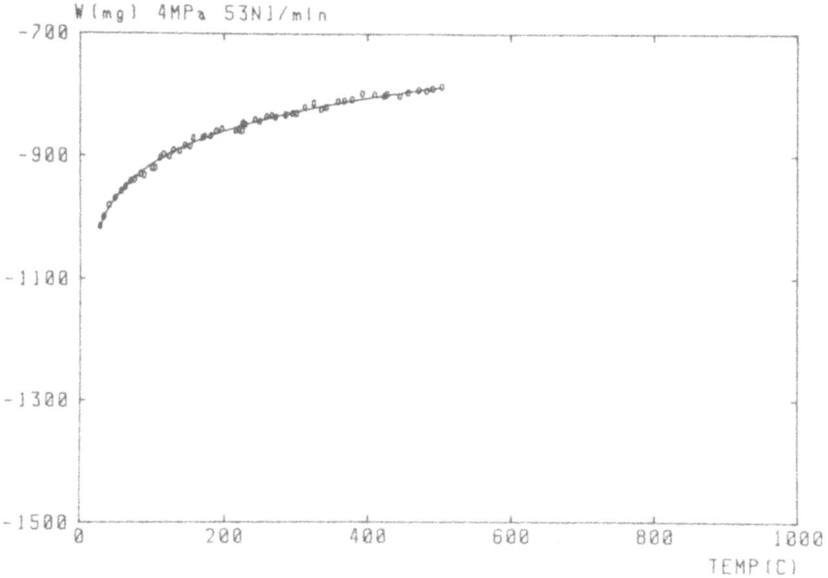

FIG. 4. Reference run with empty sample holder. ($Y = -1227{\cdot}89 + 79{\cdot}09 \ln X$; $r = 0{\cdot}9975$)

error is true density of the changing sample material. Experimentally decided densities have been made only for fresh fuel and end product.

As would be expected, in pressurized runs maintenance of a constant pressure is important because of the change in buoyancy effect on the apparent weight of the sample. This has been one of the main errors in the presented experiments, these pressure changes often being responsible for the 'bumps' that can be seen in the weight curve. A perfect pressure control yields an almost perfect curve; therefore an automatic pressure control has been installed. This is of special importance when steam is added, some tests having been ruined by steam flowing up into the balance vessel due to bad pressure control.

6. EXPERIMENTAL RESULTS

Runs have been made with samples of wood (6 g), lignin (1·2 g) and cellulose (1·2 g). The heating rate was 10 °C/min in all runs. Pure N_2 (99·995 %, O_2 10 ppm, H_2O 5 ppm) was used as purge gas with an initial room temperature velocity of approximately 1·7 cm/s at all pressures.

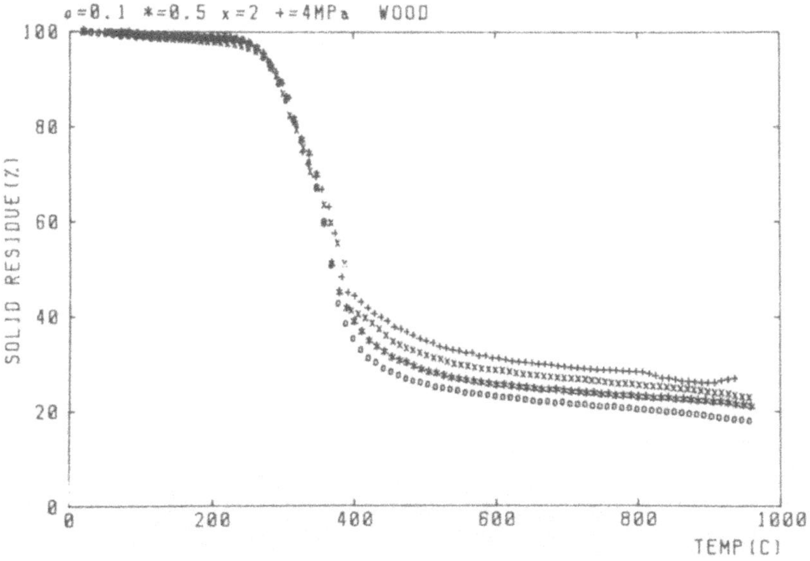

FIG. 5. TG curves for wood, pressure as parameter (0·1–4·0 MPa).

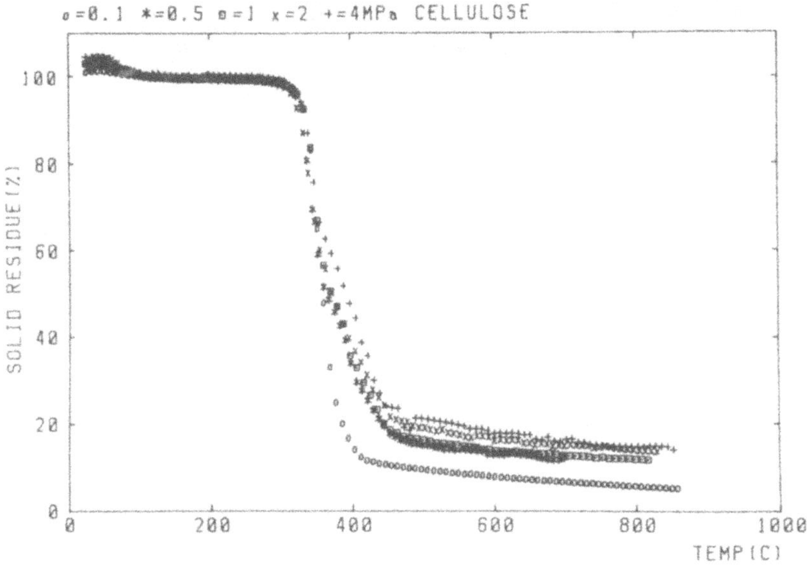

FIG. 6. TG curves for cellulose, pressure as parameter (0·1–4·0 MPa).

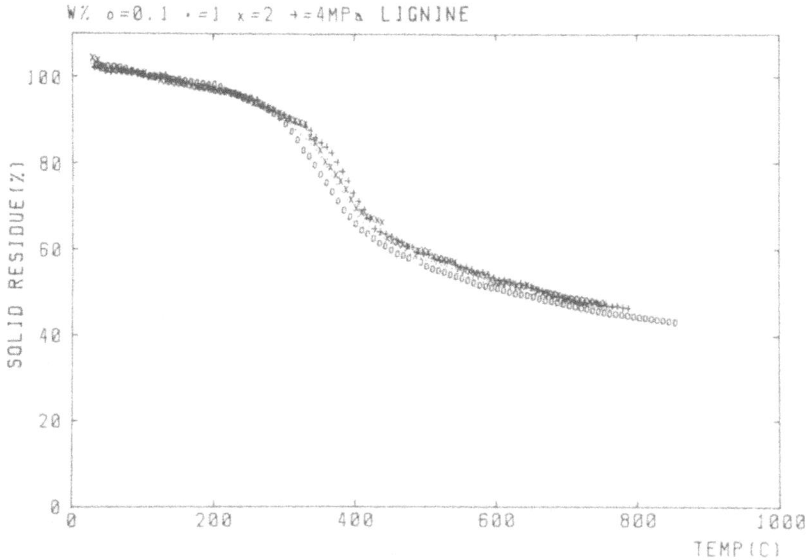

FIG. 7. TG curves for lignin, pressure as parameter.

Figures 5, 6 and 7 show results for different pressures for wood, cellulose and lignin respectively. The wood sample was a predominantly birch, cellulose filter paper and lignin of the acid type. The sample data are presented in another manner in Figs 8, 9 and 10 and clearly show the pressure effect at different temperatures. Finally a comparison between the different fuels can be made in Figs 11 and 12. One experiment had a rather poor pressure behaviour but it is still indicated in Figs 9, 11 and 12 (cellulose 3·0 MPa).

The strongest pressure dependency is observed for cellulose pyrolysis (see Figs 6, 9, 11 and 12). The char residue at high temperature (750 °C) increases from 6 % at atmospheric pressure to 15 % at 4·0 MPa. For wood a similar but weaker tendency is observed (Figs 5, 8, 10 and 11). The amount of char is increasing from 21 % at 750 °C and atmospheric pressure to 28 % at 4·0 MPa. The pressure dependency for lignin is very weak (see Figs 7, 10, 11 and 12). For all fuels the main change is taking place between 0·1 and 0·5 MPa.

Heat of reaction effects can be seen by comparison of temperature inside and outside the sample. Pyrolysis of wood at atmospheric pressure is slightly endothermic but at higher pressure it turns to exothermic behaviour. The same phenomenon, and to an even greater extent, is

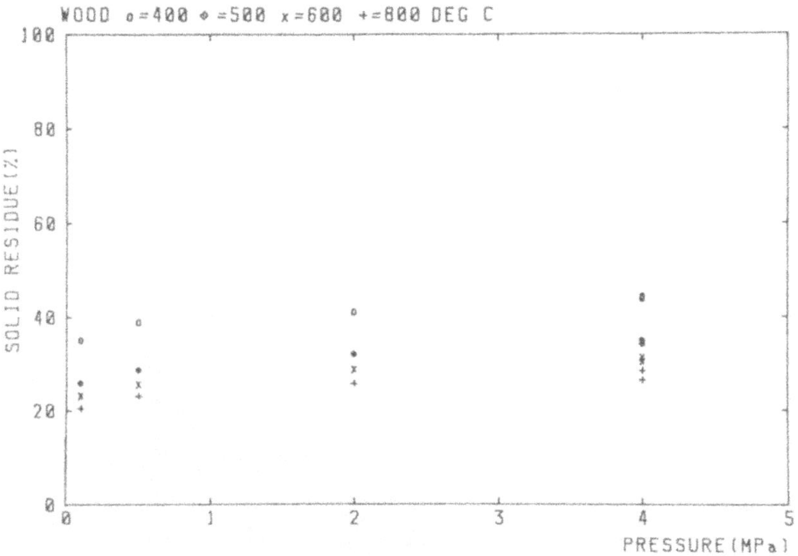

FIG. 8. Wood, solid residue relative to pressure, temperature as parameter.

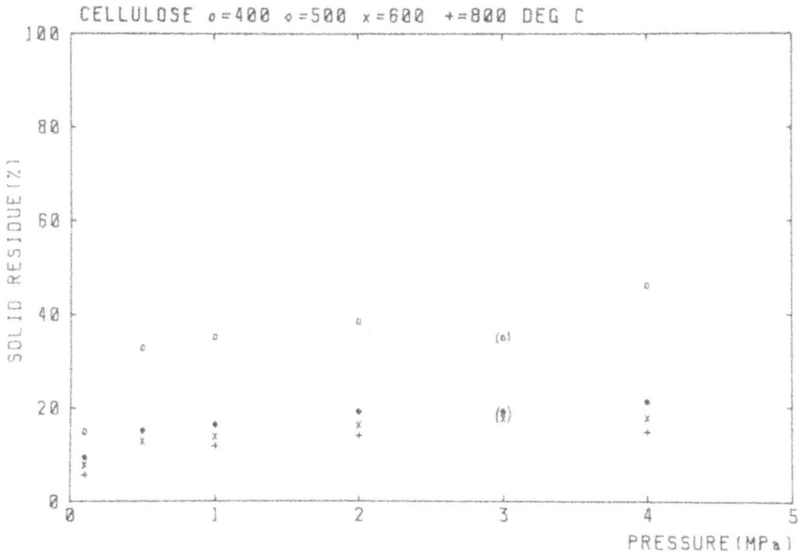

FIG. 9. Cellulose, solid residue relative to pressure, temperature as parameter.

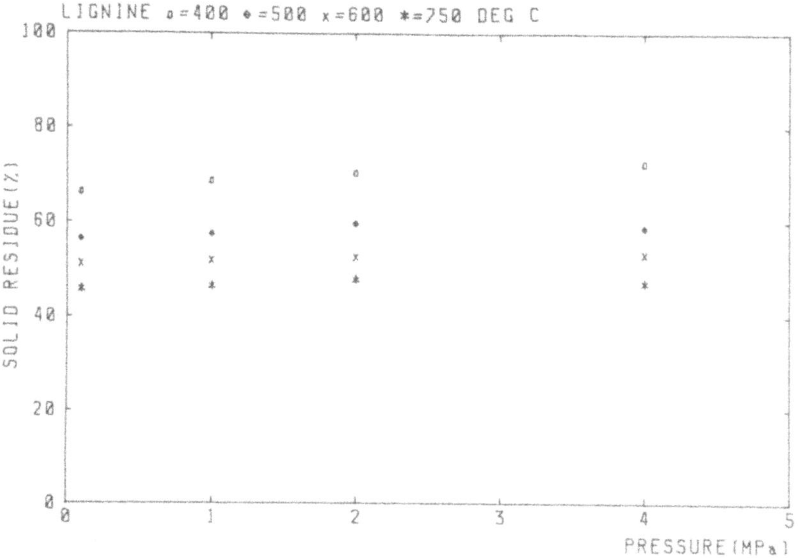

FIG. 10. Lignin, solid residue relative to pressure, temperature as parameter.

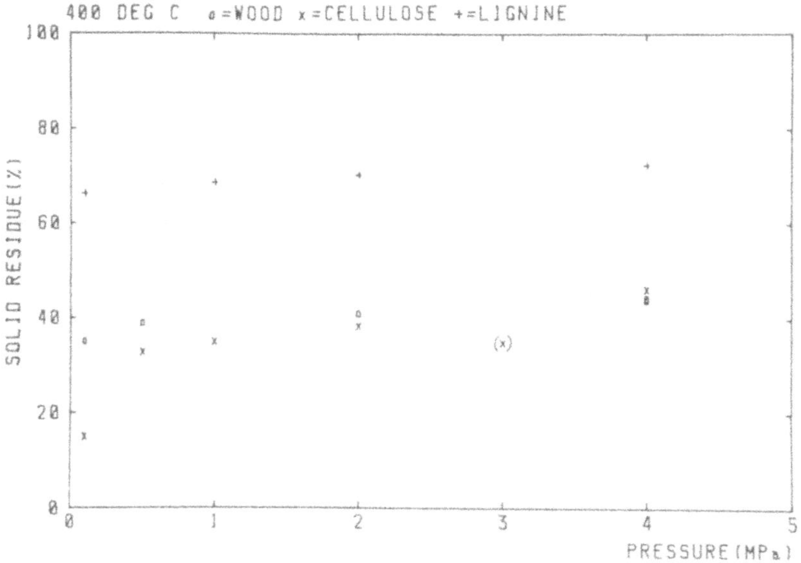

FIG. 11. Influence of pressure on pyrolysis of wood, lignin and cellulose, pyrolysis and temperature 400 °C.

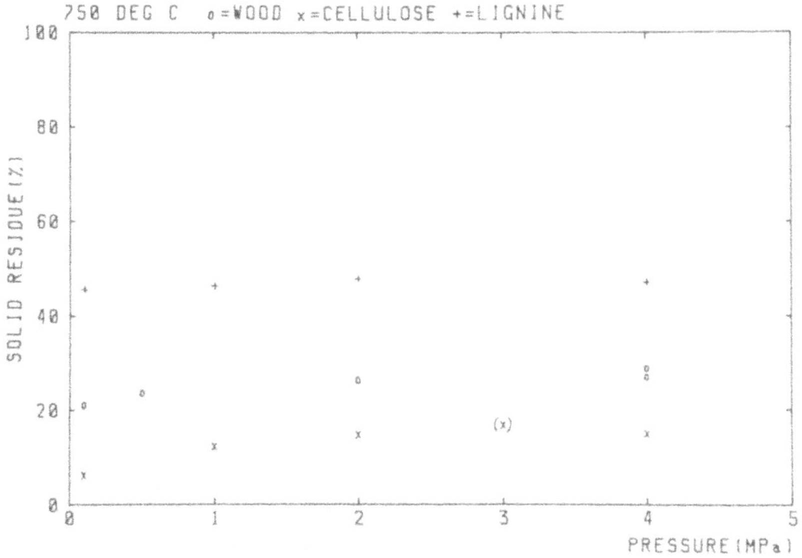

FIG. 12. Influence of pressure on pyrolysis of wood, lignin and cellulose, pyrolysis and temperature 750 °C.

observed for cellulose. Lignin pyrolysis seems to be almost thermally neutral and not pressure dependent.

7. DISCUSSION OF RESULTS

The increasing char fraction and exothermicity of the pyrolysis with pressure is in good agreement with earlier results over a smaller pressure range.[5]

The presented results show that wood behaviour at higher pressure can be roughly explained by the behaviour of its components.

The importance of increasing char fraction is clear when considering their influence on the total gasification rate achieved in a fluidized bed, e.g. oxygen blown. The presented results refer to slow heating rate, and similar tests with high heating rate are necessary to study further the pressure dependency which may be somewhat altered.

Earlier data for atmospheric pressure might indicate that the 'flash' effect can be approximately 5–10 % for wood.[7] Based on that assumption and the probable assumption that 8–10 % of the char in an oxygen blown process is combusted with oxygen or lost as ash and flyash, the influence on total amount of char to be gasified can be doubled or even trebled. The gasification rate of char in itself has a positive pressure dependency and thus the total rate may still be increasing with pressure but at a slower rate than that which can be expected from char gasification data.

The conclusion is that even the limited differences in the amount of char produced that have been shown for wood can affect the total gasification rate in a fluidized bed to a great extent.

8. FUTURE WORK

Gasification of wood char with steam and other media is just about to start. Conversion of the top flange and balance house to a design that permits rapid introduction of solids in the hot environment is considered for supplementary studies of rapid heating experiments.

ACKNOWLEDGEMENT

The work described has been funded by Studsvik Energiteknik AB itself within the programme for fundamental energy research. Bengt-Göran Espenäs have given valuable advice during the design phase and are also responsible for some of the experiments. Arthur Svensson has made the drawings and the mechanical design.

REFERENCES

1. Dobner *et al.* (1975). Modification of a thermo-balance for high pressure process studies. 68th Annual A.I.Ch.E. Meeting, November 16–20.
2. Gardner, C. N. *et al.* (1980). Thermogravimetric measurements at high pressures. National Bureau of Standards, Publ 580, May.
3. Sabrowsky, H. *et al.* (1978). Thermogravimetrie unter erhöhten Drücken. *Chem. Ing. Tech.* **50**(3), 217.
4. Rensfelt, E. *et al.* (1978). Basic gasification studies. IGT Symp. Papers. Energy from Biomass and Wastes, August 14–18, p. 465.
5. Antal, M. (1981). Effect of pressure on biomass pyrolysis and gasification. 13th Biomass Thermochemical Contractors Meeting, October.
6. Ekström, C. and Rensfelt, E. (1980). Flash pyrolysis of biomass in Sweden. Specialist Workshop on Fast Pyrolysis, SERI, Colorado.

40

Effect of Size, Shape, Density, Moisture and Furnace Wall Temperature on Burning Times of Wood Pieces

EDWARD R. HUFF

Agricultural Engineering Department, University of Maine, Orono, Maine, USA

1. INTRODUCTION

To contribute toward increasing combustion intensity in fireboxes for particulate materials, the effects of several parameters on burning time of single wood cubes were investigated in 1978 and reported in an earlier paper.[3] That paper discussed the effect of size, moisture content, furnace wall temperature and air temperature on burning time of small wood cubes.

Of the parameters tested, cube size and furnace wall temperature had the greatest effect on burning time. Moisture had a smaller effect, and air temperature had no significant effect on the total burning time. Burning time increased with the first and second order of both linear dimension and moisture content. An increase in size, air temperature or wall temperature caused the volatilization time to occupy a smaller proportion of total burning time, while an increase in moisture raised this proportion.

This paper extends the number of parameters to include the effects of density and shape on burning time of single pieces, and presents equations predicting burning time of wood pieces under various conditions.

2. DESIGN OF THE 1980 EXPERIMENT

After the 1978 tests, it seemed desirable to determine the effects of density and shape on burning time. Air temperature, varied in 1978, made so little effect that it was omitted as a variable on the 1980 tests. Moisture content was also omitted as a variable, not because it was unimportant or

TABLE I
LEVELS OF VARIABLES IN 1980 TEST

Parameter	Level	Description			
Size, $Z = \sqrt[3]{\text{volume}}$ (pine cubes only)		Dimension (cm)		Volume (cm^3)	
	1	0·500		0·125	
	2	1·000		1·000	
	3	1·145		1·5	
	4	1·50		3·375	
	5	2·00		8·0	
Shape, H (vol. = $1{\cdot}5\,cm^3$)		Name	Dimensions (cm)		ϕ_s
	1	Cube	1·145 1·145 1·145		0·806
	2	Flat	1·873 1·873 0·428		0·620
	3	Sticks	4·579 0·572 0·572		0·569
	4	Flatter	2·480 2·480 0·244		0·431
	5	Flattest	3·436 3·436 0·127		0·250
Density, D (g/cm^3)		Wood	Mean	Std. Dev.	Range
	1	Balsa	0·1406	0·036	0·085–0·231
	2	Pine	0·4244	0·032	0·374–0·511
	3	Oak	0·6567	0·047	0·522–0·751
Wall temp., W	1	797 K			
	2	1 063 K			
	3	1 345 K			

insignificant but to simplify the test procedure to allow testing in a limited time.

The parameters varied in the test were size, shape and density of the particle, and wall temperature of the furnace. However, not all combinations of these were tested. The main test consisted of a factorial design consisting of 3 densities, 5 shapes, and 3 wall temperatures in all combinations, with 3 replications, for a total of 45 combinations or 135 tests, according to Table I. In this test, volume was constant at $1{\cdot}5\,cm^3$. In an adjunct to the main test, size was varied to see if the relationship between size and burning time found in the previous tests held true for a smaller size.

These tests included 1/2 cm cubes, containing 1/8 the volume of the 1 cm cubes previously tested. This test consisted of 5 sizes and 3 wall temperatures, at only one density and shape, for a total of 15 combinations, or 45 tests, also according to Table I. Some of the cubes tested were common to both sets.

The numerical measure of shape used in this test is sphericity, described as[5]

$$\phi_s = \frac{\text{surface area of a sphere with same volume as object}}{\text{surface area of object}} = \frac{A_s}{A} \quad (1)$$

This can be put in equation form by solving for the surface area of a sphere in terms of its volume, which gives

$$A_s = \sqrt[3]{36\pi V^2} = 4{\cdot}836V^{2/3} \quad (2)$$

Putting this in eqn (1), we get

$$\phi_s = (36\pi V^2/A^3)^{1/3} = 4{\cdot}836(V^2/A^3)^{1/3} \quad (3)$$

3. TEST EQUIPMENT AND PROCEDURES

The test equipment and procedures were the same as for the 1978 tests except for testing non-dry specimens. Due to limited time, it was thought that an initial indication of the effects of density and shape could be obtained with dry specimens even though the interaction effect of these with moisture was thereby forfeited. This greatly simplified the tests.

3.1. The Furnace

The furnace used for testing was a small refractory-lined electric furnace with an on–off temperature controller. Temperature was measured by two chromel-alumel thermocouples thinly embedded in the floor and ceiling, one on one of the four furnace heating rods, one in the air stream to the furnace, and one in the room to measure ambient.

Room temperature air was blown across the specimens through a 2·66 cm ID horizontal pipe at 87 cm/s as measured by pressure drop across a sharp-edged orifice. There was about a ± 3 cm/s air velocity variation due to fan motor speed changes caused by voltage fluctuations. Also, the air flow may have been different around different size and shape pieces. These errors were not considered serious.

3.2. Preparation of Specimens

Specimens were cut to the shapes listed in Table I, to within 1·7 % of the basic dimensions so that volume would not vary from stated volume by more than 5 %. Like wood chips, both the flat specimens and sticks were cut parallel to the grain. The specimens were dried, numbered, weighed and measured. No attempt was made to homogenize density within each of the three species of wood tested. Density of small pieces cut from the same stock varied considerably, as shown in Table I.

Holes were drilled in the larger cubes, the sticks and in the flat specimens for press fit on to a 1/16 in. rod. But a 1/16 in. hole occupying nearly 8 % of the volume of a 1/2 cm cube was considered too great a loss. These were mounted by pointing the rod and sticking it into the cubes. They did not fall off in burning. Also, the 'flatter' and 'flattest' specimens (Table I) were too thin for a 1/16 in. hole through their thickness so two holes were drilled in the face for attaching to the rod with fine wires, with a flat face resting on the rod. In the sticks, holes were drilled through approximately 80 % of the length for mounting on the rods. These holes occupied 5 % of the volume of the sticks. Otherwise, the greatest volume occupied by any hole was 2·5 % of the volume of the piece.

3.2. Burning Specimens

Oven wall temperature was set using the temperatures measured by the thermocouples in the oven, in an equation that took into account the temperatures and relative areas of the walls, rods and holes.[2] The time required for the oven to reach equilibrium dictated burning cubes at one furnace temperature on a given day.

The specimens were mounted on a 1/16 in. rod by one of the three means described above. They were then thrust into the furnace and placed 4–5 cm from, and directly in front of, the air inlet tube. Time from thrusting into the oven until extinction of the flame and of the char were measured.

The larger specimens sometimes tended to crack across the grain, allowing a piece to fall off before completely burning. The piece that fell to the bottom took longer to burn. Additional specimens burned at the same test point sometimes were successful, but when all attempts failed, a probable burning time was constructed from the available data.

It made a big difference in burning time whether the smallest cubes at the lowest temperatures were kept in the air stream or brought out of the air stream to ignite. There was so little material to volatilize that the volatile gases could not accumulate in high enough concentration to ignite spontaneously in the cool air stream when radiant flux was low. They would reduce to zero size merely glowing. They took much longer to burn, probably because of the lack of radiation flux from their own flame. Therefore the pieces that failed to ignite spontaneously were momentarily removed from the air stream to ignite.

3.3. Analysis of Data

Volume, density and char burning time were computed for each specimen. An analysis of variance was performed on the density–shape–wall temperature factorial design and means were computed for main and interaction effects. Least square straight line equations were determined where appropriate.

Since the effect of size had been shown to be highly significant in the 1978 tests, no statistical analysis was performed for this parameter this time.

4. BURNING TIMES

The analysis of variance showed that the main and two-way interaction effects of density, shape and temperature on all three burning times were highly significant (1 % or better). The three-way interaction effects, except for char burning time at 2·5 %, were also highly significant. The effect of size, shape and density on burning time at each of the three test temperatures is presented, then the overall effect of temperature on burning time is given.

All equations presented were developed under the constraint that total burning time is the sum of flame time and char time, $t_t = t_f + t_c$. This is true for both the slope and intercept of all equations.

4.1. Effect of Size

Figure 1 shows the effect of size on burning time at the three temperatures tested, and Table II gives the equations of the lines. The measure of size used is Z, the cube root of the volume.

By using dimensionless groups in 1978 it was found that dividing burning time by the volume/surface area linearizes the relationship between burning time and size. The same method of linearizing the curves was used in Fig. 1 by dividing the burning times by Z, which dimensionally is identical to volume/surface area ratio.

The constants for the regression equations for the lines in Fig. 1 indicated almost the same slope for the flame time at all three temperatures. Also indicated was almost the same intercept for char time at all three temperatures. Therefore Fig. 1 and the equations were constructed under

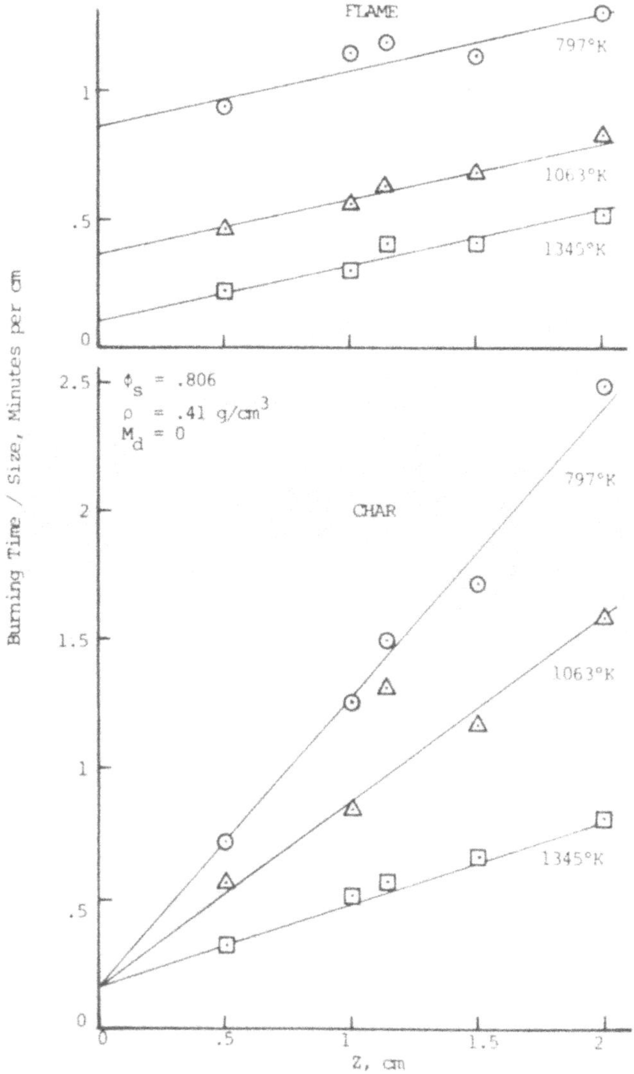

FIG. 1. Effect of size on burning time.

TABLE II

CONSTANTS FOR BURNING TIME EQUATIONS AT ZERO MOISTURE CONTENT

Parameter equation	*Type of burning time*	*Temperature (K)*					
		797		*1 063*		*1 345*	
		a	*b*	*a*	*b*	*a*	*b*
Size, Z	t_f	0·86	0·22	0·36	0·22	0·10	0·22
$t = aZ + bZ^2$	t_c	0·16	1·12	0·16	0·72	0·16	0·32
$\phi_s = 0{\cdot}806$							
$P_M = 0{\cdot}41$ g/cm^3 [a]	t_t	1·02	1·34	0·52	0·94	0·26	0·54
Shape, ϕ_s	t_f	0	1·49	0	0·71	0	0·45
$t = a + b\phi_s$	t_c	0·38	1·49	0·17	1·18	0·08	0·69
$Z = 1{\cdot}145$ cm							
$P_M = 0{\cdot}41$ g/cm^3 [a]	t_t	0·38	2·98	0·17	1·89	0·08	1·14
Density, ρ	t_f	1·79	—	0·83	—	0·52	—
$t = a\rho$	t_c	2·69	—	1·93	—	1·05	—
$Z = 1{\cdot}145$ cm							
$\phi_s = 0{\cdot}535$ [a]	t_t	4·48	—	2·76	—	1·57	—

[a] Mean values.

the constraints that at all three temperatures the flame time slopes are equal and the char burning times at zero size are equal.

The tests conducted in 1978 on 1, 1·5 and 2 cm cubes showed burning time to be a function of both the first and second powers of size. Figure 1 confirms this relationship down to 1/2 cm cubes.

Most consistent results were obtained for the flame time, and for all burning times at the highest temperature. The inconsistency at the lowest temperature was probably due largely to the difficulty in establishing flame.

4.2. Effect of Shape

Figure 2 shows the temperature-dependent effect of shape on burning time. The equations are given in Table II. An indication of the difference in burning times made by sticks vs. chips is also shown by the plots of the one stick shape tested. The sticks appear to burn somewhat faster than flat chips with the same sphericity, although the difference is small enough for sticks and chips to be incorporated in the same equations without much loss in accuracy.

In Fig. 2 the zero sphericity axis represents pieces with a volume of

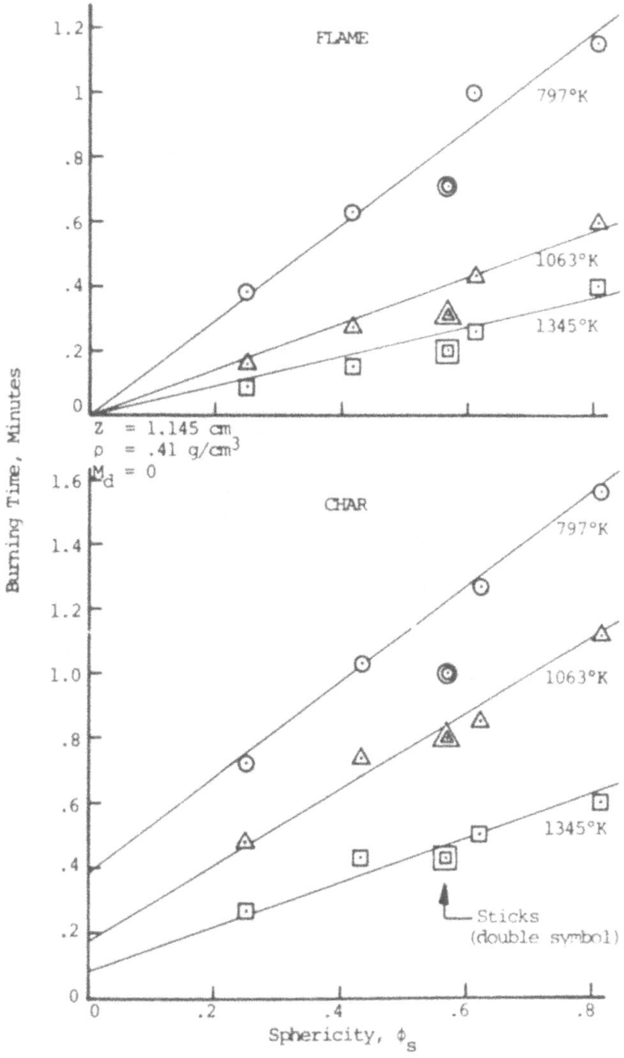

FIG. 2. Effect of shape on burning time.

$1{\cdot}5\,\mathrm{cm}^3$ but with zero thickness and infinite surface area. The regression equations for the four non-stick shapes indicated that at $\phi_s = 0$ the flame burning time was zero. If flame time closely parallels volatilization time, a flame time of zero at zero sphericity indicates instantaneous volatilization and burning.

The char-burning times are temperature dependent at $\phi_s = 0$. The indication is, therefore, that at a sphericity of zero, flame time is zero but char time is a function of temperature. This may reflect the finite time required for diffusion of oxidant to the char surface.

4.3. Effect of Density

Figure 3 shows the effect of density on burning time. All five shapes are averaged for each of the points shown. The relationship of both flame time and char burning time to density is so close to being linear that straight lines were drawn through zero at zero density and equations of the form $t = a\rho$ were assumed. The slopes are given in Table II.

4.4. Effect of Temperature

The effects of size, shape and density have been reported at all three wall temperatures. Here the effect of temperature on burning time is put in equation form.

Temperature dependency of chemical reactions usually conforms to the Arrhenius equation (4):

$$\frac{\mathrm{d}m}{\mathrm{d}t} = F\mathrm{e}^{-E/RT} \tag{4}$$

where $\mathrm{d}m/\mathrm{d}t$ is the rate of change of mass. F is the frequency factor, and is a measure of the number of effective collisions of gas molecules. It has been called a velocity constant. E is the activation energy of the material, R is the universal gas constant, and T is the absolute temperature.

The same kind of temperature dependency was assumed here with the equation taking a slightly different form:

$$t = A\,\mathrm{e}^{b/T} \tag{5}$$

where t is the burning time and A and b are constants to be determined. This assumes combustion rate to be constant enough for its inverse, burning time, to be described by the inverse of the expression in eqn (4).

Figure 4 shows flame time–temperature relations for three sets of data: all 1978 data, all 1980 data, and one specific condition from 1980. The 1980 data plot very linearly on the semi-log paper, indicating a close relation to the Arrhenius-type equation. The 1978 data plot somewhat less linearly.

All the lines of Fig. 4 have nearly the same slope. Extrapolating the lines to a $1000/T$ of zero shows a positive flame time, indicating a non-zero volatilization and burning time at infinitely high temperature.

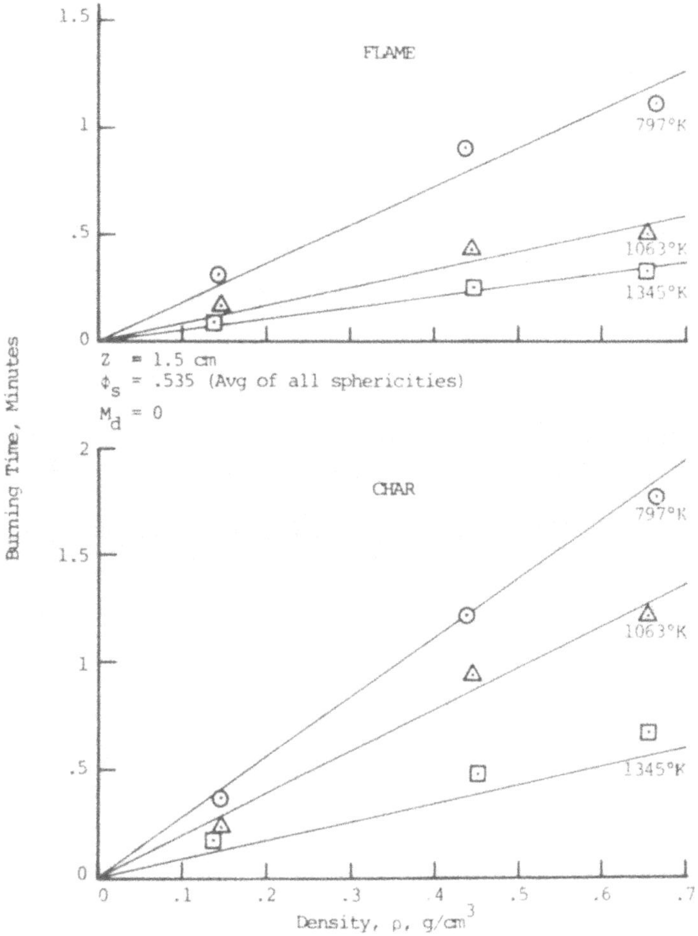

FIG. 3. Effect of density on burning time at three temperatures.

The equations for the char burning time–temperature relations for the same three sets of data are

$$t = 0{\cdot}267\,e^{1987/T} \quad \text{for 1978, all data} \tag{6}$$

$$t = 0{\cdot}195\,e^{1834/T} \quad \text{for 1980, single level} \tag{7}$$

$$t = 0{\cdot}122e^{1828/T} \quad \text{for 1980, all data} \tag{8}$$

The points, not shown on a graph, are less linear than those for flame time; lines drawn through them are concave downward. This indicated a poorer

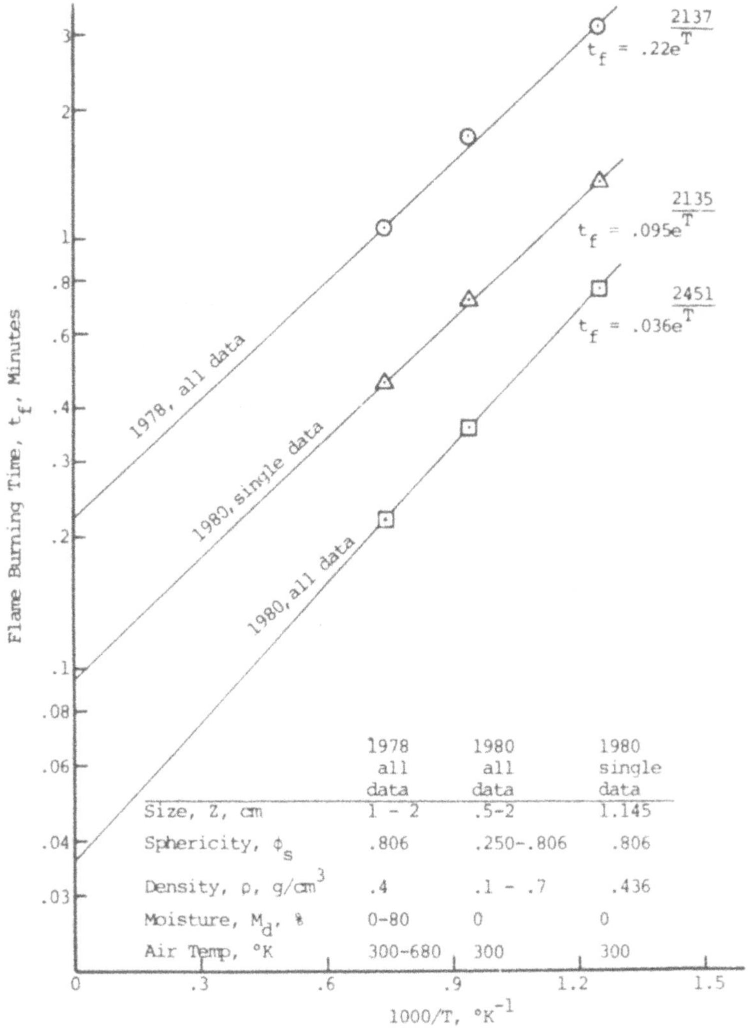

FIG. 4. Effect of temperature on flame burning time.

fit with the Arrhenius equation. This may be explained by diffusion rate as the dominant mode of control of char burning rate, rather than control of chemical reaction rate.[1,3,6]

5. PREDICTION EQUATIONS

Equations to predict burning time for any combination of size, shape, density, moisture content and furnace wall temperature were formulated using this general procedure:

(1) Develop an equation for burning time as a function of each single parameter, at a constant value of each of the other parameters. This results in five equations, one each for size, shape, density, moisture and wall temperature.
(2) Multiply these five equations together, assuming a prediction equation can be obtained from their product.
(3) Compare the resulting equation against test results and adjust the constant in the equation.

Equations for the flame time and char burning time were obtained by this method. Total burning time is the sum of these two. This method sacrifices all interaction effects on the altar of simplicity. But they are easy to calculate, and show how the parameters effect burning time.

5.1. Assembling the Single-Parameter Equations

Equations for the effect of size on flame and char burning time were obtained from Table II at 1063 K. The equation for flame time, $t_f = 0{\cdot}36Z + 0{\cdot}22Z^2$, was rearranged to read

$$t_f = 0{\cdot}36Z(1 + 0{\cdot}61Z) \tag{9}$$

Likewise, the equation for char time was rearranged to read

$$t_c = 0{\cdot}61Z(1 + 4{\cdot}5Z) \tag{10}$$

The equations for the effect of sphericity and density were obtained in like manner from Table II at 1063 K. Equations for the effect of moisture came from the 1978 data, assuming burning time to be linear with moisture content as shown in Fig. 5.

Equations for the effect of temperature are of the form of eqn (5). The constants for the '1980 single level' of Fig. 4 and eqn (7) were used for the

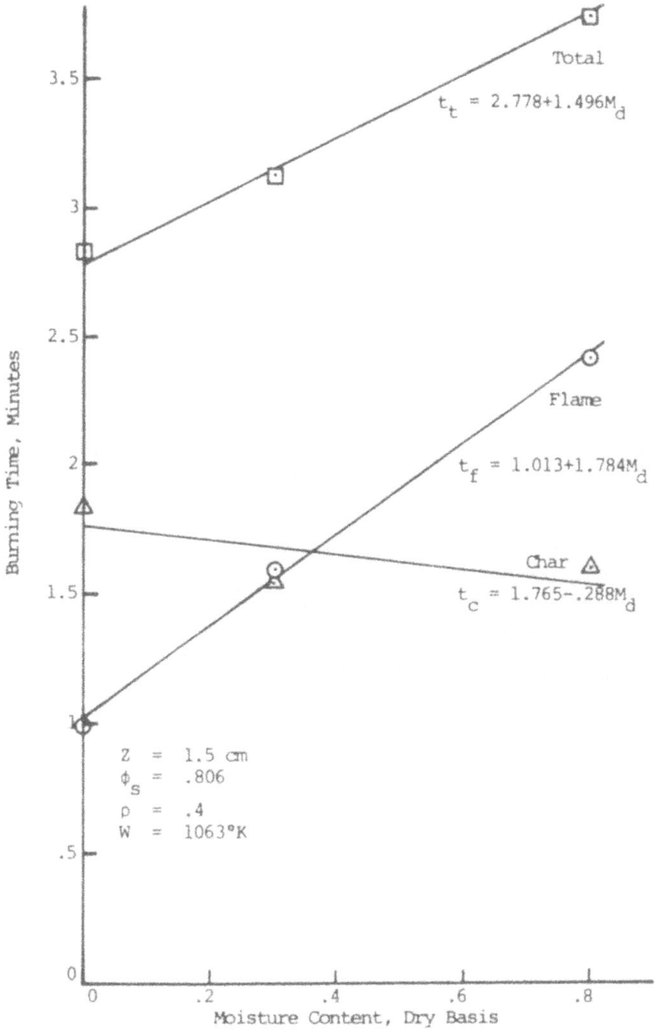

FIG. 5. Effect of moisture on burning time for 1·5 cm pine cubes at 1063 K.

constant A, and 2200 was chosen as a mean slope, b, for both flame and char burning times. This produced the equations

$$t_f = 0{\cdot}095\, e^{2200/T} \qquad \text{and} \qquad t_c = 0{\cdot}195\, e^{2200/T} \tag{11}$$

5.2. The Prediction Equation

Five equations relate size, shape and density (Table II), moisture (Fig. 5) and temperature (eqn (11)) to flame time. Five similar equations relate these parameters to char time. Multiplying together the equations for flame time and adjusting the constants to agree with measured values gives eqn (12). (The constants obtained from direct multiplication were 0·020 for t_f and 0·018 for t_c.) Doing the same for char time gives eqn (13):

$$t_f = 0{\cdot}12\rho Z(1 + 0{\cdot}61Z)\phi_s(1 + 1{\cdot}76M_d)\,e^{2200/T} \tag{12}$$

$$t_c = 0{\cdot}0064\rho Z(1 + 4{\cdot}5Z)(1 + 6{\cdot}9\phi_s)(1 - 0{\cdot}16M_d)\,e^{2200/T} \tag{13}$$

$$t_t = t_f + t_c \tag{14}$$

where t_f, t_c, t_t are flame, char burning and total burning times (min); ρ is density (g/m^3); Z is cube root of particle volume (cm); ϕ_s is sphericity (dimensionless) (range 0–1); M_d is moisture content, dry basis; and T is temperature (K). Burning time in seconds can be obtained by multiplying the constants by 60. Thus, the constant for eqn (12) is 7·2 and for eqn (13) is 0·38.

Equations (12) and (13) show the general effects of the various variables on flame and char burning time. They show, for instance, that the effects of density, shape and moisture can be considered linear but that zero moisture gives a finite burning time and that zero sphericity gives a finite char burning time. They show that the effect of size is a quadratic and that the effect of temperature is an inverse exponential.

The constants in eqns (12) and (13) were adjusted to put the means of the calculated burning times within 3% of the measured means. Then the standard deviations are 19%, 16% and 12% for the flame, char and total times, respectively (e.g. 63% of the calculated flame times are within 19% of the corresponding measured values). The greatest calculated time differences were for the flame time of the flattest oak specimens at the highest temperature (57% higher than measured) and for char burning time of pine cubes at the middle temperature (37% lower than measured).

The char burning times show a poorer temperature correction than the flame times because diffusion, the controlling factor in char burning, is not as temperature dependent as chemical reaction rate. The worst errors of the total times are not as great as those of either the flame or char burning times.

6. APPLICATION TO WOOD FURNACES

The work done here may be more readily applied to large furnaces that burn wood particles in suspension than to small furnaces where wood is burned in a pile. In suspension furnaces, wood particles are introduced near the top of a large firebox in which turbulence is maintained. The smaller particles burn completely in suspension as single particles, while the larger particles may drop on to the grate and finish burning there. Because of the constant air velocity in the tests reported here, these prediction equations may apply more closely to the particles that burn completely in suspension than those that finish burning on the grate.

To obtain equations that more closely predict burning times of wood pieces in suspension burning fireboxes, air velocity and oxygen content need to be varied in test. It may also be necessary to determine average air velocity in relation to wood pieces in a turbulent furnace, where very small pieces probably follow air currents, very large pieces drop directly to the grate, and medium sized pieces do some combination of these two. Furnace oxygen contents are lower than atmospheric and reduced oxygen contents are expected to increase burning times.

REFERENCES

1. Essenhigh, R. H. (1977). Combustion and flame propagation in coal systems: A review. Sixteenth Symposium (International) on Combustion at M.I.T., Cambridge, Mass., August 15–20. (Copyright 1977 by the Comb. Inst.), pp. 353–74.
2. Huff, E. R. (1979). Burning times of wood cubes. Unpublished Ph.D. dissertation, University of Nebraska.
3. Huff, E. R. (1980). Effect of size, moisture content, and environment temperature on burning times of wood cubes. *Agricultural Energy*, Vol. 2. Proceedings of the ASAE National Energy Symposium, Kansas City, MO., Sept. 29–Oct. 1.
4. Kanury, A. M. (1975). *Introduction to Combustion Phenomena*. Gordon & Breach, New York.
5. Kunii, Daigo and Octave Levenspiel (1969). *Fluidization Engineering*. Wiley, New York.
6. Mulcahy, M. F. R. and Smith, I. W. (1969). Kinetics of combustion of pulverized fuel: A review of theory and experiment. *Review of Pure and Applied Chemistry* **19**, 81–108.

41

Single Particle Combustion Analysis of Wood

WAYNE W. SIMMONS and KENNETH W. RAGLAND
Department of Mechanical Engineering, University of Wisconsin, Madison, Wisconsin, USA

1. INTRODUCTION

Rising energy costs have prompted many industries, and particularly wood-related industries, to examine biomass fuel sources as replacements for oil and gas. Several studies[1-4] have indicated a considerable resource in unused wood and wood waste. One factor limiting the efficient use of this resource is a lack of basic combustion data. This limitation has made industrial boiler design empirical and subject to much guesswork. Trajectory analysis of burning fuel particles in the furnace or boiler should be an important design tool; however, much of the basic data have not been available. The average fuel particle size is on the order of centimeters but may range as high as 0·3 m and as small as 100 μm.[3] Fuel particle moisture content may also vary widely. Hence, one can see that, without knowledge of mass burning, heat release rate and burn-out times under specific operating conditions, it is very difficult to size and design an efficient boiler.

The study and analysis of wood combustion is difficult due to the complex nature of wood. Only limited data are available on the process as a whole in the suspension-stoker size range. Effort has been made to examine portions of the combustion process under specialized conditions. Most work can be divided into five major topics: drying, kinetics of pyrolysis, heats of reaction during pyrolysis, char reactions, and ignition of volatile products.

Most data on drying of wood center around air drying pieces or wood piles near room temperature.[5] Some limited drying data are, however, available at elevated temperatures.[6-8] Other useful observations on the effect of moisture on combustion have been made by Williams[9] and Gardon.[10] The most extensive source of data related to wood combustion involves the kinetics of pyrolysis. There have been excellent reviews on the subject by a number of individuals.[11-13] Some of the more useful work is referenced here.[14-26] The heat of reaction during pyrolysis has been subject to considerable controversy since published values often vary widely.[27-29] Most char reaction studies have used coal as a fuel source but some work has been conducted for wood.[42,30] Ignition of the volatiles has been examined by a number of researchers.[31-39] The most detailed experimental work to date on the process as a whole was conducted by Phillips and Becker.[40] They collected data on combustion of pine sticks with application to tree fire behavior under a variety of conditions.

This present work examines the transient mass behavior of wood under conditions similar to those experienced in stoker-type furnaces. Single particle tests are conducted on sugar pine and white oak under a variety of size, temperature, gas velocity, moisture, and oxygen conditions.

2. EXPERIMENTAL APPARATUS AND PROCEDURES

A furnace has been constructed to provide a means of obtaining transient mass data during wood combustion. A cross-sectional view of the furnace is shown in Fig. 1. The free stream air enters the furnace at the bottom and flows upward through an elliptical duct, 80 × 150 mm. Suspended in this duct is a heating element (Super Kanthal) that electrically heats the free stream to the desired test temperature. The flow then passes through a flow diverter into the test section which is 50 mm in diameter. A quartz view port allows visual observation during combustion. From here the flow stream rises out of the furnace into an exhaust system. The upper exhaust section is not used in this transient mass study.

The heating element will withstand temperatures of up to about 2000 K and power levels of near 5 kW. The free stream temperature is influenced by both test flow rate and power level. The inlet air is dried prior to testing, mixed with nitrogen to achieve the desired oxygen concentration, and throttled through a laminar flow element (Hastings) to measure the flow rate. Temperature conditions in the test section are monitored using a thermocouple located just below where the fuel particle resides during testing.

The fuel particle is attached to a 0·02 mm diameter stainless steel wire which hooks to a 2 mm diameter quartz rod 0·55 m long. The quartz rod attaches to an electronic balance (Cahn 7500) suspended above the furnace. The minimum resolution of the balance during test conditions is 1 mg.

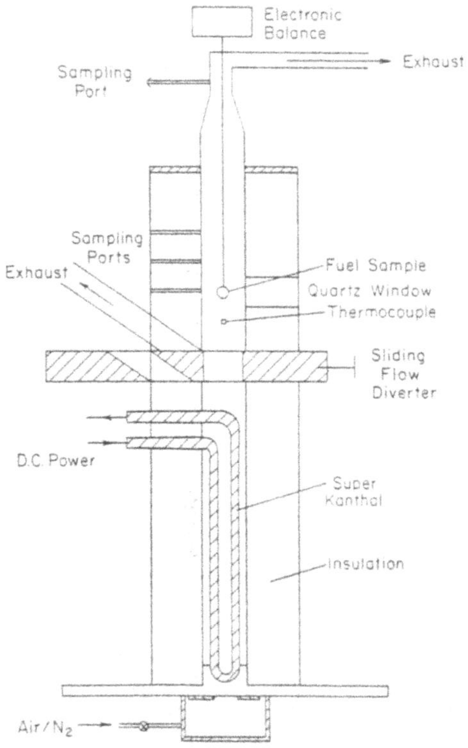

FIG. 1. Cross-sectional view of furnace.

Analog signals from the balance are digitized and stored in an Apple II+ computer for data reduction at a later time. A 12-bit A/D converter (ADA Lab) is used for this operation. Conversion time is 50 ms. To minimize noise effects, ten readings are taken and averaged for each data point.

Transient mass tests were conducted varying six parameters: temperature, Reynolds number, free stream oxygen concentration, particle size, wood type and moisture content. The ranges over which these parameters are examined is:

temperature: 900–1200 K
Reynolds number: 60–600
oxygen concentration: 0–21 %
moisture content: 0 %–fully saturated
wood type: sugar pine, white oak
particle size: 10, 20 mm cubes.

3. RESULTS AND CONCLUSIONS

The transient mass data are normalized by dividing by the initial mass of the samples. The reactivity is calculated as the first derivative of this normalized mass curve. A spline fit of the data is performed to smooth out signal fluctuations that occurred because of sample vibration during the testing. The data points that appear on the curves that follow are simply for identification purposes; the actual data were taken every 2 s. Two runs were made at every test condition to ensure a valid test. The reproducibility of the data was excellent with a maximum variation of about 5% between readings from each test.

Each of the size parameters under investigation is examined individually by varying one while holding all others fixed. The standard conditions from which perturbations are made are:

temperature: 1100 K
Reynolds number: 120
oxygen concentration: 21%
moisture: oven dried
wood type: pine
particle size: 10 mm.

Normalized mass and reactivity transient plots with varying temperature are shown in Fig. 2. The temperature effect is very clearly depicted in these graphs. On the transient mass curves there appear to be two rather distinct regions. Examining the 1200 K curve, one regime appears to begin at time = 0 and proceed through the first 20 s. The second regime extends from this point through the end of the burn. One might expect this indicates independent devolatilization and char burn periods. However, the proximate analysis indicates roughly 80% volatiles and 20% fixed carbon, yet at this transition period nearly 90% of the original mass is gone. Certainly these test conditions are different from a proximate analysis test and heating rate will affect the volatile fraction;[25] however, it still lends evidence that char and volatile combustion are occurring, at least in part, simultaneously. This conclusion is further backed up by visual observations.

As temperature increases, the initial reactivity increases strongly. At temperatures of 1000–1200 K a steep initial jump is observed. This is experienced due to virtual instantaneous ignition of the volatile gases. At 900 K an initial heating period is noted prior to ignition, and the ignition delay time for 10 mm pine particles is about 5 s. At later times, $t > 40$ s, during pure char combustion this temperature is not distinctive. Figure 3

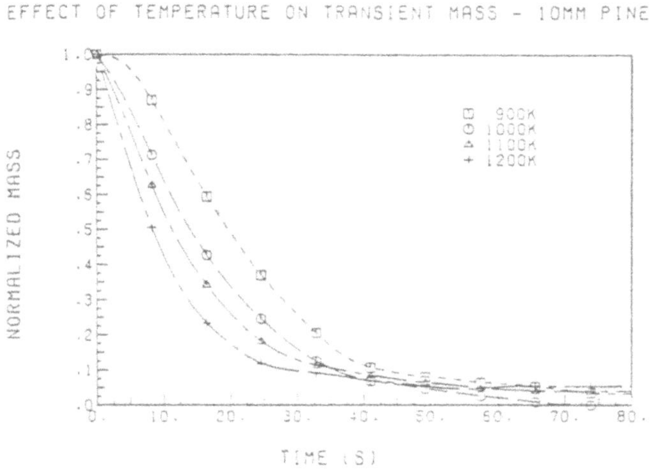

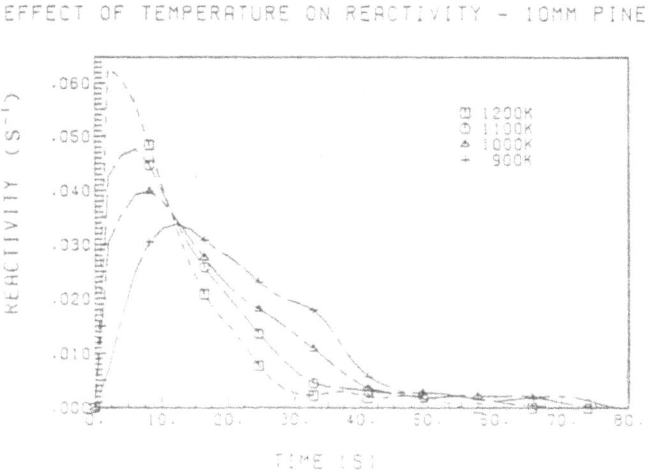

FIG. 2. Effect of temperature.

shows the reactivity plot expanded into two parts, the initial period less than 20 s and final period from 20 to 100 s. The char burn rates are similar in magnitude and duration but occur at different times in the burn.

A similar phenomena is observed when the effect of Reynolds number is examined. Figure 4 gives normalized mass and reactivity transients for 10 mm pine at 1100 K. Higher Reynolds numbers are characteristic of higher reactivities during the initial burning period. This effect again virtually disappears in the purely char burn phase as with temperature.

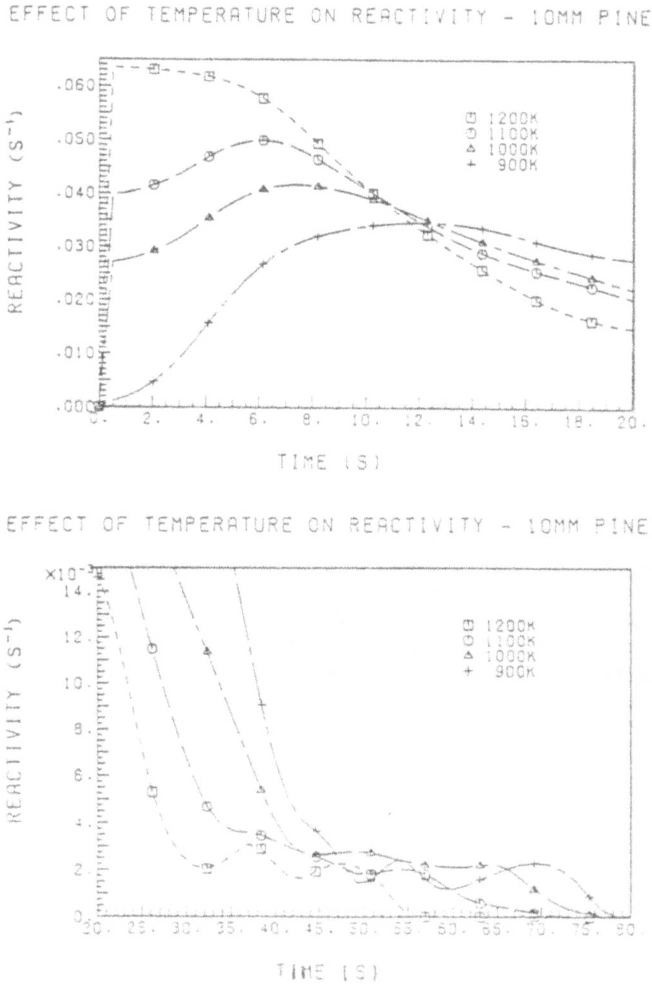

FIG. 3. Expansion of temperature curves.

These results are consistent with the temperature findings because both factors increase the heat flux, with higher values of Re affecting the heat transfer coefficient and temperature affecting the driving potential. The Reynolds number may also influence the mass transfer but this effect is examined by varying the free oxygen concentrations.

Oxygen diffusional effects are characterized by Fig. 5. The same plots of normalized mass and reactivity transients are shown under varying free stream oxygen conditions for 10 mm pine at 1100 K. It is clear from these

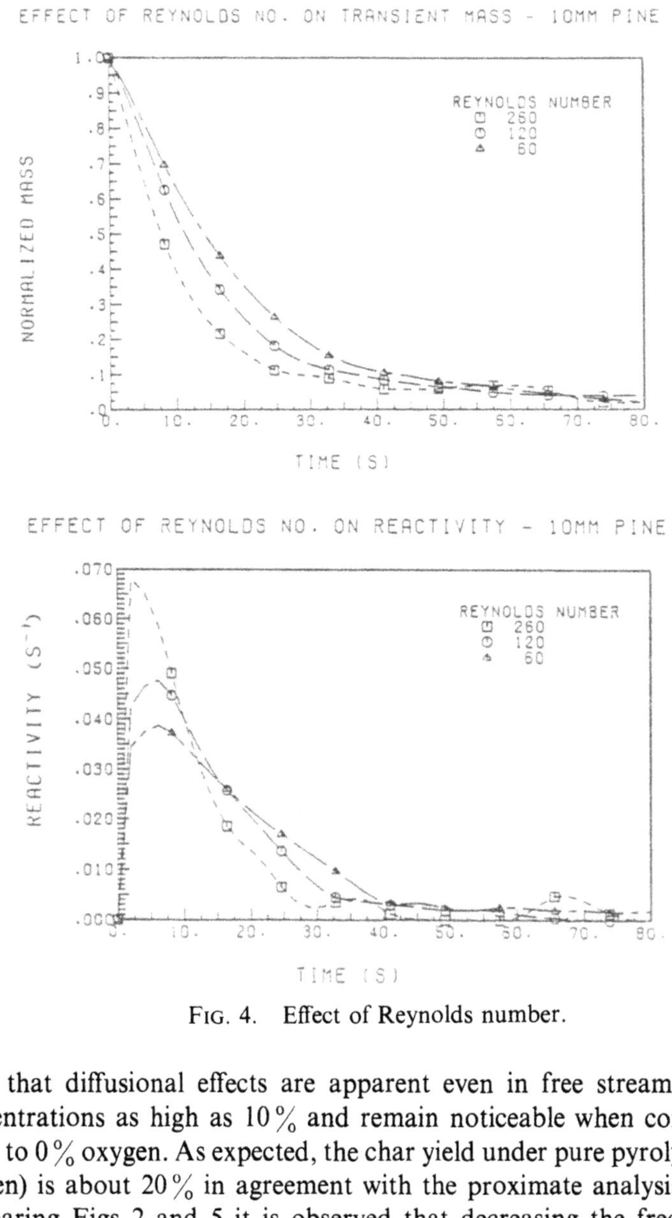

FIG. 4. Effect of Reynolds number.

plots that diffusional effects are apparent even in free stream oxygen concentrations as high as 10% and remain noticeable when comparing 2·5% to 0% oxygen. As expected, the char yield under pure pyrolysis (0% oxygen) is about 20% in agreement with the proximate analysis. When comparing Figs 2 and 5 it is observed that decreasing the free stream oxygen concentration to 10% at 1100 K is very similar to operating at 21% oxygen and 1000 K. This indicates that both heat transfer and diffusional effects are important in this region.

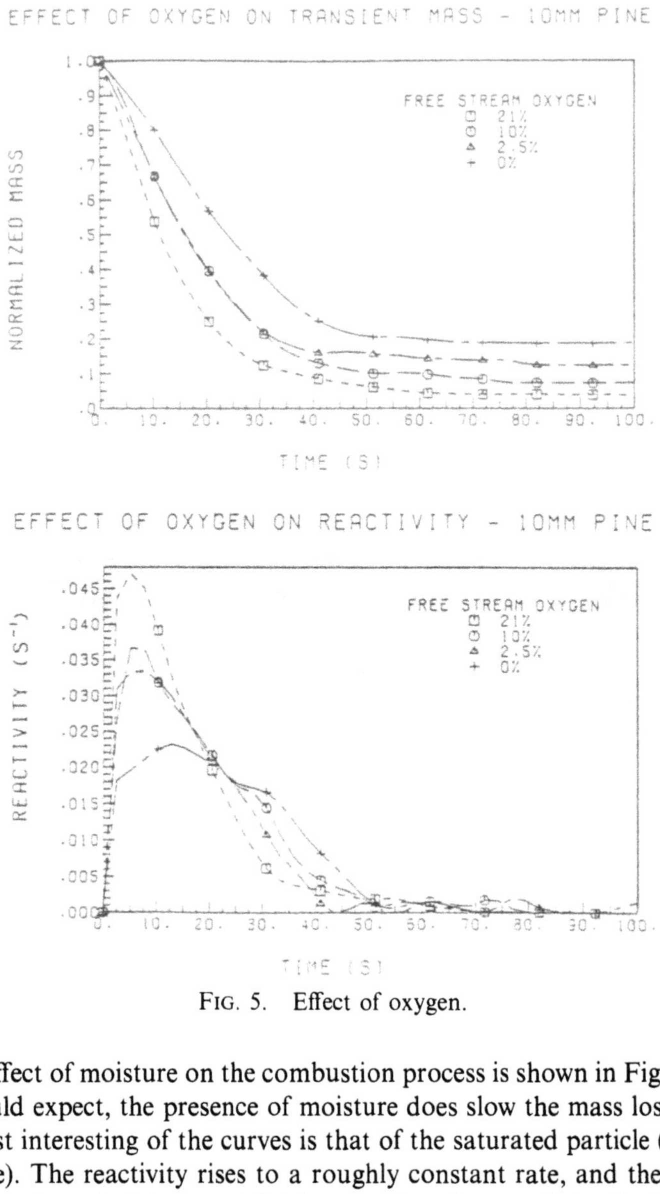

FIG. 5. Effect of oxygen.

The effect of moisture on the combustion process is shown in Fig. 6. As one would expect, the presence of moisture does slow the mass loss rate. The most interesting of the curves is that of the saturated particle (240% moisture). The reactivity rises to a roughly constant rate, and then falls steadily to the end of the burn. This is somewhat characteristic of a classical drying process. Previous researchers[8] had postulated the initial removal of 'free' water by capillary action before the major amount of combustion occurs. If this were true, one would expect the high reactivities associated

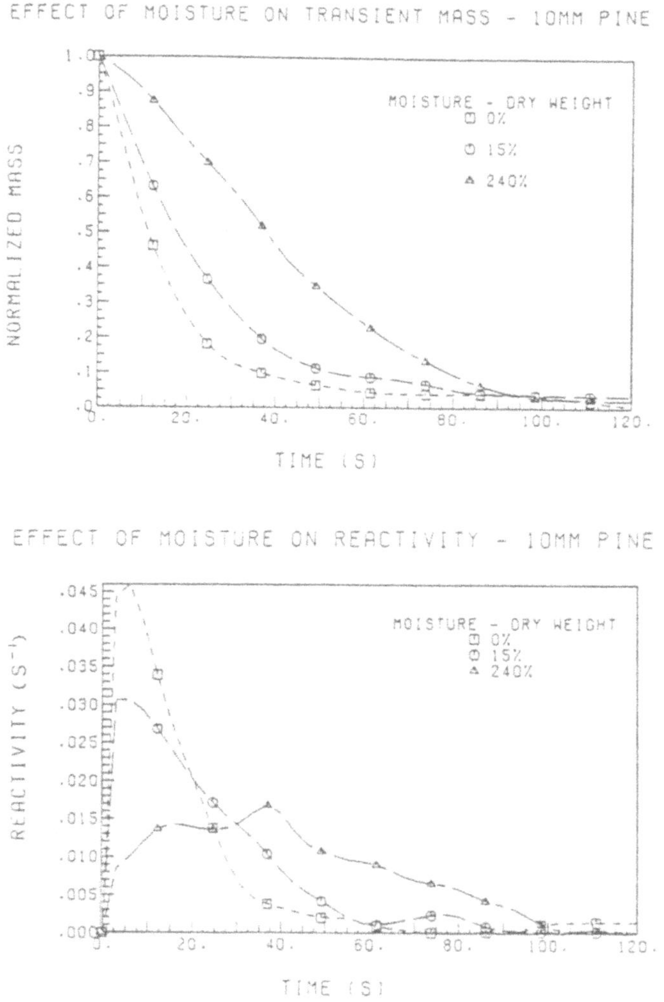

FIG. 6. Effect of moisture.

with volatile combustion to begin at 50–60 s into the test (the point where the normalized mass curve reaches about 0·3 indicating elimination of the water weight). The reactivity curve, however, shows no distinct change at this point and thus combustion and drying appear to be occurring simultaneously. This conclusion is also substantiated by visual observation of flaming during the initial 50 s of the burn.

Figure 7 shows the effect of wood type on both the transient normalized mass and reactivity for 10 mm particles at 1100 K. One can see by these

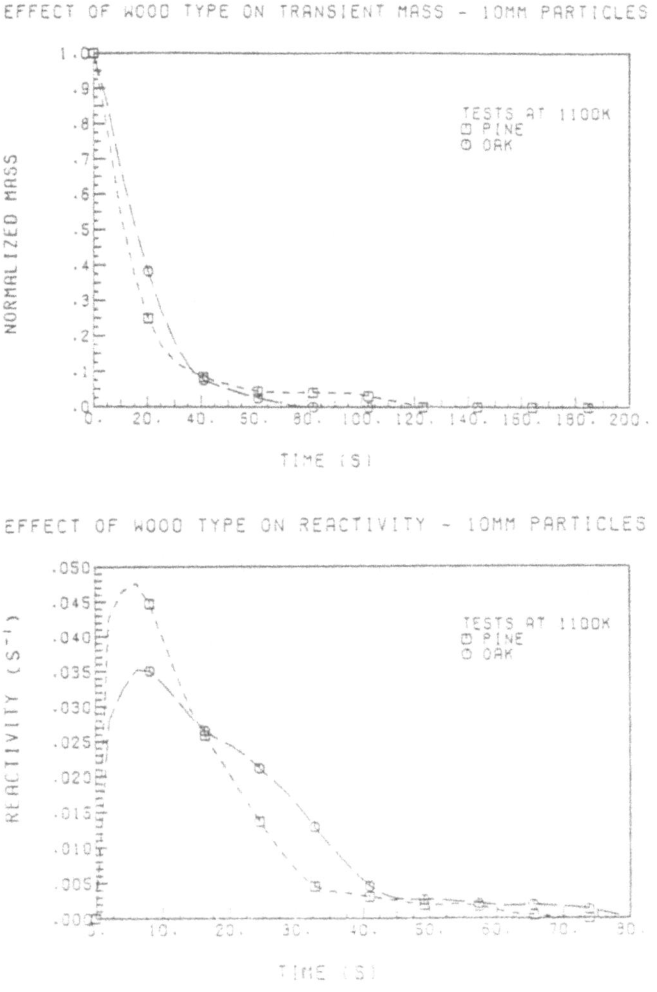

FIG. 7. Effect of wood type.

curves that pine is more reactive than the oak. This same effect is also observed in other tests with variations in parameters such as temperature, moisture and free stream oxygen concentration. This difference may be due to a higher organic extractive content in the pine versus oak. These organics will vaporize quickly and serve as an additional heat input to the particle upon vaporization. This effect also could be related to differences in thermal conductivity because the presence of extractives tends to increase conductivity. However, higher density tends toward higher conductivity

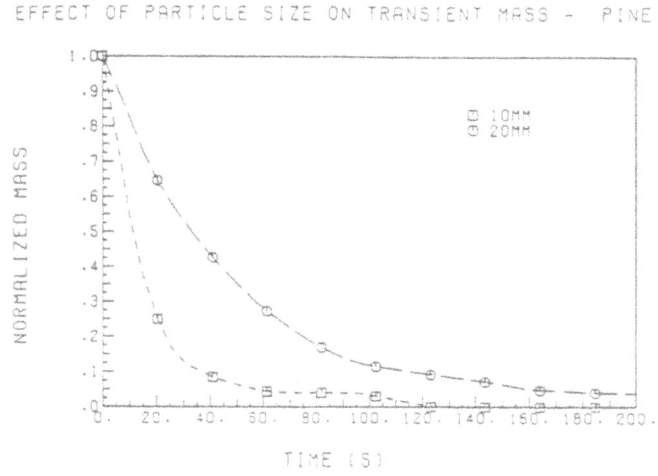

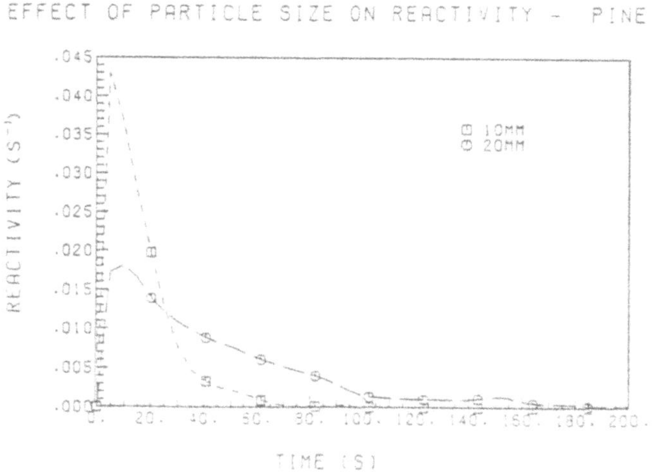

FIG. 8. Effect of particle size.

and oak has a density nearly twice that of pine. Available thermal conductivity data in fact show pine to be slightly less than oak.[41] It therefore seems the former effect of extractive content is dominating the initial higher reactivity as opposed to a thermal conductivity effect.

Interesting observations can also be made regarding the effect of particle size. Figure 8 again shows normalized mass and reactivity transients for 10 and 20 mm particles at 1100 K. The smaller particles are much more reactive. This is reasonable because the surface area-to-mass ratios for the

TABLE I
INFLUENCE OF HEAT TRANSFER PARAMETERS AND SIZE ON MAXIMUM REACTIVITY

	Temperature	
K		*max R* (s^{-1})
1 200		0·063
1 100		0·048
1 000		0·040
900		0·035
	Reynolds number	
Re		*max R* (s^{-1})
260		0·068
120		0·048
60		0·039
	Size	
	max R (s^{-1})	*max R′* ($g/cm^2/s$)
10 mm	0·044	0·003 0
20 mm	0·018	0·002 4

smaller 10 mm particles are much larger than for 20 mm particles. This same phenomenon also forces the smaller particles to be much more sensitive to change in other factors such as temperature, moisture content, Reynolds number and oxygen concentration.

The maximum reactivity for each test condition is shown in Table I. The maximum reactivities are chosen for quantitative comparison because they appear to be representative of the character of the curve. In the first grouping the effect of temperature is evident. The same is true for Reynolds number. Both indicate there is an external heat transfer effect on combustion under these conditions. This influence is predominantly seen in the volatile burn and hence is probably related to reaction rates in the boundary layer which feed back to the internal devolatilization. The next list concerning particle size indicates the importance of internal heat transfer. Reactivities are present both as (s^{-1}) and ($g/cm^2/s$) to negate effects of surface area and mass differences. In both cases the smaller

TABLE II

INFLUENCE OF FREE STREAM OXYGEN, PARTICULATE MOISTURE CONTENT AND WOOD TYPE ON MAXIMUM REACTIVITY

Free stream oxygen concentration	
% O_2	*max R* (s^{-1})
21	0·047
10	0·037
2·5	0·033
0	0·023
Particulate moisture content	
% (*dry basis*)	*max R* (s^{-1})
0	0·047
15	0·031
240	0·017
Wood type	
	max R (s^{-1})
Pine	0·047
Oak	0·032

particle is more reactive indicating the internal heat transfer also influences combustion rates.

In Table II one can see that the free stream oxygen concentration is influencing reactivity. This is a mass transfer effect where the rate of diffusion of oxygen into the boundary layer to support volatile combustion is being limited. This causes a similar feedback effect as did external heat transfer, restricting the heat flux to the particle. The moisture content of the wood has a strong effect on the maximum reactivity as evidenced by the next set of values. One sees as little as 15 % moisture reduces the maximum reactivity by one-third. The final two values indicate that wood type influences the maximum values significantly.

In conclusion, these experimental results tend to indicate that both external boundary layer and internal heat transfer are important in wood combustion under the conditions tested. The same is true for diffusion of oxygen into the boundary layer. Increases in wood moisture content of as little as 15% cause an observable change in particle reactivity. Finally, variations in wood type can influence the combustion process significantly.

ACKNOWLEDGEMENTS

This work was supported by the US Department of Agriculture, Forest Products Laboratory. Many helpful discussions with Andrew Baker and John Zerbe of FPL are gratefully acknowledged. Jeff Horning and Janet Krauze ably assisted with the experiments.

REFERENCES

1. Aton, C. L., Birchfield, J. L. and Bulpitt, W. S. (1980). Wood as an alternative fuel. ASME paper No. 80-C2/Pwr-6 presented at the Century 2 Potpourri Conference, San Francisco, CA, August 13–15.
2. Bergvall, J. K., Bullington, D. C., Gee, L. and Koss, W. (1978). Wood waste for energy study, inventory, assessment and economic analysis. State of Washington Dept. of Natural Resources, Olympia, Washington.
3. Raymond, D. R. and Slinn, R. J. (1975). Testimony by the American Paper Institute. Hearings by the Subcommittee on Energy Research, Development, and Demonstration, House Committee on Science and Technology, US Congress, September 11.
4. Schweiger, R. (1980). Power from wood. *Power* **124**, S1–S32.
5. Clay, D. (1981). Personal communication. Weyerhaeuser Company, Tacoma, WA, November.
6. Malte, P. C. and Dorri, B. (1981). The drying of fuel particles in wood fired combustion systems. Final Technical Report to Weyerhaeuser Company, Energy Division, Tacoma, WA, February.
7. Adams, T. N. Unpublished work at the University of British Columbia.
8. Blackshear, P. L. Jr and Murty, K. A. (1965). Heat and mass to, from and within cellulosic solids burning in air. *Tenth Symposium on Combustion*, The Combustion Institute, pp. 911–23.
9. Williams, C. C. (1953). Damage initiation in organic materials exposed to high intensity thermal radiation. Massachusetts Institute of Technology Fuels Research Laboratory, Technical Report No. 2, Cambridge, MA.
10. Gardon, R. (1953). Temperature obtained in wood exposed to high intensity radiation. Massachusetts Institute of Technology Fuels Research Laboratory, Technical Report No. 3, Cambridge, MA.
11. Roberts, A. F. (1970). A review of kinetic data for the pyrolysis of wood and related substances, ul. *Combustion and Flame* **14**, 261–72.
12. Soltes, E. J. and Elder, T. J. Pyrolysis. Reprint of chapter from forthcoming book, *Organic Chemicals from Biomass*, CRC Press.

13. Milne, T. (1979). Pyrolysis—The thermal behaviour of biomass below 600 °C. In *Survey of Biomass Gasification*, Vol. II. SERI Report TR-33-239, Golden, CO.
14. Browne, F. L. (1958). Theories of combustion of wood and its control. Forest Products Laboratory, US Dept. of Agriculture, Report No. 2136.
15. Stamm, A. J. (1956). Thermal degradation of wood and cellulose, ul. *Industrial Engineering Chemistry* **48**(3), 413–17.
16. Tang, W. K. (1967). US Department of Agriculture Forest Service Research Paper, FPL 71.
17. Murty, K. A. and Blackshear, P. L. (1966). *Pyrodynamics* **4**, 285–98.
18. Tinney, E. R. (1965). The combustion of wood dowels in heated air. *Tenth Symposium on Combustion*. The Combustion Institute, pp. 925–30.
19. Akita, K. (1959). *Rept. Fire Research Institute, Japan* **7**, 99.
20. Eickner, H. W. (1962). Basic research on the pyrolysis and combustion of wood. *Forest Products Journal* **12**, 194–9.
21. Kanury, A. M. (1972). Thermal decomposition kinetics of wood pyrolysis. *Combustion and Flame* **18**, 75.
22. Shafizadeh, F. (1968). Pyrolysis and combustion of cellulosic materials. *Advances in Carbohydrate Chemistry* **23**, 419.
23. Lewellen, P. C., Peters, W. C. and Howard, J. B. (1979). Cellulose pyrolysis kinetics and char formation mechanism. *Sixteenth Symposium on Combustion*. The Combustion Institute, p. 1471.
24. Wai-Chun Chan, R. and Krieger, B. B. (1981). Kinetics of dielectric-loss microwave degradation of polymers: lignin. *Journal of Applied Polymer Science* **26**, 1533–53.
25. Wiggins, D. and Krieger, B. B. (1980). Effect of heating and quenching rates on volatiles produced from combustion-level-heat-flux pyrolysis of biomass. Eighty-ninth National AIChE Meeting, Portland, OR, August 17–20.
26. Slottee, J. S. and Krieger, B. B. (1980). Effect of particle size in volatiles produced from plasma pyrolysis of lignin. Eighty-ninth National AIChE Meeting, Portland, OR, August 17–20.
27. Roberts, A. F. (1971). The heat of reaction during the pyrolysis of wood. *Combustion and Flame* **17**, 79–86.
28. Tang, W. K. and Neill, W. K. (1964). *Journal of Polymer Science*(c) **6**, 65.
29. Roberts, A. F. and Clough, G. (1963). Thermal decomposition of wood in an inert atmosphere. *Ninth Symposium on Combustion*. The Combustion Institute, p. 158.
30. Laurendeau, N. (1978). Heterogeneous kinetics of coal char gasification and kinetics. *Progress in Energy Combustion Science* **4**, 221.
31. Simms, D. L. (1960). Ignition of cellulose materials by radiation. *Combustion and Flame* **4**, 293.
32. Simms, D. L. (1963). On the pilot ignition of wood by radiation. *Combustion and Flame* **7**(3), 253.
33. Simms, D. L. and Law, M. (1967). The ignition of wet and dry wood by radiation. *Combustion and Flame* **11**, 377–88.
34. Martin, S. (1965). Diffusion-controlled ignition of cellulosic materials by intense radiant energy. *Tenth Symposium on Combustion*. The Combustion Institute, p. 877.

35. Weatherford, W. D. and Shepard, D. M. (1965). Basic studies of the mechanism of ignition of cellulosic materials. *Tenth Symposium on Combustion*. The Combustion Institute, p. 897.
36. Akita, K. (1956). *Rept. Fire Research Institute, Japan* **9**, 10–20.
37. Lawrence, E. K. (1952). Analytical study of flame initiation. Sc.M. Thesis, Dept. of Chem. Engr., Massachusetts Institute of Technology, Cambridge, MA.
38. Bamford, C. H., Crank, J. and Malan, D. H. (1945). *Proc. Cambridge Phil. Soc.* (*Part I*) **42**, 166.
39. McLean, W. J. (1980). Ignition of pulverized fuels: A comparative study. Presented at the 1820 Spring Meeting of the Western States Section of the Combustion Institute, Salt Lake City, Utah, April 5, 6.
40. Phillips, A. M. and Becker, H. A. (1982). Pyrolysis and burning of single sticks of pine in a uniform field of temperature, gas composition, and gas velocity. *Combustion and Flame* **46**, 221–51.
41. MacLean, J. D. (1941). Thermal conductivity of wood. *Heating, Piping, and Air Conditioning* **13**(6), 380–91.
42. Arthur, J. R. (1951). *Transactions Faraday Society* **45**, 164.

42

Plane Flame Furnace Combustion Studies of Pulverized Wood

JAMES J. REUTHER

Fuels & Combustion Laboratory, Pennsylvania State University, University Park, Pennsylvania, USA

GRANT G. KARSNER

Energy and Environmental Science Division, Weyerhaeuser Company, Tacoma, Washington, USA

and

SAMUEL T. JACK

Pennsylvania State University, University Park, Pennsylvania, USA

1. INTRODUCTION

The problem of fuel selection is becoming a more difficult one for the combustion engineer, primarily because of the rapidly increasing number and variety of new fuels available.[1] These manifold new fuels are being identified or developed in response to the dual world problem of energy availability and maintenance of environmental air quality.[1,2] Nations around the world will place an increased reliance on the utilization of biomass resources, particularly wood, to partially meet short term energy demands. Although the increased usage of virgin wood or wood wastes seems feasible, detailed combustion performance characteristics of pulverized wood (ignition and burnout times; pollutant emissions) as a function of fuel properties (species; moisture; particle size) and combustor operating conditions (preheat; excess air) are virtually unknown. Development of second-generation technology is necessary for increased wood utilization. Combustion research is required to supply well-defined data on biomass flames, and a fundamental understanding of pulverized-fuel combustion. Present wood-burning equipment has been designed and

TABLE I

TYPICAL COMPOSITIONS OF SELECTED CONVENTIONAL AND ALTERNATIVE FUELS

Solid fuel	*Ultimate analysis*					*Proximate analysis*			*Moisture (%)*	*Heating value (Btu/lb)*	*H/C*
	C	*H*	*N*	*S*	*O*	*Ash*	*Volatile matter*	*Fixed carbon*			
	(dry, wt %)					*(dry, wt %)*					
Bituminous coal	80·5	5·0	1·5	1·3	5·9	5·8	29·6	64·6	6·7	14 430	0·75
Anthracite coal	84·2	2·8	0·8	0·6	2·2	9·4	3·4	84·2	4·5	13 810	0·40
Lignite	64·4	4·2	1·1	0·8	20·7	8·8	41·8	49·4	41·5	11 034	0·78
Peat	52·2	5·3	1·8	0·4	30·3	10·0	67·3	22·7	72·5	9057	1·22
Bark	53·1	5·7	0·2	0·1	39·0	2·5	66·0	27·0	50	8930	1·29

installed in the absence of quantitative data on the combustion performance of this renewable biomass fuel. As a result, a number of units are now operated in non-optimal modes. Moreover, the potential exists for pulverized wood to be used as a replacement fuel for oil or gas. This represents an economically attractive option.

2. STATEMENT OF THE PROBLEM

Some qualitative information can be derived from analogous studies on pulverized *coal* combustion characteristics.[1,2] To some extent the physical (ignition and burnout times) and chemical (pollutant emission potentials) combustion behavior of a pulverized solid fuel can be estimated or predicted by its compositional properties, i.e. ultimate and proximate analyses. Table I lists the 'typical' compositions of selected conventional and alternative fuels, including wood.[1,2] The ease of ignition of a finely-sized fuel may be related to the percent volatile matter: generally, the higher the percent volatile matter, the more readily the pulverized fuel will ignite. Coupling high fuel volatilities with high moisture contents (undried fuels) can reverse this trend (lignite and peat). Burnout efficiency is a complex function of many compositional variables, with the fixed carbon playing a significant role: generally, the higher the fixed carbon, the more likely it is for carbon oxidation to be incomplete within the residence time of the combustor with resulting carbon particulate carry-over. An evaluation of a fuel's pollution potential can be made by inspection of its ultimate analysis: the higher the percent fuel-nitrogen, fuel-sulfur and mineral matter (ash), and the lower the atomic H/C ratio, the greater the tendency for the combustion of that fuel to result in higher nitrogen oxides (NOX), sulfur oxides (SOX), inorganic particulates (ROX), and smoke pollution, respectively.

Table II has been constructed from the data in Table I and the aforementioned 'rules of thumb'. It displays a comparison between a conventional solid boiler fuel (bituminous coal) and several alternative solid fuels. The qualitative information in Table II would suggest that wood has compositional properties that would make it a preferred solid fuel. This is because of its superior ease of combustion (ignition and burnout) and its low pollution potential (NOX, SOX and ROX). Although oil combustion is always easier and cleaner to conduct than pulverized coal firing, wood appears to be an attractive solid fuel alternative. One major drawback of

TABLE II

COMPARISON OF CONVENTIONAL AND ALTERNATIVE FUEL COMBUSTION CHARACTERISTICS

Solid fuel	*Ignition*	*Burnout*	*NOX* (%*N*)	*SOX* (%*S*)	*ROX* (%*Ash*)	*Smoke* (*H*/*C*)
Bituminous coal			Reference			
Anthracite coal	D	D	L	L	H	H
Lignite	D	E	L	L	H	L
Peat	D	E	H	L	H	L
Bark	E	E	L	L	L	L

D = more difficult; E = easier; L = lower potential; H = higher potential.

wood is its high moisture content (40–60 + %), which retards fuel ignition. Also, the heating value of wood is lower than that of coal and other fuels.

The 'scoreboard' data given in Table II only provide a *qualitative* analysis of the similarities and differences between pulverized coal and wood flames. The quantitative comparison is lacking. The purpose of this paper is to present comprehensive results from The Pennsylvania State University Fuels and Combustion Laboratory (PSU-FCL) program conducted to obtain pulverized wood combustion data. First results have been published previously.[2] Pulverized wood has already been found to have the shortest ignition time and lowest ignition temperature of the solid fuels listed in Tables I and II.[2]

This paper reports on a continuing program, the overall objective of which is to determine, under the well-controlled and defined simulated boiler conditions produced by the PSU-FCL plane flame furnace, the combustion characteristics of pulverized wood. These results establish the parametric relationships of wood properties (moisture: 10–28 %; mass-mean particle size: 36–612 μm; fraction of fines: 0–100 % < 100 μm) and combustor operating conditions (preheat: 500–850 °C; excess air: ±25 %), with combustion behavior (ignition time and temperature; burnout time; gaseous and solid products of combustion).

3. EXPERIMENTAL

Pulverized Douglas fir bark was fired in the PSU-FCL plane flame furnace, shown schematically in Fig. 1. The construction of the furnace permits characterization of the pre-to-post combustion regime. Despite its small

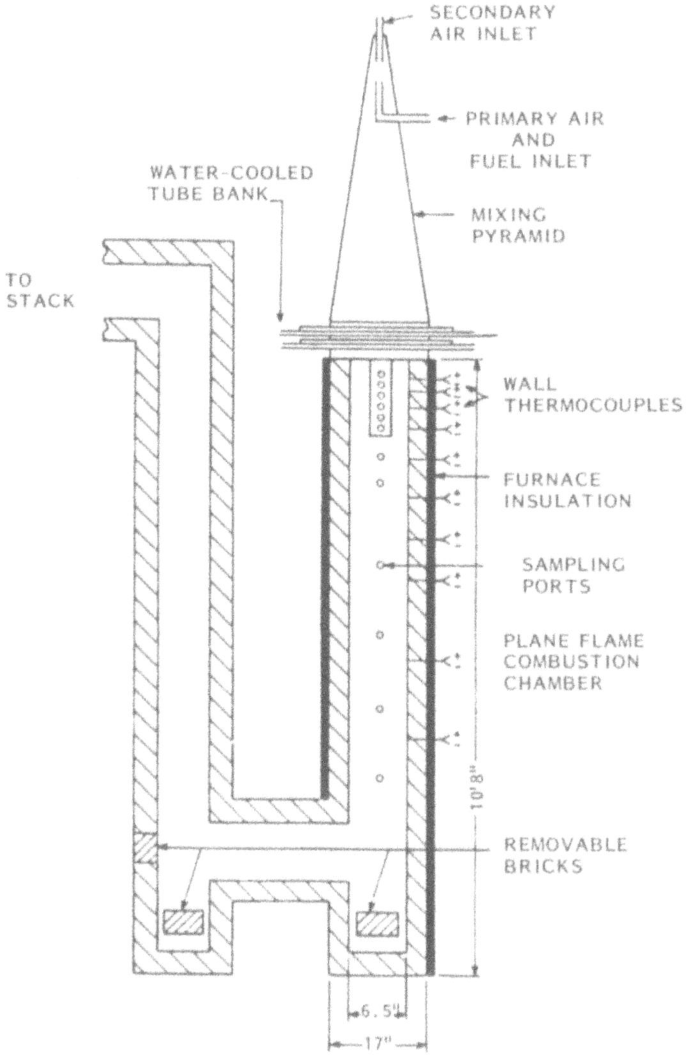

FIG. 1. Pennsylvania State University Fuels and Combustion Laboratory Plane Flame Furnace used for solid fuel combustion tests. The behavior of anthracite coal, reconstituted coal and biomass (wood) are currently being evaluated.

scale (1–2 orders of magnitude smaller than a typical utility boiler), the PSU-FCL plane flame furnace adequately simulates the time/temperature history of a full-scale boiler: heating rates of 10^4 K/s; peak gas temperature of 1400–1500 K; and residence times of 1–2 s. However, it does not recreate the turbulent aerodynamics within an industrial flame. The furnace has been designed to generate a microturbulent flame of essentially one-dimensional geometry. The flat flame structure is generated by close control of mixing of secondary air and adiabatic combustion products. To achieve such an environment, premixtures of pulverized fuel and primary air (*ca.* 18% of the required combustion air) pass through a water-cooled tube bank before flowing downward into the combustion chamber. This key furnace component helps to prevent combustion flashback into the mixing chamber (a precaution that is required for reactive fuels such as wood) and streamlines the velocity profile of the freely-falling pulverized bark/air cloud into plug-flow.

Flat wood flames stabilize at different positions downstream of the down-fired burner primarily as a result of different moisture levels and particle size distributions. Because of the 'once through' flow, all wood particles experience a nearly identical time/temperature history as they vertically traverse the combustion zone. Thermal and chemical profiles of the combustion environment can be well-controlled and resolved as a function of time.

After varying degrees of furnace-preheating, pulverized Douglas fir bark was burned in the furnace at a constant firing rate, with fuel-rich and fuel-lean stoichiometries, and with wood having narrow and wide polydisperse particle size cuts. Fuel moisture contents varied from 10 to 28 wt%. Each fuel and combustion parameter was varied independently in the current study.

The internal refractory fire-brick walls of the plane flame furnace were preheated with a natural gas/air flame to a set limit temperature profile below which a self-sustaining solid fuel flame would not develop. The plane flame furnace had been previously calibrated according to its time/temperature preheat history so that this standardized ignition procedure could be executed. For example, at 500 °C, 700 °C or 900 °C, the natural gas flame had to be fired at 0·5, 1·0 *or* 7·0 h, respectively. After qualitative experiments had established the minimum ignition temperature for a test fuel, the furnace would be preheated to that target temperature, the gas burner shut off, removed, its entry port plugged, and the desired steady-state pulverized bark feed rate established. The wood was pneumatically delivered to the furnace, after metering via a commercial

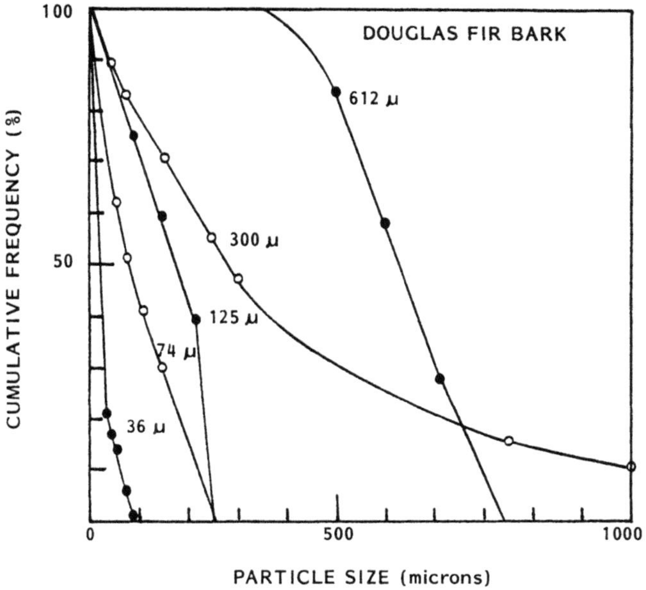

FIG. 2. Particle size distributions of pulverized Douglas fir bark burned in this program.

vibrating screw-feed system, at a rate of 28 lb/h, or $0{\cdot}25 \times 10^6$ Btu/h. The maximum deviation in feed rate from the mean value during a given run was ±5%. The wood/air stoichiometry was varied over ±25% excess air by changing the secondary air flow rate.

Current results include a study of Douglas fir bark, with an ultimate and proximate analysis such as given in Table I. The important wood fuel properties parametrically varied were particle size and moisture. Figure 2 presents the cumulative frequencies versus particle sizes for the five size distributions tested, along with their mass-mean particle sizes: 36, 74, 125, 300 and 612 μm. These distributions will be referred to by their respective mass-mean sizes. It may be expected that the variation in the percentage of fines, arbitrarily defined as those particles less than 100 μm, may be used as a correlating parameter with the combustion results. All samples were size-classified from that labeled as 300 μm.

The as-received moisture of the previously dried wood was *ca.* 10%. This moisture level was enriched to a maximum value of 28% by a very simple spray-wetting technique.[3] The method consisted of spreading the wood into a thin layer, spraying a measured quantity of water on to it, mixing, and allowing the wetted fuel to equilibrate overnight in a plastic bag. The

TABLE III

RESULTS OF PSU-FCL PLANE FLAME FURNACE COMBUSTION STUDIES OF PULVERIZED WOOD BARK

Mass-mean particle size (μm)	*Particle size range (μm)*	*Amount of fines (% < 100 μm)*	*Moisture content (%)*	*Excess air (%)*	*Ignition temperature (°C)*	*Ignition time (ms)*	*Burnout time (ms)*	*NOX (ppm)*
36	0–90	100	10	+25	500	0	30	—
	0–90	100	28	+25	500	0	30	—
74	0–250	56	10	+25	500	2	145	300
	0–250	56	28	+25	500	2	142	—
125	0–250	25	10	+25	500	3	338	290
	0–250	25	10	0	500	4	425	225
	0–250	25	10	−25	500	9	1 140	30
125	0–250	25	28	+25	500	3	342	210
	0–250	25	28	0	500	5	430	—
	0–250	25	28	−25	500	10	1 150	—
300	0–2 000+	20	10	+25	600	5	535	150
	0–2 000+	20	28	+25	600	30	540	—
612	350–850	0	10	+25	850	201	635	—

moisture content was checked before and after this preparatory procedure, and immediately before firing using standardized methods for moisture analysis.

Flame and wall temperatures and compositions of gaseous and solid products of combustion were characterized as a function of residence time (distance down the furnace centerline). Sampling of the combustion environment was performed through ports distributed along the vertical axis of the furnace (see Fig. 1). The residence time between measuring positions was determined by correcting volumetric flow rates to give local flame velocities.[4] Volumetric changes caused by molar and temperature variations were incorporated into the calculations. Probes designed and fabricated at the PSU-FCL were used to measure: (a) the internal furnace wall (thermocouples); (b) the gas (suction pyrometer) temperatures; and (c) the gaseous and solid products of combustion (various water-quenched gas and particulate sampling probes). Gaseous product concentrations were measured on-line both in the flame and the exhaust gases using non-dispersive infrared (CO_2, CO), paramagnetic (O_2) and chemiluminescent (NOX) instrumental techniques. Captured solid samples were subjected to elemental analyses.

The extent of combustion in the pulverized wood flame was monitored in two independent ways by measuring: (1) the appearance of carbon dioxide in the gaseous products, and (2) the disappearance of the elemental carbon in the char. Thus, carbon burnoff was resolved as a function of time. The independently derived gas- and solid-phase progress of combustion data agreed within $\pm 5\%$ of each other.

With these data on carbon burnoff and residence time, important events in the initiation and termination of the bark flame were defined, ignition time and burnout time, respectively. Ignition time is defined as the residence time where the most rapid temperature rise occurs. Burnout time is the time required for 100% carbon burnoff.

4. RESULTS

A representative set of time-resolved data for one series of flame tests on pulverized wood is given in Fig. 3: the internal furnace wall temperature, the flame gas temperature and the appearance of CO_2 and NOX and the disappearance of elemental carbon within the plane flame furnace.

Table III contains a listing of the results of PSU-FCL plane flame furnace combustion studies of pulverized wood obtained in this phase of

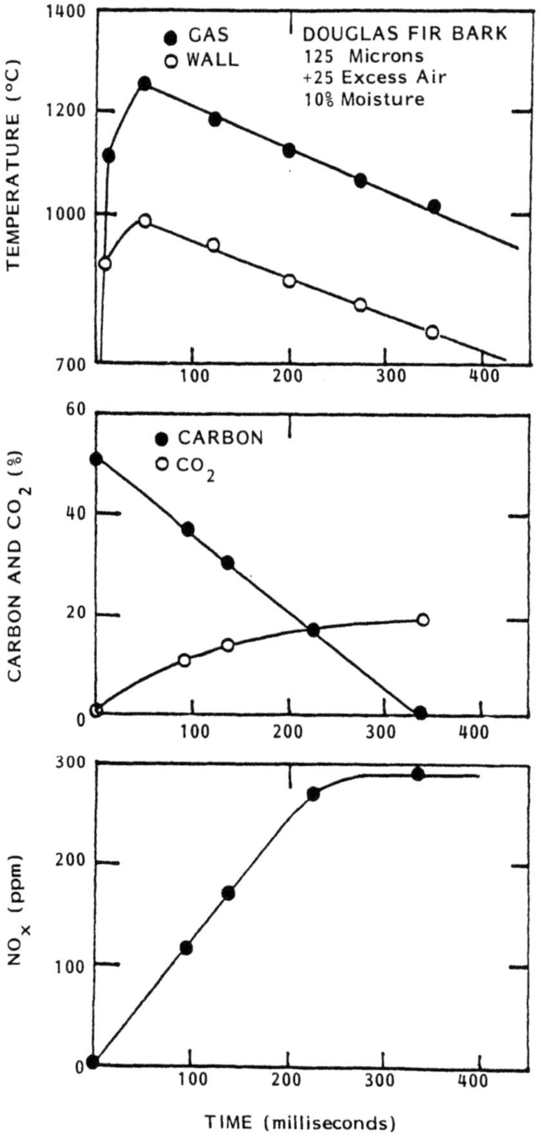

FIG. 3. Representative plane flame furnace wood bark combustion data.

testing. As the data in Table III indicate, the effects of mass-mean particle size (36–612 μm), particle size range (0–2000 + μm), percentage of fines (100–0 %), moisture content (10–28 %) and excess air (±25 %) on ignition temperature and time, burnout time and NOX pollution have been parametrically evaluated. The numerical data presented in Table III represent an average of reproducible data from two or more experiments in all cases. The uncertainties in these measurements are approximately: ignition temperature: ±3 %; ignition time: ±10 %; burnout time: ±10 %; NOX: ±5 %. The time-resolution of ignition and burnout is reported in milliseconds (10^{-3} s); NOX concentrations are given in parts per million (ppm) on a dry volume basis and corrected to 3 % oxygen in the exhaust.

5. DISCUSSION OF RESULTS

5.1. Effect of Particle Size on Ignition

The analysis of the data shown in Fig. 2 and Table III can be correlated with three size parameters: (a) the mass-mean size, (b) the particle size distribution, and (c) the amount of fines (defined as less than 100 μm). Focusing first on the overall relationship between mass-mean particle diameter and ignition temperature/time, at constant moisture and excess air, changing the mass-mean particle size distribution from 36 to 612 μm results in a 350 °C and 200 ms increase in minimum ignition temperature and time, respectively. This trend appears to be consistent with other solid fuel combustion studies. Figure 4 shows the impact of the percentage of fuel fines on ignition time. The ignition time is less than 10 ms for those fuels which contain 20 % or more fines (less than 100 μm). Further, the ignition temperature of 500 °C was independent of particle size over a 3·5-fold range in mass-mean particle size (36–74–125 μm), and ignition time was very insensitive to particle size (0-to-5 ms) over an even larger (10-fold) range in particle size (36–74–125–300 μm). This independence and insensitivity abruptly disappears between the 300 and 612 mass-mean size range where the fuel fines were removed by screening prior to the combustion test. This distribution was classified by sieving to remove all particles of sizes less than 350 μm and greater than 850 μm. Its ignition time of 201 ms is 70 times longer than for the original particle size distribution (that labeled 300 μm). These data suggest that there exists some threshold level of fuel fines less than 20 % which is required for low ignition times. We anticipate that some fraction of particles with high surface/volume ratios are required for stable ignition.

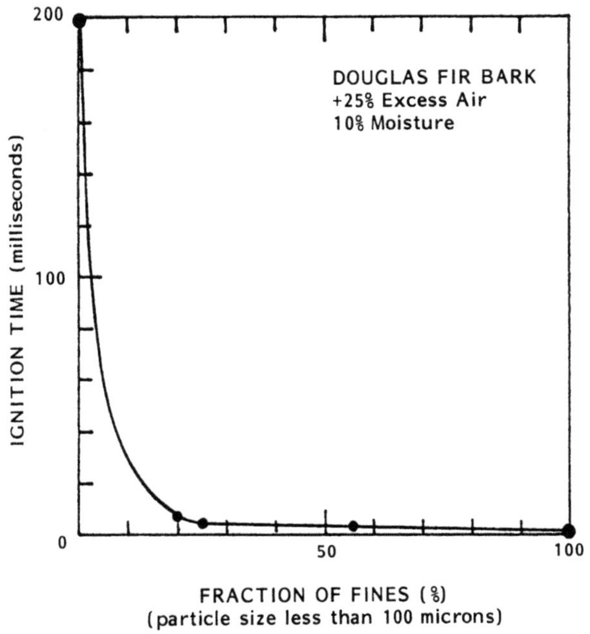

FIG. 4. Relationship between percentage fines ($<100\,\mu m$) and ignition time.

The data presented in Table III are considered internally comparative and consistent, but not absolute because ignition and burnout time determinations are device (time–temperature history) dependent. McLean,[5] in collaboration with Karsner, found at constant gas temperature and oxygen concentration that the ignition delay of time of Douglas fir bark was relatively insensitive to particle size for 60–90 μm size particles, but that it nearly doubled over the 90–200 μm range. The data obtained in this study and given in Table III tend to corroborate this behavior. Moreover, these trend results also further support the modeling conclusion of McLean that the ignition of pulverized wood is delayed in the absence of fines and may only occur as the result of non-isothermal particle surface heating of larger particles.

Most of the current understanding of pulverized fuel combustion is a consequence of several investigations on the combustion performance characteristics of pulverized coal. Comparisons between pulverized wood and coal will contrast their similarities and differences. Weight found that the ignition temperature of pulverized high-volatile bituminous coal was independent of the percentage of fines, defined as sizes less than 40 μm.[6] In his experiments, a change in ignition temperature of 100 °C was deemed insignificant. The current results indicate that pulverized wood does not possess ignition characteristics analogous to pulverized coal at the same particle size. As the percentage of 40 μm 'fines' is decreased from 80 to 18 to 10, or as the mass-mean wood size is increased from 36 to 74 to 300 μm, respectively, a 100 °C increase in ignition temperature is observed. Thus, it is concluded that bituminous coal and wood have differing reactivities and observed differences in fuel variable impacts on ignition times may be attributed to reactivity differences.

There are several possible ignition mechanisms for pulverized wood. The data suggest one interesting possibility. For example, by the time a 125 μm bark particle ignites, only *ca.* 10 % of volatiles have evolved. Wood, when rapidly heated in the plane flame furnace (10^4 K/s), may experience a 90 % or more devolatilization weight loss.[5] The unique character of wood combustion—as indicated by the rapid transient mass loss because of its high volatile evolution upon rapid heating, and the high reactivity of its char at low and high temperatures[7]—make the overlap of devolatilization and heterogeneous combustion very likely. Thus, heterogeneous ignition may be competitive with homogeneous ignition. Ignition may, indeed, occur on the solid surface at temperatures far below those required to ignite flammable mixtures of volatiles and air. In this regard, wood ignition is very similar to that for low-rank brown coals.[8]

5.2. Effect of Moisture on Ignition

Since moisture level is a dominant factor affecting the ignition time required for any pulverized fuel, the current study determined the minimum ignition temperature of pulverized wood as a function of moisture (at constant stoichiometry and differing particle sizes). The data in Table III indicate that the ignition temperature of pulverized wood is independent of fuel moisture level over the 10–28 % range for three of the individual mass-mean particle sizes studied—36, 74 and 125 μm—but that such is not the case for the fourth—300 μm. This result is in near-agreement with the modeling efforts of McLean for Douglas fir bark. The current study shows that the effect of increased moisture level (10–28 %) delays ignition 6-fold (5–30 ms, respectively). The *ca.* 1 ms rise in ignition time per percent moisture increase is comparable to that estimated by McLean.[5] It should be noted that this model considers only small size cuts whereas the experimental data reported here are for wider polydispersion. The independence of ignition temperature upon fuel moisture level at small mass-mean particle sizes is also related to the large percentage of fines. This is suspected since a small but measurable effect of increased moisture on ignition time is witnessed as the percentage of fines is reduced to 25%: under stoichiometric and sub-stoichiometric conditions (0% and −25% excess air, respectively), an increase in the moisture content of 125 μm particles from 10% to 28% increases ignition time by 10–20%. Finally, it should be noted that experiments by Douglas[9] on 74 μm sub-bituminous coal (volatile matter = 24·7%) particles revealed that ignition temperature was independent of moisture over the 0–20% range. Hence, whereas pulverized wood is different from coal with respect to ignition for similar particle sizes, it is more similar with respect to fuel moisture effects.

5.3. Effect of Excess Air on Ignition and Burnout

Combustion engineers have been concerned with balancing the amount of excess air required to ensure stable fuel ignition and complete burnout with that needed to achieve maximum thermal efficiency. Until recently, this tradeoff has usually guaranteed that pulverized fuel combustors have been operated in the fuel-lean mode.[10] With the enactment of regulations to limit the quantity of combustion-generated air pollutants that can be emitted into the atmosphere have come several strategies for pollution control by flame modification.[11,12] For NOX control, these modifications usually involve the minimization of excess air, or the operation of a combustor in two stages, where the first stage is fuel-rich.[11,12] The current study examined non-stoichiometric combustion experiments on pulverized wood to determine the impact of this pollution control strategy in anticipation of possible NOX emission problems.

Table III indicates that the penalties for burning pulverized wood at stoichiometries of less than +25% excess are a delayed ignition time and a prolonged burnout time. Specifically, a reduction of excess air from +25% to −25%, at constant gas preheat temperature, results in a factor of 3 or more increase in ignition and burnout times for the 125 μm mass-mean size distribution.

5.4. Effect of Particle Size on Burnout

When particle size effects are parametrically investigated, the main emphasis is usually concerned with their influence on pulverized fuel burnout time, once a flame has been initiated. A previous section examined the influence of fines on fuel ignition. This section deals with burnout trends, specifically with regard to the top-size of the particle distribution. The results in Table III indicate the expected increase in burnout time with increasing mass-mean particle size. However, unlike the ignition correlation presented in Fig. 4, the results from the combustion tests on the 300 and 612 μm mass-mean size distributions suggest that burnout is not exclusively a function of maximum particle size. Although the bulk size distribution is increased by a factor of 2 from 300 to 612 μm mass-mean sizes, the top-particle size has been *reduced* by more than a factor 2 (2000 to 850 μm) by prescreening. The data in Table III indicate that the whole particle size distribution (300 mass-mean size), consisting of a 15% cumulative frequency of particles larger than 850 μm (see Fig. 2), burns out almost 20% faster than the narrower distribution comprised only of particle sizes within the 350–850 μm range. Alternatively, in the presence of 20% fines (< 100 μm), the burnout times of 1000–2000 μm particles are faster than those of 850 μm particles whose distribution is void of fines. Modeling efforts are underway to better understand this interesting phenomenon. These results suggest that burnout must be related to two parameters defining size distribution and is linked to fuel ignition. The amount of fine particles in the polydispersion is the suspected driving force giving rise to this effect. There is a precedent for such behavior in the coal combustion literature.[13] Beér reported two decades ago that as the percentage of fines in a pulverized anthracite coal distribution, with a constant top-size of 250 μm, was reduced from 80% to 20%, the flame length, or burnout time, increased by 40–50%.[13]

5.5. Effect of Moisture on Burnout

The effect of moisture enrichment (from 10% to 28%) on burnout time has been studied for bulk particle sizes of 36, 74, 125 and 300 μm. In every test, the burnout time was virtually unaffected by the 3-fold increase in moisture. This behavior is not unexpected. Since the ignition times of these fuels are similar (Table III), the time–temperature histories of these small polydispersions in the furnace are similar.

5.6. Effect of Excess Air, Moisture and Particle Size on NOX

As with almost all solid fuels, pulverized wood contains small amounts of fuel-bound nitrogen (FBN), which upon combustion can be readily oxidized into NOX (where $X = 1, 2$). During the course of this investigation, the fate of this fuel-bound nitrogen was parametrically determined as a function of excess air, moisture and particle size.

Switching from a fuel-lean oxidizing flame environment to a fuel-rich reducing one is perhaps the best known NOX control strategy, not only for pulverized coal[11] but for distillate and residual oil[12] as well. As stoichiometry is changed from +25% to −25% excess air, reductions in NOX of 70–90% are common for pulverized coal and petroleum.[11,12] There is a paucity of NOX data, however, for pulverized wood flames. Results from the current study are summarized in Table III. For 125 μm mass-mean size wood, the NOX production level was 290 ppm. This represents a 95+ % conversion efficiency of FBN to NOX, which is similar to that of a (low-nitrogen) distillate petroleum oil.[12] Nitrogen enriched coal and residual oil have an FBN conversion efficiency of *ca.* 10–20% and 20–50%, respectively. Upon decreasing the combustion stoichiometry from +25% to −25% excess air, NOX from the wood flame was reduced 90%, to 30 ppm. The current study suggests that in a bark flame, almost all of the FBN is converted to NOX, which is undesirable from a *pollution production* viewpoint. However, NOX can be very effectively controlled by reduced excess air, which is very good from a *pollution control* viewpoint. Both of these effects primarily result because of the very high volatility of wood. In general, the higher the volatile matter content of the fuel, the higher the FBN-to-NOX conversion efficiency and the more effective fuel-rich combustion is in the reduction of NOX. In contrast, coal is a high-nitrogen, low-volatile fuel. Residual and distillate oils are medium- and low-nitrogen, high volatile fuels, respectively. Pulverized bark flames should pollute no more than distillate oil flames, which are relatively very clean compared to those of coal and heavy oil. This is why distillate oil is an endangered fossil fuel species, and why surrogates, such as pulverized wood, are now sought.

At a constant excess air of +25% and a bulk particle size of 125 μm, an increase in moisture from 10% to 28% resulted in a 27% decrease in NOX (from 290 to 210 ppm). Furthermore, at a constant excess air of +25% and a moisture content of 10%, decreasing the mass-mean particle size from 300 μm to 125 to 74 μm resulted in an increase in NOX from 150 to 290 to 300 ppm, respectively. These trends concerning the effects of moisture and particle size on NOX have also been observed in lignite coal flames.[11] Of the various coal ranks listed in Table I, lignite has compositional properties most like wood, so that this behavior is not too surprising. These moisture and particle size effects on fuel-bound-nitrogen-conversion-to-NOX cannot be attributed to differences in maximum flame temperature. Fuel-bound nitrogen oxidation is quite insensitive to temperature.[11] The changes in NOX cannot be explained by an acceleration of air-bound nitrogen-to-NOX conversion chemistry because the maximum temperatures in the bark flames are too low for such an atmospheric nitrogen fixation to take place.

5.7. Comparisons between Pulverized Bark and Coal or Heavy Oil Combustion

Industrial mists of atomized heavy petroleum oil (300 μm mass-mean diameter) burn out in *ca.* 300–500 ms.[1] Equi-sized pulverized bark burns out just as fast (Table III). Moreover, the 150 ppm NO_2 emitted from this bark flame corresponds to *ca.* 0·3 lb of NO_2 per million Btu fired, which is equal to the federal NO_2 emission limit for heavy oil combustion.[12] By burning as fast and as cleanly as heavy oil, pulverized wood appears to be quite an acceptable substitute for this petroleum fuel.

6. SUMMARY AND CONCLUSIONS

The combustibility data generated by The Pennsylvania State University Fuels and Combustion Laboratory plane flame furnace has allowed the decoupling of the flame initiation and termination events. The study establishes, under well-controlled simulated boiler combustion conditions, the relationships among physicochemical wood properties, combustor operating conditions, and thermochemical combustion behavior. Investigated were the effects of fuel particle size distribution, moisture and excess air on ignition and burnout times and on NOX production. Many similarities and differences were found in comparisons between pulverized wood and coal flames. Previous experiments at the PSU-FCL[1] and elsewhere[5] have already indicated that pulverized wood is more reactive than low and high rank pulverized coals. Research is continuing in this area at the PSU-FCL in order to quantify these relationships. Pulverized peat will be studied next. There are some legitimate similarities between pulverized wood and petroleum flame chemistry, namely NOX production and control. Experiments in which solid, liquid and gaseous fuels are all fired in one combustor, and the flame characteristics directly compared, are continuing at The Pennsylvania State University Fuels and Combustion Laboratory using a prototype multiphase fuel burner system.

ACKNOWLEDGEMENTS

Support for this research by the Weyerhaeuser Company is gratefully acknowledged.

REFERENCES

1. Reuther, J. J. (1982). The Pennsylvania State University Fuels and Combustion Laboratory Combustion Data Bank. *Proc. Fourth Ann. Energy Seminar* **10**, 1–14, Gannon Univ., Erie, PA.
2. Daley, R. D., Kim, H. T., Lima, J. G. H., Ogunsola, O. I., Taylor, P. H., Warchol, J. J., Withum, J. A. and Reuther, J. J. (1982). Solid, liquid, and gaseous alternative low-sulfur-fuel combustion research at The Pennsylvania State University Fuels and Combustion Laboratory. *Proc. Central States Section Combustion Symposium: The Use of High Sulfur Coal: Combustion Gasification, and Control of Emissions (Notably* SO_2*)*. Columbus, OH, CSS-CI-82-13:1-14.
3. Karsner, G. G. (1981). Private communication to Dr J. Reuther of The Pennsylvania State University Fuels and Combustion Laboratory, September 3.
4. Smith, I. (1971). Kinetics of combustion of size-graded pulverised fuels in the temperature range 1200–2270 °K. *Combustion and Flame* **17**, 303–14.
5. McLean, W. (1982). Ignition of pulverized fuels: a comparative study. *Proc. Western States Section Combustion Institute*. Salt Lake City, UT, WSS-CI-82-23.
6. Weight, R. P. (1981). Ignition temperatures of pulverized fuel. *J. Inst. Energy* **54**, 176.
7. Brunner, P. and Roberts, P. (1980). The significance of heating rate on char yield and char properties in the pyrolysis of cellulose. *Carbon* **18**, 217–24.
8. Juniper, L. A. and Wall, T. F. (1980). Combustion of particles in a large pulverized brown coal flame. *Combustion and Flame* **39**, 69–81.
9. Douglas, L. S. (1980). Ignition temperatures of pulverized fuel. *J. Inst. Energy* **53**, 151–2.
10. *Steam: Its Generation and Use* (1978). Principles of combustion. 6-1–6-22. Babcock & Wilcox.
11. Lee, J., Chen, S., Pershing, D. and Heap, M. (1979). Pollutant production in pulverized coal flames: The effect of coal characteristics. Western States Section Combustion Institute, Provo, UT, WSS-CI-79-9.
12. Turner, D., Andrews, R. and Siegmund, C. (1972). Influence of combustion modification and fuel nitrogen content on nitrogen oxides emissions from fuel oil combustion. *Combustion* **44**, 21–30.
13. Beer, J. (1964). International Flame Research Foundation: The effect of fineness and recirculation on the combustion of low-volatile pulverized coal. *J. Inst. Fuel* **37**, 286–313.

43

Analysis of Condensate and Gases Formed in Combustion of Model Wood Volatile Compounds

MARK M. THORNTON
Department of Chemistry

and

PHILIP C. MALTE
Department of Mechanical Engineering,
University of Washington, Seattle, Washington, USA

1. INTRODUCTION

Substantial attention has been focused recently on energy production from biomass and synthetic biomass-based fuels. Several studies report in detail the pyrolysis of cellulose/lignin fuels; for example, see ref. 1. However, little detailed study has been directed to the reaction of biomass pyrolysis products, or synthetic biomass-based fuels in combustion processes. In the present research, model compounds typical of the products of wood devolatilization, or pyrolysis, are burned under well-controlled, well-aerated conditions in a laboratory stirred reactor. The intermediate condensed and gas phase products are analyzed, and the amount of material remaining unburned is determined as a function of increasing temperature. Limited results are obtained for combustion of wood pyrolysis gas, which is supplied to the stirred reactor from a separate 'stick' pyrolyzer.

The model compounds selected for the present experiments represent two of the three major fractions of wood. The cellulose fraction of wood yields a typical pyrolysis compound mix which is reported by Shafizadeh[2,3] and Milne,[4] among others. DeAngelis *et al.*[5] report similar compounds as emissions from residential woodburners. Some of the compounds from the cellulose fraction are light aldehydes and ketones, such as acetaldehyde and methyl ethyl ketone, which we have found to burn readily.[6] In the present work we consider furan and furfural, interesting because of their cyclic epoxy structure and because of the aldehyde character of furfural. We burn a mono-terpene,α-pinene as a model of the extractable fraction of the wood.

The lignin fraction of wood may produce the most difficult-to-burn material. Typical lignin model compounds represent an experimental challenge because they are hard to vaporize, although later experiments will involve the combustion of a phenol–water mixture. With respect to our present experiments, however, it is interesting to note that Krieger *et al.*[7] report certain products from pyrolysis of lignin pellets by microwave heating essentially similar to some products reported here for incomplete combustion of α-pinene.

The present experiments have two objectives: (1) to separate the devolatilization and combustion processes, so that the volatiles oxidation can be studied in detail under independently controlled combustion conditions; and (2) to obtain the data on the unburned products in order to make it possible to link certain products and emissions with the distinctive fractions of the biomass or synthetic biomass-based fuel. The latter has implication for the tailoring of fuels for optimal combustion efficiency and cleanliness.

2. EXPERIMENTAL

Combustion takes place in a dual-chamber jet-stirred reactor, shown schematically in Fig. 1. The left-hand chamber is fired on premixed hydrogen/air, which produces vitiated air for mixing with the prevaporized model compounds in the right-hand chamber, termed the 'reactor'. Model compounds (α-pinene, ALFA; furan, Quaker Chemical; pentane, J. T. Baker) are run from an infusion pump into a stream of heated nitrogen where flash evaporation occurs. This mixed stream then flows into the reactor. Model compounds with higher boiling points (furfural, ALFA) are pumped under pressure into an atomizing vaporizer (Fig. 2), which works by producing a strong shearing action across a liquid sheet.[8] All carbon species in the combustion reactor are thus attributable to the additive, which has a nominal concentration of 10 000 ppmC (ppm equivalent to carbon dioxide). The amount of carbon entering the reactor is determined from the measured CO and CO_2 concentrations at the highest combustion

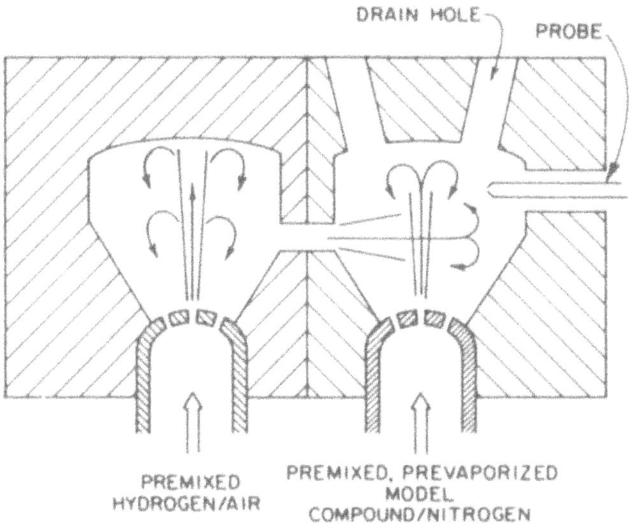

FIG. 1. Dual jet-stirred combustion reactor.

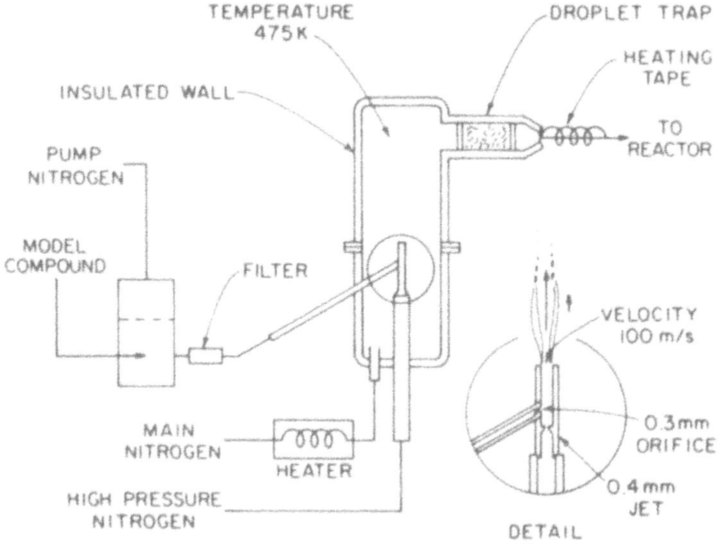

FIG. 2. Model compound vaporizer.

temperatures. Any changes in the carbon concentration are observed by timing the infusion pump syringe, or by checking the rotameter reading on the liquid pump.

Reactor pressure is 1 atm, volume is 50 cm^3 and residence time is approximately 12 ms. Reactor temperature is controlled in the range 975–1450 K, by varying the stoichiometry of the hydrogen/air flame which is always very fuel-lean. Unburned oxygen ranges from 7 to 13 mole %. Thus, the prevaporized model compounds mix into and burn with an excess of hot air, at temperatures and residence times appropriate for finite rate oxidative pyrolysis.

The reactor resembles others described previously,[9,10] although the present reactor suffers from poorer mixing and less uniformity. In the upper 70 % of the reactor, for pentane additive, the relative standard deviations of temperature and oxygen are 5 and 8 %, respectively, over 53 locations; and the relative standard deviations for carbon monoxide and carbon dioxide are 19 and 17 %, respectively, over 18 locations. However, the reactor is less uniform with respect to unburned organic compounds; the relative standard deviation for unburned pentane is 114 %. The sample probe is positioned in an 'average' location in order to compensate as much as possible for these effects.

Combustion temperature is measured by a fine platinum rhodium thermocouple coated with BeO/Y_2O_3 by the method of Kent.[11] Radiative losses from the bead to the outside environment are estimated to be 40 K at 1300 K.

Wood pyrolysis gas is produced in a separate thermal pyrolysis unit. This unit heats nitrogen in an electrical furnace, and the hot nitrogen then flows over a wood stick in an immediately adjoining, insulated, stainless steel pyrolyzer section. Each wood stick is fed into the pyrolyzer by a motorized screw drive system. The stick feed rate essentially controls the pyrolysis rate, which is *ca.* 0·01 g/s and corresponds to roughly 2500 ppmC in the combustion reactor. The pyrolysis gas produced in this manner, and carried in the N_2, passes through a short, insulated line to the reactor. The nominal gas temperatures are: 800 K pyrolyzer inlet temperature, 650 K pyrolyzer exit temperature, and 550 K reactor injection temperature. The wood sticks, 47 cm × 4 cm × 0·5 cm (38 cm useful length) with 20 % moisture content (dry basis), are prepared from billets all cut from the heartwood of a single log of Douglas fir, *Pseudotsuga menziesii*.

The amount of carbon entering the reactor is determined as above, and then relative carbon input at cooler conditions is scaled to the actual stick feed rate, a procedure which we estimate is certain to a factor of 2.

3. SAMPLING AND ANALYSIS

A clear fused quartz probe previously described[12] draws a continuous stream at *ca.* 75 torr from the combustion reactor. This probe has a sonic entry; a short, uncooled section of 7·5 cm length and 2 mm ID; and is concluded by a water cooled section. Because of the low probe pressure, modest organic concentration and relatively well mixed combustion, deposition of condensed material in the cooled section of the probe occurs only when wood gas is burned at cool combustion temperatures—a situation which requires probe rinsing. Close downstream to the probe a condensate trap at 273 K collects water and condensed organic compounds. The gas sample stream then passes through a 6·35 mm OD Teflon line to a conditioning system and on-line gas chromatograph (GC).

The conditioning system allows the gas sample stream to flow continuously at low pressure, while aliquots are isolated and pressurized (without dilution) to any selected pressure less than 2 atm for prompt separation and analysis by GC. All parts of the pressurization system are swept with helium or are maintained at 353 K to forestall condensation of water or organic compounds. This system greatly improves the usable sensitivity of the GC.

The GC is a Perkin–Elmer Sigma 1B. This instrument incorporates microprocessor control and on-line digital peak integration and data reduction. A two column (molecular sieve and porous polymer), two detector (hot wire detector—HWD, and flame ionization detector—FID) system is used for analysis. Two porous polymer columns are used alternatively: Tenax-GC, where elution of higher boiling and highly polar components is required; and Porapak-T, where greater selectivity, especially between lower boiling alkanes, alkenes and alkynes is required. On-line gas samples of 0·1 ml volume, conditioned as indicated above, are introduced at 1 atm and 423 K by means of a 10-port valve (Valco). (Columns and standard conditions used are as follows: molecular sieve 5A (Alltech): 1·82 m × 2·1 mm ID, 60/80 mesh, helium 15 ml/min, oven isothermal at 423 K; Tenax-GC (Alltech): 1·82 m × 2·1 mm ID 80/100 mesh, helium 15 ml/min, oven isothermal at 423 K or temperature programmed: 313 K hold 1 min, then 10 K/min to 513 K; Porapak-T (Supelco): 1·82 m × 2·1 mm ID, 80/100 mesh, helium 15 ml/min, oven isothermal at 423 K or temperature programmed: 323 K hold 1 min, then 10 K/min to 443 K).

Direct GC analysis by syringe injection of aqueous condensate before the tar extraction step is a quick way to estimate yields of components such as acetone and aldehydes, which are not detected in the subsequent tar analysis. Volatile polar organic compounds in solution are often a major part of the yield of condensed products, and although not identified specifically by this method, they are separated into general classes. Direct injections of the aqueous condensate are made by syringe on to the Tenax-GC column described above, and using only the FID. A quick separation is carried out isothermally at 473 K, and a more detailed separation is accomplished by temperature programming the chromatograph oven from 373 to 523 K at 10 K/min. For the present work, the results given below for these condensed compounds are based on a carbon response factor determined for propane.

In order to analyze for tars, the procedures described next are used. Residue, if any, is rinsed from the traps and probe with dichloromethane or methanol (Burdick & Jackson). This rinse is then combined with the condensate and extracted for 1 h in a Wehrli extractor with 50 ml of dichloromethane refluxing rapidly. The extract is dried over anhydrous sodium sulfate, filtered with glass wool, and rinsed with extra dichloromethane into a Kuderna-Danish concentrator and evaporated down to 1–2 ml as appropriate. All samples are kept frozen when not in use. Many of the samples darkened at first when they were exposed to light. Later, such exposure was avoided. This method has been tested and found to recover 95 % of furfural from 1:500 (vol/vol) aqueous solution.

These extracted samples are chromatographically separated by fused silica open tubular (capillary) columns directly connected to the FID. Split or splitless injections are made depending on the concentration of the sample. Some of the components in the split injections show poor peak shapes, possibly resulting from adsorption or decomposition in the injector, even though a thoroughly deactivated liner packing (silanized glass wool and 1 % SE-30 on Chromosorb W-HP) is used. Dilution and splitless injections are then used for these cases. (Chromatographic conditions are as follows: non-polar column (J & W): fused silica 30 m × 0·25 mm, SE-30 coating, helium 210 kPa, 313 K for 1 min hold, then 4 K/min to 523 K; polar column (Hewlett Packard): fused silica, 20 m × 0·25 mm, Carbowax-20M coating, helium 140 kPa, 313 K for 1 min hold, then 5 K/min to 473 K; with injector in each case at 448 K, detector at 523 K.)

Approximate Kovats retention indices are estimated for unknown components and are used only as a guide to possible identification. Assignments based on retention time are made only for compounds whose actual retention time on our system is known, since retention data are complicated by the splitless injection.[13]

Quantitative determinations of the tars are made on the basis of FID responses. Each extracted sample is spiked with 0·1 μg/μl p-bromophenol as an internal standard. Wherever possible, response factors are determined relative to p-bromophenol with our own system. Otherwise where reliable literature response factors are not available, components are assumed to have the same response factor as p-bromophenol.

Mass spectra of the tars are obtained using a Hewlett-Packard 5985 gas chromatograph/mass spectrometer/data system (GC/MS). Samples are introduced by splitless injections on fused silica open tubular columns as above. Similar columns and conditions are employed. Electron impact mass spectra at 70 eV are recorded during the whole of the useful chromatogram. The mass range scanned is 40–250 amu with 8 steps per amu. The column outlet discharges directly into the ion source which is maintained at 473 K. The transfer lines in which the column itself is carried are kept at 523 K.

Background subtraction and searches of the EPA/NIH mass spectral data base are made to confirm chromatographic assignments. Manual interpretation is also made by reference to tables[14–16] and by direct interpretation.[17]

4. GAS PHASE RESULTS

Figure 3 illustrates the completeness of combustion of the model compounds, which is similar with respect to compound type. Above 1200 K, input carbon is more than 95 % converted to carbon monoxide and carbon dioxide, and above 1300 K, limiting carbon monoxide of 3 % of input carbon is approached (close to the GC detection limit).

Two distinct modes of combustion are shown in Fig. 3. Above temperatures of about 1150 K, the combustion rate is 'fast'. In this mode the measured concentrations show carbon dioxide > carbon monoxide > unburned organic compounds. In the range of approximately 1050–1150 K, the combustion is unstable. That is, it is nearly impossible to operate the reactor in this range of temperatures. It is thought that high temperature ignition of the model compounds occurs in this range, and thus, when ignition is achieved, the temperature quickly rises to some value greater than 1150 K. On the other hand, as the reactor is cooled below about 1050 K, combustion is 'slow' and there is little heat release from the model compounds.

The limited data which have been obtained in the 1050–1200 K range show a sharp rise in the $CO/(CO+CO_2)$ ratio with decreasing

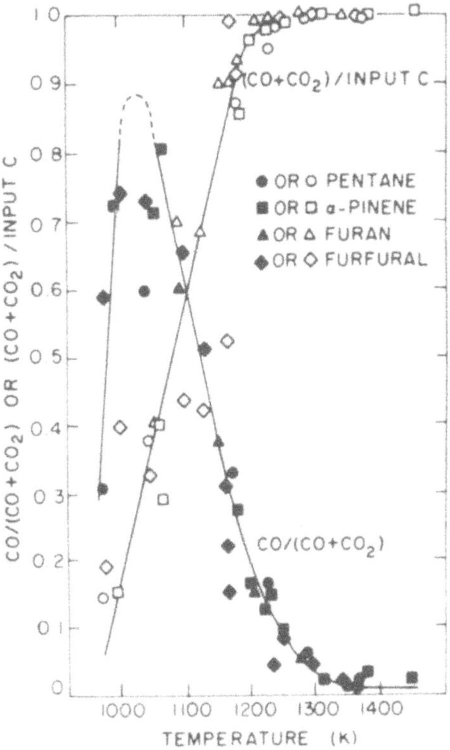

FIG. 3. Measured CO and CO_2 versus combustion temperature for four model compounds and wood pyrolysis gas.

temperature, with a maximum ratio occurring at about 1050 K. The sharp rise in this ratio is due to declining oxidation of CO to CO_2, which occurs because the OH radical concentration falls off at low temperatures. In the 'slow' combustion mode below *ca.* 1050 K, CO_2 appears to predominate over CO, although neither is produced in abundance.

Figure 4 shows total gas phase organic material as ppmC as determined from specific response factors for known hydrocarbons. Figure 5 shows the breakdown for CH_4, C_2H_4, C_2H_2 and C_3H_6 as a function of model compound type. For pentane additive, readily identifiable C_1 to C_5 hydrocarbons account for the gas phase organic sum in Fig. 4. Likewise for furan, the sum is due mainly to C_1 to C_5 compounds, including a large C_5 response. For the α-pinene additive, however, there are approximately 20 peaks ranging up to C_{10}, and including unidentified components. This indicates a much more complex behavior.

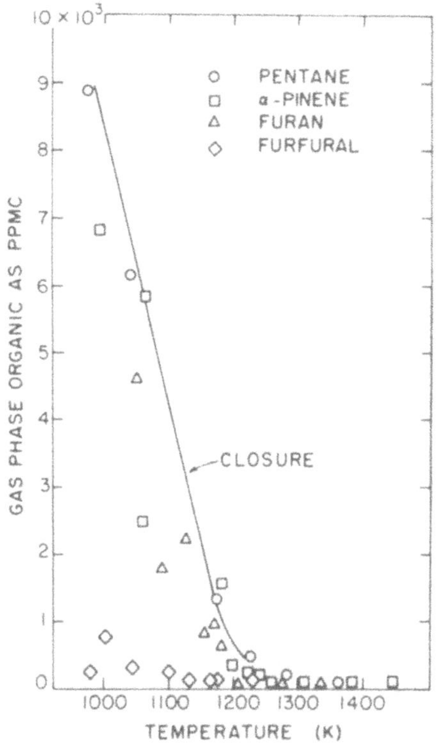

FIG. 4. Total gas phase organic material, as ppmC, versus combustion temperature for four model compounds. 'Closure' indicates a nominal carbon mass balance.

The curve labeled 'closure' in Fig. 4 indicates the amount of gas phase organic material which if added to the measured CO and CO_2 would equal the total input carbon. Closure is indicated for the pentane additive; approximate closure is indicated for the α-pinene and furan additives. Any lack of closure for the cases with α-pinene and furan is due to uncertain response factors for unidentified components, particularly those containing oxygen. Condensate material cannot account for the lack of closure, since (as shown below) for furan this is almost nil, and for α-pinene the maximum amount of condensed material is less than 0·1 % (by mass) of the starting material.

Furfural shows a poor carbon balance, due to the fact that it yields only small amounts of gas phase chromatographable material, and due to the fact that this deficiency is not completely made up by condensed material.

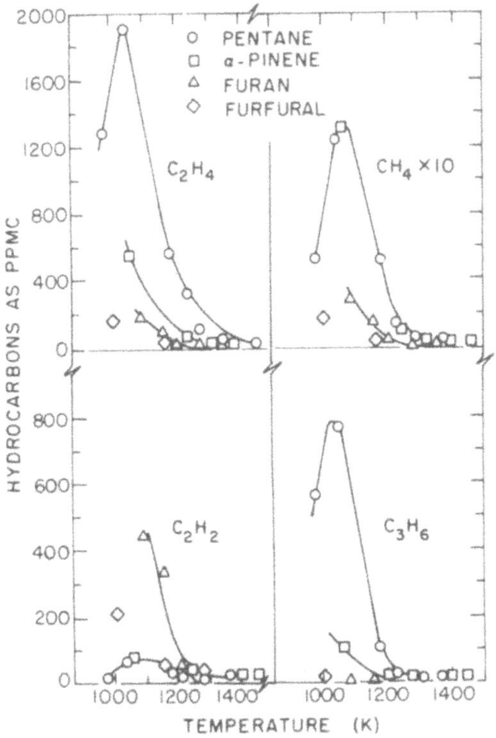

FIG. 5. Methane, ethylene, acetylene and propylene versus combustion temperature for four model compounds.

As shown below, the condensed material is mainly volatile polar organic compounds in solution, which have a maximum yield of at most 10 % of the input carbon.

The yield of light olefins and methane from the model compounds, as shown in Fig. 5, is in general agreement with straight chain hydrocarbon studies (e.g., see ref. 18). Also, acetylene is predominant in the case of furan (and possibly furfural) combustion.

5. CONDENSED PHASE RESULTS

Figure 6 shows the yield of organic material in the aqueous condensate. This includes volatile polar organic compounds in solution. Pentane and furan additives give little condensed material, whereas α-pinene and

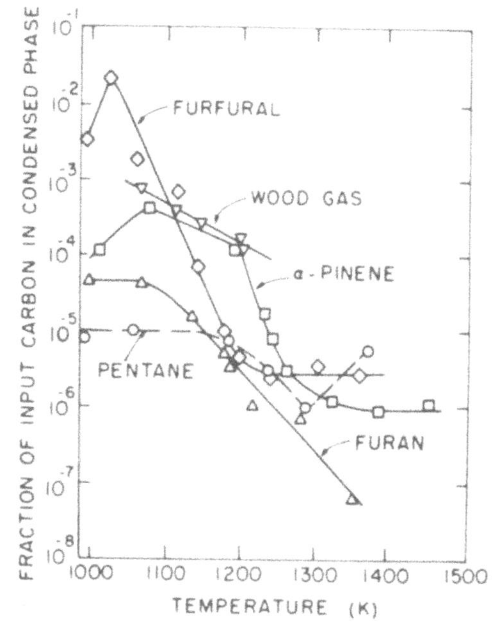

FIG. 6. Fraction of input carbon measured in aqueous condensate versus combustion temperature.

furfural give larger amounts, especially furfural at the coolest temperatures. Wood pyrolysis gas, although not studied over the whole temperature range, produces condensed material yields similar to α-pinene and furfural (at 1050 K).

The five most abundant classes of condensed compounds from α-pinene, as separated by temperature programmed runs, give the same declining trend with increasing temperature for $T > 1050$ K. This implies that relative product speciation measured at one temperature may be a good indicator of relative abundance at other temperatures. It also implies that a single oxidative pyrolysis mechanism is operating in the reactor for $T > 1050$ K.

Table I summarizes the findings of the tar analysis for α-pinene, furfural and wood gas. All three cases represent the temperature for highest total condensed (aqueous condensate and tars) product yield. Tar yields from pentane and furan are not shown since they are essentially nil. Tar weight yields are essentially similar for the α-pinene and the wood pyrolysis gas, given the uncertainties associated with the wood pyrolysis gas input rate and probe deposits (see above). However, the yield of tar from furfural is less.

TABLE I
SUMMARY OF TAR YIELDS AND COMPOSITION

Additive	*Tar yield*	*Weight percent by class*			
		Phenols	*Alcohols*	*Others*	*Unassigned*
α-Pinene	0·303 g/kg	61·9	11·3	4·7	14·9
Furfural	0·090 g/kg	5·9	2·6	39·9	51·6
Wood gas	0·175 g/kg	18·0	1·8	0·9	79·3

Tables I, III and IV show the breakdown of the tar components of α-pinene, furfural and wood gas, respectively. Along with the name and weight percent of each component is indicated the nature of the evidence used for identification. Compounds identified (tentatively) by only one method are minor components and are listed as traces. In this respect, unchecked library matching of mass spectral peaks in complicated chromatographs is unsatisfactory.

The analysis with the greatest definition is that for α-pinene (Table II).

TABLE II
α-PINENE TAR COMPOSITION

Name	*Rank*	*wt %*	*SE-30*	*C-20 M*	*MS*
Phenol	1	37·6	+	+	+
o-Cresol	2	13·7	+	+	+
Benzyl alcohol	3	10·1	+	+	+
Unassigned	4	7·8	–	–	–
2,5-Dimethylphenol	5	6·6	+	+	–
2,4-Dimethylphenol	6	5·7	+	+	–
1,4-Naphthalenediol	7	2·6	+	+	–
3,5-Dimethylphenol	8	2·4	+	+	–
Furfural	9	2·1	+	+	–
3,4-Dimethylphenol	10	2·0	+	+	–
Unassigned	11	1·4	–	–	–
Unassigned	12	1·3	–	–	–
Tetrahydrofurfuryl alcohol	13	1·2	+	+	–
3,5-Dimethylphenol	14	1·1	+	+	–
Unassigned	15	1·0	–	–	–
Other unassigned	—	3·5	–	–	–

Trace components: acenaphthene, guaiacol, 2,6-dimethoxyphenol. Combustion temperature: 1 062 K.

TABLE III

FURFURAL TAR COMPOSITION

Name	*Rank*	*wt%*	*SE-30*	*C-20 M*	*MS*
Furfural	1	39·9	+	+	+
*	2	8·0	–	–	+
Unassigned	3	7·9	–	–	–
*	4	6·6	–	–	+
Unassigned	5	5·8	–	–	–
Unassigned	6	4·3	–	–	–
Unassigned	7	4·0	–	–	–
*	8	3·7	–	–	+
p-Cresol	9	3·7	+	+	–
Unassigned	10	3·4	–	–	–
Unassigned	11	2·8	–	–	–
Unassigned	12	2·7	–	–	–
Tetrahydrofurfuryl alcohol	13	2·6	+	+	+
*	14	2·5	–	–	+
2,5-Dimethylphenol	15	2·2	+	+	–
Other unassigned	—	0·9	–	–	–

Trace components: furfuryl alcohol, acetic acid, guaiacol, catechol.
* All four peaks related to *m/e* 98·1 and possible isomers of furfuryl alcohol.
Combustion temperature: 1011 K.

TABLE IV

WOOD PYROLYSIS GAS TAR COMPOSITION

Name	*Rank*	*wt%*	*SE-30*	*C-20 M*	*MS*
1,4-Naphthalenediol	1	10·4	+	+	–
3,5-Dimethylphenol	2	2·4	+	+	–
3,4-Dimethylphenol	3	1·7	+	+	–
Benzyl alcohol	4	1·7	–	+	–
o-Cresol	5	1·5	+	+	–
m-Cresol	6	1·1	+	+	–
Furfural	7	0·7	–	+	–
2,5-Dimethylphenol	8	0·6	+	+	–
2,4-Dimethylphenol	9	0·3	+	+	–
5-Hydroxymethylfurfural	10	0·2	+	+	–
Tetrahydrofurfuryl alcohol	11	0·1	+	+	–
Unassigned		79·3	–	–	–

Traces: phenol, *p*-cresol, guaiacol, naphthalene, catechol, resorcinol, 2,6-dimethoxy phenol. Combustion temperature: 1052 K.

The predominance of phenols is the main feature of the α-pinene results. It is likely that some of the unassigned components could be matched up with dimethyl, trimethyl and tetramethyl substituted phenols for which reference material is not on hand. Reference mass spectra of these isomeric compounds resemble each other greatly, and thus little unequivocal progress can be made by mass spectrometry alone. Species such as furfural and tetrahydrofurfuryl alcohol are common to the other additive cases. The only known component having 10 or more carbon atoms is 1,4-naphthalenediol.

Unburned starting material is the main feature of the furfural results (Table III). In addition, four major components show mass spectra closely related to that of furfuryl alcohol. Other components have structural similarities to furfural. The two phenols are assigned on the basis of chromatographic data; mass spectra uncomplicated by close eluting components have not been obtained. Production of trace phenol (C_6H_6O) from combustion of furfural ($C_5H_4O_2$) implies aggregation of carbon species in the flame, or possibly in the probe.

The wood pyrolysis gas tar components which have been identified (Table IV) are consistent with those found for α–pinene. Substantially higher boiling components are also found which are relatively abundant, and many of the other unassigned peaks are likely to be phenols. The unassigned peaks are not included in the ranking.

6. SUMMARY AND CONCLUSIONS

Although the overall oxidation of the different model compounds is similar with respect to temperature, the nature of the unburned organic material is strikingly different for the different model compounds. The pentane reference compound gives very well defined experimental results, with measured CO, CO_2 and C_1 to C_5 hydrocarbons giving a carbon mass closure. The results for α-pinene are also well defined. Although the chromatograms for the gas phase material are complex, and include unidentified components, approximate carbon mass closure is obtained. Furthermore, most of the condensed tar components are identified, and are seen to be phenols. Of the model compounds examined, α-pinene gives the most tar, although this is less than 0·1 % (by mass) of the starting material. Results for furan are dominated by unburned C_5 material in the gas phase.

Furfural gives the largest amounts of condensed polar organic material, approaching 10 % of the input carbon, and its tar yield is dominated by starting material. However, carbon mass closure is not obtained, and the cause of this needs to be identified.

The tar produced by the combustion of wood pyrolysis gas contains phenols also produced by the combustion of α-pinene as well as a number of high boiling components yet to be assigned. Since the extractable materials are typically a small fraction of the wood, and since furfural produced only little phenol, the measured phenols are probably from the lignin fraction of the wood.

Detailed kinetic interpretation of these data will be reported later. However, a single global oxidative pyrolysis rate is indicated, one that holds approximately for straight chain hydrocarbons, because of the similarity in burning rate for all of the compounds examined here.

ACKNOWLEDGEMENTS

This research has been sponsored by the Solar Energy Research Institute, Golden, Colorado, through their Office of University Programs. The efforts of Mr Peter Kamber, Ms Jean Hertzberg and Mr Robert Morley at the University of Washington are sincerely appreciated.

REFERENCES

1. Anon. (1980). Specialists' Workshop on fast pyrolysis of biomass. SERI Report CP-622-1096, Solar Energy Research Institute, Golden, CO.
2. Shafizadeh, F. (1968). Pyrolysis and combustion of cellulosic materials. In *Advances in Carbohydrate Chemistry*, Vol. 23. M. L. Wolfrom and R. S. Tyson (Eds). Academic Press, New York, p. 419.
3. Shafizadeh, G. and Lai, Y. Z. (1972). Thermal degradation of 1,6-anhydro-D-glucopyranose. *J. Org. Chem.* **37**, 278.
4. Milne, T. (1979). Pyrolysis—The thermal behavior of biomass below 600 °C. In *Survey of Biomass Gasification*, Vol. II. SERI Report TR-23-239, Solar Energy Research Institute, Golden, CO.
5. DeAngelis, D. G., Ruffin, D. S. and Reznik, R. B. (1980). Preliminary characterization of emissions from wood-fired residential combustion equipment. US Environmental Protection Agency Report EPA-600/78-8-040, Washington, D.C.
6. Thornton, M. M. and Malte, P. C. (1981). Combustion rate of model wood volatiles. In *Proceedings of First International Conference on Residential Solid Fuels*. Oregon Graduate Center, Beaverton, OR.
7. Krieger, B. B., Chan, R. and Graef, M. (1978). Reactions of biomass/lignin-microwave degradation. Report, Department of Chemical Engineering, University of Washington, Seattle, WA.

8. Chigier, N. (1981). *Energy, Combustion and Environment.* McGraw-Hill, New York.
9. Malte, P. C., Schmidt, S. C. and Pratt, D. T. (1977). Hydroxyl radical and atomic oxygen concentrations in high intensity turbulent combustion. *Proceedings of the 16th Symposium (International) on Combustion.* Combustion Institute, Pittsburgh, PA, p. 145.
10. Malte, P. C., Schmidt, S. C., Kramlich, J. C., Spitzer, K., Yee, D. and Singh, S. (1981). Behavior of OH and NO in a mixing influenced jet-stirred reactor. *AIAA Journal of Energy* **5**, 193.
11. Kent, J. H. (1970). A noncatalytic coating for platinum-rhodium thermocouples. *Combustion and Flame* **14**, 279.
12. Kramlich, J. C. and Malte, P. C. (1978). Modeling and measurement of sample probe effects on pollutant gases drawn from flame zone. *Combustion Science and Technology* **18**, 91.
13. Grob, K. and Grob, K. Jr (1978). Splitless injection and the solvent effect. *High Res. Chromatography and Chrom. Comm.* **1**, 57.
14. Abrahamsson, S., Stenhagen, E. and McLafferty, F. W. (1974). *Registry of Mass Spectral Data*, Vols. 1–4. Wiley, New York.
15. Stenhagen, E., Abrahamsson, S. and McLafferty, F. W. (1969). *Atlas of Mass Spectral Data, Vols. 1–3.* Interscience, New York.
16. Cornu, A. and Massot, R. (1966). *Compilation of Mass Spectral Data.* Heyden & Son, London.
17. McLafferty, F. W. (1966). *Interpretation of Mass Spectra.* W. A. Benjamin, New York.
18. Glassman, I. (1977). *Combustion.* Academic Press, New York.
19. Rosenthal, D. (1982). Theoretical limitation of gas chromatographic mass spectrometric identification of multicomponent mixtures. *Analytical Chem.* **54**, 63.
20. Haack, L. P., Marano, R. S., Riley, T. L. and Levine, S. P. (1981). Gas chromatographic technique for the on-line evaluation of solvent emission abatement devices. *Environmental Science and Technology* **15**, 1463.
21. Singh, S., Grosshandler, W. L., Malte, P. C. and Crain, R. W. (1979). Oxides of nitrogen formed in high intensity methanol combustion. *Proceedings of 17th Symposium (International) on Combustion.* Combustion Institute, Pittsburgh, PA, p. 689.
22. Zweig, G. and Sherma, J. (1972). *Handbook of Chromatography*, Vol. I. CRC Press, Cleveland.

44

Dissolution of Peat and Wood in Tetralin Compared with Coal

CHRISTINA HÖRNELL, PEHR BJÖRNBOM, EMILIA BJÖRNBOM, AGNETA BERGSTRÖM, GERTH KARLSSON, OLA KARLSSON, BENGT OLSSON, YRJÖ SOLANTAUSTA and BÖRJE ÅHGREN

Royal Institute of Technology, Department of Chemical Technology, Stockholm, Sweden

1. INTRODUCTION

In Sweden, a country without any substantial coal resources, there is an interest in using peat and biomass as raw materials for fuel conversion processes.[1,2] The present investigation, which is sponsored by the National Swedish Board for Energy Source Development, was undertaken to obtain data for comparison of coal, peat and biomass in such direct liquefaction processes where the raw material is heated and liquefied in a slurrying vehicle.

Part of this investigation was in cooperation with VTT—the Technical Research Centre of Finland.

Several investigators have used tetralin as a model for the hydrogen donor solvent in coal liquefaction investigations. Tetralin reacts with free radicals formed by the thermal dissociation of coal. The radicals abstract hydrogen from the tetralin which is transformed into naphthalene.[3-7] A few studies were made on peat liquefaction by thermal dissolution.[8-11]

Up to a temperature of 400 °C the thermal dissociation of pure tetralin is very slow, not reaching more than a few percent conversion of tetralin in 8 h according to Hooper.[4] Neavel has shown that up to 400 °C the formation of naphthalene well corresponds to the conversion of tetralin in the presence of coal. Therefore, the naphthalene to tetralin ratio can be used to measure hydrogen transfer at least up to 400 °C.[3]

We have made some experiments above 400 °C. For these experiments the high ratio of naphthalene to tetralin may then depend not only on the hydrogenation of the solid fuel, but also on the thermal destruction of the tetralin itself. The results should then be interpreted with some reserve. In the present investigation the dissolution of coal, peat, wood and cellulose in tetralin were compared.

The objective was to obtain information about the differences between coal dissolution on the one hand, and dissolution of peat and wood on the other hand. For this purpose, we tried to eliminate all differences between coal, peat and wood experiments, except for the type of raw material.

2. EXPERIMENTAL

2.1. Raw Materials

The coal used was Polish steam coal, a typical Swedish import coal. Normal grade Finnish milled peat from Oulu was used in the experiments with peat. The biomass used was branches, including needles, from spruce (*Picea abies*). The cellulose used was Solkafloc. The elemental composition and ash content of the raw materials are given in Table I. As hydrogen-donor we used 1,2,3,4-tetrahydronaphtalin zur Synthese from Merck.

TABLE I
RAW MATERIALS

	Moisture content (%)	*C* (%)	*H* (%)	*N* (%)	*O* (%)	*S* (%)	*Ash* (%)
Coal	0·5–2	80·1	4·9	1·4	10·0	0·38	3·2
Peat	5–10	52·1	6·3	3·2	32·7	0·25	5·5
Biomass	1·5–3·5	50·0	6·1	0·6	38·5	0·1	4·7
Cellulose	7·5	42·4	6·4	0·3	49·7		1·2

2.2. Experimental Procedure

The raw material was dried at 50 °C under CO_2/vacuum. Thereafter it was ground in a Retsch ultracentrifuge under a stream of CO_2.

With a sieve with holes of 0·12 mm we obtained a stirrable slurry with a 1:3 ratio of raw material to tetralin.[12] Approximately 130 g of raw material, with 390 g of tetralin, were placed in a 1 litre autoclave from Pressure Products Industries equipped with a Magne-drive stirrer. After flushing with nitrogen it was pressurized to 5 MPa cold pressure with nitrogen.

The heating-up period for the experiments varied between 121 and

160 min, mainly depending on the chosen reaction temperature. After the desired reaction time the autoclave was left to cool by itself until 165 °C, thereafter it was water-cooled by an internal cooling-coil. During heating and cooling the Magne-drive, equipped with a propeller, rotated at 1000 rpm, and during the time at the reaction temperature at 1500 rpm. Experiments with a plexiglass model of the reactor have indicated that 1000 rpm is enough to keep the solids in suspension while at 1500 rpm gas bubbles are drawn down into the slurry.

After the experiments, the gas was collected in a gas bag. To collect all the gases the reactor was repeatedly flushed with helium while the slurry was stirred. The gases were then analysed by GC (Shimadzu GC-3 BT). The liquid phase, i.e. tetralin, naphthalene and products, was taken from the reactor by suction and accompanying solids filtered off. The liquid phase was stored in a refrigerator under nitrogen until further analysis. The solids remaining in the reactor were washed out with acetone. The solids, together with the filtrate from the tetralin phase, were then extracted with fresh acetone in a soxhlet apparatus, before drying to determine the weight of the solid residue. The tetralin phase was later analysed by GC (Aerograph Autoprep, Model A-700, from Wilkens Instr. Res. Co.) to determine the ratio between newly formed naphthalene and tetralin remaining in the phase.

The reaction temperature of the experiments varied within ± 5 °C, which has been omitted in the tables.

Conversion of the raw material was calculated in the way which is accepted in coal liquefaction processes with the following formula:[13]

$$\text{conversion} = \frac{\text{maf}_{\text{input}} - \text{maf acetone insoluble}_{\text{out}}}{\text{maf}_{\text{input}}} \times 100$$

where $\text{maf}_{\text{input}}$ is the weight of the moisture and ash-free sample in grams.

3. RESULTS AND DISCUSSIONS

3.1. Influence of the Processing Conditions on the Conversion of the Raw Material and the Consumption of the Hydrogen Donor Solvent

The results obtained from coal dissolution in tetralin are shown in Table II. It may be seen that the dissolution of coal at 350 °C is very low, and independent of the processing time in the interval from 15 to 120 min (see experiments K4 and K9). Coal conversion at 445 °C is considerable after 15 min time (experiment K6). Longer processing time has no effect on the

TABLE II

DISSOLUTION OF COAL IN TETRALIN

Exp. No.	*Processing time (min)*	*Temperature (°C)*	*Conversion (%)*	*Naphthalene / Tetralin (g/g)*	*N/T*[a] / *Conversion* × 100
K9	15	350	8	0·02	0·25
K4	120	350	13	0·02	0·15
K6	15	445	73	0·24	0·33
K7	120	445	74	0·39	0·53

[a] N/T = naphthalene/tetralin.

coal dissolution but tetralin consumption increases considerably (experiment K7). The results in our study are in good agreement with previous work on coal dissolution in tetralin.[3]

Table III illustrates the effect of temperature and processing time on the dissolution of peat in tetralin. The results obtained indicate that peat dissolves in tetralin already at 350 °C and the conversion depends both on the temperature and the duration of the process. The effect of the reaction time decreases with the increase of temperature. The latter result is in good agreement with our previous work on raw peat liquefaction, without preliminary drying.[14,15] The consumption of tetralin increases with time and temperature.

Biomass and cellulose dissolution in tetralin is considerable at 350 °C and

TABLE III

DISSOLUTION OF PEAT IN TETRALIN

Exp. No.	*Processing time (min)*	*Temperature (°C)*	*Conversion (%)*	*Naphthalene / Tetralin (g/g)*	*N/T / Conversion* × 100
T6	15	345	49	0·02	0·04
T5	120	350	79	0·04	0·05
T9	15	375	80	0·05	0·06
T11	67	375	87	0·08	0·09
T8	15	400	79	0·09	0·11
T10	67	400	91	0·12	0·13
T3	15	400	93	0·17	0·18
T4	120	445	97	0·36	0·37

TABLE IV
DISSOLUTION OF BIOMASS IN TETRALIN

Exp. No.	*Processing time (min)*	*Temperature (°C)*	*Conversion (%)*	*Naphthalene / Tetralin (g/g)*	*N/T / Conversion* × 100
B2	15	345	75	0·03	0·04
B9	120	350	85	0·07	0·08
B10	15	375	84	0·07	0·08
B11	67	375	87	0·10	0·11
B6	35	435	93	0·24	0·26
B7	120	420	93	0·28	0·30

15 min processing time (see Tables IV and V). The conversion of the solid material increases with the increase of temperature and time. Higher processing temperature and longer duration contribute mainly to the consumption of the hydrogen donor solvent.

Tables II–V show considerable difference in the results obtained in the dissolution of different solid fuels.

3.2. Comparison of the Dissolution in Tetralin of the Different Raw Materials

Figure 1 illustrates the conversion of the studied raw materials as a function of the temperature, at 15 min processing time. It may be seen that peat and biomass are much more reactive than coal. The difference is very significant at lower temperatures. Peat and biomass can be dissolved in tetralin already at 350 °C. Approximately 80 % of these raw materials are converted

TABLE V
DISSOLUTION OF CELLULOSE IN TETRALIN

Exp. No.	*Processing time (min)*	*Temperature (°C)*	*Conversion (%)*	*Naphthalene / Tetralin (g/g)*	*N/T / Conversion* × 100
C3	15	350	78	0·05	0·06
C2	120	350	86	0·08	0·09
C1	15	375	86	0·07	0·08
C6	67	375	89	0·11	0·12
C5	67	400	95	0·18	0·19

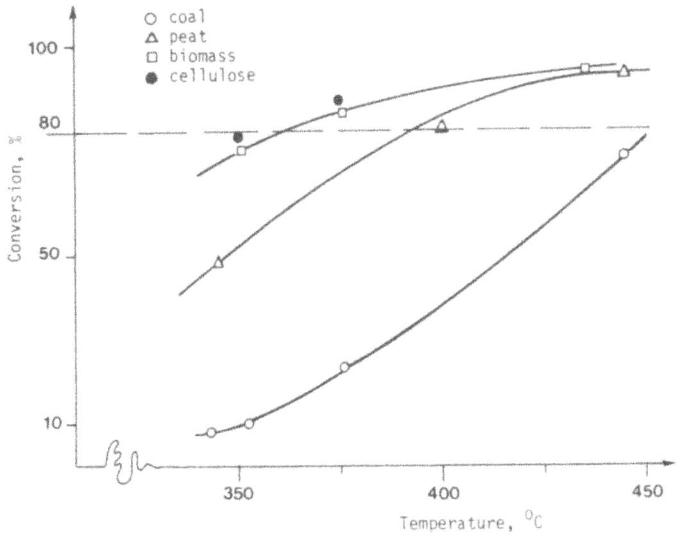

FIG. 1. Conversion of coal, peat, biomass and cellulose in 15 min thermal dissolution in tetralin.

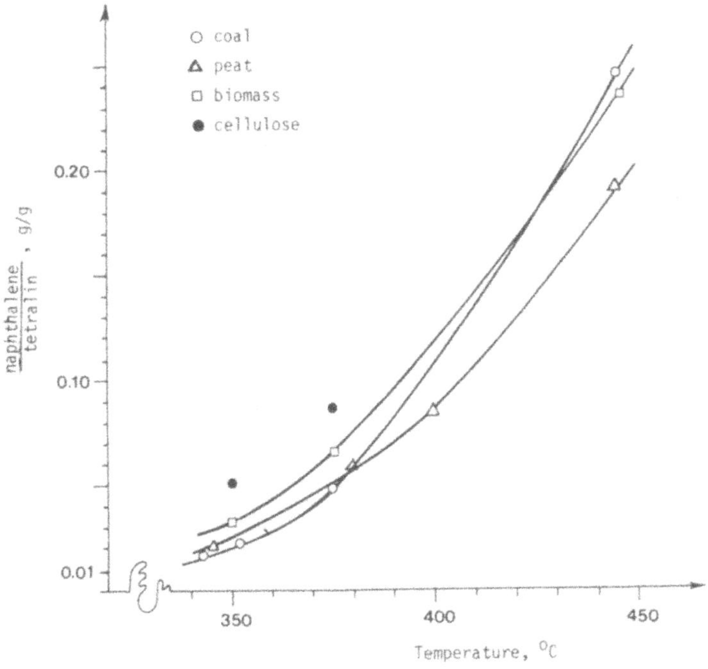

FIG. 2. Consumption of hydrogen-donor solvent in thermal dissolution of coal, peat, biomass and cellulose in tetralin; duration 15 min.

at 375 °C. The same conversion of coal can be reached at a temperature of 450 °C. Coal is practically not dissolved in tetralin at 350 °C.

The consumption of hydrogen from the hydrogen-donor solvent tetralin is of practical interest.

The conversion of tetralin into naphthalene in the presence of different raw materials is demonstrated in Fig. 2. The consumption of tetralin in the process is expressed by the ratio between the newly formed naphthalene and the remaining amount of tetralin. The processing conditions are the

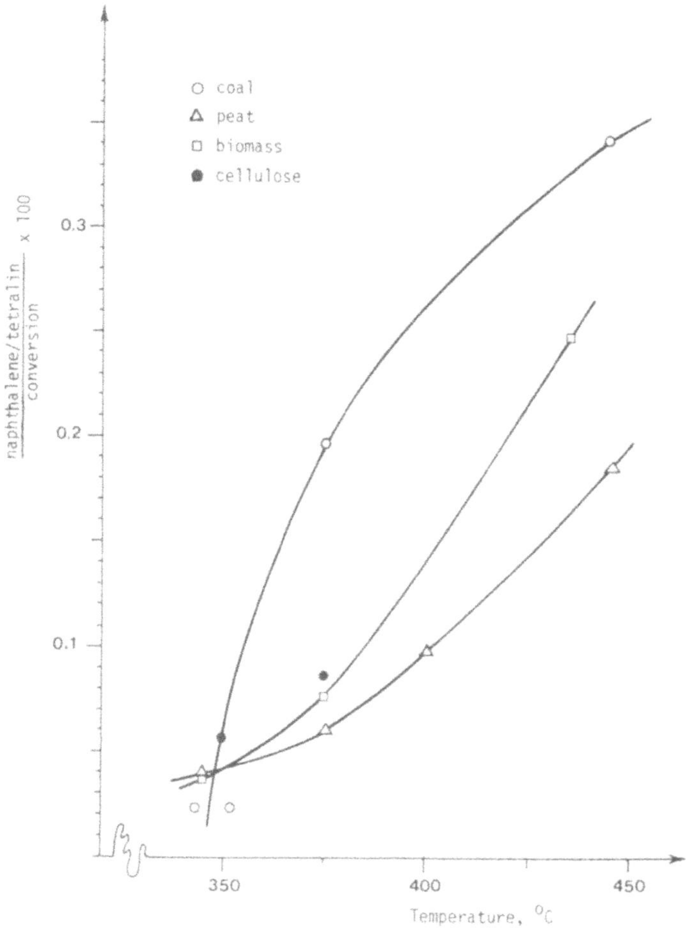

FIG. 3. Dissolution of coal, peat, biomass and cellulose in tetralin; duration 15 min.

same as in Fig. 1. The results indicate that the consumption of hydrogen from the different raw materials is nearly the same though coal conversion is much lower (see Fig. 1).

The results presented in Figs 1 and 2 are summarized in Fig. 3. It shows the connection between the conversion of the raw material and the consumption of hydrogen from tetralin. This connection is expressed by the following ratio:

$$\frac{\text{naphthalene (which is formed in the process)}}{\text{tetralin (which is left after the experiment)} \times \text{conversion of the solid fuel}}$$

Figure 3 demonstrates this ratio for different raw materials as a function of temperature. It can be seen that it increases with increase of temperature for all of the studied raw materials. The rate of increase is different, however. The consumption of tetralin in the presence of coal is much higher compared to that in the presence of peat and biomass.

The conversion of the studied solid fuels and the consumption of the hydrogen-donor solvent at longer processing times and different temperatures is shown in Tables VI and VII. The results confirm the higher reactivity of peat, biomass and cellulose compared to coal. Coal reacts at higher temperatures and requires much larger tetralin consumption than peat, biomass or cellulose. The transformation of tetralin into naphthalene is considerable at higher temperatures for all the studied raw materials.

TABLE VI
DISSOLUTION OF COAL, PEAT, BIOMASS AND CELLULOSE IN TETRALIN AFTER 67 min AT REACTION TEMPERATURE

Raw material	*Exp. No.*	*Temperature (°C)*	*Conversion (%)*	*Naphthalene/Tetralin (g/g)*	$\frac{N/T}{\text{Conversion}} \times 100$
Coal	K13	375	34	0·08	0·24
Peat	T11	375	88	0·08	0·09
Biomass	B11	375	87	0·10	0·11
Cellulose	C6	375	89	0·11	0·12
Coal	K5	395	46	0·08	0·17
Coal	K11	400	54	0·14	0·26
Peat	T10	400	91	0·12	0·13
Biomass	B4	392	86	0·11	0·13
Cellulose	C5	400	95	0·18	0·19

TABLE VII
DISSOLUTION OF COAL, PEAT, BIOMASS AND CELLULOSE IN TETRALIN AFTER 120 min AT REACTION TEMPERATURE

Raw material	*Exp. No.*	*Temperature (°C)*	*Conversion (%)*	*Naphthalene / Tetralin (g/g)*	*N/T / Conversion* × 100
Coal	K4	343	13	0·02	0·15
Peat	T5	350	79	0·04	0·05
Biomass	B9	350	85	0·07	0·08
Cellulose	C2	350	86	0·08	0·09
Coal	K7	445	74	0·39	0·53
Peat	T4	445	97	0′36	0·37
Biomass	B7	420	93	0·28	0·30

The consumption of hydrogen from the hydrogen-donor solvent is worth further discussion. Figure 4 represents the transformation of tetralin into naphthalene as a function of the conversion of the different fuels. The data are based on the results presented in Figs 1, 2 and 3 and Tables VI and VII. It may be seen that the curve obtained in the presence of coal differs essentially from those obtained in the presence of peat, biomass and cellulose.

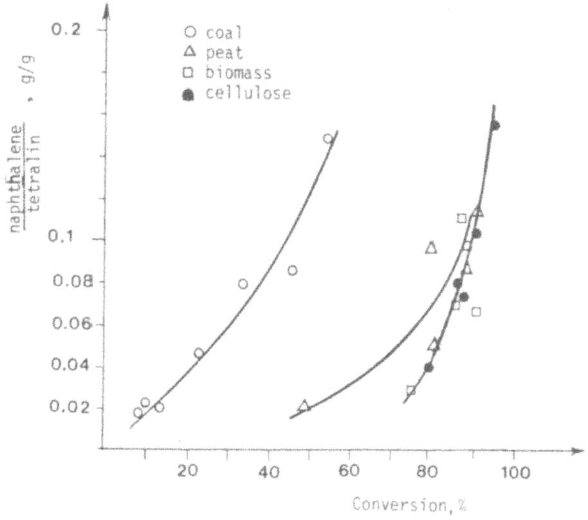

FIG. 4. Consumption of tetralin as a function of solid fuel conversion.

The conversion of coal and the consumption of hydrogen begin simultaneously. The conversion of tetralin in the presence of the younger fuels is lower and it begins after the conversion of the solid fuel has advanced considerably.

It seems that the initial conversion of peat and biomass occurs independently of the hydrogen-donor effect of the solvent. The dissolution of these fuels begins without any H consumption—no tetralin transformation into naphthalene is observed. Probably the initial conversion stage of these fuels consists mainly of thermal destruction of the fuel with evolution of H_2O and CO_2. The product of this initial conversion stage maybe resembles coal. Its further dissolution in tetralin is similar to coal dissolution; it occurs with participation of H from the solvent. The consumption of tetralin in the presence of coal is much higher, however, than that observed in the presence of peat and biomass. Probably, in the presence of coal tetralin is consumed in side reactions which do not contribute to the dissolution of the solid fuel. Figure 4 shows that peat has a somewhat intermediate position between coal and biomass. H consumption from tetralin begins earlier in peat dissolution compared to biomass dissolution. The higher the oxygen content of the initial solid fuels, the longer is the initial conversion period, in which no H consumption from the H-donor solvent occurs.

The observed phenomenon is of considerable practical importance. It indicates that the H consumption in the liquefaction processes of the younger fuels is not as high as expected, considering their high O content. Further insight into this phenomenon could be obtained by determination of the quantity of CO_2 which was evolved from the raw materials during the thermal dissolution in tetralin (see Table VIII).

The quantity of CO_2 which was formed in the experiments was calculated in percent of the initial moisture and ash-free (maf) input material. The following expression was used:

$$\frac{CO_2 \text{ evolved}}{\text{maf solid fuel input}} \times 100$$

It can be seen that the amount of gas obtained from coal is very low, while peat, biomass and cellulose evolve large quantities of CO_2. A considerable part of the O content of these young solid fuels is removed by thermal decomposition of the fuel with CO_2 evolution. This contributes to the low consumption of hydrogen-donor solvent in the thermal dissolution of peat, biomass and cellulose.

These results are in good agreement with our previous work on raw peat

TABLE VIII
EVOLUTION OF CO_2 IN THERMAL DISSOLUTION IN TETRALIN OF COAL, PEAT, BIOMASS AND CELLULOSE

Raw material	*Exp. No.*	*Processing time (min)*	*Temperature (°C)*	$\frac{CO_2\ evolved}{maf\ fuel_{in}} \times 100$ (%)
Coal	K4	120	343	0·4
Peat	T7	15	350	18·3
Biomass	B2	15	345	16
Cellulose	C3	15	350	15·4

liquefaction with CO.[15] Very low consumption of CO was observed. The yield of CO_2 obtained by the thermal decomposition of the wet peat was, however, higher. The low consumption of hydrogenating agent in younger fuels' liquefaction contributes to the technical feasibility of the process.

CONCLUSIONS

The results obtained in this work may be summarized in the following conclusions. Peat and biomass are more reactive than coal in thermal dissolution in tetralin. They react in milder conditions compared to coal and require less consumption of hydrogen-donor solvent.

ACKNOWLEDGEMENTS

The authors are indebted to the National Swedish Board for Energy Source Development and to the Energy Department of the Finnish Ministry of Trade and Industry for their financial support.

REFERENCES

1. Lindström, O. (1980). *Ambio* **IX**(6), 309–13.
2. Lindström, O. (1977). *Teknisk Tidskrift* **10**.
3. Neavel, R. C. (1976). Liquefaction of coal in hydrogen-donor and non-donor vehicles. *Fuel* **55**, 237–42.
4. Hooper, R. J. *et al.* (1979). Thermal dissociation of tetralin between 300 and 450 °C. *Fuel* **58**, 132–8.

5. Liebenberg, B. J. and Potgieter, H. G. J. (1973). The uncatalysed hydrogenation of coal. *Fuel* **52**, 130–3.
6. Potgieter, H. G. J. (1973). Kinetics of conversion of tetralin during hydrogenation of coal. *Fuel* **52**, 134–7.
7. Lasarov, L. *et al.* (1976). On thermal dissolution of coal in tetralin. *Solid Fuel Chemistry* **2**, 38–41 (in Russian).
8. Westlin, A. and Bresle, Å. (1950). High temperature extraction of peat I. *IVA* **21**, 208–19.
9. Westlin, A. and Bresle, Å. (1950). High temperature extraction of peat II. *IVA* **21**, 250–60.
10. Westlin, A. and Bresle, Å. (1951). High temperature extraction of peat III. *IVA* **22**(2), 46–58.
11. Djakova, M. K. and Melentieva, N. W. (1942). Thermal dissolution of solid fuels. Note II. Peat and lignin stage. *J. Prikl. Krim.* **15**(4), 236–47.
12. Solantausta, Y. (1981). Solvent extraction of coal and peat. Internal report from VTT (Technical Research Centre of Finland), Helsinki.
13. Gorin, E. *et al.* (1971). A synthetic fuels process. *Eighth World Petroleum Congress, Proceedings*, Vol. 4, Moscow.
14. P-Björnbom, E. *et al.* (1979). The combined dewatering and liquid phase hydrogenolysis of raw peat using carbon monoxide. *Fuel Processing Technology* **1979**(2), 161–9.
15. Björnbom, P. *et al.* (1981). Liquefaction of Swedish peats. *Fuel* **60**, 7–13.

45

Fluidodynamic Effects in the Fractional Solubilization of Biomass Leading to Liquefaction

E. CHORNET, D. EUGÈNE
Département de Génie Chimique, Université de Sherbrooke, Sherbrooke, Québec, Canada

and

R. P. OVEREND
Energy Project, National Research Council of Canada, Ottawa, Ontario, Canada

1. INTRODUCTION

From the considerable body of information on direct coal liquefaction and the experiments reported for biomass, a working hypothesis for the process of liquefaction includes the following sequential steps:

(i) 'solubilization' of the polymeric structures in biomass materials;
(ii) scission of the solubilized polymers into smaller units;
(iii) stabilization of the smaller units by either increased solvolytic action, addition of an external reducing agent (generally hydrogen) or autogenous reactions (thermally induced).

Information on the rates and products of the three steps, that is solubilization, scission and stabilization, is currently very sparse. Using pumpability of slurries of wood or peat as a criterion, then with different carrier vehicles[1-3] it should be evident that pretreatment processes contribute to steps (i) and (ii). Recent work[4-6] has shown that biomass has extremely high rates of reaction in step (ii) and that, as might be inferred from the heterogeneity of the wood polymer matrix, the products of steps (ii) and (iii) are extremely heterogeneous in composition.[7-11]

With this knowledge we came to the conclusion that a significant fraction of the biomass materials may be 'solubilized' in the preheating stages of liquefaction. Thus it was decided to investigate steps (i) and (ii) at timescales much less than reported previously using a sequential transfer autoclave system. Our programme had the following objectives.

(1) To conduct separately the preheating and the hydrogenolysis/hydrogenation stages.
(2) To study the 'transferability' of preheated wood slurries between the preheater and the hydrogenation reactor. The transfer is to be conducted through conduits and valves thus simulating the pumping action in real high pressure systems.
(3) To rapidly heat the transferred slurry up to reaction temperature thus facilitating studies at short contact times (either for hydrogenation or solvolysis).
(4) To investigate the influence of mechanical stress (shear) in the rupture (and rearrangement) of the polymeric structures into smaller and soluble fragments.
(5) To separate and quench the reaction products thus limiting the extent of secondary reactions following primary liquefaction.

2. EXPERIMENTAL

A schematic diagram of the experimental set-up used is depicted in Fig. 1. In its simplest form two autoclaves (A1 = 1 litre) and (A2 = 2 litres) are connected through a 4·8 mm ID medium pressure tubing. An actuated valve, controlled by a timer, permits the transfer of the charge, initially in A1, to A2. A suitable orifice is placed at the discharge end of the connecting tubing. Agitation and heating are independently controlled in both autoclaves.

The experimental programme reported herein was based on wood slurried in anthracene oil (A0). It was felt that the fluidity and aromaticity of the A0 would facilitate the experimentation. It was also expected that the A0 would favor the solubilization of the lignin fraction due to the hydroaromatic nature of the latter. Eventually an aromatic wood-enriched recycle oil would be produced and used in the later stages of the experimental programme.

The input material to A1 had the following characteristics:

Poplar (*Populus Tremuloides*, grown in Quebec) passing a −60 US Tyler mesh and air dried to $\simeq 10\%$ moisture content. The maf wood

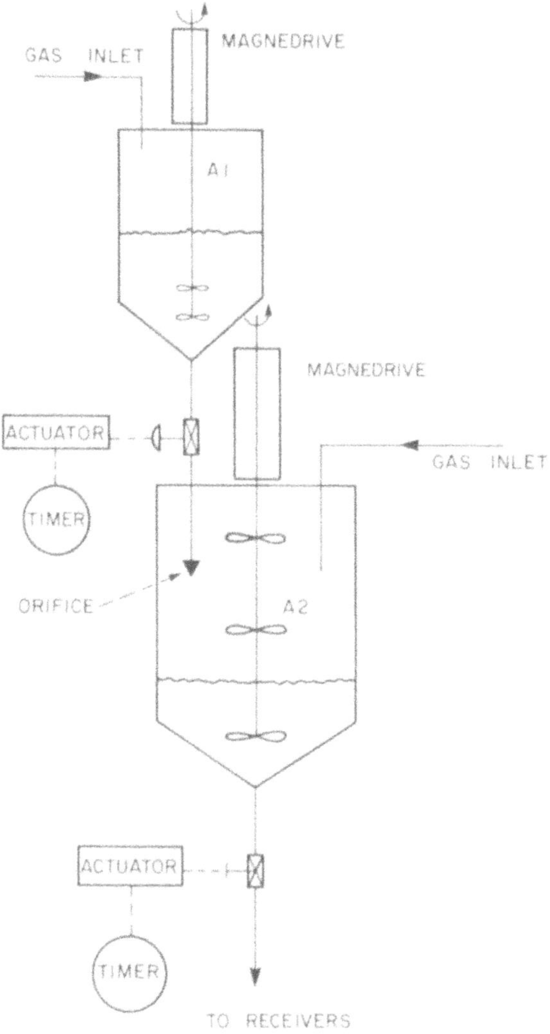

FIG. 1. UdS Cascade reactor system.

concentration in the anthracene oil suspension was comprised between 12 and 15% by weight. The experimental system (A1 and A2) was normally pressurized with hydrogen—a standard practice in our laboratory when handling anthracene oils. In these particular experiments, with no catalyst present, it was not expected that there would be hydrogenolysis or hydrogenation of the mixture in A1.

The temperature in A1 was increased from ambient at a rate of 5–6 °C/min until a maximum preheating temperature comprised between 200 and 250 °C was attained. The residence time of the slurry in A1 after attaining the target preheat temperature was varied between 0 and 60 min.

A pressure difference of 4–8 MPa was used to transfer the cellulose slurry to A2 where final solubilization takes place at reaction temperatures between 300 and 340 °C. Reactor A2 was kept under continuous internal agitation (Magnedrive) at 600 rpm. A small amount of anthracene oil was initially placed in A2 in order to facilitate heat transfer to the transferred slurry.

The residence time in A2 was varied between 5 and 12 min after which the treated slurry was discharged into cold receivers and quenched.

Simple analytical methods were used to characterize the products discharged from A2. They included:

—gas analyses via GC;
—solution of the slurry in THF (at 25 °C) to determine the following parameters:

$$\text{conversion:}\quad 1-\frac{\text{THF insoluble organics}}{\text{maf wood}}$$

$$\text{organic liquid yield:}\quad \frac{\text{THF soluble organics} - \text{solvent in}}{\text{maf wood}}$$

—H_2O determination via xylene extraction;
—elemental analysis (C, H, N) of the derived oils.

3. RESULTS

The initial experimentation established the pumpability and reactivity of the liquefaction process as follows:

—Good 'transferability' of the slurry between A1 and A2 requires a temperature T_{A1} 230–250 °C (preferably the highest one).

—Acid catalysts, to promote cleavage of the glucosidic bonds in the carbohydrates, were beneficial for the transferability and allowed satisfactory use of temperatures not higher than 240 °C.

—The maf wood/carrier oil ratio could not be raised beyond 15 wt % without causing serious problems and limitations during transferability.

—Overall conversions of 90 % could be obtained even at short residence times in A2 (5 min).

—H_2 (or CO) were found not necessary to achieve high conversions. In fact when the reaction in A2 was conducted in the presence of argon, conversions in excess of 90 % were also achieved.

—'Oil yields' (THF soluble organics) were very difficult to measure with accuracy since their computation is largely affected by only a few percentage errors in the material balances. Also H_2O determination proved to be of limited reproducibility via xylene extraction and our laboratory now only uses the Karl Fischer technique for this determination. Despite these limitations, 'oil yields' were consistently better than 50 % and in some cases achieved around 70 %.

The influence of acidity in the conversion is shown in Table I. Acid concentrations in excess of 0·1 w/o resulted in excessive 'charring' thus lowering conversion to oil.

The production of a wood-concentrated oil was carried out by successive

TABLE I

INFLUENCE OF ACIDITY IN THE CONVERSION FOLLOWING PREHEATING AT 230–255 °C, SUDDEN DISCHARGE AND 12 min REACTION TIME AT 320–335 °C

Acid		*Conversion*
CH_3COOH	1·0/1 000	96·3
HCl	1·0/1 000	90·0
HCl	1·7/1 000	88·2
HCl	1·75/1 000	87·2
HCl	2·0/1 000	87·7
HCl	2·5/1 000	79·0
HCl	3·0/1 000	80·0
H_2SO_4	5·0/1 000	72·0
HCl	5·0/1 000	71·0

The concentration of acid refers to H^+ expressed in parts per thousand weight relative to the slurry.

TABLE II

SOLVOLYSIS CHARACTERISTICS OF RECYCLE OILS CONTAINING INCREASED AMOUNTS OF WOOD-DERIVED OILS

	Oil yields from wood during solvolysis (% maf wood)	*Wood-derived liquids in recycle oil (% total oil)*
1st recycle oil (from A0 + wood)	43·6	5·6
2nd recycle oil + wood	62·0	13·0
3rd recycle oil + wood	71·5	20·5
4th recycle oil + wood	67·4	28·1

These oils are 28·1 % wood-derived and they seem excellent as carrier solvent since after 4 recycles they continue being efficient in the solvolysis.

recycle of the oil obtained with added wood. By this means an oil containing up to 28·1 % wood-derived liquids was produced. Table II shows the oil yields which were obtained during four successive recycle runs. It is to be noted that the solvolysis characteristics are not affected by the recycling thus giving support to the feasibility of a recycle operation. Elemental analysis of recycle oils (A0 plus wood-derived) are shown in Table III. A progressive enrichment in oxygen is evident from the data, although the value of 6·1 % in the 4th recycle oil (containing 28·1 % wood oil) is not excessive and could be improved in a potential upgrading operation via cracking and hydrogenation. The low hydrogen content of the oils indicates that little or no hydrogenation has taken place under these non-catalytic conditions, suggesting a solvolytic route towards solubilization.

The most interesting information obtained from this experimentation concerns, however, the fluidodynamic effects observed during the transfer

TABLE III

ELEMENTAL ANALYSIS OF RECYCLE-OILS

	$\% C$	$\% H$	$\% N$	$\% (O+S)_{by\ diff}$
Virgin anthracene oil	91·2	5·8	1·6	1·4
1st recycle (5·6 % wood oil)	90·1	6·4	1·1	2·4
2nd recycle (13·0 % wood oil)	88·6	5·9	1·2	4·3
3rd recycle (20·5 % wood oil)	88·6	5·9	1·2	4·3
4th recycle (28·1 % wood oil)	86·8	5·8	1·3	6·1

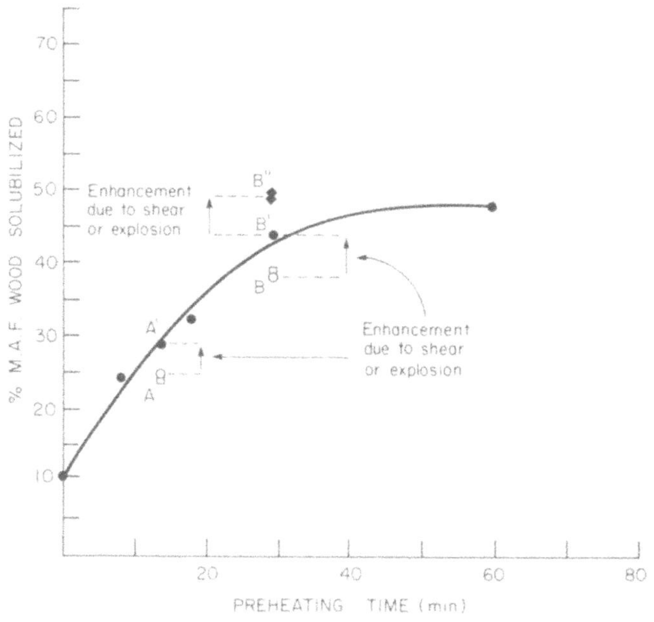

FIG. 2. Effects of preheating time and discharge through orifices in the solubilization of poplar at 250 °C in the presence of anthracene oil.

of the material from preheater A1 to a cold quench receiver A2. Figure 2 shows the conversion profiles of wood expressed as maf wood 'solubilized' following discharge from A1 to A2 after a given duration of preheating time had elapsed in A1. Inert gas (argon or nitrogen) is present in this operation and the temperature in A1 is maintained at 252–255 °C while in A2 quenching takes place as soon as the discharge is effected.

The baseline experiment is given by points A and B (note repeat experiment accuracy) in which no transfer was effected. Points A′ and B′ illustrate the effect of placing a discharge tube (60 cm × 0·48 cm ID). Point B″ is for two repeat experiments through the discharge tube with materials from experiment B′.

4. DISCUSSION

In order to increase solubilization during preheating and thus facilitate pumping at higher wood concentrations we have found that it is necessary to maintain the lignocellulosic suspension above 230 °C, and preferably around 250 °C. No catalyst seems to be required under these conditions. Residence time in the eventual preheater can be considerably shortened if recirculation and discharge through an orifice is continuously done. In fact with one recirculation it is possible to halve the time needed for achieving about 45 % solubilization in the preheating stage.

No clear evidence of any specific cause on the effect of recirculation and discharge can be inferred at this time through our limited experimentation. Yet many indications exist in the literature about polymeric changes and rupture in the presence of sudden explosion, shear forces or ultrasonics.[12,13] Considering the conditions used in our experiments it is quite likely that the lignocellulosic material achieves, at about 250 °C, a certain degree of plasticity which results in large shear forces when discharged through a fine conduit. At the low pressure side of the conduit sudden 'explosion' plus the generation of ultrasonic frequencies are possible (although not proven) resulting in polymer breaking and increased solubilization. Quantification of such effects should be a major task in the next experimental round.

The characteristics of the solids (the non-solubilized material) after passage through the orifice have not been studied in detail as yet. Optical inspection of the samples reveals, however, a very distinct morphology from the parent wood material. This parallels the change in the production of a microfibrillar-type cellulose reported recently by Turbak *et al.*[14]

5. CONCLUSIONS

The simple experimentation reported in this work leads to a number of conclusions.

- Extensive solubilization of poplar via solvolysis can be achieved at about 250 °C. Routinely more than 50 % maf wood can be solubilized using simple organic solvents (anthracene oil in this study).
- The action of shear forces is important and results in an increase in solubilization.
- The transfer of wood slurries through conduits (a measure of pumpability) has proven to be a key problem. In our experimentation we consistently found difficulties at slurry concentrations higher than 15 % solids content.
- It is simple and technically feasible to concentrate wood slurries by repeated additions of wood followed by partial solubilization via

solvolysis. This can be done by recirculating the slurry through a suitable orifice across which a suitable pressure difference is maintained. 8 MPa were used in most of our work. In this way we have achieved slurries containing 28·1 % solubilized wood.

ACKNOWLEDGEMENTS

R. P. Overend is a member of the staff of the Division of Energy of the National Research Council of Canada and is an adjunct professor with the Université de Sherbrooke.

The authors are indebted to the National Research Council and the National Science and Engineering Research Council (both from Canada) for the generous support in grants and equipment received to conduct this project.

Discussions with H. Ménard and G. Chauvette concerning analytical points are gratefully acknowledged.

REFERENCES

1. Schaleger, L. L., Yaghoubzadeh, N. and Ergun, S. (1979). Pretreatment of biomass prior to liquefaction. Presented at DOE's Thermochemical Conversion Contractor's Meeting, Rolla, Miss., November.
2. Figueroa, C. and Schaleger, L. L. (1982). LBL continuous bench-scale liquefaction unit, operation and results. Presented at 6th Annual Conference on Energy from Biomass Wastes, Lake Buena Vista, Florida, January.
3. Davis, H., Figueroa, C., Karatas, C., Kloden, D., Schaleger, L. and Yaghoubzadeh, N. (1981). Catalytic liquefaction of biomass. Presented at DOE's Thermochemical Conversion Contractor's Meeting, 493–520, October.
4. Eager, R. L., Mathews, J. F. and Pepper, J. M. (1982). Conversion of aspen poplar into liquid fuel. Proc. Specialists Meeting on Biomass Liquefaction, Saskatchewan, Can. Forestry Services (ENFOR), 1–14, February.
5. Beckman, D., Boocock, D. G. B., Mackay, D., Poretta, F. and Sherman, K. (1982). Wood liquefaction: A semi-continuous unit and fast heat-up studies. Proc. Specialists Meeting on Biomass Liquefaction, Sask., Can. Forestry Services (ENFOR), 15–30, February.
6. Chornet, E., Eugene, D. and Arsenault, R. (1982). An update on the conversion of lignocellulosics via hydrogenolysis and solvolysis at low severity conditions. Proc. Specialists Meeting on Biomass Liquefaction, Sask., Can. Forestry Services (ENFOR), 30–31, February.
7. Elliott, D. C. (1981). Bench scale research in biomass direct liquefaction. Presented at DOE's Thermochemical Conversion Contractor's Meeting, 551–575, October.

8. Elliott, D. C. (1982). Biomass liquefaction product analysis and upgrading. Proc. Specialists Meeting on Biomass Liquefaction, Sask., Can. Forestry Services (ENFOR), 157–173, February.
9. Kallury, R. K. M. R. and Tidwell, T. T. (1982). Investigation of the chemical composition of the product from high temperature hydrogenation of poplar wood. Proc. Specialists Meeting on Biomass Liquefaction, Sask., Can. Forestry Services (ENFOR), 183–190, February.
10. Schaleger, L. L. and Davis, H. G. (1982). Progress in the characterization of the products of direct liquefaction of Douglas fir. Proc. Specialists Meeting on Biomass Liquefaction, Sask., Can. Forestry Services (ENFOR), 191–204, February.
11. Heitz, M. (1982). Hydrogénolyse de matériel ligneux. Thesis (Dr. 3ième Cycle), Institut National Polytechnique de Lorraine, ENSIC, Nancy, March.
12. Winspear, S. (1975). Thermo-chemical starch conversion—the Multivisc converter. *Appita* **28**(5), 322–5.
13. De Gennes, P. G. (1974). Coil-stretch transition of dilute flexible polymers under ultrahigh velocity gradients. *The Journal of Chemical Physics* **60**(12), 5030–42.
14. Turbak, A. F., Snyder, F. W. and Sandberg, K. R. (1982). Food products containing microfibrillated cellulose. US Patent 4,341,807, July 27.

46

The Determination of the Most Important Flame Characteristics of Pellet Burners

V-P. HEISKANEN, J. HUOTARI and D. ASPLUND

Domestic Fuel Laboratory, Technical Research Centre of Finland, Jyväskylä, Finland

1. INTRODUCTION

Research directed to the combustion of peat pellets is not plentiful, because the idea of processing peat to pellets is quite new. In this work, an attempt was made to determine the most important flame characteristics of pellet burners in order to obtain some information about the combustion of pellets.

2. FUELS USED

Pellets were made of dried milled peat and had the following properties:

—moisture 15·0%
—diameter 10 mm
—length 10–25 mm
—calorimetric heat value of dry peat 22·19 MJ/kg
—composition (%):

C	46·0	N	1·1
H	4·9	ash	3·9
S	0·2	H_2O	15·0
O	28·9		

—volume weight 700–750 kg/m^3
—humification degree H6–7, on von Post scale.

Pellets with moisture 19·3 and 30·0% were tested also.

3. APPARATUS

The apparatus is composed of fuel storage, screw feeder, burner and cylindrical water-cooled combustion chamber. Combustion air is brought in one stage without preheating. The air and fuel feed devices are adjustable.

The flue gas cooler in the combustion chamber is a 1·6 m high cylinder, surrounded by a cooling water jacket. The water jacket is divided into four sections. The temperature rise and rate of water flow can be measured for every section.

The burner used in the experiments is commercial and is, therefore, not described. The schematic picture of the apparatus is shown in Fig. 4.

4. MEASUREMENTS

Measurements were carried out according to Fig. 1.

The flame temperature was measured by a suction pyrometer. The

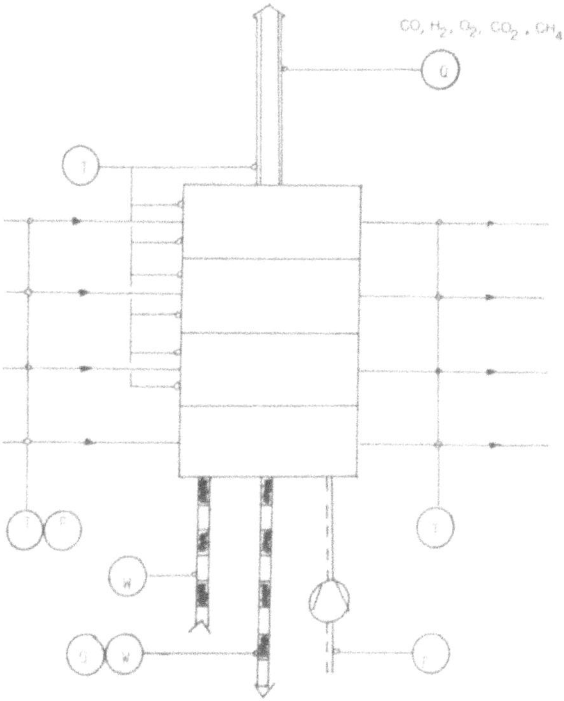

FIG. 1. Schematic picture of instrumentation.

TABLE I

MEASURED AND CALCULATED RESULTS

Test	Fuel	Diameter (mm)	Moisture (%)	Temperatures in the combustion chamber (°C) 1	2	3	4	5	6	Composition of flue gas (%) H_2	O_2	CO_2	CO	N_2	CH_4	λ	Flame length (m)	Ash (kg/h)	Ash loss (%)
1	Peat pellet	10	15·0	705	650	520	445	435	425	0·71	2·90	15·22	2·14	78·82	0·20	1·06	0·70	0·847	25
2	Peat pellet	10	15·0	705	690	560	480	455	440	—	3·04	15·46	1·95	79·40	0·16	1·09	0·70	0·740	16·1
3	Peat pellet	10	15·0	800	740	610	520	490	470	—	3·81	15·65	0·44	80·12	—	1·21	0·80	0·717	14·2
4	Peat pellet	10	15·0	850	705	625	625	545	500	—	4·64	15·12	0·34	79·91	—	1·27	0·90	0·209	4
5	Peat pellet	10	15·0	860	850	720	620	590	580	—	4·90	14·95	—	80·16	—	1·30	0·85	0·178	4·5
6	Peat pellet	10	15·0	900	900	720	650	610	590	—	5·48	14·02	—	80·10	—	1·39	0·95	0·161	3·6
7	Peat pellet	10	15·0	900	870	705	620	570	520	—	5·66	12·83	—	81·49	—	1·41	1·00	0·230	5·2
8	Peat pellet	10	15·0	890	850	680	600	525	500	—	7·79	11·46	—	80·76	—	1·63	0·80	0·578	19·6
9	Peat pellet	10	15·0	770	540	415	385	365	350	—	6·49	12·97	0·35	80·17	—	1·08	0·40	0·547	25·9
10	Peat pellet	10	15·0	630	480	420	375	370	360	—	9·45	11·14	0·20	79·21	—	1·37	0·50	0·412	23·0
11	Peat pellet	10	15·0	790	605	510	435	380	370	—	9·76	10·91	0·10	79·24	—	1·51	0·60	0·343	17·7
12	Peat pellet	10	15·0	930	695	455	380	365	360	—	10·40	10·44	—	79·16	—	1·75	0·60	0·224	9·3
13	Peat pellet	10	15·0	930	800	550	420	380	370	—	10·72	10·18	—	79·10	—	1·89	0·60	0·137	4·5
14	Peat pellet	10	15·0	930	690	480	370	360	350	—	10·01	9·24	—	80·79	—	1·97	0·60	0·126	3·1
15	Peat pellet	10	19·3	615	435	420	410	385	375	—	6·59	12·81	0·37	80·19	—	1·04	0·35	0·912	29·5
16	Peat pellet	10	19·3	810	680	550	490	470	450	—	5·56	14·16	0·15	80·15	—	1·14	0·60	0·527	16·9
17	Peat pellet	10	19·3	1 025	805	605	500	500	490	—	4·64	15·27	—	80·08	—	1·20	0·60	0·262	6·5
18	Peat pellet	10	19·3	685	660	510	440	430	410	—	6·88	12·77	0·25	80·08	—	1·24	0·50	0·518	17·2
19	Peat pellet	10	30·0	610	470	380	320	300	290	0·17	14·75	6·74	0·37	78·00	—	1·71	0·10	0·835	41·6
20	Peat pellet	10	30·0	625	490	370	340	310	290	—	11·96	9·31	0·18	78·51	—	1·79	0·10	0·368	35·2
21	Peat pellet	10	30·0	700	490	405	315	295	290	0·19	15·66	5·95	0·44	77·76	—	2·33	0·20	0·609	28·2
22	Peat pellet	10	30·0	690	470	380	320	305	290	0·16	15·39	5·94	0·43	78·11	—	2·49	0·30	0·486	22·5

TABLE II

MEASURED AND CALCULATED RESULTS

Losses						Total losses		Heat transfer to the sections (kW)				Mass flow rate of fuel (kg/h)	Efficiency (%)	Test
u_{ug}		u_{ent}		u_{cn}										
(kW)	(%)	(kW)	(%)	(kW)	(%)	(kW)	(%)	1	2	3	4			
3·06	10·4	4·42	15·1	5·88	20·1	13·36	45·6	1·62	5·55	5·12	4·90	5·46	54	1
2·29	7·8	4·58	15·6	3·79	12·9	10·66	36·4	2·16	4·85	5·62	5·60	5·46	65	2
0·46	1·6	5·55	18·9	3·34	11·4	9·35	31·9	2·04	6·01	5·64	5·59	5·46	68	3
0·40	1·4	6·53	22·3	0·94	3·2	7·87	26·9	1·46	5·01	4·94	5·34	5·46	73	4
0	0	8·06	27·5	1·06	3·6	9·1	31·1	1·84	5·80	5·81	6·37	5·46	69	5
0	0	8·55	29·2	0·85	2·9	9·40	32·1	1·77	6·07	5·81	6·53	5·46	68	6
0	0	6·93	23·6	1·22	4·2	8·15	27·8	1·66	4·46	5·81	5·62	5·46	72	7
0	0	7·58	25·9	4·61	15·7	12·19	41·6	1·43	5·54	5·18	5·46	5·46	58	8
0·22	1·4	2·33	14·4	3·36	20·8	5·90	36·5	1·19	2·99	3·02	2·64	3·01	64	9
0·15	0·9	2·75	17·0	3·00	18·5	5·89	36·5	1·27	3·26	3·22	3·60	3·01	64	10
0·08	0·5	3·08	19·1	2·30	14·2	5·45	33·8	1·32	3·52	3·06	3·05	3·01	66	11
0	0	3·37	20·8	1·20	7·4	4·57	28·3	1·11	3·27	3·16	2·80	3·01	72	12
0	0	3·87	24·0	0·58	3·6	4·45	27·6	1·16	3·56	2·75	2·97	3·01	72	13
0	0	3·86	23·9	0·40	2·5	4·26	26·4	1·50	3·98	3·99	4·40	3·01	74	14
0·41	1·4	4·57	15·6	6·94	23·7	11·92	40·7	1·31	3·55	3·49	3·45	5·82	59	15
0·17	0·6	5·70	19·4	3·98	13·0	9·85	33·6	1·50	4·73	4·63	3·86	5·82	66	16
0	0	6·25	21·3	1·53	5·2	7·78	26·5	1·46	4·95	5·07	4·81	5·82	74	17
0·30	1·0	5·33	18·2	4·05	13·8	9·68	33·0	0·73	3·46	3·07	2·97	5·82	67	18
0·41	2·6	2·35	15·1	5·11	32·9	7·86	50·7	0·65	1·63	1·27	1·86	3·5	49	19
0·15	1·0	2·47	15·9	4·32	27·8	6·94	44·7	0·72	2·17	2·11	2·33	3·5	55	20
0·65	4·2	3·10	20·0	3·46	22·3	7·21	46·5	0·67	2·04	2·11	2·33	3·50	54	21

temperature of the cooling water was measured with a mercury thermometer.

Analyses of flue gas were performed chromatographically. When measuring gas compositions in the combustion chamber, a water-cooled probe was used to prevent further combustion.

The water flow rates were measured by decanter. The air flow rate was measured by rotameter.

The fuel flow rate was known by calibration of the feeder. The calculation of λ (= measured air flow/stoichiometric air flow) was based on fed, not burnt fuel. Because of ash loss the compositions and amounts of fed and burnt fuel are not the same. The mass flow rate of ash was determined by collecting the ash produced during the test. The percentage of unburnt fuel in ash was determined.

The dust concentration in flue gas was measured by sucking gas isokinetically from the flue gas channel and filtering the solid particles.

The flame length was estimated with the measure scale.

The test period was 2 h, which enabled a stable condition to be reached. The results are presented in Tables I and II.

5. RESULTS

5.1. Heat Transfer

The calculated values of heat radiation and convection are presented in Table III. The calculation method of radiation is explained in the appendix to this paper. The convective heat flow is determined indirectly by subtracting the calculated radiation from the measured total heat transfer. The direct calculation of convection does not produce accurate values, because of unknown gas velocities and temperatures along the cylinder wall. The convection is most effective in the upper part of the cylinder, because of turbulence caused by the flue gas channel.

TABLE III

RADIATION, CONVECTION AND TOTAL HEAT TRANSFER

Test	*Radiation* (*kW*)	*Convection* (*kW*)	*Total heat transfer* (*kW*)
3	16·0	3·3	19·3
4	14·1	2·7	16·8
7	15·2	2·3	17·5

In three tests the radiative, convective and measured total heat transfer had the values shown in Table III.

The amount of radiation compared with convection is not constant. This is understandable, considering the structure of the combustion chamber, gas temperature and the rather small convection surface compared with the chamber volume.

The distribution of heat transfer to separate sections can be seen in Table II. A relatively high amount of the heat transfer to the uppermost section is because convection is most effective there. Heat flows from the cylinder cover to the uppermost section by conduction.

The heat transmission is most intensive at the distance 0·3–0·5 m, depending on the fuel flow rate, which has an effect on the flame length. When considering, for example, the results presented in Tables I and II, it will be found, that the distribution of heat transfer in test 7 is exceptional when compared with tests 1–6 and 8. The heat transfer at a distance 0·1–0·5 m is less, though the temperatures in the flame are higher with the

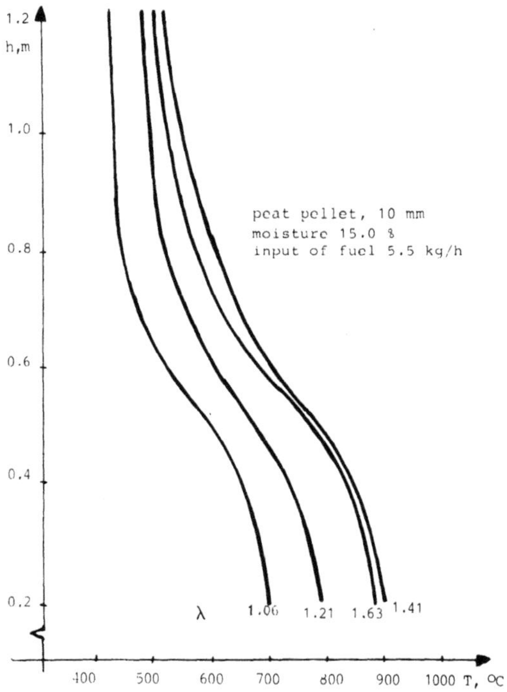

FIG. 2. Axial distribution of temperature in the combustion chamber, as parameter.

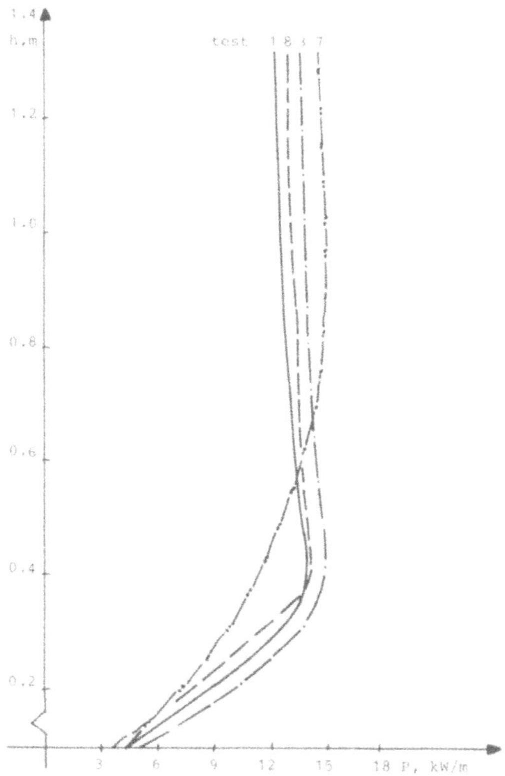

FIG. 3. Axial distribution of heat transfer.

exception of test 6. This results from the higher flame in test 7. Therefore the flowing gas mixture reaches the final composition in the later stage and the contents of radiant gases CO_2 and H_2O are lower at the same heights in the flame. Therefore, the radiative heat transfer is less at a distance 0·1–0·5 m in test 7. The axial distributions of temperature and heat transfer in four tests are presented in Figs 2 and 3.

Any clear correlation between heat flow distribution and, for example, amount of air or fuel moisture cannot be concluded but the heat flow is divided among the sections in a fairly constant ratio irrespective of the variation of the aforesaid factors. The average ratios are 1:3. 1:3:3.

5.2. Incompletely Burnt Gases

It can be noticed, when the measured compositions of the flue gases are closely considered, that carbon monoxide, hydrogen and methane are not found when the amount of air is more than 1·3–1·4 times the stoichiometric amount excluding the experiments when the pellets with a moisture of 30 % were used. Methane cannot be found when the excess air is 15–20 % of the stoichiometric amount. Disappearance of hydrogen occurs with almost the same excess of air. To prevent the presence of carbon monoxide in flue gas, the excess air must be 30–40 % of the stoichiometric ratio, as mentioned before. The carbon monoxide content is generally higher than the hydrogen or methane content, while there is more hydrogen than methane.

The reasons for the incomplete combustion of gas are the imperfect mixing between air and gas and the low temperature of the combustion chamber's water-cooled walls. For the latter reason the flame temperature is rather low because of intensive radiation to the walls. The effect can be observed by comparing the flame temperatures and gas compositions of the separate tests. The comparison indicates that lower temperatures cause higher content of incompletely burnt gases as expected. The low temperatures were also caused by significant ash loss. There is an obvious correlation between ash loss and incompletely burnt gases. The effect of unburnt fuel on the gas temperatures in the combustion chamber is caused by two factors. The composition of the gas forming part of the fuel is not the same as the composition of the fuel fed to the burner. In addition, fuel flow rate decreases and therefore λ increases, because the air flow rate is unchanged. This can be seen very clearly by considering tests 15–18. The effect of ash loss upon efficiency, the temperature at point 1 and carbon monoxide content were as presented in Table IV.

Decreasing ash loss causes higher temperatures and, furthermore, reduction of carbon monoxide content. The amount of excess air has an influence on the amount of incompletely burnt gases too but, for example, in test 18 more excess air was used than in test 17. However, the carbon monoxide content was higher because of lower temperatures.

On this basis the amount of excess air, ash loss and the loss caused by

TABLE IV

ASH LOSS, TEMPERATURE AND CARBON MONOXIDE CONTENT IN TESTS 15–18

Test	*Ash loss* (%)	*Temperature, point 1* (°C)	*CO* (%)	*Temperature, mean value* (°C)
15	29·5	615	0·37	440
18	17·2	685	0·25	523
16	16·9	810	0·15	575
17	6·5	1 025	0	654

incompletely burnt gases have such interdependence that scarce excess air causes large ash loss and this causes low flame temperatures; for this reason gases leave the combustion chamber unburnt.

When considering unburnt gases carbon monoxide causes the most significant loss in efficiency. Methane has the largest heat of reaction, 802·3 MJ/kmol, but methane content is mostly so low that the loss is not higher than 2 % even when excess air is used only scarcely. The loss caused by unburnt hydrogen is in the same order of magnitude. It can be stated generally that the loss caused by incompletely burnt gases is rather small when all losses are considered and it concerns only the case when little excess air is used. Instead of that, for example, the emission of polyaromatic hydrocarbons (PAH) has been noted to correlate with the

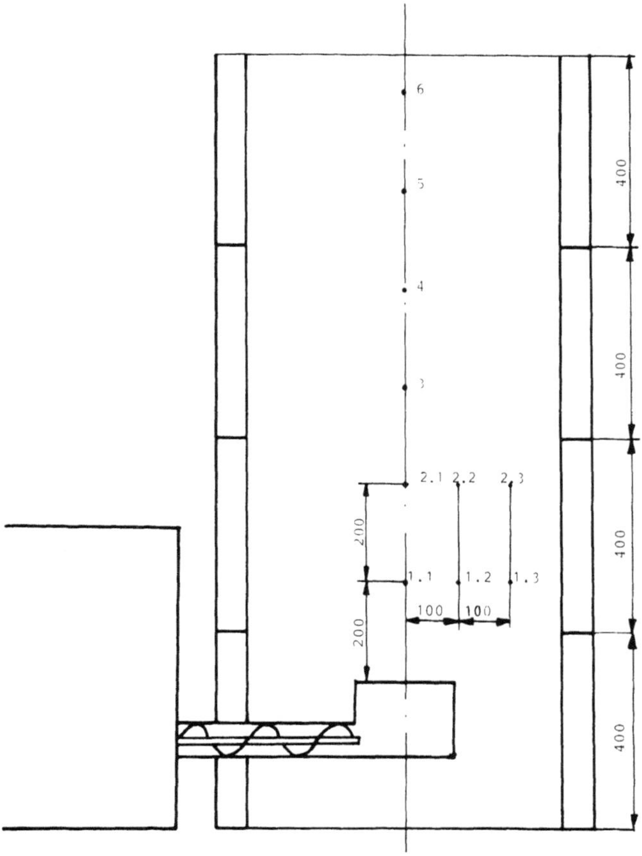

FIG. 4. Measuring points.

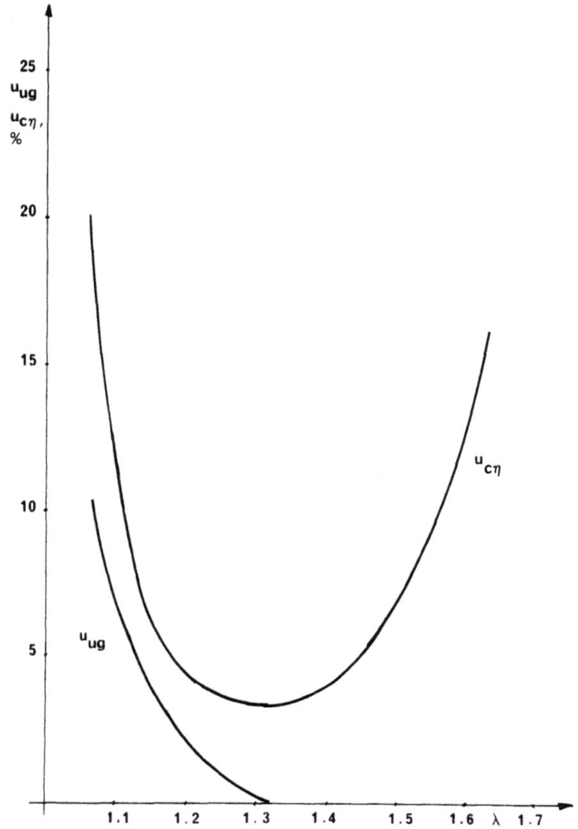

FIG. 5. The losses caused by unburnt fuel and incompletely burnt gases as a function of λ in tests 1–8. Peat pellet 10 mm, moisture 15·0 %, 5·5 kg/h.

amount of unburnt gases so that PAH content increases when the content of carbon monoxide, hydrogen and methane increase. The losses caused by incompletely burnt gases and unburnt fuel as a function of λ in tests 1–8 are presented in Fig. 5.

5.3. The Effect of Fuel Moisture and Excess Air

The fuel moisture reduces heat value and adiabatic temperature of combustion. Because combustion intensity can be supposed to depend on temperature exponentially, the increased moisture decreases reaction rates in the flame.

In tests 19–22, when the pellets with a moisture content of 30 % were used, the measured temperatures were very low. This depended partly on the effect of moisture, but also on profuse excess air. Profuse excess air was needed because proper combustion did not take place otherwise.

The flue gas contained hydrogen and carbon monoxide even though excess air was used, 1·7–2·5 times the stoichiometric amount. Though carbon monoxide and hydrogen contents were low, the absolute amounts were significant because of profuse flue gas flow.

When considering the temperatures above the burner, the temperature decreases very abruptly at a distance of 0·2–0·4 m. The flame lengths were only 0·1–0·3 m. Thus the combustion of produced gases occurs at a short distance, even though incompletely, because the low temperature results in too short a reaction zone.

The amount of air has a clear effect upon the flame length, too. When comparing the results, for example, in tests 1–8, it can be found that the flame length increases with increasing excess air, reaching a maximum when 1·45–1·50 times the stoichiometric amount of air is used. It can be noted that the flame is rather high when considering the use of the burner in a boiler.

5.4. Gas Compositions in the Flame

The gas compositions in the flame were determined by sucking gas at a high velocity using a water-cooled probe to prevent further combustion. Six samples were taken from different points during 2 h. Because there is continuous variation in the combustion, the samples cannot be used for the determination of the concentration distributions. The points where the samples were taken are shown in Fig. 4. The measured compositions and temperatures as presented in Table V.

Combustion is taking place with less than the stoichiometric amount of air finding the chemical equilibrium; the rest of the air has not yet reacted at

TABLE V

THE VALUES OF THE MEASURED COMPOSITIONS

Sample	CH_4	CO_2	H_2	O_2	N_2	CO	T (°C)
1·1	1·37	9·68	4·01	5·21	67·98	11·55	1 082
1·2	$2{\cdot}6 \times 10^{-3}$	8·12	—	9·75	81·47	0·46	890
1·3	—	1·11	—	19·08	79·63	$9{\cdot}4 \times 10^{-3}$	370
2·1	0·17	4·38	0·36	14·64	78·84	1·39	976
2·2	0·13	7·28	0·24	11·91	79·22	1·03	901
2·3	$4{\cdot}8 \times 10^{-3}$	7·96	—	17·25	74·46	0·09	375

this stage. The chemical equilibrium can be determined by the aid of a homogeneous water gas reaction:

$$CO + H_2O \rightleftarrows CO_2 + H_2$$

The theoretical compositions for samples 1·1 and 2·1 are calculated and shown in Table VI. The parameter λ' is the ratio between the amount of air used and the stoichiometric amount. First λ' was given a value smaller than 1. The theoretical composition in the measured temperature was calculated. After that an amount of air was added to the composition, such that the oxygen content was equal to the measured content, and then the nitrogen content was calculated. If the nitrogen content was lower than the measured content, λ' was supposed to be too little and vice versa. Then λ' was changed and iteration executed, until the correct composition was found. The theoretical compositions corresponding to the measured compositions of dry gas at points 1·1 and 2·1 were calculated that way. The results are in Table VI.

TABLE VI
THE VALUES OF THE CALCULATED COMPOSITIONS

			CH_4	CO_2	H_2	O_2	N_2	CO	SO_2	T (°C)
a	1·1	$\lambda' = 0{\cdot}58$	$1{\cdot}99 \times 10^{-6}$	11·07	9·05	—	64·31	15·53	$3{\cdot}65 \times 10^{-2}$	1 082
b			$1{\cdot}50 \times 10^{-6}$	8·31	6·79	5·21	67·55	11·65	$2{\cdot}74 \times 10^{-2}$	1 082
c		measured	1·37	9·68	4·01	5·21	67·98	11·55	—	1 082
a	2·1	$\lambda' = 0{\cdot}93$	$2{\cdot}11 \times 10^{-9}$	18·30	1·12	—	78·55	2·00	$2{\cdot}79 \times 10^{-2}$	975
b			$0{\cdot}64 \times 10^{-9}$	5·52	0·34	14·64	78·88	0·60	$0{\cdot}84 \times 10^{-2}$	975
c		measured	0·17	4·38	0·36	14·64	78·84	1·39	—	975

[a] Calculated composition corresponding to the chemical equilibrium without unreacted air.
[b] With unreacted air.
[c] Measured composition.

When comparing the compositions it can be seen (and the differences depend on the fact) that there is still methane in the gas, though according to theory the methane content should be insignificant at the temperatures in question. Obviously the methane reaction

$$CH_4 + H_2O \rightleftarrows CO + 3H_2$$

is happening too slowly to reach equilibrium and therefore the methane content is higher than it should be.

When considering the burner design λ' can be plotted against the height of the combustion chamber with various operational parameters. This will give further information for burner design.

6. SUMMARY

It can be noted from the results that the ash loss has a significant effect on the shape and temperature of the flame as well as on the efficiency. Gas compositions in the flame can be explained partly with the aid of a homogeneous water gas reaction.

The results presented in this paper are a part of the research which is continuing with the tests of 6 mm peat pellets and wood pellets.

REFERENCES

1. Gasstrahlung (1974). In *VDI-Wärmeatlas*, 2nd edn. VDI-Verlag, Düsseldorf, s. Kc1–Kc8.
2. Gumz, W. (1953). *Kurzes Handbuch der Brennstoff- und Feuerungstechnik*, 2nd edn. Springer-Verlag, Berlin, s. 356–357.

APPENDIX

The Calculation Method of Radiative Heat Transfer According to Ref. 1

The calculation involves only gas radiation because the distribution of dust concentration is not known. The average radiation temperature in the cylinder calculated by aid of the measured results:

$$(\bar{T}_4) = \frac{T_1^4 + T_2^4 + T_3^4 + T_4^4 + T_5^4 + T_6^4}{6}$$

T_{1-6} are the temperatures measured above the burner; the distance between the points is 0·2 m.

The mean beam length L for radiation of gas:

$$L_0 = 4V/A$$

V = the volume of the chamber (0·423 m^3)
A = the wall area of the chamber (3·443 m^2).

For the cylinder in question $h = 6r \rightarrow$

Radiation to the wall $L_0 = d$

Radiation to the middle of the bottom and cover $L_0 = d$

For the total radiation approximately $L_0 = d$. L can be found when the value for the product pL is estimated first.

CO_2: $p_{CO_2}L \approx 0{\cdot}15 \times 0{\cdot}5$ bar m $= 0{\cdot}075$ bar m

H_2O: $p_{H_2O}L \approx 0{\cdot}1 \times 0{\cdot}5$ bar m $= 0{\cdot}05$ bar m

The partial pressures of the radiant gases CO_2 and H_2O can be calculated on the basis of the air flow rate and the composition of the fuel. The emissivities ε_{CO_2} and ε_{H_2O} are taken from the diagrams shown in ref. 1 using the exact values of the products $p_{CO_2}L$ and $p_{H_2O}L$.

The emissivity of gas mixture:

$$\varepsilon_g = \varepsilon_{CO_2} + C_w \varepsilon_{H_2O} - \Delta\varepsilon$$

The absorption coefficient:

$$\alpha_g = \alpha_{CO_2} + C_w \alpha_{H_2O} - \Delta\varepsilon$$

The values of C_w and $\Delta\varepsilon$ are taken from the diagrams shown in ref. 1.

Radiative heat transfer between gas and the walls of the grey chamber ($\varepsilon_{wall} = 0{\cdot}8$) is

$$q = M_{mg}\bar{\varepsilon}_g - M_m \bar{\alpha}_g$$

where

$$\bar{\varepsilon}_g = \varepsilon^2[\varepsilon_g(L) + \rho\varepsilon_g(2L) + \rho^2\varepsilon_g(3L) + \cdots]$$
$$\alpha_g = \varepsilon^2[\alpha_g(L) + \rho\alpha_g(2L) + \rho^2\alpha_g(3L) + \cdots]$$

Total radiative heat transfer is

$$Aq = A(M_{mg}\bar{\varepsilon}_g - M_m\bar{\alpha}_g) = \phi_{rad}$$

A is the added area of the walls, bottom and cover.

47

Catalysis of Gas-Phase Reactions in Steam Gasification of Biomass

E. G. BAKER, L. K. MUDGE and W. A. WILCOX

Battelle Pacific Northwest Laboratories, Richland, Washington, USA

1. INTRODUCTION

Catalysts play a major role in conventional hydrocarbon processing but only recently have they been applied to conversion of biomass materials. Pacific Northwest Laboratory (PNL) is studying catalytic steam gasification of biomass for producing specific gas products. These studies are sponsored by the Biomass Energy Systems Division of the US Department of Energy.

Gasification of biomass with steam involves a combination of three types of reactions:

(1) Pyrolysis to produce gaseous products (CO_2, CO, CH_4 and some heavier hydrocarbons), tar, water soluble organics (acetic acid, methanol, acetone, esters and aldehydes) and char.

(2) Reaction of char from (1) with H_2O, CO_2 and H_2 to produce CO, CH_4 and CO_2.

(3) Reforming and condensation of the products of (1) and (2) to produce additional gas products, tars and char.

In studying catalytic steam gasification we have used alkali carbonates impregnated or mixed with the wood to enhance the steam char reaction (2) and supported metal catalysts to catalyze the gas phase reactions (3). This paper presents the results of laboratory studies on catalysis of the gas phase reactions.

With the proper choice of catalysts and operating conditions methanol synthesis gas, methane-rich gas, hydrogen-rich gas and ammonia synthesis gas can be produced directly by steam gasification of biomass. High carbon conversion is achieved and no condensible organics (tar, oil, etc.) are formed when the catalysts are active.

Many commercial catalysts were effective in producing specific gas products; however, most of the catalysts quickly lost their activity. This paper discusses efforts to develop long-lived catalysts for the production of methanol synthesis gas and methane-rich gas.

2. EARLY EXPERIMENTAL RESULTS

The objective of our studies has been to develop catalyst systems for the direct production of specific gases by steam gasification of biomass. Preliminary catalyst screening experiments were done in a batch reactor, but the majority of the work utilized a continuous reactor described below.

2.1. Description of Experimental Systems

Two continuous-wood-feed, fixed-catalyst-bed reactors constructed of quartz glass were used for these studies. The reactors, which were operated at atmospheric pressure, are shown in Fig. 1. Wood is fed into the reactors at 8–12 g/h and gasified in the top of the reactor. The hot gases pass through a catalyst bed in the bottom of the reactor. The space velocity over the catalyst is typically 1000–2000 cm^3/g catalyst/h. Char and ash accumulate on a support just above the catalyst bed and are vacuumed out periodically.

Analytical equipment used includes a Carle AGC-S analytical gas chromatograph for gas analysis, a Perkin Elmer 240 elemental analyzer for wood and char analysis, and a Dohrmann DC80 carbon analyzer for water analysis. An Aminco-Dietz Sor-BET surface area meter and a hydrogen chemisorption apparatus were used for catalyst characterization.

2.2. Production of Specific Gases

The primary gas phase reactions that occur during steam gasification are shown in Table I. The conditions and catalysts used for these reactions in the hydrocarbon processing industries are also included. Using this and equilibrium data on the C—H—O system as a basis we selected operating conditions and catalysts for production of methane, methanol synthesis gas ($2H_2/CO$), ammonia synthesis gas ($3H_2/N_2$) and hydrogen.

Steam gasification of biomass is endothermic. At temperatures above about 500 °C heat must be added to the gasifier to maintain the reaction temperature. While this is not a problem in laboratory studies, in a commercial scale unit the amount of heat that can be added economically is limited. Experience with our 1 ton/day process development unit and economic evaluations of larger gasifiers indicated 750 °C was about the

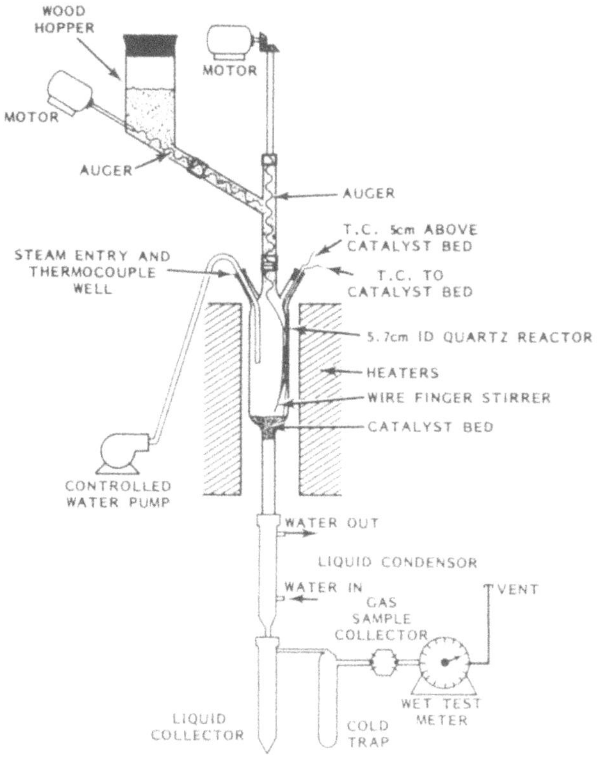

FIG. 1. Continuous wood-feed reactor.

TABLE I
GAS PHASE REACTIONS IN STEAM GASIFICATION OF BIOMASS[5,6,3]

	Temp. (°*C*)	*Press.* (*atm*)	*Catalysts*
1. Shift reaction ($CO + H_2O \leftrightarrows CO_2 + H_2$)	315–485	1–20	Metal oxides Fe, Cr, Cu, Zn
2. Methanation ($CO + 3H_2 \leftrightarrows CH_4 + H_2O$)	230–450	1–300	Ni on a support (silica, alumina, refractory)
3. Hydrocarbon reforming ($C_nH_m + nH_2O \rightarrow nCO + (n + (m/2))H_2$)	550–1 000	1–40	Ni on oxides (alumina, magnesia)
4. Cracking	250–500	2–200	silica-alumina, Ni W on silica-alumina

TABLE II

PRODUCTION OF SPECIFIC GASES FROM BIOMASS BY STEAM GASIFICATION

Product gas *Secondary catalyst*	*Methane rich gas* *Harshaw 3266* *Ni/support*	*Methanol synthesis gas* *Harshaw 3266* *Ni/support*	*Ammonia synthesis gas* *Girdler G-93 Co-Mo* *Co-Mo/Al_2O_3*	*Hydrogen* *Girdler G-3,* *Harshaw 1404* *Fe-Cr, Ni/support*
Gasification temperature (°C)	560–540	740–770	640–660	740–770
Catalyst temperature (°C)	550–560	740–760	650–670	720–750
steam/wood (g/g)	0·33	0·70	1·60	5·7
air/wood (g/g)	—	—	0·65	—
Carbon conversion to gas[a] (%)	68	90	65	78
Gas composition vol %				
H_2	29·5	53·1	50·7	64·1
CO_2	34·3	15·5	21·2	29·9
CO	10·8	28·3	7·3	5·8
CH_4	25·4	3·1	2·7	0·2
C_2^+	tr	0·1	0·3	—
N_2	—	—	17·8	—

[a] No liquid products were formed when the catalysts were active. Char is the only other product.

maximum practical upper limit. We also set 550 °C as the minimum gasification temperature. Production of methane is thermodynamically more favorable at lower temperature; however, the rate of gas production is too low to be practical.

Table II shows typical results obtained for each case at the optimum conditions determined by our studies. The gasification temperature was constrained to 550–750 °C as previously indicated and all tests were run at 1 atm pressure. No tar or condensible organics were produced in any of the cases as long as the catalyst was active; char was the only byproduct.

3. CATALYST LIFETIME STUDIES

The processes for making a methane-rich gas and a methanol synthesis gas were selected for further development. The preliminary laboratory studies identified catalysts which had the right chemical composition to promote the desired reactions, but no attempt was made to determine the activity over an extended period of time. For the processes to be economically feasible the catalysts have to be resistant to deactivation or easily regenerated.

The first sign of loss of activity in both the methane and the methanol synthesis gas case is the appearance of tars in the product gas. At this time the gas quality begins to deteriorate rapidly. We defined catalyst lifetime as the grams of wood processed per gram of catalyst prior to the appearance of tars in the product gas.

Early in the lifetime studies it became apparent that carbon deposition would be the most serious problem with respect to catalyst deactivation. Reactions which can deposit carbon include:

(1) $2CO \leftrightarrows C + CO_2$
(2) $CH_4 \leftrightarrows C + 2H_2$
(3) $C_nH_m \longrightarrow \text{polymers} \longrightarrow \text{coke} + xH_2$

Carbon can also be removed from the catalysts by the carbon–steam reaction (4):

(4) $C + H_2O \longrightarrow H_2 + CO$

Even though the equilibrated gas may show no affinity for carbon formation, in an open thermodynamic system carbon may be stable in a steady state and the accumulation of carbon may continue. The question of whether there is a net buildup of carbon is thus a kinetic one and can be influenced by the choice of catalysts.[3]

3.1. Catalysts for Methane-Rich Gas

Efforts to develop a long-lived catalyst system for the production of a methane-rich gas have not been successful. Numerous active catalysts were tested but the longest lifetime achieved was 73 g wood/g catalyst.

All of the catalysts found to be active for the methane case generally had the following properties:

(1) a nickel concentration of 25 wt % or more,
(2) a high surface area support (100 m^2/g),
(3) a nickel surface area of 30 m^2/g.

The active lifetime of the catalyst seems to be related primarily to the amount of active metal surface area on the catalyst as shown in Fig. 2. Coprecipitated catalysts containing approximately 50 wt % Ni and 10 wt % Fe on alumina had the longest lifetimes. Work at the University of Louisville indicates that the maximum metal surface area that can be achieved with coprecipitated nickel catalysts is in the range of 40–60 m^2/g.[2] Other methods of catalyst preparation result in even less surface area.

Deactivation of high surface area, dispersed metal catalysts is a

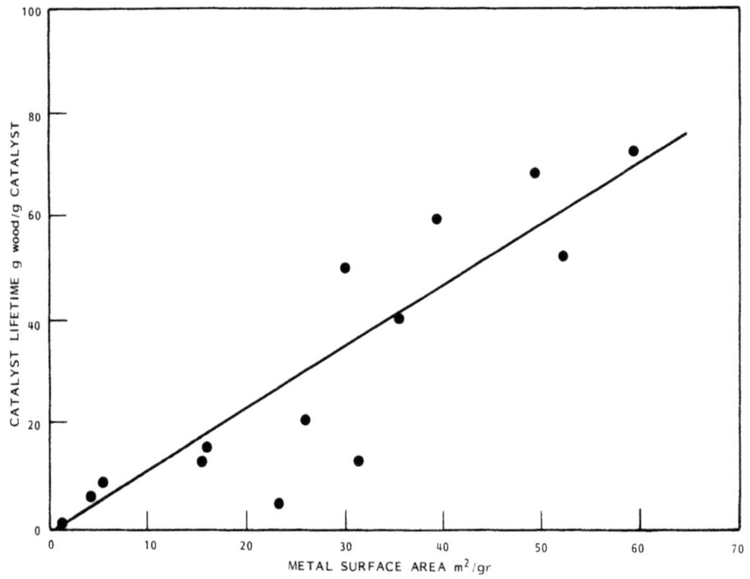

FIG. 2. Catalyst lifetime for production of methane-rich gas.

complicated phenomenon in which chemical poisoning (carbon deposition in our case) and thermal sintering may both be involved.[2] At temperatures over 500 °C, sintering, the loss of metal surface area due to growth in nickel particle size on the catalyst, may be a significant cause of deactivation. Two coprecipitated catalysts were steamed at 550 °C for 48 h and their surface area was reduced by 50–70 %. This rate of surface area loss corresponds approximately to the rate of loss of activity.

Deactivated catalysts generally had 15–30 wt % carbon on them indicating carbon deposition may also be a problem. Some catalysts were tested in both the methane case and the methanol synthesis gas case at 750 °C and in all instances had longer active lifetimes at 750 °C where sintering should be more severe than at 550 °C. It is likely that both sintering and carbon deposition contribute to catalyst deactivation at 550 °C. A better understanding of the mechanism of deactivation is necessary before a long-lived catalyst can be developed.

3.2. Catalysts for Methanol Synthesis Gas

Supported nickel catalysts are also the most effective catalysts for synthesis gas production at 750 °C. Lifetimes of up to 100 g wood/g catalyst were achieved with commercial catalysts; however, longer lifetimes are needed to make the process economically feasible. Carbon deposition is the cause of deactivation.

With the aid of W. R. Grace, Davison Technical Division, we developed a series of supported trimetallic nickel alloy catalysts for the methanol synthesis gas case which are resistant to deactivation. One of these catalysts, Ni–Co–Mo on a silica-alumina support, ran for over a year in a laboratory reactor without deactivation. A lifetime of 1470 g wood/g catalyst was achieved before the run was terminated with the catalyst still active. Table III shows the progress of this run.

At the start of a run the trimetallic catalysts are not quite as active as several commercial nickel catalysts in terms of the quantity of synthesis gas produced per gram of wood. The trimetallic catalysts maintain their activity for a much longer period of time as shown in Fig. 3.

Carbon deposition up to 25 wt % was found on deactivated catalysts. The trimetallic catalysts are unique among the active supported metal catalysts we tested for their ability to prevent accumulation of carbon and remain active. Carbon builds up on the catalyst to 4–8 wt % and then does not increase further. The catalysts are still active with this quantity of carbon on them. The carbon can be removed by steaming at 800 °C and the catalyst returns to its original activity.

TABLE III

LIFETIME STUDY WITH Ni–Co–Mo/SiO_2–Al_2O_3 CATALYST

Date	*4/2/81*	*5/8/81*	*8/13/81*	*11/17/81*	*2/22/82*	*3/17/82*	*4/8/82*
Lifetime (g wood/g catalyst)	Start	100	580	816[b]	1 197	1 358	1 469
Steam rate (g/g wood)	0·8	0·9	1·6	1·1	0·9	0·9	0·9
Gas composition							
% H_2	51·5	49·6	47·1	43·6	45·9	51·7	50·9
% CO_2	16·3	19·4	25·9	23·6	22·2	18·8	19·4
% CH_4	4·0	6·8	7·6	8·2	8·0	5·3	5·0
% CO	28·0	24·0	18·1	23·6	23·1	24·1	24·4
% C_2H_4	0·0	0·2	0·3	0·5	0·5	0·1	0·2
% C_2H_6	0·0	0·04	0·5	0·5	0·4	0·1	0·1
Carbon conversion to gas (%)[a]	91	90	95	98	98	95	90
Synthesis gas yield (nm^3/kg wood)	1·41	1·24	1·09	1·08	1·10	1·33	1·21

[a] Conversion to liquid is very small—less than 0·1 %. The rest of the carbon is converted to char.

[b] At a lifetime of 802 the reactor plugged and the catalyst had to be removed from the reactor. After being replaced in the reactor and reduced the catalyst had lost activity and a significant quantity of tar was formed. The catalyst was regenerated with steam at 800 °C and then reduced. Activity was restored to the level prior to shutdown.

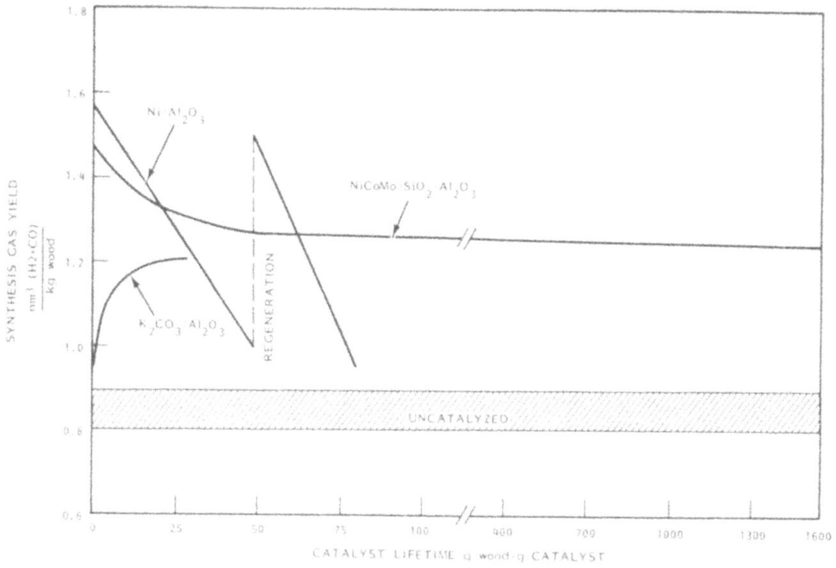

FIG. 3. Catalyst lifetime for production of methanol synthesis gas.

There are several possible mechanisms which may account for the reduction in carbon deposition on trimetallic catalysts. Nickel alloy catalysts are known to be less active but more stable with respect to carbon deposition than typical nickel catalysts in methanation and steam reforming reactions.[1,4,3] We do not have enough information to identify the mechanism by which this takes place; however, it is apparent from our studies that the trimetallic catalysts have active sites which either reduce the rate of carbon deposition or increase the rate of carbon removal compared to active sites on typical nickel catalysts.

The trimetallic catalysts can also be used to produce a hydrogen-rich gas and ammonia synthesis gas at 750 °C. Because of the similarity to the methanol synthesis gas case long lifetimes are anticipated for the hydrogen case. The use of air to produce ammonia synthesis gas does not appear to effect the catalyst; however, the oxygen is mainly consumed in the upper part of the reactor and very little (0·5 % in the gas) reaches the catalyst.

Alkali carbonates impregnated on the wood also catalyze gas phase reactions. A tar-free gas with a composition similar to that obtained with the trimetallic catalysts is produced with 17 wt % K_2CO_3 impregnated on the wood. A catalyst containing 25 wt % alkali carbonate (K or Na) on alumina also produced a tar-free gas at 750 °C, but the gas quality was somewhat poorer in terms of selectivity to H_2 and CO. Table IV compares

TABLE IV

ACTIVITY OF ALKALI CARBONATE CATALYSTS FOR GAS PHASE REACTIONS

Catalyst	*$NiCoMo/SiO_2$–Al_2O_3*	*17 wt % K_2CO_3 impregnated on wood*	*25 % Na_2CO_3 on Al_2O_3*	*None*	*Ni, Na_2CO_3/Al_2O_3*
Gas composition					
% H_2	51·7	52·4	46·7	42·5	50·6
% CO_2	18·8	21·8	24·2	27·7	21·7
% CH_4	5·3	3·2	7·9	8·8	6·7
% CO	24·1	21·4	19·7	18·3	20·2
% C_2^+	0·3	1·2	1·5	2·7	0·8
Carbon conversion to gas (%)	95	90	97	94	97
Carbon conversion to char (%)	5	10	3	3	3
Carbon conversion to liquid (%)	0	0	0	2	0
Synthesis gas yield (nm^3) (H_2 + CO)/kg feed	1·32	1·23	1·05	0·85	1·23

All runs at 750 °C and steam-to-wood rate 1/1 except uncatalyzed which was 1·4/1.

the activity of alkali carbonate catalysts with nickel catalysts and an uncatalyzed run at 750 °C.

There are two interesting points regarding the supported alkali catalysts:

(1) the activity *increased* slightly during the run (Fig. 3);
(2) there was no carbon on the catalyst at the end of the run.

Both of these indicate the catalyst may have a long or even unlimited lifetime with respect to deactivation. Addition of a small amount of nickel to a supported alkali catalyst improved the gas quality as shown in Table IV, but longer term tests are needed to see if the nickel will remain active.

4. CONCLUSIONS

By selecting the proper operating conditions and catalyzing the appropriate gas phase reactions in the gasifier it is possible to produce methanol synthesis gas, methane-rich gas, hydrogen-rich gas, and ammonia synthesis gas directly by steam gasification of biomass. Char is the only other product in all four cases as long as the catalyst remains active.

Catalyst deactivation, caused primarily by carbon deposition and/or sintering, is a problem in all four cases with the commercial nickel catalysts we tested. A series of trimetallic catalysts developed by PNL for the methanol synthesis gas case are unique among supported metal catalysts we tested for their ability to prevent accumulation of carbon and remain active for an extended period of time. These catalysts can also be used for ammonia synthesis gas and hydrogen production. A long-lived catalyst for production of a methane-rich gas has not been found.

Alkali carbonates also catalyze some gas phase reactions although they are not quite as effective as supported nickel catalysts. Supported alkali catalysts deserve further investigation because they are unique in several ways. When used for production of methanol synthesis gas at 750 °C there is no carbon on the catalyst and the catalyst activity increases as the run progresses.

REFERENCES

1. Araki, M. and Ponec, V. (1976). Methanation of carbon monoxide on nickel and nickel–copper alloys. *J. Catalysis* **44**, 439–48.
2. Reucroft, P. J. and Park, J. W. (1980). Reducibility and surface area of coprecipitated supported nickel methanation catalysts. *J. Materials for Energy Systems* **2**, 28–33.

3. Rostrup-Nielson, J. R. (1975). *Steam Reforming Catalysts*. Danish Technical Press, Copenhagen.
4. Stencil, J. M., Heinz, R. E. and Bradley, E. B. (1979). Structural changes in sulfur-resistant methanation catalyst by X-ray photoelectron spectroscopy. *J. Vacuum Science Tech.* **16**, 557–60.
5. Thomas, C. L. (1979). *Catalytic Processes and Proven Catalysts*. Academic Press, New York.
6. Venuto, P. B. and Habib, E. T. Jr (1979). *Fluid Catalytic Cracking with Zeolite Catalysts*. Marcel Dekker, New York.

48

Liquid Hydrocarbon Fuel Potential of Agricultural Materials

JAMES L. KUESTER, CARMO M. FERNANDEZ, TA-CHING WANG and GARY HEATH

Department of Chemical Engineering, Arizona State University, Tempe, Arizona, USA

1. INTRODUCTION

A project has been under development at Arizona State University (ASU) since 1975 with the goal of producing quality liquid hydrocarbon fuels from cellulosic and waste polymer materials. An indirect liquefaction approach is used, i.e. gasification to synthesis gas followed by liquefaction of the synthesis gas. The primary virtue of an indirect liquefaction approach for cellulosic type feedstocks is that the oxygen contained in the materials is easily separated. Thus the hydrocarbon liquid is free of oxygenated compounds and can therefore be tailored to match transportation fuel products currently derived from petroleum (e.g. diesel, jet fuels, high octane gasoline).

Thermochemical conversion processes (as opposed to biological or biochemical) are generally considered to be relatively insensitive to different types of cellulosic feedstocks. Thus it is assumed that a wide variety of individual or mixed materials could be fed into a process and thus achieve an enhanced economy of scale and minimize seasonal feedstock availability fluctuation problems.

One of the objectives of the ASU project has been to screen a wide variety of feedstocks to ascertain whether any performance fluctuations exist for alternative materials. This task has been coupled with the prime project objective of optimizing system performance, i.e. maximizing product yields subject to achieving acceptable product quality and system operational reliability. Thus alternative feedstock testing is 'piggybacked' on to an experimental run with the primary objective of improving system performance. The net result is that operating condition deviations will exist for alternative feedstock runs over a period of time.

This paper will review the current status and future plans for the assessment of liquid hydrocarbon fuel potential of various feedstocks. The general project strategy is depicted in Fig. 1. The procedure calls for development of gasification system performance data for various feedstocks coupled with generation of characterization data (composition, heating values, etc.) for the materials. Regression techniques are then utilized to isolate the effect of feedstock characterization data on gasification system performance. Promising results from the regression

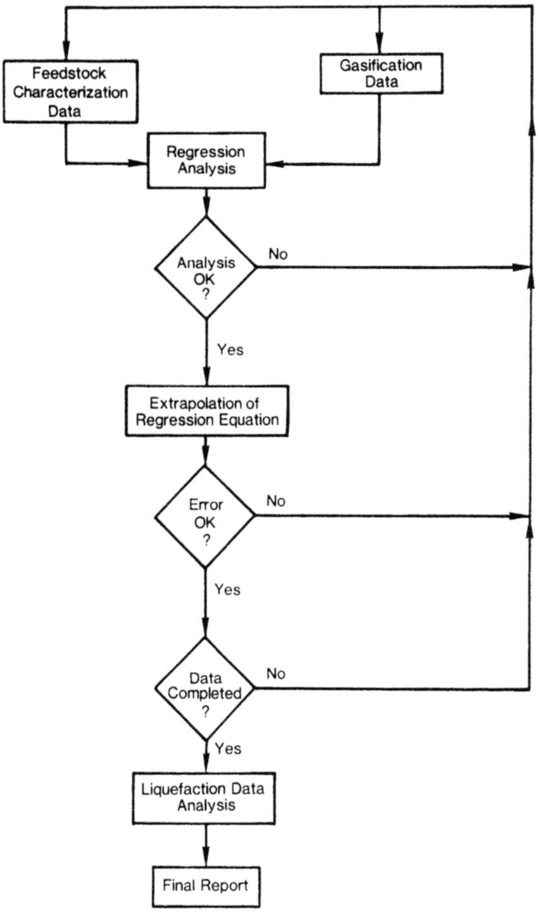

FIG. 1. Project strategy.

study would be used to predict conversion system performance from characterization data for new feedstock candidates. The liquefaction system performance is highly predictable from gasification component yield data and thus liquefaction system performance confirmation is performed for selected feedstocks only that have produced a satisfactory correlation of characterization data and gasification system performance.

2. CONVERSION SYSTEM DESCRIPTION

The existing system is laboratory scale with a capacity of approximately 10 lb/h of feedstock. Target product yields are in the 50–100 gal of diesel type fuel per ton of feedstock (dry, ash free). Continuous processing is employed. While the unit is of small scale, the processing steps and procedures are commercially realistic. Thus designed experiments can be sometimes tedious to accomplish but the resulting data should be highly reliable for scale-up purposes.

A flow schematic of the conversion system is depicted in Fig. 2. A photograph is given in Fig. 3. The gasification system is comprised of two fluidized beds with connecting circulating solid transfer loops. One fluidized bed is used as a feedstock pyrolyzer while the second bed (regenerator) operates in a combustion mode to heat the circulating solids media. Both inert solids (sand) and catalytic materials are under investigation. The fluidized bed approach allows for efficient heat transfer, continuous solids recirculation and elimination of a combustion zone in the pyrolyzer (and thus avoids gas clean-up steps). Cellulosic (biomass) feedstocks are continuously fed to the pyrolyzer and flashed to a synthesis gas consisting of paraffins, olefins, carbon monoxide, hydrogen and carbon dioxide. The gas passes though a cyclone-scrubber system to a compressor. From the compressor, the gas can be distributed to the pyrolyzer and/or liquefaction reactor. Additional gas candidates for fluidizing the pyrolyzer are steam and off gas from the downstream reactors. Studies to date indicate that the use of recycle pyrolyzer gas is not desirable for fluidizing the pyrolyzer due to the increased effective residence time with respect to the reactive gas components. The regenerator is fluidized by air and recycle gas from the pyrolyzer and/or downstream reactors. The off gas from the regenerator is passed through a cyclone-scrubber system before being vented.

The liquefaction system consists of a catalytic fluidized bed reactor to produce paraffinic fuel (diesel, jet) with an additional fixed bed catalytic

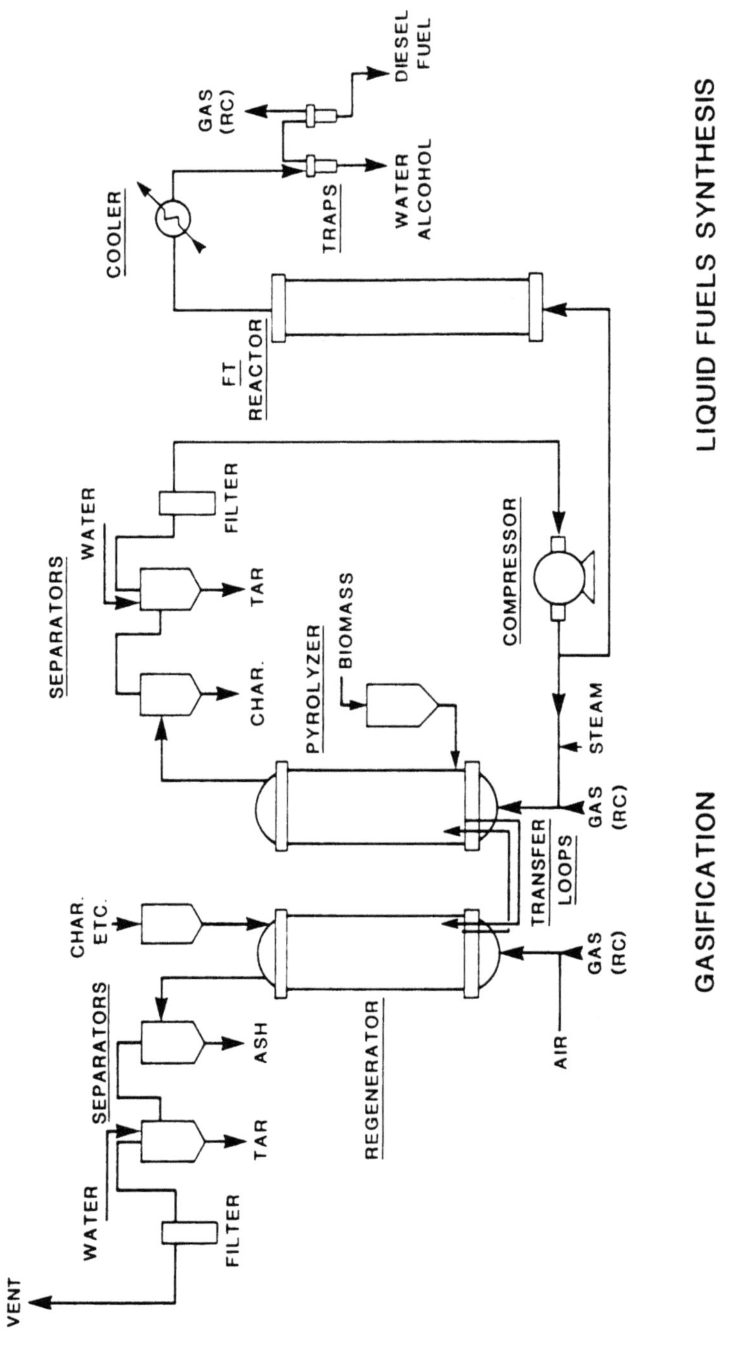

FIG. 2. Conversion system schematic.

reformer if a high octane gasoline product is desired. A fluidized bed is used to control the reactor temperature with the presence of the significant exothermic heat of reaction that is evolved. Also the fluidization allows for continuous regeneration of the catalyst if necessary. In the catalytic fluidized bed, the reactive components in the synthesis gas (olefins, carbon monoxide, hydrogen) are converted to a primary paraffinic hydrocarbon phase and a secondary alcohol–water phase. The off gas from this reactor

FIG. 3. Conversion system.

accumulates an appreciable amount of normal paraffins plus carbon dioxide. A slurry phase liquefaction reactor is also being investigated as an alternative to the fluidized bed liquefaction reactor. Potential advantages include enhanced temperature control, use of unsupported catalysts and operating condition flexibility, particularly with respect to residence time. In the reformer, structural complexity formation and cracking reactions take place to increase the octane rating of the liquid hydrocarbon compounds. The off gas is of high heating value due to the presence of C_1–C_4 normal paraffins.

Typical operating conditions for the reactors are as follows:

	Pyrolyzer	*Fluid bed liquefaction reactor*	*Reformer*
Temperature (°C)	600–800	250–300	490
Pressure (psig)	0–5	125	400
Residence time (s)	<2	18	11

Progress in the laboratory (systems development, factor studies) has been described in several publications.[1-3]

3. FEEDSTOCK CHARACTERIZATION

All materials can be characterized by elemental analysis, chemical compound analysis and various physical property measurements. For cellulosic type materials, the elemental analysis (C, H, O, N, S) is essentially constant for all materials.[4] With regard to chemical composition, the number of individual compounds present for a given bulk species is enormous. A generalized breakdown is depicted in Fig. 4. The carbohydrates account for the bulk of the material with subcategories including cellulose, hemicellulose, polysaccharides, etc. Phenolic type compounds (lignins, suberin, cutin, etc.) are also present in significant proportions. Other compound types (terpenes, acids and alcohols, proteins and inorganics) may appear in various amounts depending on the particular species. Each type, however, would generally be present at a composition of less than 5 wt % of the bulk plant material.[5]

A major laboratory effort to classify agricultural materials with regard to compound analysis and potential for energy crop production has been conducted at the USDA Northern Regional Research Center (see, for example, ref. 6). An extraction scheme developed by this laboratory is shown in Fig. 5. The various fractions can, of course, be subclassified almost without bound.

The physical properties of interest in assessing thermochemical conversion potential would include heating value, heat capacity, thermal conductivity, density, etc. These properties are all a function of composition but measurement is normally much less tedious than a composition analysis. Values have been tabulated for selected materials.[4]

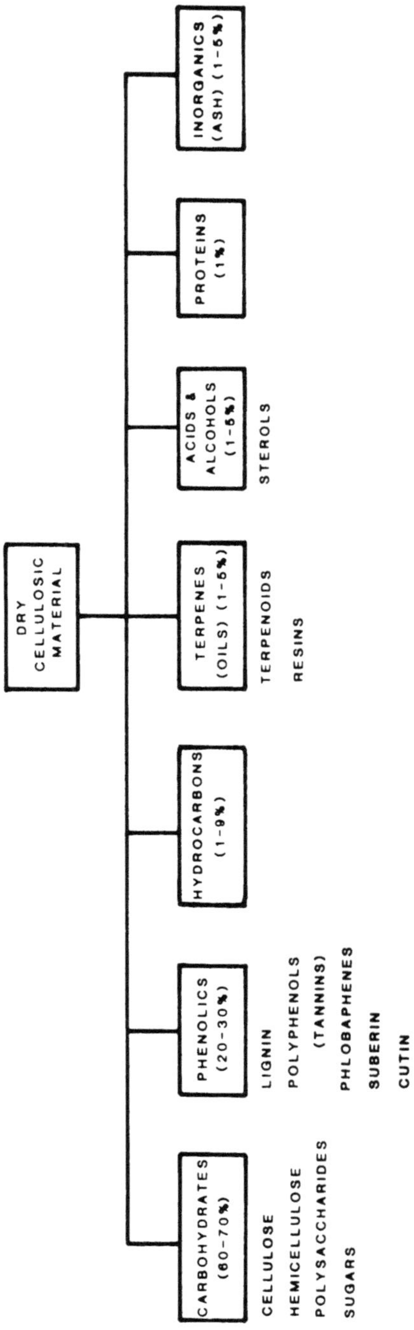

FIG. 4. Classes of compounds generally present in cellulosic materials.[5]

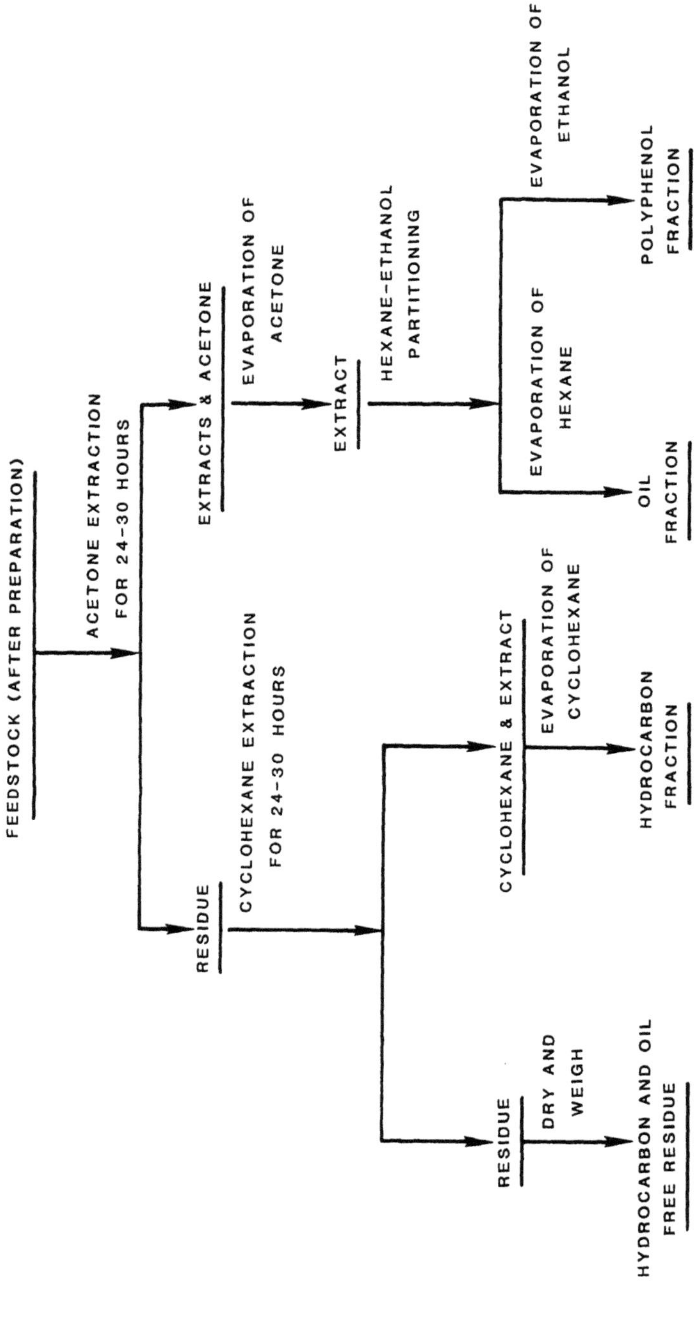

FIG. 5. Feedstock extraction procedure.[6]

TABLE I
CHARACTERIZATION TESTS

Analysis	*Reference*
Total ash	AOAC—22010 (1965)—ref. 7
Crude protein	AOAC—22011 (1965)—ref. 7
Crude fiber	AOAC—20038 (1965)—ref. 7
Heating value	ASTM—D-2382-76 (1981)—ref. 8
Lignin	ASTM—H-06-560 (1981)—ref. 8
Polyphenols	Buchanan *et al.* (1978)—ref. 9
Oils	Buchanan *et al.* (1978)—ref. 9
Hydrocarbons	Buchanan *et al.* (1978)—ref. 9
Suberin	Kolattukudy (1977), Zhuncheko and Cherkasova (1973)—refs 10, 11

The various characterization tests that have been utilized to date in the project are listed in Table I.

4. CURRENT STATUS

A list of feedstocks currently under investigation in the laboratory is given in Table II. The materials have been supplied by private industry, municipalities, other university laboratories and US and foreign government laboratories. The industrial wastes represent residues available after a primary product is removed (e.g. cotton gin trash after removal of cotton). The forest residues listed are all cut and burned in the field in the southwest United States by the US Forest Service to conserve water. The urban wastes represent preprocessed municipal refuse (Eco-Fuel II) and digested sewage sludge. The energy crop category represents various materials under investigation by the USDA and private industry with regard to cultivation for primary energy conversion purposes.

The approach to relating the characteristics of the various feedstocks to conversion system performance (see Fig. 1) has consisted of the following:

(1) broad screening of materials in the gasification system;
(2) broad screening of materials for selected characterization data;
(3) regression analysis on broad screening data;
(4) reduction of feedstock list for additional characterization and regression analysis work based on range of performance in the gasification system; and
(5) liquefaction runs on selected feedstocks.

TABLE II
FEEDSTOCKS

Industrial wastes	*Energy crops*
Sawdust	Russian thistle
Paper chips	Water hyacinth
Fir bark	Raw kelp
Hog fuel	Euphorbia lathyris
Cotton gin trash	Silver maple
Guayule bagasse	Smooth sumac
Guayule cork	Common milkweed
Candelilla bagasse	Red tartarian honeysuckle
Jojoba meal	Common elder
Lignin	Giant ragweed
Kelp residue	Field thistle
Almond hulls	Cherry eleagnus
Almond shells	Poke weed
Almond prunings	Tall boneset
Almond prunings (cured)	Rosin weed
Sugarcane bagasse	Compass plant
Wheat straw	Tall goldenrod
	Canada wild rye
Forest residues	Sassafras
Creosote bush	Coral berry
Sugar sumac	Wild bergamot
Arizona cypress	Swamp milkweed
Pringle manzanita	Grass leaved goldenrod
Wright silktassel	Sow thistle
Pointleaf manzanita	Tall bellflower
Shrub live oak	Pale Indian-plantain
Hairy mountain mahogany	Raw guayule
Utah juniper	Sweet sorghum
Piñon pine	
Greasewood	*Others*
Velvet mesquite	Corn starch
	Coal
Urban wastes	Polyethylene
Eco-Fuel II	Polypropylene
Sewage sludge	Peat
	Portuguese oak cork

The results for the gasification screening runs are shown in Table III. The responses listed are pyrolysis gas composition with a particular emphasis on the reactive components for synthesis gas usage (total olefins, H_2/CO ratio). The factors listed are heating value, ash content and extraction analysis with respect to protein, polyphenol, oil and hydrocarbon fractions (see Fig. 5) and gasification system operating conditions (fluidizing gas type and flows, temperature, residence time). The purpose of including the reactor operating conditions data was to isolate these factors from the characterization factors via inclusion in the regression analysis and/or 'blocking' the data. In general, the temperature variations and the fluidizing gas type would be considered important factors with regard to manipulating pyrolysis gas composition over the range reported. The gas flow rates (and related residence time) would be considered less significant for the ranges listed.

The response data in Table III indicate a significant range for both total olefins (1–38 mol %) and H_2/CO mole ratio (0·2–8·8) for the cellulosic materials. The synthetic polymers yielded very high olefins and H_2/CO ratios as expected. The feedstock characterization data also varied significantly with the following range exhibited for the cellulosic materials:

heating value (Btu/lb)	7400–12 700
ash (wt %)	0·1–35·9
protein (wt %)	0·1–25·3
polyphenol (wt %)	0·1–20·2
oil (wt %)	0·03–9·20
hydrocarbons (wt %)	0–10·4

Various regression fits were then performed on the data with typical results for a quadratic model listed in Table IV. Here, various combinations of three factors were tested with regard to significance on olefin and H_2/CO manipulation. In general, factor combinations that resulted in correlation coefficients of $>0{\cdot}9$ would be considered strong fits. Note that the combination of factors involving ash, heating value and polyphenol + oil + hydrocarbon gave the best fit for both olefin content and H_2/CO ratio for the data in Table III. Further analysis of this data is continuing.

A reduced list of feedstocks were selected based on the data in Table III for further characterization work and possible uniform operating condition conversion runs. The selection consisted of the following materials with the choice based on covering the range of olefin and H_2/CO response values: guayule cork, Portuguese oak cork, euphorbia lathyris, Russian thistle, shrub live oak and almond prunings. Thus, for example, the cork materials

TABLE III

SUMMARY OF FEEDSTOCK ANALYSES, OPERATING CONDITIONS AND GASIFICATION RESULTS

No.	Feedstock	Feedstock analysis[a]			Pyrolysis reactor performance								
		Heating value (Btu/lb)	Ash (wt%)	Extraction[b] fractions (wt%)	Operating conditions[c]			Composition (mole%)[d] summary					H_2/CO ratio
					Fluidizing[e] gas (lb/h)	Temp. (°F)	Res. (s)	H_2	CO	C_2H_4	Total olefins	Paraf-fins	
	Industrial wastes												
1	Sawdust	8 579	7·0	0·53, 3·17, 0·71, 0·58	23 RPG	1200	4·5	15·13	55·57	2·63	4·13	18·83	0·27
2	Papers chips	8 664	0·6	0.10, 0·38, 0·26, 0·08	25 RPG	1280	5·5	14·77	58·86	3·70	5·76	17·20	0·25
3	Fir bark	10 076	5·1	1.23, 17·2, 4·98, 0·18	22 RPG	1100	7·5	18·98	48·21	7·34	13·43	15·44	0·39
4	Fir bark				60S + 5 RPG	1250	3·2	29·24	44·41	7·30	11·61	11·24	0·65
5	Hog fuel	8 991	6·2	2·06, 6·14, 1·35, 0·22	30S + 5 RPG	1200	4·8	16·36	36·25	4·73	13·25	25·25	0·45
6	Guayule bagasse	7 838	3·9	3·7, 5·9, 2·0, 4·5	24 RPG	1450	4·2	25·33	37·00	7·75	13·60	19·46	0·68
7	Guayule bagasse				60S + 5 RPG	1170	2·5	29·93	33·15	2·71	13·70	14·85	0·90
8	Guayule cork	11 376	2·1	6·2, 6·6, 6·0, 2·3	19 RPG	1210	4·7	17·55	20·86	9·92	21·10	32·67	0·84
9	Guayule cork				19 RPG	1210	4·6	14·32	22·23	11·03	20·60	34·58	0·64
10	Guayule cork				22 RPG	1300	5·0	20·57	22·14	14·80	22·30	31·06	0·92
11	Guayule cork				9S + 5 RPG	1320	4·9	35·24	14·40	16·67	38·73	14·65	2·25
12	Candelilla bagasse	8 570	10·7	4·54, 7·88, 2·48, 0·56	45S + 5 RPG	1420	3·4	33·69	36·17	7·86	9·82	10·13	0·93
13	Jojoba meal	9 809	2·5	25.3, 13·0, 9·2, 0·1	23 RPG	1320	4·4	11·96	37·56	9·15	13·39	26·69	0·32
14	Jojoba meal				45S + 5 RPG	1240	3·0	39·12	18·09	9·49	13·21	17·65	2·16
15	Almond hulls	7 727	5·9	6·92, 3·78, 1·04, 0·56	18 RPG	1250	6·9	29·08	35·44	4·01	6·27	16·29	0·79
16	Almond shells	8 311	8·8	5·32, 2·27, 0·75, 0·13	18 RPG	1280	4·6	26·03	38·06	3·09	4·76	19·01	0·68
17	Almond prunings	8 235	2·8	2·17, 7·26, 0·59, 0·27	23 RPG	1340	5·3	22·96	44·87	6·11	13·97	16·37	0·51
18	Almond prunings				45S + 5 RPG	1280	3·5	31·21	42·39	5·74	10·67	7·81	0·74
19	Almond prunings (cured)	8 550	25·4	9·02, 6·23, 0·6, 0·38	22 RPG	1520	3·7	39·15	33·75	4·41	5·49	15·60	1·16
20	Sugarcane bagasse	9 060	8·3	7·08, 4·45, 0·82, 0·8	30S + 5 RPG	1400	4·8	28·60	36·74	7·29	10·38	15·31	0·78
21	Wheat straw	8 726	8·2	1·14, 0·28, 0·57	22 RPG	1600	4·7	30·19	48·27	2·57	2·57	14·7	0·63
	Forest residues												
22	Creosote bush	8 822	3·8	6·4, 5·7, 0·5, 0·08	22 RPG	1380	4·5	25·99	39·43	6·48	10·24	16·61	0·65
23	Sugar sumac		5·3	3·7, 16·3, 2·8, nil	22 RPG	1200	4·1	28·89	31·88	5·75	10·42	18·25	0·90
24	Arizona cypress	7 955	4·5	1·6, 9·5, 2·0, nil	22 RPG	1350	4·6	26·64	38·40	6·40	10·34	17·52	0·69
25	Pringle manzanita	8 775	2·2	2·1, 11·4, 0·6, 0·27	22 RPG	1400	4·6	24·99	40·68	6·29	11·11	16·42	0·61
26	Wright silktassel	9 265	3·2	1·8, 4·6, 2·6, 0·38	22 RPG	1320	4·7	25·64	38·69	7·30	11·89	18·32	0·66
27	Pointleaf manzanita	8 444	1·7	2·0, 10·4, 0·9, 0·33	22 PRG	1380	5·2	24·46	35·50	5·69	12·85	16·28	0·68
28	Shrub live oak	8 524	3·1	4·6, 7·4, 1·0, 0·09	22 RPG	1400	5·2	27·99	41·28	5·56	8·13	18·00	0·67

29	Hairy mountain mahogany	9076	4·2	6·0, 4·75, 2·16, 0·21	22 RPG	1380	4·2	27·61	37·84	10·26	12·83	16·91	0·72
30	Utah juniper	8589	4·2	3·4, 6·7, 1·3, 0·19	22 RPG	1300	4·5	28·83	39·54	6·56	7·33	17·86	0·73
31	Pinion pine	9060	3·3	2·6, 5·0, 3·0, 0·26	22 RPG	1420	5·3	25·82	41·78	6·33	10·00	17·56	0·61
32	Velvet Mesquite	8507	4·4	6·5, 4·8, 0·96, 0·08	21 RPG	1690	3·5	33·01	44·35	2·61	5·01	18·46	0·74
	Urban wastes												
33	Eco-Fuel II (Cambridge)	9572	14·7	2·59, 5·36, 2·16, 0·99	22 RPG	1250	4·2	20·85	48·51	5·16	8·20	19·35	0·42
34	Eco-Fuel II (Cambridge)				12S + 5 RPG	1180	4·8	23·62	45·24	7·90	13·47	13·83	0·52
	Energy crops												
35	Russian thistle	7936	15·4	9·2, 3·1, 0·8, 0·2	20 RPG	1280	4·5	26·37	36·08	3·21	4·98	17·96	0·73
36	Water hyacinth	7740	35·9	7·8, 6·1, 1·0, 0·4	19 RPG	1270	5·6	23·00	42·43	3·52	4·56	16·02	0·54
37	Raw kelp	6448	31·7	15·32, 0·82, 0·34, 0·33	22 RPG	1500	3·7	42·44	27·36	1·43	1·43	8·88	1·55
38	Euphorbia lathyris	8725	3·1	3·0, 2·9, 1·4, 0·2	31 RPG	1370	6·4	32·97	26·27	3·84	14·08	18·66	1·26
39	Euphorbia lathyris				45S + 5 RPG	1370	3·0	52·61	5·96	1·32	11·73	5·75	8·82
40	Silver maple	8838	4·7	16·3, 19·8, 2·4, 0·39	30S + 5 RPG	1080	5·0	15·89	44·64	4·39	13·53	16·99	0·36
41	Smooth sumac	8817	7·7	7·1, 20·2, 5·9, 0·21	30S + 5 RPG	1330	4·9	34·98	25·36	9·83	14·55	13·67	1·38
42	Red tartarian honeysuckle	8949	5·2	10·2, 15·8, 3·4, 1·77	45S + 5 RPG	1400	3·4	44·08	28·41	8·47	9·86	8·33	1·55
43	Giant ragweed	8461	11·4	11·4, 4·4, 8·3, 0·6	30S + 5 RPG	1420	4·1	37·14	28·49	7·79	9·95	13·29	1·30
44	Pokeweed	8396	11·5	15·5, 5·9, 3·4, 0·17	30S + 5 RPG	1260	5·0	42·29	18·20	7·46	13·42	11·24	2·32
45	Tall boneset	8913	6·5	8·6, 10·8, 5·9, 0·56	30S + 5 RPG	1140	4·7	20·32	35·15	8·22	15·29	19·44	0·58
46	Rosin weed	8264	9·4	6·2, 7·0, 2·8, 0·79	30S + 5 RPG	940	5·2	9·91	52·32	1·07	4·98	21·32	0·19
47	Tall goldenrod	9304	7·5	9·8, 5·7, 2·8, 0·92	35S + 5 RPG	1260	4·2	37·02	29·29	2·78	8·77	17·08	1·26
48	Sassafras	7431	3·1	8·9, 14·4, 5·7, 0·23	45S + 5 RPG	1190	4·0	31·87	36·00	6·84	13·98	9·17	0·88
49	Coral berry	8676	4·8	5·9, 11·1, 2·3, 0·81	30S + 5 RPG	1200	5·3	28·37	31·33	6·58	13·88	14·75	0·90
50	Wild bergamot	8674	7·2	7·22, 6·9, 2·4, 1·24	30S + 5 RPG	1220	5·2	36·85	24·14	6·01	10·57	17·11	1·53
51	Raw guayule	10820	5·2	4·3, 8·9, 2·3, 10·4	20 RPG	1310	4·6	17·28	34·98	5·54	10·61	28·58	0·49
52	Raw guayule				45S + 5 RPG	1230	3·0	42·69	14·26	7·39	11·14	17·81	2·99
	Others												
53	Corn starch	7532	0·1	0·41, 0·16, 0·03, 0·03	30S + 5 RPG	970	4·8	7·19	59·41	1·82	9·49	6·34	0·12
54	Polyethylene	19986	0·04	—, 0·03, 0·08, 0·42	19 RPG	1240	4·5	14·19	0·96	19·29	33·28	50·36	14·78
55	Polypropylene	19563	0·03	—, 0·14, nil, nil	19 RPG	1240	4·5	13·57	0·69	13·34	37·13	48·62	19·67
56	Peat	9805	7·6	8·53, 2·0, 1·65, 0·60	9S + 5 RPG	1280	5·1	45·05	18·48	4·15	7·50	12·69	2·43
57	Portuguese oak cork	12676	0·3	7·7, 3·2, 2·0, 0·1	30S + 5 RPG	1180	4·9	14·18	33·07	11·74	26·80	22·18	0·92

[a] Moisture free basis.
[b] Ash free basis; listed in following order: protein, polyphenol, oil, hydrocarbon fractions.
[c] Fluidizing media either dolomite (No. 14, 38, 39, 52) or sand (rest); operating pressure 0·7–1·4 psig.
[d] Water, N_2 free basis; balance of analysis = CO_2.
[e] S = steam, RPG = recycled pyrolysis gas.

TABLE IV

CORRELATION COEFFICIENT, R^2, AND DATA RELATIONSHIP OF REGRESSION ANALYSIS

$$Y = A_0 + A_1X_1 + A_2X_2 + A_3X_3 + A_4X_1^2 + A_5X_2^2 + A_6X_3^2 + A_7X_1X_2 + A_8X_1X_3 + A_9X_2X_3$$

Independent variables, X			*Responses, Y*	
X_1	X_2	X_3	*Olefin content*	H_2/CO
A	HV	P	0·8607	0·9622
A	HV	O	0·9085	0·9533
A	HV	POH	0·9891	0·9843
A	HV	T	0·9529	0·9053
A	P	O	0·8782	0·9578
A	P	POH	0·8957	0·9780
A	P	T	0·8396	0·9392
A	O	POH	0·8574	0·9708
A	O	T	0·9208	0·9762
A	POH	T	0·8583	0·9692
HV	P	O	0·8629	0·9327
HV	P	POH	0·9528	0·9423
HV	P	T	0·9115	0·9071
HV	O	POH	0·9333	0·9434
HV	O	T	0·7892	0·9272
HV	POH	T	0·9256	0·9454
P	O	POH	0·9325	0·5816
P	O	T	0·9032	0·6736
P	POH	T	0·8738	0·5811
O	POH	T	0·9315	0·6984
P	F	HV	0·8184	0·5802
P	F	A	0·8546	0·5936
P	F	POH	0·8276	0·6994
P	F	T	0·9478	0·7231
P	HV	A	0·7817	0·4677
P	HV	POH	0·8718	0·7441
P	HV	T	0·9145	0·5873
P	A	POH	0·8201	0·6447
P	A	T	0·8251	0·4010
P	POH	T	0·7509	0·6077
F	POH	T	0·7859	0·5095
F	A	T	0·8454	0·4835
F	A	POH	0·8049	0·6166
F	HV	POH	0·8582	0·6797
F	HV	T	0·9721	0·5088
F	HV	A	0·9498	0·4989

A = Ash; F = Fiber; HV = Heating Value; P = Protein; POH = Polyphenol + Oil + Hydrocarbon; T = Temperature, °F.

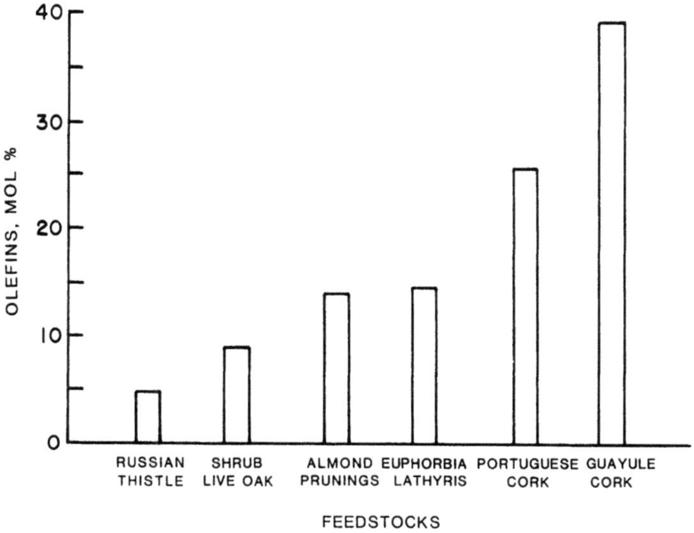

FIG. 6. Olefin response for selected feedstocks.

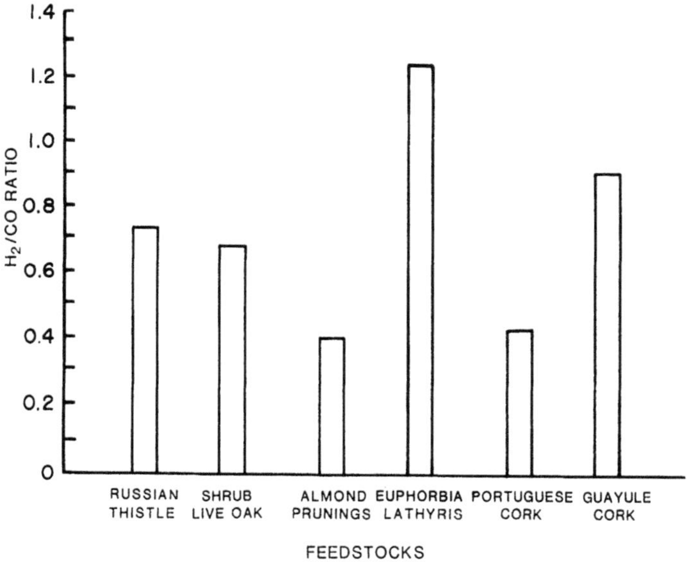

FIG. 7. H_2/CO response for selected feedstocks.

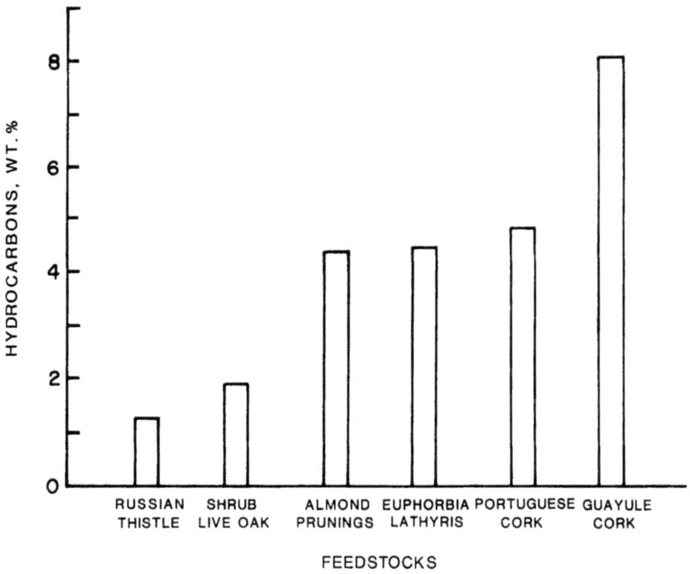

FIG. 8. Hydrocarbon content for selected feedstocks.

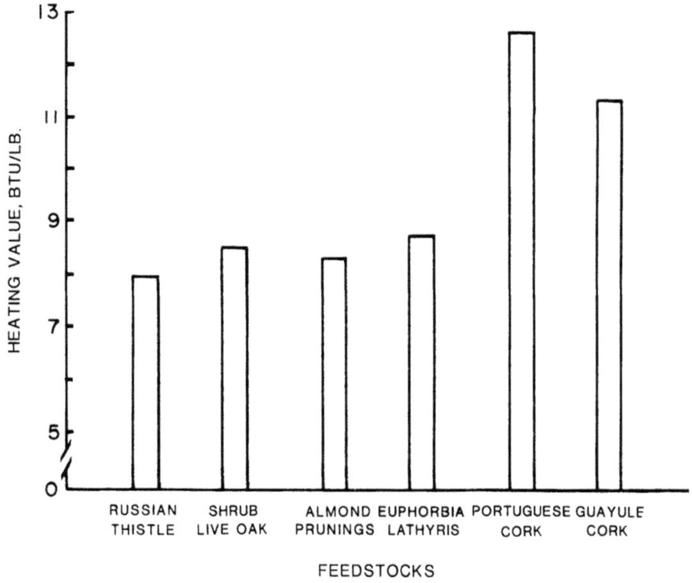

FIG. 9. Heating values for selected feedstocks.

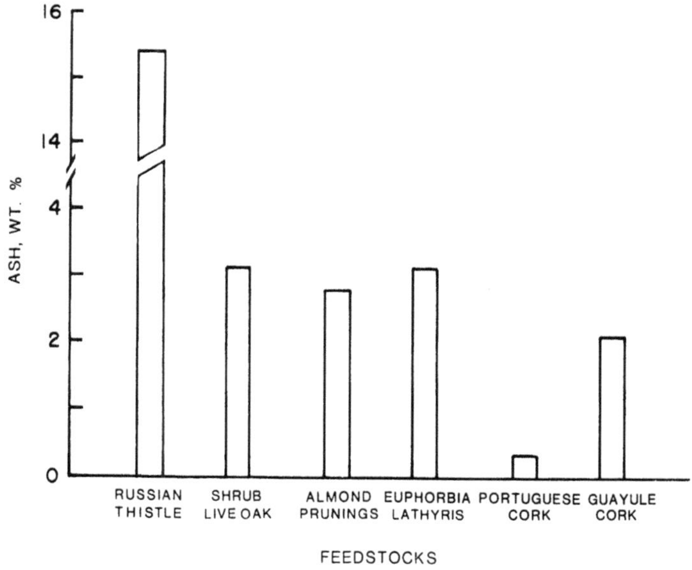

FIG. 10. Ash content for selected feedstocks.

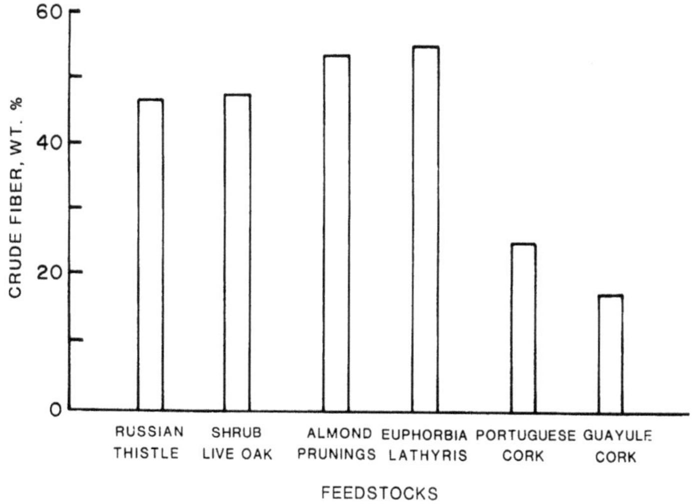

FIG. 11. Crude fiber content for selected feedstocks.

TABLE V

PROPERTIES OF FISCHER–TROPSCH PRODUCT AND COMMERCIAL FUEL OILS

	Commercial fuel oils			*Fischer–Tropsch product*	
	No. 2 diesel	*Kerosene*	*JP-4*	*Almond pruning feedstock*	*Guayule bagasse*
Specific gravity	0·8360	0·8108	0·7586	0·7902	0·7950
Gravity (API °)	37·8	43	55	47·6	46·5
Boiling point range (°F)					
evaporated at 10%	369	336	147	235	238
evaporated at 50%	458	410	302	352	414
evaporated at 90%	563	479	438	471	535
Calculated cetane index	45·9	47·8	48·3	45·3	55·7
Heating value (Btu/lb)	19383	21676	22440	19354	21043

yielded very high olefin values while euphorbia lathyris exhibited high H_2/CO ratios. Additional characterization work to ascertain the cause of this exceptional behavior included crude fiber, lignin, suberin and hydrocarbon (via heptane extraction) analysis. Response results for the selected feedstocks are grouped in Figs 6 and 7 and feedstock characterization data are shown in Figs 8–11. The visual comparison indicates that hydrocarbon content and heating value relate to olefin yields but that crude fiber and ash content are of lesser significance . None of the four factors shown appears to have a significant correlation with the H_2/CO ratio. Further characterization work is continuing on the selected samples.

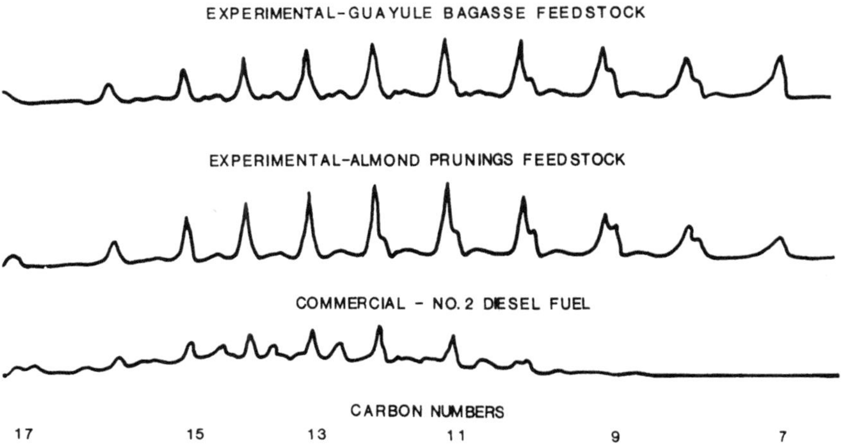

FIG. 12. Liquid fuels reactor hydrocarbon phase composition—integrated runs (gas chromatograms).

The product quality to be expected from the various feedstocks is indicated by the gas chromatograms in Fig. 12 and property data in Table V. Here, guayule bagasse (from wild stands in northern Mexico) and almond prunings (from cultivated orchards in California) were processed through the continuous indirect liquefaction system stopping at the Fischer–Tropsch reaction step. As shown, product quality is similar for both feedstocks (although perhaps botanically dissimilar) and analogous to No. 2 diesel fuel except for the low molecular weight end. External consultation has indicated that product distillation should not be required, i.e. the raw product is a candidate for transportation and/or process fuel use for existing engines and machinery.

5. FUTURE PLANS

The continuing research on the project includes the following items:

(1) Additional feedstock screening. Candidates are nominated both at the initiative of external laboratories and project personnel.
(2) Revisions in the selected feedstock list with regard to both feedstocks and characterization tests. The feedstock choice is based on response values from the conversion system. Characterization test survival is dependent on degree of correlation with the response data. New methods are introduced as new knowledge is gained.
(3) Liquefaction data on selected samples.

Clearly, in view of the finite number of cellulosic species that are available, the above approach is without bound. The intent is to reach a point where a reliable *a priori* prediction of liquid fuel yield potential for a given cellulosic species will be possible via determination of selected feedstock characteristics.

ACKNOWLEDGEMENTS

The research reported in this paper is supported by the US Department of Energy (Contract No. DE-AC02-76CS40202) and US Department of Agriculture (Grants No. 59-2043-0-2-094-0 and 59-2043-1-6-062-0). The cooperation of Dr Marvin Bagby (USDA, Northern Regional Research Center, Peoria, Illinois) and Dr Ed Davis (US Forest Service, Rocky Mountain Station) is also acknowledged.

REFERENCES

1. Kuester, J. L. (1980). Liquid hydrocarbon fuels from biomass. Chapter 8 in *Biomass as a Nonfossil Fuel Source*, D. L. Klass (Ed.). ACS Symposium Series 144, American Chemical Society, Washington, D.C.
2. Kuester, J. L. (1980). Fluidized bed pyrolysis to gases containing olefins. *Proceedings of the Specialists' Workshop on Fast Pyrolysis of Biomass*. SERI/CP-622-1096, pp. 253–70.
3. Kuester, J. L. (1980). Conversion of cellulosic wastes to liquid fuels. Chapter 15 in *Energy from Waste*, Vol. I, T. C. Frankiewicz (Ed.). Ann Arbor Science Publishers.
4. Graboski, M. and Bain, R. (1981). Properties of biomass relevant to gasification. Chapter 3 in *Biomass Gasification—Principles and Technology*. Noyes Data Corp., Park Ridge, N.J.

5. Robinson, T. (1980). *The Organic Constituents of Higher Plants.* Cordus Press, North Amherst, Massachusetts.
6. Bagby, M. O., Buchanan, R. A. and Otey, F. H. (1980). Multi-use crops and botanochemical production. Chapter 6 in *Biomass as a Nonfossil Fuel Source*, D. L. Klass (Ed.). ACS Symposium Series 144, American Chemical Society, Washington, D.C.
7. *Official Methods of Analysis, 10th edn* (1965). Association of Official Agricultural Chemists, Washington, D.C.
8. *Annual Book of ASTM Standards* (1981). American Society for Testing and Materials, Philadelphia, Pennsylvania.
9. Buchanan, R. A., Cull, I. M., Otey, F. H. and Russell, C. R. (1978). Hydrocarbon- and rubber-producing crops. *Econ. Bot.* **32**(2), 131–53.
10. Kolattukudy, P. E. (1979). *Recent Advances in Phytochemistry*, F. A. Loewus and E. Reuncles (Eds). Plenum Press, New York.
11. Zhunchenko, A. G. and Cherkasova, A. I. (1973). *Chemical Composition of Birch Bark*. Bibliography Series 181 (Suppl. 12), J. Werner and V. Pollock (Eds). Institute of Paper Chemistry, Appleton, Wisconsin.

49

Steam Gasification of Alpha Cellulose in a Fluid Bed Reactor

WALTER P. WALAWENDER, DEBORAH A. HOVELAND AND L. T. FAN
Department of Chemical Engineering, Kansas State University, Manhattan, Kansas, USA

1. INTRODUCTION

Over the past decade, a number of studies have been conducted on the fluidized bed gasification of biomass materials. Several factors influence gasifier performance, the most important being temperature. Several investigators have noted that different types of feed materials produce different results when gasified under apparently similar conditions.[1-5] Some of these investigators have postulated that the differences observed are due to differences in the cellulose content of the feed material.

Antal *et al.*[1] conducted TGA studies on various biomass materials, including cellulose, paper, hardwood, softwood and manure. Their results suggest that the higher the cellulose content of the material, the greater the extent of devolatilization. Walawender *et al.*[3] gasified various materials in a fluid bed reactor, including cane, manure and sewage sludge. They found an apparent qualitative correlation between the cellulose content of the material and the level of gas yield. This correlation indicates that the higher the cellulose content, the higher the gas yield at a given operating temperature. Beck and Wang[5] gasified sawdust and manure in a fluid bed reactor under partial oxidation conditions. They found that sawdust produced a higher yield of gas than manure under similar operating conditions and attributed the higher yield to the higher cellulose content of the sawdust. Raman *et al.*[4] reported a similar finding with different size fractions of manure. They suggested that the differences in gas yield observed were due to differences in the material make-up of the size fractions investigated.

Different types of carbonaceous residues contain different amounts of cellulose. If cellulose content is a significant factor in the gasification of biomass materials, then it is apparent that a study of the gasification characteristics of pure cellulose would aid in understanding and interpreting all biomass gasification processes.

A review of the literature shows that no work has been conducted on the gasification of cellulose in a fluid bed reactor. The closest related work is that of Antal and co-workers[1,6] who studied secondary reactions of cellulose generated volatiles in a gas phase reactor. Their experiments involved transient conditions with the resulting gas analyses representing the average over the course of each experiment. In contrast, this study examined cellulose gasification under continuous, steady state conditions with relatively constant gas residence times. The objective of the work was to generate datum lines for the gas yield, composition, heating value, energy recovery and carbon conversion under well defined conditions to serve as a bench mark for gauging biomass gasification processes.

2. EXPERIMENTAL FACILITIES AND PROCEDURE

2.1. Facilities

The bench-scale reactor used in this work is shown schematically in Fig. 1. The reactor was constructed from a 5·08 cm (2 in.) ID by 50 cm length of schedule 40 Inconel 600 pipe, fitted on top with a 10·16 cm (4 in.) ID by 15 cm length of pipe of the same material. The upper 15 cm of the reactor served as a disengaging zone. The lower 23 cm, which served as a gas distributor and preheater, was packed with aluminium oxide pellets (0·5 cm in diameter). The fluid bed section was separated from the packed bed section by means of a 0·025 mm (60 mesh) 304 stainless steel screen. The bed material was a mixture of 25% by weight limestone and 75% by weight silica sand. The bed had a static height of 8·4 cm and an expanded height of about 11 cm.

The reactor was heated by means of four pairs of semi-cylindrical electrical resistance heaters, two of which were controlled by thermostats. Steam, the sole fluidizing gas, was generated in a series of two electrical resistance furnaces and was supplied to the gas preheater section at approximately 500 K. The feed was introduced into the reactor by gravity flow through a vertical feedpipe which discharged 6·6 cm above the static bed. A Vibra Screw (Model SCR-20) screw feeder with a solid core flight screw was used to supply the feed at a uniform volumetric rate. A known volumetric flow of nitrogen was supplied to aid flow through the feedpipe and to prevent gas backflow through the feedpipe and subsequent

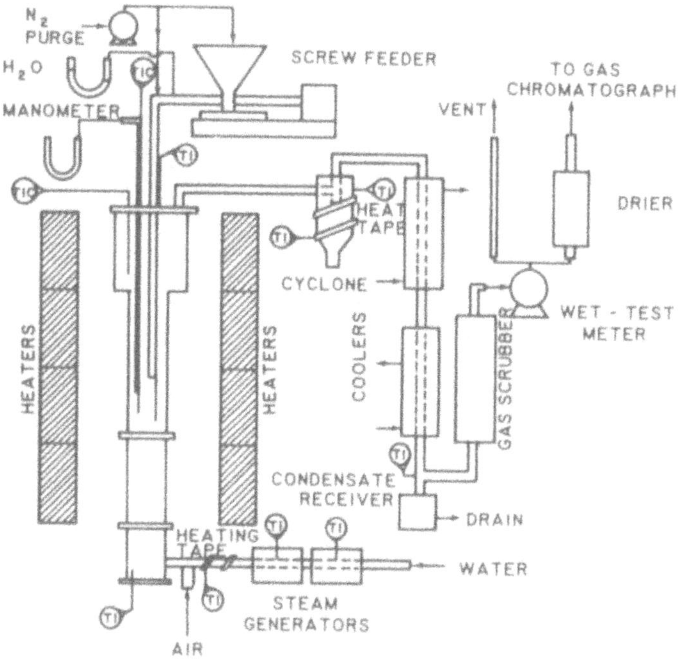

FIG. 1. Bench-scale reactor.

condensation of vapor in the feedpipe and feeder. It was also used as a tracer gas.

The off gas from the reactor was sent to a cyclone where entrained char was separated from the gas. The temperature in the cyclone was maintained above 600 K to prevent condensation of tar. The gaseous stream leaving the cyclone was cooled to approximately 300 K with two water-cooled heat exchangers in series, and the condensed steam was collected in a receiver. The resultant off gas carried with it a fine mist of condensables, which were removed by passing the gas through a packed bed of glass wool. The off gas could then vent to the atmosphere, or a portion could go through a drying column packed with Drierite ($CaSO_4$) and then to an on-line process gas chromatograph for analysis.

The temperature in the reactor was monitored by using a sliding (91 cm long) thermocouple, which measured temperatures from the bottom of the fluidized bed to the top of the disengaging zone. The sliding thermocouple was used to insure a uniform axial temperature profile throughout the reactor and freeboard. Temperatures were also monitored at other points as shown in Fig. 1. Pressure probes were used to monitor the bed and freeboard pressures.

2.2. Feedstock

The feed material used in this study was alpha (α) cellulose (rated at 99·5 % pure) obtained from the Sigma Chemical Company. The as-received α-cellulose was a fine powder which could not be fed consistently into the reactor through the screw feeder. To facilitate feeding, the α-cellulose was blended with tap water to form a paste which was pressed through a 2 mm screen to form beads. The resultant beads were dried at 323 K for 48 h and then sieved. The size fraction used was 1–2 mm. The cellulose beads contained 4·9 % moisture and 1·1 % ash. Pure α-cellulose has an empirical formula of $C_6H_{10}O_5$ and a heating value on a dry ash-free basis of 17·492 MJ/kg daf.

2.3. Procedure

For start-up of each experimental run, the system was brought to the operating temperature and the steam rate was adjusted to the desired value. The flow rate of steam required to maintain the selected superficial velocity was measured both by collecting condensate downstream of the heat exchangers and by metering the water flow into the steam generators. Nitrogen gas, used as both purge and tracer, was introduced at a rate that was about 40 % of the flow rate of the dry produced gas. The total start-up time was about 2 h from a cold start. Feeding was then started, and any change in the temperature of the bed or freeboard section due to feeding of the cellulose was corrected automatically by the thermostats. A gas sample was taken 5 min after feeding began and every 11 min thereafter. Condensate and nitrogen rates were measured over 10 min time intervals as the gas samples were being taken. A typical run lasted about 90 min with the last 30–45 min yielding steady GC readings. The feed rate was evaluated after every run by disconnecting the lower section of the feedpipe and weighing the effluent collected for three 3 min time intervals at a given setting on the feeder.

It was not possible to measure the total char produced in an experiment because of the hold-up of char in the bed. Nor was it possible to measure the total tar produced because of hold-up on the inside of the water-cooled heat exchangers and other parts of the experimental system.

2.4. Chemical Analyses

Analyses of the dry off gas were conducted with the aid of an Applied

TABLE I
REACTOR OPERATING PARAMETERS

Reactor temperature range	865 K–1060 K
Fluidizing gas	Steam
Superficial velocity	0·27 m/s–0·33 m/s
Feed rate	1·8 g/min–2·3 g/min
Cellulose particle size	1 mm–2 mm
Bed particle size:	
Sand	0·595 mm–0·297 mm (−30 + 50 mesh)
Limestone	2·82 mm–0·297 mm (−7 + 50 mesh)

Automation (Optichrom 2100) on-line process gas chromatograph. The gas components of interest included H_2, CO, CO_2, CH_4, C_2H_4, C_2H_6, C_3H_6, C_3H_8 and N_2. Moisture and ash analyses were conducted on the feed according to standard ASTM procedures.

2.5. Operating Conditions

The operating conditions for all the experimental runs are summarized in Table I. By adjusting the water rate according to the operating temperature, it was possible to maintain nearly constant gas residence times, which ranged from 1·26 to 1·54 s for all runs. The feed rate varied from 1·8 to 2·3 g/min although the feeder setting was always the same. It was believed that the feed rate variations were due to differences in the size and bulk density of the cellulose beads. The principal experimental variable was the reactor bed temperature. The freeboard section of the reactor was operated at the same temperature as the bed for each run. The superficial velocity of the fluidizing gas corresponded to about twice the minimum fluidization velocity and 7% of the terminal velocity of the feed. It also corresponded to about three times the minimum fluidization velocity of the bed material.

The conditions of the experiment were such that feed particle heat-up, in the feedpipe and the vapor space between the feedpipe and the bed surface, by both radiation and convection was not sufficient to cause devolatilization of particles prior to their entering the bed. Once in the bed heating rates and rate of particle movement in the bed were such that volatiles were produced at or near the bed surface. Further, the operating conditions were such that all of the emulsion gas was in upflow. The volatiles residence time in the freeboard was determined from the combined flow of volatiles and inerts and not the steam alone (although it made up the bulk of the flow).

3. RESULTS AND DISCUSSION

3.1. Calculations and Analyses

All the calculations were based on the average of the gas composition data obtained during the last 30–45 min of each run. From the average GC analysis and the nitrogen tracer rate for each run, the volumetric flow rate of the dry produced gas (nitrogen-free) was calculated using a nitrogen balance. The dry produced gas analysis (nitrogen-free basis) was used to evaluate the higher heating value of the dry produced gas using the standard heat of combustion for each component. The volumetric gas yield per unit mass of feed on a dry ash-free (daf) basis was calculated from the volume flow rate of the dry produced gas (at 288 K and 101·3 kPa) and the mass flow rate of the dry ash-free feed. The energy recovery was calculated as the product of the gas yield and gas heating value divided by the heat of combustion of the feed, times one hundred. It represents the percentage of the energy content of the feed converted to combustible gas. The carbon conversion was calculated as the atoms of carbon in the gas produced from a unit mass of daf feed divided by the atoms of carbon in a unit mass of daf feed. It represents the percentage of carbon in the feed that is converted to gaseous form. Because of the hold-up of char and tar in the experimental system, it was not possible to conduct an overall material balance on the bench-scale reactor. Consequently, only the yield and properties of the dry produced gas were considered in this work.

3.2. Model Development

The gasification of cellulose is in reality an extremely complex process involving numerous physical and chemical steps. Studies by Antal *et al.*[1] suggest that biomass gasification can be described by the following conceptual model: (1) pyrolysis or devolatilization from 574 K to 773 K, which produces volatile matter and char; (2) secondary reactions involving the volatiles above 773 K; (3) char gasification at temperatures greater than 1073 K.

Rensfelt *et al.*[7] have noted that char gasification does not significantly affect the gasification process below 1100 K. Because of the temperature range covered in this work, char gasification is not important. Raman *et al.*[8,9] have found that devolatilization of biomass in a fluidized bed reactor is nearly instantaneous. Hence, we are only concerned with the second step of Antal's conceptual model. The secondary reactions of gasification involve many complex reactions such as cracking as well as the water gas shift reaction.[1,9,10]

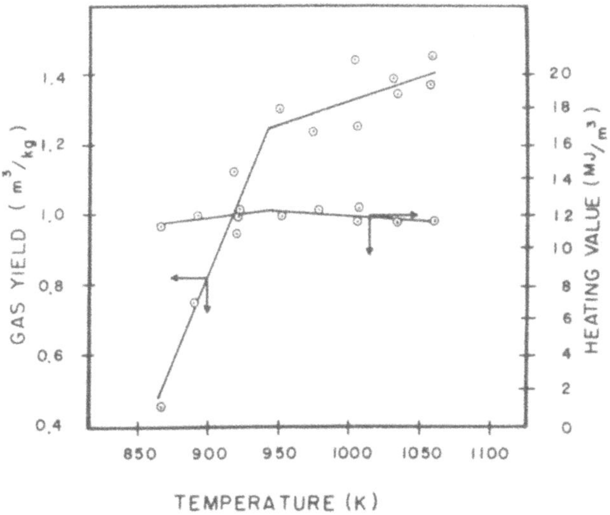

FIG. 2. Gas yield and heating value versus temperature.

A simplified mechanism is presented here based on the observations of others and a preliminary regression analysis of the gas yield, energy recovery and carbon conversion data obtained in this work. This mechanism postulates that between 865 and 1060 K the gasification of cellulose can be divided into two regimes. The first regime, up to 940 K, is controlled by cracking of volatile matter and the second, above 940 K, is dominated by the water gas shift reaction

$$CO + H_2O \rightleftarrows CO_2 + H_2$$

The transition point of $940\,K \pm 3\,K$ was first established by regression analysis of the two regimes for the gas yield, energy recovery and carbon conversion. Subsequent analysis was conducted by simultaneous regression with 940 K as the common point.

3.3. Produced Gas Yield

Figure 2 shows the volumetric yield of produced gas as a function of temperature. The lines represent the simultaneous regression analysis of the data. The gas yield varied linearly with temperature in both regimes. In the first regime the gas yield rose rapidly from $0{\cdot}48\,m^3/kg$ at 865 K to $1{\cdot}25\,m^3/kg$ at 940 K. In the second regime, above 940 K the gas yield rose slowly to $1{\cdot}41\,m^3/kg$ at 1060 K.

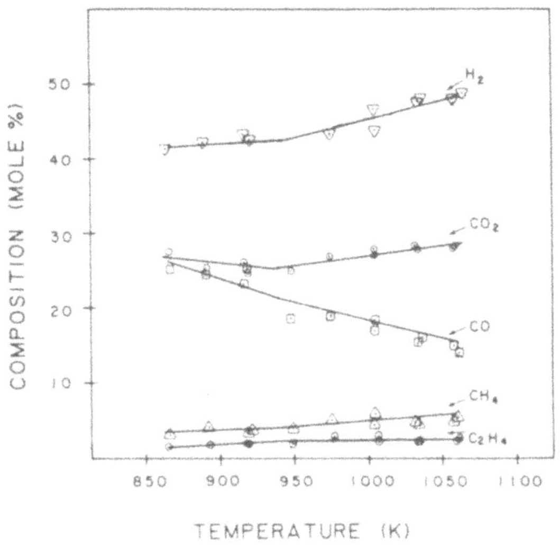

FIG. 3. Gas composition versus temperature.

3.4. Produced Gas Heating Value

The higher heating value of the produced gas as a function of temperature is also shown in Fig. 2, with the lines determined from simultaneous regression analysis of the data. The heating value increased from 11·5 MJ/m^3 at 865 K to 12·0 MJ/m^3 at 940 K and then decreased to 11·5 MJ/m^3 at 1060 K. The variation in the heating value was small and its average value was 11·8 MJ/m^3.

3.5. Produced Gas Composition

Figure 3 shows the concentrations of H_2, CO_2, CO, CH_4 and C_2H_4 as functions of temperature. The lines represent the simultaneous regression analysis of the data. The concentration of H_2 increased with temperature and ranged from 41·8% at 865 K to 48·3% at 1060 K, with a transition value of 42·6% at 940 K. Carbon dioxide ranged from 26·4% at 865 K to 25·5% at 940 K and then increased to 29·1% at 1060 K. The concentration of CO decreased with temperature and ranged from 26·4% at 865 K to 14·9% at 1060 K, with a transition value of 20·9% at 940 K. The concentration of CH_4 increased slightly with temperature and varied from 3·6% at 865 K to 5·6% at 1060 K, with a transition value of 4·3% at 940 K. The concentration of C_2H_4 also increased slightly with temperature and ranged from 1·5% at 865 K to 2·6% at 1060 K, with a transition value of 2·2% at 940 K. The minor components, C_2H_6, C_3H_6 and C_3H_8, all

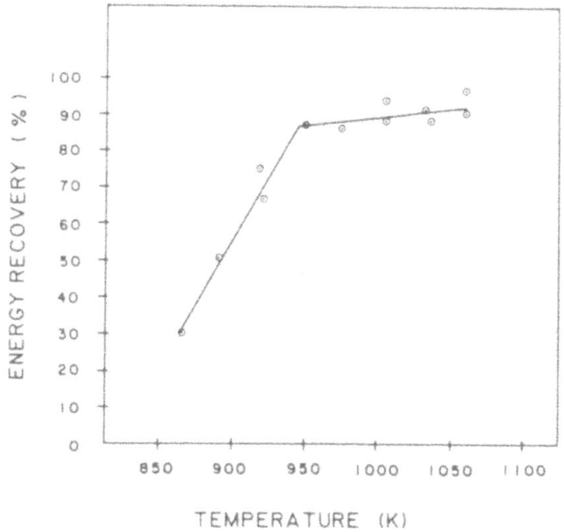

FIG. 4. Energy recovery versus temperature.

decreased in concentration with an increase in temperature. Their compositions are not plotted because combined they represent less than 3 % of the total gas yield.

3.6. Energy Recovery

The energy recovery is plotted against temperature in Fig. 4. The lines represent the simultaneous regression analysis of the data. Energy recovery was linear in both regimes. In the first regime, the energy recovery rose rapidly from 32 % at 865 K to 87 % at 940 K. Then, it rose only 5 % over the next 120 K to 92 % at 1060 K. Antal and co-workers[1,6] obtained an energy recovery of 83 % when gasifying cellulose with a gas phase reactor temperature of 973 K.

3.7. Carbon Conversion

Figure 5 shows the carbon conversion as a function of temperature. The lines represent the simultaneous regression analysis. Like the gas yield and energy recovery, the carbon conversion data are well represented with linear correlations in both regimes. In the first regime the carbon conversion rose sharply from 32 % at 865 K to 87 % at 940 K. The carbon conversion was essentially constant over the next 120 K and rose to only 89 % at 1060 K. Antal *et al.*[1,6] obtained a carbon conversion of 80 % when gasifying cellulose with a gas phase reactor temperature of 973 K.

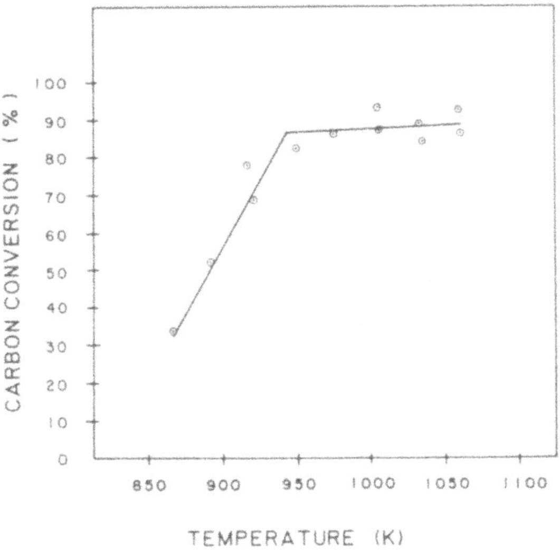

FIG. 5. Carbon conversion versus temperature.

3.8. Mechanistic Interpretations

TGA studies by Antal *et al.*[1] have shown that devolatilization of cellulose is essentially complete at 773 K with 90 % of the feed being volatiles and 10 % being char. They have further suggested that cracking and reforming of the volatile materials produced during devolatilization occurs rapidly at temperatures greater than 870 K. Additional studies by Antal[6] have indicated that reforming is probably not important and the secondary reactions are dominated by cracking reactions. Our own data also suggest that reforming is probably not important since no water consumption was noted in the first temperature regime and the concentrations of CH_4 and C_2H_4 did not decrease over the entire temperature range.

The forward step of the water gas shift reaction, which dominates the second regime in the mechanism proposed here, is faster than the reverse step below 1083 K. This is the case that existed in our experiments. Also, the reaction will tend to be driven to the right due to the excess of steam in the system. The molar concentration of steam was one to two orders of magnitude higher than the three major components of the dry produced gas.

When the water gas reaction takes place in the forward direction, the dry gas yield should increase as shown by our data. Water should be consumed as indicated by our measurements. The carbon conversion should remain unchanged. Our carbon conversion data showed only a

2 % increase over the second regime. Because of the water gas shift reaction, the heating value of the gas should decrease slightly with an increase in the temperature due to the diluent effect of CO_2. Our data indicated a slight decrease in the heating value.

It should also be pointed out that cracking volatile matter is probably not important above 940 K. Antal *et al.*[1,6] observed that the yield of tar decreases with an increase in the reaction temperature. For a reactor furnace temperature of 773 K and a gas phase reactor temperature of 973 K, they found that the tar yield was only 2 % of the initial feedstock weight, indicating that cracking was essentially complete.

3.9. Quantitative Analysis

The hypothesis that the water gas shift reaction dominates the second regime was tested quantitatively by assuming that the increase in the gas yield above 940 K was due solely to the water gas shift reaction. This assumption enables us to relate the slope of the gas yield curve, $\mathrm{d}Y/\mathrm{d}T$, to that of the mole fraction curve, $\mathrm{d}y_i/\mathrm{d}T$, for the components H_2, CO_2 and CO, over the second temperature regime.

Using the definition of mole fraction, some algebraic manipulations and differentiation, we can show for the components that are produced in the shift reaction (H_2 and CO_2)

$$\frac{\mathrm{d}y_i}{\mathrm{d}T} = \frac{1}{f}\frac{\mathrm{d}Y}{\mathrm{d}T}\frac{(n_0 - n_{i0})}{(n_0 + |\Delta n_i|)^2} \tag{1}$$

and for the component that is consumed (CO)

$$\frac{\mathrm{d}y_i}{\mathrm{d}T} = -\frac{1}{f}\frac{\mathrm{d}Y}{\mathrm{d}T}\frac{(n_0 + n_{i0})}{(n_0 + |\Delta n_i|)^2} \tag{2}$$

In eqns (1) and (2) f is the volume of 1 mole of gas at the standard conditions (i.e. 298 K and 101·3 kPa), n_{i0} is the number of moles of component i in the mixture at the reference temperature (940 K in the present work), n_0 is the total number of moles of the mixture at the reference temperature, and Δn_i is the difference between the number of moles of component i at the temperature of interest and that at the reference temperature. Absolute value signs on Δn_i have been used in eqns (1) and (2) so that calculations can be made relative to any component. Notice that the slopes of the mole fraction curves are 10^{-2} times the slopes of the mol % composition curves shown in Fig. 3.

In the water gas shift reaction 1 mole of dry gas (CO) produces 2 moles of

TABLE II

PREDICTED AND EXPERIMENTALLY DETERMINED SLOPES OF THE PRODUCED GAS MOLE FRACTION VERSUS TEMPERATURE CURVES

Component	*Average predicted slope of the produced gas mole fraction versus temperature curve between 940 K and 1060 K (1/K)*	*Experimentally determined slope of the produced gas mole fraction versus temperature curve between 940 K and 1060 K (1/K)*
H_2	$5 \cdot 63 \times 10^{-4}$	$4 \cdot 75 \times 10^{-4}$
CO_2	$7 \cdot 29 \times 10^{-4}$	$2 \cdot 91 \times 10^{-4}$
CO	$-11 \cdot 8 \times 10^{-4}$	$-5 \cdot 11 \times 10^{-4}$
	Between 950 K and 1060 K (1/K)	Between 950 K and 1060 K (1/K)
H_2	$3 \cdot 94 \times 10^{-4}$	$5 \cdot 07 \times 10^{-4}$
CO_2	$5 \cdot 12 \times 10^{-4}$	$3 \cdot 05 \times 10^{-4}$
CO	$-8 \cdot 30 \times 10^{-4}$	$-4 \cdot 89 \times 10^{-4}$
	Between 960 K and 1060 K (1/K)	Between 960 K and 1060 K (1/K)
H_2	$2 \cdot 95 \times 10^{-4}$	$5 \cdot 32 \times 10^{-3}$
CO_2	$3 \cdot 83 \times 10^{-4}$	$3 \cdot 06 \times 10^{-3}$
CO	$-6 \cdot 20 \times 10^{-4}$	$-4 \cdot 71 \times 10^{-3}$

Calculations based on regression analysis.

dry gas (H_2 and CO_2). Water is not considered as a component of the gas or a part of the gas yield since the present analysis is on a dry basis. Notice that the slope of the mole fraction curve changes with a change in temperature because Δn_i is a function of the temperature. The change in the slope is not large for the temperature range studied because the total number of moles, n_0, is much greater than the change in the number of moles of component i, Δn_i. The gas composition data should properly be represented with curves, but, to be consistent with the proposed mechanistic model, the data have been fitted with straight lines between 940 K and 1060 K.

Table II shows the average of the predicted slopes of the produced gas mole fraction versus temperature curves, calculated from eqns (1) and (2). The average values have been obtained by evaluating the slopes at 940 K and 1060 K, and taking their arithmetic averages. Table II also shows the least squares estimates of those slopes using data between 940 K and 1060 K for comparison. Agreement between the predicted value ($5 \cdot 63 \times 10^{-4}$/K) and the experimentally determined value ($4 \cdot 75 \times 10^{-4}$/K) of the slopes for H_2 is good. The values for CO and CO_2 are of the same order of magnitude, but the agreement is surprisingly good in light of the complexity of the system.

It would be expected that the transition from one regime to the other would not take place at a single temperature but rather over a range of temperatures. Consequently it should be possible to use other transition temperatures in the model that are in the neighbourhood of 940 K. The effect of variation in the transition temperature was examined by conducting simultaneous regression analysis on the gas yield with common transition temperatures ranging from 935 to 960 K. Ordinary regression analyses of the gas composition data were also conducted for compositions at or above the transition temperature. The results of these calculations are included in Table II which shows the predicted average slopes of the gas composition curves based on eqns (1) and (2) using the simultaneous regression slopes of the gas yield line for temperatures from 935 K to 960 K compared with the regression analysis slopes of the raw composition data at or above the transition temperature under consideration. Table II shows that as the transition temperature increases, the agreement for CO and CO_2 improves while the agreement for H_2 becomes poorer. However, the overall agreement for all three components improves as the transition temperature increases. This observation is consistent with Antal's observation[6] that cracking is essentially complete at 970 K.

4. CONCLUSION

Steam gasification experiments were conducted with α-cellulose in a bench-scale reactor over a temperature range of 865 K to 1060 K. The gas yield and other data indicated the existence of two distinct regimes in the gasification process which led to the postulation of a mechanistic model which describes the gasification process between 865 K and 1060 K as being composed of two distinct regimes. The first was due to the cracking of volatile matter and showed a very rapid rise in gas yield with an increase in temperature up to 940 K. The second regime was controlled by the water gas shift reaction and showed a dramatically slower rise in gas yield with an increase in temperature above 940 K. These regimes were also evident in the energy recovery and carbon conversion.

ACKNOWLEDGEMENT

This is contribution No. 82-604-A, Department of Chemical Engineering, Kansas Agricultural Experiment Station, Kansas State University, Manhattan.

REFERENCES

1. Antal, M. J., Edwards, W. E., Friedman, H. L. and Rogers, F. E. (1978). A study of the steam gasification of organic wastes. Project Report to EPA, W. W. Liberick, Project Officer.
2. Howard, J. B., Stephens, R. H., Kosstrui, H. and Ahmed, S. M. (1979). Pilot scale conversion of mixed waste to fuel, Volume I. Project to EPA, W. W. Liberick, Project Officer.
3. Walawender, W. P., Raman, K. P. and Fan, L. T. (1980). Gasification of carbonaceous materials in a fluidized bed reactor. In *Proceedings of Bio-Energy '80 World Congress and Exposition.* The Bio-Energy Council, Washington, D.C., 575–7.
4. Raman, K. P., Walawender, W. P. and Fan, L. T. (1980). Gasification of feedlot manure in a fluidized bed: The effects of superficial gas velocity and feed size fraction. Preprints, *ACS Div. Fuel Chem.* **25**(4), 233–44.
5. Beck, S. R. and Wang, M. J. (1980). Wood gasification in a fluidized bed. *Ind. Eng. Chem. Proc. Des. Dev.* **19**(2), 312–17.
6. Antal, M. J. (1979). The effects of residence time, temperature and pressure on the steam gasification of biomass. Presented at Symposium on Biomass as a Non-fossil Fuel Source. Division of Fuel Chemistry, American Chemical Society Meeting, Honolulu, April 1–6.
7. Rensfelt, E., Blomkvist, G., Ekström, C., Engström, S., Espenös, B. G. and Liinanki, L. (1978). Basic gasification studies for development of biomass medium-Btu gasification processes. In Symposium Papers Energy from Biomass and Wastes. Institute of Gas Technology, Washington, D.C., 465–94, August 14–18.
8. Raman, K. P., Walawender, W. P., Fan, L. T. and Howell, J. A. (1981). Thermogravimetric analysis of biomass. Devolatilization studies on feedlot manure. *Ind. Eng. Chem. Proc. Des. Dev.* **20**(4), 630–36.
9. Raman, K. P., Walawender, W. P., Fan, L. T. and Chang, C. C. (1981). Mathematical model for the fluid-bed gasification of biomass materials. Application to feedlot manure. *Ind. Eng. Chem. Proc. Des. Dev.* **20**(4), 686–92.
10. Schoeters, J., Maniatis, K. and Buekens, A. (1981). Fuel gas from agricultural residues in a fluidized bed reactor. In *Proceedings of 2nd World Congress of Chemical Engineering* **1**, 313–17.

50

Wood Chip Gasification in a Commercial Downdraft Gasifier

WALTER P. WALAWENDER, S. M. CHERN and L. T. FAN
Department of Chemical Engineering, Kansas State University, Manhattan, Kansas, USA

1. INTRODUCTION

Fixed bed and moving bed gasifiers for the production of low Btu gas from wood and charcoal were widely used in Sweden and other countries during the World War II era. The Swedish experience was compiled by the Swedish Academy of Engineering, and this work was recently translated by Reed and Jantzen.[1] After the war, the need for gasifiers dwindled although the Swedes continued their development efforts.

Following the Arab oil embargo, rapidly rising prices for petroleum coupled with projections of potential near-term petroleum and gas shortages prompted the search for alternative energy sources. In response to this effort, gasification technology has been resurrected and low Btu gas production became one of the major options for alternative energy in the United States. Reed *et al.*[2] reviewed the state of the art in gasification technology and summarized the status of a number of commercial and research efforts in the field.

Although the literature contains a vast amount of information on various facets of the gasification of wood and biomass in fixed bed and moving bed gasifiers, very little has been published on the complete material balance for these gasifiers. In fact, many gasifiers have been constructed with little or no quantitative information on the gasifier performance. Graham and Huffman[3] presented detailed information on the Biomass Corporation gasifier; this appears to be the only complete material balance analysis of a moving bed gasifier in the open literature.

The present work reports the material balances and performances of a recently commercialized downdraft gasifier, the Buck Rogers 'Gasafire'™. The gasifier, rated to produce 320–1600 MJ/h, was evaluated over its working capacity range and was found to be remarkably consistent in terms of gas yield, gas composition, gas heating value and other measures characterizing its performance.

2. EXPERIMENTAL FACILITIES AND PROCEDURE

2.1. Gasifier Description

Figure 1 presents a schematic representation of the commercial gasifier used in this study. It has a diameter of 0·6 m and is completely open at the top. It is unique in that there is no throat or choke plate to support the bed. A fan located downstream from the gasifier draws air through the chip bed as well as the tuyeres. A feed bin and a screw feeder (not shown in Fig. 1) supply chips to the top of the gasifier each time the bed height drops to a set level, controlled by an electric eye. The fan motor also drives the cyclone rotary valve and the 'airgitator' gearbox through a series of speed-reducing pulleys and gears. Consequently, the fan speed controls the air intake, chip feed, gas production and char output rates. The 'airgitator' rotates very slowly at 20–30 min per revolution and has several functions, listed below:

(1) It provides secondary air, through the tuyeres.
(2) It levels the feed at the top of the bed.
(3) It slowly mixes material throughout the gasifier.
(4) It generates a grinding motion in the ceramic ball bed which, in turn, reduces the char size so that it can percolate through the ball bed and finally through the grid plate.
(5) Coupled with the ball bed and the grid plate, it serves as the support for the bed.

Gas and char flow out of the gasifier through the grid plate. The fan suction draws them out of the base of the gasifier and blows them through the cyclone where the char is separated from the gas. The gas stream is then sent to a flare stack (not shown in Fig. 1) where it is incinerated.

2.2. Operating Procedure

The procedure used in operating the gasifier is outlined below.

2.2.1. (a) Start-Up

The fan was started at low speed, and propane was introduced into the flare stack and ignited. Next the feeder was manually operated to fill the gasifier just above the tuyeres. The bed was then ignited with an electric fire starter

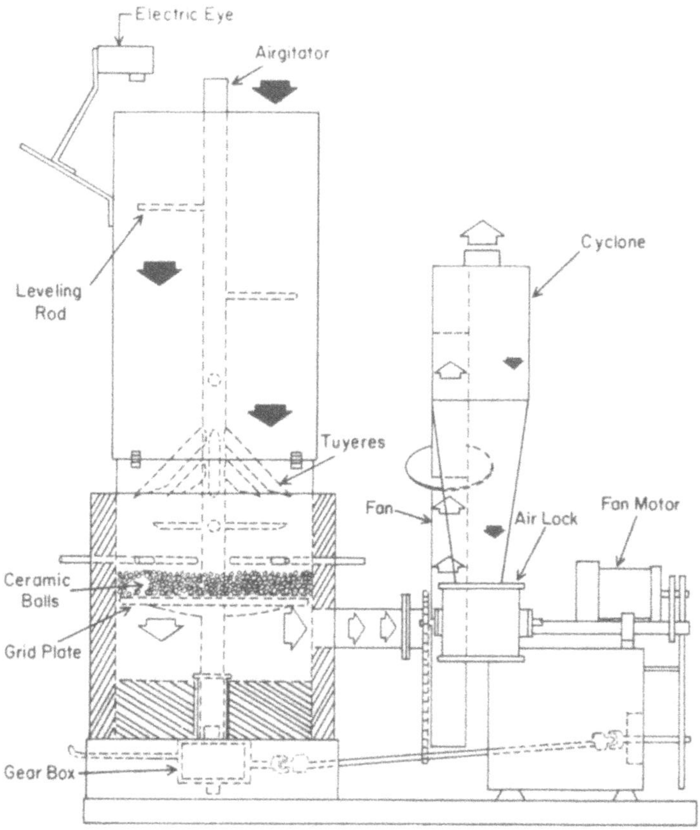

FIG. 1. Commercial gasifier.

and the fan speed was increased. The fire was allowed to develop for 15–20 min. Then chips were added to fill the gasifier and the fan speed was increased. As gas production increased, the propane rate was decreased and finally shut down. The fan speed was then gradually increased to the desired operating condition and the reactor was allowed to stabilize. This procedure normally required 2 h to complete.

2.2.2. (b) Char and Feed Measurements

Char was measured by direct collection over 30 min intervals. Chip feed was measured by a procedure to be detailed later. Both of these measurements were conducted over a 5 h period starting after the completion of start-up.

2.2.3. (c) Gas Analysis and Condensables Measurement

One hour after the char and feed measurements were started, a small gas stream was drawn for gas analysis and condensate determination. The specific procedures will be detailed later. These measurements were conducted for 2 h.

2.2.4. (d) Tracer Gas Measurement

Nitrogen tracer gas was introduced to the product gas at a known rate, and the gas analysis and condensate measurements were repeated. The specific procedures are detailed later. These measurements were conducted for 1 h.

2.2.5. (e) Post Tracer Measurements

Step (c) was again repeated without the tracer gas for a 1 h period.

2.2.6. (f) Shut-Down

The chip feed was stopped and the fan speed was increased to accelerate complete consumption of the chips. This normally required 1 h.

2.3. Supporting Measurements

The supporting measurements included moisture and ash analyses of the feed and elemental analysis of the feed. Moisture was determined by drying in an oven for 3 h at 380 K. Ash was determined in a muffle furnace by standard ASTM procedure. The elemental analysis of feed was conducted with the aid of a Perkin-Elmer Model 240b Elemental Analyzer.

2.4. Feedstock

The feed material used in this work was a mixed hardwood chip. The average chip screen size was 14 mm; the approximate size distribution is given in Table I. The chips were free of bark and were pre-dried to 6–17 %

TABLE I
CHIP SIZE DISTRIBUTION

Screen size passed (mm)	*Weight fraction*	*Typical sizes (mm)*
Sticks	0·014	10 × 75, 7 × 74, 23 × 95
>18·5	0·099	24 × 43, 21 × 32, 30 × 47
18·5	0·211	17 × 47, 20 × 25, 15 × 26
12·5	0·592	15 × 32, 12 × 17, 9 × 31
4·5	0·056	5 × 19, 3 × 41, 4 × 7
3·5	0·028	1·5 × 13, 2 × 9, 2 × 2

moisture (wet basis). The elemental analysis of the dry chips was fairly constant over the course of the experiments. The average analysis was C, 48·31 %; H, 5·94 %; O, 44·77 %; N, 0·18 %; ash, 0·80 %. The average higher heating value of the dry chips, estimated from the Dulong formula (found in basic Chemical Engineering textbooks), was 16·82 MJ/kg.

2.5. Measurement Scheme and Computations

The objective of the present work was to gather complete material balance data for the gasifier and to evaluate the gasifier performance. Because of the configuration of the gasifier and the stream flow rates involved, it was not convenient to measure all of the stream rates directly; consequently, indirect methods were devised.

The two quantities directly measured were the char output rate and the dry gas composition. Char output was determined by collecting and weighing the char over 30 min intervals for approximately a 5 h period. The gas analysis was conducted with an Applied Automation on-line process gas chromatograph (GC) which draw a continuous sample from points downstream of the cyclone. The GC had a cycle time of about 11 min and was able to detect the following components in the producer: H_2, CO_2, CO, CH_4, O_2 and N_2 along with traces of C_2H_6, C_2H_4 and C_3H_6. It was believed that the O_2 in the producer gas resulted from air leaks in the fan seal and the gas sampling system. Consequently all compositions were converted into an air-free basis. Product gas compositions were determined from the average of the pre and post trace gas analyses.

The dry product gas rate was indirectly determined with the aid of a tracer technique. Nitrogen tracer was introduced into the product gas at a known volumetric rate at a point just upstream from the fan. The gas composition was then measured with the tracer flowing over a 1 h period to insure steady state compositions. From the known tracer injection rate and the nitrogen composition with and without tracer, the dry gas volumetric flow rate could be readily evaluated. Knowing the molecular weight of the gas allowed the stream rate to be converted to a mass basis.

The dry gas flow rate, coupled with a nitrogen balance between the inlet and outlet of the gasifier, allowed the dry air input rate to be evaluated. This balance assumed that negligible nitrogen was produced in the gasification process. Knowing the temperature and relative humidity over the course of an experiment allowed the moisture input from humid air to be evaluated. Since most experiments were of 5 h duration, the moisture evaluation was based on mid-run conditions.

Condensables consisted of two parts, tar and aqueous. These were determined by taking a side draw from the main gas stream at a rate of about 0·55 m^3/h. This stream also supplied the GC with sample. The gas was first sent through a hot filter (maintained at about 420 K with a heating tape) which was packed with glass wool to remove heavy tar components. Next the gas was sent through two water-cooled condensers in series to remove most of the water. The remaining water removal was accomplished with two receivers placed in series in an ice-bath. The tarry mist was then removed by passing the stream through a tightly packed glass wool filter. Flow through the train was maintained with the aid of a 'Gast'™ compressor which provided suction. The discharge from the compressor was then sent through a wet test meter, followed by a drierite chamber and then to the GC. Two identical trains were constructed for use with and without tracer. The wet test meter readings were corrected for temperature, pressure, water of saturation and air, and tracer gas when needed. The condensates and tars were weighed and converted to a mass per unit volume dry gas basis. These results were then scaled to the main stream flow.

The chip feed to the gasifier was intermittent rather than continuous. However, chip flow through the gasifier was continuous with only the chip level fluctuating (0·3–0·5 m). Chips were delivered by a constant speed screw feeder which was activated by the chip level sensed by an electric eye. The following procedure was adopted to evaluate the chip feed rate. A timer was wired into the screw feeder circuit to determine the cumulative on-time for the feeder. This was generally recorded over a 5 h period. The feeder was calibrated by collecting and weighing the feeder discharge over 5–10 s intervals. Some fluctuations in the calibration were noticed, i.e. 0·6–0·7 kg/s. These were due to bin height variations and/or bulk density variations. The feed rate for each run was evaluated by taking the appropriate calibration factor times the cumulative on-time for that run. The above items determined the material balance for the gasifier. Material balance closure was evaluated as output/input and the results adjusted so that all closures were less than 100 %, i.e. for closure greater than 100 % the result was subtracted from 200.

The performance of the gasifier was evaluated in terms of a variety of measures determined from the gas composition, material balance data and supporting measurements. The dry gas composition gave one measure of gasifier performance and, from it, the higher heating value of the dry gas was calculated from the standard heat of combustion of the components. The gas, char and tar yields gave other measures of performance. These were calculated on the basis of a unit mass of dry feed to eliminate the influence of moisture variations. A dry ash-free basis was not used since the ash content of the dry chips was less than 1 % and not subject to much variation. Char yield was expressed as the kg of char per 100 kg of dry chips and gas yield was expressed on a volumetric basis as m^3/kg of dry chips. The volume basis used in this work was 289 K and 101·3 kPa. Tar yield was expressed in two ways, one as kg of tar per 1000 kg of dry chips and the other as parts per million in the dry gas. Also evaluated were the water output, kg of condensate per kg of dry chips, and the air input, kg of dry air per kg of dry chips. The energy output was evaluated in MJ/h and the energy yield in MJ/kg of dry chips. The cold gas efficiency was evaluated as the ratio of the energy content of the gas produced from 1 kg of dry chips to the energy content of 1 kg of dry chips. This quantity represented the fraction of the energy content of the feed that was converted to combustible gas. The mass conversion efficiency was evaluated as the ratio of dry gas output to the total input.

3. RESULTS AND DISCUSSION

3.1. Results

The material balances for runs with closures greater than 95 % are summarized in Table II. A total of 19 runs were completed, and 9 had closures greater than 95 %; the poorest closure was 80 %. Some runs had closures greater than 100 % and these were adjusted to a 100 % basis by subtracting the result from 200. The material balance closures for all runs are included in Table III.

TABLE II
MATERIAL BALANCE SUMMARY

Run	*Inputs (kg/h)*				*Outputs (kg/h)*					*Closure (%)*
	Wet chips	*Dry air*	*H_2O*	*Total*	*Dry gas*	*Char*	*Tar*	*H_2O*	*Total*	
98	32·0	43·1	0·5	75·6	66·3	0·9	0·09	7·4	74·7	98·9
910	32·0	45·2	0·5	77·7	68·0	0·9	0·14	7·4	76·4	98·3
920	35·7	62·1	0·4	98·2	92·9	1·4	0·09	7·1	101·5	96·6
101	52·4	76·9	0·8	130·1	116·9	1·7	0·09	13·3	132·0	98·6
929	52·7	74·0	1·0	127·7	113·0	1·8	0·14	12·2	127·1	99·5
106	58·1	76·8	0·5	135·4	117·0	1·2	0·09	10·8	129·1	95·4
1 119	89·1	112·1	1·0	202·2	173·8	3·0	0·27	19·7	196·8	97·3
1 117	96·2	140·4	0·7	237·3	218·9	2·5	0·18	22·9	244·5	97·0
1 221	140·5	202·1	1·1	343·7	302·1	4·1	0·54	43·3	350·0	98·2

TABLE III
PERFORMANCE SUMMARY

Run	Dry chip rate (kg/h)	Chip moisture (% wet basis)	Matl. balance closure (%)	Dry gas HHV (MJ/m^3)	Gas yield (m^3/kg DC[a])	Char yield (kg/100 kg DC)	Tar yield (kg/1 000 kg DC)	Tar in dry gas (ppm)	Dry air in (kg/kg DC)	H_2O out (kg/kg DC)	Energy yield (MJ/kg DC)	Total energy out (MJ/h)	Mass conversion efficiency (%)	Cold gas efficiency (%)
98	26·7	16·52	98·9	5·33	2·22	3·28	2·55	1 370	1·61	0·277	11·82	315	87·7	70·2
910	26·7	16·52	98·3	5·18	2·28	3·21	4·59	2 003	1·69	0·277	11·82	315	87·5	70·2
920	33·6	6·01	96·6	5·33	2·48	4·06	2·71	977	1·85	0·212	13·14	441	94·7	78·1
927	41·2	8·11	85·4	5·63	2·54	3·08	3·30	1 180	1·75	0·438	14·31	590	98·1	85·0
922	47·8	5·75	85·1	5·44	1·43	2·76	1·90	1 199	1·05	0·186	7·76	370	74·9	46·1
101	48·7	6·99	98·6	5·59	2·18	3·45	1·86	777	1·56	0·273	12·19	594	89·8	72·5
929	48·8	7·41	99·5	5·63	2·10	3·72	2·79	1 205	1·52	0·250	11·82	576	88·5	70·2
1 011	49·5	7·31	91·6	5·44	2·55	3·94	4·59	1 607	1·88	0·321	13·84	685	96·1	82·2
1 015	52·3	7·95	88·8	5·33	2·73	4·69	5·21	1 679	2·08	0·381	14·52	759	97·5	86·3
924	53·7	5·50	91·8	5·66	1·62	2·79	2·54	1 426	1·13	0·209	9·11	490	80·9	54·2
106	54·3	6·67	95·4	5·48	1·95	2·18	1·67	776	1·42	0·199	10·68	579	86·5	63·5
123	54·3	14·61	94·2	5·51	1·90	4·27	3·35	1 594	1·35	0·236	10·44	567	83·0	62·1
104	63·7	7·72	92·6	5·55	2·61	2·71	2·85	989	1·90	0·307	14·50	924	96·2	86·2
1 129	68·9	14·27	91·9	5·59	2·85	3·82	3·96	1 254	2·04	0·281	15·89	1 094	98·1	94·4
121	72·5	15·07	80·4	5·63	1·39	3·57	3·76	2 445	0·99	0·169	7·81	567	70·8	46·5
1 119	74·6	16·32	97·3	5·74	2·11	3·96	3·65	1 567	1·50	0·264	12·08	901	86·0	71·8
1 117	85·4	11·26	97·0	5·78	2·30	2·92	2·13	830	1·64	0·268	13·31	1 136	92·2	79·1
1 110	119·7	11·56	91·0	5·74	1·79	1·67	3·03	1 544	1·27	0·206	10·23	1 224	81·6	60·8
1 221	125·5	10·67	98·2	5·44	2·10	3·25	4·34	1 803	1·61	0·345	11·42	1 434	87·9	67·9
all data mean	—	—	—	5·51	2·17	3·33	3·20	1 380	1·57	0·268	11·93	—	88·3	70·9
σ	—	—	—	0·15	0·41	0·72	0·99	424	0·31	0·067	2·25	—	7·5	13·0
<95% mean	—	—	—	5·51	2·19	3·34	2·92	1 256	1·60	0·263	12·03	—	87·9	71·5
σ	—	—	—	0·19	0·14	0·54	0·99	433	0·12	0·039	0·76	—	2·1	4·5

[a] DC represents dry chips.

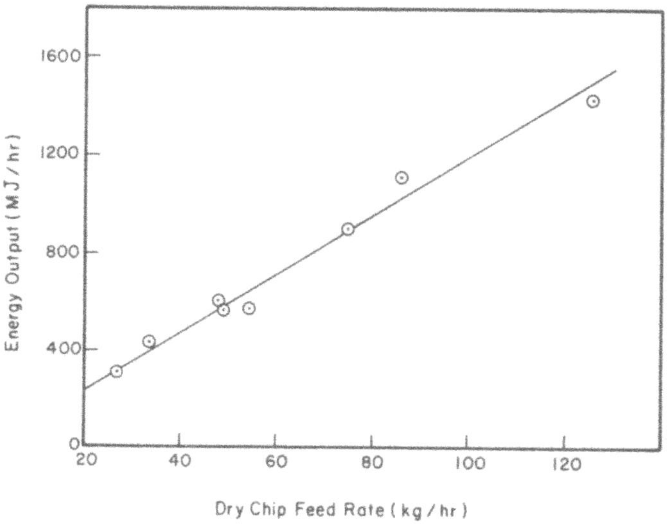

FIG. 2. Energy output versus dry chip feed rate.

Table III presents various measures of the gasifier performance as a function of the dry chip feed rate along with the moisture content of the as-received feed. Included in the table are the char yield in kg/100 kg of dry chips, the tar yield in kg/1000 kg of dry chips, the gas higher heating value, the energy yield per unit mass of dry chips, the total energy output, the water output rate in kg/kg of dry chips, the tar content of the gas in ppm, the dry air input rate in kg/kg of dry chips, the volumetric gas yield, the mass conversion efficiency and the cold gas efficiency. The means and standard deviations for the complete data set and the nine runs with closure greater than 95 % are given at the bottom of the table. As can be seen from the table, although the values fluctuated somewhat, no distinct trends were observed as the dry chip feed rate increased. The mean gas yield over the entire operating range was 2·19 m^3/kg for runs with closures greater than 95 %. Other important performance measures include the mean heating value at 5·51 MJ/m^3, the mean mass conversion efficiency at 88 %, and the mean cold gas efficiency at 72 %.

The energy yield per pound of dry chips was fairly close for the runs with closures greater than 95 %. This behavior is also illustrated by Fig. 2 in which the total energy output is plotted against the dry chip feed rate. As can be seen from the figure, a linear relationship determined by least squares analysis describes the data quite well.

Dry gas composition data as a function of dry chip feed rate for all the

TABLE IV
DRY GAS COMPOSITION (%)

Run	H_2	C_3H_6	CO_2	C_2H_4	C_2H_6	N_2	CH_4	CO
98	15·00	0·10	16·67	0·39	0·07	46·92	2·50	18·41
910	15·00	0·08	16·25	0·36	0·06	47·89	2·33	18·05
920	14·27	0·09	15·50	0·42	0·07	48·19	2·49	18·97
927	16·56	0·08	16·60	0·41	0·07	44·45	2·52	19·31
922	14·85	0·09	15·24	0·42	0·07	47·46	2·48	19·40
101	15·47	0·10	14·80	0·49	0·09	46·72	2·41	19·94
929	15·52	0·09	14·96	0·47	0·07	46·50	2·55	19·85
1 011	14·69	0·12	16·28	0·57	0·09	47·55	2·61	18·09
1 015	13·25	0·14	16·31	0·65	0·10	49·14	2·67	17·73
924	16·04	0·08	15·84	0·44	0·07	45·11	2·51	19·93
106	15·32	0·07	15·23	0·39	0·07	46·78	2·45	19·69
123	16·33	0·11	16·81	0·57	0·09	45·94	2·39	17·77
104	15·25	0·08	14·83	0·45	0·06	46·83	2·48	20·01
1 129	15·61	0·12	15·90	0·57	0·09	46·22	2·41	19·07
121	16·06	0·13	16·41	0·63	0·10	45·86	2·47	18·34
1 119	15·56	0·14	15·51	0·63	0·09	45·96	2·49	19·62
1 117	14·90	0·15	15·56	0·66	0·10	46·02	2·61	20·00
1 110	15·50	0·13	15·17	0·58	0·09	45·95	2·49	20·08
1 221	12·25	0·19	16·24	0·82	0·10	49·40	2·81	18·20

runs are summarized in Table IV. The principal components of the product gas were H_2, CO, CO_2, CH_4 and N_2 with trace amounts of C_2H_6, C_2H_4 and C_3H_8. Although the results show some fluctuation, there is no distinct trend with increasing dry chip feed rate. As noted earlier, all gas compositions were determined from the average of the pre and post trace gas analyses and are reported on an air-free basis.

3.2. Discussion

The results of this work indicate that the performance of the gasifier is remarkably consistent over a fairly wide range of operating conditions. The chip feed rate was varied over almost a five fold range from 27 to 126 kg/h with a corresponding energy output ranging from 320 to 1400 MJ/h. The consistency in performance was further reflected by the temperature measured right above the ceramic ball bed. It fluctuated only slightly from 950 to 1000 K over the course of the experiments with no distinct trends indicated as the chip feed rate was increased.

The mass conversion and cold gas efficiencies obtained in this study are compatible with the results obtained by Graham and Huffman[3] using a downdraft gasifier of comparable capacity. Using poplar chips with a 13% moisture content (wet basis), they reported a mass conversion efficiency of 87% and a cold gas efficiency of 74%. Their feed rates were comparable to the upper range examined in this work, but they made no effort to study wide ranges of feed rate. Their gas compositions were similar to ours but showed more CO (24–26%) and less CO_2 (9–10%). The differences are probably due to the differences in composition of the feedstocks used in the two studies. Furthermore, Graham and Huffman[3] reported substantially more tar than we did, i.e. of the order of 10 000 ppm. Their tar results were determined by difference whereas ours were measured values. Our measurements of tar were rather crude because some of the tar was deposited on the sample system surfaces and not measured. Even if the measurements were low by 100%, the tar observed in our study was still an order of magnitude lower than that reported by Graham and Huffman.

ACKNOWLEDGEMENT

This is contribution No. 83-14-A, Department of Chemical Engineering, Kansas Agricultural Experiment Station, Kansas State University, Manhattan. This work was supported in part by The Buck Rogers Co., Inc., Industrial Airport, KS.

REFERENCES

1. Reed, T. B. and Jantzen, D. E. (1979). Generator gas—The Swedish experience from 1939–1945. Translated from the work of the Swedish Academy of Engineering. SERI/SP-33-40.
2. Reed, T. B. *et al.* (1980). A survey of biomass gasification, Vols 2 and 3. SERI/TP-33-239. (Also published in hard cover as Reed, T. (Ed.) (1981). *Biomass Gasification.* Energy Technology Review No. 67, Noyes Data Corp., Park Ridge, NJ.)
3. Graham, R. G. and Huffman, D. R. (1981). Gasification of wood in a commercial-scale downdraft gasifier. In Symposium Papers, Energy from Biomass and Wastes V, Institute of Gas Technology, 633-50.

51

Methane Yield from Biomass Gasification at High Temperature and Pressure

L. Liinanki, N. Lindman, S.-O. Sjöberg and E. Ström
Royal Institute of Technology,
Department of Chemical Technology, Stockholm, Sweden

1. INTRODUCTION

The methane yield in biomass gasification is always of great importance. The yield must be high in fuel gas production and it has to be small in synthesis gas production for methanol or ammonia. A fundamental understanding of the methane producing and consuming reactions are therefore a prerequisite for optimal design of biomass gasifiers. This article gives a new contribution to this understanding. Results are presented and discussed from an investigation where these reactions are studied over a broad range of pressures and where the influence of biomass char and ash has been eliminated.

The interest was focused on the homogeneous gas phase reactions taking place after the primary pyrolysis and char gasification. In order to be able to study influence of tar and oxygen addition on the methane yield three types of experiments were carried out:

—reforming of peat product gas with oxygen addition;
—reforming of a synthetic, tar-free, gas without oxygen;
—reforming of a synthetic, tar-free, gas with oxygen.

Most of the experiments were of the first type and it was from these that the major results were achieved. The other types were performed to make it possible to separate different special effects.

2. EXPERIMENTS

2.1. Experimental Apparatus

The experimental system that was used for studies of gasification of wood and peat is illustrated in Fig. 1. It consists of a fluidized bed followed by a

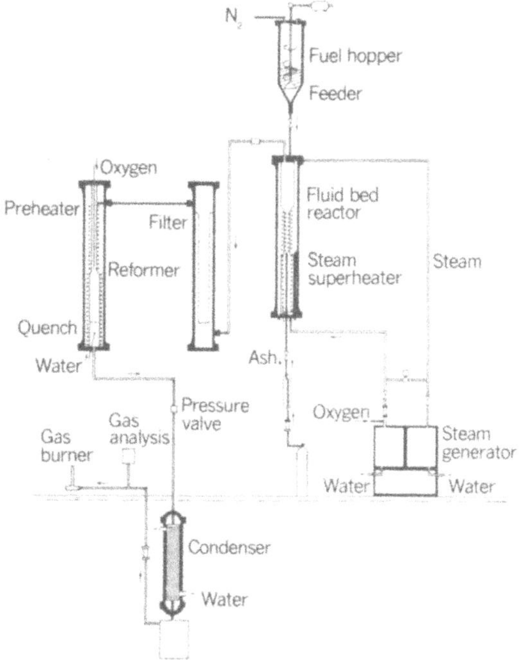

FIG. 1. Experimental gasification unit.

high temperature filter and a secondary reformer. The bubbling fluidized bed is 10 cm in diameter and 60 cm high. A mixture of steam and oxygen is used as gasification and fluidization agents. Steam is produced in an electrically heated steam generator and is superheated up to the gasification temperature before it is introduced to the gasification zone. The gas is distributed at the bottom of the fluidized bed through a perforated cone. Oxygen is mixed with steam before reaching the gas distributor. The reaction zone can be heated by electrical coils outside the tube wall when the gasifier is working without oxygen.

The fuel is fed from the fuel hopper by a rotating feeder to a falling tube which ends in the fluidized bed. The capacity is about 10 kg fuel/h. The gasifier is designed to operate at 750–900 °C.

The dust collecting equipment consists of three filter bags, with a total area of 1 m^2, hanging inside a pressure vessel. The filter bags are made of felted inconel fibres and have a working temperature of about 500 °C.

The gas reformer is a ceramic lined pressure vessel with an inside steel tube. Outside the tube there are electrical coils for heating of the reformer.

The upper part of the reformer can be used to superheat the gas by injecting oxygen in the gas stream. The reformer can work at temperatures up to 1100 °C. In the bottom of the reformer the gas is quenched with water to a temperature of about 300 °C. The experimental system is designed for 3 MPa.

For gas analysis there is an on-line gas chromatograph, with automatic sampling every 4 min, for analysis of H_2, CO, CO_2, CH_4 and C_2H_n. The equipment is also supplied with a tar sampling system at the inlet and outlet of the reformer.

2.2. Experimental Procedure

In the experiments with peat and oxygen addition the gasifier and the reformer were electrically heated to 800 °C before feeding was started. The steam rate was adjusted so that the linear velocity in the fluid bed would be about 10–15 cm/s. The system was then pressurized to the desired level in the range of 0·3–2·0 MPa. After about 30 min of feeding, when the char bed was sufficiently high, oxygen was fed into the gasifier and into the reformer. The oxygen flow rate was adjusted to give the proper temperature. Samples of tar water were taken when temperatures and pressure were stable. In some experiments gas samples were taken at the inlet of the reformer.

In experiments with synthetic tar-free gas, mixtures of H_2, CH_4 and CO_2 were added to the steam at the inlet of the reformer. The empty gasifier was maintained at temperature of about 500 °C for superheating of the steam. The gas composition at the inlet was:

$$55\%\ H_2,\ 15\%\ CH_4,\ 30\%\ CO_2$$

2.3. Experimental Results with Peat/Oxygen

The raw gas from the fluidized bed has a high content of tars, methane and C_2-hydrocarbons. Factors that influence the product distribution are

—gasification temperature,
—oxygen input,
—gasification pressure.

Increasing gasification temperature gives decreasing yields of methane and C_2-hydrocarbons. The oxygen input influences the temperature at the bottom of the fluidized bed. A high oxygen/fuel ratio gives a high temperature in the gasifier which decreases the methane and C_2-hydrocarbon yields. In the described experiments the gasification temperature was about 750 °C and the oxygen/fuel ratio between 0·2 and 0·25 kg O_2/kg peat (dry matter).

Increasing pressure results in a significant increase in yield of methane in the raw gas from the gasifier (Fig. 2). The methane yield, α_{CH_4}, is defined as the relative amount of carbon in the gas present as methane.

$$\alpha_{CH_4} = \frac{X_{CH_4}}{X_{CO} + X_{CO_2} + X_{CH_4} + 2X_{C_2H_n}}$$

$$X_{CX} = \text{mole fraction}$$

The tar and methane rich gas is introduced in the reformer at about 400 °C. The gas is heated electrically to about 800 °C and then by injecting oxygen the gas temperature is raised up to 1100 °C. Figure 2 shows the methane yield from the reformer as a function of temperature and pressure.

At low temperatures (< 800 °C) there is an increasing yield of methane and the α_{CH_4} value reaches a maximum at about 800 °C and then decreases with increasing temperature. That means that there is a net production of methane at low temperatures. The yield of tars decreased considerably at reforming temperatures up to 800 °C and was very low at 1100 °C (< 1 %). The yield of C_2-hydrocarbons is almost nil at temperatures over 900 °C.

The methane yield in the outlet is also dependent on the total pressure. Higher pressure results in increased yield of methane. In all experiments the

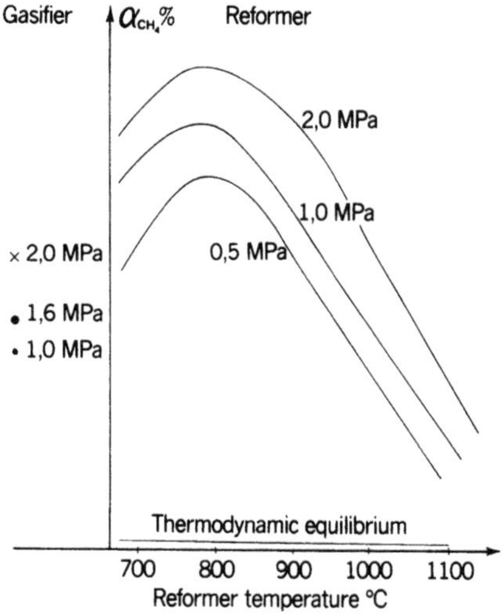

FIG. 2. Thermal reforming of methane from peat gasification.

methane yield was much higher than predicted from equilibrium for the methane reforming reaction.

$$CH_4 + H_2O \rightleftharpoons CO + 3H_2 \quad (1)$$

The other base reaction for decomposition of methane is

$$CH_4 \rightleftharpoons C(s) + 2H_2 \quad (2)$$

Other reactions which occur in the reformer are the water–gas shift reaction (3) and the steam–carbon reaction (4):

$$CO + H_2O \rightleftharpoons CO_2 + H_2 \quad (3)$$

$$C(s) + H_2O \rightleftharpoons CO + H_2 \quad (4)$$

Figure 3 shows an equilibrium index I for the reformer product gas. This index I is defined as

$$I = [\log_{10} K(\text{theoretical}) - \log_{10} K(\text{experimental})]_{T=T}$$

where K(theoretical) = thermodynamic equilibrium, K(experimental) = experimental partial pressure ratio (carbon = graphite).

Four conclusions can be drawn from the figure.

—There is a very strong chemical potential for steam reforming of methane via reaction (1).
—The water–gas shift equilibrium is reached.
—There is no potential for carbon deposition via the Boudouard reaction.
—There is a strong potential for carbon deposition via reaction (2).

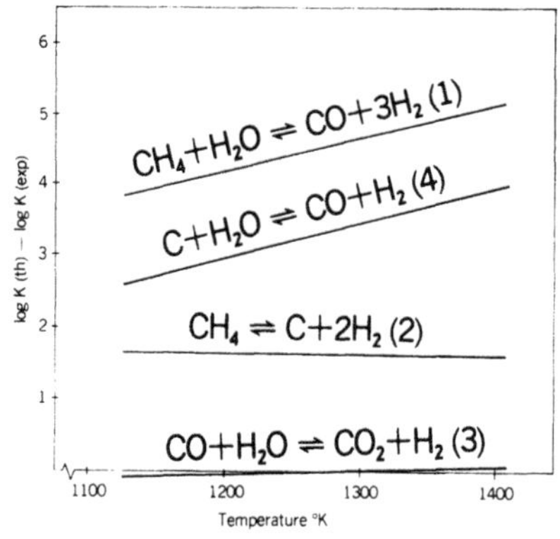

FIG. 3. Equilibrium index *vs.* reformer temperature.

Substantial amounts of soot were also found in the reformer after every experiment. The formed soot contained, besides carbon, small amounts of hydrogen (0·6 %), oxygen (1·8 %), nitrogen (0·7 %) and sulphur (0·4 %).

3. FUNDAMENTAL PROBLEMS

The experimental results are not easily interpreted in fundamental mechanistical terms. The curves of the reformer tube give rise to some problems:

—Why is the methane yield, or concentration, always so very much higher than what is predicted from thermodynamics?
—Why is this high yield so very weakly dependent on mean residence time in the reformer?
—Why do the curves always pass through a maximum? The gas entering the reformer is totally free from solid carbon.
—Why does higher pressure always result in higher yields of methane?

This article does not present a definite answer to these problems but suggests some hypotheses.

4. LITERATURE ON METHANE YIELD

Not very much has been published on the methane yield from biomass gasification. A short review is given here of some articles on the matter. The region of interest is that of 700–1100 °C and with gas residence times in the order of seconds to tens of seconds.

It is natural to start with the assumption that the methane yield from gasification is the sum of producing and consuming reactions. This starting point does not, however, seem to be common in the literature even though different investigators are obviously interested in the reactions of the same molecules under similar conditions. This is most striking when reading literature on methane yield in coal gasification. The methane producing reactions are mostly discussed in articles on SNG from coal. The proposed mechanisms show a considerable variation but some of them may be applicable also to biomass gasification.

Blackwood is one of the few authors who uses biomass char in his investigations.[1] The rate of methane formation was investigated with hydrogen pressures up to 4·0 MPa and at temperatures in the range 650–870 °C.

A first order rate equation is proposed and a typical rate for the hydrogenation of coconut or wood char is given at partial pressure of hydrogen 2·0 MPa and 830 °C: $k_0 = 3{\cdot}0 \times 10^{-5}$ mole methane/(min g carbon). Blackwood is also able to correlate methane yield to oxygen content in the char. No oxygen gives a methane production rate equal to zero.

In a later article Blackwood and McCarthy present a discussion of the mechanisms of the methane-producing reactions.[2] It was concluded that the hydrogenation is a two-stage process where the first step is a rapid hydrogenation of volatiles and the second is a slow hydrogenation of the fixed carbon. The rate of the first step is proportional to $P^n_{H_2}$ where n is less than unity, whereas the rate of the second reaction is proportional to P_{H_2}. The major part of the methane yield is normally produced in the first step.

Squires presents an investigation of steam–oxygen gasification of coal in a fluid bed and discusses the methane yield.[3] A very strong covariance is observed between the methane yield and the hydrogen partial pressure. The results are interpreted by an assumption that the methane yield is explained by the thermodynamical conditions in the fluid bed. A hypothesis of a 'quasi-equilibrium' is then proposed as the methane yield is so very much higher than could be expected from the true equilibrium:

$$C(\text{graphite}) + 2H_2 \rightleftharpoons CH_4$$

Squires' hypothesis implies that hydrogen and methane are in equilibrium in the presence of a solid carbon with a very much higher activity than graphite or even amorphous coal. An activity of up to 20 has to be assumed.[4]

Johnson gives an extensive review of the literature on methane production in coal gasification and proposes also a new mechanism.[4] The methane (and the ethane) is assumed to originate from three independent reactions:

—thermally activated decomposition of coal during the initial devolatilization,
—gas-phase hydrogenation of volatiles,
—rapid-rate hydrogenation of the remaining char (semichar).

The methane yield is also in Johnson's model very much dependent on the hydrogen partial pressure.

None of these investigators seems to consider the methane consuming reactions to have any influence, which may lead to oversimplifications. The following authors have treated the consuming reactions.

Kassel studied methane decomposition in the temperature region 700–850 °C and found a first order rate equation with a rate constant $k_0 = 5 \times 10^{11}$ 1/s and activation energy 330 kJ/mole.[5] He also stated that the reaction rate is very greatly retarded by hydrogen and when the partial pressure of hydrogen is high a 'false-equilibrium' is observed with a 'constant' $P_{H_2}^{1 \cdot 5}/P_{CH_4}$.

Gordon studied the decomposition at temperatures between 1000 °C and 1100 °C and found that the reaction rate was unaffected by the presence of steam.[6]

Skinner and Ruehrwein studied the pyrolysis and oxidation of methane in the temperature range 925 °C to 1525 °C and found that the reaction rate was of first order and had an activation energy of 420 kJ/mole.[7] The presence of oxygen accelerated the decomposition by at least one order of magnitude.

Minkoff and Tipper give a comprehensive review of the literature and present a temperature dependence for the first order reaction rate constant which can be expressed as $k_0 = 1 \cdot 06 \times 10^{14}$ 1/s and with an activation energy of 400 kJ/mole.[8] The relation is given for the temperature region 1150 °C to 2050 °C.

Dugwell and Foster study the decomposition of methane in the presence of oxygen and combustion products at temperatures between 1080 °C and 1450 °C.[9] They also found that the decomposition rate is one to two orders of magnitude faster when oxygen or combustion products are present compared to results obtained without those. Dugwell and Foster also discuss the results from Skinner and Ruehrwein and explain the very much higher decomposition rates by the effect of radicals formed in the combustion reactions.

It is clear that the decomposition reactions must be accounted for in normal cases of biomass or coal gasification if the decomposition rates given by Dugwell and Foster and Skinner and Ruehrwein are valid under the prevailing conditions. This is then very plausible as gasification often is performed with air or oxygen addition to compensate for the endothermal heat of reaction.

5. HYPOTHESES

The following section gives a description of the hypotheses, and in the next section attempts are made to verify them.

(a) The solids free gas contains substantial amounts of 'tar', i.e. large oxygen-rich molecules. This tar undergoes themal decomposition to soot. The tar and/or soot is consumed in two reactions. It is either steam reformed to CO, CO_2 and H_2 or hydrogenated to methane. This latter reaction may be of a similar nature to the so-called rapid-rate methane formation, reviewed by Johnson.[4] The methane is also, simultaneously, decomposed thermally to yield graphite and reacted with steam to form CO and H_2. The methane yield is thus controlled by the following, under the prevailing conditions irreversible, reactions:

$$\text{Tar} + H_2 \rightarrow CH_4 \tag{1}$$

$$\text{Tar} + H_2O \rightarrow CO + H_2 + CO_2 \tag{2}$$

$$CH_4 \rightarrow C(s) + 2H_2 \tag{3}$$

$$CH_4 + H_2O \rightarrow CO + 3H_2 \tag{4}$$

(b) Reactions (1) and (2) take place in the entrance of the reformer and the tar is totally consumed after a short distance. Reaction (4) is very slow. Reaction 3 does not yield graphite but a char with a higher activity than graphite or 'usual' amorphous coal. The char deposited is then in equilibrium with methane and hydrogen. This 'pseudo-equilibrium' resembles theories by Squires[3] which also are discussed by Johnson.[4]

(c) The results are not to be regarded as obtained by high temperature reforming but as results from gases which are at equilibrium at a much lower temperature. This is the same as to say that the equilibria

$$CO_2 + H_2 \rightleftharpoons CO + H_2O$$

$$CH_4 + H_2O \rightleftharpoons CO + 3H_2$$

cause the methane concentration to increase when the gas is cooled to room temperature and that the composition is 'frozen' at rather low temperature.

(d) The reformer is heated electrically but also by adding oxygen to the product gas. Methane is then combusted in the flame to a higher degree than hydrogen and carbon monoxide.

(e) The gas velocity in the reformer is so low that it is to be regarded as a tubular reactor with laminar flow or, at high mean residence times, even as a tank reactor. This explains why increased residence time does not give a much higher conversion as the degree of backmix increases, the reactant is diluted and as a consequence the reaction rate is decreased.

6. DISCUSSION OF THE HYPOTHESES

(a) The first hypothesis can give an explanation which is in qualitative agreement with the experimental results. The lower limit for the influence of the methane consuming reactions can be estimated from the following discussion.

The reformer is regarded as a tubular reactor with plug flow. This implies that the conversion X is given by

$$X = 1 - \exp(-kt)$$

for a first order reaction. k is the rate constant in 1/s and t is the residence time. From this the necessary time for a specific degree of conversion t_x is given as $t_x = \log(1-x)/k$. Table I gives some values for t_x at different temperatures for some of the rate constants discussed above. The values in the last column have been calculated from a curve for simultaneous combustion given by Dugwell and Foster.[9]

TABLE I

t_x AT DIFFERENT TEMPERATURES (s)

Temperature (°C)	*Conversion* (%)	*Kassel*[5]	*Minkoff and Tipper*[8]	*Skinner and Ruehrwein*[7]
1 100	5	0·4	0·6	0·04
	50	5	9	0·5
	95	20	40	2
1 000	5	4	10	0·3
	50	50	100	4
	95	200	600	20
900	5	50	200	4
	50	700	3 000	50
	95	3 000	10 000	200

The results from our experiments with synthetic tar-free gas without oxygen addition are well correlated to the constant given by Minkoff and Tipper. The results with synthetic gas with oxygen are not in agreement with the very small t_x values given by Skinner and Ruehrwein. A rate constant with a smaller activation energy seems to be more correct under our conditions.

The maximum in the methane yield curve at 800 °C can thus only partly be explained by the methane decomposition. The competition between the methane producing and the non-methane producing tar reactions therefore has to be the strongest factor.

This is also supported by the observation that the increase in methane yield with temperature is almost zero around 1000 °C when no oxygen is added to tar containing product gas. (See, for example, Ekström and Rensfelt.[13]) The methane decomposition cannot account for this phenomenon as can be seen in the table for t_x.

The relative independence of mean residence time which was observed is easily explained by this hypothesis by assuming that the tar is very rapidly consumed in the reformer. The reactions taking place 'downstream' do not then affect the methane yield very much, at least not at lower temperatures and especially not when no oxygen is added.

(b) The second hypothesis includes the same methane forming reactions as the first. These reactions are assumed to consume its reactants tar and possibly soot after a very short period. The methane is then decomposed to some solid carbon in reversible reactions. The methane partial pressure is then controlled by this pseudo-equilibrium. This hypothesis is supported by a striking covariance between the partial pressures of hydrogen and methane. A very good agreement is achieved when the value of $\varphi = \log(P_{H_2}^{1\cdot3}/P_{CH_4})$ is plotted against $1/T$ for all experiments. A best fit to the results gives a straight line with the temperature dependence according to

$$K = 3{\cdot}0 \times 10^4 \exp(-1{\cdot}17 \times 10^4/T)$$

which is shown in Fig. 4.

The value of φ seems to be independent of mean residence times in the range 1–25 s, and of the total pressure in the range 0·4–2 MPa. The values of

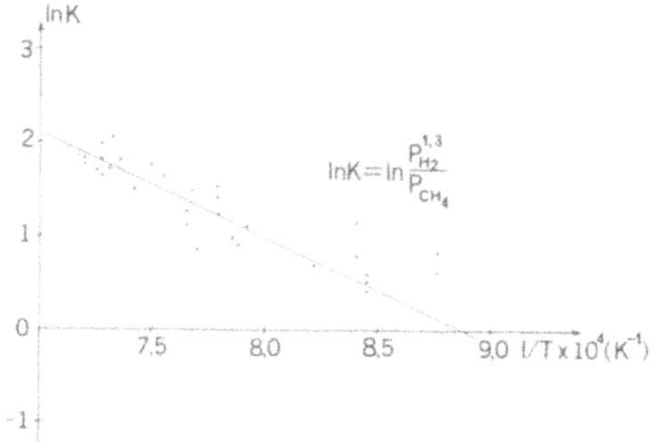

FIG. 4.

φ obtained from experiments with tar-free product gas do not show this behaviour. This may be regarded as an indication that solid deposits from tar decomposition serve as 'seeds' for the above-mentioned methane decomposition.

(c) The hypothesis of the frozen equilibrium implies a methane content which is independent of the maximum reformer temperature as the exit temperature of the gas, before quenching, is not covariating with the maximum temperature. A strong influence of the maximum temperature is observed and for that reason this theory cannot explain the experimental results.

(d) Combustion, as the most important methane consuming reaction, can explain the high methane yield and the weak dependence on residence time. The maximum of the methane yield can be explained as in hypothesis (a). However, a selective combustion of methane is very unlikely in the presence of hydrogen because of its lower activation energy for combustion and higher flame velocity.

(e) The influence of backmixing cannot be neglected as the Reynolds number is so low, especially for longer residence times. The flow is always laminar, even at the shorter residence times, with Reynolds number in the range 1800–370. This can, in combination with hypothesis (a), explain the weak dependence on residence time and that residual methane content is higher than expected.

7. COMPARISON WITH OTHER PUBLISHED RESULTS

From the discussion of the hypotheses it can be seen that both the first and the second hypothesis can give satisfactory explanations to the results obtained at RIT. It may be suspected that the results from this investigation are specific to the equipment used. A comparison with published data obtained in similar units does not support that suspicion. The data are from experiments with peat in a Koppers–Totzek, a Russian Winkler-type and an IGT fluid bed.

The experiments with the Koppers–Totzek gasifier are presented by Leppämäki[10] using milled fuel, particle size 0·1 mm. This gasifier is of a suspension type and the residence time is very short and the gasification temperature high, 1100–1200 °C. It operates at atmospheric pressure. The Winkler-type gasifier experiments were carried out at The Peat Institute in Minsk.[12] The gasification temperature was 720–760 °C and in a secondary

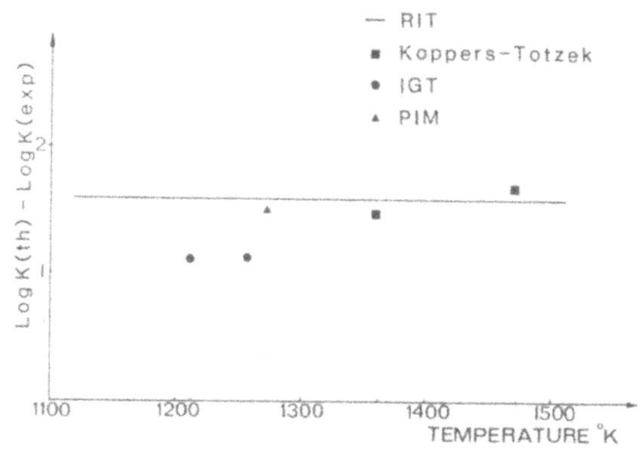

FIG. 5. Equilibrium index for product gas from different gasifiers.

blast 950–1015 °C. The IGT peat gasifier is a fluid bed operated at 940–985 °C and 3 MPa.[11]

Figure 5 shows the equilibrium index, *I*, for the methane decomposition to graphite for these gasifiers. The results from our experiments regarding methane yield do not appear to be significantly different from the others.

ACKNOWLEDGEMENT

The authors wish to acknowledge Professor O. Lindström for his encouraging support and Mr E. Rensfelt by whom this work was initiated. Grant support from the Swedish National Board for Energy Source Development is gratefully acknowledged.

REFERENCES

1. Blackwood, J. D. (1959). *Aust. J. Chem.* **12**, 14.
2. Blackwood, J. D. and McCarthy, D. J. (1966). *Aust. J. Chem.* **19**, 797.
3. Squires, A. M. (1961). *Transact. Inst. Chem. Eng. (London)* **39**, 3.
4. Johnson, J. L. (1974). *Kinetics of Coal Gasification*. Wiley, New York.
5. Kassel, L. S. (1932). *JACS* **54**, 3949.
6. Gordon, A. S. (1948). *JACS*, **70**, 395.
7. Skinner, G. B. and Ruehrwein, R. A. (1959). *J. Phys. Chem.*, **63**, 1736.
8. Minkoff, G. J. and Tipper, C. F. H. (1962). *Chemistry of Combustion Reactions*. Butterworths, London.
9. Dugwell, D. R. and Foster, P. J. (1973). *Carbon* **11**, 455.

10. Leppämäki, E. *et al.* (1976). Gasification of peat—a literature review. Technical Research Centre of Finland, Report.
11. Institute of Gas Technology (1979). Experimental program for the development of peat gasification. Interim Report No. 6: Hydrogasification and Fluid Bed Test. Chicago, March.
12. Gamburg, I. I. *et al.* (1959). *Trudi Instituta Torfa* 7. The Collected Works of the Peat Institute, Minsk.
13. Ekström, C. and Rensfelt, E. (1980). Flash pyrolysis of biomass in Sweden. Specialists' Workshop on Fast Pyrolysis of Biomass, SERI, Colorado.

52

Low Temperature Gasification of Biomass Under Pressure

DOUGLAS C. ELLIOTT and L. JOHN SEALOCK, JR
Battelle Pacific Northwest Laboratories, Richland, Washington, USA

1. INTRODUCTION

Previous studies of the influence of catalysts on the thermal gasification of biomass have shown that alkali carbonates are among the most active catalysts for the gasification of biomass with steam.[1-3] Most of these studies have investigated gasification at temperatures above 550 °C and have concentrated on gasification of the char formed after initial volatilization. When considering single reactor production of high-Btu gas from biomass, or any other organic substrate, one is faced with the contrasting reaction conditions favoring gasification and methane formation. The basic dilemma is the choice between a high temperature, low pressure reaction system which favors the breakdown of biomass to gases, and a low temperature, high pressure system which favors methane formation. In this article, we report laboratory investigations of the influence of sodium carbonate in the presence of a supported nickel metal catalyst with the intent of maximizing methane production from biomass at low temperature in a pressurized gasification system.

2. EXPERIMENTAL

A unique autoclave system illustrated in Fig. 1 was used to study low temperature (250–450 °C) gasification of biomass. The pressurized reactor vessel, manufactured by Autoclave Engineers, is a 1-liter autoclave equipped with a magnetic-coupled stirrer. The vessel was operated with standard heating and cooling and stirring equipment provided with the unit, except that the stirring baffle was removed in order to fit a stainless steel liner into

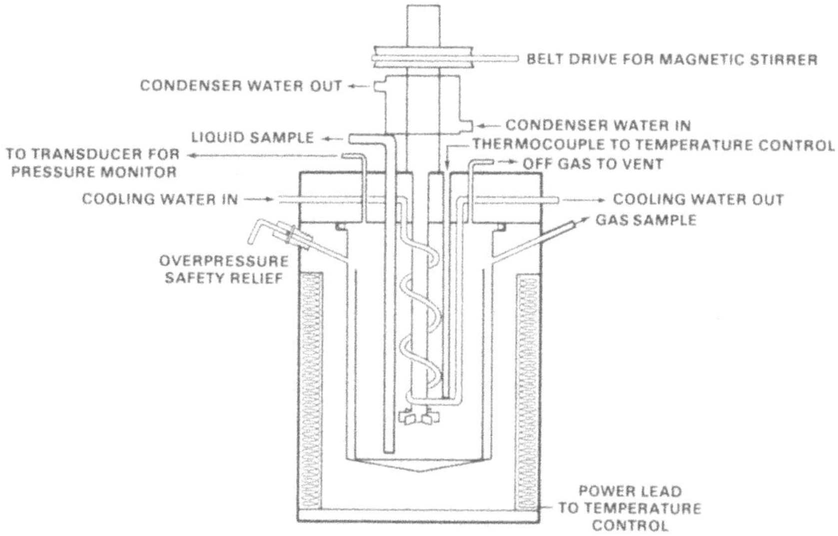

FIG. 1. One-liter autoclave modified for multiple sampling.

the reactor. The use of the liner facilitated product recovery and reactor cleanup, and contained the alkali to avoid contact between the reactants and the pressure vessel. All surfaces exposed to the reaction environment were constructed of 316 stainless steel.

The multiple sampling system used a body wall port for access to the top of the reactor for gas sampling. The standard 1/8-in. dip tube sample system was capped, and a 1/4-in. dip tube was inserted through the solids injection port in the cover of the autoclave. This larger dip tube allowed for slurry sample removal from the bottom of the reactor without plugging, which is a problem when dealing with fibrous biomass. The autoclave and sample removal system were operated remotely after the reactor was batch charged. The sample bulb system was isolated from the reactor system by an air-controlled, high pressure needle valve (Fig. 2). When the reactor reached the proper conditions and a sample was required, the air-controlled valve was opened and the sample flowed through the two open valves into the sample bulb, which was monitored via the pressure gauge. After the sample was withdrawn, which required approximately 1 s, the valves were closed, the sample bulb was disconnected at the tubing union, and another sample bulb was connected. Five samples were typically withdrawn during experiments which lasted 45–60 min at reaction temperature.

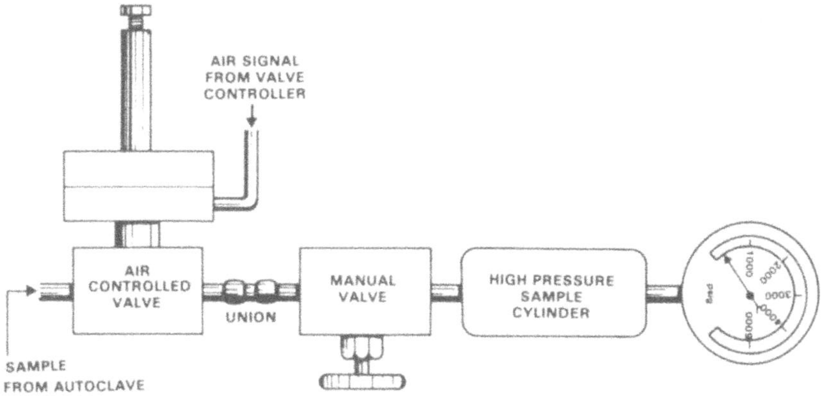

FIG. 2. Sampling bulb system.

Gasification of Douglas fir wood flour and wood components were investigated in the presence and absence of the alkali carbonate and nickel catalysts. A typical catalytic run was conducted as follows: 30·0 g of the feedstock and 5·0 g of the nickel-supported tablets were added to a solution of 300 ml of water, and anhydrous alkali carbonate was added to form a 0·321 molar solution. The resulting mixture was placed into the body of the autoclave inside the stainless steel liner. Once the top was securely fastened and the stirring initiated, the reactor was purged and then pressurized to 300 psig with nitrogen. The first sample was drawn from the top or bottom of the reactor when the predetermined temperature was attained. Additional samples were drawn at 15 min intervals thereafter. The general procedure and sampling for uncatalyzed degradation of the two feedstocks was identical except that the alkali carbonate catalysts and the nickel tablets were eliminated.

All gas compositions were determined using a Carle AGC-S gas chromatograph set up for refinery gas analysis. Integration was performed by a Spectra-Physics SP4000 central processor. The feedstocks used were: (1) Douglas fir wood flour produced by drying and milling paper-grade wood chips to −40 mesh; (2) cellulose purchased as α-cellulose from Baker Chemical Company; (3) holocellulose isolated at the University of Montana under the direction of Dr Fred Shafizadeh; and (4) lignin commercially available from Westvaco Corporation as Indulin ATR, a neutralized softwood kraft lignin.

The sodium carbonate gasification catalyst used in the gasification experiments was analyzed reagent grade or better. The nickel catalyst used was Ni-1404 from the Harshaw Chemical Company. It is a supported metal catalyst consisting of 68 % nickel, reduced and stabilized, on a proprietary support (1/8 in. tablets).

3. RESULTS AND DISCUSSION

Results of gasification studies are presented in Figs 3, 4, 5, and 6. Carbon conversion is defined as the percentage of carbon in the feedstock determined to be present in the product gas as carbon oxides or as C_1 or C_2 hydrocarbons. The gas production measurement was based on a calculation of gas production as a function of declining nitrogen concentration in the reactor as biomass was converted to gas. In order to make this calculation, an initial measured quantity of nitrogen was injected into the reactor at the beginning of the test as a cover gas. As the gasification process proceeded the initial amount of nitrogen was steadily

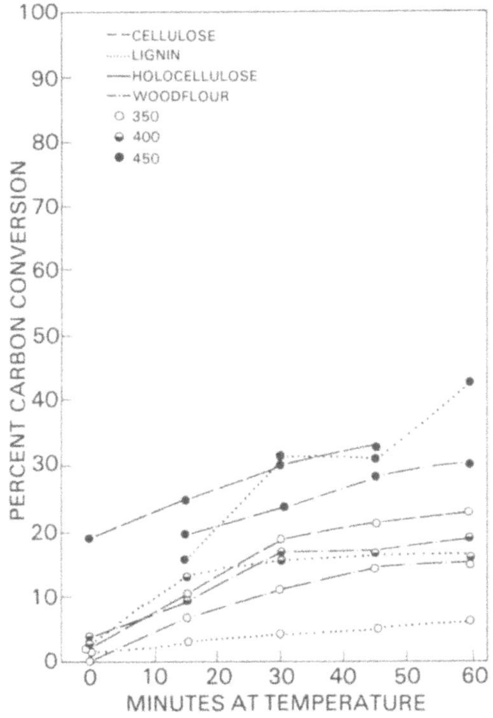

FIG. 3. Uncatalyzed biomass gasification. Carbon conversion to gases, not corrected for gas solubility in aqueous solution at ambient conditions.

diluted from 100% concentration by the product gases formed during gasification of the biomass. This measured dilution along with the concentration of the individual product gases was the basis for the calculation of the mass of product gas produced.

Little differentiation could be made for uncatalyzed gasification of wood and wood components at the 450 °C end of the study. However, lignin exhibited a dramatically low rate of conversion at lower temperatures. This is shown in Fig. 3 and serves as the baseline case for the catalyst studies.

The addition of sodium carbonate to the system increases the rate of carbon conversion for all feedstocks, at all temperatures, except for lignin at 450 °C. The catalytic effect seems to be greatest for the lignin components at the lower temperature, 350 °C, while the cellulose components are affected more at the higher temperature, 450 °C. The data from the sodium catalyzed tests are depicted in Fig. 4.

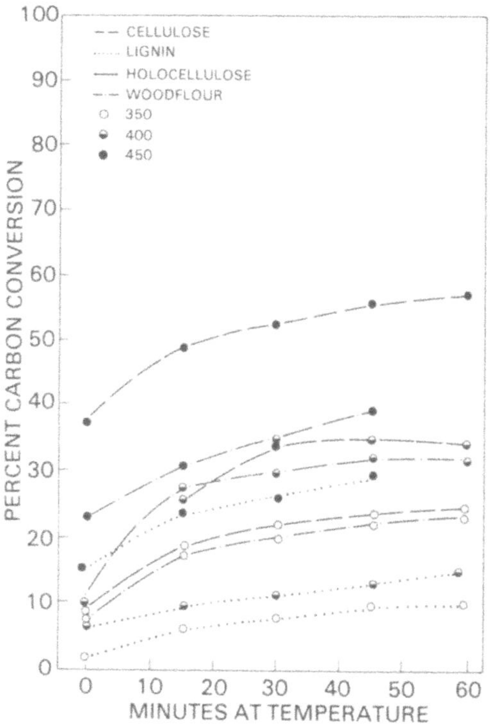

FIG. 4. Biomass gasification catalyzed with sodium carbonate. Carbon conversion to gases, not corrected for gas solubility in aqueous solution at ambient conditions.

The sodium carbonate catalyst increased carbon conversion for nearly all combinations of feedstocks and reaction temperatures; however, no similar trends are apparent for the nickel catalyst cases. As with sodium carbonate, nickel improved carbon conversion for lignin at the low temperature, but decreased carbon conversion to gas at 450 °C relative to the uncatalyzed case. At the higher temperatures there was substantial improvement in the rate of carbon conversion for the cellulose and wood flour in the presence of the nickel. As shown in Fig. 5 50% and greater conversion was achieved in these tests after only 45 min.

The combination of these two catalysts resulted in a dramatic increase in rates of carbon conversion as shown in Fig. 6. Conversions of 90% or greater are indicated for the cellulose and wood flour after 1 h at 450 °C. The ordering of the four feedstocks, lignin, Douglas fir wood flour, holocellulose and cellulose, from lowest to highest rate of conversion was consistent at all three temperatures. The great increase in activity with the

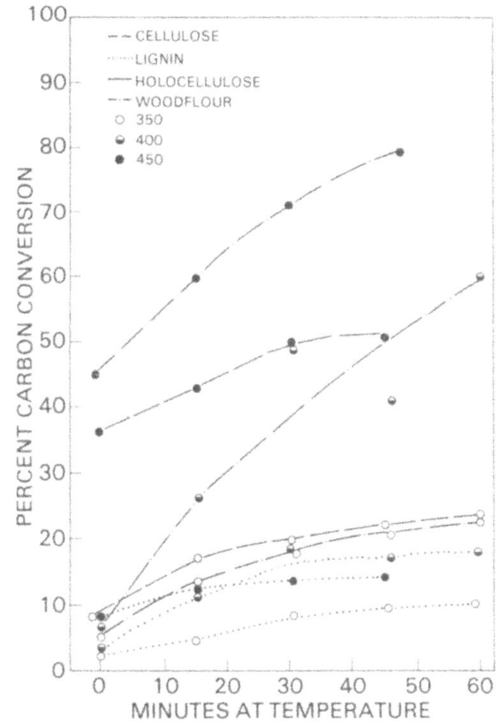

FIG. 5. Biomass gasification catalyzed with nickel. Carbon conversion to gases, not corrected for gas solubility in aqueous solution at ambient conditions.

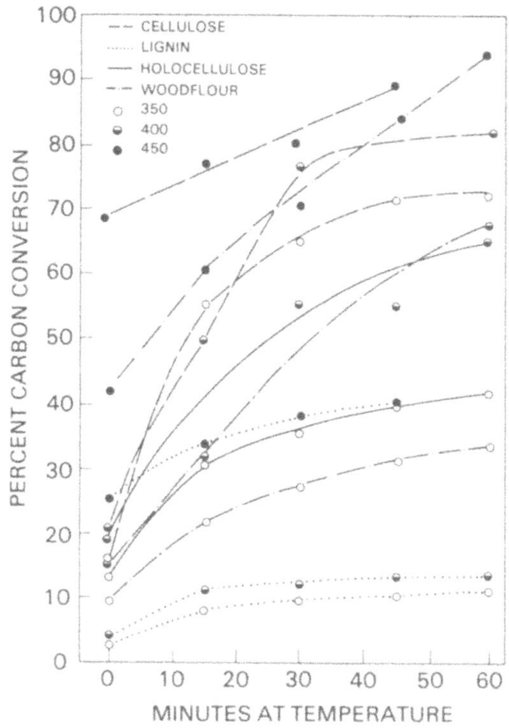

FIG. 6. Biomass gasification catalyzed with nickel and sodium carbonate. Carbon conversion to gases, not corrected for gas solubility in aqueous solution at ambient conditions.

combined catalysts could be due to a number of reasons. It may be merely a fortuitous combination wherein the non-gaseous product produced preferentially with one catalyst is easily converted to gas in the presence of the second catalyst. Alternatively, the alkali, in addition to its gasifying effect, may serve as a 'promoter' with the nickel catalyst causing the nickel to preferentially catalyze gas-forming reactions.

3.1. Gas Product Composition

Comparisons of gas product compositions for wood and wood component catalytic gasification are presented in Table I. It is a listing of gas produced during a 1 h experiment with a feedstock-to-water ratio of 1 to 10. Gas samples were removed every 15 min and the calculated final mass of gas is reported in the table. The autoclave feedstock loading was less at 450 °C

TABLE I

FINAL PRODUCT GAS COMPOSITION CALCULATED IN GRAMS

Feedstock	*Catalyst*[a]	*350°C*				*400°C*				*450°C*			
		CH_4	CO	$C{=}C/C_2$	H_2/CO_2	CH_4	CO	$C{=}C/C_2$	H_2/CO_2	CH_4	CO	$C{=}C/C_2$	H_2/CO_2
Cellulose	O	0·062	0·406	0·048[c]	0·010/4·992	0·187[b]	0·742	0·170[c]	0·021/5·541	0·506[b]	0·405	0·036/0·209	0·107/6·932
	Na	0·013	0·061	0·014/0·009	0·264/10·203	0·076[b]	0·053	0·081[c]	0·215/8·386	0·866[b]	0·072	0·088/0·462	0·377/12·84
	Ni	0·577	0·427	0·016/0·064	0·069/7·940	1·515[b]	0·256	0·238[c]	0·394/12·73	1·429[b]	0·205	0·004/0·150	0·295/10·47
	Na/Ni	1·792	0·048	0·029/0·284	0·671/18·27	3·74	0·036	0·000/0·407	0·602/20·13	2·940[b]	0·087	0·006/0·325	0·508/15·44
Holocellulose	Na/Ni	1·263	0·054	0·001/0·120	0·417/14·90	3·598	0·047	0·008/0·402	0·653/21·73	3·703	0·042	0·009/0·507	0·495/22·537
Lignin	O	0·219	0·171	0·010/0·030	0·010/2·639	1·202	0·324	0·017/0·156	0·044/5·962	1·975[b]	0·102	0·021/0·379	0·246/11·12
	Na	0·148	0·002	0·005/0·012	0·091/4·776	0·392	0·000	0·009/0·095	0·211/7·071	1·175[b]	0·018	0·018/0·330	0·281/8·095
	Ni	0·438	0·109	0·010/0·034	0·081/4·223	1·097	0·100	0·013/0·139	0·115/7·457	0·668[b]	0·094	0·006/0·101	0·109/3·387
	Na/Ni	0·251	0·005	0·005/0·022	0·131/5·869	0·468	0·000	0·006/0·091	0·184/6·378	1·559[b]	0·029	0·018/0·343	0·357/11·07
Wood flour	O	0·318	0·506	0·034/0·042	0·034/6·333	0·455	0·154	0·026/0·114	0·070/7·875	0·428[b]	0·231	0·014/0·137	0·057/3·953
	Na	0·138	0·069	0·018/0·028	0·321/10·96	0·433	0·034	0·050/0·225	0·440/14·00	0·870[b]	0·042	0·068/0·409	0·252/9·714
	Ni	0·704	0·389	0·018/0·061	0·155/8·628	4·101	0·197	0·010/0·310	0·582/18·54	4·169[b]	0·046	0·002/0·146	0·403/14·70
	Na/Ni	1·037	0·039	0·004/0·095	0·450/13·86	3·753	0·025	0·001/0·449	0·446/23·36	3·65[b]	0·038	0·003/0·330	0·534/16·98

[a] O = no catalyst added, Na = 10·2 g sodium carbonate added, Ni = 5 g of supported nickel metal catalyst added.
[b] Experiments performed with 20 g of feedstock charge, all others used 30 g of feedstock.
[c] Total C_2 hydrocarbons reported as ethane.

due to the pressure limitations of our equipment. Ratios of reactant masses were held constant in all experiments.

Considering first the effects of catalysts on cellulose, uncatalyzed gasification leads to carbon dioxide (CO_2) as the primary gas product with a significant amount of carbon monoxide (CO). Methane (CH_4) and hydrogen (H_2) are lesser products produced on a nearly equimolar basis. Increasing the temperature increases the production of all the gases, with the increase in CH_4, ethane (C_2) and ethylene (C=C), being much greater than for the other gases. The addition of sodium carbonate to the system results in a dramatic reduction in CO with an even more dramatic increase in H_2. Carbon dioxide production also increases with the addition of the sodium carbonate catalyst. This is apparently due to the combined effect of increased gasification and a strong water–gas shift reaction. Hydrocarbon concentrations are greatly reduced at 350 °C, less affected at 400 °C, and increased at 450 °C.

Introduction of nickel into the system causes an increase in CH_4 production resulting from a similar decrease in CO. The combination of nickel and sodium carbonate results in the highest amount of methane from cellulose. The combined catalysts result in a substantial reduction in CO yield but the reduction is similar in magnitude to that with sodium carbonate alone. Also, a hydrogenation effect on the C=C is noted, based on the near total disappearance of C=C and an increase in C_2 in the product gas. The introduction of nickel into the system also results in an increase in hydrogen product gas although the magnitude of this change is not consistent relative to temperature. A clear trend can be noted for increased CO_2 production relative to the introduction of nickel. Thus, it appears nickel is also catalyzing some gasification reactions.

In uncatalyzed lignin gasification, CO_2 is the primary product as was found with cellulose. The amount is only one-half of the CO_2 production from cellulose at 350 °C and increases over the temperature range so that there is actually greater production at 450 °C than from cellulose. Methane is the secondary product and is actually produced in larger quantities than from cellulose. On a molar basis H_2 production is nearly as great as CH_4 production at 450 °C. Increasing temperature tends to increase the production of all the product gases, except that CO remains nearly constant. The increase in C_2 production is much greater than for C=C. The introduction of sodium carbonate to the system decreases the hydrocarbon production. Carbon monoxide is reduced to immeasurably small levels in nearly all cases. The combined effect of this CO reduction and a coincidental increase in hydrogen again suggests a water–gas shift effect. The lack of a clear trend for CO_2 production again muddies this conclusion; however, CO_2 solubility in alkali is suspected as the confounding element. The addition of nickel increases CH_4 production at 350 °C while causing a decrease in all hydrocarbons at 450 °C. As might be expected, the effect on CO production is exactly the opposite. The effect of nickel on H_2 and CO_2 production from lignin follows no clear trends.

As was seen with lignin and cellulose, the primary gas product from Douglas fir wood flour is CO_2. Methane, CO and H_2 are lesser products produced in nearly equimolar concentration. Increasing temperature decreases the CO and CO_2 production while H_2 and CH_4 increase slightly. In the uncatalyzed case, total production of C_2 hydrocarbons does not change much; however, the mole ratio increases to 10:1 from 1:1 for C_2 to C=C. The addition of sodium carbonate causes decreases in CH_4 and CO yields. Nickel causes a great increase in CH_4 production. The addition of sodium carbonate with the nickel catalyst actually reduces the methane production except at the low temperature end of our study. With Douglas fir wood flour as the feedstock nickel appears to increase both H_2 and CO_2 production.

3.2. Methane Production

Methane production from biomass feedstocks with various catalyst systems is shown in Fig. 7. These data are for experiments which lasted 60 min at temperature. The graph actually depicts a composite temperature and pressure effect, since in this sealed system the pressure varied with the temperature. Typical pressures varied from 2400–2600 psig at 350 °C to 4600–5000 psig at 450 °C.

Gas composition varies significantly with feedstocks. Lignin is a major source of CH_4 by simple thermal degradation, but the use of catalysts with cellulose can result in even higher yields of CH_4. The fact that CO_2 is the primary product from all three feedstocks may result from a combined effect of the water–gas shift reaction under the strong influence of the high concentration of water as well as the thermal splitting of CO_2 and CO from the biomass. The most significant effect of the alkali catalyst is the reduction of CO, CH_4 and the two carbon hydrocarbons, apparently through a combination of water–gas shift and catalysis of alternate gas-forming reaction routes. An increase in hydrogen production was also noted. Nickel causes an increase in CH_4 and a decrease in CO as well as a shift of C=C to C_2. The highest methane production was found with the wood flour feedstock catalyzed with nickel only at 450 °C.

Actual product gas composition is an important consideration when

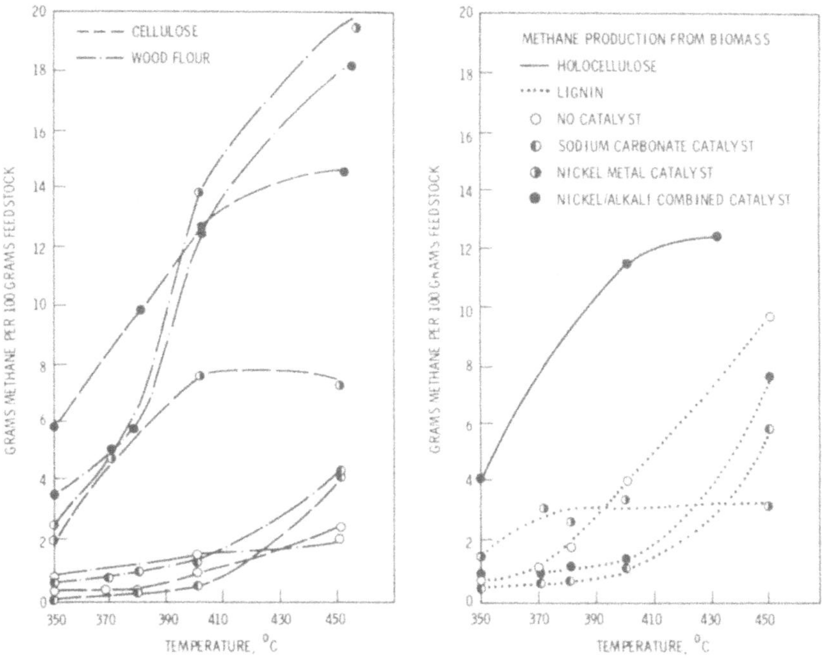

FIG. 7. Methane production from biomass in a pressurized aqueous reactor system.

comparing biomass components. Figure 8 presents the final gas composition after 1 h at various temperatures for each of the four biomass feedstocks studied. These particular sets of data were chosen because they represent the highest concentration of methane produced in our test system. The combined catalyst of sodium carbonate and nickel produced the highest methane concentration from cellulose and wood flour. In the lignin case, the nickel alone produced the highest concentration of methane at all temperatures, although the use of no catalyst at 450 °C resulted in the highest actual production of methane. In the sodium catalyzed cases presented in Fig. 8 it can be seen that the carbon monoxide level is approximately 0·1 % at all temperatures. The hydrogen content is high—generally about 30 %. This is probably the effect on equilibrium due to the relatively vast amount of water in the reactor system. It is also interesting to note that the methane content levels off or even drops above 400 °C. This also appears to be an equilibrium limitation as a result of the increasing temperature.

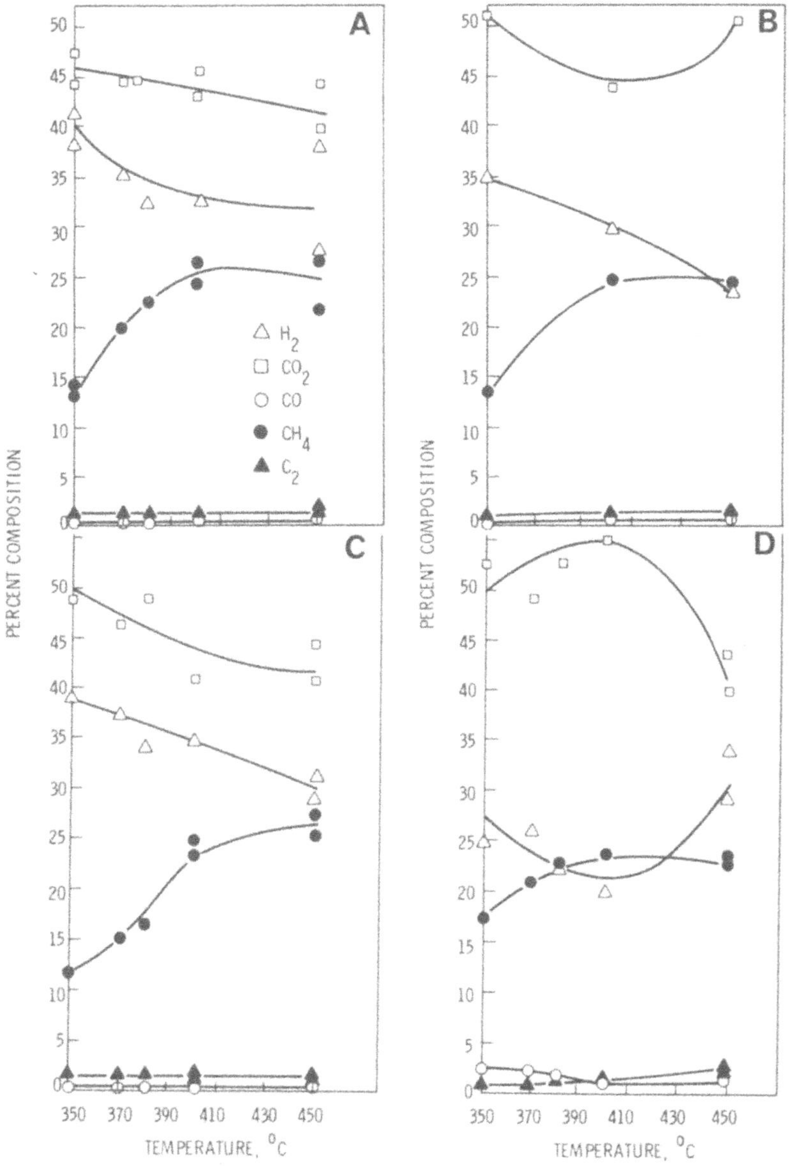

FIG. 8. Gas compositions from catalytic biomass gasification. (A) Cellulose/ Ni–Na_2CO_3; (B) holocellulose/Ni–Na_2CO_3; (C) wood flour/Ni–Na_2CO_3; (D) lignin/Ni only.

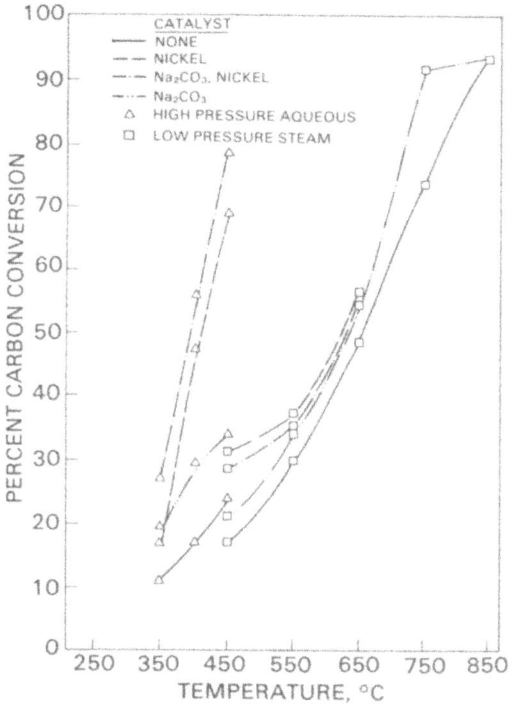

FIG. 9. Comparison of wood flour gasification in the low temperature and high temperature reactor systems.

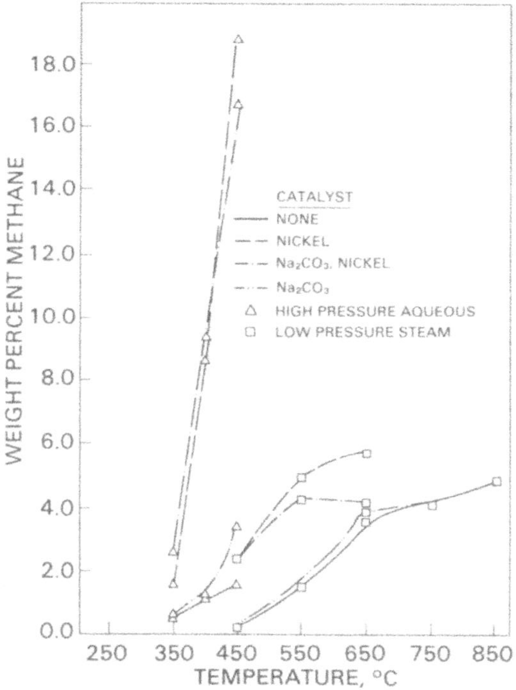

FIG. 10. Comparison of methane production from wood flour in the low temperature and the high temperature reactor systems.

3.3. Comparisons of Low and High Temperature Gasification

The data in Fig. 9 are from the experiments reported here compared with results from an atmospheric pressure, high temperature steam gasification reactor reported elsewhere in detail.[4] Though it was expected that the high pressure water system at the low temperature could conceivably accentuate methane production, the increase in carbon conversion to gases relative to an unpressurized system was unexpected. Of the differences between the two reaction environments, vapor residence time and system pressure are the obvious causes of the differences in gas production. Pressure effects have been discussed earlier; the vapor residence time affects the extent of breakdown of the tars and condensibles produced during volatilization. In the high temperature, atmospheric pressure reactor these materials were swept from the vessel shortly after they were produced, while in the low temperature, pressurized reactor all products were sealed in the reaction environment for the total time of the experiment. This increased gas residence time in the presence of the nickel catalyst along with the high pressure system also allowed for greater methane production as shown in Fig. 10.

ACKNOWLEDGEMENTS

The authors are appreciative of the financial support for this work granted to Battelle Pacific Northwest Laboratories by the Gas Research Institute, Basic Research Department. We are grateful to Dr John Cox, Manager Organic Chemistry, and Dr Charles Nelson, Project Scientist, Coal Chemistry for their support and interest in our efforts. We also wish to thank Dave Ortiz for his valuable contributions in conducting the laboratory experiments and Richard Hallen for his consultations regarding data interpretations.

REFERENCES

1. Mudge, L. K., Sealock, L. J., Jr and Weber, S. L. (1979). *J. Anal. Appl. Pyrolysis* **1**, 165–75.
2. Ross, R. A. and Fong, P. (1981). *Conserv. Recycling* **4**(1), 15–28.
3. Sealock, L. J., Jr, Elliott, D. C., Weber, S. L. and Robertus, R. J. (1981). Kinetics and catalysis of producing synthetic gases from biomass. Available from NTIS No. PB81-217614.
4. Sealock, L. J., Jr, Elliott, D. C., Hallen, R. T., Barrows, R. D. and Weber, S. L. (1981). Kinetics and catalysis of producing synthetic gases from biomass. Available from Gas Research Institute No. GRI-80/0116.

53

Biomass Gasification Reaction Velocities

T. B. Reed and M. Markson
Solar Energy Research Institute, Golden, Colorado, USA

1. INTRODUCTION

The controlled propagation of flames is an important foundation of our modern technology—from the humble gas flames in our stoves and the combustion of gasoline in our cars, to the powdered coal flames in our largest power plants.

We take it for granted that the flames in these devices are stable, but the control of some of these flames is only a few decades old. The control of gaseous flames is particularly simple; that of liquid fuel flames somewhat more complex. However, the control of flames of solids is much more difficult and less well understood. Yet we are approaching the end of our supplies of low cost gas and liquid fuels, and will more and more need to generate and control flames in the more abundant lower cost solid fuels—biomass, coal and peat.

2. BACKGROUND

2.1. Static and Dynamic Flame Propagation

A 'flame' (Fig. 1) is a self-propagating reaction in which, once ignited, the unburned reactants are heated by the flame front and encounter reactive species; they, in turn, react to continuously propagate the reaction into the unburned reactants, while the reaction products and the resulting heat are carried away behind the advancing front.[1] The sharp rise in temperature at the reaction front causes heat to flow back into the unburned gases and reactive species to diffuse into the reactants at rates which determine the 'flame velocity', V_r. Compared to gasifiers, gas flames are very well

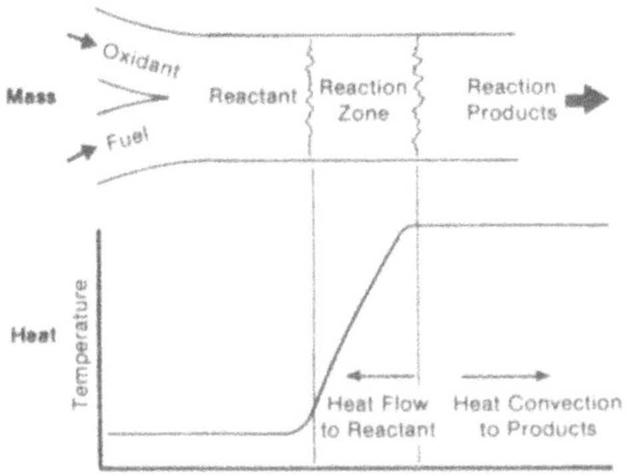

FIG. 1. Heat and mass transfer in a self propagating reaction (flame).

understood and provide a useful model for understanding the reaction propagation in gasifiers.

If the flame front shown in Fig. 1 is stationary in space, the flame is 'static' (or stable); if the flame front moves, the flame is 'dynamic'. In many continuous processes (cooking, welding) we use static flames. An explosion is a dynamic propagation. The art and science of combustion deals in part with understanding the reaction propagation mechanisms and conversion of uncontrolled dynamic reactions into static reactions. (While Fig. 1 shows a reaction front moving in one dimension, the reaction can also propagate in two and three dimensions.)

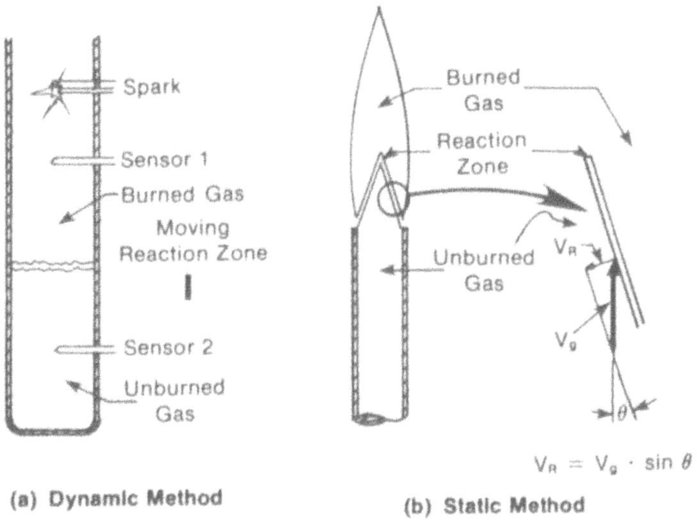

FIG. 2. Measurement of flame propagation velocity.

The 'flame velocity' of gases is sometimes studied in a reaction tube similar to that shown in Fig. 2(a). The tube is filled with a mixture of fuel and air; the mixture is ignited with a spark and the velocity of the front between two temperature sensors is recorded.

Conceptually, this dynamic moving flame front could be converted to a static front if the gases at the left entered the tube at a velocity equal to the flame velocity. In practice, it would be very difficult to maintain the flame at exactly one point.

3. REACTION STABILIZATION

The simple Bunsen burner of Fig. 2(b) automatically stabilizes the flame front over a wide range of gas velocities. The unburned reactants leave the burner tube at a velocity, V_g, higher than the flame velocity. The flame front then assumes a conical shape with the side of the cone at an angle θ to the burner axis. The reaction moves into the unburned reactants at a velocity, V_g, just sufficient to equal the forward motion of the gases. As shown in the vector diagram, the condition for stable burning is:

$$V_f = V_g \sin \phi$$

(Complex interactions near the rim of the burner make analysis at this point more complex.)

In general, the stabilization of flame fronts is not as simple and elegant as that of the Bunsen flame described above, but the principles are similar, i.e. means must be provided so that heat and reactive species can flow into the unignited mixture; the means must then be provided to limit the extent of this flow in space to produce a positionally stabilized reaction. We will discuss the mechanisms of propagation in solids relevant to gasifiers and the means to stabilize the moving front.

4. REACTION PROPAGATION IN GASIFIERS

There are a variety of gasifiers for coal and biomass in which the solid fuel is converted to a gaseous fuel on a continuous basis using partial oxidation or pyrolysis.[2,3] In all cases steady state operation depends on controlling the rate of propagation of the gasification reactions. In fluidized beds and suspension gasification, there is continuous relative motion of the discrete solid particles, while in fixed bed gasifiers there is much slower stratified

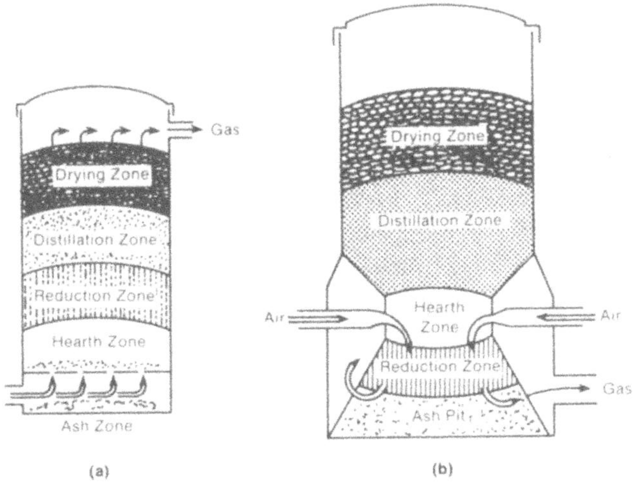

FIG. 3. (a) Updraft gasifier; (b) conventional downdraft gasifier.

motion of the solid bed as a whole, generally downward by gravity. The gas can then pass either in the opposite direction to the fuel (updraft or counterflow gasifiers) (Fig. 3a), or in the same direction (downdraft or coflow gasifiers) (Fig. 3b). While fixed bed coal gasifiers are updraft, biomass gasifiers can be either updraft or downdraft, depending on the purpose for which the gas is to be used.

Updraft gasifiers can use very wet wood, but because the pyrolysis products are immediately cooled, the gas contains up to 30 % of tar and oil. This gas is usually used in boilers close-coupled to the gasifier to prevent condensation. In the downdraft gasifier, the pyrolysis occurs before the char gasification so that the volatile materials partially burn in the incoming air and then pass through the hot char. The gases then typically contain 200–2000 ppm of tar and are sufficiently clean to be used to operate engines for shaft power, for chemical synthesis, or for pipeline use.

5. GASIFIER MODELLING

A modest effort has gone into modelling the action of gasifiers in order to be able to better predict behavior under varied conditions.[4–6] In 1978, a program at SERI was initiated to build an updraft (coflow) gasifier (Fig. 3a) and collect data on temperatures and gas compositions that would be useful for modelling the updraft gasifier.[6] The updraft gasifier was chosen

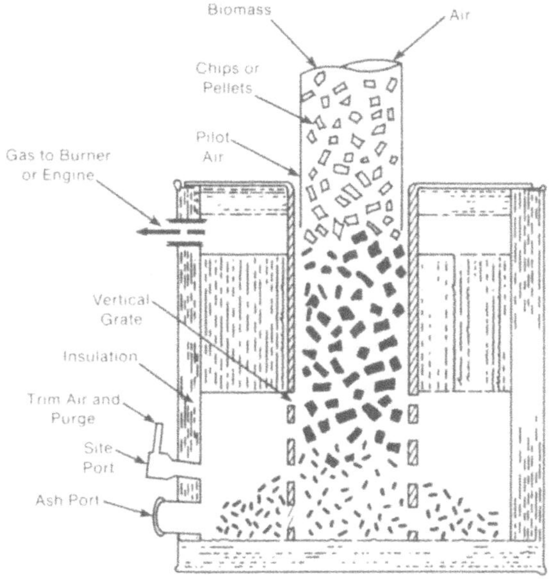

FIG. 4. 10 hp air stratified downdraft gasifier.

because the reactions can be clearly stratified, reducing the flow to one dimension, and greatly simplifying the mathematics and physical measurements. However, updraft gasifiers produce large amounts of tars and oils and this makes analysis and data gathering difficult.

The 'downdraft' (or coflow) gasifiers (Fig. 3b) used widely during World War II produce very little tar and oil and are much more practical in small sizes for producing clean gas. Air (oxygen) is introduced at nozzles or tuyères which form a complex reaction zone not amenable to scale-up or simple analysis.[7] We have modified the downdraft principle to build a high pressure oxygen gasifier that makes very clean gas for methanol synthesis.[8]

We have now developed several 'stratified' downdraft gasifiers (for instance, shown in Fig. 4) that give approximately one-dimensional flow. In operation, air or oxygen and biomass both enter the open top and pass together down to the reaction zone. One advantage of this arrangement is that the temperature at the reaction zone is uniform across the gasifier rather than having hot spots at the air nozzles. A second advantage is that the gasifier can be easily scaled. Another advantage is that it is very easy to introduce the feed at the open top and no seals or lock hoppers are required at atmospheric pressure. A very important advantage in studying the reaction velocity is that it is easy to introduce sampling probes and thermocouples through the open top to measure the position, composition and temperatures of the reaction zone. However, it introduces a problem of stabilizing the position of the reaction zone. We will discuss methods of stabilizing this zone below.

6. EXPERIMENTAL

In order to study the stability of the reaction zone position, a 'transparent stratified downdraft gasifier' (shown in Fig. 5a) was built. A 5·4 cm inside diameter quartz tube 30 cm long was cemented with silicone rubber into a 2 in. pipe coupling and attached to a heat exchanger, a blower, an orifice meter and flare as shown in Fig. 5(b). The orifice was calibrated using a positive displacement meter.

Because of the relatively small tube diameter, the measurements were limited to relatively small charcoal and biomass sizes, typically 1 cm in largest dimension.

In operation, charcoal is loaded into the tube to a height of about 10 cm, the gas flow is set, and the charcoal is ignited. After a stable hot bed of coals is established, a charge of charcoal or biomass is added on top of the hot coals, and the position of the fuel level and the incandescent interface are recorded as a function of time.

In some cases fibrous insulation 1·4 cm thick was added around the quartz tube to reduce heat losses. A 5 mm slot was cut in the insulation to permit observation of the reaction zone.

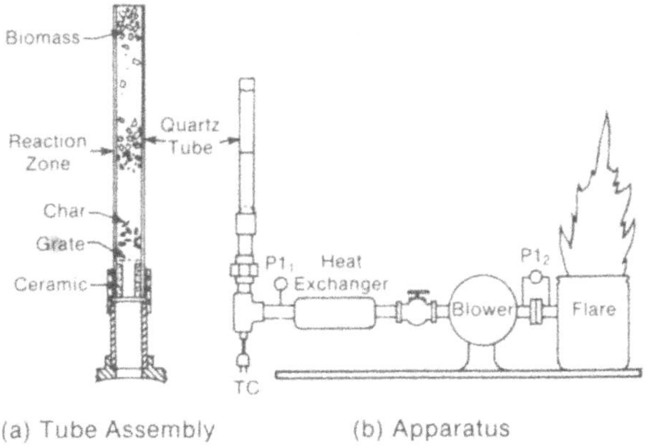

FIG. 5. SERI transparent downdraft gasifier.

7. RESULTS

The results of a particular run will be discussed in detail, and most of the comments then apply to the other figures and results.

The quartz tube (Fig. 6c) with insulation in place was filled to 25 cm above the grate with hardwood charcoal, the blower was started and the throttle valve set to give a flow of 2·1 scfm. The charcoal was ignited with a propane torch and when the bed was glowing uniformly biomass was added to fill the tube to about 40 cm. The temperature shown diagrammatically in Fig. 6(a) is a composite of several measurements made with thermocouples and an optical pyrometer. Actual temperatures at each point can vary widely around those shown, depending on the exact point of measurement and method of measurement. Temperatures will be discussed more fully in succeeding papers.

Observations were made approximately every minute of the position of the top of the fuel bed, the top of the incandescent 'flaming pyrolysis' bed, and the bottom of the incandescent char bed, and the results are shown in Fig. 6(b). The figure shows that it was necessary to add fresh chips every 3–4 min. Initially, the lower parts of the char bed were cold, but the downward heat flow caused the visibly glowing interface to progress down to the grate in about 10 min after which there was no apparent change in the bed temperatures. Chips were added continuously for 30 min, at which point the blower was extinguished and the experiment terminated. Note that during that period the top of the flaming pyrolysis zone moved down about 4 cm and that the last 10 min of operation were very close to a 'steady state' condition.

The total weight of chips fed over a 30 min period was 0·623 kg (1·37 lb), an input of 1·23 kg/h (2·74 lb/h) or about 23 156 kJ/h (21 970 Btu/h), assuming the chips had a heat of combustion of 18 550 kJ/kg (8000 Btu/lb). The energy content of the gas was not measured in this experiment but in other measurements we find the heat content of the gas varies between 3460 and 5190 kJ/m^3 (100 and 150 Btu/scf). The gas flow was 0·055 N m^3/min (2·1 scfm) for this experiment, corresponding to 13 280–19 920 kJ/h (12 600 to 18 900 Btu/h) energy content of the gas. Thus the cold gas efficiency of the gasifier (energy in gas/energy in fuel) probably lies in the range of 57–86 %.

While the position of the reaction zone in Fig. 6 is almost constant over a long period, this is not necessarily observed in all cases. The variation of zone position with gas flow is shown in Fig. 7 in an uninsulated tube for gas flows of 0·026, 0·055 and 0·0894 N m^3/min (1·0, 2·1 and 3·4 scfm). For this

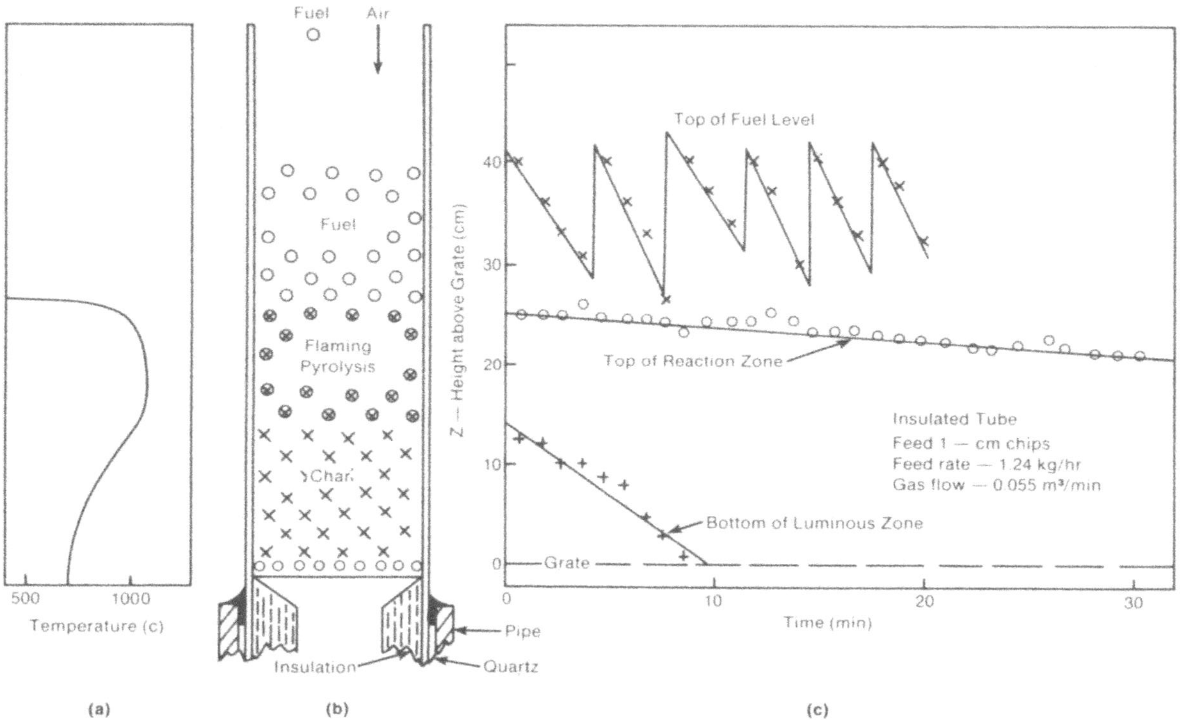

FIG. 6. (a) Temperature (schematic) and (b) position of reaction velocity zones in (c) transparent downdraft gasifier.

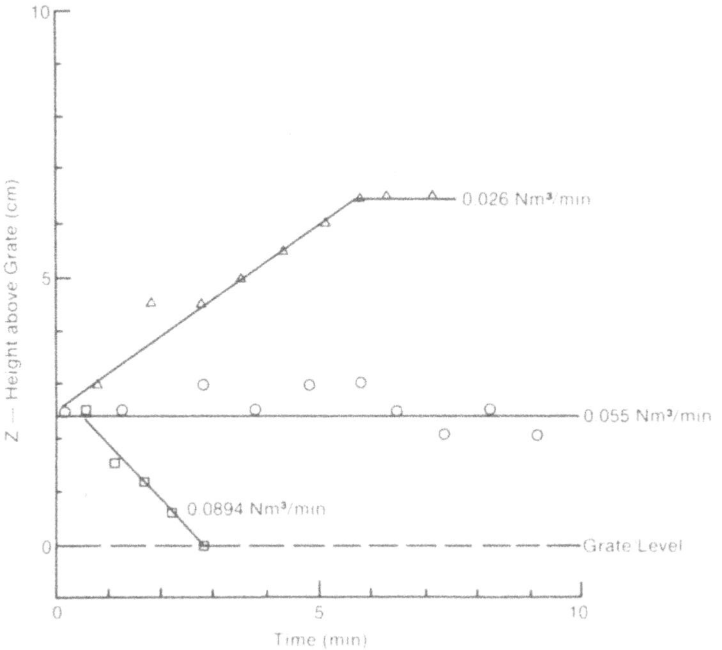

FIG. 7. Variation of zone position with time for various gas flows in uninsulated tube.

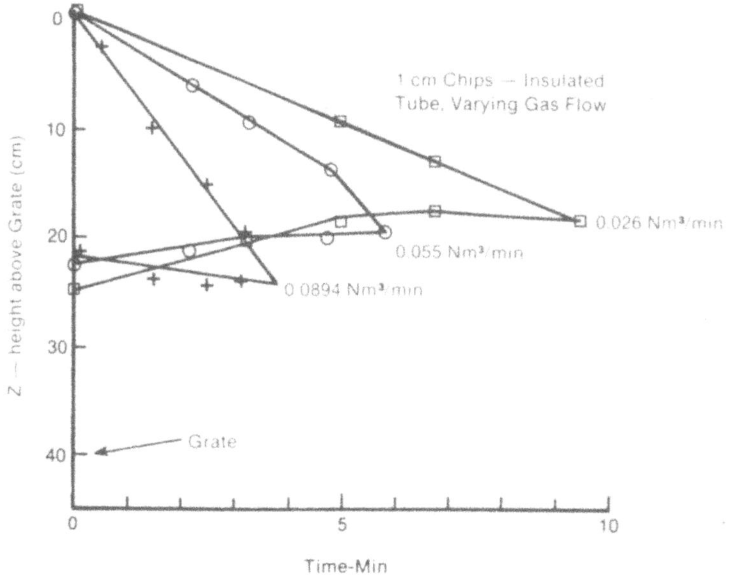

FIG. 8. Variation of reaction zone position with gas flow in insulated tube.

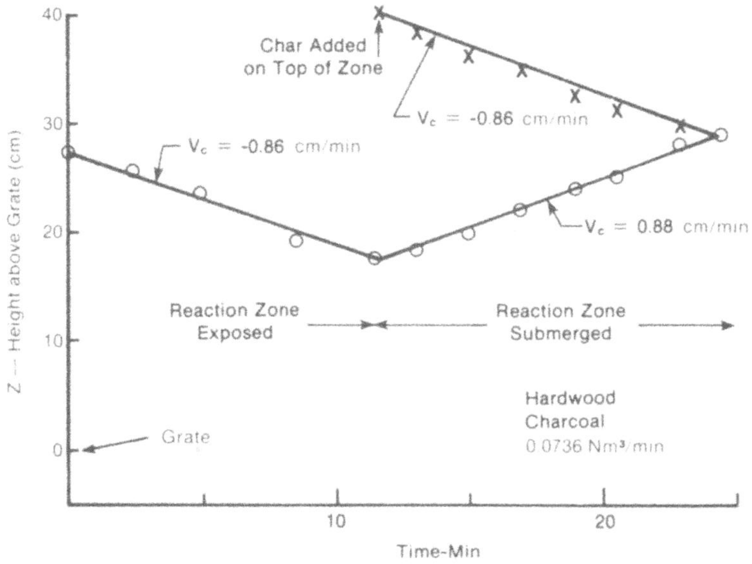

FIG. 9. Reaction zone position for charcoal gasification, exposed and submerged.

highest flow, the zone position moves downward until it reaches the grate at which time the gasification is extinguished (solid square). Figure 8 shows similar results for an insulated tube. Note that in this case the zone does not descend as fast at the highest gas flow and ascends faster at the lowest flow.

The gasification rate of hardwood charcoal is shown in Fig. 9. For the first 10 min of operation the charcoal was burning at the top surface. Then fresh cold charcoal was added to fill the tube. Note that the rate of descent of the top of the bed is similar in both cases because this represents the rate of oxidation of charcoal. However, note that in the second case the reaction interface moves up toward the air at approximately the same rate as the fuel descends. (The similarity of rate of consumption and of propagation is probably a coincidence.)

8. STEADY AND UNSTEADY REACTION RATES

As air is drawn down over the pyrolyzing biomass, the gases and volatiles burn in a process we call 'flaming pyrolysis' (see Fig. 10), providing the heat necessary to sustain pyrolysis. About 85–90% of the original mass is converted to gas in this flaming pyrolysis zone. The oxygen in the incoming air is consumed completely early in this zone, and the gases leaving the zone

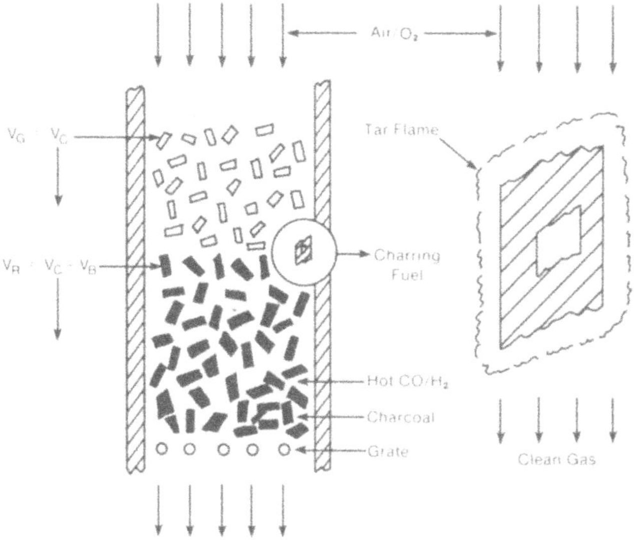

FIG. 10. Flaming pyrolysis of single particle and in gasifier showing zone motion.

are primarily CO_2, H_2O, CO, H_2 and CH_4. In this flaming pyrolysis zone gases and vapors stream from the surface at calculated velocities of about 100 cm/min, effectively preventing any reaction with the charcoal. When pyrolysis is complete, the CO_2 and H_2O resulting from flaming pyrolysis consume the fresh charcoal and are further reduced to CO and H_2.

The motion of the top of the gasifier bed, V_g, and the biomass/char reaction interface, V_r, shown in Fig. 10, can be used to measure the biomass and char gasification rates. We assume for simplicity that the char has the same volume as the biomass from which it is produced. (In fact, there is some shrinkage.) Then the observed motion of the top surface, V_g, is also the velocity of char gasification, V_c.

$$V_g = V_c \tag{1}$$

The velocity of biomass consumption, V_b, is also the rate of char production by pyrolysis. The velocity of motion of the biomass/char reaction interface, V_r, is given by

$$V_r = V_c - V_b \tag{2}$$

The velocities of motion of the interfaces are proportional to the volumetric and mass conversion rates in a tube of constant radius R, and

$$V = \mathring{v}/\pi R^2 = \mathring{m}/\rho\pi R^2 \tag{3}$$

where $\mathring{v}$ and $\mathring{m}$ are the volumetric and mass conversion rates respectively.

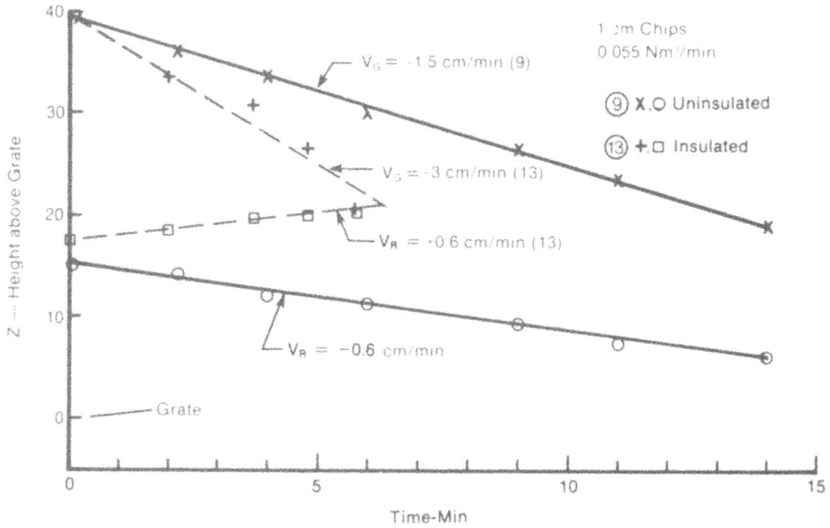

FIG. 11. Insulation effect on zone position. Various forms of interface stabilization.

At steady state, the reaction zone is stationary, $V_r = 0$, and the char volume remains constant, so that $V_b = V_c$, i.e. biomass consumption = char production = char consumption. However, when the reaction zone moves up or down as shown in Figs 7–9 and 11, the rates are no longer equal. The solution of eqn (1) for the runs in Fig. 11 gives the values of biomass and char conversion shown in Table I.

Note that in run 9 with the uninsulated tube char is being consumed (circles). Since the char requires more air for conversion to gas, the biomass is consumed at a slower rate (crosses). In run 13 there is a net production of char (squares) so that biomass is being consumed at a faster rate to make gas and to make char (plusses).

TABLE I

CHAR AND BIOMASS CONVERSION RATES IN QUARTZ GASIFIER DERIVED FROM INTERFACE MOTION

Run	*Condition*	*Char and gasification velocity (V_g and V_c) (cm/min)*	*Reaction zone velocity (V_R) (cm/min)*	*Biomass pyrolysis rate (V_R) (cm/min)*
9	Uninsulated	−1·5	−0·6	0·9
13	Insulated	−3·0	+0·6	3·6

9. PROPAGATION MECHANISMS

In the downdraft gasifier most of the heat produced in flaming pyrolysis is carried down by mass convection. How is it possible then to propagate the reaction into the incoming charcoal or biomass? Several mechanisms exist for this reverse heat flow.

Radiation from hot charcoal is a very effective mechanism for heating the next higher layer of charcoal. The propagation rate in charcoal is shown in Fig. 9 where it is seen that the rate of upward propagation is about twice the rate of consumption.

In biomass, radiation is also an effective reverse heat transfer mechanism, but another, possibly more important, mechanism also exists. As the incoming biomass is pyrolyzed, the gases and vapours mix with the incoming air and form a combustible mixture. Depending on the stoichiometry of the gases, a flame will then propagate upward into these gases at flame velocities of 10–100 cm/s, greatly exceeding the downward velocity of the biomass. Heat conduction in the reactor walls and natural convection may also play lesser roles.

10. MODES OF STABILIZATION

Four modes of gasifier interface stabilization are shown in Fig. 12. If pyrolysis proceeds more slowly than gasification, the interface will move down and can be extinguished when it reaches the bottom of the char bed. However, we observe that the interface can move toward or away from the grate (as shown in Figs 7, 8 and 12a) but can then be stabilized a fixed distance away. We call this mode 'grate stabilization' and believe that as the char zone shrinks, the rate of char gasification shrinks until it equals the rate of production, resulting in a stable interface, but reduced gas quality, due to reduced distance available for CO_2 and H_2O reduction.

If the char consumption and biomass consumption are exactly equal, the zone can be stable anywhere in the tube. In general, this is not a stable situation since small changes in any parameter can cause the zone to drift. However, the zone can also be stabilized at intermediate levels in the bed by reducing insulation, by agitation or by introducing some of the air at the level desired (tuyère stabilized).

If pyrolysis rates exceed gasification rates (Fig. 12c), the reaction zone will climb indefinitely until it reaches the top of the bed. At this point it is 'top stabilized', and its position is controlled by feed rate. Radiation from

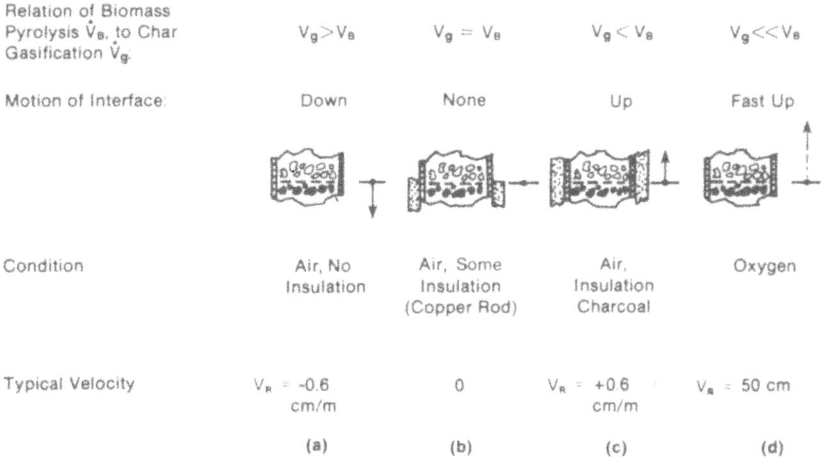

FIG. 12. Interface motion observed in stratified downdraft gasifiers.

the top of such a bed is a significant heat loss and also stabilizes the zone at this point. Finally, if oxygen is used instead of air for gasification, the pyrolysis rate greatly exceeds the gasification rate and the reaction zone immediately becomes top stabilized (Fig. 12d).

11. CONCLUSIONS

The data presented here are part of a much larger study on the mechanisms and modelling of the 'stratified downdraft gasifier'. In fact, the relatively slow motion of the reaction zone relative to the overall biomass feed rate is a second order effect, requiring an advanced state of understanding to predict the exact balance between the pyrolysis reaction and the char gasification reaction. This paper presents data on the reaction zone velocity and points qualitatively to factors that control this balance without attempting a quantitative analysis at this time. We expect to incorporate these data in a more comprehensive paper to be issued later.

ACKNOWLEDGEMENTS

We would like to thank our colleagues for helpful discussions on the experimental and theoretical aspects of this paper. We would also like to acknowledge the support of the Alcohol Fuels Office of the US Department of Energy for aspects of this work related to the SERI Oxygen Gasifier.

REFERENCES

1. Gaydon, A. G. and Wolfhard, H. G. (1960). *Flames—Their Structure, Radiation and Temperature*. Chapman & Hall, London.
2. Solar Energy Research Institute (1980). *A Survey of Biomass Gasification*, Vol. I–III, T. B. Reed (Ed.). SERI/TR-33-239. Now published in hard cover as Reed, T. (Ed.) (1981). *Biomass Gasification: Principles and Technology*. Energy Technology Review No. 67, Noyes Data Corp., Park Ridge, NJ.
3. Kaupp, A. and Goss, J. R. (1981). State-of-the-art for small (2–50 kW) gas producer-engine systems. Final Report to USDA, Forest Service. Contract No. 53-319R-0-141.
4. Bueckens, A. G. and Schoeters, J. G. (1982). This book, Chapter 35, pp. 619–89.
5. Bacon, D. W. *et al.* (1982). Annotated bibliography of gasifier modelling. Dept. of Chemical Engineering, Queen's University (EN for Project C-197), Kingston, Ontario, K7L 3N6.
6. Desrosiers, R. (1982). Fundamental air-gasification engineering parameters. SERI/PR-234-1470, Golden, CO.
7. Groenveld, M. D. and Van Swaaij, W. P. M. (1980). The design of co-current, moving bed gasifiers fueled by biomass. *Thermal Conversion of Solid Biomass Wastes*, J. L. Jones and S. B. Radding (Eds). ACS Symposium Series No. 130, American Chemical Society, Washington, D.C., p. 333.
8. Reed, T. B. and Markson, M. (1982). The SERI high pressure oxygen gasifier. Presented at the Biomass-to-Methanol Specialists Workshop, Tamarron, Colorado, 3–5 March 1982. SERI/TP-234-1455R. Golden, CO.

54

Biomass Liquefaction: An Overview

ESTEBAN CHORNET
Département de Génie Chimique, Université de Sherbrooke, Sherbrooke, Québec, Canada

and

RALPH P. OVEREND
Energy Project, National Research Council of Canada, Ottawa, Ontario, Canada

1. INTRODUCTION

The conversion of biomass materials into oils has a precise objective:

> To transform a carbonaceous solid material which is originally difficult to handle, bulky and of low energy concentration, into pumpable oils having physico-chemical characteristics which permit storage, transferability through pumping systems, and further use either in direct combustion furnaces or as feedstocks for hydrotreatment leading to specific fuels and chemicals.

The above definition can also be applied to coal, lignite and peat. Since the primary product is a 'liquid oil' the term liquefaction has been universally accepted to represent such a conversion route.

In engineering terms a generalized conceptual liquefaction process is depicted in Fig. 1. It involves:

—preparation of the feedstock at a convenient moisture content and size range (typically $<0{\cdot}5$ mm);
—slurrying the feedstock within a liquid carrier (oil recycle cut, specific solvent or simply an aqueous system);
—pumping the slurry through heat exchangers where temperature near reaction conditions is rapidly established;
—addition of a reducing gas (H_2 or H_2/CO at ≤ 30 MPa) with a fraction of the gas added to the preheater;

—main reaction carried out in an appropriate converter, i.e. reactor system;
—product separation via a multistage let-down system where vapor–liquid equilibria is normally used to separate light, middle and heavy condensates from non-condensable gases;
—solid–liquid separation conducted with the heavy condensate as feedstock—this step normally involves distillation, centrifugation and/or extraction;
—in the case of organic liquid carriers the recovery of a particular cut or the specific solvent is needed to complete the recycle loop.

In most liquefaction approaches catalysts and/or additives are either introduced initially to the slurry or they can be present in the converter in fluidized mode. Catalysts can have various functions, the most common being depolymerization, cracking, hydrogenolysis, hydrogenation and deoxygenation.

The liquefaction of carbonaceous materials takes place through a complex sequence of structural and chemical changes which involve:

—solvolysis resulting in micellar-like substructures;
—depolymerization to smaller and soluble molecules;
—thermal decomposition leading to new molecular rearrangements through dehydration, decarboxylation, C–O and C–C bond ruptures;
—hydrogenolysis in the presence of hydrogen;
—hydrogenation of functional groups.

The extent of the above reactions will vary according to the original material, the severity of the liquefaction conditions and the presence of interacting solvents and catalysts. Current evidence from coal liquefaction suggests that in the presence of organic solvents the dissolution step (through solvolysis and partial depolymerization) is rapidly achieved[1] and can be strongly influenced by plasticity effects.

The main purpose of liquefaction is to increase the H/C ratio of the product oil relative to that present in the original feedstock. A decrease of the O/C ratio is also necessary to achieve hydrocarbon-like products. Addition of H_2 is thus needed to increase the H/C ratio and to decrease the O content mainly through the formation of H_2O. In biomass materials O removal occurs via internal dehydration and decarboxylation reactions occurring during the initial pyrolytic stages.

Dehydration can contribute for 46 % removal of the initial O content in *Populus tremuloides* with decarboxylation contributing for an additional

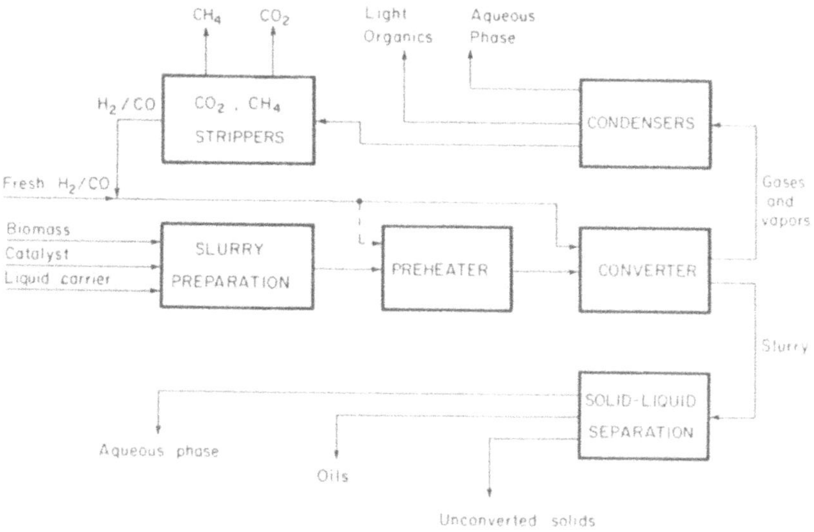

FIG. 1. Generalized biomass liquefaction flow diagram.

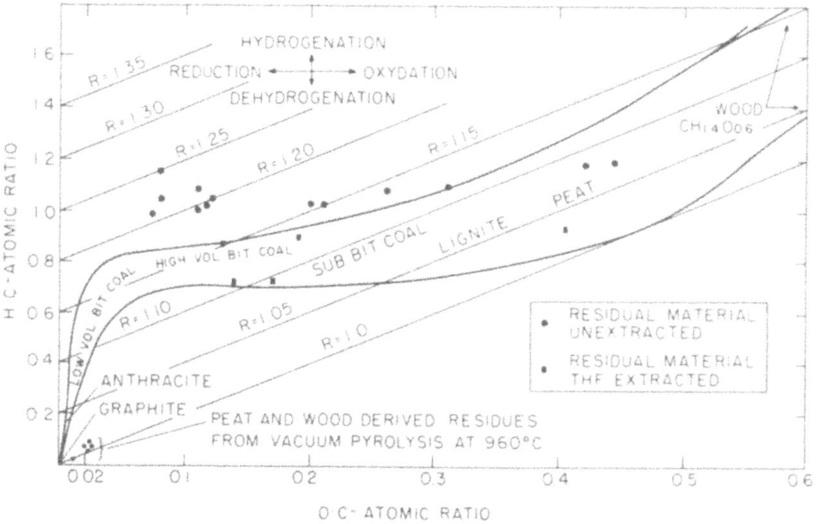

FIG. 2. H:C *vs.* O:C ratios during the course of peat liquefaction (after ref. 4). Note the close association between liquefaction unconverted residues and the coalification band. *R* values express the amount of molecular oxygen needed to fully oxidize the carbon and hydrogen present.

10%. Internal dehydration also results in a considerable decrease of the available H in the initial biomass. About 43% of the initial hydrogen is thus lost in the case of *Populus*.

Figure 2 illustrates another general feature during liquefaction of low ranking materials. The present example concerns peat.[4] The data shown suggest that liquefaction of peat closely follows the coalification band as if condensation reactions yield coal-like precursors during the liquefaction sequence. Thus depolymerization and cracking towards low molecular weight compounds competes with condensation reactions leading to new polymeric structures which will require energy and extensive H addition to achieve hydrocarbon-like products having H/C atomic ratios near 2·0.

In what follows we will present a fundamental analysis of liquefaction applied to lignocellulosics. Comparisons with coal will be initially made to illustrate the distinct nature of biomass materials, their advantages and limitations.

2. COAL, PEAT AND LIGNOCELLULOSIC BIOMASS: GENERAL BEHAVIOUR DURING LIQUEFACTION

Since liquefaction technology was developed for coal and applied particularly to high rank coals (mainly high volatile bituminous), an extension to lower rank feedstocks was considered possible and historically tried since the early beginnings of coal liquefaction. Such direct extrapolation of the coal technology might have been inappropriate in view of the fundamental differences between coal, peat and lignocellulosics.

It is a known fact that coalification induces important structural changes in the polymeric chains present in peat. As well, biochemical processes alter considerably the lignocellulosic polymers which eventually become humic substances. Thus the structural polymers present in coal, peat and lignocellulosics, although related to one another, have undergone considerable changes which result in distinct physical and chemical behaviour during conversion. A universal liquefaction approach thus seems highly questionable and altogether unwarranted.

Coal is structurally formed by isotropic polymeric structures known as macerals.[5,6] These macerals show striking differences during liquefaction.[7–9] The *liptinitic components* (derived from spores, waxes and bitumens initially present in peat and also, most likely, in the original lignocellulosics) are thermally melted and solubilized in the vehicle oil. The accepted picture is that they melt inside the coal matrix and via passage through the pores they are solubilized by the vehicle oil. Another maceral group, and most important, is *vitrinite*. When heated, it may undergo a plastic stage which can result in extensive liquid formation if available hydrogen is present to stabilize the polymeric fragments resulting from thermal decomposition reactions. If hydrogen is not available, recondensation reactions may lead to coke precursors that are refractory to liquefaction. It is generally accepted that vitrinite materials derive from huminitic components initially present in peat.[4]

The third group of coal macerals are *inertinites*. They cannot be converted into oils and are extremely stable under liquefaction and even gasification conditions. They end up as unconverted residue during liquefaction.

From a process viewpoint, the selection of a particular coal as feedstock for liquefaction is an essential choice since obviously there is little interest in processing coal having a large fraction of inertinitic components. The choice of a coal should then be based on the fractions of exinitic and vitrinitic components present.

The plastic stage associated with vitrinite changes during coal liquefaction is not yet well understood but it seems a necessary prerequisite for its eventual transformation into oils.[2]

Humic substances (i.e. peat) are the parent constituents of most of our coals, the other group being constituted by sapropel-derived coals. The link between peat and coal is made through extensive geothermal changes in the earth's crust. Thus coal macerals should be linked to thermally modified components present in humic materials. This link is not obvious for chemists or engineers but has been established by petrologists and geochemists. In a recent publication[4] it has been shown that peat, when treated under the presence of CO/H_2, undergoes significant modifications showing huminite/vitrinite reflectance patterns which eventually change to vitroplast (a precursor of the plastic stage already alluded to). The formation of oils (or coke) follows with increasing temperatures. It is significant that the overall behaviour during peat liquefaction is quite similar to the changes observed during coal liquefaction thus suggesting a possible path for coal genesis from peat. Furthermore, temperatures about 50–75 °C lower than those needed for coal liquefaction seem to be sufficient to ensure very high conversions ($>80\%$) and oil yields ($\sim 50\%$ measured as benzene solubles).

Liquefaction of *lignocellulosics* is indeed much less known than peat or coal conversion. To our knowledge no systematic study of reflectance patterns has been conducted on liquefaction residues. This is perhaps due to the high 'reactivity' of biomass materials which directs the overall effort in most laboratories towards near-complete conversions. Yet there is incomplete knowledge about the polymeric patterns during the early stages of liquefaction of lignocellulosics. An effort in this direction could eventually present a unified view of this complex field.

Lignocellulosics can be viewed as a mixture of essentially three types of polymers: *hemicelluloses* (polymers of xylose, mannose and galactose mainly), *cellulose* (a regular sequence of glucose units linked as β1,4-D-pyranose) and *lignin* (a complex mixture of highly substituted aryl-phenyl groups). The above polymers are found with small amounts of secondary components (waxes, resins, terpenes, etc.). The basic interest has been in studying liquefaction as being uniquely dictated by the properties of the recognizable chemical entities indicated above. Thus, in principle, the course of liquefaction can be predicted by a study of the individual polymers undergoing a complex sequence of thermally induced reactions.

Such a chemical approach to liquefaction is not possible with peat and coal since the structures of the polymers (macerals) are not known with any certainty in the case of coal and are only partially known for peat. In the later case, Schnitzer and Khan[10] have proposed a central aromatic core at which are attached, via hydrogen bonding, carbohydrates and peptides. Metals are also present as complexes and oxygenated functional groups account for the extremely high hydrophilicity of the entire mass. The central aromatic core is supposed to be constituted by a sequence of phenolic rings with substantial ether linkages forming the so-called humic substances, which are likely to adopt a coiled spacial conformation.

3. RHEOLOGICAL CONSIDERATIONS

The structural differences between coal, peat and lignocellulosics also dictate important engineering properties of the materials. Rheological behaviour during liquefaction is a key area which can control the pumpability of slurries as well as flow regimes in preheaters and converters. Viscous energy dissipation can be of paramount importance in the plastic stages of coal, peat and even lignocellulosics liquefaction since the control of mass and heat transfer will be conditioned by the fluidodynamic properties of the media. The concept of 'mechano-chemistry' describes in simple terms the significance of rheological changes during thermal conversion in general. Its application to liquefaction of biomass is indeed new and it is proposed in this paper based on the fact that lignocellulosics undergo important structural changes when heated.[11]

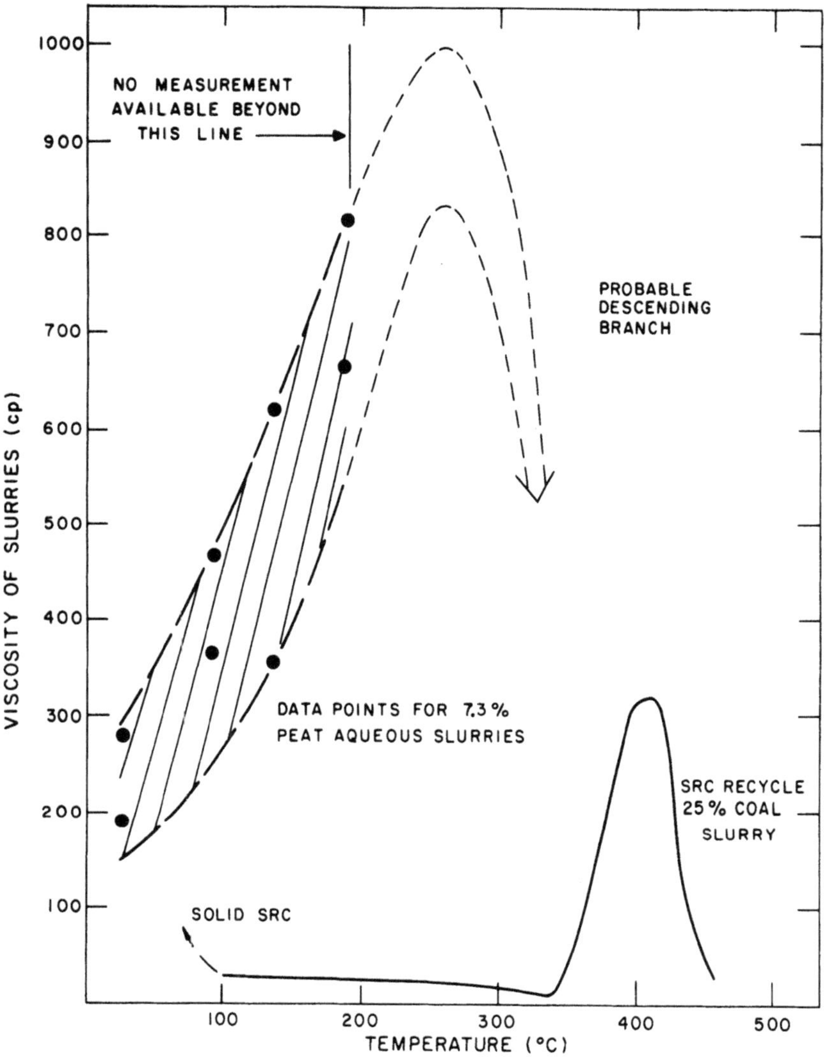

FIG. 3. Viscosity profiles for coal and peat slurries.[6]

Their behaviour when slurried could condition the extent and the quality of the liquefaction products. In Fig. 3 we show typical viscosity–temperature profiles for coal and peat slurries.[6] Viscosity profiles for lignocellulosics are not known beyond 165 °C.[12] Thus the expected viscosity maximum possibly due to development of plasticity cannot be yet ascertained. Enhancement by swelling of the cell walls in the presence of solvents having either electron-donor or proton-acceptor capabilities should also result in viscosity increases. A pressing need thus exists for viscosity data at high temperatures (180–240 °C).

Why are these rheological considerations of fundamental significance? There are three main reasons:

—*Pumpability* of slurries and their mixing patterns will depend primarily on the fluidodynamics of the suspension.

—Flow characteristics can affect considerably the *chemical kinetics* and can furthermore control the availability of hydrogen needed to stabilize free radicals created in the thermal decomposition stages of the liquefaction.

—The polymeric structures present in lignocellulosics will (or can) eventually undergo directional orientations in high velocity flows and high shear rates. Furthermore, if abrupt changes in direction of flow fields are present, very large shear forces can be generated fragmenting the long polymeric chains into smaller fractions thus reducing their chain lengths and accelerating their solvolysis. The latter is also facilitated by the availability of key sites to solvolytic agents due to polymer stretching and eventual breakdown.

The above concepts, although well known in extrusion engineering, have been unfortunately absent in the area of biomass liquefaction until recently. In fact, the approaches developed for coal liquefaction have been directly extended to biomass in most laboratories despite the significant structural and rheological differences between coal and biomass.

The discussion herein presented suggests that a different type of technology is needed for biomass liquefaction. This new technology will have to consider the unique rheology of lignocellulosics at different stages of plasticity, and the distinct yet recognizable polymeric constituents of biomass which permit initial identification, through chemistry, of monomers and derivatives which should be present in the liquefied product.

4. ULTRASTRUCTURE RELATED TO WOOD LIQUEFACTION

As previously indicated, lignocellulosics are essentially a mixture of three polymeric families. Their spacial arrangement reflects the original structure of the parent living cells. The three-dimensional organization of the different cell units is known as ultrastructure.[13]

Lignocellulosics could be considered as an ensemble of cells interlinked through openings (pits) and rays. The cells are held together by what is known as the middle lamella (ML) which surrounds the different cell layers. The latter are divided into:[14] primary wall (P), outer layer of the secondary wall (S_1), middle layer of the secondary wall (S_2), inner layer of the secondary wall (S_3) and warty layer (W). These layers differ from one another both with respect to their structural arrangement and their chemical composition.

The middle lamella has as its main function to bind the cells together. Chemically it is mainly composed of lignin and hemicelluloses, with lignin being the dominant molecular species. Its thickness is between 0·2 and 1·0 μm. The primary wall is a thin layer, 0·1–0·2 μm thick, consisting of cellulose, hemicelluloses, pectin and protein embedded in lignin. The ensemble ML-P is often referred to as compound middle lamella. Its lignin content is high, but because the compound layer is thin only 20–25 % of the total lignin is located in this layer.[13] In the presence of reagents capable of inducing swelling, the primary wall is easily peeled off thus exposing the secondary wall where cellulose is predominantly found. The important point is that unless the compound middle lamella is 'disengaged' from the secondary wall, accessibility to cellulose chains is hindered. In a pure pyrolytic context (i.e. carbonization), thermally induced rapid breakdown of the cellulosic chains results in intermediate compounds which are sterically hindered within the rigid lignin-rich structure of the compound middle lamella since lignin decomposes at higher temperatures than cellulose.[15] The net effect is the random recombination of the intermediate compounds leading to tar and char formation.

The secondary wall is made of lamella arranged in almost parallel microfibrils. Lignin and residual hemicelluloses bind, probably via hydrogen bonding, the different microfibrils. The latter are the complex structural units resulting from the interlinking of about 36 parallel cellulose molecules. Microfibrils can thus be considered as bundles of cellulose molecules aggregated together as a consequence of the micellar configurations present in the parent living cell. Within the microfibrils the cellulose molecules form ordered regions (crystalline) and less ordered (amorphous) regions. An extensive hydrogen bonding network gives a high stability to the microfibrils which build up fibrils and eventually cellulose fibers.[16,17] The thickness of the secondary wall is between 1 and 5 μm. It is to be noticed that, in the distinct layers of the secondary wall, microfibrils are oriented in different directions with respect to the cell axis. This should result in unique rheological and mechanical properties of importance when pumping or extruding wood suspensions.

Liquefaction, and pyrolytic processes in general, aim at the destruction of the ultrastructure and the conversion of the polymeric units into smaller fragments. Due to the multilayer spatial arrangement and the compactness of the microfibrilar network in the secondary wall, diffusional limitations should be expected to influence both the breakdown of the polymeric units and the recombination of the fragmented radicals. Such effects will be particularly important in the temperature range 160–450 °C where most liquefactions are performed. At higher temperatures (>600 °C) diffusional limitations are responsible for the formation of char and tar (i.e. during carbonization and gasification).

In view of the geometric limitations imposed by the ultrastructural architecture of lignocellulosics a necessary task in wood liquefaction is to find methods of orderly disruption of the ultrastructure network. The interaction with selective solvents is probably the obvious way of conducting this task. In fact, the pulp and paper industry follows exactly the same route with the objective of liberating the cellulose fibers from the other now unwanted molecules. In thermal liquefaction the objective could be reformulated as follows: 'solubilize the different macromolecules and depolymerize or crack them into smaller fragments of potential interest as chemicals or fuel substitutes'.

Solubilization of lignocellulosics can be accomplished either integrally or sequentially. Each mode of operation will lead to a colloidal suspension where macromolecules will form micellar configurations. Sequential solubilization has as its objective the fractionation of lignocellulosics into its different polymeric families. Conceptually, these families (hemicelluloses, lignin and cellulose) can be separated from one another due to their distinct physicochemical characteristics. Integral liquefaction aims at total solubilization of lignocellulosics into either aqueous or organic media. In what follows our discussion will focus on the integral mode of operating only.

5. CHEMISTRY RELATED TO LIQUEFACTION

Aspen-poplar, a northern climate biomass prototype species for liquefaction, shows a percent weight composition of 50·9 α-cellulose, 21·8 hemicelluloses, 18·7 lignin, 3·5 acetyl, 4·7 extractives and 0·4 ash. Assuming that extractives are easily solubilized during liquefaction, aspen-poplar and biomass in general will essentially react as a 'mixture of carbohydrates and lignin'.

Carbohydrates in biomass are either linear polymers of glucose (cellulose) or polymers of a variety of sugar monomers (pentoses and hexoses) yielding heterogeneous polymers known as hemicelluloses. Cellulose in wood is composed of about 10 000 glucose units arranged through H-bonding in compact microfibrils. Some H-bonding between cellulose and lignin also exists.[13] Hemicelluloses are polymers containing between 200 and 1000 sugar units, highly branched among themselves and to the lignin polymers. Ether bonding is the favored branching link.

Lignin is now considered as a complex mixture of polymeric formed from three precursors: *p*-coumaryl, coniferyl and sinapyl alcohols. The polymers possess a spherical arrangement with $MW \simeq 20\,000$. Preponderant linkages are β alkyl-4 aryl ether, α alkyl-4 aryl ether together with β alkyl-5 aryl C–C bond, and biphenyl 3–5 C–C bond.

Chemical events during liquefaction involve two main categories of reactions:

(a) substrate–solvent interactions (i.e. hydrolysis or solvolysis);
(b) thermal decomposition sequences.

5.1. Hydrolysis/Solvolysis

In *aqueous media* (acidic or basic), the chemistry of liquefaction will be similar to that of pulping processes (125–180 °C) at least in the initial stages of the liquefaction processes. In laboratory work normally conducted using slow heating autoclaves the analogy between pulping and liquefaction is quite justifiable. A mixture of hydrolytic products mainly derived from hemicelluloses and lignin will be readily obtained in the initial stages of the liquefaction.

In *acid media*, monosaccharides are rapidly formed from depolymerization of hemicelluloses. The composition of the cooking liquor and particularly its pH controls the selectivity of the hemicellulose depolymerization.[19,20] The monosaccharides formed will be subsequently hydrolysed to furfural and derivative acids when the temperature is increased during liquefaction. Further transformation will take place through thermal decomposition sequences.

Lignin undergoes extensive cleavage in acid media. During acid sulfite pulping α-hydroxyl and α-ether groups are cleaved readily forming benzylium ions.[21] Two pathways are competitive for the benzylium ions during pulping: sulfonation and condensation via carbonium ion–nucleophilic site interaction.[22,23] During liquefaction in aqueous acid media similar effects as those discussed for pulping are expected to occur, with sulfonation being replaced by addition of the conjugated base of the acid. Cellulose is only mildly attacked during acid pulping due to the low temperatures used. During liquefaction, when temperature is brought up beyond 230 °C, rapid acid hydrolysis of the glycosidic bonds takes place. Extensive kinetic studies (see ref. 24) show that depolymerization to glucose units is rapidly achieved in a few seconds at temperatures around 250 °C. Secondary hydrolysis products are also formed if acidity conditions are preserved. Thus uncontrolled acid hydrolysis at high temperatures will lead to a complex mixture of aldehydes and acids which will subsequently react at higher temperatures via pyrolytic processes (and hydrogenolysis if H_2 and a catalyst are present).

As the previous analysis has shown, the acid approach as a pretreatment for subsequent thermal liquefaction could have potential interest if conditions are adjusted in such a way that maximum production of monosaccharides and monomeric lignin precursors could be obtained in an initial mild pretreatment operation. The solubilized stream could theoretically be fractionated (via solvent extraction) into a sugar solution and a phenolic-rich organic phase. These fractions could be further treated into products of interest either as chemicals or fuels.

The cellulosic residue derived from the mild pretreatment could, in turn, be acid hydrolysed at higher temperatures and the monomeric sugars hydrogenated to a mixture of polyols.[25] Limitations of the proposed approach are mainly due to rheological constraints: that is, difficulties in processing biomass slurries which would result in high space–time yields of concentrated identifiable products.

In *basic media* hydrolysis reactions of carbohydrates, and polysaccharides in general, are slower than in the presence of acids. Temperature, however, is of importance and reaction rates are significant beyond 140 °C where the cleavage of glycosidic bonds becomes important. Reactions of carbohydrates in alkaline media are known as 'peeling' and 'stopping' reactions. According to Sjöström[26] a variety of acids can be produced during the normal cooking times at maximum temperatures (1–3 h at 155–175 °C). The following acids are representative of the degradation products: metasaccharinic, 2-C-methylgliceric, 3-deoxy-2-C-hydroxymethylpentonic (glucoisosaccharinic), lactic, 2,5-dihydroxypentanoic, xyloisosaccharinic, and 2-hydroxybutanoic. During alkaline cooking up to 30 % of the total carbohydrate is lost. Further degradation takes place at higher temperatures when crystalline cellulose begins to be attacked by the alkali. Delignification in the presence of alkali leads to cleavage of ether linkages promoted by OH^- groups. According to

Gierer[23] β alkyl-4 aryl ether bonds are cleaved preferentially; α-ether bonds (as in phenylcoumaran) lead to stilbene structures; methoxyl groups leave as methanol in the presence of OH^-; condensation reactions also occur resulting in diarylmethane structures; and finally chromophoric groups (unsaturated structures) like quinones are produced via oxidation.

Beyond 200 °C and at high OH^- concentrations (> 10 % alkali-to-wood weight) solubilization of the residual lignin–carbohydrate complex can be achieved as reported in a review by Wenzl.[24] In general a complex mixture of phenolic derivatives, acids and gaseous hydrocarbons, CO_2, CO and H_2 accompanies the production of a 'bitumen' which is likely to be a sum of condensation products obtained via both unoccupied C-5 sites produced during hydrolysis of lignin and free radicals induced by thermal excitations. The 'bitumen' thus obtained can be readily hydrogenated using Mo-based catalysts at 400 °C and about 17 MPa operating H_2 pressure.[27]

In *organic media* the chemistry of liquefaction will depend upon the nature of the solvent–substrate interactions. Solvation, as the first step towards liquefaction, will take place via the coupling electron-donor–electron-acceptor between solvent and substrate (essentially carbohydrate and lignin). As in the case of aqueous systems, good penetration of the solvent within the microfibrilar structure of the cellulosic chains is essential to achieve good solvation.

Since cellulose is the main component of biomass, liquefaction solvents should be chosen on the basis of their ability to interact with cellulose leading to or facilitating its solubilization. A voluminous literature is available on cellulose solvents.[28] However, solvents of industrial interest for liquefaction are limited to the following. Paraffinic once-through oils (kerosene and the like) having little or no solvolytic or solvation action as measured, for instance, by their swelling ability of cellulose.[29] The role of this type of solvent will be essentially limited to that of a carrier fluid within which the lignocellulosic suspension can undergo thermal decomposition reactions with or without hydrogenolysis/hydrogenation depending upon the catalytic combination used.

Solvents which can be recycled and thus probably limited to those derived from either carbohydrates or lignin. In this category we find phenols and phenolic derivatives, simple alcohols and polyalcohols.

Phenolic compounds in the presence of water are known to liquefy aspen sawdust at 350 °C.[30] Although thermal decomposition reactions do occur prior to this temperature the solvent pair phenol:water readily stabilizes reaction intermediates in the presence of simple acid–base catalysts ($ZnCl_2$). The presence of H_2 is not necessary to achieve significant conversions according to the previous authors. Early work at the US Bureau of Mines[31] showed that the use of a reducing agent, CO in their case, resulted in alkylation of cresols at temperatures of 275 °C. Solvent losses are thus expected, hence the need to use a solvent which can be potentially produced from the process. The same authors provided quantitative information on the efficient role of H-donor solvents (*n*-butanol) at moderate temperatures (250–300 °C). The presence of Na_2CO_3 as a catalyst is necessary to 'promote' the role of alcohols as H-donor.

Pure organic recycle solvents of aromatic nature have been used in the operation of the PERC process at Albany.[32] In our current work at Sherbrooke using creosote oils (of naphthenic nature) we readily solubilize about 40–50 % of the biomass at temperatures of 230–240 °C.

Direct interaction between the solvent and the aspen-poplar substrate can take place together with thermal decomposition reactions. The net effect is the total solubilization of hemicelluloses and lignin.

Little is known about the chemistry of the organic solvent interaction with lignocellulosic substrates at the temperatures leading to solubilization of cellulose (250–400 °C). Most selective seem to be phenols which, according to Miller and Fellows, result in a narrow range of products, anisole and xanthene being the dominant molecular species.

5.2. Thermal Decomposition Reactions (Pyrolysis)

The hydrolytic (or solvolytic) reactions described above compete, from the very onset of liquefaction, with 'pure pyrolytic events'. In a recent review, Antal[33] has made an exhaustive survey of the chemistry, kinetics and product distribution related to pyrolysis of cellulosic and non-cellulosic carbohydrates. The reader is referred to Antal's article as an up-to-date version of the present knowledge and unknowns in pyrolysis.

It is to be noticed that the transglycosylation reaction can be explained through either a free-radical mechanism[34,35] or via concerted displacements.[36] Carbonium-ion mechanisms can be responsible for the formation of volatile carbonyl compounds or condensation products from cellulose.[37] Let us also mention that Shafizadeh[38] suggested that the changes in conformation (concerted displacement) required for transglycosylation would require hydrogen bond breaking (thermally) and 'flexibility' in the macromolecule. The latter effect could be achieved, according to Shafizadeh, at or near to the glass transition point.

Thermal decomposition of lignin becomes significant at temperatures beyond 280 °C.[39] Prior to this temperature lignin degradation is due mainly to autohydrolysis reactions with acids generated by the decomposition of hemicelluloses.[40,41] Two interrelated problems are present in all the studies concerning thermal decomposition of lignins: their initial nature and the rôle of thermal softening.[11,42]

Assuming that hemicelluloses decompose in the 150–210 °C range, the lignin initially associated with the hemicelluloses fraction will find itself in a 'free state' (unbound). Since an upper thermal softening point of 205 °C has been reported by Goring,[11] it is likely that structural rearrangements take place within this temperature limit leading to increased mobility of the 'free lignin'. Autohydrolysis should thus be enhanced by a significant decrease in diffusional constraints when the polymer is in its mobile and glassy state. Partial breakdown of the lignin structure should thus begin at about 210 °C. With increasing temperatures thermally induced electronic excitations should generate enough free radicals to initiate subsequent thermal decompositions leading to some volatiles and above all char formation. The influence of steam formed during dehydration is still unclear but it is known[43] that steam can facilitate ether linkage breakdown under conditions of polymer stress ('explosive decompression'). Little is known about the thermal degradation of the lignin bound to cellulose but in view of the previous comments it is likely that this lignin is less reactive to low temperature degradation by autohydrolysis, mainly due to diffusional limitations. We could thus envisage that cellulose-bound lignin will undergo free-radical induced degradation and will generate considerable amounts of char during pyrolysis.

Having analyzed the hydrolytic/solvolytic and the thermal decomposition sequences of possible interest in liquefaction we could ask the following question: could we consider alternate chemical reaction sequences in liquefaction? A literature survey will rapidly specify *hydrogenolysis* as being of importance.

Two main objectives have been pursued in hydrogenolysis: (a) conversion of the entire lignocellulosic mass into hydrocarbon-like oils, and (b) depolymerization of a variety of lignin residues from either acid hydrolysis or pulping processes. Thorough phenomenological reviews are available.[24,44] The interest in a reducing medium (hydrogen) is threefold: (1) hydrogenate or deoxygenate terminal O-containing functional groups; (2) limit the extent of condensation reactions induced by free radicals; and (3) hydrocrack the lignin as well as the tar material formed in the early stages of the thermal treatment.

5.2.1. Hydrogenation–Deoxygenation Reactions at Low Temperatures

It has already been mentioned that the hydrogenation of sugars at mild temperatures is known to produce poly-alcohols (hexitols and xylitol) which undergo further transformation into mixtures of glycerol, ethylene glycol and propylene glycol.[45] Further development of this concept was pursued in the Soviet Union where acidification of wood (carbohydrates) was followed by simultaneous hydrogenation at 6–8 MPa, 160–170 °C using transition metals as catalysts.[46,47] Hydrogenation of the unconverted lignin was conducted in basic media (NaOH) and 280–320 °C under H_2 at 5–7 MPa to yield phenols and derivatives.

Hydrogenolysis of wood at mild conditions (< 200 °C) under basic media has long been a well recognized technique to conduct structural studies on lignin.[48,49] The depolymerization sequence involves according to Pepper and Steck:[50] (a) primary hydrogenolysis, in which the lignin polymers are freed from other plant constituents; (b) secondary hydrogenolysis, by which partial degradation yields phenyl propanoid derivatives; (c) cracking, or C–C bond cleavage, with loss of hydroxyl as H_2O and methoxyl as methyl alcohol; (d) ring saturation to produce cyclohexyl derivatives. A large fraction (30–35 %) of ill-defined ether soluble oils accompany the monomeric compounds obtained. Basic aqueous suspensions and ethanol–water or dioxane–water were used for the three studies.

5.2.2. Stabilization of Free Radicals

It has been previously mentioned that free radicals are involved in thermal decomposition of cellulose and probably lignin beyond 250 °C. The intermediate radicals formed at higher temperatures will have a random tendency to form condensed macromolecules leading to tar and eventually char. If H_2 can be split via catalytic action it is conceivable that the atomic species formed (H) could stabilize free radical intermediates and thus prevent unwanted condensation. This approach is largely theoretical and encounters a difficult problem: that of H-transfer from the catalyst to free radical intermediates. If the catalyst is a solid particle such a transfer will not be possible since free radical intermediates will be mostly protected by the still present ultrastructure of biomass. One way of circumventing this problem will be to use soluble catalysts together with defibrilated cellulosic material. This will ensure direct contact between the catalyst and the intermediate radicals either from cellulose or its associated lignin. Soluble salts of transition metals (particularly Ni) should be the prime candidates for such an H-transfer cycle.

An alternative to the soluble catalyst approach will consider H-donor solvents which can be thermally excited in the temperature region considered (250–350 °C). In aqueous media *n*-butanol has proven to be an effective donor in the presence of Na_2CO_3.[31] Phenols and derivatives (i.e. cresols) achieve high solubilization of lignocellulosics,[31,30] but it is unclear whether they act as H-donors in hydrogenation or as solvolytic agents through H-bonding interactions. Tetraline could be used as the prototype H-donor for model studies.[2,51]

A convincing argument for the need to stabilize H-transfer was presented by Appell and co-workers.[31] In their work with CO and H_2O they tried to interpret mechanistically the competing reactions leading to condensation versus stabilizing reactions using H produced via what is known as the CO-formate. Their argument can be formulated as follows.

In the 250–400 °C range, aldol condensation between two carbohydrate molecules leads to high molecular weight undefined viscous material. This aldol condensation occurs between a carbonyl group on one molecule and two 'active' hydrogens on another molecule with elimination of H_2O and formation of char. How does the CO act in the presence of Na_2CO_3? The sequence is supposed to take place as follows:

(i) $Na_2CO_3 + 2CO + H_2O \longrightarrow 2HCO_2Na + CO_2$

(ii)

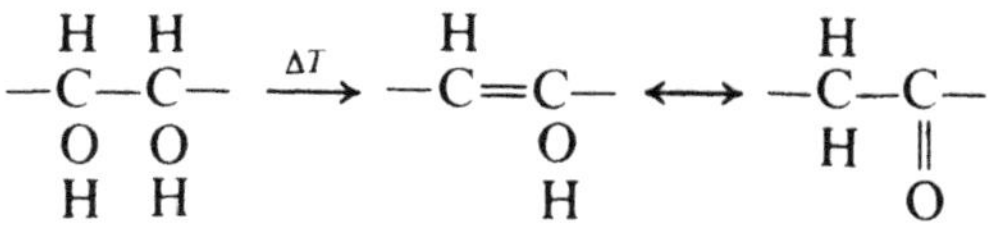

representing dehydration between vicinal OH groups in a carbohydrate

(iii)

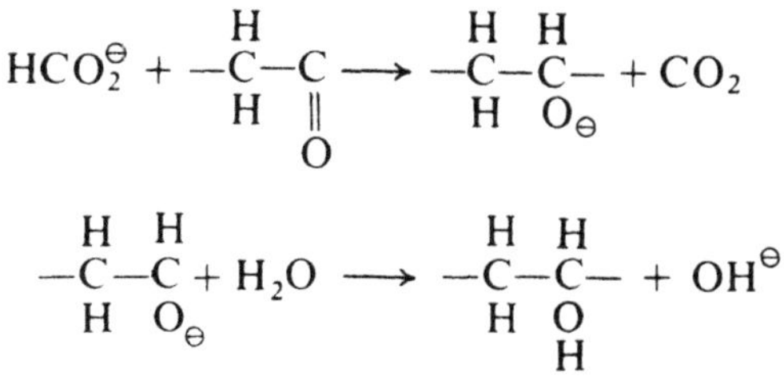

(iv) $OH^{\ominus} + CO \longrightarrow HCO_2^{\ominus}$

5.2.3. Hydrocracking of High Molecular Weight Compounds

Even in the presence of catalytic agents integral thermal liquefaction of lignocellulosics produces a complex mixture of relatively low molecular weight compounds together with a more viscous oily material (i.e. tar) derived both from the carbohydrates and the lignin. These fractions are highly oxygenated; typically 10–25% by weight of 'oil' produced. The interest in lowering the O content of these oils is twofold: to produce a hydrocarbon-like material, and to decrease its viscosity. Thus solubilization of lignocellulosics obtained at temperatures up to 350 °C is normally followed by hydrocracking at higher temperatures. This is only possible if the liquefaction is conducted in an organic solvent carrier having a critical point lower than the reaction temperature which is normally between 380 and 430 °C. In aqueous systems, the tar-like material must be separated from water and water-solubles and subjected to hydrocracking in a separate vessel.

Mechanistic studies on the hydrocracking of biomass derived tar are based on simple determinations of conversion versus time. Kaufman *et al.*[52] developed the following model:

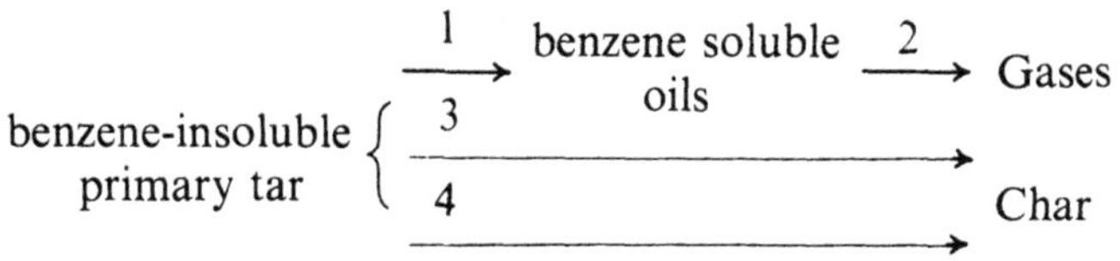

From their data obtained both in a batch reactor and a CSTR system, they concluded that hydrocracking of tar (or pyrobitumen according to the authors) preferentially followed the reactions in series 1 + 2, with reaction 3 being kinetically non-significant. With Ni crystallites as catalyst (prepared *in situ* from $Ni((OH)_2)$, Gupta *et al.*[53] found an $E_{ACT,1} = 71{\cdot}7$ kJ/mol and $k_1/k_2 = 0{\cdot}7$ at temperatures between 343 and 482 °C and at H_2 partial pressures higher than 5·5 MPa.

Two main functions are required in hydrocracking reactions of biomass derived oils: C–C cleavage and splitting C=O and C–OH bonds. Further to these functions a suitable catalyst should have a hydrogenating ability to increase the H/C ratio. A final function of the ideal catalyst system is that it should recognize the presence of two different primary heavy products: those derived from carbohydrate condensation reactions and those derived from lignin.

Although a wealth of information is available concerning hydrocracking/hydrogenation of substituted aromatic compounds, little is known about suitable catalysts for carbohydrate condensation products. Three directions are possible:

(a) Catalytic preparations based on hydrorefining approaches and directly applicable to lignin derivatives. It is well established that hydrorefining is best achieved through a combination of group VI_B

and VIII_B metals on acidic supports. A thorough review has recently been made by Kaliaguine.[54] This author speculates on the need for a new formulation of the classical hydrorefining catalysts and questions the need for sulfiding which is a common feature in the petroleum industry. CoMo and NiW on silica–alumina supports have been chosen by Elliott[55] to upgrade wood-derived oils from the Albany facility.

Combinations of Fe and/or Ni have been developed to hydrocrack lignin: the Noguchi process followed a number of years of selective catalyst development.[56] In this process, a mixture of Fe, Ni and Zn prepared together with $Cu(OH)_2$ has been used to transform desulfonated lignin into 55–65% distillable phenolic derivatives by hydrotreating at about 425 °C with a recycle oil;[57] the HRI Lignol process uses an iron-on-alumina catalyst in an ebullated bed to yield products boiling at 290 °C or lower temperature.[58]

Catalysts based on CuO/Cr_2O_3 have yielded 40% monophenols upon hydrogenolysis of lignin.[59,60]

(b) Catalytic preparations based on general knowledge about deoxygenation of O-rich tar fractions. By far the best known metals are Ni, Co and Fe prepared in a variety of forms which reduce to their metallic form under the severe reduction conditions present in the liquefaction or upgrading reactor. At 380–425 °C, Ni introduced as $Ni(OH)_2$, $NiCO_3$ or Ni-kieselguhr was used by Gupta *et al.*[53] to convert cellulose into oils. It is likely that at these conditions Ni acts essentially in the catalysis of H_2 dissociation, thus becoming available to thermal decomposition fragments.

The use of Ni in basic aqueous systems has been implemented by Boocock *et al.*[61] who showed relatively low oil yields (25–30% of dry poplar) and a significant H_2 consumption (5·8 g/50 g of wood) suggesting that at the temperatures used (about 350 °C) H_2 dissociation achieves better dehydration than thermal decomposition or basic catalysed hydrolysis mechanisms.

Clearly, better selectivity and recoverability are needed to improve this line of catalysts. Perhaps Fe should be the obvious candidate given the inherent problems associated with the recovery of the catalyst.

(c) Catalytic preparations aiming at dehydration of low molecular weight fractions obtained through hydrocracking (the latter with or without additional catalyst).

> Molecular sieves and, in particular, H-ZSM5 have been the main focus of interest in view of their capacity to dehydrate model compounds (i.e. alcohols) and also promote oxygen and carbon conversion to CO_2.[54] Direct application of this concept has been made by Eager *et al.*[62] who treated in the vapor phase about 85% (by weight) the primary oil from liquefaction (i.e. proto-oil) in a fixed bed of H-ZSM5 at temperatures between 450 and 650 °C. Substituted aromatics, C_6–C_{10}, and polycyclic compounds (naphthalene, methylnaphthalene) were present in the products. Current efforts by these authors emphasize the use of dual function catalysts where the H-ZSM5 will act preceded by a cracking preparation resistant to coke formation (i.e. CuO/Cr_2O_3).

As a summary of the hydrogenolysis of lignocellulosics we consider that the application of catalysis will be necessary if the liquefaction of biomass should yield a narrow distribution of families of products. One option would direct primary liquefaction towards the production of a proto-oil having as main components two types of families: low molecular weight polyalcohols and simple aromatics mainly constituted by phenol derivatives. Such an approach seems to be possible from the analysis presented in this section. An alternative option will force rapid thermal decomposition (325–360 °C) resulting in an ill-defined heavy oil-like material (tar) which undergoes subsequent hydrocracking at higher temperatures (380–430 °C) using 'selective hydrorefining preparations' formulated for highly oxygenated compounds. Such catalytic preparations will be complex and need to include increased functional selectivity with respect to those used in the existing conventional hydrorefining processes.

6. A COMPREHENSIVE MODEL FOR LIQUEFACTION

Based on the structural and chemical analysis presented in the preceding sections, a tentative model for liquefaction of biomass can be drawn. Table I suggests alternate options for liquefaction. It also shows associated main events which, in our model, take place in a sequential manner. Although this model is indeed simplified, particularly in regard to the distribution of products, it paints a comprehensive portrait of the complex liquefaction events.

Structural changes are needed to comminute biomass, to improve the rheology of wood flour suspensions, and to open the cellulosic aggregates

TABLE I

THE SEQUENCE OF LIQUEFACTION EVENTS

Reaction	*Technique*	*Products*		
Ultrastructure changes	Thermal stress Mechanical stress Solvolytic action	Hemicellulose → Solubles Cellulose → Activated Lignin → Modified		
		HC	C	L
Chemical degradation	1. Hydrolysis with catalysts H^+	Sugars, aldehydes	Sugars, aldehydes	'Acid' lignin
	OH^-	Acids	Acids	Partially soluble
	2. Solvolysis in presence of H· or electron donor	Solubilized oligomers, sugars, acids	Solvation, swelling, depolymerized	Solvolysis
	3. Neutral aqueous and steam treatment	Solubilized sugars, acids, aldehydes	Swelling, dehydration in amorphous regions	Attack by auto hydrolysis products
	4. Mixed aqueous/solvent systems	Solubilized	Swelling and degradation	Solubilized
Thermal decomposition 200 °C–360 °C	1. Na_2CO_3, CO, OH^- catalyzed high pressures	Mixture of oxygenates derived from above + thermal cracking		
	2. H· transfer low pressure			
	3. Autohydrolytic decomposition can be assisted by shear			
	4. H· transfer at high pressures			
Hydrocracking 380 °C–430 °C	H_2 + catalyst Catalytic cracking De-oxygenation Hydrogenolysis Hydrogenation		Proto-oil	

HC = hemicellulose; C = cellulose; L = lignin.

(microfibrils) to reagents. Steam explosion techniques (Mason-derived approach[63]) or high shear fields as generated by screw-moving devices with or without steam addition (STAKE-derived approach, 1980; White *et al.*;[12] Eager *et al.*[62]) resulting in solubilization of hemicelluloses and the freeing of a large fraction of the lignin together with substantial defibrilation. The latter facilitates subsequent accessibility to active sites in cellulose (i.e. glycosidic bonds). The material thus prepared also possesses distinct rheological properties which enhance the pumpability of concentrated wood suspensions ($>15\%$ solids). A similar beneficial rheological effect has been found to take place upon mild hydrolysis (H^+) of wood suspensions at 170–180 °C using a colloidal mill as reported by Figueroa *et al.*[64] High shear fields can also be generated by rapid passage of wood suspensions in organic solvents through nozzles.[65] In this case increased pumpability is also obtained through solubilization and defibrilation.

Chemical degradation involves hydrolytic (H^+ or OH^- approaches) and solvolytic action resulting in concerted displacements of the molecules and chemical bond rupture and/or rearrangement. These reactions take place within the 150–250 °C range in the presence of diluted acids and can be pursued to higher temperatures in concentrated alkaline media, thus overlapping with thermal decomposition reactions of carbohydrates and lignin.

Solvolytic cleavage requires either H-donor or an electron-donor capacity in the solvent together with catalytic hydrogenolysis if depolymerization of lignin is the objective.

Hemicelluloses in their entirety and lignin in a fractional way can be readily solubilized by chemical degradation even in aqueous systems where the acids formed from hemicelluloses participate in the so-called autohydrolysis reactions. As indicated in the previous section, chemical degradation of hemicelluloses is rapidly followed by thermal degradation of intermediates at temperatures lower than 250 °C.

Thermal decomposition reactions are of importance beyond 250 °C and affect preferentially the cellulose and, with increase in temperature and enhanced solvolytic action, the lignin derivatives are also transformed. It is important to notice that H-transfer will play an important role in stabilizing free radical intermediates either via H_2 dissociation and transfer through the solvent or the CO-formate mechanism as previously discussed.

Since condensation reactions are not easy to prevent, a mixture of heavy molecular weight 'condensed oils' is frequently observed in all liquefaction work. Its hydrorefining through catalytic action is a challenging program for applied chemistry in view of the heterogeneity of the feedstock and its high O-content. A number of approaches have been discussed in the preceding section. The available information on the hydrogenolysis of lignin derived materials[57,58] shows potential for the upgrading route with feedstocks derived from lignin only. The applicability of the same catalyst preparations to tar derived from carbohydrates is not yet proven.

7. STATUS OF LIQUEFACTION WORK

Historical reviews related to the phenomenological aspects of biomass liquefaction are available.[24,31,44] The routes described in Table I have been exhaustively explored at laboratory levels by many research groups.

Batch autoclave studies are the universal method of conducting preliminary exploratory work in liquefaction. Three main problems are present in batch autoclave work: long heat-up times ($\sim$5–8 °C/min) leading to increasing probabilities for secondary reactions; erratic gas–liquid–solid mixing regimes resulting in kinetic segregation effects and incorrect gas/slurry ratios not representative of continuous converters. Thus, a lot of caution is needed while interpreting either kinetic data or product distribution from autoclave runs. In fact, the product distribution might very well be a consequence of extensive secondary reactions. Recent work by Beckman *et al.*[66] shows consistent acetone-soluble oil yields comprised between 30 % and 50 % of charged biomass (dry basis) within the initial minute of reaction using fast heat-up tubing bombs. No significant changes could be discerned from the data with and without catalyst. These results obtained in a neutral aqueous system would suggest that initial thermal degradation is indeed a fast kinetic process compared to subsequent chemical interactions and further thermal decompositions. Recent Canadian autoclave work on biomass liquefaction of poplar shows confirmation of the general trends and yields obtained through many decades of laboratory experimentation: Boocock *et al.*[67] reported acetone-soluble yields comprised between 34 and 53 % using Raney-Ni and $NiCO_3$ in aqueous solutions; Eager *et al.*[68] reported 44–50 % oil yields also in an aqueous system using Na_2CO_3 and CO (note that at reaction conditions, 320–360 °C, the pH changes from initially basic, pH = 11, to slightly acidic at the end of the run); Chornet *et al.*[69] reported THF-soluble yields comprised between 42 and 71 % using anthracene oil as solvent at temperatures of about 425 °C. The oils were in all cases highly oxygenated.

Of more interest than the autoclave work are attempts to liquefy biomass

TABLE II

SUMMARY OF AVAILABLE INFORMATION ON CONTINUOUS BIOMASS LIQUEFACTION (THROUGH SEPTEMBER, 1982)

Type of process	*Research group*	*Pretreatment*	*Liquefaction conditions*					*Product yields (based on % dry feedstock*				*H_2 consumption g/100 g feedstock*	*Main references*
			Temp. (°C)	*P (MPc)*	*Reactor type*	*Residence time (min)*	*Catalyst*	*Conversion*	*Oil yield*	*Gas*	*H_2O*		
Solvolytic (mineral oil once-through)	Worcester Polytechnic Institute (WPI) 1 kg/h	Ball milling newsprint ≤0·2 mm to form a slurry of 20% solids	450	7·2 (H_2)	CSTR	15	Ni introduced as $Ni(OH)_2$ 0·22 g/100 g cellulose	81	46	~30	5	5·3	52 44 53
Solvolytic (recycle oil)	Pittsburgh Energy Research Center process (PERC) Albany, Oregon, facility (25 kg/h)	Grinding Douglas fir ≤35 mesh to form a 7% wood slurry	370 (preheated) and 330 (reactor)	20·7 1·5CO:1H_2	PFR	19	Na_2CO_3 5·0 g/100 g wood	n.a.	53 (TR 12)	n.a.	n.a.	n.a.	32
Hydrolytic (H^+ pretreatment) (OH^- during liquefaction)	Lawrence Berkeley Laboratory—LBL Univ. of California (1·8 kg/h)	Grinding Douglas fir after pH hydrolysis of chips at pH = 1·9, 185 °C for 45 min. Slurry formed at about 25% maf wood in aqueous phase	330–360	17–24 CO:H_2 at variable ratios	PFR	5–60 (normally 20)	Na_2CO_3 8·3 g/100 g wood	96–99	25–33 (MeOH/CCl_4 used as solvent) (Runs no. 8 and 11)	n.a.	n.a.	n.a.	64 79 70 71
Hydrolytic (OH^-)	Univ. of Saskatchewan (0·2 kg/h wood meal). Extrusion device	Grinding aspen chips in hammer mill and mixing flour with water to produce meal (H_2O/wood = 2:1)	360	15–21 CO	Extrusion	n.a.	Na_2CO_3 5% in H_2O at 10 ml/min	95–99	36–44 CCl_4 used as solvent	n.a.	n.a.	n.a.	62

in small continuous units. Problems studied are associated with feedstock preparation, pumpability and rheology of wood suspensions, heat and mass transfer effects as well as reactor configuration and let-down systems for temperature and pressure reduction. Table II lists a summary of the work conducted to date in continuous biomass liquefaction.

Research at WPI focused on the liquefaction of cellulosic newsprint, using mineral oil (paraffinic) as carrier solvent in a once-through mode. Powdered newsprint was prepared via ball milling: 99·6 % of the powdered material was $<841\ \mu m$, 87·1 % $<177\ \mu m$, 39·2 % $<149\ \mu m$, and 8·2 % $<74\ \mu m$. No account of changes in the ultrastructure of cellulose during grinding are reported by the authors. Our current experience in grinding cellulose suggests that temperatures near the thermal decomposition range are achieved when cellulose is milled below $500\ \mu m$. It is thus likely that WPI work was conducted on thermally modified cellulose. Oil yields obtained are reproducible and no major difficulties are reported by the authors in the operation of the unit.

Solvolysis with wood recycle oil is the basis for the analysis of run TR 12 conducted in the PERC mode at the Albany facility. This process development unit was designed based on the work of Appell *et al.*[31] Pumpability problems limited the use of the facility to a maximum concentration of 8 % dry solids. The summary of results presented in Table II represents a characteristic yield of the process in the recycle mode.

In order to increase the concentration of solids being pumped researchers at Lawrence Berkeley Laboratory exposed wood chips (typically 25 % by weight) to a mild prehydrolysis at pH = 1·9 and 185 °C for 45 min. The mixture was milled using a disc refiner (or colloidal mill) and liquefied in basic aqueous media using various CO/H_2 ratios. Although conversions are consistently high (96–99 %), oil yields (defined as methanol–chloroform solubles) are in the 25–33 % range. A significant amount of water-soluble material is thus obtained.

Initial continuous work at the University of Saskatchewan has been centered around an extrusion device in which aspen meal is progressively converted at temperatures of about 360 °C. Alkaline conditions are insured by the presence of Na_2CO_3 which acts coupled to CO.[62]

Alternate efforts in continuous biomass liquefaction are being started by other research groups. Thus Beckman *et al.*,[66] Chornet *et al.*[65] and Björnbom (1982) have given initial, preliminary reports on the design and construction of small continuous units whose commissioning and operations should proceed through 1983. As well, peat and biomass liquefaction on a continuous bench-scale are to be conducted at the

Technical University of Claustal. Initial reports concern data obtained on batch systems.[72]

In order to lower the pressure conditions which represent a significant economic limitation for eventual commercial implementation, efforts have been directed towards processes working near atmospheric pressure conditions. At the University of Tübingen[73] the liquefaction of sewage sludge initially at 70–80 % moisture is conducted in a tubular reactor in which the sludge moves along by means of an auger while the temperature is raised to 400 °C. Although undoubtedly water will evaporate under these conditions, thermal decomposition of the sludge is rapid enough to yield 20–27 % condensable oils while char yields amount to about 60 % on the dry sludge fed. Catalytic action of the mineral matter present in the sludge is suggested by the authors.

A few important design specifications have been provided by the continuous operations:

—Pumpability of wood suspensions requires either very fine comminution (WPI case: 87 % of cellulosics are smaller than 0·18 mm) or else chemical pretreatment to hydrolyze the hemicelluloses fraction (LBL approach). Alternate approaches of obtaining 20 % initial solids concentrations are needed to provide new options for concentrated slurry pumping. This is necessary if thermal efficiencies of biomass liquefaction are to be comparable to those obtained with coal where up to 40–45 % initial ground coal is slurried in a recycle oil. Since the residence times needed for biomass liquefaction are about one-half (and even one-third) those needed for coal liquefaction, simple arithmetic shows that for similar space–time yields (based on oils produced) biomass will require to be processed at concentrations of $\geq 20\,\%$.

Since concentrated wood slurries show unique rheological properties[12] which will decisively influence heat transfer in preheaters (and reactor), it is necessary that an extended effort is devoted to the development of a rheological data base at the temperatures involved during pretreatment and liquefaction. Such information will help to assess the feasibility of extrusion devices as alternate options for liquefaction either as high pressure feeders or as the mixer–reactor itself.

—Design specifications for preheaters and reactors are based on engineering 'intuition'. Heat transfer coefficients and mixing patterns will need to be developed in order to implement process modifications based on engineering parameters.

—Since primary conversion is readily achieved at very short contact times at reaction temperature,[66,72] new efforts should emphasize these new findings in the design of continuous units. Short residence time reactors involving three-phase systems represent a worthwhile engineering challenge.

—Corrosion characteristics and equipment lifetime, particularly valves and pumps, are not known. An assessment of the engineering behaviour of potential alloys will be required to establish a comparative data base from which engineering decisions can be made. Tendencies to develop moderate and low pressure processes should be stressed during primary liquefaction particularly in non-neutral media.

8. CHARACTERIZATION OF LIQUEFACTION PROTO-OIL

The quality of the primary oil produced in liquefaction (i.e. the proto-oil) is of importance for any decision-making concerning the further use of the oil either directly as fuel (blended with other heavy oils) or as a feedstock for potential upgrading. The following discussion presents a review of the existing information:

8.1. How are Proto-Oil Yields Measured?

In continuous solvolytic processes the proto-oil produced is simply added to the initial carrier solvent assuming that the latter does not suffer alteration during liquefaction. Thus, in the WPI approach the total filtered oil produced contains the initial paraffinic material and by difference an oil yield is calculated. In the PERC approach the product oil is weighed as a function of time and, knowing the recycle oil input rate, an oil yield is thus directly obtained.

In aqueous systems a two-phase mixture is produced. Thus decantation and separation of the two phases (aqueous and organic) are required prior to filtering and extracting the organic phase to obtain the oil yield. The extractive medium is indeed variable among the many research groups. This results in scattered and difficult-to-compare information concerning yields. Common solvent systems for extraction are:

—chloroform	(LBL)
—methanol–chloroform	(University of Saskatchewan)
—acetone	(University of Toronto)
—benzene	(University of Claustal)
—THF	(University of Sherbrooke)

8.2. Properties of the Proto-Oil

Of paramount importance is the elemental analysis since it provides direct access to the O-content of the oil and will thus dictate (through stoichiometry) the amount of H_2 needed to upgrade it. Table III shows the composition of proto-oil derived from continuous runs at the conditions previously specified in Table II.

TABLE III
PROTO-OIL ELEMENTAL COMPOSITION (% WEIGHT DRY BIOMASS)

	WPI (*cellulosic newsprint*)	*PERC* (*Douglas fir*)	*LBL* (*Douglas fir*)	*University of Saskatchewan* (*Aspen*)
C	n.a.	78·9	76·0	70·2
H	n.a.	8·5	7·2	7·2
O	~13·7	12·3	16·8	22·6

From Table III it can be concluded that de-oxygenation is significantly better achieved in solvolytic processes than in hydrolytic approaches for similar process conditions (PERC *vs.* LBL). It is to be noticed that PERC runs TR-8 and TR-9 provided oils with 7% and 8% oxygen respectively.

The viscosity of the proto-oil is of basic importance for its further use. Thorough viscosity data is available for the proto-oil derived from the PERC-type runs (TR-12). Water-free proto-oil showed a range of 222 centipoise at 104 °C to 1838 centipoise at 77 °C. The presence of water in the proto-oil considerably affects the viscosity, which at 99 °C is of 805 centipoise with 1·1% water and of 88 centipoise at 4·7% water.

Distillation of the TR-12 proto-oil obtained from the PERC mode provides the following information: IBP: 171 °C with 65% of the oil distilling below 427 °C with no abrupt transitions at any temperature range. Although there have been attempts to distil the biomass liquefaction oils following hydrocarbon-like distillation cuts, in view of the aromatic nature of the oils more appropriate fractionations should be followed. Thus Goheen[57] has suggested the following fractions:

IBP up to 170 °C: light ends and neutrals
170 °C–240 °C: monophenol fraction
240 °C–260 °C: catechol fraction
260 °C– : pasting or recycle tar

A detailed description of distillation profiles of PERC derived oils is

TABLE IV

SEQUENTIAL ELUTION SOLVENT CHROMATOGRAPHY APPLIED TO PROTO-OIL DERIVED FROM BIOMASS LIQUEFACTION[75]

Sequential solvents	*Predominant molecular species*	*Average weight fractions of proto-oil*	*Average O%*	M_n	M_W
1. Hexane	Saturated hydrocarbons	0–6	—	143	158
2. 15% benzene hexane	Aromatic hydrocarbons				
3. Chloroform	Polar aromatics	1–10	11	172	210
4. 4% ether-chloroform	Monophenols	5–30	16	173	211
5. 3% ethanol-ether	Highly functionalized phenols	35–55	21	210	286
6. Methanol	Polyphenols	10–25	23	350	604
7. 3% ethanol-chloroform		1–3	46	181	195
8. 3% ethanol-tetrahydrofuran	Unidentified	5–20	n.a.	687	892
9. Acetic acid (or pyridine)	Unidentified	0–10	n.a.	n.a.	n.a.
10. Non-eluted	Unidentified	—	—	—	—

presented by Elliott.[74] Vacuum distillation at 100 mm Hg (ASTM-D-1160) is used to separate the proto-oils into five to seven fractions which are individually analysed via GCMS. According to these analyses the mixture is a blend of carrier oil with derived phenolic and oxygenated cyclic components. Between 30 and 50% of the oil can thus be routinely characterized. A spectral library has been set up as a reference guide.

Characterization techniques via solvent extraction are based on thermodynamic affinity between the solvent and particular products in the proto-oil. Adaptation of this approach to biomass derived liquids is based on the work initially conducted by the MOBIL Group.[1] In a recent paper, Schaleger and Davis[75] have described the solvent sequence used to fractionate the proto-oil. Confirmation of the previous GCMS findings reported by Elliott[74] is apparent from the results of Table IV from where we ascertain that polar aromatics and phenolic derivatives constitute the majority of the oil.

Molecular weight determinations of the SESC fractions are also included in Table IV both in number and molecular weight averages. A modification of the gel permeation technique using inexpensive and simple packings (Bio-Beads) is reported by the authors.[75]

Since about 25% of the biomass material is transformed into water soluble organics, the chemical characterization is of prime importance in aqueous-based liquefaction systems (i.e. LBL). High pressure ion exclusion chromatography has been used by Schaleger and Davis[75] to determine main water-soluble organics. Formic, acetic and glycolic acids account for about 30% of all water-soluble organics and they represent 90% of the total acid content of the aqueous phase. Levulinic acid has also been detected. The pH of the aqueous solution is slightly acidic.[64]

As a summary, the analytical characterization of biomass liquefaction proto-oil confirms the polyphenolic nature of the oil. The number of phenolic structures condensed together to form polyphenols is between 3 and 6 phenolic rings. These phenolic structures cannot quantitatively be justified as being derived from lignin only. Cyclization of carbohydrate derivatives must contribute to the overall phenol-like character of the proto-oil.

9. CONCLUSIONS

Our analysis of the biomass liquefaction has brought about a number of considerations which we believe are of potential interest for future developments in this area. Major points are:

(i) Biomass being composed of identifiable polymeric constituents offers distinct options for conversion which are not possible with fossil fuels (e.g. coal). In particular, the possibility of fractionation and individual utilization of the polymers should be considered. Thus carbohydrates could be depolymerized and converted into simple alcohols via mild hydrogenation techniques. Lignin hydrogenolysis could complement the fractionation approach. The chemicals involved in fractionation/hydrogenolysis/hydrogenation are reasonably understood. The constraints of this route are mainly caused by the complexity and unfavourable economics associated with existing fractionation approaches. Efforts in this area are indeed necessary.

(ii) Biomass conversion via integral liquefaction can be chemically approached via a sequential analysis involving structural re-arrangements at low temperatures, chemical degradation via acid–base catalysis or electron-donor–acceptor interactions with specific solvents, thermal decomposition reactions and hydrogenolysis/hydrogenation at higher temperatures. The product obtained (i.e. proto-oil) with current technological limitations (moderate heat fluxes and relatively long residence times) is a complex mixture of polyphenols with varying degrees of O-containing functional groups. The high molecular weight condensed products resemble coal liquefaction products, particularly when CO/H_2 are the reducing media used for liquefaction.

(iii) New pretreatment approaches and recent short residence time liquefaction results suggest new avenues for integral biomass liquefaction. These new directions are based on the ultrastructure modifications introduced by various pretreatment methods which open up the cell structure and wood and facilitate solvent–substrate interactions, particularly during thermal decomposition. Also the pretreatment methods facilitate the pumpability of concentrated wood suspensions. This is required if significant space–time yields are to be achieved.

(iv) A lack of information concerning the transport properties of biomass suspensions (rheology and heat transfer) has limited the technological development of small continuous systems necessary to achieve a consistent and reproducible data base. Such a task is urgently needed for engineering design purposes.

(v) Upgrading of the proto-oil is not a straightforward extension of conventional hydrorefining processes. The significant oxygen content of proto-oil (typically in the range 10–20%) and the tendency of polyphenolic structures to break the ether-linkages to rapidly form condensed aromatics leading to char makes the catalytic approach a difficult challenge. Reformulated hydrorefining catalysts of specific pore sizes like those developed for heavy bitumen hydrocracking might be needed.

(vi) The environmental constraints derived from the polycyclic nature of the proto-oil will doubtless require careful handling and might impose additional economic burdens on an eventual commercial liquefaction process. Yet the almost complete absence of polyaromatic hydrocarbons in the proto-oil shows considerable promise. More complete analytical characterizations are needed to evaluate properly the potential hazards of the proto-oil or its derived substitute products.

(vii) Although material balances are not completely known in the continuous systems so far tested the accumulated data suggest that 40–50% proto-oil yields are feasible. Minimization of H_2 or H_2/CO consumption is required and efforts in catalysis should specifically be addressed to a better utilization of the necessary reducing gas.

ACKNOWLEDGEMENTS

The authors are indebted to the National Research Council, the ENFOR Program, the National Science and Engineering Research Council of Canada and the Programme de Formation des Chercheurs et Action Concertée du Québec for financial assistance throughout our liquefaction program. Special thanks are acknowledged to our colleagues Dr W. Kalkreuth, Dr M. Heitz and Mr D. Eugène for many useful discussions.

REFERENCES

1. Whitehurst, D. D., Mitchell, T. O. and Farcasiu, M. (1980). *Coal Liquefaction—The Chemistry and Technology of Thermal Processes*. Academic Press, New York, p. 29.
2. Neavel, R. C. (1982). *Coal Science*. Academic Press, New York, Ch. 1, p. 1.
3. Roy, C., de Caumia, B. and Chornet, E. (1982). Liquids from biomass by vacuum pyrolysis. *Proceedings of the Specialists Meeting on Biomass*

Liquefaction, Sponsored by the ENFOR Program. R. L. Eager, J. F. Matthews and J. M. Pepper (Eds). University of Saskatchewan, Saskatoon, 57.

4. Kalkreuth, W. and Chornet, E. (1982). *Fuel Processing Technol.*, **6**, 93.
5. Anon. (1963). *International Handbook of Coal Petrology*, 2nd edn. CNRS, Paris (addendum to 2nd edn, 1971).
6. Anon. (1981). Feasibility study of the dewatering of peat using a high velocity, short residence time reactor. Report submitted by the Sandwell Beak Research Group to the National Research Council of Canada, DSS File No. 10SX.31155-9-2666.
7. Wiser, W. (1968). *Fuel* **47**, 475.
8. Neavel, R. C. (1975). *Fuel* **55**, 237.
9. Squires, A. M. (1978). *Appl. Energy* **4**, 161.
10. Schnitzer, M. and Khan, S. U. (1972). *Humic Substances in the Environment*. Marcel Dekker, New York. p. 192.
11. Goring, D. A. I. (1963). *Pulp and Paper Magazine Canada* **64**, 12T, 517.
12. White, D. H., Homaidan, A., Chehab, M., Schott, N. R., Wolf, D. and Lezzar, A. (1981). Modified extruder for feeding fine-ground cellulosic slurries to pressure systems. *Proc. Biomass Conversion Contractors Meeting*, DOE, Washington (October).
13. Sjöström, E. (1981). *Wood Chemistry: Fundamentals and Applications*. Academic Press, New York.
14. Côté, W. A., Jr (1967). *Wood Ultrastructure*. University of Washington Press, Seattle.
15. Beall, F. C. and Eickner, H. W. (1970). Thermal degradation of wood components: A review of the literature. USDA Forest Service Research Paper, Serial No. FPL 130.
16. Kolpak, F. J., Weih, M. and Blackwell, J. (1978). *Polymer* **19**, 123.
17. Gardner, K. H. and Blackwell, J. (1974). *Biochim. Biophys. Acta* **373**, 232.
18. Adler, E. (1977). *Wood Sci. Technol.* **11**, 169.
19. Sjöström, E., Haglund, P. and Janson, J. (1962). *Sven. Papperstindn.* **65**, 855.
20. Janson, J. and Sjöström, E. (1964). *Sven. Papperstindn.* **67**, 764.
21. Gellerstedt, G. and Gierer, J. (1971). *Sven. Papperstindn.* **74**, 117.
22. Gellerstedt, G. (1976). *Sven. Papperstindn.* **79**, 537.
23. Gierer, J. (1970). *Sven. Papperstindn.* **73**, 571.
24. Wenzl, H. F. J. (1970). *The Chemical Technology of Wood*. Academic Press, New York.
25. Balandin, A. A., Vasyunina, N. A., Chepigo, S. V. and Barysheva, G. S. (1959). *Doklady Akad. Nauk SSSR* **128**, 941.
26. Sjöström, E. (1977). *Tappi* **60**(9), 152.
27. Heinemann, H. *Petroleum Refiner* **33**, 161.
28. Turbak, A. F., Hammer, R. B., Davies, R. E. and Hergert, H. L. (1980). *ChemTech*. Jan, 51.
29. Rowell, R. M. and Hart, S. V. (1981). Penetration of solvents into heartwood. Final Report, Forest Products Laboratory, USDA Forest Service Project No. 3212-5-76-3.
30. Miller, I. J. and Fellows, S. K. (1981). *Nature* **289**, 398.
31. Appell, H. R., Fu, Y. C., Illig, E. G., Steffgen, F. W. and Miller, R. D. (1975). Conversion of cellulosic wastes to oil. RI No. 8013, US Bureau of Mines.

32. Berry, W. L. and Thigpen, P. L. (1981). Operation of the biomass liquefaction facility at Albany, Oregon. *Proceedings of the 13th Biomass Thermochemical Conversion Contractor's Meeting*, Oct. 27–30, US Department of Energy, 521–50.
33. Antal, M. (1982). *Adv. in Solar Energy* **1**, 61.
34. Golova, O. P. (1975). *Russian Chemical Reviews* **44**, 687.
35. Madorsky, S. L. (1964). *Thermal Degradation of Organic Polymers*. Wiley-Interscience, New York.
36. Kilzer, F. J. and Broido, A. (1975). *Pyrodynamics* **2**, 151.
37. Byrne, G. A., Gardiner, D. and Holmes, F. H. (1966). *J. Appl. Chem.* **16**, 81.
38. Shafizadeh, F. (1980). In *Proceedings of the Specialists Workshop on Fast Pyrolysis of Biomass*, SERI/CP-622-1096.
39. Kudo, K. and Yoshida, E. (1957). *J. Jap. Wood Res. Soc.* **3**(4), 125.
40. Lora, J. H. and Wayman, M. (1978). *Tappi* **61**, 48.
41. Kollmann, F. and Fengel, D. (1965). *Holz als Roh-und Werkstoff* **23**, 461.
42. Salmen, N. L. and Back, E. L. (1977, 1980). *Tappi* **60**(12), 137; **63**(6), 117.
43. Marchessault, R. H., Coulombe, S., Morikawa, H. and Robert, D. (1980). *Can. J. Chem.* **60**, 2372.
44. Kaufman, J. A. and Weiss, A. H. (1975). *Solid waste conversion: Cellulose liquefaction*. NTIS, PB 239 509, Report No. EPA-670/2-75-031.
45. Degussa (1941). German Patent No. 892,590.
46. Vasyunina, N. A., Balandin, A. A. and Mamatov, Y. (1963). *Kinetika i Kataliz*. **4**(1), 156; *ibid.* **4**(3), 443.
47. Vasyunnia, N. A., Pogosov, Y. L., Balandin, A. A., Chepigo, S. W. and Barysheva, G. S. (1965). USSR Patent No. 168,666.
48. Brewer, C. P., Cooke, L. M. and Hibbert, H. (1948). *J. Am. Chem. Soc.*, **70**, 57.
49. Pepper, J. M. and Hibbert, H. (1948). *J. Am. Chem. Soc.* **70**, 67.
50. Pepper, J. M. and Steck, W. (1963). *Can. J. Chem.* **41**, 2867.
51. Connors, W. J. (1980). *Holzforschung* **34**, 29.
52. Kaufman, J. A., Gupta, D. V., Szatkowski, T. S. and Weiss, A. H. (1974). *Chem. Ing. Tech.* **14**, 609.
53. Gupta, D. V., Kranich, W. L. and Weiss, A. H. (1976). *Ind. Eng. Chem. Process Des. Dev.* **15**(2), 256.
54. Kaliaguine, S. (1982). Upgrading pyrolytic oils from wood, *Proc. Specialists Meeting on Wood Liquefaction*, Toronto, ENFOR Program, 75.
55. Elliott, D. C. (1982). Biomass liquefaction product analysis and upgrading. *Proc. Specialists Meeting on Wood Liquefaction*, Saskatoon, ENFOR Program, 157.
56. Oshima, M., Maeda, Y. and Kashima, H. (1963). US Patent 3,105,095.
57. Goheen, D. W. (1966). *Lignin Structure and Reactions*, R. F. Gould (Ed.). *Adv. in chem. Series* **59**, 205.
58. Huibers, D. T. A. and Jones, M. W. (1980). *Can. J. Chem. Eng.* **58**, 718.
59. Sakakibara, A. and Araki, T. (1961). *J. Japan Wood Res. Soc.* **7**, 1.
60. Coscia, C. J., Schubert, W. J. and Nord, F. F. (1961). *Tappi* **44**, 360.
61. Boocock, D. G. B., Mackay, D., Franco, H. and Lee, P. (1980). *Can. J. Chem. Eng.* **58**, 466.
62. Eager, R. L., Matthews, J. F. and Pepper, J. M. (1982). Conversion of aspen

poplar into liquid fuels. *Proc. Specialists Meeting on Biomass Liquefaction*, Toronto, ENFOR Program, 1.

63. Mason, W. H. (1938). US Patent 2,140,189. See also Delong, E. A. (1981). Can. Patent 1,096,374.
64. Figueroa, C., Schaleger, L. L. and Davis, H. G. (1981). LBL continuous bench scale liquefaction unit, operation and results. Presented at 6th Annual Conference on Energy from Biomass Wastes, Lake Buena Vista, Florida, Jan. 25–29, File No. LBL-13709, Univ. of California.
65. Chornet, E., Eugene, D. and Arsenault, R. (1982). An update on the conversion of lignocellulosic materials via hydrogenolysis and solvolysis at low severity conditions. *Proc. Specialists Meeting on Biomass Liquefaction*, Toronto, ENFOR Program, 31.
66. Beckman, D., Boocock, D. G. B., Mackay, D., Poretta, F. and Sherman, K. (1982). Wood liquefaction: A semicontinuous unit and fast heat-up studies. *Proc. Specialists Meeting on Biomass Liquefaction*, Saskatoon, ENFOR Program, 15–30.
67. Boocock, D. G. B., Jenkins, A., Poretta, F. and Mackay, D. (1981). Experimental work at the University of Toronto. *Proc. Specialists Meeting on Biomass Liquefaction*, Toronto, ENFOR Program, 11.
68. Eager, D., Matthews, J. F. and Pepper, J. M. (1981). Experimental work at the University of Saskatchewan. *Proc. Specialists Meeting on Biomass Liquefaction*, Toronto, ENFOR Program, 27.
69. Chornet, E., Arsenault, R., Roy, C. and Eugene, D. (1981). Experimental work at the Université de Sherbrooke. *Proc. Specialists Meeting on Biomass Liquefaction*, Toronto, ENFOR Program, 41.
70. Ergun, S. (1980). Biomass liquefaction efforts in the United States. Report LBL-10456, Feb.
71. Schaleger, L. L., Yaghoubzadeh, N. and Ergun, S. (1979). Pretreatment of biomass prior to liquefaction. Report LBL-10093, Nov.
72. El-Saied, H. and Oelert, H. (1980). *Cellulose Chem. Technol.* **14**, 507.
73. Bayer, E. and Kutubuddin, M. (1982). *Korrespondenz Abwasser* **29**, 377.
74. Elliott, D. C. (1981). Bench scale research in biomass direct liquefaction. *Proc. Biomass Conversion Contractors Meeting*, DOE, Washington, Oct., 551.
75. Schaleger, L. L. and Davis, H. G. (1982). Progress in the characterization of the products of direct liquefaction of Douglas fir. Report LBL-14017. *Proc. Specialists Meeting on Biomass Liquefaction*, Toronto, ENFOR Program, 191.
76. Basch, A. and Lewin, M. (1973). *J. Polym. Sci*. **11**, 3071.
77. Broido, A. and Yow, H. (1977). *J. Appl. Polym. Sci.* **21**, 1677.
78. Broido, A. and Nelson, M. A. (1975). *Comb. & Flame* **24**, 263.
79. Davis, H., Ergun, S., Figueroa, C., Karatas, C., Kloden, D., Schaleger, L. L. and Yaghoubzadeh, N. (1981). Research and development activities on direct liquefaction technology. Quarterly Report LBL-12537, Jan–Mar.
80. Domburgs, G., Sergeeva, V. N., Kalnins, A. I. and Kiselis, O. (1966). *Latvijas PSR Zinat. Akad. Vestis* **12**, 52.
81. Furneaux, R. H. and Shafizadeh, F. (1979). *Carbohyd. Res.* **74**, 354.
82. Golova, O. P. and Krylova, R. G. (1957). *Dokl. Akad. Nauk. SSR* **116**, 419.

83. Sergeeva, V. N. and Vaivads, A. (1954). *Latvijas PSR Zinat. Akad. Vestis* **86**(9), 103.
84. Shafizadeh, F. and Bradbury, A. G. W. (1979). *J. Appl. Polym. Sci.* **23**, 1431.
85. Shafizadeh, F., McGinnis, G. D. and Philpot, C. W. (1972). *Carbohyd. Res.* **25**, 23.
86. Stevenson, F. J. (1982). *Humus Chemistry: Genesis, Composition, Reactions.* Wiley-Interscience, New York, p. 258.
87. Wayman, M. (1980). Alcohol from cellulosics: The autohydrolysis-extraction process. *Proc. IV Int. Symp. on Alcohol Fuels Technology*, Guaruja, Brazil, 79.
88. Woodley, F. A. (1971). *J. Appl. Polym. Sci.* **15**, 835.

55

Analysis and Comparison of Products from Wood Liquefaction

DOUGLAS C. ELLIOTT
Battelle Pacific Northwest Laboratories, Richland, Washington, USA

1. INTRODUCTION

This paper describes the results of the analytical effort dealing with the products from the United States Department of Energy, Biomass Liquefaction Experimental Facility at Albany, Oregon. This facility, which was operated under contract to DOE by Wheelabrator Cleanfuel produced approximately 80 barrels of wood-derived oil in the period from August 1979 to March 1981.[1,2] The oil was produced by two variations of the basic CO-steam process.[3] One variation, the PERC process, involved recycle of a portion of the product so that a wood-flour-in-oil slurry system was used. The other variation, the LBL process, was a once-through process wherein the wood was prehydrolyzed and then processed in a water slurry mode. Both variations operated at about 340 °C and 3000 psig with aqueous sodium carbonate as the catalyst and CO/H_2 as the cover gas. Both processes were tested in a number of reactor configurations. The actual plant operation time is therefore broken into a number of test runs; products from runs 7 through 12 will be discussed in this paper. A detailed description of the equipment configuration and processing conditions for each test run is given in refs 1 and 2.

2. WHOLE OIL ANALYSES

The product as produced at Albany was drawn into 50-gallon drums for storage. This material was in turn pumped into 1-gallon steel containers for shipment to Pacific Northwest Laboratory (PNL) for analysis. The samples were stored at room conditions in these containers upon receipt at

PNL. Representative end-product samples were received and analyzed for TR7, 8, 9, 11 and 12. In addition, specially processed oils including flash tank condensate oil, trap out tray distillate oil, and distillate bottoms material from TR8 and the preheated distillate feed from TR9 were also received and analyzed.

A broad range of techniques was used for analysis of these products. Initial samples were drawn for a battery of tests including:

—elemental analysis for carbon, hydrogen, nitrogen and oxygen by direct measurement using a Perkin-Elmer 240 analyzer;
—moisture analysis by ASTM D-95 xylene distillation;
—solids determination by filtration of hot acetic acid insoluble material from product oil solutions;
—heat of combustion by adiabatic bomb calorimeter; and
—vacuum fractional distillation by ASTM D-1160 procedure.

Table I provides a summary of this data for each of the process oils analyzed. The differences in the analyses suggest a difference in processing severity. The high oxygen content and low volatility of TR11 indicates that a lesser amount of reaction occurred such that significantly more oxygen remained combined in the product oil. The low solids contents of TR7 and TR11 are indicative of the single-pass LBL process. The high moisture contents of the LBL oils magnifies their high oxygen content, but elemental ratios are fairly similar otherwise.

Due to the highly polar quality of the chemical constituents of wood-derived oil, there is a substantial degree of water solubility in the oil. Our

TABLE I
SUMMARY ANALYSES OF ALBANY PRODUCT OILS

Analysis	*TR7*	*TR8*	*TR9*	*TR11*	*TR12*
Elemental analysis (%):					
Carbon	72·3	82·0	76·6	67·2	76·1
Hydrogen	8·6	8·8	8·2	6·8	10·5
Oxygen	17·6	9·2	14·1	25·1	15·1
Nitrogen	0·2	0·5	0	0	—
Moisture (ASTM D-95)	8·5	3·1	8·9	13·5	7·3
Solids (HOAC insoluble)	1·0	10·8	17·8	2·2	3·7
HHV, Btu/lb	14 500	15 800	15 000	12 400	14 200
Percent distillable (D1 160, 10 mm Hg)	68	56·1	46	42·6	46

results suggest that 7–8 wt % of water will dissolve in wood-derived oil at room temperature (higher solubility is noted at higher temperature). The high moisture contents of the samples in Table I are a result of this solubility. It also appears that small droplets of water can be trapped in the more viscous products as suggested by the high moisture content of TR11. This trapped water tends to separate from the oil upon standing. This is shown by the separation of water phases from oil samples sent to us from Albany which were homogeneous before shipment. The inclusion of water in the oil samples causes problems from both a process viewpoint as well as an analytical viewpoint. Comparison of data in Table II to that in Table I shows the effect on elemental analysis which can be attributed to moisture.

TABLE II
ANALYSES CALCULATED FOR A DRY BASIS

Element	*TR7*	*TR8*	*TR9*	*TR11*	*TR12*
C	80·2	84·2	85·1	78·5	80·6
H	8·5	8·7	8·0	6·2	10·3
O	11·1	6·6	6·9	15·3	9·1
N	0·2	0·5	0	0	0
HHV	15 800	16 300	16 500	14 300	15 300
H/C	1·26	1·23	1·12	0·94	1·52

We can see that a range of products can be produced from the same biomass feedstock by modifying reactor conditions and configuration only slightly. By proper operation, this type of processing scheme should be able to routinely produce a product containing 10 % or less oxygen, having a hydrogen-to-carbon atomic ratio of approximately 1·2 and a heating value in the range of 16 000 Btu/lb.

2.1. Moisture Effect on Fluidity

The moisture content has a major effect on the fluid properties of the oil derived from wood. In our study of the pour point of the Albany oils,[4] we have determined that there is a strong correlation between moisture level and the pour point of the product. The effect of water on the pour point correlates with the earlier disclosure[1] of water content effect on product oil viscosity. In summary, reported results showed that with a given oil product, the viscosity remained nearly constant as the water content was reduced from 14 % to 3 % by weight; however, drying to less than 3 % resulted in a dramatic increase in the viscosity of the product oil.

In our own work we used a simplified version of the ASTM pour point test (D-97). The pour point is often used as the minimum temperature for storage of a fuel oil to be pumped to a firing chamber. A determination of pour point was made for wood oil products as shown in Table III. These results show a wide range in pour point associated with reaction condition variations, distillate fractionation and water dilution. Typical whole-oil products have a pour point around room temperature, but can range

TABLE III
WOOD PRODUCT OIL POUR POINTS

Product tested	*Pour point*	*Modified oil pour point*
TR7 (3% moisture)	23 °C	
TR7 (water and light oil removed by distillation)	36 °C	with 10% H_2O added, 20 °C
TR7 (20% distillation residue)	58 °C	with 5% H_2O added, 43 °C with 7% H_2O added, 40 °C
TR8 (1% moisture)	24 °C	
TR8 (40% distillate)	– 19 °C	
TR8 (40% distillation residue)	> 100 °C (estimated at 120 °C)	
TR11 (13·5% moisture)	60 °C	
TR12 (7·3% moisture)	27 °C	

higher due to non-optimal reaction conditions as in TR11 (see Table IV for conditions). Distillates show potential for use as light fuel oils as suggested by low pour point (– 19 °C for TR8 distillate). The effect of adding water to the oil product can be seen in the several cases listed. It is difficult to put into words the effect on the casual observer of this change in pour point; such a change can actually make the difference between the perception of a 'solid' product and a 'liquid' product. In the area of utilization the possibility apparently exists for the use of water as a cutting agent to improve the fluid properties of an unpumpable product oil or to adjust the fluidity of product oil after the removal of light organic constituents for other uses.

2.2. Gas Chromatography–Mass Spectrometry

An extensive gas chromatography–mass spectrometry (GC/MS) study comparing the five Albany product oils has been completed. The GC/MS was used to identify over 100 individual chemical components in the products; then a similar GC separation with a flame ionization detector was

TABLE IV

GROUP SUMMARY OF GC/MS DATA FOR ALBANY OILS

Residence time	*6 h*	*100 min*	*? min*	*11 min*	*18 min*
Temperature	*350°C*	*335°C*	*338°C*	*338°C*	*330°C*
Process	*LBL*	*PERC*	*PERC*	*LBL*	*PERC*
Sample	*TR7D-Drum*	*TR8-504*	*TR9-302*	*TR11-116*	*TR12-BBL729*
Chemical group	*No. 1*				
Water (by D-95)	8·5	3·1	8·9	13·5	7·3
C_2–C_5 acids	—	—	—	—	0·5 (4)
C_5–C_6 alcohols	—	0·1 (2)	0·1 (2)	—	—
HiMW guaiacols	—	—	—	0·4 (2)	—
HiMW oxygenate	—	—	—	0·8 (3)	0·1 (1)
Guaiacols	1·6 (3)	0·1 (1)	0·1 (1)	2·3 (6)	0·7 (3)
Cyclopentenones	2·1 (10)	0·1 (2)	0·2 (2)	0·7 (9)	0·3 (2)
C_8–C_9 (=) cyclic ketones	2·0 (15)	—	—	0·5 (8)	0·3 (2)
Cyclopentanones	1·3 (11)	0·3 (9)	0·4 (9)	0·4 (7)	0·4 (4)
C_7–C_9 cyclic ketones	0·7 (9)	0·2 (5)	0·4 (5)	0·1 (4)	0·3 (4)
Unsaturated phenols	0·7 (2)	0·6 (2)	0·5 (2)	—	0·5 (2)
Phenols	3·3 (22)	6·5 (40)	6·8 (40)	0·6 (8)	4·5 (24)
Dihydroxybenzenes	4·7 (9)	—	—	0·2 (2)	3·8 (13)
Methylnaphthols	2·8 (6)	0·6 (2)	0·4 (2)	—	—
Dihydroindenes	—	0·1 (2)	0·1 (2)	—	—
C_8 alkylbenzenes	—	0·6 (3)	0·5 (3)	—	—
PAH	1·0 (3)	23·6 (37)	14·2 (37)	—	0·6 (1)
Benzofurans	—	0·9 (3)	0·7 (3)	0·1 (1)	—
Other heterocyclics	—	1·5 (2)	0·9 (2)	—	—
No. of peaks identified	90	110	110	50	61
No. of total peaks	360	380	390	290	—
Quantity identified	28·4%	38·2%	34·1%	19·6%	19·7%
Total chromatographed	77·4%	82·7%	61·4%	25·5%	51·7%

Data listed as percent and numbers in parentheses indicate number of individual chemicals.

used to quantify those components which were identified. These analyses show some very distinct differences between the products. They also indicate that the product is hardly a random grab-bag of materials, but shows specificity for isomers and compound groups. A quantitative summary of these results is presented in Table IV. The complete table of data for all components identified is included in ref. 5.

Table IV provides some interesting insights into the oil products. The most imposing difference between the products is the large amount of polynuclear aromatic hydrocarbons (PAH) found in TR8 and TR9 products. These would apppear to be remnants of the anthracene oil carrier used in these tests. They are not likely to be products from wood since they are lacking almost entirely from the LBL oil products and also from the long-term oil slurry run, TR12. Also, the relative concentrations of the various PAHs is indicative of the anthracene oil used at Albany. This result immediately raises questions about the mass-balance calculations performed by the facility operator for TR8 and TR9, as these oils were considered to be 97% and 94% wood-derived material respectively.

Some mechanistic speculations can also be put forth based on these product compositions. Considering the reaction temperature, residence time and extent of recycle (LBL vs. PERÇ), one can conclude that the TR11 oil is the least severely processed product analyzed here. This is supported by the earlier reports of high oxygen content, high pour point and low volatility, and is reconfirmed by the low amount of product which can be chromatographed (see Table IV). The total oil chromatographed, i.e. the amount of oil which passes through the chromatography column to the detector, can be used as a measure of oil quality since less stable and highly polar compounds as well as low volatility components will not pass through the column and be detected. We conclude that guaiacols are early major products in wood liquefaction since they are the major fraction of TR11. Only the ortho-methoxy phenols (guaiacols) were found in wood-derived oil, suggesting the material is specifically derived from the coniferous lignin. The lack of naphthols indicates that these are later products resulting from condensation. The phenolics appear to be an intermediate product. More highly oxygenated products result, therefore, from less severe processing conditions.

Progressing to the next more severely processed product, TR7, we see that the cyclic ketones as well as the phenolics have become more numerous. Naphthols and dihydroxybenzenes are also noted in significant quantities. The lack of low molecular weight alcohols and acids in the TR7 and TR11 products could be explained by their high solubility in the aqueous byproduct stream. However, the acids have been found only in the TR12 product which was produced at a lower temperature. This may demonstrate the temperature dependence of the formation of the acids as suggested by our earlier work[16] on the temperature dependence of the pH of the product material of biomass liquefaction. In essence, we conclude that the acids are less likely to be formed at higher temperatures and the range between 300 and 350 °C is critical in this connection.

Considering the TR8 and TR9 oils, we see that they are nearly identical except for the quantity of anthracene oil impurity. Although the single-cycle conditions were not so severe in these tests as TR7, the recycle mode of operation has apparently changed these oils to a higher degree. The amount of saturated cyclic ketones has been reduced to a significant degree while the unsaturated cyclic ketones are almost completely gone. The amount of guaiacol has been further reduced while the amount of phenolics increased. The lack of dihydroxybenzenes and the reduced amount of naphthols is unexplained. The reduced amount of naphthol may be due to the low temperature as suggested by the TR12 result as well.

The TR12 oil, although subjected to the greatest amount of recycle (albeit at less severe conditions), appears to be a less polished product compared to TR8 or 9. The concentrations of cyclic ketones appear to be higher especially in the unsaturated case. There are also less phenolics and more guaiacols. The amount of naphthols is essentially non-existent while the dihydroxybenzenes are present in sizable quantities. Comparisons of the amount of TR12 oil chromatographed are misleading unless compared on an anthracene oil-free basis, in which case the TR12 is nearly equivalent to TR9.

As mentioned earlier, the only methoxy phenolic isomer was guaiacol. All alkyl substitutions on the guaiacol were at the paraposition. Naphthols appeared to contain only the 1-hydroxy group and only methylated naphthols were detected. A wide variety of methylated isomers of phenol were detected also, but the para-isomer was the only isomer of ethyl phenol detected. 2-methyl-3-pentanone was the only non-cyclic ketone identified. Nearly all isomers of methyl and dimethyl cyclopentanone and cyclopentenone were detected. Other more highly substituted isomers could not be distinguished with certainty from cyclohexanones or other larger cyclic ketones. Diene, alkylbenzene and indan identifications reported by us and others earlier are now believed to be incorrect. Although the computer library matches of the mass spectra in many cases appeared good, close analysis of these spectra now show that the true components are cyclopentanones, cyclopentenones and alkenyl phenols.[6]

As a result of this GC/MS analytical work, an on-site mass spectra library has been developed at PNL. Identification of the components of a complex mixture by gas chromatography/mass spectrometry can be accomplished most rapidly by automated comparison of the mass spectra obtained on the GC/MS to the spectra in a suitable library. For this reason, most commercial GC/MS instruments include computer systems and software for generating libraries and performing automated library searches. In the case of wood liquefaction products, however, we found the library search procedures to be of limited value because the standard libraries lacked many of the compounds that occur in the liquefaction products. In addition, the identification of a number of major compounds was hindered by the inclusion of spectra of poor quality in the libraries. Therefore, we have begun to compile a library of mass spectra of compounds found in wood liquefaction products. The spectra are carefully screened for accuracy and, in particular, for the absence of impurities. GC retention time information (Kovat's Indices) is also included in this library since it provides independent confirmation of the identity of an unknown. Kovat's Indices are also more reliable than mass spectra for identification of specific isomers in many cases.

2.3. Product Oil Solution Extraction

Caustic extraction and solvent separation have been used to separate Albany products into neutral, acid and insoluble fractions. Specific results of these separations are presented in Table V. The separation was made by reacting the wood-derived product with excess aqueous 5% NaOH. The solution was then extracted with methylene chloride. The solvent was stripped from the organic layer to yield the neutral fraction. The aqueous phase was then reacidified with dilute HCl and re-extracted with methylene chloride. The solvent was stripped from the organic phase to yield the acid fraction. The insolubles were filtered from the aqueous phase and dried. This particular scheme of extraction and separation was selected after several alternate techniques failed to provide adequate phase separation.

The information gained from these separations is used in comparing the products from the various process test runs and also provides an additional measure of product quality. The results indicate that a basic difference exists between products from the LBL and PERC processes as practised at Albany. The LBL product has both a higher acid content (expected to be mostly phenols) and a higher insoluble content. This type of separation could also have application from a process point of view; for example, a separation of the neutral fraction for direct use could be followed by

TABLE V
EXTRACTION AND SOLVENT SEPARATION OF WOOD-DERIVED PRODUCTS

Products	Neutral[a]	Acid[b]	Insoluble[c]	Total recovery[d]
TR7	33·1	35·0	32·0	87·5
TR8	75·9	12·9	11·3	103·4
TR9	64·9	21·2	13·8	92·4
TR11	19·3	22·6	58·1	93·7
TR12	50·3	25·3	24·4	92·0

[a] Mass percent non-reactive with NaOH(aq), $MeCl_2$ extractable.
[b] Mass percent reactive with NaOH(aq), $MeCl_2$ extractable in acidified form.
[c] Mass percent insoluble in $MeCl_2$ in acid or base solution.
[d] Mass percentage of initial product which was recovered in the three fractions.

retreatment of the acid and insoluble fractions in the CO–steam system, with hydrogen, or in a methylation step.

2.4. Distillate Fractionation

An important means of fractionating the whole oil for further analysis has been by vacuum distillation. The technique of choice has been to use the ASTM-D1160 equipment with a modified receiver which allows distillate fractions to be removed from the system. We have earlier reported detailed results from TR7[7,8] and TR8[9] distillations. A comparative graph showing the distillation curves for the five whole oils is given in Fig. 1. The two LBL oils seem to bracket the grouping of PERC oils, with TR7 being most volatile and TR11 being least volatile. The analyses of fractions recovered during these distillations was undertaken in much the same manner as the whole oils. Tables VI, VII and VIII give the data which have not been published previously for TR9, TR11, and TR12.

In addition to the whole oil products, there were a number of oil fractions which were produced at Albany as a result of different product handling measures. Fractions which were analyzed at PNL included a condensate oil (TR8C condensate) from TR8 which separated from the condensate water recovered from the pressure letdown step, the distillate trap-out tray oil (TR8D trapoil) and bottom material (TR8D bottoms) from the TR8 vacuum flash distillation test, and TR9 oil (TR9-432) recovered following

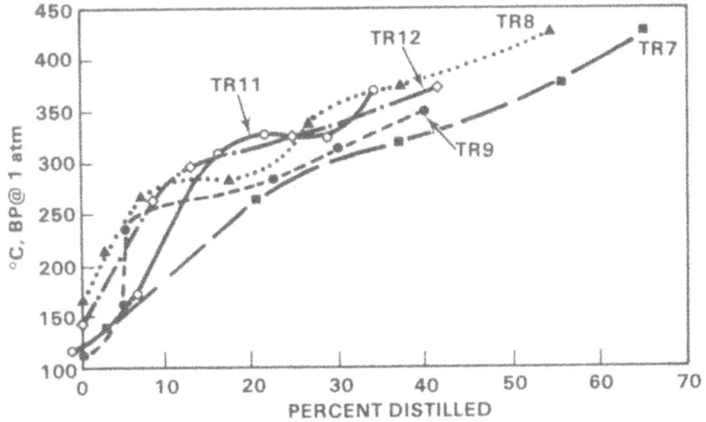

FIG. 1. Atmospheric distillation curves of dry Albany oils.

TABLE VI
VACUUM DISTILLATION FRACTION ANALYSIS, TR9

(a) Distillation data

Fraction	*Actual amount*	*Relative amount*	*BP @ atm*	*BP @ 10 mm*
No. 1	8 ml/12 ml	5%/8%	112–160 °C	7–46 °C
No. 2	24 ml	16%	238–277 °C	107–136 °C
No. 3	10·5 ml	7%	277–303 °C	136–160 °C
No. 4	14 ml	9%	303–349 °C	160–197 °C
Resid.	80·17 g	54%	(final pot temp. 276 °C)	

(b) Fraction analyses

Fraction	*C*	*H*	*O*	*N*	*Atomic H/C*
No. 1 (top)	78·8	12·0	8·5	0·2	1·8
No. 2	78·0	10·4	8·6	0·3	1·6
No. 3	80·1	9·4	9·6	0·2	1·4
No. 4	80·9	9·7	9·4	9·4	1·4
Resid.	82·5	7·4	9·1	0	1·1

TABLE VII

VACUUM DISTILLATION FRACTION ANALYSIS, TR11

(*a*) *Distillation data*

Fraction	*Actual amount*	*Relative amount*	*BP @ atm*	*BP @ 10 mm*
No. 1	2 ml/30 ml	0·9 %/10·6 %	108 °C	9 °C
No. 2	< 1 ml/19 ml	—/8·1 %	170 °C	52 °C
No. 3	4 ml	1·7 %	—	—
No. 4	16 ml	6·8 %	304 °C	162·4 °C
No. 5	10 ml	4·3 %	321 °C	175·0 °C
No. 6	14 ml	6·0 %	325 °C	178·3 °C
No. 7	10 ml	4·3 %	366 °C	214·7 °C
Resid.	161·80 g	57·4 %	(final pot temp. 370 °C)	

(*b*) *Fraction analyses*

Fraction	*C*	*H*	*O*	*N*	*Atomic H/C*	*HHV Btu/lb*
No. 1 (top)	75·2	9·7	13·1	0	1·5	—
No. 2 (bottom)	70·6	8·7	17·5	0	1·5	13 700
No. 3	71·9	8·7	16·5	0·2	1·4	14 600
No. 4	74·5	8·4	16·0	0	1·3	15 100
No. 5	75·7	8·3	15·4	0·4	1·3	14 700
No. 6	75·5	8·0	16·6	0	1·3	14 600
No. 7	73·7	7·9	17·8	0	1·3	14 200
Resid.	79·9	5·3	14·6	0	0·8	14 400

an attempt to dry the oil by distillation prior to vacuum distillation. These oils were analyzed in much the same way as the whole oils, and the results are shown in Table IX. The results are surprisingly similar for the four oils with the greatest differences being in the volatility and solubility of the oils. All the products are low in sulfur and nitrogen. These differences and similarities suggest to us that the oils are alike chemically and the difference lies primarily in the molecular weight ranges of each of the products. This thesis is confirmed in part by our GC/MS analyses which showed the same chemical groups in all the product materials with changes in proportions within the various groups providing the difference between the products.

The similarity of the fractions is also a function of the sophistication of the separation equipment. For example, the TR8C condensate oil was separated more in the manner of a steam distillation in the pressure letdown step rather than a light oil stripping with multiple-plate separation. As a result a substantial portion (~ 20 %) of this oil was made up of anthracene

TABLE VIII

VACUUM DISTILLATION FRACTION ANALYSIS, TR12

(*a*) *Distillation data*

Fraction	*Actual amount*	*Relative amount*	*BP @ atm*	*BP @ 10 mm*
No. 1	1·5 ml/20 ml	0·5%/7%	145 °C	33 °C
No. 2	19 ml	7%	263 °C	127 °C
No. 3	17 ml	6%	296 °C	153 °C
No. 4	24 ml	10%	324 °C	179 °C
No. 5	35 ml	15%	382 °C	228 °C
Resid.	147·83 g	54%	(final pot temp. 310 °C)	

(*b*) *Fraction analyses*

Fraction	*C*	*H*	*O*	*N*	*Atomic H/C*	*HHV Btu/lb*
No. 1 (top)	24·3	13·9	10·8	0·5	2·2	16 500
No. 2	72·8	12·2	17·2	0·1	2·0	11 400
No. 3	78·4	11·5	12·9	0·5	1·7	19 600
No. 4	77·2	10·3	15·0	0·7	1·6	15 300
No. 5	80·0	9·3	9·8	0	1·4	—
Resid.	76·7	6·9	10·8	0	1·1	—

TABLE IX

SUMMARY ANALYSES OF ALBANY PRODUCT FRACTIONS

Analysis	*TR8C*	*TR8D trap*	*TR8D bottom*	*TR9 432*
Elemental analysis (%):				
Carbon	78·1	84·6	83·9	82·9
Hydrogen	10·7	8·7	7·1	7·8
Oxygen	7·9	6·2	5·6	11·5
Nitrogen	0·8	0·5	0·1	0
Sulfur	< 1/2 ppm	< 1/2 ppm	< 1/2 ppm	—
Moisture (ASTM D-95)	6·9	0·2	0	0·2
Solids (HOAC insoluble)	0·4	1·8	33·2	18·0
HHV, Btu/lb	16 200	16 600	16 000	15 900
Percent distillable (D1 160, 10 mm Hg)	94	94	26·6	58

oil components (high boiling, polycyclic aromatics) in addition to the C_1–C_7 alcohols, C_5–C_7 ketones, single-ring hydrocarbons (aromatic and aliphatic) and the ever-present phenolics. The vacuum distillation unit at Albany which produced the TR8D trap oil and bottom material was in essence a single plate still. Wood oil–water mixtures are subject to foaming under vacuum distillation conditions and this was a problem in the Albany unit. The ratio of overhead to bottoms was also limited by the high viscosity of the bottoms material.

2.5. Other Special Analyses

The original goal of the DOE Biomass Liquefaction Program was to produce a storable, pumpable fuel from biomass. The product as produced at Albany is pumpable as demonstrated by the operation of the plant. The fact that the product can serve as a boiler fuel in both its whole oil form and as a distillate oil has also been demonstrated by a series of boiler firing tests undertaken at the Pittsburgh Energy Technology Center.[1,9,10] The stability of the oil product under various storage conditions has been tested in our laboratory.

Samples of TR7 distillate were recovered from long-term storage to determine effects of air, light and heat at 90 °C. Initial inspection indicated that a color change occurred under all conditions after a year's time but to a lesser degree with the exclusion of air. A slight increase in viscosity appeared to correlate in a qualitative manner with the color change. Water droplets were noted in the sample stored for 48 weeks with exposure to all three elements. However, no solid precipitate or gas pressure buildup was noticeable in any of the samples. Elemental analysis showed some minor change. Quantitative GC/MS analysis of the samples, similar to the analyses reported for the whole oil samples, also showed only minor differences. The same 69 chemical components were identifiable in each of three samples with no significant quantitative change in any component. These three samples were stored with air, light and 90 °C heat for various periods of time from zero time to 6 months to 1 year with the balance of the year's time for the first two samples spent in a dark refrigerator sealed under argon. The amount of sample identified in the three samples varied from 31·5 % at zero time to 31·6 % after 6 months to 29·9 % after a year; these differences did not indicate a significant trend. A second measure, the quantity of material which passed through the chromatograph to the detector, varied from 84·1 % at zero time to 93·2 % at 6 months to 87·0 % after a year; again no significant trend was evident. These results showed that a wood oil distillate representing 55·9 % of whole LBL oil could be stored at moderately destabilizing conditions for long periods of time with only minor degradation. Additional high temperature stability tests of TR8D trap oil at approximately 200 °C under atmospheric reflux in excess of 12 hours also showed no gross degradation of the oil product.[5] These studies confirmed the storable nature of wood-oil distillates.

Additional tests have been made to provide comparisons of wood-derived oil to other liquid fuels. Southern Petroleum Laboratories, Inc. under contract to Rust Engineers[1,11] performed a series of typical petroleum analyses on the TR7 oil and found it to be chemically very different from crude petroleum. From our own laboratories we have reported results of Ammes Assays on TR7 oils[8] and TR12 oils which showed that wood oil has a level of activity similar to crude petroleum and much lower than coal tar, coal-derived liquids or shale oil. The Ammes Assay measures mutagenic tendency of materials as a preliminary test of carcinogenic potential.

Due to the oxygenated, polar character of wood oil, its corrosion potential has been suggested.[12,13] The belief that wood-derived oil would be corrosive is in part an extrapolation from published descriptions of pyrolysis products. Recent publications[14] from Texas A&M now attribute the corrosive character of the Tech-Air pyrolysis oils to the organic acids in these products and not the phenolics. Corrosion literature[15] indicates that phenol, cresols, creosote, cresylic acid and wood tar are not considered severely corrosive. In comparison with petroleum, these substances are less corrosive to most materials of construction except concrete. The major corrosion problem with petroleum products is with copper alloys due to the sulfur impurities in petroleum. These copper alloys are essentially unaffected by phenolics. In addition, discussions with phenol users (resin producers) and cresylic acid producers (cresylic acid is a mixed phenolic) indicate that although these materials require special handling requirements to protect workers from contacting the chemicals, special materials of construction in the plants are not required. Mild steel is used throughout these plants and phenolics are shipped and stored in mild steel containers. Corrosion of these materials by phenolics is not considered significant; however, pumps and valves are often equipped with stainless steel trim. Therefore, the corrosive character of phenol is limited to its chemical reactivity with organic materials with the primary concern to chemical manufacturers being worker exposure and the corrosive effects on clothing. This corrosive character decreases with alkyl substitution of the aromatic ring such that pure butyl phenol is not considered corrosive and chemical grade octyl phenol is designated by the Department of Transportation as non-corrosive. The highly substituted character of the biomass oil phenolics would make these materials fairly impotent corrosive agents.

3. CONCLUSION

Results of the analyses show that the oil product from wood can be produced with a range of properties depending on the processing conditions. All experiments have used Douglas fir wood chips as the feedstock in the system. Oxygen content is reduced from 42% in the wood flour to the range of 10–15% in the oil product. Generally, the oxygen content appears to vary in direct proportion to the yield of oil product. The oil product is relatively viscous with a pour point at about room temperature. Heats of combustion for the products range around 16 000 Btu/lb on a dry basis. The products are from one-half to two-thirds distillable under vacuum. The product components number in the hundreds. Of the over 100 compounds and isomers identified and quantified by GC/MS, the major compound groups appear to be methylated cyclopentanones, methylated cyclopentenones, guaiacol and alkylated guaiacols, phenol and alkylated phenolics, and methylated naphthols.

REFERENCES

1. Rust International Corporation (1982). Final report, an investigation of liquefaction of wood at the Biomass Liquefaction Facility, Albany, Oregon. Prepared under DOE contract DE-AC01-78ET-23032.
2. Thigpen, P. L. and Berry, W. L., Jr (1982). Liquid fuels from wood by continuous operation of the Albany, Oregon Biomass Liquefaction Facility. Symposium Papers from Energy from Biomass and Wastes VI, Lake Buena Vista, FL., Jan. 25–29, p. 1057.
3. Elliott, D. C. (1982). Thermochemical production of liquids from biomass. In *Solar World Forum, Solar Technology in the Eighties*, D. O. Hall and J. Morton (Eds), Vol. II. Pergamon Press, Oxford, pp. 1286–91.
4. Elliott, D. C. (1981). Bench scale research in biomass direct liquefaction. In *Proceedings of the 13th Biomass Thermochemical Conversion Contractor's Meeting*. Arlington, VA., October, pp. 551–75, available NTIS No. CONF-8110115.
5. Elliott, D. C. (1982). Bench scale research in biomass direct liquefaction. In *Proceedings of the 14th Biomass Thermochemical Conversion Contractor's Meeting*. Arlington, VA, June, available NTIS No. CONF-820685.

6. Schirmer, R. E., Pahl, R. T. and Elliott, D. C. (1982). Analysis of a thermochemically derived wood oil. Submitted for publication May.
7. Elliott, D. C.,1980). *Am. Chem. Soc., Div. Fuel Chem. Prprt.* **25**(4), 257–63.
8. Elliott, D. C. (1981). Process development for direct liquefaction of biomass. From *Fuels from Biomass and Wastes*, D. L. Klass and G. H. Emert (Eds). Ann Arbor Science Publishers, Inc., Ann Arbor.
9. Elliott, D. C. (1981). *Biotech. & Bioeng.* Symp. No. 11, 187–98.
10. Ekmann, J. M. and Snedden, D. (1981). Final report, test report on the combustion of PERC and LBL wood oils. Pittsburgh Energy Technology Center.
11. Winfrey, J. C. (1979). A laboratory evaluation of biomass oil (TR-7-136). Lab Report No. 77782, Southern Petroleum Laboratories, Houston, TX.
12. Wilhelm, D. J. and Stallings, J. W. (1980). Assessment of the biomass liquefaction facility in Albany, Oregon, and related programs. Report to US DOE on contract No. DE-AC03-80CA83008, June 16.
13. Molton, P. M., Miller, R. K., Russell, J. A. and Donovan, J. M. (1981). Effects of reaction conditions on the aqueous thermochemical conversion of biomass to oil. ACS Symposium Series No. 144, pp. 137–62.
14. Lin, S.-C. and Soltes, E. J. (1981). Hydrocarbons via hydrogen treatment of pine pyrolytic oil. Presented at the 181st Nat'l. Amer. Chem. Soc. Meeting, Atlanta, GA, March 29–April 3.
15. Radbald, E. (1951). *Corrosion Guide.* Elsevier Publishing Co., Inc., New York; Polar, J. P. A guide to corrosion resistance. Climax Molybdenum Co.
16. Elliott, D. C. (1981). Bench scale research in biomass liquefaction in support of the Albany, Oregon, experimental facility. Presented before the 12th Biomass Thermochemical Conversion Contractor's Meeting, Washington, D.C., available NTIS No. PNL-SA-9430.

56

Characterization of Peat and Biomass Liquids

OLA KARLSSON and PEHR BJÖRNBOM
Royal Institute of Technology, Department of Chemical Technology, Stockholm, Sweden

1. INTRODUCTION

The work presented in this paper is a part of the Oil from Peat and Biomass project sponsored by the National Swedish Board for Energy Source Development.

At the present the goals of the project are:

(1) To study the liquefaction of different solid materials such as peats, biomasses and coals at varying reaction conditions in batch autoclaves. Both wet liquefaction with and without carbonate catalyst in the presence of carbon monoxide and dry liquefaction with tetralin as a hydrogen donor solvent are being studied.
(2) To build and run a bench-scale continuous reactor.
(3) To develop suitable analytical methods for the liquefaction products and byproducts and to evaluate the data obtained.

This paper will deal with the third part.[1]

The predominant need in the field of analysis of liquefaction products and reporting of analytical data is perhaps the need for standardization. This standardization is necessary in order to be able to compare and exchange results and conclusions between different research groups.

Many of the ASTM standard methods for petroleum are not applicable on these liquids: in some cases due to the difference between chemical composition of petroleum and liquefaction products, and in some cases due to the fact that when small batch reactors are used the amount of product is simply not enough.

The products from a liquefaction process can be separated and analysed as follows:

(1) Crude oil (liquefaction product).
   (a) Rapid, simple and preferably standardized routine characterization with the main purpose being to compare the outfall of separate experiments.
   (b) Analysis to meet the demands from the petroleum and catalyst industries in order to facilitate research on upgrading of liquefaction products.
   (c) Detection and quantification of compounds and classes of compounds critical to health and environment and/or of pure scientific interest.

(2) Water and water solubles.
   (a) Analysis to facilitate waste water treatment.
   (b) Analysis to make recovery of water solubles possible.

(3) Gases.
   Quantification of reaction and product gases.

(4) Solid residue.
   Elementary analysis etc.

Most of our efforts so far have been made on the characterization of the crude oil. The methods reported here are gradient elution chromatography (GEC) and distillation.[2]

A basic assumption in this investigation was that the products from liquefaction of wood and peat would give the same fractions in GEC as would coal liquids. This assumption is justified by the fact that coal liquids contain a very wide variety of compounds consisting of carbon, hydrogen, oxygen, nitrogen and sulphur.

Our experimental results support this view. Our products from the liquefaction of wood, peat and lignite were all fractionated into the same fractions. Further analysis of these fractions confirmed that they correspond to the fractions obtained from coal liquids according to the literature.[3]

2. EXPERIMENTAL

Gradient elution chromatography was performed on a preparative scale with alumina prepared according to Brockman, as stationary phase. The solvents were hexane, dichloromethane, tetrahydrofurane and methanol.

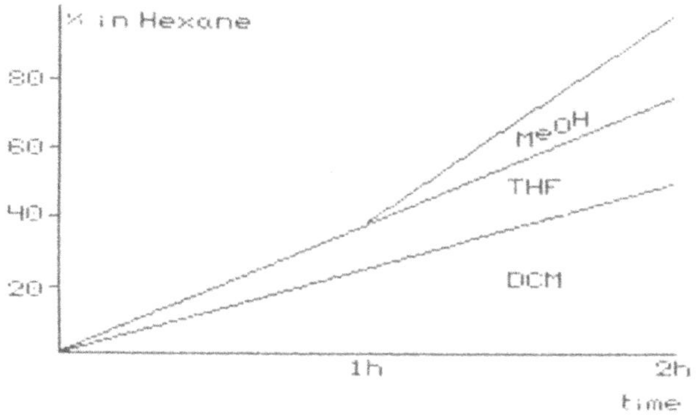

FIG. 1. Solvent system for GEC.

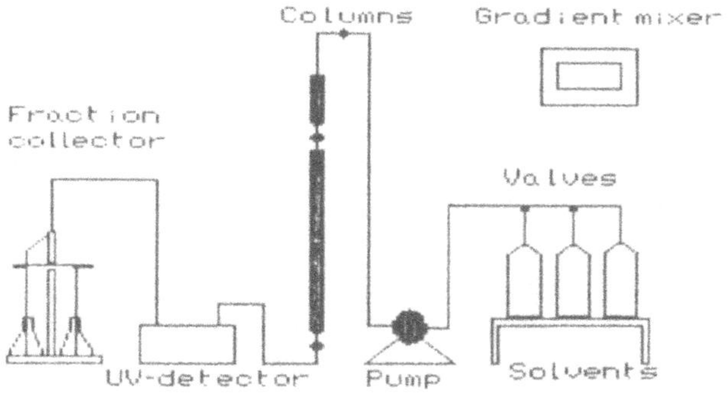

FIG. 2. Gradient elution chromatography.

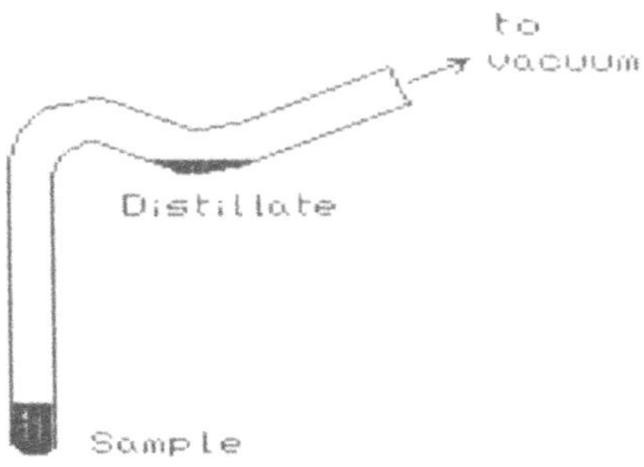

FIG. 3. Micro distillation equipment.

The gradient was controlled by an LKB-Beckman Ultrograd gradient mixer, as shown in Fig. 1. The total flow rate was 15 ml/s and the elution time was 2 h plus purging time.

Four fractions were detected by a variable UV/Vis detector and collected by a fraction collector.

The GEC equipment is shown in Fig. 2. The solvents were evaporated off in a rotavaporator to constant weight.

The fractions were analysed employing UV and IR spectroscopy, GC, GC/MS, elementary analysis and a micro distillation apparatus as shown in Fig. 3. The temperature gradient was 3 °C/min.

3. RESULTS AND DISCUSSION

A typical chromatogram is shown in Fig. 4. The wavelengths used for detection were 210, 254, 300, 341, 405 and 435 nm.

The first fraction consisted, according to elementary analysis data, mainly of hydrocarbons (H/C: 1·9, O/C: 0·005, N: 1 %, S: 0·1 %—typical values for a biomass product). IR and UV data indicated that the main type of components were branched saturated hydrocarbons. However, some aromaticity could occur, possibly due to highly substituted aromatic rings of alkane character.

Distillation data indicated that approximately 2/3 of the first fraction distilled under 370 °C. IR, UV, density and RI analysis confirmed that the distillate only consisted of highly branched and cyclic saturates. GC and GC/MS data indicated that more than 100 isomers of C10–C22 saturates occurred, none of them in a concentration exceeding 1 % of the fraction.

The molecules containing heteroatoms and aromatics were concentrated

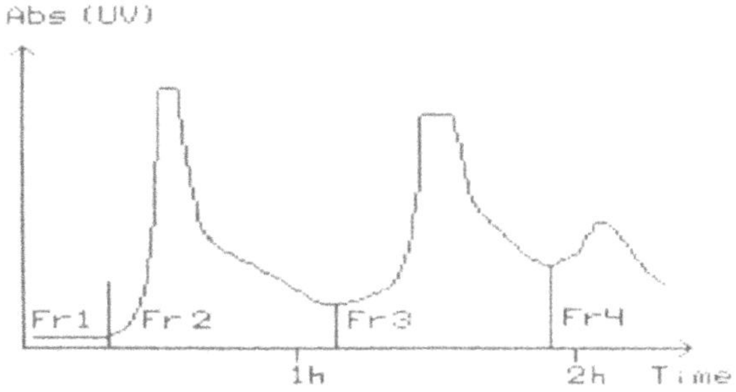

FIG. 4. Typical chromatogram—GEC.

in the distillation residue. The nitrogen occurred as amines and the oxygen as ethers according to IR spectra.

Depending on the raw material and the reaction conditions the yield of the first fraction varied between 5 and 20 wt % of our liquefaction product. For crude petroleum rich in asphalt or lubricating oils, which were analysed for comparison, the yield of the first fraction was around 37 wt % and for crude petroleum rich in fuel or petrol approximately 50 wt %.

The second fraction was highly aromatic according to IR, UV and elementary analysis (H/C: 1·7, O/C: 0·07, N: 1·9 %, S: 0·17 %—typical values for a biomass product). It contained more oxygen, nitrogen and sulphur than the first fraction. IR data showed that the oxygen occurred mostly as ketone or ester groups and the nitrogen as amines and pyridines. However, there was no trace of phenolic oxygen.

The corresponding fraction from coal liquids appeared to have much less oxygen than when it was obtained from peat or biomass.

Distillation data indicated that approximately 2/3 of the second fraction was distillable up to 370 °C. IR and elementary analysis showed that many of the oxygen-containing compounds were relatively low boiling. GC confirmed that substituted mono-, di- and triaromatics were dominant in the distillate. The nitrogen was concentrated in the distillation residue.

Depending on the raw material and the reaction conditions, the yield of the second fraction varied between 5 and 35 wt % of our liquefaction product. For crude petroleum rich in lubricating oils the yield was 45 wt %; for crude petroleum rich in asphalt or petrol, approximately 35 wt %; and for crude petroleum rich in fuel, 13 wt %.

The third fraction was phenolic or resinous in its character according to IR and UV spectra. Elementary analysis data showed that the heteroatoms were concentrated in the third fraction, when compared to liquefaction product as a whole (H/C: 1·5, O/C: 0·12, N: 2·2 %, S: 0·15 %—typical values for the third fraction for products from biomass, H/C: 1·4, O/C: 0·12, N: 1·6 %, S: 0·09 %—typical values for a biomass-originating product as a whole).

Products from biomass had more low boiling phenolic material than products from peat according to distillation data. GC/MS and IR spectra showed that ketone and alcohol groups were also present. The only compounds exceeding 1 wt % in concentration were phenol and substituted phenols such as cresols.

The yield of the third fraction varied between 10 and 45 wt %. For all crude petroleums tested the yield was less than 15 wt %.

The fourth fraction was much like the third according to IR, UV and elementary analysis data. However, very little of the fourth fraction was distillable under 370 °C.

The yield of the fourth fraction varied between 2 and 20 wt % for our products and between 2 and 3 wt % for all crude petroleums tested.

The non-eluted material was found, through comparison with the results from extraction with hexane in a Soxhlet apparatus, to consist of asphaltenes, preasphaltenes and inorganic material.

The yield of the non-eluted fraction varied between 6 and 65 wt % for our products, depending on the reaction conditions and the raw material. For all crude petroleums tested the percentage of the non-elutables was less than 20 %.

More than four fractions could easily be obtained by modifying the gradient. Four fractions were, however, assumed to give adequate information when the purpose was to characterize.

The reproducibility of the GEC method was found to be better than 7 %, which was quite enough for this purpose.

3.1. Other Methods for Characterization

It is essential that the set of characterization methods is carefully selected, otherwise the results can be completely misleading. For example, moisture and viscosity are not independent of each other and to determine the viscosity without checking the moisture could be very serious.[4]

It is also essential that all analyses are performed as soon as possible after the liquefaction experiment is terminated and that all samples are protected from air, light and heat. An example of a crude oil stored in a refrigerator without proper protection against air is shown in Fig. 5.

In order to fully appreciate the analytical results the isolation method for the crude oil must be known. In Fig. 6 an example of the effect of the solvent

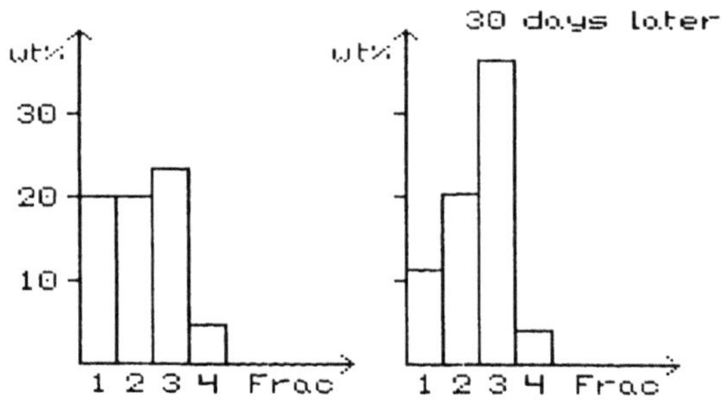

FIG. 5. Effect of storage—GEC fractions.

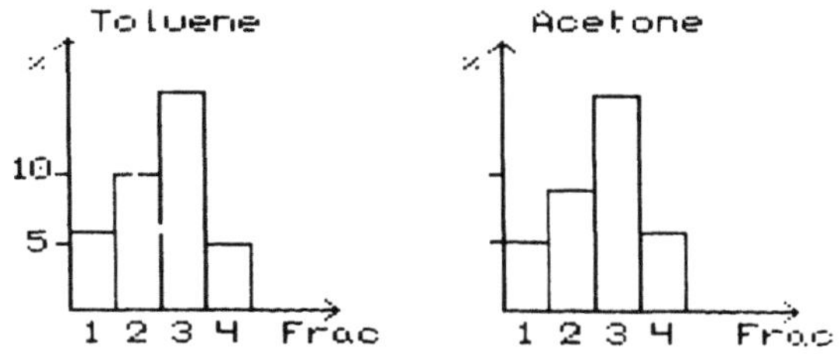

FIG. 6. Effect of solvent. The product yield was 40 and 50 wt % respectively.

on the composition of the crude oil, under separation of the solid residue from the liquefaction product, is shown.

Some of the chemical properties that are of interest and that are practically available are listed in Table I, together with the most common analytical methods.

As this information often is not enough to properly characterize the liquefaction product, it is necessary to determine some other properties that can reflect the chemical properties on which it is not practical or even possible to obtain information. Some of these other properties are listed in Table II, together with the most common analytical methods.

TABLE I

IMPORTANT CHEMICAL PROPERTIES

Mean molecular weight—Osmometry, MS
Types of compounds—Chromatography (GEC, HPLC)
Elemental composition—Elementary analysis
Moisture—Fischer titration
Salt, mineral matter

TABLE II

OTHER IMPORTANT PROPERTIES

Viscosity
Density
Carbon residue—Ramsbottom carbon residue (Conradson)
Ash
Aromaticity—NMR
Pour—and softpoint
Distillability
Heating value
Thermo analysis—Thermo balance

These properties are important as these types of liquefaction products, before upgrading, often contain a significant amount of large molecules for which little or no chemical information is available.

Which of these properties/methods that are the most important can be discussed and must be discussed in order to establish some form of standardization for characterization of peat and biomass liquids.

REFERENCES

1. Björnbom, P. *et al.* (1982). Oil from peat and biomass. Final report for the fifth stage, Royal Institute of Technology, Dept. of Chemical Technology, Stockholm, Sweden.
2. Karlsson, O. (1980). Unpublished paper. Royal Institute of Technology, Dept. of Chemical Technology, Stockholm, Sweden, Analytical report.
3. Callen, R. B., Bendoraitis, J. G., Simpson, C. A. and Voltz, S. E. (1976). Upgrading coal liquids to gas turbine fuels. *Ind. Eng. Chem. Prod. Res. Dev.* **15**(41), 222–33.
4. Thigpen, B. (1981). Liquid fuels from wood by continuous operation of the Albany, Oregon biomass liquefaction facility. The Rust Engineering Company, Birmingham, Alabama.

57

The Products of Direct Liquefaction of Biomass

HUBERT G. DAVIS, MICHAEL A. EAMES, CARLOS FIGUEROA, RAYMOND R. GANSLEY, LARRY L. SCHALEGER and DAVID W. WATT

Energy and Environment Division, Lawrence Berkeley Laboratory, University of California, Berkeley, California, USA

1. INTRODUCTION

Wood is converted to a dense liquid fuel by thermal treatment at 300–360 °C under a pressure of water and reducing gas in the presence of aqueous alkali. The product includes both an oil phase and an aqueous phase rich in dissolved organics. Previous work[1] showed that the oil phase has little in common with petroleum fuels, and the standard methods of petroleum chemistry are therefore largely inapplicable. Thus there existed a need for new methods for the fractionation of biomass oils into constituent chemical classes for the purpose of assessing comparative product quality and evaluating the best end uses and upgrading options.

Three general approaches to the problem of the chemistry of wood oil include solvent extraction, gas chromatography–mass spectroscopy (GC–MS), and adsorption chromatography. Solvent extraction schemes are in common use in the characterization of synfuels including those derived from biomass. The 'oil' fraction of coal fluids is typically defined, for example, as that portion extractable into pentane.[2] By this definition, there is no 'oil' in wood oil since the solubility of wood-derived fuels in pentane is virtually nil. Thus wood oil is more commonly taken as the acetone, chloroform or benzene-soluble product, and would, in the parlance of coal fluid chemistry, be classified as a mixture of asphaltenes and preasphaltenes.

An approach based on a combination of both solubility and acid–base properties of biomass oils is somewhat more useful. For example, a major part of wood oil can be extracted from chloroform into aqueous sodium hydroxide but not bicarbonate, revealing the predominantly phenolic character of these materials. In practice, however, extraction techniques have proved to be of limited utility. They are time consuming and irreproducible owing to opacity of the two phases, difficulties with emulsions and incompleteness of extractability of high molecular weight phenolics into aqueous caustic.

A good deal of information regarding the structural characteristics has been obtained by GC–MS.[3,4] The major components include methoxyphenols, cyclopentenones, cycloalkanones, alkylphenols, dihydroxybenzenes and methylnaphthols.[3] Many of these types of compounds are also found in the aqueous phase along with C1–C6 carboxylic and dicarboxylic acids.[4,5] However, these identifications pertain only to the 20% or so of the constituents of wood oil that have yielded to GC–MS analysis. Further progress along these lines will be difficult for two reasons. First, even with the high resolution afforded by capillary GC, there is the general problem of overlapping peaks that plagues analysis of extremely complex mixtures. Elliott has observed as many as 360 distinct peaks in a typical wood oil sample.[3] The second limitation is that 30% or more of wood oil is usually too involatile to be detectable by GC. Although pyrolysis GC might find application in this connection,[6] this technique has as yet not yielded significant information about wood oils.

There seemed to be a need, therefore, for an approach that would fill the gaps left by GC–MS for the characterization of whole biomass oils. The ideal method would be fairly simple, rapid, reproducible, and capable of separating these materials into classes according to chemical functionality. Adsorption chromatography, although widely used in other branches of fuel chemistry, had not been applied to this problem and so seemed a logical starting point. Farcasiu[7] had already developed SESC (sequential elution by solvents-chromatography) for separating coal fluids into distinct chemical classes. The method seemed even better suited for the separation of biomass oils which, unlike coal products, have no nitrogenous fraction to deal with. We therefore chose SESC as the basis for the development of such a method.

2. RESULTS AND DISCUSSION

2.1. Method Development

The original SESC technique required only slight modification for the purpose at hand. Since wood oil is virtually devoid of aliphatic hydrocarbons, the first step, elution with hexane, was omitted. Whatever amount of this fraction, F1, does exist is eluted along with the aromatic hydrocarbons by 15% benzene–hexane in the present technique. This hydrocarbon fraction is designated F1,2. Also, we have found acetic acid to be superior to pyridine for removing the last wood oil residues from the column. The complete sequence of solvents is shown in Table I.

A further improvement was the use of the low pressure preparative liquid or 'flash' chromatographic technique,[8,9] which employs small, irregularly-shaped particles of silica gel (40 μm) and low positive pressures (20 psig) to afford speedier analysis and improved resolution with reduced consumption of solvent.

One sample of wood oil, TR-7, produced from Douglas fir at DOE's Albany, Oregon liquefaction test facility by LBL's water-based process, was fractionated four times including once with an anhydrous sample, in order to check repeatability and the effect of water. The presence of 8% of water in the sample did not alter the results (Table I).

TABLE I
SESC FRACTIONATION OF TR-7 OIL

Fraction	*Solvent*	*Amount (%) Std. Dev.*
F1,2	15% benzene–hexane	3·5 ± 0·4
F3	Chloroform	16·8 ± 2·3
F4	6% ether–chloroform	44·1 ± 3·7
F5	4% ethanol–ether	21·1 ± 3·0
F6	Methanol	6·5 ± 1·3
F7	4% ethanol–$CHCl_3$	1·0 ± 0·4
F8	4% EtOH–THF	2·5 ± 1·7
F9	HOAc	4·2 ± 0·6

2.2. Characterization of SESC Fractions

Several dozen biomass-derived oil samples have now been subjected to SESC as modified for wood oil. Information regarding the chemistry of various fractions has been provided by gel permeation chromatography (GPC) for molecular weight distributions, GC–MS for specific compound identification, infrared spectroscopy, elemental analysis and solubility classification.

In Table II we show the ranges of the percentages found in each fraction of the oils tested, typical average molecular weights, the typical percentage of oxygen in the fraction and an average molecular formula.

Molecular weight distributions of fractions and whole oil samples were obtained by gel permeation chromatography. The predominant band in

TABLE II

CHARACTERIZATION OF WOOD OIL SESC FRACTIONS

Fraction	*Wt%*	$\bar{M}_n^a$	$\bar{M}_w^b$	*% O*	*Average molecular formula*
1,2	0–6	140	160	5	$C_{11}H_{14}O_{0\cdot5}$
3	1–20	170	210	11	$C_{13}H_{15}O_{1\cdot3}$
4	5–45	170	210	16	$C_{12}H_{14}O_2$
5	5–55	210	290	21	$C_{15}H_{16\cdot5}O_{3\cdot3}$
6	5–35	350	600	23	$C_{30}H_{30}O_7$
7	1–3	180	210	46	—
8	2–20	690	900	—	—
9	0–10	—	—	—	—

[a] Number-average molecular weight.
[b] Weight-average molecular weight.

the infrared spectra of most fractions was that due to hydroxyl stretching, except that this band is absent in F1, 2 and weak in F3. Other bands included the carbonyl stretching band (F3–F6 and F8) and those characteristic of the aromatic ring.

Fractions 1–5 are liquid while 6, 8 and 9 are solid. Fraction 7, always minor, is usually solid. Capillary gas chromatography indicates that the distillable components reside exclusively in fractions 1–5 with the suggestion that F5 also contains less volatile matter. Fractions 4–6 dissolve rapidly and almost completely in 5% aqueous sodium hydroxide.

Fraction 1, 2 contains aromatic hydrocarbons. GC–MS analysis of one F1,2 sample gave the following tentative identifications: alkylated benzene derivatives (C_8H_{10}, C_9H_{12}, $C_{10}H_{14}$ and $C_{11}H_{16}$ species); substituted indanes or alkenylbenzenes (C_9H_{10}, $C_{10}H_{12}$ and $C_{11}H_{14}$); 2-methylnaphthalene, biphenyl and a tetramethylphenanthrene isomer.

According to Farcasiu,[7] the constituents of F3 include non-hydroxylic polar aromatics and oxygen heterocycles. Anisole, a variety of cycloalkanones, a butyrophenone (1-(p-methoxyphenyl)-1-butanone) and a hindered phenol (2,4,6-trimethylphenol) were found by GC–MS in this study. Standards were used to show that anisole, guaiacol and eugenol also elute with F3.

The present work reinforces the conclusion that F4 consists largely of monophenols.[7] Phenol and 5-indanol elute cleanly with 6% ether–chloroform. Further studies with model compounds are in progress.

Fractions 5 and 6 are also phenolic in nature as evidenced by solubility in aqueous caustic, infrared spectra and elemental analyses. It appears from

TLC data[7] that 1-naphthol, 2,3-naphthalenediol, salicylic acid and chrysin, $C_{15}H_8O_2(OH)_2$, would be found in F5. We find that 3,4-bis(p-hydroxyphenyl)-3,4-hexanediol appears in F5. The average molecular formula can be fairly represented by the structure

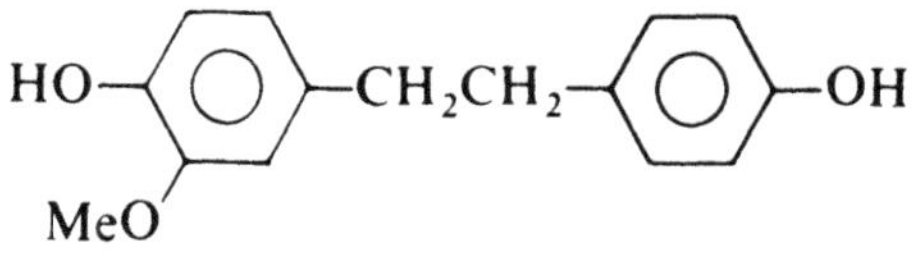

a fragment conceivably derivable from softwood lignin. Although there is probably considerable overlap between F4 and F5, it can be inferred that monophenols prevail in the former and phenolic dimers (e.g. bis-phenols) in the latter.

Methanol marks the approximate line between asphaltenes (benzene-soluble pentane insolubles) and preasphaltenes in SESC. The sixth fraction elutes sharply as a dark brown band with little or no contamination by the tail end of F5. This material is involatile, highly polar and phenolic in nature. The major components of SRC-derived F6 have been termed 'highly functional molecules' having greater than 10 wt % heteroatoms. In the case of wood oil, F6 would seem to be polyphenolic material, perhaps partially degraded lignin.

2.3. SESC Profiles of Wood Oils

SESC profiles of eight biomass oils are shown in Table III. These runs are described in Table IV, along with additional product information. The oils

TABLE III
SESC PROFILES OF SELECTED WOOD OILS

Fraction	*Oil:*							
	TR-7	*TR-10*	*TR-12*	*CL-8*	*CL-10*	*CL-11B*	*CL-12B*	*CL-21*[a]
1,2	4	1	4	1	3	4	3	5
3	17	7	14	3	18	16	15	15
4	44	19	14	6	8	20	25	40
5	21	40	38	44	50	41	39	20
6	6	21	18	31	14	12	12	13
7,8,9	8	11	12	15	6	7	5	7
1–4	65	27	32	10	29	40	43	60
4–6	71	80	70	81	72	73	76	73

[a] From aspen. All others were from Douglas fir.

TABLE IV
OPERATING CONDITIONS AND ANALYTICAL RESULTS FOR SELECTED RUNS[a]

	TR-7	*TR-10*	*TR-12*[b]	*CL-8*	*CL-10*	*CL-11B*	*CL-12B*	*CL-21*
Temperature (°C)	~350	~350	330	330	350	360	360	350
Steady-state yield (%)	~25	~26	53	40	39	41	32	43
Reducing gas	H/CO	H/CO	H/CO	H/CO	CO	CO	H	H
Residence time (h)	5	—	0·3	0·4	0·35	0·3	0·3	0·3
% Oxygen	14	17	13	18·5	16	16	16	13·5
$\bar{M}_n$	220	290	240	280	270	250	250	180
$\bar{M}_w$	310	470	370	360	340	340	340	270

[a] Data for TR runs taken from ref. 13.
[b] PERC mode. All other runs were done by the LBL process.

identified as TR-7, TR-10 and TR-12 were produced at the US Department of Energy's (DOE) process development unit at Albany, Oregon. TR-7 and TR-10 were made by a process developed at LBL in which wood chips are prehydrolyzed under mildly acidic conditions, neutralized, mechanically refined and introduced into a one-pass reactor system as a 15–25 % slurry of soluble and insoluble wood solids in water.[10,11] TR-12 oil was the product of a process developed from the original work at the Bureau of Mines' Pittsburgh Energy Research Center in the late 1960s–early 1970s.[12] This is a recycle process which introduces a slurry of wood flour in anthracene oil in the presence of aqueous sodium carbonate. Over a number of cycles the anthracene oil used for start-up is replaced by wood oil. The TR-12 used in the present study was judged to be 99 % wood-derived.[13]

The CL oils were produced in a 2 kg h^{-1} continuous liquefaction unit (CLU) designed, built and operated at LBL.[14] They were made by the LBL process using 18–20 % aqueous slurries of prehydrolyzed wood. Except for CL-21 from aspen (*Populus tremuloides*), all of the oils of Table III were derived from Douglas fir.

2.4. Quality of Wood Oils

Desirable process goals for biomass liquefaction can now be cast in terms of these SESC profiles of oil quality. From the standpoint of fuel quality the ideal wood oil would consist entirely of fractions 1–3. However, from the results shown in Table IV (4–21 % in F1–F3) this appears to be unrealistic within the context of the current generation of liquefaction processes. A more practical process goal would be to maximize phenolic monomers (F4) at the expense of F5 and F6. Although a reliable index of reaction severity is not available, this does appear to be the most significant factor affecting the proportion of F1–4 in the case of Douglas fir. The effect of reaction severity is most evident in comparisons of TR-7 (65 %) with TR-10 (27 %), and of CL-8 (10 %) with CL-10, 11B or 12B (29–43 %). In general it appears that the CLU oils are somewhat underreacted. This is probably due to the fact that residence times in the CLU were of the order of 20 min, a constraint imposed by the operating characteristics of the system and the lack of oil recycle.

Several additional points about oil quality may be noted. We reported previously that the nature of the reducing gas did not appear to affect the course of biomass liquefaction, at least not under the conditions employed in runs CL-10 and 11.[1] Although work by others suggest the possibility of a unique role for carbon monoxide as a reducing agent, perhaps acting by the putative formate ion mechanism,[12,15,16] the results of CLU operation suggest that wood liquefaction is largely thermal in nature and that any advantages to be gained by using carbon monoxide over hydrogen are minor at best.

SESC also provides a very approximate measure of total phenolics in the sum of fractions 4–6. These totals vary from 70 to 81 % for the oils listed in Table III. If an average steady state oil yield of 40 % for CLU runs 8, 10, 11 and 12 is assumed, then the yields of phenolics based on dry wood are 32, 29, 29 and 30 %, respectively. These numbers compare closely with the lignin content of Douglas fir, 28 %. It would be erroneous, however, to assume that lignin is the exclusive source of phenolic material. In the first place, the liquefaction of pure cellulose also gives rise to considerable phenolic product.[17] Second, the aqueous phase of wood liquefaction by the LBL process also contains significant quantities of substances that probably originate with lignin.[18] These include dihydroxybenzenes such as 4-methylcatechol and hydroxyalkyl derivatives of guaiacol such as vanillyl alcohol. Finally, we have seen that certain hindered phenols and simple guaiacols populate F3 rather than F4.

2.5. Effect of Feedstock

The results of CL-21 underscore the importance of choice of feedstock. Under conditions of comparable reaction severity, aspen gives rise to a product of quality superior to that from Douglas fir and in slightly higher yield. The total phenolics amount to 31 % based on dry feedstock. This compares to a lignin content of approximately 17–18 %. That aspen is more reactive than coniferous woods is well established in other technologies. A variety of additional biomass feedstocks are under examination as candidates for liquefaction in these laboratories.[19]

2.6. Product Upgrading by Catalytic Hydrogenation

Catalytic hydrogenation offers a promising option for upgrading biomass oils. In one experiment SESC was used to evaluate the effectiveness of copper chromite catalyst. Neat, anhydrous TR-7 oil slurried with finely divided catalyst (2 %) was reacted for 50 min at 400 °C under a hydrogen pressure of 2000 psig in a stirred autoclave. The product was subjected to SESC with the following results (dry TR-7 oil basis):

Fraction:	F1,2	F3	F4	F5	F6	F7–9
Wt %:	7	22	47	4	1	8

It appears that catalytic hydrogenation was very effective in the depolymerization of higher molecular weight phenols and that monophenol was converted to less polar materials. Further studies with model substances are in progress.

In the temperature range of 180–240 °C or higher, biomass samples react with recycled oil product to form a homogeneous solution or mixture.[20] This is a specific form of solvolysis.[21] The crude solvolytic product must be further treated to recover the recycle oil and net product (about 50 % by weight of the dry biomass feed). Experiments such as the above with TR-7 and preliminary experiments with solvolytic products indicate that catalytic hydrogenation is the most promising way to effect this.

From the average molecular weights given inTable II we can estimate the boiling ranges of SESC fractions 1 to 5. Very approximately these are the following:

Fraction	Approximate boiling range (°C at 1 atm)
1,2	120–260
3	120–300
4	170–360
5	260–430

When we consider that low boilers ($< \sim 120$ °C) are inevitably lost in the oil sample recovery and/or the SESC work-up, it is clear that oils with 20 % or more in SESC fractions 1–3 and 50 % or more in fractions 1–4 should be easily distillable at atmospheric pressure to give over 20 % volatiles. This is all that is necessary to provide the right proportion of net product and recycle vehicle.

2.7. Summary

Several methods of characterizing products of biomass liquefaction including GC–MS, wet analysis, infrared and various forms of liquid chromatography have been tried. Of these the SESC sequential elution technique has been particularly helpful in separating whole product oils into chemically distinguishable fractions.

Variations in the distribution of SESC fractions have been used to show effects of changing conditions and changing feedstocks in biomass liquefaction. They have also been used to follow the effects of catalytic hydrogenation.

Liquefaction of aspen results in higher yields than liquefaction of Douglas fir. The aspen oil product is richer in SESC fractions 1–4, is more fluid and has lower oxygen content.

Catalytic hydrogenation is effective in increasing the percentage of oil in the more volatile SESC fractions 1–4. It appears to be the key to successful development of an oil recycle solvolysis process.

3. EXPERIMENTAL

'Flash' chromatography employed commercially available 'silica gel for flash chromatography', and columns and fittings. Generally the ratio of sample to adsorbent was 1:40. Sample was preadsorbed on to silica gel by rotary evaporation of a solution of wood oil in methylene chloride; the resulting semisolid was then quantitatively transferred to the top of a column packed with the aid of hexane. Solvents were reagent grade or, in the case of chloroform, ether and tetrahydrofuran, unstabilized HPLC grade.

After elution of each fraction, solvent was removed by rotary evaporation at water pump pressure and 50–60 °C, and the weight of the residue was determined. Typical recoveries were in the range of 100–115 %, probably owing to the impossibility of completely removing solvent. The SESC results reported in Table III are normalized to 100 %. With a few samples from runs at low severity we had significantly less than 100 % recovery.

Gel permeation chromatography employed two columns in series packed with 200–400 mesh styrene–divinylbenzene copolymer beads with 8 % and 12 % cross-linkage.[22] The solvent was tetrahydrofuran. Molecular weight calibration curves were constructed using a series of 12 polystyrene and phenolic standards. The high molecular weight exclusion limit was 700–800 daltons. The calibration was checked frequently with nordihydroguaiaretic acid (mass 302) or phenolphthalein (318). The solvent, tetrahydrofuran, was pumped at 1·2 ml min^{-1} with UV detection at 254 or

280 nm. Generally little or no elutable material having molecular weights above the exclusion limit of the system was observed. Molecular weight calculations were performed with the aid of commercially available computer software.

ACKNOWLEDGEMENTS

The contributions of the following individuals and groups are gratefully acknowledged: Pat Bronnenberg, Mark Featherstone, Dieter Kloden, Mohindar Singh, Nasser Yaghoubzadeh and Rafael Zamora; Pierre-Yves Lamy of the University of Technology, Compiegne, France; Larry Taylor of Virginia Polytechnic Institute and State University; and the Rust Engineering Corp.

This work was supported by the Assistant Secretary for Conservation and Renewable Energy, Biomass Energy Technology Division of the US Department of Energy under Contract No. DE-AC03-76SF00098 and funded through Battelle Pacific Northwest Laboratory, Lead Laboratory for the US Department of Energy, Biomass Energy Technology Division, Thermochemical Conversion Program under Contract No. DE-AC06-RL0-1830.

REFERENCES

1. Schaleger, L. L., Figueroa, C. and Davis, H. G. (1982). *Biotechnol. Bioeng. Symp.* **12**.
2. Mima, M. J., Schultz, H. and McKinstry, W. E. (1978). In *Analytical Methods for Coal and Coal Products*, Vol. I, C. Karr, Jr (Ed.). Academic Press, New York, Chapter 19.
3. Elliott, D. C. (1982). Paper presented at the 14th Biomass Thermochemical Contractors' Meeting, Arlington, Virginia, June 23–24. NTIS CONF-820685.
4. Davis, H., Djafar, R., Ergun, S., Figueroa, C., Karatas, C., Schaleger, L., Seth, M., Wrathall, J., Yaghoubzadeh, N. and Yu, G. (1980). Lawrence Berkeley Laboratory Report, LBID-272, Aug. LBID-314, Nov.
5. Davis, H. G., Kloden, D. J. and Schaleger, L. L. (1981). *Biotechnol. Bioeng. Symp.* **11**, 151.
6. Gallegos, E. J. (1979). In *Chromatography in Petroleum Analysis*, K. H. Altgelt and T. H. Gouw (Eds). Marcel Dekker, New York, Chapter 8.
7. Farcasiu, M. (1977). *Fuel* **56**, 9.
8. Still, W. C., Kahn, M. and Mitra, A. (1978). *J. Org. Chem.* **43**, 2923.
9. Crane, L. J., Zief, M. and Horvath, J. (1981). *Amer. Lab.* June.

10. Davis, H. G., Figueroa, C., Karatas, C., Kloden, D. J., Schaleger, L. L. and Yaghoubzadeh, N. (1981). Paper presented at the 13th Biomass Thermochemical Contractors' Meeting, Washington, D.C., Oct. LBL-13449.
11. Ergun, S., Schaleger, L. L. and Seth, M. (1980). In *Energy from Waste*, Vol. 1, T. C. Frankiewicz (Ed.). Ann Arbor Sci. Publ. Inc., Ann Arbor, Mich., Chapter 13.
12. Appell, H. R., Fu, Y. C., Friedman, S., Yavorsky, P. M. and Wender, I. (1971). Bureau of Mines Report of Investigations 7560, US Dept. Interior, Washington, D.C.; Appell, H. R. (1977). In *Fuels from Waste*, L. L. Anderson and D. A. Tillman (Eds). Academic Press, New York, pp. 121–40.
13. Thigpen, P. L. and Berry, W. L., Jr (1982). In Symposium Papers, Energy from Biomass and Wastes VI. Institute of Gas Technology, Lake Buena Vista, Florida, Jan. 25–29, Chapter 41.
14. Figueroa, C., Schaleger, L. L. and Davis, H. G. (1982). *ibid.*, Chapter 42.
15. Schuchardt, U. and Matos, F. (1981). *Fuel* **61**, 106.
16. Tran, A. D. and Rogers, D. Z. (1981). Examination of alternative catalysts for biomass direct liquefaction. Final Report, D.O.E. Contract B-92646-A-Q, June.
17. Molton, P. M., Miller, R. K., Donovan, J. M. and Demmitt, T. F. (1979). *Carbohyd. Res.* **75**, 199.
18. Schaleger, L. L. Unpublished work.
19. Gansley, R. and Vermeulen, T. Unpublished work.
20. Davis, H. G., Figueroa, C., Karatas, C., Kloden, D. J., Schaleger, L. L. and Yaghoubzadeh, N. (1981). Quarterly Research Report, LBID-426, July.
21. Yu, S.-M. (1982). Solvolytic liquefaction of wood under mild conditions. Ph.D. Thesis, Univ. of Calif., Berkeley.
22. Hansler, D. W., Hellgeth, J. W., McNair, H. M. and Taylor, L. T. (1979). *J. Chromatogr. Sci.* **17**, 617.

58

Formation of Aromatic Compounds from Condensation Reactions of Cellulose Degradation Products—II

D. A. NELSON, J. A. RUSSELL and P. M. MOLTON
Battelle Pacific Northwest Laboratories, Richland, Washington, USA

1. INTRODUCTION

Considerable interest has developed over the last decade concerning renewable alternatives for petroleum-based fuels. A likely candidate for such renewable fuels may be produced by the liquefaction of biomass under aqueous, alkaline conditions.

Unfortunately, from the standpoint of understanding liquefaction process chemistry, biomass is a mixture of cellulose, hemicellulose, lignin and other components, depending much upon the source of the material. Each component can contribute a plethora of compounds to the liquefaction product. To avoid variable and complex mixtures this research was initially directed toward the liquefaction of a single component, cellulose. This has helped facilitate compound identification and allowed insight into the formation mechanisms of the liquefaction product.

Prior liquefaction research performed with cellulose has identified 86 compounds in the product oil.[1,2] These compounds can be assigned to various groups including aliphatic and cyclic hydrocarbons, aliphatic and cyclic alcohols, aliphatic and cyclic ketones, furans, alkylbenzenes and phenolics. The presence of the aromatic components, except for the furans, offers a generally unrecognized area for exploration.

The comprehensive work of Richards *et al.*[3-6] concerning cellulose and related polysaccharide degradation in aqueous alkali indicated conversion of these materials through a series of saccharinic acid derivatives to formic, acetic, lactic and glycolic acids. The alkali treatment of monomeric saccharides provides similar conversions to saccharinic acid and acyclic fragments.[7-9] Acetol, pyruvaldehyde, pyruvic acid and biacetyl were also detected in the products, albeit in low quantity.[9] Furans, cyclopentanediones, 3-hydroxy-2,2-butanone and cyclopentenones have been isolated after similar alkaline treatment of monosaccharides.[11,12]

Phenols were reported quite early as products from the alkali treatment of monosaccharides.[13] More recent research has shown that alkyl-p-benzoquinones and catechols were formed when sucrose was refluxed in basic solution.[14] However, only catechols were formed under basic conditions with glucose or fructose. Several alkylbenzenediols (hydroquinones), alkylacetophenones and catechol have been isolated from the alkaline degradation of xylose and glucose.[15]

It has been generally assumed that lignin would be the source of aromatic products (minus furans) in the biomass liquefaction oils. However, our research and that of others cited above indicate that a varied selection of aromatic compounds may be prepared under alkaline conditions from monosaccharides and cellulose. The object of this research was to identify further aromatic components from the cellulose liquefaction oil and determine the conditions necessary for aromatic compound formation. Such information may ultimately make it possible to alter liquefaction conditions to favor products with more desirable characteristics.

2. EXPERIMENTAL

Melting points were determined on a Thomas–Hoover capillary melting point apparatus and are uncorrected. Nuclear magnetic resonance spectra (^{1}H and ^{13}C) were recorded with a Varian FT-80 spectrophotometer. Gas chromatographic analyses were obtained with a Hewlett-Packard 5880A instrument. Two capillary columns were quite useful for product elution: SE-54 and Superox, both on fused silica (2 mm × 30 m). Column temperature requirements for these columns were 50 °C isothermal for 10 min, programmed to 250 °C at 3 °C/min. Preparative GC was performed with a Perkin-Elmer model 900 equipped with an SE-30 column (4 mm × 1·8 m). A Hewlett-Packard 5992B GC/MS coupled with an HP 18947A GC/MS capillary interface was also used. The labelled compounds 3-hydroxy-2-2-butanone-2-^{13}C and 3-hydroxy-2-butanone-1-^{13}C were obtained from Merck, Sharp and Dohme Canada Ltd., Quebec.

3. LIQUEFACTION PROCEDURE

For our work with cellulose, a 300 ml capacity autoclave was used. To this was added 30 g cellulose (Solka-Floc®), 1·92 g sodium carbonate and 120 g distilled water. The autoclave was sealed and flushed with nitrogen for 5 min. The temperature of the autoclave was then raised to the required level (heating time 80 min to reach 290 °C), maintained at that temperature for 60 min, and cooled (cooling time 3·5 h). After cooling, the gas was vented and the autoclave opened. The aqueous phase was decanted, while the oil and char remaining in the vessel were extracted with acetone. Usually about 3 g of acetone insoluble material is obtained. The solvent from the acetone soluble fraction was removed under a nitrogen stream. This oil was then extracted with dichloromethane. Generally, 8 g of dichloromethane solubles and 7 g of insolubles are obtained in this manner. Extraction of the aqueous phase with dichloromethane provided about 0·5 g of oil. The pH of the aqueous phase was approximately 5.

4. ACETYLATION OF CELLULOSE OIL FRACTIONS

Acetylation of various liquefaction products was performed using *in situ* generated ketene.[16] For example, 1·0 g of cellulose oil (dichloromethane soluble material, fraction II) was dissolved in 50 ml dichloromethane. To this was added 2·6 g triethylamine. The solution was heated to reflux, and 2·0 g acetylchloride was added dropwise. The reflux was maintained for 2 h. The solution was cooled and washed with water. The dichloromethane solution was dried over anhydrous sodium sulfate. After filtration, the solvent was removed *in vacuo*.

5. METHYLATION OF CELLULOSE OIL FRACTIONS

A modified phase-transfer catalysis procedure[17] was used for fractions soluble in dichloromethane. For example, 3·0 g of cellulose oil (fraction II) was dissolved in 100 ml dichloromethane. To this was added 20 ml of 40 wt % sodium hydroxide solution and 0·1 g tetrabutylammonium hydroxide. Dimethylsulfate (10 g) was added dropwise, with stirring, to the

® Trademark of Brown Co., New York.

immiscible layers. The temperature during addition was held at 5 °C. The mixture was then stirred for 4 days at 25 °C. Aqueous ammonia was added to destroy unreacted dimethylsulfate. The dichloromethane layer was separated and water washed. After drying, the solvent was removed *in vacuo*.

A second methylation procedure[18] was applied to the cellulose oil (dichloromethane insoluble fraction III). Since this fraction was not soluble in common non-polar solvents, 6·0 g of solid material was dissolved in 100 ml dimethylsulfoxide. To the solution was added 37·7 g barium oxide and 35·2 g barium hydroxide. The mixture was stirred and 44 g dimethylsulfate was added. An exotherm occurred after approximately 4 h. After 4 days, aqueous ammonia was added. A very persistent emulsion was overwhelmed by continuous dichloromethane extraction. The solvent was removed *in vacuo*. Only 0·3 g of a tan oil was recovered.

6. HYDROLYSIS OF CELLULOSE OIL FRACTION

Due to its intractibility, fraction III was subjected to acid hydrolysis. Thus, 20 g of this solid was added to 250 ml 6 M hydrochloric acid and refluxed for 4 h. The hot solution was filtered, and 19·5 g solid was recovered. The aqueous acid was lyophilized to 0·3 g of a white solid. This material was silylated with hexamethyldisilazane and trimethylchlorosilane in pyridine.[19]

The unreacted fraction III was resubjected to acid hydrolysis using refluxing 6 M sulfuric acid for 4 h. However, only 0·2 g of a tan solid was recovered after a laborious procedure.[20] It was not pursued further.

7. ALKALINE TREATMENT OF INTERMEDIATES

The liquefaction procedure was applied to acetoin and biacetyl. To a 180 ml capacity autoclave was added 11·5 g of an 85 % acetoin solution in water, 0·219 g sodium carbonate and 5·75 g distilled water. The biacetyl reaction was performed with 9·8 g of this intermediate. The autoclave was sealed and flushed with nitrogen for 5 min. The temperature of the autoclave was then raised to the required level (heating time 70 mm to reach 290 °C), maintained at that temperature for 60 min, and cooled (cooling time 3 h). After cooling, the gas was vented and the autoclave opened. The mixture was filtered to remove 2,5-dimethylhydroquinone (1·4 g) present during acetoin reactions. This solid was washed with water and dichloromethane. The filtrate and washes were extracted with dichloromethane. This extract was dried over anhydrous sodium sulfate. The solvent was removed *in vacuo*.

Similar reactions were performed under atmospheric conditions. Either 11·5 g of 85% acetoin or 9·8 g biacetyl was added to 100 ml 0·1 M sodium hydroxide. The solution was sparged with nitrogen and kept under the inert gas during heating. Reflux was generally maintained for 1 h. The aqueous solution was acidified to pH 5 and extracted with dichloromethane. The solvent was removed under vacuum.

8. 2,5-DIMETHYLHYDROQUINONE

This compound was isolated in 18% yield from the 290 °C autoclave reaction of acetoin with base. It could be readily purified by sublimation, MP 223–224 °C. ^{13}C NMR (CD_3OD) δ 149·03 (aryl-1), 123·43 (aryl-2), 118·30 (aryl-3) and 15·87 (methyl) with TMS as internal reference. Mass spectrum, *m/e* (relative intensity) 138 (base peak), 123 (41·5), 121 (9·1), 109 (7·1) and 95 (11·1).

Anal. Calcd for $C_8H_{10}O_2$: C, 69·56; H, 7·25; O, 23·19. Found: C, 69·51; H, 7·33; O, 23·29.

9. 2,5-DIMETHYLBENZOQUINONE

This compound was prepared in quantitative yield by dissolving 1·0 g of 2,5-dimethylhydroquinone in ether and adding to it 1·0 g sodium sulfate and 2·0 g silver oxide. The mixture was stirred overnight, then filtered. The solvent was removed to give yellow needle crystals, MP 124–126 °C. ^{13}C NMR ($CDCl_3$) δ 187·89 (cyclohexadienyl-1), 145·82 (cyclohexadienyl-2), 133·42 (cyclohexadienyl-3) and 15·35 (methyl).

10. RESULTS AND DISCUSSION

Although we have performed a great deal of research into the identification of cellulose liquefaction products by GC/MS,[1] an annoying observation has become evident with this technique. Not all solvent extracted components can be eluted through the gas chromatographic columns under standard conditions. Residue remaining within the glass-lined injection port indicated that less than 50% of the acetone soluble components were volatile at 250 °C. The dichloromethane extraction (Fig. 1) reduced the injection port residue. Approximately 70% of fraction II could be eluted at 250 °C. The dichloromethane extract of the decanted aqueous layer (I) provided 90% GC elution. Although the solvent extraction process improved vapor phase elution, it did not reduce the complexity of the fractions. For instance, over 150 components were detected by GC in fraction II.

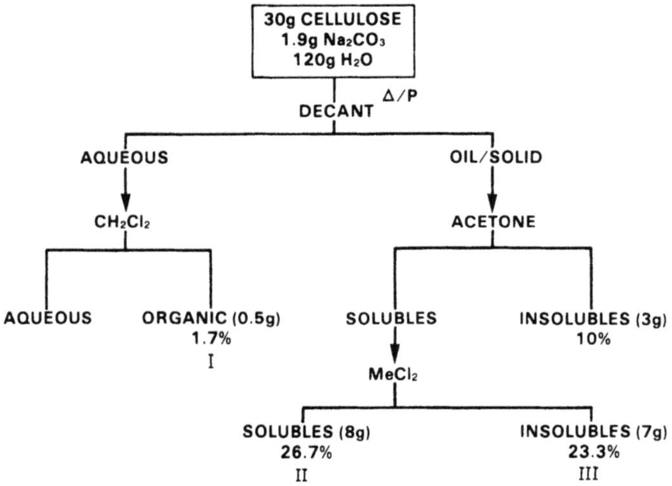

FIG. 1. Extraction procedure for cellulose liquefaction products.

Acetylation was used as a further attempt to improve the volatility of fraction II. This also allowed us to confirm the identification of some phenols previously reported in cellulose oil.[1] We were able to identify phenol, 4-methylphenol and 2-ethylphenol, as well as 3-propylphenol, by comparing the retention times of the respective acetylated and unacetylated compounds. However, 2-methoxyphenol (guaiacol) could not be confirmed in this manner.

Elution of acetylated fraction II through an alumina column using a sequential solvent system (benzene, dichloromethane, acetonitrile and methanol) did provide further component separation. However, most components were present in quite small amounts, except for several compounds with lengthy GC retention times. The identification of one of those compounds will be discussed later.

Methylation of fraction II did improve the volatility of its components, i.e. more than 90% of II was eluted through the GC columns based on injection port residue. This derivatized fraction was eluted through silica gel with the sequential solvent system. By this method, the methylated derivatives of compounds **1–3** were identified in the methanol eluent by GC/MS (Fig. 2). Catechol and 3,4-dihydroxyacetophenone have been previously reported as products of the aqueous alkaline degradation of glucose and xylose.[15] Catechol was also a product of similar reactions with fructose and sucrose.[14]

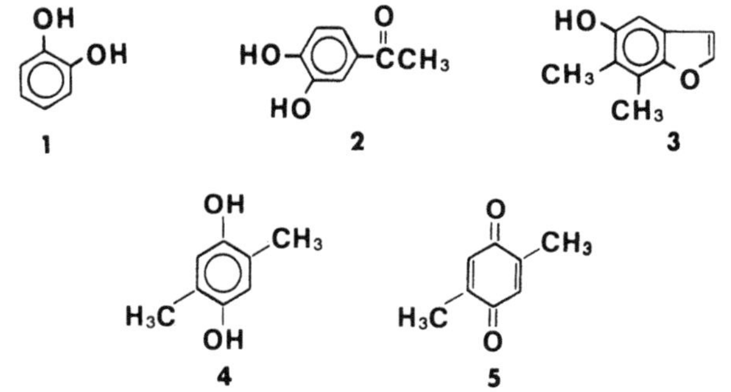

FIG. 2. Compounds identified in cellulose oil.

Fraction III, a solid at 25 °C was not responsive to separation techniques. An attempt was made to methylate III.[18] However, only about 1·0 wt % of III could be converted to a GC elutable material. None of the components separated by GC were readily identifiable. Fraction III was subjected to hydrolysis conditions with hydrochloric acid followed by sulfuric acid. Less than 1 % of III was soluble in either aqueous acid after 4 h of reflux. The hydrolyzate remaining in the hydrochloric acid solution was lyophilized, silylated and examined by gas chromatography. No identifiable trimethylsilyl monosaccharides were observed.

Four components were observed in the acetylated fraction II with GC retention times greater than 60 min. One of these was also present in unacetylated II, albeit in small quantity (*ca.* 0·1 %). However, if the acetone soluble fraction, after solvent removal, was extracted with ether, the quantity of that single component could be enhanced to more than 0·8 %. Preparative gas chromatography was applied to the ether extract. After several passes, a single crystalline component (MP 223–224 °C) was isolated. The compound was identified as 2,5-dimethylhydroquinone (**4**).

Further confirmation was achieved by oxidizing this material to 2,5-dimethylbenzoquinone (**5**). Reexamination of the ether extract by GC indicated that less than 0·1 % of **5** was present.

The presence of **5** was noted after alkali treatment of sucrose.[14] Quite early work had indicated that both **4** and **5** could be prepared by refluxing biacetyl in alkaline solution.[21] However, yield information was only given for the product of 2,3-pentadione (10 % 2,5-diethylbenzoquinone). Our experience has been that yields of **4** and **5** from biacetyl were more in the 1–5 % range depending upon reaction time. Even lower amounts of **4** and **5** were detected when biacetyl was exposed to aqueous sodium carbonate at 100 atm and 290 °C.

Biacetyl was present in ether extracts of the aqueous solution decanted after cellulose liquefaction. The biacetyl content was approximately 0·2 %. Acetoin or 3-hydroxy-2-butanone was also present (0·1 %) in this ether extract. Acetoin has been previously identified as a product of alkali treated fructose.[10] Interestingly enough, both **4** and **5** were detected in low amounts (1–5 %) after acetoin had been refluxed in caustic solution from 1 to 24 h. However, under liquefaction conditions, acetoin yielded 18 % of **4** and approximately 0·1 % of **5**. This indicated that the conversion of acetoin to **4** was more efficient than biacetyl at high temperature and pressure.

Since the liquefaction conditions had been designed for cellulose, the temperature requirements for acetoin conversion were examined. These are presented in Table I.

The yield of **4** was quite dependent upon temperature since little of it was

TABLE I

YIELDS OF ACETOIN PRODUCTS FROM AUTOCLAVE EXPERIMENTS

Temperature (°C)	*Product* **4**[a] (%)	*Product* **5** (%)
50	<0·01	<0·01
100	<0·01	<0·01
150	1·4	0·03
200	8·7	0·04
250	11·7	0·07
290	18·6	0·10

[a] Crystalline **4** was isolated from reactions performed at and above 150 °C. The other values were determined by GC of the reaction mixtures.

detected in reaction mixtures exposed to less than 150 °C. In all cases **5** was present in quite low yield (0·1 %). In each of the reaction mixtures, the aqueous phase had a pH of about 5. The amount of sodium carbonate was doubled in one acetoin experiment performed at 290 °C to maintain a higher pH value. However, the yield of **4** was severely decreased to 1 % instead of the expected 18 %. This may suggest that acidic conditions are necessary later in the formation of **4**. Atmospheric reactions of acetoin in refluxing acid did not produce **4**.

Since the autoclave reactions were performed with a high ratio of acetoin to water (to enhance isolation of **4**), an experiment was performed to more reflect the cellulose conditions. Thus, 30 g acetoin, 1·92 g sodium carbonate and 120 g water were exposed to 290 °C for 1 h. The yield of **4** was 18 %. Therefore, the ratio of acetoin to water does not appear to be a factor in the preparation of **4**.

Von Euler and Hasselquist[22] presented a mechanism for the formation of 2,5-dimethylhydroquinone from biacetyl (Fig. 3). This involved an

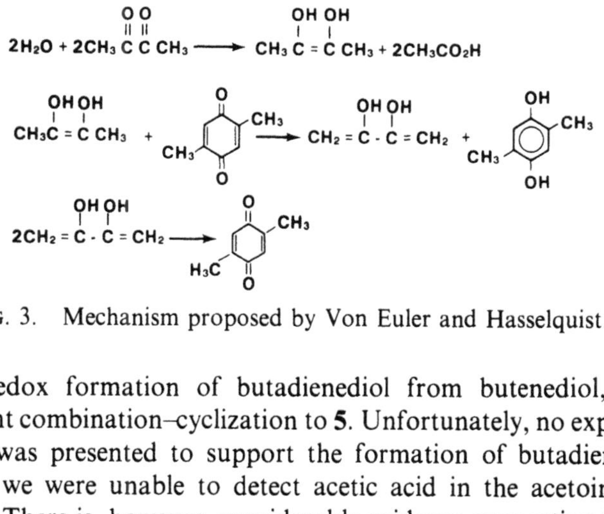

FIG. 3. Mechanism proposed by Von Euler and Hasselquist.[23]

elegant redox formation of butadienediol from butenediol, and the subsequent combination–cyclization to **5**. Unfortunately, no experimental evidence was presented to support the formation of butadienediol. In addition, we were unable to detect acetic acid in the acetoin reaction products. There is, however, considerable evidence supporting formation of enediolates from biacetyl in alkaline solution.[23,24] In alkaline solution, acetoin has also been shown to form the bivalent enediol anion corresponding to biacetyl.[25] This reaction was reversible upon neutralization.

We find the reduction–oxidation mechanism that results in the formation of **4** to be attractive, since it suggests that **5** is an intermediate with **4** as the final product. This could explain the very low amounts of **5** in the autoclave experiments with acetoin. On the other hand, it is well known

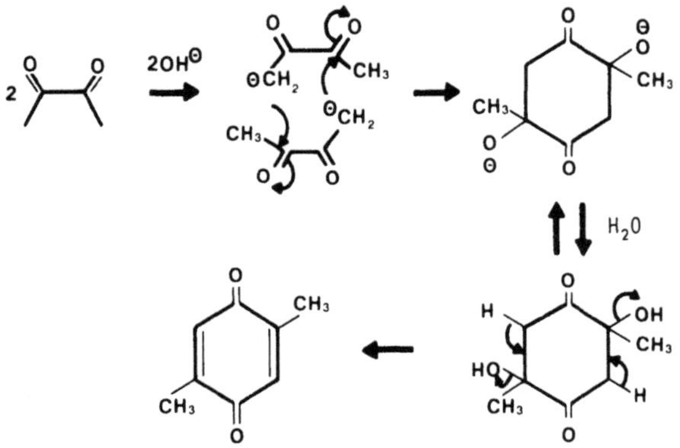

FIG. 4. An alternate mechanism for the formation of **5**.

that hydroquinones are reducing agents.[26] Thus, small amounts of **5** could be formed as a byproduct. As an example, samples of **4** (retained for 4 months) were resublimed and gave both **4** and **5** as products.

An alternative scheme for the formation of **5** from biacetyl is presented in Fig. 4. Such a mechanism depends upon aldol condensations and would not generate acetic acid. In addition, the oxidative conversion of acetoin to biacetyl could be linked to the reduction of **5** to **4**.

Experiments performed at 290 °C with ^{13}C-labelled acetoin suggested that two molecules of biacetyl or enediolate combined to form **4** and **5**. The results are presented in Fig. 5. The ^{13}C-NMR spectra of **4** and **5**, formed by acetoin-1-^{13}C or acetoin-2-^{13}C, indicated equal distribution of ^{13}C-label in four respective positions. This suggests a symmetrical precursor prior to an aldol condensation–cyclization.

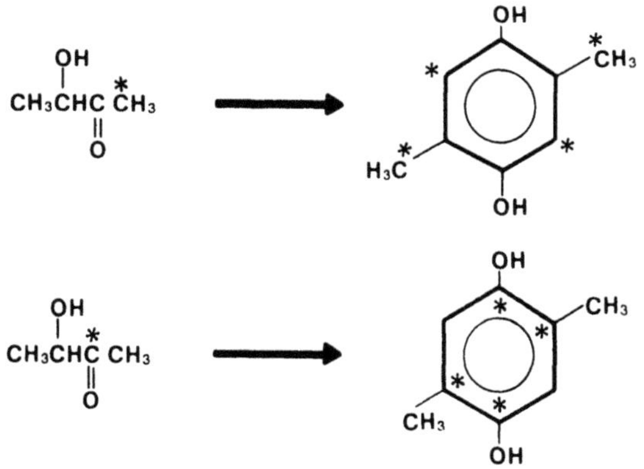

FIG. 5. Results of 290 °C alkaline exposure of ^{13}C-labelled acetoin.

The increasing efficiency of **4** formation from acetoin with increasing temperature might indicate the presence of a radical or radical-ion intermediate. In order to test this possibility the following experiment was performed. Water and acetoin were thoroughly degassed (freeze–thaw under vacuum) and combined with base in an inert gas glovebox. The autoclave was also assembled in the glovebox. All transfer lines were well swept with nitrogen prior to heating the autoclave to 290 °C. An 18 % yield of **4** was obtained even with the exclusion of oxygen. We should note that the color of the product oil and **4** was greatly reduced. Presumably, chromophoric components such as semiquinones were reduced.

11. CONCLUSIONS

Alkaline liquefaction of cellulose has provided a varied selection of aromatic compounds among the complex mixture of other products. During this research six aromatic components were identified using a variety of derivatization and instrumental techniques. Of these six components, 2,5-dimethylhydroquinone and 2,5-dimethylbenzoquinone were noteworthy since they could be formed from the condensation of acetoin or biacetyl, both of which were identified as cellulose liquefaction products.

A reduction–oxidation mechanism was proposed involving formation of dimethylbenzoquinone from biacetyl (or the enediolate), reduction of the benzoquinone to the hydroquinone with concurrent oxidation of acetoin to biacetyl. Experiments conducted with acetoin under liquefaction conditions supported acetoin as the precursor for the hydroquinone. Yet ^{13}C-labelling experiments suggested involvement of biacetyl. The proposed mechanism tended to satisfy these and other conditions.

Knowledge of the components in cellulose oil and the mechanisms of formation may allow control and selection of specific products. A reduction of the number of components may also decrease potential noxious products, and permit such biomass oils to become acceptable as fuel or chemical sources.

ACKNOWLEDGEMENTS

We thank S. D. Landsman for GC/MS identifications and Dr J. A. Franz for ^{13}C-NMR spectra. We gratefully acknowledge support from the US

Department of Energy, Division of Chemical Sciences, Office of Basic Energy Sciences.

REFERENCES

1. Molton, P. M., Miller, R. K., Russell, J. A. and Donovan, J. M. (1981). In *Biomass as a Nonfossil Fuel Source*, D. L. Klass (Ed.). ACS Symposium Series No. 144, Washington, D.C., pp. 137–62.
2. Russell, J. A., Miller, R. K. and Molton, P. M. (1983). *Biomass*, 43.
3. Richards, G. N. and Sephton, H. H. (1957). *J. Chem. Soc.* 4492.
4. Machell, G. and Richards, G. N. (1957). *J. Chem. Soc.* 4500.
5. Machell, G. and Richards, G. N. (1958). *J. Chem. Soc.* 1199.
6. O'Meara, D. and Richards, G. N. (1958). *J. Chem. Soc.* 1204.
7. Feather, M. S. and Harris, J. F. (1973). *Advan. Carbohyd. Chem. Biochem.* **28**, 161.
8. Sowden, J. C. (1957). *Advan. Carbohyd. Chem.* **12**, 35.
9. Pigman, W. and Anet, E. F. L. J. (1972). In *The Carbohydrates, Chemistry and Biochemistry*, W. Pigman and D. Horton (Eds), Vol. A1. Academic Press, New York, pp. 165–94.
10. Shaw, P. E., Tatum, J. N. and Berry, R. E. (1968). *J. Agr. Food Chem.* **16**, 979.
11. Giaturco, M. A., Giammarino, A. S. and Pitcher, R. G. (1963). *Tetrahedron* **19**, 2051.
12. Enkvist, T., Alfredsson, B., Merikallio, M., Paakkonen, P. and Jarvela, O. (1954). *Acta Chem. Scand.* **8**, 51.
13. Hoppe-Seyler, F. (1871). *Ber.* **4**, 346.
14. Kato, Hiromichi, Mizushima, M., Kurata, T. and Fujimaki, M. (1973). *Agr. Biol. Chem.* **37**, 2677.
15. Forsskahl, I., Popoff, T. and Theander, O. (1976). *Carbohyd. Res.* **48**, 13.
16. Brady, W. T., Liddell, H. G. and Vaughn, W. L. (1966). *J. Org. Chem.* **31**, 626.
17. Merz, A. (1973). *Angew. Chem. Internat. Edit.* **12**, 846.
18. Wallenfels, K., Bechtler, G., Kuhn, R., Trischmann, H. and Egge, H. (1962). *Angew. Chem. Internat. Edit.* **2**, 515.
19. Sweeley, C. C. and Chang, T. C. L. (1961). *Anal. Chem.* **33**, 1860.
20. Whistler, R. L. and Richards, G. N. (1958). *J. Amer. Chem. Soc.* **80**, 4888.
21. Von Pechmann, H. (1888). *Ber.* **21**, 1411.
22. Von Euler, H. and Hasselquist, H. (1949). *Arkiv Kemi* **1**, 325.
23. Von Euler, H. and Hasselquist, H. (1949). *Arkiv Kemi, Mineral. Geol.* **26A**, 1.
24. Von Euler, H. and Hasselquist, H. (1950). *Rec. Trav. Chim.* **69**, 402.
25. Petuely, F. and Meixner, N. (1953). *Monatsh.* **84**, 1061.
26. Fieser, L. and Fieser, M. (1961). *Advanced Organic Chemistry*. Reinhold Publishing Corp., New York, p. 848.

59

Liquefaction of Aspen Poplar to Produce an Oil and Chemicals

R. L. EAGER, J. F. MATHEWS and J. M. PEPPER
University of Saskatchewan, Saskatoon, Saskatchewan, Canada

1. INTRODUCTION

In a period when fossil hydrocarbons are likely to become scarcer, methods to convert renewable biomass to useful fuels and chemical feedstocks are increasingly attractive. Such methods would permit the conversion of the very large solar energy reserves trapped as lignocellulosic materials into a form which would have the advantages of being initially sulfur-free and readily portable. All such biomass is composed of two vastly chemically different components: the carbohydrate-based holocellulose (cellulose and hemicelluloses) and the aromatic based lignin, which together have a much lower heating value than fossil fuel due to their high oxygen content. The production of a fuel compatible with natural oil must therefore involve significant oxygen removal. Any one such process may not be suitable for conversion of both the holocellulose and the lignin and an optimization of reaction conditions may be required.

One such process is that based on the early research[1] at the US Bureau of Mines. These workers had shown that a variety of ligno-cellulosic materials (wood products, agricultural and civic wastes) could be converted, in part, into a fuel oil by reaction with carbon monoxide and water at elevated temperatures in the presence of sodium carbonate which acts as a catalyst.

It was of interest to determine if a typical hardwood such as aspen (*Populus tremuloides*—Michx), of which Canada has large, virtually unused resources, could be treated similarly to produce a useable oil.

This paper summarizes the results of work at the University of Saskatchewan on the liquefaction of aspen poplar to produce an oil. (The word 'oil' used to describe the viscous dark coloured product does not imply that this material is the same as fossil crude oil; there are significant differences which will be mentioned later.)

The work to be summarized includes:

(a) Initial runs using a batch reactor to elucidate the effect of the various operating parameters on the liquefaction process.
(b) Runs using a semi-continuous reactor to study open-system behavior and to permit use of much shorter reaction times than could be achieved with the batch reactor.
(c) Analysis of the liquefaction products in order to determine if industrially useful chemicals were produced and also to obtain some clues to the mechanism of the reaction.

2. EXPERIMENTAL

2.1. Raw Materials

The wood used in this study was mature, 8 in. in diameter at shoulder height, aspen poplar (*Populus tremuloides*—Michx). Air dried wood, ~6% moisture, was chipped and then ground with a Wiley mill to pass a 1 mm screen. The ground meal was uniformly mixed before use.

2.2. Apparatus and Procedure

2.2.1. Batch Reactor

The reactor was an AMINCO high pressure shaking assembly with a $1{\cdot}4 \times 10^{-3}\,m^3$ reaction vessel. Temperatures were measured with an iron-constant thermocouple and controlled to ± 5 K by a Brown Electric Pyrometer. Pressures were measured to an accuracy of $\pm 2\%$ by a Validyne pressure transducer and modulator, and displayed as a digital readout.

The following procedure was followed for the reactions. Wood meal (100 g dry basis) and a solution containing the desired quantities of catalyst and water were added to the reaction vessel. The wood meal and the solution were thoroughly mixed using a large spatula. With the reactor in the shaking assembly, carbon monoxide (Liquid Carbonics, 99·5% pure) was used first to flush out the reactor and then to pressurize it to the selected starting pressure. The shaking assembly and the electric heater were started; a time interval of approximately 2·5 h was required to bring the reactor to the desired operating temperature. To prevent overshooting the desired final reaction temperature 40 min of this time was required for the last 30 K. The operating pressures ranged from a low of 22·4 MPa corresponding to low reaction temperature (593 K), low initial carbon monoxide pressure (5·17 MPa) and low water-to-wood ratio (1 : 1) to a high of 45·5 MPa (633 K, 8·60 MPa and 2·5:1). After the required reaction time (the nominal reaction time for all runs was 1 h, unless specified otherwise) the heating was stopped but shaking continued for a further 1·5 h. A 20 min cooling period was required for the reactor temperature to drop to 30 K below the reaction temperature. Because of the time required for heating and cooling it is estimated that the reactor contents were at an effective reaction temperature for approximately 24 min longer than the nominal reaction time. After the reactor had cooled to room temperature, which required approximately 4 h, a sample of the gas phase was taken and the remaining gas vented to the atmosphere. After cooling to room temperature the pressure was higher than the initial cold pressure. This was due to the shift reaction with an additional contribution from carbon and oxygen being removed from the wood as carbon dioxide.

The liquid products consisted of an aqueous phase and an oil phase. The aqueous phase was decanted and weighed. It was initially colorless but upon exposure to air it rapidly underwent a series of colour changes through yellow to a reddish brown and finally to a deep blue black color. As much as possible of the oil phase was removed by pouring it from the reactor followed by scraping the walls with a spatula. Over 90% of the oil was recovered in this manner. This portion of the product was used for carbon and hydrogen analysis. The oil still adhering to the walls of the reactor was recovered by dissolving in acetone. The two oil portions were combined and the total mass used to calculate the percentage yield of oil, based on the mass of wood used.

2.2.2. *Semi-Continuous Reactor*

(*a*) *Early design.* The concept upon which the semi-continuous reactor was designed involved a small hopper to hold the wood meal, a screw-type conveyor which was to carry the reactants through a tubular reactor section whose temperature could be controlled by a series of heating units surrounding the reactor tube, and a receiver system to collect the liquefaction products. The entire system would be sealed, filled with carbon monoxide to some initial pressure which could be later controlled and released through an off-gas vent. To this end an apparatus was designed and built and is illustrated schematically in Fig. 1. The insert depicts that part of the equipment which was originally built and operated but was subsequently modified, as shown in the full diagram, for reasons that are discussed later.

The apparatus was machined from 304 stainless steel and consisted of three main sections. The hopper (4200 ml capacity, 10·5 cm ID, 15 cm OD

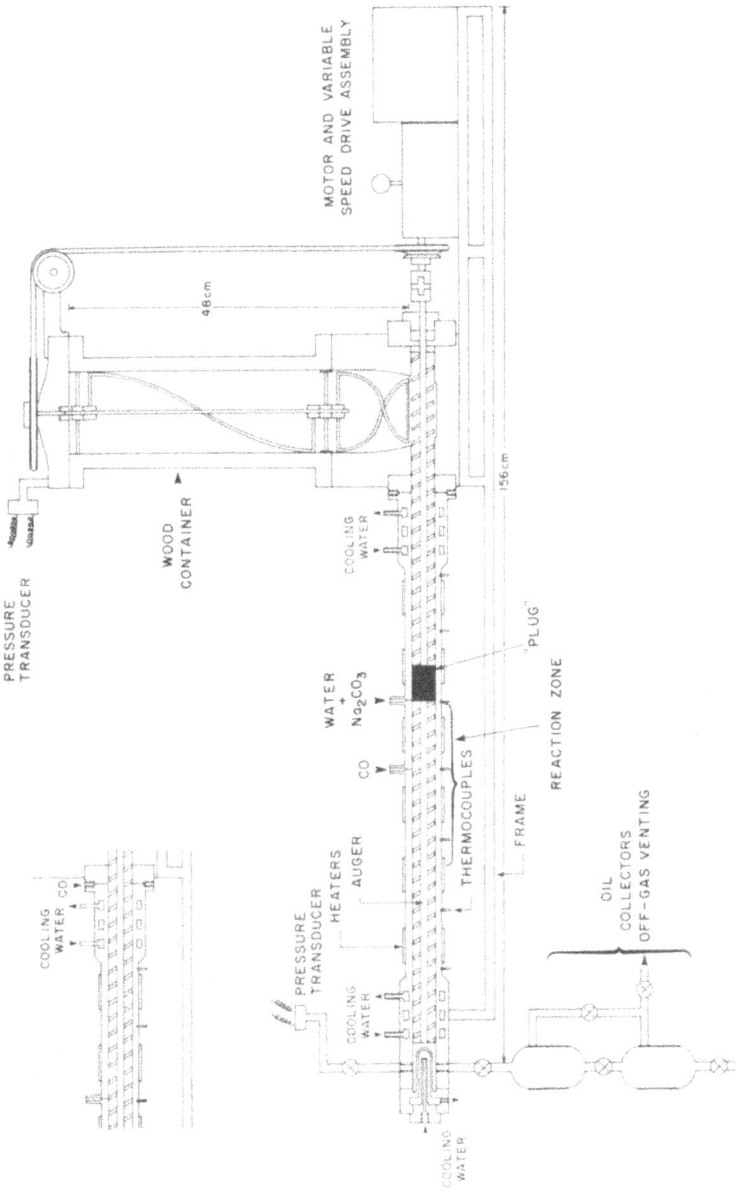

FIG. 1. Small-scale semi-continuous reactor, modified design. Insert shows early design.

and 48 cm high) was designed to hold approximately 500 g of wood meal in the form of a slurry with the water and sodium carbonate. The removable top permitted easy access. To insure downward feeding on to the horizontal auger an S-shaped stirrer was included. It was driven initially by a belt through a series of pulleys but later connected by a Master Craft angle drive unit with steel gears to the same motor/power block/gear arrangement (Zero-Max Industries, Inc.) that drove the auger. The reactor tube (92 cm long, 1·59 cm ID, 3·18 cm OD) was bolted on to the lower section of the hopper unit such that the auger could be inserted from the hopper end and rotated by the motor. Both ends of the reactor tube section were fitted with units for circulating cooling water. Several ports along the reactor tube permitted entries for gas or other reactants. Two collector vessels were arranged with suitable valves so that products could be collected periodically by maintaining the upper collector under pressure while releasing the lower one to atmospheric pressure. The reactor section was heated by means of eight separate 350 W heating units (D-M-E Company band heaters), each of which was controlled by a variable autotransformer. Iron-constant thermocouples were inserted into wells located along the reactor tube and the temperatures monitored using a multipoint channel recorder (Tracor Westronics, Model M11E). The heating units were wrapped with Cera blanket insulation (Johns-Manville). Pressure transducers (Validyne Engineering Corporation, Model DP 22-70) connected to digital readouts (Validyne Engineering Corporation, Carrier Demodulator Model CD 101-3) monitored the gaseous pressure. The auger drive shaft was sealed with Chevron type Teflon gaskets (Garlock XV Chevron) and the thrust washer was machined from Oilite Bronze (Perth Metal Industries). The gaskets used to seal the lid on the hopper, the hopper to the base, and the cap at the exit end of the reactor tube were made from composite asbestos material (Garlock 7021). Teflon gaskets were used on the ports.

Since the unit was to be used in some circumstances as a closed system it was assembled and cold pressure-tested with water to 69 MPa and to 54 MPa with the reactor tube heated to 673 K. The complete unit was assembled in a room with all controls and monitoring done through a wall which had been lined inside with 3·2 mm ($\frac{1}{8}$ in) steel plate.

The screw feeder (auger) was made from 304 stainless steel, was 88 cm long with a 1·4 cm pitch and was machined to fit closely the inner bore of the reactor tube. The original auger (type A, Fig. 2) had a square cut spiral which resulted in extensive binding and thus forced a shutdown. Two other augers were constructed each with a full radius bottom. One (type B) had a

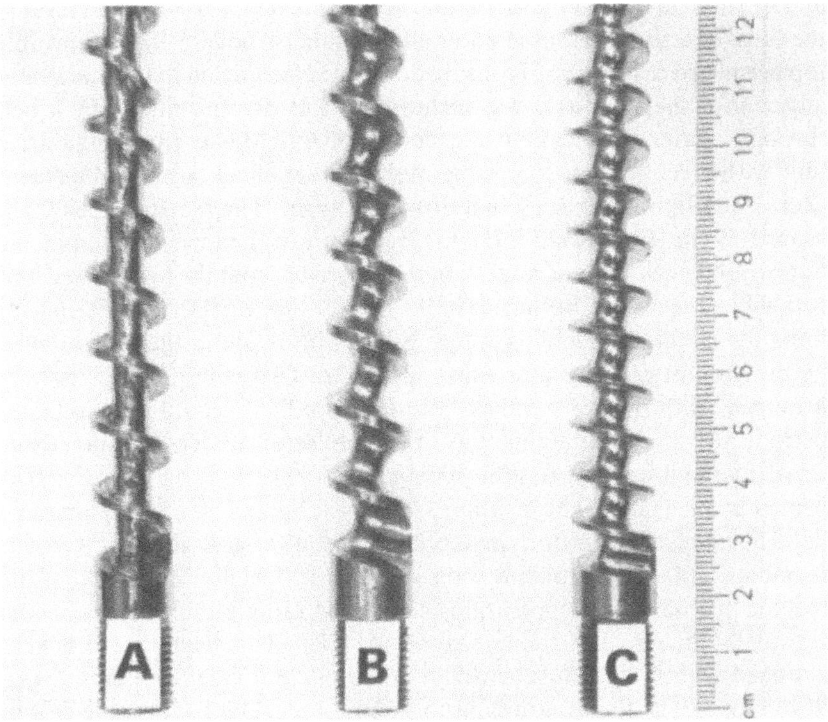

FIG. 2. Auger types used in semi-continuous reactor.

1·4 cm pitch and the other (type C) had a 0·7 cm pitch. Limited tests made with type B indicated problems would be encountered with water backing up into the hopper. Type C proved to be most satisfactory and was used in all the runs described below.

For the early runs using this semi-continuous reactor the following typical procedure was used. A slurry of aspen wood meal (*Populus tremuloides*—Michx) (2 mm mesh, 4% moisture) (150 g) with water (400 ml) containing sodium carbonate (20 g) was added to the hopper. After flushing the apparatus with carbon monoxide, the vent was closed and the apparatus pressurized with carbon monoxide using an Aminco (10 000 psi Single-End) Diaphragm-Type Compressor, to the required pressure (e.g. 27·6 MPa). Heating was begun and when the temperature of the centre portion of the reactor zone reached ~633 K the motor was started to rotate the auger at a speed of 1 rpm to advance the wood slurry into and through the reactor zone. At the conclusion of a run, the apparatus was allowed to cool overnight, the excess gas pressure released, the bulk of the product removed from the collectors and the residue washed out with acetone. The auger was withdrawn from the motor end and examined for evidence of reaction (i.e. unchanged meal, oil, char and tar).

For this paper oil is defined as that product composed of the water-insoluble fraction which is recovered by decantation plus the acetone-soluble residue. Reported yields are calculated using the weight of this oil and the weight of wood meal converted (initial weight less that of wood remaining in the hopper and on the auger). The water content of the oil was determined by first separating and removing a portion of the water by centrifugation at 2000 rpm and then analyzing the residue for water using the Karl-Fisher procedure. A typical oil analysed 15–17% water.

As a result of several runs made using this general procedure but varying the parameters of auger speed, temperature profile along the reactor and pressure several important observations were made:

(a) The principle that a continuously fed wood substrate could be converted in part into an oil was established. The resulting product was shown to be similar in yield and quality to that produced earlier by the batch process.

(b) Problems were experienced in feeding the wood slurry. As the process continued steam produced in the reactor portion would condense in the cold hopper and dilute the slurry to the point that it would no longer be fed by the auger.

(c) Loss of water (by condensation) from the reactor modified the reaction conditions and led to poorer conversions. The amount of water had been shown to be an important parameter.[2]

(d) Serious erosion and/or corrosion occurred on several spirals of the auger at the point where reaction was being initiated.

(e) The conversion occurred very rapidly as evidenced by an examination of the withdrawn auger showing the meal-to-oil reaction taking place over an approximate 2 cm length representing only about 3 min residence time at ~633 K.

(f) Char formation downstream from the obvious point of initial reaction suggested a lack of water present, possibly as a result of the back condensation (see b and c).

(*b*) *Modified design.* In an attempt to overcome some of the problems referred to above the apparatus was modified as shown by the full diagram in Fig. 1. The aqueous solution of the catalyst (sodium carbonate) was now pumped directly into the reactor under pressure at a point where the temperature had been raised to 633 K. To do this a high pressure liquid pump (Eldex precision metering pump, Model B-100-S) was installed. The rate of flow of the sodium carbonate solution was controlled by adjusting the stroke volume of the pump. In this way, the wood meal was now introduced through the hopper as the air-dried form which was easier for the auger to feed than was the slurry form.

The carbon monoxide inlet was changed from the original position immediately after the wood hopper to a point located 17·8 cm downstream from the aqueous solution injection port. This permitted the continual addition, under pressure, of carbon monoxide into the reaction zone. The carbon monoxide was pumped intermittently from a storage cylinder to maintain a constant reactor pressure. Pressure transducers were installed to record pressures in the hopper and near the end of the reactor zone.

Under these conditions, provided the auger speed was 12 rpm it was found that a wood meal 'plug' formed in the reactor just prior to the aqueous carbonate injection port (Fig. 1). (The term 'plug' is used to designate a state that the unreacted wood meal assumes on the auger over a region of 4 cm. The plug represents a dynamic system in which a steady state is reached between the arrival of fresh wood meal at the upstream end and the reaction to form liquid products and gases at the downstream end. Water vapor from the downstream end of the reactor tube diffuses only part of the way through the plug. Under proper sets of conditions the plug is firm enough to support pressure gradients up to 10 MPa.) At auger speeds of 12 rpm a plug did not always form. At steady state conditions the plug remained even though the wood meal was continuously being fed into the reactor and converted to oil and water-soluble products. Such a plug may be seen when the auger is withdrawn at the end of a run. Figure 3 shows a 15 cm length of such an auger with the wood meal plug. The incoming wood meal is initially compressed (light area) before the liquefaction reaction begins (dark area).

An important result of the modified design is a significant reduction in the extent of erosion and/or corrosion occurring on the auger.

2.2.3. *Product Analysis*

The higher heats of combustion of the oils were determined with a Parr oxygen bomb adiabatic calorimeter, Model 1221. The gas samples were analyzed using an F and M temperature programmed gas chromatograph, Model 720, equipped with a thermal conductivity detector. A 3·2 mm × 4·57 m copper column packed with 60–120 mesh silica gel was used.[3] Argon was used as the carrier gas, with a flow rate of 18 ml/min. For

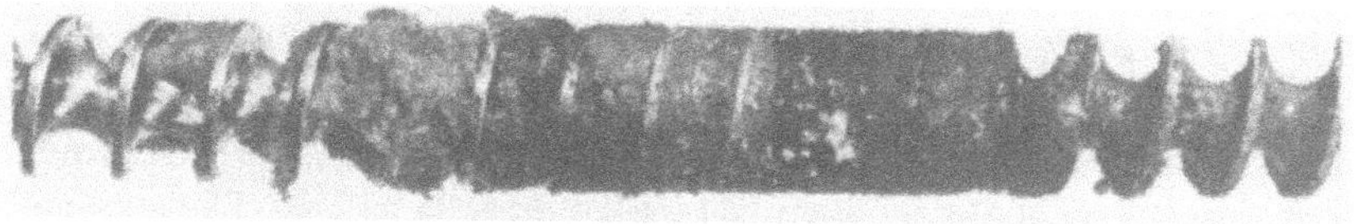

FIG. 3. Section of auger showing 'plug' and reaction zone.

each analysis the oven temperature was initially held for 5 min at 323 K in order to detect hydrogen, carbon monoxide and methane. The temperature was then increased at a rate of 30 K per minute to 423 K to obtain a satisfactory resolution of the carbon dioxide peak.

3. RESULTS AND DISCUSSION

3.1. Batch Reactor Studies

3.1.1. Factorial Design Experiments

The effect of the process variables—temperature of reaction, initial pressure of carbon monoxide (cold reactor), concentration of sodium carbonate (wt % based on dry wood) and water content (g H_2O/g dry wood)—were determined using two 2^3 factorial experiments. The factorial points are shown in Table I. The response factors that were experimentally measured were percentage yield [(g oil/g wood) × 100], and percentage (C + H) of the product [(gC + gH)/g oil) × 100]. The standard deviations of these response factors were calculated from the center points and other replicate runs (df = 16) to be 1·08 % and 2·00 % respectively.

The mathematical product of these two response factors gives a third factor which will be called the carbon and hydrogen recovery (CHR) factor [CHR = (g(C + H)/g oil) × (g oil/g wood) × 100] because it is a measure of the mass of carbon and hydrogen which can be recovered as part of an oil product from a given mass of original dry wood. Since the heat of

TABLE I
DESIGN OF 2^3 FACTORIALS

I(a)[a]	−	+	*I(b)*[b]	−	+
Temperature (K)	593	633	Temperature (K)	593	633
P_{CO}(MPa)	5·17	8·60	P_{CO}(MPa)	5·17	8·60
Catalyst (wt %)	7·5	12·5	g H_2O/g dry wood	1·0	2·5

[a] g H_2O/g dry wood = 5·00.
[b] wt % catalyst = 12·5.

combustion of the oil is a function of the percentage (C + H) (a linear function as will be shown below), the CHR also indirectly measures the heat energy or fuel value obtained from a given mass of wood. This latter would be of importance if the oil were to be used directly as fuel The standard deviation for CHR was calculated (df = 16) to be 1·01 %.

A statistical analysis of the data derived from the first 2^3 factorial experiment (Table Ia) showed no significant effect ($\alpha = 0{\cdot}05$) of either temperature or initial carbon monoxide pressure on any of the response factors. The percentage of catalyst had a negative effect ($\alpha = 0{\cdot}05$) on yield, a positive effect ($\alpha = 0{\cdot}01$) on percentage (C + H) and no effect ($\alpha = 0{\cdot}05$) on the CHR. No significant interaction or curvature was found. The effects of catalyst were relatively minor: a change in yield from 37·5 % to 34·8 % and a change in (C + H) from 77·5% to 82·4% for a change in catalyst concentration from 7·5 % to 12·5 wt %. The second 2^3 factorial experiment (Table Ib) showed no significant effect of initial CO pressure on any of the response factors. Temperature appeared to have a small positive effect ($\alpha = 0{\cdot}05$) on percentage (C + H) but on neither of the other two response factors. The water-to-wood ratio had a significant ($\alpha = 0{\cdot}01$) negative effect on both percentage yield and CHR. The percentage yield values are shown in Table II.

The most important variable appears to be the ratio of water to wood. Table II also shows the results of runs made at a ratio of 0·5:1. The yield is inversely proportional to the ratio. The relationship between ratio and yield is linear from a ratio of 5:1 to a ratio of 1:1, with no statistically significant effect of temperature or initial carbon monoxide pressure. For the high temperature (633 K) runs this linearity continues to a ratio of

TABLE II
PERCENTAGE YIELD OF OIL[a]

		Water:wood			
P_{CO} *(MPa)*	*Temp. (K)*	*5:1*	*2·5:1*	*1:1*	*0·5:1*
5·17	593	35·3	40·7	43·6	55·9
8·60	593	34·9	42·8	44·5	54·1
5·17	633	34·3	40·1	44·0	47·6
8·60	633	30·0	41·7	43·7	45·8
Average		33·6	41·3	44·0	[b]

[a] wt % catalyst = 12·5.
[b] Since there is an obvious temperature effect here, an average is not meaningful.

0·5:1. However, the runs at low temperature (593 K) show a sharp upward trend as the ratio is dropped to 0·5:1. There is at present no satisfactory explanation for this latter phenomenon.

Water certainly serves as a solvent for the catalyst initially and probably distributes it throughout the wood particles. That water need not be present as a liquid phase at reaction temperatures was demonstrated in the runs in which the ratios of water to dry wood were 1:1 and 0·5:1. With those amounts of water present in the reactor the vapor specific volumes were greater than the specific saturated vapor volumes and so a liquid phase as such could not have been present.

A 2^3 factorial experiment was made using the design of Table I(b), but replacing sodium carbonate with potassium carbonate. There were no significant differences with respect to either levels of the response factors or sensitivity to process variables caused by the use of potassium carbonate. Therefore it can be concluded that, if the cation has a role to play in the reaction, both Na^+ and K^+ are equally suitable.

3.1.2. Effect of Reaction Temperature

Although temperature appeared to have no significant effect in the range 593–633 K it has been shown by Eager *et al.*[4] that temperatures below about 533 K result in a char product rather than an oil or a tar. Qualitative observations also indicate that the oil produced at 633 K is somewhat less viscous than that produced at 593 K. This last observation may be of some importance if the product is to be used directly as a fuel oil. However, if the product is to be further upgraded, a more viscous oil or even a tar which could be melted would probably be a suitable intermediate.

3.1.3. Effect of Reaction Time

Because of the large thermal inertia of the high pressure reaction vessel it was difficult to study the kinetics of the reaction. A series of four runs was made for nominal reaction times ranging from 0 to 1·0 h. In all cases the product was a fluid oil; hence it may be deduced that the reaction time at 633 K is less than 24 min (see experimental section).

3.1.4. Effect of Catalyst Type—Interaction with Carbon Monoxide

It is unclear at present what roles the catalyst and the carbon monoxide play in the reaction. In order to shed some light on this subject several sets of exploratory runs were made using different catalysts and different combinations of catalyst concentration and initial carbon monoxide pressures. Table III shows the results of a set of runs studying the effects of

TABLE III
CATALYST–CO INTERACTIONS[a]

Initial P_{CO} *(MPa)*	*%* Na_2CO_3	*%NaOH*	*Product*	*%Yield*	*%(C+H)*	*CHR*
0[b]	0	0	black fluffy powder	36·2	70·2	25·4
3·45	0	0	fine black powder	35·5	71·5	25·4
0[b]	10·0	0	tarry lump	36·0	77·1	27·8
0[b]	0	7·6	sticky tar	35·0	73·2	25·6
3·45	10·0	0	oil	31·2	79·6	24·8
3·45	0	10·0	oil	33·7	79·8	26·9

[a] 100 g wood; 633 K; 500 g H_2O.
[b] Initial $P_{N_2} = 3{\cdot}45$ MPa.

the presence and absence of catalyst and carbon monoxide as well as the use of sodium hydroxide instead of sodium carbonate.

In the cases where carbon monoxide was not present, nitrogen at the same initial pressure was substituted in order to keep the total pressure the same. An inspection of the response factors and the qualitative product descriptions shows that sodium hydroxide is a suitable catalyst and that, if a tar which could be melted is a satisfactory product (see discussion under 'Effect of reaction temperature'), carbon monoxide is probably not necessary. This latter point has considerable implication on the operation and economic feasibility of any commercial plant based on this reaction.

Since it was clear that some catalyst is required for a tar or oil product, it was then desirable to determine the minimum catalyst concentration to produce this result. A series of 5 runs was made in which the loading of sodium carbonate was varied from 2 to 10 wt % (based on dry wood). From this series it was apparent that a minimum loading of 7 g sodium carbonate per 100 g dry wood was required to produce oil or tar without solid char particles being present. The role of the catalyst is still a matter for conjecture.

3.1.5. Effect of Using Hydrogen–Carbon Monoxide Mixtures

It may be economically attractive to use synthesis gas (carbon monoxide and hydrogen) instead of pure carbon monoxide in the initial gas charge to the reactor. A series of runs using dry wood was made in which the total initial gas pressure was 8·60 MPa but with H_2:CO ratios of 2:1, 1:1, 1:3 and 0:1. Reaction temperatures and water contents were also varied for the 1:1 ratio. Table IV summarizes the conditions and results.

TABLE IV
EFFECT OF USING H_2:CO MIXTURES[a]

			Product oil			
Initial H_2:CO	*Temperature (K)*	*Water:wood*	*%Yield*	*%C*	*%H*	*CHR*
3:1	593	2·5:1	39·5	69·7	7·2	30·4
1:1	593	1:1	41·2	72·1	7·4	32·8
1:1	593	2·5:1	37·5	71·3	6·8	29·3
1:1	633	1:1	40·2	73·8	7·0	32·5
1:1	633	2·5:1	36·2	74·8	8·4	30·1
1:3	593	2·5:1	40·2	68·9	7·0	30·5
0:1	593	2·5:1	42·8	68·7	7·3	32·5

[a] Common reaction parameters: 12·5 wt % Na_2CO_3, total initial gas pressure 8·60 MPa, reaction time (nominal) 1 h.

A comparison of the yields and CHR factors with those obtained with corresponding conditions, but using pure carbon monoxide, shows that replacing part of the carbon monoxide with hydrogen does not have any significant effect on these response factors. Qualitatively it was observed that the oil obtained from the runs where hydrogen was present initially was slightly more viscous than that produced when carbon monoxide alone was used.

3.1.6. Green Wood–Sodium Carbonate Catalyst

Since freshly cut (green) wood might be used as the supply for a wood–oil conversion plant, a study was made using freshly cut aspen wood as the raw material in a 2^3 factorial experiment with the same points as Table I(b). No centrepoints were run. The green wood ($\sim$30 % moisture) was ground to pass a 2 mm screen in the Wiley mill because the 1 mm screen tended to plug. Recovered oil yields were on the average 5 percentage points higher than those for the corresponding dry wood set, possibly due to an easier penetration of the sodium carbonate into the wood pores. The effects of the variables on the response factors were not significantly different to those for the dry wood.

3.1.7. Recycling the Aqueous Phase

Recycling the aqueous phase should reduce the cost of catalyst in a plant operation. A series of seven runs was made at 633 K in which the initial reactor charge was 10 g sodium carbonate, 500 g water and 100 g dry wood. In each succeeding run the reaction mixture was 100 g of fresh dry wood,

TABLE V
RECYCLE OF AQUEOUS PHASE[a]

	%*Yield*	%(*C* + *H*)	*CHR*
Initial[b]	32·0	79·0	25·3
1st recycle[c]	36·0	76·5	27·5
2nd recycle[c]	37·7	77·1	29·1
3rd recycle[c]	39·7	80·0	31·8
4th recycle[c]	38·2	75·2	28·7
5th recycle[c]	38·9	78·2	30·4
6th recycle[c]	37·4	78·9	29·5
Average of recycle runs	38·0	77·7	29·5

[a] Common reaction parameters: reaction time (nominal) = 1 h, $T = 633$ K, P_{CO} (initial) = 3·45 MPa, wood = 100 g.
[b] Water = 500 g, Na_2CO_3 = 10 g,
[c] Make-up water = 100 g, aqueous phase from previous run = 400 g, Na_2CO_3 = 2 g.

400 g aqueous phase from the preceding run and 100 g of additional water containing 2 g of sodium carbonate. The initial carbon monoxide pressure in all runs was 3·4 MPa. Table V shows the results of this series of runs. The average percentage yield of the recycle runs was 38·0, significantly higher than that of the initial run (32·0).

The average percentage (C + H) of the recycle runs was not appreciably different to that of the initial run and hence the average CHR for the recycle runs (29·5) was significantly higher than that for the initial run (25·3). It may be that the increased yield of oil product resulted from a secondary conversion of the initially water-soluble organic products or their increased solution in the oil phase due to partial saturation of the aqueous phase from previous cycles. Alternatively, because of the presence of these compounds in the recycle medium, a greater direct conversion to oil may have occurred.

3.1.8. Material Balance, Carbon Recovery, and Heat of Combustion

Table VI shows an overall mass balance and a carbon balance for a typical run at 633 K and 1 h nominal time.

The overall material balance is closed within 0·8 % and the overall carbon balance within 6·4 %.

Analysis of the residual gas phase showed 23·4 mol % hydrogen, 1·6 mol % carbon monoxide and 75·0 mol % carbon dioxide. These values were relatively constant (±5 %) for approximately 200 runs made using various combinations of values of the experimental parameters. The

TABLE VI
MATERIAL BALANCES

	Material	*Weight (g)*	*wt %C*	*Weight of C (g)*
Feed:	Total	642[a]		62·9
Products:	Oil phase	33·0	72·4	23·9
	Ether extractable oxygenated hydrocarbons[b]	19·5	49	9·6
	Solid residue	12·5	15	1·9
	H_2	1·2	—	—
	CO	1·1	42·9	0·47
	CO_2	84·5	27·3	23·0
	Water (aqueous phase)	485	v. small	
	Total	637		58·9

[a] 100 g wood, 10 g Na_2CO_3, 500 g water, 32 g CO.
[b] Ref. 4.

hydrogen and part of the carbon dioxide are no doubt due to the shift reaction. The remainder of the carbon dioxide probably originates directly by decomposition of the wood or as a result of oxidation of carbon monoxide by oxygen from the wood.

The recovery of the carbon of the original wood in the form of oil is ~47 % and, in the form of liquid oxygenated hydrocarbons, ~20 %. The recoveries of the mass of the original wood as oil and as oxygenated hydrocarbons are ~33 % and ~20 % respectively.

The measured higher heats of combustion of the oils ranged from 32·2 to 36·0 MJ/kg. These values can be compared with ~19·8 MJ/kg for wood and 43·2 MJ/kg for a Bunker C-type fuel. The heats of combustion of the oils were found to correlate with the percentage (carbon + hydrogen) in the sample. The regression equation is $\Delta H\,(\mathrm{MJ/kg}) = 3{\cdot}6 + 0{\cdot}394\,[\%(\mathrm{C}+\mathrm{H})]$. The correlation coefficient is 0·862 with 11 degrees of freedom and is significant at $\alpha = 0{\cdot}01$. The standard error of estimate for ΔH is 0·62 MJ/kg.

4. SUMMARY AND CONCLUSIONS: BATCH REACTOR STUDIES

Previously dried wood may be converted into an oil in yields up to approximately 50 % by weight based on the original wood.

The product oils, which may be rendered fluid on mild warming, have a carbon plus hydrogen content of 80% and a higher heat of combustion ranging from 32·2 to 26·0 MJ/kg.

Within the ranges of the parameters studied, temperature (593–633 K), nominal reaction times (0–1 h), water-to-wood ratio (0·5:1–5:1), nature of catalyst (Na_2CO_3, K_2CO_3, and NaOH), amount of catalyst (7·0–12·5 wt %) and initial hydrogen-to-carbon monoxide ratios (2:1–0:1), the most important variable appears to be the water-to-wood ratio.

Oil yields from freshly cut (green) wood are about five percentage points higher than those from dried wood.

Total carbon recovery in the form of oil and water-soluble oxygenated products represents approximately two-thirds of the carbon in the original wood.

There is some evidence that in alkaline media carbon monoxide may not be necessary to effect oil production.

A series of seven runs, for each of which 80% of the aqueous phase came from the previous run, led to successful oil production in each case.

Replacing part of the carbon monoxide with hydrogen does not have any significant effect.

5. SEMI-CONTINUOUS REACTOR STUDIES

Although the operation of the early designed reactor was satisfactory as it showed that wood meal could be converted rapidly into products, including an oil similar to that obtained using a batch process, the modified form had a significant advantage. It was recognized that the formation of the 'plug' permitted the establishment of a pressure differential between the reactor side (higher pressure) and the hopper side (lower pressure) (see Table VII). This 'plug' prevented steam generated in the hot reaction zone from returning to the hopper, condensing and ultimately stopping the injection of wood meal. To check the reproducibility of the performance of the modified reactor a series of seven runs were performed under identical conditions. In each case aspen wood meal (500 g) was loaded into the hopper. The sealed system was pressurized with carbon monoxide to an initial pressure of 27·6 MPa and heated to 633 K. A solution of sodium carbonate (5%) was introduced at a rate of 240 ml/h and the wood meal fed in at a rate of 200 g/h using an auger speed of 12–14 rpm. The pressure was allowed to build up on the receiving side and controlled at 34 MPa by the release of gas. After 2·25 h the run was stopped, the reactants cooled

TABLE VII
CONVERSION OF ASPEN WOOD TO OIL[a]

Pressure (MPa)				
Hopper	*Controlled*	*Oil yield*[b] (%)	*Char yield*[b] (%)	*Description of oil*
27·3	33·8	41[c]	8·8[c]	Fluid
17·2	20·7	42	2·6	Fluid, greasy
17·2	20·7	44	0·7	Fluid, greasy
13·8	17·2	42	0·8	Slightly viscous
10·3	13·8	36	5·9	Very viscous
10·3	13·8	37	4·7	Moderately viscous[d]

[a] Wood meal 500 g; auger speed 12–14 rpm; sodium carbonate solution 5% at 8 ml/min; duration of run 2·25 h; temperature 633 K.
[b] Percentage yield based on wood meal converted (see text).
[c] Data are averages of seven runs made to prepare sample for analysis.
[d] Sodium carbonate solution 5% at 10 ml/min.

overnight, excess gas released, and the oil collected under an atmosphere of nitrogen and stored at 279 K. During the course of each run a pressure differential of 9·6 MPa was developed. Oil was produced in an average yield, based on original air-dried wood meal, of 41·5% ± 2·5%.

With the recognition that the modified semi-continuous reactor was reliable in that the plug formation was readily reproducible and the yield and quality of the oil equal to any previously produced, if not slightly more fluid, studies were made on the effects of changing reaction parameters.

5.1. Effect of Reaction Pressure

Unlike the studies using a closed-system batch reactor in which the resulting reaction pressure was related primarily to the temperature, it was now possible using the semi-continuous reactor to control the pressure in the system. A series of runs was made to determine if a satisfactory conversion could be achieved at significantly reduced pressures. The results of these runs, involving the reduction of the reaction pressure from 34 MPa to 14 MPa, are given in Table VII. It may be seen that only a slight decrease in the yield of oil (44 to 37%) results. There is a noticeable decrease in the fluidity of the oil as the reaction pressure decreases. Depending on the end use of the oil the viscosity may not be important since increased fluidity readily occurs on warming.

5.2. Effect of Duration of Run on Char Formation

It was observed that in all runs a char fraction was found at the exit end of

TABLE VIII
EFFECT OF DURATION OF RUN ON CHAR FORMATION

Duration of run (h)	*Char (g)*
2·3	3·5
2·0	10·5
2·3	9·0
0·5	6·0
0·75	11·0

Wood meal 500 g, auger speed 12 rpm; initial hopper pressure 13·8 MPa; controlled reaction pressure 17·2 MPa; sodium carbonate solution (5 %) introduced at 8 ml/min; temperature 633 K.

the reactor tube. Previous experiments had shown that this material could not be subsequently converted to an oil and hence it would appear to be a product resulting directly from the wood. A series of runs was made at different durations which would mean different amounts of wood meal admitted. These results are given in Table VIII. The relatively small amount of char that is formed appears to be the result of start-up operations. Once a steady state has been achieved in the reactor there is no evidence of further char build-up and hence disposal of the char would not become a problem in continuous operation.

5.3. Effect of Sodium Carbonate Concentration

A series of runs was also made to study the effect of the sodium carbonate

TABLE IX
EFFECT OF SODIUM CARBONATE CONCENTRATION[a]

Na_2CO_3[b] (%)	*Oil yield*[c] (%)	Char yield[c] (%)	*Description of oil*
18·8	37	4·7	Moderately viscous oil
9·1	40	5·4	Moderately viscous oil
5·0	41	4·5	Moderately viscous oil
0·0	32·5	15·6	Viscous tar containing particles of char

[a] Wood meal 500 g; auger speed 12–14 rpm; initial hopper pressure of CO 10·3 MPa; controlled reactor pressure 13·8 MPa (2000 psig); sodium carbonate solution introduced at 10 ml/min; duration of run 2·25 h; temperature 633 K.
[b] Na_2CO_3 used based on wood meal converted.
[c] Percentage yield based on wood meal converted (see text).

concentration. These results are given in Table IX. The presence of sodium carbonate appears to be necessary to prevent char formation; however, a satisfactory product is obtained when 5% (on a dry wood basis) is used.

6. COMPARISON OF SEMI-CONTINUOUS vs. BATCH REACTORS FOR PRODUCTION OF OIL

The operation of the small-scale semi-continuous reactor described herein has been shown to be successful and to have several advantages over that of the batch reactor. There is no recognizable difference in the yield of oil produced but since it has been shown that char production is essentially a start-up product only, the subsequent oil produced would be of a superior fluid quality. The required residence time at an elevated temperature has been shown to be short (<1 min) which is an important factor in design of scaled-up operations. Of greatest significance is the realization that, even though the same high temperature of reaction (633 K) can be maintained in the short reaction zone, significantly lower pressures ($\sim$14 MPa vs. $\sim$35 MPa) can be used compared to those required in a batch operation.

Although the yield of oil is similar from both processes, some differences have been noted in the chemical composition. As determined by ^{13}C-NMR. the percentage of the carbon content which is of an aromatic or unsaturated type is much higher ($\sim$56 vs. $\sim$37%) for the oil from the semi-continuous process than that from the batch process. There is also evidence for less degradation in the former case, which may be directly related to the shorter residence time at elevated temperatures. A detailed study of the chemical differences between the oils is under investigation and will be reported at a later date.

At this time it is difficult to predict the type of use such biomass-derived oils may have in the future. Their direct use as a fuel or as raw material for upgrading to a higher quality gasoline fraction and/or to potentially useful chemicals may thus be different for batch or continuously produced products, depending on the original chemical compositions of each.

7. SUMMARY AND CONCLUSIONS: SEMI-CONTINUOUS REACTOR STUDIES

A small-scale, semi-continuous reactor is described whereby air-dried aspen wood meal can be converted into an oil in yields comparable to those obtained in a previously described batch process.

The success of this reactor is based upon the utilization of a screw-type feed mechanism whereby a 'plug' of wood meal is maintained between wood supply and reaction zone permitting a pressure differential to be maintained between the two. The recognition that the formation of a plug of wood meal permits the establishment of a significant pressure differential between the wood hopper and the reaction zone raises the distinct possibility that a fully continuous reactor can be designed to operate satisfactorily.

Conversion of wood into oil and other products is rapid and the semi-continuous reactor permits the use of a short residence time at the reaction temperature.

The recognition that the reactor can operate satisfactorily at significantly lower pressures than those established in the closed batch system suggests that a more favorable situation exists for a commercial plant than was formerly assumed.

Preliminary chemical examination of the oil from the semi-continuous reactor reveals some significant differences in chemical composition over that obtained using the batch process (e.g. a higher percentage of aromatic and/or unsaturated carbon.

Extended use of the modified procedure whereby air-dried wood meal is introduced directly has caused little corrosion and/or erosion damage to the auger.

ACKNOWLEDGEMENTS

The authors wish to acknowledge the financial support provided by the Canadian Forestry Service, Department of Environment (ENFOR Programme). The technical assistance provided by Messrs K. Thoms and V. Gummadi is appreciated. The assistance of the College of Engineering machine shop personnel, and in particular Mr G. Dyck, with the design and construction of the apparatus is gratefully acknowledged. The donation of the Cera blanket by Fergus Industries of Saskatoon is appreciated.

REFERENCES

1. Appell, H. R., Fu, Y. C., Friedman, S., Yavorsky, P. M. and Wender, I. (1971). Converting organic wastes to oil. US Bureau of Mines Report of Investigation No. 7560.

2. Eager, R. L., Mathews, J. F. and Pepper, J. M. (1982). Liquefaction of aspen poplar wood. *Can. J. Chem. Eng.* **60**, 289–94.
3. Jeffery, P. G. and Kipping, P. J. (1964). *Gas Analysis by Gas Chromatography.* Macmillan, New York, p. 94.
4. Eager, R. L., Mathews, J. F., Pepper, J. M. and Zohdi, H. (1981). Studies on the products resulting from the conversion of aspen poplar to an oil. *Can. J. Chem.* **59**, 2191–8.
5. Appell, H. R. (1977). The production of oil from wood waste. In *Fuels from Waste.* Academic Press, New York, Chapter VIII.

60

Thermal Degradation of Cellulose in Alkali

J. M. KROCHTA, J. S. HUDSON, C. W. DRAKE,
T. R. MON and A. E. PAVLATH

*Western Regional Research Center,
US Department of Agriculture,
Berkeley, California, USA*

1. INTRODUCTION

Thermochemical liquefaction involves conversion of biomass at high temperatures and pressures to liquid products. These reactions occur in an aqueous or organic carrier liquid, often in the presence of other reactants and a catalyst. A wide range of lignocellulosic feedstocks are possible, and reactions are relatively rapid.

One thermochemical conversion approach for producing liquid products which has been explored is reaction of biomass material catalyzed by the alkali salt sodium carbonate in an aqueous medium or in an organic medium with water present. So far, most research on this thermochemical liquefaction approach has focused on urban refuse, wood or peat feedstocks.[1-12] The product of interest has been a black, tarry material usually defined as 'oil' by its solubility in acetone or benzene. These 'oils' have improved bulk density and fuel value compared to the feedstocks. However, they are poorly defined, contain considerable oxygen, and are corrosive and quite viscous. Some work has been done with cellulose to define the reaction mechanisms leading to more desirable liquid products, but a great deal remains to be done.[13-18]

While studies on thermochemical conversion of cellulosic materials have focused on the production of acetone- or benzene-soluble 'oils', much less attention has been directed to the amount of other products formed in the conversion. These include water-soluble product, carbon dioxide and water. The objective of this study was to determine the effect of temperature, time and alkali concentration on the thermal degradation of cellulose to all the conversion products. The intent was to establish better the effect of important variables on the overall material and carbon balances for the degradation process and perhaps clarify the overall chemical reactions and mechanisms involved.

Of particular interest in the investigation was the water-soluble product formed over the range of variables investigated. In the instance where acetone- or benzene-soluble product is optimized, some water-soluble product is also produced. Questions arise as to the amount, composition and utilization of this water-soluble material. In the best case, the water-soluble product would contain components of value as fuel or chemicals.

The work of Eager *et al.*[9] indicated the production of certain alcohols, ketones and carboxylic acids in the water-soluble product from cellulose and from aspen poplar at reaction conditions found to optimize the 'oil' phase in the thermochemical conversion of aspen poplar. Davis *et al.*[10] found significant production of formic, acetic and glycolic acids, along with larger carboxylic acids, phenols and neutral organics in the water phase at reaction conditions selected to optimize 'oil' yield from Douglas fir. Other work[19–21] has established the existence of many water-soluble compounds among the alkaline degradation products of glucose, cellulose and other saccharides at low temperatures (25–200 °C), including the saccharinic acids and formic, acetic, glycolic and lactic acids. Our work looked at the effect of a range of reaction conditions on the production of several of the volatile compounds from cellulose by using gas chromatography.

One possible scenario involves the total conversion of cellulosic materials to water-soluble product and the separation of the components into useful compounds. Therefore, those conditions which gave large conversion of the cellulose to water-soluble product were of great interest. Alcohols and ketones produced in significant amounts would be useful as fuels or chemicals. Carboxylic acids could be used directly as chemicals. They could also be decarboxylated or esterified to produce fuels or chemicals.

Both sodium carbonate and sodium hydroxide were used in the investigation. It was thought that sodium hydroxide might be more effective because it can swell various forms of biomass considerably compared to their initial volume. The swelling effect has the potential of making the cellulose, hemicellulose and lignin more accessible and therefore more reactive. Further, the products of reactions might diffuse more easily from the site of reaction, thus decreasing inhibition of further primary conversion and decreasing likelihood of undesirable secondary and tertiary reactions.

2. EXPERIMENTAL PROCEDURES

Highly purified, finely powdered cellulose (Cellulay-Cellulose, United States Biochemical Corp., Cleveland, Ohio) was used. Most reactions were run with a 20% by weight slurry of the cellulose in water. However, the larger concentrations of sodium hydroxide studied produced such swelling of cellulose that only a 10% cellulose slurry was used. Dry pyrolysis experiments were also made for comparison.

Reactions were performed in magnetically stirred Autoclave Engineers' 1 liter or 300 cm^3 autoclaves using either glass or Hastelloy C liners (Autoclave Engineers, Erie, Penn.). Most reactions were condicted in nitrogen atmosphere 210 and 360 °C, with the majority of experiments at 305 °C. Pressure builds to 9308–11 377 kPa (1350–1650 psig) at this temperature. Most reaction times at temperature have been 10 min, with additional runs for 60 min. However, the time to get from 200 °C, when reaction begins, to 305 °C was 20–40 min. The time required for cooling from 305 to 200 °C was 2–4 min.

After cooling to room temperature, a gas sample was taken from the autoclave and analyzed by gas chromatography for H_2, O_2, CO, CO_2 and CH_4. Gas chromatographic analysis of the gas sample was performed using a 2·5 mm × 2·1 m stainless steel column packed with Spherocarb 100/120 (Analabs, Inc., No. Haven, Conn.), with helium as the carrier gas at a flow rate of 15 cm^3/min. Injector temperature was 175 °C and detector temperature was 165 °C. Column temperature was programmed to rise at 8 °C/min to 150 °C after 50 °C for 7·5 min. The final reactor pressure at room temperature and the reactor gas space then allowed determination of the amount of each gas present. The liner with reaction products was next removed and the solid product filtered from any free water. The solid product was then extracted of water solubles with several consecutive rinses with water, and the combined water was analyzed for total solids and for carbon content. Total solids determination was made by drying at 40 °C, first in a forced air oven and then in a vacuum oven. The extracted, solid material was also dried in a forced air oven and then in a vacuum oven, both set at 40 °C.

The resulting dried product was extracted with acetone in a Soxhlet extractor for 16 h followed by removal of the acetone by drying at 100 °C. The resulting acetone-insoluble fraction was defined as 'residue'. Finally, the acetone-soluble and residue materials were analyzed for carbon and hydrogen content.

In several cases, the Soxhlet extractions were repeated with ethanol replacing the acetone to assess the solubility and composition of the resulting products. In each case, the results were quite similar to those obtained with acetone.

The yield of water soluble material produced in the thermochemical conversion reaction was calculated as follows:

$$\text{water-solubles (WS) yield} = 100 \times \frac{\text{water solubles (WS)}}{\text{feed (moisture and ash free)}}$$

A similar calculation was performed to calculate acetone-soluble (AS), carbon dioxide (CO_2) and residue (R) yield.

The yield of water produced in the conversion reaction was determined by calculating the feed material weight not accounted for by water-soluble, acetone-soluble, carbon dioxide and residue products and subtracting from it the material lost in the handling of reactants and products. This loss weight was estimated by calculating the carbon not accounted for by the water-soluble, acetone-soluble, carbon dioxide and residue products and dividing this value by the carbon content of the water-soluble product. This latter value was believed to be close to the carbon content of the lost material.

Thus, the water yield was calculated as follows:

$$\text{water } (H_2O) \text{ yield} = 100 \times \frac{\text{feed} - (\text{WS} + \text{AS} + CO_2 + \text{R}) - \dfrac{\text{lost carbon (C)}}{\text{wt fract. C in WS}}}{\text{feed (moisture and ash free)}}$$

When sodium hydroxide or sodium carbonate was added to the reaction, an adjustment had to be made in the mass balance. This involved simply assuming that the sodium ended up in the water extract. Carbonate was assumed to convert entirely to CO_2 at the conditions of the experiments (final $pH \leq 5$). Solubility of CO_2 in the reaction solution was assumed negligible.

Gas chromatographic analyses of the water phase first involved separation into two fractions. This was done by simple distillation up to approximately 101 °C, leaving a dark brown, viscous residue. The distillate was collected and analyzed for alcohols and ketones. The residue was solubilized in water, acidified to $pH = 3$, filtered, and then analyzed for carboxylic acids. Gas chromatographic analysis of these aqueous fractions was performed using a 2·1 mm × 1·83 m nickel column packed with Chromosorb 101-80/100 (Johns Manville, Denver, Colo.), with nitrogen passing over formic acid as the carrier gas at a flow rate of 30 cm^3/min. Injector temperature was 200 °C, column temperature was 180 °C, and detector temperature was 250 °C.

3. RESULTS

3.1. Effect of Temperature without Alkali

The effect of temperature on the thermal degradation of cellulose in water without alkali to products consisting of water solubles, acetone solubles, carbon dioxide, water and residue is shown in Fig. 1. Production of other gases besides carbon dioxide was extremely small. The temperatures examined cover most of the range of interest in the thermochemical conversion of cellulosics. Reaction time at temperature was 10 min.

It appears that at approximately 200 °C degradation of cellulose is initiated. The amount of water solubles produced goes through a maximum of about 24 % yield at approximately 240 °C, and then drops quickly to less than 10 %. The yield of acetone solubles rises to approximately 11 % at 260 °C and then falls to below 10 % also. The production of carbon dioxide rises slowly and then levels off at approximately 10 % yield.

The amount of residue drops quickly to about 33 % at 260 °C and then

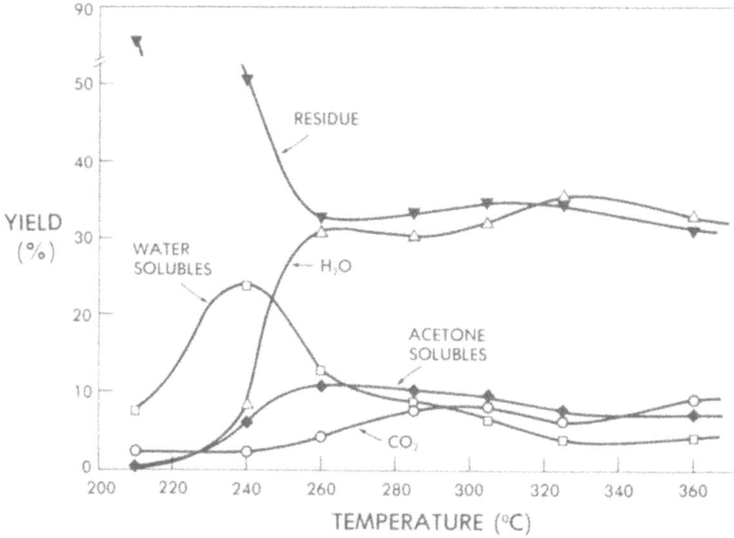

FIG. 1. Effect of temperature on cellulose thermochemical conversion in water (20 % cellulose slurry in water, 10 min reaction).

TABLE I

EFFECT OF TEMPERATURE ON COMPOSITION OF CELLULOSE THERMOCHEMICAL DEGRADATION PRODUCTS (20 % CELLULOSE SLURRY IN WATER, 10 MIN)

Temperature (°C)	*Products*						
	Water solubles	*Acetone solubles*			*Residue*		
	%C	*%C*	*%H*	*%O*	*%C*	*%H*	*%O*
210	56	67	8	25	43	6	51
240	50	66	6	28	46	6	48
260	67	66	5	29	68	5	27
285	66	67	5	28	70	5	25
305	70	70	6	24	71	5	24
325	71	71	6	23	73	5	22
360	71	71	6	23	73	5	22

levels off as temperature increases. The amount of water produced in the cellulose conversion reaction increases quickly between 210 and 260 °C, and levels off at about 33 % yield.

Thus, the amount of residue never dips below 30 %, and residue and water are the main products of cellulose conversion between 260 and 360 °C. Water solubles are at a maximum at about 240 °C, but the residue yield at this point is over 50 %.

Table I gives an idea of the composition of the products. Unfortunately, none of the products is very hydrocarbon-like, and they all possess large oxygen contents. At low conversion temperature, the residue and water solubles compositions are quite close to that of cellulose. At higher temperature, the water solubles, acetone solubles and residue have similar composition.

One might then conclude by looking back at Fig. 1 that water solubles are the first product of thermochemical degradation of cellulose in water, and that these eventually dehydrate and polymerize to form acetone solubles, which in turn polymerize further to residue material. Thus, at low temperatures the residue appears to be unconverted cellulose, whereas at higher temperatures the residue more closely resembles acetone solubles.

Dry pyrolysis of cellulose at 295 °C for 10 min produced very small yields of water solubles (0·6 %) and acetone solubles (1·7 %) and a large yield of residue (44·8 %) compared to reaction of cellulose in water at similar conditions. Interestingly, the yields of water (28·7 %) and carbon dioxide (11·2 %) were similar to those obtained with cellulose in water. Thus, the

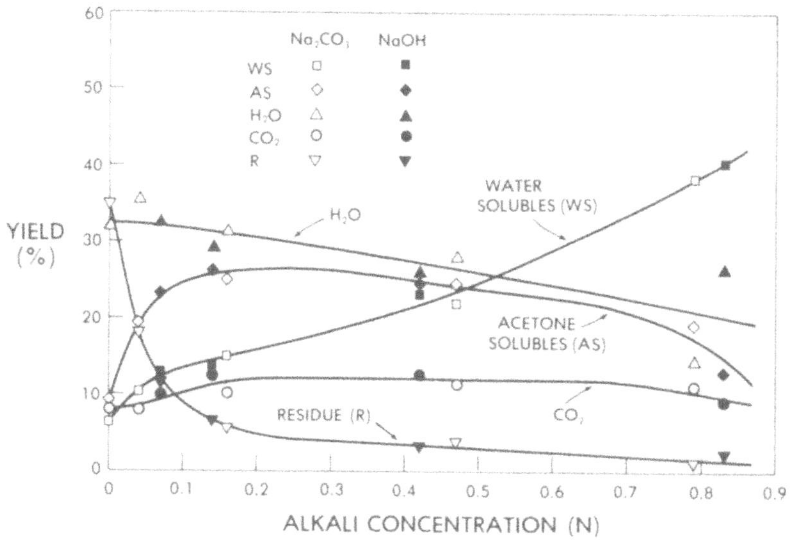

FIG. 2. Effect of alkali solution concentration on cellulose thermochemical conversion (20 % cellulose slurry in solution, 305 °C, 10 min reaction).

major effect of performing cellulose thermochemical degradation experiments in water may be some inhibition of the polymerization of initially water-soluble compounds to acetone-soluble compounds and, in turn, the inhibition of the polymerization of acetone solubles to residue material. The inhibition may occur by solubilization and dilution of the water solubles.

There was very little effect of increasing reaction time from 10 min to 1 h.

3.2. Effect of Alkali

Figure 2 shows the effect of sodium hydroxide and sodium carbonate on the thermochemical degradation of cellulose at 305 °C. This temperature is approximately in the middle of the range of temperatures which have been of interest in the thermochemical conversion of cellulosics. The gram-equivalent weights used for sodium hydroxide and sodium carbonate were, respectively, 40 g and 53 g. The effect of alkali is significant, greatly decreasing residue yield while substantially increasing yields of water solubles and acetone solubles. At the same time, production of water is reduced. Interestingly, the yield of water solubles continues to increase as alkali concentration increases, while the yield of acetone solubles increases to approximately 26 % and then begins to decrease.

TABLE II

EFFECT OF ALKALI CONCENTRATION ON COMPOSITION OF CELLULOSE THERMOCHEMICAL DEGRADATION PRODUCTS (20% CELLULOSE SLURRY IN SOLUTION, 305 °C, 10 MIN)

Alkali conc.	*Products*						
	Water solubles	*Acetone solubles*			*Residue*		
	%C	*%C*	*%H*	*%O*	*%C*	*%H*	*%O*
0	70	70	6	24	71	5	24
0·04 N Na_2CO_3	78	73	6	21	75	5	20
0·07 N NaOH	71	73	6	21	76	5	19
0·14 N NaOH	67	74	7	19	75	5	19
0·16 N Na_2CO_3	70	74	6	20	77	5	18
0·42 N NaOH	58	75	7	18	76	5	19
0·47 N Na_2CO_3	62	75	7	18	77	5	18
0·79 N Na_2CO_3	48	75	7	18	76	5	19
0·83 N NaOH	58	74	7	19	76	5	19

Unfortunately, sodium hydroxide does not appear to possess any advantage over sodium carbonate. Thus, the cellulose swelling advantage which sodium hydroxide possesses compared to sodium carbonate does not translate into any different result in thermochemical conversion reactions at the conditions studied.

The effect of alkali concentration on the composition of cellulose thermochemical conversion products is given in Table II. In water and in dilute alkali solutions, the composition of the water solubles, acetone solubles and residue are similar. This supports the belief that these materials differ by degree of polymerization. In more concentrated solutions of alkali, the composition of the acetone-soluble and residue material remains approximately the same as the amounts of these materials decline. The carbon content of the water solubles declines, reflecting less production of water.

Again, increasing reaction time from 10 min to 1 h had little effect on the products.

Table III shows the effect of higher concentrations of alkali compared to the lower alkali concentrations of Fig. 2. The result of higher alkali concentrations is eventually to eliminate the production of residue. After passing through a maximum, production of acetone solubles is also eliminated. The yield of water solubles steadily increases as alkali concentration increases, so that eventually they are nearly the sole product.

TABLE III

EFFECT OF ALKALI CONCENTRATION ON THERMOCHEMICAL DEGRADATION OF CELLULOSE (305 °C, 10 MIN)

Cellulose slurry (%)	*Alkali conc.*	*Yield*[a] *Acetone solubles* (%)	*Yield*[a] *Residue* (%)
20	0	9·1	34·8
20	0·14 N NaOH	26·3	6·5
20	0·42 N NaOH	24·7	2·9
20	0·83 N NaOH	13·1	2·1
10	0·42 N NaOH	4·2	0
10	0·91 N NaOH	0	0
10	1·03 N Na_2CO_3	0	0

[a] Based on maf feed.

3.3. Gas Chromatographic Analyses of Water Solubles

Figure 3 shows the yields of the several alcohols and ketones detected by gas chromatography in the distillate obtained by distilling up to approximately 101 °C the water phase from several cellulose conversion experiments. These experiments were performed with 20 % cellulose slurries in 0·47 N Na_2CO_3 for 10 min at temperature. At these conditions approximately equal amounts of water solubles and acetone solubles (20–25 % yields) and small amounts of residue (3–5 % yields) are produced at 305 and 360 °C. An equivalent amount of water solubles yield (20 %), small acetone solubles yield (3 %) and large residue yield (65 %) are produced at 240 °C. After distilling the water solubles to 101 °C, a dark brown, viscous solution remains as residue. Figure 4 shows the yield of the several carboxylic acids detected by gas chromatography in the distillation residue after dilution with water and acidifying to pH = 3 and filtering.

In all cases, the yields increase with increasing temperature. The amounts of the alcohols, ketones and carboxylic acids determined at 360 °C are several times the yields at 240 and 305 °C, and represent a total of approximately 8 % of the original weight of cellulose. The yield of acetic acid is the largest at 3·6 %.

Figures 5 and 6 show the yields of alcohols, ketones and carboxylic acids determined in the water phase after reaction of a 20 % cellulose slurry in 2·36 N Na_2CO_3 for 1 h. At these conditions, the sum total of acetone solubles and residue produced at 305 and 360 °C is quite small (10–12 %).

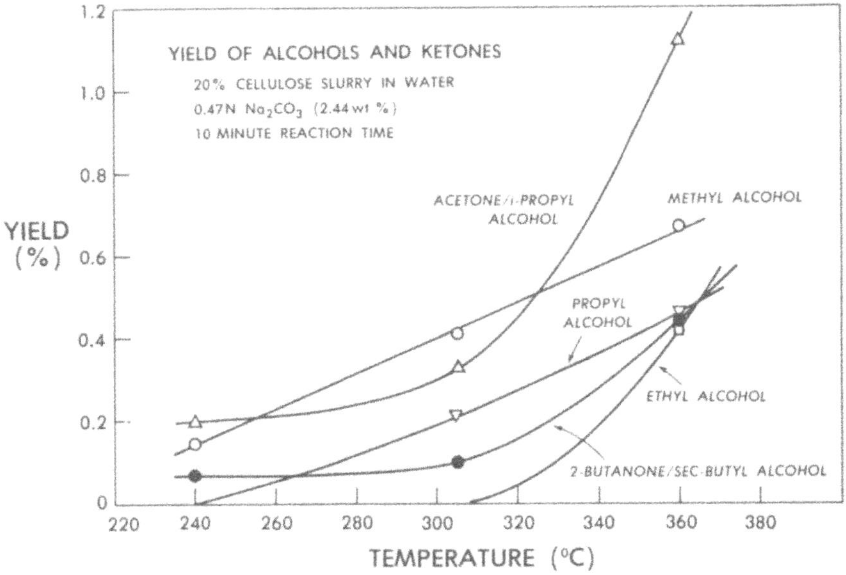

FIG. 3. Yields of certain alcohols and ketones from thermochemical conversion of 20% cellulose slurry in 0·47 N Na_2CO_3 (2·44 wt %, 10 min reaction).

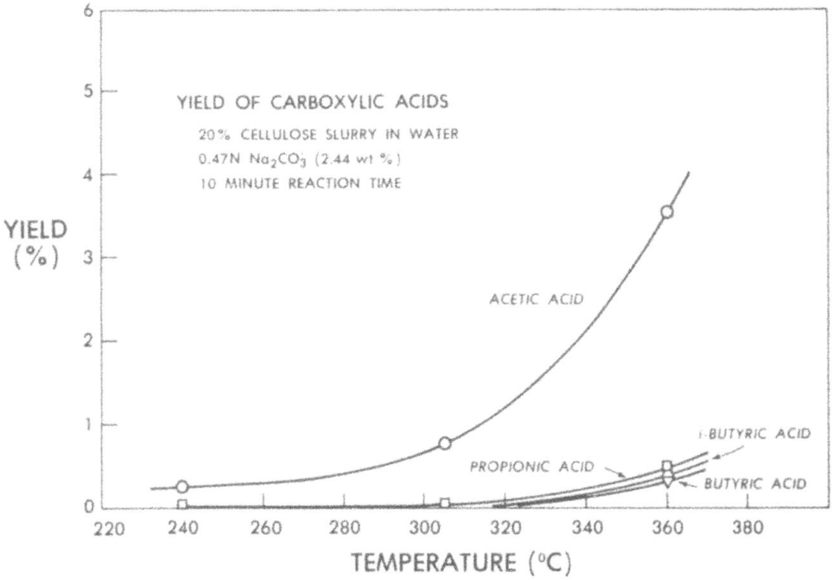

FIG. 4. Yields of certain carboxylic acids from thermochemical conversion of 20% cellulose slurry in 0·47 N Na_2CO_3 (2.44 wt %, 10 min reaction).

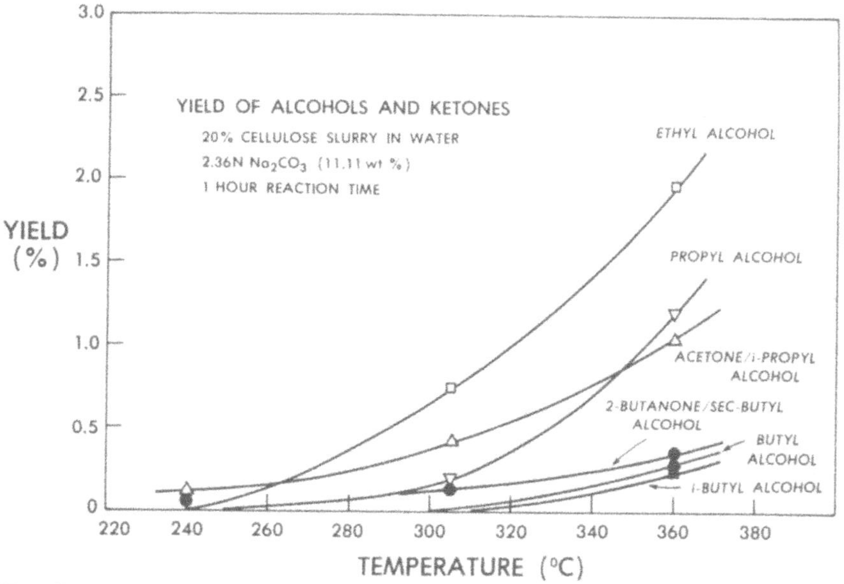

FIG. 5. Yields of certain alcohols and ketones from thermochemical conversion of 20% cellulose slurry in 2·36 N Na_2CO_3 (11·11 wt%, 1 h reaction).

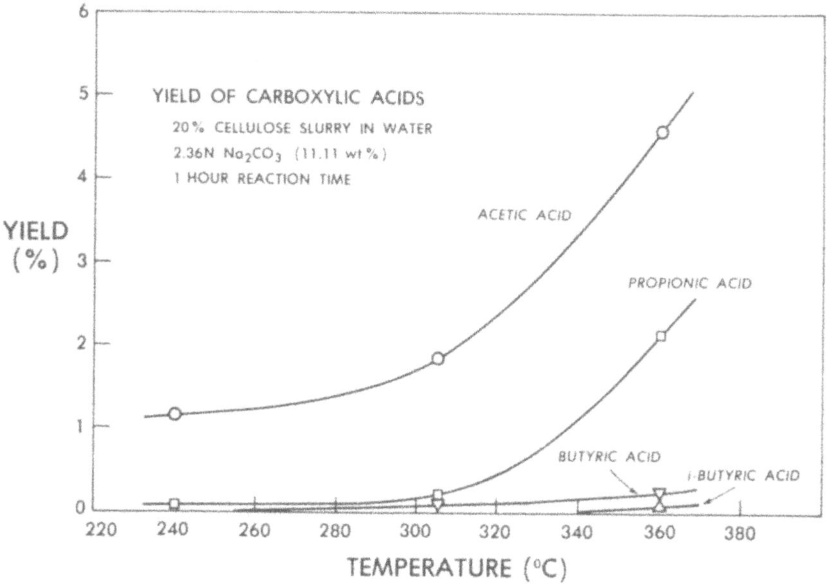

FIG. 6. Yields of certain carboxylic acids from thermochemical conversion of 20% cellulose slurry in 2·36 N Na_2CO_3 (11·11 wt%, 1 h reaction).

At 240 °C, a large residue yield is obtained (37 %) with the remainder being mainly water soluble product. Again, the amounts of the alcohols, ketones and carboxylic acids determined at 360 °C are several times the yields at 240 and 305 °C. They represent a total of approximately 12 % of the original cellulose weight. Yields of ethyl alcohol and acetic acid are, respectively, 2·0 and 4·6 %.

4. SUMMARY AND CONCLUSIONS

Thermochemical degradation of cellulose in aqueous solution is a complex process which produces a complex mixture of products. These products can be identified as water solubles, acetone solubles, insoluble residue, carbon dioxide and water. Addition of alkali to the reaction system increases the yields of water-soluble and acetone-soluble products and decreases the production of residue. At high concentrations of alkali, the production of water-soluble material dominates and yields of acetone solubles and residue approach zero.

The water-soluble products obtained by thermochemical degradation of cellulose at several temperatures in different concentrations of alkali were analyzed for several simple alcohols, ketones and carboxylic acids by gas chromatography. At reaction conditions of 2·36 N Na_2CO_3, 360 °C and 1 h reaction time, the following yields were obtained from a 20 % cellulose slurry (based on moisture and ash free cellulose): acetic acid: 4·6 %, propionic acid: 2·2 %, ethyl alcohol: 2·0 %, *n*-propyl alcohol: 1·2 %, and acetone/i-propyl alcohol: 1·1 %. The remaining compounds were found in yields less than 1 %.

The complete solubilization of cellulosic material in alkali solutions may produce valuable compounds in significant yields. Further work should be done to analyze the large percentage of water soluble material still unidentified.

REFERENCES

1. Appell, H. R., Wender, I. and Miller, R. D. (1970). Conversion of urban refuse to oil. Bureau of Mines Solid Waste Program, Technical Progress Report-25, US Dept. of Interior.
2. Appel, H. R., Fu, Y. C., Friedman, S., Yavorsky, P. M. and Wender, I. (1971). Converting organic wastes to oil—a replenishable energy source. Bureau of Mines Report of Investigations 7560, US Dept. of Interior.

3. Appell, H. R., Fu, Y. C., Illig, E. G., Steffgen, F. W. and Miller, R. D. (1975). Conversion of cellulosic wastes to oil. Bureau of Mines Report of Investigations 8013, US Dept. of Interior.
4. Elliot, D. C. and Walkup, P. C. (1977). Bench scale research in thermochemical conversion of biomass to liquids in support of the Albany, Oregon, experimental facility. Battelle Pacific Northwest Laboratories for US Dept. of Energy, Report TID-28415.
5. Elliot, D. C. and Giacoletto, G. M. (1979). Bench scale research in biomass liquefaction in support of the Albany, Oregon, experimental facility. Third Annual Biomass Energy Systems Conference, Golden, Colorado.
6. Elliot, D. C. (1980). Bench scale research in biomass liquefaction by the CO–steam process. *Canadian Journal of Chemical Engineering* **58**, 720–34.
7. Elliot, D. C. (1980). Process development for biomass liquefaction. *ACS Div. Fuel Chem. Preprints* **25**(4), 257–63.
8. Chornet, E., Fonseca, R., Grandbois, M. and Brown, R. P. (1980). The conversion of peat and humic materials into liquid fuels using carbon monoxide. *Canadian Journal of Chemical Engineering* **58**, 713–7.
9. Eager, R. L., Matthews, J. F., Pepper, J. M. and Zohdi, H. (1981). Studies on the products resulting from the conversion of aspen poplar to an oil. *Canadian Journal of Chemistry* **59**(14), 2191–8.
10. Davis, H. G., Figueroa, C. and Schaleger, L. L. (1982). Liquid fuels by direct liquefaction of biomass. Lawrence Berkeley Laboratory, LBL-14155.
11. Schaleger, L. L., Figuero, C. and Davis, H. G. (1982). Direct liquefaction of biomass: results from operation of continuous bench scale unit in liquefaction of water slurries of Douglas fir wood. Lawrence Berkeley Laboratory, LBL-14019.
12. Davis, H., Figueroa, C. and Schaleger, L. (1982). Research and development activities on direct liquefaction technology. Lawrence Berkeley Laboratory, LBL-14595.
13. Molton, P. M., Demmitt, T. F., Donovan, J. M. and Miller, R. K. (1978). Mechanism of conversion of cellulosic wastes to liquid fuels in alkaline solution. Energy from Biomass and Wastes—Symposium Papers. Institute of Gas Technology, Chicago, Il.
14. Ostermann, R. D., Bishop, K. A. and Rosson, R. F. (1980). Kinetics of the thermochemical conversion of cellulose to oil in aqueous alkaline solution. Energy from Biomass and Wastes IV. Institute of Gas Technology, Chicago, Il.
15. Miller, R. K., Molton, P. M. and Russell, J. A. (1980). Thermal degradation of cellulose in alkali. Battelle Pacific Northwest Laboratory, PNL-SA-8603.
16. Russell, J. A., Molton, P. M. and Landsman, S. D. (1980). Chemical comparisons of liquid fuel produced by thermochemical liquefaction of various biomass materials. Battelle Pacific Northwest Laboratory. PNL-SA-8602.
17. Donovan, J. M., Molton, P. M. and Demmitt, T. F. (1981). Effects of pressure, temperature, pH, and carbon dioxide on oil yields from cellulose liquefaction. *Fuel* **60**, 898–902.
18. Molton, P. M., Miller, R. K., Russell, J. A. and Donovan, J. M. (1981). Effects of reaction conditions on the aqueous thermochemical conversion of biomass to oil. In *Biomass as a Nonfossil Fuel Source*, D. T. Klass (Ed.). ACS

Symposium Series 144. American Chemical Society, Washington, D.C., pp. 137–62.

19. Molton, P. M. and Demmitt, T. F. (1978). A literature survey of intermediate products formed during the thermal aqueous degradation of cellulose. *Polym.-Plast. Technol. Eng.* **11**(2), 127–57.
20. Alfredsson, B. and Samuelson, O. (1968). Hydroxyacids formed by alkali treatment of hydrocellulose. *Sven. Papperstidn.* **71**, 679.
21. Sjostrom, E. (1977). The behavior of wood polysaccharides during alkaline pulping processes. *Tappi* **60**, 151.

61

The Electrochemical Conversion of Biomass-Derived Compounds

James H. P. Utley
Department of Chemistry, Queen Mary College, London, UK

It is timely to describe the application of electrochemical methods for the transformation of organic compounds; the method is powerful and non-thermal and its possible advantages are being increasingly explored for the conversion of biomass-derived organic compounds. Specific problems concerned with the conversion of biomass will be discussed but first it is necessary to describe the reactive intermediates which result from electron transfer, to explain the link with biomass problems, and to outline the requirements for operation on a large scale.

1. REACTIVE INTERMEDIATES VIA ELECTRON TRANSFER

The early work of Faraday[1] and Kolbe[2] established that fatty acids could usefully be oxidized to hydrocarbons but Haber in his classical examination[3] of the controlled potential reduction of nitrobenzene showed that the electrochemical method could be selective (Scheme 1), i.e. the electrode was a reagent of 'tuneable' redox power. More recently it has been shown that the reactive intermediates formed at anodes and cathodes are those already familiar to organic chemists, namely radicals,

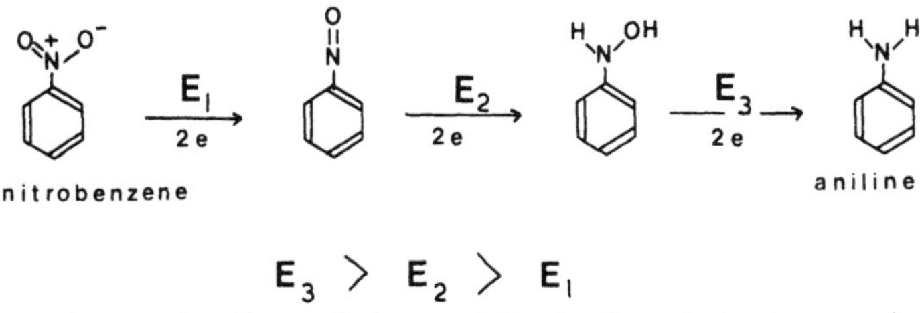

SCHEME 1. Controlled potential reduction of nitrobenzene.[3]

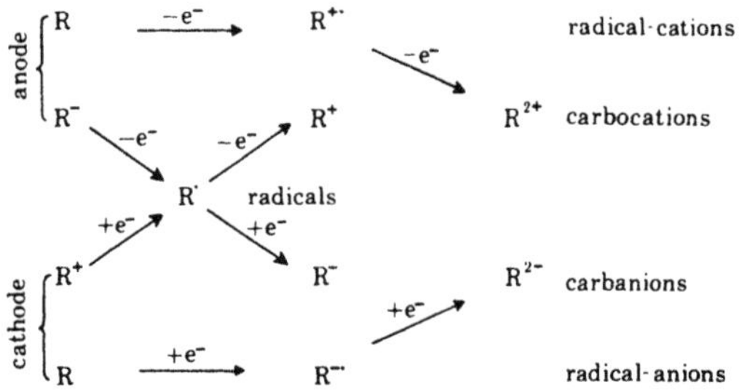

SCHEME 2. Organic reactive intermediates produced at electrodes.

carbocations and carbanions. In addition radical-anions and radical-cations, and occasionally dianions and dications, are formed by electron transfer (Scheme 2). Such intermediates can undergo rapid chemical reaction; the overall rate of the reactions may be controlled by the rate at which electron transfer occurs or by the rate of transport of starting materials to the electrode surface or products away from the surface.

Consideration of the energetics of electron transfer is a convenient way of appreciating the power of the electrochemical methods. Although Scheme 3 depicts the especially relevant case of cathodic cleavage, the expression for the rate of electron transfer applies equally to anodic reactions. The availability of practical solvent and electrode systems dictates operation within the range $+3$ to -3 V and for a one-electron transfer ($z = 1$) and a typical value for the transfer coefficient ($\beta = 0{\cdot}5$) the free energy of activation ($\Delta G^{\neq}$) may be modified by βzEF which is *ca.* 290 kJ mol^{-1}. A rough calculation shows this to be equivalent to a change in rate of electron transfer, over a 6 V range, of 10^{14}, i.e. equivalent to thermal activation by several thousands of degrees. The practical requirements for harnessing this oxidizing and reducing power are that electrode materials are inert to oxidation or reduction, or both (e.g. platinum, various forms of carbon, mercury, lead, lead dioxide, titanium clad with inert oxides or metals) and

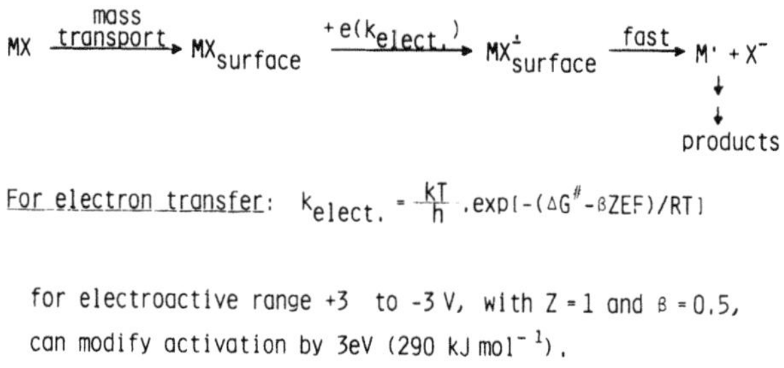

SCHEME 3. Energetics and rates.

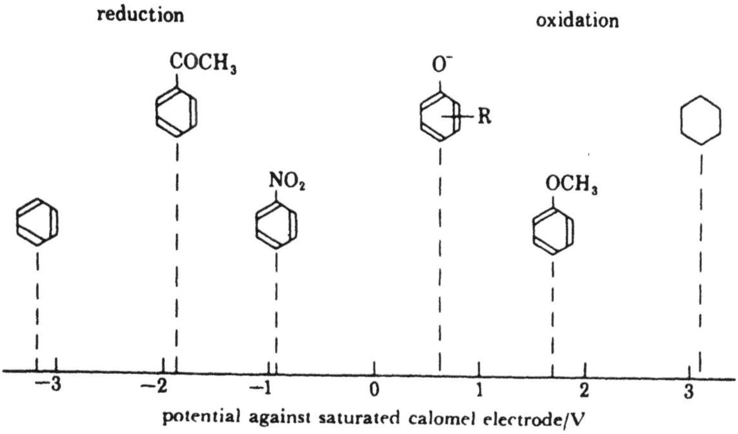

FIG. 1. A scale of redox power.

that solvents are reasonable conductors which permit a good electroactive range, e.g. dipolar aprotic solvents such as acetonitrile, or N,N-dimethylformamide. In the important case of the Monsanto adiponitrile process (see below) aqueous solution is used but a hydrophobic cathode environment is established by the use of a tetraalkylammonium supporting electrolyte. Supporting electrolytes are usually required to aid conductivity. To maintain a good electroactive range at least one ion should be difficult to oxidize or reduce, depending on which is the working electrode. The practical requirements outlined above will be amply illustrated in the examples to be discussed.

An organic chemist's view of the power of electro-oxidation and electro-reduction is given in Fig. 1; at the anodic extreme carbon–carbon and carbon–hydrogen bonds may be cleaved and at the cathodic limit solvated electron reductions are possible. The relative order of ease of oxidation or reduction (a few examples are given) comes as no surprise to the physical-organic chemist and is largely predictable according to substituent effect theory.

2. THE LINK WITH BIOMASS PROBLEMS

Although there is much uncertainty concerning timescales it is generally accepted that eventually fossil sources of chemicals (petrochemicals, coal-based chemicals) will be supplemented or replaced by chemicals derived from renewable resources (biomass). In considering the technology to be

Starting materials available from biomass?

[Many C_4/C_5, polyfunctional, highly oxygenated]

↓

Conversion into fuels (de-oxygenation) or
petrochemical substituted (often higher M.Wt., polyfunctional)?

[Involves reductive cleavage and dimerisation]

↓

What are good chemical methods for the above, capable of
operation on a large scale?

[ELECTROCHEMICAL]

SCHEME 4. A strategy for the conversion of biomass.

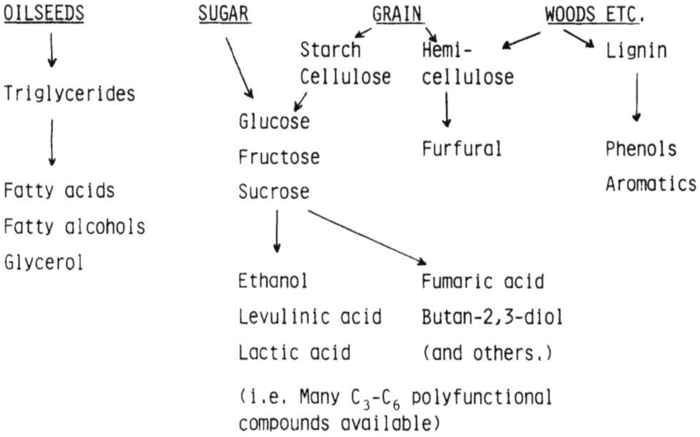

SCHEME 5. Chemicals from biomass.

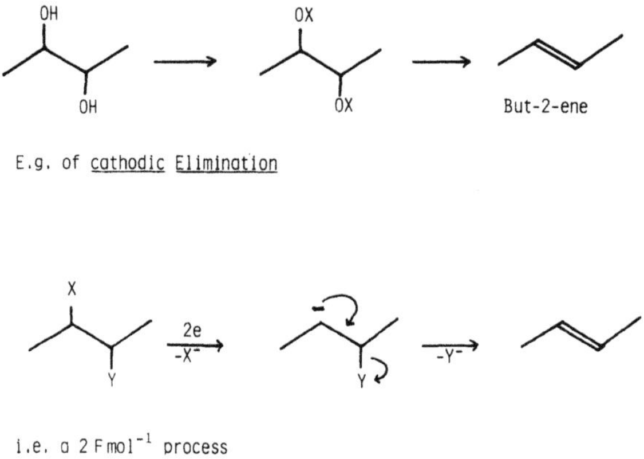

SCHEME 6. Transformations of butan-2,3-diol.

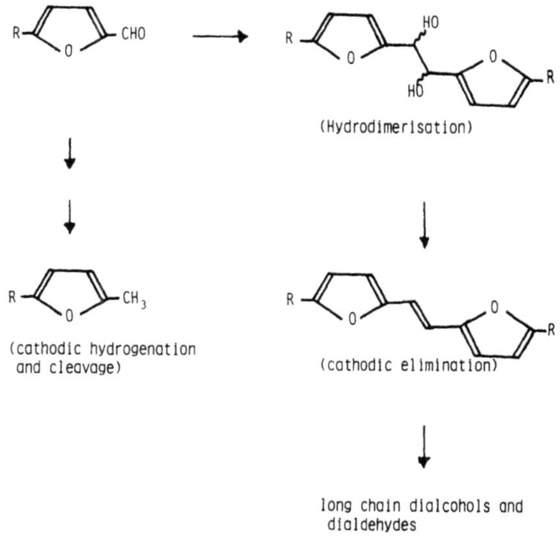

SCHEME 7. Transformations of furfural.

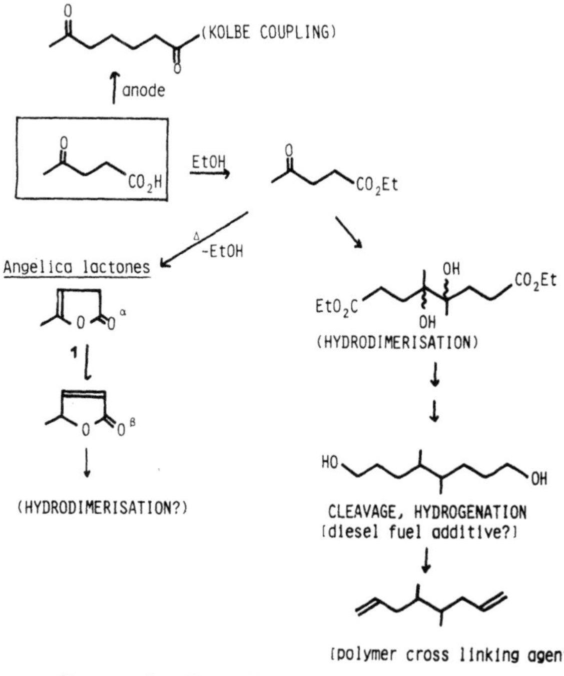

SCHEME 8. Transformations of levulinic acid.

applied to this challenge electrochemistry presents itself according to Scheme 4. Of the statements included in Scheme 4 we must consider further the nature of the likely starting materials, the contention that electrochemical processes may be operated on a large scale, and the conversions (reductive cleavage, dimerization) that appear to be particularly useful. The composition of biomass and the availability of (cheap) starting materials has been well reviewed[4] and is summarized in Scheme 5.

In the first instance it is likely that cathodic reactions will be the more important; biomass-derived starting materials are highly oxygenated and therefore amenable to reduction. Anodic reactions, especially cleavage, might be of importance for the degradation of such polymers as lignin which is a potentially rich source of phenols and aromatic carboxylic acids and aldehydes. For three important substrates, butan-2,3-diol, furfural and levulinic acid, possibly useful transformations which may be carried out cathodically are outlined in Schemes 6–8. But on what scale could such electrochemical processes be operated?

3. SCALE-UP OF ELECTRO-ORGANIC REACTIONS

Operation at controlled potential, as first envisaged by Haber,[3] was tedious until the development of the potentiostat and it must remain impracticable for large scale electrolysis at high currents. One of the first requirements for scale-up is therefore to arrange conditions so that the selectivity of controlled potential operation is retained in controlled current operation. The situation is outlined in Fig. 2 in terms of voltammetric curves. The

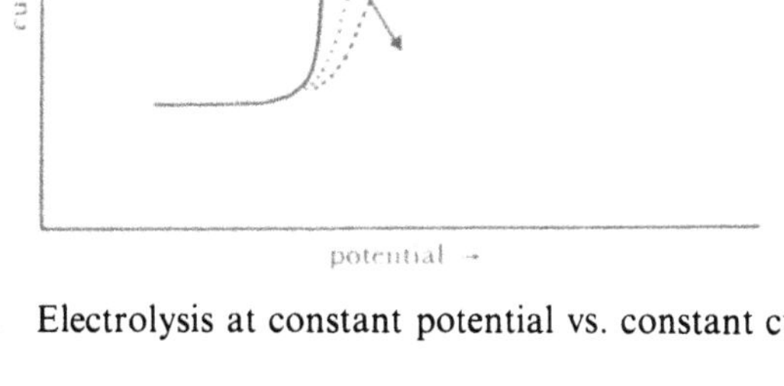

FIG. 2. Electrolysis at constant potential vs. constant current.

current is proportional to the concentration of electroactive substrate and to the number of electrons being transferred. The limit which may be reached is also a function of the rate of mass transport to the electrode. A potentiostat is a device which will allow the current to vary so as to maintain a constant potential, e.g. E_1, at the working electrode *vis à vis* a nearby reference electrode such as the saturated calomel electrode or a silver/silver halide couple. Consequently, as the concentration of electroactive compound falls, the current falls. Without potentiostatic control the potential would shift in an attempt to maintain current and might well reach, say, E_2 at which another function in the molecule would be reduced/oxidized. For large scale reactions, however, it is possible to control the concentration of electroactive starting material. This enables operation of a cell to be maintained in the region of E_1, i.e. the concentration is kept at a level which ensures that the current does not drop below i_1. Operation at controlled current requires cheaper and less sophisticated equipment than operation at controlled potential.

The Monsanto adiponitrile process provided great impetus to organic electrochemistry and key features of it are outlined in Scheme 9. It is an example of a cathodic hydrodimerization reaction and its development from a laboratory discovery by M. M. Baizer to a process running at *ca.* 300 000 tonnes yr^{-1} contains many lessons.[5] Chief among these are the manner in which the reaction was made efficient in aqueous solution and the reduction of the cell voltage (to be distinguished from the electrode potential, see below). In aqueous solution with alkali metal salts as supporting electrolyte the main product is that of hydrogenation, i.e. propionitrile, rather than that of hydrodimerization. This hurdle was

$$\underset{\text{(AN)}}{2CH_2=CHCN} \xrightarrow{2e,\ 2H^+} \underset{\text{(ADN)}}{NC(CH_2)_4CN}$$

Key conditions

(1) High concentration of AN in aqueous electrolyte - use of $Et_4\overset{+}{N}EtSO_4^-$ (TEAES) crucial.

(2) Detailed design of parallel plate cells important - mass transfer rates.

(3) Optimum current density 0.45 A cm^{-2}, cell volts 11.65 in pilot plant (14,500 tonnes yr^{-1}).

(4) Long life of membrane.

SCHEME 9. The Monsanto adiponitrile process.

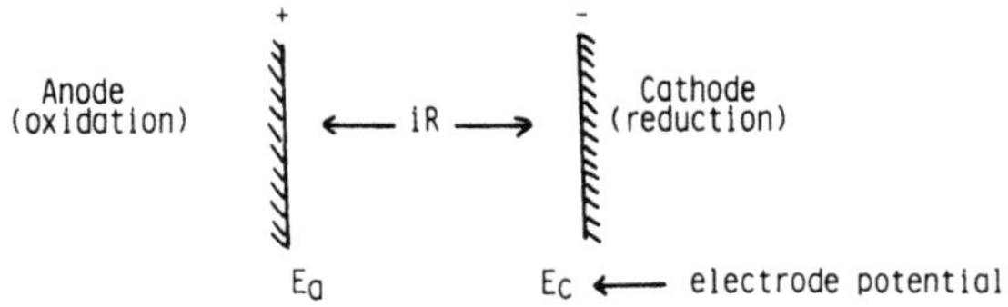

Cell voltage (V) $= E_a + E_c + iR$

Power usage (kWh . kg^{-1}) $= \dfrac{V \times F \times \text{number of } F.mol^{-1}}{\text{product M.Wt.} \times \text{current efficiency}}$

FIG. 3. The reactor (cell).

overcome by the use of tetraalkylammonium salts, e.g. tetraethylammonium ethyl sulfate, which acted as supporting electrolyte, as solubilizing agent to provide a high concentration of acrylonitrile, and, via the concentration of tetraalkylammonium ions at the cathode, as providing a hydrophobic environment which encouraged dimerization of acrylonitrile radical-anions rather than their protonation. Recent developments have been in the engineering of the process; the cell-dividing membrane has been dispensed with resulting in a smaller electrode separation and a reduction of the cell voltage to *ca.* 3·8 V.

The simplest form of reactor/cell is shown in Fig. 3. The parallel plate arrangement depicted also serves to make the distinction between electrode potential and cell voltage; from the expression for power usage it is easy to calculate that for the adiponitrile process a change of cell voltage from 11·65 V to 3·84 V brings about a threefold drop in power usage, to 2·4 kWh kg^{-1}. There are several excellent accounts[6] of the variety of cells which may operate on a large scale. Features common to all good designs are low cell voltages, versatility (useful for a variety of electro-oxidations and reductions), and cheap construction with long-lived electrode materials and membranes. Provision for continuous operation is also highly desirable. Considerable experience may be gained by experiments on a laboratory scale using cells constructed from glass, steel or plastics such as polypropylene.

An obvious requirement for the practical electrochemical conversion of biomass emerges. The likely starting materials will be present in dilute aqueous solution; therefore processing in flow cells, with selective reduction/oxidation of one component of a mixture, will be advantageous. Products which may easily be separated should be aimed for, e.g. gases or high molecular weight water insoluble compounds. From the equation in

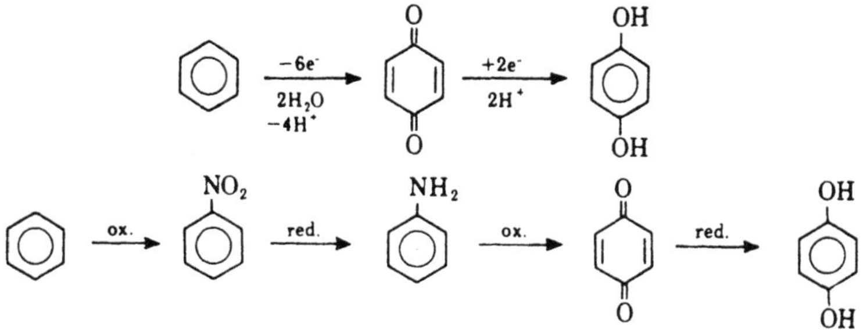

SCHEME 10. The elimination of steps; a conventional vs. electrochemical route. Reproduced from Reference 7 by permission of the publisher, The Royal Society, London.

Fig. 3 it is clear that high molecular weight products require proportionately less power consumption.

Finally, in this section, it is appropriate to mention some other perceived advantages of electrosynthesis. Electro-organic synthesis usually proceeds via *umpolung*, i.e. an electron-rich, nucleophilic, species such as an aromatic hydrocarbon is converted by electron transfer into a radical-cation which is electron deficient and electrophilic. Similarly, formation of radical-anions also involves *umpolung*. Consequently chemical steps, necessary in conventional routes, may be circumvented. An example is given in Scheme 10 which compares a conventional and an electrochemical process.[7] Another possible advantage is that as no chemical redox reagent is employed (these may be toxic, hazardous, expensive, or all three), there is no spent reagent to dispose of. It is possible therefore that some environmental advantage may accrue. A more esoteric feature, of particular relevance to the production of fine chemicals and pharmaceuticals, is that because electrochemical methods are non-thermal, and chemical conditions may be mild, isomerization and further reactions of sensitive products may be avoided.

4. APPLICATION TO PARTICULAR 'BIOMASS PROBLEMS'

4.1. The Cleavage of Carbon–Oxygen Bonds

The reductive cleavage of carbon–oxygen bonds has been long known and the systematic investigation[8] of the cleavage of benzylic ethers and acetates has received particular attention. Cleavage of hydroxyl from carbon is, however, especially problematical. The hydroxyl group has poor leaving group ability; the example given in Scheme 11 illustrates this. At very

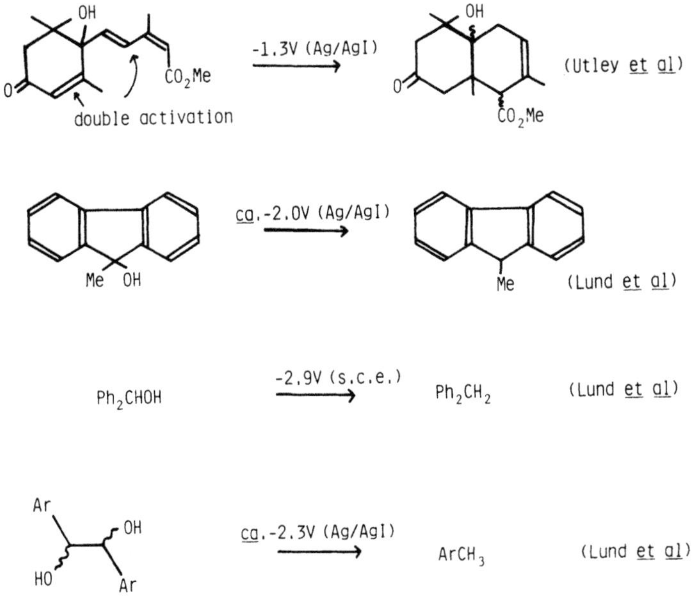

SCHEME 11. Hydroxyl as a leaving group in electro-reduction.

negative potentials hydroxyl can be removed[9] from diphenylmethyl and 9-methylfluorenyl, positions normally to be regarded as activated for cleavage. In another example (a stilbenediol) hydroxyl is lost but the carbon–carbon bond also cleaves.[9] In the case of the methyl ester of abscisic acid, reduction[10] proceeds at an acceptably low potential but hydroxyl does not depart from the radical-anion which prefers in this case to isomerize and cyclize intramolecularly.

The problem is analysed further in Scheme 12, the conclusion being that cheap derivatives of hydroxyl compounds are required which, preferably, reduce at low potentials.

$$MX \xrightarrow[(E_{red})]{e} MX^{\dot{-}} \longrightarrow M^{\cdot} + X^-$$

(1) Low potential reduction requires activation of M (the electrophore)

(2) Cleavage requires X^- to be a "good leaving group"

But. OH^- is a "poor leaving group" ∴ prior conversion necessary.

i.e. $MOH \xrightarrow[Step]{Chem.} MOX \xrightarrow{e} \longrightarrow M^{\cdot} + \bar{O}X$, where $\bar{O}X$ is a "good leaving group".

SCHEME 12. The cathodic cleavage of hydroxyl.

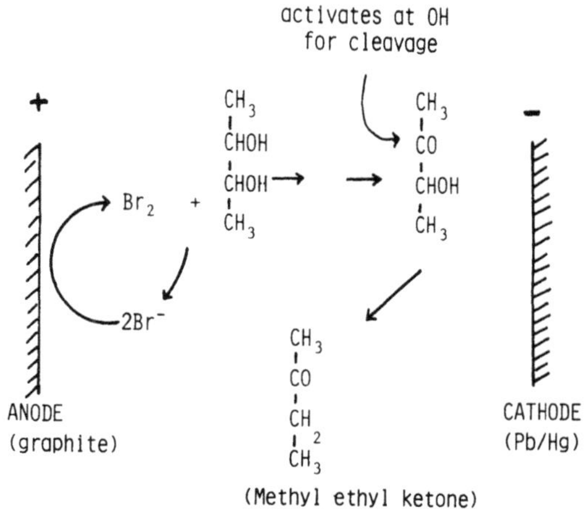

SCHEME 13. Conversion of butan-2,3-diol; a paired synthesis. Best results at pH 7, 3 mA cm^{-2} and 25 % conversion; 50–60 % current efficiency; product composition acetoin (20 %): MEK (80 %): butan-2,3-dione (nil).

One elegant approach to this involves the use of a 'paired synthesis', a method which attempts usefully to employ the reactions at both electrodes.[11] Such a method for the conversion of butan-2,3-diol into methyl ethyl ketone is outlined in Scheme 13. In essence the anode is used to produce bromine which, chemically, oxidizes one of the hydroxyl groups. The carbonyl group in the resulting acetoin ($CH_3COCH(OH)CH_3$) activates the molecule toward electro-reduction with subsequent loss of hydroxyl.

The bromine oxidation in Scheme 13 is an example of an 'indirect' use of electrochemistry, i.e. a redox reagent is generated or regenerated electrochemically. This important method is further exemplified in Scheme 14 which shows how an electrochemical cell may be used to sustain a concentration of hydriodic acid, the classical reagent for removal of hydroxyl from carbohydrates. Production of protons at the anode is balanced, surprisingly, by reduction of mercuric iodide at the cathode. The involvement of organomercurials, as in Scheme 14, has been demonstrated for the reduction of allylic alcohols.[12] A suitable choice of membrane and operating temperature is required to allow diffusion of protons to the cathode compartment with suitably rapid chemical reaction between the carbohydrate and hydriodic acid. In this manner moderate yields of hexenes and other compounds have been obtained[12] from a variety of carbohydrates.

Cathodic elimination has been exemplified in Scheme 6; in some cases it

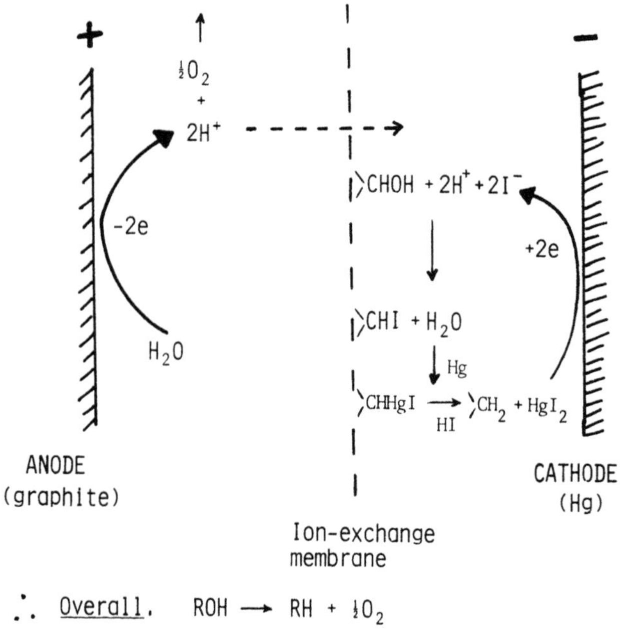

SCHEME 14. Indirect reduction of carbohydrates.

is possible to intercept the carbanion intermediate[13] but in most cases the second, elimination, step is very fast. There is a major problem associated with the application of this reaction. The desired product is the alkene, yet these are usually more easily reduced than the starting materials.

As with cleavage of a single group the substituents X and Y (Scheme 6) should be good anionic leaving groups, e.g. halide, *p*-toluene sulfonate, methanesulfonate etc. The latter substituents allow cathodic elimination

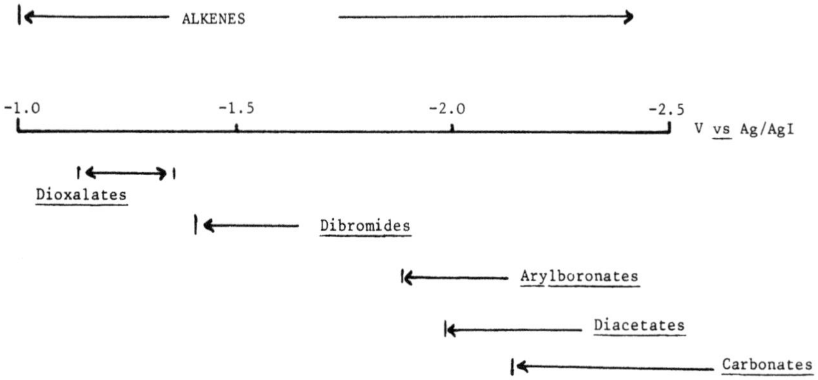

FIG. 4. Potentials for reductive elimination.

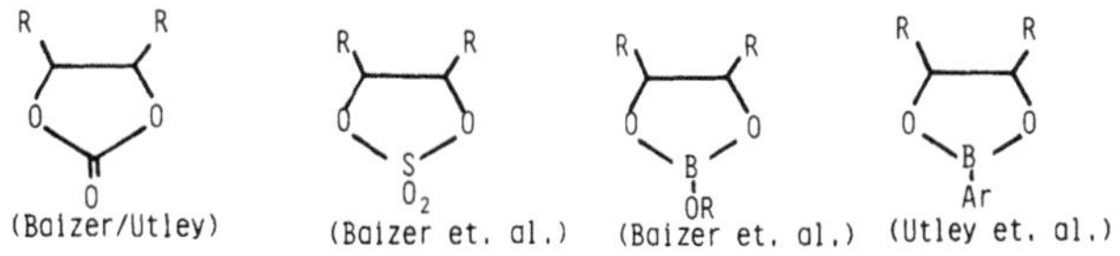

	(Baizer/Utley)	(Baizer et. al.)	(Baizer et. al.)	(Utley et. al.)
Preparation	Easy	Difficult	(Easy)	Easy (expensive)
Reduction	Difficult (E_{red} > -2.3V (Ag/AgI))	Moderately difficult	Difficult	Moderately easy (Ep, *ca* -2V)
Products	Alkenes/Alkanes	Alkene	Hydrolysis	Dimeric dianion and slow hydrolysis, alkenes/alkanes in some cases.

SCHEME 15. Cyclic derivatives.

from aliphatic systems but at very negative potentials. For instance, the dimethanesulfonate of butan-2,3-diol is reduced at $\geq -2{\cdot}3$ V (vs. Ag/AgI) and may be converted[14] at a lead cathode into a mixture of *cis* and *trans* but-2-ene in *ca.* 30% yield. In this case the alkene survives further reduction because a flow cell is used which rapidly removes the gaseous product from the electrolyte.

A comparison of the potentials required for cleavage of a variety of vicinal substituents is made in Fig. 4. The overlap with potentials for alkene production is clearly seen.

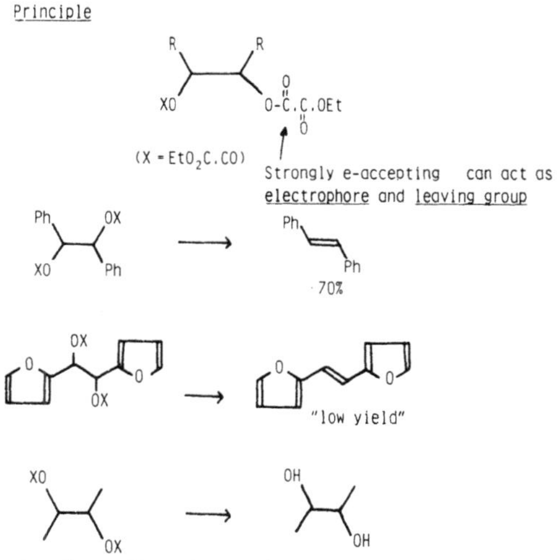

SCHEME 16. The reduction of oxalates. Features: (1) low reduction potentials; $-1{\cdot}2$ V (Ag/AgI); (2) cleavage pattern of radical-anions different for aromatic and aliphatic systems—why?

$$2HC(=O)H \xrightarrow[2H^+]{2e} HOCH_2CH_2OH$$

(as formalin) (E.G.)

i.e. formally a hydrodimerisation but seems much more complex!

Key results

(1) Graphite cathode, KOAc (1M), 0.17 A cm^{-2}, EG (96% current yield)

(2) C_3, C_4, C_5 Products also formed

(3) Current flows at -1.6 V vs. s.c.e.

SCHEME 17. Formaldehyde to ethylene glycol.

Cyclic derivatives of diols, especially carbonates,[15] boronates[15] and sulfates,[11] have attracted attention as substrates for cathodic elimination. The results of this approach are summarized in Scheme 15. The carbonates[15] show some promise although in the cases examined much of the alkene product was further reduced.

Conversion of hydroxyl into oxalate esters[16] provides derivatives in which the reducible function (the *electrophore*) is also the leaving group. In this case reduction potentials are low and further reduction of alkene is avoided. The method works well[16] for production of conjugated alkenes but, so far, good conditions for the desired cleavage of aliphatic oxalates have not been found.[14] This approach is summarized in Scheme 16.

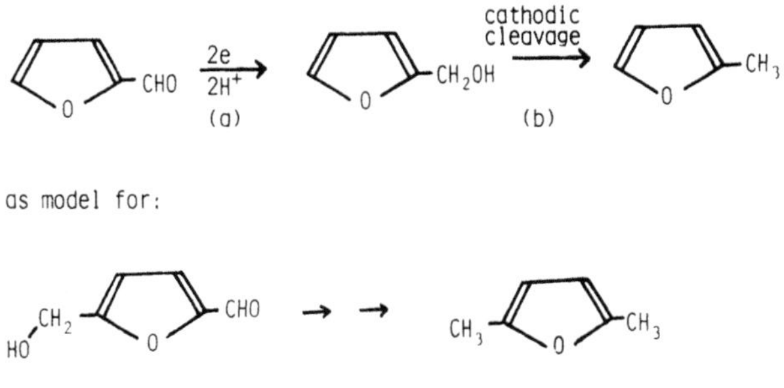

Conclusion. Step (a) feasible; step (b) not achieved by direct reduction.

SCHEME 18. Reduction of furfurals.

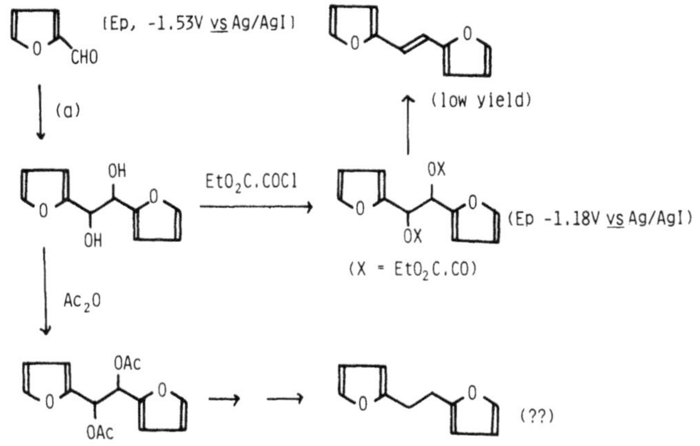

SCHEME 19. Hydrodimerization of furfural. Step (a) Divided cell, c.p.e., Hg cathode, MeCN-Bu NOAc, HOAc, yield 67 %; or parallel plate undivided flow cell, constant current 0·02 A cm^{-2}, Pb cathode, 50 % aqueous MeOH-NaH_2PO_4 (0·2 M) – Me_4NBr (0·2 M), yield 50 %.

The electrohydrodimerization of carbonyl compounds is likely to be important in the context of biomass chemistry. Preliminary experiments indicate[17] that, although mechanistically complex, the electrochemical conversion of formaldehyde into ethylene glycol could be workable (Scheme 17).

Furfural and 4-hydroxymethylfurfural have long been regarded as possibly important renewable feedstocks. The conversions in Scheme 18 have been examined[18] but for the reasons outlined in Section 4 direct conversion into methylfurans has not been achieved. The methylfurans are good target molecules, however; 2-methylfuran has an octane rating of 86 and might be of some use as a fuel additive.

The conversion of furfural into higher molecular weight derivatives is easier to achieve, electrochemically, and recent results[19] are summarized in Scheme 19.

ACKNOWLEDGEMENT

I acknowledge the support and encouragement of the Solar Energy Research Institute, Golden, Colorado, and especially that of Dr Helena Chum, Senior Electrochemist. Financial support for part of the work described was via SERI sub-contract No. HK-1-1136-1.

REFERENCES

1. Faraday, M. (1834). *Pogg. Ann.* **33**, 438.
2. Kolbe, H. (1849). *Annalen* **69**, 257.
3. Haber, F. (1898). *Z. Elektrochem.* **4**, 506.
4. Chum, H. L. and Osteryoung, R. A. (1981). Survey of the electrochemistry of some biomass derived compounds. SERI/TR-332-417, Solar Energy Research Institute, Golden, Co.; Lipinsky, E. S. (1981). *Science* **212**, 1465.
5. Baizer, M. M. (1980). US Patent 3,193,581. *J. Appl. Electrochem.* **10**, 285; Danly, D. E. (1981). *Hydrocarbon Processing* **161**; Baizer, M. M. (1980). *Chemtech.* 161.
6. Baizer, M. M. (Ed.) (1973). *Organic Electrochemistry.* M. Dekker Inc., New York; Weinberg, N. W. (Ed.) (1982). *Techniques for Electro-organic Synthesis*, Part III. Wiley-Interscience, New York; Goodridge, F. (1981). *Phil. Trans. R. Soc. Lond.* **A302**, 275.
7. Jansson, R. E. W. (1981). *Phil. Trans. R. Soc. Lond.* **A302**, 285.
8. Coleman, J. P., Naser-ud-din, Gilde, H. G., Utley, J. H. P., Weedon, B. C. L. and Eberson, L. (1973). *J. Chem. Soc. Perkin Trans.* **2**, 1903.
9. Lund, H., Michel, M. A., Mousset, G. and Simonet, J. (1975). *Electrochimica Acta* **20**, 143; Lund, H., Doupeux, H., Michel, M. A., Mousset, G. and Simonet, J. (1974). *Electrochimica Acta* **19**, 629.
10. Terem, B. and Utley, J. H. P. (1979). *Electrochimica Acta* **24**, 1081.
11. Baizer, M. M. (1982). Electrochemical Society Meeting, Montreal, Abstract No. 395; Baizer, M. M., Nonaka, T., Park, K., Saito, Y. and Nobe, K. (1982). Innovative concepts in applications of organic electrochemistry. UCLA School of Engineering and Applied Science, UCLA—Ch.E.—821.
12. Lund, T. and Lund, H. To be published.
13. Adams, C., Kamkar, N. M. and Utley, J. H. P. (1979). *J. Chem. Soc. Perkin Trans.* **2**, 1767.
14. Ellis, K. J. and Utley, J. H. P. (1982). Electrochemistry of biomass-derived organic compounds. Prepared for Solar Energy Research Institute, Golden, Co. under subcontract No. HK-1-1136-1.
15. Riley, J. H., Sopher, D. W., Utley, J. H. P. and Walton, D. J. (1982). *J. Chem. Research* (*S*) 326; *idem. ibid.* (M) 3355.
16. Sopher, D. W. and Utley, J. H. P. (1981). *J. Chem. Soc. Chem. Commun.* 134.
17. Weinberg, N. L. and Chum, H. L. (1982). Electrochemical Society Meeting, Montreal, Abstract No. 589.
18. St. John, M. R., Furgala, A. J., Solka, B. H. and Samuells, A. F. (1982). Liquid fuels from biomass electrolysis. Prepared for Solar Energy Research Institute, Golden, Co. under contract No. XB-1-9039-2.
19. Nazar-ul-Islam and Utley, J. H. P. Unpublished results.

62

Electrochemistry of Lignin Materials and Derived Compounds

HELENA L. CHUM
Solar Energy Research Institute,
Golden, Colorado, USA

and

D. W. SOPHER and H. A. SCHROEDER
Colorado State University, Fort Collins,
Colorado, USA

1. INTRODUCTION

One of the broad classes of renewable biomass materials of interest for conversion either to liquid fuels or to chemical feedstocks are the underutilized low-grade hardwoods such as aspen and other poplar species, southern hardwoods, etc.[1] Such lignocellulosic materials consist of three major biopolymers—cellulose, hemicelluloses and lignins—and a small amount of extractives. The quantities of the various components are a function of the species.[2] The chemical and physical properties of the individual polymers, mainly the lignin fraction, are such that their response to chemical or microbiological treatments is variable. In order to simplify the conversion of biomass to chemicals and fuels, the separation (pretreatment) of the major polymeric species is necessary, except where thermal processes such as gasification or pyrolysis are concerned. The separation of the lignin component from wood in the manufacture of paper has long been known and practised in industry[4] by a variety of methods. In the context of the conversion of biomass into fuels and chemicals, a number of pretreatments have been considered.[3,5] Such treatments can lead to low-molecular-weight cellulosic and hemicellulosic polymers that can be more easily converted into alcohols or other chemicals by chemical or microbiological processes. At the same time, a modified lignin fraction is recovered. Unlike other natural polymers lignin cannot be degraded to give structurally intact precursors, in this case phenyl propane monomers. Therefore, each pretreatment will generate lignin materials of different chemical and physical properties. Lignins are the major renewable source of phenolic compounds and those that have low molecular weight and are chemically reactive may be of future industrial importance. A more detailed knowledge of the relationship between their structural features and their physical and chemical properties is a prerequisite for commercial application, whether as a source of phenolic compounds or as a replacement of phenol in thermosetting resins.[6] The utilization of these phenolic compounds is of importance for an overall favorable economics of utilization of biomass to fuels and chemicals.

This paper addresses one specific type of processing of lignin polymers, electrochemical treatment, to increase the reactivity of the polymer or to generate low-molecular-weight phenolic compounds. The relevant literature on lignin electrochemistry will be reviewed as well as the research that we have been carrying out at the Solar Energy Research Institute since 1981. The reader is referred to the paper on the Electrochemical Conversion of Biomass-Derived Compounds by J. Utley in this book for an overview of the electrochemical reactions and methodology (Chapter 61). Synthetic work rather than mechanistic will be the subject of this paper. A recent survey of the electrochemistry of biomass and derived compounds has been compiled by one of us.[7]

2. ELECTROCHEMISTRY OF LIGNINS

The electrochemical processes of lignin materials are a complex sequence of electrochemical reactions, often combined with chemical reactions (mainly in alkaline media). These reactions are a marked function of the solvent and supporting electrolyte system, of the nature of the electrode material, and of the electrode potential, among other variables. In addition to simple transformations involving functional groups, such as

$$RCH_2OH \underset{\text{reduction}}{\overset{\text{oxidation}}{\rightleftharpoons}} RCHO \underset{\text{reduction}}{\overset{\text{oxidation}}{\rightleftharpoons}} RCOOH$$

which can be anodic (oxidative) or cathodic (reductive) processes, a number of other reactions (anodic and cathodic) are possible: substitution, addition, coupling, and bond cleavage, usually followed by the formation of new bonds. In addition, there are cases in which both anodic and cathodic processes generate useful products—a paired or matched synthesis.[8]

2.1. Electrochemical Delignification

This pretreatment utilizes the fact that in a divided cell the catholyte becomes more alkaline and the anolyte acidic. The generation of bases and acids then promotes hydrolytic processes. In 1954, Sychev and Petrova[9] proposed a method of delignification of agricultural wastes to be used in Russian collective farms, in order to transform the wastes into suitable animal feed, without having to supply chemicals (bases and acids, or bleach) for delignification, but using electricity instead. A divided cell was employed with a refractory cement as a separator. Graphite electrodes were employed and sodium chloride was the supporting electrolyte (3 %). The anolyte contained the same electrolyte and also sodium sulfate. The cathode chamber was filled with coarsely chopped winter rye straw or cotton husks. This scheme allowed the delignification of the plant material to take place by the base, with concomitant production of hydrogen. In the anode compartment chlorine, oxygen and acid were generated. After electrolysis the 40 % delignified material was washed with acidified water and dried—the resulting material was a suitable animal feed.

A subsequent treatment was also proposed. It consisted in placing the material after treatment in the cathode compartment, in the anode compartment, to be neutralized and bleached. The resulting pulp was colorless. The byproducts hydrogen and chlorine were valuable for the local industries and agriculture. In addition, hydrolysis of the hemicellulosic fraction yielded simple sugars such as xylose (and possibly xylitol by reduction), which could also be used. There are not enough data to estimate the power consumption of the process; the reported estimates seem quite low.

A more recent patent explores the acid hydrolysis of biomass performed in the anode compartment of a divided cell, which makes the resulting carbohydrate fractions much more suitable for enzymatic hydrolysis to glucose and other sugars. In the cathode compartment hydrogen was generated.[10]

2.2. Electrochemical Processes Applied to Liquors Derived from Pulping Processes

The use of the acid generated in the anode compartment of a divided cell has been proposed to precipitate lignins from spent pulping liquors.[11,12]

Neutralization and coagulation by formation of complexes (with the cations generated by oxidation of the metallic anodes)[11] were used. Extensive lignin oxidation was also observed. The coupling of anodic oxidation of spent sulfite liquor with electrodialysis to reduce the pH through a cation-selective membrane was also tested.[13] Reasonable current densities (50–200 mA/cm^2) at 90 °C were employed.

Several attempts have been made to utilize electrolysis of waste liquors (black and sulfite) to increase pulping yields. Most of the modified lignins were separated by filtration or flotation, and the pollution was decreased.[14]

2.3. Electrochemical Oxidation of Lignins

2.3.1. Halogenation Reactions

In the wood and wood pulp industry, where chlorination processes play an important role (bleaching), the halogenation reactions were extensively studied.[15] To a much lesser extent, electrochemical chlorination has also been studied in connection with bleaching, pulp recovery and waste liquor treatments.[16] More systematic studies have been performed of the changes in lignins as a function of the electrolyte composition during the electrochemical bleaching of pulp.[17] In addition, the resulting chlorinated lignins were also investigated as a function of the electrolysis time.[18] Dioxane lignin was chlorinated and extensively degraded oxidatively by electrolysis in NaCl (3 %) at constant current (12 mA/cm^2).[19]

Pershina investigated potential uses of chemically and electrochemically modified hydrolysis lignins (from wood saccharification plants).[20] Electrochemical chlorination was one example, in addition to chemical modification generating nitrile- and nitrolignins. Ultrasound purification of the hydrolysis lignins was successful; it was used in the preparation of bakelite-type molding powders.

Electrochemical bromination and iodination reactions were also carried out. Kraft lignin and lignosulfonic acid were brominated by electrolysis in aqueous HBr ($\sim$14 %) and HCl ($\sim$7 %) with an average current density of 50 mA/cm^2 on a Ti anode.[21] Milled wood lignin and hydrolysis lignin electrochemical iodination have been compared.[22] At pH 10, the hydrolysis lignin can attain 21 % iodine content in short electrolytic experiments. Longer reaction times decrease the iodine content of the resulting lignins, presumably by degradation reactions caused by the accumulation of iodate and hypoiodate. The reactivity of the milled wood lignin was much smaller than that of the hydrolysis lignin. Indirect bromination reactions of hydrolysis lignin were also investigated in aqueous solutions.[23] In addition to bromination, demethoxylation and lignin degradation also took place. Polybasic phenols produced by demethoxylation were oxidized to quinone structures. In aqueous solution, hypobromite and bromate caused the lignin degradation to low molecular weight products (~90%) in the form of amorphous gels. In DMSO extensive bromination (~30% Br content) took place with little degradation.

2.3.2. Exhaustive Electrolyses

Butanol lignin in NaOH was electrolytically oxidized on a lead dioxide anode with a current density of 20 mA/cm^2 for 48 h.[24] Such an exhaustive electrolysis degraded the lignin to low-molecular-weight compounds such as methylethylketone (23%), acetone (15%, acetic acid (15%), aromatic carboxylic acids (24%), oxalic acid (3%) and isobutylmethylketone (2%).

2.3.3. Electrolytically Modified Lignins

Kovalenko and coworkers[24-30] have also attempted to modify electrochemically the properties of the hydrolysis lignins from wood saccharification plants. In many cases, the resulting modified polymers were more reactive in applications such as filler, hardener, and vulcanizing agent in epoxy resins. Such chemical property changes were achieved by increasing the content of phenolic OH groups, and by decreasing the methoxy content by demethylation or demethoxylation reactions. At the same time, oxidative cleavages of C–C and C–O bonds led to the formation of low-molecular-weight carbonyl- and carboxyl-containing compounds. The medium chosen for most of these electrolyses was an aqueous alkaline solution. These conditions were suitable to dissolve the acid-hydrolysis hardwood lignins of low molecular weight. In order to dissolve the corresponding higher molecular weight softwood lignins, it was necessary to subject them to an alkaline treatment (180 °C, 10 atm). Under these conditions, some degradation (and recondensation) reactions occur, which are similar to those observed in alkaline pulping,[31] with hydrolysis of the β-O-4 alkyl aryl ether bonds of both phenolic and non-phenolic units. The materials submitted to electrolyses were not extensively characterized in many of these investigations.

The influence of the anode material was assessed in some cases.[26,28,29] The highest yields of products of mild oxidation such as vanillin and vanillic acid were obtained at Pt and Ni electrodes for the softwood lignin investigated. Such yields were comparable to those obtained in nitrobenzene oxidation reactions of the corresponding lignins. On Cu and C electrodes more drastic oxidation reactions occurred, with the highest contents of phenolic OH groups being produced.

The highest current efficiencies for oxidation of the hardwood hydrolysis lignins were in the range of 50% with ruthenium oxide anodes. The efficiency increased with temperature in the 25–75 °C range. Graphite electrodes did not perform as well as ruthenium oxide, but were better than nickel and platinum electrodes.[26] The current densities employed in these investigations were in the range 20–50 mA/cm^2.

Electrolysis of suspensions of softwood hydrolysis lignins in 5% KOH caused increased solubilization of material in water compared to that of the low-temperature alkaline treatment alone.[32] Electrolysis in acid medium did not cause solubilization.

Investigations of the anodic oxidations of lignosulfonic acids were performed at constant current in 2·5 M NaOH on Pt electrodes. Increased carboxylic acid and phenolic OH contents were observed similarly to the above-mentioned studies.[33] Decarboxylation of the resulting carboxylic acids was observed if the current densities were > 32 mA/cm^2.

Electrolysis of hydrolysis lignins from cotton hulls submitted to partial solubilization by alkaline treatment in NaOH also produced materials with increased carboxylic acid content (decreased pH from 13·5 to 9).[34] The resulting lighter brown lignin had an increased number of easily reducible functionalities as detected by polarographic techniques.

A more recent investigation of the anodic oxidation of kraft lignin was performed in 0·1 M KCl aqueous solutions. The reaction was shown to be a complex multistage process, and the product distribution depended on the applied anode potential. Oxidations at 0·625 V yielded lignins of increased molecular weight due to the predominance of oxidative coupling reactions at that potential. Increased potential to 0·8 V led to a decrease in molecular weight due to oxidative degradation reactions.[35]

2.4. Electrochemical Reduction of Lignins

Electrolyses on lead cathodes (30 mA/cm^2) were performed on Klason lignin suspensions in alkaline solutions (1 M NaOH or LiOH). The solubilization of the lignin after electrolyses at 55 °C was ~70% compared to 60% of the hydrolytic treatment alone.[35]

Electrochemical solvated-electron reductions were performed on dioxane and kraft lignins in aqueous ammonia solutions as a function of the concentration of ammonia.[36] At higher concentrations of ammonia degradation of the polymer into monomers and dimers was observed. Ether linkages of the lignin were cleaved. These results were similar to those of Shorygina *et al.*[37] when reducing lignins with sodium in liquid ammonia. In Shorygina's experiments a loss of about $\frac{1}{3}$ of the methoxy groups was reported. Freundenberg and coworkers[38] treated spruce lignin in potassium and potassium amide in liquid ammonia and observed a loss of $\frac{2}{3}$ of the methoxy groups. These Birch-type[39] reductions have been investigated for a number of model compounds of lignins.[40,41]

The polarographic investigation of a variety of lignin model compounds was applied to the analysis of composition of functional groups of lignins. For instance, in dimethylsulfoxide, irreversible waves were observed and the half-wave potential was shown to be a function of the functional group present in the model compound. The ease of reduction is as follows:

quinone > aldehyde (aromatic) > ketone (aromatic)
> olefin conjugated with aromatic ring[42]

These polarographic studies continued with the investigations of the sprucewood lignin isolated by Pepper's method, which were compared to polarograms of dioxane and Bjorkman lignins. The sprucewood lignin exhibited six peaks over a wide potential range. One of them was associated with adsorption, whereas the others were faradaic, and corresponded primarily to the reduction of carbonyl groups.[43] The order of ease of reduction was proposed as: quinone > carbonyl groups hydrogen-bonded to free-phenolic groups > free carbonyl groups. The number of reduction waves, their nature and the potential at which they occur were shown to differ from one type of lignin preparation to the other.

Electrochemical reduction of nitrolignins was investigated and shown to depend on the electrolyte pH and on the cathode material.[44] Zinc electrodes in acid solutions were very effective (90 % reduction) whereas lead cathodes were the least effective electrodes.

We have characterized the ethanol-extracted explosively depressurized aspen (*Populus tremuloides*) lignin (EEEDAL) with respect to its functional group composition and electrochemical reduction reactions on mercury cathodes in methanol/tetraethylammonium perchlorate supporting electrolyte.[45] The preparation of the lignin was similar to that described by Marchessault,[46] and included a carbon tetrachloride extraction procedure to remove the wood-extractives which are also extracted by ethanol. The explosively depressurized aspen wood was supplied by Iotech Corp., and underwent some air oxidation upon storage. The statistical representation of this irregular polymer with respect to functional groups is

$$C_{9\cdot 0}H_{6\cdot 7}O_{1\cdot 2}(O_{CO})_{0\cdot 3}(OH)_{1\cdot 2}(OCH_3)_{1\cdot 2}$$

A fractionation procedure was developed in order to recover a large fraction of the lignin by differential solubility: precipitation in aqueous acid of an acid-insoluble fraction, and extraction of the soluble material in ethyl acetate, yielding the acid-soluble fraction. The acid-insoluble material had a 10% smaller phenolic OH content than the original material, and a similar aliphatic OH content. The acid-soluble lignin had a higher phenolic content ($\sim 0{\cdot}9/C_9$), and a similar aliphatic OH content. The acid-insoluble lignin had a higher double-bond equivalent number, and appeared to be more condensed than the acid-soluble fraction. After electrolysis on a mercury pool cathode at a constant potential of $-2{\cdot}6$ V vs. Ag/AgCl (about 60–90 min, initial current density ~ 150 mA/cm^2) the amount of the low-molecular-weight acid-soluble material increases. This material has a similar phenolic OH content and a higher alcoholic content than the corresponding material isolated after fractionation only. The main electrochemical reactions are carbonyl reduction to hydroxy groups, and cathodic cleavage of β-O-4 alkyl aryl ether bonds, which are activated for reduction by the presence of adjacent α-carbonyl groups. Similar activation was observed by Evstigneev *et al.*[42] in the reduction of dimeric model compounds containing α-carbonyl groups and –O–4 ether bonds. In our investigations, lignin reduction (mostly by the reactions mentioned above) consumes $\frac{2}{3}$ of the charge; the other $\frac{1}{3}$ of the charge was used to produce hydrogen, through solvent discharge. These reactions were carried out at pH 12. The molecular weight distribution of all these lignin fractions, as well as lignins from other pretreatments, were investigated by high-performance size-exclusion chromatography.[47] The fractionated acid-soluble material clearly shows paucidispersity (monomers, dimers, trimers and tetramers can be distinguished in the polydisperse material). The acid-insoluble materials, fractionated or electrolyzed, are polydisperse, and have about twice weight-average molecular weight than the corresponding acid-soluble materials. Our research continues completing the characterization of the materials by evaluation of the proportion of C–C and C–O bonds present in these materials,[48] as well as the suitability of these materials for phenol replacement in thermosetting resins.[49] The derivative prepared by reaction of EEEDAL and formaldehyde has been synthesized and partially characterized spectroscopically. This derivative was shown to replace successfully 25% of phenol in a phenol–formaldehyde resin.[49] We are presently testing the other electrochemically modified polymeric materials and also investigating anodic modifications of EEEDAL[50] as well as the electrochemical reactions of other lignins obtained by other pretreatments such as organosolv, anaerobic digestion and acid hydrolysis.

ACKNOWLEDGEMENTS

The authors gratefully acknowledge the support of the Biomass Energy Systems Division and of the Office of Alcohol Fuels of the Department of Energy. Thanks are also due to Iotech Corp. for the sample of wood.

REFERENCES

1. Goldstein, I. S., Holley, D. L. and Deal, E. L. (1978). *For. Prod. J.* **28**, 53; Sarkanen, K. V. (1976). *Science* **191**, 773; Goldstein, I. S. (1977). In Proc. Conf. Chemical Feedstocks Alternatives, F. J. Antwerpen (Ed.). Am. Inst. Chem. Engineers, 61; Sarkanen, K. V. (1977), *ibid.* 74.
2. Browning, B. L. (Ed.) (1975). *The Chemistry of Wood.* R. E. Krieger Publishing Co., Huntington, New York; Sjostrom, E. (1981). *Wood Chemistry.* Academic Press, New York.
3. (a) Goldstein, I. S. (Ed.) (1981). *Organic Chemicals from Biomass.* CRC Press, Boca Raton; (b) St.-Pierre, L. E. and Brown, G. R. (Eds) (1980). *Future Sources of Organic Raw Materials—Chemrawn-I.* Pergamon Press, Oxford.
4. See, for instance, Rydholm, S. A. (1964). *Pulping Processes.* Interscience, New York.
5. Allen, B. R., Cousin, M. J. and Pierce, G. E. (1980). Pretreatment methods for the degradation of lignin. Final Report to the National Science Foundation. Batelle Columbus Laboratories, Columbus; Klass, D. L. (Ed.) (1981). Biomass as a nonfossil fuel source. ACS Symposium Series **144**.
6. Glasser, W. G. (1981). *For. Prod. J.* **31**, 24; Falkehag, S. I. (1975). *Appl. Polym. Symp.* **28**, 247; Herrick, F. W. and Hergert, H. L. (1977). The utilization of chemicals from wood: retrospect and prospect. In *The Structure, Biosynthesis, and Degradation of Wood*, Recent Advances in Phytochemistry, Loweus, F. A. and Runeckles, V. C. (Eds). Plenum Press, New York **11**, 443; Kringstadt, K. The challenge of lignin. In ref. 3(b), 627.
7. Chum, H. L. and Osteryoung, R. A. (1982). Survey of the electrochemistry of some biomass-derived compounds. SERI/TR-332-417 (available from NTIS), Solar Energy Research Institute, Golden, CO.
8. Baizer, M. M. (Ed.) (1973). *Organic Electrochemistry.* Marcel Dekker, Inc., New York; Weiberg, N. L. (Ed.) (1974, 1975, 1982). *Technique for Electroorganic Synthesis.* Wiley-Interscience, New York, Parts I, II and III.
9. Sychev, N. A. and Petrova, V. N. (1954). *Zhur. Prklad, Khim.* **27**, 461.
10. Eskamani, A. and Dernar, H. D. (1982). US Patent, 4,341,609.
11. Babkin, V. A., Serbol'skii, E. N., Medvedeva, S. A., Anisimova, M. I. and Isakova, N. V. (1979). *Khim. Drev.* 94.
12. Venderevskii, A. D., Semenov, V. L., Syrovatko, V. N. and Pershina, T. P. (1976). USSR Patent, 504,847.
13. Mazitov, L. A., Kuznetsova, G. N., Stepanova, O. A., Smirnova, V. A. and Kirsanov, V. A. (1975). USSR Patent, 493,530.

14. Examples: Shinn, D. L. (1945). *Pulp & Paper Ind.* **19**, 66; Santiago, E. (1971). *Tappi* **54**, 1641; Hara, T. (1971). Japan Patent, 71 32,442; Ichiki, M. and Ishii, M. (1974). US Patent, 3,793,174 (CA **78**: 75628e); Malkov, Yu. A., Nepenin, Yu. N., Kipriyanova, V. I. and Burov, V. A. (1976). *Bum. Prom-st.* 16 (CA **84**: 107330u).
15. Dence, C. W. (1971). In *Lignins: Occurrence, Formation, Structure and Reactions*, K. V. Sarkanen and C. H. Ludwig (Eds). Wiley, New York, p. 373.
16. Examples: Kashihara, S. (1972). Japan Patent, 72 38,881 (CA **80**: 72252h); Ohta, M. (1974). Japan Patent, 74 132,857 (CA **83**: 47933k); Vulchev, V., Khristova, E., Nankov, N., Kamenov, G. and Tunkusheva, D. (1974). *Tseluloza Khartiya* **5**, 8 (IPC AN 45-3780); Buczylo, E. (1975). *Przeglad Wlok* **29**, 164 (IPC An 46-4522); Selivanov, V. G., Svitel'skii, V. P., Ryumin, V. G. and Samborskii, N. V. (1976). *Bumazh. Prom.* 24 (IPC AN 47-9381); Davydov, V. D., Demin, V. A. and Bogomolov, B. D. (1976). *Izv, VUZ, Lesnoi Zh.* **19**, 159 (IPC AN 48-397); Anisimova, M. I. and Babkin, V. A. (1978). *Khim. Drev.* 69; Serdobol'skii, E. N., Anisimova, M. I., Babkin, V. A. and Permyakova, G. N. (1978). *Khim. Drev.* 41.
17. Demin, V. A., Davydov, V. D. and Bogomolov, B. D. *Tesizy Dokl.—Vses. Konf. Khim. Ispol'z. Lignina, 6th*, V. N. Sergeeva (Ed.). Riga, USSR: Zinatne (CA **86**: 191508e).
18. Pershina, L. A. (1967). *Itogi Issled. Khim. 50 Let, 1917–1967* 77 (CA **71**: 23014n).
19. Coskuner, U. (1973). *Diss. Abstr.* **33**, 12: 5793B.
20. Demin, V. A., Davydov, V. D. and Bogomolov, B. D. (1979). *Khim. Drev.* 41.
21. Zeigerson, E., Gileadi, E. and Bloch, M. R. (1974). Israeli Patent, 40, 198 (CA **83**: 12600x).
22. Shalimov, V. N., Kovalenko, E. I., Kovalenko, N. A. and Smirnov, V. A. (1980). *Zh. Prikl. Khim.* **53**, 1284.
23. Kovalenko, E. I., Shalimov, V. N. and Kovalenko, N. A. (1980). *ibid.* **53**, 850.
24. Bailey, A. J. and Brooks, H. M. (1946). *J. Am. Chem. Soc.* **68**, 445.
25. Kovalenko, E. I., Raskin, M. N., Smirnov, V. A. and Shifron, E. M. (1973). *Izv. Sev.-Kavkaz. nauch. tsentra vyssh. shkoly. Serv. tkhn.* 80 (CA **83**: 30162k).
26. Kovalenko, E. I., Shifron, E. M. and Raskin, M. N. (1973). *Novosti Elektrokhim. Org. Soedin. Tezisy Dokl. Vses. Soveshch. Elektrokhim. Org. Soedin., 8th.* L. G. Feoktistov (Ed.). Riga, USSR: Zinatne, 24 (CA **83**: 62159f).
27. Kovalenko, E. I., Koshutin, V. I., Smirnov, V. A. and Shifron, E. M. (1976). USSR Patent, 512,222 (CA **85**: 47685x).
28. Kovalenko, E. I., Raskin, M. N. and Smirnov, V. A. (1976). *Tr. Novocherk. Politekh. Inst.* **320**, 69 (CA **83**: 154596c).
29. Kovalenko, E. I., Smirnov, V. A. and Shalimov, V. N. (1977). *Zh. Prikl. Khim.* (Leningrad) **50**, 1741.
30. Smirnov, V. A., Zverev, V. A., Kovalenko, E. I., Mil'man, V. I. and Raskin, M. N. (1975). *Izv. Sev.- Kavkaz. nauch. tsentra vyssh. shkoly. serv. tekh. n.* **3**, 89 (IPC AN 47-7179).
31. Gierer, J. (1982). *Holzforschung* **36**, 43, 55.
32. Forostyan, Yu. N., Forostyan, E. I. and Soroka, I. I. (1976). *Khim. Prir. Soedin.* **408** (CA **85**: 110292j).
33. Davydov, V. D., Tedoradze, G. A., Tysyachnaya, G. Ya. and Razmanova, I.

A. (1976). *Tesizy Dokl.-Vses. Konf. Khim. Ispol'z Lignina, 6th.* V. N. Sergeeva (Ed.). Riga, USSR: Zinatne; 122 (CA **86**: 191511a).

34. Gorokhov, G. I. and Krayanskii, O. B. (1971). *Tr. Krasnodar. Politekh. Inst.* 180 (IPC AN 43-9416).
35. Itoh, H., Ishii, T. and Takeya, G. (1968). *Hokkaido Daigaku Kogakubu Kenkyu Hokoku* 141 (CA **70**: 79292y).
36. Chupka, E. I., Lipskia, A. A. and Nikitin, V. M. (1976). *Khim. Drev.* 72.
37. Shorygina, N. N. and Kefeli, N. Ya. (1947). *J. Gen. Chem. USSR* **17**, 2058; **18**, 528; Shorygina, N. N., Kefeli, N. Ya. and Semechkina, A. F. (1950). *Khur. Obschei. Khim.* **20**, 1199; Shorygina, N. N. and Semechina, A. F. (1949). *ibid.* **19**, 1101.
38. Freudenberg, K., Flickinger, E., Sobek, A. and Klink, F. (1939). *Chem. Ber.* **72**, 217.
39. Birch, A. J. (1947). *J. Chem. Soc.* 102.
40. Pernelman, P. A. and Dence, C. W. (1974). *Acta Chem. Scand.* **B28**, 453.
41. Freudenberg, K., Lautsch, W. and Piazolo, G. (1941). *Chem. Ber.* **74**, 1879.
42. Evstigneev, E. I., Bronov, L. V. and Nikitin, V. M. (1979). *Khim. Drev.* 71.
43. Evstigneev, E. I., Bronov, L. V. and Nikitin, V. M. (1979). *ibid.* 82.
44. Gogotov, A. F. and Chupka, E. I. (1979). *ibid.* 53.
45. Chum, H. L., Ratcliff, M., Schroeder, H. A. and Sopher, D. W. (in press). Electrochemistry of biomass-derived materials. I. Characterization, fractionation, and electrochemical reduction reactions of ethanol-extracted explosively-depressurized aspen lignin; Sopher, D. W., Schroeder, H. A. and Chum, H. L. (1982). In Electrochemistry applied to biomass. SERI/PR-622-1321 (available from NTIS), Solar Energy Research Institute, Golden, CO.
46. Marchessault, R. H. and St.-Pierre, J. Ref. 3(b), 613.
47. Himmel, M. E., Oh, K., Sopher, D. W. and Chum, H. L. (in press). High-performance size-exclusion chromatography of low molecular-weight lignins and model compounds. *J. Chromatography*; Chum, H. L., Johnson, D. K., Oh, K. and Himmel, M. E. (in press). The molecular weight distribution of hydrothermally degraded aspen (*Populus tremuloides*) and mixed hardwoods by GPC.
48. Chum, H. L. and Glasser, W. G. Unpublished results.
49. Chum, H. L., Wallace, K. and Schroeder, H. A. Unpublished results.
50. Johnson, D. and Chum, H. L. Unpublished results.

63

The Electrolysis of Lignosite at a Reticulated Vitreous Carbon Electrode

VICTOR R. KOCH
Covalent Associates, Inc., Woburn, Massachusetts, USA

1. INTRODUCTION

Lignin is a three-dimensional, randomly linked macromolecule comprising methoxylated phenylpropane units. While lignins from grasses, softwoods and hardwoods differ in terms of the level of phenyl group linkages and degree of methoxylation, the principal structural features involve ether linkages α and β to the phenylpropane aromatic ring. These C–O bonds comprise approximately 65% of monomer–monomer bonding and include alkyl-O-aryl bonds linking phenylpropane units, alicyclic ethers of the tetrahydrofuran type and diaryl-ethers.[1] By selectively cracking these bonds one could fragment lignin into mono- and diaryl phenols which could serve as an alternative source of phenolic feedstocks or high-value speciality chemicals.

While phenolics have been previously obtained from lignin by various hydrogenation techniques,[2-4] poor selectivity and the high cost of hydrogen have precluded commercialization. Rather than employing the rigorous reaction conditions of petrochemical hydrocracking technology, we have instead utilized the infinitely variable power of a potentiostated working electrode at ambient temperature. Thus, our approach to electrowinning phenolics from lignin involves a simple electrochemical cell, an Li^+-containing supporting electrolyte in either DMF or H_2O, and a high surface area reticulated vitreous carbon electrode. Sulfite lignin (lignosite) was the substrate employed in our experiments. Lignosite is a flash-dried powder of Ca-based acid sulfite spent liquor which is supplied by Georgia-Pacific Corporation. This material contains, as a major component, the Ca salts of softwood lignin sulfonates along with minor contributions by pentose sugars.[5]

Previous electrochemical work with lignin has recently been reviewed by Chum and Osteryoung.[6] Most of this research has been carried out by Japanese or Russian workers and detailed information regarding the experimental particulars is sparse. In those cases where preparative electroreductions were carried out, a galvanostatic (constant-current) technique was employed and the resulting products were poorly characterized.

The most recent electrochemical study of lignin has been conducted by Chum and coworkers at SERI.[7] Using lignin extracted with ethanol from steam-exploded aspen, a number of potentiostatic reductions at a mercury cathode were carried out. They found that the reaction products contained reduced carbonyl functionalities as well as a lower methoxyl content compared to starting material.

2. EXPERIMENTAL

2.1. General

Infrared (IR) spectra were recorded on a Perkin Elmer Model 237 Spectrophotometer. Optical densities (OD) were measured with a Bausch & Lomb Spectronic 20. The solubility of various lignins was measured in H_2O and DMF by saturating a known volume of solvent with a known weight of lignin. After stirring for 30 min, the excess lignin was filtered, dried and weighed. All electrochemical processes and solubility studies were accomplished at 27 °C.

2.2. Materials

Lignosite was obtained from Georgia-Pacific Corporation (Bellingham, WA) through our consultant, Professor H. A. Schroeder of Colorado State University. Softwood kraft lignin was supplied by Westvaco Corporation (Charleston, SC). Samples of Klason lignin (spruce and aspen) were obtained from Professor Schroeder. All of these samples were used as received. Dimethylformamide (DMF) (distilled-in-glass) was obtained from Burdick & Jackson and dried over 4 Å molecular sieves (Fisher). $LiClO_4$ (anhydrous) was obtained from Fisher and used without further purification.

2.3. Electrochemical Cell and Instrumentation

A three-electrode, two-compartment cell (Fig. 1) was used for cyclic voltammograms (CV) and for preparative electrolyses of lignosite. All

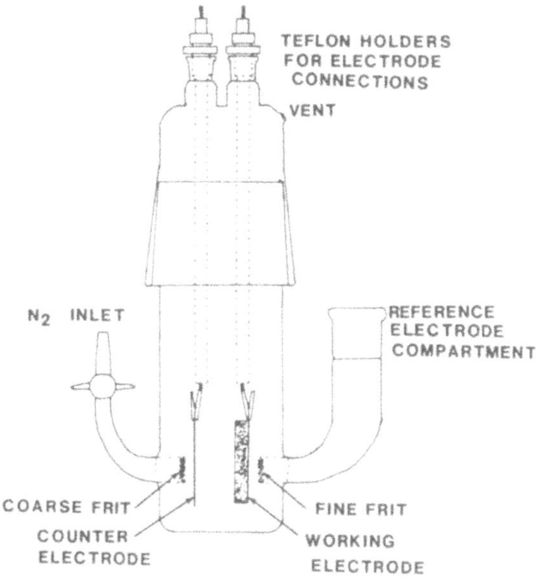

FIG. 1. Electrochemical cell (Pyrex) used for cyclic voltammetry studies and preparative electrolyses. Capacity of central compartment is 125 ml.

potentials are quoted against a saturated calomel reference electrode (SCE) (Fisher) separated from the central compartment by a fine glass frit. The electrolyte, 100 ml of either 1 M $LiClO_4$/DMF, LiOH/H_2O or LiCl/H_2O, was always deaerated for 20 min with N_2 prior to beginning any electrochemistry. The working electrode for CV consisted of a 3 mm dia. vitreous carbon disk (Tokai) sealed in Pyrex and polished to a mirror-like finish with a 0·4 μm alumina/H_2O slurry. For preparative electrolyses, reticulated vitreous carbon (Fluorocarbon, No. 100S) of approximately 97% void space was used. The immersed portion had dimensions of 3 cm × 3 cm × 1·3 cm. The counter electrode consisted of 3 cm × 3 cm × 10 mil Ni shim stock (Roblinger) wrapped with Celgard 2400 separator material. Approximately 100 ml electrolyte was used in all cases.

Cyclic voltammetry was accomplished with a BAS Model CV-1B signal generator and potentiostat interfaced with a Houston Omnigraphic 2000 X-Y recorder.

Controlled potential electrolysis was achieved with an Amel Model 551 potentiostat. The amount of charge passed during an electrolysis was measured with an Acromag coulometer. Reaction times varied from 1 to 4 h.

2.4. Electroreduction Procedure

After the electrolyte was purged with N_2, the cell was equipped with a Teflon spin bar and placed over a magnetic stirrer. Background currents were monitored starting at the rest potential, typically 0·08 V, and moving in the negative direction until electrolyte decomposition was reached: −1·50 V for H_2O and −3·00 V for DMF. Lignosite was then added with continued stirring and N_2 perfusion. The applied potential was increased from rest to the desired value and the time noted. Over the course of the electrolysis the contents of the cell, originally yellow-brown, turned a dark red-brown in color.

At the end of a preparative run, the cell contents were poured into 100 ml dilute HCl (10% v/v) and then extracted with either $CHCl_3$ or Et_2O (2 × 100 ml). The organics were combined, dried over Na_2SO_4, and the solvent removed under reduced pressure. A sweet-smelling, yellow oil (acid to litmus) usually remained.

The dark brown aqueous fraction was further treated by saturating the solution with LiCl at 0 °C. This resulted in the formation of a fine light tan precipitate and a brown oil which floated to the top. The oil was adsorbed on filter paper and solidified when allowed to dry at room temperature. The light tan precipitate was removed by filtration and dried at 40 °C. The LiCl-saturated filtrate was then extracted with 2 × 50 ml EtOAc which, after solvent removal, afforded a small additional quantity of yellow oil. Material balances by weight of recovered products and unreacted lignosite ranged from 85 to 94%. Small amounts of residual lignosite and/or reaction products remained dissolved in the aqueous layer as evidenced by the persistent brown coloration and subsequent analysis for phenolics.

A control experiment was conducted wherein 2 g lignosite was added to deaerated 1 M LiCl/H_2O in the electrochemical cell in the *absence* of any applied potential. After 2 h of stirring, the cell contents were worked up as usual. The organic fraction yielded ~20 mg of a red solid whose phenolic content was similar to that of lignosite itself (see below).

2.5. Analysis for Phenolics

Owing to the potential variety of the lignin reduction products, a direct photometric method which quantitatively assesses the concentration of a mixture of substituted phenols was employed.[6,7] This technique relies on the reaction of phenols with 4-aminoantipyrine at pH 10 in the presence of $K_3Fe(CN)_6$ to form a deep red antipyrine dye. Phenol itself is employed as the standard for this colorimetric procedure. Thus any color produced by the reaction of other phenolics with 4-aminoantipyrine is reported as phenol. Because substitution on the aromatic ring generally reduces response, the measured concentration represents the minimum concentration of phenolic compounds present. All determinations were carried out in Bausch & Lomb test tubes at 510 nm.

The direct photometric method covers the phenol concentration range of 0·0 to 50 mg/l with a sensitivity of 1 mg/l. Measured ODs for the lignins and lignosite reduction products were converted to percent phenolics by weight.

Additional analysis of the liquid product fraction was conducted at the Olin Research Center (New Haven, CT). Thus, high pressure liquid chromatography (HPLC) was conducted on a Varian instrument employing a Waters Associates Styragel column (pore exclusion volume of 100 Å). Identification of the major components of the liquid fraction was accomplished with a Finnegan GC/MS.

3. RESULTS AND DISCUSSION

3.1. Lignin Solubilities

The solubility of lignin in a given solvent is an important parameter in electrochemical synthesis. Indeed, the rate of an electrochemical reaction is directly proportional to (among other things) the concentration of the depolarizer. We measured the solubilities of four lignins in DMF (since this solvent has an excellent negative electrochemical window) and also in H_2O (since Chum and co-workers have recently reported on the electrochemistry of hardwood lignin in this medium).[7]

Table I presents the solubility obtained at 27 °C. As expected, the Ca^{2+} salt of lignin sulfonate was very soluble in water and marginally soluble in

TABLE I

SOLUBILITIES OF VARIOUS LIGNINS IN WATER AND IN DIMETHYLFORMAMIDE AT 27 °C

Lignin	*Solubility (g/100 ml)*	
	H_2O	*DMF*
Lignosite	23·3	0·5
Kraft	1·5	13·8
Klason (aspen)	~0·02	0·3
Klason (spruce)	~0·02	0·1

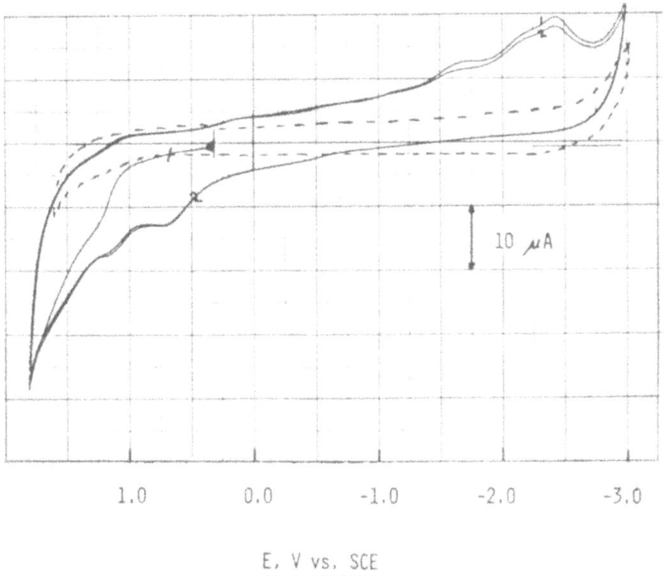

FIG. 2. Cyclic voltammogram of background (----), and 0·5 mg/ml lignosite (-----) in 1 M $LiClO_4$/DMF_2; $v = 100$ mV/s; $A = 0{\cdot}07\,cm^2$.

DMF. Kraft lignin from Westvaco Corporation has good solubility in DMF but is marginally soluble in H_2O. This particular sample of kraft lignin has an average degree of polymerization of 2200 and possesses 0·001 carboxylate/phenylpropane unit of lignin. Thus kraft lignin behaves more like an organic than the lignosite salt. Each of the Klason lignins was poorly soluble in DMF and virtually insoluble in H_2O. This reflects the highly condensed nature of this material obtained from the hydrolysis of woody biomass with 72% H_2SO_4. From a synthetic electrochemical perspective, either lignosite in H_2O or kraft lignin in DMF would be good solute/solvent combinations.

3.2. Electrochemistry of Lignins in DMF

The electrochemistry of lignosite in $LiClO_4$/DMF was first investigated since earlier in-house CV experiments revealed that the presence of Li^+ shifted lignosite reduction waves to more anodic (less energetic) potentials compared to $n Bu_4NClO_4$/DMF. This suggested that one could electrochemically access reactive functionalities (electrophores) under less stringent reaction conditions.

Figure 2 shows two successive CV scans of lignosite in 1 M $LiClO_4$/DMF. The dashed line represents a background scan without lignosite. We note one peak at −2·42 V with shoulders at −1·65 and −2·15 V. One new anodic wave at 0·65 V is also evident—a consequence of scanning through the peak at −2.42 V. Peak currents were determined as a function of increasing sweep rate, v, from 50 to 1000 mV/s. When these currents were plotted against $v^{1/2}$, a straight line was obtained, indicating that the electron transfer process was diffusion controlled and not due to adsorption on the electrode surface.

Controlled potential electrolysis of lignosite in 1 M $LiClO_4$/DMF was carried out at −2·00, −2·25, and −2·50 V. In each case the electrolysis did not go to completion due to passivation of the reticulated vitreous carbon cathode by lignosite reaction products. This manifested itself in terms of a rapid decrease in current with concomitant formation of a tenacious yellow film on the electrode. No more than 200 coulombs (~2 mF) of charge were passed. If we assume that the average molecular weight ($\overline{MW}$) of lignosite is 2000 amu,[5] only 0·5 Faradays/mole of electricity was passed.

Lignosite electrolysis products in $LiClO_4$/DMF comprised small amounts (20–40 mg, 1–2 % by weight) of a yellow oil which appeared to be a mixture of hydroxyaromatic-substituted carboxylic acids according to infrared spectroscopy. Poor yields were endemic to electrolyses in the DMF solvent, which most likely took part in electrode passivation.

3.3. Electrochemistry of Lignosite in H_2O

Because of the difficulties encountered in DMF media, we explored the electrochemistry of lignosite in H_2O and found several remarkable results. To begin with, three CV curves are reproduced in Fig. 3: background $LiCl/H_2O$ electrolyte; with lignosite; and after potentiostatic electrolysis at −1·20 V. In curve B, shoulders at −0·5 and −1·3 V are evident. A plot of i vs $v^{1/2}$ for the first wave was not linear, indicating an adsorption process. The second wave was too close to background to get useful analytical information. Finally curve C clearly shows that the two electroactive species in lignosite have been converted to other products which are not electroreactive in this potential region.

A variety of preparative electrolyses were conducted in H_2O-based electrolytes and are summarized in Table II. No electrode passivation was observed and, indeed, cathodic currents after dropping slowly at first remained constant due to our proximity to background processes (water-splitting). Regardless of salt and length of electrolysis, the same yellow oil was obtained in varying amounts.

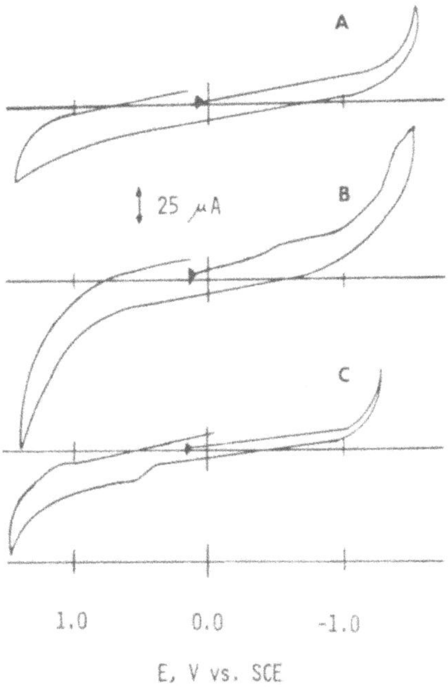

FIG. 3. Cyclic voltammetry curves at vitreous carbon in 1 M $LiCl/H_2O$. (A) Background; (B) 20 mg/ml lignosite; (C) after potentiostatic reduction of lignosite; all curves at $v = 100\,mV/s$, and $A = 0{\cdot}07\,cm^2$.

3.4. Generation of Phenolics via Electrolysis

We have employed a simple colorimetric technique which directly measures the substituted and unsubstituted phenol content of as-received lignins and lignosite electrolysis products. The example is solubilized in an aqueous solution at pH 10, then reacted with 4-aminoantipyrine in the presence of $K_3Fe(CN)_6$. A red antipyrene dye results, the intensity of which is directly proportional to phenol concentration. Since phenol itself is used as a standard and since substituted phenols reduce the intensity of the dye, the values obtained are minimum concentrations of phenolics.

The phenol contents of three lignins as measured by this technique are presented in Table III. Both lignosite and kraft lignin are essentially equivalent with the more highly condensed Klason lignin, having about half of the phenol content compared to the other two. The low molecular weight kraft lignin has, according to Westvaco, about 15% phenolics by weight, which is close to what we measure. Moreover, the phenolic content of soluble lignin sulfonates of molecular weights from 4000 to 40 000 amu is

TABLE II

SUMMARY OF LIGNOSITE ELECTROLYSES UNDER VARIOUS CONDITIONS

Amount lignosite (g)	*Electrolysis medium*	*Applied potential, V vs. SCE*	*Faradays/ mole*[a]	*Ether soluble products (mg)*
0·5	1 M $LiClO_4$/DMF	−2·25	0·5	40 mg yellow oil (20-1)
2·0	1 M $LiOH/H_2O$	−1·50	6·1	140 mg yellow oil (23-1)
2·0	1 M $LiCl/H_2O$	−1·20	21·6	194 mg yellow oil (25-1); 150 mg pungent yellow oil (29-1)
2·0	1 M $LiCl/H_2O$	−1·20	14·5	100 mg yellow oil (30-3)[b]

[a] Assuming $\overline{MW} = 2000$ amu.

[b] When added to solid products (30-1 and 30-2, 1·78 g) isolated from the aqueous fraction, the material balance for this run becomes 94 %.

17–37 %.[10] It would appear that we are measuring the lower value since our lignosite sample has an average molecular weight of less than 15 000 amu.

Table IV lists lignosite electrolysis products and the calculated phenol content. We first note that the red solid isolated from the control experiment contains 16 % phenolics by weight, similar to that of lignosite itself. The other products range in phenolic content from 43 % for product 25-1 to a high of 87 % for product 23-1. We stress that these values may not reflect absolute phenolic content. Their main usefulness lies in demonstrating the *change* in phenolic content of lignosite before and after electrolysis.

TABLE III

PHENOL CONTENT OF AS-RECEIVED LIGNINS

Sample	*Optical density at 510 nm*	*Phenol concentration (mg/ml)*	*Calculated phenol content of lignin, and by weight*
Phenol standard	0·092	0·100	—
Lignosite	0·120	0·130	13
Kraft	0·136	0·150	15
Klason (spruce)	0·071	0·077	7·7

TABLE IV

PHENOL CONTENT OF LIGNOSITE AND LIGNOSITE ELECTROLYSIS PRODUCTS

Material	*Applied potential, V vs. SCE*	*Calculated phenol content (% by wt)*
Lignosite (as received)	—	13
Red Solid (20-2)	—	16
Yellow oil (23-1)	−1·50	87
Yellow oil (25-1)	−1·20	43
Tan solid (30-1)	−1·20	52
Solidified brown oil (30-2)	−1·20	79
Brown aqueous solution (30-4)	−1·20	65

3.5. Structural Elucidation of Electrolysis Products

Samples of liquid products obtained from electrolyses conducted in aqueous media were subjected to HPLC and GC/MS analysis. When injected into a Waters Associates Styragel column, two major peaks each possessing molecular weights less than 300 amu were found. These two peaks represented 87 % of the injected sample and, in conjunction with GC/MS data, were assigned to mono- and diaromatic species.

When these liquid products were silylated and analyzed by GC/MS, only three major products were identified by computer library searches made on the background subtracted mass spectra: vanillic acid (I); 3-methoxy-4-hydroxy-phenylacetic acid (II); and what appears to be a dimer of 3-methoxy-4-hydroxy-phenylacetone (III).

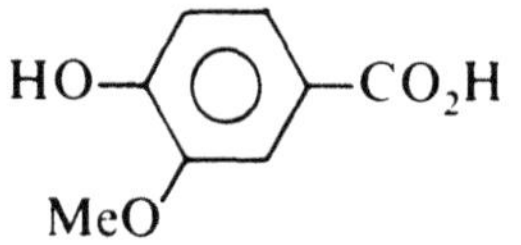

I

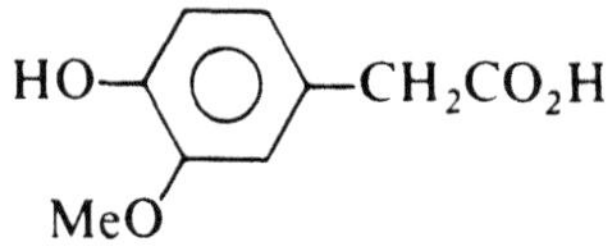

II

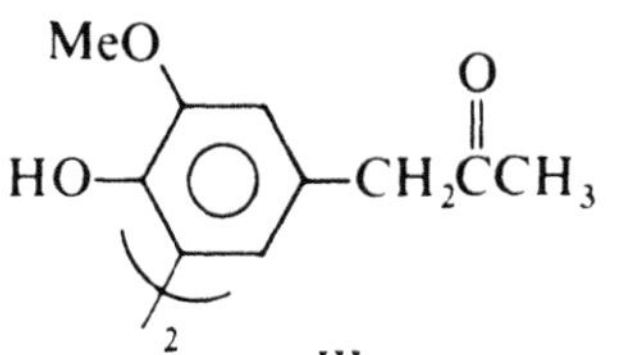

III

The integrated yields of these compounds are 15, 16 and 29 % respectively. These three compounds when combined represent an isolated yield of 12 % based on added lignosite. While vanillic acid has been obtained from the oxidative alkaline hydrolysis of lignin sulfonates, the other two compounds have, to our knowledge, not been reported in lignin degradation chemistry.

ACKNOWLEDGEMENTS

This work was supported by the National Science Foundation. We thank Professor H. A. Schroeder (Colorado State University) for providing lignin samples and for useful discussions of the results. We also thank Dr R. E. Maizell (Olin Research Center) for arranging analytical services.

REFERENCES

1. *Lignins: Occurrence, Formation, Structure, and Reactions* (1971), K. V. Sarkanen and C. H. Ludwig (Eds). Wiley-Interscience, New York.
2. Prahacs, S., Barclay, H. G. and Bahtia, S. P. (1971). *Pulp Pap. Mag. Can.* **72**, T 199.
3. Enquist, T., Turunen, J. and Ashorn, T. (1962). *Tappi* **45**, 128.
4. *Biomass Digest* (1980). Vol. 2, p. 1.
5. Lo, C. F., Schroeder, H. A., Laver, M. L. and Aft, H. (1971). *Tappi* **54**, 1731.
6. Chum, H. L. and Osteryoung, R. A. (1981). Survey of the electrochemistry of some biomass-derived compounds. SERI Report No. TR-322-417, August.
7. Electrochemistry applied to biomass, Annual Report (1981), H. L. Chum (Ed.). SERI Report No. PR-622-1321, September.
8. Adapted ASTM D-1783-Standard as published in *ASTM Book of Standards* (1968), Part 23.
9. Eisenstaedt, E. (1938). *J. Org. Chem.* **3**, 153.
10. Ref. 1, p. 613.

64

Specification of Standard Biomass Materials for Thermochemical Research

G. M. BARTON, Chairman

INTRODUCTION BY THE EDITORS

An important event of the Estes Park meeting was the first attempt to arrive at a specification for standard biomass materials for thermochemical research. Prior to the conference, the National Research Council of Canada retained a wood chemist, Mr George Barton, to survey a number of institutions and their researchers (listed in Appendix A) as to their viewpoints on the need for what could loosely be described as 'standard wood'. The various standards of TAPPI and ASTM were reviewed and along with the survey results were used as a foundation for a round table panel presentation and discussion.

Appendix B details the questions asked and the responses of the attendees at Estes Park to this issue.

What follows are the synopses of the discussion of the expert panel:

Mr George Barton	—Chairman
Dr Charles Nelson	—Gas Research Institute, Chicago, Illinois
Dr Fred Shafizadeh	—Wood Chemistry Department, University of Montana, Missoula, Montana
Dr Wolfgang Glasser	—Dept. of Forest Products & Dept. of Chemical Engineering, Virginia Polytechnic Institute and State University, Blacksburg, Virginia
Dr Herbert Schroeder	—Dept. of Forest and Wood Sciences, Colorado State University, Fort Collins, Colorado

along with a special intervention by Dr Hillis and the points raised by the delegates.

Appendix C is a list of North American laboratories and organizations which are capable of providing assistance in the area of standard wood materials.

Mr Barton kindly prepared the contribution that follows which distils his impressions and findings into a series of recommendations for researchers that we, the editors of this book, endorse and hope that both researchers and referees of journal articles will adhere to—at least until the issue of standard reference materials is better resolved.

SYNOPSIS

Prior to the introduction of individual Panel Speakers, a very short review of wood's variability and its importance to thermochemical biomass conversion was made. It was acknowledged that Dr Hillis's chapter 'Wood and Biomass Ultrastructure' provides an excellent introduction to wood's variability, a theme which was reinforced by several speakers throughout the remainder of the Conference.

A slide comparing wood components from a typical hardwood and softwood grown in a moderate climate was shown. Both qualitative and quantitative differences among lignin, hemicelluloses and extractives were discussed. For example, lignin is not only higher (24–33 %) in softwoods compared to hardwoods (19–28 %), but is chemically different. Softwood lignin is derived from phenyl propane groups containing one methoxyl group, whereas hardwood lignin is derived from phenyl propane groups containing two methoxyl groups. It was emphasized that this is not merely an academic consideration since the easier chemical pulping and animal digestibility of hardwoods is due to these differences. In the case of hemicelluloses, hardwoods contain higher amounts (19–26 %) compared to softwoods (12–18 %). More importantly, hardwood hemicelluloses consist of hardwood xylans, which are not fermentable, while the galactoglucomannans of the softwoods are readily fermentable. Finally, it was emphasized that although hardwoods generally have lower amounts of extractives than softwoods, the important differences are seen in the variety of unique chemicals found in the softwoods. Reference was made to only two of these, namely dihydroquercetin in Douglas fir, an extractive which inhibits sulphite pulping of this species and the tropolones in western red cedar which chelate iron and interfere with enzyme reactions.

The Panel Speakers were then introduced and spoke about their individual concerns. Questions from the conference delegates were deferred until all panelists had spoken.

Dr Charles Nelson reviewed some of the background in arriving at the concept of a 'standard coal' and subsequent problems associated with implementing the recommendations of the coal workshop. It would appear that although the majority of scientists surveyed favoured standard coal samples, they were reluctant to work on coal samples outside their immediate vicinity. It seemed that political and financial considerations were of higher priority than scientific ones. Dr Nelson concluded by saying that even if 'standard coal' samples were available without charge few scientists would be willing to use them.

Dr Fred Shafizadeh emphasized the importance of defining research samples, especially in regard to accurate species identification, heartwood–sapwood delineation and particle size. On the other hand he deplored unnecessary repetition of analytical studies to obtain data on species already adequately covered in the literature. Dr Shafizadeh concluded by stressing the desirability of being able to procure 'standard' samples of wood components such as lignin and hemicelluloses on a fee basis.

Dr Wolfgang Glasser's main concern was lignin variability and the need for more precise definition of lignin research samples. By means of selected slides he reviewed modern techniques and analytical procedures used at the Polytechnic Institute to define and characterize lignin. He concluded by listing a variety of lignin preparations such as synthetic lignin, both normal and radioactive, organosolve and milled wood lignin, which could be provided by his Institute on a fee basis. Additionally, he offered a comprehensive analysis of commercial lignins also on a fee basis.

Dr. Herbert A. Schroeder reviewed the survey results and commented on those species nominated for standard hardwoods and softwoods. He pointed out that aspen was not a typical hardwood, since it pulped easier than other hardwoods and also had a higher animal digestibility. He also reminded the delegates that Douglas fir from the coast behaved differently from interior fir, in terms of permeability. He also cautioned scientists to be careful in correctly identifying the oak species. White oak has low permeability and is used for cooperage, whereas red oak is very permeable due to its different anatomical structure.

Although this concluded the formal Panel presentation, *Dr. Hillis* requested a short time to comment on standard wood. Dr Hillis vigorously supported the dual concept of better sample definition and standardization of research samples and added his own special concern of adequate storage of research samples. He reminded the Conference that wood is not a static substance and in fact is subject to rapid chemical and biochemical change unless it is stored in the dark at low temperatures.

QUESTION PERIOD SYNOPSIS

Michael Antal, Jr, University of Hawaii, thanked the Panel for presenting a balanced picture of the 'standard fuels concept'. In particular, he was impressed that the Panel recognized the probability that even if standard fuel samples were made available, scientists would continue to give priority to those samples in their own 'back yards' because of political and financial considerations. Dr Antal strongly supported the need to validate experimental results through a common 'standard fuel'. He nominated Avicel as a standard for cellulose and Indulin AT for lignin. He concluded by recommending that a good report of the future would include one experiment with a 'standard fuel' in order that other scientists could compare it to their own research effort.

David Beckman, University of Toronto, expressed concern over the choice of 40–60 mesh size for standard samples. He felt that commercial processes would have difficulty achieving these small sizes and that it would be easier to relate to experiments with standard samples if they were larger. He also asked Dr Schroeder of the Panel to elaborate on his statement about the difficulty of scaling up the results of standard wood on a commercial basis? *Dr Schroeder* replied that a standard wood sample would probably consist of clear, normal wood whereas material used for commercial processes would inevitably consist of limbs and tops containing a large amount of juvenile wood, as well as slabs and edgings containing bark. Thus chemical composition, moisture content and heat content (Joules) would be significantly different from the standard sample.

Barbara Krieger, University of Washington, strongly recommended that the comments of the Panel Discussion be included in the *Proceedings*. She also recommended that scientists report both the density and moisture content of their samples in properties such as thermal conductivity and heat capacity and thereby make rough calculations as to temperature differences in order that other researchers could look at the thermal heat flux differences and differences in product distribution. She further recommended that single particle sample moisture content and density be reported as well as bulk moisture content and density, since both of these data will aid anyone modelling gasification or combustion processes.

Edward Lipinsky, Battelle, suggested that materials other than hardwoods or softwoods should be considered for discussion. One of these is corn which amounts to over a hundred billion tons per year. He pointed out, however, that corn is a tricky raw material from which to obtain standard samples since there are great differences in chemical composition between green corn stalks and those that have been left in the fields for a few days. In particular, there is a very sharp reduction in hemicelluloses. He suggested that corn sampling should be done on the advice of experts in agronomy and agriculture. In terms of wood, he nominated red oak since it made inferior paper and was likely to become a weed species in pine forests. His last comment was to change the emphasis from a standard sample to a 'representative sample' which should be so well characterized that everyone will know what it is.

Perry L. Blackshear, Jr, University of Minnesota, in keeping with the suggestion of Ms Krieger of density and moisture data, wished to add the additional measurement of fall velocity of chips in the moist state, dry state and charred state. If the rate at which the particle in free fall would heat in a controlled temperature environment were known, it would be possible for any furnace manufacturer to develop the appropriate specifications. His main point, in conclusion, was that we standardize our specifications rather than standardize the wood.

Dr H. A. Schroeder, University of Colorado wished to clarify the important differences between 'free' and 'bound' water in wood, because the subject of moisture content had been raised several times during the discussion. He explained that free water is found in the cell cavity and can be moved reasonably rapidly from wood whereas bound water is in the cell wall and its rate of movement is of the order of one hundred to one thousand times slower than free water. He concluded by emphasizing that any heat calculation involving bound water must take into account not only heat of vaporization but also the heat of wetting which at oven-dry conditions could be 260 calories per gram of water.

APPENDIX A

1. Forintek Canada Corp., Western Laboratory, 6620 N.W. Marine Drive, Vancouver, B. C. V6T 1X2. Telephone: (604) 224-3221.

   Persons contacted:
   Dr R. M. Kellogg, Manager Wood Science Department; Dr E. P. Swan, Research Scientist; and Dr J. F. G. Mackay, Research Scientist.

2. Solar Energy Research Institute, Thermochemical and Electrochemical Research Branch, Solar Fuels and Chemicals Division, 1617 Cole Boulevard, Golden, Colorado 80401. Telephone: (383) 321-1440.

Persons contacted:
Dr Thomas A. Milne, Chief Thermochemical and Electrochemical Research Branch, Solar Fuels and Chemicals Division; and Dr Helena Li Chum, Research Scientist, Bio/Chem Conversion Branch.

3. Pulp and Paper Research Institute of Canada, 570 St. John's Boulevard, Point Claire, Quebec, Canada H9R 3J9. Telephone: (514) 697-4110.

   Persons contacted:
   Dr M. Kamran Azarniouch, Head Energy Research Section; Dr Cyril Heitner, Manager Physical Analysis Section; Dr Bruce I. Fleming, Head Wood Chemistry Section; and Mr Steven Prahacs, Director Environmental and Energy Research Division.

4. Domtar Inc. Research Centre, Sortie 40, Trans-Canada Highway West, Senneville, Quebec, Canada H9X 3L7. Telephone: (514) 457-6810.

   Person contacted:
   Mr H. D. Lightfoot, Group Leader Design Engineering and Business Analysis.

5. Forintek Canada Corporation, Eastern Laboratory, 800 Montreal Road, Ottawa, Ontario, Canada K1G 3Z5. Telephone: (613) 744-0963.

   Persons contacted:
   Dr Harold H. Brownell, Research Scientist; Dr Norman Levitin, Research Scientist; Mr C. T. Keith, Head Wood Anatomy Section; and Mr D. R. Huffman, Research Scientist in Wood Energy.

6. National Science Foundation, Renewable Materials Engineering Program, Division of Chemical and Process Engineering, 1800 G. Street, N.W., Washington, D.C. 20550. Telephone: (202) 357-9606.

   Person contacted:
   Dr Oskar R. Zaborsky, Program Director, Renewable Materials Engineering Division of Chemical and Process Engineering.

7. USDA-Forest Service, Forest Products Laboratory, PO Box 5130, Madison, Wisconsin 53705.

   Persons contacted:
   Dr Andrew Baker, Research Scientist in Energy and Chemical

Production Section; Drs Jack Rowe, Duane Zinkel and T. Kent Kirk, Research Scientists in the Chemistry and Paper Research Section.

8. University of Montana, Missoula Wood Chemistry Laboratory, Missoula, Montana 59812. Telephone: (406) 243-6212.

   Person contacted:
   Dr Fred Shafizadeh, Director of the Wood Chemistry Laboratory.

9. Pacific Forest Research Centre, Department of the Environment, 506 West Burnside Road, Victoria, B.C., Canada V8Z 1M5. Telephone: (604) 388-3811.

   Person contacted:
   Dr John Manville, Research Scientist.

APPENDIX B

Survey Responses on 'Standard Wood'* for Fundamentals of Thermochemical Biomass Conversion Workshop, Estes Park, Colorado, October 18–22, 1982

What species of wood (common and Latin names) or commercial sample (e.g. Westvaco Lignin) are you currently using?
Many reported interest in wood components, commercial cellulose and lignin samples.

In what physical form is it required (e.g. solid, 60 mesh, powder, etc.)?
In the first place was powder, followed by ground wood (40–80 mesh). Also some interest in wholewood, chips and sawdust.

What components of wood are you most interested in (e.g. cellulose, lignin, etc.)?
Equally interested in cellulose and lignin with hemicellulose a close third.

Are you also interested in bark or foliage?
Although nearly a third of the replies showed no interest in either bark or foliage, almost a half were interested in bark and a quarter in foliage.

* Includes wood, wood components such as cellulose, hemicelluloses and lignin as well as bark and foliage.

What standard methods of analyses are you using (e.g. Lignin, ASTM 1106-560)? Please indicate any variations from the standard method (e.g. different mesh size).
This question was poorly answered. Of those commenting, ASTM was most frequently mentioned with some references to TAPPI and CPPA. Elemental and non-standard methods also.

From the following short list of North American hardwoods, please rate your preference for a 'standard species', using numbers 1 to 4.
1. Aspen (Populus tremuloides)
2. Yellow birch (Betula alleghaniensis)
3. White oak (Quercus alba)
4. White ash (Fraxinus americana)
Others, including those not indigenous to North America.
Nil response.

Please do the same from the following list of North American softwoods.
1. Douglas fir (Pseudotsuga menziesii)
2. Loblolly pine (Pinus taeda)
3. White spruce (Picea glauca)
4. Western hemlock (Tsuga heterophylla)
Others, including those not indigenous to North America.
Nil response.

How important is the concept of 'standard wood' to your current research? To your future research?
The majority of comments were 'very important, important for the future, important or not applicable'. A few answered 'not important'.

(1) Is there a need to provide 'standard wood' from the shelf? (2) Should the wood be stored under nitrogen in chip form or should it be available as whole wood, including bark?
(1) Majority answered 'yes'. (2) Majority favoured whole wood including bark.

Should components of 'standard wood', prepared by ASTM or TAPPI methods, be available from the shelf?
The majority answered 'yes'; however, a few thought that preparation of wood components was the researcher's responsibility.

Should a genetic bank from 'standard species' be established?
Equally divided among yes, maybe, no opinion and no.

APPENDIX C

List of Names and Laboratories* Interested in Providing Information, Analytical Services and Standard Wood Components

Wood Identification
Forintek Canada Corp., Eastern Forest Products Laboratory, 800 Montreal Road, Ottawa, Ontario K1G 3Z5, Telephone (613) 744-0963. Mr C. T. Keith, Head Wood Anatomy Section, Wood Science Department.

Forintek Canada Corp., Western Laboratory, 6620 N.W. Marine Drive, Vancouver, B.C. V6T 1X2, Telephone (604) 224-3221. Dr R. M. Kellogg, Manager Wood Science Department.

USDA-Forest Service, Forest Products Laboratory, P.O. Box 5130, Madison, Wisconsin 53705, Telephone (608) 264-5600. Dr Regis Miller, Project Leader, Centre for Wood Anatomy Research.

Lignin
Virginia Polytechnic Institute and State University, Department of Forest Products and Department of Chemical Engineering, 210 Cheatham Hall, Blacksburg, Virginia 24061, Telephone (703) 961-7673. Dr Wolfgang Glasser, Professor.

Forintek Canada Corp., Western Laboratory, Dr Eric P. Swan, Research Scientist (Address and telephone number as above).

Hemicelluloses
University of Montana, Wood Chemistry Department, Missoula, Montana 59812, Telephone (406) 243-6212.

B.C. Research, 3650 Wesbrook Mall, Vancouver, B.C. V6S 2L2, Telephone (604) 224-4331. Dr John Howard, Research Scientist, Chemical Technology Division.

* Since many of these services are on a free basis, preliminary questions should be referred to the individuals directly.

65

Definition of Biomass Samples Involving Wood, Bark and Foliage

G. M. BARTON
530 Bayview Road, Lion's Bay, British Columbia, Canada

1. INTRODUCTION

In recent years there has been a growing interest in using wood, bark and foliage for chemical biomass conversion studies.[1] The reasons are understandable since forests represent one of the largest sources of renewable biomass still available to mankind. Also, the Forest Product Industries provide a constant, collected source of potential thermochemical conversion material such as tops, limbs, bark and foliage not required for lumber or pulp. This potential will increase dramatically if plans to introduce whole tree logging materialize. Unfortunately, many scientists have been attracted to this potential biomass who are unfamiliar with the wide variation among and between tree species. To many wood is wood and they make little attempt to define the sample on which valuable scientific research is done. Borrowing a sentence from the Basic Coal Sciences Project Advisory Report[2] and substituting wood for coal, the following statement emerges and describes the current situation concisely: 'Considerable basic research has been done on a wide range of wood samples for various purposes, yet much of this previous research cannot be correlated since little, if any comparisons can be drawn from the samples used'. The Estes Park conference unanimously endorsed the need for reviewers and authors to be aware of tree variability and to define all research samples in such a way that other scientists may be able to calibrate and correlate their own research results. Accordingly, this paper discusses tree variability and recommends that, as a preliminary measure to Standard Samples, researchers and authors use a checklist based on classical analytical wood chemistry to define research samples.

2. PRINCIPAL FACTORS OF SAMPLE VARIATION IN WOOD, BARK AND FOLIAGE

Samples of wood, bark and foliage originating from different trees, or even from different locations from the same tree, may vary considerably in anatomical structure and in physical, chemical and mechanical properties. Only the more important variables will be discussed in this report; for more detail it is recommended that the reader consult several fine text books on the subject, two of which are readily available.[3,4] The purpose here is to identify some of the more important variables in order that scientists will recognize those factors which influence the results of their own studies and which will cause their colleagues problems in duplicating and extending the original research study. The following headings discuss the more important of these variables beginning with the most important.

2.1. Hardwood–Softwood Differences

Trees can be classified into two main types: angiosperms—usually deciduous, also known as broad-leafed trees or hardwoods, and gymnosperms, i.e. coniferous or evergreens and softwoods. In this article the terms 'hardwood' and 'softwood' will be used.

One of the important chemical differences between softwoods and hardwoods is the difference, both qualitatively and quantitatively, in their lignins. Softwood lignin is derived from phenyl propane units carrying one methoxyl group and constitutes 24–33 % of the wood. Hardwood lignin, on the other hand, is derived from phenyl propane groups carrying two methoxyls and constitutes only 19–28 % of the wood. These differences are not merely academic since both easier chemical pulping and greater animal digestibility of the hardwoods is due to these chemical differences. Obviously, experiments done on lignin from either hardwoods or softwoods must be so specified in order to make future research studies relevant.

Equally important, chemical differences exist between the hemicelluloses of softwoods and hardwoods. Softwoods contain two major hemicelluloses—softwood galactoglucomannan (approximately 12–18 % of the wood) and softwood xylan (approximately 7–14 % of the wood). In contrast, all hardwoods are believed to contain only one major hemicellulose, commonly called hardwood xylan. This material usually constitutes 19–26 % of the wood, although white birch (*Betula papyrifera*) contains an exceptional 35 %.

In terms of bark utilization it is unlikely that bark would be important in the hardwoods since it is very thin resulting in a low bark-to-wood ratio. An exception to this might be found in poplar bark from which Pearl[5] isolated potentially important pharmaceuticals.

Although hardwood foliage has the disadvantage of being seasonal, certain species such as poplar can be hybridized to produce high protein contents of up to 19·5 %. This high protein content, together with poplar's rapid growth potential when grown as a crop, suggests a future in animal feeding.

2.2. Species Differences

The basis for any meaningful research studies by the analytical school of wood chemists was the proper identification of the species used in the experiment. Generally, this was easy in the case of the living tree due to clues provided by foliage and bark, but more difficult in the case of isolated wood samples. Fortunately, a definitive answer can nearly always be obtained from microscopic wood identification techniques developed by wood technologists. One has only to compare the chemical differences among the Canadian species listed in Table 5.1 of ref. 3 to appreciate the necessity for this measure. Even more dramatic differences can be seen by comparing extractive differences. (Extractives can be defined as substances present in wood, not an integral part of the cellular structure, that can be removed by solvents such as hot or cold water, benzene, alcohol, ether or others that do not react chemically with wood components.) Western red cedar (*Thuja plicata*), for example, contains tropolones,[6] which are potent fungicides, thus making it a preferred species where durability is required. Douglas fir (*Pseudotsuga menziesii*), on the other hand, has no tropolones but does contain an extractive, dihydroquercetin,[7] which interferes with the sulfite process of pulping. Since, with few exceptions, each species contains its own unique blend of organic chemicals, these extractives could be viewed as 'personality chemicals'[8] which give each species its distinctive chemical and some utilizational properties. If, however, the concern is only with elemental analysis, then wood is relatively constant in composition (carbon 50 %, oxygen 44 %, hydrogen 6 % and nitrogen 0·1 %) irrespective of species.

Because bark is more heterogeneous than wood and its cellulose and lignin are difficult to isolate in clearly defined fractions, a summative analysis is rarely attempted. A range in values for softwood bark is greater than for wood, for example holocellulose 30–40 %, lignin 17–42 % on an extractive-free basis. As in the case for wood, bark extractives are species dependent and generally higher than in the adjacent wood.

The chemical composition of foliage also varies with species but is more dependent on the content of included twigs and branches. The following composition of foliage preparations from softwoods indicates a range of 32–36 % for cellulose, 33–37 % for lignin, 8–11 % for protein and 9–11 % for extractives. Foliage contains chemicals associated with photosynthesis and the living cell which are not generally present in wood and bark. Of the extractives, the essential oils are the most important and are species dependent; western red cedar (*Thuja plicata*), for example, contains a very high proportion of thujone[9] (80–85 %) not found in other western softwoods.

2.3. Sapwood–Heartwood Differences

When first formed, layers of wood are light-colored and conduct sap and store food. These layers are called sapwood and are found in different thicknesses next to the bark. After a number of years this sapwood undergoes physical and chemical changes, ceases to conduct sap, and becomes heartwood. In many species it is easy to see the dividing line between sapwood and heartwood in freshly cut material, but in some species, such as western hemlock (*Tsuga heterophylla*) for example, this is difficult. In some cases chemical enhancement can be useful to indicate where the boundary line occurs.[10] Also the sapwood of most species contains a higher moisture content than the adjacent heartwood.

As heartwood forms, its cell walls become encrusted causing a significant reduction in permeability. A good example of this is found in white oak (*Quercus alba*) heartwood and accounts for this species' wide acceptance for barrel staves. In some species such as interior Douglas fir (*Pseudotsuga menziesii*), the low permeability of its heartwood is a disadvantage since it resists the penetration of preservative treating solutions for protection against decay. There are often marked differences in decay resistance between sapwood and heartwood of the same species. An excellent example of this is to be found in western red cedar (*Thuja plicata*). The sapwood of this species contains none of the protective fungicidal tropolones found in abundance in the heartwood, only a few cells removed. Thus, cedar poles must have their sapwood removed before being placed into service. If, however, these poles are to be treated with commercial preservatives, then the sapwood is left on since it is more permeable than the heartwood.

These heartwood–sapwood differences were generally recognized by the analytical school of wood chemists and sampling was confined, for the most part, well into the heartwood zone to avoid the problems just discussed.

2.4. Age Differences

Depending on the species, during the first 5–20 years of growth, wood found in the tree must be considered juvenile wood. After this period and until the tree dies, the wood formed is referred to as mature wood and possesses characteristics which are considered normal for the species. There are some notable differences between juvenile and mature wood. In general, for example, juvenile wood tends to have a lower cellulose and a higher lignin content. Because of this and other anatomical reasons, juvenile wood is not as suitable for structural purposes or where the shrinking and warping of sawn lumber is important. Thus, in order to define the wood used for research purposes, it would be desirable to confine the sample to mature wood.

2.5. Other Differences

In addition to those differences already discussed there are others which should be known and avoided if possible.

2.5.1. Site Differences

A number of mineral constituents are necessary for tree growth. Depending on the site, a variety of inorganic constituents in the soil in differing concentrations will be taken up by the tree's root system and ultimately will be found in the mature wood, bark and foliage. Thus, the same species grown in different locations could have a higher or lower concentration of these constituents. Since some inorganic constituents, particularly potassium, sodium, calcium and magnesium, can influence thermochemical conversion studies, it is important to do a qualitative as well as a quantitative analysis on the ash sample.

2.5.2. Reaction Wood

Reaction wood is an abnormal type of wood that occurs in the stems of crooked or leaning trees. In softwoods it is called compression wood, in hardwoods it is known as tension wood. Compression wood is noted for its higher density and contains a lower cellulose content and a higher lignin and galactose content. Tension wood varies in density, depending on the species, but contains higher cellulose levels, lower lignin and lower amounts of xylose and normal wood.

2.5.3. Pitch Pockets

Pitch pockets are areas of excessive high resin content occurring occasionally in softwoods. The range in size for pitch pockets is quite great and may be large enough, in the case of Douglas fir (*Pseudotsuga menziesii*), to contain several litres of resin.

2.5.4. Brash Wood

Brash wood may be defined as weak wood with low resistance to impact. It is sometimes associated with softwoods with exceptionally wide growth rings. Deterioration of wood by decay is another cause of brashness and such wood should be eliminated from the research sample.

2.5.5. Decayed Wood

Wood-decaying fungi can be classified according to their preference to either the lignin or the cellulose portion of the wood. In either case the proportion of these vital wood components will have been altered to render the sample non-typical and should be eliminated.

3. DEFINITION OF SAMPLES FOR RESEARCH

From the foregoing discussion it is apparent that sampling wood, bark and foliage for chemical biomass conversion studies is subject to many variations. In an effort to minimize these variations a checklist described in Table I is recommended. This list, based on sampling techniques developed by wood chemists, ensured a high degree of correlation of research results. Although most of the headings are self-explanatory some further

TABLE I
CHECKLIST FOR SAMPLE DEFINITION

1. *Species* (checked by a botanist or wood technologist)
2. *Site* (geographic location, including elevation)
3. *Date sample obtained*
4. *Vertical distance above ground that the sample was obtained* (traditionally this was breast height, 4·5 ft. or 1·37 m)
5. *Diameter of cross-section*
6. *Thickness of cross-section*
7. *Age of sample* (number of annual rings from pith to periphery)
8. *Extent of heartwood* (distance from pith to heartwood/sapwood boundary)
9. *Extent of sapwood* (distance of heartwood/sapwood boundary to periphery)
10. *Condition of sample* (with special reference to presence of knots, pitch pockets, decay, compression or juvenile wood, etc.)
11. *Sample storage conditions* (temperature, absence of light, inert atmosphere, etc.)

comments may be desirable. Species identification, for example, especially in isolated wood samples, can be difficult and should be confirmed by a wood anatomist or botanist. In rare cases, such as in one of the southern pines (*Pinus tauda*), it is only possible to identify the sample if it has been removed from a professionally identified living tree. Also, if bark from any species is being sampled, the distance above the ground the sample is taken should be noted since bark composition varies widely in terms of its vertical position in the tree. Distinction should also be made between inner bark, which is light brown in color, and outer bark which is dark brown. Once again the reason for this is that these barks have a different chemical composition. Finally, although foliage tends to have fewer differences because of its similar role in photosynthesis, seasonal variations are important and should be noted and standardized. Softwood foliage, for example, has higher carotene[11] amounts in the spring and summer than during the winter months.

4. CONCLUSION

It must be concluded upon examination of factors such as species, site, age and vertical and horizontal location that wood and bark samples vary in chemical composition among trees and even within the same tree. Foliage samples also vary in chemical composition especially as a result of seasonal changes.

5. RECOMMENDATIONS

Until standard wood, bark and foliage samples are readily available to scientists involved in chemical biomass conversion studies, it is recommended that samples be carefully defined according to species, site, age, date, condition and vertical and horizontal location within the tree. It is also recommended that reviewers of biomass papers be alerted to the problems of poorly defined research samples and encourage authors to include adequate experimental details in their publications.

REFERENCES

1. Goldstein, I. S. (1981). Biomass availability and utility for chemicals. *Organic Chemicals from Biomass.* CRC Press Inc., Boca Raton, Florida, USA, Chapter 1.

2. Basic Coal Sciences Project Advisory Committee Report (1980). Prepared for Gas Research Institute by Science Applications Inc., Energy Utilization Group, McLean, Virginia, USA.
3. Mullins, E. J. and McKnight, T. S. (Eds) (1981). *Canadian Woods, Their Properties and Uses*, 3rd edn. University of Toronto Press in cooperation with the Canadian Forestry Service, Environment Canada, Ottawa, Canada.
4. Panscin, A. J. and de Zeeuw, C. (1980). *Textbook of Wood Technology, Structure, Identification, Properties and Uses of the Commercial Woods of the United States and Canada*, 4th edn. McGraw-Hill, New York, USA.
5. Pearl, I. H. (1969). Water extractives of American Populus pulpwood species bark: A review. *Tappi* **52**, 428.
6. Barton, G. M. and Macdonald, B. F. (1971). The chemistry and utilization of western red cedar. Department of Fisheries and Forestry, Canadian Forest Service Publication No. 1023, 31 pp., Vancouver, B.C.
7. Pew, J. C. (1948). A flavanone from Douglas fir heartwood. *J. Am. Chem. Soc.* **70**, 3031–4.
8. Barton, G. M. (1973). Wood chemistry of western conifers, questions and answers. Information Report VP-X-106, 28 pp., Department of the Environment, Canadian Forestry Service, Western Forest Products Laboratory, Vancouver, B.C.
9. Barton, G. M. (1979). Chemicals from trees, outlook for the future. Technical Report No. 4. 12 pp., Forintek Canada Corporation, Western Forest Products Laboratory, Vancouver, B.C.
10. Barton, G. M. (1973). Chemical colour tests for Canadian woods. *Can. For. Ind.* **93**, 57–62.
11. Barton, G. M. and Macdonald, B. F. (1978). A new look at foliage chemicals. *Tappi* **61**, 45–8.

Index

www.ingramcontent.com/pod-product-compliance
Ingram Content Group UK Ltd.
Pitfield, Milton Keynes, MK11 3LW, UK
UKHW021901220926
475870UK00004B/122
* 9 7 8 9 4 0 1 0 8 6 8 5 1 *